2018 China Semiconductor Technology International Conference (CSTIC 2018)

Shanghai, China
11 – 12 March 2018

IEEE Catalog Number: CFP1860Y-POD
ISBN: 978-1-5386-5309-8

**Copyright © 2018 by the Institute of Electrical and Electronics Engineers, Inc.
All Rights Reserved**

Copyright and Reprint Permissions: Abstracting is permitted with credit to the source. Libraries are permitted to photocopy beyond the limit of U.S. copyright law for private use of patrons those articles in this volume that carry a code at the bottom of the first page, provided the per-copy fee indicated in the code is paid through Copyright Clearance Center, 222 Rosewood Drive, Danvers, MA 01923.

For other copying, reprint or republication permission, write to IEEE Copyrights Manager, IEEE Service Center, 445 Hoes Lane, Piscataway, NJ 08854. All rights reserved.

*** This is a print representation of what appears in the IEEE Digital Library. Some format issues inherent in the e-media version may also appear in this print version.

IEEE Catalog Number: CFP1860Y-POD
ISBN (Print-On-Demand):   978-1-5386-5309-8
ISBN (Online):   978-1-5386-5308-1

Additional Copies of This Publication Are Available From:

Curran Associates, Inc
57 Morehouse Lane
Red Hook, NY 12571 USA
Phone: (845) 758-0400
Fax: (845) 758-2633
E-mail: curran@proceedings.com
Web: www.proceedings.com

Table of Contents

Preface

Chapter I - Device Engineering and Memory Technology

Active-Performance Benchmark for Advanced 3D-CMOS Devices* (1-9) — 1
Hitoshi Wakabayashi, Eisuke Anju and Iriya Muneta
School of Engineering, Tokyo Institute of Technology, Yokohama, Japan

Introduction of 95 nm SPOCULL Technology* (1-64) — 5
Honglin Zeng, Liang Wu, Tianhua Dong, Lan Jin, Yulei Jiang, Yi Peng Chan, Leong Tee Koh, Jui Lin Lu, Min-Hwa Chi andTzu-Yin Chiu
Semiconductor Manufacturing International Corporation, Shanghai, China

Opposite Trends Between Digital and Analog Performance for Different TFET Technologies* (1-41) — 9
P.G.D. Agopian[1,2], C. Bordallo[1], J.A. Martino[1], R. Rooyackers[3], E. Simoen[3], N. Collaert[3] and C. Claeys[4]
[1]LSI/PSI/USP, University of Sao Paulo, Brazil
[2]Sao Paulo State University (UNESP), Campus Sao Joao da Boa Vista, Brazil
[3]imec, Leuven, Belgium
[4]E.E. Department, KU Leuven, Belgium

III-V GaAs and InP HBT Devices for 4G & 5G Wireless Applications* (1-1) — 13
C.R. Bolognesi, Wei Quan, Akshay M. Arabhavi, Diego Marti and Olivier Ostinelli
Millimeter-Wave Electronics Group (MWE), ETH-Zürich, Switzerland

Nonvolatile Memory Outlook: Technology Driven or Application Driven* (1-8) — 17
Jing Li
Department of Electrical and Computer Engineering, University of Wisconsin-Madison, Wisconsin, USA

MJT-Based Nonvolatile Logic LSI for Ultra-Low Power and Highly Dependable Computing* (1-54) — 19
Masanori Natsui, Tetsuo Endoh, Hideo Ohno and Takahiro Hanyu
Tohoku University, Japan

Do We Have to Worry About Extended Defects in High-Mobility Materials?* (1-55) — 23
Eddy Simoen[1], Po-Chun Hsu[2], Liang He[3], Yves Mols[1], Bernadette Kunert[1], Robert Langer[1], Niamh Waldron[1], Geert Eneman[1], Nadine Collaert[1], Mark Heyns[1,2] and Cor Claeys[4]
[1]imec, Leuven, Belgium
[2] Department MTM, KU Leuven, Belgium
[3]School of Advanced Materials and Nanotechnology, Xidian University, China
[4] E.E. Department, KU Leuven, Belgium

Predictive As-Grown-Generation Model for NBTI of Advanced CMOS Devices and — 27

Circuits* (1-13)
Zhigang Ji, Rui Gao and Jianfu Zhang
School of Engineering, Liverpool John Moores University, Liverpool UK

Resistive Switching and Synaptic Plasticity in HfO$_2$-Based Memristors With Single-Layer and Bilayer Structures (1-39)   31
Qingxi Duan, Liying Xu, Jiadi Zhu, Xinhao Sun, Yuchao Yang and Ru Huang
Institute of Microelectronics, Peking University, Beijing, China

Smart IC Technologies for Smart Devices in IoT Applications* (1-24)   34
Min-Hwa Chi
Semiconductor Manufacturing International Corporation, Shanghai, China

Synthesis of MoS$_2$ Via In-Situ Sulfurize Sputtering Mo (1-58)   38
Wei Junqing, Li Yue, Wang Fang, Zhang Zichao, Feng Yulin, Li Yi, Zhang Kailiang
School of Electronics Information Engineering, Tianjin Key Laboratory of Film Electronic & Communication Devices, Tianjin University of Technology, Tianjin, China

Improvement of 28HKMG NIO Device HCI by Implant Scheme and Sequence Optimize Doping Profile and E-Field (1-18)   42
Jiajia Tao, Qingchun Zhang, Jie Zhou, Yongsheng Yang, Byunghak Lee, Falong Zhou, TzuChiang Yu
Semiconductor Manufacturing International Corporation, Shanghai, China

7 μm Pitch deep Trench Super Junction Process Development (1-20)   45
Zhimin Zhu, Fei Wang and J.K. Shen
Shanghai Huahong Grace Semiconductor manufacturing Corporation, Shanghai, China

A Study of 28nm LDMOS Layout Optimization of 28HKMG for RF Application (1-21)   48
Guiying Ma, WeiJia Xu and Byunghak Lee
Semiconductor Manufacturing International Corporation, Shanghai, China

A Study of 28HKMG DLCMOS Linear Drain Current Degradation Induced By Hot Carrier Injection (1-22)   51
Guiying Ma, Hao Sun, Ling Tang, Yongsheng Yang, Xuejie Shi, Byunghak Lee, Miao Liao and Tzuchiang Yu
Semiconductor Manufacturing International Corporation, Shanghai, China

Investigation and Three Implementations For Low Power Self-Aligned 1.5-T SONOS Flash Device (1-26)   54
Zhaozhao Xu[1,2], Donghua Liu[1,2], Wei Xiong[2], Jun Hu[2], Ziquan Fang[2], Wenting Duan[2], Hualun Chen[2] and Wensheng Qian[2]
[1] State Key Laboratory of Functional Materials for Informatics, Shanghai Institute of Microsystems and Information Technology, Chinese Academy of Sciences, Shanghai, China
[2] Shanghai Huahong Grace Semiconductor manufacturing Corporation, Shanghai, China

Substrate Current Improvement for a 25V N-Type LDMOS (1-27)   59

Ziquan Fang[1], Donghua Liu[1,2] and Wensheng Qian[2]
[1]*Shanghai Huahong Grace Semiconductor Manufacturing Corporation, Shanghai, China*
[2]*State Key Laboratory of ASIC and System, School of Microelectronics, Fudan University, Shanghai, China*

Improvement of Bond Pad Crystal Defect By New Aluminum Pad Film Stack (1-28) 63
Junhong Zhao, Zhengying Wei, Yansheng Wang, Wei Zhou, Chang Sun, Jun Qian, Yufei Peng and Ran Huang
Huali Microelectronics Corporation, Shanghai, China

A Technique for Fabricating Low Voltage Charge-Coupled MOSFET with 0.9 μm Pitch Size (1-30) 67
L. Shi, R.X. Fan, J.Z. Miao, Y.P. Wang, J.K. Shen and L. Yao
Shanghai Huahong Grace Semiconductor manufacturing Corporation, Shanghai, China

28 nm High Density SRAM Bit Cell Design and Manufacture Study (1-31) 70
Meili Hu, Lijun Song and Lei Wu
Semiconductor Manufacturing International Corporation, Shanghai, China

A Compact Model of Analog RRAM For Neuromorphic Computing System Design (1-32) 73
Wenqiang Zhang, Hauqiang Wu, Peng Yao, Bin Gao and He Qian
Institute of Microelectronics, Tsinghua University, Beijing, China

Reversed Trapezoid Trench Profile Formation with Silicon Nitride Confinement (1-34) 76
Shi-liang Ji, Qiu-hua Han and Hai-Yang Zhang
Semiconductor Manufacturing International Corporation, Shanghai, China

An Improved Yield Model For Embedded Flash (1-35) 79
Yun Xu and Yuxiang Zhang
Shanghai Huahong Grace Semiconductor Manufacturing Corporation, China

A Novel SCR ESD Protection Structure For RF Power Amplifier (1-37) 82
Chunguang Wang, Zexue Liu, Junhua Liu and Huailin Liao
Key Laboratory of Microelectronic Devices and Circuits (MOE), Institute of Microelectronics, Peking University, Beijing , China

Effect of Wafer Edge Cut on Testing and Yield (1-40) 85
Yuxiang Zhang, Qin Huang and Yun Xu
Shanghai Huahong Grace Semiconductor Manufacturing Corporation, Shanghai, China

Efficient Multi-Bit SRAMs Using Nanostructures Field-Effect Transistors (Nano-FETs) (1-42) 88
Bander Saman
School of Engineering, Taif University, Taif, Kingdom of Saudi Arabia

TiSi$_2$ Formation and Mechanism Through Ultra-Thin Al$_2$O$_3$ Intermediation (1-47) 92
Peilin Hao, Ming Li, Bingxin Zhang, Siyang Gao, Xia An and Ru Huang
Key Laboratory of Microelectronic Devices and Circuits, Institute of Microelectronics, Peking University, Beijing, China

A New Method for Traveling Memory Address Based on Vector Function 95

Qingwen Zhang, Yun Xu ,YuXiang Zhang, Zhimin Zeng, Mingming Li, Yuan Wu
Shanghai Huahong Grace Semiconductor Manufacture Corporation, Shanghai, China

First-Principles Study on Ge$_{1-X}$Sn$_x$Si Core-Shell Nanowire Transistors(1-49)   98
Zeguo Gu, Feng Xu, Bin Gao, Huaqiang Wu and He Qian
Institute of Microelectronics, Tsinghua University, Beijing, China

A Physical Current Model for Junction-Modulated Tunneling Field-Effect Transistor with Steep Switching Behavior (1-60)   102
Zhu Lv, Qianqian Huang, Yang Zhao, Cheng Chen, Runsheng Wang and Ru Huang
Key Laboratory of Microelectronic Devices and Circuits (MOE), Institute of Microelectronics Peking University, Beijing, China

Benchmarking of Multi-Finger Schottky-Barrier Tunnel FET for Ultra-Low Power Applications (1-61)   105
Jiadi Zhu, Qianqian Huang, Lingyi Guo, Libo Yang, Cheng Chen, Le Ye and Ru Huang
Key Laboratory of Microelectronic Devices and Circuits (MOE), Institute of Microelectronics, Peking University, Beijing, China

Study on Microscopic Model of Resistive Switching in Amorphous Tantalum Pentoxide from First-Principle Calculations (1-62)   109
Lin Bao[1], Yichen Fang[1], Zongwei Wang[1], Jian Kang[1], Yuchao Yang[1], Jintong Xu[2], Yimao Cai[1] and Ru Huang[1]
[1]*Institute of Microelectronics, Peking University, Beijing, China*
[2]*Key Laboratory of Infrared Imaging Materials and Detectors, Shanghai Institute of Technical Physics, Chinese Academy of Sciences, Shanghai, China*

Performance Evaluation of Tunneling Field Effect Transistor on Terahertz Detection (1-63)   112
Q. Yang[1], J. Zhang[1], C. Zhu[1], X. Lin[2] , F. Yan[1] and X. Ji[1]
[1]*Institute of the Electronic Science and Engineering, Nanjing University, China*
[2]*Beijing Institute of Space Mechanics and Electricity, China*

Study on NBTI Improvement of HfO$_2$-based 14 nm P-type FinFET with Post High-k Deposition Thermal Treatment (1-65)   115
Wen Wang[1,2], Yongmin Zhao[1,2], Danniel Feng[2], Hao Jiang[2] and Jun Luo[1,3]
[1]*School of Microelectronics, University of Chinese Academy of Sciences (UCAS), Beijing, China*
[2]*Semiconductor Manufacturing International Corporation (SMIC), Shanghai, China*
[3]*Institute of Microelectronics, Chinese Academy of Sciences (IMECAS)*

Simulation for the Feasibility of High-Mobility Channel in 3D NAND Memory (1-74)   118
Zhaozhao Hou[1,2], Jiaxin Yao[1,2], Zhenhua Wu[1] and Huaxiang Yin[1,2]
[1]*Key Laboratory of Microelectronics Devices and Integrated Technology, Institute of Microelectronics of Chinese Academy of Sciences, Beijing, China*
[2]*University of Chinese Academy of Sciences, Beijing, China*

Self-aligned Metallic Source and Drain Fin-on-Insulator FinFETs with Excellent Short Channel Effects down to 20 nm Gate Length (1-76)   121
Qingzhu Zhang[1,2], Junjie Li[1], Hailing Tu[2], Huaxiang Yin[1,4], Jiang Yan[3], Lingkuan Meng[1], Jiaxin Yao[1], Guilei Wang[1], Zhijun Cao[1], Yudong Li[1], Zhaohao Zhang[1], Zhenhua Wu[1], Feng Wei[2], Hongbin Zhao[2], Jiangfeng Gao[1], Xiaobin He[1], Qifeng Jiang[1], Wenjuan Xiong[1],

Jinjuan Xiang[1], Zhangyu Zhou[1], Yihong Lu[1], Gaobo Xu[1], Kun Luo[1], Yu Pan[1], Renren Xu[1], Jie Gu[1], Chaozhao Hou[1], Junfeng Li[1] and Wenwu Wang[1]

[1]*Key Laboratory of Microelectronics Devices & Integrated Technology, Institute of Microelectronics of Chinese Academy of Sciences (IMECAS), Beijing, China*
[2]*State Key Laboratory of Advanced Materials for Smart Sensing, General Research Institute for Nonferrous Metals, Beijing, China*
[3]*College of Electronic and Information Engineering, North China University of Technology, Beijing, China*
[4]*University of Chinese Academy of Sciences, Beijing, China*

Engineering Resistive Switching Behavior in Tao$_x$ Based Memristive Devices for Non-Von Neuman Computing Applications (1-77)   124

Jingxian Li[1,2], Teng Zhang[1], Qingxi Duan[1], Lidong Li[2], Yuchao Yang[1] and Ru Huang[1]

[1]*Key Laboratory of Microelectronic Devices and Circuits (MOE), Institute of Microelectronics, Peking University, Beijing, China*
[2]*State Key Lab for Advanced Metals and Materials, School of Materials Science and Engineering, University of Science and Technology Beijing, China*

Chapter II – Lithography and Patterning

Resist Model Setup for Negative Tone Development at 14nm Node* (2-11)   127

Lijun Zhao[1,2], Lisong Dong[1], Libin Zhang[1], Yayi Wei[1,2] and Tianchun Ye[1,2]

[1]*Key Laboratory of Microelectronic Devices & Integrated Technology, Institute of Microelectronics of Chinese Academy of Sciences, Beijing, China*
[2]*University of Chinese Academy of Sciences, Beijing, China*

Source and Mask Co-Optimization Based on Depth Learning Methods (2-16)   131

Fei Peng and Yijiang Shen
School of Automation, Guangdong University of Technology, Guangzhou, China

An Optical Proximity Model for Negative Toned Developing Photoresists (2-36)   134

Qiang Wu
Semiconductor Manufacturing International Corporation Technology Research and Development, Shanghai, China

Lens Thermal Behavior Based Exposure Process Control for Flat Panel Display Lithographic Tools (2-12)   138

Zhiyong Yang, Hao Jiang and Shiyuan Liu
State Key Laboratory of Digital Manufacturing Equipment and Technology, Huazhong University of Science and Technology, Wuhan, China

High Power LPP-EUV Source with Long Collector Mirror Lifetime for High Volume Semiconductor Manufacturing (2-19)   141

Hakaru Mizoguchi, Hiroaki Nakarai, Tamotsu Abe, Krzysztof M Nowak, Yasufumi Kawasuji, Hiroshi Tanaka, Yukio Watanabe, Tsukasa Hori, Takeshi Kodama, Yutaka Shiraishi, Tatsuya Yanagida, Georg Soumagne, Tsuyoshi Yamada, Taku Yamazaki and Takashi Saitou

Gigaphoton Inc, Kanagawa, Japan
Multi-beam Mask Writer MBM-1000* (2-23)   148

Hiroshi Matsumoto, Hideo Inoue, Hiroshi Yamashita and Kenji Otoshi
NuFlare Technology, Inc., Yokohama, Kanagawa, Japan

Robust Overlay Metrology by Mueller Matrix Ellipsometry with a Differential Calculus (2-20)   151
Xiuguo Chen and Shiyuan Liu
State Key Lab of Digital Manufacturing Equipment and Technology, Huazhong University of Science and Technology, Wuhan, China

Study on Spin-On Hardmask for Quad-Layer Application (2-15)   155
Yushin Park, Sanghak Lim, Seungwook Shin, Jinhyung Kim, Yunjun Kim, Seunghyun Kim, Jaebum Lim, Sung Hwan Kim, Hyeonil Jung, Yumi Heo, Sunyoung Yang and Jeong Yun Yu
Development group, Electron Materials Division, Samsung SDI, Suwon, Korea

Comparative Analysis of Resist Model Stability in Negative Tone Development Process (2-28)   158
Hua Cai, Hanmo Gong and Qian Ren
Semiconductor Manufacturing International Corporation, Shanghai, China

Advanced Photoresist and Material Development in China* (2-22)   163
Ran Ruicheng, Mao Guoping and Sun Yousong
Jiangsu Hantop Photo-Material Company, Pizhou Economic Zone, Xuzhou, China

A Study of Multi-Factor Influence on Resist LWR Performance (2-39)   167
Dongxu-Yang and Lei-Ye
Semiconductor Manufacturing International Corporation, Shanghai, China

Ultra-Fast Directly Self-Assembly Materials for Sub-5nm Lithographic Patterning (2-18)   170
Xuemiao Li, Chenxu Wang and Hai Deng
Department of Macromolecular Science, State Key Laboratory of Molecular Engineering of Polymers, Fudan University, Shanghai, China

Study of Unique Pseudo Buried Layer in 0.18 μm SiGe BICMOS Process (2-9)   173
Donghua Liu[1,2], Xi Chen[2] and David Wei Zhang[1]
[1]*State Key Laboratory of ASIC and System, School of Microelectronics, Fudan University, Shanghai, China*
[2]*HuaHong Grace Semiconductor Manufacturing Corporation, Shanghai, China*

Lithography Process Optimization for Emitter Window In SiGe-HBT Device (2-10)   178
Donghua Liu[1,2], Ziquan Fang[2] and Zhaozhao Xu[2,3]
[1]*State Key Laboratory of ASIC and System, School of Microelectronics, Fudan University, Shanghai, China*
[2]*HuaHong Grace Semiconductor Manufacturing Corporation, Shanghai, China*
[3]*State Key Laboratory of Functional Materials for Informatics, Shanghai Institute of Microsystem and Information Technology, Chinese Academy of Sciences, Shanghai, China*

Study on Planarization Performance of Spin On Hardmask (2-14)   183
Huichan Yun, Jinhyung Kim, Yoona Kim, Seulgi Jeong, Sanghak Lim and Jeong Yun Yu
Development group, Electron Materials Division, Samsung SDI, Suwon, Korea

A Novel Lithographic Material Constructed by Polymerizable Liquid Crystal Molecule (2-30)   186

Hui Cao and Hai Deng
Department of Macromolecular Science, State Key Laboratory of Molecular Engineering of Polymers, Fudan University Shanghai, China

A Study of Photoresist Residue Defect Induced by Substrate Surface Condition (2-32)   189
Zhou Fang and Chang Liu
Semiconductor Manufacturing International Corporation, Shanghai, China

Challenges and Solutions of Few-Line Patterns with Freeform Illumination in Immersion Lithography (2-34)   192
Peipei Liang, Zhifeng Gan and Qiang Wu
Semiconductor Manufacturing International Corporation, Shanghai, China

Research of Lithograph Process of Polyimide Photo Resist for Passivation Thick Films (2-37)   195
Zhao Wang and Ke Feng
CRRC Yongji Electric Co., Ltd, Xi'an, China

The Study on the Diffraction Spectrum of Adding S-BAR during the Source Mask Optimization (SMO) Process (2-41)   198
Miao Xia, Song Bai and Qin Wang
Semiconductor Manufacturing International Corporation, Shanghai, China

Nanometer Scale Mixed Patterns Formed by Self-aligned Spacer Image Transfer and Optical Lithography Technology (2-46)   202
Zhaohao Zhang[1], Qingzhu Zhang[1,2], Junjie Li[1], Xiaobin He[1] and Huaxiang Yin[1,3]
[1]Key Laboratory of Microelectronics Devices & Integrated Technology, Institute of Microelectronics of Chinese Academy of Sciences (IMECAS), Beijing, China
[2]State Key Laboratory of Advanced Materials for Smart Sensing, General Research Institute for Nonferrous Metals, Beijing, China
[3]University of Chinese Academy of Sciences, Beijing, China

Chapter III – Dry & Wet Etch and Cleaning

The Study of STI Etching Micro-Loading in Reactive Ion Etch (RIE) (3-57)   205
Yunhe Dong, Zhongwei Jiang and Yahui Huang
Beijing NAURA Microelectronics Equipment Co., Ltd, Beijing, China

Patterning Technology Options for Future Scaling* (3-7)   208
Kenichi Oyama
Tokyo Electron Technology Solutions Limited, Nirasaki City, Yamanashi, Japan

Dummy Gate Critical Dimension Uniformity Improvement for 14nm and Beyond (3-23)   210
Jie Luo, Dongping Zhang and Haiyang_Zhang
Semiconductor Manufacturing International Corporation, Shanghai, China

Zero Lag Dispense to Increase the Etching Uniformity in a Single Wafer Wet Cleaner (3-58)   213

Wei Liu, Xiaoyan Liu and Yi Wu
Beijing NAURA Microelectronics Equipment Co., Ltd. Beijing Tianzhu Export Processing Zone, ShunYi District, China

Study of Downstream CF4 Contained Plasma Process Impact on Chamber Condition (3-13)   **217**
Yali Fu[1,2], Shawming Ma[3], Yi Wang[1], Ken Wang[4], Mingjie Jiao[2], Nancy Zhang[2], Vijay Vaniapura[3], Hai-Au Phan-Vu[3] and Bob Elliston[3]
[1]Peking University, School of software and microelectronics, Beijing, China
[2]Mattson Technology, China Field Process Group, Beijing, China
[3]Mattson Technology, Plasma Product Group, Fremont, California, USA
[4]SMNC Etch Department, Beijing, China

Advanced Techniques for Enhanced Ion Beam Etch Uniformity (3-36)   **221**
Peter Goglia[1], Hari Hegde[1] and Yannick Pilloux[2]
[1]Plasma-Therm, LLC, Fremont, CA, USA
[2]Plasma-Therm, LLC, St. Petersburg, FL, USA

The Development of Novel STI Spacer in an Epi Free SiGe BICMOS Process (3-14)   **224**
Donghua Liu[1,2], Ziquan Fang[2] and Wensheng Qian[2]
[1]State Key Laboratory of ASIC and System, School of Microelectronics, Fudan University, Shanghai, China
[2]HuaHong Grace Semiconductor Manufacturing Corporation, Shanghai, China

Fabrication of Silicon Nanopores with Tunable Shapes and Sizes (3-15)   **227**
Qi Chen, Yifan Wang and Zewen Liu
Institute of Microelectronics, Tsinghua University, Beijing, China

Study of Plug Etch Back Process Matching on Faraday Shielded ICP and Traditional CCP Plasma Etch Chambers (3-18)   **230**
Yali Fu[1,2], Shawming Ma[3], Yi Wang[1], Jammy Xiao[4], MH Kim[3], Nancy Zhang[2], Mingjie Jiao[2], Frank Wang[3], Tammy Lee[3]
[1]Peking University, School of Software and Microelectronics, Beijing, China
[2]Mattson Technology, China field process Group, Beijing, China
[3]Mattson Technology, Plasma Product Group, Fremont, California, USA
[4]SMNC ETCH department, Beijing, China

De-Wetting Model Study on Spinning Substrate – Challenges for Low Chemical Consumption (3-21)   **333**
David Mui[1], Mark Kawaguchi[1] and Christian Haigermoser[2]
[1]Lam Research Corp., Fremont, USA
[2]Lam Research Service (Shanghai) Co., Ltd., Shanghai, China

Advanced Material for Process Performance in Plasma Process (3-26)   **236**
Xingjian Chen, Tuqiang Ni and Shenjian Liu
Advanced Micro-Fabrication Equipment Inc

Germanium Compatible Photoresist Strippers and Residue Cleaners for 5nm and Below Technology Nodes (3-27)   **240**
Chien-Pin Sherman Hsu
Avantor, Tai-Yuan St., Chu-Bei, Hsinchu, Taiwan, China

Mega-Sonic Effect on Sub-Resolution Assist Features During Photomask Cleaning (3-40) — 242
Sophia Xue, Irene Shi, Alan Li, Schafer Qin, Eric Tian, Max Lu and Eric Guo
Semiconductor Manufacturing International Corporation, Shanghai, China

Fast Model for Positive Tone Development (PTD) Source Mask Optimization (SMO) Calibrated by Dose Tuning (3-52) — 245
Song Bai, Xuan Li and Manhua Shen
Semiconductor Manufacturing International Corporation, Shanghai, China

Analysis of Reversed Depth Micro-Loading Based on Pulsed Plasma (3-55) — 248
Guodong Chen, Liming Yao, Zhongwei Jiang and Yahui Huang
Beijing NAURA Microelectronics Equipment Co.,Ltd Beijing, China

A Study of Non-Bosch Deep Trench Etch — 251
Wang Jing, Nie Miao, Jiang Zhongwei and Huang Yahui
Beijing NAURA Microelectronics Equipment Co. Ltd

Chapter IV – Thin Film, Plating and Process Integration

Investigation on the Leakage of Triple Split-Gate Flash Device and Its Improved Solution (4-19) — 253
Zigui Cao, Lingyue Zhang, Yan Sun and Buchun Su
Huahong Grace Semiconductor Manufacturing Corporation, Shanghai, China

Hybrid Integration of FDSOI And Si Bulk For CMOS Process (4-23) — 257
Yang Song, Kang Ye and Changfeng Wang
Huali Microelectronics Corporation, Shanghai, China

Patterning Challenges in 193i-Based Tip To Tip in N5 Interconnects* (4-40) — 260
Basoene Briggs, Janko Versluijs, Juergen Bömmels, Christopher J. Wilson, Zsolt Tőkei, Arindam Mallik and Job Soethoudt[#]
imec, Leuven, Belgium
[#]KU Leuven, Leuven, Belgium

Overview of ALD Applications for Advanced CMOS Technology* (4-33) — 264
X. Shi , C. Li, H. Ji, H. Qin, W. Zhang, W. Xia and P. Ding
Beijing NAURA Microelectronics Equipment Co. Ltd, Beijing, China

Conformal SiGe Selective Epitaxial Growth for Advanced CMOS Technology (4-29) — 268
Yiqun Liu, Lan Jin, Kunshan Song, Qiong Wu, Youfeng He and Yonggen He
Semiconductor Manufacturing International Corporation, Shanghai, China

Effect of Idle Time on the Property of Ultra Low-k Dielectric Film (4-31) — 271
Wenrong Hou, Jack Chen, Neil Li, Eric Li, Xiaorong Wang, Ming Feng, Xiaowei Shu and Jianqiang Liu
Semiconductor Manufacturing North China (Beijing) Corporation (SMNC), Beijing, China

In-situ Plasma Monitoring of PECVD a-Si:H(i)/ a-Si:H (n) Surface Passivation for — 274

Heterojunction Solar Cells Application (4-39)
Yu-Lin Hsieh[1], Li-Han Kau[2], Hung-Jui Huang[2], Chien-Chieh Lee[3], Yiin-Kuen Fuh[1] and Tomi T. Li[1]
[1]*Department of Mechanical Engineering, National Central University, Taiwan, China*
[2]*Institute of Opto-mechatronics Engineering, National Central University, Taiwan, China*
[3]*Optical Science Center, National Central University, Taiwan, China*

PVD Systems for Advanced Packaging Applications* (4-25)   279
Jinguo Zhen, Kuanmao Wang, Wei Xia, Hougong Wang and Peijun Ding
Beijing NAURA Microelectronics Equipment Co. Ltd, Beijing, China

The Proposal and Application of the Hole Smoluchowski Effect to Explain the Current-Voltage Characteristics of High-k MIM Capacitors (4-27)   283
W.S. Lau
Zhejiang University, Department of Information Science and Electronic Engineering, Hangzhou, China

Engineering Implication of the Correlation Between the Leakage Current in High-k Dielectric Materials and the Electronic Defect States Detected by Zero-Bias Thermally Stimulated Current Spectroscopy (4-28)   286
W.S. Lau
Zhejiang University, Dept Information Science & Electronic Engineering, Hangzhou, China

Optimization of Backside Metal Deposition in Power IC Process for Suppression of Wafer Warpage and Film Peeling Issue (4-1)   289
Wei Zhang, Weijie Liu and Bingxin Ma
Semiconductor Manufacturing International Corporation, Tianjin, China

A Technique for Improving Contact Filling with Aluminum (4-6)   300
Chunling Liu, J.K. Shen, Xingjie Wang and Zhihui Ji
Huahong Grace Semiconductor Manufacturing Corporation Shanghai, China

Emitter-Base Short Issue Study and Improvement in a Low Cost and High Performance 0.18 μm SiGe BICMOS Process (4-7)   303
Donghua Liu[1,2], Zhaozhao Xu[2,3] and Wensheng Qian[2]
[1]*State Key Laboratory of ASIC and System, School of Microelectronics, Fudan University, Shanghai, China*
[2]*HuaHong Grace Semiconductor Manufacturing Corporation, Shanghai, China*
[3]*State Key Laboratory of Functional Materials for Informatics, Shanghai Institute of Microsystem and Information Technology, Chinese Academy of Sciences, Shanghai, China*

Advanced Mass Flow Controllers (MFC) with Ethercat Communication Protocol and Embedded Self Diagnostics (4-15)   307
Kevin Findleton, Berwin Banares and Mohamed Saleem
Brooks Instrument, Hatfield, PA, USA

The Study of TiN Residues Formation Mechanism and Removal Solution on Bondpad Surface in PI/PA Mask Combined Process (4-26)   311
Xu Jie, Huang Chong, Li Zhiguo and Ding Tongguo
Huahong Grace Semiconductor Manufacturing Corporation, Shanghai, China

Management of TEOS-CVD Process Tool Exhausts in 3D-NAND Manufacturing   315

Andrew A. Chambers
Edwards, Clevedon, UK

Hydrogen Based High Strain Si:P Epitaxy Process Development and Optimization (4-35)   318
Jin Yang
Applied Materials China, Shanghai, China

Chapter V – CMP and Post-Polish Cleaning

New CMP Processes Development and Challenges for 7nm and Beyond (5-14)   321
Haigou Huang[1], Dinesh Koli[1], John H Zhang[2], Stan Tsai[2], Taifong Chao[1], Yuanfang Lu[1], Hong Jin Kim[1] and Qiang Fang[1]
[1]*Advanced Technology Development, GLOBALFOUNDRIES, Malta, NY, USA*
[2]*GLOBALFOUNDRIES, Albany Nanotechnology Center, Albany, NY 12203, USA*

STI Scratch Defects Reduction by Using Solid Pad in 1x Technology Node (5-22)   326
Kuang-Wei Chen[1], Yen-Ting Chen[1], Chun-Fu Chen[1], Tuung Luoh[1], Lin-Wuu Yang[1], Ta-Hone Yang[1], Kuang-Chao Chen[1], Yuchi Tay[2], Yi Ming Chen[2], Johnson Ton[2] and Yen Pin Lu[2]
[1]*Macronix International Co., Ltd., Technology Development Center, Hsinchu, Taiwan*
[2]*Thomas West Inc., CA, USA*

Low Scratch High Throughput Auto Stop Slurry (5-12)   329
Jinfeng Wang, Juyeon (Jay) Chang and Jeongdeog (Jeidy) Koh
Cabot Microelectronics Corporation, Aurora, IL, USA

Characterization of Lanthanide Elements Doped Ceria Nanoparticles and Its Performance in Chemical Mechanical Polishing as Novel Abrasive Particles (5-23)   332
Jie Cheng[1], Yang Li[2], Xinchun Lu[1]
[1]*State Key Lab of Tribology, Tsinghua University Tianjin Hwatsing Technology Company Limited, China*
[2]*Department of Chemistry, Tsinghua University, China*

Investigation of Effects of Pattern Structures Arrangement on Chemical Mechanical Polishing Process (5-16)   336
Lixiao Wu[1], Sookap Hahn[2] and Changfeng Yan[1]
[1]*School of Mechanical & Electronical Engineering, Lanzhou University of Technology, Lanzhou, China,*
[2]*SKW Associates, Santa Clara, CA, USA*

Colloidal Silica: Chemistry, Properties and Adaptations for Electronic Polishing Applications* (5-5)   339
Francois Batllo, Yulia Tataurova and Michael Kamrath
Nalco Water: An Ecolab Company, Naperville, IL, USA

Novel CMP Pads by Special Structural Design (5-18)   343
Yijie Luo, Min Liu, Weitung Wang and Jun Yan
Hubei Dinghui Microelectronics Materials Co., Ltd, Wuhan, China

Achieving Highly Accurate Profile Control by Applying a Multi-Zone Polish Head for   346

200mmThin-Film SOI CMP (5-20)
Mike Liu[1], Jamie Leighton[2] and Mike Rosa[2].
[1]*Applied Global Services, Applied Materials, Xi'an, Shaanxi, China*
[2]*Applied Global Services, Applied Materials, Santa Clara, California, USA*

CMP Pad Surface Uniformity Optimization after Polish (5-1)   349
Ying Emily Lu, Wei William Guo and Jenny Wu
Corporate Q&R Center, Semiconductor Manufacturing International Corporation, Shenzhen, Guangdong, China

Chapter VI – Metrology, Reliability and Testing

Methods for Testing Signals Considering Jitter Transfer Function (6-10)   353
Kenichi Nagatani
Advantest (China) Co., Ltd. Shanghai, China

The Monitoring Evolutionary Algorithm in Device Production Testing (6-17)   356
Kun Xu[1], Songmiao Wang[1], Wei Xu[1] and Qin Wang[2]
[1]*System Engineering Department, Advantest (China) Co., Ltd, Shanghai, China*
[2]*Department of Micro/Nano Electronics, Shanghai Jiaotong University, Shanghai, China*

Achieving Mastery in Chip Manufacturing (6-31)   359
Cyril-Patrick Fernandez and Jim Jensen
Fuping Semi Yield Masters, Fuping, China
Advanced Fab Technology, Dallas, Texas, USA

Application of an Enhanced Failure Bitmap Method for Memory Fault Isolation (6-9)   362
Zhimin Zeng, Yun Xu, Qingwen Zhang, Mingming Li, Yutian Zhang and Xiaojun Xu, Xiao Zou
Shanghai Huahong Grace Semiconductor Manufacture Corporation, Shanghai, China

Advanced Process Equipment Matching Methodology in Semiconductor Manufacturing (6-3)   366
Ziqian Javaer Liu, Hongtao HT Qian and Mengyang Elaine Liu
Semiconductor Manufacturing International Corporation, Shanghai, China

Applications of Advanced Techniques of Transmission Electron Microscope in Characterization of Semiconductor Devices* (6-25)   370
Jinghong Li
Globalfoundries, Hopewell Junction, NY, USA

The Detection and Investigation of Polyline Leakage by Electron-Beam Inspection (6-2)   374
Rongwei Fan, Hunglin Chen, Kai Wang, Kun Cai, Yin Long, Qiliang Ni and Xiaofang Gu
Shanghai Huali Microelectronics Corporation, Shanghai, China

Legacy Profiler Capacity/Utilization Improvement with Automation for High Volume Baw Production (6-14)   377
Yanghua He, Dario Nappa, Hector Nevarez and Michael Lube
Process Engineering, Qorvo, Inc., Richardson, Texas, USA

Monitoring Critical Process Steps in 3D NAND and Advanced RF Using Picosecond   381

Ultrasonic Metrology (6-16)
Johnny Dai[1], Robin Mair[1], Kwansoon Park[2], Xinglin Zeng[1], Priya Mukundhan[16-1], Cheolkyu Kim[2] and Tim Kryman[1]
[1]Rudolph Technologies, Budd Lake, NJ, USA
[2]Rudolph Technologies, Sungnam-si, Gyunggi-do, Korea

The Total Inspection Solutions of Extreme Tiny Defect in Beol Advanced Semiconductor Process (6-12)   384
Xingdi Zhang, Hunglin Chen, Yin Long, Qiliang Ni, Rongwei Fan and Kai Wang
Shanghai Huali Microelectronics Corporation, Shanghai, China

The Detection and Investigation of Tungsten-Plug Voids by Electron-Beam Inspection (6-1)   387
Rongwei Fan, Hunglin Chen, Kai Wang, Zuoyue Liu, Yin Long, Qiliang Ni and Xiaofang Gu
Shanghai Huali Microelectronics Corporation, Shanghai, China

Probing and Manipulating the Interfacial Defects of InGaAs Dual-Layer Metal Oxides at the Atomic Scale (6-8)   390
Xing Wu[1,4], Chen Luo[1], Peng Hao[2], Tao Sun[2], Runsheng Wang[2], Chaolun Wang[1], Zhigao Hu[1], Yawei Li[1], Jian Zhang[1], Gennadi Bersuker[3], Litao Sun[4] and Kinleong Pey[5]
[1]Department of Electrical Engineering, East China Normal University, Shanghai, China
[2]Institute of Microelectronics, Peking University, China
[3]The Aerospace Corporation, Los Angeles, CA, USA
[4]SEU-FEI Nano-Pico Center Key Laboratory of MEMS of Ministry of Education School of Electronic Science and Engineering, Southeast University, Nanjing, China
[5]Singapore University of Technology and Design, Singapore

Reliability Engineering: Help Enable Technology Scaling* (6-23)   393
S. Pae, HC. Sagong, M. Jin, T. Jeong, J. Kim, I. Baick, H. Shim, J. Park, H. Kim, Y.C. Choi, and S. Shin
Foundry Business, Samsung Electronics, Gi-Heung City, Korea

Changeable Electromigration Failure Mode in Wide Cu Interconnects (6-11)   398
Hui Zheng, Binfeng Yin, Leigang Chen, Ke Zhou and Chinte Kuo
Shanghai Huali Microelectronics Corporation, Shanghai, China

Impact of TID Radiation on Hot-Carrier Effects in 65nm Bulk Si NMOSFETs (6-15)   401
Zhexuan Ren[1], Xia An[1], Jianing Wang[2], Xing Zhang[1] and Ru Huang[1]
[1]Institute of Microelectronics, Peking University, Beijing, China
[2]Faculty of Information Technology, Beijing University of Technology, Beijing, China

Reliability Comparison Between Al Process and Cu Process Based on 0.11μm Technology Node (6-27)   404
Xiang Fu Zhao
Semiconductor Manufacturing International Corporation, Shanghai, China

Assessing the Accuracy of Statistical Properties Extracted From a Limited Number of Device Under Test for Time Dependent Variations* (6-33)   406
J. F. Zhang, M. Duan, Z. Ji and W. D. Zhang
Dept Electronics and Electrical Engineering, John Moores University, Liverpool, UK

Advanced N-Channel LDMOS with Ultralow Specific On-resistance by 0.18 μm Epitaxial BCD Technology (6-4) ... 410
Yao Yao[1], Linhui Hu[2], Gangning Wang[2], Shanon Pu[2], Min-Zhi Lin[1], Zhiyuan Ye[1] and Peng-Fei Wang[1]
[1]*School of Microelectronics, Fudan University, Shanghai, China*
[2]*Semiconductor Manufacturing International Corporation, Shanghai, China*

Reliability Study of Novel Gate Oxide Process Control Method (6-32) ... 413
Guangpeng Zeng[1], Weidong Geng[1], Yi Shih Lin[2]
[1]*Key Laboratory of Photo-Electronics Thin Film Devices and Technique of Tianjin, Key Laboratory Opto-electronic Information Science and Technology, Institute of Photo-electronics Thin Film Devices and Technique, Nankai University, Tianjin, China*
[2]*Semiconductor Manufacturing International Corporation, Shanghai, China*

Research On Ultra-Low R_{DSON} Wafer Measurement (6-7) ... 416
Lei Wang
Shanghai Huahong Grace Semiconductor Manufacturing Corporation, Shanghai, China

A Method of Contact Level PVC Enhancement (6-18) ... 419
Xiaojun Xu, Jinjin Xie, Yutian Zhang, Zhimin Zeng and Yun Xu
Shanghai Huahong Grace Semiconductor Manufacture Corporation, Shanghai, China

Study on Optimization Method of Flash Cell Analog Measurement (6-19) ... 423
Wu Yuan, Zhang Qingwen and Li Mingming
Shanghai Huahong Grace Semiconductor Manufacture Corporation, Shanghai, China

Thickness and Concentration Control for Silicon-Germanium (SiGe)Films (6-20) ... 426
Liying Wu, Xiaofeng Wang and Wei Li
Semiconductor Manufacturing International Corporation, Shanghai, China

ULK Optimization for Cu/ULK(k=2.5) BEOL Interconnect TDDB and EM Improvement at 28/14nm Technology Node and Beyond (6-29) ... 430
Xiaodong Zou, Tao Dou, Zheyuan Tong, Fenglian Li and Duohui Bei
Semiconductor Manufacturing International, Shanghai, China

Chapter VII – Packaging and Assembly

Fan-Out Wafer-Level Packaging for 3D IC Heterogeneous Integration** (7-21) ... 435
John H Lau
ASM Pacific Technology, Hong Kong, China

Improvement of Package Warpage Through Substrate and EMC Optimization* (7-16) ... 441
Ken Lee[1], Min Sung Kim[1], Jaesung Kim[1], Sangkyun Kim[2], Donghwan Lee[2] and Kyunghag Jung[2]
[1]*Simmtech, Co., Ltd., Heungdeok-gu, Cheongju, Chungcheongbuk-do, Korea*
[2]*Samsung SDI, Giheung-gu, Yongin-si, Gyeonggi-do, Korea*

Emerging Fine Line Panel Level Fan Out Technology* (7-23) ... 445

David Fang, Michael Hsu, CC Chang, KW Chung, Alex Liu, Irving Lin and Daniel Fann
Powertech Technology Inc., Hsinchu, Taiwan

Process Development of Five- and Six-Side Molded WLCSP* (7-15)   449
Shuying Ma, Teng Wang, Zhiyi Xiao and Daquan Yu
Huantian Technology (Kunshan) Electronics Co., Ltd. Kunshan, Jiangsu Province, China

The Signal Integrity Simulation of Sip Package using TSV Interposer Technology* (7-14)   452
Ge Sima, Jian Xu and Peng Sun
The National Center for Advanced Packaging (NCAP China), WuXi, China

Effective EMI Shielding for Semiconductor Packages Through Novel Conformal Coating* (7-12)   455
Xinpei Cao, Andrew Sun, Dan Maslyk, Junbo Gao, Qizhuo Zhuo and Jinu Choi
Henkel Electronic Materials, LLC. Irvine, California USA

Latest Developments of Molding Compound Material for Power Semiconductors (7-2)   460
Yusuke Tanaka and Itaru Watanabe
Sumitomo Bakelite (Suzhou) Co., Ltd., Suzhou, China

Thermal and Mechanical Performance for Different Package Design of Ultra Thin 8 Die Stacked Flash Packages (7-6)   463
Hao Liu, Qian Wang, Tianhan Wu, Xiaolin Bai and Fangfang Guo
Engineering Department, Ramaxel Technology Co., LTD, Dong Guan, China

Wafer Warpage Control by Epoxy Molding Compounds for Wafer Level Package (7-11)   467
Kihyeok Kwon, Joo Young Chung, Donghwan Lee and Sang Kyun Kim
Samsung SDI, Gyeonggi-do, Korea

Advanced Panel Level Die Attach* (7-19)   470
Hugo Pristauz
Besi Austria GmbH Radfeld, Tyrol, Austria

Laser Release Technology for Wafer Level Packaging (7-3)   473
Dongshun Bai, Xiao Liu, Hong Zhang, Qi Wu, Ram Trichur, Rama Puligadda and Tony Flaim
Brewer Science, Rolla, MO, USA

Capillary Wedge Bonding Technology for Stacked Die Packages (7-7)   476
Hao Liu[1]*, Maopeng Zhou[1], Xiansheng Duan[1], Jiantao Lin[1], Fangfang Guo[1] and Naoki Sekine[2]
[1]Engineering department, Ramaxel Technology Ltd, Dongguan, China
[2]Global Engineering Support Management Dept, Shinkawa Ltd, Tokyo, Japan

Study on the High Reliability Performance and High Thermal Conductivity Epoxy Molding Compound (7-8)   480
Xingming Cheng[1], Wei Tan[1], Zhen Wang[2], Hongjie Liu[1] and YangYang Duan[1]
[1]Jiangsu HHCK Advanced Materials Co., Ltd
[2]Lianyungang Branch of Traditional Chinese Medicine, Jiangsu Union Technical Institute, Lianyungang, Jiangsu, China

Highly Functionalized EMC Package Materials for Fingerprint Sensors (7-9)   483
Junghwa Kim, Junwoo Lee, Yongyeop Park, Jiah Yoon, Jiyoon Cho, Kyoungchul Bae,
Donghwan Lee and Sangkyun Kim
Samsung SDI, Giheung-gu, Yongin-si, Gyeonggi-do, Korea

Structuring Redistribution Layers Down to 2 Micron Line Spacing with Laser Direct   486
Ablation* (7-22)
Dirk Müller[1], Jan Brune[2] and Rainer Pätzel[2]
[1]*Coherent Inc., Santa Clara, CA, USA*
[2]*Coherent Laser Systems GmbH & Co. KG, Göttingen, Germany*

New Application Method for Package Level EMI Shield Coating (7-10)   489
Stuart Erickson
Ultrasonic Systems, Inc., Haverhill, MA, USA

Thin Die Stacking Technologies for 3D Memory Packages (7-20)   492
Jie Wu[1], Oranna Yauw[1], Andrew Tan[1], Horst Clauberg[2] and Daniel Buergi[2]
[1]*Kulicke & Soffa Pte Ltd, Singapore*
[2]*Kulicke & Soffa Industries Inc., Fort Washington, PA, US*

Electroplated Nanotwin Copper for Fine Line RDL (7-17)   495
Stream Chung, Yao-Tsong Chen and Zong-Cyuan Chen
Chemleader Corporation, Hsinchu, Taiwan

Method to Reduce Metal Filament on Die With Backside Metal (7-1)   498
Bo Peng, Deguang Zheng and Yue Yu
Wafer Test Department, NXP semiconductor (China) Ltd., Tianjin, China

Silane Oligomer in Epoxy Molding Compound (7-5)   501
Hongjie Liu[1], Zhen Wang[2], Wei Tan[1], Lanxia Li[1], Xingming Cheng[1], Xiaojuan Jiang[1] and
Liang Cui[1]
[1]*JiangSu HHCK Advanced materials Co., Ltd*
[2]*Lianyungang Branch of Traditional Chinese Medicine，Jiangsu Union Technical Institute
Lianyungang, Jiangsu, China*

Chapter VIII – MEMS, Sensors and Emerging Semiconductor Technologies

Nanostructures for Smart Systems* (8-17)   504
Joerg Martin[1], Ray Saupe[1], Joern Langenickel[2], Martin Moebius[2], Kathleen Heinrich[1],
Alexander Weiß[1] and Thomas Otto[1, 2]
[1]*Fraunhofer Institute for Electronic Nano Systems, Chemnitz, Germany*
[2]*Center for Microtechnologies, University of Technology, Chemnitz, Germany*

RF MEMS Resonant Devices for Wireless Communication* (8-9)   508
Q. Yuan[1,2], X. Kan[1,2,3], Z.J. Chen[1,2,3], J. L. Yang[1,2,3], J.C. Zhao[4], L. Sun[4] and F. H. Yang[1,2]
[1]*Institute of Semiconductors, Chinese Academy of Sciences, Beijing, China*
[2]*University of Chinese Academy of Science, Beijing, China*
[3]*State Key Laboratory of Transducer Technology, Shanghai, China*
[4]*Jiangsu Key Laboratory of ASIC Design, Nantong University, Nantong, China*

Study of the Mesa Etched Tri-Gate InAs HEMTs with Extremely Low SS for Low-Power Logic Applications (8-29)   512
Yueh-Chin Lin[1], Jing-Neng Yao[2], Hisang-Hua Hsu[3], Ying-Chieh Wong[1], Chi-Yi Huang[4], Heng Tung Hsu[5], Hiroshi Iwai[5] and Edward Yi Chang[1,2,5]
[1]*Department of Materials Science and Engineering, NCTU, Hsinchu, Taiwan*
[2]*Department of Electronics Engineering, NCTU, Hsinchu, Taiwan, China*
[3]*Institute of Imaging and Biomedical Photonics, NCTU, Tainan, Taiwan, China*
[4]*Institute of Lighting and Energy Photonics, NCTU, Tainan, Taiwan, China*
[5]*International College of Semiconductor Technology, NCTU, Hsinchu, Taiwan, China*

Next Generation of WLCSP Contacting Technologies for 250 Micron Pitch and Below* (8-31)   515
Bert Brost and Valts Treibergs
Xcerra Corporation, St. Paul, Minnesota, USA

Photovoltaic Properties of Lateral Ultra-Thin Si p-i-n Structure (8-24)   519
Suguru Tatsunokuchi[1], I. Muneta[1], T. Hoshii[1], H. Wakabayashi[1], K. Tsutsui[2], H. Iwai[2] and K. Kakushima[1]
[1]*School of Engineering, Tokyo Institute of Technology, Yokohama, Japan*
[2]*Institute of Innovative Research, Tokyo Institute of Technology, Yokohama, Japan*

Study on Sensitivity of the Bilateral Ultra-Thick Silicon Sensors as Neutron Dosimeter (8-18)   522
Y.H. Huang, M. Yu, RY. Yin, Z. Zhang, X.Y. Zha, Z.Y. Zhu, J.L. Liu, M.J. Wang, J.Y. Wang and Y.F. Jin
National Key Laboratory of Nano/Micro Fabrication Technology, Key Laboratory of Microelectronic Devices and Circuits (MOE), Institute of Microelectronics, Peking University, Beijing, China

A Novel Three Observation-Windows Measurement Scheme for SPAD Fluorescence Lifetime Imaging Detector (8-20)   525
Ding Li[1], Zhong Wu[1], Yue Xu[1,2] and Tingchen Zhao[1]
[1]*College of electronic and optical engineering & College of microelectronics in Nanjing University of Posts and Telecommunications, Nanjing, Jiangsu Province, China*
[2]*National and Local Joint Engineering Laboratory of RF integration & Micro-Assembly Technology, Nanjing, Jiangsu Province, China*

An Improved Behavioral Simulation Model for Afterpulsing Phenomenon (8-21)   528
Tingchen Zhao[1], Feiyang Sun[1], Yue Xu[1,2], Bin Li[1] and Ding Li[1]
[1]*College of Electronic and Optical Engineering & College of Microelectronics, Nanjing University of Posts and Telecommunications, Nanjing, Jiangsu Province, China* 2*National and Local Joint Engineering Laboratory of RF Integration and Micro-assembly Technology, Nanjing, Jiangsu Province, China*

Chapter IX – Design Automation of Circuit and Systems

Cellular Neural Network (CENN) FPGA Implementation using Multi-Level Optimization (9-5)   531
Zhongyang Liu[1], Shaoheng Luo[1], Xiaowei Xu[2] and Cheng Zhuo[1]
[1]*Zhejiang University, Hangzhou, China*
[2]*University of Notre Dame, Notre Dame, Indiana, USA*

Quadrature Amplitude Modulated Backscatter for 2.4 GHz Self-Powered Chips (9-16) 534

Luyao Zhang, Hao Zhang, Zhengkun Shen, Mingxiao He, Xiucheng Hao, Enbin Gong, Le Ye and Huailin Liao

Laboratory of Microelectronic Devices and Circuits (MOE), Institute of Microelectronics, Peking University Beijing, China

A Measurement Method for the Chip-to-Board Transmission of Radio Frequency Waves in the Power Distribution Networks with Simple Embedded Circuits (9-11) 537

Lan Luo, Dihu Chen and Tao Su

School of Electronics and Information Technology, Sun Yat-sen University, Guangzhou China

Novel Approaches to Circuit Timing (9-21)** 540

Ulf Schlichtmann

Chair of Electronic Design Automation Department of Electrical and Computer Engineering, Technical University of Munich, Munich, Germany

Electrical and Thermal Characterization of SiC Power MOSFET* (9-23) 544

Takashi Sato[1], Kazuki Oishi[1], Masayuki Hiromoto[1] and Michihiro Shintani[2]

[1]Graduate School of Informatics, Kyoto University, Kyoto, Japan
[2]Graduate School of Information Science, Nara Institute of Science and Technology, Nara, Japan

MTTF-Aware Design Methodology for Adaptive Voltage Scaling* (9-22) 548

Masanori Hashimoto and Yutaka Masuda

Dept. Information Systems Engineering, Osaka University, Suita, Japan

Analysis of Affecting Factors in Neutron Interactions with Gate Oxide in CMOS Transistors (9-31) 552

C.-Z. Chena[1] and David Y. Hu[2]

[1]Qualchip Technologies, Inc., Wuxi, Jiangsu, China
[2]MetroSilicon Microsystems, Kunshan, Jiangsu, China

An Improved S-Box of Lightweight Block Cipher Roadrunner for Hardware Optimization (9-4) 555

Juhua Liu[1], Wei Li[2] and Guoqiang Bai[3]

[1]Institute of Microelectronics, Tsinghua University, Beijing, China
[2]Institute of Microelectronics, Tsinghua University, Beijing, China
[3]Tsinghua National Laboratory for Information Science and Technology, Beijing, China

Ceramic Interconnect Bridge for Heterogeneous Integration and Multiple Chip Packaging* (9-30) 559

Boping Wu,

Huawei Research, Beijing, China

Auto-Generation of Pipelined Hardware Designs for Polar Encoder* (9-35) 562

Zhiwei Zhong[1,2], Xiaohu You[2] and Chuan Zhang[1,2]

[1]Lab of Efficient Architectures for Digital-communication and Signal-processing (LEADS)
[2]National Mobile Communications Research Laboratory, Southeast University, Nanjing, China

An Accelerator-Aware Microarchitecture Simulator for Design Space Exploration* (9-24)   566
Di Gao and Cheng Zhuo
Key Lab. of Advanced Micro/Nano Electric Devices & Smart Systems of Zhejiang College of Information Science & Electronics engineering, Zhejiang University, Hangzhou, China

High-Speed Implementation of SM2 Based on Fast Modulus Inverse Algorithm (9-17)   570
Wei Li[1], Juhua Liu[1] and Guoqiang Bai[2]
[1]*Institute of Microelectronics, Tsinghua University, Beijing, China*
[2]*National Laboratory for Information Science and Technology, Tsinghua University, Beijing, China*

Technology Mediated Tutorial on RISC-V CPU Core Implementation and Sign-Off using Revolutionary EDA Management System (EMS) - VSDFLOW (9-32)   573
Kunal Promode Ghosh and Anagha K Ghosh
VLSI System Design Corporation Pvt. Ltd., Bangalore, India

An Improved Leakage-Driven Runtime Decap Modulation Algorithm for Microprocessors (9-15)   576
Leilei Wang and Pingqiang Zhou
School of Information Science and Technology, ShanghaiTech University, Shanghai, China

Implementation of Heart Rate Detection Algorithm Based on a Low Power Chip (9-2)   579
Yingying Wu, Hao Chen, Zhongmin Lin, Zirong Chen and Xin-an Wang
The Key Lab of Integrated Microsystems Peking University Shenzhen Graduate School, Shenzhen, China

Design And Implementation Of A Low-Complexity R-peak Detection Algorithm (9-3)   582
Zirong Chen, Zhongmin Lin, Hao Chen, Xin-an Wang and Yingying Wu
The Key Lab of Integrated Microsystems Peking University Shenzhen Graduate School, Shenzhen, China

An Application-Learnable Neuromorphic Frequency Synthesizer with Random Vector Neural Network (9-7)   585
Lizhao Gao, Tingbing Ouyang, Chao Zhang, Jiangtao Gu and Bo Wang
Key Lab of IMS, School of ECE, Peking University Shenzhen Graduate School Shenzhen, China

Empowering Edge Mining o Smartphones with Reconfigurable Fabrics (9-8)   588
Zeyu Yan, Xiaowei Xu, Guangyu Yu and Hu Yu
School of Optical and Electronic Information, HUST, Wuhan, China

Accelerating Earth Movers Distance with Instruction Set Extension for Image Retrieval (9-9)   591
Guangyu Yu, Xiaowei Xu, Zeyu Yan and Hu Yu
School of Optical and Electronic Information, HUST, Wuhan, China

A Universal Implementation of Cardiovascular Disease Surveillance Based on HRV (9-10)   594
Yufan Feng, Ying Zhang, Xiaole Cui and Xin-an Wang
The Key Lab of Integrated Microsystems Peking University Shenzhen Graduate School, Shenzhen, China

A Low-Offset Current-Mode CMOS Vertical Hall Sensor Microsystem with Four-   597

Phase Spinning Current Technique (9-12)
Xingxing Hu[1], Yue Xu[1,2] and Jun Xu[1]
[1]*College of electronic and optical engineering & College of microelectronics in Nanjing University of Posts and Telecommunications, Nanjing, Jiangsu Province, China*
[2]*National and Local Joint Engineering Laboratory of RF Integration and Micro-assembly Technology, Nanjing, China*

A Novel Gas Sensor Signal Drift Adjustment Method Based on Controlled Measurement (9-18)
Chang-Yong Chiu[1] and Zhun Zhang[2]
[1]*School of Communication & Information Engineering, Shanghai University, Shanghai, China*
[2]*College of Optoelectronic Engineering, Shenzhen University, Shenzhen, China*

China Semiconductor Technology International Conference 2018 (CSTIC 2018)

Editors:

Cor Claeys
KU Leuven
Leuven, Belgium

Ru Huang
Peking University
Beijing, China

Hanming Wu
Semiconductor Manufacturing International Corporation - SMIC
Shanghai, China

Qinghuang Lin
IBM Thomas J. Watson Research Center
Yorktown Heights, NY, USA

Steve Liang
Jiangsu Changjiang Electronics Technology Co. Ltd
Wuxi, China

Peilin Song
IBM Thomas J. Watson Research Center
Yorktown Heights, NY, USA

Zhen Guo
Intel
Santa Clara, USA

Kafai Lai
IBM Research Division
Hopewell Junction, NY, USA

Ying Zhang
Applied Materials
Santa Clara, CA, USA

Yuchun Wang
Anji Microelectronics
Shanghai, China

Yiyu Shi
Missouri University of Science and Technology
Columbia, MO, USA

Hsiang-Lang Lung
Macronix International, Ltd.
Hsinchu, Taiwan, China

PREFACE

This issue contains a selection of the accepted papers presented at *China Semiconductor Technology International Conference 2018 (CSTIC 2018)*, March 11-12, 2018 in Shanghai, China. The memory stick version of these papers serves as the CSTIC 2018 conference proceedings and is distributed to all registered conference attendees. After reviewing a selection of the presentations will be considered for publication in IEEE Xplore.

CSTIC is the largest and the most comprehensive annual industrial semiconductor technology conference in China. It aims to provide a platform for executives, managers, engineers and researchers from around the world to exchange the latest developments in semiconductor technology and manufacturing and related fields. It also offers an opportunity for those who are interested in investing and collaboration opportunities in the semiconductor industry in Asia, particularly in China.

CSTIC covers all the aspects of semiconductor technology and manufacturing, including circuit design, system integration, devices, materials, patterning (lithography and etching), processes, integration, testing, reliability, device physics and manufacturing as well as emerging semiconductor technologies, including clean energy such as light emitting diodes (LEDs), III-V semiconductors, sensors and micro-electromechanical systems (MEMS).

CSTIC 2018, organized by Semiconductor Equipment and Material International (SEMI), imec, and The Integrated circuit Materials Industry Technology Innovation Alliance (ICMTIA) and technically sponsored by the IEEE Electron Devices Society relies on a long time tradition, which started in 2001. The original International Semiconductor Technology Conference (ISTC) merged in 2009 to become CSTIC, aiming for a broad international representation and increased paper submissions from around the world. For CSTIC 2018 the papers came from all major semiconductor manufacturing regions in the world, including Austria, Belgium, Brazil, China, Germany, India, Japan, Korea, Malaysia, Poland, Saudi Arabia, Singapore, Switzerland, Taiwan, United Kingdom and the United States of America. About 231 papers have been selected for oral presentations and approximate 137 papers for poster presentations after careful reviews by the conference organizing committee.

In total 174 papers are included in these Proceedings after peer reviews. They represent a snapshot of the recent developments in semiconductor technology and manufacturing in the world. In particular, they offer a glimpse into the state-of-the-art of semiconductor technology and manufacturing in China. These papers are divided into nine (9) chapters according to the nine symposia of CSTIC 2018:

• Device Engineering and Memory Technology
• Lithography and Patterning
• Dry & Wet Etch and Cleaning
• Thin Film, Plating and Process Integration
• Chemical-Mechanical Polishing (CMP) and Post-Polish Cleaning
• Metrology, Reliability and Testing
• Packaging and Assembly
• MEMS, Sensors and Emerging Semiconductor Technologies
• Design and Automation of Circuits and Systems

These Proceedings are very valuable to engineers and researchers in the fast-moving and growing semiconductor industry. It will give readers a clear understanding of the status of semiconductor technology and manufacturing in China. Furthermore it will also serve as a useful reference for those who are interested in nanofabrication, micro- and nano-fluidics, micro- and nano-photonics, organic electronics, bio-chips, light emitting diodes (LEDs) and other clean energy technologies.

We thank the invited speakers and the authors, particularly the conference plenary speakers, Chenming Hu Professor at Berkeley University, USA, Dr. PR (Chidi) Chidambaram, Vice President QCT Process Technology & Foundry Engineering Qualcomm Inc, USA, Dr. Kevin Zhang, Vice-President Foundry Business Development, TSMC, Taiwan and Dr. Zhiyong Ma, Vice President Technology and Manufacturing Intel, USA, for their valuable contributions to CSTIC 2018. We also thank the more than 120 organizing committee members, particularly the symposium chairs, for their dedication and hard work to help improve the quality and to broaden the reach of CSTIC. These committee members are experts in their respective fields of semiconductor technology and are from well-known companies or prestigious institutions. They all have demanding day jobs, yet they have volunteered to help organizing this conference and to critically review papers presented in these Proceedings. Their contributions were crucial for the success of the conference. We are also indebted to the financial support from the sponsors of CSTIC 2018. Finally, we extend our sincere thanks to SEMI for their tireless efforts and their meticulous organizational skills to help organize CSTIC 2018 and to assemble and publish these CSTIC 2018 proceedings.

Ru Huang, General Chair CSTIC 2018
Peking University, Beijing, China

Cor Claeys, Co-Chair, CSTIC 2018
KU Leuven, Leuven, Belgium

CSTIC 2018 Organizing Committee

10 March 2018, Shanghai, China

ACTIVE-PERFORMANCE BENCHMARK FOR ADVANCED 3D-CMOS DEVICES

Hitoshi Wakabayashi, Eisuke Anju and Iriya Muneta

School of Engineering, Tokyo Institute of Technology, Yokohama 226-8502, Japan

E-mail: wakabayashi.h.ab[at]m.titech.ac.jp

ABSTRACT

Active performances are discussed by using benchmark for advanced 3D-CMOS devices. FinFETs as the 3D-CMOS devices have simultaneously improved the indexes of speed and active power, because of not only a gate length scaling and velocity enhancement but also a W_{eff} / $W_{footprint}$ enhancement to drive both gate and parasitic capacitances. To further enhance the performances, it is required to reduce the supply voltage with improving those factors.

INTRODUCTION

CMOS devices for advanced logic LSIs in could/fog and IoT-edge applications have been being scaled using not only a dimension scaling and but also higher velocity, thin equivalent oxide thickness (EOT) and large effective device width per footprint (W_{eff} / $W_{footprint}$) [1-35]. Especially, in order to enhance the W_{eff} / $W_{footprint}$, a FinFET technology has been applied for advanced 3D-CMOS devices [24-35] to drive both gate and parasitic capacitances. However, it tends to be difficult to reduce an active power of device itself per footprint, due to the variability due to performance degradation through an effective device width. In order to evaluate this difficulty, a performance (speed), power, area, cost and temperature (PPACT) have been discussed [36,37].

In this paper, the speed and active power are discussed for advanced 3D-CMOS devices using benchmark data.

BENCHMARK FOR SCALING

Prior to a discussion on the speed and active power, a scaling behavior of technology node is shown in Figure 1 [1-35]. One dimensional (1D) scaling along with x 0.7 per two years is being continued even below 10 nm, as compared to the ITRS table along with x 0.7 per three years. Furthermore, 7-nm technology node has been published earlier than our prediction [32,33]. In the future in 2030, the technology node is going to be 1-nm technology node, which corresponds to the almost just two sets of a lattice constant in silicon. As everybody understands, the technology node has no meaning in technical consideration. Therefore a cell size of 6T-SRAM depending on the years is plotted in Figure 2, to discuss on a physical 2D area scaling [1-35]. The scaling has been being scaled along with 0.7^2 - 0.8^2 per two years. In the future in 2030, the cell area of 6T-SRAM is going to be scaled as approximately

10^{-3} μm^2, which is almost the same as the size of ribosome.

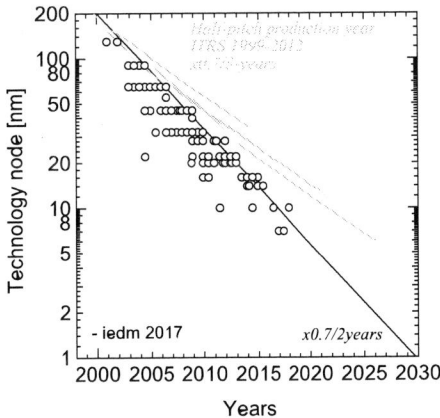

Figure 1: Technology-node dependence on published years for advanced CMOS devices

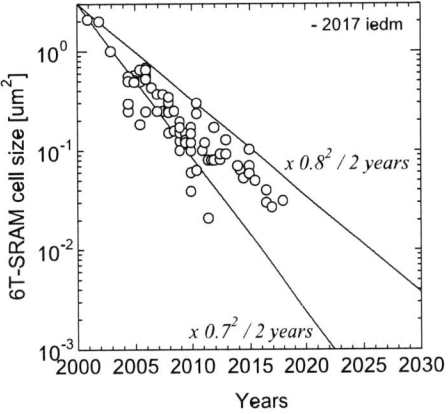

Figure 2: 6T-SRAM cell size dependence on published years

In advanced CMOS devices employing three-dimensional (3D) structure such as a FinFET, a value of W_{eff} / $W_{footprint}$ is enhanced to improve a drivability of transistor per footprint area. Estimated W_{eff} / $W_{footprint}$

dependence on the years is shown in Figure 3. This value increases with an increase in the years along with 1/0.7 per 2 years to enhance the drivability per footprint area to drive both gate and parasitic capacitances. To further enhance the value of $W_{eff}/W_{footprint}$, the structures of lateral stacked-nanowires and monolithic transistors are promising candidates [38-41].

dependence on the power density is shown in Figure 4. The FinFET devices among 22-10-nm technology-nodes have achieved a valuable progress because of gate length scaling and higher velocity caused by FinFET technology [27,28,30,33,34].

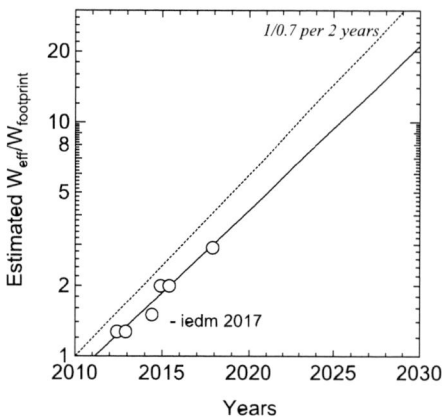

Figure 3: Estimated W_eff/W_footprint dependence on years for 3D-CMOS devices such as FinFET

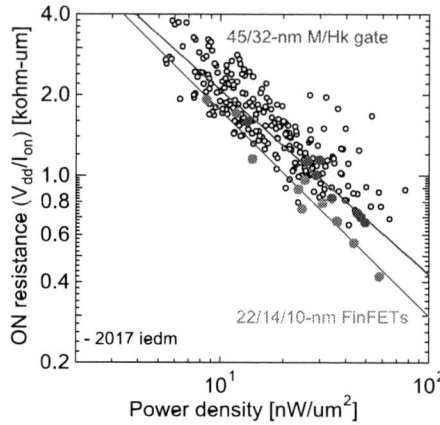

Figure 4: Active-performance benchmark between ON resistance and power density for 3D-CMOS devices

ACTIVE-PERFORMANCE BENCHMARK

In order to evaluate the active performance of CMOS devices, the speed and active power are estimated as follows. Firstly, to estimate the speed, a delay time of 3D CMOS devices were calculated by using

$$\tau_g = (C_i L_g W_{eff}/W_{footprint} + C_{para}) V_{dd}/I_{on} \qquad (1).$$

Here, τ_g, C_i, L_g, W_{eff}, $W_{footprint}$, C_{para}, V_{dd} and I_{on} are respectively the delay time of transistor, insulator capacitance per footprint area, effective device width, footprint device width, parasitic capacitance per footprint device width, supply voltage and on current per footprint device width, which are calculated as a harmonic mean. In the nanometer scaled 3D-devices, it is reasonable to consider on the relationship of $C_i L_g W_{eff}/W_{footprint} << C_{para}$. Therefore in this paper, ON resistance of V_{dd}/I_{on} is discussed as an index of the speed.

Secondly, an active power of transistor per footprint area, P_{active}, is considered to be

$$P_{active} = 1/\tau_g (C_i L_g W_{eff}/W_{footprint} + C_{para}) V_{dd}^2/L_g \qquad (2).$$

In this paper, a power density per footprint area of $V_{dd} I_{on}/L_g$ is used as an index of active power. Eventually as an active-performance benchmark, the ON resistance

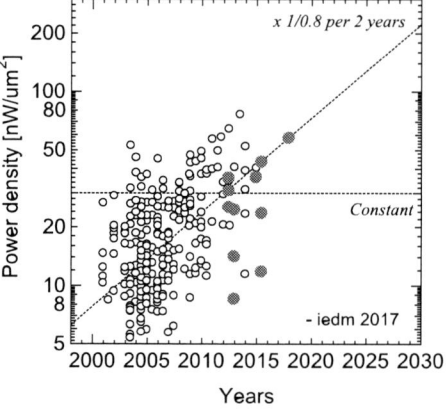

Figure 5: Power-density dependence on years for 3D-CMOS devices

However due to the difficulty to reduce the supply voltage, the power density per footprint area is concerned along with scaling, as shown in Figure 5. In this figure, the distribution of vertical axis is mainly caused by the variation of multi threshold voltages. If we see the trend of behavior for especially red dots, the power densities gradually increase along with 1/0.8 per two years, approximately, even a constant behavior is expected. This

978-1-5386-5309-8/18 $31.00 © 2018 IEEE

might be due to the difficulty to reduce the supply voltage maintaining the on current [42].

NEAR-FUTURE DIRECTION

In order to further reduce the supply voltage with higher velocity, the technologies such as small band-gap channel material and other operation principles are strongly required [43]. For example of III-V nanowire, ON resistance dependence on power density is shown in Figure 6 [44]. An InGaAs nanowire is superior to the FinFETs, however it is still concerned on a yield of products, due to higher dislocation densities with approximately 10^5 cm^{-2} in the channel on silicon substrate. Furthermore even in III-V channels, a self-heating is one of the most important issue in the 3D-CMOS devices, such as slim Fin, stacked nanowires, and monolithic devices [45,46].

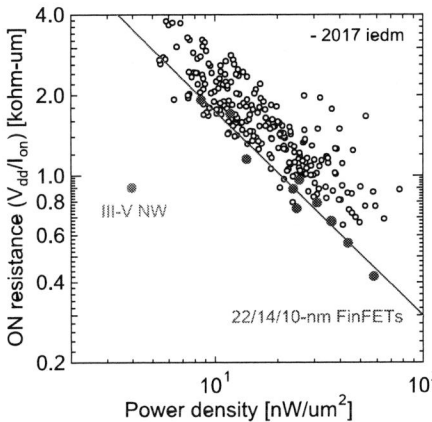

Figure 6: On resistance dependence on power density including 3D-III-V nMOSFETs

CONCLUSIONS

The benchmarks of speed and active power were discussed for advanced 3D-CMOS devices, as the discussion on active performance. For further progress, the reduction of supply voltage is strongly required simultaneously with both the enhancement of velocity and less self-heating effect.

REFERENCES

[1] Gordon E. Moore: Electronics, 38, No. 8, April 19, p. 114, 1965.

[2] Robert H. Dennard, et al.: in IEEE J. Solid-State Circuits, SC-9, p. 256, 1974.

[3] S. M. Sze: Physics of Semiconductor Devices, 2nd Edition, John Wiley & Sons, 1981.

[4] Hon-Sum Philip Wong, et al.: in Proceedings of the

IEEE, 87, No. 4, p. 537, 1999.

[5] Dimitri A. Antoniadis, et al.: "MOSFET performance scaling: Limitations and future options," IEDM, p. 253, 2008.

[6] Yuan Taur and Tak H. Ning: Fundamentals of Modern VLSI Devices, Second Edition, Cambridge University Press, 2009.

[7] http://www.itrs.net/

[8] Hitoshi Wakabayashi, et al.: "A Dual-Metal Gate CMOS Technology Using Nitrogen-Concentration-Controlled TiNx Film, IEEE T-ED, VOL. 48, NO. 10, OCTOBER 2001, p. 2363.

[9] Hitoshi Wakabayashi, et al.: "Characteristics and Modeling of Sub-10-nm Planar Bulk CMOS Devices Fabricated by Lateral Source/Drain Junction Control," IEEE T-ED, VOL. 53, NO. 9, SEPTEMBER 2006, p. 1961.

[10] Y. Tateshita, et al.: "High-Performance and Low-Power CMOS Device Technologies Featuring Metal/High-k Gate Stacks with Uniaxial Strained Silicon Channels on (100) and (110) Substrates," IEDM, p. 63, 2006.

[11] K. Mistry, et al.: "A 45nm Logic Technology with High-k+Metal Gate Transistors, Strained Silicon, 9 Cu Interconnect Layers, 193nm Dry Patterning, and 100% Pb-free Packaging," IEDM, p. 247, 2007.

[12] C. Auth, et al.: "45nm High-k + Metal Gate Strain-Enhanced Transistors," Symposium on VLSI Technology, p. 128, 2008.

[13] S. Natarajan, et al.: "A 32nm logic technology featuring 2nd-generation high-k + metal-gate transistors, enhanced channel strain and 0.171 μm^2 SRAM cell size in a 291Mb array," IEDM, p. 941, 2008.

[14] C.-H. Jan, et al.: "A 45nm low power system-on-chip technology with dual gate (logic and I/O) high-k/metal gate strained silicon transistors," IEDM, p. 637, 2008.

[15] Mark Bohr: "The New Era of Scaling in an SoC World," ISSCC, p. 23, 2009.

[16] Satoru Mayuzumi, Hitoshi Wakabayashi, et al.: "High-Performance Metal/High-k n- and p-MOSFETs With Top-Cut Dual Stress Liners Using Gate-Last Damascene Process on (100) Substrates," IEEE T-ED, VOL. 56, NO. 4, p. 620, APRIL 2009.

[17] Satoru Mayuzumi, Hitoshi Wakabayashi, et al.: "Channel-Stress Enhancement Characteristics for Scaled pMOSFETs by Using Damascene Gate With Top-Cut Compressive Stress Liner and eSiGe," IEEE T-ED, VOL. 56, NO. 11, NOVEMBER 2009 , p. 2778.

[18] C.-H. Jan, et al.: "A 32nm SoC platform technology with 2nd generation high-k/metal gate transistors optimized for ultra low power, high performance, and high density product applications," IEDM, p.

978-1-5386-5309-8/18 $31.00 © 2018 IEEE

647, 2009.

[19] P. Packan, *et al.*: "High performance 32nm logic technology featuring 2nd generation high-k + metal gate transistors," IEDM, p. 659, 2009.

[20] Satoru Mayuzumi, Hitoshi Wakabayashi, *et al.*: "Effects of Channel Width on Stress Enhancement for Damascene-Gate nFETs With Top-Cut Tensile-Stress Liner," IEEE EDL, VOL. 31, NO. 1, JANUARY 2010, p. 65.

[21] Satoru Mayuzumi, Hitoshi Wakabayashi, *et al.*: "Mobility and Velocity Enhancement Effects of High Uniaxial Stress on Si (100) and (110) Substrates for Short-Channel pFETs," IEEE T-ED, VOL. 57, NO. 6, JUNE 2010 p. 1295.

[22] C.-H. Jan, *et al.*: "RF CMOS Technology Scaling in High-k/Metal Gate Era for RF SoC (System-on-Chip) Applications," IEDM, p. 604, 2010.

[23] Hitoshi Wakabayashi, *et al.*: "Progress and prospects of silicon transistors based on junction technologies," 2013 International Workshop on Junction Technology (IWJT), pp. 98-103.

[24] Digh Hisamoto, *et al.*: "A fully depleted lean-channel transistor (DELTA)-a novel vertical ultra thin SOI MOSFET," IEDM, p. 833, 1989.

[25] Digh Hisamoto, *et al.*: "FinFET - A Self-Aligned Double-Gate MOSFET Scalable to 20 nm," IEEE Transactions on Electron Devices, 47, No. 12, p. 2320, 2000.

[26] Xuejue Huang et al.: "Sub-50 nm P-channel FinFET," IEEE Transactions on Electron Devices, 48, No. 5, p. 880, 2001.

[27] C. Auth, *et al.*: "A 22nm high performance and low-power CMOS technology featuring fully-depleted tri-gate transistors, self-aligned contacts and high density MIM capacitors," Symp. on VLSI Tech., p. 131, 2012.

[28] C.-H. Jan, *et al.*: "A 22nm SoC platform technology featuring 3-D tri-gate and high-k/metal gate, optimized for ultra low power, high performance and high density SoC applications," IEDM, p. 44, 2012.

[29] S. Natarajan, *et al.*: "A 14nm Logic Technology Featuring 2nd-Generation FinFET Transistors, Air-Gapped Interconnects, Self-Aligned Double Patterning and a 0.0588 μm^2 SRAM cell size," IEDM, p. 941, 2014.

[30] C. Auth, *et al.*: "A 10nm High Performance and Low-Power CMOS Technology Featuring 3rd Generation FinFET Transistors, Self-Aligned Quad Patterning, Contact over Active Gate and Cobalt Local Interconnects," IEDM, p. 673, 2017.

[31] B. Sell, *et al.*: "22FFL: A High Performance and Ultra Low Power FinFET Technology for Mobile and RF Applications," IEDM, p. 685, 2017.

[32] T. Song, *et al.*, "A 7nm FinFET SRAM Macro Using EUV Lithography for Peripheral Repair Analysis,"

ISSCC 2017, 12.2.

[33] Daewon Ha, *et al.*, "Highly Manufacturable 7nm FinFET Technology featuring EUV lithography for Low Power and High Performance Applications," 2017 Symposium on VLSI Technology, T6-1.

[34] Shien-Yang Wu, *et al.*, "Demonstration of a sub-0.03 μm^2 High Density 6-T SRAM with Scaled Bulk FinFETs for Mobile SOC Applications Beyond 10nm Node," 2016 Symposium on VLSI Technology, p. 978.

[35] Cliff Hou, *et al.*, "A Smart Design Paradigm for Smart Chips," ISSCC 2017, 1.1.

[36] https://irds.ieee.org

[37] https://eps.ieee.org/technology/heterogeneous-integration-roadmap.html

[38] J.-P. Colinge, *et al.*: "Silicon-on-insulator 'gate-all-around' MOS device," in IEDM, p. 137, 1990.

[39] N. Loubet, *et al.*, "Stacked Nanosheet Gate-All-Around Transistor to Enable Scaling Beyond FinFET," 2017 Symposium on VLSI Technology, T17-5, T230.

[40] S. Barraud, T. Ernst, *et al.*, "Performance and Design Considerations for Gate-All-Around Stacked-NanoWires FETs," IEDM, p. 677, 2017.

[41] N. Macha, *et al.*, "Ultra high density 3D SRAM cell design in Stacked Horizontal Nanowire (SN3D) fabric,"IEEE/ACM Int'l Symp. On Nanoscale Architectures, pp. 155-161, 2017.

[42] Hitoshi Wakabayashi, Dimitri A. Antoniadis, *et al.*: "Supply-Voltage Optimization for Below-70-nm Technology-Node MOSFETs," IEEE Transactions on Semiconductor Manufacturing, Vol. 15, No. 2, MAY 2002, p. 151.

[43] Ian A. Young, *et al.*, "Principles and trends in quantum nano-electronics and nano-magnetics for beyond-CMOS computing," 2017 47th European Solid-State Device Research Conference (ESSDERC), pp. 1 - 5.

[44] Cezar B. Zota, *et al.*, "InGaAs Nanowire MOSFETs with I_{ON} = 555 $\mu A/\mu m$ at I_{OFF} = 100 nA/μm and V_{DD} = 0.5 V," 2016 Symposium on VLSI Technology, p. 162.

[45] Eisuke Anju, Iriya Muneta, Hitoshi Wakabayashi, *et al.*, "Self-Heating-Effect-Free p/n-Stacked-NW on Bulk-FinFETs and 6T-SRAM Layout," IEEE/EDS, Electron Devices Technology and Manufacturing (ETDM) Conference 2018, to be presented.

[46] Kentaro Matsuura, Iriya Muneta, Hitoshi Wakabayashi, *et al.*, "Chip-Level-Integrated nMISFETs with Sputter-Deposited-MoS$_2$ Thin Channel Passivated by Al$_2$O$_3$ Film and TiN Top Gate," ETDM 2018, to be presented.

978-1-5386-5309-8/18 $31.00 © 2018 IEEE

Introduction of 95nm SPOCULL Technology

Honglin Zeng[1,], Liang Wu[1], Tianhua Dong[1], Lan Jin[1], Yulei Jiang[1], Yi Peng Chan[1], Leong Tee Koh[1], Jui Lin Lu, Min-Hwa Chi[1], Tzu-Yin Chiu*

[1]Semiconductor Manufacturing International Corporation (SMIC)

Shanghai, 201203, China

*Corresponding author, Honglin_Zeng@smics.com

ABSTRACT

SPOCULL is an innovative technology successfully developed by adding a 2nd poly (Poly2, or P2) layer into a mature logic technology as local interconnection (M0), not only the chip area can be reduced significantly but also transistor performance is enhanced. The SPOCULL technology is successfully implemented on 0.13μm CMOS node with logic gate density (~2x than standard 0.13μm LL technology) and small SRAM cell (size ~0.739μm²); this is equivalent to typically 95nm node (thus referred to as 95SPOCULL). Furthermore, the parasitic capacitance and device leakage of transistor is also reduced significantly, e.g. ~30% leakage reduction with same speed of logic standard cells. Also, the P2 layer enables the one-time-programmable (OTP) and multiple-time-programmable (MTP) capability as embedded into the technology platform for MCU. The SPOCULL can also be easily extended to 0.18μm node similarly.

Keywords— SPOCULL, gate density, low leakage, interconnect, ULP

INTRODUCTION

CMOS devices have been scaled for more than 40 years follow Moore's law in order to achieve higher density, high performance, and lower power consumption. Developing more–than-Moore (MtM) technology is a great opportunity for 8-inch fab to face the advanced technology challenges. [1-3]. Currently, MtM focuses on CMOS logic platform integrated with diversified functions, e.g. Analog/Mixed-signal, RF, sensors, 3D IC, power control etc. [4-6]

Per Chiu,et.al.'s innovation [7-10] based on a self-aligned poly-silicon electrode technology [11-12], an enhanced logic platform technology is developed, This new platform technology not only reduces the contact size and parasitic capacitance of source/drain, but also results in shallower junction with lower junction leakage. The advantages of self-aligned poly-silicon appear more significant at advanced CMOS node (e.g. 0.13μm) for future internet of things (IoT) applications.

SPOCULL technology (Figure 1) is optimized based on self-aligned poly-silicon electrode technology with significant advantages by adding a 2nd poly-Si (highly-doped) layer. It serves as local interconnect (referred to as M0) for achieving smaller contact (CT),

also results in formation of shallower junction with significantly lower leakage current by a few x. There are more device features demonstrated, e.g. with the 2nd poly, it is easily implementing a double-poly non-volatile memory (NVM), poly-insulator-poly (PIP) capacitor, and also the "field plate" for high voltage transistors.

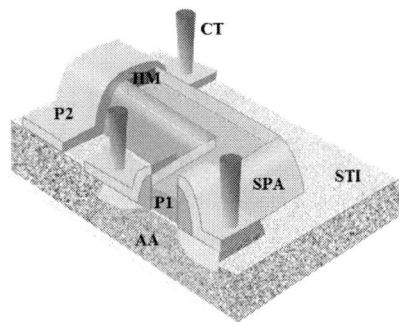

Figure 1: The P2 (M0) layer serves as local interconnect.

SPOCULL technology is currently demonstrated on several CMOS nodes, such as 95SPOCULL based on 0.13μm LL (0.13LL), 60nm SPOCULL (on 90nm node), and 120nm SPOCULL (on 0.18μm) as platform for wide applications. There are high voltage (HV) and ultra-low power (ULP) versions implemented on 95SPOCULL (as referred to as 95HV and 95ULP). Single device in 95SPOCULL can shrink ~50% with the contact (CT) to source/drain (S/D) connection (by using the highly doped P2 local interconnect). 95SPOCULL also results in the lowest leakage comparing with similar technology nodes. Further, 95SPOCULL can result in ~2x gate density than 0.13 LL (with same gate length), i.e. STD cells size reduced ~50% (with comparable performance). The SRAM bit-cell is small (0.739μm²) with ultra-low leakage (<1pA/cell); it is probably the smallest among available technologies available in 8-inch foundry. The introduction of highly doped M0 can provide embedded poly-insulator-poly (PIP) capacitor as well as double-poly non-volatile memory (eNVM) for embedded applications for MCU. Currently, there are quite a few new tape out (NTO) in Fab and one in risk production.

95SPOCULL ARCHITECTURE

Device Architecture

The introduction of highly doped poly-silicon layer

978-1-5386-5309-8/18 $31.00 © 2018 IEEE

can provide more features in device structure, e.g. the connection of Gate and S/D can be readily through the local interconnect without narrow down process windows. As shown in Figure 2, the device parameters (e.g. SA/SB, i.e. the distance from Gate to STI edge) can be reduced significantly due to the local interconnection (M0) and the CT landing partially on M0 atop STI area. Thus, the transistor dimension based on 95SPOCULL technology is mainly dominated by the gate length and SA/SB parameters; and the transistor size (also SRAM cell and logic standard cells) can be possibly reduced greatly by ~50% than 0.13 LL.

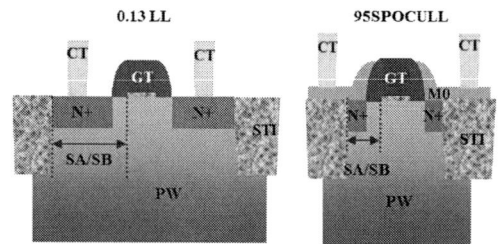

Figure 2: The innovative device architecture of 95SPOCULL and comparison with traditional 0.13 LL.

Local Interconnection

The local interconnection based on 95SPOCULL platform has additional features. As in Figure 3, typical inverter needs to connect the S/D of N/PMOS through CT and M1; however, the M0 readily available on 95SPOCULL can connect the S/D directly as a flexible and convenient metal routing scheme. The optimized local interconnection on 95SPOCULL can reduce the chip size significantly and reduce the amount of contacts and metal layers with enhanced yield.

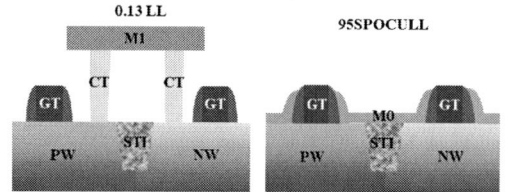

Figure 3: Comparison of 95SPOCULL platform (with local interconnection M0) vs traditional 0.13μm technology (no M0).

Standard Cells and SRAM Bit-Cell

With the M0 local interconnection and device architecture, 95SPOCULL can achieve gate density ~2X than 0.13LL (at same gate length), and as in Figure 4, the gate density about same as the conventional 90nm node. Further, 95SPOCULL can achieve the smallest SRAM bit cell ($0.739\mu m^2$) in 8-inch fab with M0 interconnection. It equals to conventional 80nm node as in Figure. 5. These STD cells, SRAM bit-cell will result in smaller circuit with good performance and chip size.

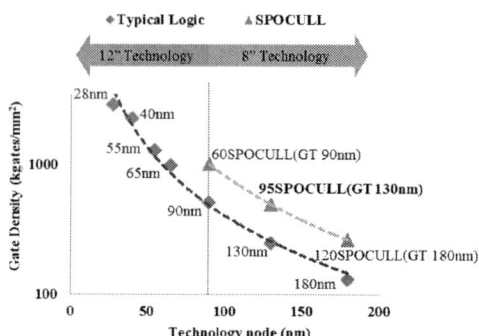

Figure 4: Industry benchmark of gate density of typical logic technology node and 95SPOCULL

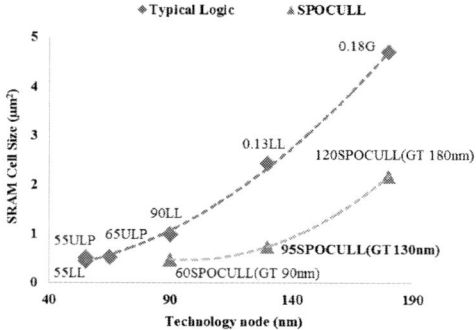

Figure 5: Industry benchmark of SRAM bit cell of typical logic technology node and 95SPOCULL

95SPOCULL PERMANCE

Device Leakage Reduction

95SPOCULL include two separate versions, i.e. 95HV and 95ULP. As in Figure 6, the leakage performance ($<1pA/\mu m$) on 95SPOCULL is the lowest among 0.13LL, 0.13μm LG (0.13LG), and 55nm ULP (55ULP), and is highly preferable for IoT applications. Power consumption of 95ULP is about 30%~40% of 0.13LL.

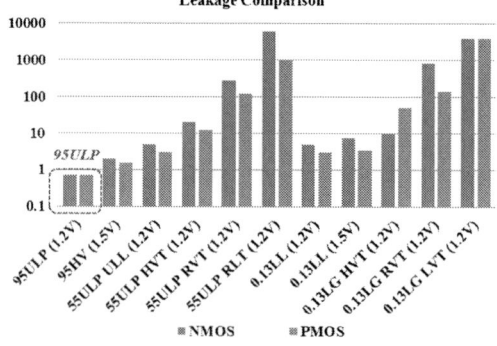

Figure 6: Device leakage comparison among 55ULP,

978-1-5386-5309-8/18 $31.00 © 2018 IEEE

0.13LL, 0.13LG, and 95ULP (the smallest).

Standard Cell Speed

As in Figure 7, the standard (STD) cells based on 95SPOCULL have comparable performance with 0.13LL, but only ~50% chip size for typical STD cells. Moreover, we illustrate the speed and leakage performance of three typical STD cells. 95SPOCULL cell leakage is about 60% lower than 0.13LL with comparable speed performance.

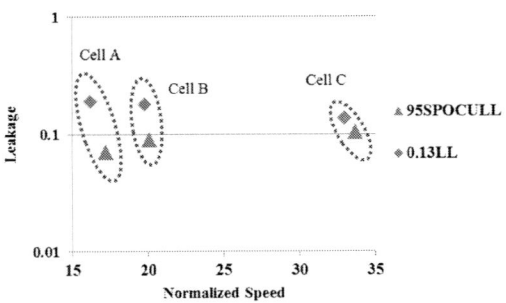

Figure 7: Speed v.s. leakage performance comparason

SRAM Performance

SRAM bit-cell on 95SPOCULL platform is also developed from 0.13LL with SPOCULL device architecture. As in Figure 8, SRAM cells performance of Iread v.s. Istandby is compared among 95SPOCULL, 0.13LG, 0.13LL, 90nm LL (90LL), 65nm ULP (65ULP), 65nm LL (65LL), 55nm LL (55LL), and 55nm ULP (55ULP). The SRAM cells based on 95ULP and 95HV can result in competitive SRAM cell size but also the lowest leakage (95ULP).

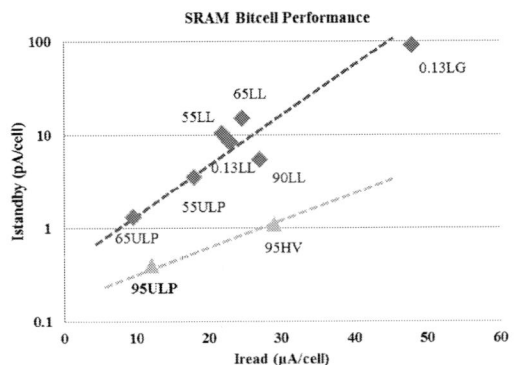

Figure 8: SRAM bitcell Iread v.s. Istandby performance comparason for 95SPOCULL, 0.13LG, 0.13LL, 90LL, 65ULP, 65LL, 55LL, and 55ULP.

Taking SRAM chip as an example as shown in Figure 9, chip performance are comparable between 95ULP SRAM and 0.11μm LL (0.11LL) SRAM, however 95ULP SRAM chip leakage is about one order

lower, Moreover, the leakage performance is even beffer at high temperature (85 ℃) due to the benefit of SPOCULL shallower junction. Besides, the chip size of 95ULP is 40%~50% smaller.

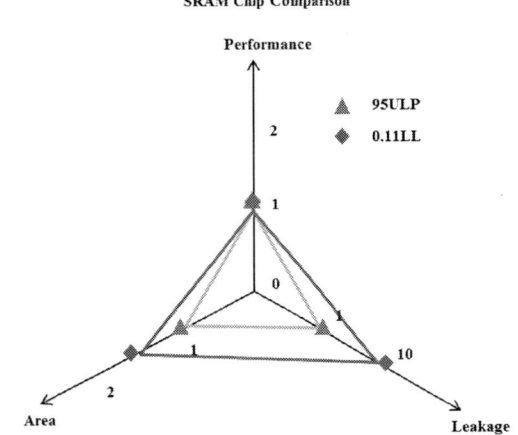

Figure 9: normalized performance, leakage, and area of 95ULP SRAM comparison with 0.11LL SRAM

Build-in MTP

The highly doped M0 on 95SPOCULL can also provide poly-insulator-poly (PIP) capacitor. Furthermore, the M0 (highly doped poly2) can be used as the control gate (CG) of multiple time programmable (MTP) embedded non-volatile-memory (eNVM). As in Figure 10, cross-section of SPOCULL and its process flow is compatible with basic SPOCULL logic flow. The functional MTP illustrated here (a p-type cell) can provide enhanced capability of NVM.

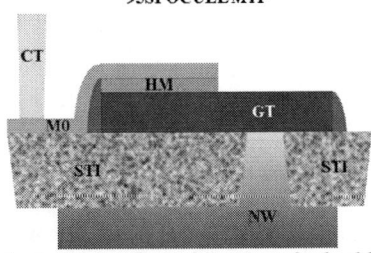

Figure 10: A sketch of cross-section of a build-in p-type MTP cell structure on 95SPOCULL.

The p-type MTP based on 95ULP has competitive bit-cell size using SPOCULL device architecture, but also show good performance of program (PGM) and erase (ERS). The data retention of the p-type MTP on 95ULP can meet stringent requirement. As shown in Figure 11a, cell Vth margin is enough (window >4.5V) after 50K cycling. The bit distribution vs Vcg bias of the p-type

978-1-5386-5309-8/18 $31.00 © 2018 IEEE

MTP also show good performance even after 10K cycling as in Figure 11b. data retention is acceptable after high temperature (150C, 168h) bake. Meanwhile, p-type MTP on 95ULP has good margin after PGM disturb. The error correction code (ECC) function can be added to further enhance the yield and reliability performance of MTP. An n-type MTP cell is also possible and performed similarly, but not illustrated here.

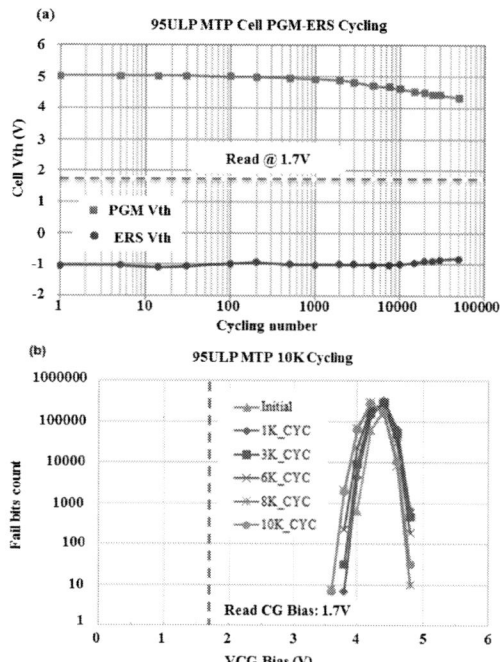

Figure 11: (a) MTP cell on 95ULP cycling time with PGM-ERS Vth;(b) MTP bit distribution vs Vcg bias after different cycling.

CONCLUSIONS

95SPOCULL is successfully developed based on 0.13LL with high voltage (HV) and ultra-low power (ULP) versions. The size of single device can be reduced ~50% related to contact to source/drain (S/D) connection through the additional highly doped P2 (as local interconnection M0). The leakage current is also reduced to lowest among adjacent conventional technology nodes. 95SPOCULL can double the logic gate density (~2x) vs 0.13LL (at same gate length). STD cell size is also reduced ~50% with comparable speed. The SRAM cell size ($0.739\mu m^2$) is among the smallest in nodes in 8-inch foundry with ultra-low leakage (<1pA/cell) as well. SRAM chip level leakage of 95ULP is one order less than 0.11LL with same performance, and 95ULP leakage performance is much beffer than 0.11LL at high temperature due to shallower junction. Meanwhile SRAM chip size of 95ULP is also 40%~50% smaller than 0.11LL.

The highly doped Poly2 (or M0) can provide PIP capacitor as well as double-poly MTP with good Program/Erase and data retention. Also, ECC function can provide additional MTP reliability. Currently there is one product in risk production and multiple new product tape-out (NTO's).

REFERENCES
[1] R. R. Schaller, "Moore's law: past, present and future," *IEEE spectrum,* vol. 34, pp. 52-59, 1997.
[2] S. E. Thompson and S. Parthasarathy, "Moore's law: the future of Si microelectronics," *Materials today,* vol. 9, pp. 20-25, 2006.
[3] R. G. Dreslinski, M. Wieckowski, D. Blaauw, D. Sylvester, and T. Mudge, "Near-threshold computing: Reclaiming moore's law through energy efficient integrated circuits," *Proceedings of the IEEE,* vol. 98, pp. 253-266, 2010.
[4] M. M. Waldrop, "More than moore," *Nature,* vol. 530, pp. 144-148, 2016.
[5] L. Clavelier, C. Deguet, L. Di Cioccio, E. Augendre, A. Brugere, P. Gueguen*, et al.*, "Engineered substrates for future more moore and more than moore integrated devices," in *Electron Devices Meeting (IEDM), 2010 IEEE International*, 2010, pp. 2.6. 1-2.6. 4.
[6] W. Arden, M. Brillouët, P. Cogez, M. Graef, B. Huizing, and R. Mahnkopf, "More-than-Moore white paper," *Version,* vol. 2, p. 14, 2010.
[7] T.-Y. Chiu, G. M. Chin, R. C. Hanson, M. Y. Lau, K. F. Lee, M. D. Morris*, et al.*, "Process for manufacturing semiconductor BICMOS device," ed: Google Patents, 1989.
[8] G. M. Chin, T.-Y. Chiu, T.-Y. M. Liu, and A. M. Voshchenkov, "Process for fabricating a bipolar transistor with a self-aligned contact," ed: Google Patents, 1990.
[9] T.-Y. Chiu, G. M. Chin, R. C. Hanson, M. Y. Lau, K. F. Lee, M. D. Morris*, et al.*, "Process for manufacturing semiconductor BICMOS device," ed: Google Patents, 1988.
[10] T.-Y. Chiu, G. M. Chin, M. Y. Lau, R. C. Hanson, M. Morris, K. Lee*, et al.*, "The design and characterization of nonoverlapping super self-aligned BiCMOS technology," *IEEE Transactions on Electron Devices,* vol. 38, pp. 141-150, 1991.
[11] T.-Y. Chiu, G. M. Chin, M. Y. Lau, R. C. Hanson, M. D. Morris, K. F. Lee*, et al.*, "A high speed super self-aligned bipolar-CMOS technology," in *Electron Devices Meeting, 1987 International*, 1987, pp. 24-27.
[12] K. Moerschel, T. Chiu, W. Possanza, K. Lau, R. Swartz, R. Mantz*, et al.*, "BEST: a BiCMOS-compatible super-self-aligned ECL technology," in *Custom Integrated Circuits Conference, 1990., Proceedings of the IEEE 1990*, 1990, pp. 18.3/1-18.3/4.

OPPOSITE TRENDS BETWEEN DIGITAL AND ANALOG PERFORMANCE FOR DIFFERENT TFET TECHNOLOGIES

P.G.D. Agopian[1,2], C. Bordallo[1], J.A. Martino[1],
R. Rooyackers[3], E. Simoen[3], N. Collaert[3] and C. Claeys[4]

[1] LSI/PSI/USP, University of Sao Paulo, Brazil
[2] Sao Paulo State University (UNESP), Campus Sao Joao da Boa Vista, Brazil
[3] Imec, Leuven, Belgium
[4] E.E. Dept, KU Leuven, Belgium
*Corresponding Author's Email: paula.agopian@sjbv.unesp.br

ABSTRACT

Different Tunnel-FET technologies are analyzed in terms of digital and analog figures of merit. The digital figure of merit used was the subthreshold swing (SS), while the analog parameter was the intrinsic voltage gain (A_V). In the early technologies based on silicon TFET devices, the SS was much higher than the ideal behavior. However, the Av was very good, reaching a value up to 80 dB. The opposite trends were observed for up to date technologies based on III-V materials, where the SS finally reaches values down to 60 mV/dec while the A_V degrades to 32 dB. The explanation is related to the predominant conduction mechanism. In the III-V TFETs, Band to Band (B2B) Tunneling is the predominant mechanism, which is more sensible to the drain electric field, increasing the output conductance and degrading the A_V. In the silicon based TFETs the Trap-Assisted-Tunneling (TAT) is the predominant mechanism, which is less dependent on the drain electric field, resulting in a better A_V.

Keywords—TFET; geometries; new materials; digital and analog performance

INTRODUCTION

The technological advances of the semiconductor industry are focused on better performance but it is important to remember that lower power consumption is one of the major concerns. In this scenario, new devices as Tunnel Field Effect Transistors (TFETs) have been proposed due to their faster switching capability (SS below 60mV/dec at room temperature). However, some disadvantages for silicon transistors have been reported, like the low on-state current (I_{ON}) due to its large and indirect bandgap (E_G) and the non-ideal SS for experimental research devices due to the high TAT component.

The use of different materials with lower bandgap, as Si_xGe_{1-x} alloys [1-4] and III-V materials [5, 6], are a widely studied approach to enhance I_{ON}. Besides the studies of the switching characteristics, some researches about the analog performance of TFETs have shown promising results [10-14].

In this work the evolution of the TFET technology is presented and the discussion about the trends of digital and analog performance is initiated.

DISCUSSION

Figure 1 presents a comparison between silicon MOSFET and TFET technologies fabricated with Gate-All-Around (GAA) (left figure) and FinFET structures (right figure). Considering the Si homo-junction TFETs with MOSFET transistors fabricated with a TFET process scheme (using a drain underlap), for both the GAA-structure [5, 7] and the FinFET [8], the on-state current (I_{ON}) is very low compared with the MOSFET counterpart. Beside this, the subthreshold swing is not so good as idealized (lower than 60mV/dec) and the off-state current (I_{OFF}) is high due to the ambipolar effect.

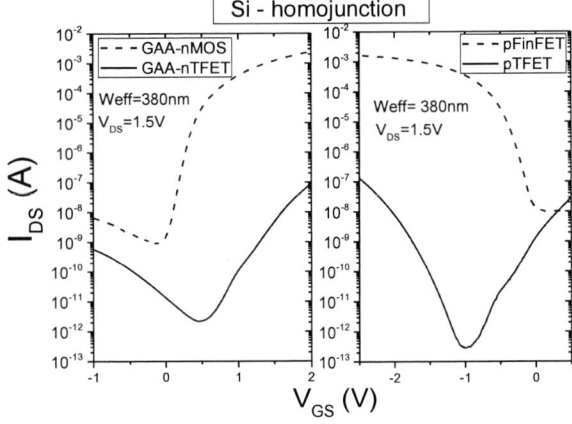

Fig.1. Comparison between MOSFET and TFET for Si-homo-junction devices (FinFET and GAA structures).

Aiming to improve these characteristics, a lot of research has been performed using different materials and different geometries focusing on reaching an ideal switch and improving the I_{ON} current level as presented in figure 2.

Figure 2A shows the transfer characteristic of TFET transistors for a GAA-structure, but in this case changing the source composition from Si 100% (homojunction device) to 100% germanium (hetero-junction transistor). A gate to drain underlap architecture is used in these GAA structures in order to minimize the ambipolar effect and consequently improve the I_{OFF}. Increasing the germanium amount at the source, the effective bandgap is reduced, increasing the band-to-band tunneling and improving the I_{ON} and switching behavior [9].

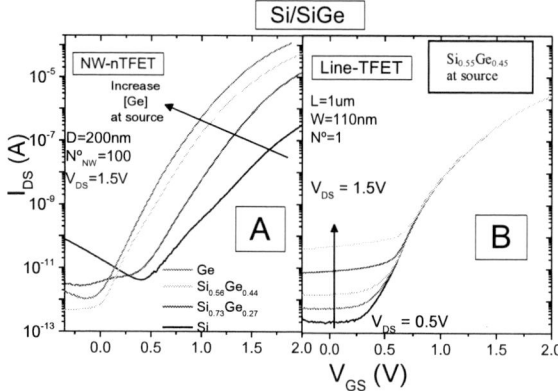

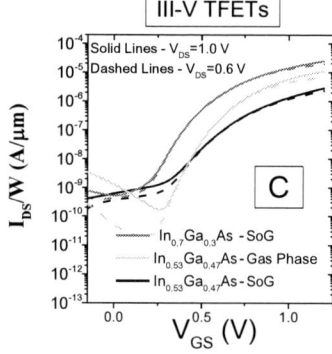

Fig. 2. Comparison between hetero-junction TFET technologies: GAA-TFETs (A), Line-TFETs (B) and planar III-V TFETs (C).

Figure 2B presents the same transfer characteristics for a planar Line-TFET for different drain bias. The line TFET structure has the channel above the source (composed by $Si_{0.55}Ge_{0.45}$) preferential without channel overlap, which enriches the tunneling rate between bands,

improving even more de SS behavior. In addition, the smaller drain bias reduces I_{OFF} without harming I_{ON}.

In figure 2C, the III-V TFET drain current (I_{DS}) is presented for different indium amounts and different doping processes. It is possible to observe that when considering the spin on glass (SoG) process, a higher indium amount results in a better switching behavior. When the Gas Phase Zn Diffusion process is considered, even for a smaller indium percentage a better I_{DS} behavior is obtained.

Considering the evolution of the TFET devices presented above, when the subthreshold swing is compared (figure 3), it is possible to observe that the smaller the bandgap (from silicon to III-V materials) the better is the SS behavior because the band-to-band tunneling becomes more predominant. For TFETs fabricated with III-V materials, the SS value surpasses 60mV/dec, which is the traditional theoretical limit for MOSFET transistors.

Although the TFET devices have been designed to meet the need for faster switching, these transistors have also shown to be promising in analog applications [10 - 13].

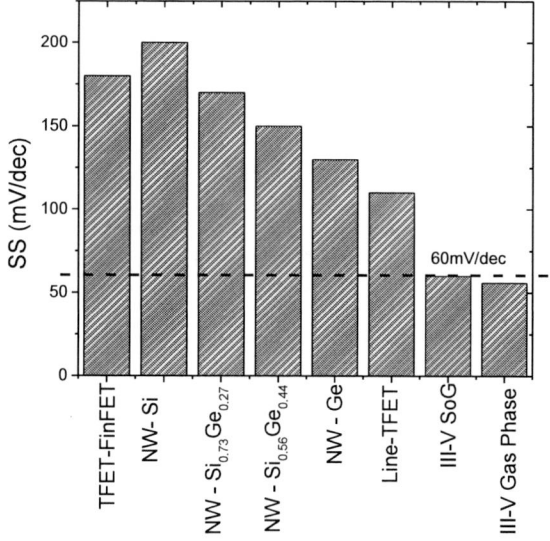

Fig.3. Subthreshold swing values as a function of technology evolution.

In order to understand the analog behavior of the tunneling transistors, it is necessary to keep in mind two statements:
- It is important to remember that the influence of the drain electric field on TFET operation is smaller than that shown for MOS transistors.
- As the TFET structures are modified to reach a high

switching speed, it is necessary to know which is the transport mechanism that dominates the operation of each TFET structure.

Comparing the tunneling transport mechanisms it is known that the Trap-Assisted Tunneling is quite dependent on gate voltage and almost independent on the drain bias, while the band-to-band (B2B) tunneling is more susceptible to the drain electric field.

In order to analyze how the transconductance (gm) and output conductance (g_D) behave with the dominant transport mechanism, the same GAA-TFET structures were used, varying only the Ge composition in the source, from pure Si (0%) to pure Ge (100%), as can be seen in figure 4.

Increasing the germanium amount at the source, the bandgap becomes narrower resulting in a higher B2B dominance on the TFET operation. This B2B dominance improves the I_{ON} current and consequently the transconductance. However, with the increasing germanium percentage at the source, the output conductance also increases due to the higher dependence of B2B on the drain electric field. This behavior occurs for all analyzed temperatures.

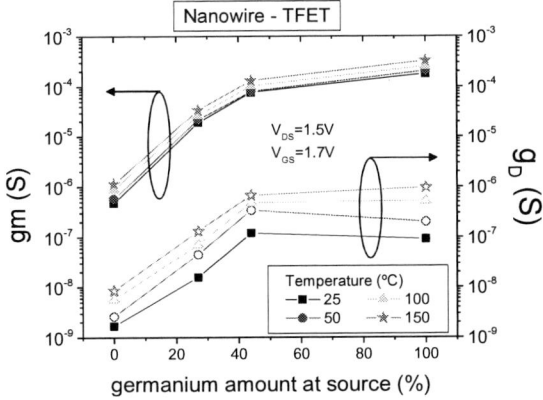

Fig. 4. *Experimental transconductance and output conductance for nanowires (GAA) TFETs varying the germanium amount at the source.*

Figure 5 shows the experimental intrinsic voltage gain (A_V) for MOSFET and TFET structures with different materials. The A_V values were calculated by the transconductance over output conductance ratio ($|A_V| \approx gm/g_D$).

Although the gm increases as the B2B becomes more dominant, the g_D is degraded in this direction. Considering the competition between these two factors (gm and g_D), the best situation for analog application occurs when the device is biased near to B2B conduction because the gm is improved and g_D was not too degraded by the influence of

the drain voltage (V_{DS}).

In summary, even if for digital applications, the performance becomes better as the transistors moved toward the III-V materials, for analog application a better A_V response was obtained for Si TFET-FinFETs. This opposite trend can be explained by the difference between the transport mechanism (trap assisted tunneling or band-to-band tunneling) that dominates for each TFET structure. As the band-to-band (B2B) tunneling becomes more important, the more abrupt becomes the SS, i.e., the better becomes the device switching. On the other hand, the more dependent on B2B for the device behavior, the greater is the dependence of I_{DS} on the applied drain voltage, which degrades the output conductance and in turn reduces the intrinsic voltage gain.

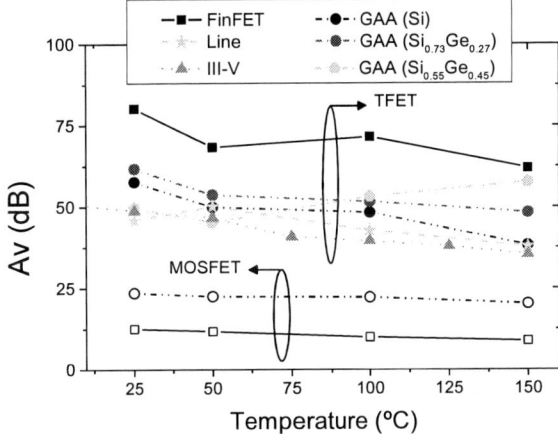

Fig.5. *Intrinsic voltage gain versus temperature for different TFET technologies.*

CONCLUSIONS

As the TFET structures are optimized to ensure a higher on-state current and a lower TAT influence to achieve a high switching speed, the transport mechanism becomes predominantly B2B tunneling which results in an increase of gm, which is good for the analog behavior. However, it also causes a g_D increase what tends to degrade the analog performance.

While to reach the desired digital performance (subthreshold swing smaller than 60mV/dec at room temperature) the TFET conduction must be strongly dependent on B2B and weakly dependent on TAT, however, the devices operating with an important TAT influence present a very low output conductance and consequently a better analog response.

978-1-5386-5309-8/18 $31.00 © 2018 IEEE

ACKNOWLEDGMENT

The authors would like to thank CNPq and FAPESP for the financial support during the execution of this work and the support from the imec's Logic Device Program and its Core Partners.

REFERENCES

[1] A. Verhulst, W. G. Vandenberghe, K. Maex, G. Groeseneken, J. Appl. Phys., 104(6), 2008, 064514.

[2] D. Leonelli,; A. Vandooren, R. Rooyackers, S. De Gendt, M. M. Heyns, G. Groeseneken, Solid-State Electronics, 65-66, 2011, pp.28-32.

[3] A. Vandooren, D. Leonelli, R. Rooyackers, A. Hikavyy, K. Devriendt, M. Demand, R. Loo, G. Groeseneken, C. Huyghebaert, Solid-State Electronics, 83, 2013, pp 50- 55.

[4] M. Schmidt; A. Schäfer, R. A. Minamisawa, D. Buca, S. Trellenkamp, J.M. Hartmann, Qing-Tai Zhao, S. Mantl, IEEE Electron Device Lett., vol. 35, no. 7, 2014, pp. 669-702.

[5] P. G. D. Agopian, M. D. V. Martino, S. D. dos Santos, F. S. Neves; J. A. Martino, A. Vandooren, R. Rooyackers, E. Simoen, A. Thean, C. Claeys, IEEE Trans. Electron Devices, vol. 62, no.1, 2015, pp. 16-22.

[6] A. Mallik, A. Chattopadhyay, IEEE Trans. on Electron Devices, vol. 59, no. 4, 2012, pp. 888-894.

[7] C. M. Bordallo, V. B. Sivieri, J. A. Martino, P. G. D. Agopian, R. Rooyackers, A. Vandooren, E. Simoen, Voon-YewThean, C. Claeys, IEEE Trans. Electron Devices, vol. 63, no. 7, 2016, pp. 2930-2935.

[8] P. G. D. Agopian, J. A. Martino, R. Rooyackers, A. Vandooren, E. Simoen, C. Claeys, IEEE Trans. Electron Devices, vol. 60, no. 8 , 2013, pp. 2493-2497.

[9] A. Alian, Y. Mols, C. M. Bordallo, D. Verreck, A.Verhulst, A. Vandooren, R. Rooyackers, P. G. D. Agopian, J. A. Martino, A. Thean, D. Lin, D. Mocuta, N. Collaert, Appl. Phys. Lett., vol. 109, 2016, pp. 243502.

[10] M. D. V. Martino, F. S. Neves, P. G. D. Agopian, J. A. Martino, A. Vandooren, R. Rooyackers, E. Simoen, A. Thean, C. Claeys, Solid-State Electronics, vol. 112, 2015, pp. 51-55.

[11] C. Bordallo, J. A. Martino, P. G. D. Agopian, A. Alian, Y. Mols, R. Rooyackers, A. Vandooren, A. Verhulst, Q. Smets, E. Simoen, C. Claeys, N. Collaert, Semicond. Sci. Technol., vol. 31, 2016, p. 124001.

[12] C. C. M. Bordallo, J. A. Martino, P. G. D. Agopian, A. Alian, Y. Mols, R. Rooyackers, A. Vandooren, A. S. Verhulst, E. Simoen,C. Claeys, N. Collaert, IEEE Trans. Electron Devices, vol. 64, no. 9 , 2017, pp. 3595-3600.

[13] P. G. D. Agopian, J. A. Martino, A. Vandooren, R. Rooyackers, E. Simoen, A. Thean C. Claeys, Solid-State Electronics, vol. 128, 2017, pp. 43-47.

III-V GAAS AND INP HBT DEVICES FOR 4G & 5G WIRELESS APPLICATIONS

C.R. Bolognesi,* Wei Quan, Akshay M. Arabhavi, Diego Marti, and Olivier Ostinelli

Millimeter-Wave Electronics Group (MWE), ETH-Zürich, CH-8092, Switzerland

*Corresponding Author's Email: colombo@ieee.org

ABSTRACT

With 5G implementation details still remaining to be defined, it is already clear that the corresponding handset power amplifiers (PAs) will require high-linearity, high-efficiency performance. In this context, the present paper first examines the PA landscape of available candidate technologies and then shows that InP/GaAsSb Double Heterojunction Bipolar Transistors (DHBTs) offer performance differentiating features. Indeed, InP/GaAsSb DHBTs exhibit weaker C_{BC} variations than other bipolar technologies, a key differentiator of nonlinear performance in bipolar transistors. Some drawbacks of InP DHBT technology are addressed with potential anticipated pathways for solutions.

INTRODUCTION

The exponentially increasing appetite for broadband mobile internet access is partly fueling interest in the development of 5G networks. Whereas 4G networks already do provide fast broadband capabilities, their latency (> 40 ms) and quality-of-service (QoS) limitations represent roadblocks for certain applications. Anticipated developments such as autonomous (self-driving) vehicles, telemedicine and tactile remote control of objects will require low round-trip latency (1 ms) not possible in current 4G networks.

In terms of QoS, higher carrier frequencies generally offer increased bandwidths: consider that the 28 GHz band comes with a bandwidth of 0.85 GHz, that 1.6 GHz is available around 37 GHz, and that a comparably huge 14 GHz is associated with the V-band around 60 GHz. Frequency reuse becomes easier at higher frequencies, as are beam-forming techniques, both facilitating factors for QoS improvements. Whereas current 5G demonstrator systems operate near 28 GHz for practical reasons (for example [1]), one can anticipate an inexorable move to higher frequencies because of the above considerations.

While earlier networks such as 2G relied on constant envelope signals, power amplifier linearity has become increasingly important in modern systems because of the increasing peak-to-average-power-ratio (PAPR) of the complex waveforms used to achieve higher data rates (*e.g.* 16-QAM). As operating frequencies rise, so do linearity requirements —an unfortunate coupled requirement situation because scaling generally tends to make devices less ideal. Furthermore, because of the wide signal dynamic range associated with the high PAPR, PA transistors are operated with up to 10 dB power back off (PBO) for linearity, and thus well below their peak power-added efficiency (PAE) operating conditions.

With clear performance advantages over CMOS and SiGe [2], GaAs HBTs have dominated the handset PA market since linear performance capability became a key requirement for 3G and later networks. However, GaAs HBTs feature a $f_T \times BV_{CEO}$ product of 650 GHz-V which likely will prove insufficient for the anticipated performance requirements of 5G systems at 30 GHz and above. In what follows, we therefore briefly consider technological alternatives before focusing on "Type-II" InP/GaAsSb DHBTs [3] for the remainder of the paper.

HANDSET PA LANDSCAPE

A. Silicon-Based MOSFETs (CMOS, SOI)

MOS-based transistors can feature maximum oscillation frequencies of the order of $f_{MAX} = 300$ GHz which make them capable of providing power amplification at least up to 60 GHz if one considers a 5× frequency ratio. However, an important limitation of CMOS technology is a rather severe shortcoming in power output and linearity already below 30 GHz which can only become more detrimental at higher frequencies. For example, 5G 28 nm bulk CMOS PA operating at 28 GHz with a peak PAE 36%, deliver 10-11% with the 9.5 dB PBO required to maintain enough linearity for 16-QAM waveforms [1].

CMOS is an established technology carrying a lot of forward momentum thanks to very refined device/circuit modeling capabilities and a broad design expertise —as can be gauged from the extensive transistor geometry tuning in discussed in [1]. It therefore cannot be precluded that 5G requirements might evolve toward very small cells and low output powers tailored to CMOS performance characteristics —in other words, it might be that CMOS *de facto* will mold the evolution of 5G requirements.

B. SiGe HBTs

SiGe HBTs have developed excellent bandwidth capability and also benefit from the established silicon microelectronics industrial infrastructure, extensive modeling capabilities and a vast pool of design expertise [4]. Transistor cutoff frequencies well in excess of 300 GHz are readily available from several manufacturers, albeit with lower breakdown voltages than in III-V HBT technologies. Circuit techniques such as transistor stacking and/or above BV_{CEO} operation has been exploited to extend the output power capabilities of these devices. For example, "above BV_{CEO} cascodes" were shown to operate reliably albeit with significant non-linearities associated with carrier multiplication [5]. More recently, Pottrain *et al.* [6] showed that impact ionization in the collector of advanced SiGe HBTs sets the fundamental limitation for output power in that technology.

The main nonlinearity source in SiGe HBTs is related to impact ionization at low current levels and nonlinear base-collector feedback C_{BC} at higher currents. Buisman *et al.* has shown a stronger C_{BC} variation (and nonlinearities) in SiGe when compared to GaAs HBTs [7].

C. GaN HEMTs

Whereas GaN mm-wave HEMTs are largely foreseen as serious candidates for 5G base station PAs at 28 GHz and above, 20 nm gate GaN HEMTs on SiC substrates were recently suggested for use in handset PAs in light of impressive drain efficiencies when operated at low (for GaN) drain voltages V_{DS} = 2-3 V [8]. The idea of using deeply scaled wide bandgap in high-efficiency low-voltage handset applications is intriguing but may face practical obstacles in terms of the large scale manufacturability of deep nanometer gate HEMTs by electron lithography and SiC substrate costs (which could be mitigated if ported to high-resistivity substrates). Furthermore, the reported load-pull characteristics at 2 V betray significant nonlinearities and whereas the peak PAE is roughly 59%, it rapidly drops to ~5% with a 10 dB PBO [8].

D. InP DHBTs

Zampardi *et al.* recently evaluated "Type-I" chirp superlattice collector (CSL) InP/GaInAs DHBTs for envelope tracking (ET) applications where the devices showed some promise because of their rapid turn on and good operating characteristics down to low collector voltages [9]. As potential hurdles for InP DHBT technology the authors cited InP substrate costs and an unknown cost structure. The commercial availability of 6-inch InP substrates should alleviate substrate cost concerns: the situation of InP DHBTs may resemble that for GaAs at the beginning of 3G. That notwithstanding, SmartCut™ like lifting and transfer of InP DHBT epitaxial layers onto GaAs carrier wafers (which could possibly incorporate E/D-mode HEMTs), and subsequent InP substrate reuse (as in silicon SOI) could provide an alternative route to substrate InP cost mitigation [10,11].

In our opinion a more serious drawback of InP DHBTs (and indeed for all III-V technologies) lies with the relative lack of sophistication of device models for circuit simulation and design, coupled with a comparably limited pool of design experience in the community. Together, these factors play out to somewhat disappointing circuit demonstrations compared to raw device performance metrics.

As an alternative to "Type-I" GanAs-based DHBTs, "Type-II" InP/GaAsSb DHBTs demonstrated excellent large-signal properties in load-pull at 40 GHz with peak PAE values of 62% in non-power optimized transistors with f_T/f_{MAX} = 350/300 GHz [12]. Since then, InP/GaAsSb DHBTs are routinely produced with f_{MAX} > 600 GHz [3] and research prototypes have reached 1.2 THz as determined by conventional de-embedding methods [4].

Also of direct interest, Xu *et al.* previously compared the linearity of "Type-I" [InP/GaInAs] and "Type-II" [(Al,In)P/GaAsSb] and found GaAsSb based devices to have superior DC and linearity characteristics compared to foundry supplied GaInAs DHBTs [13]. In that work, the authors associated the drop in linear performance at intermediate currents to the chirp superlattice (CSL) grading used in the base/collector region grading between GaInAs and InP [13].

INP/GAASSB DHBTS

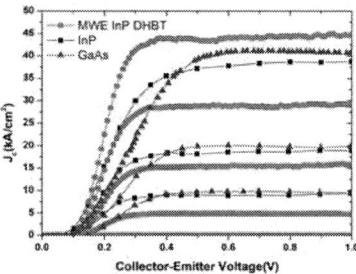

Figure 1: Static I-V characteristics of InP/GaAsSb DHBT (MWE, in red) superposed with InP/GaInAs DHBT and GaAs HBT characteristics from [9].

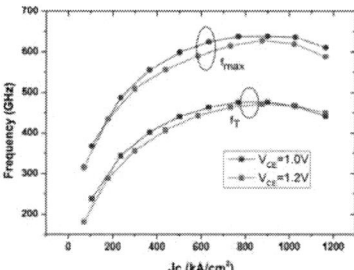

Figure 2: RF characteristics of 20 nm base InP/GaAsSb DHBT at two V_{CE} biases. The emitter contact area is 0.2 × 5 μm².

Figure 1 above shows the low-current region of the I_C-V_{CE} characteristics of an InP/GaAsSb DHBT superposed to the measurements for the Type-I InP/GaInAs DHBT and GaAs HBT described in [9]. The InP/GaAsSb DHBT clearly shows lowest saturation voltages and ON resistance values R_{ON}.

The present $0.2 \times 5\ \mu m^2$ InP/GaAsSb DHBTs were developed for high-speed performance and offer a relatively low breakdown voltage $BV_{CEO} = 5V$ with $f_T/f_{MAX} = 480/640$ GHz. Fig. 3 shows the low-current density details of the f_T/f_{MAX} dependence on bias conditions. Clearly, cutoff frequencies vary only weakly above $V_{CE} = 0.6$ V. These transistor were tested under similar conditions to those in the works of [9] and [13] to facilitate comparisons and determine whether they offer any potential interesting performance features for 5G handset PA applications.

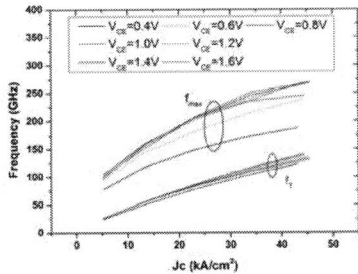

Figure 3: Low-current detail of f_T/f_{MAX} for InP/GaAsSb DHBT. The cutoff frequencies show weak variations with voltage even at low V_{CE} (df_T/dJ_C shows little dependence on V_{CE}). The emitter contact area is $0.2 \times 5\ \mu m^2$.

Figures 4-5 show RF-knee measurements indicating a weak dependency of MSG/MAG on bias conditions at 28 and 40 GHz superposed with the data of [9] for InP/GaInAs DHBTs with a CSL collector grading. It is apparent that the present InP/GaAsSb DHBTs outperform InP/GaInAs devices at low collector voltages, and show weaker gain variations with V_{CE}.

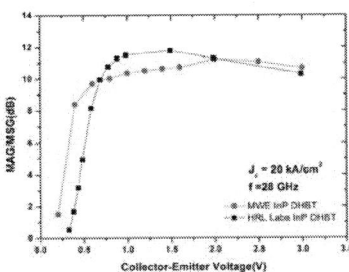

Figure 4: 28 GHz RF knee measurements superposed with the InP/GaInAs DHBTs characterized in [9].

Figure 6 shows the extracted base/collector capacitance C_{BC} at $f = 5$ GHz for our InP/GaAsSb DHBTs and the InP/GaInAs DHBTs characterized [9]. Whereas the "Type-I" InP/GaInAs DHBTs show a sharply increasing C_{BC} below $V_{CE} = 1$ V (as do GaAs HBTs [9]),

our "Type-II" devices only show a notable C_{BC} rise below 0.4 V. The modulation of C_{BC} was identified as the key differentiator of nonlinear behavior across bipolar transistor technologies [7].

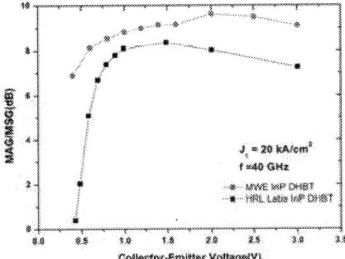

Figure 5: 40 GHz RF knee measurements superposed with the InP/GaInAs DHBTs characterized in [9].

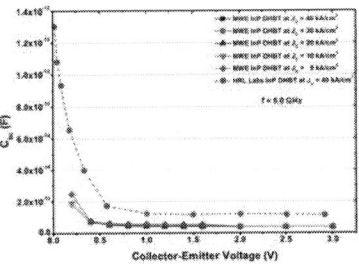

Figure 6: Bias dependence of C_{BC} in InP/GaAsSb DHBT extracted at 5 GHz (MWE labels). The values extracted for InP/GaInAs DHBTs in [9] are included for comparison (HRL label).

Finally, Fig. 7 shows the results of two-tone intermodulation measurements at 18 GHz with a 4 MHz tone separation as a function of collector current density. It is immediately apparent that the third order intercept point IP3 varies only weakly with current density while the associated power gain is essentially independent of J_C over a broad range. Because nonlinearities in bipolar transistors are primarily related to variations in C_{BC} [7], these linearity results are consistent with the weak bias dependence shown above.

DISCUSSION

The RF knee roll-off at low V_{CE} in bipolar transistors occurs when the base-collector junction begins to experience forward biases. At the same time, C_{BC} increases strongly and nonlinear behavior worsens. Consideration of the InP/GaAsSb DHBT band diagram in Fig. 8 shows two reasons behind their excellent performance: *a)* no current blocking occurs in the conduction band when the reverse collector-base bias decreases or as current density rises (as otherwise seen in "Type-I" InP/GaInAs DHBTs); and *b*) even under forward base/collector biases, no holes can be injected from the *p+* base into the *n−* InP collector due to the massive valence band discontinuity between GaAsSb and InP. This band alignment results in a strong suppression of saturation minority charge storage because only few electrons can be injected into the *p+* GaAsSb base.

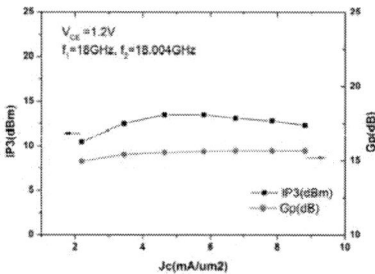

Figure 7: 18 GHz intermodulation measurement (4 MHz tone separation) for InP/GaAsSb DHBT. IP3 and the associated power gain are very flat as a function of collector current density. The emitter contact area is 0.2 × 10 μm².

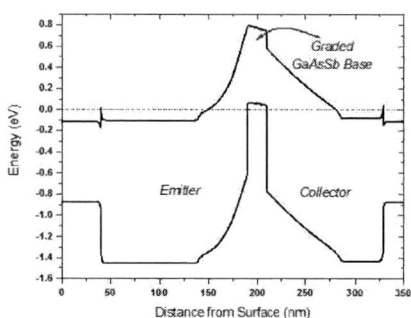

Figure 8: Equilibrium band diagram for "Type-II" InP/GaAsSb DHBT. There is no collector current blocking at high J_C and/or low V_{CE} nor collector hole saturation charge in saturation mode [4].

SUMMARY

Power amplifiers for 5G handsets will require high-linearity and high-efficiency characteristics. The present work briefly reviewed the features of several technologies of interest and characterized "Type-II" InP/GaAsSb DHBTs within a framework compatible with the work of others [7, 13] to comparatively assess InP/GaAsSb DHBTs in contrast to various alternatives.

DC current-voltage characteristics as well as RF knee plots show that InP/GaAsSb DHBTs not only offer superior performance at low V_{CE} (making them attractive for ET applications), but also display weaker MSG/MAG

variations with bias suggestive of good linearity properties. As nonlinearities in bipolar transistors tend to be dominated by C_{BC} variations, the bias dependence of C_{BC} was characterized down to very low V_{CE} values. As expected, C_{BC} in InP/GaAsSb DHBTs is far less bias dependent than in "Type-I" InP/GaInAs DHBTs and GaAs HBTs [9]. As could be anticipated from the above, InP/GaAsSb DHBTs showed nearly bias-independent IP3 and associated gain characteristics in two-tone intermodulation distortion measurements around 18 GHz. Considering that Buisman *et al.* showed that GaAs HBTs show significantly weaker C_{BC} variation than SiGe HBTs [7], InP/GaAsSb DHBTs appear to be very well-suited for 5G handset PA applications because of a remarkably weak C_{BC} modulation following from their unique band structure. More advanced testing methods such as non 50 Ω IMD and ACPR measurements are warranted.

ACKNOWLEDGEMENTS

Prof. Renato Negra from RWTH Aachen is thanked for clarifications on linearity requirements evolution in broadband networks. Mr. Hansruedi Benedicter is thanked for his help with microwave characterization.

REFERENCES

[1] S. Shakib *et al. Tech. Digest IEEE ISSCC 2016*, San Francisco CA, Jan. 31-Feb. 4 2016, pp. 352-354.

[2] H.S. Bennett *et al. IEEE Trans. Electron Dev.*, vol. 7, 2005, pp. 1235-1257.

[3] M. Alexandrova *et al. IEEE Electron Dev. Lett.*, vol. 35, 2014, pp. 1218-1220.

[4] P. Chevalier *et al. Proc. of the IEEE*, vol. 105, 2017, pp. 1035-1050.

[5] C.M. Grens *et al. IEEE Trans. Dev. and Materials Reliability.* vol. 9, 2009, pp. 431-439.

[6] A. Pottrain *et al. Proc. 2011 European Microwave Conference (EuMIC)*, Manchester UK, Oct. 10-11 2011, p. 12437268.

[7] K. Buisman *et al. Proc. 2013 European Microwave Conference (EuMIC)*, Nuremberg, Germany, Oct. 6-10 2013, p. 259-262.

[8] M. Micovic *et al. Proc. 2016 IEEE Int. Electron Dev. Meeting (IEDM)*, San Francisco, Dec. 3-7 2016, pp. 3.3.1-4.

[9] P. Zampardi *et al. Proc. 2016 IEEE Compound Semicond. IC Symp. (CSICS)*, Austin TX, Oct. 23-26 2016, p. 16499247.

[10] W. Chen *et al. Electrochemical and Solid-State Lett.* vol. 12, 2009, pp. H149-H150.

[11] N.Daix *et al. APL Materials.* vol. 2, 2014, p. 086104.

[12] V. Teppati *et al. IEEE Electron Dev. Lett.*, vol. 32, 2011, pp. 886-888.

[13] H. Xu *et al. Proc. 2012 Compound Manufacturing Tech. Conf. (CS MANTECH)*, Boston MA, Apr. 23-26 2012, paper 6b.3.

NONVOLATILE MEMORY OUTLOOK: TECHNOLOGY DRIVEN OR APPLICATION DRIVEN

JING LI

Department of Electrical and Computer Engineering
University of Wisconsin-Madison, Madison, WI 53706, USA
jli@ece.wisc.edu

The last decade has seen significant progress in emerging nonvolatile memory (NVM) technologies including Spin Torque Transfer RAM (STT RAM), phase change memory (PCM) and resistive RAM (RRAM). Until now, the key industry players have all demonstrated Gb-scale capacity in advanced technology nodes, including 1Gb PCM at 45nm by Micron [1], 8Gb PCM at 20nm by Samsung [2], 32Gb RRAM at 24nm by Toshiba/Sandisk [3], 16Gb conductive bridge (CBRAM) at 27nm by Micron/Sony [4], 128Gb 3D XPoint technology by Micron/Intel [5] and 1Gb STT-RAM with DDR4-compatible interface by Everspin [6]. Even with successful commercialization, the insertion of these technologies to exiting computer system as a direct drop-in replacement seems not being effective. The fundamental reasons for that are 1) Technically, the inherent nature of these technologies does not align well with either main memory or persistent storage in terms of cost-per-bit, latency, power, endurance, and retention. 2) Economically, besides more investment to existing memory manufacturing facilities for producing these new technologies, it is difficult to convince end-users to switch to a new technology as long as they can still use DRAM or Flash for the same purpose, unless significant benefits are provided.

Looking back into history, the interaction between computer architecture and IC technology has been complicated and bi-directional. On one hand, the availability of emerging technology will affect the decisions system architects would make, while the advances in computer architecture also influences the viability of different technologies. On the other hand, emerging big data applications, such as, analytics, recognition, search, machine translation, natural language processing and others, also provide new drivers and pose new challenges for future architecture design. The same principle is applied to the case of emerging nonvolatile memories. It is challenging for any of the emerging NVMs to take over the dominant mature market of DRAM or Flash. However, we envision that to enable a wide adoption of these NVM technologies, a potentially viable path is to explore non-traditional usage models or new paradigms beyond traditional memory applications, for instance, in-memory processing (IMP). We believe that the emerging NVMs may become an enabling technology for IMP.

The exploratory PIM architectures reported in literature can be broadly divided into the several types: 1) One type is to utilize the inherent dot-product capability of the crossbar structure to accelerate matrix multiplication, which is a key computational kernel in a wide range of applications including deep learning, optimization, etc. Representative work includes PRIME [7], ISAAC [8], and memristive Boltzmann machine [9]. By augmenting RRAM crossbar design with various digital or analog circuits in the periphery, these architectures can realize different accelerator functions that are built atop matrix multiplication. 2) Another type is to implement a neuromorphic system, which exploits the analog nature of NVM array to implement synaptic network in order to mimic the fuzzy, fault-tolerant and stochastic computation of the human brain, without sacrificing its space or energy efficiency. 3) The third type is associative processer (AP), also known as non-volatile Content addressable memory (nv-CAM) or Ternary content addressable memory (nv-TCAM), which supports associative search to locate data records by content rather than address. We demonstrated the very first large-scale PCM TCAM chip and prove the feasibility of implementing

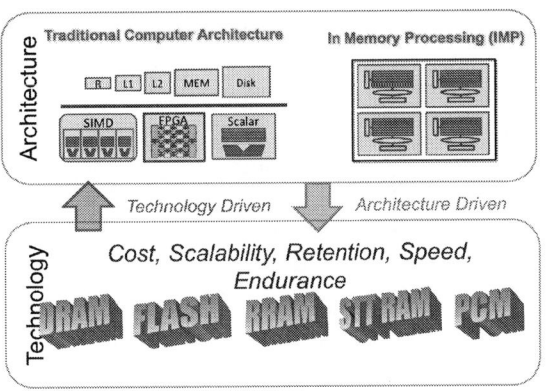

Figure 1: Technology-Architecture interaction:

978-1-5386-5309-8/18 $31.00 © 2018 IEEE

technology driven or architecture driven

in-memory processing using emerging NVM in a cost-effective manner [10]. Other representative work includes RRAM-based TCAM [11], AC-DIMM [12] 4) The fourth type is reconfigurable architecture (RA). Representative work includes non-volatile field programmable gate array (nv-FPGA [13]) that utilizes NVM cells to replace the SRAM cells in Look-Up Tables (LUTs) and Block RAMs (BRAMs), or uses NVM cells as switches in the connection blocks and switch blocks. Recently, an interesting reconfigurable in-memory computing architecture that combines the best advantages of TCAM and FPGA [14,15].

Both AP (Type-3) and RA (Type-4) show great promise in implementing the concept of in-memory processing without necessarily incurring high cost. Specifically, they do not need expensive mixed-signal circuits (A/D, D/A) as Type-1 and Type-2 and thus, their adoption barrier is lower than Type-1 and Type-2.

Among all the work, we would like to specifically highlight an interesting reconfigurable in-memory computing architecture (Type-4) developed by us. It shares some similarities to FPGA in morphable dataflow architecture but also radically differs from it by providing: 1) flexible on-chip storage, 2) flexible routing resources, and 3) enhanced hardware security. For the first time, it exploits a continuum of PIM capabilities across the whole spectrum, ranging from 0% (pure data storage) to 100% (pure compute engine), or intermediate states in between (partial storage and partial computation), as illustrated in Error! Reference source not found.. Such superior programmability blurs the boundary between computation and storage. We believe it may open up rich research opportunities in driving new reconfigurable architecture, design tools, and developing new data-intensive applications, which were not generally considered to be suitable for FPGA-like accelerations.

Conversely, these IMP architectures further drive technology innovation. For instance, existing NVM technologies has limited ON/OFF resistance ratio (except for CBRAM), typically on the order of 100x, in contrast to $>10^5$x of MOSFET. It is not a problem for either memory or storage applications but can be a severe reliability concern for certain types of IMP. Let us take PCM as an example, to achieve high ON/OFF ratio, optimization must be made towards improving essential property such as resistivity contrast between amorphous and crystalline, whereas the other properties including endurance (chemical stability), programming speed (crystallization speed) and write power (melting/crystallization temperature, thermal conductivity), etc. are all secondary concerns. Interestingly, these are much relaxed constraints to meet as compared to the case of DRAM (unlimited endurance and ns-scale programming speed) or NAND

flash ($4F^2$ cell size as well as MLC capability).

To summarize, we see an interesting interplay between emerging nonvolatile memory technologies and novel IMP architectures. As our community embarks upon the post-Moore's Law era and looks towards a post-CMOS technology, these NVM technologies have been the front-runner in the race of development and commercialization among various other technology options. The close interaction of architecture and technology will open up new opportunities and become the keys for future semiconductor and computer research.

REFERENCES

[1] C. Villa, et al. A 45nm 1gb 1.8v phase-change memory. ISSCC, 2010

[2] Y. Choi, et al. A 20nm 1.8v 8gb pram with 40mb/s program bandwidth. ISSCC, 2012

[3] T. Y. Liu, et al. A 130.7mm2 2-layer 32gb reram memory device in 24nm technology. ISSCC, 2013

[4] Micron Technology, Inc. Breakthrough Nonvolatile Memory Technology.

[5] Fackenthal, Richard, et al. 19.7 A 16Gb ReRAM with 200MB/s write and 1GB/s read in 27nm technology, ISSCC 2014

[6] Everspin, Inc. Everspin Announces Sampling of the World's First 1-Gigabit MRAM Product

[7] Ping Chi, et al. PRIME: A Novel Processing-in-Memory Architecture for Neural Network Computation in ReRAM-Based Main Memory, ISCA 2016

[8] Ali Shafiee, et al. ISAAC: A Convolutional Neural Network Accelerator with In-Situ Analog Arithmetic in Crossbars, ISCA 2016

[9] Mahdi Nazm Bojnordi, et al. Memristive Boltzmann machine: A hardware accelerator for combinatorial optimization and deep learning, HPCA 2016

[10] Jing Li, et al. 1 Mb 0.41 μm^2 2T-2R Cell Nonvolatile TCAM With Two-Bit Encoding and Clocked Self-Referenced Sensing, JSSC 2014

[11] Qing Guo, et al. A Resistive TCAM Accelerator for Data-Intensive Computing, MICRO 2011

[12] Qing Guo, et al. AC-DIMM: Associative Computing with STT-MRAM, ISCA 2013

[13] J. Cong and B. Xiao. mrfpga: A novel fpga architecture with memristor-based reconfiguration. In NANOARCH, pages 1–8, June 2011.

[14] Yue Zha and Jing Li, Reconfigurable In-Memory Computing with Resistive Memory Crossbar, ICCAD 2017

[15] Yue Zha and Jing Li, RRAM-based Reconfigurable In-Memory Computing Architecture with Hybrid Routing, ICCAD 2016

978-1-5386-5309-8/18 $31.00 © 2018 IEEE

MTJ-BASED NONVOLATILE LOGIC LSI FOR ULTRA LOW-POWER AND HIGHLY DEPENDABLE COMPUTING

Masanori Natsui[1]*, Tetsuo Endoh[1], Hideo Ohno[1], and Takahiro Hanyu[1]

[1]Tohoku University, Japan

*Email: natsui@riec.tohoku.ac.jp

ABSTRACT

A novel logic-LSI architecture, "nonvolatile logic-in-memory (NV-LIM) architecture," where nonvolatile storage elements are distributed over a logic-circuit plane, is proposed as a promising candidate to overcome performance, power and reliability wall on the present CMOS-only-based logic LSIs. Some concrete design examples based on the NV-LIM architecture using MTJ device are demonstrated and their usefulness is discussed in comparison with the corresponding CMOS-only-based realization.

INTRODUCTION

In the Internet of Things (IoT) era, it is strongly necessary to realize a computer architecture that is compatible with ultra-low-power and high-performance computing power. However, in the present CMOS-only-based VLSI, the communication bottleneck between memory and logic modules inside a VLSI chip, as well as increasing standby power dissipation and variation in the device characteristic, limits to solving the above problems [1]. In a conventional logic-LSI architecture, logic and memory modules are separately implemented together and these modules are connected each other through global interconnections. Even if the device feature size is scaled down in accordance with the semiconductor technology roadmap, the global interconnections are relatively getting longer, which results in the increase in longer delay and higher power dissipation due to signal propagation between the interconnections. In addition, since on-chip memory modules are "volatile," they always consume the static power to maintain the stored data.

On the other hand, several emerging storage devices are getting developed to overcome the weak points of conventional semiconductor memories such as dynamic random-access memory (DRAM) and static random-access memory (SRAM). In particular, a magnetoresistive random-access memory (MRAM) is one of the emerging memories that is currently drawing the most attention in the memory business. Spin-transfer torque (STT) MRAM promises speed and reliability comparable to that of SRAM along with the "non-volatility" of flash [2,3]. Since magnetic tunnel junction (MTJ) device, the key element of MRAM, is easily distributed over a logic-circuit plane by using a three-dimensional (3D) stack structure,

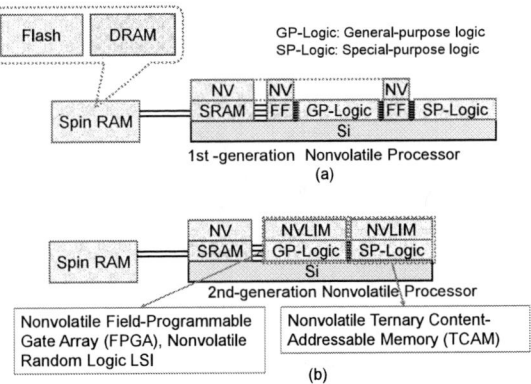

Figure 1: Combination of power-gating and nonvolatile logic techniques; (a) Conventional CPU without power gating, (b) Conventional CPU with power gating, (c) NV-LIM CPU with power gating.

performance degradation due to intra-chip global wires could be drastically mitigated, which leads to a high-performance, ultra-low-power and highly reliable (or highly resilient) logic LSIs.

One of the most useful methods to cut off leakage power is to use power gating. If the power gating is applied in the conventional logic LSI, a certain amount of standby power can be eliminated, but two additional operations, "back-up" and "boost-up" procedures, must be performed before and after applying the power gating, respectively, which may limit the situation where the power-gating technique can be applied. In contrast, the use of non-volatility makes it possible to apply power gating without performing the above operations, which ideally eliminates the wasted power dissipation.

Figure 1(a) shows nonvolatile VLSI processor architecture, where high-density and high-speed MRAMs and nonvolatile flip-flops are used to simply realize nonvolatile logic LSIs [4,5]. When a part of the nonvolatile on-chip memory is merged into logic-circuit modules as shown in Fig. 1(b), the performance of the nonvolatile logic LSI can be improved. The use of spintronics-based nonvolatile logic-in-memory (NV-LIM) architecture makes not only performance improved, but

978-1-5386-5309-8/18 $31.00 © 2018 IEEE

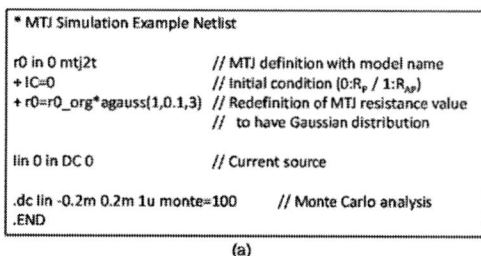

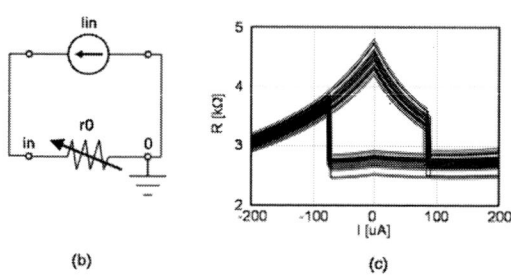

Figure 2: Monte-Carlo simulation of the MTJ device characteristic by a SPICE simulator with a built-in MTJ device model: (a) example netlist, (b) corresponding circuit schematic, (c) simulated waveform.

also reliability enhanced in the future logic LSIs.

In the following section, we introduce a nonvolatile motion vector prediction unit designed by a sophisticated design environment for large-scale MTJ/MOS-hybrid LSI as a concrete design example, as well as a circuit technique for maximizing the performance of embedded MRAM in a microcontroller unit for IoT sensor nodes. As another possibility of NV-LIM application, we also describe an application example for improving the reliability of LSI.

NONVOLATILE MOTION VECTOR PREDICTION UNIT

Various types of general/special-purpose circuits have been reported for demonstrating the impact of embedding nonvolatile memory devices into a logic circuit plane [6-16]. To design practical-scale MTJ/MOS-hybrid logic LSI and broaden the scope of its application, it is necessary to establish a sophisticated design environment that can precisely handle the behavior of the MTJ device while retaining sufficient compatibility with a conventional design environment for CMOS-based VLSI. For this purpose, we have developed a design flow for MTJ/MOS-hybrid LSI by combining de-facto standard EDA tools and new supplementary design tools for the precise simulation

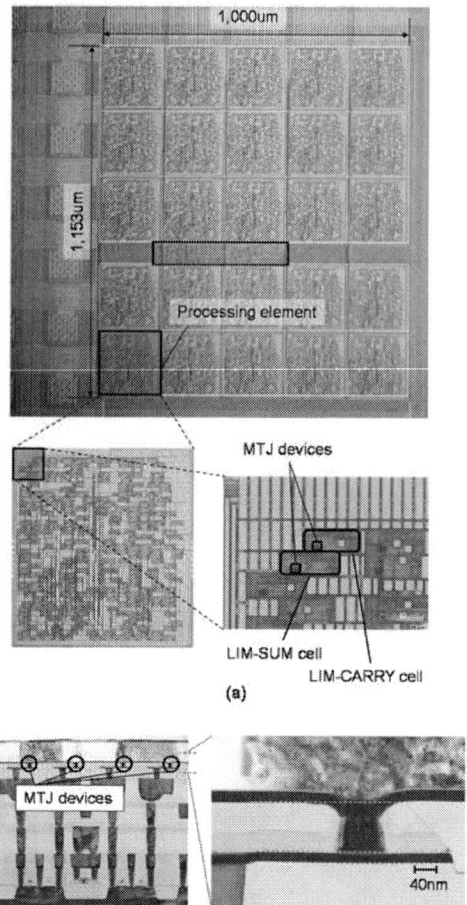

Figure 3: Fabricated nonvolatile motion-vector prediction unit: (a) micrograph after fourth metal process, (b) cross-sectional SEM image of fabricated MTJ devices.

of MTJ device characteristics [17-19]. By utilizing the proposed flow, various MTJ-based NV-LIM circuits can be designed and simulated with HDL, and the corresponding layout, including MOS and MTJ/MOS-hybrid cells, can be automatically synthesized. Moreover, the operation verification including the timing analysis can be accomplished by the de-facto standard EDA tools.

Figure 2 shows an example of the Monte-Carlo simulation of the MTJ device characteristic by a circuit simulator with a built-in MTJ device model (ns-spice_mtj). As shown in Fig. 2 (a), the simulator enables the definition of an MTJ device by a line of text including the input and

978-1-5386-5309-8/18 $31.00 © 2018 IEEE

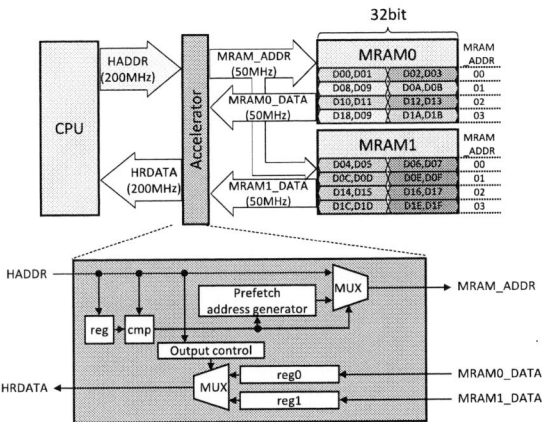

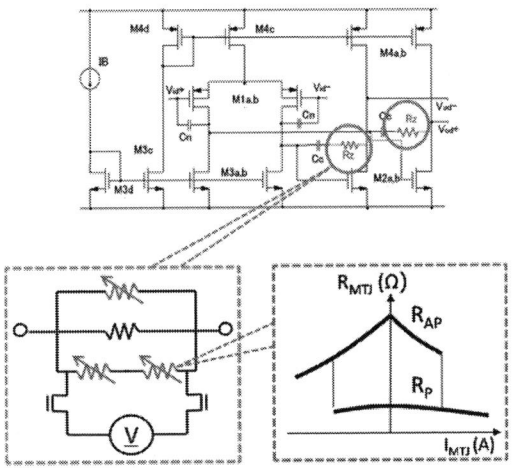

Figure 4: A proposed architecture. An accelerator circuit is embedded between a CPU and an STT-MRAM, and the memory access is done in a multiplexing manner.

Figure 6: Process variation-resilient operational transconductance amplifier.

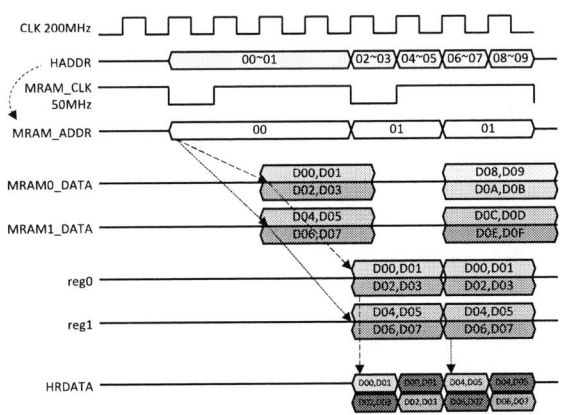

Figure 5: Detailed example data flow in the proposed architecture in case of of 16-bit instructions allocated at consecutive memory addresses.

output terminal names, model name, and other device-dependent parameters in the same manner as the other built-in devices such as transistors. The behavior of the circuit schematic can be simulated as shown in Fig. 2 (b) by applying the netlist to ns-spice_mtj, the circuit simulation result can be obtained as shown in Fig. 2 (c).

Figure 3 shows a test chip photomicrograph of the nonvolatile motion-vector prediction unit using a 90-nm

MTJ/MOS process on a 300-mm wafer fabrication line [18,19]. This chip is the world's first fabrication example fully designed by the proposed design environment for NV-LIM LSI. The circuit is designed by a combination of primitive cells in the MTJ/MOS-hybrid call library and those in the conventional CMOS cell library. Because the intermediate data can be maintained in the nonvolatile memory, the power supply for each domain can be precisely controlled over each operation cycle, without regard for data retention. In case that the power supply of each processing element was controlled independently cycle by cycle, its leakage power and total power were reduced to 26.0% and 48.3%, respectively, compared with those of the conventional CMOS-based design.

MEMORY ACCESS CONTROLLER FOR ENERGY-EFFICIENT NV-MCU

In realizing a nonvolatile microcontroller unit (MCU) for sensor nodes in Internet-of-Things (IoT) applications, it is important to solve the data-transfer bottleneck between the central processing unit (CPU) and the nonvolatile memory constituting the MCU. As one circuit-oriented approach to solving this problem, we propose a memory access minimization technique for MRAM-embedded nonvolatile MCUs [20,21]. Figure 4 shows a proposed architecture with the proposed memory access controller. In addition to multiplexing and prefetching of memory access, the proposed technique realizes efficient instruction fetch by eliminating redundant memory access while considering the code length of the instruction to be fetched and the transition of the memory address to be

978-1-5386-5309-8/18 $31.00 © 2018 IEEE

accessed as shown in Fig.5. As a result, the performance of the MCU can be improved while relaxing the performance requirement for the embedded MRAM, and compact and low-power implementation can be performed as compared with the conventional cache-based one. Through the evaluation using a system consisting of a general purpose 32-bit CPU and embedded MRAM, we can confirm that the proposed technique increases the peak efficiency of the system up to 3.71 times, while a 2.29-fold area reduction is achieved compared with the cache-based one.

DEPENDABLE CIRCUIT DESIGN USING MTJ DEVICE

As another possibility of the NV-LIM architecture, we can utilize the non-volatility of the MTJ device to design highly-reliable logic-circuit LSIs [22-24]. By using nonvolatile memory devices embedded into logic circuitry as programmable memory elements, we can compensate the performance degradation due to process variation after fabrication. For example, Fig. 6 shows a process variation-resilient operational transconductance amplifier (OTA) with MTJ-based multi-level variable resistors. The proposed compensation circuitry can be distributed over the circuit plane with quite a small overhead compared with conventional circuit conditioning methods because of the non-volatility and 3D stackability of MTJ devices, which leads to fine-grained circuit function optimization per circuit building block. Simulation results demonstrated that the proposed circuitry realizes flexibility for circuit performance tuning, which is capable of designing a process-variation-resilient analog circuit.

SUMMARY

NV-LIM architecture has a potential capability of realizing high-performance, ultra-low-power and highly resilient logic LSIs. As a future prospect, we are currently considering the possibility of realizing a new-type MTJ/MOS-hybrid logic LSI that enables to simulate the intelligent behavior in the brain such as massively-parallel processing, autonomous decentralized control, and learning and recognition function. By utilizing the knowledge obtained by our former results, we will create a path toward a new paradigm LSI beyond the brain.

ACKNOWLEDGEMENTS

The authors thank Y. Takako of Focal Agency for excellent technical assistance. Part of this research was supported by the JSPS ImPACT Program, R&D for Next-Generation IT of MEXT, Japan, and JSPS KAKENHI Grant Number 16KT0187.

REFERENCES

[1] T. Hanyu, et al., IEEE IEDM, pp.28.2.1-28.2.3, Dec. 2014.
[2] S. Ikeda, et al., IEEE Trans. Electron Devices, vol.54, no.5, pp.991-1002, 2007.
[3] R. Courtland, IEEE Spectrum, pp.11-12, 2014.
[4] N. Sakimura, et al., IEEE ISSCC, pp.184-185, Feb. 2014.
[5] H. Koike, et al., IEEE ASSCC 2013, pp.317-320, Nov. 2013.
[6] D. Suzuki, et al., J. Appl. Phys. vol. 115, 17B742, 2014.
[7] D. Suzuki, et al., IEICE ELEX, vol.10, no.23, 20130772, Dec. 2013.
[8] D. Suzuki, et al., IEEE Trans. Magn. vol.50, no.11, pp.3402104, 2014.
[9] D. Suzuki, et al., JJAP, vol.52, no.4, pp.04CM04, 2013.
[10] D. Suzuki, et al., JJAP, vol. 53, no.4, pp.04EM03, 2014.
[11] D. Suzuki, et al., IEICE ELEX, vol.11, no.13, pp.20140296, 2014.
[12] D. Suzuki, et al., Symp. VLSI Circuits, pp.172-173, June 2015.
[13] S. Matsunaga, et al., Symp. VLSI Circuits, pp.106-107, June 2013.
[14] S. Matsunaga, et al., JJAP, vol.51, no.2, pp.02BM06, 2012.
[15] S. Matsunaga, et al., IEICE ELEX, vol. 11, no.10, pp.20140297, 2014.
[16] S. Matsunaga, et al., Symp. VLSI Circuits, pp.298-299, June 2011.
[17] N. Sakimura, et al., IEEE ISCAS, pp. 1971-1974, May 2012.
[18] M. Natsui, et al., IEEE JSSC, vol.50, no.2, pp.476-489, 2015.
[19] M. Natsui, et al., IEEE ISSCC, pp.194-195, Feb. 2013.
[20] M. Natsui and T. Hanyu, Ext. Abstr. Int. Conf. Solid State Devices and Materials, pp. 977-978, 2017.
[21] M. Natsui and T. Hanyu, JJAP, 2018 (in press).
[22] M. Natsui and T. Hanyu, J. Multiple-Valued Logic and Soft Computing, Vol.21, No.5-6, pp.597-608, 2013.
[23] M. Natsui et al., ISMVL2012, pp.214-219, May 2012.
[24] Y. Kim et al., ISCAS2012, pp.2705-2708, May 2012.

DO WE HAVE TO WORRY ABOUT EXTENDED DEFECTS IN HIGH-MOBILITY MATERIALS?

Eddy Simoen[1]*, Po-Chun Hsu[2], Liang He[3], Yves Mols[1], Bernadette Kunert[1], Robert Langer[1], Niamh Waldron[1], Geert Eneman[1]. Nadine Collaert[1], Mark Heyns[1,2] and Cor Claeys[4]

[1]Imec, Kapeldreef 75, B-3001 Leuven, Belgium

[2] Dept. MTM, KU Leuven,

[3]School of Advanced Materials and Nanotechnology, Xidian University, 710126 Xi'an, Shaanxi, China

[4]Dept. Electrical Engineering, KU Leuven, Kasteelpark 10, B-3001 Leuven, Belgium

*Corresponding Author's Email: eddy.simoen@imec.be

ABSTRACT

An overview is given on how to tackle the question of the electrical activity of extended defects which are inevitably present in hetero-epitaxial III-V layers on silicon. Analysis methods are described which rely on simple device structures containing a specific type of extended defect (here, threading dislocations). Applying the same methods to real scaled FinFETs is rather challenging. Instead Generation-Recombination noise spectroscopy provides data that can be compared with other more standard techniques, like Deep-Level Transient Spectroscopy (DLTS) in order to identify the presence of an electrically active extended defect in the channel material.

INTRODUCTION

The use of hetero-epitaxy may offer novel functionalities to silicon, e.g., the integration of III-V materials to a CMOS platform enables to exploit the advantages that a direct band gap and high mobility can offer. However, one of the main issues to overcome is the presence of extended defects in the hetero-epitaxial layer, which form due to relaxation of the mechanical stress, associated with the lattice and thermal misfit between two different materials. This can lead to misfit and threading dislocations (TDs) (Fig. 1), stacking faults in different (111) glide planes or anti-phase boundaries, when a polar semiconductor is grown on a non-polar one.

Ideally, one would like to avoid the presence of such extended defects as they represent a local change in the crystal structure and hence in the band structure of the material. Huge efforts have been undertaken over the years to improve and optimize crystal growth in order to reduce the defect density and researchers have come up with some clever solutions, like the use of buffer layers or relying on the so-called aspect ratio trapping (ART), where inclined dislocations are stopped at the oxide sidewall in narrow trenches, so that the top layer is in principle defect free. However, this principle only works for TDs, while there is always a finite chance for the presence of extended defects close to the device channel.

The question is: can we tolerate the presence of such defects, can we put a number on the maximum density we would like to have and, in general, can we predict what their impact will be on the device performance?

It is the aim of this work to address these issues for the case of n-type InGaAs, which could be of potential interest for the fabrication of high-mobility nFinFETs, Tunnel-FETs or the integration of optical interconnects. The problem is tackled from two different angles. First, one can try to collect basic information on the impact of a specific type of extended defect – here, the focus is on TDs – on the electrical parameters of InGaAs (generation and recombination lifetime), using simple device structures, like a p-n junction diode or a Schottky barrier. The same structures can then be used in a more fundamental study of the electrical properties of TDs (the activation energy E_T; the capture cross section σ_n and the trap concentration N_T), using Deep-Level Transient Spectroscopy (DLTS) [1],[2]. Alternatively, one can try to assess the impact of a dislocation on individual transistors. This may be achieved for example by Generation-Recombination (GR) noise spectroscopy [3],[4].

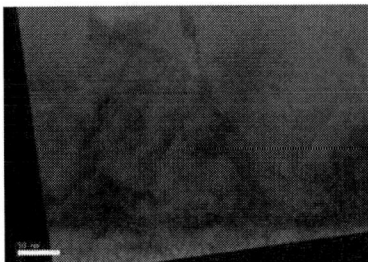

Figure 1: Transmission Electron Microscope (TEM) cross-section of threading and misfit dislocations in an $In_{0.53}Ga_{0.47}As$ layer on a silicon substrate.

BASIC STUDIES

Impact on diode characteristics

It has been shown before that p-n junction diodes are very useful tools to study the impact of defects on materials' properties [5]: the change in the reverse current is inversely proportional to the generation lifetime τ_G, while from the forward (diffusion) current the recombination lifetime τ_R can be derived. The ratio of the two provides a good estimate of the energy level of the responsible defects. Applying this to the case of In$_{0.53}$Ga$_{0.47}$As, epitaxial layers have been grown by Metalorganic Chemical Vapor Deposition (MOCVD) on a semi-insulating (SI) GaAs or an n-type InP substrate. The latter case is a lattice-matched reference with a low TD density (TDD). A bottom Si-doped InP layer is used for ohmic contact formation, while the diode is formed by a 1 μm thick undoped n-type and a Zn-doped p$^+$ InGaAs layer. Mesa structures have been etched and Mo/Al metallization is performed by lift-off. The schematic lay-out of the diodes is shown in Fig. 2, together with a cross-section Scanning Electron Microscope (SEM) image. The TDD has been evaluated by X-ray Diffraction (XRD).

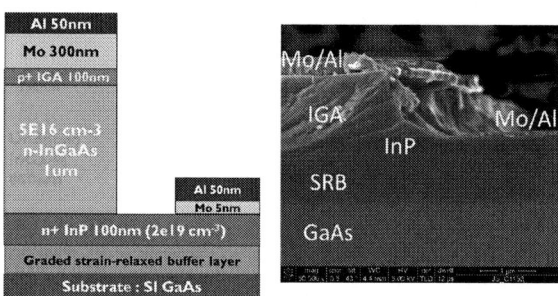

Figure 2: Schematic lay-out of the p$^+$n diode structure and cross-section SEM image after processing.

Current-voltage *(I-V)* and Capacitance-Voltage *(C-V)* measurements have been performed on large-area diodes, to derive the area current density like in Fig. 3. One can clearly see that the reverse current density increases with TDD, which can be ascribed to the impact of the dislocations on the generation lifetime τ_G. The extracted values are represented in Fig. 4, showing a reduction from about 1 μs for low TDD to about 10^{-8} s for a density of 10^9 cm^{-2}. Likewise, the recombination lifetime evolves from 0.2 μs to a few ns over the same range of TDD. From the dashed lines in Fig. 4, one can come to the conclusion that for a TDD below roughly 5×10^6 cm^{-2}, the lifetimes will be dominated by other grown-in or processing-induced defects. Beyond this limit, the TDs will dominate the lifetime properties. A similar result has been obtained in the past for the base layer in GaAs solar cells [6]. In that case, the minority carrier lifetime from Time-Resolved Photoluminescence could be modeled by [7]:

$$\tau_R^{-1} = \tau_{max}^{-1} + \pi^3 D_{min} TDD/4 \qquad (1)$$

with D_{min} the diffusion coefficient for minority carriers.

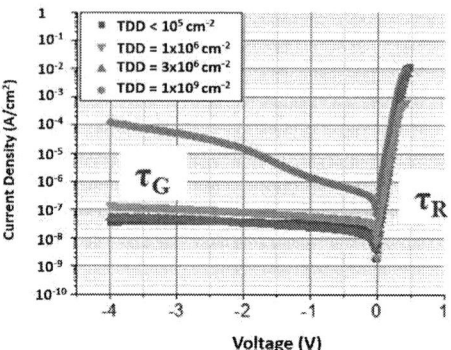

Figure 3: Current density versus bias for a set of In$_{0.53}$Ga$_{0.47}$As p$^+$n junction diodes with different TDD.

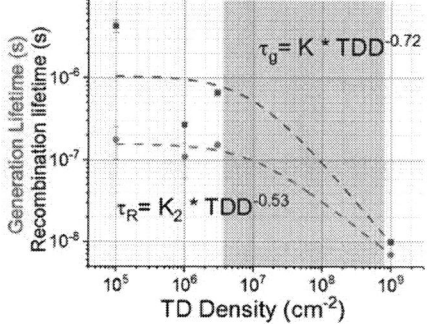

Figure 4: Generation (black square) and recombination lifetime (red dot) extracted from the diode I-V as a function of the TDD. The dashed curves serve mainly as a guide to the eye.

From the empirical trends in Fig. 4 rather a power law dependence is found with exponent between -0.53 (τ_R) and -0.72 (τ_G). However, more data points in the TDD range between 5×10^6 cm^{-2} and 10^9 cm^{-2} are required to determine more accurately the relationship. A square root dependence has been found in the past for the case of TDs in SiGe p$^+$n diodes, which has been ascribed to the fact that dislocations do not behave like a classical Shockley-Read-Hall (SRH) center [8]. One of the differences is that dislocations introduce a one-dimensional density-of-states instead of a discrete energy level in the band gap. These states cannot be considered independent of each other, so that upon charging a dislocation, a potential barrier builds up, yielding non-SRH capture. Currently, InGaAs material with a wider variety of TDD is being studied to further refine the obtained results.

The ratio of τ_G/τ_R yields a trap level in the range of 0.42 eV, slightly off-center of the band gap. As will be seen below, this is in agreement with the activation energy

978-1-5386-5309-8/18 $31.00 © 2018 IEEE   24

derived from DLTS measurements.

DLTS on diodes

The advantage of using a p^+n diode is that it enables the use of a forward-bias injection pulse during DLTS in order to address the minority carrier hole traps in the n-type depletion layer. The diodes have been designed such that the spectral information comes from the undoped n-layer. Spectra obtained on a diode with a TDD of 10^9 cm^{-2} are represented in Fig. 5. Three main peaks are observed, labeled E1 (electron trap) and H1 and H2 (hole trap). E1 has an activation energy $E_C-E_T = 0.45$eV and is thought to be associated with the TDs. This is among others supported by the large width of the peak, discernable in the bottom spectrum of Fig. 5. The origin of the hole traps is less clear for the moment; while H1 could be dislocation-related, H2 is most likely originating from surface states along the edges of the mesa diodes. An Al$_2$O$_3$ surface passivation layer is normally applied, which, however, degrades with time.

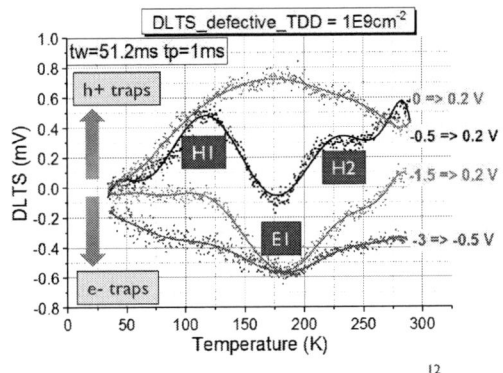

Figure 5: DLT-spectra for an InGaAs-on-GaAs p^+n junction with a TDD of 10^9 cm^{-2}. The bias pulses are indicated in the figure. A pulse frequency tw=51.2 ms and a pulse duration tp=1 ms has been utilized.

Our data for the activation energy for TDs in In$_x$Ga$_{1-x}$As is compared with the literature in Fig. 6.

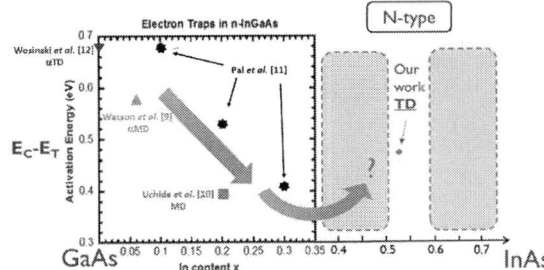

Figure 6: Evolution of the DLTS activation energy of the electron traps associated with TDs in In$_x$Ga$_{1-x}$As as a function of the In composition. The data are from Watson et al. [9], Uchida et al. [10], Pal et al. [11] and Wosiński et al. [12].

While for lower x, the E_C-E_T value drops more or less linearly, suggesting that the dislocation states are pinned to the top of the valence band for low In content alloys, the data obtained here indicate a reversal of this trend for $x>0.5$.

IMPACT ON TRANSISTORS

One should first of all realize that for a TDD in the range of 10^9 cm^{-2} there is only a limited probability that a small-area transistor contains a defect. Dislocations thus form in the first instance another source of device variability, whereby both the threshold voltage, the inversion-layer mobility and the density of interface states can be impacted by its presence. The larger the device area, the higher the probability for the presence of a TD.

While DLTS is the standard technique for the detection of electrically active defects in semiconductor materials, it usually requires a rather large-area device and is difficult to apply to minimum-size transistors [13]. One method which can be applied to any size of device is GR-noise spectroscopy. This relies on the GR noise spectrum, with a Lorentzian shape as a function of the frequency f. An example is shown in Fig. 7, corresponding with a 5 µm long In$_{0.3}$Ga$_{0.7}$As nFinFET with a fin width of 20 nm and a fin height of 25 nm (4 fins) [14]. The gate stack consists in this case of 1 nm Al$_2$O$_3$, 3 nm HfO$_2$ and 10 nm TiN gate.

What is represented is the product of $f \times S_I$, with S_I the noise power spectral density (PSD) as a function of f. The maximum position corresponds with the characteristic frequency of the GR center $f_0=1/(2\pi\tau_0)$. As can be seen in Fig. 7, f_0 is not changing with the gate voltage (or with the current through the transistor), which indicates that the GR centers are in the semiconductor channel and not in the gate oxide [15]. In that case, the characteristic GR center time constant τ_0 can be studied as a function of the temperature T. According to Eq. (2), τ_0 is thermally activated and can be represented in an Arrhenius diagram [15]:

$$\ln(\tau_0 T^2)=(E_C-E_T)/kT + \ln[h^3/(4k^2\sigma_n(6\pi^3 M_c \cdot m_e^{*3/2} \cdot m_h^{*1/2})^{0.5}]$$
(2)

In Eq. (2), E_C is the conduction band minimum, k is Boltzmann's constant. h is Planck's constant, M_c the number of equivalent maxima in the conduction band, m_e^* and m_h^* are the conduction effective mass for electrons and for holes, respectively. The trap activation energy can be derived from the slope of the Arrhenius plot, while the capture cross section is calculated from the intercept.

The noise spectra have been measured as a function of temperature, keeping the bias conditions fixed and stepping T by 25 K between 300 and 400 K. The resulting Arrhenius plots obtained on several 5 µm long nFinFETs are represented in Fig. 8. From the slopes of the linear fits, different activation energies in the range of 0.33 eV to 0.45 eV have been derived [14]. In order to identify the GR centers, one can compare with literature data, mainly obtained by DLTS analysis. Comparing with Fig. 6, an E_T value around 0.40 eV may correspond to the presence of a dislocation in the $In_{0.3}Ga_{0.7}As$ fin. Further proof could then come from a Transmission Electron Microscopy inspection of the specific device.

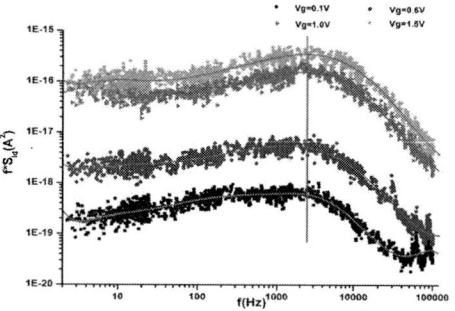

Figure 7: Normalized PSD as a function of the frequency corresponding with different gate voltages V_g and a fixed drain voltage (0.5 V).

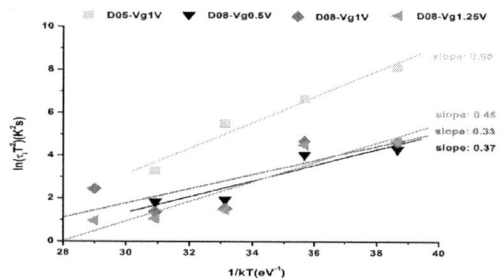

Figure 8: Arrhenius plot obtained for several L=5 μm $In_{0.3}Ga_{0.7}As$ nFinFETs at constant drain and gate voltage, while varying the temperature between 300 and 400 K. The slope represents the activation energy of the different GR centers.

CONCLUSIONS

A methodology has been described to determine the electrical activity of grown-in induced extended defects in high-mobility channel materials on silicon. In first instance, simple device structures can be fabricated containing a specific type of extended defect. From the change in the forward and reverse current density one may extract a relationship between the generation/ recombination lifetime and the defect density. More in-depth characterization can be achieved by performing DLTS analysis on the same diode structure, when properly designed. For the case of scaled transistors, it is much more difficult to establish whether an extended defect is present or not. Information can be indirectly gained by applying GR noise spectroscopy. DLTS results then serve as a benchmark to possibly identify the observed GR centers in the high-mobility channel material.

ACKNOWLEDGEMENTS

This work has been performed in the frame of imec's Core Partner Program on III-V High Mobility Devices.

REFERENCES

[1] D. V. Lang, *J. Appl. Phys.*, vol. 45, 1974, pp. 3023-3032.

[2] D. K. Schroder, *Semiconductor Material and Device Characterization*. Wiley, New York, 1998

[3] N.B. Lukyanchikova, *Noise and Fluctuations Control in Electronic Devices*, A. Balandin (Ed.), American Scientific, Riverside, CA, 2002.

[4] D. C. Murray, A. G. R. Evans and J. C. Carter, *IEEE Trans. Electron Devices*, vol. 38, 1991, pp. 407-416.

[5] E. Simoen, C. Claeys and J. Vanhellemont, *Defect and Diffusion Forum*, vol. 261-262, 2007, pp. 1-24.

[6] C. L. Andre, J. J. Boeckl, D. M. Wilt, A. J. Pitera, M. L. Lee, E. A. Fitzgerald, B. M. Keyes and S. A. Ringel, *Appl. Phys. Lett*, vol. 84, 2004, pp. 3447-3449.

[7] M. Yamaguchi, A. Yamamoto and Y. Itoh, *J. Appl. Phys.*, vol. 59, 1986, pp. 1751-1753.

[8] G. Eneman, E. Simoen, R. Delhougne, P. Verheyen, V. Simons, R. Loo, M. Caymax, C. Claeys, W. Vandervorst and K. De Meyer, *J. Electrochem. Soc.* Vol. 153, 2006, pp. G379-G384.

[9] G. P. Watson, D. G. Ast, T. J. Anderson, B. Pathangey and Y. Hayakawa, *J. Appl. Phys*, vol. 71, 1992, pp. 3399-3407.

[10] Y. Uchida, H. Kakibayashi and S. Goto, *J. Appl. Phys.*, vol. 74, no, 1993, pp. 6720-6725.

[11] D. Pal, E. Gombia, R. Mosca, A. Bosacchi and S. Franchi, *J. Appl. Phys.*, vol. 84, 1998, pp. 2965-2967.

[12] T. Wosiński, *J. Appl. Phys.*, vol. 65, 1989, pp. 1566-1570.

[13] E. Simoen, J. Lauwaert and H. Vrielinck, *Semiconductors and Semimetals*, Eds. L. Romano, V. Privitera and C. Jagadish, vol. 91, pp. 205-250, Elsevier, 2015.

[14] L. He, E. Simoen, C. Claeys, H. Chen, D. D. Guo, L. N. Hu and Y. Qin, *Proc. of ICNF 2017*, Vilnius (Lithuania), 20-23 June, 2017, IEEE Explore doi: 10.1109/ICNF.2017.7985987.

[15] V. Grassi, C. F. Colombo, and D. V. Camin, *IEEE Trans. Electron Devices*, vol. 48, 2001, pp. 2899-2905.

PREDICTIVE AS-GROWN-GENERATION MODEL FOR NBTI OF ADVANCED CMOS DEVICES AND CIRCUITS

Zhigang Ji[*], Rui Gao, and Jianfu Zhang

School of Engineering, Liverpool John Moores University, Liverpool, UK

*Corresponding Author's Email: z.ji@ljmu.ac.uk

ABSTRACT

Increasing pressure on the quest towards low power applications with stringent reliability demands requires narrowing the gap between NBTI models and real circuit operation for design optimization. In this work, we reviewed the recent-proposed As-grown-Generation (AG) model. By separating different types of defects through experiment, we demonstrated that it can make reliable prediction beyond the measurement window. Compared with other existing models, the proposed model delivered its original mission: Reliably predicting long term aging at low Vg based on a model extracted from Vg-accelerated short tests. Its simple formula makes it readily implementable for circuit level simulation.

INTRODUCTION

Negative bias temperature instability (NBTI) is a well-known reliability issue which can has impact on CMOS integrated circuits [1]. Although firstly reported in 1966 [2], it is only during the last ten years that NBTI became the most serious reliability issue due to the increase of the gate electric field and chip operating temperature with scaling and the routinely introduction of nitrogen adding to the gate dielectric [3]. Modeling the negative bias temperature instability (NBTI) is essential for optimizing circuit design. Several models have been proposed [4-6]. They are usually physical-based and are validated by fitting with the test data. The drawbacks are two-fold: a) There is still no agreement on physical mechanism yet; b) By increasing the number of parameters, the apparent success of the model could be just the consequence of good fitting of existing data and thus have little use in predicting the unknown circuit operating condition. There is a pressing need for a NBTI model to reliably predicting long term NBTI at low use-bias, based on a model extracted from Vg-accelerated short tests.

In this work, a defect-based As-grown-Generation (A-G) model is proposed to model the NBTI-induced degradation. Through separating different types of defects directly from the measurement, the framework shows good predictive capability for long term degradation under any given voltages. The work is arranged as follows: The details about the devices is described in Section II. Then in Section III, we will show how different types of defects can be experimentally separated. Based on the extracted kinetics, the A-G model will be proposed and validated in section IV. Finally, it comes to the conclusion of the paper.

DEVICES AND EXPERIMENT

pFinFETs fabricated using a Hf-based high-k oxide and a metal gate are used to demonstrate the proposed model. The EOT is sub-1-nm with TiN metal gate [7]. Fast measurement with the Id-Vg measured within 3μs is used [8]. The threshold voltage degradation is monitored by sensing at a constant Id of 500nA*W/L around threshold voltage. Unless specified, the temperature is 125 °C.

Although NBTI is considered as electric-field driven phenomenon [3], the tests were usually performed under the constant Vg. The underlying assumption is that the total degradation, ΔVth, is much smaller than the applied voltage. **Fig.1** compared the NBTI degradation under the constant-Vg and constant-E condition, wherein, constant-E is approximately achieved by consistently increasing stress bias, Vgst, by the amount of ΔVth, which is measured in the last step. Clear difference can be observed between the two, indicating that the electric filed has been affected by the ΔVth. In this work, unless specified, the constant-E test will be applied.

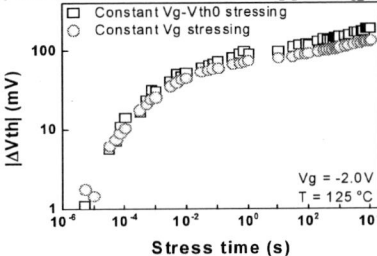

Fig. 1 The comparison between the constant Vg and constant Eox paradigms for NBTI.

As-GROWN-GENERATION (A-G) FRAMEWROK

The pre-existing defects and the generated defects have been considered as two main types of defects contributing in NBTI. The former are those already exist in the dielectric, while the latter require much higher voltage to create in the first instance. To model NBTI correctly, it is important to characterize and understand the kinetics of each defect, as will be elaborated below.

i. Generated defects (GD)

Clear experiments, such as SILC [9] and charge pumping [10] strongly support the existence of GD. It is usually extracted by arbitrarily taking the last point at the end of the recovery trace [11], making it sensitive to the measurement conditions such as. discharging time, tdisch, and the discharge voltage, Vgdisch. To understand this phenomenon, the energy distribution of defects [12] are

978-1-5386-5309-8/18 $31.00 © 2018 IEEE

extracted from the device in fresh and after 1ks stress, as shown in **Fig.2a**. Subtracting the fresh from the stressed one gives the distribution of GD component, as shown in **Fig.2b**. Clearly, GDs are located both within the Si bandgap and beyond Si conduction band. In favor of the energy level, those GD within the bandgap can be charged and discharged repeatedly, so that they are referred to as cyclic positive charges (CPC). On the other hand, the defects above Si Ec are more difficult to neutralize and they are called as Anti-Neutralization positive charges (ANPC), as shown in **Fig.2c**.

Fig.3a&d shows ANPC reduces against either Tdisch or Vgdisch. However, the same amount of the increase can be observed in CPC in **Fig. 3b&e**. This can be understood that the discharged ANPC can be refilled and thus be accounted into CPC. The total GD, i.e. GD=CPC+ANPC in **Fig.3c&f**, is independent of Tdisch and Vgdisch. Based on the understanding, the Stress-Discharging-Recharging (SDR) procedure can be used to capture the kinetics of the entirety of GD, as shown in **Fig.4** [13]: After stressing under Vgst for a certain time, the measurement starts by applying an opposite polarity of Vgdisch to accelerate the discharge of pre-existing traps. Then the low use-bias under real operation Vgch ($|Vgch| < |Vgst|$) was applied to refill all the traps that contribute under use condition. By subtracting the measured degradation with the same recharging step but on fresh device, GD can be obtained. **Fig. 5** shows the measured GD kinetics under different overdrive voltage, Vgov. GD follows power law relationship against time and can be well described with the classical power law using Eqn (1) for long term prediction. Wherein, g0, m and n are the pre-factor, voltage and time exponent respectively. The time exponent is 0.2.

$$GD = g_0 \cdot V_{gov}{}^m \cdot t^n \quad (1)$$

ii. Pre-existing defects

Pre-existing defects originate from the fabrication imperfectness. their charging and discharging kinetics will not be affected by the device's stress history. Therefore, they can be characterized using heavily stressed devices.

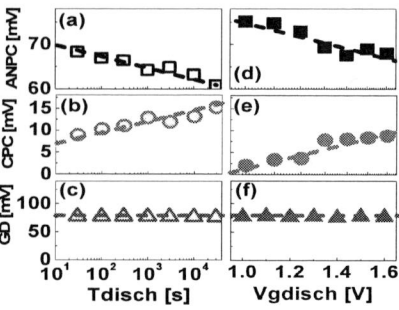

Fig. 3 Dependence on discharge conditions for **(a&d)** ANPC, **(b&e)** CPC and **(c&f)** GD=ANPC+CPC. The discharging voltage is +1.6V. Wherein, ANPC equals to Vth(end of discharge) – Vth(Fresh), CPC equals to Vth(end of recharge) – Vth(end of discharge) and GD is the sum of ANPC and CPC.

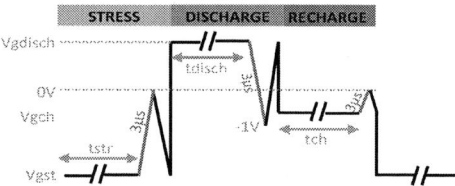

Fig. 4 The test follows a stress-discharge-recharge sequence. At the end of each step, ΔVth was monitored from a corresponding Id~Vg, taking from the 3μs pulse edge.

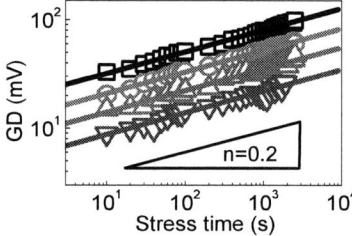

Fig.5 Generation kinetics of GD. The degradation follows power law with Vgov and time.

To understand these pre-existing traps, the energy profile is extracted after the device charging up under the bias, Vgch [14]. As shown in **Fig.6**, when Vgch is low, the profiles after filling at different Vgch overlap well. However, when Vgch increases further, they deviate from each other. This can be considered that two different types of hole traps in the gate dielectric, as illustrated in **Fig. 7a&b**: one type can capture a hole without changing its energy level. While another type, after capturing one hole, its energy level shift upwards from the ground state to the charged state, resulting in the increased ΔVth in **Fig.6** at a given Vgdisch-ΔVth, after filling at higher Vgch. One simple method can be applied to separate these two types of traps by exploiting the parallel region of the distributions extracted for two consecutive Vgch levels, as illustrated in **Fig. 8a**: starting with the distribution trace at the lowest Vgch in which only AT traps are involved, the

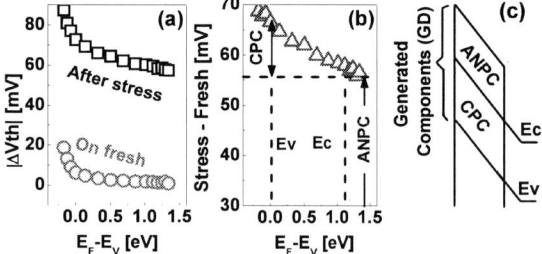

Fig. 2 **(a)** Comparison of the energy profiles before and after stress. **(b)** The profile for GD extracted substracting two curves in (a). **(c)** Illustration of the energy range for the cyclic positive charges (CPC) and the anti-neutralization positive charges (ANPC).

978-1-5386-5309-8/18 $31.00 © 2018 IEEE   28

discharging trace under the next charging level, Vgch+ΔVgch, is shifted down to align these two curves within the tail region from the charging level. By following this procedure up to the highest Vgch, the distribution of saturated AT traps can be extracted from the experimental data and fitted empirically with Eqn(2), as shown in **Fig. 8b**. EAD traps can then be extracted by subtracting AT from the total, as shown in **Figs. 9**.

$$Saturated\_AT = p_1 \cdot \exp(p_2 \cdot V_{gov}) \qquad (2)$$

The charging kinetics of EAD under different Vgov are shown in **Fig. 10**. EAD follows power law relationship in time and voltage and can be modelled with Eqn (3). It is important to note that although both EAD and GD follow power law relationship, they have different time exponent and thus need to be modelled separately. **Fig. 10a** shows the charging kinetics for AT under different Vgov, they all quickly reach saturation. When they are normalized against its saturated value, as shown in **Fig. 10b**, they follow the same stretched exponent kinetics as described by Eqn (4).

$$EAD = g_2 \cdot V_{gov}^{m2} \cdot t^{n2} \qquad (3)$$

$$AT = Saturated\_AT \cdot [1 - \exp(-\frac{t_{AT\_ch}}{\tau})^{\gamma}] \qquad (4)$$

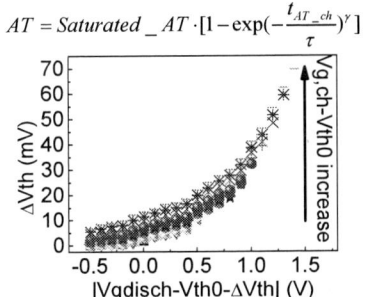

Fig. 6. Energy distribution of the Pre-existing trap after charging under different levels, $V_{g,ch}$ for 1000 s at 125°C.

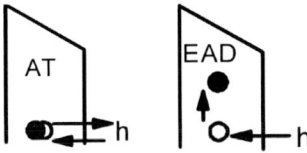

Fig. 7. Illustration of charging (a) AT and (b) EAD traps. AT do not change energy level while EAD will after charging.

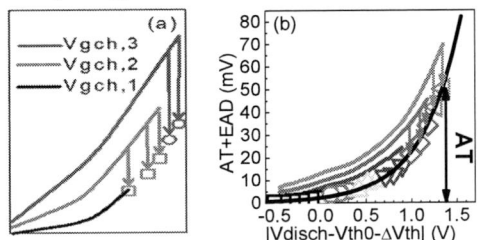

Fig. 8 (a) Illustration on extracting AT. **(b)** The method is applied to real data.

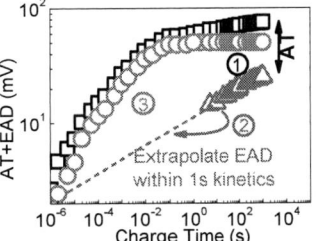

Fig. 9 EAD kinetics by subtracting total AT from total degradation. AT kinetics is obtained by subtracting EAD kinetics from the total.

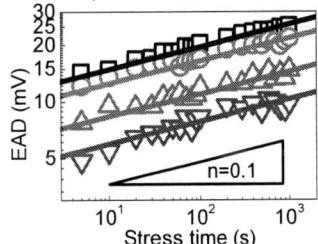

Fig.10 The charging kinetics of EAD: The degradation follows power law with Vgov and time.

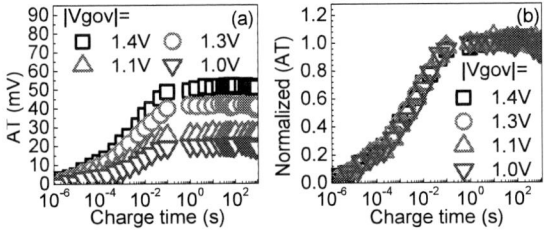

Fig. 11 (a) AT charging kinetics under different Vgov, it saturated in short time. **(b)** When they are normalized against its saturated value, it follows the same stretched exponent kinetics

It is widely accepted that charging and discharging of pre-existing traps dominates during AC operation. Therefore, the discharging kinetics is required, which can be measured at the end of charging under each Vgov. The discharging of the total pre-existing traps under different charging time or voltages are shown in **Fig.12a&b**. They can be scaled following universal recovery curve as shown in **Fig.12c** and described with Eqn (5).

$$f(t_{disch}) = (1 + B \cdot t_{disch}^{\beta})^{-1} \qquad (5)$$

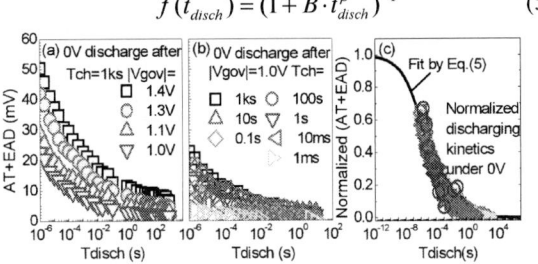

Fig. 12 Discharging kinetics of the pre-existing defect after different charge voltage **(a)** and charging time **(b)**. **(c)** They can be normalized with universal recovery trace as described in Eqn (5).

Model Prediction and Validation

Based on the knowledge of the defects involving in the NBTI degradation, an A-G simulator can be built on the Eqns (1) - (5) [15]. By solving these coupled equations, the NBTI degradation under both DC and AC operation mode can be predicted. To be practical useful, the model must has the predictive capability, i.e. the model parameters collected from short-term high stress conditions should be able to be used to predict the long-term degradation kinetics at devices' low operating condition. One demonstration is shown in **Fig. 13**. Four V_g-accelerated short DC stresses were firstly carried out to extract the A-G model, which then is used to predict the both DC and AC NBTI under use-V_{g_ov}. The good agreement between the prediction and the measurement can be observed, delivering the original mission of NBTI modelling.

It is intuitive to understand the contribution of each defect under operating condition. As shown in **Fig. 14**. For DC NBTI stress, it is obvious that there are more EADs compared with GD at the initial stage. However, due to the higher time exponent, GD increase much faster than EAD and plays more important role in the long term. For AC stress, since GD has negligible discharge at 0V while EAD discharges a lot, in a much earlier term GD has already surpassed EAD compared to DC case. AT is relative small compared with the other two components.

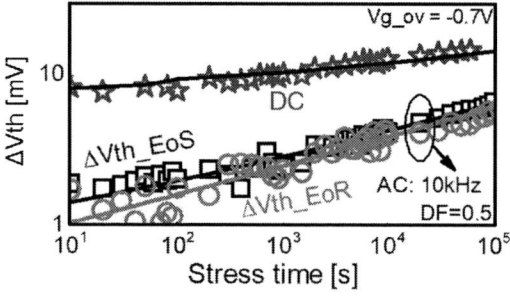

Fig. 13 Demonstration of the prediction capability for both DC and AC NBTI. The frequency and DF dependent after stressing for 1ks is compared between the prediction (lines) and test data (symbols).

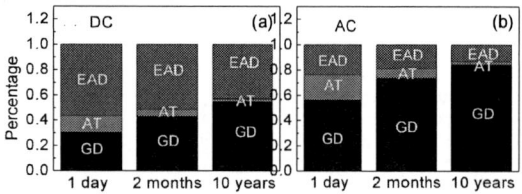

Fig. 14 Kinetics of different components against stress time under DC (a) and 1kHz (b) NBTI stress. The normalized contribution of different components under different stress time is given in (c) DC

ACKNOWLEDGEMENTS

The authors thank B. Kaczer, D. Linten, and G. Groeseneken of IMEC, Belgium, for fruitful discussions and supply of test samples. This work is supported by the EPSRC of UK under the grant no. EP/L010607/1.

REFERENCES

[1] W. Wang, S. Yang, S. Bhardwaj, S. Vrudhula, F. Liu, and Y. Cao, "The Impact of NBTI Effect on Combinational Circuit: Modeling, Simulation, and Analysis," IEEE Trans. VLSI, p.173, 2010.

[2] B. E. Deal, M. Sklar, A. Grove, and E. Snow, "Characteristics of the Surface - State Charge (Qss) of Thermally Oxidized Silicon," J. Electrochem. Soc., p. 266, 1967.

[3] D. K. Schroder, "Negative bias temperature Instabilty: What do we understand?" Micro. Rel., p.841, 2007.

[4] S. Deora, V. D. Maheta and S. Mahapatra., "NBTI Lifetime Prediction in SiON p-MOSFETs by H/H2 Reaction Diffusion(RD) and Dispersive Hole Trapping Model," in *Proc. IRPS*, p. 1105, 2010.

[5] V. Huard, C. Parthasarathy, N. Rallet, C. Guerin, M. Mammase, D. Barge, and C. Ouvrard, "New characterization and modeling approach for NBTI degradation from transistor to product level," in *IEDM.*, p. 797, 2007,

[6] T. Grasser, B. Kaczer, W. Goes, H. Reisinger, T. Aichinger, P. Hehenberger, P. J. Wagner, F. Schanovsky, J. Franco, M. Luque and M. Nelhiebel, "The paradigm shift in understanding the bias temperature instability: From reaction–diffusion to switching oxide traps," IEEE Trans. Electron Devices, pp. 3652, 2011.

[7] L, Ragnarsson, Z. Li, J. Tseng, T. Schram, E. Rohr, M. J. Cho, T. Kauerauf, T. Conard, Y. Okuno, B. Parvais, P. Absil, S. Biesemans and T. Y. Hoffmann, "Ultra low-EOT gate-first and gate-last high performance cmos achieved by gate-electrode optimization", IEDM p.1, 2009.

[8] Z. Ji, L. Lin, J. F. Zhang, B. Kaczer, and G. Groeseneken, "NBTI Lifetime Prediction and Kinetics at Operation Bias Based on Ultrafast Pulse Measurement," IEEE Trans. Electron. Dev. p. 228, 2010.

[9] J. Q. Yang, M. Masuduzzman, J. F. Kang and M. A. Alam, "SILC-Based Reassignment of Trapping and Trap Generation Regimes of Positive Bias Temperature Instability," in Proc. IRPS, p.196, 2011.

[10] L. Lin, Z. Ji, J. F. Zhang, W. Zhang, B. Kaczer, S. De Gendt and G. Groeseneken, "A Single Pulse Charge Pumping Technique for Fast Measurements of Interface States," IEEE Trans. Electron. Dev., p.1490, 2011.

[11] X. Wang, S.-H. Song, A. Paul, and C. H. Kim, "Fast characterization of PBTI and NBTI induced frequency shifts under a realistic recovery bias using a ring oscillator based circuit," in Proc. IRPS, p. 6B.2.1, 2014.

[12] Z. Ji, S. Hatta, J. F. Zhang, J. Ma, W. Zhang, N. Soin, B. Kaczer, S. De Gendt, and G. Groeseneken, "Negative bias temperature instability lifetime prediction: Problems and solutions," in IEDM Tech. Dig., p. 15.6.1, 2013.

[13] R. Gao, A. B. Manut, Z. Ji, J. Ma, M. Duan, J. F. Zhang, J. Franco, S. W. M. Hatta, W. Zhang, B. Kaczer, D. Vigar, D. Linten, and G. Groeseneken, "Reliable time exponents for long term prediction of negative bias temperature instability by extrapolation," IEEE Trans. Electron. Dev., p.1490, 2017.

[14] P. P. Ren, R. Gao, Z. Ji, H. Arimura, J. F. Zhang, R. Wang, R. Huang, M. Duan, W. Zhang, J. Franco, S. Sioncke, D. Cott, J. Mitard, L. Witters, H. Mertens, B. Kaczer, A. Mocuta, N. Collaert, D. Linten, A. V.-Y. Thean, and G. Groeseneken, "Understanding charge traps for optimizing Si-passivated Ge nMOSFETs," VLSI-T, p.214 2015.

[15] R. Gao, Z. Ji, S. M. Hatta, J. F. Zhang, J. Franco, M. Duan, W. Zhang, B. Kaczer, S. De Gendt, D. Linten and G. Groeseneken, "Predictive, defect-centric As-grown-Generation (A-G) model for device/circuit co-design with nanoscale CMOS technology," VLSI-T p.355, 2016.

Resistive switching and synaptic plasticity in HfO₂-based memristors with single-layer and bilayer structures

Qingxi Duan, Liying Xu, Jiadi Zhu, Xinhao Sun, Yuchao Yang[*], and Ru Huang

Key Laboratory of Microelectronic Devices and Circuits (MOE),
Institute of Microelectronics, Peking University,
Beijing 100871, China
Email: yuchaoyang@pku.edu.cn

ABSTRACT

Here we report a device size and structure dependent study on the resisitive switching and synaptic plasticity in HfO₂-based memristive devices. We found that the on/off ratio of resistive switching constantly increases as device size decreases, and bilayer Pt/TaO$_x$/HfO₂/Pt devices generally exhibit more uniform switching characteristics and lower threshold voltages compared with single-layer Pt/HfO₂/Pt devices. Long-term potentiation and depression behaviors have been emulated using such bilayer devices, suggesting potential applications for neuromorphic hardware.

Keywords—memristor, synaptic plasticity, conducting filament, oxygen vacancy

INTRODUCTION

As emerging nanoionic devices with high scalability and rich dynamics, memristors have acquired a lot of attention due to their potential applications in memory logic[1], nonvolatile memory[2], and neuromorphic computing etc[3]. In neuromorphic systems, memristors are envisioned to implement the functionality of synapses in a compact way and therefore enable learning and memory in resultant neural networks. A number of synaptic functions have been emulated using memristors till now, such as spike timing dependent plasticity[4], paired pulse facilitation[5], heterosynaptic plasticity[6]. It is of critical significance to develop scalable nanodevices with the capability of emulating synaptic plasticity and similar dynamics in order to truly simulate the functionality of the human brain.

In this work, we study the resistive switching behavior and synaptic plasticity in Pt/HfO₂/Pt and Pt/TaO$_x$/HfO₂/Pt devices The results showed that the on/off ratio constantly increases as device size decreases, and bilayer Pt/TaO$_x$/HfO₂/Pt devices generally exhibit more uniform switching characteristics and lower threshold voltages compared with single-layer Pt/HfO₂/Pt devices. The long-term potentiation (LTP) and long-term depression (LTD) implemented using Pt/TaO$_x$/HfO₂/Pt devices indicate potential application of these device as electronic synapses.

EXPERIMENTAL

All devices studied in this work were fabricated on SiO₂/Si substrates. Both the bottom and top electrodes were patterned by photo lithography, and subsequently 30 nm thick Pt film with 5 nm thick Ti adhesion layer was deposited by e-beam evaporation and adopted as the electrodes after lift-off processes Devices with a capacitor structure and four different sizes (300×300 μm², 200×200 μm², 100×100 μm², 50×50 μm²) and crossbar structure (2×2 μm²) were both fabricated 5 nm thick HfO₂ film was deposited by magnetron sputtering using a HfO₂ target and serves as the switching material in Pt/HfO₂/Pt devices. A TaO$_x$ layer was further deposited on top of HfO₂ by direct current (DC) reactive sputtering using a Ta metal target in Ar/O₂ gas mixture (the oxygen partial pressure was 3%) at 400 °C, forming a switching bilayer in Pt/TaO$_x$/HfO₂/Pt devices.

RESULTS AND DISCUSSION

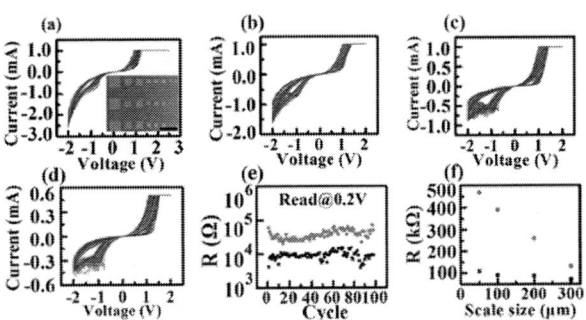

Fig. 1 *I–V* curves in 100 cycles of measurements from Pt/HfO₂/Pt devices with a size of (a) 300×300 μm², (b) 200×200 μm², (c) 100×100 μm² and (d) 50×50 μm². The inset in (a) shows scanning electron microscopy image of the devices. Scale bar: 500 μm. (e) Endurance property of Pt/HfO₂/Pt devices with a size of 100×100 μm². (f) Dependence of on- and off-state resistance on device size.

Fig 1(a-d) shows the I-V curves of Pt/HfO₂/Pt single-layer devices with four different sizes, indicating reliable bipolar switching characteristics, as also shown by the endurance property of 100×100 μm² device as an example. Besides, it can be observed from Fig 1(a-d) that the maximum current during reset decreases as the device area decreases. Fig 1(f) further shows the dependence of on- and off-state resistance values of Pt/HfO₂/Pt devices on cell size. While off-state resistance

---

978-1-5386-5309-8/18 $31.00 © 2018 IEEE

increases as cell size decreases, on-state resistance shows weak dependence on the size of devices, thus suggesting that the resistive switching is induced by formation of conducting filaments. The above effects have resulted in a larger on/off ratio when device size decreases, as shown in Fig 1(f).

The improved switching characteristics as device size decreases have suggested us to further scale down the devices. We have therefore replaced the capacitor structure of Pt/HfO$_2$/Pt single-layer devices with a crossbar structure to allow much smaller cell size, as shown in Fig 2(a). Indeed, the *I-V* characteristics and endurance measurements have shown even larger on/off ratio in Pt/HfO$_2$/Pt devices with a size of 2×2 μm^2. However, it could be noticed that the Pt/HfO$_2$/Pt devices exhibit relatively high switching voltages up to 4 V and large variations in the threshold voltages (Fig 2(b)). Fig 2(c) further indicates existence of notable cycle-to-cycle variation in off-state resistance.

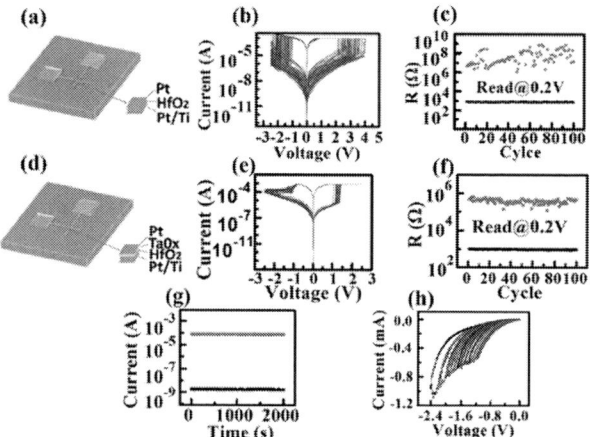

Fig. 2 Pt/HfO$_2$/Pt and Pt/TaO$_x$/HfO$_2$/Pt crossbar devices. (a) Schematic of the single-layer crossbar structure. (b) *I-V* sweeps repeated for 100 cycles. (c) Endurance characteristics of the Pt/HfO$_2$/Pt crossbar devices with a size of 2×2 μm^2. (d) Schematic of the bilayer device structure. (e) *I-V* sweeps repeated for 100 cycles. (f) Endurance characteristics of the Pt/TaO$_x$/HfO$_2$/Pt crossbar devices with a size of 2×2 μm^2. (g) Retention characteristics of the Pt/TaO$_x$/HfO$_2$/Pt crossbar devices. (h) Analog switching achieved by gradual reset.

In order to further optimize the device uniformity, we have introduced a TaO$_x$ layer with substantial oxygen vacancy (V$_O$) concentration on top of HfO$_2$ to form a TaO$_x$/HfO$_2$ bilayer structure, as schematically shown in Fig 2(d). It has been reported that such bilayer structure is favorable for improving the switching uniformity of valence change based memristive devices[7, 8]. Indeed, Fig 2(e-f) shows significantly improved device uniformity and decreased switching voltages, and Fig 2(g) also verifies that the resistive switching in Pt/TaO$_x$/HfO$_2$/Pt devices is nonvolatile. More importantly, the introduction of the oxygen-deficient TaO$_x$ layer has enabled the exchange of V$_{OS}$ between the TaO$_x$ and HfO$_2$ layers and subsequent formation of conducting filaments mainly in the more insulating HfO$_2$ layer, serving as the switching mechanism in these devices. This well-defined oxygen exchange dynamics are significantly different from the distribution of V$_{OS}$ in single-layer HfO$_2$, leading to more

gradual switching characteristics in Pt/TaO$_x$/HfO$_2$/Pt devices, especially in the reset process (Fig 2(e)). We can therefore take advantage of such gradual switching behavior in reset and utilize this to achieve analog resistive switching, as shown in Fig 2(h). Such analog switching is essential for the implementation of synaptic plasticity, as will be discussed in more detail afterwards.

One of the important requirements for synaptic devices is that their weights are incrementally adjustable in order to enable online learning capabilities. The above results obtained from Pt/HfO$_2$/Pt single-layer devices and Pt/TaO$_x$/HfO$_2$/Pt bilayer devices have shown that the Pt/TaO$_x$/HfO$_2$/Pt devices have better analog switching characteristics that can be used for implementation of synaptic potentiation and depression processes. Fig 3(a) shows the adjustment of device resistance by a consecutive sequence of positive and negative voltage sweeps, suggesting continuous potentiation and depression, respectively. Fig 3(b) further shows the modulation of device conductance by electric pulses. The measurement includes 500 identical potentiation pulses with an amplitude of 1.2 V and width of 100 ns, while the interval between neighboring pulses is 20 ms. These are followed by the 500 identical depression pulses with an amplitude of -1.5 V and width of 100 ns, and the interval between neighboring pulses is once again 20 ms. As can be seen in Fig 3(b), LTP and LTD behaviors analogous to biological systems can be implemented using the Pt/TaO$_x$/HfO$_2$/Pt bilayer devices.

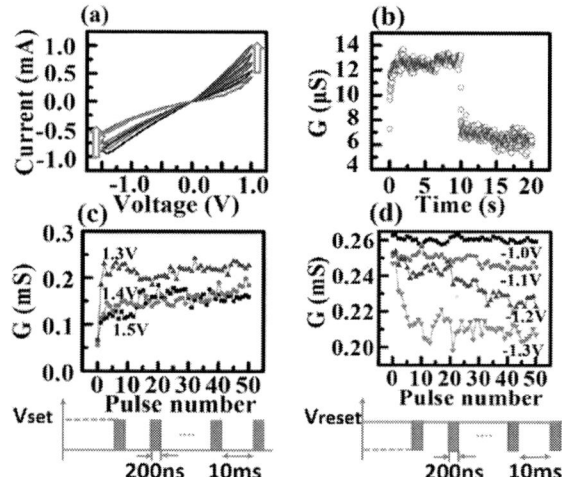

Fig. 3 (a) Continuous conductance modulation of the device by repeated voltage sweeps in the same range. (b) Potentiation and depression behaviors by applying identical pulses (potentiation: 1.2 V, 100 ns; depression: −1.5 V, 100 ns). (c-d) Potentiation/depression processes of the device by pulse tests with varied amplitudes. The pulse width was fixed at 200 ns and the interval between neighboring pulses was 10 ms.

The amount of conductance change in each step can be tuned by the amplitude of the applied pulses. As shown in Fig 3(c) and 3(d), when the pulse width was fixed at 200 ns and the pulse interval was fixed at 10 ms, a higher pulse amplitude usually leads to a larger tuning step during LTP and LTD processes. This is in agreement with the electric field-driven V$_{OS}$ migration and filament formation mechanism. It can be seen that for each potentiation or depression process the resistance changes gradually and finally saturates after a certain number of pulse stimulations. The amplitude of the LTP and LTD pulses also effectively changes the conductance level where the saturation takes place, leading to generally larger tuning ratio when the pulse amplitude is increased, as can be seen in Fig 3(c-d). The aboverealization of LTP and LTD plasticity and the tunability indicate the potential of the Pt/TaO$_x$/HfO$_2$/Pt devices for applications as electronic synapses in emerging neural computing systems. However, further optimization might be needed for optimizing the linearity of weight adjustment in these devices[9].

The above resistive switching behavior and synaptic plasticity of the Pt/TaO$_x$/HfO$_2$/Pt devices can be interpreted by the migration of oxygen vacancies and formation/dissolution of conducting filaments, as schematically illustrated in Fig 4. In the Pt/TaO$_x$/HfO$_2$/Pt bilayer device, the initial state of the device is illustrated in Fig. 4 where the HfO$_2$ layer is essentially depleted of V_O so the overall resistance of the device is high. When a positive voltage is applied on the top electrode, V_O migration into and the formation of conducting paths in the HfO$_2$ layer lead to the transitions from high resistance states (HRS) to low resistance states (LRS). However, when a negative voltage is applied on the top electrode, V_O migrate backward from the HfO$_2$ layer into the TaO$_x$ layer, so that the filament is broken and the device switches to high resistance state (HRS). Such physical processes therefore underpin the resistive switching and synaptic plasticity in the Pt/TaO$_x$/HfO$_2$/Pt devices.

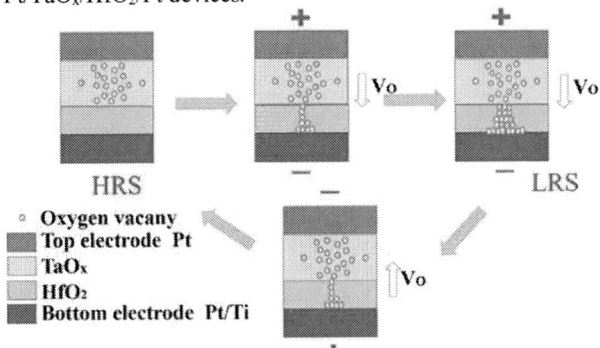

Fig. 4 Schematic illustration of the resistive switching process in the Pt/TaO$_x$/HfO$_2$/Pt devices

CONCLUSION

In this work, we have studied the resistive switching and synaptic properties of HfO$_2$-based memristors with different device sizes and dielectric structures, namely, single-layer and bilayer structures. We found that the on/off ratio of resistive switching constantly increases as device size decreases, and bilayer Pt/TaO$_x$/HfO$_2$/Pt devices generally exhibit more uniform switching characteristics and lower threshold voltages compared with single-layer Pt/HfO$_2$/Pt devices. More importantly, the Pt/TaO$_x$/HfO$_2$/Pt devices also showed superior analog switching characteristics, which were subsequently utilized to emulate the potentiation and depression processes by both DC and pulse stimuli. The implementation of LTP and LTD plasticity indicates the potential for applications as electronic synapses, and the resistive switching mechanism of the Pt/TaO$_x$/HfO$_2$/Pt devices can be well understood by the migration of oxygen vacancies and formation/dissolution of oxygen-deficient conducting filaments.

ACKNOWLEDGEMENTS

This work was supported by National Key R&D Program of China (2017YFA0207600), Beijing Municipal Science & Technology Commission Program (Z161100000216148) and National Natural Science Foundation of China (61674006, 61421005). Y.Y. acknowledges support from the "1000 Youth Talents Program" of China.

REFERENCES

[1] Borghetti, J., Snider, G. S., Kuekes, P. J., Yang, J. J., Stewart, D. R., & Williams, R. S. Nature, vol. 464, 2010, pp. 873-876.

[2] Lee, M. J., Chang, B. L., Lee, D., Lee, S. R., Man, C., & Ji, H. H., et al. Nature Materials, vol. 10, 2011, pp. 625-630.

[3] Upadhyay, N. K., Joshi, S., & Yang, J. J. Science China Information Sciences. vol. 59, 2016, pp. 061401-061426.

[4] Jo, S. H., Chang, T., Ebong, I., Bhadviya, B. B., Mazumder, P & Wei, L. Nano Letters. vol. 10, 2010, pp. 1297–1301.

[5] Du, C., Ma, W., Chang, T., Sheridan, P., & Lu, W. D. Advanced Function Materials. vol. 7, 2015, pp. 4290–4299.

[6] Yang, Y., Yin, M., Yu, Z., Wang, Z., Zhang, T., & Cai, Y., et al. Advanced Electronic Materials, vol. 3, 2017, p. 170003.

[7] Yang, Y., Choi, S., & Lu, W. Nano Letters, vol. 13, 2013, pp. 2908-2915.

[8] Li, J., Duan, Q., Zhang, T., Yin, M., Sun, X., Cai, Y., Li, Li., Yang, Y. & Huang, R. RSC Advances, vol. 7, 2017, pp. 43132 – 43140.

[9] Wang, Z., Yin, M., Zhang, T., Cai, Y., Wang, Y., & Yang, Y., et al. Nanoscale, vol. 8, 2016, pp. 14015-14022.

Smart IC Technologies for Smart Devices in IoT applications

Min-hwa Chi

Technology Development, SMIC, Shanghai, 201203, China

Email: min-hwa_chi@smics.com

ABSTRACT

The Internet of Things (IoT) as a driving force for future business needs massive capabilities of sensing, computing, and transmitting with analysis, learning, and decision-making. The IoT hardware/software includes key functions (i.e. sensing, computing and transmitting) in hierachical systems with wide range of performance and low power. Thus, state-of-art foundry needs to provide balanced and wide portfolio of technologies on logic platforms (i.e. CMOS, SOI, FinFET, IPs) with various specialty functions (e.g. NVM, power management, CIS, MEMS, RF) and capability of 3DIC/SiP for future smart devices in IoT era.

INTRODUCTION

The Internet of Things (IoT) paradigm is revolutionizing our live with capability of connection of anything and anyone for quality analysis, deep learning, and decision-making. The IoT Hardware/software realizes key functions (i.e. sensing, computing and transmitting) in hierachical systems with wide range of performance and low power. Modern foundry leads to advanced manufacturing technologies with platforms of logic and specialty functions as well as 3DIC/SiP capability for future IoT. Promising IC technologies for IoT are overviewed in this paper. After a brief introduction of IoT, we describe state-of-art CMOS devices (FinFET and FDSOI) for ultra-low-power (ULP) IC, design methodology (Near-threshold and Neuromorphic circuits) for low-power, RFIC, Power Management IC (PMIC) and wireless-power-transfer (WPT) technologies. Then, CMOS compatible NVM technologies at FEOL (e.g. OTP and Ferroelectric FET) and BEOL (e.g. RRAM, PCRAM, NRAM), and 3D embedded-NVM are described. The magnetic memory (MRAM) and Logic can be integrated at BEOL together with CMOS as a scheme of 3D monolithic IC. The 3D-NAND technology can be extended to form 3D-RRAM and 3D-PCRAM. Various MEMS sensors and CMOS Image Sensor (CIS) can be integrated at BEOL with CMOS. The through-Si-Via (TSV) technology enables 3D stack-up capability for 3DIC/SIP. The 3D-DRAM high-bandwidth memory (HBM) is a recent success based on the stack-up of DRAM chips through TSV. In short, the state-of-art IC foundry is continuously progressing with enhanced capability in sensing, transmitting, and computing as well as capability in 3DIC/SIP, IP/design, software/firmware development in order to serve as "virtual-IDM".

INTERNET-OF-THINGS (IoT)

The Internet of Things (IoT) as a driving force for future business connects any person/identity or machine (i.e. "things") together in network with massive capabilities for analysis/learning and decision-making. The IoT hardware/software [1] includes key functions of sensing, computing and transmitting with hierachical structures and performance in wide range, i.e. from ULP to high performance, from wearable/mobile devices to embedded components in systems. Thus, modern foundry needs to provide balanced technologies with wide portfolio on platforms of logic and specialty functions with capability of 3DIC/SIP for future IoT applications **(Figure 1)**.

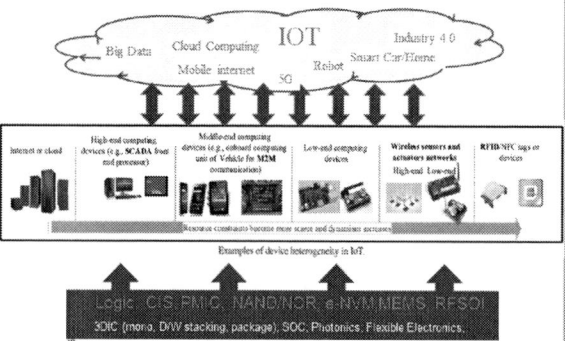

Figure 1: Foundry technologies provide a balanced portfolio in wide range for supporting future IOT applications.

ULP DEVICE AND DESIGN

Sustainable IoT systems **[2-3]** may need to operate CMOS ICs or 3DIC in ultra-low-power (ULP) as well as energy harvesting capability for computing, sensing, and communications. The 28nm bulk CMOS technology at ultra-low-voltage (ULV) mode potentially can meet these challenges for low-power digital signal processing, power management and RF front-end circuits. The FinFET on bulk and FDSOI technologies **[4-5]** are more attractive for operations at ULV or ULP as related to the better gate control on channel and less variability in device parameters (vs planar transistors on bulk).

Near-threshold circuit (**NTC**) design has been considered as promising to achieve energy efficient for ULP; however, the performance is seriously degraded **[6]**. Fortunately, the advancement of 3DIC technology (for fast and energy efficient signal transmission) as well as neural circuit (for non von Neumann circuit) can enhance its performance significantly. Thus, a combination of NTC designs, neuromorphic circuit (for

978-1-5386-5309-8/18 $31.00 © 2018 IEEE

effective computing), and 3DICs appears good optimization for both high performance and ULP.

Various types of neuromorphic computing circuits [7-8] can be implemented by analog circuits, NVM (E2PROM, Flash, RRAM, PCRAM, MRAM), or elements with memory (mem-ristor, mem-capacitor, and mem-inductor). As each solution has its own trade-offs, there is no commonly accepted best one. Neuromorphic computing based on spiking neural networks (SNN) can realize computations with extremely low energy **(Figure 2)**.

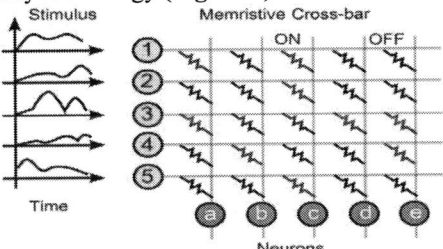

Figure 2: Schematic of a cross-bar array based SNN. The analog currents（as input）are integrated by neurons [7].

RFIC, PMIC, AND WPT

Currently, RFICs are widely based on CMOS platform. A 28nm low-power CMOS [9] on bulk can integrate low-noise amplifier (LNA), power amplifier (PA), and digital baseband together. A 130nm RFSOI CMOS [10] can integrate RF-switch, LNA, and PA together as well as CMOS for digital signal processing. The higher voltage LDMOS (lateral-diffusion MOSFET) and EDMOS (extended-drain MOSFET) can be integrated modularly together [11] on CMOS platforms for RF power amplifier as well as DC power management IC (PMIC).

Wireless power transfer (WPT) [12] is a key function for future smart devices in IoT, e.g. in battery-less sensors, passive RFID, wireless sensor network, machine-to-machine solutions, implantable medical devices, etc. The power for wireless transmission can be from harvesting from surrounding energy (e.g. electromagnetic energy, lights, vibration, temperature, etc.), or by "beamed" wireless energy (e.g. the "near" field beam from inductive/capacitive coupling or the "far" field beam of RF). The key component is the receiving antenna (Rx coils) integrated in chips for receiving energy. One WPT scheme [13] for 3DICs in SIP is proposed to use the "near" field energy from the PCB substrate to the chips through the magnetic resonance coupling with significantly reducing power interconnections and power supply currents and dissipation.

EMBEDDED-NVM AT FEOL

The NVM technologies are recently demonstrated with full FEOL compatibility in 14nm FinFET as OTP

(Figure 3) [14], Ferroelectric FET (FeRAM) [15] and 2-transistor RRAM **(Figure4)** [16] as illustrated below.

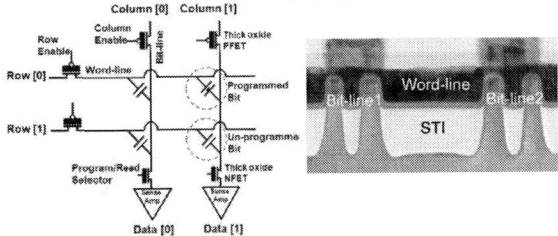

Figure 3: Schematic of the OTP memory array. The anti-fuse cell is programmed by rupturing the interfacial layer in the gate dielectric stack [14].

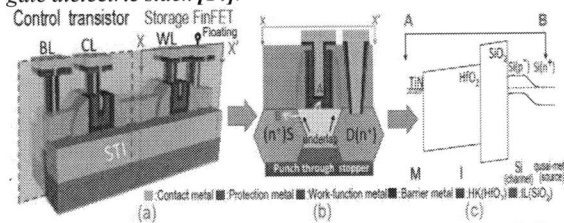

Figure 4: (a) Two identical FinFETs in series as an emb-flash cell, (b) the cross-section of the storage FinFET from A to B (gate to drain), and (c) its band diagram [16].

EMBEDDED-NVM AT BEOL

The NVM technologies are progressed toward full compatibility at BEOL **(Figure 5)**, e.g. Resistive-RAM (RRAM) [17], Phase-Change-RAM (PCRAM, or PCM) [18], and Nanotube-RAM (NRAM) [19]. It is particularly interesting, as they can serve not only for the embedded applications, but also capable of stacking-up into 3D array for high density and high performance systems. *Firstly*, the advanced RRAM and PCRAM cells can be formed within contact, via, or MIM layers with simpler process (e.g. 2-3 extra masks) in BEOL. By applying suitable electrical pulses the resistance is switched between high and low states (with orders of magnitude difference) by forming or rupturing conductive paths in RRAM or phase change in PCRAM. *Secondly*, the **NRAM** is based on electromechanical switch using spin-on Carbon nanotube (CNT) material with excellent characteristics (e.g. >10^{11} endurance, low power and fast array program). *Thirdly*, various emb-NVM at BEOL can be stacked-up by repeating the process (i.e. 3D stacking-up of horizontal memory array) by using access devices (or selectors), e.g. Ovonic Threshold Switch (OTS) in Intel/Micron's 3XPoint technology [20] and C-nanotube FET [21].

Figure 5: Various NVM technologies, e.g. RRAM [17], PCRAM [18], and NRAM [19] at BEOL.

978-1-5386-5309-8/18 $31.00 © 2018 IEEE

MRAM AND MAGNETIC LOGIC

The magnetic-tunnel-junction (MTJ) based MRAM [22-23] is a promising NVM technology by its almost infinite endurance, fast switching/sensing. There is also magnetic spin-based device [24-25] formed at BEOL for nonvolatile and reconfigurable logic circuits. The combination of spin-logic of BEOL and CMOS logic at FEOL can enhance logic function with simpler circuits and smaller foot-print.

The magnetic tunnel junction (MTJ), typically 2 ferro-magnetic layers sandwiched with a thin insulator barrier, shows bi-stable tunneling magneto-resistance (TMR) as memory element in MRAM. The TMR is resulted from the magnetic states of the "free" ferro-magnetic layers with respect to the "pinned" layer in spin polarity (parallel or anti-parallel). The MTJ with CoFeB/MgO/CoFeB structure can achieve high TMR ratio (~5x). The typical 1T-1MTJ MRAM cell can be switched by magnetic field or spin-torque-transfer (STT). The STT-MRAM has demonstrated recently with low write current ($<10^6$ A/cm^2 in 10ns pulse), good retention (>10 years), small cell size, fast read (~30ns), and endurance (~10^{14}). This approach is actively pursued in industry for replacing DRAM and/or SRAM in future. Recently, a 1Gb MRAM array with perpendicular-MTJ (pMTJ) is demonstrated **(Figure 6)** **[22]** at 28nm logic node.

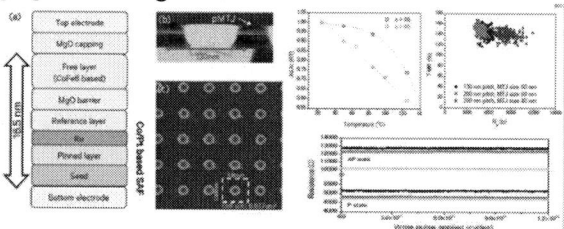

Figure 6: A 1Gb MRAM array with pMTJ is demonstrated [22] with bit-cell size of 22F² at 28nm logic node.

The all-spin-logic (ASL) device **[24]** is of interest due to its fabrication at BEOL **[25]** in addition to those features of non-volatility, high density, reconfigurable, lower device count per gate, and good scalability. The ASL device **(Figure 7)** consists of I/O nano-magnets for storing digital information, connected by a channel (typically Cu, Al, or Graphene) for transferring spin information, an isolation layer for separating devices, and an interface between the nano-magnet and channel for injecting spin-polarized electrons. The spin injection, diffusion, and STT switching in metallic lateral-spin-valve (LSV) structure are basic mechanisms for logic operation with non-volatility.

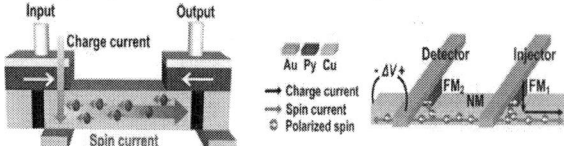

Figure 7: ASL buffer circuit with the effective spin current illustrated assuming positive bias at the input terminal [24].

MONOLITHIC 3DIC

The monolithic 3DIC integration including emerging NVM (e.g. RRAM, PCRAM, MRAM, C-nanotube RAM, …etc.) as well as logic circuits (e.g. magnetic spin-logic, C-nanotube FET, etc.) at BEOL together with CMOS at FEOL promises revolutionary digital systems with massive connectivity and data communication with high security, high speed, and low-power. The monolithic 3DIC has advantages of precision alignment of contacts at gate or transistor levels with performance gain (e.g. wire length reduction) and independent optimization of stacked devices **[26].** The monolithic 3DIC with its high contact density is also a powerful solution for heterogeneous co-integration with high density of via's. The main process flow is illustrated in **Figure 8**, with top transistors processed at low temperature with similar structure as FDSOI **[27].**

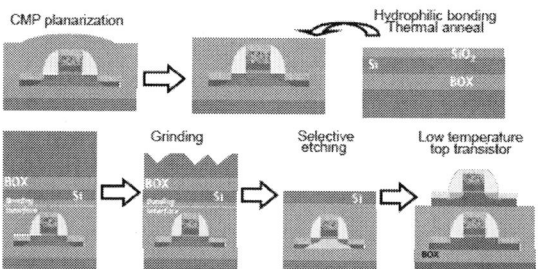

Figure 8: Main process flow for monolithic 3DIC [26].

VERTICAL 3D-NAND AND 3D-RRAM

The state-of-art 3D stacking-up NAND vertically (3D-VNAND) **(Figure 9)** **[28]**, already revolutionized the world for solid-state data (SSD) storage, is integrated by multiple cells integrated in vertical manner (i.e. 3D stacking-up of memory array vertically) and already in mass production by Samsung since 2014. Intel/Micron is also ramping up 3D-VNAND in 2016. Furthermore, 3D-RRAM **[29]** **(Figure 10)** and 3D-PCRAM based on similar architecture as 3D-VNAND are also in progress with simpler process.

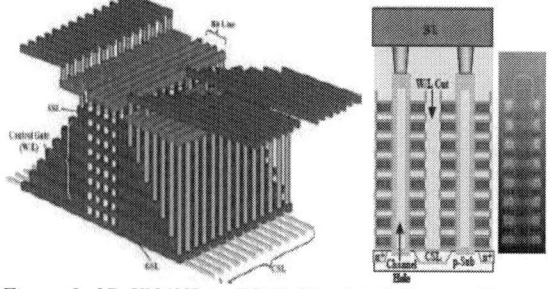

Figure 9: 3D-VNAND in TCAT (Terabit Cell Array Transistor) architecture with 3D stacking up and cross-section view [28].

978-1-5386-5309-8/18 $31.00 © 2018 IEEE

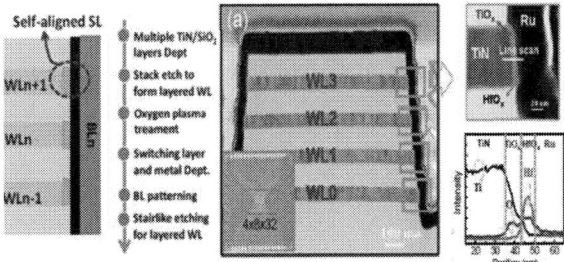

Fig. 10: Schematic of self-aligned VRRAM and integration flow. The multi-layer WLs were formed by TiN with surface oxidized by O plasma. Also shown the cross-section TEM and image of TiN/TiOx/HfO2/Ru layers with EDX spectrum.[29]

MEMS AND CMOS IMAGER

Modern MEMS technology [30–31] is driven by smartphones, e.g. microphone, fingerprint sensor, accelerometer, gyroscope, RF switch, energy harvester, etc. Modern MEMS sensor ICs are fabricated on CMOS platform **(Figure 11)** [31] with low-noise mixed signal circuitry, low-power data processing, computing, and connectivity.

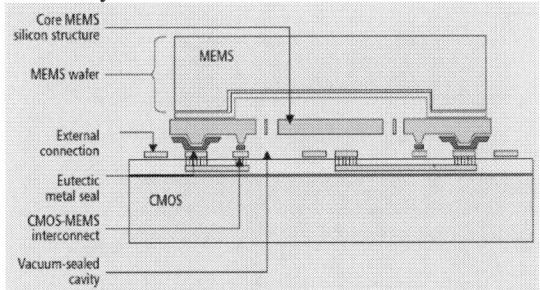

Figure 11: Silicon MEMS/CMOS integration technology [31].

CMOS imager sensor (CIS) [32] is already a "must" in smartphones, and also increasingly serving for 3D imaging, object recognition, drone, robot, medical, self-driving auto-mobile, …etc. The sensing pixels can be front-side-illuminated (FSI) or back-side-illuminated (BSI) by wafer bonding techniques **(Figure 12)**. Modern CIS chip can be integrated on CMOS platform with expanded functions (e.g. communication, computing and memory) through 3DIC technology.

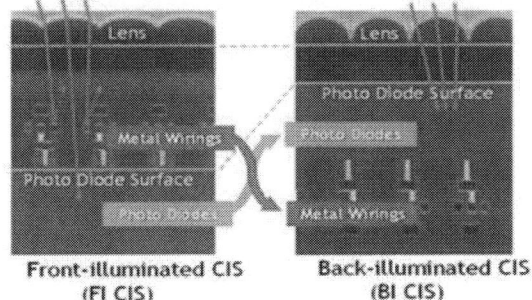

Figure 12: Illustration of front-side-illuminated (FSI) and back-side-illuminated (BSI) imagers [32].

TSV-BASED 3DIC/SIP AND 3D-DRAM

The conventional 3DIC technology [33] as based on chip-to-chip or wafer-to-wafer stacking by using through-Si-via (TSV) technique or based on system-in-package (SIP) can offer significant advantages in low power, high speed, data security, etc. The 3D-DRAM high-bandwidth memory (HBM) [34] is a recent success based on stacking-up DRAM chips by TSV, e.g. SK Hynix's 4/8Gb stackable DRAM. 3DIC integration is an effective way to extend the Moore's law into "More-than-Moore" era.

CONCLUSIONS

As the Moore's law slowing down, the IoT becomes new driving force. The state-of-art IC foundry industry continuously progresses rapidly with enhanced capability in sensing, transmitting, and computing; as well as capability of 3DIC/SIP, IPs and design services, software/firmware in order to perform as "virtual-IDM or System house" for the coming IoT world.

REFERENCES

[1] M.A.Razzaque, et.al., IEEE Internet of Things, J. Vo. 3, No. 1, p.70, 2016. [2] D. Bol, G. de Streel, D.Flandre, S3S, 2015. [3] A.Sharma, et.al., IEDM, p.147, 2016. [4] O.Weber, ICICDT, p.1-3, 2017. [5] T.Dry,T.Letavic; VLSI-T, p.T164, 2017. [6] V.M. van Santen, et.al., IEEE Circuits & Systems I: Reg. Papers, p.1-14, 2017. [7] B.Bajendran, et.al., IEEE J. on Emerging and Selected Topics in Circuits and Systems, v6, No. 2,p.198, 2016. [8] G.Volanis, et.al., IEEE Design and Test, p.91, 2016. [9] D.Fritsche ,et.al., IEEE Trans. Microwave Theory and Techniques, v63, No.6, p.1910, 2015. [10] R.V.Purakh1,et.al., IEEE RFIC Symp. p.47, 2015. [11] Jun Cai, et.al., ISPSD, p.196, 2009. [12] COST Action IC1301, IEEE Micro Magazine, p.56, June, 2017. [13] J. Song, et.al., IEEE WPTC, p.1, 2017. [14] Y.Liu, et.al. VLSI-T, p.40, 2014. [15] M.Trentzsch, et.al., IEDM, p.295, 2016. [16] E.R.Hsieh,et.al., VLSI-T, p.T73, 2017. [17] H-S.P.Wong,et.al., IEEE Proc. v100, No.6, p.1951, 2012. [18] H.Y.Cheng, et.al., IEDM, p.56 ,2015. [19] S.Ning, et.al., IEEE ED, p.2837, 2015. [20] D. C. Kau, et.al., IEDM, p.617, 2009. [21] C.Ahn, et.al., IEEE ED, v62, No.7, p2197, 2015. [22] C.Park, et.al., IEDM, p.664 (2015). [23] W.S.Zhao, et.al.,Intl. Conf. on VLSI and SOC, p.37, 2011. [24] J.Kim, et.al., IEEE Proc., v.103, No.1, p.106, 2015. [25] Azad Naeemi, et.al., DAC, 2014. [26] C.Fenouillet-Beranger,et.al., ICICDT, 2016. [27] G.Ghibaudo, ESSDERC, p.53,2016. [28] J. Jang, et.al., VLSI-T, p.192,2009. [29] Xiaoxin Xu1, et.al., VLSI-T,2016. [30] G.Lammel, MEMS, p.61, 2015. [31] D.K.Schaeffer, IEEE Com. Magazine, p.100, April, 2013. [32] T.Nomoto, et.al., VLSI Circuits, 2016. [33] J.lau, Keynote at IEEE Japan ICEP,2011. [34] H. Jun, et.al., IEEE Design&Test, Jan/Feb, p.16-25,2017.

SYNTHESIS OF MoS₂ VIA IN SITU SULFURIZE SPUTTERING Mo

Wei Junqing, Li Yue, Wang Fang*, Zhang Zhichao, Feng Yulin, Li Yi, Zhang Kailiang*
School of Electronics Information Engineering, Tianjin Key Laboratory of Film Electronic & Communication Devices, Tianjin University of Technology, Tianjin, China, 300384
86-22-60214229 and fwang75@163.com; kailiang_zhang@163.com

ABSTRACT

Transition metal dichalcogenides (TMDS) have attract more attention, but it is still a tricky issue to synthesize large area MoS2 nanofilms. Here a two-step process method was used to produce MoS2 nanofilms with large scale. Pre-studies have proved that the crystallinity of sputtered Mo precusors have great effect on MoS2 synthesis. By using varing deposition powers the diffierent crystallinities of Mo films can be obtained. Then magnetron sputtered Mo thin-films were used as Mo precusor to synthesize MoS2 nanofilm in high tempreture. By controlling the sputtering power of Mo, a certain crystal orientation of Mo thin film can be deposited which can enhance the Raman spectrum and improve the quality of MoS2 nanofilms.

INTRODUCTION

Physical vapor deposition (PVD) combined with CVD have been aroused great interest to prepare large area of MoS2 nanofilms.1 The two-step process method which means using pre-sputtered Mo or MoO3 or MoS2 as precusor and sulfide powder as another precusor in high tempreture under Ar atmosphere. Compared to CVD , two-step methods are not limited by some factors such as the amount of precursor reactants, gas flow rate and distance of diffierent precursors.There are few studies focus on that the crystal orientation of pre-sputtered precursor Mo can affect the synthesized MoS2 nanofilms2-4.

In this work, pre-sputtered Mo thin films were deposited under diffierent powers and all the pre-deposited Mo thin film were sulfurized with the same conditions. We found that the diffierences between Mo thin films deposited under varing powers have great effects on MoS2 nanofilms which can be proved by Raman spectrum. By optimizing the deposition power, highly crystal quality and great surface topography of Mo thin films were deposited. Also by adjusting the amount of sulfur powder and reaction tempreture, the MoS2 naonfilms displayed high crystalline quality as well as electrical transmission characteristics. All results displayed in our work proved that large scale and high quality of TMDS nanofilms can be synthesized by control the pre-deposited precusors.

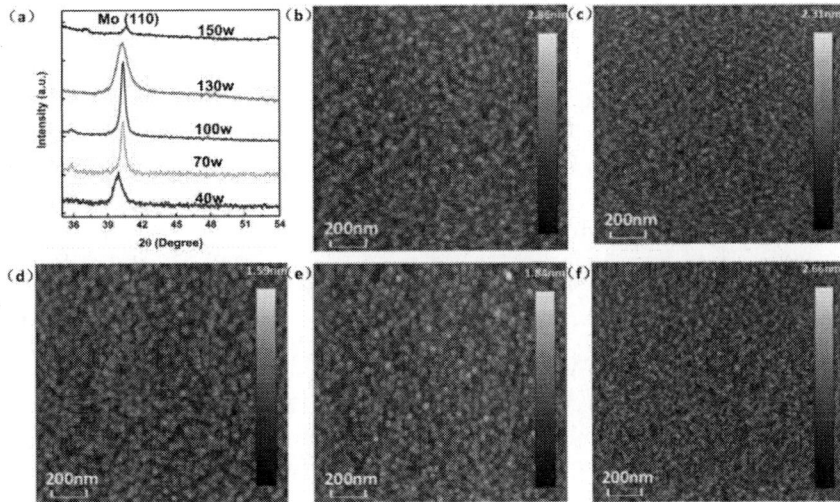

Fig1. Sputtering Mo thin film (a)XRD of Mo thin films under diffierent powes; AFM of Mo thin films(b)40w (c)70w (d)100w(e)130w (f)150w

EXPERIMENTAL DETAILS

The silicon (290nm SiO_2/p+silicon) substrates were cleaned in acetone and isoprophy alcohol(IPA) solution under ultrasound environment for 15 minutes. Then the substrates rinsed with deionized(DI) water following by drying with nitrogen and baking.

Fabrication of Mo thin film

The Mo thin films were sputtered onto SiO_2/Si substrate by magnetron sputtering highly pure Mo target in Ar atmosphere. Different DC powers were used to acquire Mo films with diffierent intenses of (110) peaks. The same thicknesses(~3nm) of Mo thin flims were acquired by modulate depositing time.

Fabrication of MoS2 nanofilm

Pre-sputterred Mo were placed at midstream of CVD chamber, then it was heated to 1000℃. Highly pure sulfur power(99.99%) was placed at upstream of CVD chamber and heated to 150℃. The Ar gas flow of 65sccm was used to carry the sulfur vapor to midstream. The amount of S powder(~120mg) and reaction time(~30min).

Fabrication of devices

All electrodes were patterned by using e-beam lithography (EBL ,Raith e_LINE plus). Following by developing and fixing, Ti(20nm) and Au(100nm) were evaporated by Electron beam evaporation(EB500) and magnetron sputtering.

CHARACTERIZATION

Topography and thickness of Mo thin films and post-sulfurized MoS_2 nanofilms were extracted by optical microscopy(OM, Olympus) and atomic force microscopy(AFM, Aglient 5600LS). Crystallographic orientation of pre-deposited Mo thin films was measured by X-ray diffraction(XRD, D/MAX-2500). And XRD patterns were recorded with 2θ ranged from 5°to 80°and a wavelength of λ=1.5418Å. Raman spectrum(ThermoFisher DXR) of MoS_2 were recorded by a laser power of 1mW and a wavelength of 532nm and a spot size of 1.2μm to confirm the crystal quality, thickness of as-synthesized MoS_2 nanofilms.

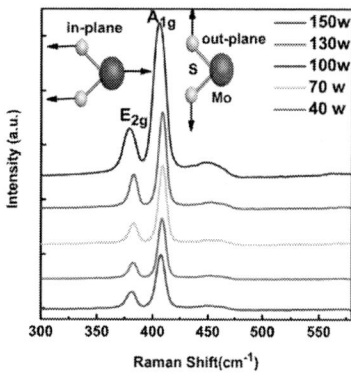

Fig2. Raman spectrum of MoS_2 nanofilms

RESULTS AND DISCUSSION

Mo thin films were deposited onto 290 nm SiO_2/Si substrate at room temperature(RT) with 28 sccm Ar gas flow and a distance of 6.5cm between target and substrates. The XRD patterns of the Mo thin films deposited in varing powers were exhibited in Figure1(a).

The most intense peak appeared at $2\theta=40.515°$ corresponding to Mo(110) preferred orientation[5]. In Figure1(a), different diffraction peaks displayed that the deposited Mo thin films have different crystal quality under five varing DC powers with an operating pressure of 0.8Pa. This phenomenon can be illustrated by Eout and Ein that are the sputtering ejected particles from the Mo target and the kinetic energy of Mo atoms or clusters when they arrive at the substrates. Higher sputtering power can produce larger Eout and Ein, thus give the Mo atoms adequent energy to form their preferred orientation and lead to more intense peak. The relation of Eout and Ein showed in (1)[6], where L~0.65cm is the distance between target and substrate, σ is the cross section for momentum transfer collision with background gas atom,k is the Boltzmann constant, P is the working pressure, T is the tempreture.

$$Ein=Eoutexp(-\frac{\sigma \rho L}{kT}) \qquad (1)$$

AFM measurements of Mo thin films were exhibited in Figure1(b)-(f), the topographies of Mo thin films under different deposition powers

were obvious.

The Raman intensities of A_{1g} and E_{2g} become stronger when the depositing power increase from 40w to 100w, but further increase the power, the intense of A_{1g} and E_{2g} become weaker which can be proved in Figure 2 and the Raman spectrum(100w) can greatly conform the XRD results mentioned above.

The pre-sputtered Mo thin film(power 100w and pressure 0.8 Pa) was put in the midstream of the chamber and 150 mg sulfur powder in the upstream and then pump the chamber under $5×10^{-4}$ Pa to preclude the influence of atmosphere. Along with injecting 65 sccm Ar as the carrier gas and the midstream was heated from room tempreture(RT) to 1000℃ at 15℃ min^{-1} and kept for 20 min at 1000℃ to ensure the Mo thin films vulcanize completely.[9-11]

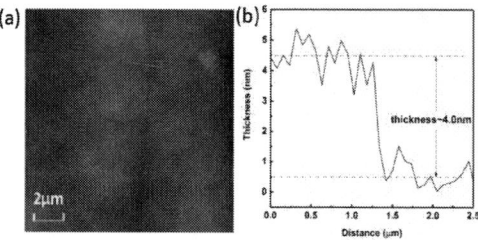

Fig3. Synthesized MoS_2 nanofilm (a)AFM measurement; (b) the thickness of MoS_2 nanofilm was ~4.0nm

were synthesized which were measured by AFM in Figure 3a. Figure 3(b) showed the thickness of synthesized MoS_2 nanofilm is ~4nm.

CONCLUSION

This study gave the effects of the crystallinity of sputtering Mo thin flims on synthesizing MoS_2 nanofilms by two-step process method for the first time and via optimizing the sputtering condition of precursor Mo thin films can improve the quality of synthesized MoS_2 nanofilms. The sputtering power(100w) can obtain the best crystallinity of Mo thin films(110) and enhance the Raman spectrum of synthesized MoS_2 nanofilms. By controlling the sputtering power of Mo and sulfurizing process can we acquire high quality of MoS_2 nanofilms which greatly promote the application of MoS_2 in Electronic devices and optoelectronic devices.

REFERENCES

[1] Song X, Guo Z, Zhang Q, et al.. Small, 2017.

[2] Shahzad R, Kim T W, Kang S W. Thin Solid Films, 2016.

[3] Chelvanathan P, Shahahmadi S A, Arith F, et al.Thin Solid Films, 2017, 638: 213-219.

[4] Fàbrega L, Fernández-Martínez I, Parra-Borderías M, et al.IEEE Transactions on applied superconductivity, 2009, 19(6): 3779-3785.

[5] Zoppi G, Beattie N S, et al. Journal of materials science, 2011, 46(14): 4913-4921.

[6] Keller B D, Bertuch A, et al. Chemistry of Materials, 2017, 29(5): 2024-2032.

[7] Feng Y, Zhang K, Wang F, et al.. 2015.

[8] Li Y, Zhang K, Wang F, et al.. ACS applied materials & interfaces, 2017, 9(41): 36009-36016.

[9] Tao J, Chai J, Lu X, et al.Nanoscale, 2015, 7(6): 2497-2503.

[10] Bai H, Ma J, Wang F, et al. Solar Energy, 2018, 159: 88-96.

[11] Sun J, Li X, Guo W, et al. Crystals, 2017, 7(7): 198.

[12] Kim Y, Kim A R, Zhao G, et al. ACS applied materials & interfaces, 2017, 9(42): 37146-37153.

[13] Late D J, Liu B, Matte H S S R, et al. ACS nano, 2012, 6(6): 5635-5641.

[14] Feng Y, Zhang K, Li H, et al. Nanotechnology, 2017, 28(29): 295705.

Improvement of 28HKMG NIO device HCI by Implant Scheme and Sequence Optimize Doping Profile and E-field

Jiajia Tao*, Qingchun Zhang, Jie Zhou, Yongsheng Yang, Byunghak Lee, Falong Zhou, TzuChiang Yu

LTP-2 Center, Technology R&D Management, SMIC
18 Zhangjiang Road, Pudong New Area, Shanghai, P.R. China 201203
E-mail: Jeremy_Tao@smics.com

ABSTRACT

Hot-carrier injection (HCI) is a challenging work of every logic device generation. In this paper, there are two kinds of directions to improve NIO HCI based on 28nm HKMG logic process. The first one is the implant scheme optimization. In this paper implant species combination, tilt and twist angles were studied. The second one is different implant sequences which also have significant impact the doping profile and e-field. In this paper, we studied how different IO LDD implant positions (after LDD SPIKE anneal, after SiGe process, and after main spacer process) impacted substrate current and HCI performance. Extensive TCAD simulations and experiments revealed the optimized implant scheme combining with implant sequence can greatly improve substrate current ~46% and HCI life time ~3.4X. The reduced substrate and improved HCI life time are attributed to two mechanisms 1) less junction impact ionization with dopant profile optimization 2) moving impact ionization point away from Si/SiO2 interface.

INTRODUCTION

Hot Carrier Injection (HCI) degradation is an important reliability concern for advanced technology nodes, which have a high Ion driving capability, and NIO device Hot Carrier Injection (HCI) always was a worst case in every logic device generation. Generally, substrate current (Isub) is a good indicator of HCI degradation. The important factors to minimize hot carrier generation and injection are to reduce the electric field magnitude and to make impact ionization region deep into silicon, which can result in the lower possibility of hot carriers reaching the Si/SiO2 interface. [1-4]

TCAD SIMULATIONS AND DISCUSSION

For 28nm HKMG logic technology, 2.5V IO NMOS suffer high Isub and low HCI life time issue. These works focus on two directions: (a) implant species combination, tilt and twist angles, (b) different implant sequences which also have significant impact the doping profile and e-field. Both of them can greatly improve IO NMOS Isub and HCI performance by TCAD simulations and Silicon experiments results.

Implant species combination

In this paper, firstly of all check different implant species combination impact on doping profile by TCAD simulation as Fig.1 show. There can be found Pure Phosphorus diffuse better than Phosphorus and Arsenic combination and make more gradient doping profile. That also can be verified by E-Field under maximal substrate current (Isub) X-cut by TCAD as Fig.2 show, Pure Phosphorus E-Field ~15% lowers than Phosphorus and Arsenic combination.

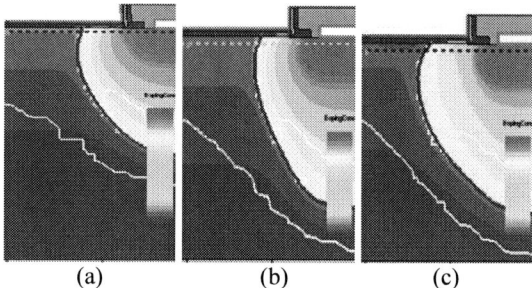

Fig.1: Doping Profile by different Implant species combination. (a) Phosphorus and Arsenic (b) Pure Phosphorus (c) Phosphorus and Nitrogen

Secondly, After Phosphorus implant, we add Nitrogen implant after it, which founds the doping profile, got better again as fig.1 showed, and From Fig.2, we can find the E-field reduced ~30% more. From TCAD model show Nitrogen implant will make interstices in the junction and the Phosphorus will diffuse again by future steps' thermal. So we can find the E-field distribute range also larger than others, that's mean Phosphorus distribute more uniform.

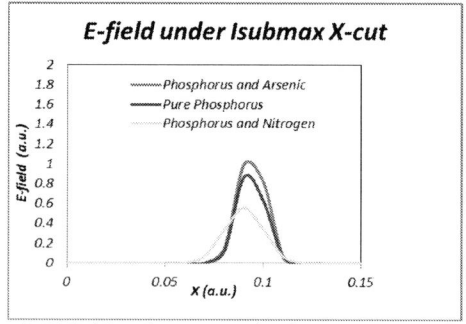

Fig.2: E-field under maximal substrate current (Isub) X-cut by TCAD for different implant species combination

Implant tilt and twist angles

For implant tilts and twist angles optimize, there are many combinations. In this paper, we will only introduce one special combination. It is tilt big angle combine twist big angle, as Fig.3 (b) shows the junction doping profile will be changed much than (a) tilt small angle and twist small angle. In TCAD model showed there is a special channeling effect when tilt big angle combine twist big angle. So the Phosphorus will diffuse quickly follow this channeling effect.

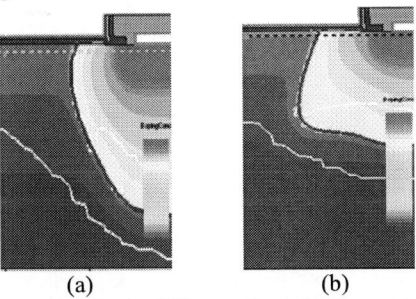

(a)   (b)

Fig.3: Doping Profile by different tilt and twist angles. (a) Small Tilt and Twist angle (b) Big Tilt and Twist angle

From E-field distribution as Fig.4 showed, E-field reduces ~ 15% and the E-field maximum location more far away from gate edge. That's also verified the special channeling effect make doping profile changed better.

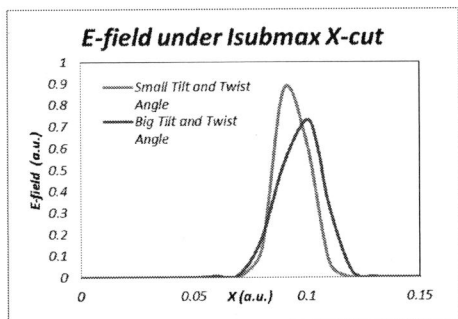

Fig.4: E-field under maximal substrate current (Isub) X-cut by TCAD for different implant tilts and twists angles

Implant Sequence

Generally, optimized doping profile to improve E-filed distributions, one way reduces maximal E-filed number, another way makes maximal E-filed location far away from gate edge also can benefit for HCI.

TABLE I.

Spacer thickness for different location		
Control (a.u.)	After SiGe loop (a.u.)	After Main Spacer (a.u.)
1	2.3	3.5

For push the maximal E-filed location away from the gate edge, we changed implant sequence to 1) after LDD Spike anneal loop, 2) after SiGe loop, the spacer thickness will thicker due to dummy poly 3) after Main Spacer (before S/D implant) which spacer thicker more. Table I shows the spacer thickness differences for different imp location.

Firstly, we changed implant sequence to after LDD spike anneal loop, which hopes that skip the high temperature thermal step left the junction defect, and then use below SiGe loop middle temperature thermal steps make Phosphorus diffused more, and doping profile as Fig.5 (a) showed. From Fig.6 we can find Maximal E-field number still can be reduce ~11%, and the E-filed distribution range also shows larger than control group which also can verify the model.

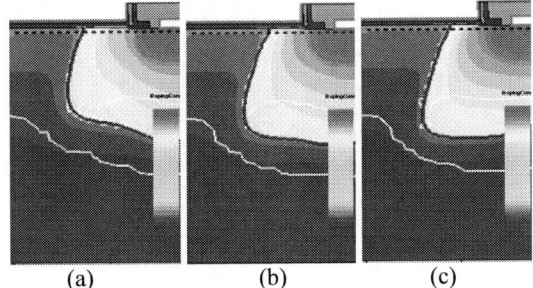

(a)   (b)   (c)

Fig.5: Doping Profile by different Implant species combination. (a) After LDD SPIKE (b) After SiGe loop (c) After Main SPACER

Secondly, we tried put LDD implant after SiGe loop which spacer thickness ~2.3X thicker than routine loop. That easy to know the maximal E-filed should be far away from gate Edge. By TCAD simulation confirmed the maximal E-field location changed much as Fig.6 showed. But from Fig.6 we can find the maximal E-filed much higher than control group. From TCAD model, that shows when recover device on same VT, after SiGe loop group need much more Phosphorus dosage due to less thermal steps make Phosphorus diffuse less.

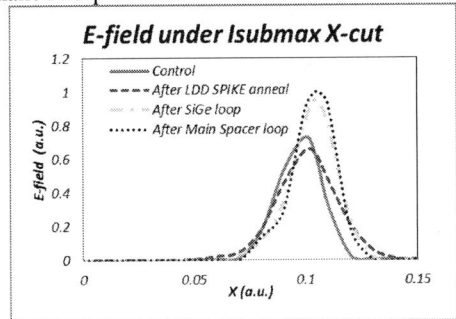

Fig.6: E-field under maximal substrate current (Isub) X-cut by TCAD for different implant sequence

The same results can be found on the third group which we put LDD implant after Main Spacer loop which spacer thickness ~3.5X thicker than routine loop.

RESULTS AND SUMMARY

After TCAD members help, we selected some good directions to do wafer splits. Below plots we can find splits conditions and also showed results by Isub/Id and HCI life time.

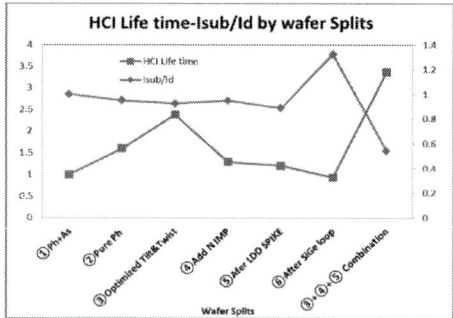

Fig.7: E-field under maximal substrate current (Isub) X-cut by TCAD for different implant sequence

The Silicon results almost match TCAD results, and finally Isub/Id reduced ~46% and 2.5V NIO HCI life time improved ~3.4X by combine 1) pure phosphorus, 2) Big Tilt and Twist angle, 3) Add Nitrogen implant, 4) After LDD SPIKE together.

REFERENCES

[1] E. Takeda, C.Y. Yang, and A. Miura-Hamada, Hot Carrier Effects in MOS Devices, Academic Press, San Diego, 1995; A. Acovic, G. La Rosa, and Y.C. Sun, "A Review of Hot- Carrier Degradation Mechanisms in MOSFETs," Microelectron. Reliab. 36, 845–869, July/Aug. 1996.

[2] E. Takeda, lEE Proceedings I - Solid-State and Electron Devices, vol. 131, pp. 153-162, 1984.

[3] Chen Min-Liang, Leung Chung-Wai, W. T. Cochran, IEEE Transactions on Electron Devices, vol. 35, pp. 22 10-2220, 1988.

[4] long Mun Park, Martin Knaipp, Hubert Enichlmair, et al. 2012 24th International Symposium on Power Semiconductor Devices and ICs, pp. 189-192, 2012

7μm Pitch Deep Trench Super Junction Process Development

Zhimin Zhu*, Fei Wang, J.K. Shen

Shanghai Huahong Grace Semiconductor Manufacturing Corporation

Shanghai, China

*Corresponding Author's Email:zhimin.zhu@hhgrace.com

ABSTRACT

With the rapid development of Super-junction(SJ) process in MOSFET,it is more and more widely used in power switcher area, and the performance of on resistance and switch speed became lower and faster[1].The structure becomes more and more compacted. The pitch of SJ structure is becoming more and more narrow to fit the developing trend, thus more gross dies can be got and the cost will be reduced. HuaHong Grace(HHGRACE) has developed its own deep trench SJ MOSFET platform, and has optimzied the process to make the platform keeping up with the technology developing trend. For the SJ MOSFET platform, it has gone through 3 periods: Generation 1.0(16μm pitch@RSP 24mohm·cm^2), 2.0(11μm pitch@RSP 18mohm·cm^2) and 3.0(9μm pitch@RSP 11mohm·cm^2). Now HHGRACE is developing the 7μm pitch SJ process technology, and it can get RSP 10mohm·cm^2 now. This paper describes the deep trench SJ developing history and new direction,also the achievements during the process optimization for 7μm pitch deep trench SJ structure.

INTRODUCTION

SJ MOSFET Developing Trend

The Super-junction (SJ) structure is an innovative breakthrough due to its ability to realize dramatically lower RSP (Resistance per square) [2]. With the rapid development of SJ MOSFET (Metal-Oxide -Semiconductor Field-Effect Transistors),it is more and more widely used in power switcher area,and the performance of on resistance and switch speed have become lower and faster.The structure is becoming more and more compacted. Figure 1 shows the developing trend.The pitch of SJ structure is becoming more and more narrow to gct lower RSP and more gross dies.

There are two methods to realize SJ MOSFET fabrication. One is by multiple layers epitaxy(EPI) growth as Fig2 shown, such as Infineon ,Fuji Elctric and ST. It can have better control accuracy by using implant process,but the cycle time is too long and the cost is higher than deep trench SJ MOSFET[3].

The other is by deep trench(DT) and single epitaxy growth such as HHGRACE , it has shorter cycle time and lower cost.Figure 3 shows deep trench SJ MOSFET structure [4] and the current status of DT SJ MOSFET in HHGRACE.

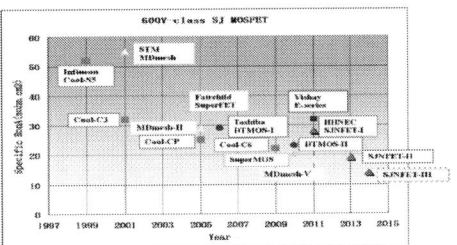

Figure 1: 600V SJ MOSFET Developing trend for Ron·A(RSP) specification

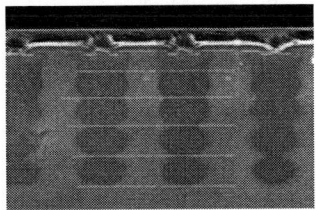

Figure 2: Multiple-EPI Growth Super-junction SEM

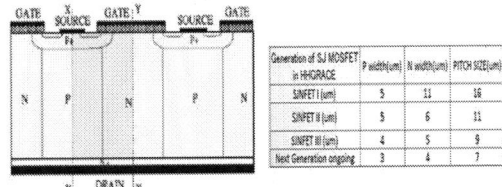

Figure 3:Cross section of Deep Trench SJ MOSFET and HHGRACE's SJ MOSFET Status

SJ MOSFET Developing Status in HHGRACE

HHGRACE has developed its own deep trench SJ MOSFET platform with 8 inch wafers fabout above 220Kpcs since 2011, and it has optimized the process to make the platform keeping up with the technique developing trend and speed.

Figure 4 shows the developing history of HHGRACE's SJ MOSFET.It has gone through 3 periods:

Generation 1.0(16μm pitch@RSP 24mohm·cm^2);

Generation 2.0(11μm pitch@RSP 18mohm·cm^2);

Generation 3.0(9μm pitch@RSP 11mohm·cm^2).

The next generation of SJ MOSFET's pitch size direction will be 7μm or less than 7μm. Now we have studied 7μm pitch with planar gate SJ process and found it available in fabrication.The 7μm pitch device might get RSP below 10mohm·cm^2,which is attractive.7μm pitch

978-1-5386-5309-8/18 $31.00 © 2018 IEEE

with trench gate SJ MOSFET device developing is also ongoing.

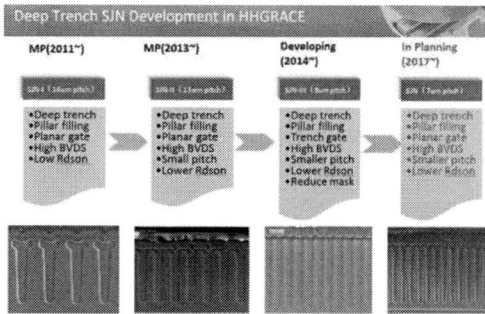

Figure 4: Development of HHGRACE's DT SJ MOSFET platform

HHGRACE'S 7μm PITCH SJ MOSFET

There are 3 main difficulties on the way of fabricating 7μm pitch size's SJ devices.

The first difficulty is N EPI control. With pitch size decreasing,the P-column turns shallower, and thus N EPI Resistance(Rs) will turn smaller and smaller to keep the P/N matching balance, and it will get lower BVdss as Figure 5 shown. Traditional N EPI cannot meet the target of BV and Rdson when pitch size is less than 9μm,so we designed a type of new EPI for SJ with pitch size less than 9μm. This new type of N EPI makes BVdss higher about 30V than traditional EPI, however Rdson keeps low level. There is no need to increase BV by increasing trench depth again.This type of N EPI has been widely used in 9μm pitch products since 2017 in HHGRACE and it can support 7μm pitch design very well.

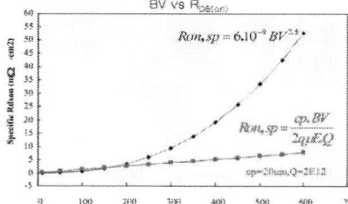

Figure 5: SJ product RSP VS BV correlation

The 2nd difficulty is trench etching control.We have mature and stable process of deep trench etching, it can get stable deep trench profile with the desired and tapered angle.

The on-resistance reduction tendency requires lateral pitch narrowing(high SJ aspect ratio)[5]，and trench aspect ratio has changed from 1:7 to over 1:12.Our trench etching tools and process can provide accurate trench depth control after recipe tuning, and the profile is stable with good uniformity within wafer.This makes the 7μm pitch process much easier to control.

The 3rd and the biggest difficulty is trench epi filling process control.It's very difficult to fill the trench well without EPI void since the pitch size turns smaller and smaller with no change in trench depth. After a lot of study and efforts, tuning the recipes,we can fill in the 7μm pitch trench perfect without void or defect now, and the uniformity and current leakage keep the same level as last version.

Figure 6 shows the post wright etch 30seconds SEM comparison after trench epi filling improvement. We can see that there is no void after epi filling process tuning. Figure7 shows the post wright etch 30seconds SEM comparison after CMP(Chemical and Mechanical Polishing) process and the former epi void issue has been resolved after process tuning. After process optimization, we can keep the high efficency without impacting the throughput of EPI filling tools.

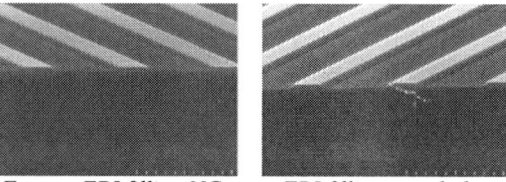

Former EPI filling NG   EPI filling tuned ok
Figure 6: Wright Etch 30s After Trench Filling

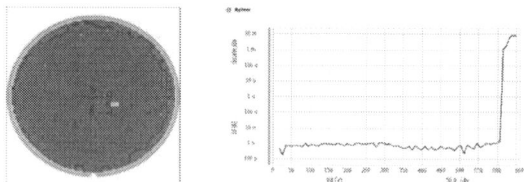

Former EPI filling NG   EPI filling tuned ok
Figure 7: Wright Etch 30s After CMP@7 μm Pitch

We have a lot of trying to optimize the EPI filling for small pitch trench. Because of the high aspect ratio, it is easy to seal the trench top without bottom filling good enough. After controling the epi filling speed and distribution, we got good result of little trench filling, with perfect lattice filling and defect free.Figure 8 shows we got good performance in BV and leakage and it indicates 7μm pitch SJ MOSFET can go ahead without any technical obstacles in fabrication.

Figure 8: 7μm pitch SJ BV distribution and leakage performance

978-1-5386-5309-8/18 $31.00 © 2018 IEEE   46

CONCLUSION

Now we can provide 7μm pitch SJ process to keep the advantage of products in both RSP and Gross dies.we can accomplish the key process control and provide device wafers with RSP@10mohm·cm^2@7μm pitch.There is no technical obstacle now for future device fabrication. The Direct Current(DC) parameters of devices can meet the target (shown in Figure 8)and the process can have international competence in SJ MOSFET fabrication.

We will keep studying and optimizing the process for small pitch (<=7 μ m) SJ to get better performance for customers ,and try to make the process more stable and easier to control.

REFERENCES

[1] Shunsuke Katoh，Eiji Shimada, Takayuki Yoshihira, Akihiro Oyama,Proceedings of ISPSD 2015,May 10-14,Hong Kong, pp. 93.

[2] Yasuhiko Oonishi, Superjunction MOSFET, Vol. 56 No. 2 FUJI ELECTRIC REVIEW.

[3] B. Jayant Baliga, Advanced Power MOSFET Concepts,2010,Chapter 7.1, pp. 324-325.

[4] Pravin N. Kondekar，Journal of the Korean Physical Society, Vol. 44, No. 6, June 2004, pp. 1565.

[5] Wataru Saito, Proceedings of ISPSD 2015,May 10-14,Hong Kong, pp. 125-128.

A STUDY OF LDMOS LAYOUT OPTIMIZATION OF 28HKMG FOR RF APPLICATION

Guiying Ma, WeiJia Xu, Byunghak Lee
Technology R&D Center, SMIC
18 Zhangjiang Road, Pudong New Area, Shanghai, P.R. China 201203
E-mail: Connie_Ma@smics.com

ABSTRACT

LDMOS (Laterally Diffused Metal Oxide Semiconductor) applications get extensively in high voltage and smart power management. In this paper, various LDMOS layout design was studied to get high f_T (Cut off frequency) and high breakdown voltage by TCAD simulation based on 28nm HKMG logic process, TCAD simulations reveal that layout and implant optimization can get f_T (Cut off frequency) > 50GHz with BV_{DS} (breakdown voltage) > 5V performance on 1.8V IO NLDMOS.

Keywords: LDMOS, f_T, breakdown.

INTRODUCTION

LDMOS is a very attractive device for high-voltage (HV) integrated circuit applications. Such as the flat panel display, motor drive, RF communication, and power management integrated circuit [1]-[3]. Many concepts for integrating high-speed and low on-resistance DMOS and CMOS have been proposed [4, 5]. RF systems are currently expanding to both higher powers (at watt level) to provide enough access, bandwidth and power for different RF applications, ranging from car radar, secure car access to power amplifiers in 5G systems[6,7]. In any of these systems, a good RF transistor with sufficiently high breakdown voltage is needed for an effective amplification of high-frequency signal. For an advanced 28nm CMOS process, high frequency and high breakdown LDMOS need to study also.

In this paper, we propose novel 3.3V NLDMOS structures with high frequency and high breakdown voltage, which can be fabricated using the same process flow optimized for 28nm HKMG logic technology. Technology Computer Aided Design (TCAD) analysis result clearly shows the compare result of breakdown voltage and frequency performance among conventional LDMOS structure with STI isolation on drain area and novel structures. New structure with drain area layout optimization and implant optimization can get better f_T and breakdown voltage performance than conventional structure and satisfy power amplifier electrical performance requirement: f_T >50GHz and BV_{DS} >5V.

DEVICE STRUCTURES

Fig1(a) is conventional NLDMOS structure with STI isolation on drain area，and Fig1(b) is conventional NLDMOS schematic cross section with layout variables, Lc, Ls, Lp and Lw, the layout parameter are aligned with paper[8]. Lp is defined as N-drift region and poly gate overlap size, Lc means effective gate length, Lw is defined as N-drift region and Pwell overlap or space and Ls means STI width.

Fig.2(a) is new NLDMOS structure1 with silicide block layer (SAB) isolation on drain area to refer paper [9], and Fig.2(b) is new structure2 with dummy gate isolation on drain area to refer paper [10].

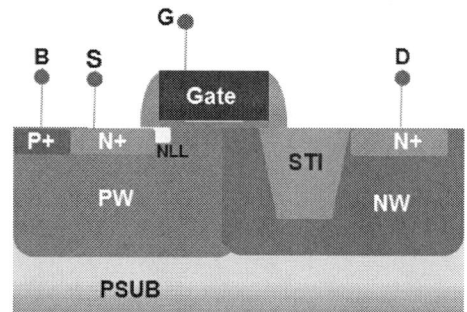

Fig1 (a): conventional structure with STI isolation on drain area.

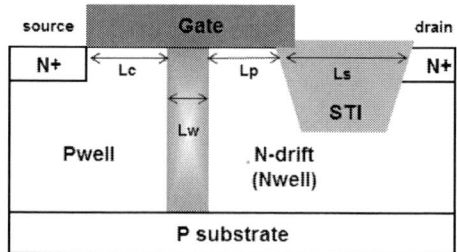

Fig1 (b): The schematic cross section of NLDMOS with layout parameters Lc, Lp, Lw and Ls

978-1-5386-5309-8/18 $31.00 © 2018 IEEE

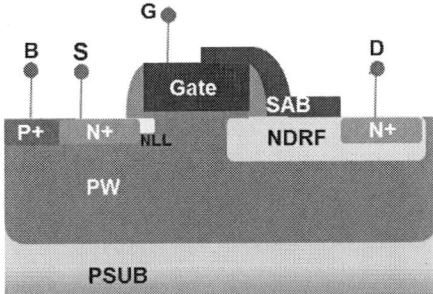

Fig2 (a): New structure1 with SAB isolation

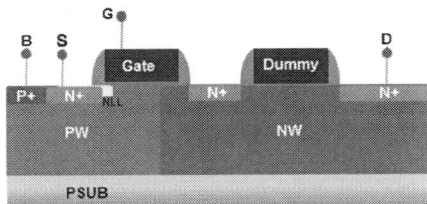

Fig2 (b): New structure2 with dummy gate isolation on drain area.

KEY PROCESS INFORMATION

Starting material is 12", EPI 2um P type silicon substrate with resistivity of 8-12 Ω .cm.

LDMOS was integrated into the normal 28nm HKMG logic process and no additional mask for the conventional NLDMOS structure and new structure2. New structure1 with silicide block (SAB) layer need one additional drain area LDD mask and implant loop to achieve higher BV_{DS} (breakdown voltage) and higher f_T (Cut off frequency) performance than conventional structure and new structure2.

Fig3 is 28nm HKMG general flow process. The step that is circled by red line is specially used for NLDMOS new structure1.

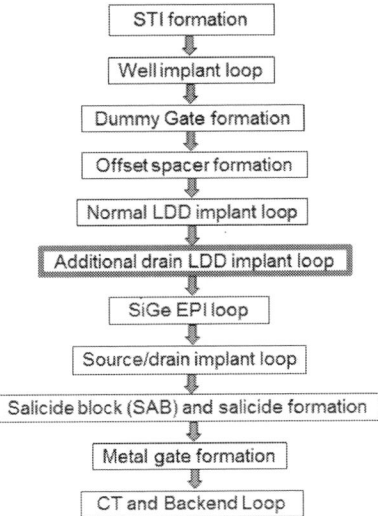

Fig3: 28nm HKMG general flow process.

TCAD SIMULATION RESULT

TCAD did f_T frequency and BV_{DS} breakdown voltage simulation based on the three NLDMOS structure. Fig.4(a) is conventional structure (with STI isolation) implant doping profile and electric field distribution; Fig.4(b) and Fig.4(c) are new structure1(with SAB isolation) and new structure2(with dummy gate isolation) implant doping profile and electric field distribution, respectively.

The conventional NLDMOS structure can get higher breakdown voltage, but f_T can only reach 30GHz with baseline size layout and baseline process, because of high on-resistance and high overlap capacitance with STI isolation. With Lp and Lc reduction optimization, f_T can be close to 50GHz, but breakdown voltage is lower than 5V from TCAD simulation (table1) due to short channel effect. Fig5 show that 28HKMG silicon data also indicate breakdown voltage is lower than 5V while Lc (effective gate length) is smaller (f_T 50GHz related Lc). so the structure has high risk for short channel effect, another disadvantage is that NW and PW boundary violation (NW/PW PR profile, NW/PW overlay) easily bring NLDMOS device performance violation.

The new structure1 (with SAB isolation) has best f_T performance, high breakdown voltage(>6V) and lowest Rdson performance with asymmetric lightly doped drain (LDD) implant between source and drain area, source area can share LDD with normal IO NMOS LDD implant, but drain area need special drain LDD implant as drift area doping. The implant has no process variation impact on NLDMOS performance since it is self-alignment with offset spacer formation after dummy gate process. But the structure disadvantage is that process need one additional mask for drain area LDD implant loop.

The new structure2 (with dummy gate isolation) can get better f_T performance since it has lower Rdson (on-resistance) by layout optimization. But breakdown voltage is a little lower than new structure1. The new structure2's advantage is that it does not need any special process but the disadvantage is that NLDMOS performance has big violation induced by NW/PW boundary violation (NW/PW PR profile, NW/PW overlay) and same with the conventional structure.

978-1-5386-5309-8/18 $31.00 © 2018 IEEE   49

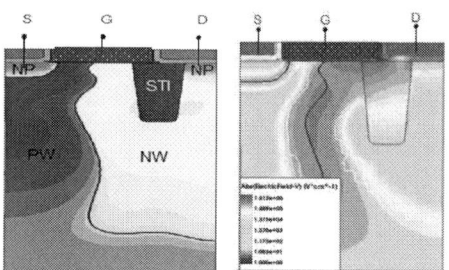

Fig4(a) conventional structure (with STI isolation) electric field distribution

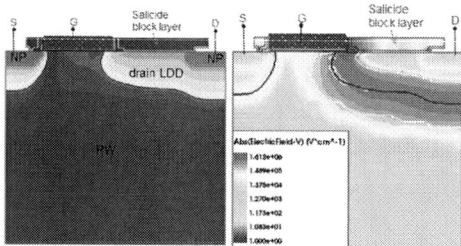

Fig4(b) new structure1(with SAB isolation) electric field distribution

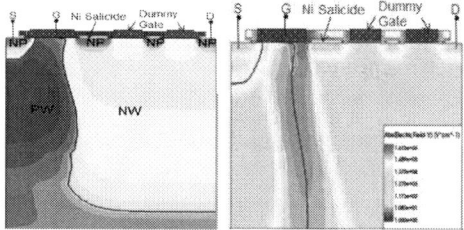

Fig4(c): new structure2 (with dummy gate isolation) electric field distribution

Table1: TCAD simulation result

Structure Type	BV_{DS} (V)	Ft (GHz)	Re_Y21 (S/um)	Im_Y11/ 2πf (F/um)	VTLIN (V)	IDLIN (μA/um)	Rdson (mOhm*mm^2)
Conventional with STI (baseline)	best	low	high	high	high	low	high
Conventional with STI (optimization)	low	good	high	half of baseline	double of baseline	a little higher than baseline	a little lower than baseline
New structure1 with SAB	good	good	low	low	medium	high	low
New Structure2 with dummy Gate	good	good	medium	medium	low	medium	medium

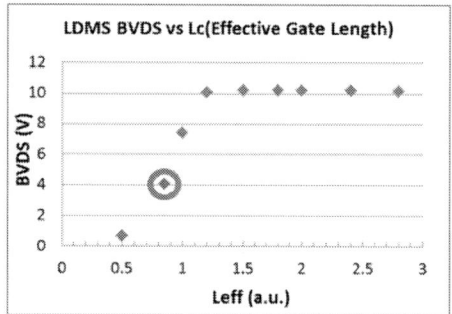

Fig5. Conventional LDMOS BV_{DS} (breakdown voltage) vs Lc Trend (28nm HKMG silicon data)

CONCLUSION

Based on TCAD simulation, new structure1 with SAB isolation NLDMOS can get the best balance among f_T, Rdson and BV_{DS} performance, and the NLDMOS violation can get best performance without NW/PW boundary violation impact, the disadvantage is that need one self-alignment LDD mask and special implant loop on drain area.

REFERENCES

[1] A.Giry, G.Tant, etc, "A monolithic watt-level SOI LDMOS linear power amplifier with through silicon via for 4G Cellular applications," in Proc. Power Amplifier. Wireless Radio Appl.,2013,pp.19~21.

[2] H.Chang, J.Jang, etc, "Advanced 0.13um smart power technology from 7V to 70V." in Proc. 24th ISPSD, Bruges, Belgium, Jun.2012, pp.217-220.

[3] T.Matsudai, etc, "0.13um CMOS/DMOS platform technology with novel 8V/9V LDMOS for low voltage high-frequency DC-DC converters," in Proc. 22ed ISPSD. Hiroshima, Japan, Jun. 2010, pp.315-318.

[4] W.Huang, etc, "5.5V Zero-Channel Power MOSFETs with Ron, SP of 1.0mΩ.mm2 for Portable Power Management Applications," Proc.of ISPSD 2009,pp.319-322.

[5] D.Muller, etc,"High-Performance 15-V Novel LDMOS Transistor Architecture in a 0.25um BiCMOS Process for RF-Power Applications," IEEE trans. Electron Devices, vol54, no 4, pp. 861-868.

[6] B.Bangerter et al.,IEEE communications Magazine,pp.90-96,2014.

[7] T.Johansson et al. IEEE Transactions on Microwave Theory a techniques, vol.62, pp.111-124, 2014.

[8] RuoYuan Li, Yongsheng Yang, Fang Chen, etc. A Study of 28nm LDMOS HCI Improvement By Layout Optimization. 2017 CSTIC, Pages: 1 – 4.

[9] Ming Li; Shaoqiang Zhang; Parthasarathy Shyam and Raj Verma Purakh, An optimized isolated 5V EDMOS in 55nm LPx platform for use in Power Amplifier applications, pp.107-109, 2016.

[10] Thanh Viet Dinh, Jan Sonsky, Jan Claes, Oliver Dieball, Guido T.Sasse and Celine Detcheverry, Novel 5V-EDMOS transistor with a record fmax of 450GHz in a baseline 40nm CMOS technology, pp.25.5.1-25.5.4, IEDM,2017.

A STUDY OF 28NM LDMOS LINEAR DRAIN CURRENT DEGRADATION INDUCED BY HOT CARRIER INJECTION

Guiying Ma, Hao Sun, Ling Tang, Yongsheng Yang, Xuejie Shi, Byunghak Lee, Miao Liao, Tzuchiang Yu

Technology R&D Center, SMIC

18 Zhangjiang Road, Pudong New Area, Shanghai, P.R. China 201203

E-mail: Connie_Ma@smics.com

ABSTRACT

LDMOS (Laterally Diffused Metal Oxide Semiconductor) applications get extensively in high voltage and smart power management. However, high operational drain voltage makes LDMOS devices highly vulnerable to the damage caused by HCI (hot-carrier injection). In this paper, shallow trench isolation (STI) based LDMOS linear drain current (Idlin) degradation induced by HCI (hot carrier injection) was systematically studied with the various layout parameters of NLDMOS (n-channel LDMOS) and well implant condition using 28nm Poly/SiON logic process, extensive TCAD (Technology Computer Aided Design) simulations and experiments revealed that layout, OPC (optical proximity correction), well implant optimization can greatly improve NLDMOS linear drain current degradation induced by HCI. The physical mechanism behinds the results should be that the impact ionization region has been driven further away from the STI edge and Si/SiO2 interface with a reduction in magnitude, which can reduce Idlin shift induced by HCI stress and make sure NLDMOS HCI pass industry general criteria of Idlin shift.

Keywords: LDMOS, Idlin shift, HCI.

INTRODUCTION

LDMOS transistors are widely used as integrated high-voltage switches, smart power technologies due to its low on-resistance (Rdson) and it is easily integrated within existing technologies. STI based LDMOS transistor is a device with STI structure in the drift region, where the STI is used to reduce the possibility of gate oxide breakdown at high drain biases [1].

As the technology is scaled down, the optimization of LDMOS becomes more challenging in terms of obtaining comparable HCI performance with the previous nodes. Scaling down of devices have resulted in higher electric field and worsen the HCI due to high concentration of well implant and field peak is closer to Si/SiO2 interface. In addition, high operational drain and gate biases make the LDMOS device vulnerable to the damage caused by HCI.

The reliability in STI based LDMOS devices have drawn much attention [2-7]. However, there is still lack of understanding and investigation available for STI-based LDMOS devices linear drain current degradation induced by HCI on advanced technology. The paper focused on NLDMOS linear drain current degradation induced by HCI with various layout parameters, OPC and well implant condition based on 28nm Poly/SiON logic process. TCAD simulation clearly verified experiment result and established the degradation mechanism.

EXPERIMENTAL

The LDMOS with STI on drain region structure (Fig1.a) used in the study is fabricated using 28nm Poly/SiON logic process with 1.8V/2.5V IO gate-OX. The Fig1.(b) is NLDMOS schematic cross section picture with layout variables, Lc, Ls, Lp and Lw, the layout parameter are aligned with paper[2]. Lp is defined as N-drift region and poly gate overlap size, Lc means effective gate length, Lw is defined as N-drift region and Pwell overlap, and Ls means STI width.

Fig2 is 28nm Poly/SiON general flow process. After STI process, the poly gate with gate oxide and offset spacer were formed, followed by the selective deposition of in situ boron-doped SiGe on PMOS source/drain. The nickel silicide/contact (CT) loop and Cu backend are applied to complete the process.

The typical operation condition of the device is 5V for VD and 1.8V/2.5V for VG. During the HCI stress, devices are stressed under high VD (i.e. 1.35X~1.55X Vcc) =6.75~7.75V with Vg @ISUBMAX (i.e. low VG) conditions while the source and substrate are at ground potential. Device parameters Idlin (Linear drain current) under VDS=0.1V are monitored during stress. The hot carrier reliability of NLDMOS is evaluated for various layout parameters structure and well implant condition. For TCAD simulation, a 2-dimensional device simulation is performed to analyze the degradation mechanism. All measurements are done at room temperature.

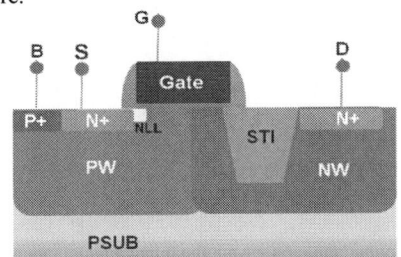

Fig.1(a). The Schematic cross-section of STI based NLDMOS.

978-1-5386-5309-8/18 $31.00 © 2018 IEEE

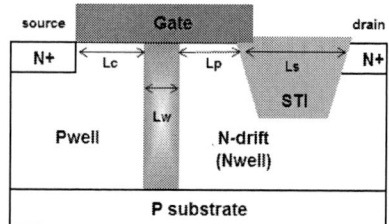

Fig.1(b). The schematic cross section of NLDMOS with parameters Lc, Lp, Lw and Ls.

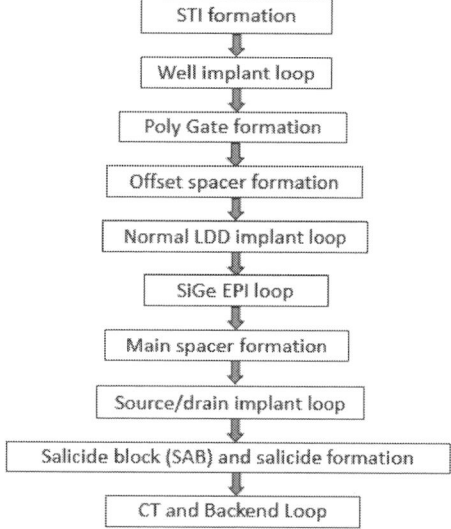

Fig2: 28nm Poly/SiON general flow process.

RESULT AND DISCUSSION

A. Layout dependence of HCI degradation

Lp and Ls as N-drift & poly gate overlap and STI width, are critical layout parameter for LDMOS key electrical characteristics, such as HCI, the device on-state resistance and breakdown voltage. This paper did wide Lp combined wide Ls split, Fig.3(a) on-state resistor (Rdson) and Fig.3(b) breakdown (BVDS) indicate that big Lp/Ls can improve BVDS but degrade Rdson performance, and it can be well understood since layout pitch (source to drain space) gets big. Fig.4 indicated that big Lw can improve BVDS and degrade Rdson performance also. Fig5 Idlin shift curve along with stress time (all Idlin shift value is normalized by baseline) show only optimized Lp/Ls can't improve Idlin shift performance, but optimized Lp/Ls (big size) combined optimized Lw (big size) can obviously reduce Idlin shift (Fig5 purple line) and make sure HCI pass industry general criteria of Idlin shift (AC 10year). The results can be clearly explained by TCAD simulations. Fig.7, Fig8 and Fig9 are NLDMOS impact ionization simulated by TCAD, which exhibits that the baseline impact ionization peak locates at N-drift and Lp region. When Lp and Ls increase, the impact ionization has been driving further away from

the STI edge. When Lw (NW/PW overlap) increase, the impact ionization has been driving further away from Si/SiO2 interface with a reduction in magnitude as observed in Fig.8. When big Lp/Ls combines big Lw, the impact ionization has been driving further away from STI edge and Si/SiO2 interface at the same time in Fig.9, which has implied a lower rate of hot carrier generation.

Although optimized Lp/Ls and Lw would degrade Rdson performance (Fig3, Fig4), Rdson bench mark trend (Fig.6) shows the Rdson (with big Lp/Ls/Lw) performance can match 40nm and other technology performance.

B. NW implant dependence of HCI degradation

This paper also studied NW implant impact on Idlin degradation caused by HCI, when draft area reduce half of NW implant dose and enhance 10% NW implant energy, combine optimized Lw, Idlin shift along HCI stress time can reduce also (Fig5 red line) and make sure marginally pass HCI with industry general criteria of Idlin shift (AC 10year). TCAD simulation can explain it that Fig.10 indicates the condition impact ionization is further away from STI edge and Si/SiO2 interface also. But the condition need special mask for drift area implant. For process and cost concern, the paper suggests optimizing layout Lp/Ls size and Lw size to achieve HCI pass performance.

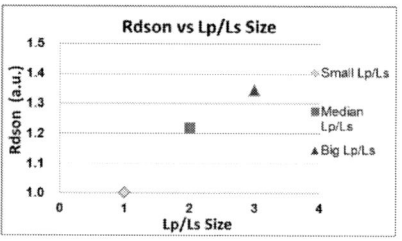

Fig3.(a). Rdson trend by Lp/Ls size

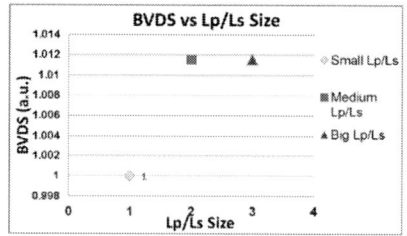

Fig3.(b). BVDS trend by Lp/Ls size

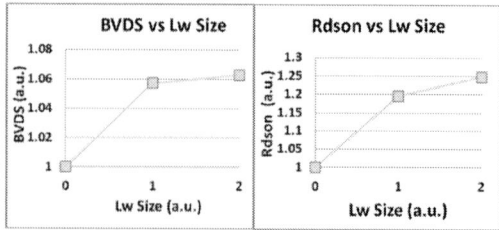

Fig4. BVDS/Rdson Trend by Lw Size

978-1-5386-5309-8/18 $31.00 © 2018 IEEE

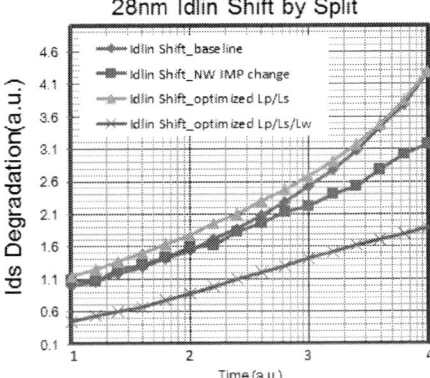

Fig5. Idlin shift by HCI stress time

SUMMARY

In this paper, LDMOS linear drain current degradation induced by hot carrier injection was systematically studied with the various layout parameters of NLDMOS and well implant condition using 28nm Poly/SiON logic process, extensive TCAD simulations and experiments reveal that layout, OPC, well implant optimization can greatly improve NLDMOS linear drain current degradation induced by HCI. The optimized Lp/Ls size combined optimized Lw size layout can make sure NLDMOS HCI pass industry general criteria of Idlin shift (AC 10year) that impact ionization has been driven further away from the STI edge and Si/SiO2 interface with a reduction in magnitude. NW implant optimization combined optimized Lw size also can make NLDMOS HCI marginal pass, but for process and cost concern, only layout optimization to improve HCI is suggested.

The layout implemented on 28nm HKMG process and 1.8V/2.5V NLDMOS HCI pass industry general criteria of Idlin shift (AC 10year) also.

REFERENCE

[1] Moens, P., & Van den Bosch, G. "Characterization of total safe operating area of lateral DMOS transistors." IEEE T.Device Mat. Re., bol. 6, no.3, pp.349-357,2006.

[2] Ruo Yuan Li, Yongsheng Yang, Fang Chen, etc. A Study of 28nm LDMOS HCI Improvement By Layout Optimization. 2017 CSTIC, Pages: 1 –4.

[3] S.Poli, S.Reggiani, G.Baccarani, etc. "Full Understanding of Hot-Carrier-Induced Degradation in STI-based LDMOS transistors in the Impact-Ionization Operation Regime." In Power Semiconductor Devices and ICs (ISPSD), 2011 IEEE 23rd international Symposium on, pp.152-155.IEEE, 2011.

[4] A.F.M.Alimin, H.H. Hizamul-din, S.F.W.M.Hatta, and N.Soin."The influence of Shallow Trench Isolation Angle on Hot Carrier Effect of STI-Based LDMOS Transistors." RSM of 2017 IEEE, pp.248-251. IEEE, 2017.

[5] Wu, D.J., D.Maji, J.R.Shih, Y-H.Lee, C.C.Liu, Y.H.Huang,R.Ranjan, "Understanding of hot-carrier-injection induced IDLIN kink effect in sub-100nm HV NLDMOS." In Reliability Physics Symposium (IRPS), 2012 IEEE International, pp.3B-4.IEEE, 2014.

[6] S. Haynie, A. Gabrys, T. Kwon, et al. 2010 22nd International Symposium on Power Semiconductor Devices & IC's (ISPSD), pp.241-244, 2013.

[7] Huixiong Zheng, Huaqiang Wu, Bin Wang, Electron Devices and Solid-State Circuits (EDSSC), 2014 IEEE International Conference, pp.1-2, 2014.

Fig6. Rdson bench mark trend

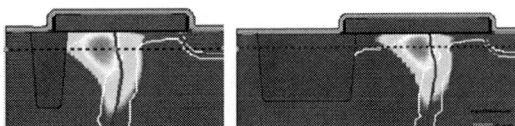

Fig7(a). small Lp/Ls   Fig7(b). big Lp/Ls

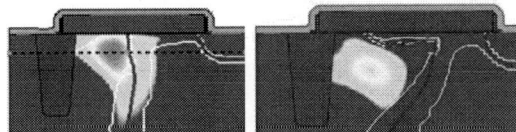

Fig.8(a). Lw=0nm   Fig8(b). big Lw size

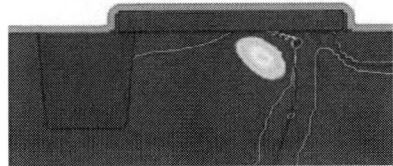

Fig.9. big Lp/Ls combined big Lw

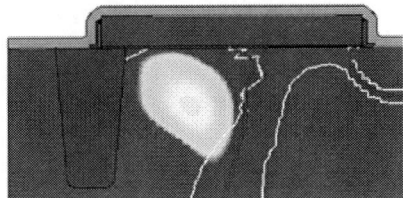

Fig.10.NW implant optimization combined Medium Lw

978-1-5386-5309-8/18 $31.00 © 2018 IEEE   53

Investigation and Three Implementations for Low Power Self-Aligned 1.5-T SONOS Flash Device

Zhaozhao Xu[1,2,*], Donghua Liu[2,3], Wei Xiong[2], Jun Hu[2], Ziquan Fang[2], Wenting Duan[2], Hualun Chen[2], Wensheng Qian[2]

[1]State Key Laboratory of Functional Materials for Informatics, Shanghai Institute of Microsystem and Information Technology, Chinese Academy of Sciences, Shanghai 200050, China
[2]ShanghaiHuahong Grace Semiconductor Manufacturing Corporation, Shanghai 201203, China
[3]State Key Laboratory of Functional Materials for Informatics, Shanghai Institute of Microsystem and Information Technology, Chinese Academy of Sciences, Shanghai 200050, China
*Corresponding author. Tel.: +86-18721794823. E-mail addresses:Xuzhaozhao_816@163.com (Z. Xu)

Biography

Zhaozhao Xu Now is a Ph.D Candidate of Shanghai Institute of Microsystem and Information Technology. His research interests focuses on embedded flash and PMIC.

Abstract

Two novel implementations (N+&P+-Polysilicon Memory/Memory1 and Memory-First/Memory2) and a conventional way (Memory-Last/Memory3) for low power self-aligned 1.5-T SONOS are demonstrated in this paper. The process flows steps of those devices are presented and characteristic of them are compared and investigated. It was founded that electrical characteristics of Memory1/Memory3 is much superior to that of Memory2. The degeneration of characteristic of Memory2 is mainly ascribed to the repeated doped channel and source/drain regions. Moreover, Memory1 and Memory3 have sampler process condition and better uniformity of ONO layers compared with Memory2, which is imperative to flash reliability. Compared to the Memory3 and Memory2, better device channel uniformity of Memory1 would be beneficial to improve the product yield.

Key Words—1.5-T SONOS; low power; self-aligned; reliability; uniformity; degeneration of characteristic

Introduction

Low-power, fast speed and highly reliable embedded memories are required for consumer, industrial, and automotive products. 1.5-T/Split-gate flash memories can realize the low-power and fast speed employed source side injection (SSI) for program. Compared to the 2-T structure, Split-gate structure is more popular for sub-90 process for its less chip area penalty due to removal of common source between selected gate (SG) and memory gate (MG) and its self-aligned structure. In [1], self-aligned architectures with "memory last'' configuration, in which the memory gate is a poly-Si spacer, are studied.

With the shrinking of device size, the structure of device is also evolved to improve their performance and to decrease the chip area penalty. More and more structures are proposed. To memory device, 3-D structures are employed in NAND flash to increase the memory density. However, planar 2-D structure is still popular for embedded system application. As device dimensions shrink, the short-channel effect (SCE) is more and more serious. Thus, split-gate structure is more popular for sub-90 nm node nor flash, especially in floating-gate (FG) memory [2, 3]. Split-gate structure offers higher suppression of SCE due to back-to-back gate structure. The split-gate structures are also employed in nitride-based charge-trapped flash [4] due to its higher shrinking capability recently. 40-nm node as well as 28-nm node split-gate MONOS flash had been demonstrated in [5, 6]. Furthermore, FinFET Split-Gate MONOS for high-speed and highly-reliable embedded flash in 16/14 nm-node and beyond also has been demonstrated [7]. In all of above mentioned literatures, either SSI or CHEI is employed for program, which can not meet the requirement of reliability for some high security products such as financial IC cards. Furthermore, the P/E power of SSI is also should be reduced for some special ultra-low power products because P/E with FN/FN is required for extending the life of battery. Therefore, three implementations for 1.5-T SONOS with ultra-low power FN/FN for P/E, in this paper, have been studied.

Devices and Experiments

For low power application, nomally-on SONOS device was chosen. This device has a postive V_{TH} after program and a negative V_{TH} after erase. Therefore, MG can be grounded during the read. Three implementations for low power self-aligned 1.5-T SONOS are illustrated in Fig.1.

The first novel implementation, Memory1, is formed with a "memory last" configuration cell with same lightly n-type channel doping for both MG and SG and a p-type doped polysilicon for SG while n-type doped polysilicon for MG. After the first implanted lightly n-type doping for adjusting the V_{TH} of SG, gate oxide of SG is formed followed by deposit and pattern of first p-type doped polysilicon. Then, second self-aligned lighlty n-type doping is employed to form a nomally-on SONOS device, as illustrated in Fig. 1 (a). Finally, ONO layers and second polysilicon are deposited and etched to form MG gate. Logic polysilicon gate is shared with the second polysilicon. In this way, a n-type buried-channel SG transistor is formed.

The second novel implementation, Memory2, is a counterpart of Memory1 and Memory3 and MG is formed with a "memory first" configuration, which is different form conventional process [1, 8, 9]. After the first implanted lightly

978-1-5386-5309-8/18 $31.00 © 2018 IEEE

n-type doping for adjusting the V_{TH} of MG, ONO layers of MG are formed followed by the deposit and pattern of first n-type doped polysilicon. After the removal of the ONO, V_{TH} adjust of SG is performed. After the formation of SG gate oxide which is shared with the high voltage logic transistor, second polysilicon is then deposited and etched to form SG gate and other logic transistor gate.

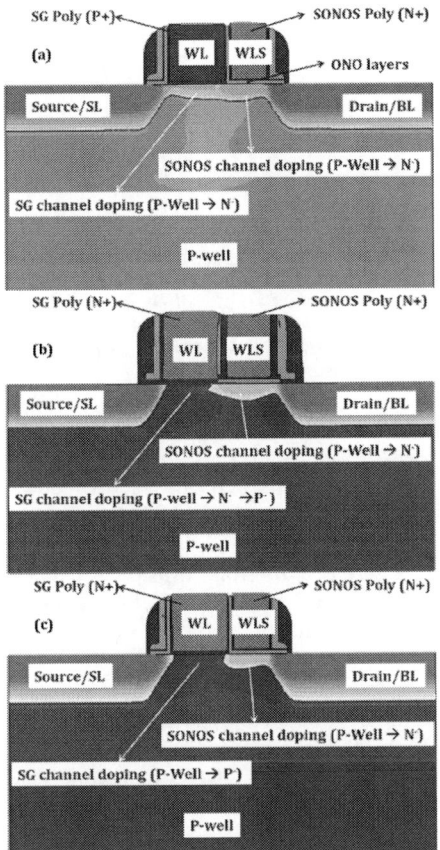

Fig. 1. Schematic cross-sectional drawings of 1.5-T SONOS device structure. (a) N$^+$&P$^+$-Polysilicon Memory/Memory1 with P$^+$ polysilicon for SG transistor and N$^+$ polysilicon for MG transistor. (b) Memory-First/Memory2. (c) Memory-Last/Memory3.

Memory3 is also formed with a "memory last" configuration, in which the SG is processed first (deposited and patterned). Then, self-aligned lighlty n-type doping is employed to form a nomally-on SONOS device, as illustrated in Fig. 1 (c). The MG is then formed [9]. This implementation has been demonstrated with capability of shirinking down to 20 nm. With the metal gate, FinFET Split-Gate MONOS with "memory last" configration for high-speed and highly-reliable embedded flash in 16/14 nm-node and beyond also has been demonstrated [7].

The drawing length of SG and MG are 0.17 μm and 0.14 μm, respectively. MG and SG V_{TH} are determined from I_D-V_G curve using max transconductance definition. V_{TH}(MG) repsents the MG transistor threshold voltage obtained under the biasing condition V_D=0.1 V, and V_B=V_{SL}=0 V with overdrive V_{WL} =2.0 V. V_{TH}(SG) repsents the SG transistor threshold voltage obtained under the biasing condition V_D=0.1 V, and V_B=V_{SL}=0 V with overdrive V_{WLS}=1.0 V (Cell is in erase state).

The read is performed with grounded MG and overdrived SG with V_{WL}=2.0 V.

FN is employed for program and erase (P/E) for its ultra-low power consumption. The array configration for 1.5-T SONOS is same as that of 2-T SONOS flash, which has demonstrated in [10]. The biases (V_{WL}/V_{WLS}/V_D/V_B/V_{SL}) of 2.0/7.5/-4.0/-4.0 V/float and 2.0/-4.0/7.5/7.5 V/float are applied for progam and erase, respectively. Therefore, program and erase are performed through the whole channel region with FN tunneling. Employed FN for both program and erase enables higher reliability, which meets the requirement of high security products due to less induced oxide damage. In addtion to the low power P/E, the read operation in that array is also low power by ground the MG and biased the SG with the standard on-chip voltage, which avoid the activation of charge pump circuit for higher voltage.

Results and discussion

A. Programming and erasing characteristics

The programming and erasing characteristics of 1.5-T devices are shown in Fig. 2. It can be noted that "memory last" 1.5-T devices show nearly same programming characteristic. Aditionally, Memory2 with "memory first" configuration shows poorer programming performance cmopared with Memory1 and Memory3, which employ "memory last" configuration. It indicated that "memory last" can improve the quality of the ONO layer by avoiding the impact of additional thermal buget. However, Memroy1 and Memory3 also show a smaller V_{TH} window compared with the 2-T device. This is due to poorer ONO layer unifromity resulted from the back-to-back gate structure. Furthermore, erasing characteristic of 1.5-T devices are comparable to that of 2-T device.

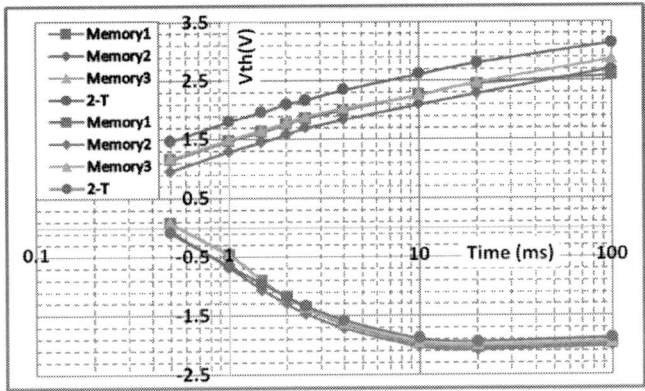

Fig. 2. Programming and erasing characteristics of 1.5-T devices. The 2-T device is also shown as a reference. The drawing channel length of SG and MG are 0.17 μm and 0.14 μm, respectively.

B. On-state current

For embeded application, the on-state current of flash cell is critical for circuit design. The MG is grounded during the read. The on-state current of these threes devices are compared and shown in Fig. 3, with the V_{TH}(SG) and V_{TH}(MG) of 1.0 V and -1.0 V, respectively. It is shown that Memory2 has a

978-1-5386-5309-8/18 $31.00 © 2018 IEEE

significantly lower on-state read current due to the degraded SG channel mobility and longer total effective channel length. Memory3 has a slightly lower on-state current due to the longer total effective channel length, compared with the 2-T SONOS. However, higher on-state current is achieved in Memory1 even with a longer total effective channel length and degraded SG channle mobility compared to the 2-T SONOS.

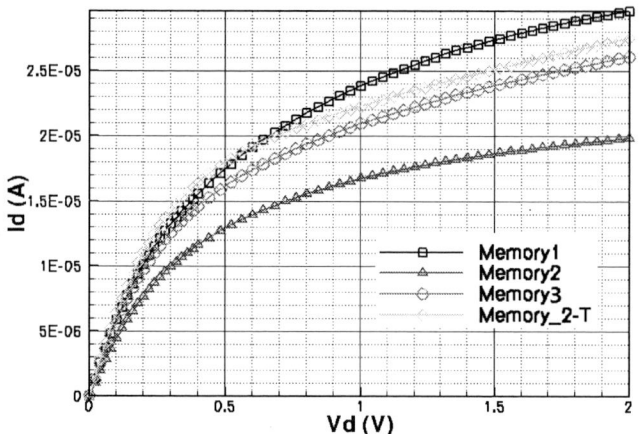

Fig. 3. I_D-V_D curves of 1.5-T devices with read bias condition. The 2-T device is also shown as a reference. The drawing length of SG and MG are 0.17 μm and 0.14 μm, respectively.

The SG channel mobility of these 1.5-T SONOS devices are illustrated in the Fig. 4. It is clear that mobility of Memory2 is dramitically degraded due to the repeated doped channel. Thus, the on-state current for read operation is reduced, which may increases the complexity of circuit design. It can be observed that SG channel electron mobility of Memeory1 is also degraded compared with Memory3 or 2-T. Thus, the increase of the on-state current in Memory1 is attributed to the wider conduction path under the SG channel, as illustrated in Fig. 5. Although the current density of Memory1 is smaller than that of Memory3 at the channel/gate oxide interface, the Memory1 has a threefold increase in the current conduction path compared with the Memory3. This can effectively help to increase the on-state current even with a degraded electron mobility in the SG channel of Memory1.

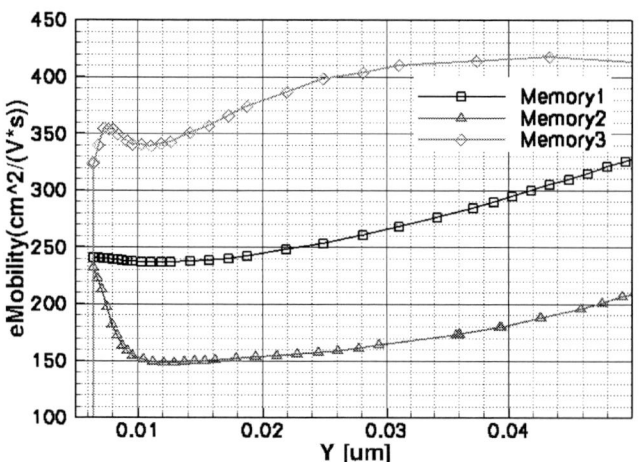

Fig. 4. SG channel center electron mobility of 1.5-T devices along the vertical direction. 2-T SONOS device has the same mobility as the Memeory3.

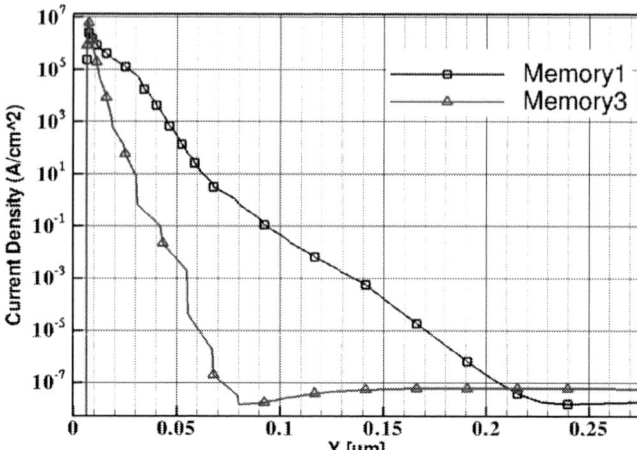

Fig. 5. The current density of Memory1 and Memory3 at the SG channel center along the vertical direction.

With the same drawing channel length of SG and MG, the current of 2-T device is larger compared to Memory2 and Memory3 of 1.5-T devices. With 2-T structure, there is common source between the SG and MG, which reduces the total effective channel length with the light doped drain (LDD) implantation. Therefore, the total effective channel lenth (SG and MG) of the entire 2-T SONOS device with four LDD extensions, is shorter than that of 1.5-T SONOS devices which only have two LDD extensions with back-to-back gate structure.

The Fig. 6 shows the read charaterasitcs of 1.5-T devices and 2-T device with the same total effective channel length of 0.252 μm according to the TCAD simulation. The V_{TH}(MG) and V_{TH}(SG) are -1.0 V and 1.0 V, respectively. The on-state current of Memory1 are obviously higher than the 2-T SONOS due to wider vertical conduction path and a higher current is also obtained in Memory3 due to the removal of common source between SG and MG, compared to 2-T device. However, the on-state current of Memory2 is still significantly smaller than that of 2-T SONOS with the same effective length. The shrinked length of the Memory2 can not increase the on-state current due to the more degraded electron mobility, as illustrated in Fig. 7. Higher repeated dose, which further degrades the mobility, is required to obtained the same threshold voltage when device is shrinked. That is, implementation of Memory2 is not a best choice for advanced process such as 45 nm or 28 nm node. This issue of repeated channel doping is avoided in both Memory1 and Memory3. Therefore, the current of Memory1 and Memory3 are increased with the shirinked total effective channel length. Moreover, Memory1 demonstrated a larger on-state current than that of both Memory3 and 2-T SONOS device with the same total effective channel length. This is attibuted to wider current conduction path under the SG.

978-1-5386-5309-8/18 $31.00 © 2018 IEEE   56

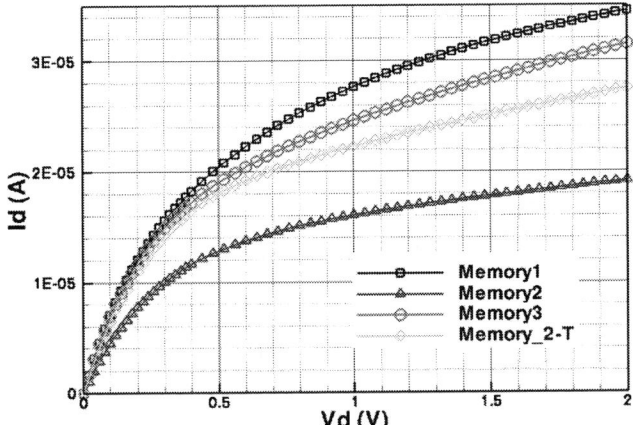

Fig. 6. I_D-V_D curves of shirked 1.5-T devices with read bias condition. The 2-T and 1.5-T SONOS devices have the same total effective channel length of 0.252 μm.

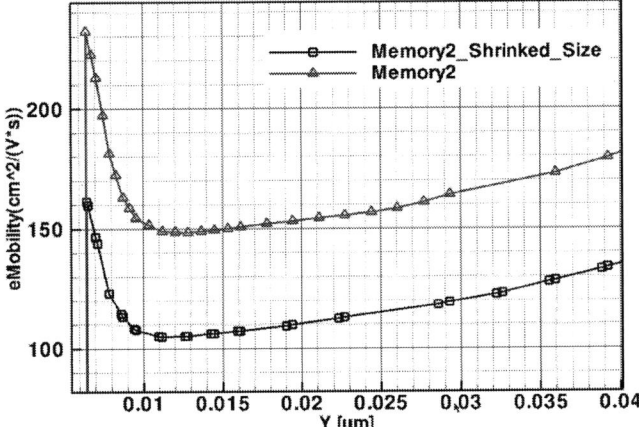

Fig. 7. SG channel electron mobility of Memory2 with two channel size. Memory2 with shirnked size induces a lower mobility.

One solution for higher on-state current is applying higher voltage to the gate of SG. This solution can improve the on-state current of Memory2 with cmpromsing the requirement of higher voltage generation. This may require the activation of the charge pump circuit during the read operation, which induces more power loss.

The read on-current as a function of the WL voltage are shown in Fig. 8(a). Noted that the on-current of Memory2 can increase by 50% at V_{WL}=3.0 V compared with V_{WL}=2.0 V and that the on-current of Memory1 can increase by 64% at V_{WL}=3.0 V compared with V_{WL}=2.0 V. However, it can be observed that higher V_{WL} seems not a good solution for the Memory2 due to the gradually staurated I_D-V_{WL} curve in the high V_{WL} region due to further degraded interface mobility with higher gate voltage. The gradually saturated I_D-V_{WL} curve due to the decreased interface mobility is retarded in the Memory3 for non-repeat doped channel. In Memory1, gradually saturated I_D-V_{WL} curve is further retarded due to both non-repeat doped channel and wider current conduction path. Memory1 with buried-channel has a part of current conduction path which is away from the silicon/oxide interface where the mobility is degraded significantly due to higher electric field in the gate oxide as the gate voltage increases. Therefore, the increase of

gate voltage results in less impact on the SG channel mobility of Memory1. It is also noted that, the leakage of the un-selected and erased Memory1 is silightly larger than that of 2-T cell. And, as illustrated in Fig. 8(b), Memeory2 and Memory3 have lower current leakage. However, the leakage of Memory1 can be reduced by adjusting the buried channel profile through the channel imlant. By the way, Memory1 is fully equivalent to Memory3 if the metal gate process is employed.

In summary, increasing WL voltage induces more improvement of on-state current for Memory1, compared with that of Memory2, Memory3, and 2-T device. In addition to convential Memory3, Memory1 is also a good choice for low power products which have lower standard on-chip voltage. Due to higher power comsuption is introduced with higher SG gate voltage for flash read operation, this solution should be concerned within the ultra-low power application.

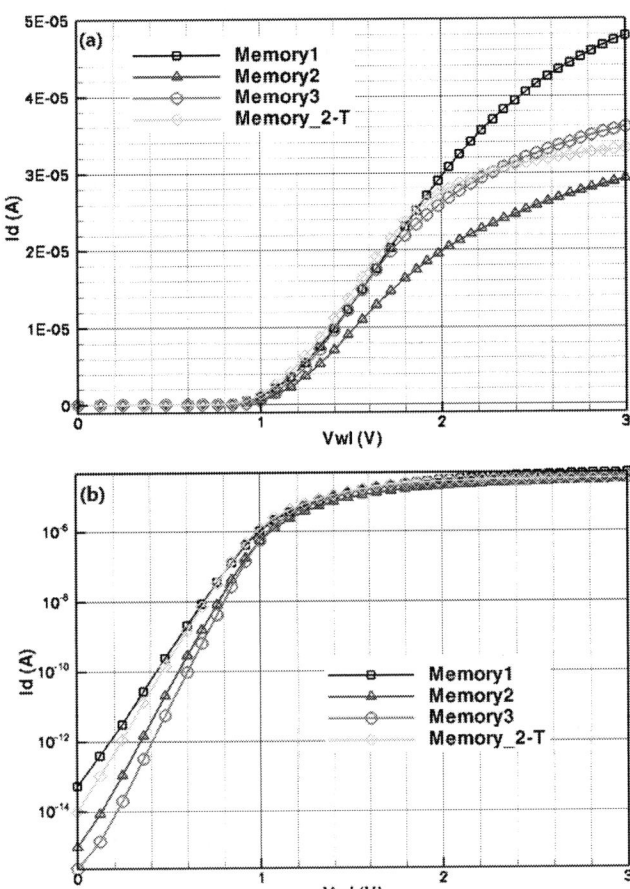

Fig. 8. Read on-current as a function of V_{WL}. The read bias are applied and that 1.5-T and 2-T devices have same drawing channel length.

Conclusion

Three implementations of self-aligned 1.5-T SONOS for low-power are compared and studied in this paper. It is shown that programming characteristic of 1.5-T devices are slightly degraded due to poor quality and uniformity of ONO layers. It also indicated that highest on-state current is achieved in novel Memory1 device due to the buried-channel, compared to

978-1-5386-5309-8/18 $31.00 © 2018 IEEE

Memory2, Memory3, and 2-T device. Wider current conduction path is achieved in Memory1, which results in higher on-state current. Memory2 with repeat doped channel shows a poorest on-state current performance although its perfect leakage suppression, which verified that Memory2 is not a good option for advanced process nodes. Conventional 1.5-T device, Memory3, shows the lowest leakage current and higher on-state read current.

References

[1] C. Charpin-Nicolle, A. D. Luca, A. Persico, C. Tallaron, F. Aussenac, R. Kies, G. Molas, L. Masoero, O. Cueto, and B. D. Salvo, "Review Article: A technological and electrical study of self-aligned charge-trap split-gate memory devices," Microelectronic Engineering, vol. 118, no. 4, pp. 15-19, 2014.

[2] S. K. Saha, "Design Considerations for Sub-90-nm Split-Gate Flash-Memory Cells," IEEE Transactions on Electron Devices, vol. 54, no. 11, pp. 3049-3055, 2007.

[3] L. Fang, J. Gu, B. Zhang, W. R. Kong, and S. C. Zou, "A Highly Reliable 2-Bits/Cell Split-Gate Flash Memory Cell With a New Program-Disturbs Immune Array Configuration," IEEE Transactions on Electron Devices, vol. 61, no. 7, pp. 2350-2356, 2014.

[4] K. T. Chang, W. M. Chen, C. Swift, and J. M. Higman, "A new SONOS memory using source-side injection for programming," Electron Device Letters IEEE, vol. 19, no. 7, pp. 253-255, 1998.

[5] T. Kono, T. Ito, T. Tsuruda, T. Nishiyama, T. Nagasawa, T. Ogawa, Y. Kawashima, H. Hidaka, and T. Yamauchi, "40-nm Embedded Split-Gate MONOS (SG-MONOS) Flash Macros for Automotive With 160-MHz Random Access for Code and Endurance Over 10 M Cycles for Data at the Junction Temperature of 170 $^{\ci}$C," IEEE Journal of Solid-State Circuits, vol. 49, no. 1, pp. 154-166, 2013.

[6] Y. Taito, T. Kono, M. Nakano, T. Saito, T. Ito, K. Noguchi, H. Hidaka, and T. Yamauchi, "A 28 nm Embedded Split-Gate MONOS (SG-MONOS) Flash Macro for Automotive Achieving 6.4 GB/s Read Throughput by 200 MHz No-Wait Read Operation and 2.0 MB/s Write Throughput at Tj of 170 C," IEEE Journal of Solid-State Circuits, vol. 51, no. 1, pp. 213-221, 2015.

[7] S. Tsuda, Y. Kawashima, K. Sonoda, A. Yoshitomi, T. Mihara, S. Narumi, M. Inoue, S. Muranaka, T. Maruyama, and T. Yamashita, "First demonstration of FinFET split-gate MONOS for high-speed and highly-reliable embedded flash in 16/14nm-node and beyond." pp. 11.1.1-11.1.4.

[8] Y. Tsuji, M. Terai, S. Fujieda, T. Syo, T. Saito, and K. Ando, "Lateral Profile of Trapped Charges in Split-Gate SONOS Memory," IEEE Transactions on Electron Devices, vol. 57, no. 2, pp. 466-473, 2010.

[9] L. Masoero, G. Molas, F. Brun, M. Gely, J. P. Colonna, V. D. Marca, O. Cueto, E. Nowak, A. D. Luca, and P. Brianceau, "Scalability of split-gate charge trap memories down to 20nm for low-power embedded memories," Electron Devices Meeting .iedm.technical Digest.international, vol. 47, no. 10, pp. 9.5.1-9.5.4, 2011.

[10] Z. Xu, W. Qian, H. Chen, W. Xiong, J. Hu, D. Liu, W. Duan, W. Kong, W. Na, and S. Zou, "Investigation and process optimization of SONOS cell's drain disturb in 2-transistor structure flash arrays," Solid-State Electronics, vol. 129, pp. 44-51, 3//, 2017.

SUBSTRATE CURRENT IMPROVEMENT FOR A 25V N-TYPE LDMOS

Ziquan Fang[1], Donghua Liu[1,2], Wensheng Qian[2]*

[1] HuaHong Grace Semiconductor Manufacturing Corporation, Shanghai 201206, China

[2] State Key Laboratory of ASIC and System, School of Microelectronics, Fudan University, Shanghai 200433, China

* Corresponding Author's Email: Ziquan.Fang@hhgrace.com

ABSTRACT

Substrate current (I_{sub}) is known as a critical indicator for device hot carrier degradation behavior.

This paper presents the result of I_{sub} improvement for a 25V NLDMOS transistor (operation voltage $V_g=V_d=25V$) design in 700V BCD process. The impact of accumulation region size, field oxide size and drift region doping concentration are discussed and optimized with assistant of TCAD simulation.

The optimized device has a great improvement on HCI reliability and also meets the I_{on}, BV, R_{sp} requirements.

INTRODUCTION

Lateral DMOS are nowadays more widely used in power applications than vertical DMOS as they can be easily integrated in conventional CMOS process. However, hot carrier degradation is a big concern for LDMOS as its operational voltage is normally high (12-100V) [1] and the current flows at Si/SiO_2 surface which makes LDMOS easier to be attacked by hot carrier injection and trapping [2].

Unlike conventional LV MOS transistor only has single peak of I_{sub}, LDMOS normally has two peaks of I_{sub}. The mechanism of the first peak is similar to LV MOS, while the second peak can be explained by the strong impact ionization near drain side which is induced by Kirk effect [3][4]. A number of studies have been done show that the second peak happen at higher V_g produces a greater degradation than first peak happen at lower V_g, also the second peak may trigger the parasitic triode resulting in device snapback [5][6].

In this paper, the N_{drift} region which is the key part of a LDMOS is carefully optimized to reduce both peaks of I_{sub}, while other electrical characteristics like specific on resistance (R_{sp}), BV and I_{on} are kept to meet the device requirements.

DEVICE DESCRIPTION

Figure 1 shows the cross section of the N-type LDMOS device discussed in this paper. The device is with a field oxide in the drift region. The device is fabricated in a 700V BCD process with the gate oxide thickness of 800A. The target operation voltage for both drain and gate is 25V. The required off-state breakdown voltage is over 40V with R_{sp} around 30mΩ·mm² (measured at $V_g=25V$

and $V_d=0.1V$). The critical layout parameters are indicated in Figure 1, LCH is channel region defined by P_{well} implant, LA is accumulation region defined by N_{drift} implant, and PF plus PA are field oxide size, in which PF is the length of poly plate. A highly doped NBL implant layer isolates the device from p-substrate to ensure the device can be used in high side circuit.

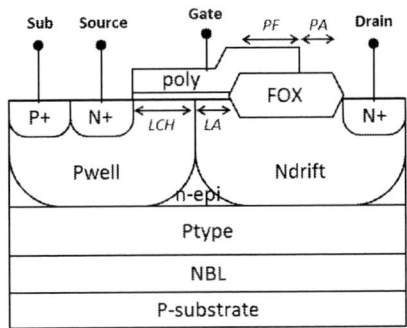

Figure 1: The cross section of N-type LDMOS transistor

Figure 2 shows the substrate current as a function of V_g before optimization. The first I_{sub} peak occurs at $V_g=6V$, and the second I_{sub} peak shows up when $V_g>17V$. The solid line is measured I_{sub}-V_g curve, while the dash line is TCAD simulation data by using Sentaurus workbench. The simulation result has been calibrated with silicon data for further analysis and optimization.

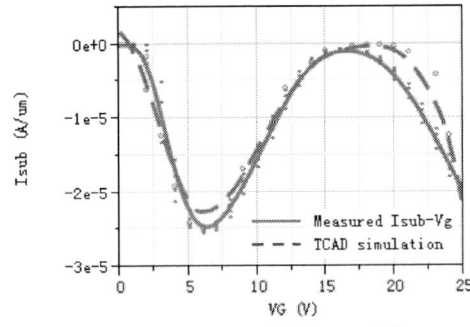

Figure 2: Substrate current as a function of V_g

Figure 3a shows the impact ionization rate at first I_{sub} peak when $V_g = 6V$, the strong impact ionization happens near bird's beak in the channel region, the hot carrier induced damage at this region will cause the channel mobility degradation, which leads to R_{on} and I_{on} degradation [6].

Figure 3b shows the impact ionization rate at second I_{sub} peak when $V_g = 25V$, the strong impact ionization region shifts towards drain side due to Kirk effect, both channel and N_{drift} region suffer from the hot carrier induced damage. R_{on} and I_{on} will have greater degradation at high V_g as the damage location under FOX has weaker gate control [7] and also the damage causes drift region resistance increase significantly [8][9].

The optimization discussed in next section aim to eliminate the second I_{sub} peak and reduce the first one.

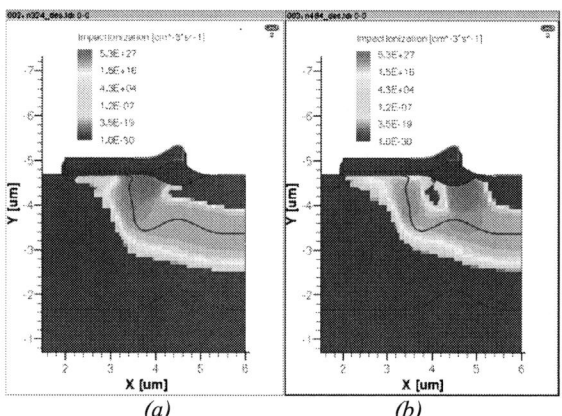

(a)   *(b)*

Figure 3: (a) Impact ionization rate at $V_g = 6V$, $V_d = 25V$ (b) Impact ionization rate at $V_g = 25V$, $V_d = 25V$

RESULTS AND DISCUSSION

N_{drift} doping

The whole N_{drift} region is composed of phosphorus implant steps with multiple implant depth (X_j). Figure 4 shows the I_{sub}-V_g curves under various 700kev phosphorus implant dose. As the Kirk effect occurs at high current densities when the current charge density exceeds the charge density in the depletion region, increasing depletion region doping is the most effective and straight forward way to eliminate the Kirk effect. From Figure 4, it is clear that the second peak of I_{sub} reduces significantly with 700kev dosage increase. However, as higher N_{drift} doping also reduces the drift region Rs, the lateral electric field under low Vg bias at LA area also increases, which cause the first peak of I_{sub} increases with higher 700kev dosage.

Figure 5 shows the I_{sub}-V_g curves under various 250kev phosphorus implant dose. At this implant energy, phosphorus dopant cannot reach the area below FOX, so it has minimum impact to Kirk effect. From figure 5, we can see the second peak of I_{sub} has almost no response to 250kev dosage change. However, the accumulation area concentration is dominant by this implant energy, higher

doping leads to higher I_{sub} first peak due to more impact ionization at bird's beak area at low V_g.

Therefore, by increasing the N_{drift} implant dosage at deeper X_j, while reducing the dosage at shallower X_j, both peaks of I_{sub} can be improved.

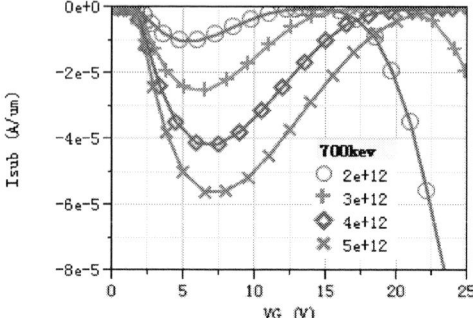

Figure 4: Simulated I_{sub}-V_g curves under various 700kev implant dose

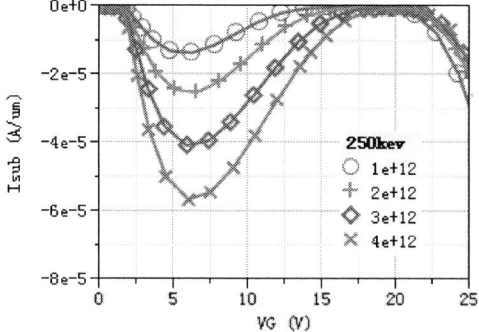

Figure 5: Simulated I_{sub}-V_g curves under various 250kev implant dose

LA size

Figure 6 shows the I_{sub}-V_g curves under various LA size. I_{sub} peak at low Vg bias first increase then decrease with larger LA size. Small LA is not preferred due to below reasons.

Firstly, at small LA, the impact ionization region extends from bird's beak to underneath FOX at low V_g, the I_{sub} peak at low V_g cannot reflect both damage at bird's beak and underneath FOX area. Therefore, although the I_{sub} value is low with small LA, actual HCI performance is even worse. Secondly, when LA is small, the device suffers from quasi-saturation phenomenon [9]. The drain current is more limited by the carrier velocity saturation in N_{drift} region due to high impact ionization rate under FOX. Instead, V_g has less control to the I_d, thus I_d increase slowly with higher V_g. Figure 7 are the I_d-V_d curves comparison for LA = 0.1um and 0.7um.

Compare to N_{drift} doping, LA size has less impact to

978-1-5386-5309-8/18 $31.00 © 2018 IEEE   60

I_{sub}, it is more adjusted to balance the trade-off between I_{on} and BV.

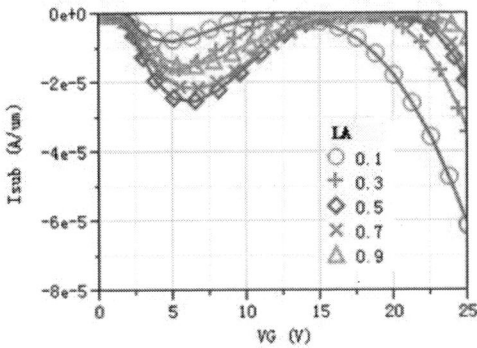

Figure 6: Simulated I_{sub}-V_g curves under various LA size

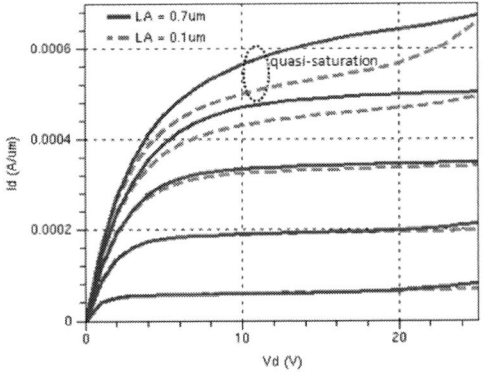

Figure 7: Simulated I_d-V_g curves with LA = 0.1 & 0.7um

PF & PA size

Figure 8 shows the I_{sub}-V_g curves under various PF+PA (FOX) size. The size of FOX also has less impact to I_{sub} compare to N_{drift} doping. PF/PA ratio has no impact to I_{sub}. Instead, the ratio usually can be moderated for BV optimization.

Compare to the impact to I_{sub}, FOX size has more significant impact on BV, I_{on} and R_{sp}. Longer FOX gives higher BV due to wider depletion width, but much lower I_{on} again due to quasi-saturation effect [10] and higher R_{sp}. Therefore, PF+PA should keep as small as possible as long as BV can meet the design requirement.

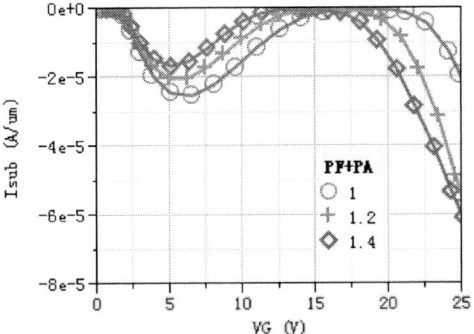

Figure 8: Simulated I_{sub}-V_g curves under various PA+PA

Optimization result

As a result of above works, Figure 9 shows the I_{sub} performance before and after optimization, the optimized device has eliminated I_{sub} second peak tail and first peak has ~16% improvement. Figure 10 shows the I_{on} performance before and after optimization, the device I_{on} performance also improved ~12%.

Figure 11 shows the I_{on} degradation% as a function of stress time (stress @ I_{submax}). The optimized device passed the HCI reliability test spec of DC $T_{0.1\%}$>0.2yrs (@1.1V_d, RT, @ I_{on} shift 10%).

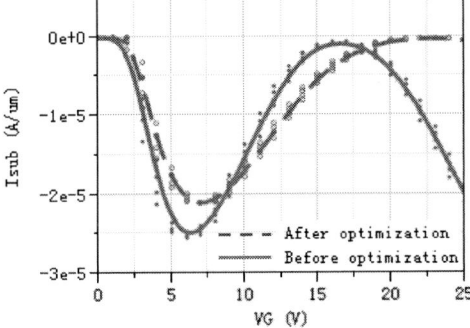

Figure 9: Measured I_{sub}-V_g curves before and after optimization

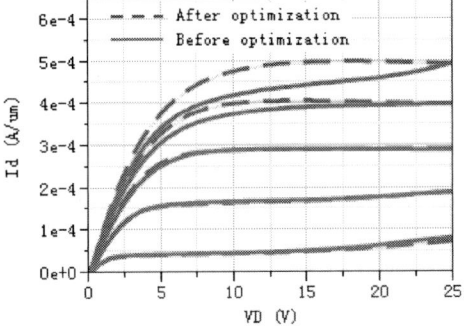

Figure 10: Measured I_{on} curves before and after optimization

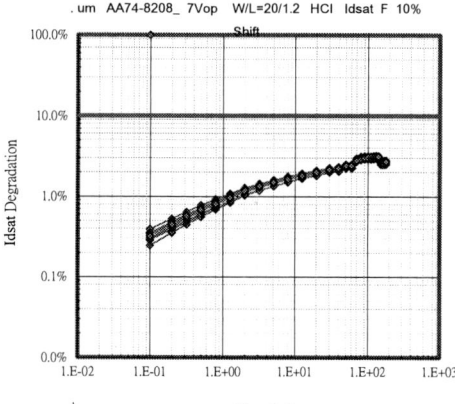

Figure 11: I_{on} degradation as a function of stress time (stress @ I_{submax})

CONCLUSION

A 25V N-type LDMOS device has been studied in this paper for its substrate current improvement. TCAD simulation has been used to understand the I_{sub} double peak phenomenon and also for I_{sub} improvement through N_{drift} region optimization.

The study shows that: to optimize the I_{sub} peak, firstly, by adjusting the N_{drift} implant concentration at different X_j, both peaks of I_{sub} can be improved. Secondly, LA size cannot be too small due to HCI and I_{on} concern, while PF+PA size should keep as small as possible as long as BV can meet the design requirement. The last but not the least, unlike conventional LV MOS device, I_{sub} is not the only indicator for LDMOS device HCI degradation behavior. The location of impact ionization is also a critical factor which must be taken into consideration for device degradation judgement.

ACKNOWLEDGEMENTS

The authors would like to thank all members of device group and high voltage group of HHgrace for their great support on this work.

REFERENCES

[1] Jone F. Chen, J. R. Lee, Kuo-Ming Wu, Tsung-Yi Huang, and C. M. Liu, "Effect of drift-region concentration on hot-carrier-induced R_{on} degradation in nLDMOS transistors," *IEEE Electron Device Letters*, vol. 29, no. 7, pp. 771–774, July 2008.

[2] P. Moens, G. Van den Bosch, and G. Groeseneken, "Competing hot carrier degradation mechanisms in lateral n-type DMOS transistors," *in Proc. IRPS*, 2003, pp. 214–221.

[3] S. K. Lee, C. J. Kim, J. H. Kim, Y. C. Choi, H. S. Kang, and C. S. Song, "Optimization of safe-operating-area using two peaks of body-current in submicron LDMOS transistors," *International Symposium on Power Semiconductor Devices and ICs (ISPSD)*, 2001, pp. 287-290.

[4] J. F. Chen, K. M. Wu, K. W. Lin et al., "Hot-carrier reliability in submicron-meter 40V LDMOS transistors with thick gate oxide," *IEEE Intl. Reli. Phy. Symp.*, 2005, pp. 560–564.

[5] Jun Wang, Rui Li, Yemin Dong, Xin Zou, Li Shao, and W.T. Shiau, "Substrate current characterization and optimization of high voltage LDMOS transistors," *Solid-State Electronics*, vol. 52, 2008, pp. 886–891.

[6] K. M. Wua, J. F. Chen, Y. K. Su, J. R. Lee, and K. W. Lin "Effects of gate bias on hot-carrier reliability in drain extended metal-oxide semiconductor transistors," *Applied Physics Letters*, 2006, 89:18.

[7] J.Hui and j. Moll, "Submicrometer device design for hot-electron reliability and performance," *IEEE Electron Device Letters*, vol.6, no.7, pp.350-352, Jul 1985.

[8] Y. Rey-Tauriac, J. Badoc, B. Reynard, R.A. Bianchi, D. Lachenal, A. Bravaix, "Hot-carrier reliability of 20V MOS transistors in 0.13μm CMOS technology," *Microelectronics Reliability*, vol. 45, 2005, pp. 1349-1354.

[9] S. Shahabuddin, N. Soin, K.K. Goh, Y. Abdul Wahab, H. Hussin, "Voltage dependences of parameter drifts in hot carrier degradation for n-channel LDMOS transistors," *Microelectronic Engineering*, vol. 109, 2013, pp. 101–104.

[10] Jone F. Chen, Shiang-Yu Chen, Kuo-Ming Wu, J. R. Shih, and Kenneth Wu, "Convergence of hot-carrier-induced saturation region drain current and on-resistance degradation in drain extended MOS transistors," *IEEE Transactions on Electron Devices*, vol. 56, no. 11, pp. 2843-2847, Nov 2009.

IMPROVEMENT OF BOND PAD CRYSTAL DEFECT BY NEW

ALUMINUM PAD FILM STACK

Junhong Zhao[1]*, Zhengying Wei[1], Yansheng Wang[1], Wei Zhou[1], Chang Sun[1], Jun Qian[1], Yufei Peng[1], Ran Huang[1]

[1]Haili Microelectronics Corporation, Shanghai201203, China

*E-mail addresses: zhaojunhong@hlmc. cn

ABSTRACT

With the rapid development of very large scale integration (VLSI) and continuous scaling in the metal oxide semiconductor field effect transistor (MOSFET), pad corrosion in the aluminum (Al) pad surface has become practical concern in the semiconductor industry. This paper presents a new method to improve the pad corrosion on Al pad surface by using new Al/Ti/TiN film stack. The effects of different Al film stacks on the Al pad corrosion have been investigated. The experiment results show that the Al/Ti/TiN film stack could improve bond pad corrosion effectively comparing to Al/SiON film stack. Wafers processed with new Al film stack were stored up to 28 days and display no pad crystal (PDCY) defects on bond pad surfaces.

INTRODUCTION

Aluminum (Al) bond pad on semiconductor chips plays an important role in integrated circuit (IC) device reliability and yield [1, 2], which is mainly to connect chip and outside circuits at assembly processes. Al bond pads quality is an important index that needs to be monitored carefully. However, defect like fluorine (F)-induced corrosion [3] has been a major issue in Al pad fabrication, which will affect Al bond pad quality and causes discolored or non-stick on pads [4].

At pad fabrication, the F on bond pad surface is inevitably introduced due to the need for the use of F-containing chemicals or gases during the passivation etch process [5]. Such F residue on pad surface can react with Al under the presence of moisture to form pad corrosion defects, which are usually called pad crystal (PDCY) defects [6]. Since the formation of PDCY defect is considered to be related with the existence of F ion and moisture around the Al pad, controlling F and moisture concentrations are the common method to eliminate this type of pad defect. In [7], Ahn used UV/ozone cleaning to prevent the Al pad corrosion. Fu [8] solves the pad corrosions by optimizing the pad fabrication process. Although many methods have been reported, it is still very important to explore a more effective method to reduce PDCY defects on pad surface.

In this paper, we provide a new method by using the Al/Ti/TiN film stack to replace the conventional Al/SiON layer during the Al pad fabrication process to reduce the pad corrosion. The effect of various factors such as Al film stack, wafer counts on FOUP and stored times on the formation of PDCY defects have been studied. It was found that the Al/Ti/TiN film stack could effectively improve PDCY defects compared to Al/SiON film stack.

EXPERIMENT

To improve the PDCY defects on pad surface, many split conditions such as different Al film layers stack and stored times are listed in Table 1. Firstly, two types of Al film stacks were deposited on 2.8μm thickness Al film separately. Subsequently, the bond pads are opened to expose Al by removing the passivation layers on top. Finally, wafers with different Al pad structures are stored in front opening unified pod

(FOUP) for different times to investigate the formation of PDCY defects on pad surface. In addition, different piece of wafers (4~25 pcs) is placed in five FOUP to observe the suffer PDCY rate. The optical microscope (OM) and the scan electron microscope (SEM) were used for defect morphology observation. Surface analysis was performed using energy dispersive X-ray (EDX) on samples.

Table 1 Experiment item to deal with PDCY defect issue

Splits	Old film stack	New film stack
Layout	400A SiON 28K Al	300A TiN 50A Ti 28K Al
PVD	Al + SiON 400A	Al + Ti/TiN 350A
Cover Etch	Recipe bodies are the same (ME 65s and OE 70s)	
After Final anneal	Stored for 4, 12, 18 and 28 days	

RESULTS AND DISCUSSION

Fig.1 (a) shows OM image of the bond pad of a wafer prepared by old film stack stored for 12 days. One can clearly see that there was a circular shape defect covered on the bonding pad. From its magnified SEM micrograph (Fig.1 (b)), it can be seen that the defect is a leaf-like feature [9] with the size is about 13μm. The EDX spectrum shows F was detected at the defect area, which confirmed that the PDCY defect is due to fluorine-induced Al bond pad corrosion. However, for the Al bond pad prepared by new film stack, no corrosion found after stored for 12 days as shown in Fig.1 (c-d). This result indicates that the new film stack has the advantage of reducing the risk of the pad corrosion.

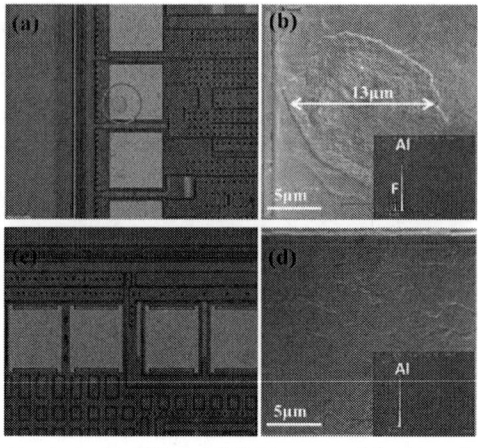

Fig.1. (a, b) OM and SEM images of the Al pad prepared by old film stack; (c, d) OM and SEM image of Al pad prepared by new film stack; Inset in (b, d): the corresponding of EDX analysis

In order to illustrate the mechanism of new Al film stack to prevent the pad corrosion issue, we compare the Al pad fabrication process through different Al film stack. Fig.2 shows the schematic diagram for Al bond pad prepared by using different Al film stack. Since the Ti/TiN layer is harder than SiON, higher etch rate can be achieved in SiON layer. It was suspected that there was higher F remaining on the bonding pads with only SiON film after bonding pads etching and cleaning. Subsequently, the remained F will easily react with Al when the bonding pad absorbed moisture, then forming the PDCY defects on the Al bond pads. Thus, the main reason for the new film stack layer to prevent PDCY defects is to reduce the content of F on pad surface.

978-1-5386-5309-8/18 $31.00 © 2018 IEEE

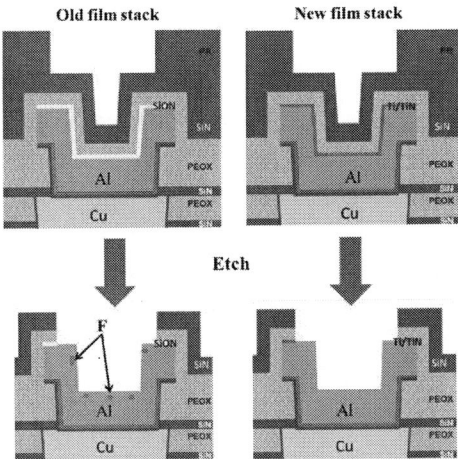

Fig.2. Schematic diagram for Al bonding pad process using different Al film stack

In our prior studies, the PDCY defects were hard to recur when there are few wafers on one FOUP. To further evaluate the impact of two different Al film stacks on PDCY defects, the relationship between the PDCY defects and wafer counts on one FOUP is shown in Fig.3. We found that the wafer with Al old film stack surfer PDCY rate increased from 25% to 92% when the pieces of wafer on FOUP increased from 4 to 25 pcs after stored for 12 days. Whereas, no PDCY defect was found in the wafer with Al new film stack no matter how the wafer counts on FOUP changes.

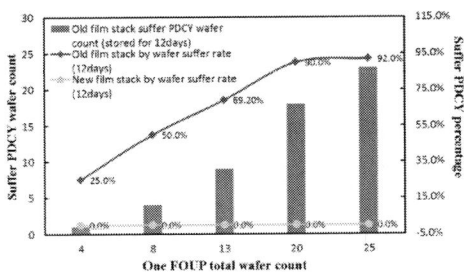

Fig.3. Suffer PDCY wafer count depending on the wafer count in one FOUP

Fig.4 shows the suffer PDCY rate of wafers with Al pad prepared by the new film stack and old film stack after stored for different days. It can easily observe that the Al pad with new film

stack did not suffer PDCY defects even after stored 28 days. However, for the old film stack wafer, the PDCY defects were detected on the surface of Al pad when only stored for 4 days. Moreover, the PDCY suffer rate of old film stack wafers is increased as the increasing storage days.

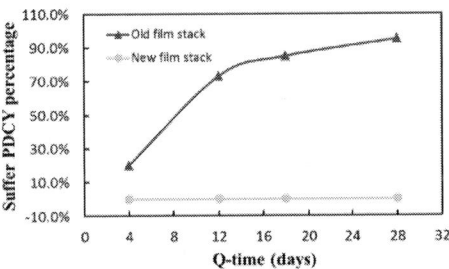

Fig.4. Suffer PDCY rate on wafers with Al pad prepared by the new film stack and old film stack in various Q-times

To evaluate if the changing of Al film stack impacted the yield of the wafer, the yield loss of major fail bin in different wafers prepared by Al old film stack and new film stack are shown in Fig.5. As can be seen in Fig.5, the yield loss of major fail bin in Al new film stack are similar with that of old film stack. This result suggests that new Al film stack can effectively prevent the Al pad corrosion.

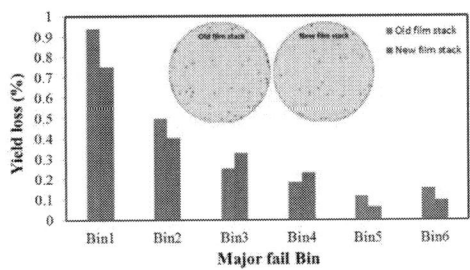

Fig.5. Yield loss of major fail bin in the wafers prepared by the new film stack and old film stack

CONCLUSIONS

New Al/Ti/TiN film stack experiment was designed to verify and improve F-induced corrosion on Al bond pad. The experiment

978-1-5386-5309-8/18 $31.00 © 2018 IEEE

results show the Al/Ti/TiN film layers can effectively reduce the F remaining concentration on the bond surface compared to the Al/SiON layers due to the lower etch rate of Ti/TiN at the same etch conditions. For the wafers were processed with new Al film stack, were stored up to 28 days and showed no PDCY defects on bond pad surfaces. In addition, the yield results show the new Al film stack did not affect yield. Therefore, the new Al film stack is an effective method to improve the F-induced corrosion on pad surfaces.

REFERENCES

[1] Y. N. Hua, "A rapid chemical diagnostic method to detect silicon oxide and TiN residues on Aluminum bondpads in wafer fabrication," *Semiconductor Electronics, 2002. Proceedings*, pp. 177-181, 2002

[2] Y. N. Hua, S. Redkar, C. K. Lau, et al, "A study on non-stick Aluminum bondpads due to Fluorine contamination using SEM, EDX, TEM, IC, Auger, XPS and TOF-SIMS techniques," *International Symposium for Testing and Failure Analysis 2002*, pp. 495-504, 2002

[3] Y. N. Hua, Z. X. Xing, X. M. Li, "Characterization studies of fluorine-induced corrosion crystal defects on microchip Al bondpads using X-ray photoelectron spectroscopy," *Proceedings of the 21th International Symposium on the Physical and Failure Analysis of Integrated Circuits*, pp. 90-93, 2014

[4] Y. N. Hua, "A study on Al bondpad grain boundaries and galvanic corrosion in wafer fabrication," *International Conference on Semiconductor Electronics*, pp. 83-85, 2004

[5] J. F. Graves and W. Gurany, Reliability effects of fluorine contamination of aluminum bonding pads on semiconductors chips, *Microelectronics Reliability*, vol. 24, pp. 805-806. 1984

[6] M. Li, W. T. K. Chien, Q. Q. Yu, et al, "A study on the relationship between pad surface fluorine concentration and the formation of pad corrosion defect," *Proceedings of the 20th IEEE International Symposium on the Physical and Failure Analysis of Integrated Circuits*, pp. 311-315, 2013

[7] S-Ho. Ahn, T-Je. Cho, Y-Soo Kim, et al, "Prevention of aluminum pad corrosion by UV/ozone cleaning," *Proceedings 46th Electronic Components and Technology Conference*, pp. 107-112, 1996

[8] G. C. Fu, L. Ni, X. M. Zhang, "A new method to prevent aluminum pad corrosion," *International Workshop on Junction Technology*, pp. 1-4, 2014

[9] R. J. Qi, S. Q. Duan, M. Li, et al, "Study on a leaf-like bonding pad defect," *Proceedings of the 20th IEEE International Symposium on the Physical and Failure Analysis of Integrated Circuits*, pp. 481-484, 2013

A TECHNIQUE FOR FABRICATING LOW VOLTAGE CHARGE-COUPLED MOSFET WITH 0.9UM PITCH SIZE

L. Shi*, R.X. Fan, J.Z. Miao, Y.P. Wang, J.K. Shen and L. Yao

Department of Engineering, Shanghai Huahong Grace Semiconductor Manufacturing Corporation,
Shanghai, China
*Corresponding Author's Email: Lei.Shi@HHgrace.com

ABSTRACT

The low voltage Charge-Coupled MOSFET with deep-submicron technology is fabricated with extra thick trench oxide process. The extra process module consists of thick oxide growth in trench, resist coating, resist expose to control thick oxide height in trench which would finally affect the break down voltage of the device. The effects of trench length, thickness of the thick trench oxide layer and cell pitch size on the breakdown voltage were discussed in this work. This low voltage Charge-Coupled MOSFET with deep-submicron technology provides an approach to achieve the significantly smaller specific on-resistance at the same break down voltage that obtained on the conventional trench MOSFET by using a greater doping concentration in the drift region.

INTRODUCTION

The rapid development of power electronics technology and the proposal of Smart Grid put forward higher requirements on power electronic devices. Informationization, digitization, automation and interaction are the goals of development, and advanced power electronic technology is an important technology toolbox. As a kind of important power electronics device, Power MOSFETs are an important class of discrete devices used in a variety of power conversion applications, because they have low conduction power loss, high input impedance, and high switching speed [1]. The increasing popularity of power MOSFETs has led to greater efforts to try to reduce their specific on-resistance, to minimize die sizes, and to reduce cost, especially for low voltages of less than 50V. To optimize the power handling capability for a specified break down voltage rating and on-state resistance, it is essential to increase the cell packing density of a power semiconductor device and use an optimized drift layer design [2]. In addition to an increased cell packing density, the MOS gates formed along the etched silicon regions result in a reduced on-state resistance per unit cell because they lack the JFET pinching resistance normally present in a silicon DMOSFET structure. The more improvements are typically achieved by increasing the trench density (i.e. the channel density) through decreasing the pitch size of the trench MOSFET. Several attempts to push the limits based on general process have been reported resulting in a

smallest pitch size of 0.8μm [3]. The reported device structure made by the conventional fabrication methods have some limitations on further reducing the cell size. A novel fabrication process of trench technology has been reported to achieve a pitch size down to 0.6μm fabricated with an unconventional implant topology [4]. Further, to reduce on-state resistance, the power MOSFETs can be fabricated using wide band gap materials such as silicon carbide and gallium nitride. The emerging technologies like GaN and SiC are still at a starting stage and will take few more years to mature [5]. However, in low voltage and medium voltage with low power requirements, the silicon devices with deep-submicron technology are best suited since they are easy to fabricate and relatively cost effective when compared to compound semiconductors and unconventional process. Hence the optimization of silicon structures and novel device structures using silicon by the general process is need of the day [6]. The power Charge-Coupled MOSFET structure, based up on utilizing an electrode embedded within oxide coated deep trenches as a part of the drift region, can be proposed to achieve high performance by altering the electric field distribution from the triangular shape in a one-dimensional case to a rectangular shape for the two dimensional case. Several papers have been published on this device structure and their fabrications [7]. However, the reported fabrications may be not suitable for deep-submicron technology.

In this work, the low voltage Charge-Coupled MOSFET with deep-submicron technology is fabricated by using only a few additional steps in the standard process sequence. The effects of some factors on the breakdown voltage also are discussed.

DEVICE FABRICATION

The proposed device, the low voltage Charge-Coupled MOSFET with deep-submicron technology, is shown in Figure 1 along with the conventional trench MOSFET, for comparison. As shown in the figure, the Charge-Coupled MOSFET uses highly doped N-type polysilicon as gate material. The height of thick oxide in trench may be easily controlled as it is independent of the lithography conditions. The gate oxide on the upper side-walls of the trenches is grown by thermal oxidation.

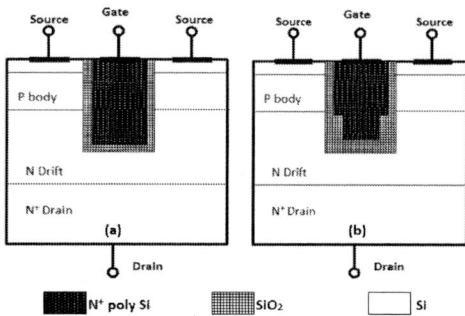

Figure 1: Schematic cross-sectional view of the (a) conventional trench MOSFET and the (b) proposed Charge-Coupled MOSFET

The proposed device can be fabricated using only a few additional steps in the standard process sequence, as shown in figure 2(a). The proposed fabrication process follows the standard process of conventional trench MOSFET fabrication until the trench opening using hard mask and thick oxide deposition in trench. Then, photo resist is coating with the desired thickness and filled in the deep trench, as shown in figure 2(b). In the next step, the desired photo resist height in the trench is achieved by the controlled expose energy and focus, as shown in figure 2 (c). The thick oxide is wet etched from the upper part of the trench which is not filled by photo resist, then, photo resist is completely striped, as shown in figure 2(d). The remaining steps are similar to the standard process flow, wherein the gate oxide on the upper side-walls of the trenches is grown by thermal oxidation, the trench is refilled with highly doped N-type polysilicon and an inter-metal dielectric is then deposited followed by etching contact windows for the N+ source and P-base regions, as shown in figure 2 (e).

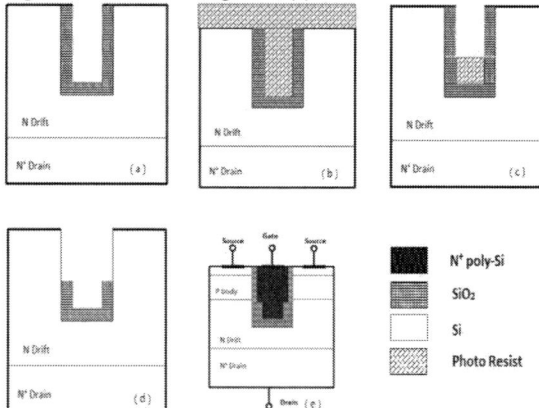

Figure 2: Proposed fabrication process steps for the Charge-Coupled MOSFET

RESULTS AND DISCUSSION

Figure 3 shows the cross-sectional SEM photograph of fabricated Charge-Coupled MOSFET with 0.9μm cell pitch by the above process. It is observed that thick oxide formed at the trench bottom, while gate oxide is gown at the upper trench. And, the thick oxide height can be adjusted by the different expose energy and focus in the progress of lithography. The bottom trench profile is magnified, as shown in figure 4.

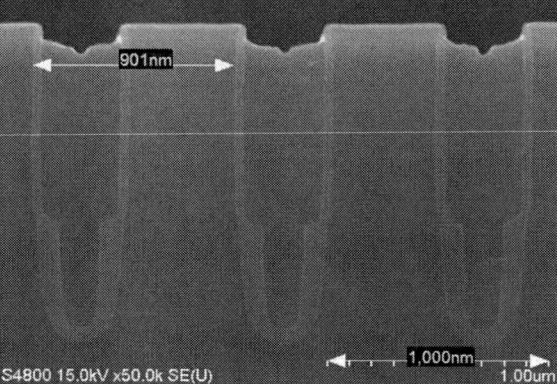

Figure 3: cross-sectional SEM photograph of fabricated Charge-Coupled MOSFET with 0.9μm cell pitch

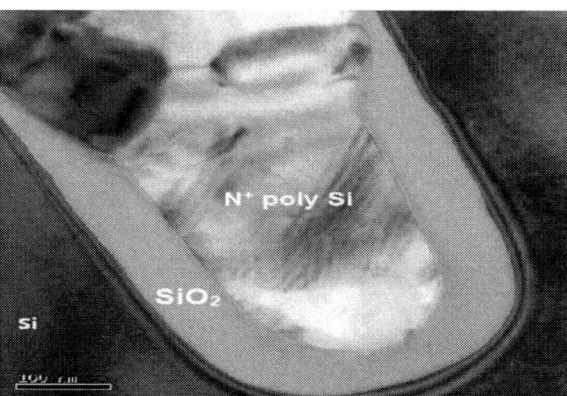

Figure 4: The TEM of the bottom trench profile

The effects of trench depth, thickness of the thick trench oxide layer and cell pitch size on the breakdown voltage are discussed by the performed 2-D numerical simulations with this structure.

As shown in figure 5, when the depth of trench is 1.5μm and the thick oxide height is 0.5μm. It can be observed that the breakdown voltage increases with decreasing cell unit. Smaller cell pitch size resulting in a slightly larger critical electric field for breakdown. For a drift length of 0.5μm, the breakdown voltage for the charge-coupled device structure is predicted to be about 30.2V for a pitch size of 0.9μm.

978-1-5386-5309-8/18 $31.00 © 2018 IEEE

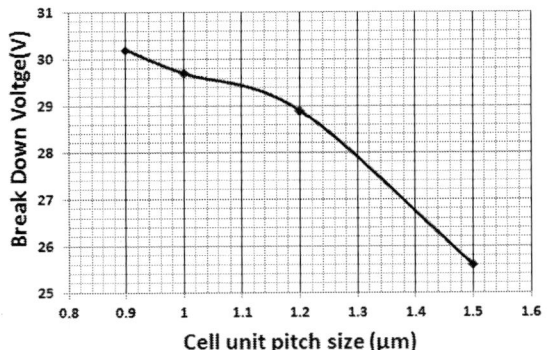

Figure 5: Dependences of Break down Voltage on cell unit pitch size

The breakdown voltage for the charge-coupled MOSFET is also dependent on the thickness of the thick oxide in the trenches and the drift region length which defined by the thick oxide height, as shown in figure 6. The peaks of electric field occur at the junction and the bottom of the trenches, respectively. A high electric field occurs at the junction when the oxide thickness is large while a high electric field occurs at the bottom of the trenches when the oxide thickness is reduced. From this figure, the optimum oxide thickness is 0.1μm. The breakdown voltage of 30.4V for proposed devices can be observed when the trench depth is 2.0μm, based on the simulated epitaxial layer thickness of 3.3μm.

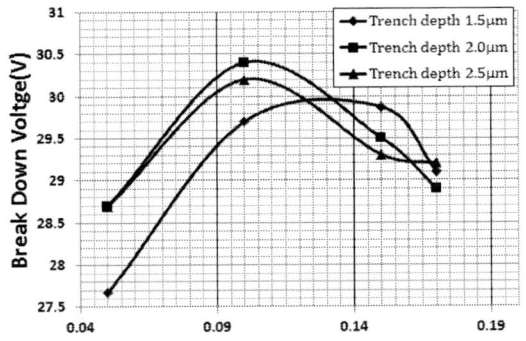

Figure 6: Dependences of Break down Voltage on Trench depth and Trench Bottom Oxide thickness

In the Fab TEG measurement, the break down voltage of 28.3V could be confirmed at approximately 1.3 times higher than that conventional one, while using the same substrate, as shown in figure 7. On the hand, the device has a good leakage performance. The specific on-resistance for the fabricated device of 0.9μm cell pitch is comparable with the conventional one whose break down voltage is 21V. So, this low voltage Charge-Coupled MOSFET with deep-submicron technology provides an approach to achieve the significantly smaller specific on-resistance at the same break down voltage that obtained on the conventional trench MOSFET by using a greater doping concentration in the drift region.

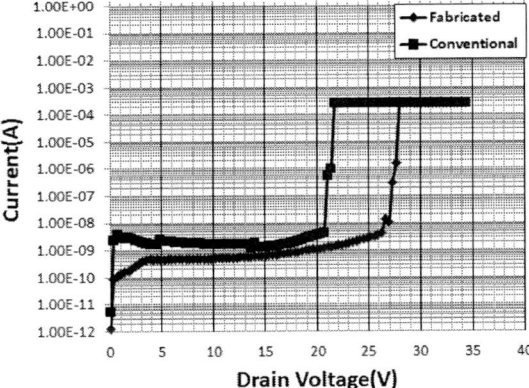

Figure 7: Measured I-V characteristics for the conventional and the fabricated device of 0.9μm cell pitch.

CONCLUSION

The low voltage Charge-Coupled MOSFET with 0.9μm cell pitch is observed that thick Oxide formed at the trench bottom, while gate oxide was gown at the upper trench. And, the thick oxide height can be adjusted by the expose energy and focus in the progress of lithography. Moreover, we demonstrated the break down voltage of 28.3V at the 1.3 times higher than the conventional one by TEG measurement. This low voltage Charge-Coupled MOSFET with deep-submicron technology provides an approach to achieve the significantly smaller specific on-resistance at the same break down voltage that obtained on the conventional trench MOSFET by using a greater doping concentration in the drift region.

REFERENCES

1. B.J. Baliga. *IEEE Trans. on Electron Devices*, vol. 38, no. 7, 1991, pp.1568-1575.
2. S. Sandeep, Rama Komaragiri, *J Mater Sci: Mater Electron*, vol.26, 2015. pp.6692-6698
3. Ono S et al. *proceedings of ISPSD2003*, Cambridge, April14-17, 2003,pp.28-31
4. P. Goarin et al. *Solid-State Electronics*, vol. 51, 2007.pp.1589 – 1595
5. Jongmn Baek et al. *Materials Science and Engineering*, Vol. B97, 2003, pp.123-128.
6. P. Goarin, G.E.J. Koops, R. van Dalen, C. Le Cam and J. Saby, *Proceedings of ISPSD2007*,May 27-30, 2007 Jeju, 2007. pp. 61-64.
7. M.A. Gajda, et al, *Proceedings of ISPSD2006*, Jun4-8, Naple, 2006.pp.112

28NM HIGH DENSITY SRAM BIT CELL DESIGN AND MANUFACTURE STUDY

Meili Hu[1]*, Lijun Song[2], and Lei Wu[2],

Division of Design Technology Co-optimization, SMIC ATD, Shanghai, 201203, China
Division of Logic and Device Eng, SMIC TD, Shanghai, 201203, China
*Corresponding Author's Email: Melissa_Hu@smics.com

ABSTRACT

SRAM is a basic component in logic circuit. To qualify 28nm technology, $0.127cm^2$ high-density SRAM is chosen as a factor. To get good nominal and VMIN yield, some key processes need to focus on. For examples, Active Area formation, Gate dielectric process and MG (metal gate) boundary. This paper will give study result for these three key steps to improve SRAM yield.

Keywords— SRAM; 28nm; AA formation; MG; VMIN yield

INTRODUCTION

SRAM is a base component in logic circuit. And it is also a necessary part to evaluate the technology manufacturability. 28nm technology generation uses $0.127cm^2$ high-density SRAM to qualify the process.

SRAM yield can be improved not only hard defects but also soft failures. This paper will discuss some process sensitivities from SRAM bit cell design point view.

Key Process Discussion

A. AA (Active area) width design

SRAM AA dimension is determined by some factors. The first one is SRAM SNM margin. That means AA domains PU/PD/PG Vt and Idsat. For 28nm node, SiGe process is new process and is key for PU performance. In order to get good profile and better process control, AA width should be big. However AA width increasing results in AA space reduction because SRAM bit cell dimension is fixed. Another reason that AA width is focus on is that SiGe profile and stress added to channel can be controlled by AA width. Below figure (Fig 1) shows SRAM VTMM of different types of SRAM. From Fig1, VTMM sensitivity with PU AA width is stronger than PD/PG. There is SiGe process effect. But AA width gives a big influence on SiGe profile and growth.

SRAM AA space determines SRAM isolation. For design point view, AA space is no big room to enlarge. So N-well/P-well and N+/P+ boundary are a way to improve isolation. After several experiments, only N+ CD is a key point to improve SRAM array Istandby.

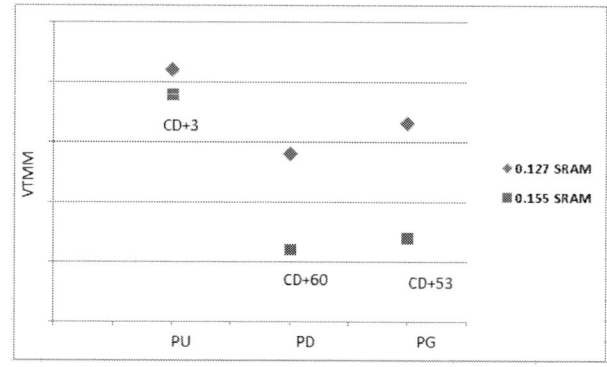

Fig 1: SRAM VTMM of different types of SRAM

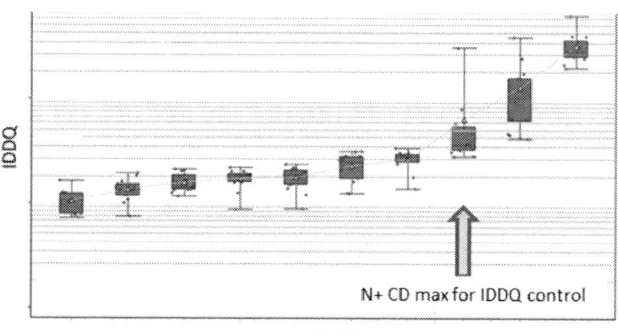

Fig 2: SRAM N+ CD vs SRAM array Istandby

SRAM speed and leakage is a trade-off for AA dimension as well as core device. But isolation is special factor that need to consider reducing overall SRAM array leakage. AA CD increase can reduce VTMM to the target while controlling SRAM array leakage.

B. GT Dielectric development

28nm Gate film stack is very complicated. Not only work function metal but also interfacial layer has multiple-layers. To meet device Vt target, interfacial layer includes OX (gate oxide) and HK film (HfO2). HfO2 film deposition and thermal treatment are very import to control oxygen vacancy. As the dielectric process maintains enough low Vt and passes device reliability, SRAM PU VTMM can be reduced 20% (Fig 3).

978-1-5386-5309-8/18 $31.00 © 2018 IEEE

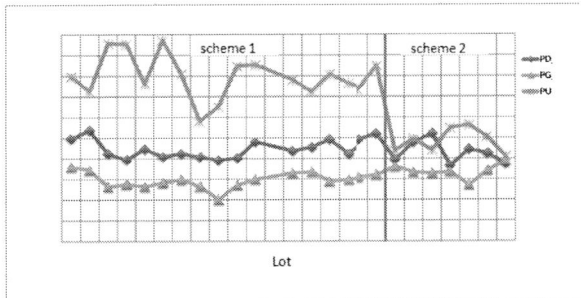

Fig 3: Dielectric scheme vs SRAM VTMM

To realize manufacturing, adding cap layer is a must-be. Cap layer choice is very limited because of production line contamination concern and device work function. Below figure (Fig 4) shows different cap layer impact for VTMM.

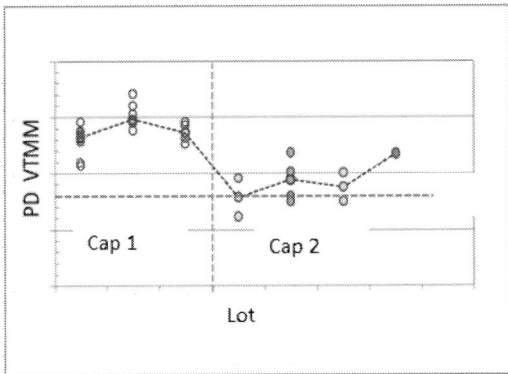

Fig 4: (a)

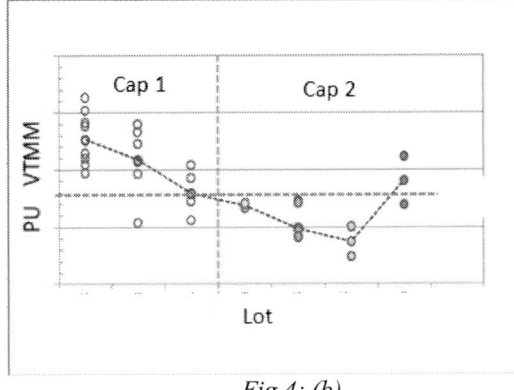

Fig 4: (b)

Fig 4a: Gate OX Cap1 vs SRAM PU VTMM
Fig 4b: Gate OX Cap2 vs SRAM PD VTMM

There is big influence for VTMM by cap layer process. Material selection need to base on work function consideration and process simplification.

C. MG process development

Gate film stack is fixed by work function. Then for SRAM

VTMM, the bigger is gate CD, the better MG (metal gate) gap filling capability is. To get bigger process widow for contact to MG short, gate CD enlargement is limited. So in order to get good SRAM VTMM, STI step height, dummy poly profile and MG N/P boundary should be optimized. Fig 5 is a schematic for three factors that mentions above.

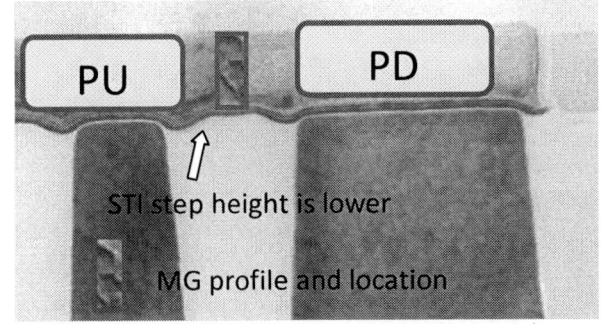

Fig 5 MG location and profile schematic

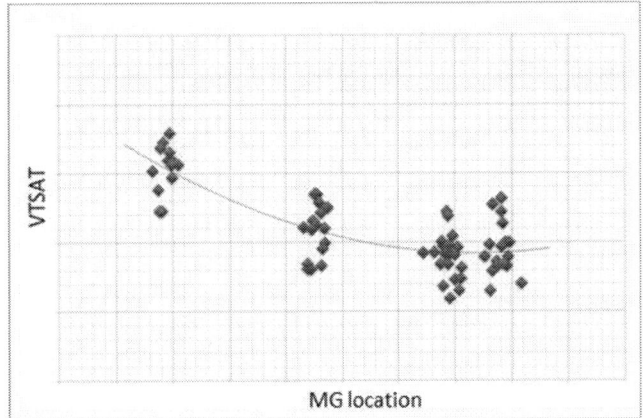

Fig 6: MG location vs SRAM Vtsat

From Fig 6, SRAM VT is very sensitive to MG location (space to PU AA). In a SRAM bit cell, MG location is determined by SRAM PU/PG VT target and process control capabilities. Thus MG location is fixed, MG lithography overlay can double PG-PG Vt difference in a SRAM bit cell. So MG overlay and CD control can't give an enough margin for process. MG film stack is key point. Thinking core device MG film stack, Fig6 show current MG film stack can reduce SRAM Vtsat sensitivity.

CONCLUSION

This paper only shows study on some key processes for SRAM design and yield improvement. AA width and space can domain PU performance and cell isolation. Gate dielectric layer, MG film stack and MG boundary optimization can get good VTMM. In addition SiGe loop and SRAM implantation are another way to optimization SRAM sweet spot and get better VTMM. Contact loop is a killer for production. So these processes should all be taken care.

REFERENCE

[1] Gregory Chen et al., "Yield-Driven Near-threshold SRAM Design" IEEE Transactions on VLSI VOL.18, No. 11, pp.1590-1598, 11 2010

[2] Axel Nackaerts et al. "Lithography and yield sensitivity analysis of SRAM scaling for the 32nm node" Proc. Of SPIE Vol. 6521

[3] Masakazu Goto et al., "The study of Mobility-Tinv Trade-off in Deeply Scaled High-k/Metal Gate Devices and Scaling Design Guideline for 22nm- node Generation" 2009 Symposium on VLSI Technology, pp. 214-215

A COMPACT MODEL OF ANALOG RRAM FOR NEUROMORPHIC COMPUTING SYSTEM DESIGN

Wenqiang Zhang[1], Huaqiang Wu[1]*, Peng Yao[1], Bin Gao[1], and He Qian[1]
[1] Institute of Microelectronics, Tsinghua University, Beijing 100084, China
*Corresponding Author's Email: wuhq@tsinghua.edu.cn

ABSTRACT

The emerging memory such as resistive random access memory (RRAM) is considered as a promising solution to emulate the synapse for neuromorphic computing. It is still difficult to evaluate the performance and optimize the design of RRAM-based neuromorphic system due to the intrinsical device characters. Here, the bidirectional analog and multi-level compact models are proposed to cover the tuning fluctuations for on-chip training and inference processes. Both the analog and multi-level simulated results are experimentally verified on 1-Kbit RRAM array.

INTRODUCTION

Deep learning has produced extremely promising results in image recognition, speech recognition, natural language processing and video tracking [1]. The training and inference of a large-scale deep neural network (~10^9 synapses) consume too much energy (more than 10GJ) due to the physical separation of processing and memory [2]. RRAM-based neuromorphic computing can integrate computation with memory to support high performance acceleration efficiently. Neuromorphic computing with RRAM has shown the ability to recognize binary characters [3] and faces [4]. These demonstrations indicate that RRAM can be integrated in the on-chip neuromorphic system design.

Recently, some RRAM-based architecture designs such as PRIME [5] and ISAAC [6] have been proposed. These designs employ RRAM array as the ideal analog or multi-level weights to achieve matrix-vector multiplication. However, due to the imperfect switching characters of RRAM, such as variations, fluctuation, nonlinearity, *etc.*, these system designs may not perform as expected. The nonlinearity and variation of analog switching will reduce the performance of online training. The conductance distribution of each level will influence the results of offline inference. Hence, to enable evaluation and optimization in the on-chip implementation of RRAM-based neuromorphic system, the analog and multi-level switching models are needed.

Here, we present the bidirectional analog and multi-level compact models, which are based on the measured data. Accordingly, both set process and reset process can perform the turning fluctuation in the analog compact model, while each level is represented as a conductance distribution in the multi-level compact model.

EXPERIMENT

The fabricated 128-by-8 RRAM array used in the measurements consists of 1024 RRAM cells, as illustrated in Fig. 1a. The RRAM element is integrated on top of the 130nm metal–oxide–semiconductor (MOS) transistor, which is known as "1T1R" cell (Fig. 1b). To realize the better bi-directional analog switching behavior, a stack of TiN/TEL/HfOx/TiN is employed for each RRAM element. Compared to conventional metal/HfOx RRAM, the local temperature could increase significantly due to the lower thermal conductivity of thermal enhanced layer (TEL) [7]. The increasing local temperature can help to form multiple conductive filaments in TEL/HfOx RRAM. Thus, reliable bi-directional analog switching behavior can be obtained for TEL/HfOx RRAM.

The parallel programing schedules can accelerate the on-chip training process or inference-only process for neuromorphic computing. Then, as shown in Fig. 1c and 1d, both set and reset schedules are parallel in the measurements. In the set process, the WL voltage pulses are applied row by row on the fabricated array through WLs, the SL voltages are fixed to 0V and the set voltage pulses are applied to BLs of relative programed cells. The reset process is similar to the set process, except that the SLs and BLs of the relative programed cells are applied with reset voltage pulses and 0V respectively.

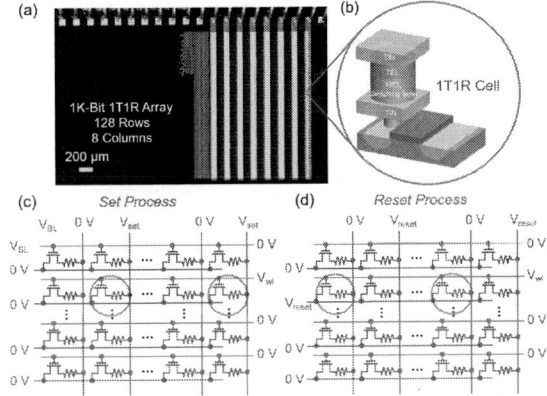

Figure 1: (a) The 1K-Bit RRAM array and (b) relative cell structure, (c) set schedule and (d) reset schedule of the RRAM in the array.

RESULTS AND DISSCUSION
Bidirectional Analog Behavior

In the on-chip training RRAM based neuromorphic

system, the conductance of devices will be updated during the training process. The bidirectional analog model should include the tuning fluctuations when the conductance of RRAM is updated. Fig. 2a shows the typical bidirectional analog behavior in 1-Kbit array. The set and reset voltage are 2.1V and 1.9 V respectively. After a set/reset voltage pulse is applied to the device, the conductance is updated by

$$G_{next} = G_{preious} + \Delta G,$$

where $G_{preious}$ and G_{next} are conductance of the RRAM before and after set/reset operation respectively, ΔG can be generated by adding a random conductance sampled from the uniform distribution to a base conductance. As shown in Fig. 2b, a compact model can cover the nonlinear and tuning fluctuations of conductance in the set and reset process.

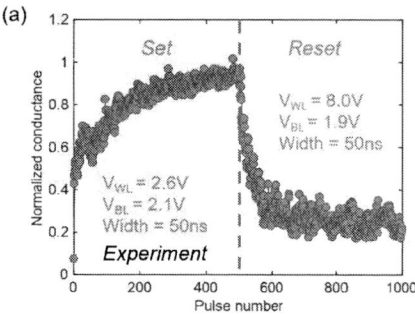

(a)

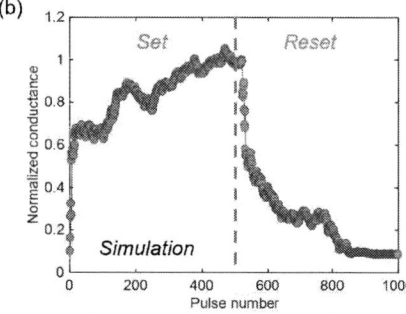

(b)

Figure 2: (a) The typical measured bidirectional analog switch behavior under identical pulses in 1-Kbit array; (b) The tuning fluctuations of the proposed analog compact model. The conductance is normalized by $0.1/G_{min}$. The red and blue circle dots indicate the set process and reset process respectively.

Multi-level Digital Behavior

For the inference-only neuromorphic computing system, the optimized weights are mapped to the conductance of RRAM arrays and the system only perform the forward phase. The multi-level digital behavior should include the write noise in mapping phase and read noise in forward phase. Fig. 3a shows the measured 16-level conductance distributions in 1-Kbit

array. The conductance distribution in each level is fitted by a normal distribution. The mean and relative standard deviation (SD) of each conductance distribution are listed in Table 1. Then a multi-level compact model can be generated by fitting the relation between the mean and SD:

$$SD = -6.0 \times 10^{-4} \cdot Mean^2 + 6.2 \times 10^{-2} \times Mean + 7.2 \times 10^{-1}.$$

Fig. 3b shows the simulated 16-level conductance distributions of the proposed multi-level compact model.

TABLE I.　FITTING PARAMETERS

Level	1	2	3	4	5	7
Mean(μS)	2.47	8.17	14.86	21.44	28.70	35.05
SD(μS)	0.51	0.98	1.87	2.07	2.27	2.02
Level	7	8	9	10	11	12
Mean(μS)	41.29	48.19	54.60	61.02	67.68	74.56
SD(μS)	2.44	2.59	1.77	2.23	1.71	1.89
Level	12	14	15	16		
Mean(μS)	80.71	87.37	94.37	102.00		
SD(μS)	2.00	1.54	1.22	0.86		

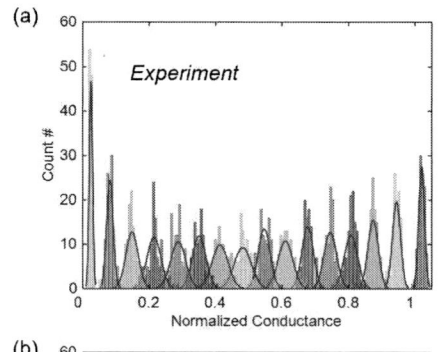

(a)

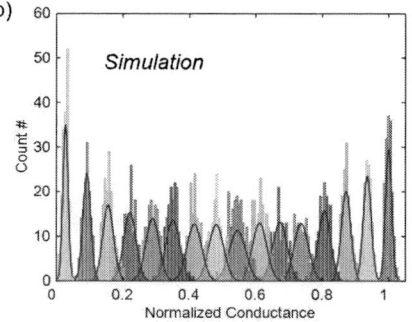

(b)

Figure 3: (a) The measured 16-level conductance distributions in 1-Kbit array; (b) The simulated conductance distributions of the proposed multi-level compact model. The conductance is normalized by $1/G_{level=max}$.

System-level Simulation

To verify the above proposed models, the accuracy of neuromorphic system is evaluated. Fig. 4a illustrates the accuracy evolution of a two-layer full-connected (FC) neural network on MNIST dataset in the on-chip training phase. The weights of the neural network are mapped to the conductance of RRAM arrays. The details of neural network structure can be found in [7]. During the training process, the conductance is updated by the direction of the gradient in the backpropagation.

For the inference-only system, we evaluate the influence of weight bits and noise on the accuracy of a three-layer (784-500-100-10) FC network. As shown in Fig. 4b, when the noise become larger, the accuracy of neural network will drop. As the weight precision become lower, the accuracy lost will be larger. Thus, it is important to optimize the noise of devices and choose the proper weight precision in the neuromorphic system design.

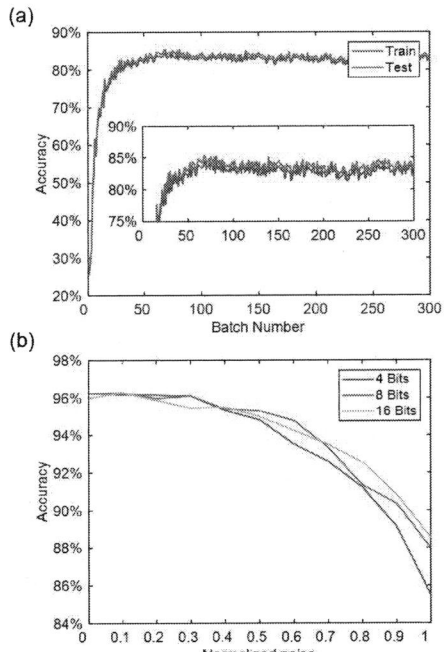

Figure 4: (a) The evolution of accuracy in on-chip learning with respected to mini-batch on MNIST dataset; (b) The accuracies of inference-only system with different weight precision and noise. The noise is normalized by the max noise of proposed compact model.

CONCLUSION

In this paper, we proposed the bidirectional analog and multi-level compact models to cover the tuning fluctuations for on-chip training and inference processes. Both the analog and multi-level simulated results are experimentally verified on 1-Kbit RRAM array. The system-level simulation also shows the above compact model can be integrated to the evaluation of the neuromorphic computing system design.

ACKNOWLEDGEMENTS

This work is supported in part by MOST (2016YFA0201801), Beijing Innovation Center for Future Chip (ICFC), Beijing Municipal Science and Technology Project (D161100001716002), and NSFC (61674087, 61674089, 61674092, 61076115)

REFERENCES

[1] Y. Lecun, Y. Bengio, and G. Hinton, "Deep Learning," Nature, vol. 521, pp. 436-444, 2015.

[2] S. B. Eryilmaz, K. Duygu, S. Yu, and H. S. P. wong, "Device and system level design considerations for analog-non-volatile-memory based neuromorphic architectures." In Electron Devices Meeting (IEDM), 2015 IEEE International. IEEE, 2015.

[3] P. Yao, H. Wu, B. Gao, S. B. Eryilmaz, X. Huang, W. Zhang, N. Deng, L. Shi, H. S. P. Wong and H. Qian, "Face classification using electronic synapses." Nature Communications 8, 2017.

[4] P. Mirko, M. B. Farnood, B.D. Hoskins, G. C. Adam, K. K. Likharev, and D. B. Strukov, "Training and operation of an integrated neuromorphic network based on metal-oxide memristors." Nature, vol. 521, pp. 61-64, 2015.

[5] C. Ping, S. Li, C. Xu, T. Zhang, J. Zhao, Y. Liu, Y. Wang, and Y. Xie, "Prime: A novel processing-in-memory architecture for neural network computation in ReRAM-based main memory." Proceedings of the 43rd International Symposium on Computer Architecture. IEEE Press, 2016.

[6] A. Shafiee, A. Nag, N. Muralimanohar, R. Balasubramonian, J. P. Strachan, M. Hu, R. S. Williams, and V. Srikumar, "ISAAC: A convolutional neural network accelerator with *in-situ* analog arithmetic in crossbars." Proceedings of the 43rd International Symposium on Computer Architecture. IEEE Press, 2016.

[7] P. Yao, H. Wu, B. Gao, W. Wu, Q. Zhang, W. Zhang, H. S. P. Wong, S. Yu and H. Qian, " Device and circuit optimization of RRAM for Neuromorphic computing." In Electron Devices Meeting (IEDM), 2017. IEEE International. IEEE, 2017.

Reversed Trapezoid Trench Profile Formation with Silicon Nitride Confinement

Shi-liang Ji, Qiu-hua Han, and Hai-Yang Zhang*
Semiconductor Manufacturing International Corporation, No.18 Zhang Jiang Rd., Pudong
New Area, Shanghai, 201203, P.R.China
*Corresponding Author's Email: *Steven_Z@smics.com*

ABSTRACT

Gate last approach has started to appear in the high performance applications since 45nm technology node. Such approach will require metal gate seamless gap-fill becomes more critical and more challenging as the semiconductor industry moves to 28nm and beyond. This work introduces a breakthrough scheme for gap-fill improvement and seam suppression by profile treatment process. In this novel scheme, we deliver two treatment processes to get reversed trapezoid trench profile; this profile can solve gap-fill over-hang issue and make metal gate deposit inside the structure to achieve "seamless" gap-fill. Both two kind of process have little spacer SiN loss.

INTRODUCTION

The technology of HKMG has led to the continuity of device scaling and enabled the extension of Moore's Law towards 28 nm and beyond nodes. The gate-last approach has been widely adopted since it can better control the transistor Vt (threshold voltage) and yield better electrical performance. [1-2] However, this approach requires more complex process engineering than gate-first approach. It has challenged every process of metal gate loop, especially in the metal gap-fill. Over-hang will block metal gap-fill and form voids, which will increase Rs and degrade Vt as figure1.

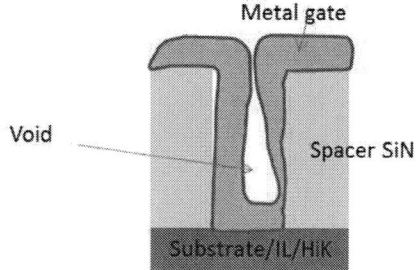

Fig1. Overhang will induce metal gate void.

As well known, enlarging trench top CD (critical dimension) to form the reversed trapezoid trench profile can improve this issue. However, dummy gate was limited

by spacer SiN (nitride), additional SiN etch step could induce lower gate height and more ILD (Interlayer dielectric) loss. In this manuscript, we used two schemes, with proper gas composition and sequence, to get a larger process window to achieve the reversed trapezoid trench profile, larger trench top CD, precise oxide loss control. All processes are more effective without sacrificing gate height. At the same time, this scheme can better control the boundary profile.[3-5]

In this manuscript, we insert different nitride treatment step to enlarge dummy gate top CD with little spacer nitride loss. Traditional F base nitride treatment and remote plasma treatment were induced for this scheme.

RESULTS

We delivered two schemes to form reversed trapezoid trench profile at dummy poly gate removal step. The process flows are as below figure 2.

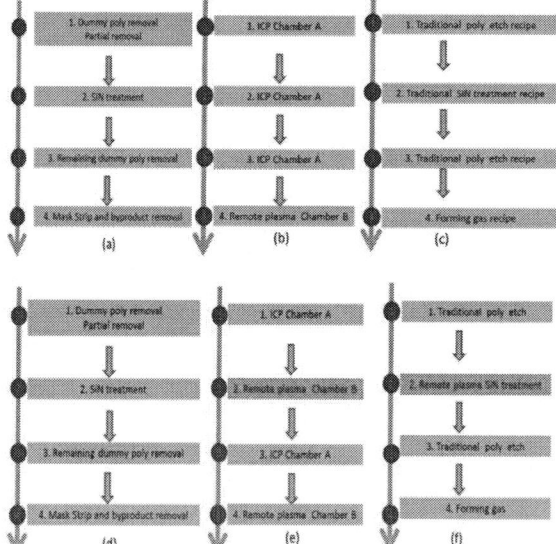

Fig 2. Two schemes to form reversed trapezoid trench profile at dummy poly gate removal step. (a), scheme 1 flow; (b), scheme 1 chamber sequence; (c), scheme 1 etch recipe design; (d), scheme 2 flow; (e), scheme 2 chamber sequence; (f), scheme 2 etch recipe design.

978-1-5386-5309-8/18 $31.00 © 2018 IEEE

Both two schemes are the same flow; the key steps are partition dummy poly removal and trim spacer SiN. Based on device requirements; we can decide dummy poly gate removal depth and SiN trim thickness. Firstly, we process dummy poly gate removal. Then, in the trim step, we adopt two approaches to enlarge top CD. The first is traditional SiN treatment approach, etch spacer SiN with F-base gas; the second is remote plasma trim spacer SiN, this approach need process in another chamber. Thirdly, remove all remaining dummy poly with traditional poly etch recipe. Finally strip PR and byproduct with forming gas in another chamber.

We get typical across section images as figure3 with two different schemes. As we known, perfect trench profile after poly removal could reduce the gap-fill difficulty and deliver better device performance. For the dummy poly gate removal process, too much profile pull-back is not acceptable, which increase over-hang impact as figure3 a. This requires anisotropic dry etching. On the other hand, n/p MOS boundary connects pull-up area and pull-down area. In anisotropic dry etching, the generated polymer/byproduct could deposit on this boundary sidewall. This could result in worse n/p MOS boundary contact. So in traditional ICP (inductively coupled plasma) etch, we must increase over etch amount to cover this conflict. However, isotropy dry etching can overcome the boundary interface with remote plasma.

Through, we can meet two aspects at the same time under conventional continuous wave plasma. From Fig3b, we can see, reversed trapezoid trench profile was achieved by enhancing anisotropic dry etching. Certainly, such contradiction could not be decoupled even with precise control of each etching parameter to gain perfect reversed trapezoid trench profile and clean n/p MOS boundary interface at the same time. This also indicates the process window is quite marginal. Remote plasma approach gives us a surprise. We used remote plasma to trim spacer SiN after poly-Si recess step of dry etching process. This step will clean a lot of byproduct/polymer because of high gas flow and low pressure. We can presume: in remote plasma, as the polymer/by-products will be pumped out easily during process, n/p MOS boundary can be residue free. With proper process time and gas base, there'll be a large process window to achieve reversed trapezoid trench profile with clean n/p boundary as shown in Fig 3c.

At the same time, traditional dummy poly etch has weak clean corner ability because of larger atom size (figure3 d); in this case, the trench corner residue will be kept. To solve this issue, NF_3/H_2 gas combination could be attempted. We can find trench corner residue has been cleaned, but the device test show PID (plasma induce damage) happened. The reason is H has smaller atom size and attacks dielectric layer more easily.[6] Predictably, remote plasma can benefit both corner residue cleaning and PID improvement as remote plasma process has a very low level averaged Te which is one solution to PID issue.

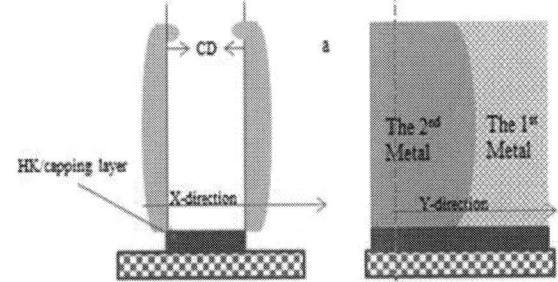

Figure 3 (a). Bowing profile with scheme1.

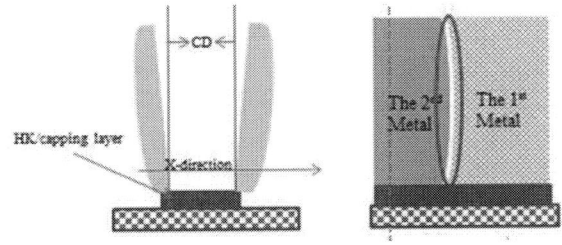

Figure 3 (b). Better reversed trapezoid trench profile with scheme 1, some sample suffer byproduct residue issue.

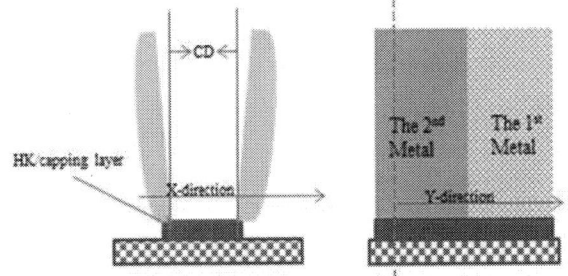

Figure3(c).Reversed trapezoid profile with scheme 2.

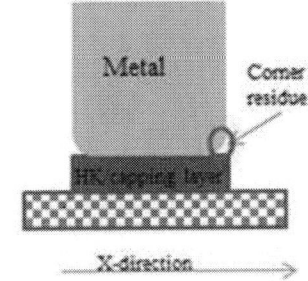

Figure3(d). Corner residue with scheme 1

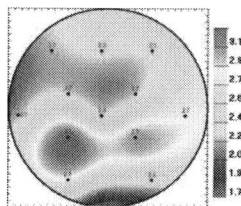 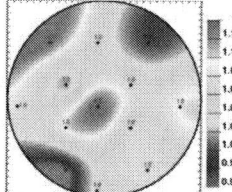

Mean : 2.4 N
3-sigma: 0.1 N
Range(U%): 0.1 N

Mean : 1.0 N
3-sigma: 0.1 N
Range(U%): 0.2 N

Figure5. ILD loss: Traditional etch process (left); remote plasma etch process (right).

Another benefit of scheme is less ILD loss with the stable capping layer loss. Too much ILD loss will degrade electric performance and lead to the overheating problem. In addition, ILD loss has N/PMOS bridge risk because of metal residue after CMP (chemical mechanical polishing). And the capping layer is also very critical to the adjustment of gate work function and control leakage. Therefore, PET (post etch treatment) will play a crucial role. Both capping layer surface interface and more loss could induce Vt shift. In remote etch scheme, we can deliver softer plasma to reduce the capping layer loss without sacrificing the ability of interface treatment. At the same time, the ILD loss amount will reduce because of soft process to remove byproduct or polymer. Remote plasma will reduce the plasma attack to ILD. The results were shown Fig 4. But the remote plasma scheme disadvantage is lower WPH, it will increase cost.

CONCLUSIONS

In this manuscript, we deliver two treatment processes to get reversed trapezoid trench profile; this profile can solve gap-fill over-hang issue and make metal gate deposit inside the structure to achieve "seamless" gap-fill. At the same time, we investigated the process difficulty under conventional etch approach, and introduced remote plasma schemes application in this process. One more step has been applied in the remote plasma chamber. The results showed that remote plasma scheme can better balance the requirements of different features and enlarge CD.

ACKNOWLEDGEMENTS

The authors would like to greatly appreciate the team members in the Etch department of Technology Development and Research Center in SMIC for their continuous help and useful discussion.

REFERENCES

[1] Naoki Tega, Hiroshi Miki, Zhibin Ren, Christoper P. D'Emic, Yu Zhu, David J. Frank, Michael A. Guillorn, Dae-Gyu Park, Wilfried Haensch, and Kazuyoshi Torii, *IEEE, IRPS11(2011)*.

[2] Hyohyun Nam, Changhwan Shin. *IEEE ELECTRON DEVICE LETTERS*, VOL. 34, NO. 4, APRIL (2013).

[3] Samer Banna, Ankur Agarwal, *J. Vac. Sci. Technol. A* 30(4), Jul/Aug (2012).

[4] Demetre J Economou, *J. Phys. D: Appl. Phys.* 47, (2014).

[5] Ruixuan Huang, Xiao-Ying Meng, Qiu-Hua Han, Hai-Yang Zhang, *SPIE*, (2015).

[6] Jun-Qing Zhou, Hai-Yang Zhang, Wu Suna, Xing-Peng Wang, Min-Da Hu. Fan Li, Li-Ya Fu, Shih-Mou Chang, Kwok-Fung Lee, *ECS Trans.* 34(1), 445, (2011).

An improved yield model for embedded flash

Yun Xu, Yuxiang Zhang

Shanghai Huahong Grace Semiconductor Manufacturing Corporation.

Postcode: 201206

yun.xu@hhgrace.com

Abstract

The product yield of IC fabrication (FAB) comes from the two contributions of Yline (Line Yield) and Yw (Wafer Yield), in which Yw is determined by Ys*Yr. Yr is generally shown as a random distribution of failure. We can use the Yr of the existing products to get the D0 that represents the control level of the process particle/defect through a Yr-D0 model. The applicability and accuracy of the Yr-D0 model is the key to yield prediction and FAB line particle improvement. This paper first introduces several commonly used yield models (Yr-D0), mainly describes an improved yield model in practical applications, and finally gives the effect of this model.

Introduction

The statistical average of the wafer yield is determined by the following formula

$$Yw=Ys*Yr \qquad (1)$$

In which, Ys is the systematic-limited yield and Yr is the defect-limited yield. Yr comes from die failure caused by a particle in the processing microenvironment or a random defect during wafer processing. Unlike the failure of Ys with characteristic graphics, Yr is generally shown to be a random distribution failure in the wafer, but sometimes it also appears a certain graphic tendency on the wafer map.

A product's Yr can represent the particle control level during wafer fabrication. However, the FAB owner and client cannot know the Yr of the product before deciding to put into production. So it's very necessary to create an index representing the control level of the line particle. D0 (Defect Density, the average number of faults per unit area), is such a characterization index.

In production line, we usually use the Yr of the product to calculate D0. The relationship between Yr and D0 can be expressed by the random failure yield model (Yr-D0 Model). The D0, derived from an effective yield model, should be able to accurately represent the particle control level of the production line. And through reliable D0 values, we can accurately predict the yield of a new product.

Yr extraction method and Yr-D0 Models in common use

Yr extraction method

After getting enough yield data (Yw) of a new product, we first need to make a reasonable separation of Yw. We can use some yield analysis software to discriminate and separate Ys and Yr. We also can, from a statistical point of view, separate Ys and Yr relatively quickly as below,

$$Ymax- Yr = 3*\sqrt{Yr * (1 - Yr)/GD} \text{ (3 sigma)} \qquad (2)$$

In which, Ymax is the highest yield (yield of wafer) of the statistical object, and GD is the number of gross die. We can get the Yr value according to the formula (2).

Particle distribution and Bose-Einstein formula

After getting Yr, the corresponding yield model can be constructed by different particle distribution assumptions (Yr-D0). The widely used distribution models include Poisson distribution, Binomial distribution, exponential distribution, and so on. For these distribution models, we can assume that all of the distribution types conform to a certain distribution function f (D), and then there are:

$$Yr = \int_0^\infty e^{-D0A} f(D) dD \qquad (3)$$

Many yield models based on different distribution types can be obtained from the formula (3). In particular, when the distribution of particles conforms to an exponential distribution:

$$f(D) = \frac{1}{D0} e^{\frac{D}{D0}} \qquad (4)$$

According to (3) (4) we can get the index distribution yield model:

$$Yr = \frac{1}{1+AD0} \qquad (5)$$

Considering the manufacturing complexity of different process platform, it needs to introduce factor n (Process Complexity), which reflects the complexity of the process. Now that we get the most commonly used Bose-Einstein formula:

$$Yr = \left(\frac{1}{1+AD0}\right)^n \qquad (6)$$

The D0 at this time represents the defect density when the complexity is 1(Sometimes thought as one photo layer). This normalization makes a comparison of the defect density of different products and different manufacturing lines. In fact, others yield models also can get the corresponding D0 with n of 1 by using the n power method.

Problems in practical application and model improvement

Choosing an appropriate yield model, we can get the D0 value that represents the particle control level of the production line. Through some calculations, we find that when the area of chip is far less than wafer area (for example, A/Aw<0.001, that is GD>1000), the differences in D0 obtained from the various yield models have been very small. But in practical applications, we find that even in the same process platform, the D0 obtained by the yield model formula for different products is often very different. This leads D0 could not to accurately and steadily characterize the defect density of the production line. As more and more embedded chips

are applied, simple Logic products are less and less. After careful study, we found that, the memory in the embedded chip has a great change in the proportion of the area within the different products. These differences originate from the products themselves, such as the memory area occupation ratio, the number of gross die and so on, resulting in a huge difference in D0. This requires that we need to make the necessary amendments to these formulas to accurately and steadily reflect the D0 level of the production line.

Based on the Bose-Einstein formula, we try to embody the influence of the gross die and the memory area ratio in this formula. The formula of the improved yield model is as follows:

$$Yr = 1/ (1+A* (M*D0)^G)^n \qquad (7)$$
$$D0 = (((Yr^{-1/n}-1)/A)^{(1/G)})/M \qquad (8)$$

A: Chip Area (in2, excluding scribe line);

D0: Defect Density (inch2 per layer);

n: Process Complexity (number of critical layer)

M: Memory size effect

G: Gross Die effect

The introduction of M factor (Memory size Effect) is easier to understand. Because the storage unit is generally the most critical dimension in the chip layout, the storage unit is often the most sensitive to the process particles. Different storage unit area ratio can result in different wafer yield performance, so it is necessary to weigh the area occupation ratio of the storage unit.

The purpose of introducing G factor (Gross Die Effect) is to give a reasonable die size correction value through the actual fitting of different platforms-product D0.

The specific fitting method is as following.

We fit the index G according to the target D0. The function curve G= A*GDPW-B is fitted according to the relationship between the exponential G and the GDPW. The index G is calculated and verified according to the function, and the D0 obtained by the formula is calculated and verified by the exponential G.

978-1-5386-5309-8/18 $31.00 © 2018 IEEE

An example of yield model amendment

We selected the 0.35μm, 0.25μm, 0.13μm three process platforms Embedded Flash products. First, the following results are obtained according to the Bose-Einstein formula which is not modified.

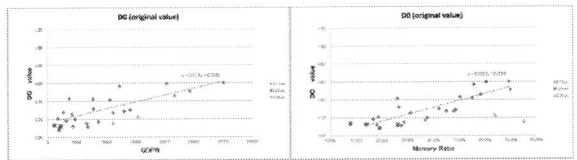

Fig1-1: Do-GDPW   *Fig1-2: Do-Memory Ratio*

We found that the same platform, the different Memory area occupation ratio and the different gross die (GDPW) products of the D0 have a huge difference. In short: the greater the gross die, the greater the D0; the greater the proportion of the memory area, the greater the D0. The fitting line in our ideal should be a flat line, indicating that the D0 of each product should be approximately equal to the same process platform. So we first add the M factor and set the related step coefficients as below,

The M (Memory Size Effect) is set as follows:
By Memory area/Chip area < 10%, D0 = D0_Yr
By Memory area/Chip area > = 10% D0 = D0_Yr/M
M=Memory account for x10

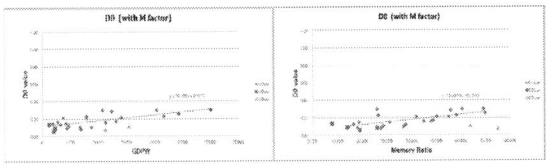

Fig2-1: Do-GDPW with M   *Fig2-2: Do-Memory Ratio with M*

With the addition of the M factor, we found that the fitting lines have tended to be flat. But there is 1 to 2 times the difference between different die size products. So we continue to introduce the G factor. The G factor is set as follows:

G= A*GDPW-B, in which A=0.00004, B=-0.94,

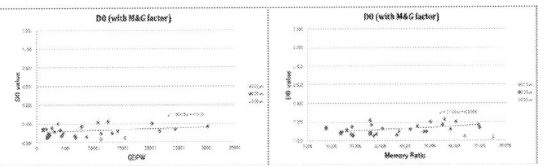

Fig3-1 :Do-GDPW with M&G   *Fig3-2: Do-Memory Ratio with M&G*

We found that after adding the G factor, the D0 fitting line was close to flat. The D0 of the same process platform is at the same level, and the D0 of different process platforms also shows certain differences, which conform to the basic concept. We have achieved the goal of improving the yield model.

Summary

The yield model is a practical requirement in semiconductor manufacturing. But there is no definite industry standard. Due to the complexity of the process and the variety of equipment, the distribution of particles in the process of manufacturing is very complex. It is difficult to cover all of them by using single distribution mode. Moreover, even the same process platform of the same factory, because of the complexity of products, they often get very different defect density although use the same defect distribution model. So we need to make the necessary correction to the yield model according to product particularity, so that the D0 of different products of the same process platform in the same factory tends to be equal. Of course, a specific yield model does not necessarily apply to other manufacturing plants. We need to consider this when comparing two companies' D0.

Reference

[1] Robert C. Leachman，IEOR 130, Methods of Manufacturing Improvement 2006

[2] A Novel Filtering Method to Extract Three Critical Yield Loss Components (Gross, Repeated, and Random) FIMER， Kiyotaka Imai, Member, IEEE, and Toru Kaga, Member, IEEE

A NOVEL SCR ESD PROTECTION STRUCTURE FOR RF POWER AMPLIFIER

Chunguang Wang, Zexue Liu, Junhua Liu* and Huailin Liao*

Key Laboratory of Microelectronic Devices and Circuits (MOE), Institute of Microelectronics,
Peking University, Beijing 100871, CHINA
*Corresponding Author's Email: junhua.liu@pku.edu.cn, liaohl@pku.edu.cn

ABSTRACT

In this paper, a novel silicon-controlled rectifier (SCR) structure is proposed to protect a radio-frequency (RF) power amplifier (PA) from electrostatic-discharge (ESD) damages. With a distinguished feature of an imbedded Deep N-well (DNW) area beneath the P-well/N-well regions, the novel SCR is demonstrated to be superior to the conventional SCR in terms of relatively low trigger voltage (13.3V) and high holding voltage (7V). The new device can be realized in SMIC 0.18μm RF technology without requiring extra layout area. The operational mechanism of the new device structure is investigated, and the effect of the additional DNW region on the SCR's I-V characteristics is analyzed by TCAD simulation.

INTRODUCTION

Rapid development of radio-frequency (RF) integrated circuits (IC) realized in CMOS technologies has occurred recently, because of the advantages of small feature size and lower cost. Nevertheless, The RF circuits realized in CMOS technologies are susceptible to electrostatic discharge (ESD) damages, which turns into the major reliability concerns for RFIC. Therefore, appropriate ESD protection design should be applied at all I/O pads in RFIC against ESD damages. Illustrated in Figure 1, a general scheme of on-chip ESD protection for RFIC consists of a power-rail ESD clamp circuit connected from V_{DD} to V_{SS}, ESD protection device connected from I/O to V_{DD} and ESD protection device connected from V_{SS} to I/O [1].

Power amplifier (PA) is one of the most important building blocks in RF Frontend Circuits. The output terminals of PA are usually connected to drain/source terminal of the MOSFET, which can be susceptible to ESD damages. For some PAs, the signal swing at output pad may be two to three times higher than the supply voltage. Therefore, the PA needs large-swing-tolerant ESD protection structure to protect its output pad [2]. The conventional double-diode ESD protection structure does not meet the requirement of large swing tolerance at PA output pads. Silicon-Controlled rectifier (SCR) has been reported to be an attractive ESD protection device for RF PA due to its low parasitic capacitance, high current handling capacity and area efficiency [3]. However, the conventional SCR structure has drawbacks such as, relatively high trigger voltage (V_{t1}), low holding voltage

(V_h) and large turn-on time in CDM event. The high trigger voltage can make the SCR device fail to turn on and internal circuitry may be damaged by ESD pulse. The low holding voltage can result in the latch-up problem if power supply voltage is higher than the holding voltage [4]. Different SCR-based structures and circuits have been presented to improve the trigger voltage and holding voltage [5]-[8]. However, all these solutions require an external trigger circuity which consumes extra silicon area and enlarges the parasitic capacitance.

In this paper, a new SCR device is proposed, which is characterized by the Deep N-well (DNW) region inserted beneath the P-well/N-well. This novel structure can be realized in the standard CMOS RF process without adding extra silicon area. The relevant TCAD simulation results show that, the new SCR has a lower trigger voltage and higher holding voltage than that of the conventional one.

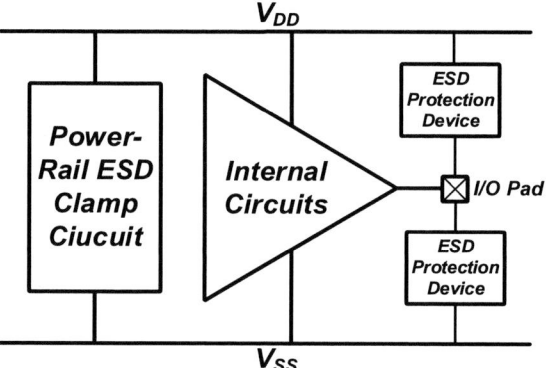

Figure 1: A general scheme of on-chip ESD protection for RF ICs

DEVICE STRUCTURE

The conventional SCR is depicted in Figure 2(a), and its equivalent circuit consists of a PNP BJT and a NPN BJT. Before being triggered, SCR device equals to a resistor with high resistance. When the SCR device is triggered, avalanche breakdown occurs at the diode between N-well/P-well and the current increases dramatically. As a result, the potential drop along the base-emitter contact area of the transistor is large enough to forward bias the junction, so large carriers are injected into the P-well/N-well areas and a low resistance path is formed due to the conduction modulation. Consequently,

978-1-5386-5309-8/18 $31.00 © 2018 IEEE

the SCR is clamped to a low holding voltage.

Compared with the conventional SCR, the new SCR, shown in Figure 2(b), has an extra DNW region inserted under the N-well/P-well regions. The doping density of DNW ($\sim 10^{20}$cm^{-3}) is much larger than the doping density of N-well ($\sim 10^{17}$cm^{-3}). Since the depletion-region width of the DNW/P-well junction is much smaller than that of the N-well/P-well junction under the same reverse-biased voltage, we can expect that the former can break down at a smaller voltage than the latter. Thus, the avalanche breakdown occurs at the DNW/P-well junction in the new SCR device rather than the N-well/P-well junction in the conventional SCR. This can result in a decreased trigger voltage. On the other hand, the DNW is expected to extend the current paths and make the effective base region wider. This can result in an increased holding potential.

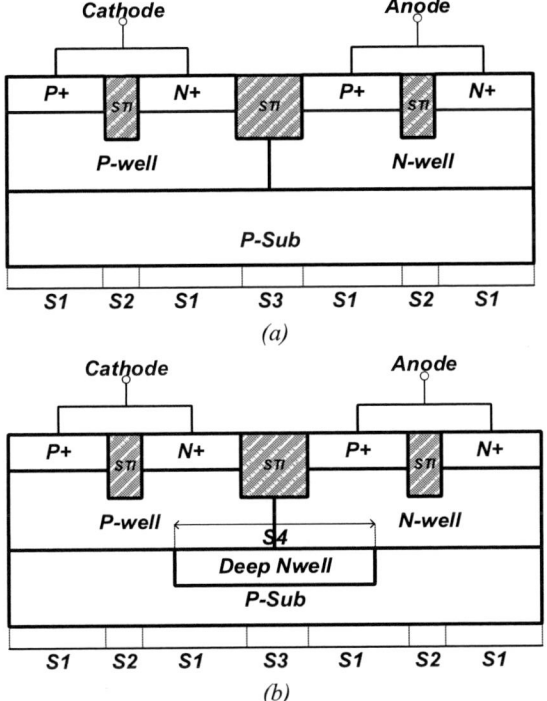

Figure 2: Device Cross-sectional view of (a) the conventional SCR (b) new SCR with DNW

Simulation Results and Discussion

To develop a deep understanding of the physical mechanism of the DNW region on the trigger voltage and holding voltage, 2-D device simulation with Sentaurus is performed. Device physical models, such as Fermi, avalanche, Shockley-Read-Hall, SRH Auger recombination and thermodynamic are used for device simulation in high-current injection condition. The device simulation is utilized to estimate the junction breakdowns and current density distributions of the SCR device during the ESD stress.

Figure 3 shows the electric field contours at the triggering point in the conventional SCR device and the new SCR device with DNW. As we expected, the avalanche breakdown takes place at different locations. As shown in Figure 3(a), avalanche breakdown occurs between N-well and P-well in the conventional SCR device. Illustrated in Figure 3(b), avalanche breakdown occurs between DNW and P-well in the new SCR device.

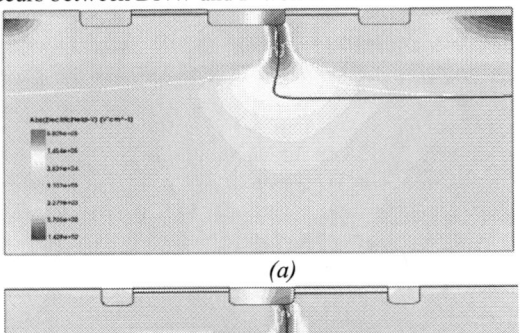

(a)

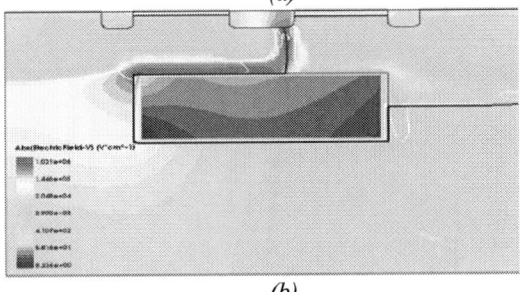

(b)

Figure 3: Simulated electric field at the triggering point of (a) the conventional SCR and (b) the new SCR with DNW

Figure 4 shows the current density contours of the conventional SCR and the new SCR, respectively, at the triggering point. Without the DNW region, as in the case of conventional SCR, the electron current (from the cathode to the anode) flows predominately through the N+/P-well/N-well/P+ path., as shown in Figure 4(a). In the new SCR, a large portion of the electron routes through a N+/P-well/DNW/N-well/P+ second path due to the presence of the DNW region, as shown in Figure 4(b). This second path is longer than the first one, thus making the effective base region width of the BJT in the new SCR wider than that in the conventional counterpart. As a result, the holding voltage increases.

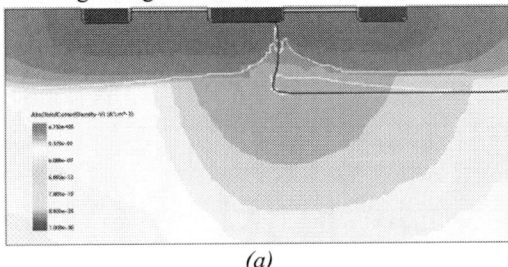

(a)

978-1-5386-5309-8/18 $31.00 © 2018 IEEE   83

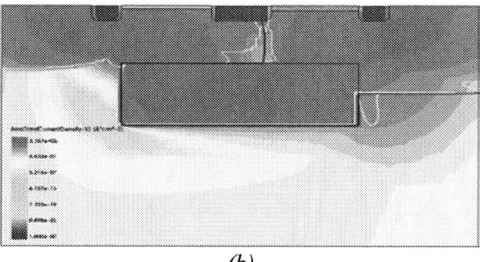

(b)

Figure 4: Simulated current density contours after triggering of (a) conventional SCR and (b) new SCR with DNW

The simulated I-V characteristics of the new SCR and conventional SCR are shown in Figure 5. The trigger voltage of the new SCR with DNW is 13.3V while the trigger voltage of the conventional SCR device is 16.3V. The holding voltage of the new SCR with DNW is 7V while the holding voltage of the conventional SCR is 4.9V. The design parameters and I-V simulation results are listed in TABLE I. The use of the DNW region achieves the low trigger voltage and high holding voltage according to TCAD simulation results.

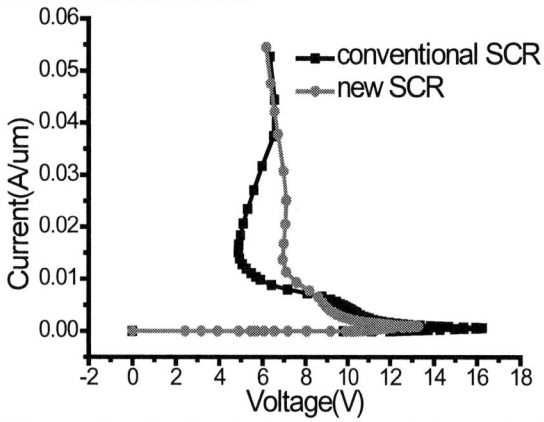

Figure 5: Simulated I-V characteristics of the conventional SCR and the new SCR with DNW

TABLE I
DESIGN PARAMETERS AND SIMULATION RESULTS

	New SCR	Conventional SCR
S_1(μm)	1.5	1.5
S_2(μm)	0.5	0.5
S_3(μm)	1	1
S_4(μm)	4	0
V_{t1}(V)	13.3	16.3
V_{hold}(V)	7	4.9

CONCLUSION

A new SCR structure with a relatively high holding voltage and low trigger voltage has been proposed, analyzed and verified by TCAD simulation. The improved ESD performance of the new SCR device resulted from additional DNW regions which reduce the avalanche breakdown voltage and extend the current path. Compared with the conventional SCR structure, the new proposed SCR structure achieves better ESD protection performance for the RF PA.

ACKNOWLEDGEMENTS

The authors would like to thank Mr. Lizhong Zhang, Mr. Yuancheng Yang and Mr. Cheng Chen at Peking University for their valuable suggestions on ESD device simulation.

REFERENCES

[1] Ming-Dou Ker, Chun-Yu Lin and Guo-Xuan Meng, "ESD protection design for fully integrated CMOS RF power amplifiers with waffle-structured SCR," 2008 IEEE International Symposium on Circuits and Systems, Seattle, WA, 2008, pp. 1292-1295.

[2] C. Y. Lin et al., "Large-Swing-Tolerant ESD Protection Circuit for Gigahertz Power Amplifier in a 65-nm CMOS Process," in IEEE Transactions on Microwave Theory and Techniques, vol. 61, no. 2, pp. 914-921, Feb. 2013.

[3] M.-D. Ker and K. C. Hsu, "Overview of on-chip electrostatic discharge protection design with SCR-based devices in CMOS integrated circuits," IEEE Trans. Device Mater. Rel., vol. 5, no. 2, pp. 235–249, Jun. 2005

[4] A Jian Liu et al., "Design and Analysis of Low-Voltage Low-Parasitic ESD Protection for RF ICs in CMOS," in IEEE Journal of Solid-State Circuits, Vol. 46, No. 5, pp. 1100-1110, 2011.

[5] A. Chatterjee and T. Polgreen, "A low-voltage triggering SCR for on-chip ESD protection at output and input pads," IEEE Electron Device Lett., vol. 12, no. 1, pp. 21–22, Jan. 1991.

[6] J. Zeng et al., "Design and analysis of an area-efficient high holding voltage ESD protection device," IEEE Trans. Electron Devices, vol. 62, no. 2, pp. 606–614, Feb. 2015.

[7] C.-Y. Lin, Y.-H. Wu, and M.-D. Ker, "Low-leakage and lowtrigger-voltage SCR device for ESD protection in 28-nm highk metal gate CMOS process," IEEE Electron Device Lett., vol. 37, no. 11, pp. 1387–1390, Nov. 2016.

[8] F. Ma, Y. Han, S. Dong, et al., "Improved low-voltage-triggered SCR structure for RF-ESD protection," IEEE Electron Device Lett., vol. 34, no. 8, pp. 1050–1052, Aug. 2013.

EFFECT OF WAFER EDGE CUT ON TESTING AND YIELD

Yuxiang Zhang[*], Qin Huang, Yun Xu

Shanghai Huahong Grace Semiconductor Manufacturing Corporation, Shanghai 201206, China

*Corresponding Author's Email: yuxiang.zhang@hhgrace.com

ABSTRACT

As the diameter of the silicon wafer grows larger and larger, meanwhile the line width of device becomes smaller and smaller, the number of effective chips on the edge of silicon wafer increases in multiples. In addition, with the increase of the process complexity, the layered problem of stack film on wafer edge, especially on ugly dice (incomplete dice), is becoming more and more serious. Therefore, improving the wafer edge process becomes more and more important to enhance the yield and test stability. This paper starts with several failure analysis cases, through logical inference and physical analysis, finally finds the edge cut issue from front-end manufacture process which causes the back-end testing failure. Several kinds of the layered problem caused by edge cut issue are discussed in this paper and some corresponding suggestions for improvement are put forward.

Keywords — wafer edge cut, layered problem, ugly dice, test yield, failure analysis

INTRODUCTION

Parallel testing is generally used to improve the test efficiency. Sharing power supply for multi device under test (DUT) is generally adopted because of the limitation of the number of the test channels by parallel testing. Therefore, effective dice (complete dice) and ugly dice (incomplete dice) may share power supply on wafer edge. Due to the special formation of the layer stack on the edge of the wafer, ugly dice often produce unexpected large current. That will induce the failure of normal effective dice, which share power supply with ugly dice, finally affect the wafer yield. In addition, unexpected large current from ugly dice will cause abnormal loss of the needle, which will shorten the life time of the probe card. Serious current even cause the test machine fault down. Let's study the relationship between wafer edge process and the test yield through the following three failure analysis cases.

FAILURE ANALYSIS

Case 1

Serious issue was found around wafer edge by back end inspection on an early product A. Something black like coke with large area on wafer edge was found by optical microscope (OM), and mass burst site was disclosed through cross section analysis (as shown in Figure 1).

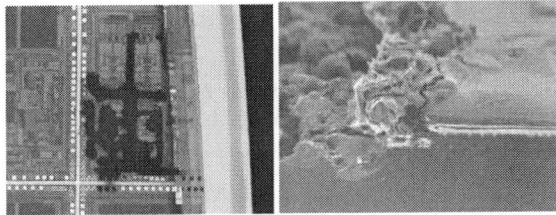

Figure 1: OM and SEM photo for arcing issue

So, what cause such serious mass burst? It's arcing by metal line discharge! During wafer manufacturing process, if metal line is exposed surrounding the wafer edge as an antenna, the strong current will suddenly introduce into the chip inside while plasma etching of the passivation layer. No discharged path for the strong current in short time result in the mass burst. As an improvement strategy, the edge cut of the passivation layer is reduced, so that the top metal layer is completely covered by the passivation layer around the wafer edge (as shown in Figure 2). Without exposed metal line, there is no abnormal tip discharge, and the issue is solved.

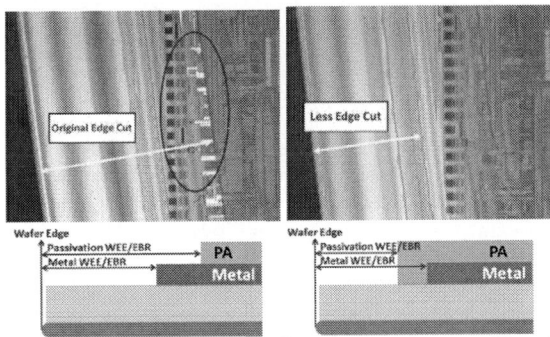

Figure 2: Countermeasure for arcing issue on product A

Edge cut for lithography usually has two ways, EBR (Edge Bead Removing) and WEE (Wafer Edge Exposure). The accuracy of WEE is higher than that of EBR.[1][2][3]

Case 2

In another case, serious short failure on product B was found at some of the peripheral dice of wafer while yield testing. Needle of the probe card was loss rapidly. White bright circle zone could be found around the wafer edge. Some black dot on this zone was caught by optical microscope. The location of the black dot was exactly where the power bonding pad of ugly die was located (as shown in Figure 3).

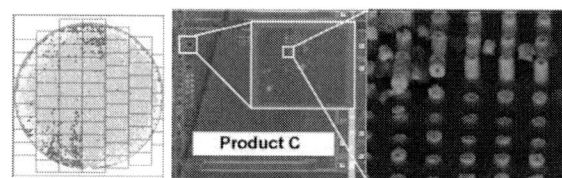

Figure 3: White circle and black dot by OM on product B

Because of the parallel testing and the limitation of the number of the test channels, the multi DUT share power operation is generally adopted. Unexpected current from ugly dice will induce the failure of the normal complete dice, which share power with ugly dice, finally affect the testing yield.

By FIB observing from three different positions at wafer edge and the analysis of the surface component, we found that the white bright circle at wafer edge was tungsten (W) residue. It caused the circuit short during test and burnt at the location of the power bonding pad in the ugly dice (as shown in Figure 4).

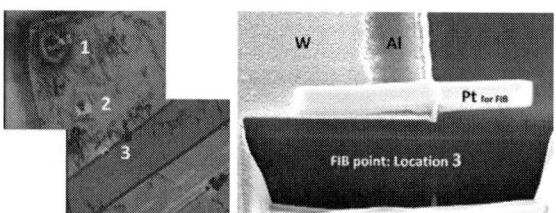

Figure 4: OM and FIB photo for product B

So, how does the tungsten residue happen? It is also caused by the wafer edge cut issue.

In this case, in addition to the edge cut of lithography, it also involves the ring shadow edge cut of tungsten deposition. Normally, the edge cut of via layer is less than that of tungsten deposition, so the edge step between two thin film layers is clean. If the edge cut of tungsten is less than that of via layer, then tungsten residue occurs (as shown in Figure 5). We find that the ring shadow edge cut from one of the tungsten deposition equipment is different from the others due to different equipment type. It causes the tungsten residue on wafers by this equipment.

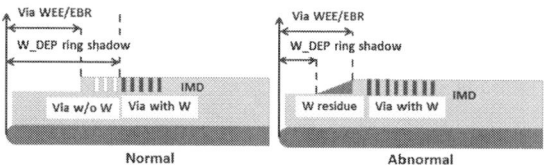

Figure 5: Formation mechanism of tungsten residue
This issue is solved after this type of the tungsten

deposition equipment is forbidden for this product.

Case 3

By touch O/S failure was found on product C, more serious at wafer edge. The needle tip was extraordinary dirty by inspection. We found the punctured bonding pad on ugly die by wafer edge at the same time. By FIB, we confirmed that no surface metal layer covered on the pad area of ugly dice, so the Tungsten plugs were exposed, spaded up and adsorbed on the tip of the needle while testing. Test failure started from these pad, and Tungsten plugs were constantly brought to the pad of next touch, affecting the subsequent DUT test results (as shown in Figure 6).

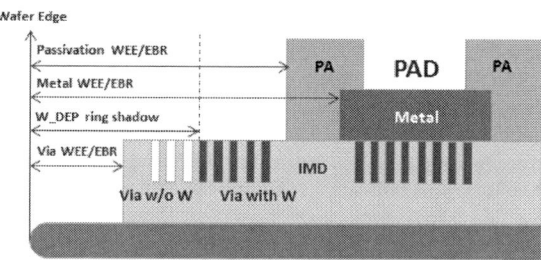

Figure 6: Binmap, OM and FIB photo for product C

Let's study the thin film stacking profile of these ugly dice on wafer edge (as shown in Figure 7). Theoretically, we can reduce the edge cut of top metal layer, meanwhile increase the edge cut of via layer below, so that we can make more pad tungsten plug be covered by metal layer. But in fact, less edge cut of metal layer brings great risks, such as arcing and metal peeling. Then can we reduce the edge cut of passivation layer and make the pad tungsten plug surface be covered with more passivation? The answer is also negative, because the passivation layer in bumping pad area is open.

Figure 7: Schematic of the thin film stack on wafer edge

In the course of the investigation, we also found that the product D, with same process as product C, did not have O/S failure issue like product C. The different point was the PAD layout. Via layout under top metal of product C was full array, while that of product D was peripheral ring shape. So there were no tungsten plugs in the needle touching area on product D (as shown in Figure 8).

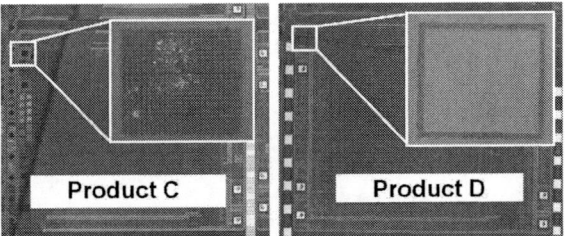

Figure 8: Bonding pad layout difference between product C and Product D

Pad layout with full array via is generally used in Pad-On-Circuit (POC) structure. It can increase the bearing capacity of pad to protect the circuit under the pad while needle touch. So the pad layout is restricted to the structure of the IC circuit. In case of non POC, in order to reduce the test problem, peripheral ring shape configuration of pad is preferred.

CONCLUSION

Summarizing the above several cases, we can find that the wafer edge thin film structure formed by front end process not only affects the subsequent processes, but also impacts on wafer test and final yield. It's difficult to solve these problems fundamentally by optimizing the test technology. The key to this problem is to properly handle the edge cut at wafer edge.

We should thoroughly cover top metal layer by passivation layer at wafer edge (edge cut of top metal layer should be more than that of passivation layer), so that arcing from exposed metal line can be avoided.

We should prevent the tungsten from depositing on the step between thin film layers (edge cut of tungsten deposition should be more than that of via), so as to prevent power pin from shorting by tungsten residue.

We should balance the solution between thin films edge cut optimization and bonding pad layout scheme optimization, so as to avoid the exposed tungsten plugs which affect the lifetime of probe card as well as the yield of the wafer.

ACKNOWLEDGEMENTS

The author would like to thank Mr. Zigui Cao, Mr. Yuanyuan Zhu and Mrs. Xiuying Gu for the technical discussion.

REFERENCES

[1] Hong Xiao, "Introduction to Semiconductor Manufacturing Technology", Prentice Hall (2000)

[2] Ankun Mao, "Study on the fluctuation of photolithography on the edge of silicon wafer in IC manufacturing", Fu Dan University (2012)

[3] Xiaofeng Yuan, Qiang Zhang, Jing'an Hao, "Wafer edge treatment in lithographic process for peeling defect reduction", IEEE/CSTIC (2017)

978-1-5386-5309-8/18 $31.00 © 2018 IEEE

EFFICIENT MULTI-BIT SRAMS USING NANOSTRUCTURES FIELD-EFFECT TRANSISTORS (NANO-FETS)

Bander Saman[1]

[1]School of Engineering, Taif University, Taif 21974, Kingdom of Saudi Arabia

*Corresponding Author's Email: bander.saman@uconn.edu

ABSTRACT

This paper presents emerging quantum dot (QDs) and quantum wells (QWs) field-effect transistors (FETs) where nanotechnology is integrated for 2-bit static random access memory (SRAM) cells and multiple valued logic (MVL) circuits. As feature sizes have started to approach sub-7 nm regime, several issues have begun to make further miniaturization difficult. We present n-channel QDs gate/channel and n-channel QWs spatial wave-function switching FETs. Unlike conventional MOS-FET (metal-oxide semiconductor field-effect transistor) which can process 1-bit at a time, the QDs and wells FETs can process 2-bits and potentially implement multi-valued logic and reduce device count.

INTRODUCTION

The semiconductor industry follows Moore's Law, which predicts the size of MOS-FET devices to shrink by 50% every 18-24 months [1,2]. When the devices sizes have started to approach sub-7 nm regime, several issues have begun to make further miniaturization difficult [3]. In this paper, the nano-FETs are introduced to implement MVL and SRAMs in future. The nano-FETs are n-type FET having quantum dots gate (QDG-FET), quantum dots channel (QDC-FET), and quantum wells transistor which is called the spatial wave-function switching field-effect transistor (SWS-FET).

n-type QDG-FETs exhibit multi-states due to change in threshold voltage (V_{TH}) as a result of stored charge in the QDs in the gate region [4]. n-type QDC-FETs produce multiple states in their transfer characteristics as the charges flow through the mini-band structure formed an array of QDs in the channel region of the FET [5].

On the other hand, n-type SWS-FET has more number of states in its transfer characteristic based on the switching of charge carriers from one channel to other channel of the device [6,7,8].

THE DEVICES STRUCTURES

Quantum Dots n-type Field-Effect Transistors (n-QDs-FET)

Figure 1 shows the cross-sectional schematic of n-QDs-FET. It is similar to n-type FET with two Silicon Oxide (SiOx) cladded Si QDs layers in gate and p-type Si substrate using self-assembly process [7]. Figure 2 shows the transmission electron microscopy (TEM) image of four SiOx-cladded Si QDs layers in the gate and channel.

In QDs-FET, a shift in the threshold (ΔV_{TH}) depends on transfer of charges to quantum dot layer, where $\Delta V_{TH}=0$ in conventional FETs [1]. In four states n-QDs-FET, the V_{TH} changes linearly when the gate voltage (V_G) increases through a range of voltages V_{g1} to V_{g2} and V_{g3} to V_{g4}, the range of voltages can be adjusted by the size of the dots [1].

Figure 3 shows the four-state characteristics of a fabricated n-QDs-FET, where the vertical axis is the current that flows between the drain and source (I_{DS}), and the horizontal axis is the voltage between the gate and source (V_{GS}).

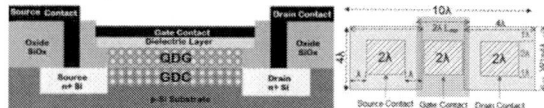

Figure 1. n-QDs-FET Cross-sectional.

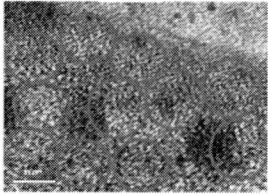

Figure 2. TEM image QDs [7].

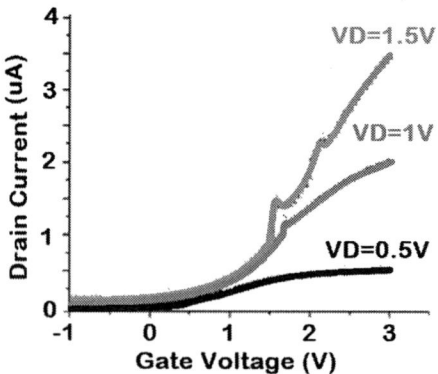

Figure 3. Experimental n-QDs-FET I_{DS}-V_{GS} characteristics [7].

Spatial Wave-function Switching n-type Field-Effect Transistor (n-SWS-FET)

n-SWS-FET is shown in Figure 4 where the Si QWs and Silicon-Germanium (SiGe) barriers layers are epitaxially grown on p-Si. Here, two Si QWs (W_1 upper well, W_2 lower well) are sandwiched between SiGe barriers. The current is transported between the source and drain through upper and lower Si QWs [8,9].

The channel current in a two quantum well (W_2 and W_1) n-SWS-FET is as a function of gate voltage (V_G). Once the V_G is increased above threshold voltage of W_2 (V_{TH2}), the electrons appear in W_2, and as $V_G >$ threshold voltage of $W1$ (V_{TH1}), the carriers start transferring to $W1$. In a small gate voltage range, current flows in both lower and upper quantum wells. Finally, as V_G is further increased, carriers completely transfer from W_2 to $W1$ and the current flows only W_1.

Figure 5 presents I_{DS} V_{DS} characteristic of a fabricated QD-n-SWS-FET.

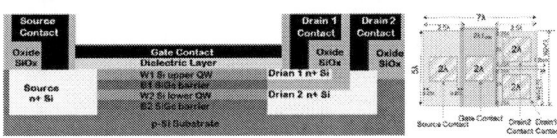

Figure 4. Two well n-SWS-FET Cross-sectional.

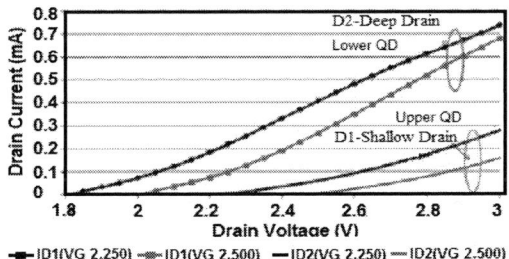

Figure 5. Experimental QD-n-SWS-FET I_{DS}-V_{DS} characteristics [9].

THE DEVICES MODLING

The Four-state n-QDs-FETs and The four QWs n-SWS-FETs have been modelled using Berkeley Short-channel IGFET Model (BSIM) and Analog Behavior Model (ABM) in Cadence-OrCAD CIS simulators. The model is suitable for transient analysis.

The Simulation of I_{DS}-V_{GS} characteristics for four-state n-QDs-FET is presented in Figure 6 [6-10]. The device parameters are shown in Table I. The simulation result shows the four regions of the transfer characteristics as (OFF-00, MOD1-01, MOD2-10, and ON-11) [11,12].

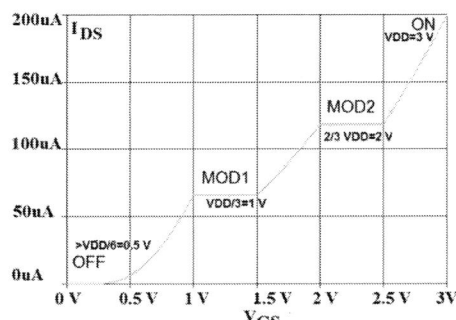

Figure 6. The simulation I_{DS}-V_{GS} of QDs-FET [10].

TABLE I.   N-QDs-FET DEVICE PARAMETERS

Parameter	Value	Unit
L	1	μm
W	3	μm
V_{DD}	3	V
V_{TH}	0.5	V
Vg1	1	V
Vg2	1.5	V
Vg3	2	V
Vg4	2.5	V

The four-state quaternary signal can be inverted by using one n-QDs-FET and one p-channel FET as pseudo n-MOS inverter design style. Figure 7 shows a quaternary inverter circuit and simulation. The input single is regenerated simply by inverting the output of the quaternary inverter using additional four states inverter, the regenerating input signal is showing in Figure 7. As result, a four-state n-QDs-FET is suitable for quaternary logic with fewer devices and Si area.

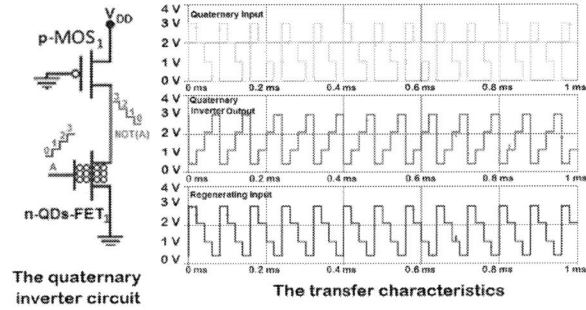

Figure 7. Circuit and simulations of the quaternary inverter logic.

978-1-5386-5309-8/18 $31.00 © 2018 IEEE

Figure 8 shows n- SWS-FET I_{DS}-V_{GS} characteristics Cadence simulation of two QWs (up) and four QWs (down), the two QWs n- SWS-FET device parameters are shown in Tables II. The simulation results show excellent performance on switching and output currents.

TABLE II.   TWO QWs N- SWS-FET DEVICE PARAMETERS

Parameter	Value	Unit
L	20	nm
W1	50	nm
W2	100	nm
V_{DD}	1	V
V_{TH1}	0.3 V	V
V_{TH2}	0.2 V	V

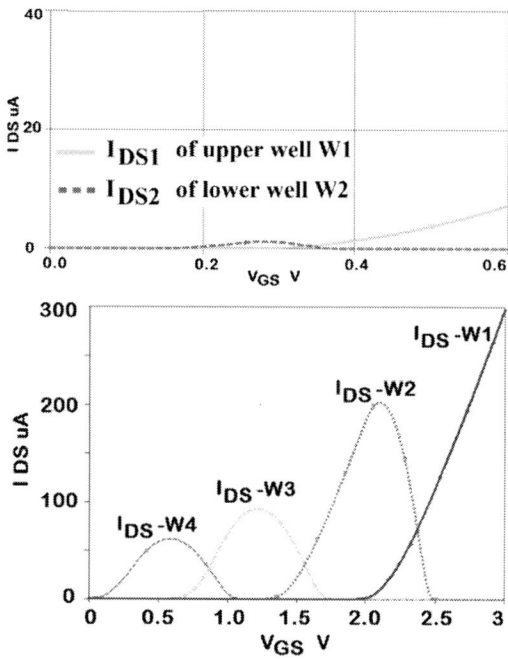

Figure 8: The simulation I_{DS}-V_{GS} of two QWs n-SWS-FET.

SRAM CIRCUIT AND SIMULATION

A Conventional 1-bit SRAM consists of two conventional-coupled COMS inverter for the memory implementation. The circuit has two coupled inverter and two n-type FET access transistors.

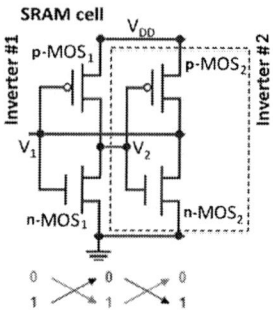

Figure 9. Conventional CMOS 1 bit SRAM cell.

The same concept of the conventional 1-bit SRAM CMOS SARM circuit is used to make 2-bit SRAM by replacing the conventional coupled COMS inverter with two n-QDs-FET inverters as showing in Figure 10.

Figure 11 shows a 2-bit SRAM cell using four Si/SiGa QWs n-SWS-FET. The sources (S_1, S_2, S_3, S_4) are connected to different levels power supply. The four levels data (Quaternary) is stored by using the loop in SWS-FET and switching properties of the four QWs n-SWS-FETs.

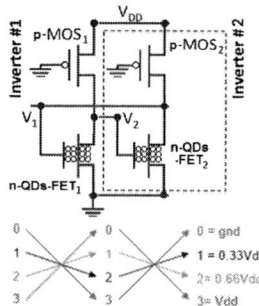

Figure 10. n-QDs-FET 2-bit SRAM cell

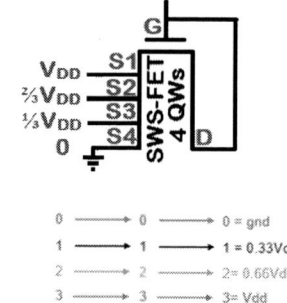

Figure 11. Four-QWs n-SWS-FETs 2-bit SRAM cell.

In 2-bit SRAM cell using QDs and four-QWs FETs, the data line is a quaternary input signal and the word line is a binary signal that enabling and disabling the write operation.

Figure 12 shows the simulations of 2-bit SRAM using QDs and four-QWs FETs. The simulation of the both cell are identical and matching the truth table of 2-bit SRAM (Table III).

Where, the QDs-FET device in Figure 12 has the parameters as Table II. Moreover the QWs n-SWS-FET SRAM circuit shown in Figure 12 has device parameters as L=1 µm, W1=1.25 µm, W2=2.5 µm, W3=3.755 µm, W4=5 µm, V_{DD}=3 V, V_{TH1} =2.5 V, V_{TH2} =2.1 V, V_{TH3} =1.18 V, V_{TH4} =0.3V.

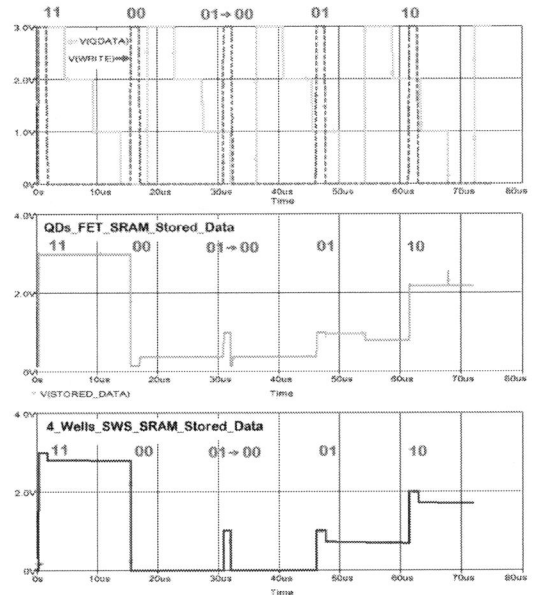

Figure 11. The simulation of 2-bit SRAM using QDs and four-QWs -FET.

TABLE III.   THE TRUTH TABLE OF 2 BIT SRAM

Write Signal	Data Signal	Stored Data
0	> ⅙ V_{DD} = Logic 0 (00)	Same State
0	⅓ V_{DD} = Logic 1(01)	Same State
0	⅔V_{DD} = Logic 2 (10)	Same State
0	⅔V_{DD} = Logic 3 (11)	Same State
1	Logic 0 (00)	> ⅙ V_{DD} Logic 0
1	Logic 1 (01)	⅓ V_{DD} Logic 1
1	Logic 2 (10)	⅔V_{DD} Logic 2
1	Logic 3 (11)	⅔V_{DD} Logic 3

CONCLUSIONS

In summary, we have demonstrated the single/multi-bit static random access memory (SRAM) based on FETs incorporating QDs and QWs. The number of the circuit element is reduced significantly compared to conventional CMOS circuit (50% in QDs, and 87.5% in 4QWs)

ACKNOWLEDGEMENTS

We thank Dr. M. Alsharef (ECE Dept. Head-Taif University), Dr. Faquir Jain (University of Connecticut), Dr. E. Heller (Synopsys Inc. NY USA) and Eng. K. Alomari (SEMC- Jeddah KSA) for comments that greatly improved the manuscript.

REFERENCES

[1] P. Gogna, M. Lingalugari, J. Chandy, E. Heller, E-S. Hasaneen and F. Jain. *Int. Journal of VLSI design & Communication Systems*, vol.3, No.5, 11 (2012).

[2] Moore, G. *Solid-State Circuits Newsletter, IEEE*, 20(3), 33-35.

[3] Kim, Austin, Baauw, Mudge, Flautner, Hu, Irwin, Kandemir, and Narayanan. *Computer*, vol. 36, no. 12, 2003, pp. 68–75.

[4] Karmakar, S., Chandy, J., Gogna, A., & Jain, M. *Journal of Electronic Materials*, 41(8), 2184-2192.

[5] Jain, Karmakar, Chan, Suarez, Gogna, Chandy, and Heller. *Journal of Electronic Materials*, vol. 41, no. 10, 2012, pp. 2775–2784.

[6] Bander Saman, P. Gogna, El-Sayed Hasaneen, J. Chandy, E. Heller, and F. C. Jain. *International Journal of High Speed Electronics and Systems*, vol. 26, No. 3 (2017) 1740009.

[7] Karmakar, Supriya, and Faquir Jain. Silicon, vol. 8, no. 3, 2016, pp. 369–379.

[8] Bander Saman, P. Mirdha, M. Lingalugari, P. Gogna and F. C. Jain. *International Journal of High Speed Electronics and Systems*, Vol. 24, Nos. 3 & 4 (2015) 1550008.

[9] B. Saman, P. Gogna, E-S. Hasaneen, J. Chandy, and E. Heller, and F. Jain. *Proceedings of International Semiconductor Device Research Symposium (ISDR) 2016*, Maryland, December 7-9, 2016, FA2-06.

[10] B. Saman, Ahmed Aziz, K. Alomari, E. Heller, and F.C Jain. *Proceedings of 26th Annual Connecticut Symposium on Microelectronics & Optoelectronics CMOC*, Storrs April 5,2017.

[11] Jain, Chan, Suarez, Lingalugari, Kondo, Gogna, Miller, Chandy, and Heller. Journal of Electronic Materials, 42(11), 3191-3202.

[12] Jain, Lingalugari, Kondo, Mirdha, Suarez, Chandy, & Heller. (2016). Journal of Electronic Materials, 45(11), 5663-5670.

TISI2 FORMATION AND MECHANISM THROUGH ULTRA-THIN AL2O3 INTERMEDIATION

Peilin Hao, Ming Li*, Bingxin Zhang, Siyang Gao, Xia An and Ru Huang

Key Laboratory of Microelectronic Devices and Circuits, Institute of Microelectronics, Peking University, Beijing 100871, China
* Email: liming.ime@pku.edu.cn

ABSTRACT

In this paper, $TiSi_2$ formation through ultra-thin Al_2O_3 intermediation is experimentally demonstrated and the mechanism for solid-state phase reaction is studied. By the intermediation of ultra-thin buffer layer, a thinner layer of amorphous Ti-silicide was obtained under the same annealing condition. It's found that the Si atomic diffusion process is almost retarded by the buffering layer while Ti atoms can still diffuse into Si substrate through the oxygen exchange effect between TiO_2 and Al_2O_3. Due to lack of nucleation centers in the $Ti/Al_2O_3/Si$ system, only amorphous C49 phase $TiSi_2$ was formed with a Ti-rich silicide phase formed at the top surface. In the normal Ti/Si system, however, C54 phase can be obtained at high temperature.

INTRODUCTION

With the feature size of the devices keeps diminishing, the degradation of metal-oxide-semiconductor field effect transistors (MOSFETs) performance has become difficult to overcome. Among all the non-ideal effects, the parasitic resistance, especially the source-drain parasitic series resistance, seriously restrict the increase of current driving capability [1]. Therefore, the metal silicide, which has low schottky barrier and low resistivity to the substrate, is widely used because of their great importance to maintain the performance. According to the ITRS roadmap, at 14nm technology node, the thickness of silicide layer needs to be lower than 12.5nm to avoid silicon over-consumption. As for the 7/5nm technology node, the contact resistivity needs to be reduced to 1E-9 Ω•cm^2 [2]. Owing to its various advantages, including good adhesion to substrate and SiO_2, low resistivity (about 15$\mu\Omega$•cm) and less Ti diffusion into the substrate, Ti has been the main metal alternative to nickel at technology nodes beyond 14nm [3]. The present literatures demonstrate that Ti films deposited on Si will transform to $TiSi_2$ at 600°C annealing temperature and into high conductive C54 phase at a higher temperature [4-7]. However, during $TiSi_2$ formation, Si acts as the active diffusion source so that a considerable silicon will be consumed [8]. In extremely scaled device, such silicon consumption will cause serious reliability issue and performance degradation. Therefore, how to control the solid-state phase reaction process of $TiSi_2$ becomes a key challenge in the modern technology.

In this paper, an oxide-mediated $TiSi_2$ formation process is proposed and experimentally demonstrated. Ti silicide formation through Al_2O_3 intermediation shows extraordinary diffusion behavior. The atomic diffusion and reaction mechanisms are deeply investigated to provide a theoretical insight into the key factors controlling the silicide thickness and structure.

EXPERIMENT

$TiSi_2$ was formed in both Ti/Si and $Ti/Al_2O_3/Si$ systems in this paper. The p-type Si(100) substrates with resistivity of 25.5-42.5Ω•cm were used as the starting wafers. After being cleaned by diluted hydrofluoric acid (HF) and hydrochloric acid (HCl), 10Å Al_2O_3 was deposited on the Si substrate by plasma enhanced atomic-layer deposition (PEALD). Next, 200Å Ti was deposited by physical vapor deposition (PVD) immediately. Then, the solid-phase reaction process was conducted with rapid thermal process (RTP). The annealing temperature ranges from 700-900°C and the annealing time was set to 30s. The unreacted Ti was removed by a mixture of sulfuric acid and hydrogen peroxide with the ratio of 4:1 (H_2SO_4:H_2O_2=4:1). The lattice structure and product thickness were examined using transmission electron microscope (TEM). Raman spectroscopy was used to identify the phase composition and energy dispersive spectrum analysis (EDS) was applied to detect relevant element content and distribution.

RESULTS AND DISCUSSION

Fig.1 shows the TEM images of two samples prepared in Ti/Si and $Ti/Al_2O_3/Si$ system respectively. From Fig.1 (a), it can be seen that 26-nm amorphous Ti silicide is formed in the Ti/Si system with smooth interface between silicon substrate. In the $Ti/Al_2O_3/Si$ system, however, layering phenomenon is observed as shown in Fig.1 (b). A 20-nm thick amorphous Ti silicide is formed at the silicon substrate while another 14-nm susceptible Ti-rich silicide or Ti/Si oxide mixture is observed at the top. Thinner silicide formation indicates possibly limited diffusion and reaction mechanism. Only at low temperature below 500°C, such Ti-rich phase can be observed in Ti/Si deposition system due to not enough silicon diffusion into Ti layer. It also suggests that silicon diffusion from substrate could be effectively retarded by the buffering Al_2O_3 layer.

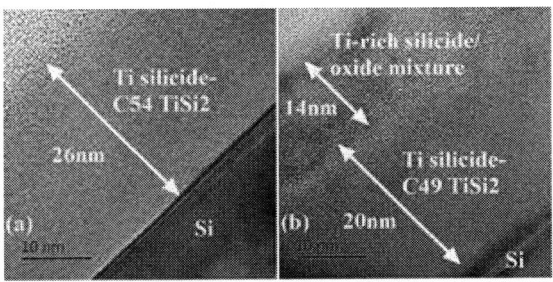

Figure. 1. TEM images of all systems Ti silicide formed by (a) Ti/Si system (b) Ti/Al$_2$O$_3$/Si system

To confirm the phase in Fig.1, Raman spectroscope is carried out with 514nm wavelength laser source. As shown in Fig.2 (a), the final product of Ti/Si system is clearly C54 TiSi$_2$, which indicates that the Ti-Si reaction has been completed. In Fig.2 (b), two optoelectronic peaks are observed near 275cm^{-1} and 300cm^{-1} in the final product of Ti/Al$_2$O$_3$/Si system. The first one is usually referred as C49 TiSi$_2$ and the latter is Ti$_5$Si$_4$ phase. The incomplete silicidation process under such high temperature is very extraordinary

out in Fig.3. From Fig.3 (a), almost Ti:Si=1:2 composition ratio is observed in the silicide layer but slight oxygen contamination is also observed, which is possibly due to the surface oxidation of TiSi$_2$. In Fig.3 (b), however, very clear layering phenomenon is observed. At the top surface, a Ti-rich layer exists, containing considerable oxygen composition. As compared to Fig.3 (a), which is the normal Ti/Si system, the abnormally high oxygen composition cannot be originated from native oxidation. Thus, it is strongly supposed to come from the inserted Al$_2$O$_3$ buffering layer. From the hump on Ti profile in Fig.3 (b), it can be inferred that Ti diffusion into substrate has been effectively retarded. On the other hand, Si diffusion is also stopped by the buffering Al$_2$O$_3$. Meanwhile, Al is almost not detected in the final product of Ti/Al$_2$O$_3$/Si system possibly because it has been dissolved as Ti grabs oxygen from Al$_2$O$_3$ layer. As figured out from Fig.3, the atomic diffusion source in the oxide intermediation system has been changed from Si to Ti. It means that Ti silicidation can been controlled by the limited diffusion ratio, depending on annealing temperature and time. Meanwhile, the silicon consumption can also be reduced.

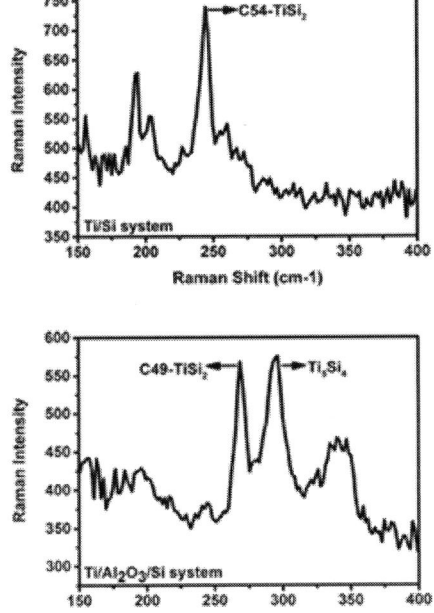

Fig. 2. Raman spectroscopy analysis of phase composition (a) Ti/Si system (b) Ti/Al$_2$O$_3$/Si system

To make clear of the composition distribution in the two different systems, EDS line scanning results are given

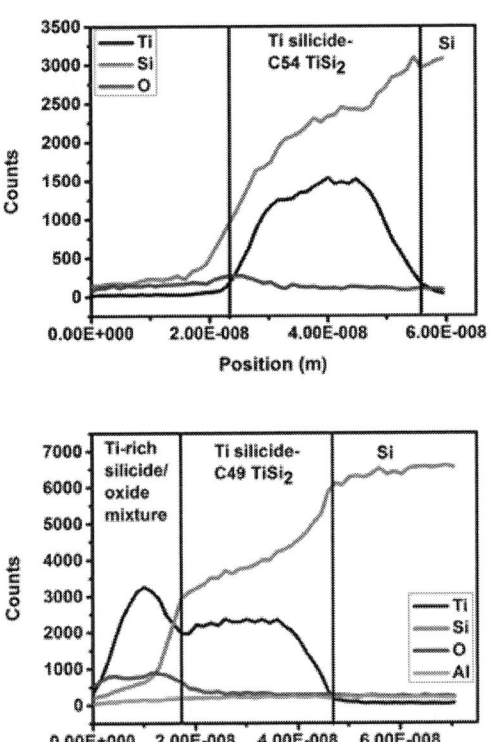

Fig. 3. EDS analysis of element content (a) Ti/Si system (b) Ti/Al$_2$O$_3$/Si system

Systematically we can conclude the atomic diffusion and solid-phase reaction mechanisms in the Al2O3-mediated Ti silicide formation process as the following three steps:

1) Atomic diffusion: In usual Ti/Si system, silicon acts as the active diffusion source so that the silicide reaction can been completed in a short time under high temperature such as 900°C. In Ti/Al$_2$O$_3$/Si system, due to the diffusion barrier of Al$_2$O$_3$, the silicon is hard to diffuse into Ti layer. At the same time, due to the oxygen scavenging effect of Ti, oxygen in Al$_2$O$_3$ will be concentrated in the Ti layer, and as an exchange effect, Ti "diffuses" into substrate through Al$_2$O$_3$ layer.

2) Solid-phase reaction: With more and more Ti atoms penetrating the buffering layer, Si atoms will react with Ti atoms to form silicide. At the early stage, unstable Ti-rich phase is formed. With annealing process lasting, TiSi$_2$ is firstly formed near the interface between substrate since there are enough supply of Si. In the buffering layer and above it, however, only Ti-rich phase can be formed since very few Si atoms can reach such position. The TiSi$_2$ thickness is then limited by the amount of Ti capable to penetrate the buffering layer.

3) C54 phase transformation: In normal Ti/Si system, C49 TiSi$_2$ will be transformed into low resistivity C54 phase under temperature higher than 700°C. However, in Ti/Al$_2$O$_3$/Si system, with annealing process lasting, the Ti-rich layer conversion into C49 will be preferred other than C54 formation. If no more Si atoms diffuse into the top Ti-rich layer during the whole annealing process, no C54 TiSi$_2$ will be formed at the end of process just as observed in this experiment. On the other hand, due to TiO$_2$ formation during Ti penetration through Al$_2$O$_3$, Ti will become more immobilized so that C49 phase layer cannot be extended any more. As a result, the thickness of C49 TiSi$_2$ will be limited by the Ti penetration through Al$_2$O$_3$, depending on the Al$_2$O$_3$ process itself.

SUMMARY

In this paper, TiSi$_2$ formation through Al$_2$O$_3$ intermediation is proposed and experimentally demonstrated. A deep insight into the atomic diffusion and solid-state reaction is provided to clarify the special silicide formation mechanism in Ti/Al$_2$O$_3$/Si system. It's supposed that the buffering layer can mediate Ti diffusion into substrate and almost fully retard Si diffusion. As a result, the thickness of TiSi$_2$ is almost limited by the buffering layer deposition process itself. On the other hand, C49 phase will be preferred in the Al$_2$O$_3$ mediated TiSi$_2$ process.

ACKNOWLEDGEMENTS

This work is supported in part by 863 Project (2015AA016501) and Cooperative Project with SMIC.

REFERENCES

[1] R. H. Dennard, F. H. Gaensslen, H. N. Yu, V. L. Rideout, E. Bassous, and A. R. LeBlanc, *IEEE J. Solid-State Circuits*, vol. 9, no. 5, pp. 256-268, 1974.

[2] C. N. Ni, et al. *IEEE Symposium on VLSI Technology*, 2016, pp. 70-71.

[3] P. Adusumilli, et al. *IEEE Symposium on VLSI Technology*, 2016, pp. 68-69.

[4] B. W. Bower and J. W. Mayer, *Appl. Phys. Lett.*, vol. 20, no. 9, pp. 359-361, 1972.

[5] W. K. Chu, S. S. Lau, J. W. Mayer, H. Muller, and K. N. Tu, *Thin Solid Films*, vol. 25, no. 2, pp. 393-402, 1975.

[6] H. Kato and Y. Nakamura, *Thin Solid Films*, vol. 34, no. 1, pp. 135-138, 1976.

[7] S. P. Murarka and D. B. Fraser, *J. Appl. Phys.*, vol. 51, no. 1, pp. 342-349, 1980.

[8] G. G. Bentini, R. Nipoti, A. Armigliato, M. Berti, A. V. Drigo, and C. Cohen, *J. Appl. Phys.*, vol. 57, no. 2, pp. 270-275, 1985.

A NEW METHOD FOR TRAVELING MEMORY ADDRESS BASED ON VECTOR FUNCTION

Qingwen Zhang, Yun Xu ,YuXiang Zhang, Zhimin Zeng, Mingming Li, Yuan Wu

Product Engineering Division, Shanghai Huahong Grace Semiconductor Manufacture Corporation

No.1188 Chuanqiao Road, Pudong New Area, Shanghai, China

Corresponding Author's Email: qingwen.zhang@hhgrace.com

ABSTRACT

MBIST (Memory Built-In Self-Test) design can reduce the cost of the test by improving the efficiency of test and program development. In many cases, MBIST design will pay more attention to test efficiency and less consideration for the convenience of EFA (Electrical Failure Analysis). Correspondingly, MBIST design the function of internal comparison with address automatic accumulation module, but this module is often not open to output function which bring inconvenience to EFA. Memory EFA relies heavily on bitmap function, that means we must send memory address serially and cumulatively. For some Graphic Interface Testers, the bitmap is highly dependent on the storage depth of VECTOR which increases exponentially when address width added. This article describes a method that traveling all the addresses of a large memory based on smaller VECTOR space.

Key words: Vector; EFA; MBIST; Bitmap

BACKGROUND

With the development of semiconductor technology, more and more functions can be integrated into single chip. SoC (System-on-Chip) is a system that integrates all the functions needed to integrate microelectronic applications on a single chip. The basic architecture of SoC usually includes MCU, Memory, Digital/Analog-Module. In these modules, the defects are most likely to occur in memory for its large size and complex process. Before shipment these defective chips must be sorted by the test.

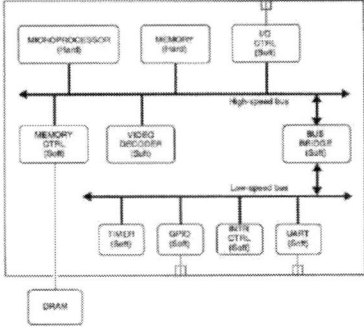

Figure 1: Example Structure of Soc

Generally, integrated circuit testing includes CP(chip probing) and FT(final test).In both test, the screening of memory occupies most of the test time. So reduce the memory test time is most important in test. CP test is closely after the chip manufacturing, so it plays an important role in sorting. The number of Memory bit is huge, so the ASIC designer develop MBIST(Memory Built-In-Self-Test) module for CP test use. MBIST is a dedicated circuit module used to speed up die screening.

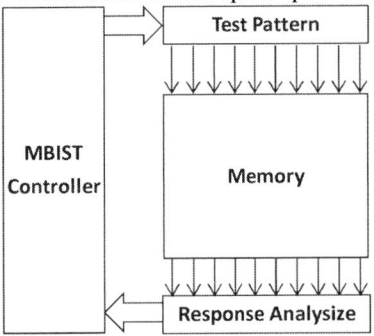

Figure 2: Example Structure of MBIST

In chip manufacture, another issue is equally important. The semiconductor process is highly complex, and the product yield is highly sensitive to the process parameters. It is directly related to how to quickly find the corresponding process of low yield event, and the recovery of production as soon as possible. EFA(Electrical Failure Analysis)engineers play an important role in the process of searching the process. The EFA engineer also develops the EFA program based on MBIST function. Unlike test engineers, EFA engineers usually focus not on testing speed, but on how to locate process problems through effective experiments. By using the special test platform (for example, MS5205) and matching the appropriate programming, each bit information of the memory can be displayed on a Bitmap. Bitmap is one of the most important tools for EFA engineers, which plays a vital role in positioning process defects. A good test shall be fast sorting and efficient to find the process of low yield.

Unfortunately, many times, MBIST design will pay more attention to test efficiency and less consideration of the convenience of EFA Electrical Failure Analysis. A

978-1-5386-5309-8/18 $31.00 © 2018 IEEE

common problem is: to improve test efficiency, the MBIST can do comparison of all bit internally, with a circuit module which has the address auto-accumulated function and output a pass/fail information. But the IO pad can't output all bit information automatically for this module is not be designed to output use.

Such design brings great inconvenience to the Bitmap program development: EFA engineer must spend a lot of energy on bitmap program development because read out memory bit information serially is difficult. For EFA experiment is very flexible, the tester often be Graphic Interface. For such type of tester, the input of address is generally based on the Vector Function. When the address width is increased, the Vector needed is increased exponentially with traditional method. If there is no proper programming method optimized, the address description needed length is possible to beyond the storage limitation of Vector.

In our work, we often encounter such problem, and finally set up an improved bitmap program development method, which greatly reduces the dependence on Vector storage depth

TRANDITIONAL METHOD

In general, for serial program development, if we want to traverse all addresses, the Vector based Pattern programming process is:
1. Translate the address into binary format and put it together in the right sequence. Copy the binary address into Vector file.
2. According to protocol format, insert the Vector to right position of Pattern

We illustrate this process by a specific example:
If a memory's address width is 18,the Vector can be simply expressed as"00 0000 0000 0000 0000   00 0000 0000 0000 0001   00 0000 0000 0000 0010   00 0000 0000 0000 0011   …   11 1111 1111 1111 1111"

Traditionally, the corresponding Pattern shall be executed as Figure3:

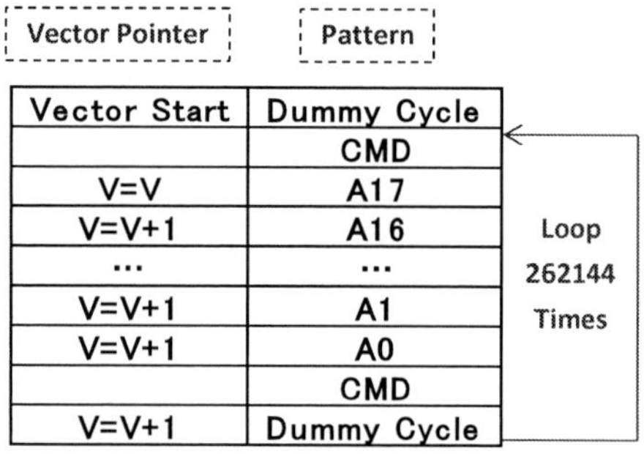

Vector Start	Dummy Cycle	
	CMD	
V=V	A17	
V=V+1	A16	Loop
…	…	262144
V=V+1	A1	Times
V=V+1	A0	
	CMD	
V=V+1	Dummy Cycle	

Figure 3: Traditional Method Schematic Diagram

The traditional method requires the tester Vector have the storage depth of 4718592(18*2^18) bits, which is not satisfied with some testers.

DIVIDE ADDRESS METHOD

Based on the similar problems encountered in bitmap program development, we explore a Divide Address Method which is not particularly dependent on storage depth of Vector. Simply ,the address is divided into two sub-address, each sub-address corresponding to the independent Vector, independent register. With the appropriate register pointer, we can achieve the aim of traversing all address of large memory size with small Vector storage.

The following examples give us a more intuitive understanding of the process and effect of the this method. Suppose:
1. Storage depth of Vector is 65536.
2. A memory's address width is 18 bits.
We divide the 18 bits address into two parts, a 9 bits sub address Address-A and anther 9 bits sub address Address-B. The corresponding sub-Vector is VA and VB, so VA and VB both can be simply expressed as"0 0000 0000   0 0000 0001   0 0000 0010   …   1 1111 1111". The Pattern after using the new method is executed according to Figure 4:

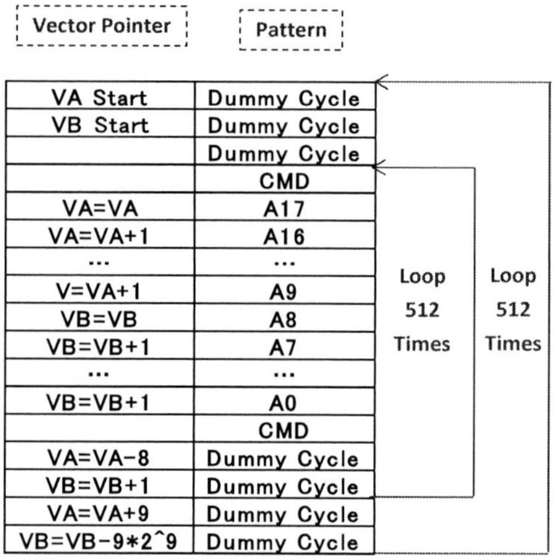

VA Start	Dummy Cycle		
VB Start	Dummy Cycle		
	Dummy Cycle		
	CMD		
VA=VA	A17		
VA=VA+1	A16		
…	…	Loop	Loop
V=VA+1	A9	512	512
VB=VB	A8	Times	Times
VB=VB+1	A7		
…	…		
VB=VB+1	A0		
	CMD		
VA=VA−8	Dummy Cycle		
VB=VB+1	Dummy Cycle		
VA=VA+9	Dummy Cycle		
VB=VB−9*2^9	Dummy Cycle		

Figure 4: Divide Address Method Schematic Diagram

From the schematic diagram we can find out the length of needed Vector is only 9216 bits (9*2^9+9*2^9) in Divide Address Method, which is greatly reduced.

CONCLUSION

The Divide Address Method can greatly reduce the length of necessary Vector. Using this method, we have

978-1-5386-5309-8/18 $31.00 © 2018 IEEE

succeeded in solving two related problems, and we are sure it will play a greater role in the future.

ACKNOWLEDGEMENTS

We would like to thank EFA team for their rich knowledge accumulation and experience, and thanks our division manager Yun Xu, without his encouraging this paper could not be written.

REFERENCES

[1] *Yoshiyuki Nakamura1,Jacob Savir, Hideo Fujiwara1. BIST Pretest of ICs: Risks and Benefits[C].VLSI Test Symposium, 2006. Proceedings.24th IEEE:6 pp.-149.*

[2] *Abhijit Ghosh，Srinivas Devadas，Riehard Newton. Sequential Test Generation At the Register-Transfer and LogieLevels, IEEE0738-IO0x/90/006/05802006: 31-42.*

FIRST-PRINCIPLES STUDY ON Ge$_{1-x}$Sn$_x$-Si CORE-SHELL NANOWIRE TRANSISTORS

Zeguo Gu, Feng Xu, Bin Gao[*], Huaqiang Wu, and He Qian

Institute of Microelectronics, Tsinghua University, Beijing 100084, China

[*]Corresponding Author's Email: gaob1@tsinghua.edu.cn

ABSTRACT

The compositional dependence of electronic band structure in relaxed and biaxially strained Ge$_{1-x}$Sn$_x$ alloy is investigated with the first principles method. Based on energy band dispersion along the [100] crystal orientation, hole effective mass is extracted via parabolic line fit. Calculation results indicate Sn composition dependence of the hole effective mass for relaxed alloy is much less pronounced than biaxially strained alloy. And then the electronic structure of Ge$_{1-x}$Sn$_x$-Si core-shell nanowire along the [110] direction is studied from first-principles calculation.

INTRODUCTION

The use of Ge$_{1-x}$Sn$_x$ alloys as FET channel material has attracted much attention, due to the higher carrier mobility than Ge [1-2] and the direct bandgap transition with a low Sn fraction [3]. These merits make Ge$_{1-x}$Sn$_x$ alloys as a promising channel material for high performance CMOS logic [4] and low power consumption electronic-photonic integrated circuits [5]. On the other hand, to meet the continued scaling down challenge, gate-all-around nanowire transistors [6] become a competitive candidate. In particular, Ge core Si shell nanowire was implemented in 2002 [7]. Difference of energy gap and lattice mismatch between Ge and Si result in a significant boosting in carrier mobility. In this perspective, p-FET performance is expected to be enhanced further by introducing Sn fraction into Ge core, since lattice mismatch between Ge$_{1-x}$Sn$_x$ alloy and Si shell is larger, which can introduce larger compressive strain to boost hole mobility.

In this paper, impacts of Sn fraction and biaxial strain on hole mobility of Ge$_{1-x}$Sn$_x$ alloy are studied by first-principles calculation in the framework of density functional theory (DFT). In further calculation a Ge$_{1-x}$Sn$_x$-Si core-shell nanowire is constructed and charge density distribution is acquired.

METHOD

In band structure calculation of Ge$_{1-x}$Sn$_x$ alloy, the electronic interaction is described within generalized gradient approximation (GGA) in the form of the Revised Perdew-Burke-Ernzerhof (PBE) [8-9] exchange-correlation functional. The electron-ion interaction is represented by the norm conserving pseudopotentials [10].

Table I. Calculation Calibration via Measurement of Carrier Effective Mass

Effective Mass/ m_0	Si		Ge	
	Calculation	Measure	Calculation	Measure
m_l	0.970	0.970	1.595	1.580
m_t	0.203	0.190	0.078	0.082
$m_{hh[111]}$	0.739	0.620	0.410	0.436
$m_{lh[111]}$	0.163	0.152	0.057	0.042
$m_{hh[100]}$	0.312	0.400	0.201	0.244
$m_{lh[100]}$	0.190	0.175	0.053	0.046
$m_{hh[110]}$	0.768	0.555	0.379	0.368
$m_{lh[110]}$	0.154	0.156	0.056	0.043

In order to guarantee the accuracy and convergence, the plane wave basis set energy cutoff is set as 1140eV and a $3 \times 3 \times 3$ Monkhorst-Pack grid [11] of k-points is applied over Brillouin zone integrations. To split heavy hole and light hole band, spin-orbit coupling effect is included. Firstly, we calibrate band structure calculation by electronic and hole effective mass of pure Ge and Si. Calculation result is listed in Table I, that matches the cyclotron resonance measurement [12].

In order to model Sn composition of different concentration, a supercell with 16 atoms composed of $1 \times 1 \times 2$ cubic diamond unit cells is constructed. 0-4 Ge atoms are replaced randomly by Sn atoms, to model given Sn composition of x = 0%, 6.25%, 12.5%, 18.75% and 25%. Although it is reported that equilibrium lattice constant is deviated from compositionally averaged value, the nominal bowing parameter is 0.0468Å [14]. In this work Sn composition is below 25% and bowing value is very small, close to be neglected. So in relaxed alloys calculation, lattice constant of different Sn composition is determined simply by Vegard's law. Considering the structure relaxation has an important influence on the final energy band structure [15], geometry optimization is performed with residual atomic forces less than 0.01eV/Å after Sn atoms substituting.

For biaxial strained Ge$_{1-x}$Sn$_x$ alloys grown on (001) Ge buffers, the lateral lattice constant $a_\parallel$ equals that of Ge. The in-plane compressive strain is $\varepsilon_\parallel = (a_\parallel - a_0)/a_0$, where $a_0(x)$ is the lattice constant of relaxed Ge$_{1-x}$Sn$_x$ alloys. Simultaneously out of plane tensile strain $\varepsilon_\perp$ is caused by in plane strain, their relations are determined by Poisson's ratio $\varepsilon_\perp/\varepsilon_\parallel = -2C_{12}/C_{11}$ [16]. C_{12} and C_{11} are elastic constants of $Ge_{1-x}Sn_x$ alloy. Moreover out of plane strain can be represented as $\varepsilon_\perp = (a_\perp - a_0)/a_0$, where $a_\perp(x)$ is the lattice constant in the perpendicular direction.

Next a $Ge_{1-x}Sn_x$-Si core-shell nanowire along the [110] direction is constructed in the supercell with about 3nm vacuum in lateral direction to avoid interaction between neighboring nanowires. In order to study the intrinsic property, the nanowire surfaces are passivated with H atoms. The nanowire diameter reaches beyond 4nm with about 470 atoms in one supercell. The plane-wave cutoff energy is taken as 400 eV and $1 \times 1 \times 3$ Monkhorst-Pack k-point grids is performed. HOMO (the highest occupied molecular orbital) [13] charge density distribution is obtained.

REASULTS AND DISCUSSION

Relaxation $Ge_{1-x}Sn_x$ Alloy

Electron energy band structure of 12.5% Sn composition along several special path in the Brillouin zone is shown in Fig. 1. We choose the same high symmetry points as the work in references [17]. The valence band top is taken to be the energy zero. It is known that band gap is often underestimated by DFT calculation. We set scissors as 0.5eV to correct the band gap.

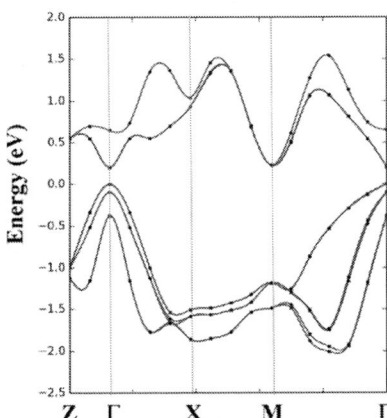

Figure 1: Calculated band structure for relaxation $Ge_{1-x}Sn_x$ alloy with 12.5% Sn composition

From Fig. 1, we can find $Ge_{1-x}Sn_x$ alloy undergoes indirect to direct band gap transition at about $x = 12.5\%$. Focusing on the valence band edge, the degeneracy of the Heavy-Hole (HH) and Light-Hole (LH) band at the Γ point is lifted. And the LH band lies underneath the HH band. The broken double level degeneration results from the broken tetrahedral symmetry in the supercell owing to the displacement of Ge and Sn atoms during the structure relaxation.

Based on the calculated band structure of $Ge_{1-x}Sn_x$ alloys, hole effective masses along [100] direction are

extracted via parabolic line fit. The HH and LH effective masses versus Sn composition are plotted in Fig. 2.

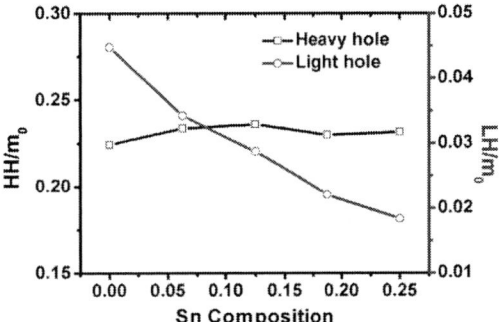

Figure 2: HH and LH effective mass of relaxation $Ge_{1-x}Sn_x$ alloy along [100] direction versus Sn composition

From Fig.2, we observe LH effective mass decreases with Sn composition increasing. But HH effective mass remains almost unchanged. The result is consistent with the work based on the empirical pseudopotential method [18]. Because DOS effective mass of HH is much larger than LH, the majority of holes occupy the HH band, and the increase in mobility due to the decrease of LH effective mass is negligible for relaxation $Ge_{1-x}Sn_x$ alloy.

Biaxial Compressive Strained $Ge_{1-x}Sn_x$ Alloy

Considering $Ge_{1-x}Sn_x$ alloys grown on (001) Ge buffers, strain tensor can be represented with Sn composition x. We performed first principles calculation with corresponding lattice constant. The variation with Sn fraction of the hole effective mass along [100] direction is given in Fig. 3.

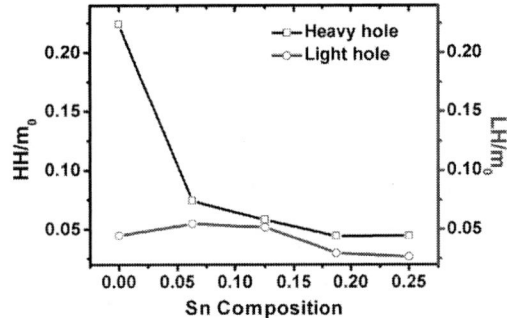

Figure 3: HH and LH effective mass of biaxial strained $Ge_{1-x}Sn_x$ alloy versus Sn composition

The higher the Sn composition, the strain effect is more significant. For the HH band, an initial steep decrease arises with the increased strain. Besides strain induced energy split between HH and LH band will result in the reduction of inter valley scattering. And then enhance of hole mobility is expected.

$Ge_{1-x}Sn_x$-Si Core-Shell Nanowire

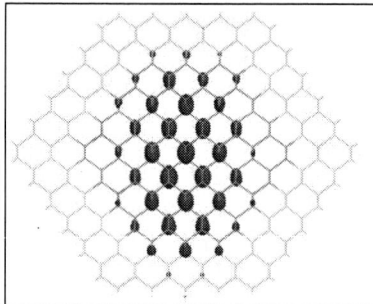

Figure 4: HOMO charge density distribution cross section for the H-passivated $Ge_{1-x}Sn_x$-Si core-shell nanowire along the [110] direction. Ge, Sn, Si atoms are respectively marked with the color of green, gray and yellow.

From the contour plots of electronic wave function near the Fermi level as shown in Fig. 4, HOMOS are localized on Ge or Sn atoms, which illustrates valence-band maximum energy state is located in the core. The result demonstrates that the type-II band offset [19] consists in $Ge_{1-x}Sn_x$-Si core shell and hole are confined in the quantum potential well of the core.

CONCLUSION

The band structure of relaxation $Ge_{1-x}Sn_x$ alloy and $Ge_{1-x}Sn_x$ layer under biaxial compressive strain in (001) plane is calculated via first principles method. Hole effective mass along [100] orientation is extracted. The analysis indicates that the obvious reduction of hole effective mass benefits from compressive strain. Moreover, $Ge_{1-x}Sn_x$-Si core-shell nanowire along the [110] direction is studied in first principles and HOMO charge density distribution reveals hole are confined in the core, so higher hole mobility is expected.

ACKNOWLEDGMENTS

The authors would like to thank Prof. Huanglong Li with the Tsinghua University for the support of first principles calculation and valuable discussion. This work is supported in part by the 863 Program (2015AA016501), Beijing Municipal Science and Technology Project (D161100001716002), and National Natural Science Foundation of China (61474072).

REFERENCES

[1] J. D. Sau and M. L. Cohen, "Possibility of increased mobility in Ge-Sn alloy system," *Phys. Rev. B* 75, 045208 (2007).

[2] O. Nakatsuka, N. Tsutsui, Y. Shimura, S. Takeuchi, A. Sakai and S. Zaima, "Mobility behavior of $Ge_{1-x}Sn_x$ layers grown on silicon-on-insulator substrate," *Jpn. J. Appl. Phys.* 49, 04DA10 (2010).

[3] H. Pérez Ladrón de Guevarai, A. G. Rodriguez, H. Navarro-Contreras, and M. A. Vidal, "Nonlinear behavior of the energy gap in Ge_xSn_{1-x} alloys at 4K," *Appl. Phys. Lett.* 91, 161909 (2007).

[4] G. Han, S. Su, C. Zhan, Q. Zhou, Y. Yang, L. Wang, P. Guo, W. Wei, C. P. Wong, Z. X. Shen, B. Cheng and Y. C. Yeo, "High-mobility GeSn P-channel MOSFETS featuring metallic source/drain and sub-370 ^{0}C process modules," in *Proc. IEDM*, pp. 402-406, 2011.

[5] S. Wirths, D. Buca and S. Mantl, "Si-Ge-Sn alloys: from growth to applications," *Prog. Cryst. Growth and Ch.* 62, pp. 1-39, 2016.

[6] Wei Lu, Ping Xie, and C. M. Lieber, "Nanowire transistor performance limits and applications," *IEEE Trans. Electron Devices* 55, 2859 (2008).

[7] L. J. Lauhon, M. S. Gudiksen, D. Wang, and C. M. Lieber, "Epitaxial core-shell and core-multishell nanowire heterostructures," *Nature* 420, pp. 57-61, 2002.

[8] J. P. Perdew, M. Ernzerhof, and K. Burke, "Generalized gradient approximation made simple," *J. Chem. Phys.* 105, 9982 (1996).

[9] B. Hammer, L. B, Hansen, and J. K. Norskov, "Improved adsorption energetics within density-functional theory using revised Perdew-Burke-Ernzerhof functionals," *Phys. Rev. B* 59, 7413 (1999).

[10] D. R. Hamann, M. Schluter and C. Chiang, "Norm-Conserving Pseudopotentials," *Phys. Rev. Lett.*, 43, 1494-1497 (1979).

[11] H. J. Monkhorst and J. D. Pack, "Special points for Brillouin-zone integrations," *Phys. Rev. B* 13, 5188 (1976).

[12] N. Naka, K. Fukai, Y. Handa and I. Akimoto, "Direct measurement via cyclotron resonance of the carrier effective masses in pristine diamond," *Phys. Rev. B* 88, 035205 (2013).

[13] R. Pekoz and J. Y. Raty, "From bare Ge nanowire to Ge/Si core/shell nanowires: A first-principles study," *Phys. Rev. B* 80, 155432 (2009).

[14] R. Beeler, R. Roucka, A. Chizmeshya, J. Kouvetakis and J. Menendez, "Nonlinear structure-composition relationships in the $Ge_{1-y}Sn_y$/Si(100) (y<0.15) system," *Phys. Rev. B* 84, 035204 (2011).

[15] M. P. Polak, P. Scharoch and R. Kudrawiec, "The electronic band structure of $Ge_{1-x}Sn_x$ in the full composition range: indirect, direct, and inverted gaps

regimes, band offsets, and the Burstein-Moss effect," *J. Phys. D: Appl. Phys.* 50, 195103 (2017).

[16] W. Huang, B. Cheng, C. Xue and Z. Liu, "Comparative studies of band structures for biaxial (100)-, (110)-, and (111)-strained GeSn: A first-principles calculation with GGA+U approach," *J. Appl. Phys.* 118, 165704 (2015).

[17] S. Gupta, R. Chen, B. Kope, H. Lin, B. Yang, A. Nainani, Y. Nishi, J. Harris and K. Saraswat, "GeSn Technology: Extending the Ge electronics roadmap," *Electron Devices Meeting* IEEE, 2012: 16.6.1-16.6.4.

[18] K. Low, Y. Yang, G. Han, W. Fan and Y. Yeo, "Electronic band structure and effective mass parameters of $Ge_{1-x}Sn_x$ alloy," *J. Appl. Phys.* 112, 103715 (2012).

[19] M. E. Pistol and C. E. Pryor, "Band structure of core-shell semiconductor nanowires," *Phys. Rev. B* 78, 115319 (2008).

A PHYSICAL CURRENT MODEL FOR JUNCTION-MODULATED TUNNELING FIELD-EFFECT TRANSISTOR WITH STEEP SWITCHING BEHAVIOR

Zhu Lv, Qianqian Huang*, Yang Zhao, Cheng Chen, Runsheng Wang and Ru Huang*

Key Laboratory of Microelectronic Devices and Circuits (MOE), Institute of Microelectronics
Peking University, Beijing 100871, China
*Corresponding Author's Email: ruhuang@pku.edu.cn, hqq@pku.edu.cn

ABSTRACT

Compared with conventional tunneling field-effect transistor (TFET), the new junction-modulated TFET (JTFET) can reliably and effectively achieve much steeper switching behavior and higher ON current by only changing the gate layout configuration. To facilitate the JTFET circuit simulation, in this work, we establish a physical tunneling current model of JTFET based on the analytical surface potential modeling. The modeled results with different device parameters are in good agreement with the simulated results by Sentaurus TCAD tools, showing the validity of proposed model.

INTRODUCTION

Recently, TFET has attracted much attention as one of the most promising candidates for low power applications [1]. Because of the band-to-band tunneling (BTBT) mechanism, TFET can break the fundamental subthreshold swing (SS) limitation of conventional MOSFET (60 mV/dec at room temperature). Additionally, low off-state current (I_{off}) can be achieved in silicon TFET. However, the on-state current (I_{on}) of the conventional silicon TFET is restricted by the poor tunneling probability of silicon. To tackle this problem, narrow bandgap materials are taken into consideration [2]. Although I_{on} can be enhanced to some extent, SS gets worse and I_{off} gets higher. Therefore, different from the materials perspective, we propose a new junction-modulated TFET (JTFET) structure recently [3], which can significantly improve SS while producing higher I_{on} and maintaining low I_{off} compared with TFET. The key point of the junction-modulated effect is to introduce the self-depleted region. JTFET has been experimentally demonstrated with the excellent performance with 36 mV/dec SS at room temperature [3].

To evaluate the circuit performance of JTFET with the junction-modulated effect, the accurate analytical current model is necessary. Although conventional TFET compact models have been proposed by some research groups [4][5], these models ignore the inversion charge in the channel and can not be applicable to the new TFET structures. We have recently proposed an analytical model for conventional TFET [6], which obtains the tunneling current based on the surface potential and can be easily extended to other TFETs. Based on our proposed model of

conventional TFET and further modeling the surface potential in self-depleted region, this study establishes the tunneling current model of JTFET for the first time.

In this work, an analytical surface potential model of bulk-Si JTFET controlled by both the gate and drain voltages is established first. Based on the surface potential, the tunneling current of JTFET can be calculated. The validity of proposed model are strongly supported by the simulation data from Sentaurus TCAD.

THE STRUCTURE OF JTFET AND SELF-DEPLETED REGION

In this work, an n-type JTFET on silicon substrate is studied. Compared with conventional TFET, JTFET has a striped gate extended into the P+ source region, leading to a larger tunneling area as shown in Figure 1 (a). Here, the width of active area is 70 nm and the finger gate width (W_G) is set as 30 nm. The oxide layer thickness (t_{ox}) is 2 nm and the effective silicon layer thickness (t_{Sieff}) is 20 nm.

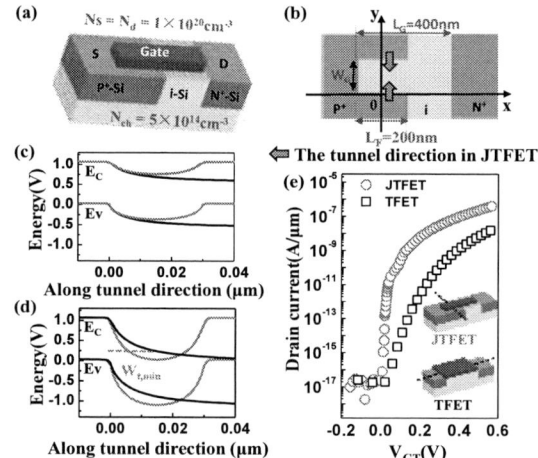

Figure 1: (a) Schematic view and (b) surface top view of JTFET; simulated surface energy band along the tunnel directions of JTFET and TFET in the off-state (c) and on-state (d); (e) simulated transfer characteristics of JTFET and TFET. ($V_{GT} = V_{GS} - V_{BTBT}$, V_{BTBT} defines as the gate voltage when devices turn on).

Besides, the $L_F \times W_G$ region under the gate is fully depleted by the build-in potential of the tunnel junctions in

978-1-5386-5309-8/18 $31.00 © 2018 IEEE

JTFET, which is called the junction-modulated effect. Because of the introduced self-depleted region, JTFET obtains higher conduction band energy and lower surface potential in the channel compared to conventional TFET (Figure 1 (c)), which results in sharper tunnel junction when devices turn on, as Figure 1 (d) shows. Moreover, narrower W_G makes stronger self-depletion effect, leading to a steeper tunneling junction when BTBT happens. Thus, JTFET can obtain much steeper SS and higher I_{ON} than conventional TFET, as plotted in Figure 1 (e). It can be seen that the key to construct the current model of JTFET is to establish the surface potential model in self-depleted region.

MODEL DERIVATION

Surface Potential Modeling

An accurate expression of the surface potential along tunnel junctions is established first. Since the device structure is symmetrical in the direction of $y = W_G/2$, the surface potential in self-depleted region is also symmetrical in this direction. Here, we define the surface potential along tunnel direction as $\varphi_{s\text{-}self}(y)$ for JTFET and $\varphi_{s0}(y)$ for conventional TFET. For their formulas, we can use the boundary conditions of the surface potential to solve the well-known Poisson equation, as follow

$$\frac{\partial^2 \varphi_{s\text{-}self}(y)}{\partial y^2} - \frac{\varphi_{s\text{-}self}(y) - (V_{GS} - V_{FB})}{\lambda^2} = \frac{qN_{ch}}{\varepsilon_{si}} \quad (1)$$

$$\lambda = \sqrt{\varepsilon_{si} t_{Sieff} t_{ox}/\varepsilon_{ox}} \quad (2)$$

where V_{GS} is the applied gate voltage, V_{FB} is the flat band voltage. However, the boundary conditions of the surface potential are different between TFET and JTFET due to their different structures. As for TFET, the boundary conditions of the surface potential can be formulated as

$$\varphi_{s0}(y)\big|_{y=0} = V_{s0} \quad (3)$$

$$\varphi_{s0}(y)\big|_{y=y_{sc}} = \varphi_{ch}(V_{DS}, V_{GS}) \quad (4)$$

$$\frac{\partial \varphi_{s0}(y)}{\partial y}\bigg|_{y=y_{sc}} = 0 \quad (5)$$

where V_{s0} is the surface potential in the source region, V_{DS} is the applied drain voltage, φ_{ch} is the channel surface potential, and y_{sc} is the width of depleted region in the channel. φ_{ch} and y_{sc} are quoted from our previous work [6]. Finally, $\varphi_{s0}(y)$ can be formulated as

$$\varphi_{s0}(y) = \left(\varphi_{ch} - V_{GS} - V_{FB} - \frac{qN_{ch}}{\varepsilon_{si}}\lambda^2\right) \cdot \cosh\left(\frac{y - y_{sc}}{\lambda}\right) \quad (6)$$

However, for JTFET, the boundary conditions of the surface potential are different from TFET owing to the existence of self-depleted region, as follow

$$\varphi_{s\text{-}self}(y)\big|_{y=0} = V_{s0} \quad (7)$$

$$\frac{\partial \varphi_{s\text{-}self}(y)}{\partial y}\bigg|_{y=\frac{W_G}{2}} = 0 \quad (8)$$

The structure of self-depleted region causes a decrease in the surface potential of JTFET. We define the surface potential difference ($\varphi_{s\text{-}self}(y) - \varphi_{s0}(y)$) as $\Delta\varphi(y)$. Noted that the key to solve $\varphi_{s\text{-}self}(y)$ is to solve $\Delta\varphi(y)$. The boundary conditions of $\Delta\varphi(y)$ can be expressed as

$$\frac{\partial \Delta\varphi(y)}{\partial y}\bigg|_{y=\frac{W_G}{2}} = -\frac{\left(\varphi_{ch} - V_{GS} - V_{FB} - \frac{qN_{ch}}{\varepsilon_{si}}\lambda^2\right)}{\lambda}$$
$$\cdot \sinh\left(\frac{\frac{W_G}{2} - y_{sc}}{\lambda}\right) \quad (9)$$

$$\Delta\varphi(y)\big|_{y=0} = 0 \quad (10)$$

Besides, $\Delta\varphi(y)$ can be linearly approximated as the TCAD simulations shown in Figure 2. Based on the linear approximation and their boundary conditions, we can get the exact formula of $\Delta\varphi(y)$, as follow

$$\Delta\varphi(y) = -\alpha \frac{\left(\varphi_{ch} - V_{GS} - V_{FB} - \frac{qN_{ch}}{\varepsilon_{si}}\lambda^2\right)}{\lambda} \cdot \sinh\left(\frac{\frac{W_G}{2} - y_{sc}}{\lambda}\right)y \quad (11)$$

where $\alpha = (0.05 + 0.01 \times W_G \times 1\mu m^{-1})$ is the correction factor.

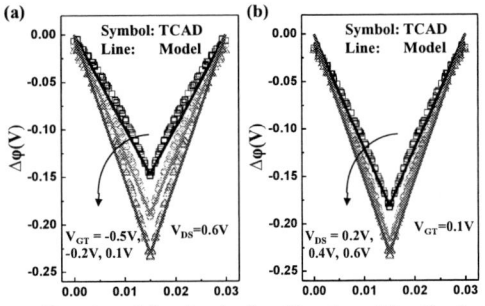

Figure 2: (a) $\Delta\varphi(y)$ along tunnel direction as a function of gate voltage, (b) $\Delta\varphi(y)$ along tunnel direction as a function of drain voltage.

As shown in Figure 2, the model results are in good agreement with the simulations. Therefore, we can obtain the expression of $\varphi_{s\text{-}self}(y)$ in JTFET. The model results of surface potential are in consistent with the simulations as plotted in Figure 3. Compared to conventional TFET, the model accurately describes the junction-modulated effect with lower surface potential in self-depleted region of JTFET.

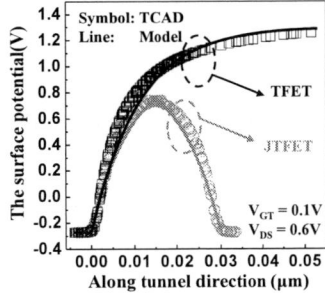

978-1-5386-5309-8/18 $31.00 © 2018 IEEE   103

Figure 3: The modeled and simulated surface potential along tunnel direction of JTFET and conventional TFET at the same gate voltage.

Likewise, the model results of $\varphi_{s\text{-}self}(y)$ under different gate voltages are also in good agreement with TCAD simulations, as plotted in Figure 4. With the gate voltage increasing, the surface potential enhances.

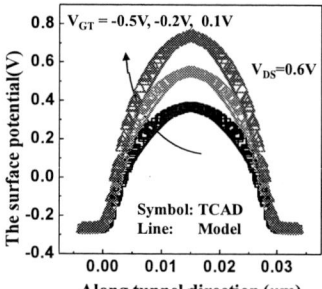

Figure 4: The modeled and simulated surface potential along tunnel direction with the variation of gate voltage.

In addition, the extent of the junction-modulated effect strongly depends on the finger gate width W_G. Narrower W_G makes stronger self-depletion, leading to the lower surface potential in the channel. As shown in Figure 4, the model results of $\varphi_{s\text{-}self}(y)$ with the variation of W_G are also in consistent with the TCAD simulations.

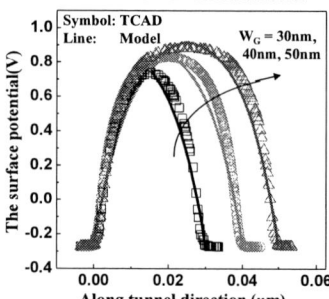

Figure 5: The modeled and simulated surface potential along tunnel direction with the variation of the finger width for $V_{DS} = 0.6V$ and $V_{GT} = 0.1V$.

Drain Current Derivation

The tunnel width (W_t), which is the key calculation factor of the tunneling current, can be derived from the surface potential in JTFET. The tunneling path is defined as the physical equal energy path between the conduction and valence band in self-depleted region. The shortest tunneling path corresponds to the largest tunneling probability, making the dominant contribution to the tunneling current. Here, we define $W_{t,min}$ as the tunnel width in the shortest tunneling path. $W_{t,min}$, where the surface potential $\varphi_{s\text{-}self}(y)$ increases by E_g/q above the source potential V_{s0} [7], can be expressed as

$$W_{t,min}(V_{DS}, V_{GS}) = y(V_{s0} + \frac{E_g}{q}) - y(V_{s0}) \quad (12)$$

Based on Kane's calculations for BTBT probability [8] and the modeled surface potential in self-depleted region, the tunneling current can be calculated as follows

$$I_{tunnel} = t_{Sieff} \cdot 2L_F \cdot E_g \cdot \frac{A_{Kane}}{B_{Kane}} \cdot \frac{1}{W_{t,min}^2} \cdot \exp\left[-B_{Kane}\left(\frac{E_g}{q}\right)^{1/2} \cdot W_{t,min}\right] \quad (13)$$

Figure 6 shows the modeled and simulated transfer characteristics. With W_G getting narrower, the switching behavior becomes steeper. The modeled results with different W_G are in good agreement with the simulated results, showing the validity of proposed model.

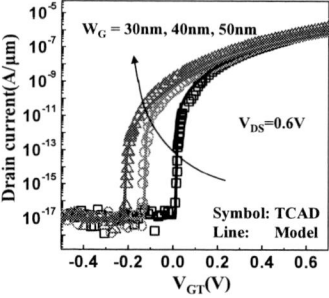

Figure 6: The modeled and simulated transfer characteristics of JTFETs with different finger width.

CONCLUSIONS

In this paper, an analytical surface potential model of Si JTFET with junction-modulated effect is proposed and verified. Based on the surface potential, the tunnel current can be accurately calculated. The results of the proposed model are in good agreement with the TCAD simulations, showing the validity of the model.

ACKNOWLEDGEMENTS

This work was supported in part by the NSFC (61421005 and 61604006) and 863 Project (2015AA016501).

REFERENCES

[1] A. M. Ionescu, et al. *Nature*, vol. 479, pp. 329-337, 2011.

[2] F. Mayer, et al. *IEEE International Electron Devices Meeting (IEDM)*, 2008, pp.1-5.

[3] Q. Huang, et al. *IEEE International Electron Devices Meeting (IEDM)*, 2012, pp.8.5.1-8.5.4.

[4] L. Zhang, et al. *IEEE Trans. Electron Devices*, vol. 59, no. 12, pp. 3217-3223, 2012.

[5] Y. Yang, et al. *IEEE Electron Device Lett.*, vol. 31, no. 7, pp. 752-754, 2010.

[6] C. Wu, et al. *IEEE Trans. Electron Devices*, vol. 61, no. 8, pp. 2690-2696, 2014.

[7] C. Wang, et al. *Science China Information Sciences*, vol. 58, pp. 1-8, 2014.

[8] E. O. Kane, *J. Phys. Chem. Solids*, vol. 12, no. 2, pp. 181-188, 1959.

BENCHMARKING OF MULTI-FINGER SCHOTTKY-BARRIER TUNNEL FET FOR ULTRA-LOW POWER APPLICATIONS

Jiadi Zhu, Qianqian Huang[*], Lingyi Guo, Libo Yang, Cheng Chen, Le Ye and Ru Huang[*]

Key Laboratory of Microelectronic Devices and Circuits (MOE), Institute of Microelectronics,
Peking University, Beijing 100871, CHINA
Phone: 86-10-62757761, Fax: 86-10-62757761, [*]E-mail: ruhuang@pku.edu.cn; hqq@pku.edu.cn

ABSTRACT

In this work, the power, performance and area of novel multi-Finger Schottky-Barrier Tunnel FET (mFSB-TFET) are benchmarked with standard MOSFET and low-power MOSFET from device and circuit level. Under the same area penalty, mFSB-TFET circuits could obtain comparable delay with significantly lower static and total power consumption by orders of magnitude than standard MOSFET circuits. The advantages are more significant at ultra-low working voltage, even compared with low-power MOSFET circuits, indicating great practical potential of mFSB-TFET for ultra-low power application.

INTRODUCTION

As the scaling of MOSFET continues, circuit power consumption dramatically increases and has become a big problem [1]. The subthreshold slope (SS) limitation of MOSFETs, which is 60mV/dec at room temperature, has made it impossible for MOSFETs to reduce its power consumption by continuously decreasing V_{DD} [2]. Silicon tunnel FET (TFET) is capable of obtaining ultra-steep SS due to its band-to-band tunneling (BTBT) mechanism and I_{OFF} several decades lower than MOSFETs. However, it suffers from small I_{ON}, which would cause large circuit delay and may limit its applications [3],[4].

To deal with such a problem, a new multi-Finger Schottky-Barrier Tunnel FET (mFSB-TFET) (Fig.1a) is proposed in our previous works [5]-[9]. By introducing comb-shaped gate and Schottky injection mechanism with self-depletion effect, the novel TFET can obtain about 3 decades higher I_{ON} and much steeper SS than conventional TFETs, experimentally demonstrating outstanding device characteristics [6]. However, to evaluate the practical potential of mFSB-TFET for ultra-low power application, a comprehensive device and circuit comparison among mFSB-TFETs, standard MOSFETs and low-power MOSFETs (high V_{th}) is also necessary.

In this paper, we benchmarked the power, performance and area of mFSB-TFET with standard MOSFET and low-power MOSFET from device and circuit level for the first time. We also discussed the trade-off in power and

performance caused by gate geometry parameters in mFSB-TFET. Results illustrate mFSB- TFET circuits

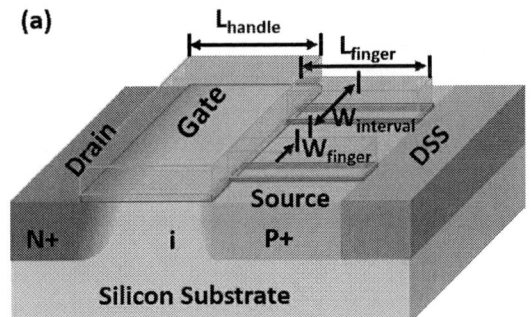

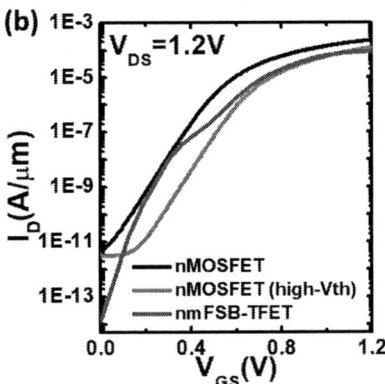

Fig.1. (a) Schematic of multi-Finger Schottky Barrier Tunnel FET (mFSB-TFET). (b) Transfer characteristics comparison among mFSB-TFET, standard MOSFET and low-power MOSFET (high-V_{th}).

could possess more than 3 decades lower static power, more than one-decade lower total power and the same delay level compared with MOSFET circuits under the same area penalty. Such advantage in power and performance is even more significant at ultra-low working voltage, demonstrating great potential of the mFSB-TFET for ultra-low power circuits.

DEVICE STRUCTURE AND SIMULATI-ON APPROACH

The mFSB-TFET has a comb-shaped gate with multi-fingers (Fig.1a). A dopant-segregated Schottky (DSS) region and several heavily doped P+ regions act as the source regions for Schottky injection and BTBT mechanism respectively. N+ pockets are also introduced in the source tunnel junction for more abrupt junction.

The simulation of mFSB-TFET is carried out using *Synopsys Sentaurus TCAD simulator*. Related parameters are calibrated with experimental data [5]. A Look-up-Table model is developed based on the results for further circuit simulation with HSPICE in this work. Standard MOSFET and low-power MOSFET (high V_{th}) of the same footprint under the same technology node are simulated for the mFSB-TFET benchmark.

PERFORMANCE, POWER AND AREA BENCHMARK

A transfer characteristic comparison among mFSB-TFET, standard MOSFET and low-power MOSFET is demonstrated in Fig.1b. The I_{ON} of mFSB-TFET is slightly lower than MOSFET due to the smaller Schottky junction width than the gate width of MOSFET with the same device area, but can be comparable with low-power MOSFET. Benefiting from the self-depletion effect, I_{OFF} can be more than two decades lower than MOSFETs. Such advantage still exists even compared with low-power MOSFETs. The mFSB-TFET also possesses steeper SS due to the BTBT mechanism. Accordingly, by separating the ON-OFF-transition mechanism, mFSB-TFET could achieve outstanding device characteristics.

Since the source and drain doping types are different in mFSB-TFET, the circuit area of mFSB-TFET might be a little larger than the MOSFET even with the same device footprint. Therefore, the circuit performance and power benchmark of mFSB-TFET needs to take the circuit area into consideration. Taking NAND circuit as an example, with the same circuit area, the static power of the mFSB-TFET circuit could be about three decades lower than that of standard-MOSFET circuit (Fig.2). The total power of

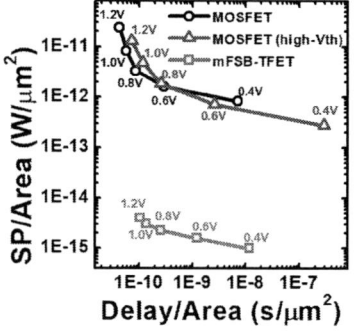

Fig.2. Comparison of static power (SP) and delay at different V_{DD} of NAND circuits based on mFSB-TFET, standard MOSFET and low-power MOSFET respectively.

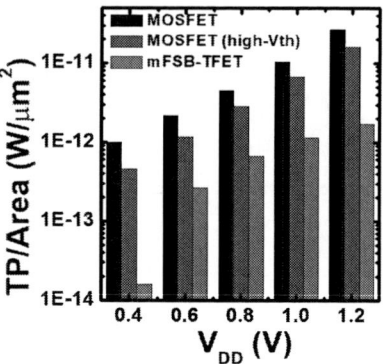

Fig.3. Comparison of total power (TP) at different V_{DD} of NAND circuits based on mFSB-TFET, standard MOSFET and low-power MOSFET respectively.

the mFSB-TFET circuit could also be more than one decade lower than that of standard-MOSFET circuit (Fig.3). The power advantages are due to the lower I_{OFF} and smaller gate capacitance of the mFSB-TFET [10]. As shown in Fig.2, the mFSB-TFET circuit also possesses similar delay level as standard MOSFET circuit. Moreover, for the smaller working voltage, the ultra-steep SS of mFSB-TFET ensures even more significant advantages in delay and total power consumption. As shown in Fig.2 and Fig.3, at operation voltage of 0.4V, the total power of mFSB-TFET circuit would be almost two decades lower than standard MOSFET circuits, and the circuit delay would be significantly smaller than low-power MOSFET circuit at the same time.

GEOMETRY IMPACTS

Comb-shaped gate geometry parameters could influence the device characteristics of mFSB-TFET and might also influence its circuit behavior. We compared the impacts on I_{ON}, I_{OFF}, and V_{TH}, when finger length (L_{finger}), handle length (L_{handle}), finger width (W_{finger}) and finger interval ($W_{interval}$) changes respectively (Table I). Results indicate W_{finger} has the dominant influence on I_{OFF} and V_{TH} of the device, and can be considered as the key geometric parameter. Hence, we further studied the circuit behavior

TABLE I
GATE GEOMETRY IMPACTS ON THE
PERFORMANCE OF mFSB-TFET

Parameter	Relative Changes ($\Delta X/ X$)	$\Delta I_{ON}/I_{ON}$	$\Delta I_{OFF}/I_{OFF}$	$\Delta V_{TH}/V_{TH}$

L_{finger}	31.6%	11.8%	-5.94%	0.722%
L_{handle}	33.3%	36.1%	-22.5%	0.144%
W_{finger}	33.3%	12.0%	192%	-9.2%
$W_{interval}$	33.3%	-7.23%	13.2%	1.46%

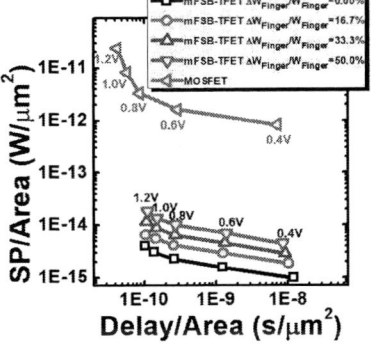

Fig.4. Comparison of static power (SP) and delay at different V_{DD} of NAND circuits based on mFSB-TFET with different finger widths and standard MOSFET.

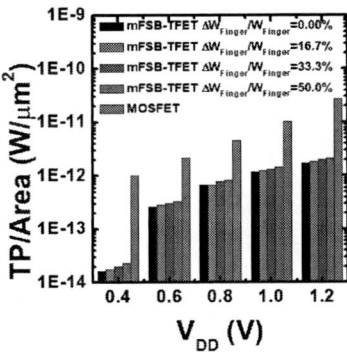

Fig.5. Comparison of total power (TP) at different V_{DD} of NAND circuits based on mFSB-TFET with different finger widths and standard MOSFET.

of mFSB-TFET with different W_{finger}, in order to figure out the impacts on the power consumption and performance of mFSB-TFET circuit.

When W_{finger} increases, both static and total power would increase (Fig.4&5), which is because larger W_{finger} results in weaker self-depletion, namely higher I_{OFF}. In the meanwhile, as shown in Fig.4, the circuit delay would decrease due to the increase in Schottky current. In this case, the mFSB-TFET circuit has comparable delay with MOSFET circuit and can still outweigh MOSFET circuits significantly in power consumption due to its much lower I_{OFF} and smaller gate capacitance. Such superiority is more notable at ultra-low working voltage. The static and total power consumption of mFSB-TFET circuit can both be about two decades lower than MOSFET circuit at 0.4V V_{DD}, demonstrating its outstanding potential for ultra-low power circuits.

CONCLUSION

This paper benchmarked the power, performance and area of a new multi-Finger Schottky Barrier Tunnel FET (mFSB-TFET) with standard MOSFET and low-power MOSFET from device and circuit level. Under the same circuit area, the power consumption of mFSB-TFET circuits could be much lower than low-power MOSFET circuits with comparable delay even for ultra-low voltage, indicating that the mFSB-TFET is a promising candidate for ultra-low power applications.

ACKNOWLEDGEMENTS

This work was supported in part by NSFC under Grant 61421005 and Grant 61604006 and in part by the 863 Project under Grant 2015AA016501

REFERENCES

[1] M. Mitchell Waldrop, "The Semiconductor Industry Will Soon Abandon Its Pursuit of Moore's Law. Now Things Could Get a Lot More Interesting." *Nature*, vol. 530, 2016, pp.144-147.

[2] A. M. Ionescu, *et al.*, "Tunnel field-effect transistors as energy efficient electronic switches," *Nature*, vol. 479, 2011, pp. 329–337

[3] R. Pandey, *et al.*, "Opportunities and challenges of tunnel FETs," *IEEE Trans. Circuits Syst. I*, vol. 63, 2016, pp. 2128–2138.

[4] J. Zhu, *et al.*, "Design and Simulation of a Novel Graded-Channel Heterojunction Tunnel FET with High I_{ON}/I_{OFF} Ratio and Steep Swing," *IEEE Electron Device Lett.*, vol. 38, 2017, pp. 1200-1203.

[5] Q. Huang, *et al.*, "A novel Si tunnel FET with 36mV/dec subthreshold slope based on junction depleted-modulation through striped gate configuration," *IEEE Inter. Electron Devices Meeting (IEDM)*, San Francisco, CA, USA, Dec. 10-13, 2012, pp.8.5.1-8.5.4.

[6] R. Huang, *et al.*, "High Performance Tunnel Field-effect Transistor By Gate and Source Engineering," *Nanotech.*, vol. 50, 2014, pp.505201.

[7] Q. Huang, *et al.*, "Comprehensive Performance Re-assessment of TFETs with a Novel Design by Gate and Source Engineering from Deivce/Circuit Perspective," *IEEE Inter. Electron Devices Meeting (IEDM)*, San Francisco, CA, USA, Dec. 15-17, pp.13.3.1-13.3.4.

[8] Q. Huang, *et al.*, "Self-Depleted T-gate Schottky Barrier Tunneling FET with Low Average Subthreshold Slope and High I_{ON}/I_{OFF} by Gate configuration and Barrier Modulation." *IEEE Inter. Electron Devices Meeting (IEDM)*, Washington, DC, USA, Dec. 5-7, 2011. pp.16.2.1-16.2.4,

[9] R. Huang, *et al.*, "Multi-Finger Schottky-Barrier Tunneling FET with Hybrid Operation Mechanism for Steep Transition and High On Current (invited)," *IEEE Inter. Conf. on Electron Devices and Solid-State Circuits (EDSSC)*, Singapore, India, Jun. 1-4, 2015. pp. 61-62.

[10] Q. Huang, *et al.*, "Device Physics and Design of T-Gate Schottky Barrier Tunnel FET with Adaptive Operation

Mechanism," *Semicond. Sci. and Tech.*, vol. 29, pp.095013, 2014.

STUDY ON MICROSCOPIC MODEL OF RESISTIVE SWITCHING IN AMORPHOUS TANTALUM PENTOXIDE FROM FIRST-PRINCIPLE CALCULATIONS

Lin Bao[1], Yichen Fang[1], Zongwei Wang[1], Jian Kang[1], Yuchao Yang[1], Jintong Xu[2], Yimao Cai[1]* and Ru Huang[1]

[1] Institute of Microelectronics, Peking University, 100871 Beijing, China

[2] Key Laboratory of Infrared Imaging Materials and Detectors, Shanghai Institute of Technical Physics, Chinese Academy of Sciences, Shanghai 200083, China

*Corresponding Author's Email: caiyimao@pku.edu.cn

ABSTRACT

As one of the emerging memory device, Ta_2O_5-based RRAM has shown various outstanding features, such as low power consumption, high reliability and excellent scalability. In Ta_2O_5 RRAM, the formation and the migration of oxygen vacancies are crucial to the resistive switching characteristics, reliability and uniformity of the device. In this paper, we used the first-principle simulation to study the properties of oxygen vacancies including forming energy and migration barrier in amorphous tantalum pentoxide. The β-Ta_2O_5 supercell was built and annealing algorithm was used for generating amorphous Ta_2O_5 in Material Studio. The band structure and DOS of both materials were studied by Vienna Ab initio simulation program (VASP). We also calculated the formation energy and migration barrier of oxygen vacancies, and discussed the mechanism of conductive path formation in amorphous Ta_2O_5. Finally, we proposed a new microscopic model of resistive switching mechanism in amorphous Ta_2O_5 RRAM device.

INTRODUCTION

In recent years, thanks to the good endurance, uniformity, and retention, TaO_x-based RRAM has been considered as a promising emerging nonvolatile memory. It is widely accepted that the formation and migration of oxygen vacancies determine the resistive-switching of Ta_2O_5 RRAM devices, so investigating the properties of oxygen vacancies are essential to understand the underlying mechanism of resistive switching of the devices and hence helpful to enable high performance RRAM design. In literatures, some researchers used first principle simulation to analyze the feature of the crystal Ta_2O_5 material. However, the function layer of Ta_2O_5 RRAM is usually in amorphous phase, which may lead to obvious difference in the properties of oxygen vacancies. In this paper, we studied main parameters of amorphous Ta_2O_5 related to properties of oxygen vacancies, such as DOS, band structure, formation energy and migration barrier. At last, we proposed a microscopic model of resistive-switching in amorphous Ta_2O_5 based RRAM device.

One critical challenge for simulating the resistive switching property in amorphous Ta_2O_5 is that the structure information of low-temperature Ta_2O_5 (L-Ta_2O_5) is in controversial [4]. In our work, we chose β-Ta_2O_5 as the basic research object which has an orthorhombic structure with lattice parameters a=6.217Å, b=3.677 Å, and c=7.794 Å, sp.gr. *Pccm*, as shown in Figure1. (a) [5]. Then we built amorphous Ta_2O_5 with annealing algorithm as shown in Figure1. (b). With these structures, we calculated electronic structures of Ta_2O_5, the formation energy and migration barrier of oxygen vacancies.

DOS AND BAND SRUCTURE

Figure2. (a) and (b) show the electronic of states and the band structure in FBZ of β-Ta_2O_5 and amorphous Ta_2O_5 respectively. We used PAW_PBE pseudopotentials with a plane-wave cutoff of 300eV in the simulation.

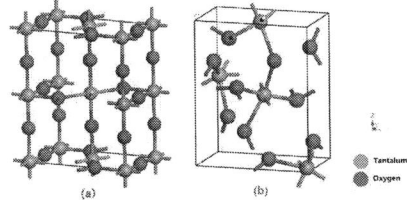

Figure 1: (a) β-Ta_2O_5 cell and (b)amorphous Ta_2O_5 cell

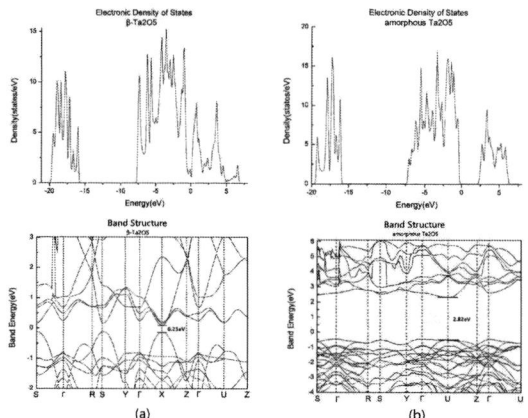

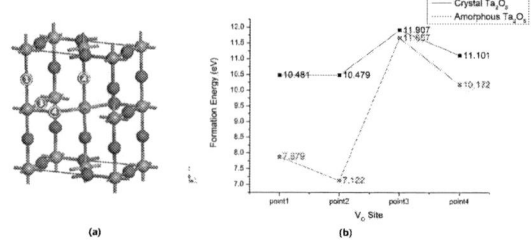

Figure 2: DOS and band structure of (a) β -Ta$_2$O$_5$ and (b) amorphous Ta$_2$O$_5$

Because of the inherent drawback of density functional theory (DFT), the bandgap was underestimated [6]. Here we concentrate on the changes of bandgap rather than the matching of experimental value. From the Figure2. (a), we can see that β -Ta$_2$O$_5$ has a direct bandgap of 0.23eV. This value is consistent with [7] and [8]. As shown in Figure2. (b), amorphous Ta$_2$O$_5$ also has direct bandgap, but the magnitude is larger than β-Ta$_2$O$_5$, reaches 2.82eV. The changes of bandgap indicate the changes of conduction. From band structure, we can speculate that the β -Ta$_2$O$_5$ is more conductive than amorphous Ta$_2$O$_5$, so that one reason of resistive-switching may be the changes between two phases of Ta$_2$O$_5$.

FORMATION ENERGY OF OXYGEN VACANCIES

we calculated the formation energy of oxygen vacancies in β-Ta$_2$O$_5$ and amorphous Ta$_2$O$_5$. As we can see from Figure 3. (a), there are four spatial positions of oxygen in β-Ta$_2$O$_5$, which denoted as point1 to point4 respectively. Because of the symmetry destruction in amorphous Ta$_2$O$_5$, it is hard to calculate formation energy at all spatial positions. We chose four spatial position randomly in amorphous Ta$_2$O$_5$ to calculate the formation energy and the average formation energy. In view of the macroscopic characteristics of materials are the average of the microscopic features, it is meaningful to use statistical approach to study amorphous Ta$_2$O$_5$.

Figure3(a)four spatial positions of oxygen in β -Ta$_2$O$_5$, (b)formation energy at four spatial position in β -Ta$_2$O$_5$ and in amorphous Ta$_2$O$_5$

From Figure 3 (b), it is easy to see that in β-Ta$_2$O$_5$, there are little difference of formation energy between four spatial position, the average vacancies formation energy is 10.992eV, so we can draw a conclusion that in the β-Ta$_2$O$_5$ oxygen vacancies are generated randomly. In the amorphous Ta$_2$O$_5$, the average vacancies formation energy is 9.208eV. Obviously, because of the destruction of crystal lattice, it is easier to generate oxygen vacancies in the amorphous Ta$_2$O$_5$ than in the β-Ta$_2$O$_5$. High vacancies density will decrease the average distance between the defects so that increase Poole-Frenkel effect in the material, and improve the body conductance of the amorphous Ta$_2$O$_5$. In view of that β-Ta$_2$O$_5$ has higher conductance than amorphous Ta$_2$O$_5$, we guess that the formation of β-Ta$_2$O$_5$ will decrease the resistance of the amorphous Ta$_2$O$_5$ layer.

MIGRATION BARRIER OF OXYGEN VACANCIES

The formation of conductive filament is the result of the migration of oxygen vacancies, so it is necessary to study the migration of vacancies for the understanding of resistive switching. By using linear synchronous transit (LST) method, we calculated the migration barriers between four spatial positions in β-Ta$_2$O$_5$. These data summarized in TABLE I.

In the TABLE I, symbol ∞ denotes that migration barrier higher than 7.5eV, and we deem that the barrier is so high that vacancies cannot pass over. Because of the significant asymmetry of the migration barriers, although oxygen vacancies are generated randomly in the β-Ta$_2$O$_5$, the quantity of vacancies at four spatial points is different. The vacancies generated at point2 and point3 are easy to move to point1 or point4, on the contrary, the vacancies at point1 and point4 are hard to move away. From these data, we can conclude that the most probably migration path in the β -Ta$_2$O$_5$ is point4 to point3 to point4, the highest barrier of this path is 2.31eV. In consideration of the LST method will overestimate barrier height, we speculate the highest migration barrier of the path is between 1eV to 2eV.

TABLE I. MIGRATION BARRIER BETWEEN FOUR SPATIAL POINTS

Start / End	Point1	Point2	Point3	Point4
Point1	∞	2.09eV	0.35eV	3.41eV
Point2	7.29eV	∞	2.77eV	6.05eV
Point3	4.15eV	1.82eV	∞	2.31eV
Point4	2.57eV	0.01eV	0.09eV	∞

Because of the destruction of crystal lattice, there are a lot of defects in amorphous Ta_2O_5 such as dislocations and interfaces. The migration barriers along the interfaces are decreased so that the oxygen vacancies will move along the interfaces easily. So we can guess that in amorphous Ta_2O_5, the deviation of migration barrier is larger than that in β-Ta_2O_5, and the dislocations and interfaces will form possible migration paths of oxygen vacancies.

RESISTIVE SWITCHING MODEL BASED ON LOCALIZED CRYSTALIZATION

From the band structure, we drew a conclusion that the β-Ta_2O_5 is more conductive than the amorphous Ta_2O_5. In fact, the formation of localized β-Ta_2O_5 is the main reason of the decay of tantalum capacitor [9]. For the reason that the β-Ta_2O_5 is the low temperature phase of tantalum pentoxide, it is easy to generate localized β-Ta_2O_5 microcrystals in amorphous Ta_2O_5 layer. On one hand, these microcrystals increase the body conductance of the amorphous Ta_2O_5 layer; on the other hand, localized β-Ta_2O_5 introduces a lot of interfaces, and these interfaces accelerate the migration of oxygen vacancies. These factors result in the increase of conducting current, so that generate Joule heat which accelerates the crystallization of Ta_2O_5. This positive feedback procedure in amorphous Ta_2O_5 result in the formation of conductive filaments so that decrease the resistance of device.

CONCLUSION

In summary, by using first-principle-calculation, we investigated the microscopic model of resistive-switching in amorphous Ta_2O_5 based RRAM device. The DOS and band structure of β-Ta_2O_5 and amorphous Ta_2O_5 were obtained and compared through simulation as well as the formation energy and migration barrier were studied, which play important role in determine the migration path in the β-Ta_2O_5. As a result, a new resistive switching microscopic model was put forward, which takes into account the effect of localized β-Ta_2O_5 microcrystals in resistive switching behavior of amorphous Ta_2O_5 based RRAM device. These localized crystallizations not only increased the body conductance of the amorphous Ta_2O_5 layer, but also introduced more interfaces and strains in the material which would accelerate the formation and migration of oxygen vacancies.

ACKNOWLEDGEMENTS

This work was supported in part by National Natural Science Foundation of China (no. 61421005, 61376087, and 61574007), the National High Technology Research and Development Program of China (no. 2011AA010401 and 2011AA010402), Key Laboratory of Infrared Imaging Materials and Detectors, Shanghai Institute of Technical Physics, Chinese Academy of Sciences (IIMDKFJJ-14-08).

REFERENCES

[1] S Cho, C Yun, S Tappertzhofen, A Kursumovic, S Lee et.al. *Nature Communications*, 2016, 7:12373

[2] J Lee, E Kioupakis, W Lu. *Aps March Meeting*, 2016

[3] J Lee, W Lu, E Kioupakis. *Applied Physics Letters*, 2014, 118 (20) :269

[4] Sung Hyun Jo, Kuk-Hwan Kim, and Wei Lu. *Nano Lett*, 2009, 9:870-874.

[5] L.A. Aleshina, S.V. Loginova. *Crystallography Reports*, 2002, 47(3): 415-419.

[6] K Ohno, K Esfarjani, Y Kawazoe. *Computational Materials Science*. Springer Berlin Heidelberg, 1999, 4 (95) :373–382.

[7] R. Nashed, W.M. Hassan, Y. Ismail, N.K. Allam. *Physical Chemistry Chemical Physics*, 2003, 15(5):1352-7.

[8] T. Gu, Z. Wang, T. Tada, and S. Watanabe. *Journal of Applied Physics*, 2009, 106(10):262907.

[9] G Sethi, M Olszta, J Li et.al. *Conference on Electrical Insulation & Dielectric Phenomena*, 2008: 815-818.

PERFORMANCE EVALUATION OF TUNNELING FIELD EFFECT TRANSISTOR ON TERAHERTZ DETECTION

Q. Yang[1], J. Zhang[1], C. Zhu[1], X. Lin[2], F. Yan[1] and X. Ji[1]*

[1]Institute of the electronic Science and Engineering, Nanjing University, China

[2] Beijing Institute of space mechanics and electricity, China

*Corresponding Author's Email: xji@nju.edu.cn

ABSTRACT

Tunneling field-effect transistor (TFET) with integrated circuit manufacture process is capable of rectifying high frequency radiations in THz region. As an efficient device to detect Terahertz signal. In this paper, we demonstrate a Si-based TFET device for THz imaging. The static characteristics and THz performance are presented by using technology computer-aided design simulation. For an optimally designed TFETs with SS of 40.3mV/dec, it has a high responsivity of 4300V/W to a 1THz radiation. Compared to the traditional CMOS Terahertz detector, TFET-based one has faster response rate and higher detect efficiency, suggesting TFET a promising device for the low-power Terahertz imaging.

INTRODUCTION

CMOS terahertz (THz) detectors based on Si-MOSFETs have multiple advantages including a low noise equivalent power (NEP) comparable to other room temperature THz detectors. However, conventional Si-MOSFETs has a relatively low frequency response at the room temperature due to the channel carriers driven by drift and diffusion mechanisms. Tunneling field-effect transistor (TFET) operated by band-to-band tunneling (BTBT) have the plausible feature of the high voltage response [1-3]. Besides, as the tunneling time of an electron is as short as 1e-15s, TFETs has a theoretical response quick enough for relatively higher frequency signals. There have been a lot of work on TFET design, simulation, fabrication and RF application. In this paper, we investigate the Terahertz response of Si-based TFET with PIN channel doping profile by technology computer-aided design (TCAD) tools to understand the performance of TFETs as a THz detector.

TFET Sensing Principle Analysis:

Fig 1 shows the cross-sectional view of the TFETs. An optimal set of parameters are determined for PIN TFETs with gate oxide thickness T_{OX} of 0.8nm, source doping concentration of 5e19cm^{-3} and junction depth of 40nm. The structure with the certain circumstances can achieve a subthreshold swing (SS) of 40.3 mV/dec. When the THz radiation (with its polarization vector in the paper plane) is incident on the source terminal of the TFET device, the device under the gate voltage rectify the signals and generate a DC source–drain which could be detected.

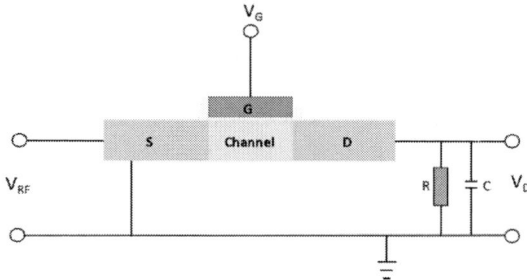

Fig 1: The cross-sectional view of TFETs for THz imaging.

The THz signal detection principles are as follows. The tunneling current has a relationship with the electric field E in the junction between source and channel as follows:

$$I_t = \frac{\sqrt{2m^*}q^3 E}{4\pi^3 \hbar^2} \frac{V_{appl}}{E_g^{1/2}} \exp(-\frac{4\sqrt{2m^*}}{3q\hbar E}E_g^{3/2}) \qquad (1)$$

here m* is effective electron mass, V_{appl} is the voltage applying to tunneling junction , ℏ is Planck constant, E_g is the forbidden band width. Using the second-band Taylor expansion, Eq. (1) can be rewritten as follows:

$$I_t = \frac{\sqrt{2m^*}q^3 E}{4\pi^3 \hbar^2} \frac{V_{appl}}{E_g^{1/2}}[M + N(E - E_0) + Q(E - E_0)^2] \qquad (2)$$

where

$$M = \exp(-\frac{4\sqrt{2m^*}E_g^{3/2}}{3q\hbar E_0}),$$

$$N = \exp(-\frac{4\sqrt{2m^*}E_g^{3/2}}{3q\hbar E_0}) * \frac{4\sqrt{2m^*}E_g^{3/2}}{3q\hbar E_0^2},$$

$$Q = \frac{4\sqrt{2m^*}E_g^{3/2}}{3q\hbar E_0^3} \exp(-\frac{4\sqrt{2m^*}E_g^{3/2}}{3q\hbar E_0})(\frac{2\sqrt{2m^*}E_g^{3/2}}{3q\hbar E_0} - 1)$$

when signal V_{RF} input in the source of TFET device, shown in Fig. 1, there will induce a junction electric shift in the channel:

$$E - E_0 = \frac{V_{RF}}{d} \tag{3}$$

here d is the junction width, so we can present the rectified DC signal as:

$$I = V_{RF}^2 \times \frac{\sqrt{2m^*}q^3}{4\pi^3\hbar^2} \times \frac{V_{app1}}{2d^2 E_g^{1/2}}[\frac{V_{app1} \times Q}{d} + N] \tag{4}$$

DC signals could be detected in the drain terminal.

Simulation results and discussion:

TCAD simulations were operated by Silvaco Atlas tool to understand the THz response of TFETs. In the simulation, the proper structure parameters of devices are selected and are listed in Table I. The gate length, width and thickness of TFETs are 50nm, 1μm and 40nm respectively, the gate oxide thickness is selected as 0.8nm while the doping concentrations of Source (P$^+$), Substrate (I) and Drain (N) terminals is 5e19cm^{-3}, 1e16cm^{-3} and 1e19cm^{-3} respectively. A traditional MOSFET structure is also listed for comparison.

Table I. Parameters of PIN-TFET and MOSFET used for detecting Terahertz signals.

	PIN-TFETs	MOSFETs
Gate Length	50nm	0.18μm
Gate Width	1μm	1μm
Channel Thickness	40nm	60nm
Oxide Thickness	0.8nm	2nm
Source Doping	5e19cm^{-3}	5e19cm^{-3}
Drain Doping	1e19cm^{-3}	1e19cm^{-3}
Channel Doping	1e16cm^{-3}	1e16cm^{-3}

The simulated I_d-V_g curves of two devices are shown as Fig 2. For PIN-TFETs, SS is 40.3mV/dec, V_{th} is 0.9V and I_{ON} is 1e-6A while for MOSFETs, those are 82.3mV/dec, 0.5V and 1e-4A.

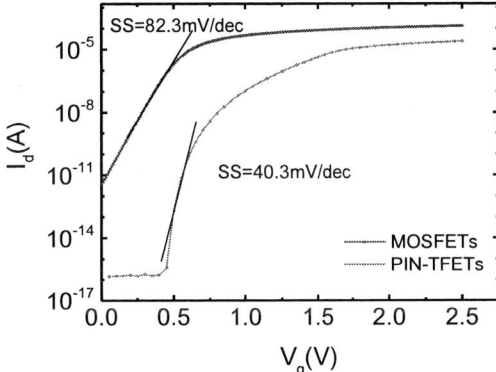

Fig 2: Id-Vg curves for MOSFETs and PIN-TFETs.

Fig 3 shows the voltage response R_v under the different gate bias of TFETs. When $V_g < V_{th}$, the width of barrier between channel and p$^+$-section is broad which means the tunneling probability is very small and no voltage response achieves as well. When V_g increase to V_{th}, the width of barrier becomes narrow so that a large amount of electron tunnels from valence band to conduction band resulting in the R_v boosts. As the Vg continues to increase, the channel resistor reduces and the R_v decreases rapidly.

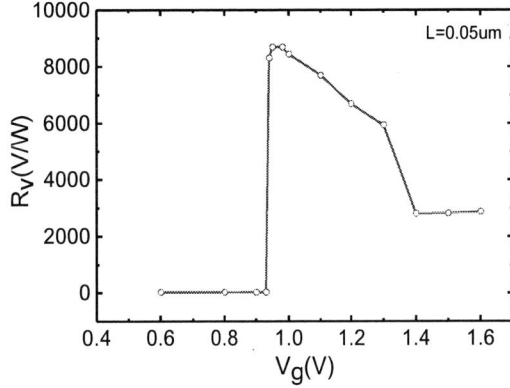

Fig 3: The voltage response with gate voltage V_g for PIN-TFETs under the input signal at 0.5THz.

Fig 4 shows the maximum voltage response R_v in the THz range from 0.5THz to 1.5THz for TFETs with L=50nm. It is seen that R_v of PIN-TFET decreases as the frequency of the imported signal increases. The R_v at 1THz signal is only half of that at 0.5THz signal. Even so, R_v at 1.5THz still have a value of 2.7kV/W. Compared to conventional Si-MOSFETs, R_v decreases much slower with frequency in the range of 0.5THz~1.5THz.

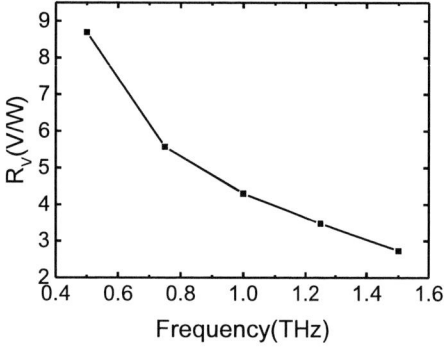

Fig 4: R_v of PIN-TFETs for signals in the frequency range from 0.5THz to1.5THz.

Fig 5 shows a comparison of the voltage response to a 0.5THz signal between PIN-TFETs and MOSFETs. The response time of the PIN-TFETs is about 1e-11s which is much faster than that of MOSFETs which means it is much fast for a PIN-TFET detector to detect Terahertz signals than a traditional MOSFET-based THz detector.

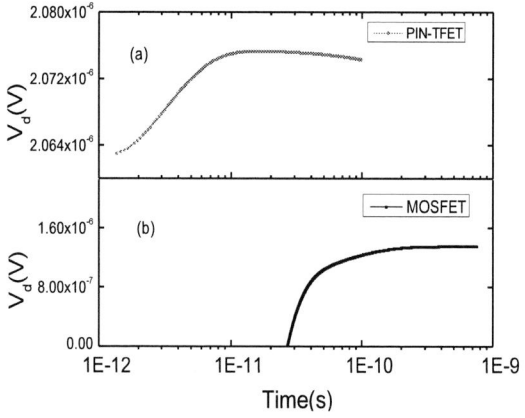

Fig 5: The rectified drain voltage with time for (a) PIN TFETs and (b) MOSFETs.

Conclusion

A Terahertz detector based on PIN-TFETs is proposed and its characteristics are analyzed through numerical and analytical modeling. TCAD simulation results show that PIN-TFETs has a quicker switching response rate and a higher voltage response to Terahertz signals compared to these for MOSFETs. For an optimally designed TFETs with SS of 40.3mV/dec, it has a high responsivity of 4300V/W to a 1THz radiation. For practical applicants, in order to achieve higher voltage response, the length of Gate should be as short as possible. Also, designing new structures of TFETs or optimizing the parameters further may achieve improvements on the voltage response and response rate.

Acknowledgements

This work was supported by the National Key R&D Program of China (No. 2016YFA0202102), CAST Project, China (No. 08201601), National Key R&D Program of China (No. 2016YFB-0402403), and the National Science Foundation for Young Scholars of China (No.61404072).

References

[1] W. Y. Choi, B. G. Park, J. D. Lee and T. J. K. Liu. *IEEE Electron Device Lett.*, vol. 28, 2007, pp. 743-745.

[2] K. K. Bhuwalka, J. Schultze and I. Eisele. *IEEE Electron Devices*, 2005, vol. 52, pp. 1541-1547.

[3] R. Narang, M. Saxena, R. S. Gupta and M. Gupta. *Journal of Semiconductor Technology and Science*, vol. 13, 2013, pp. 224-236.

Study on NBTI Improvement of HfO$_2$-based 14 nm P-type FinFET with Post High-k Deposition Thermal Treatment

Wen Wang[1,2], Yongmin Zhao[1,2], Danniel Feng[2], Hao Jiang[2], Jun Luo[1,3]

[1] School of Microelectronics, University of Chinese Academy of Sciences (UCAS), Beijing, China
[2] Semiconductor Manufacturing International Corporation (SMIC), Shanghai, China
[3] Institute of Microelectronics, Chinese Academy of Sciences (IMECAS)
Correspondent author E-mail: Keira_Wang@smics.com

Abstract

Negative Bias Temperature Instability (NBTI) of P-type FinFET device remains a major device reliability concern in sub-16/14 nm nodes. In this paper, the impact of HKMG stack process on the NBTI performance is the focus. Especially, post high-k deposition thermal treatment including PDA and PCA process is utilized to improve the NBTI performance of HfO$_2$-based P-type FinFET device. The intrinsic mechanism for the reliability improvement using different annealing temperature is disclosed. The intrinsic correlation between annealing temperature and reliability improvement is also investigated.

Results show that both PDA and PCA processes can lead to improved NBTI performance due to the restoration of Si/IL interface trap states, as well as the IL regrowth. As the increasing of annealing temperature, the improvement is enhanced. Furthermore, NBTI performance is more sensible to the variation of PDA temperature than PCA temperature. Finally, by optimizing the joint PDA & PCA thermal treatment processes, the NBTI performance is improved significantly.

Keywords—FinFET; reliability; NBTI; PDA; PCA.

Introduction

FinFET has become dominant device architecture in 14 nm node and beyond, and high-k metal-gate (HKMG) is applied for its advantages on increasing drive current, reducing gate leakage current as well as source-drain leakage current [1]. However there are also challenges such as thermal instability of metal gate and poor reliability of HK dielectric.

Negative Bias Temperature Instability (NBTI) remains a major reliability concern for P-type FinFET device [2]. Compared with its planar counterparts with only <100> crystal interface orientation in the substrate, FinFET device with fully depleted channel has two kinds of crystal interface orientations, which for fin top it is <100> and for fin sidewalls it is <110>. It is widely accepted that interface trap generation and holes trapping in the pre-existing defects of the HK/IL interface is the main cause for NBTI [3]-[4]. As the interface state density of <110> interface is larger than <100>, this fact may lead to more severe NBTI degradation for FinFET structure [5].

HKMG process loop has significant impact on device reliability. A dual-layer insulator is applied in HKMG stack, several angstroms of thermal SiO$_2$ is grown before HK deposition to alleviate lattice mismatch between HK and Si substrate which may lead to severe charge trapping. Efforts should be made to develop the quality of IL and HK as well as the interface in order to improve device reliability [6].

In this work, thermal treatment process to improve NBTI performance is manifested, including post high-K deposition anneal (PDA) and post high-K capping anneal (PCA).

Experiments

A schematic FinFET device structure investigated in this work is shown in Fig.1, it is fabricated in typical gate last FinFET process flow. First, fin formation is completed with self-aligned-double-patterning (SADP) followed by shallow trench isolation (STI) process and dummy gate formation. Then stress engineering including SiGe epitaxy and sourse/drain dopant is processed. After which is the dummy gate removal and HKMG loop, including SiO$_2$ IL, HfO2 dielectric and work function metal layers deposition. Finally back end of line (BEOL) loops are done.

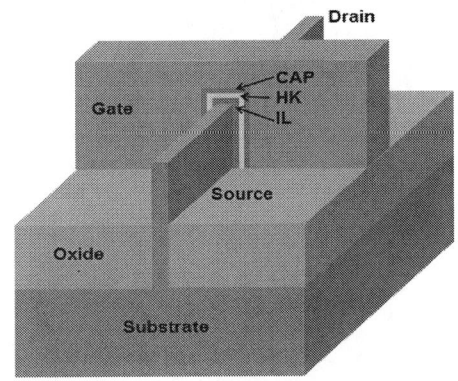

Figure 1: Schematic of FinFET device structure.

During insulator deposition process, we have 4 groups of devices: (A) using normal low temperature thermal treatment after HfO$_2$ deposition; (B) using only high temperature PDA process, annealing is applied with different temperature after HfO$_2$ deposition; (C) using only high temperature PCA process, annealing is applied with different temperature after HfO$_2$ TiN capping layer deposition; (D) using high temperature PDA & PCA combined process.

The device NBTI performance is tested with a measure-stress-measure (MSM) method with the temperature kept at 125℃, a negative bias is applied at the gate, and the source/drain are grounded.

Results and Discussion

As PMOS NBTI will lead to an increase of threshold voltage and gate leakage current, a decrease of drive current and transconductance, Vth is an important parameter to characterize NBTI performance. Fig.2. demonstrated the normalized Vth shift of devices using different thermal treatment process. Devices with normal low temperature thermal treatment process have worst Vth shift, devices with either high temperature PDA or high temperature PCA process have much less Vth shift, while devices with high temperature PDA & PCA combined process have the most favorable Vth shift. This indicates that PDA and PCA process can both improve NBTI performance of P-type FinFET significantly, and PDA & PCA combined process can even improve NBTI performance better.

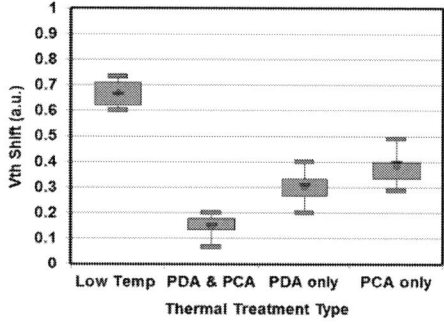

Figure 2: Normalized Vth shift of devices with normal low temperature thermal treatment, high temperature PDA / PCA only and high temperature PDA&PCA combined process.

Note that annealing temperature can have a strong impact on device reliability performance, Fig.3. shows the normalized device △ Vth degradation versus stress time curve of devices under different PDA temperature. Devices with higher PDA temperature have more favorable Vth shift, this implies that higher annealing temperature is more effective in NBTI improvement.

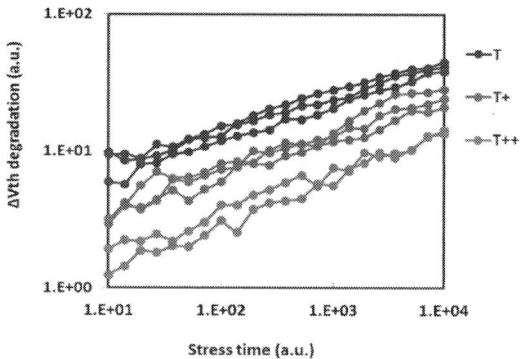

Figure 3: Normalized △ Vth degradation versus stress time of devices with PDA at temperature T / T+ / T++

PDA & PCA combined process optimizing experiment results are presented in Fig.4. PDA annealing includes high temperature T1 and increased high temperature T1+A, PCA annealing process includes high temperature T2 and increased high temperature T2+B. It is evident that PDA T1+A combining PCA T2+B process has the best NBTI performance, meaning that when using the same PDA annealing temperature T1+A, NBTI performance gains when PCA annealing temperature increase from T2 to T2+B; when using the same PCA annealing temperature T2+B, NBTI performance gains when PDA annealing temperature rise from T1 to T1+. While B is 2 times the size of A, We can get the conclusion that with the annealing temperature increasing of either PDA or PCA or both processes will result in a more stable reliability performance, what's rmore, NBTI improvement is more sensible to PDA temperature variation than PCA temperature variation.

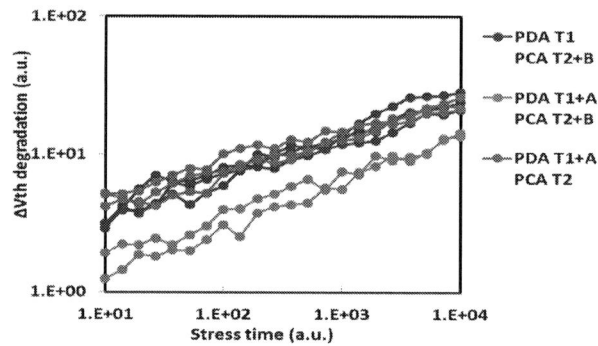

Figure 4: Normalized △ Vth degradation versus stress time of devices with PDA & PCA combined process at different annealing temperature.

It is shared that NBTI mainly results from trap generation and charge (hole) trapping which are strongly related to interface trap states (Nit), it is reasonable to assume that interface trap states of HK stack is reduced during post High K thermal treatment. We use charge pumping measurement to characterize the Si/IL interface trap states. Fig.5. illustrates the normalized results of the measurement, it is evident that Nit density of devices with both high temperature PDA and high temperature PCA is much smaller than devices with normal low temperature thermal treatment process, this result is in accordance with the NBTI results showed above.

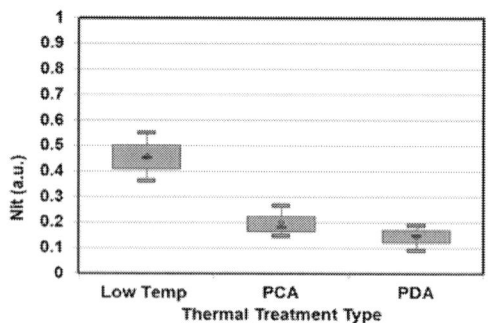

Figure 5: Normalized Nit of SiO₂/IL interface of devices with normal low temperature thermal treatment / high temperature PDA / high temperature PCA process

978-1-5386-5309-8/18 $31.00 © 2018 IEEE   116

To further investigate the influence of PDA annealing temperature on high K insulator stack and device leakage current performance, we applied MOSCAP capacitance-voltage (CV) measurement, the extracted the normalized Ig versus Tinv curve is as bellow. We can get from Fig.5. that with the increasing of PDA annealing temperature, Tinv gets thicker and Ig decreases. This indicates that PDA can cause IL(SiO2) regrowth, which can benefit device reliability but may be unfavorable to device static electrical performance, higher annealing temperature can contribute to this phenomenon.

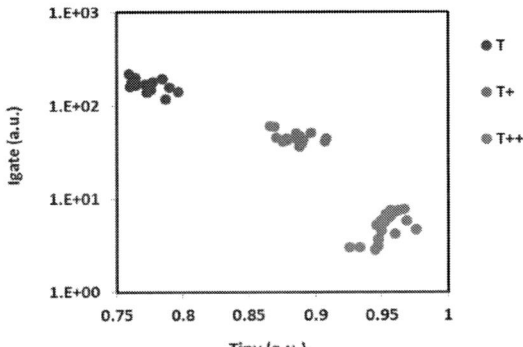

Figure 6: Normalized Ig versus Tinv curves of devices with PDA at temperature T / T+ / T++

Device reliability study is an intricate project that relate to every aspect of the device design and fabrication process. When trying to improve NBTI performance, we also study on the influence of HK stack process on device static electrical performance. The normalized universal curves of devices with different PDA annealing temperature are compared in Fig.7. Devices with lower temperature PDA process have superior static electrical performance, and devices' uni-curves degrade by higher temperature PDA. One feasible mechanism may be that higher annealing temperature can cause IL (SiO2) to regrow, and the regrowth is not well-controlled in sub-nano meter level, which may result in the degradation of device static electrical performance. Thus, further HK stack process optimization will be needed to make the tradeoff between reliability and device static electrical performance.

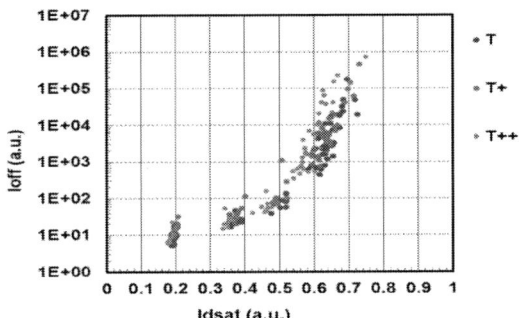

Figure 7: Normalized universal curve of the devices with different PDA annealing temperature.

Conclusion

In this work, the impact of HKMG stack process on P-type FinFET reliability performance is investigated. In detail, post high K deposition thermal treatment including PDA, PCA and the influence of annealing temperature on NBTI performance are revealed. Results show that both PDA and PCA process can lead to improved NBTI performance effectively, and the joint effort of PDA & PCA processes is more beneficial to improve the NBTI performance. Furthermore, reliability improvement is enhanced with the increase of annealing temperature, and is more sensible to the temperature variation of PDA process than that of PCA process. The mechanism is that during PDA and PCA annealing process, Si/IL interface trap states are restored and PDA process can cause SiO2 IL regrowth. Charge pumping measurement and CV measurement confirmed this inference. However, high annealing temperature in HK stack process may be harmful to device static electrical performance due to the regrowth of SiO2 interlayer. Further efforts need to be made to optimize the HK stack process.

References

[1] Xu J, Wang A, He J, Jing X, Zhang Z, Zhang B, et al. 14nm metal gate film stack development and challenges[C]// Semiconductor Technology International Conference. IEEE, 2017:1-3.

[2] Parihar N, Southwick R G, Sharma U, Wang M and Stathis J H, Mahapatra S. Comparison of DC and AC NBTI kinetics in RMG Si and SiGe p-FinFETs[C]// Reliability Physics Symposium. IEEE, 2017:2D-4.1-2D-4.7.

[3] Mukhopadhyay S, Goel N and Mahapatra S. A Comparative Study of NBTI and PBTI Using Different Experimental Techniques[J]. IEEE Transactions on Electron Devices, 2016, 63(10):4038-4045.

[4] He X, Xie X, Li Y, Zhang S, Shen Z, Zhou F, et al. Improved NBTI characteristic of HKMG FinFET with thermal oxidized interlayer and post interlayer anneal[C]// China Semiconductor Technology International Conference. IEEE, 2016:1-3.

[5] Lee Y H, Liao P J, Joshi K and Huang D S. Circuit-based reliability consideration in FinFET technology[C]// IEEE, International Symposium on the Physical and Failure Analysis of Integrated Circuits. IEEE, 2017:1-7.

[6] Joshi K, Hung S, Mukhopadhyay S, Chaudhary V, Nanaware N, Rajamohnan B, et al. HKMG process impact on N, P BTI: Role of thermal IL scaling, IL/HK integration and post HK nitridation[C]// Reliability Physics Symposium. IEEE, 2013:4C.2.1-4C.2.10.

[7] Desai S, Mukhopadhyay S, Goel N, Nanaware N, Jose B, Joshi K, et al. A comprehensive AC / DC NBTI model: Stress, recovery, frequency, duty cycle and process dependence[C]// Reliability Physics Symposium. IEEE, 2013:XT.2.1-XT.2.11.

[8] Linder B P, Dasgupta A, Ando T, Cartier E, Kwon U, Southwick R, et al. Process optimizations for NBTI/PBTI for future replacement metal gate technologies[C]// Reliability Physics Symposium. IEEE, 2016.

SIMULATION FOR THE FEASIBILITY OF HIGH-MOBILITY CHANNEL IN 3D NAND MEMORY

Zhaozhao Hou[1,2], Jiaxin Yao[1,2], Zhenhua Wu[1], and Huaxiang Yin[1,2]*

[1] Key Laboratory of Microelectronics Devices and Integrated Technology, Institute of Microelectronics of Chinese Academy of Sciences, Beijing 100029, People's Republic of China
[2] University of Chinese Academy of Sciences, Beijing 100049, People's Republic of China
*Corresponding Author's Email: yinhuaxiang@ime.ac.cn

ABSTRACT

Vertical 3D NAND flash memory devices, featuring poly-Silicon (poly-Si) channel, have been anticipated to replace conventional planar NAND flash because of the difficult scaling below 15-nm regime due to the increased noise and cell-to-cell interference. However, the conduction inside a long poly-Si channel is significantly degraded due to the scattering at grain boundaries (GB) and interface defects, leading to a reduced and unstable drive-current (I_d) with a broad distribution. In addition, the I_d required for reading operations decreases with the increasing number of stacked memory layers, resulting in poly-Si unsustainable for long-term scaling. The purpose of this paper is to study alternative channel materials with electron mobility higher than poly-Si, as a feasible approach to enable further scaling for future 3D NAND generations. A series of Sentaurus TCAD simulations are performed to clarify the necessity for the adoption of high-mobility channel materials in 3D NAND.

INTRODUCTION

In the past few years, 3D NAND flash memory has been attractive for low-cost and high-density memory. Vertical stackable 3D NAND flash is currently in mass production as replacement of planar NAND flash beyond the 15-nm node. Several approaches have been proposed to vertically stack 3D NAND cells, such as bit-cost scalable (BiCS), terabit cell array transistors (TCAT), and stacked memory array transistor (SMArT) [1].

However, owing to the difficulties of epitaxially growth for single-crystal channel in a multi-layer 3D structure, the industrial schemes typically have poly-Si as the channel material. Thus, the current 3D NAND devices are mainly based on a vertical poly-Si channel transistor [2]. Nevertheless, poly-Si conduction is inhibited by scattering events at grain boundaries and charged defects, thus leading to degradation of device performance such as lower drive-current (I_d) for reading operations and larger variability of threshold voltage (V_{th}) [3-4]. Furthermore, I_d is decreased by increasing the number of stacked layers, causing poly-Si unsustainable for the growth of storage density. Fig. 1 shows the reduction trend of cell current as the increasing word line (WL) stacks. When the stacked WL in 3D NAND flash reaches 64 layers and beyond, the increase in the total channel length limits the stacks of

vertical NAND flash memory devices due to decreased cell current for sensing scheme.

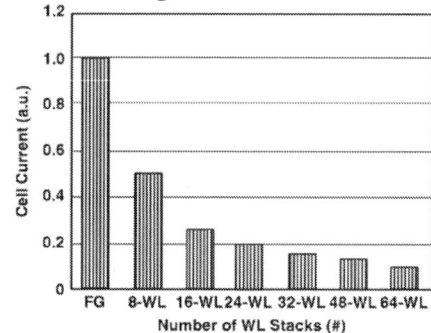

Fig. 1. Trend of cell current as word line (WL) stacks [5]

In the current 3D NAND channel, due to the constrained geometry, the poly-Si grain size is on the order of nanometers. The transition region between two grains is a grain boundary which acts as a defect sites to trap carriers [6]. These trap sites create a space-charge region between two crystallites, which then impedes the flow of charge carriers from one grain to the other, as shown in Fig. 2. The current conduction depends on both the grain size and the crystal orientation.

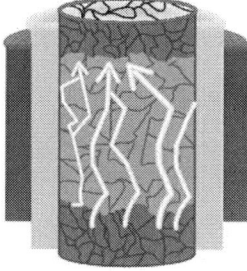

Fig. 2. Schematic view of current conduction in 3D NAND with poly-Si channel [7]

To improve memory performance, material development to enhance the drive-current/mobility of the poly-Si channel should be considered concretely. A possible solution to boost the drive-current was previously reported by J. G. Lisoni et al., and it includes engineering the poly-Si grain size and curing the defects at interface and grain boundaries via laser thermal annealing [8]. On

978-1-5386-5309-8/18 $31.00 © 2018 IEEE   118

the other hand, strong improvement in I_d has also been demonstrated by E. Capogreco et al. with the use of epitaxially grown SiGe, Si and $In_xGa_{1-x}As$ owing to their higher electron/hole mobility and tunable energy bandgap [9-10]. Meanwhile, the grain boundaries are greatly reduced in these epitaxially grown materials.

Furthermore, the drive-current (I_d) for reading operations in 3D NAND can be estimated from the following equation:

$$I_d = g_m [V_{read} - V_{th} - V_d/2] \quad (1)$$

The transconductance g_m in the above equation is related with the electron mobility of the channel material. Therefore, alternative channel materials with electron mobility higher than poly-Si can be considered as a promising solution to boost read current.

METHODOLOGY

We have implemented a series of computer-aided design (TCAD) simulations to verify the feasibility of high-mobility channel materials for 3D NAND. We created the structure of the 3D NAND flash memory by using the Sentaurus sprocess tool. Fig. 3 shows a schematic structure of vertical NAND flash memory for simulation, where a string consists of a channel, Oxide-Nitride-Oxide (ONO) stack with Gate-All-Around (GAA) structure, and poly-Si control gate. The simulation model follows the same integration sequences of a BiCS NAND flow: the gate patterning is done after the memory hole formation. Furthermore, as it mimics a string, it renders the study for the effects of the spaces between the gates and to control the junctions by means of the selectors.

The gates are arranged periodically along the vertical strings, and the gate length L_g and inter-poly dielectric (IPD) length L_{IPD} are 50 and 30 nm, respectively. The tunneling oxide, trap charge nitride, and blocking oxide layer are 3, 7, and 6 nm, respectively. A vertical channel hole with a 50-nm diameter is etched through the stack and filled with $Si_{0.7}Ge_{0.3}$ or poly-Si. Here, a P-type channel with uniform Boron concentration of $5 \times 10^{17}/cm^3$ is used for simulation. Furthermore, P^+ poly-silicon is used as gate material to reduce back tunneling and to achieve the required threshold voltage with Boron doping concentration $1 \times 10^{20}/cm^3$. The source and drain are N^+ doped at a concentration of $10^{20}/cm^3$.

The simulated 3D NAND structures consist of 3, 6, 32, and 64 gates, respectively. The device structures used for simulation are shown in Fig. 4 (a) and 4 (b). The side gates, named top select gate (TSG) and bottom select gate (BSG) are used as selectors, while the central gate represents the memory cell. Critical structure dimensions are listed in Table I.

TABLE I. STRUCTURE PARAMETERS OF SIMULATION DEVICE

Parameter	Value
Gate length	50 nm
Inter-poly dielectric length (L_{IPD})	30 nm
Channel hole diameter	50 nm
Tunneling oxide	3 nm
Trap charge nitride	7 nm
Blocking oxide	6 nm
Number of stacked memory layers (N_{layer})	3/6/32/64

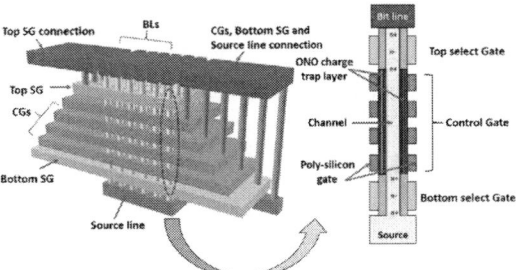

Fig. 3. Schematic structure of the 3D NAND flash

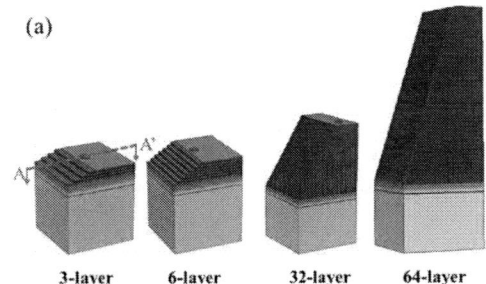

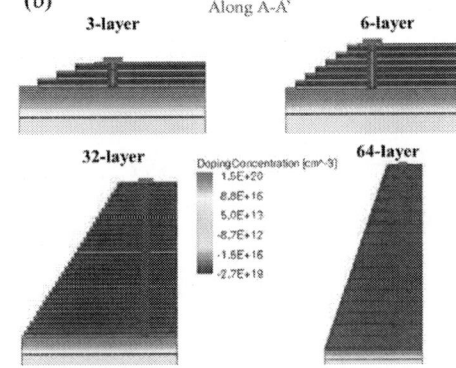

Fig. 4. Schematics of the simulated (a) structures and (b) the cross-sections for 3D NAND devices with various stacked layers

A schematic circuit diagram of the 3D NAND string during the read operation is shown in Fig. 5. The simulation was performed by sweeping gate of the

978-1-5386-5309-8/18 $31.00 © 2018 IEEE   119

memory cell under read operation while keeping 1 V at the drain side (V_d) and 2 V (V_{pass}) at the gates for the unselected memory cell.

Sentaurus sdevice numerically solves Poisson for the carrier density and the electric field and the drift-diffusion model for the currents on a meshed finite element structure representing the device under investigation.

RESULTS AND DISCUSSION

Fig. 6 (a) and (b) show the typical I_d-V_g characteristics obtained from the simulation results. It is obvious that as the number of stacked memory layers increases, the reduction of drive-current (I_d) for 3D NAND with poly-Si channel is much more significant than that for counterpart with $Si_{0.7}Ge_{0.3}$ channel.

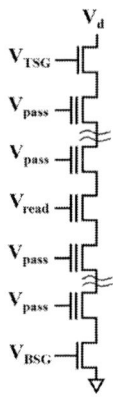

Fig. 5. A schematic circuit diagram of the vertical NAND string during the read operation

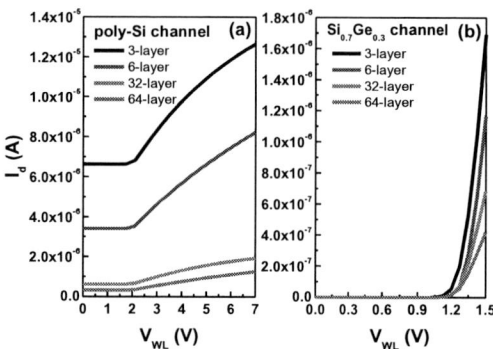

Fig. 6. Typical I_d-V_g characteristics for 3D NAND with (a) poly-Si channel and (b) $Si_{0.7}Ge_{0.3}$ channel under different stacked layers

Fig. 7 compares the percentage of I_d reduction for 3D NAND with $Si_{0.7}Ge_{0.3}$ channel and poly-Si channel as a function of stacked layers (N_{layer}). I_d reduction is about 90% for 64-layer NAND with poly-Si channel, which is much higher than that for counterpart with $Si_{0.7}Ge_{0.3}$ channel (74% reduction). Therefore, the advantage of

high-mobility channel material is clearly observed and confirmed from the simulation results.

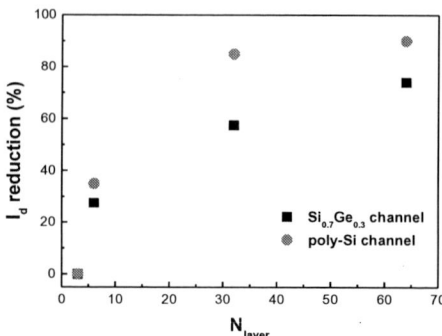

Fig. 7. Percentage of I_d reduction as a function of stacked layers (N_{layer}) for 3D NAND with poly-Si and $Si_{0.7}Ge_{0.3}$ channel

CONCLUSION

We have demonstrated that the high-mobility channel materials have the potential to give the highest possible current performance in vertical 3D NAND memory, based on simulation results. It has been confirmed that the drive-current (I_d) reduction for 3D NAND with high-mobility channel materials is much lower than that for counterpart with poly-Si channel as the number of memory stack layer increases. Thereafter, the introduction of epitaxially grown high-mobility channel materials can be regarded as an effective and practical solution to enable further scaling for future 3D NAND generations.

ACKNOWLEDGEMENTS

This work is supported by the National Science and Technology Major Project of China under Grant No 2013ZX02303007.

REFERENCES

[1] E. Capogreco et al., *IEEE Trans. Electron Dev.*, vol. 64, no. 1, p. 130, 2017.
[2] K. H. Lee et al., *Microelectro. Eng.*, vol. 147, p. 45, 2015.
[3] L. Breuil et al., *Proc. IEEE IMW*, p. 81, 2016.
[4] U. Celano et al., *VLSI Symp. Tech. Dig.*, p. 192, 2016.
[5] E. S. Choi et al., *Proc. IEEE IEDM Dig.*, p. 211, 2012.
[6] H. J. Yang et al., *J. Nanosci. Nanotechnol*, vol. 17, p. 2628, 2017.
[7] M. Toledano-Luque, et al., *Proc. IEEE IEDM Dig.*, p. 203, 2012.
[8] G. Lisoni et al., *VLSI Symp. Tech. Dig.*, p. 24, 2014.
[9] E. Capogreco et al., *Proc. IEEE IEDM Dig.*, p. 40, 2015.
[10] E. Capogreco et al., *Proc. IEEE IMW*, p. 109, 2015.

978-1-5386-5309-8/18 $31.00 © 2018 IEEE

Self-aligned Metallic Source and Drain Fin-on-Insulator FinFETs with Excellent Short Channel Effects Immunity down to 20 nm Gate Length

Qingzhu Zhang[1,2], Junjie Li[2], Hailing Tu[1], Huaxiang Yi[2,4], Jiang Yan[3], Lingkuan Meng[2], Jiaxin Yao[2], Guilei Wang[2], Zhijun Cao[2], Yudong Li[2], Zhaohao Zhang[2], Zhenhua Wu[2], Feng Wei[1], Hongbin Zhao[1], Jiangfeng Gao[2], Xiaobin He[2], Qifeng Jiang[2], Wenjuan Xiong[2], Jinjuan Xiang[2], Zhangyu Zhou[2], Yihong Lu[2], Gaobo Xu[2], Kun luo[2], Yu Pan[2], Renren Xu[2], Jie Gu[2], Chaozhao Hou[2], Junfeng Li[2], and Wenwu Wang[2]

[1] State Key Laboratory of Advanced Materials for Smart Sensing, General Research Institute for Nonferrous Metals, Beijing 100088, China; [2] Key Laboratory of Microelectronics Devices & Integrated Technology, Institute of Microelectronics of Chinese Academy of Sciences (IMECAS), Beijing, 100029, China;[3] College of Electronic and Information Engineering, North China University of Technology, Beijing, 100144; [4] University of Chinese Academy of Sciences, Beijing, China, 100049.

Tel: +8610-18010047883, *E-mail: yinhuaxiang@ime.ac.cn; zhangqigzhu@ime.ac.cn

ABSTRACT

In this paper, a fin-on-insulator (FOI) FinFET structure with a metallic source and drain (MSD) and HKMG-last processes was investigated extensively. FOI FinFETs demonstrated ten times of leakage current reduction compared with conventional bulk FinFET using p-n junction. In contrast to bulk FinFETs, MSD FOI FinFETs demonstrate a greatly improved control of short channel effects (SCEs) (~47% DIBL and ~32% SS reductions for 20-nm-Lg PMOS) compared to conventional bulk FinFET. The drive ability of 20-nm-Lg FOI transistor is increased about 30 times with full Ni(Pt) silicide process, up to 547 $\mu A/\mu m$ for NMOS and 324 $\mu A/\mu m$ for PMOS at V_{DD} = 0.8 V. However, the excellent controls of SCEs and channel leakage were also obtained. No obvious increase of leakage happens in FOI FinFETs using fully Ni(5%Pt) silicide process.

.

Keywords— Fin-on-insulator (FOI); Ni(Pt) silicide; fully silicide; leakage; metallic source and drain (MSD)

INTRODUCTION

Bulk silicon FinFET with 3D architecture has been applied to the advanced (16-/14-nm generation) CMOS technologies today [1-2]. However, how to restrain SCEs and bulk leakage is extremely difficult for 7 nm node and beyond by punch through stop (PTS) doping under the fin channel. Silicon on insulator (SOI) FinFET shows great advantages e.g. better leakage control, higher operation speed and lower power consumption compared to bulk FinFET. However it suffers from lacking of design flexibility, high-cost starting wafers and fin tilting problems at extreme thin fin thickness. Using bulk Si to fabricate SOI liked FinFET can combine both of advantages, which is a perfect candidate for future advance device. Many previous works have achieved the structure by thermal oxidation the fin bottom or bottom oxidation through STI process [3-4]. In this paper we realized FOI FinFET using a more simple and effective method. First, notched fin arrays were formed by advanced reactive ion etching (RIE) with mixed anisotropic, plasma passivation and isotropic etching.

Second, a 10 nm liner oxidation process by rapid thermal oxidation (RTO) with well controlled uniformity is carried out to isolate top fin structure and the bottom Si substrate [5].

Besides the serious problems of poor gate control ability, bad values of SS and DIBL and high consuming power with the device continuing scaling, the large parasitic resistance is another important factor limiting driving ability of 3D structure device greatly due to small sizes of fins or nanowires. The parasitic resistance basically consists of contacted resistances (R_cs) and sheet resistances (R_{sh}s). Conventional selective epitaxy and MSD are two effective methods to decrease the parasitic resistance, and a large bulk leakage is usually generated with MSD process due to excessive diffusion of metal atoms or silicide piping defects. Therefore, selective epitaxy process has been used in 22 nm and 14-16 nm FinFET mass productions to reduce the SD parasitic resistance by enlarging the sizes of SD. However, it suffers from great disadvantages e.g. complex integrated process, high cost, inhomogeneity wafer to wafer, and the high quality epitaxy could not easy to achieved with the continue scaling of fin pith width. As the FOI FinFET structure is totally isolated by bottom oxide, similar to SOI, this great advantage makes it immune to silicide induced leakage. Therefore, the option of the MSD process maybe the most effective and low cost way to decrease parasitic resistance and promote device driving abilities.

Conventional NiSi silicide is particularly a favorable candidate due to its low R_cs (10–18/squre), low Si consumption (consumption of 1.83-nm Si per nanometer of Ni, yielding 2.21-nm NiSi), and low thermal budget formation (~500℃). However, Ni silicide is very easy to accumulation at very low temperature (~600℃) and the diffusion of Ni atoms is very fast along SD defects (also namely silicide piping defects). Ni with a small mount Pt is verified the most effective way to produce a robust silicide with very low resistivity, improve silicide thermal stability, and reduce the trends of silicide piping defects. In this paper, the MSD was formed with conventional two-step rapid thermal annealing (RTA) using Ni(5%Pt) on 3D FOI FinFET and bulk FinFET structures, without selective epitaxy process. The related device

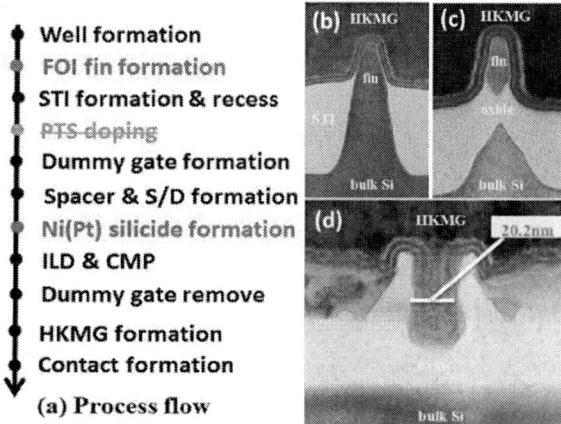

(a) Process flow

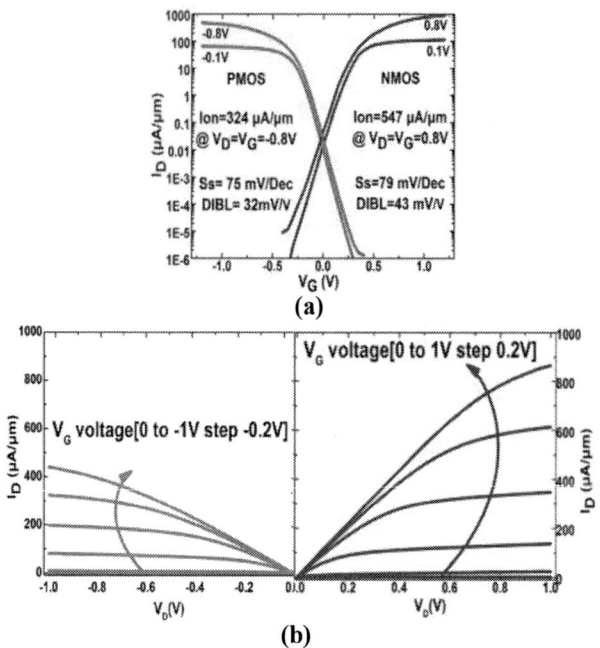

Fig. 1. (a) The fabrication process flow of FOI FinFETs. (b) the cross sectional view of fabricated FOI Fin in gate region (Fin height ~28nm, Fin top width ~7nm) and (c) the cross section view of fabricated FOI Fin in source and drain region (Fin height ~35 nm, Fin top width ~7 nm).

characteristics and the fully MSD effect on FOI FinFET and FinFET are studied by experiment.

EXPERIMENTAL PROCEDURE

The FOI FinFETs were fabricated on 200 mm Si (100) wafers and the process integration flow was based on conventional FinFET fabrication, as is briefly shown in **Fig.1 (a)**. Several processes which are different from FinFET fabrication are signed in colored. The ultra-small SOI-like fin structures were patterned by self-aligned the spacer image transfer (SIT) method. First, notched fin arrays were formed by advanced RIE with anisotropic etch, plasma passivation and isotropic etch. Then a 10 nm liner oxidation process by rapid thermal oxidation (RTO) process with well controlled uniformity was carried out to isolate the top fin structure and the bottom Si substrate. The detailed information of FOI fin structure formation has been shown in our previous work in detail [5]. The adjacent fins were isolated by deposited HARP STI oxide and a chemical mechanical polish (CMP) process was implemented to achieve a flat surface. The height of STI region was controlled accurately by diluted-HF (DHF) cleaning process to prevent fin from tilting. A punch through stop (PTS) doping process was skipped due to the fully isolation by bottom oxide of the FOI fin, as is shown in **Fig.1 (b)**. In the following steps, the dummy gate, spacers, source/drain (S/D) doping, activation anneal were implemented using regular integration process of bulk FinFETs [9]. A series of gate lengths (L_gs) ranging from 20 nm to 500 nm were achieved by mixed lithography technology. The nanometer patterns were formed by electron beam and large patterning was generated by I-line lithography. Self-aligned Ni(5%Pt) silicide with traditional two-step RTA process was carried out after 20 nm-thick spacers and S/D formation. FinFETs and FOI FinFET both with and without silicide process were fabricated on 200 mm Si (100) wafers. The Si fins in S/D region were almost fully consumed by Ni(5%Pt) for FOI FinFET, as is shown in **Fig.1**

Fig. 2. (a) CMOS I_D-V_G curves of 20-nm-Lg MSD FOI FinFET, excellent device parameters are achieved with fully MSD, (b) CMOS output characteristics of 20-nm-Lg MSD FOI FinFET, relative high driving abilities are achieved with fully MSD

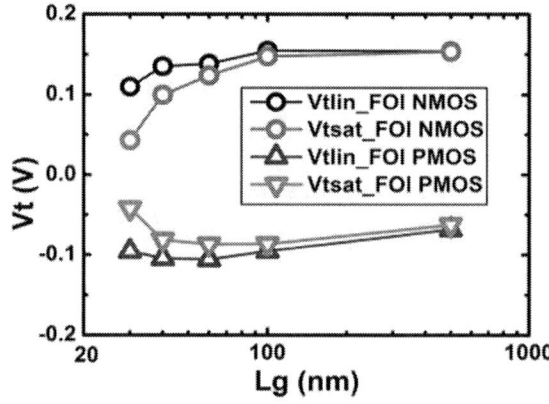

Fig. 3. Threshold voltage vs. gate length for conventional FOI n- and p-FinFETs.

(d), which is helpful to greatly reduced SD parasitic resistance. A 2 nm-thick high-k dielectric (HfO$_2$) was deposited using ALD process upon the formation of approximately 8-Å-thick interfacial layer (IL). NMOS with ALD TiAlC/TiN and PMOS with ALD TiN/TaN/TiN layers were subsequently formed on the formed high-k layer. The SD metal-contact processes and metallization for device formation were finished in following steps. The final device architectures were carried out by transmission electron microscope (TEM) and the electrical characterization was performed with a HP4156C precision semiconductor parameter analyzer.

RESULTS AND DISCUSSION

The I_D-V_G transfer and I_D-V_D output characteristics curves of CMOS FOI FinFETs with 20 nm physical gate lengths are

978-1-5386-5309-8/18 $31.00 © 2018 IEEE

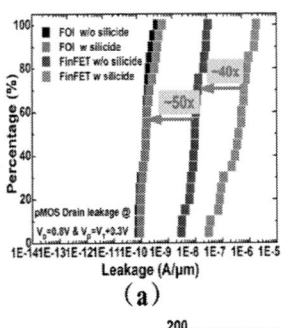

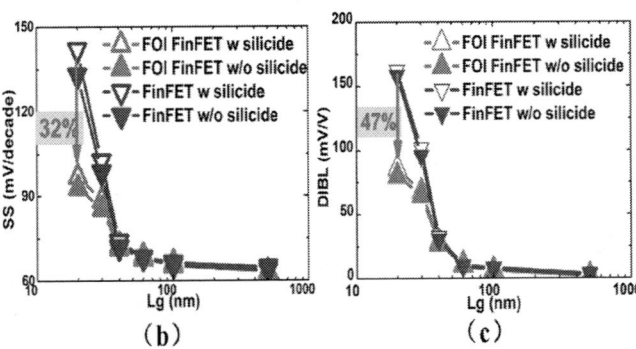

Fig. *4. (a) Junction leakage results for p-type FOI FinFET and bulk FinFET, no obvious leakage increase for FOI FinFET with MSD, (b) DIBL varying as a function of Lg in PMOS, 47% DIBL improving on that in 20nmLg bulk FinFET achieved,(c) Ss varying as a function of Lg in PMOS, 32% Ss improving on that in 20nmLg bulk FinFET achieved.*

shown in **Fig.2 (a)** and **(b),** respectively. In contrast to previous report [4], I_{ON} of the fabricated FOI transistors are increased about 30 times, up to 547 $\mu A/\mu m$ for NMOS and 324$\mu A/\mu m$ for PMOS at V_{DD} = 0.8 V. Meanwhile, subthreshold parameters (DIBL = 32 mV/V and SS = 76 mV/dec for PMOS and DIBL = 43 mV/V and SS= 79 mV/dec for NMOS) are achieved. Besides that, the new fully isolated device can reach a better I_{ON}/I_{OFF} ratio (I_{ON}/I_{OFF} >10^6) due to self-allied fully insulator isolation and a good channel control of similar Ω-shaped multi-gate structure. The results show the great driving performance improvement and better SCEs control than bulk FinFETs with a little process changes, which means this insulator isolated structure and fully MSD process are very suitable for future scaled device, especially for advanced low power and extra low power applications.

The curves of the threshold voltages (V_t-lin and V_t-sat) vs. different L_gs are plotted in **Fig. 3** for conventional MSD FOI *n*- and *p*-FinFETs. As can be seen from the picture, A good controls of SCEs and channel leakage are achieved, which means that FOI FinFET has a very strong gate control ability to the extra-thin channel.

As is shown in **Fig.3 (a),** FOI FinFETs demonstrate ten times of leakage current reduction compared to conventional bulk FinFET with *p-n* junction isolation. It demonstrates that the fully insulator isolation is a very good solution for leakage reduction. There are no obvious leakage increasing happens in FOI FinFET with conventional fully Ni(Pt) silicide process, however an ~40 times leakage increase was demonstrated on bulk FinFETs by experiment. The large leakage of bulk

FinFET is mainly caused by Ni piping defect or the formed silicide go through p-n junction at SD region. In contrast to bulk FinFET, MSD FOI FinFET demonstrates a greatly improved of SCE control. ~47% DIBL and ~32% SS reductions for 20-nm-Lg PMOS, as is shown in **Fig.3 (b)** and **Fig.3 (c)**, respectively. The ultra-thin fin thickness (top width ~7nm and maximum width is ~15nm) and Ω-shape multi-gate structure for FOI FinFET may be the most important factors for a better gate control. Therefore, the great DIBL and SS reductions in values are achieved compared to bulk FinFET.

SUMMARY

In summary, advanced fully silicidation processes performed on isolated bulk-fin to form fully MSD were successfully demonstrated for FOI FinFETs. The great drive current improvement, perfect control of the junction leakage and the immunity to SCEs of the *n*- and *p*- FOI transistors were achieved at the same time. We conclude that the fully MSD is a valuable and robust process method for the practical application of novel devices in future device scaling for fully isolated structure. The proposed FOI FinFET also provides a potential solution for future low power CMOS and the very sensitive sensor application.

ACKNOWLEDGEMENT

This work was supported by the Ministry of Science and Technology of China under Contract 2013ZX02303 through the 14-16nm Technology Program, the National Key Research and Development Program of China (2016YFA0301701).

REFERENCES

[1] Jan C H, Bhattacharya U, Brain R, et al. "A 22nm SoC platform technology featuring 3-D tri-gate and high-k/metal gate, optimized for ultra low power, high performance and high density SoC applications" in Proc. IEEE VLSI, Jun. 2014, pp. 136-137, doi: 10.1109/IEDM.2012.6478969

[2] Seo K I, Haran B, Gupta D, et al. "A 10nm platform technology for low power and high performance application featuring FINFET devices with multi workfunction gate stack on bulk and SOI", in Proc. IEEE IEDM, Dec. 2014, pp. 1-2, doi: 10.1109/VLSIT.2014.6894342.

[3] K. Cheng, S. Seo, J. Faltermeier, D. Lu, "Bottom Oxidation through STI (BOTS)- A Novel Approach to Fabricate Dielectric Isolated FinFETs on Bulk Substrates" in Proc. IEEE VLSI, Jun. 2012 , 48 (11) :3.1.1-3.1.4 doi: 10.1109/VLSIT.2014.6894390.

[4] Q. Zhang, H. Yin, J. Luo, H. Yang, L. Meng, Y. Li, Z. Wu, Y. Zhang, Y. Zhang, C. Qin, J. Li, J. Gao, G. Wang, W. Xiong, J. Xiang, Z. Zhou, S. Mao, G. Xu, J. Liu, Y. Qu, T. Yang, J. Li, Q. Xu, J. Yan, H. Zhu, C. Zhao, and T. Ye. "FOI FinFET with ultra-low parasitic resistance enabled by fully metallic source and drain formation on isolated bulk-fin", in Proc. IEEE IEDM, Dec. 2016, pp. 452-455, doi: 10.1109/IEDM.2016.7838438.

[5] X. Ma, H. Yin, P. Hong and W. Xu "Self-Aligned Fin-On-Oxide (FOO) FinFETs for Improved SCE Immunity and Multi-VTH Operation on Si Substrate". ECS Solid State Letters, Vol. 4, No. 4, pp. Q13-Q16, Jan., 2015, doi: 10.1149/2.0071504ssl.

ENGINEERING RESISTIVE SWITCHING BEHAVIOR IN TAO_X BASED MEMRISTIVE DEVICES FOR NON-VON NEUMAN COMPUTING APPLICATIONS

Jingxian Li[1,2], Teng Zhang[1], Qingxi Duan[1], Lidong Li[2*], and Yuchao Yang[1*], and Ru Huang[1]

[1]Key Laboratory of Microelectronic Devices and Circuits (MOE), Institute of Microelectronics, Peking University, Beijing 100871, China
[2]State Key Lab for Advanced Metals and Materials, School of Materials Science and Engineering, University of Science and Technology Beijing, Beijing 100083, China.
*Corresponding Author's Email: yuchaoyang@pku.edu.cn, lidong@mater.ustb.edu.cn

ABSTRACT

Memristive devices are nowadays widely investigated for non-von Neuman computing, such as neuromorphic and in-memory computing, where memristors with incremental switching behavior are favorable for neuromorphic hardware while binary resistance state with high on/off ratio is desirable for the in-memory logic. Here we report the engineering of the resistive switching behavior in tantalum oxide based memristive devices by introducing an ion diffusion limiting layer Al_2O_3 at the TiN/TaO_x interface in order to satisfy the different requirements of various applications. This study could provide clues to the optimization of memristive devices for future neuromorphic and logic applications.

Keywords—memristive; analog; neuromorphic; in-memory logic; TaO_x; Al_2O_3

INTRODUCTION

Memristors (or memristive devices) are 2-terminal nanoionic devices [1] that are considered as promising building blocks for future computing systems, due to their low power consumption, high scalability and compatibility with conventional complementary metal oxide semiconductor (CMOS) technology [2], which have been actively exploited for the non-von Neumann computing applications, such as neuromorphic computing [3, 4] and in-memory computing [5, 6]. To be used in neuromorphic hardware, the memristors shall need to show incrementally tunable resistance, while the binary resistance is desirable for the in-memory logic. We have reported a novel approach to engineering the conductance

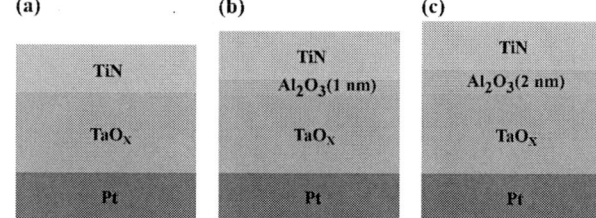

Fig. 1. Schematic of three device structures.

modulation linearity in TaO_x based memristors by insertion of a SiO_2 diffusion limiting layer (DLL) to homogenize the filament growth/dissolution rate [7]. However, SiO_2 is still limited by its relatively low activation energy for oxygen ion migration (0.92–1.17 eV). Here we optimize the device performance by using Al_2O_3 as the DLL material since Al_2O_3 have even higher activation energy of ~6.5 eV for oxygen diffusion [8] and preferably limits the ion diffusion speed and therefore reduces the number of oxygen ions/vacancies that participate in the initial, abrupt filament growth/dissolution process. Meanwhile, the present devices are able to increase the on/off ratio of the binary resistance and thus are also desirable for in-memory logic. Typical Boolean logic functions including OR and AND have been experimentally demonstrated using these devices, making them promising candidates for future non-von Neuman computing.

EXPERIMENTAL

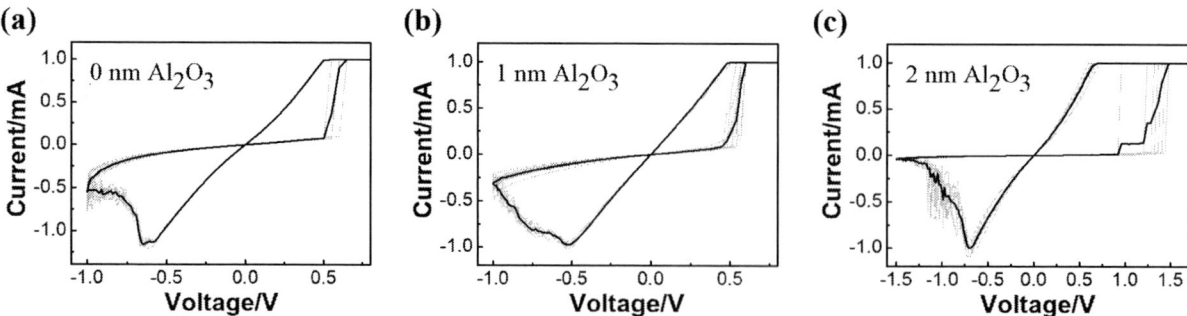

Fig. 2. 10 consecutive *I–V* sweeps (gray) and averaged *I–V* curve (black) of TaO_x based memristors with a Al_2O_3 layer thickness of 0, 1 and 2 nm (from a to c).

Three types of memristors were fabricated and studied in this work as illustrated in Fig. 1. TiN/Al$_2$O$_3$/TaO$_x$/Pt devices (2 × 2 μm^2 in size) with a typical crossbar structure were fabricated by conventional photo-lithography, film deposition, and lift-off processes. The TaO$_x$ layer with a thickness of ~30 nm was deposited by reactive sputtering utilizing an Ar : O$_2$ ratio of 3 : 2 and a RF power of 500 W. In order to limit the oxygen diffusion and thus control the filament growth/dissolution rate at the interface, a Al$_2$O$_3$ layer with varied thicknesses of 1 and 2 nm was grown by atom layer deposition following TaO$_x$ sputtering to form a TiN/Al$_2$O$_3$/TaO$_x$/Pt structure. In all device structures, a Pt capping layer was always coated on top of the TiN electrode for protection purpose. Electrical measurements including direct current (DC) and pulse tests were performed using an Agilent B1500A semiconductor parameter analyzer with a probe station.

RESULTS AND DISCUSSION

Fig. 2 shows 10 consecutive current-voltage *(I-V)* sweeps (gray) from the TiN/Al$_2$O$_3$/TaO$_x$/Pt devices after forming, and the black line represents the averaged *I-V* curve. The set (reset) of the device was achieved by a positive (negative) voltage sweep on the TiN top electrode, in agreement with the fact that the TiN electrode serves as a reservoir of oxygen vacancies, and the exchange of oxygen between TiN and TaO$_x$ defines the switching polarity of these devices. When compared with TiN/TaO$_x$/Pt device (Fig. 2a), memristor with 1 nm Al$_2$O$_3$ layer (Fig. 2b) shows a more uniform resistive switching behavior and both the set and reset processes present more gradual transitions. It is expected that the Al$_2$O$_3$ limits the ion diffusion speed and thus reduces the number of oxygen ions/vacancies that participate in the initial, abrupt filament growth/dissolution process, accounting for the more gradual switching that will also be shown in Fig. 3 afterwards. However, a thicker Al$_2$O$_3$ of 2 nm (Fig. 2c) significantly increases the set voltage from 0.5 to 1.5 V. When the Al$_2$O$_3$ becomes thicker, the effect of limiting diffusion is stronger, therefore resulting in the more difficult set process. This makes the TiN/Al$_2$O$_3$(2 nm)/TaO$_x$/Pt devices less favorable compared with TiN/Al$_2$O$_3$(1 nm)/TaO$_x$/Pt devices.

Fig. 3 further demonstrates that improved linearity in weight modulation was indeed achieved in TiN/Al$_2$O$_3$/TaO$_x$/Pt devices with Al$_2$O$_3$ thicknesses of 1 nm, via application of 300 identical pulse stimuli in each potentiation or depression processes. Both the DC and pulse measurements have identified the excellent resistive behavior in device with Al$_2$O$_3$, thereby highlighting the role of DLL in effectively regulating the dynamic rate of filament growth/dissolution.

In addition to neuromorphic computing, an alternative approach to addressing the von Neumann bottleneck is to perform logic operations using emerging nonvolatile memory devices. Compared with neuromorphic computing in-memory logic generally has different requirements on device performance, i.e. binary resistive switching behavior in memristive devices since high and low resistive states can correspond to the two logic states "1" and "0". To be used in nonvolatile logic computing with crossbar arrays, the memristors need to show high on/off ratio due to the device-to-device variations, therefore, the TaO$_x$ based devices with 2 nm Al$_2$O$_3$ are desirable for logic applications (Fig. 2c). Fig. 4 shows the accomplishment of OR and AND Boolean logic functions using TiN/Al$_2$O$_3$ (2 nm)/TaO$_x$/Pt devices. The logic operation has been suggested by Linn et al. to implement Boolean logic in at most three sequential cycles [9]. The electric potentials applied to the electrodes act as logic inputs *p* and *q*, while the resistance state of the device acts as the logic output *s*. In the present scenario, the high potential applied to the electrode and the low resistance state of the device is designated as logic "1", while the low potential applied to the electrode and the high resistance state is designated as logic "0". An initialization step is always needed to set the devices into state "1", followed by two write steps, as shown in Figs. 4a and 4c. The pulse measurements in Figs. 4b and 4d have clearly demonstrated the experimental implementations of OR and AND functions using a single device, which could be promising for energy-efficient computing applications.

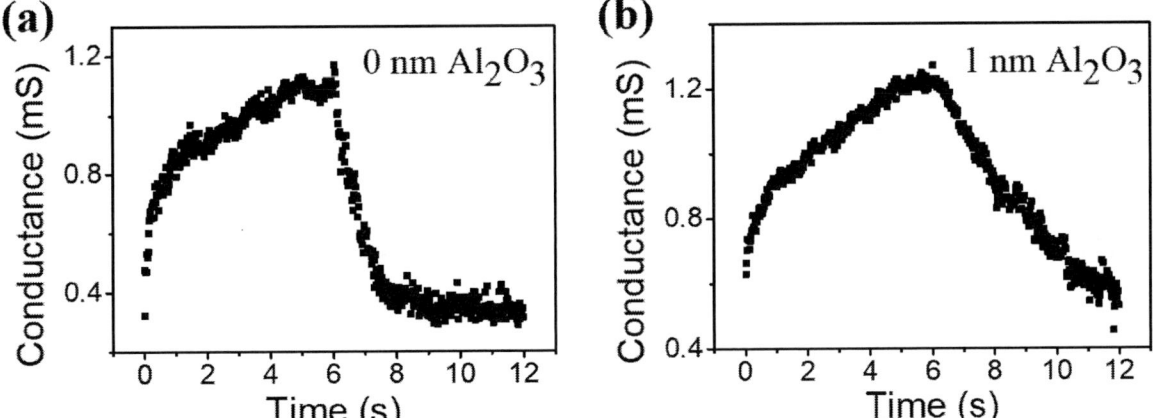

Fig. 3. Conductance modulation processes by application of 300 identical potentiation pulses and 300 identical depression pulses for TiN/Al$_2$O$_3$/TaO$_x$/Pt devices with a Al$_2$O$_3$ thickness of (a) 0 nm, (b) 1 nm. The pulse parameters used in the measurements were (a) potentiation: 0.85 V/100 ns, depression: −0.95 V/100 ns; (b) potentiation: 1.02 V/100 ns, depression: −1.03 V/100 ns.

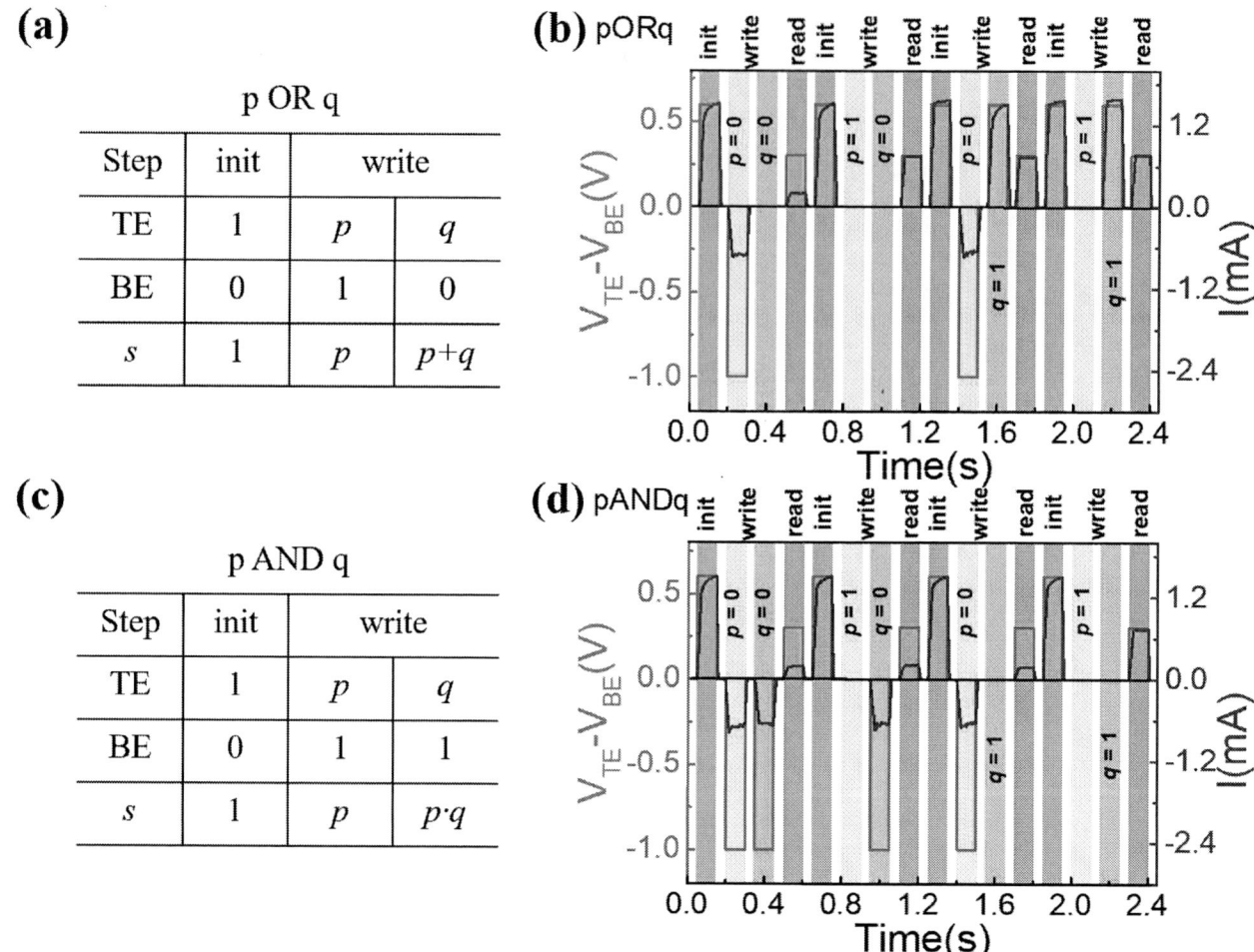

Fig. 4. Logic operation of (a) OR function, (c) AND function. Pulse measurements of (b) OR function, (d) AND function in TiN/Al$_2$O$_3$ (2 nm)/TaO$_x$/Pt devices

CONCLUSION

In this study, we have fabricated TiN/Al$_2$O$_3$/TaO$_x$/Pt devices with high switching linearity and high on/off ratio that can be used for neuromorphic and in-memory logic applications. Experimental implementations of OR and AND functions using a single device are demonstrated. This study could provide clues to the optimization of memristive devices for non-von Neuman computing applications.

ACKNOWLEDGEMENTS

This work was supported by National Key R&D Program of China (2017YFA0207600), Beijing Municipal Science & Technology Commission Program (Z161100000216148) and National Natural Science Foundation of China (61674006, 61421005). Y. Y. acknowledges support from the "1000 Youth Talents Program" of China.

REFERENCES

[1] Y. Yang, X. Zhang, L. Qin, Q. Zeng, X. Qiu and R. Huang, "Probing nanoscale oxygen ion motion in memristive systems," Nat. Commun., vol. 8, p. 15173, 2017.

[2] D. B. Strukov, G. S. Snider, D. R. Stewart, and R. S. Williams, "The missing memristor found," Nature, vol. 453, pp. 80–83, 2008.

[3] M. Prezioso, F. Merrikh-Bayat, B. D. Hoskins, G. C. Adam, K. K. Likharev, and D. B. Strukov, "Training andoperation of an integrated neuromorphic network based on metal-oxide memristors" Nature, vol. 521, pp. 61–64, 2015.

[4] N. K. Upadhyay, S. Joshi, and J. J. Yang, "Synaptic electronics and neuromorphic computing" Sci. China Inform. Sci., vol. 59, pp. 061404:1-061404:26, 2016.

[5] J. Borghetti, G. S. Snider, P. J. Kuekes, J. J. Yang, D. R. Stewart, and R. S.Williams, "'Memristive' switches enable 'stateful' logic operations via material implication," Nature, vol. 464, pp. 873–876, 2010.

[6] Y. Yang, M. Yin, Z. Yu, Z. Wang, T. Zhang, Y. Cai, W. D. Lu, and R. Huang, "Multifunctional nanoionic devices enabling simultaneous heterosynaptic plasticity and efficient in-memory Boolean logic" Adv. Electron. Mater., vol. 3, p. 1700032, 2017.

[7] Z. Wang, M. Yin, T. Zhang, Y. Cai, Y. Wang, Y. Yang and R. Huang, "Engineering incremental resistive switching in TaOx based memristors for brain-inspired computing," Nanoscale, vol. 8, pp. 14015-14022, 2016.

[8] A. H. Heuer, "Oxygen and aluminum diffusion in α-Al2O3: How much do we really understand?," J. Eur. Ceram. Soc., vol. 28, pp. 1495-1507, 2008.

[9] E. Linn, R. Rosezin, S. Tappertzhofen, U. Böttger, R. Waser, "Beyond von Neumann--logic operations in passive crossbar arrays alongside memory operations," Nanotechnology, vol. 23, p. 305205, 2012.

RESIST MODEL SETUP FOR NEGATIVE TONE DEVELOPMENT AT 14NM NODE

Lijun Zhao[1,2], Lisong Dong[1], Libin Zhang[1], Yayi Wei[1,2], Tianchun Ye[1,2]*

[1]Key Laboratory of Microelectronic Devices & Integrated Technology, Institute of Microelectronics of Chinese Academy of Sciences, Beijing 100029, China
[2]University of Chinese Academy of Sciences, Beijing 100029, China
*Corresponding author: weiyayi@ime.ac.cn

ABSTRACT

The shrinkage effect of negative tone development (NTD) process is the primary cause of the mismatch between simulation data and experimental data. Therefore, the resist model is strongly required in the lithography simulation for NTD process. A practical flow of building a resist model is presented for 14nm NTD process, including test patterns design, model calibration and model verification. The test patterns are designed according to the optical parameters and resist parameters. Then, we use experimental data to calibrate and verify the model. The results show the simulation data have the same trend with experimental data, which indicates that the resist model has the capacity of capturing the resist shrinkage effect of NTD process.

INTRODUCTION

As the semiconductor technology node has been shrinking down to 14nm and below, achieving smaller critical dimension (CD) becomes more challenge in lithography. Various techniques have been developed to meet the requirements of advanced node. One of these techniques is NTD which uses inverse development compared to positive tone develop (PTD) [1-2]. By using bright-field mask and NTD process, lithography can achieve higher image contrast, lower Line Edge Roughness (LER) and larger process window [3].

It is known that a prominent characteristic of NTD process is the resist shrinkage induced by the de-protection reaction during the post-exposure bake (PEB) [4]. The resist shrinkage effect contains vertical shrinkage effect, which is also called the resist thickness loss, and the horizontal shrinkage effect. In addition to the common acid and base diffusion and reaction mechanism for general chemically amplified resist, there are the following special effects in the NTD process [5]: resist shrinkage effect, developer depletion and diffusion, wafer CD jump induced by sub-resolution assistance feature (SRAF) effects.

Some papers have illustrated the mechanism [6-11]. Here is summary of the common point. The reason of resist shrinkage is that the evaporation of the intermediate product at PEB stage can lead to the change of free volume. The formation of shrinkage can be described as following. Acid molecules are created by photo-acid generators

(PAG) in the exposure step. Then de-protection reactions are catalyzed by the acid molecules during the PEB step. The de-protected molecules are volatile and can easily escape from the resist once they diffuse to the film top. The escaped volatile molecules leave behind vacancies or holes that begin to collapse, which eventually leads to the resist volume shrinkage [4].

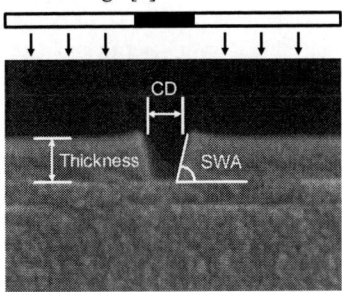

Figure 1: Experimental result of resist shrinkage

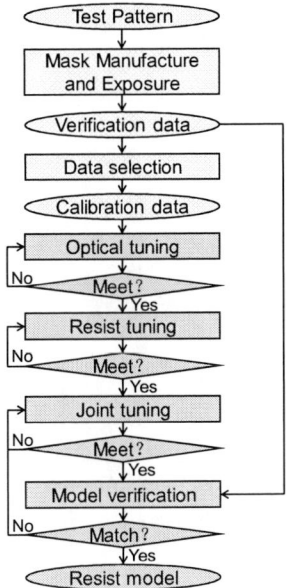

Figure 2: The flow of model calibration and verification

The experimental result of resist shrinkage is shown

in Figure 1. The CD, resist thickness and side wall angle (SWA) depend on the volume shrinkage that reserved after the development. For the dense line/space (L/S) patterns, the density of patterns is close to 50%. But for the isolated L/S patterns, the density of patterns is only about 10%. The volume of reserved resist of isolated patterns is larger than the dense patterns, so the degree of shrinkage of isolated patterns is much higher than the dense patterns. From above results, the side wall angle of isolated patterns is larger than dense patterns.

To truly reflect the impact of such physical and chemical reaction on the accuracy of lithography simulation, a resist model for NTD process is established in this paper. The whole flow is depicted in Figure 2, which includes test patterns design, model calibration and model verification.

In this paper, the section 2 provides the design flow of test patterns based on the parameters of aerial image. The flows of model calibration and verification are introduced in section 3. Finally, section 4 concludes this paper.

TEST PATTERNS DESIGN

To describe the characteristic of real layout as much as possible, the patterns for verification contain base type and complex type. The base type of patterns includes the 1D patterns, the head-to-head (HTH) patterns and head-to-line (HTL) patterns, which is shown as Figure 3. And the complex type of patterns contains H bar, broken H bar, L shape, U shape and some of real chip layout, which is not shown in this paper.

The size of verification patterns should cover the capability of the process and the initial design rules. For example, the actual limit pitch of immersion lithography is about 80nm, so the minimum pitch of L/S patterns should be 80nm*0.95=76nm.

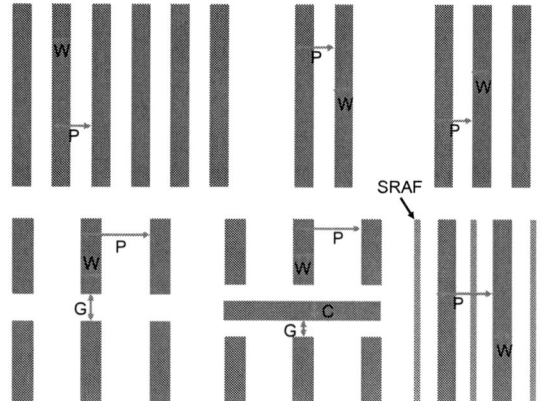

Figure 3: Base type of test patterns

For resist models, the fewer count of patterns for calibration selected from verification patterns, the lower cost of calculation. To maximize the coverage of

calibration patterns and minimize the cost, we focus on pattern type selection and pattern size selection. For type selection, the base type of patterns should be selected as the calibration patterns. For size selection, we select calibration patterns from verification patterns based on the parameter space that they occupied.

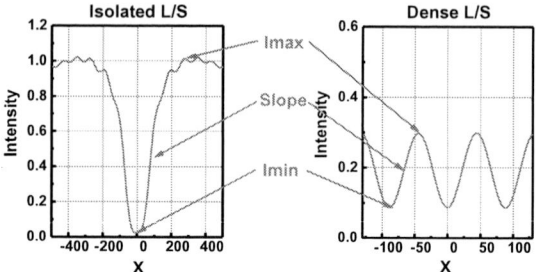

Figure 4: The parameters of aerial image: Imin, Imax, Slope

Figure 4 gives the intensity of aerial image of the dense L/S patterns and isolated L/S patterns. From the aerial image, certain image parameters can be extracted: Minimum intensity (Imin) and maximum intensity (Imax), the slope of the aerial image (Slope) [12-13].

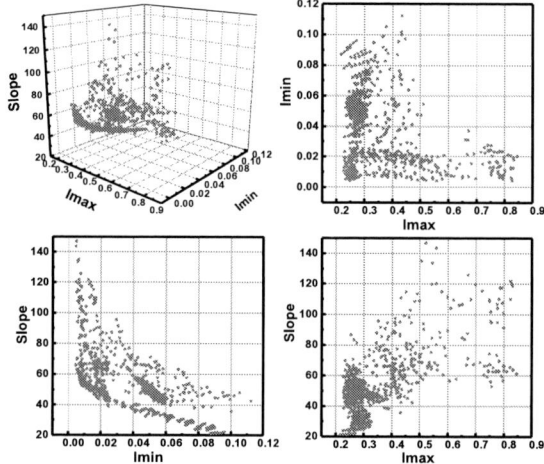

Figure 5: Parameter space of test patterns for model verification

We use the optical model to simulate the aerial image of the patterns for verification. The result is shown as Figure 5, including 3D parameter space, Imin/Imax cross-section, Slope/Imin cross-section and Slope/Imax cross-section. Each point represents an individual calibration structure's location within the aerial image parameter space.

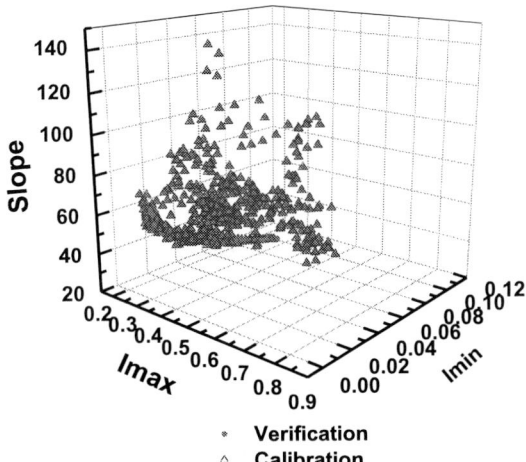

Figure 6: Parameter space of test patterns for model calibration and verification

Base on the parameter space of verification, we select the calibration patterns. First, these points are categorized into one of 18000 individual grid volumes throughout the 3D model space. Then, each grid volume occupied by the verification patterns is chosen as the candidate grid volume. In each candidate, at least one pattern should be chosen as the representative patterns for calibration. For the dense grid volume, more representative patterns need to be considered. Finally, the selected patterns are simulated and marked in the parameter space, which is shown as Figure 6. From the result, the calibration patterns cover the verification patterns in the dimension of parameters and the count of calibration patterns is about 60% of the verification patterns.

MODEL CALIBRATION AND VERIFICATION

Based on the designed test patterns and the corresponding measured data, the gauge file which include name, coordinate, drawn CD, measured CD and weight is created.

For calibration flow, there are three steps. First, we tune the optical parameters in simple resist model. Then, we fix the optical model and add related resist model terms to acquire a more complex resist model. Finally, we jointly fine tune the optical model and resist model.

In calibration flow, there are five key points can evidently affect the accuracy of model:

1) As the input data of model calibration, the measured data should be effective. For some patterns beyond the process limit, the profiles of these patterns are rough and the measured CDs have a large deviation. Therefore, these patterns should be added to the verification patterns. The other ineffective information is the noise from measurement. In order to improve the accuracy of measurement data, one point should be measured at least three times. And the large variance should be checked according to the SEM image.

2) In the selection of model form, use the simple form as an initial and add resist terms step by step. The initial model form should contain the diffusion terms and neutralization terms, which characterize the latent image at PEB stage without resist deformation. Then, the shrinkage term is added to the model form to reflect the deformation at PEB and development stages.

3) The constraints of model term coefficient decide the direction of optimization. As the optimizing engine has multiple solutions, a constraint lack of physical significance maybe gains an unreasonable result with small RMS.

4) The weights of patterns and conditions should be resized according to the calibration result. Due to the smaller noise in the measurement of 1D patterns, increasing the weights of these patterns will be benefit to catch the shrinkage effect and improve the accuracy of resist model. Here we set the weight of 1D patterns is 3 and the weight of 2D patterns is 1. For the robust process, the probability of small process variation is higher than the one of large process variation. In addition, the weight of nominal condition is 5 and the weight of off condition is 1.

5) The model terms should be divided into some groups, such as optical terms, diffusion terms, neutralization terms and development terms.

The error distribution in model calibration is shown as Figure 7. The model error distribution of 1D and 2D patterns are in the range of [-2.5nm, 2.5nm] and [-5nm, 5nm] respectively.

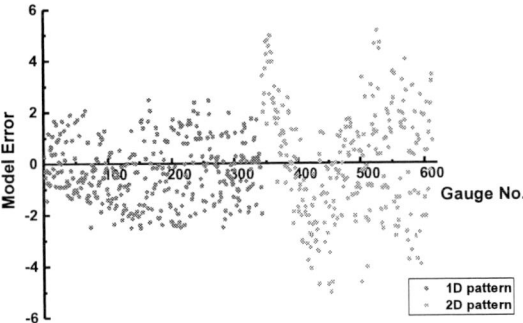

Figure 7: Model error distribution of 1D and 2D patterns

The bossung curves of the simulation result and experimental data have same trend. The max error between model_CD and wafer_CD of pattern CD86P800 is 2nm under all process condition, which is shown in Figure 8.

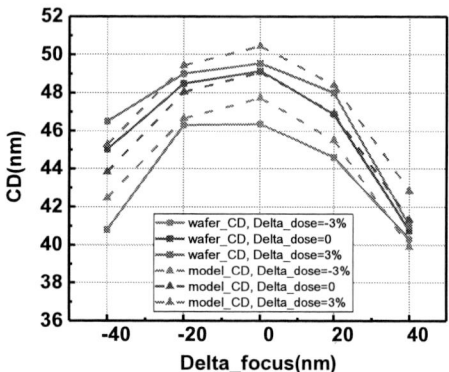

Figure 8: Bossung Curve of model_CD and wafer_CD

Finally, we use the calibrated model to check the patterns for verification. The resist bias trend of trench after adding SRAF is depicted in Figure 9. A coincidence between the model_CD and wafer_CD of these patterns can be easily observed. The resist bias is defined as model CD or wafer CD subtracted by the aerial image CD (AI_CD). The resist bias is about 6nm to 10nm, which reflect the degree of resist shrinkage and the resist model can capture the resist deformation.

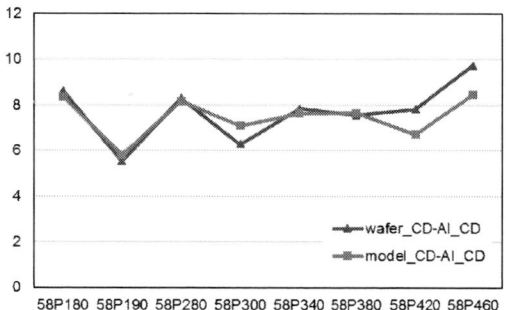

Figure 9: Resist bias trend of trench after adding SRAF

The result of model error distribution is represented in Figure 10. The standard deviation of 1D patterns is about 1nm to 2nm, and the one of 2D patterns is about 2nm to 3nm.

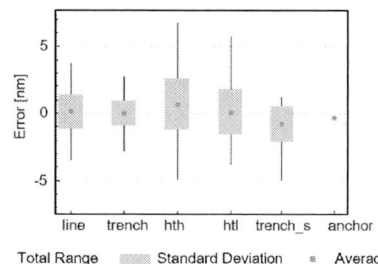

Figure 10: Model Error Distribution

CONCLUSION

This paper introduces a practical flow of establishing a resist model for 14nm NTD process, including test patterns design, model calibration and model verification. Through using the parameter space, the calibration patterns are selected effectively from the verification patterns. Then, the experimental data is used to calibrate and verify the model. The model error distribution of 1D patterns is ±2.5nm, which indicates that the resist model has the capacity of capturing the resist shrinkage effect of NTD process. For the linewidth offset effect compared to PTD, we will study and add fitting parameters in the future.

ACKNOWLEDGEMENTS

This work is supported by the National Science and Technology Major Project of China (Grant Nos. 2016ZX02301001, 2017ZX02315001003) and the National Natural Science Foundation of China (Grant No. 61604172). The authors would like to thank the Integrated Circuits Advanced Process Center and Key Laboratory of Microelectronics Devices & Integrated Technology of IMECAS.

REFERENCES

[1] G. Landie, Y. Xu, S. Burns, K. Yoshimoto, et al., *Proc. SPIE,* vol. 7972, 2011, pp. 797206.

[2] S.A. Robertson, M. Reilly, J.J. Biafore, M.D. Smith, et al., *Proc. SPIE,* vol. 7972, 2011, pp. 79720Y.

[3] L. Van Look, J. Bekaert, V. Truffert, V. Wiaux, et al., *Proc. SPIE,* vol. 7640, 2010, pp. 764011.

[4] P. Liu, L. Zheng, M. Ma, Q. Zhao, et al., *Proc. SPIE,* vol. 9779, 2016, pp. 97790Y.

[5] C.M. Hu, F. Lo, E. Yang, T. Yang, et al., *Proc. SPIE,* vol. 9780, 2016, pp. 978018.

[6] J. Bao, C.J. Spanos and A.R. Romano, *Proc. SPIE,* vol. 3743, 1999, pp. 16-24.

[7] C. Fang, M.D. Smith, S. Robertson, J.J. Biafore, et al., *Journal of Photopolymer Science and Technology,* vol. 27, 2014, pp. 53-59.

[8] W. Gao, U. Klostermann, T. Mülders, T. Schmoeller, et al., *Proc. SPIE,* vol. 7973, 2011, pp. 79732W.

[9] N. Jakatdar, J.W. Bao and C.J. Spanos, *Proc. SPIE,* vol. 3678, 1999, pp. 275-282.

[10] B. Küchler, T. Mülders, H. Taoka, W. Gao, et al., *Proc. SPIE,* vol. 10147, 2017, pp. 101470F.

[11] T. Mülders, H.-J. Stock, B. Küchler, U. Klostermann, et al., *Proc. SPIE,* vol. 10146, 2017, pp. 101460M.

[12] M. Niehoff, V.K. Singh, S. Shang and O. Toublan, *Proc. SPIE,* vol. 6156, 2006, pp. 615619.

[13] K. Patterson, Y. Trouiller, K. Lucas, J. Belledent, et al., *Proc. SPIE,* vol. 5756, 2005, pp. 294.

Source and Mask Co-Optimization Based on Depth Learning Methods

Fei Peng[1], Yijiang Shen[*]

[1]School of Automation, Guangdong University of Technology, Guangzhou 510006. China
[*]Corresponding Author's Email: yjshen@gdut.edu.cn

ABSTRACT

In this paper, we investigate the source and mask co-optimization (SMO) in optical lithography based on depth learning methods. We first describe the polarization state of light propagation in the wafer image formation using an analytical vector imaging model together with a stratified medium model; then we apply the depth learning algorithms including Adagrad and Adam methods to formulate simultaneous source-mask co-optimization framework; improved performance in terms of pattern fidelity is merited by the simulation results.

INTRODUCTION

In semiconductor manufacturing, optical lithography is a critical step transferring desired circuit patterns from a mask onto a semiconductor wafer using light. The resolution of the printed pattern is limited by the incident light wavelength, the numerical aperture (NA) and proportionality factor k_1 [1].

Source and mask co-optimization (SMO), which inverts the procedure of lithography process without the constraint of the topology of the original design [2], is a widely used resolution enhancement technique (RET) by modifying the source and the mask simultaneously to improve the printability. SMO methods including gradient-based methods with process robustness [3], augmented Lagrangian methods [4] are based on scalar imaging models which are only accurate for NA less than 0.4. Peng et al. summarized the above developments in high-NA projection optics into a uniform and consistent formulation with further simplification [5], based on which Ma et al. developed pixelated gradient-based SMO using vector imaging model [6], followed by the level-set based optimization framework by Shen [7, 8, 9] which describes the vector nature of image formation in hyper-NA lithography systems, however, the polarization of optical propagation in the wafer is not accounted for, compromising the accuracy of the wafer image formation .

In this paper, we develop a SMO optimization framework based on a vector imaging model, further improving the accuracy of the lithography process by embedding the wafer imaging formation as the reflection and refraction of electromagnetic waves in stratified medium [10]; subsequently, depth learning methods including Adagrad [11] and Adam [12] are exploited to jointly optimize the source and the mask.

VECTOR IMAGING MODEL

A schematic of image of formation based on vector model and the reflection from and through the stratified media model are given in Figures 1 and 2, respectively. Consider a monochromatic wave propagating in the direction $\vec{k}$, where $\vec{k} = (\alpha, \beta, \gamma)^T$ is the direction cosine with $\gamma = \sqrt{1 - \alpha^2 - \beta^2}$, the mapping between the polarization coordinate system and the spatial coordinate system is [13]

$$E'(\alpha_s, \beta_s) = \begin{bmatrix} E_x \\ E_y \\ E_z \end{bmatrix} = T \begin{bmatrix} E_\perp \\ E_\| \end{bmatrix}, \quad (1)$$

where $E'(\alpha_s, \beta_s)$ with E_x, E_y, E_z being the electric field components along the x, y, and z axes is the electric field in spatial coordinate emanated from a point source (α_s, β_s); $E_\perp, E_\|$ are the electric field components along the $e_\perp$ (TM) and $e_\|$ (TE) axes. Let the mask be denoted as a scalar matrix $M \in R^{N \times N}$, the mask near field can be calculated as [13]

$$E = E'(\alpha_s, \beta_s) \odot B \odot M, \quad (3)$$

where $\odot$ is the entry-by-entry multiplication. B is the mask diffraction matrix defined by the constant scattering coefficient assumption (CSCA) [5,6] as

$$B(m, n) = e^{\frac{j2\pi\beta_s m + j2\pi\alpha_s m}{\lambda} \times pixel}, m, n = 0, 1, \ldots N - 1, \quad (4)$$

where $pixel$ is the pixel size on the mask side, and more accurate scattering coefficients can be applied to (3) to approximate the mask near field with improved accuracy.

According to the [6, 14, 15], the aerial image intensity and the printed pattern are computed as

$$I = \frac{1}{J_s} \sum_{(\alpha_s, \beta_s)} J(\alpha_s, \beta_s) \sum_{p=x,y,z} \left\| H_p^{\alpha_s\beta_s} (B^{\alpha_s\beta_s} \odot M) \right\|_2^2, \quad (5)$$

and

$$Z = sig(I) = [1 + e^{-a(I-t_r)}], \quad (6)$$

where $J(\alpha_s, \beta_s)$ is point source intensity from a point source (α_s, β_s) and $J_s = \sum_{(\alpha_s, \beta_s)} J(\alpha_s, \beta_s)$ embeds the sum of all point source intensities, $H_p^{\alpha_s\beta_s}$, $B^{\alpha_s\beta_s}$ are both functions of (α_s, β_s), and $H_p^{\alpha_s\beta_s}$, p = x, y, z are referred as the equivalent filters of the x, y, z components and can be calculated as

$$H_p^{\alpha_s\beta_s} = \frac{2\pi}{n_w R} \mathcal{F}^{-1} \left\{ \sqrt{\frac{\gamma}{\gamma}} V \odot U \odot E \right\}, \quad (7)$$

where $V(m, n)$ characterizes the rotating factor in a hyper-NA system incorporating the stratified media model [10]

978-1-5386-5309-8/18 $31.00 © 2018 IEEE

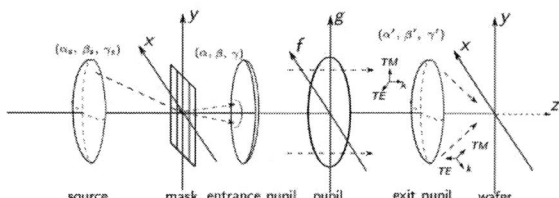

Figure 1: Schematic of vector model

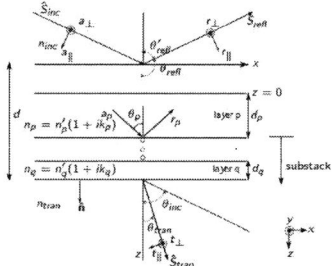

Figure 2: Reflection from and transmission through a stratified medium

$$V(\alpha, \beta, \gamma) = \begin{bmatrix} F_\perp & F_\parallel^{xy} & 0 & 0 & 0 \\ 0 & 0 & F_\perp & F_\parallel^{xy} & 0 \\ 0 & 0 & 0 & 0 & F_\parallel^z \end{bmatrix} \begin{bmatrix} \frac{\beta^2}{1-\gamma^2} & \frac{-\alpha\beta}{1-\gamma^2} \\ \frac{\alpha^2\gamma}{1-\gamma^2} & \frac{\alpha\beta\gamma}{1-\gamma^2} & 0 \\ \frac{-\alpha\beta}{1-\gamma^2} & \frac{\alpha^2}{1-\gamma^2} & \alpha \\ \frac{\alpha\beta\gamma}{1-\gamma^2} & \frac{\beta^2\gamma}{1-\gamma^2} & \beta \\ -\alpha & -\beta & \gamma \end{bmatrix}, \quad (8)$$

where (α, β, γ) is the wave propagates at the back end of the exit pupil and $F_\perp$, $F_\parallel^{xy}$, $F_\parallel^z$ are the ratios of the photoresist electric field amplitude to the incident field amplitude for TE polarization and x, y, z components in TM polarization where details can be referred to [1].

SOURCE AND MASK COOPTIMIZATION FRAMEWORK

The object of SMO is to find the optimal source S and mask M which minimizes the designed distance between the target pattern and printed pattern which is normally defined by the l2 norm, leading to the minimization of

$$F = \|Z - M\|_2^2. \quad (9)$$

Lifting the bound-constraint of $M \in [0,1], S \in [0,1]$ with parameter transformations $J = \frac{1+\cos\theta}{2}$ and $M = \frac{1+\cos\omega}{2}$, the gradients of the cost function F with respect to the source pattern θ and mask pattern ω are calculated as

$$\nabla F_J =$$
$$\frac{a}{J_s^2}\left(\frac{-J_s \sin\theta}{2} + \frac{\sin\theta}{2}\sum_{(\alpha_s,\beta_s)} J(\alpha_s,\beta_s)\right) \odot$$
$$\sum_{m,n=1}^N \sum_{p=x,y,z}\left\|E_p^{wafer}\right\|_2^2 \odot (Z-M) \odot Z \odot (1-Z), \quad (10)$$

$$\nabla F_M = \frac{-a\sin}{2J_s}\sum_{(\alpha_s,\beta_s)} J(\alpha_s,\beta_s)\sum_{p=x,y,z}((H_p^{real}+H_p^{img})\otimes(M \odot B))) \odot (Z-M) \odot I_{output} \odot (1-Z). \quad (11)$$

ALORITHM 1. ADAM

> **Stochastic optimization:** g_t^2 indicates the element-wise square $g_t \odot g_t$, and t denote the time-step
>
> **Require:** α' (learning rate)
> **Require:** $\beta_1, \beta_2 \epsilon [0,1]$: Exponential decay rates for the moment estimates.
> **Require:** f(θ'): Stochastic objective function.
> **Require:** θ': Initial parameter vector
> $m_0 \leftarrow 0$ (Initialize 1st moment vector)
> $v_0 \leftarrow 0$ (Initialize 2st moment vector)
> $t \leftarrow 0$ (Initialize time-step)
>
> While θ_t' not converged do
> $t \leftarrow t + 1$
> $g_t \leftarrow \nabla_{\theta'} f_t(\theta_{t-1}')$
> $m_t \leftarrow \beta_1 m_{t-1} + (1-\beta_1)g_t$
> $v_t \leftarrow \beta_2 v_{t-1} + (1-\beta_2)g_t^2$
> $\hat{m}_t \leftarrow m_t/(1-(\beta_1)^t)$
> $\hat{v}_t \leftarrow v_t/(1-(\beta_2)^t)$
> $\theta_t' \leftarrow \theta_{t-1}' - \alpha'\hat{m}_t/(\sqrt{\hat{v}_t} + \varepsilon)$
> end

In particular, illumination symmetry is required to avoid pattern placement error and is therefore guaranteed by,

$$\overline{\nabla}F_J = \frac{1}{2}\left[\frac{\nabla F_J + Q\nabla F_J}{2} + \frac{\nabla F_J + Q\nabla F_J}{2}Q\right], \quad (12)$$

where ∇F_M, Q, $\overline{\nabla}F_J$ are gradient of source, flipping matrix and averaged gradient. The updating procedure using the Adam algorithm is given in ALORITHM 1, where without loss of generality, the gradients of the cost functions ∇F_J and ∇F_M are denoted as g_t. In ALORITHM 1, self-adapted step-sizes are calculated accordingly, showing more robustness than conventional steep-descent methods, which minimizes the cost function rapidly without regard for step sizes. The Adagrad algorithm, similar to the Adam algorithm, outperforms the latter with faster convergence rate in the early stages, which make a better performance with source optimization. Consequently, the Adagrad and Adam methods are applied to modify the source pattern and the mask pattern in the simulations, respectively.

NUMERICAL RESULTS

The learning rates of Adam and Adagrad are empirically set as 0.1 and 0.3, respectively. System parameters used for the simulations are: λ = 193nm, NA = 1.35, spatial resolution Δx = 4 nm/pixel, a = 80 and $t_r = 0.25$ using to describe sigmoid function and the system is initially illuminated by an annular source with partial coherent factor $\sigma_{in} = 0.6$ and $\sigma_{out} = 0.9$. The parameters of the wafer stack are given in Table 1 [10].

TABLE 1: WAFER STACK PARAMETERS

Layer	Index	Thickness
Incident medium	(1.45, 0)	
Top anti-reflection	(1.55, 0)	35nm
Photoresist	(1.8, 0.02)	100nm
Bottom anti-reflection	(1.72, 0.33)	87nm
Substrate	(0.883, 2.788)	

978-1-5386-5309-8/18 $31.00 © 2018 IEEE

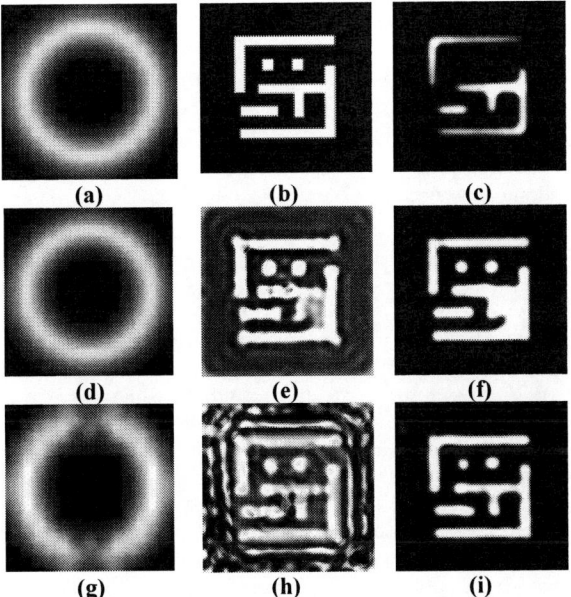

(a) **(b)** **(c)**

(d) **(e)** **(f)**

(g) **(h)** **(i)**

Figure 3: Simulation results with MO and SMO approaches after 50 iterations. (a) and (b) are the initial source and target patterns, respectively, with wafer image given in (c). (d) the initial source pattern (e) synthesized mask using MO, (f) is the corresponding wafer image with (e) as input. (g) and (h) are the synthesized source and synthesized mask using SMO, with the wafer image shown in (i).

In Figure 3, the desired pattern is given in (b) with image size 257×257; (e) is the synthesized mask pattern calculated by MO after 50 iterations, the synthesized mask pattern and the synthesized source pattern are shown in (g) and (h) using the proposed SMO 44 iterations. From the corresponding wafer images in (c), (f) and (i) with input in (b), (e) and (h) illuminated by the original annular source in (a), (d) and the synthesized source in (g), it is observed that the proposed SMO methods greatly improves pattern fidelity.

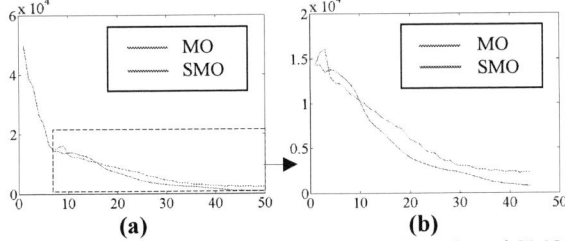

(a) **(b)**

Figure 4: Pattern Error (PAE) for the MO and SMO approaches. In (a), the first 7 iterations applies the MO to achieve an initial mask pattern with PAE below 15000, and (b) the PAE drifts of MO and SMO in iterations 8-50 improving pattern fidelity from PAE to 2286 and 756 respectively.

Figure 4 (a) illustrates the whole process of the proposed SMO approach, where an initial mask pattern for the SMO is synthesized using only the MO in the first 7 iterations, and pattern fidelity of 756 using the proposed approach and 2286 of the MO after 50 iterations is presented in (b), showing significant PAE improvement and much faster convergence.

CONCLUSIONS

A vector imaging model based SMO approach is formulated with the stratified medium model describing the reflection and refraction of electromagnetic fields in the wafer after synthesized source and mask are computed efficiently using the Adam and Adagrad methods.

ACKNOWLEDGEMENTS

The work in this paper was partially supported by Natural Science Foundation of Guangdong Province, China (2016A030313709, 2015A030310290) and Guangzhou Science and Technology Project, China (201607010180).

REFERENCES

[1] A. K. Wong, *Resolution Enhancement Techniques in Optical Lithography.* Washington: SPIE Press, 2001.

[2] L. Pang, Y. Liu and A. Dan, *Proceedings of the SPIE 6607*, May 15, 660739.

[3] N. Jia and E. Lam, *Opt. Express.*, vol. 19,2011, pp. 19384-19398.

[4] J. Li and E. Lam, *Ecs Transactions.*, vol. 52, 2013, pp. 163-168.

[5] T. Pistor, A. Neureuther, and R. Socha, *Proceedings of the SPIE 4000*, 2000, pp. 228-237

[6] X. Ma, Y. Li and L. Dong, *J. Opt. Soc Am.*, vol. 29,2012, pp. 1300-12.

[7] Y. Shen, N. Wong and E. Lam, *Opt. Express.*, vol. 19, 2011, pp. 5511-5521.

[8] Y. Shen, *proceeding of CSTIC 2017*, March 12-13, 2017, pp. 1-3

[9] Y. Shen, *Opt. Express.*, vol. 25, 2017, pp. 21775-21785.

[10] A. K. Wong, *Optical Imaging in Projection Microlithography.* Washington: SPIE, 2005.

[11] J. Duchi, E. Hazan, and Y. Singer. *J. JMLR.*, vol. 12, 2011, pp.257-269.

[12] DP. Kingma, and J. Ba, *J. CoRR.*, vol. abs/1412.6980, 2014.

[13] D. Peng, P. Hu, V. Tolani, and T. Dam, *Proceeding of the SPIE 7640*, 76402Y (2010).

[14] X. Ma, C. Han, Y. Li, L. Dong and GR Arce, *J. Opt. Soc Am.*, vol. 30,2013, pp. 112-23.

[15] N. Jia and E. Y. Lam, *proceeding of CSTIC 2010*, Sep 26-29, 2010, pp. 4173-4176

An Optical Proximity Model for Negative Toned Developing Photoresists

Qiang Wu

Semiconductor Manufacturing International Corporation
Technology Research and Development
18 Zhangjiang Road, Pudong New Area
Shanghai, P.R.C. 201203
+86-13585860701, ken_wu@smics.com

ABSTRACT

At 14 nm technological node and beyond, negative toned developing (NTD) has become needed process for trench and contact/via photolithographic processes owning to its higher utilization of light energy compared with positive toned developing (PTD) process. However, the efficiency gain for the NTD process only happens in semi-dense to isolated pitches where more light can pass through the mask. For dense pitches, such as the pitches ranging from 76-100 nm, the NTD process has no advantage compared to the PTD process. This is because in the PTD process, the photochemical reaction is only needed to weaken the targeted photoresist volume so that the targeted area can collapse and be removed by the developer to form photoresist image. In the case of negative toned developing method, however, the targeted photoresist portion has to be fully reacted, or its dissolution properties in the solvent has to be fully changed in order to remain after solvent developing. As a confirmation of the model, after developing, the profile of the NTD resists will shrink in thickness and size. This indicates that, for the same amount of light, there will be more chemical amplification in NTD photoresist than in PTD photoresist. We will present a model based on the above idea.

I INTRODUCTION

As photolithography has provided the biggest contribution to the linewidth shrink in semiconductor manufacturing process. Shown in equation (1), the resolution is limited by three factors, the wavelength, the numerical aperture NA, and the K_1 factor, or the ways of illumination.

$$CD(\text{half pitch}) = K_1 \frac{\lambda}{NA} \tag{1}$$

Over the past 30 years, linewidth shrink has been done through the shrink in wavelength, the increase of numerical aperture, and the off-axis illumination angle. With the introduction of copper back-end technology of the 130 nm technology node, due to the difficulty in finding an acceptable etching process for the copper metal layer, the copper has to be done through the so-called

damascene process. In this case, the photolithography has to print trenches instead of lines which is used in the aluminum process.

Although from aerial image, the resolution limit for the lines and trenches are the same. For the minimum pitch, the lines and spaces are usually equivalent. A photoresist that can print very dense lines also prints the same dense trenches with the same image contrast. A simulated aerial image is shown in Figure 1[1]. However, for the semi-dense patterns of the backend, the area occupied by the trenches or vias is usually less than 50%, which means that the average transmission rate of the entire mask is usually <50%. A similar simulation is done for a pitch of 200 nm, shown in Figure 2(a).

As we know that the photoresist, especially the chemically amplified photoresist can be basically described with a threshold model, due to their high dissolution contrast with the exposure energy. From the point of view of the photoresists, for the same exposed part, the pattern shown in Figure 2(a), also named the bright field pattern, certainly receives more exposure per unit area than the pattern shown in Figure 2(b). For the backend trench/via lithography, the Figure 2(b) represents the positive toned developing (PTD) process, and Figure 2(a) represents the negative toned developing (NTD) process.

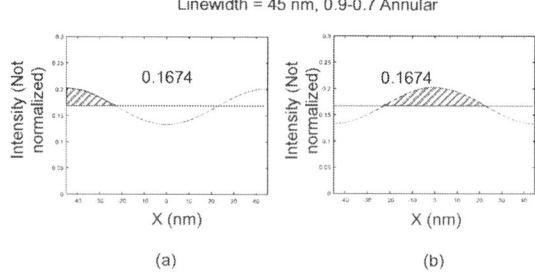

(a)   *(b)*

Figure 1: aerial image simulated with 90 nm thick photoresist with zero acid diffusion and 1.35NA with no polarization for a 45 nm (a) line (b) space at 90 nm pitch. The partial coherence is 0.9-0.7 annular with 9 x 9 segmentation in the 1st quadrant of the illumination pupil.

The mask is a binary with no Mask 3D scattering considered. The threshold is 0.1674 for both (a) and (b).

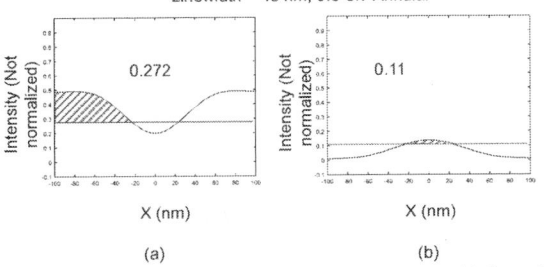

(a)   (b)

Figure 2: aerial image simulated for a 45 nm (a) line (b) space at 200 nm pitch with the same exposure condition as in Figure 1. The threshold is 0.272 for (a) and 0.11 for (b).

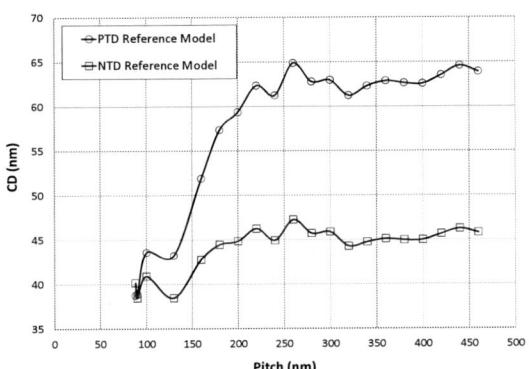

Figure 3: 2 simulated linewidth through pitch curves using a set of post-OPC mask values (for the NTD process). The simulations are done with a pure optical model with 10 nm Gaussian diffusion length and a photoresist model, respectively, under 1.35NA and an illumination condition resembling a cross quadrupole 0.9-0.7 CQ60 degrees with ASML Brion's SMO software.

From many years of the use of the chemically amplified resist, we know that the chemical amplification is done through the simultaneous reaction-diffusion of the photoacid, which catalytically trigger the deprotection of the photoresist material. The diffusion of photoacid, however, will cause the aerial image contrast to degrade significantly and too much photoacid diffusion will cause linewidth roughness (LWR) to increase. In case of the positive toned developing, the deprotection of the photoresist polymer does not need to be 100% done since we only need to weaken and tear apart the exposed area so that it can be removed by the liquid phase developing process. What we found during the use of the chemically amplified resist is that the aerial image degradation by the photoacid diffusion can be described very well by a simple Gaussian diffusion. In case of NTD, we found the following:

(1) The MEF for the densest patterns, such as the anchoring pitch in the 1-dimensional dense lines/spaces is larger than that of the PTD.

(2) The semi-dense and isolated trenches can be printed with about 15-20 nm smaller linewidth with no degradation of image intensity gradient, or the exposure latitude, as described in Figure 3.

(3) Comparing with the same PTD resist, the NTD's resolution is worse. E.g., the PTD can print 80 nm line and spaces, but the NTD resist we have can not.

For the NTD, since it is the exposed area that remains after the developing process, the deprotection, or dissolution rate change is also through the chemical amplification. Unlike the PTD, the deprotected regions made by individual photoacids have to grow to some minimum size such that they can join one another to avoid being removed by the developing process, as illustrated by Figure 4. This suggests that the diffusion length of the photoacid in a NTD photoresist will be longer compared to the same resolution class PTD photoresist.

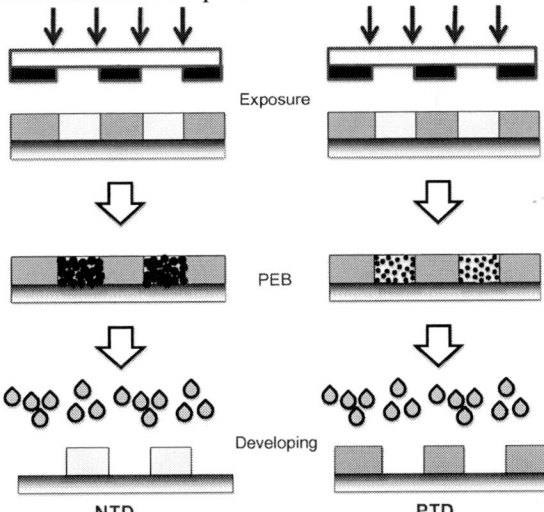

Figure 4: Schematic diagram of the negative toned developing and the positive toned developing processes.

II MICROSCOPIC MODEL OF PHOTOACID SCATTERING AND DIFFUSION

As found from our evaluation of our best 193 nm immersion photoresists, the photoacid diffusion length is found to be around 5 nm for the PTD and 10 nm for the NTD. Longer diffusion length can explain higher MEF and worse resolution. It can also explain the fact that the deprotection areas made by individual photoacids must join one another. The NTD's 2 times longer diffusion can have 8 times higher deprotection volume compared to the

978-1-5386-5309-8/18 $31.00 © 2018 IEEE   135

PTD. In NTD, the deprotected pockets of the polymer will look like packed spheres with equal radius, as illustrated in Figures 5(a)-5(b). The total exposed area can be found to be 1.35-1.91 times than the total area occupied by the spheres, respectively, for the hexagonal and cubic geometries. That is to say, the saturation in deprotection is about 1.35-1.91 times the NTD reaction threshold. In reality, due to the existence of random distribution of the photoacids and random deprotection reaction, the number may be larger.

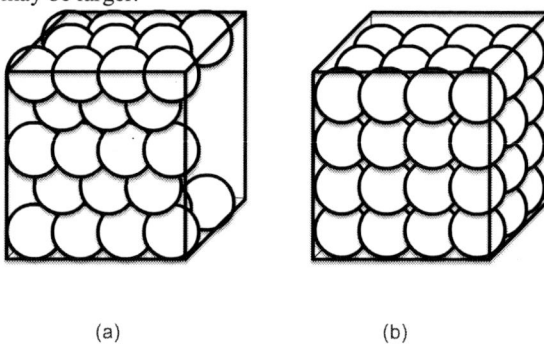

(a) (b)

Figure 5: Schematic sketches of minimum deprotected pockets inside an exposed (a) 3D hexagonal volume; (b) 3D cubic volume.

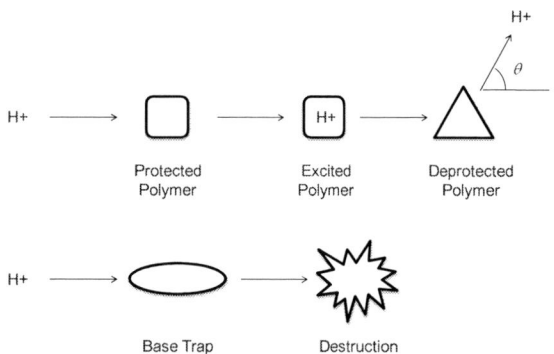

Figure 6: Schematic sketches of 2 processes that can change the diffusion of a photoacid.

For an individual photoacid, there are two scattering centers that can change its diffusion: the deprotection centers and the base quencher traps, as depicted by Figure 6. In case of the former, the photoacid is absorbed by the polymer temporarily and, after triggering the catalytic deprotection reaction, the photoacid is released with a random direction angle with respect to the incident angle. In case of a base quencher, the photoacid is destroyed forever.

For the deprotection reaction, assuming a polymer with molecular weight of 5000, which consists of mostly carbon and hydrogen, such as the poly (methyl

methacrylates) with adamantyl groups, lactone groups, and some carboxylic acids as pendant groups. If each carbon atom carries 2 hydrogen atoms, and we model a polymer molecule as a twisted chain with chain segment length of 0.154 nm (the single Carbon-Carbon chain length), there are a total of 357 segments. If we treat the segment length as some unit cell diameter, and the molecule is densely twisted and packed into a sphere, the radius of the molecule can be calculated as follows:

$$R(\text{number of segments}) = \sqrt[3]{\frac{3 \times 357}{4\pi}} = 4.4$$

(2)

The diameter of the molecule is 0.154 nm $2R$=1.36 nm.

This says that during the course of a photoacid that diffuses for 5-10 nm before it is finally quenched, it may travel across many polymer molecules, each with a diameter of 1.36 nm! Assuming nearly all the exposed polymer molecules will be deprotected, as necessary for NTD, in 1-dimension, an extra 5 nm diffusion will mean it will traverse about 4 deprotection sites. As depicted by Figure 7, in order for the photoacid to diffuse 5 extra nm, in 1-dimensional space, there is only 1/8 the probability in going for 4 molecular sites in one direction before it is quenched! For 3D, the probability can be smaller!

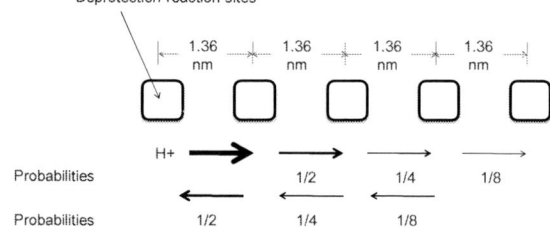

Figure 7: Schematic diagram of a photoacid that will diffuse for 5 extra nm in 1-dimensional space.

When the deprotection sites are saturated, the photoacid may diffuse up to 8 times more distance before it is quenched in 1D diffusion. In Figure 8, we have performed a simulation with an additional variable diffusion convolution for the 10 nm diffused image intensity above the 1.6 x threshold. The diffusion length for the saturated area is 60 nm and 10 nm for the non-saturated area. Equation (3) has shown our proposed model for the variable acid diffusion, where $a_{\text{non-saturating}}$=10 nm, a_{sat}=60 nm, I_{Trunc} is the truncated aerial image intensity illustrated in Figure 8, $level_{sat}$=1.6$threshold$, and the n is an empirical exponent and is set to equal 1 in our model, which means that the diffusion length is roughly proportional to the already deprotected site density within the exposed volume.

$$a = a_{\text{non-saturating}} + I_{\text{Trunc}}(x)^n / level_{sat}^n \left(a_{sat} - a_{\text{non-saturating}} \right)$$

(3)

Figure 8 indicates that the final aerial image intensity profile is pushed by the large acid diffusion from the deprotection saturated areas, or the exposed areas. However, after they enter the non-saturating area, they meet with quickly increasing deprotection sites and the diffusion length will be quickly reduced. This will create a sharper image gradient, which helps maintain or even improve image contrast. This is observed in wafer exposure experiment. The overall effect has been reported previously [2].

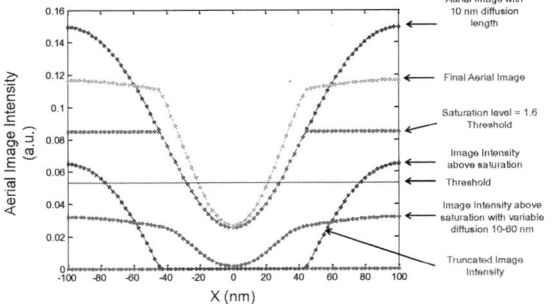

Figure 8: Simulated aerial image at 200 nm pitch with energy anchored at 90 nm pitch. The imaging threshold = 0.053.

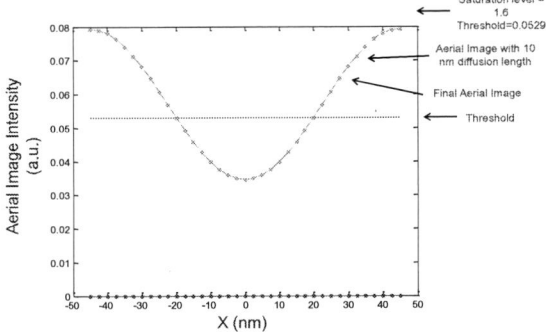

Figure 9: Simulated aerial image at 90 nm pitch. The maximum image intensity dose not go above the saturation threshold, which is 1.6x the photolithography threshold (threshold=0.053): 0.085.

But for the dense pitch, the saturation will not happen since all aerial images are sinusoidal with low contrast, any saturation will degrade image contrast. Not surprisingly, we found that the saturation does not happen at the dense pitches, such as the 90 nm pitch, as described by Figure 9, since the maximum image intensity is below the saturation threshold

III LINEWIDTH THROUGH PITCH SIMULATION COMPARISON WITH COMMERCIAL SOFTWARE

We have performed a simulation comparison with that of the commercial software (Brion), which has been verified by wafer data.

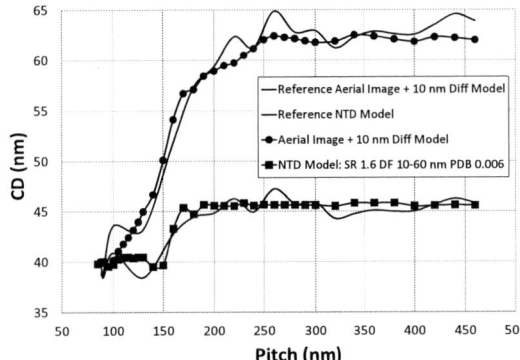

Figure 10: Simulation of the linewidth through pitch with our proposed model with the commercial model. The simulation also uses 0.6% photo-decomposable base (PDB) concentration.

The comparison data are displayed in Figure 10 and the agreement is very good. The information we get from the fit is that the model can accurately characterize the NTD exposure-developing effect with 3 new physical parameters. What we learn is that the photoacids can diffuse much longer distance in areas where the deprotection has reached completion, or saturation.

CONCLUSION

We have proposed a physical model for the negatively toned developing (NTD) process with three more physical parameters in addition to the original photolithography imaging model based on partially coherent imaging with mask 3D scattering effect and 10 nm Gaussian diffusion. The agreement with commercial software is very good.

ACKNOWLEDGEMENTS

We would like to thank SMIC TD colleagues and management for the support of this work.

REFERENCES

[1] Qiang Wu, "Key points in 14 nm photolithographic process development, challenges and process window capability", Proc. CSTIC 2017.

[2] S. S. Mehta, C. Higgins, V. Chauhan, S. Pal, H. P. Koh, J. R. Fakhoury, S. Gao, L. Subramany, S. Iqbal, J. Bumhwan, P. Morrison , C. Karanikas, Y. Wei, D. R Cho, "Investigation of Trench and Contact Hole Shrink Mechanism in the Negative Tone Develop Process", Proc. SPIE 8682, 23, 2013.

LENS THERMAL BEHAVIOR BASED EXPOSURE PROCESS CONTROL FOR FLAT PANEL DISPLAY LITHOGRAPHIC TOOLS

*Zhiyong Yang, Hao Jiang, Shiyuan Liu**

State Key Laboratory of Digital Manufacturing Equipment and Technology, Huazhong University of Science and Technology, Wuhan 430074, China
*Corresponding Author's Email: shyliu@hust.edu.cn

ABSTRACT

This paper presents a comprehensive approach to control flat panel display (FPD) lithographic tool exposure process based on projection lens thermal behavior. Lithographic tool can maintain enough process window and high throughput for large DOSE exposure on this control method. In some FPD lithographic process layers, the exposure tool would generate huge thermal aberration, such as focal plane and magnification drift, because of the large DOSE. This control method includes several technics such as high accuracy patterning quality prediction, high accuracy thermal aberration prediction, aberration correction and equipment process adaptive management. In exposure process, the lithographic tool can adjust the internal parameters and working sequences in real time to maintain the aberration level and tact time. Based on this method, the FPD lithographic tool is applicable for high resolution and high DOSE processes with high manufacturing efficiency.

INTRODUCTION

Projection lithographic tool is the core equipment for FPD manufacturing. A typical FPD lithographic process includes more than 10 layers. First several layers are key layers that the patterning resolution is 2μm line/space and the typical energy is 30mJ/cm^2. However, the last layer is the organic film exposure that the resolution is lower than 10μm line/space but the exposure energy is higher than 300mJ/cm^2. In such processes, lithographic tool should adapt to different conditions including both high resolution medium exposure energy and medium resolution but very high exposure energy.

In generally, a cool projection lens has perfect optical performance that the wavefront aberration should be smaller than nanometer level and the grid distortion should be much smaller than overlay and focus requirements. However, the thermal effect of projection lens significantly affects the image quality in high energy exposure. Due to the absorption of ultraviolet light energy, the lens elements become deformed with index changing. Obvious drifts or declines of optical performance, such as focal plane, magnification, distortion and high order aberrations, would appear.

Optical aberration of projection lens dominates the patterning quality. Previous studies have already shown that the optical aberrations could result in a series of serious problems such as overlay error, critical dimension variance, pattern asymmetry, defocus and image blur, which are unacceptable in FPD lithographic process.

Since active elements are adopted, the projection lens of IC lithographic tool can self-adjust the wavefront aberration and magnification, and therefore most of the thermal effect could be accurately corrected. However, FPD projection lens has a quite different structure for large area exposure, in which only magnification and focal plane could be adjusted. It is necessary to develop a new control strategy to eliminate the strong thermal effects and high order thermal aberration in FPD projection lens.

This paper proposes an approach to control the exposure process of FPD lithographic tool based on projection lens thermal behavior, including the considerations such as high precision patterning quality prediction, high precision thermal aberration prediction, aberration correction and equipment process adaptive management. Firstly, we simulate the pattering result that includes resolution and overlay at different optical aberration levels. Then, we calculate the thermal behavior of projection lens and calibrate to several functions that describe the aberrations variant with time. In this step, we should also consider the typical mask patterns and resolutions in FPD manufacturing process, because different pattern has different thermal behaviors and different aberration tolerances. The first two steps are finished off-line. Furthermore, we also design some adjustable elements for projection lens to correct the low order thermal aberrations. Finally, the control system of lithographic tool would adjust the internal parameters and working sequences based on all result achieved in the first three steps.

In comparison to traditional thermal aberration control method, two novel techniques for FPD manufacturing process are specially introduced. First one, we define and modify the aberration tolerance based on the requirement of typical pattern resolution in a real case. And the second one is that an adaptive control strategy has been introduced in exposure process. In this paper, process control refers to the internal functions of lithographic tool could be adjusted based on the real time thermal status. For example, when the thermal aberration is too high to be corrected by the adjustable elements, control system would tune the exposure duty cycle and add some automatic calibration processes between dies. While the thermal aberration is acceptable, control system would cancel some auxiliary process to improve the throughput.

Based on this control strategy, lithographic tool can maintain sufficient process window and high throughput. Especially for organic film process, the lithographic tool obtains high manufacturing efficiency with huge thermal aberration corrected for patterning quality.

LENS THEMAL BEHAVIOR

In generally, UV light is commonly used in FPD lithographic tool for higher resolution. However, that is the nature behavior that the glass material of optical lens has certain absorption of UV light. During the exposure process, the UV light will be converted to thermal energy and the temperature of glass will rise. Despite the use of some technical means, such as gas purging and edge cooling, the internal temperature of glass is raised and the index of refraction also changes at the same time [1].

The optical aberration determines the lithographic patterning quality. In projection lens of lithographic tool, the surface shape, element distance and material refraction index are the main three factors that contribute to optical aberration. As shown in Fig 1, thermal lens would generate additional aberration which causes by refraction index changing.

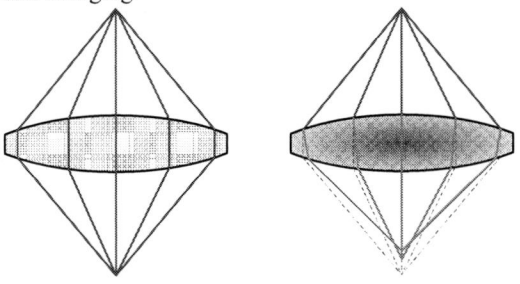

Cool Lens Imaging   Thermal Lens Imaging

Figure 1: Thermal lens imaging

The exposure duty cycle of FPD lithographic tool is about 20%~30%. In another time, the equipment performs auxiliary actions such as plates exchanging, alignment and focus mapping. In exposure process, the thermal aberration of projection lens raises continuously with the material absorbing UV energy; in auxiliary process, the aberration of projection lens decrease with the glass material cooling. Therefore, two typical thermal behaviors can be defined as Fig 2. The first behavior curve represents the thermal aberration variance in continuous heating process. In this process, the thermal aberration would increase rapidly at the beginning and be saturated as the lens reaches the thermal equilibrium state. Similarly, the second behavior represents the thermal aberration variance in continuous cooling process from thermal equilibrium state to complete cooling state. The thermal aberration decreases rapidly at beginning of this process and remains steady near the cooling state.

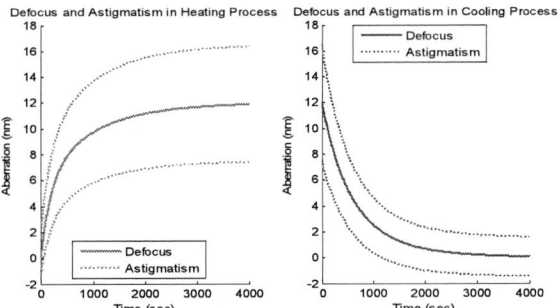

Figure 2: Typical thermal aberration, variant with time

The following two formulas express the thermal behavior of lens. Equation (1) indicates the aberration variance in heating process that A_{u1} and A_{u2} are the scaling factor of thermal aberration, the τ_{u1} and τ_{u2} are the time factor of variance, t_u is the heating time. Equation (2) indicates the cooling process and the parameters are similar meanings.

$$TA_u(t_u) = A_{u1} \cdot (1 - exp(-t_u/\tau_{u1})) + A_{u2} \cdot (1 - exp(-t_u/\tau_{u2})) \quad (1)$$

$$TA_d(t_d) = A_{d1} \cdot (-exp(-t_d/\tau_{d1})) + A_{d2} \cdot (-exp(-t_d/\tau_{d2})) \quad (2)$$

The variance of thermal aberration in complete FPD production process can be expressed as a continuous fragment of the two behavior curves. Therefore, the parameters of these two behavior curves need to be calculated [3] and calibrated accurately.

PATTERNING QUALITY

Resolution, overlay accuracy and total pitch accuracy are the main specs of FPD lithographic patterning quality. As shown in Fig3, optical aberrations such as defocus, field curvature, and astigmatism will result in a decrease in the resolution of patterning, while magnification error and distortion will significantly cause the overlay error and total pitch error.

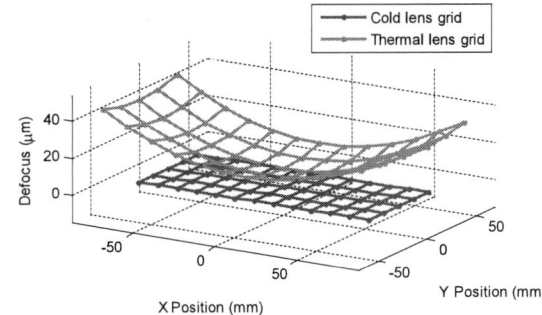

Figure 3: Focus, magnification and distortion variance, caused by thermal wavefront aberration

978-1-5386-5309-8/18 $31.00 © 2018 IEEE

These typical types of thermal aberrations must be controlled within a certain range to ensure FPD lithographic patterning quality. It is important to emphasize that different process layers require different exposure DOSE for FPD lithography, and the patterning resolution and overlay tolerance are not same. The influence of optical aberration on patterning quality can be expressed accurately by linear or quadratic model [4]. Based on the calculation results, the quantization rules of thermal aberration tolerance, shown as Table 1, can be established for different typical process to control the thermal aberration in the exposure process and the WFE is RMS value of high order aberrations.

TABLE I.   IMAGING QUALITY RULE FOR TYPICAL PROCESS

Typical Process	Image quality rule for thermal aberration		
	Defocus	*Mag*	*WFE*
POLY/M1, 2μm 30mJ/cm²	10μm	0.3ppm	5nm
CH, 3μm 30mJ/cm²	20μm	0.3ppm	5nm
ORG, 5μm 300mJ/cm²	60μm	2ppm	20nm

LITHOGRAPHIC TOOL CONTROL

The compensation of thermal aberrations usually adopts the feed-forward control method. Firstly, the accurately calibrated model is used to calculate the current thermal aberration based on the accumulated working status, and then the aberration is adjusted by using the compensation devices in lithographic tool and projection lens.

Precise calibration of heating and cooling behavior curves is a prerequisite for accurate control of thermal aberrations. During this off-line calibration process, reticle alignment sensor measures the position of imaging mark for a defined period and traverses the entire heating and cooling process.

In IC lithography, all thermal aberrations must be compensated precisely, so the projection lens of IC lithographic tool equipped with sophisticated wavefront adjusting device to achieve this function [2]. Subject to the optical and mechanical structure, the projection lens of FPD lithographic tool can only compensate a part of aberrations, such as defocus and magnification. The other part of aberrations, such as field curvature, astigmatism and distortion, cannot be compensated.

For this reason, FPD lithographic tool uses the following strategies to control the thermal aberrations. For compensable aberrations, the device performs the tuning according to the real-time calculating result directly. For non-compensable aberrations, lithographic tool adjust the exposure duty cycle according to the process recipe and patterning quality tolerance, shown as Fig 4.

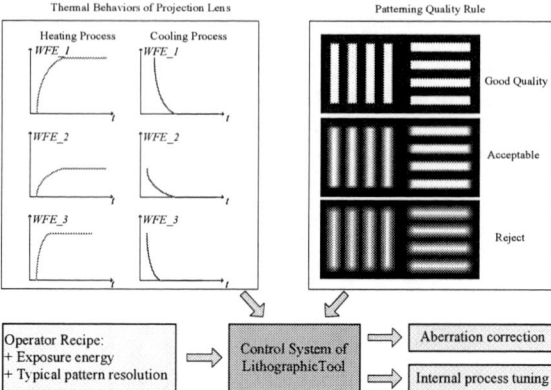

Figure 4: Lithographic tool control process, based on lens thermal behavior and patterning rule

Fig 5 represents the control result that the fragments of heating and cooling curve will be optimized combination, so that the non-compensable aberrations do not exceed the limitation of tolerance during the entire production process.

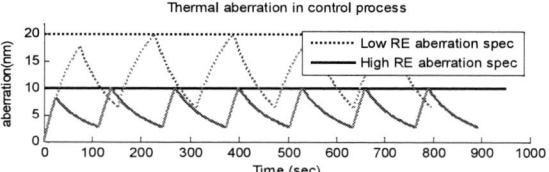

Figure 5: Thermal aberrations control result in FPD lithographic tool

CONCLUTION

Thermal aberration in FPD lithographic tool has serious impacts for patterning quality. Due to the limitation of optical and mechanical structure, some novel approaches must be taken to solve this problem. Aberration tolerances setting for different operation condition, as well as the adaptive adjustment of duty cycle, are well suited for thermal aberration control in all different types of FPD lithographic process.

REFERENCES

[1] Grace H. Ho, "Lens-heating-induced focus drift of I-line step and scan: correction and control in a manufacturing environment," Proc. SPIE 4344, August 2001

[2] Yasuhiro Ohmura, "High-order aberration control during exposure for leading-edge lithography projection optics," Proc. SPIE 9780, 2016

[3] Xinfeng Yu, "Computational method for simulation of thermal load distribution in a lithographic lens," Applied Optics 55(15) 4186-4191, 2016

[4] Zhiyong Yang, "CDU linear model based on aerial image principal components," Proc. SPIE 7640, March 2010

978-1-5386-5309-8/18 $31.00 © 2018 IEEE

High Power LPP-EUV Source with Long Collector Mirror Lifetime for High Volume Semiconductor Manufacturing

Hakaru Mizoguchi, Hiroaki Nakarai, Tamotsu Abe, Krzysztof M Nowak, Yasufumi Kawasuji, Hiroshi Tanaka, Yukio Watanabe, Tsukasa Hori, Takeshi Kodama, Yutaka Shiraishi, Tatsuya Yanagida, Georg Soumagne, Tsuyoshi Yamada, Taku Yamazaki and Takashi Saitou

Gigaphoton Inc. Hiratsuka facility: 3-25-1 Shinomiya Hiratsuka Kanagawa,254-8567, JAPAN
*Corresponding Author's Email: hakaru_mizoguchi@gigaphoton.com

ABSTRACT

We have been developing CO_2-Sn-LPP EUV light source which is the most promising solution as the 13.5nm high power light source for HVM EUVL. Unique and original technologies such as; combination of pulsed CO_2 laser and Sn droplets, dual wavelength laser pulses shooting and mitigation with magnetic field have been developed in Gigaphoton Inc.. We have developed first practical source for HVM; "GL200E" [17] in 2014. We have proved high average power CO_2 laser more than 20kW at output power cooperate with Mitsubishi electric cooperation[16]. Pilot#1 is up running and its demonstrates HVM capability; EUV power recorded at111W average (117W in burst stabilized, 95% duty) with 5% conversion efficiency for 22hours operation in October 2016[21]. Recently we have demonstrated, EUV power recorded at113W in burst stabilized (85W in average, 75% duty), with 5% conversion efficiency during 143hours operation. Also the Pilot#1 system recorded 64% availability and idle time was 25%. Availability is potentially achievable at 89% (2weeks average), also superior magnetic mitigation has demonstrated promising mirror degradation rate (= -0.5%/Gp) above 100W level operation with dummy mirror test[22] . Very low degradation (= - 0.4%/Gp) of actual collector mirror reflectance has been demonstrated above 100W level operation (in burst) with magnetic mitigation EUV source.

1. Introduction

Extreme ultraviolet (EUV) light source has been being developed together with a scanning exposure tool. As the tool with 10 W EUV light source, ASML shipped " -demo tool" in 2007[1] and Nikon shipped EUV-1 in 2008[2]. ASML has been developing -tool; NXE-3100 in the beginning of 2011 with 100 W EUV light source.[3)4] Requirement of the EUV exposure tool is now in the β-tool; NXE3300 (for high volume manufacturing (HVM))[5]. Several machines are already shipped in 2013. The required EUV power is 250 W clean power (after purifying infrared (IR) and deep ultra violet (DUV) spectra) at intermediate focus (IF). However demonstrated power level is still around 80 W[6)7]. In this paper we introduce latest EUV source more than 100W level EUV power demonstration by Pilot #1 system; 250W EUV source for HVM of Gigaphoton .

2. LPP EUV light source system and key components
2.1 System concept

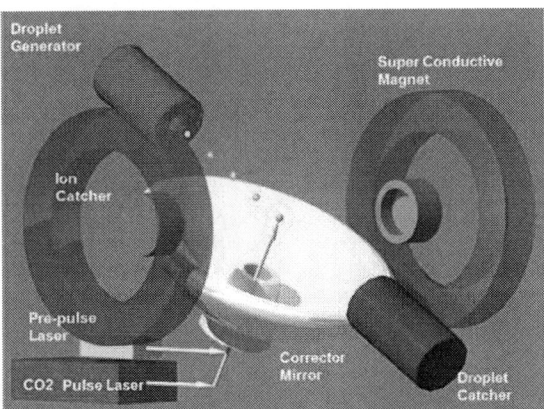

Fig. 1 The concept of Gigaphoton HVM EUV light source.

Since 2002, we have been developing the carbon dioxide (CO_2) laser produced Tin (Sn) plasma (CO_2-Sn-LPP) EUV light source which is the most promising solution as the 13.5 nm high power (>200 W) light source for HVM EUV lithography (EUVL) We have chosen the LPP-EUV method because of its high efficiency, power scalability, and spatial freedom around plasma. Our group has proposed several unique original technologies[8)9]. The theoretical[10] and experimental[11] data have clearly demonstrated, combination of CO_2 laser and Sn plasma realize high conversion efficiency (CE) from driver. The conceptual Stracture of Gigaphoton HVM EUV light source is shown in Fig.1. And procedure of Sn debris mitigation with the magnetic field is shown in Fig. 2. At first step Sn droplet target is irradiated with pre-pulse laser. The Sn droplet is crushed to sub-micron mists. The mists are expanded in time. At second step after certain delay time the mists cloud is expanded and heated by pulsed CO_2 laser beam. The cloud is converted to high temperature plasma. Sn ions have several number of charges. During recombination process Sn plasma emits 13.5 nm EUV light. Most of the Sn ions can be trapped by the magnetic field by Larmor movement. To prevent the collector mirror from being contaminated, Sn plasma needs to be trapped before being deposited on the collector mirror. Residues of the plasma after emitting EUV light are eventually scattered inside the vessel. To enhance EUV energy and to maximize Sn debris mitigation, number of Sn ions should be maximized in these laser heating processes

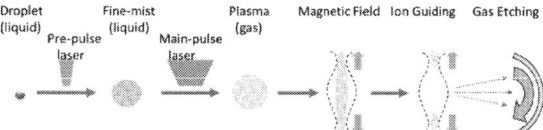

Fig. 2 EUV emission & Sn mitigation process concept.

2.2 Pre-pulse technology[12)13)]

CE of 4.7 % with the 20 μm in diameter droplet has been demonstrated by optimizing the pre-pulse laser conditions as shown in Fig. 3 (red dot) with small experimental device. These basic studies have contributed to the development of the high-power production machine and to the basic design for further EUV power scaling together with theoretical calculations. This phenomena is explainable with difference of pre-pulse expansion mechanism of ns-pulse and ps-pulse. This high CE technology enables 250 W EUV source with 20 kW CO_2 laser.

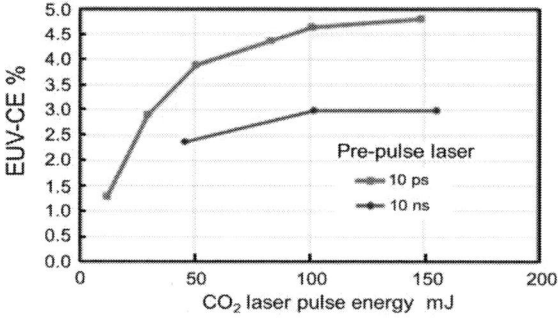

Fig.3 Conversion efficiency vs. CO_2 laser energy

The amount and the distribution of the Sn neutral atoms after the pre-pulse laser irradiation in a certain magnetic field were observed with the Laser Induced Fluorescence (LIF) method. On the other hand, in case of 10 ps pre-pulse ionization rate is very high even in case of low CO_2 pulse energy (Fig4).

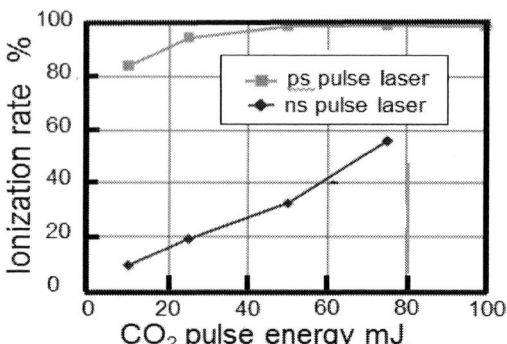

Fig.4 Ionization ratio vs. CO_2 laser energy.

2.3 Droplet generator & Magnetic-mitigation

Our Sn debris mitigation concept with the magnetic field is simple. Because EUV light is emitted from the Sn plasma, which is mainly composed of Sn ions and electron, almost all the Sn ions can be trapped in the strong magnetic field. Also, some neutral atoms can be guided and trapped by charge exchange with ions. In reality, however, not all the Sn atoms and ions can be trapped in the magnetic field. (Fig.5).[13)] We have investigated behaviors under various conditions to optimize Sn debris mitigation parameters in the compact EUV generation tool (patented).

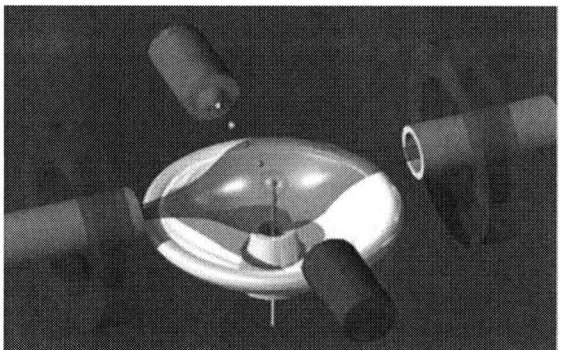

Fig.5 Collector mirror and mitigation system.

2.4 Driver laser System development

To realize our EUV light source, we are constructed three devices; proto #1, #2 and Pilot #1. Major difference between three systems is CO_2 laser power and output angle, other specification is essentially same (Table 1). We have been develop system technology and component test at proto #1 since 2011[13)]. The new CO_2 laser amplifier has been developed by co-operate with Mitsubishi electric supported by NEDO[15)16)]. We have started the construction of proto #2 system in 2H of 2013. In 2015 we succeeded to demonstrate 20 kW output power with combination of pre-amplifier by Mitsubishi electric (Fig.6), and other CO_2 laser by the Trump at proto #2 test bench[17]. The operation has started beginning of 2014 (Fig.7). This system achieved 20kW with 15ns pulse duration, repetition rate was 100kHz. We have demonstrated 256W EUV emission in burst operation with 95% duty cycle by proto#2 device[19]

Table.1 Target specification of Gigaphoton EUV sources

		Proto#1 Proof of Concept	Proto#2 Key Technology	Pilot#1 HVM Ready
Target Performance	EUV Power	25W	>100W	250W
	CE	3 %	4.0 %	5.0 %
	Pulse Rate	100kHz	100kHz	100kHz
	Output Angle	Horizontal	62°upper	62°upper
	Availability	~1 week	~1 week	>75%
Technology	Droplet Generator	20 - 25μm	< 20μm	< 20μm
	CO_2 Laser	5kW	20kW	27kW
	Pre-pulse Laser	picosecond	picosecond	picosecond
	Collector Mirror Lifetime	Used as development platform	10 days	> 3 months

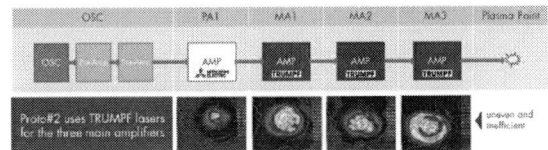

Fig.6 Driver laser system configuration of Proto#2.

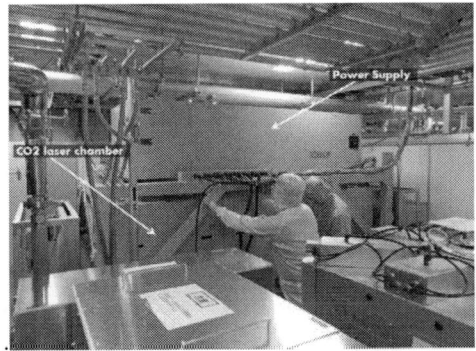

Fig.7 CO_2 pre-amplifier installation byMitsubishi-electric.

3. 250W HVM LPP-EUV Source[19)20]
3.1 Pilot #1 system construction and configuration

The Pilot#1 HVM EUV light source is consist of driver laser system, beam transfer system and EUV chamber system. We are planning to use the four amplifiers which are produced by Mitsubishi electric. At present the Pilot#1 system is constructed, the picture of construction

is shown in Fig.8.

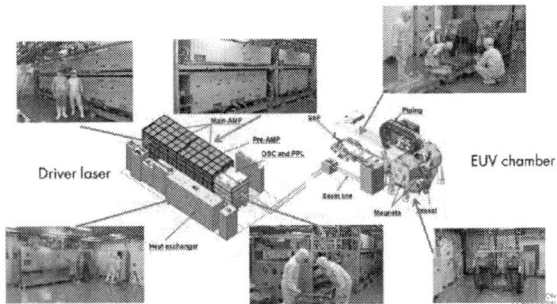

Fig. 8 Pictures of Pilot#1 system construction

The specification of Proto #1, Proto#2 and Pilot #1 are shown in Table 1. Also outlook of the whole system is shown in Fig. 9. In 2016 we developed 27kW CO_2 driver laser system for Pilot #1 system. (15ns pulse duration, repetition rate was 100kHz) composed with one pre-amplifier and three main amplifier produced by Mitsubishi electric. Configuration is shown in Fig. 10, the beam profile is much better than Proto #2 CO_2 driver laser system[18]. Also picture of laser installation is shown in Fig.11.

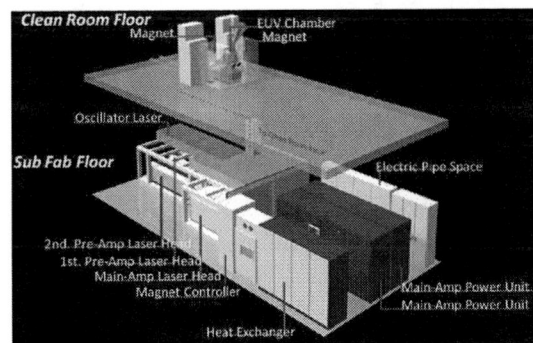

Fig.9 GL200E-Pilot#1

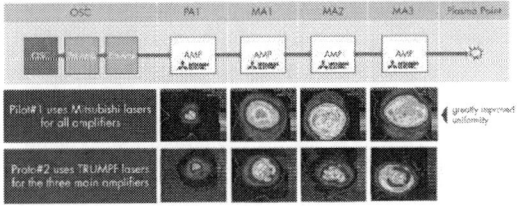

Fig.10 Driver laser system configuration of Pilot#1.

Fig.11 Pre-amplifier + 3x main amplifier installation by Mitsubishi Electric.

3.2 High Ce experiment and actual experiment in high power device

Difference of Driver CO_2 laser beam profile of Proto #2 system and Pilot #1 is shown in fig.12. Beam profile of Pilot #1 system is much symmetrical and homogeneous. This difference make difference of conversion efficiency. Pilot #1 system achieved significant progress of conversion efficiency between 5.0% and 5.5%. We expect this homogeneity coming from difference of excitation scheme of CO_2 laser.

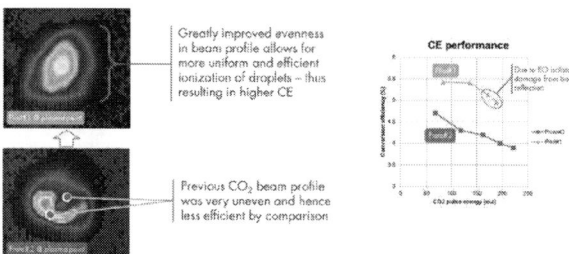

Fig.12 Driver laser beam profile and conversion efficiency

3.3 Latest operation data of Pilot#1 system

Since September 2016, we started the operation of Pilot #1 system. One example of the operation data of the Pilot #1 system is shown in Table 2 and Fig.13. EUV in band power is 117W, 95% duty cycle, 50kHz (111W average power) operation time is 22hours. CO_2 laser power is only 9.5kW, because of 5% high conversion operation. Also dose stability data is shown in Fig.12 during 22 hours. This tata shows very good stability. The measured data shows 92.3% die yield (Fig.14). Pilot#1 is up running and its demonstrates HVM capability in October 2016[21].

Table 2. Pilot #1 Operation parameters

	Performance
Conversion Efficiency	**4.9%**
Power (average at IF)	111W
Duty Cycle	**95%**
Power (in band)	**117W**
Operation Pulse Number	3.8Bpls
Operation Time	**22hrs**
Dose Margin	25%
CO_2 Power	**9.5kW**
Pulse Rate	50kHz
Pulse Duration	~10ns
Availability (4 week average)	17%

Note

Although CE and average power show relatively good results, availability continues to be a gating issue. The biggest cause of this problem is the back reflection in the CO_2 laser.

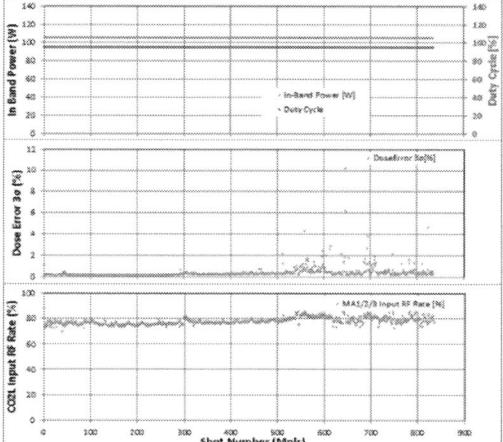

Fig.13 EUV emission data

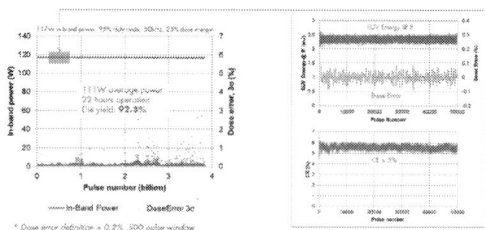

Fig.14 Dose stability data

3.4 Mitigation data of collector mirror

Picture of actual collector mirror is shown in Fig. 15. Diameter is around 400mm, weight is 22kg. Collector reflectivity is >48%. The shape of the mirror is ellipsoid and one focal point is plasma point, the other point is correspond to intermediate focus point. The surface is coated with 50 pare of Mo/Si layers. The surface is coated by capping layer.

Fig. 15 Picture of actual collector mirror

After EUV plasma is created, EUV light is collected by multilayer mirror. Fig.16 shows the schematic of this new filter.[14] However EUV plasma emits not only EUV light, but also UV light, visible light and infrared light respectively. These light components are called "Out of band light" (Fig.17). Reflected IR light from the multilayer makes interference pattern at focal plane (patented). Only IR light is absorbed by aperture stop.

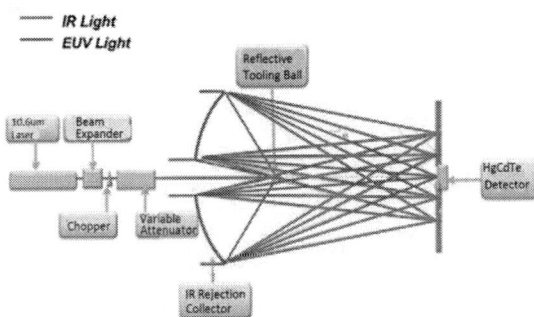

Fig. 16 Schematic of new type filter

Thin mitigation simulation on the corrector mirror of Pilot #1 system shows significant improvement of cleanness from Proto #2 system (Fig.18), by optimized design of hydrogen gas flow on the corrector mirror. This data promise very long lifetime of the corrector mirror of Pilot #1 system. Also capping layer has important role to avoid deposition of thin. Fig.19 shows significant

deference between capping layer materials (A,B,C). The "material A" shows blistering of the capping layer. The "material B" shows no blistering, but thick deposition of thin. The "material C" shows very good performance of Tin deposition. Therefore we chose "material C" for capping layer.

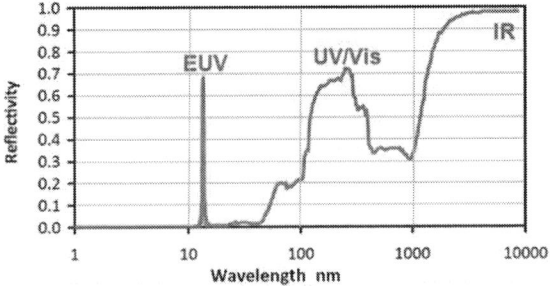

Fig.17 Out of Band spectrum

Fig.18 Simulation of thin deposition rate on corrector mirror

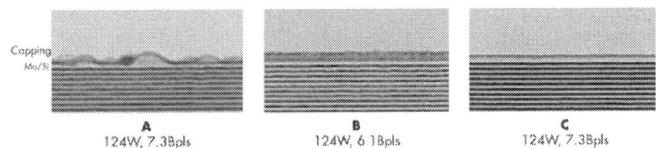

Fig. 19 Capping layer comparison (material A,B,C)

Recently we have demonstrated, EUV power recorded at113W in burst stabilized (85W in average, 75% duty), with 5% conversion efficiency during 143hours operation. Also the Pilot#1 system recorded 64% availability and idle time was 25%. Availability is potentially achievable at 89% (2weeks average), also superior magnetic mitigation has demonstrated promising mirror degradation rate (= -0.4%/Gp) above 100W level operation.[22] Fig.20 shows the degradation of reflectance of actual corrector mirror with "material C" capping layer. Blue color picture shows far field pattern of reflected light. This data shows very low degradation rate is realized at actual EUV source operation.

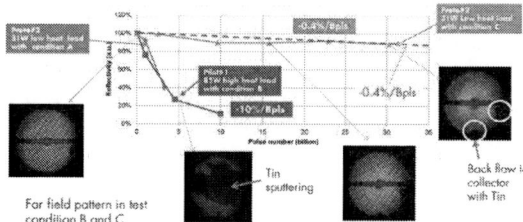

Fig.20 Reflectivity measurement of actual corrector mirror

4. Conclusion

We have developed Pilot #1 system for HVM EUV lithography. We demonstrated 117W EUV power (I/F clean in burst), 50 kHz, 22 hours stable operation. This system is using 27 kW CO_2 laser amplifier system consist of pre-amplifier and 3x main amplifiers by Mitsubishi electric in Japan. Furthermore, very low degradation (= -0.4%/Gp) of actual collector mirror reflectance has been demonstrated above 100W level operation (in burst) with magnetic mitigation EUV source.

Next step of our challenge is extend the lifetime test further and increase the power up to 250W level. We will ship Pilot system to EUV lithography market for less than 10nm node semiconductor mass manufacturing.

5. Acknowledgement

This work was partly supported by the New Energy and Industrial Technology Development Organization (NEDO), Japan. We acknowledge their continuous support.

We acknowledge to following researchers and organizations; Plasma simulation is supported by Dr. Jun Sunahara in Osaka University. Plasma diagnostics is supported by Dr. Kentaro Tomita, Prof. Kiichiro Uchino and others in Kyushu University . Laser engineering is supported by Dr. Akira Endo in HiLase Project (Prague). Also collector mirror suppliers – especially RIT. Also CO_2 laser amplifier development is supported by Mitsubishi electric CO_2 laser amp. develop. team: Dr. Yoichi Tanino, Dr. Junichi Nishimae, Dr. Shuichi Fujikawa and others. Authors are very sorry to miss Dr. Yoichi Tanino in Mitsubishi electric corporation with sudden death on 1st February in 2014. We appreciate his extreme great job of CO_2 amplifier development in very short period and pray for the soul of him.

6. References

[1] Noreen Harned, et. al.: 'EUV Lithography with the Alpha Demo Tools: status and challenges', Proc. SPIE 6517 (2007) 6517-06.

[2] Takaharu Miura et. al.,: 'Nikon EUVL development progress update' Proc. SPIE 6921 (2008) 6921-0M.

[3] Christian Wagner, et. al.: 'EUV into production with ASML's NXE platform', Proc. SPIE 7636 (2010) 7636-1H .

[4] Christian Wagner, et. al.: 'Performance validation of ASML's NXE:3100', Proc. SPIE 7969 (2011) 7969-49.

[5] R. Peeters, S. Lok, et.al.: "ASML's NXE platform performance and volume Introduction" Extreme Ultraviolet (EUV) Lithography IV, Proc. SPIE 8679 (2013) [8679-50]

[6] Jack J.H. Chen, TSMC: "Progress on enabling EUV lithography for high volume manufacturing" 2015 EUVL Symposium (5-7 October 2015, Maastricht, Netherlands)

[7] Mark Phillips, Intel Corporation "EUVL readiness for 7nm" 2015 EUVL Symposium (5-7 October 2015, Maastricht, Netherlands)

[8] Akira Endo, et. al.: 'Laser produced EUV light source development for HVM', Proc. SPIE 6517 (2007) .

[9] Hakaru Mizoguchi, et. al.: '100W 1st Generation Laser-Produced Plasma light source system for HVM EUV lithography', Proc. SPIE 7969 (2011) 7969-08.

[10] K. Nishihara, et. al., EUV Sources for Lithography, Chap. 11, ed. V. Bakshi, SPIE, Bellingham, 2005.

[11]H. Tanaka, et. al.: 'Comparative study on emission characteristics of extreme ultraviolet radiation from CO_2 and Nd:YAG laser-produced tin plasmas', Appl. Phys. Lett. 87, (2005) 041503 .

[12] Tatsuya Yanagida, et. al.: 'Characterization and optimization of tin particle mitigation and EUV conversion efficiency in a laser produced plasma EUV light source', Proc. SPIE 7969 (2011) 7969-100.

[13] Junichi Fujimoto, et. al.: 'Development of the reliable 20 kW class pulsed carbon dioxide laser system for LPP EUV light source', Proc. SPIE 7969 (2011) 7969-99.

[14] RIGAKU technical display: 'IR Rejection Collector Optic Manufacturing Process' (Proc. of International Symposium on Extreme Ultraviolet Lithography 2013, Oct.6-10.2013,Toyama)

[15] Yoichi Tanino: 'A proposal for an EUV light source using transverse flow CO_2 lasers' (International Symposium on Extreme Ultraviolet Lithography 2012, Oct.1-4.2012, Brussel) 1016.

[16] Krzysztof M Nowak, Yoichi Tanino et.al.: 'EUV driver CO_2 laser system using multi-line nano-second pulse high-stability master oscillator for Gigaphoton's EUV LPP system'(Proc. of International Symposium on Extreme Ultraviolet Lithography 2013, Oct.6-10.2013, Toyama).

[17] Hakaru Mizoguchi, et al: "Update of one hundred watt HVM LPP-EUV source performance" 2015 EUVL Symposium (5-7 October 2015, Maastricht, Netherlands)

[18] Koji Yasui, Naoyuki Nakamura, Jun-ichi Nishimae, Masashi Naruse, Kazuo Sugihara, Masato Matsubara, "Stable and scalable CO2 laser drivers for high-volume-manufacturing extreme ultraviolet lithography applications" 2016 EUVL symposium (24-26,

Oct.2016, Hiroshima, Japan)

[19] Hakaru Mizoguchi, Hiroaki Nakarai, Tamotsu Abe, Krzysztof M. Nowak, Yasufumi Kawasuji, Hiroshi Tanaka, Yukio Watanabe, Tsukasa Hori, Takeshi Kodama, Yutaka Shiraishi, Tatsuya Yanagida, Tsuyoshi Yamada, Taku Yamazaki, Shinji Okazaki, Takashi Saitou:"Performance of new high-power HVM LPP-EUV source" Extreme Ultraviolet (EUV) Lithography VII, Proc. SPIE9776 (2016)

[20] H. Mizoguchi "Development of 250W EUV Light Source for HVM Lithography", 2016 EUVL symposium (24-26, Oct.2016, Hiroshima, Japan)

[21] Hakaru Mizoguchi et al.:" Performance of 250W High Power HVM LPP-EUV Source", Proc. SPIE 10143, Extreme Ultraviolet (EUV) Lithography VIII (2017)

[22] Hakaru Mizoguchi, et al: " High Power HVM LPP-EUV Source with Long Collector Mirror Lifetime", EUVL Workshop 2017, (Berkeley, 12-15, June, 2017)

Multi-beam mask writer MBM-1000

Hiroshi Matsumoto*, Hideo Inoue, Hiroshi Yamashita, and Kenji Otoshi

NuFlare Technology, Inc., 8-1 Shinsugita, Isogo, Yokohama, Kanagawa 235-8522, Japan

*Corresponding Author's Email: matsumoto.hiroshi@nuflare.co.jp

ABSTRACT

Multi-beam mask writer MBM-1000 has been developed for N5. It is designed to accomplish high resolution with 10 nm beam and high throughput with 300 Gbps blanking aperture array (BAA) and inline real-time data path. It is equipped with function of pixel level dose correction (PLDC) to improve patterning resolution, which is method specific to multi-beam writer. Design concept of MBM-1000 is described. Writing test is carried out to demonstrate performance of MBM-1000 and PLDC function.

INTRODUCTION

Trend of shot count increases and lower sensitivity resist continues toward N5 and beyond. Multi-beam writer is faster than variable shaped beam (VSB) writer at large figure count since beam current usable for exposure is constant independent of pattern size; VSB writer, on the other hand, uses reduced shot size with small beam current to expose fine pattern. Therefore, multi-beam writer is expected to be inserted for HVM production in a few years [1]. We developed multi-beam mask writer MBM-1000 for N5. It is designed to accomplish higher throughput than a single-beam VSB writer EBM-9500 [2] at shot count higher than 500 G/pass.

OVERVIEW OF MBM-1000

Figure 1 shows basic design of MBM-1000 and flow of data processing. MBM-1000 uses multi-beam system with single cathode and multi-beam aperture array accommodated in single column [3,4]. Shaping aperture array (SAA) is illuminated by e-beam from cathode to formulate beamlets, which are individually blanked or unblanked by blanking aperture array (BAA). Beamlets are collectively positioned on the target by common deflectors. Key parameters of MBM-1000 are shown in Table 1.

Concept of MBM-1000 is to accomplish good patterning resolution by using 50 kV single-acceleration, 10 nm beam size and 10 bit dose resolution. Single-stage acceleration is selected not to elongate effective beam path and to minimize the effect of electromagnetic noise on beam.

To accomplish high throughput, 512×512 blanking cells are implemented in BAA chip, which is maximum array size available on single LSI chip. BAA chip has data ratio of 300 Gbps. Current density is adjusted to 2 A/cm^2 to have total beam current of 500 nA with 10 nm beam as standard specifications. Beam resolution with more beam current will be investigated to accomplish higher throughput.

Mask blank is exposed stripe by stripe. Stripe size is a key parameter for stage speed and number of stage turn. It is defined by beam deflection field size in the case of VSB writer, and by beam array size in the case of multi-beam writer. MBM-1000 uses beam array size of 82 um to have stripe size compatible with EBM-9500. Large beam array allows to use acceptable stage speed.

Air bearing stage is introduced to MBM-1000. It is more reliable than frictional drive stage because wearing of mechanism structure is absent. It is suitable for writing mode of constant stage-speed, which is standard mode of multi-beam writer. In the case of VSB writer, frictional drive stage with higher acceleration capability is preferable as writing mode with variable stage-speed is necessary in terms of writing speed. This is because shot count fluctuates depending on pattern.

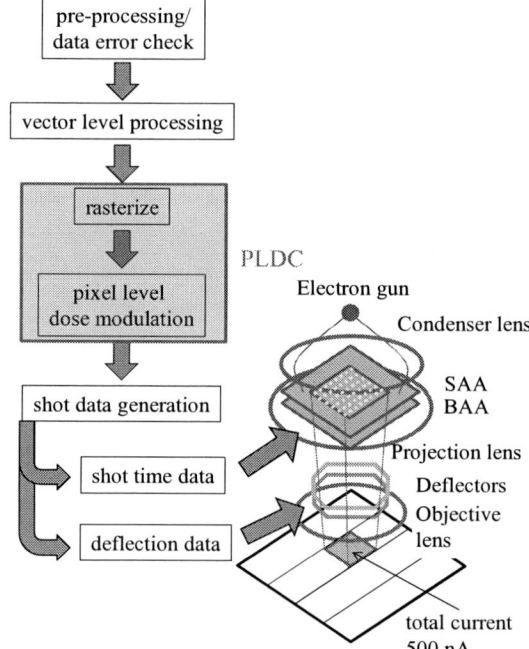

Figure 1: Overview of MBM-1000 writing system. Beamlets are generated by shaping aperture array (SAA) and controlled by blanking aperture array (BAA), along with common deflector. Data is generated by inline and real-time processing and sent to BAA at the data ratio of 300 Gbps.

978-1-5386-5309-8/18 $31.00 © 2018 IEEE

Table 1. Configuration of MBM-1000.

Item	Spec	Remarks
Accel. voltage	50 kV	Single stage acceleration for beam stability and tool serviceability
Cathode	2 A/cm^2	Adjusted to have total current of 500 nA with 10 nm beam size.
Beam blur	Smaller than EBM	Realized by optics design
Beam size	10 nm	Standard size optimized for patterning resolution.
Exposure field size	82 um□/ 512×512 beamlets	Compatible to deflection field size of EBM-9500 in order not to increase stage speed.
Shot time resolution	Variable (standard is 10 bit/pass	Accurate CD/position control by avoiding dose quantization error. Bit/pass selectable to balance writing accuracy and TPT.
Stage	Air bearing stage	Performs good with constant speed mode used for MBM.
Data format	MBF, VSB12i, OASIS. MASK	MBF is designed for MBM-1000 and supports any-angle figures and polygon figures.
Correction	Added EUV-PEC and PLDC	All corrections done inline. Correction functions of PEC/LEC/FEC/GMC/CEC for EBM inherited to MBM.

DATA PATH OF MBM-1000

Polygon and any-angle patterns are supported by MBF format which is newly developed for MBM-1000.

MBM-1000 inherits correction function from EBM-9500 except thermal effect correction (TEC), as resist heating is insignificant at current density of 2 A/cm^2. It has new inline corrections of pixel level dose correction (PLDC) for better patterning resolution and linearity. EUV-PEC function is also added to address 1 um range backscattering occurring at multilayer of EUV substrate.

Data path of MBM-1000 is based on inline processing to minimize mask turnaround time (TAT), as shown in Fig. 1. Main portion of data processing and correction calculation are performed in parallel to writing, in inline and real-time manner to utilize write time for processing time.

Writing data is prepared in MBF format by offline data conversion from user's CAD data. When writing job is registered, pre-write processing is invoked. This process performs data transfer from storage unit to high-speed disk array in writing control unit. It also performs data format check and preprocessing for global correction. Maps of pattern density and dose density are generated and global corrections are carried out by using these maps. Global correction includes loading effect correction (LEC), fogging effect correction (FEC) and grid-matching correction (GMC), and charge-effect correction (CEC). Pre-write process is automatically invoked and carried out in preceding mask writing, and thus it does not impact TAT when writing masks sequentially. In-writing processing comprises rasterizing step and dose correction step. In rasterizing step, vector data are reconstructed and divided into stripes, and then pattern density map is generated for each stripe. Then dose is modulated for local CD corrections, such as PEC and EUV-PEC.

PLDC is also carried out. This function performs dose modulation pixel by pixel to improve dose profile at pattern edge [5]. It brings improved pattern fidelity of curvilinear patterns, better resolution with wider process margin, and corrected linearity. It performs mask process simulation on fine mesh smaller than beam blur to calculate dose contour and corrected dose. Hence, processing of PLDC requires enormous computation power, and it employs GPU cluster system.

After dose modulation, shot data and deflection data are transferred to BAA and deflection voltage amplifiers via hardware system.

WRITING PERFORMANCE

Performance of MBM-1000 with 300 Gbps BAA and PLDC function was evaluated by writing test.

LCD of 1.3 nm (3σ) is obtained with high sensitivity pCAR resist as shown in Fig. 2. Image placement across full-mask area (141 mm×141 mm) is 1.0 nm (3σ) for horizontal and 1.1 nm (3σ) for vertical, as shown in Fig. 3. Patterning resolution is better than that of EBM-9500, as already reported [6].

Complex OPC pattern with curvy figure is important application of multi-beam writer. It demands long write time and large shot count for VSB writer, but multi-beam writer can write it without write time penalty.

Example of writing result of complex OPC pattern is shown in Fig. 4. Curvy pattern with 80 nm contact hole is printed. Dot pattern of 30 nm diameter is printed, but with smaller CD than design. The size difference could be reduced when PLDC function was used to enhance exposure dose of hole pattern.

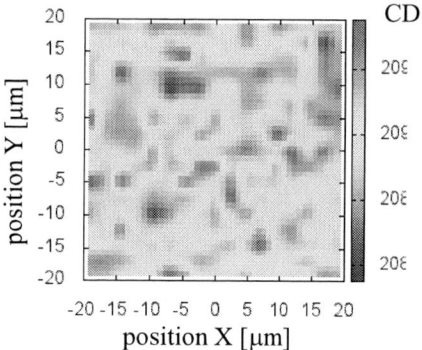

Figure 2: Local CD uniformity of 160 nmhp 1:1 line and space pattern in horizontal direction. Deviation was 1.3 nm (3σ) for horizontal line and 1.2 nm (3σ) for vertical line, with pCAR resist of FUJIFILM PRL-009 (30 μC/cm2).

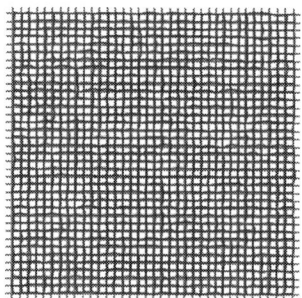

Figure 3: Global image placement measured as overlay of 1st and 2nd write pass. Horizontal deviation was 1.0 nm (3σ) and vertical 1.1 nm (3σ).

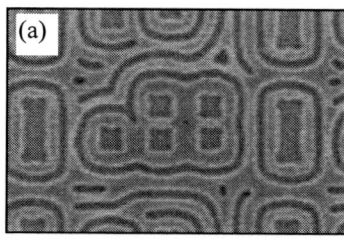

Figure 4: (a) Complex OPC pattern printed on low sensitivity pCAR resist (140 μC/cm2), compared with (b) CAD design in MBF format.

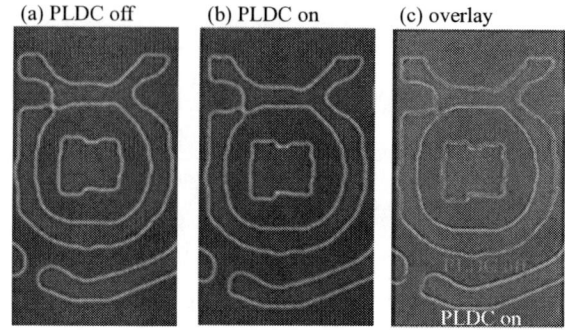

Figure 5: Curvilinear pattern printed with resolution enhancement by PLDC on low sensitivity pCAR resist (140 μC/cm2). Main pattern is 4X of 80 nm contact hole. Cr image with (a) PLDC off and (b) on are overlaid in (c).

Example of writing result with PLDC is shown in Fig. 5. Dose modulation for dose contrast enhancement resulted in higher pattern fidelity as seen especially in the shape of square corner.

SUMMARY

Writing test demonstrated performance of MBM-1000, which is designed for N5 and beyond. 300 Gbps BAA offers high throughput without penalty in writing accuracy. PLDC makes capability of multi-beam writer for curvilinear pattern more effective.

Tool tuning with total current of 500 nA (standard configuration) is being continued to meet N5 requirement. Writing test with larger beam current will be carried out to optimize writing accuracy and throughput.

REFERENCES

[1] M. Chandramouli, F. Abboud, N. Wilcox, A. Sowers, A. and D. Cole, *Proc. SPIE* 8522, 85221K (2012).

[2] H. Matsui, T. Kamikubo, S. Nakahashi, H. Nomura, N. Nakayamada, M. Suganuma, Y. Kato, J. Yashima, V. Katsap, K. Saito, R. Kobayashi, N. Miyamoto, M. Ogasawara, *Proc. SPIE* 9985, 998508 (2016).

[3] H. Yasuda, S. Arai, J. Kai, Y. Ooae, T. Abe, Y. Takahashi, S. Hueki, S. Maruyama, S. Sago, and K. Betsui, *Jpn. J. App. Phys.* 32(12B), 6012 (1993).

[4] J. Klein, H. Loeschner, E. Platzgummer, *Proc. SPIE* 7970, 79700C (2011).

[5] H. Zable, H. Matsumoto, K. Yasui, R. Ueba, N. Nakayamada, N. Shirali, Y. Masuda, R. Pearman, A. Fujimura, *Proc. SPIE* 10454, 104540D-1 (2017).

[6] H. Matsumoto, H. Inoue, H. Yamashita, H. Morita, S. Hirose, M. Ogasawara, H. Yamada and K. Hattori, *Proc. SPIE* 9984, 998405 (2016).

ROBSUT OVERLAY METROLOGY BY MUELLER MATRIX ELLIPSOMETRY WITH A DIFFERENTIAL CALCULUS

Xiuguo Chen[*] and Shiyuan Liu

State Key Lab of Digital Manufacturing Equipment and Technology, Huazhong University of
Science and Technology, Wuhan 430074, China
*Email: xiuguochen@hust.edu.cn

ABSTRACT

The differential Mueller matrix calculus is introduced to investigate Mueller matrices of double-patterned gratings with overlay displacements. We find and demonstrate that the linear birefringence and dichroism, LB′ and LD′, along the ±45° axes obtained from the differential calculus show a linear response to the overlay displacement and are zero when the overlay displacement is absent at any conical mounting. The better property of LB′ and LD′ than the Mueller matrix elements of the off-diagonal blocks in the presence of overlay displacement reveals that LB′ or LD′ can be a more robust indicator for diffraction-based overlay metrology.

INTRODUCTION

Overlay control is of vital importance to good device performances in semiconductor manufacturing, especially to the emerging double patterning or multi-patterning lithography techniques. Over the past years, the diffraction based overlay (DBO) metrology techniques have been developed to address the challenges posed by the demand for tighter overlay control to the traditional image based overlay metrology, of which the DBO metrology using Mueller matrix ellipsometry (MME) is a good candidate due to the much more information provided by the 16 elements of a 4×4 Mueller matrix [1]. Current DBO metrology using MME is typically based on a property that the elements from the two 2×2 off-diagonal blocks of the Mueller matrix are zero when the overlay displacement is absent, otherwise will deviate from zero and respond to the overlay displacement linearly. However, the above property of the off-diagonal block elements of the Mueller matrix is essentially only valid at the azimuthal angle of $\phi = 90°$, i.e., with the plane of incidence parallel to grating lines. Furthermore, it is also necessary to accurately align the sample at $\phi = 90°$ when performing measurement, since even a minor offset from $\phi = 90°$ will lead to a large measurement error. This makes the DBO metrology using MME lack of robustness. In this work, we applied the differential Mueller matrix calculus to investigate the Mueller matrices of double-patterned grating structures with overlay displacements. We found that the elementary optical properties LB′ and LD′ obtained from the differential calculus exhibited a better property than Mueller matrix elements of off-diagonal blocks in the presence of overlay displacements [2]. The

simulation results also demonstrated that LB′ and LD′ can be adopted as a more robust indicator than the off-diagonal block elements of the Mueller matrix for DBO metrology.

THEORY

In the differential Mueller matrix calculus, the differential Mueller matrix **m** relates the Mueller matrix **M** to its spatial derivative along the light propagation direction z as

$$d\mathbf{M}/dz = \mathbf{m}\mathbf{M}. \qquad (1)$$

The solution of Eq. (1) can be achieved by taking the logarithm of **M** if **m** does not depend on z, that is [3, 4],

$$\mathbf{L} = \ln(\mathbf{M}) = \mathbf{L}_m + \mathbf{L}_u, \qquad (2)$$

where **L** is the accumulated differential Mueller matrix and $\mathbf{L} = l\mathbf{m}$, with l being the optical path length in the medium. $\mathbf{L}_m = (\mathbf{L} - \mathbf{G}\mathbf{L}^T\mathbf{G})/2$ and $\mathbf{L}_u = (\mathbf{L} + \mathbf{G}\mathbf{L}^T\mathbf{G})/2$ are the G-antisymmetric and G-symmetric terms of **L**, respectively. Here, **G** is the Minkowski metric and $\mathbf{G} = \text{diag}(1, -1, -1, -1)$. The superscript "T" denotes the matrix transpose. If **M** is a nondepolarizing Mueller matrix, we will have $\mathbf{L}_u = 0$ within the experimental error, provided that the isotropic absorption has been subtracted beforehand from the diagonal elements of **L**. The term $\mathbf{L}_m$ is given by [5]

$$\mathbf{L}_m = \begin{bmatrix} 0 & -LD & -LD' & CD \\ -LD & 0 & CB & LB' \\ -LD' & -CB & 0 & -LB \\ CD & -LB' & LB & 0 \end{bmatrix}, \qquad (3)$$

where LB and LD refer to the linear birefringence and dichroism along the p-s axes of the reference frame, LB′ and LD′ refer to the linear birefringence and dichroism along ±45° axes, CB and CD refer to the circular birefringence and dichroism, respectively.

Considering the electromagnetic reciprocity and the reflection symmetry of the double-patterned grating in Fig. 1 relative to a plane perpendicular to grating lines, we can restrict the azimuthal angle ϕ to the range of 0 to 90°. In addition, we assume that the grating is composed of only reciprocal materials and the investigated Mueller matrices are exclusively associated with the zeroth-order diffracted light. We can derive that [2]

$$L(-\delta, \phi) = L(\delta, \phi), \qquad (4a)$$

$$L'(-\delta, \phi) = -L'(\delta, \phi), \qquad (4b)$$

$$C(-\delta, \phi) = C(\delta, \phi), \qquad (4c)$$

where δ is the overlay displacement, L = LB − iLD, L′ = LB′ − iLD′ and C = CB − iCD, which represent the linear anisotropy along the p-s and ±45° axes as well as the circular anisotropy, respectively. According to Eq. (4), we can further derive the following properties that are useful in the overlay measurement.

– *Property 1*: Among the six elementary optical properties (LB, LD, LB′, LD′, CB, and CD), only LB′ and LD′ are sensitive to the direction of overlay displacement. When $\delta = 0$, we have LB′ = LD′ = 0 for any azimuthal configuration. However, when $\delta \neq 0$, we have LB′ ≠ 0 or LD′ ≠ 0 for any azimuthal configuration except $\phi = 0°$.

– *Property 2*: When $\phi = 0°$, we have LB′ = LD′ = CB = CD = 0 for any overlay displacement; and when $\phi = 90°$, we have CB = CD = 0 for any overlay displacement.

The Property 1 indicates that we can judge whether or not the grating sample has an overlay displacement from the absolute value of LB′ or LD′ and meanwhile we can distinguish the direction of overlay displacement from the sign of LB′ or LD′. According to electromagnetic reciprocity as well as Properties 1 and 2, we can obtain the relation for any element of the two 2×2 off-diagonal blocks of the Mueller matrix at $\phi = 90°$ when $\delta \neq 0$, that is,

$$m_{ij}(-\delta) = -m_{ij}(\delta) \neq 0. \qquad (5)$$

The relation in Eq. (5) was the starting point for most of the reported literature handling overlay measurement based on Mueller matrix formalism [6-8]. However, we should note that the above relation is only valid at $\phi = 90°$.

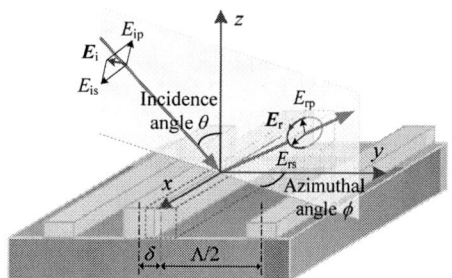

Figure 1: Representation of polarized light incidence upon a double-patterned grating structure with an overlay displacement of δ.

According to Property 1, we can adopt LB′ (or LD′) as the indicator for overlay monitoring. In actual measurements, the sample Mueller matrices are typically collected in a spectral range. Therefore, the LB′ extracted from the collected Mueller matrices will also vary in a spectral range. For overlay measurement, it is necessary to define a scalar indicator that has a linear relation with respect to the overlay displacement within a small range.

In this work, we define the scalar overlay indicator as

$$\gamma_{LB'} = \sum_{i=1}^{N} \omega_i LB'_i, \qquad (6)$$

where N denotes the number of spectral points and ω_i is the weight associated with LB'_i at the ith spectral point. Two approaches can be used to calculate the weight ω_i, namely, namely the mean weighting approach with $\omega_i \equiv 1$ and the principal-weighting (PC) approach based on principal component analysis with $\sum_{i=1}^{N} \omega_i^2 = 1$ [2]. With the scalar overlay indicator, we can design the DBO target to realize the overlay measurement. Figure 2 illustrates the layout of the DBO target, where d (or $-d$) is a known value representing the designed shift and ε is the actual overlay error induced in the manufacturing processes. According to Fig. 2, we know that the total overlay displacement will be $\delta = d + \varepsilon$ (or $-d + \varepsilon$). Based on the linear assumption between the scalar indicator $\gamma_{LB'}$ and the overlay displacement δ, we can obtain the actual overlay error ε by

$$\varepsilon = \frac{\gamma_{LB'}(d+\varepsilon) + \gamma_{LB'}(-d+\varepsilon)}{\gamma_{LB'}(d+\varepsilon) - \gamma_{LB'}(-d+\varepsilon)} d. \qquad (7)$$

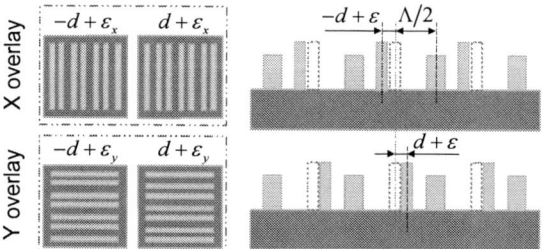

Figure 2: Layout of the DBO target, which consists of two cells per direction used to measure the overlay errors ε_x and ε_y along the X and Y directions, respectively. In the DBO target, d is a designed shift, and Λ denotes the pitch of the doubled pattered structure. Without loss of generality, ε_x and ε_y are denoted as ε for simplicity.

RESULTS AND DISCUSSION

Figure 3 depicts the schematic of the investigated double-patterned grating in the simulation, which has a pitch of $\Lambda = 660$ nm. The profile of Resist1 is characterized by top CD $w_1 = 120$ nm, line height $h_1 = 50$ nm, and sidewall angle $\alpha_1 = 87°$, and the profile of Resist2 is characterized by top CD $w_2 = 100$ nm, line height $h_2 = 140$ nm, and sidewall angle $\alpha_2 = 87°$. The thickness of the BARC (bottom anti-reflective coating) layer is $t = 30$ nm. The total overlay displacement is denoted as δ, which should be the sum of the designed shift d and the actual overlay error ε, as illustrated in Fig. 2. In the simulation, the Mueller matrices are calculated using rigorous coupled-wave analysis [9, 10] in the spectral range of 250 to 800 nm with an increment of 5 nm and by fixing the incidence angle at $\theta = 65°$. The Mueller matrices at different azimuthal angles varied from 0 to 90° will be examined.

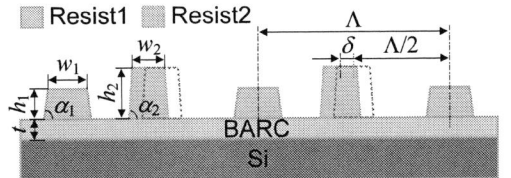

Figure 3: Schematic of the investigated double-patterned grating.

Figure 4 presents the spectra of the six elementary optical properties extracted from the simulated Mueller matrices at azimuthal angles varied from 0° to 90° with an increment of 30°. We can observe from Fig. 4 that when $\phi = 0°$, LB′ = LD′ = CB = CD = 0, and when $\phi \neq 0°$, only LB′ and LD′ exhibit sensitivity to the direction of overlay displacement while other optical properties are at most sensitive to the absolute value of the overlay displacement. In addition, when $\phi = 90°$, CB = CD = 0 even if the overlay displacement $\delta \neq 0$. However, when $\phi \neq 90°$ and $\phi \neq 0°$, CB and CD will not be equal to zero any more.

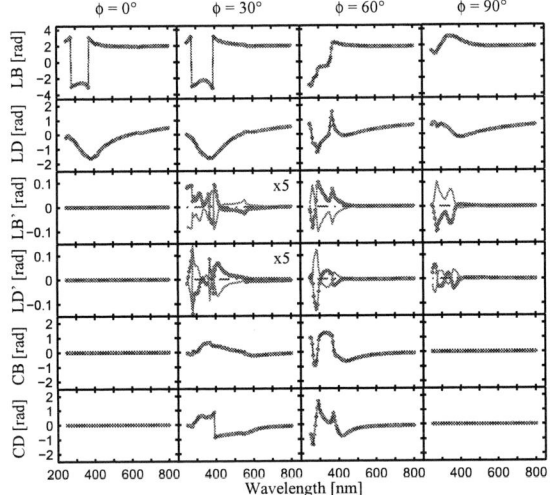

Figure 4: The spectra of the elementary optical properties LB, LD, LB′, LD′, CB, and CD extracted from the simulated Mueller matrices. The green solid curves with open circles and the red solid curves are for an overlay displacement of $\delta = -10$ nm and of $\delta = 10$ nm, respectively. The blue dashed-dotted curves correspond to $\delta = 0$. The spectra of LB′ and LD′ obtained at $\phi = 30°$ are multiplied by a factor of 5 for clarity.

Next, we try to measure the overlay error. In the following simulation, we fixed the designed shift d at 20 nm. We varied the overlay errors from 2 to 30 nm with an increment of 2 nm and compared the measured overlay

errors by Eq. (7) with the input (given) overlay errors. Random noise was also added to the simulated Mueller matrices, which was generated using a signal dependent noise model given in our previous work [11]. The generated noise was dependent on both the sample under test and the measurement configurations (combination of the wavelength and the incidence and azimuthal angles). We used $\gamma_{LB'}$ and $\gamma_{m_{13}}$ ($\gamma_{m_{13}} = \sum_{i=1}^{N} \omega_i m_{13,i}$, here we took the Mueller matrix element m_{13} from the 2×2 off-diagonal blocks as an example) to measure overlay errors at $\phi = 60°$ and $\phi = 90°$ to make a comparison, and both $\gamma_{LB'}$ and $\gamma_{m_{13}}$ were calculated using the PC and mean weighting approaches, respectively.

Figure 5 presents the measurement results by $\gamma_{LB'}$ and $\gamma_{m_{13}}$ at $\phi = 60°$ and $\phi = 90°$, respectively. As can be observed from Fig. 5, except the case of $\gamma_{m_{13}}$ using the mean weighting approach at $\phi = 90°$, the overlay indicators can achieve good measurement results in most cases. As shown in Figs. 5(a) and 5(d), when using $\gamma_{LB'}$ to measure overlay errors at $\phi = 60°$, both the PC and mean weighting approaches can lead to good measurement results. The PC and mean weighting approaches can only achieve good measurement results for the overlay errors less than about 24 nm, when using $\gamma_{LB'}$ to measure overlay errors at $\phi = 90°$, as illustrated in Figs. 5(b) and 5(e). If we want to measure overlay errors using $\gamma_{m_{13}}$ at $\phi = 90°$, Figs. 5(c) and 5(f) suggest that we can only use the PC weighting approach.

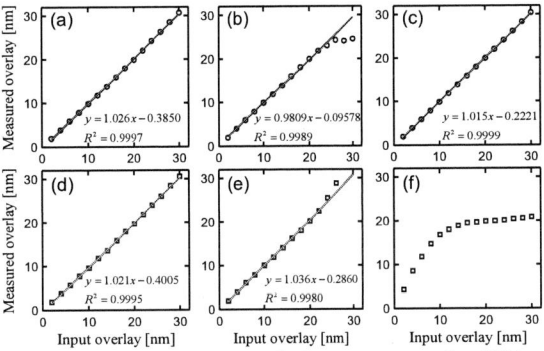

Figure 5: The measurement results by using $\gamma_{LB'}$ at $\phi = 60°$ (a, d) and $\phi = 90°$ (b, e), as well as by using $\gamma_{m_{13}}$ at $\phi = 90°$ (c, f). (a), (b) and (c) correspond to the results by $\gamma_{LB'}$ and $\gamma_{m_{13}}$ using the PC weighting approach. (d), (e) and (f) correspond to the results by $\gamma_{LB'}$ and $\gamma_{m_{13}}$ using the mean weighting approach. The inserted equations associated each subfigure are the fitted linear equation and the coefficient of determination.

Since the PC weighting approach exhibits a better linear performance than the mean weighting approach even at the azimuthal configuration of $\phi = 60°$, we want to examine whether or not we could achieve good measurement results using $\gamma_{m_{13}}$ based on the PC weighting approach at $\phi = 60°$, just as the indicator $\gamma_{LB'}$ did in Fig. 5(a). In the test, we fixed d at 10 nm and assumed that the input overlay error $\varepsilon = 5$ nm. We used $\gamma_{m_{13}}$ based on the PC weighting approach to obtain the measured overlay error ε' at $\phi = 60°$. It was found that the measured overlay error $\varepsilon' = -24.560$ nm, which was obviously wrong! We further performed the test to examine whether or not we could achieve good measurement results using $\gamma_{m_{13}}$ based on the PC weighting approach at the azimuthal configurations that slightly deviate from 90°. The results shown in Fig. 6 indicate that even a small offset of 0.1° from $\phi = 90°$ will lead to a large measurement error of nearly 1 nm (a relative measurement error of nearly 20%). We then replaced the original indicator with a corrected one $\gamma'_{m_{13}} = \sum_{i=1}^{N} \omega_i \Delta m_{13,i}$ to subtract the nonzero $m_{13,i}(\delta = 0)$ when δ is absent from the corresponding value $m_{13,i}(\delta \neq 0)$ at $\delta \neq 0$, namely $\Delta m_{13,i} = m_{13,i}(\delta \neq 0) - m_{13,i}(\delta = 0)$. As can be observed from Fig. 6, the corrected indicator leads to a much better measurement results than the original one. However, we should note that the corrected indicator involves the calculation of the off-diagonal block element value of the Mueller matrix at $\delta = 0$, which is unrealizable based on the DBO target given in Fig. 2. It thus suggests that if we want to use the off-diagonal block elements of the Mueller matrix as the overlay indicator, we can only perform overlay measurement at $\phi = 90°$. Moreover, the azimuthal angle should be accurately aligned; otherwise large measurement error will be generated.

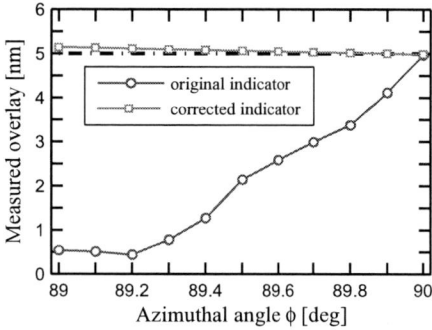

Figure 6: The measured overlay errors at different azimuthal angles using the original indicator $\gamma_{m_{13}}$ and the corrected indicator $\gamma'_{m_{13}}$, respectively. The black dash-dot line corresponds to the input overlay error.

CONCLUSIONS

(1) The overlay indicator $\gamma_{m_{13}}$ can only be used at $\phi = 90°$, and moreover even at $\phi = 90°$, only the PC weighting approach can achieve good measurement results while the mean weighting approach that was commonly adopted in the reported literature will lead to inaccurate results.

(2) When using $\gamma_{m_{13}}$ to measure overlay error based on the PC weighting approach at $\phi = 90°$, it is required to accurately align the azimuthal angle since a minor offset of 0.1° from $\phi = 90°$ will lead to a large relative measurement error of nearly 20%.

(3) The reason why the overlay indicator $\gamma_{m_{13}}$ can only be used at $\phi = 90°$ was attributed to that the Mueller matrix elements of the 2×2 off-diagonal blocks will not be equal to zero any more at a general conical mounting ($0° < \phi < 90°$) even when $\delta = 0$.

(4) The overlay indicator $\gamma_{LB'}$ can achieve good measurement results using both PC and mean weighting approaches even at a general conical mounting, since when $\delta = 0$, we have LB' = 0 for any azimuthal configuration, while when $\delta \neq 0$, we always have LB' $\neq 0$ for any conical mounting.

REFERENCES

[1] S. Liu, X. Chen, and C. Zhang, *Thin Solid Films*, vol. 584, 2014, pp. 176-195.

[2] X. Chen, H. Gu, H. Jiang, C. Zhang, and S. Liu, *Opt. Express*, vol. 25, 2017, pp. 8491-8510.

[3] R. Ossikovski, *Opt. Lett.*, vol. 36, 2011, pp. 2330-2332.

[4] N. Ortega-Quijano and J. L. Arce-Diego, *Opt. Lett.*, vol. 36, 2011, pp. 1942-1944.

[5] O. Arteaga and B. Kahr, *Opt. Lett.*, vol. 38, 2013, pp. 1134-1136.

[6] Y. N. Kim, J. S. Paek, S. Rabello, S. Lee, J. Hu, Z. Liu, Y. Hao, and W. McGahan, *Opt. Express*, vol. 17, 2009, pp. 21336-21343.

[7] J. Li, Y. Liu, P. Dasari, J. Hu, N. Smith O. Kritsun, and C. Volkman, *Proc. SPIE*, vol. 7638, 2010, pp. 76382C.

[8] C. Fallet, T. Novikova, M. Foldyna, S. Manhas, B. H. Ibrahim, A. De Martino, C. Vannuffel, and C. Constancias, *J. Micro/Nanolith. MEMS MOEMS*, vol. 10, 2011, pp. 033017.

[9] M. G. Moharam, E. B. Grann, D. A. Pommet, and T. K. Gaylord, *J. Opt. Soc. Am. A*, vol. 12, 1995, pp. 1068–1076.

[10] L. Li, *J. Opt. Soc. Am. A*, vol. 13, 1996, pp. 1024-1035.

[11] X. Chen, S. Liu, H. Gu, and C. Zhang, *Thin Solid Films*, vol. 571, 2014, pp. 653-659.

STUDY ON SPIN-ON HARDMASK FOR QUAD-LAYER APPLICATION

Yushin Park[*], Sanghak Lim, Seungwook Shin, Jinhyung Kim, Yunjun Kim, Seunghyun Kim, Jaebum Lim, Sung Hwan Kim, Hyeonil Jung, Yumi Heo, Sunyoung Yang, Jeong Yun Yu

Development group, Electron Materials Division, Samsung SDI, Suwon, Korea 443-370

*Corresponding Author's Email: yushin.park@samsung.com

ABSTRACT

Quad-layer and tri-layer systems have been widely used as stack integration schemes for fine patterning. Compared to tri-layer system, the quad-layer system includes an additional layer of silicon oxynitride (SiON), which requires higher curing temperature (300~400℃) for its deposition. However, the quad-layer system has several advantages in patterning process. The quad-layer system is more efficient in reflectivity control including pitch dependency than tri-layer system. Besides, it demonstrates excellent pattern fidelity after etch without wiggling. In this paper, we present the benefits of the quad-layer system, and demonstrate well-defined sub-20 nm patterns obtained with carbon-based spin-on hard mask (C-SOH) at high temperature in quad-layer application by multiple patterning process.

INTRODUCTION

As the density of the integrated circuits increases and the node size of the device shrinks, the aspect ratio of the pattern structure gets higher, so that the thinner photoresist is required for the photolithographic process. However, it is difficult to transfer the pattern into the substrate at the required depth with thin photoresist. Therefore, it is necessary to use multi-layer resist system of Si-based and C-based layers with high etch selectivity between PR and substrate [1-3].

A set of amorphous carbon layer (ACL) and SiON layer is the most commonly used system because of its high etch-selectivity and physical strength. Due to the benefit of spin-coating process, however, C-SOH has replaced ACL in both tri- and quad-layer systems. The stacks consist of substrate/C-SOH/Si-antireflective coating (Si-ARC)/photoresist (PR) for tri-layer and substrate/C-SOH/SiON/bottom anti-reflective coating

(BARC)/PR for quad-layer systems. In both systems, C-SOH must have high etch resistance and selectivity for good pattern transfer into sufficient depth.

Especially, quad-layer system is a beneficial stack strategy for fine patterning. It is possible to control the reflectivity more efficiently in quad-layer system with BARC and SiON layers than tri-layer system with Si-ARC only. Furthermore, it reduces wiggling phenomena and demonstrates high pattern fidelity after etch. In quad-layer system, however, C-SOH layer must have good thermal stability as well as high etch resistance throughout high temperature treatment for SiON deposition on the next step.

RESULTS AND DISCUSSION

Due to the process difference between tri- and quad-layer systems, C-SOH candidates for each tri- and quad-layer application require different physical/chemical properties. Various physical/chemical properties of several C-SOH candidates for the tri- and quad-layer applications are summarized in Table I. Because tri-layer system has fewer layers for anti-reflective function than quad-layer system, it is important to choose C-SOH which has adequate refractive index (n) / extinction coefficient (k) for tri-layer application [4, 5]. The reflectivity between PR and under layers is calculated for tri- and quad-layer system by PROLITH[TM] to investigate the effect of the optical properties (n/k values) of C-SOH. The reflectivity maps are shown in Figure 1 with the investigated C-SOH candidates of yellow dots. Depending on the stack thickness and illumination conditions, the reflectivity in tri-layer system could be changed. However, it demonstrates that quad-layer system has a much wider optical window for C-SOH candidates than tri-layer system, which indicates that the reflectivity

TABLE I. THE PHYSICAL/CHEMICAL PROPERTIES OF SAMSUNG SDI C-SOH CANDIDATES

Samples	A		B		C		D		E	F
Baking Temperature (°C)	240	400	240	400	240	400	240	400	400	400
Carbon Content (%)	83	73	83	77	84	76	79	71	79	88
Etch Resistance (%, compared to ACL)	70	56	71	65	72	64	67	56	63	73
Coating Uniformity (%)	1.6	2.3	1.5	3.1	0.8	1.6	0.7	1.6	2.0	0.8

978-1-5386-5309-8/18 $31.00 © 2018 IEEE

is unlikely to be affected by the optical properties (n/k values) of the C-SOH candidates.

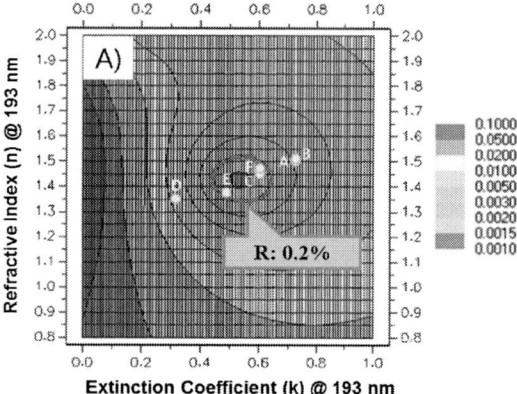

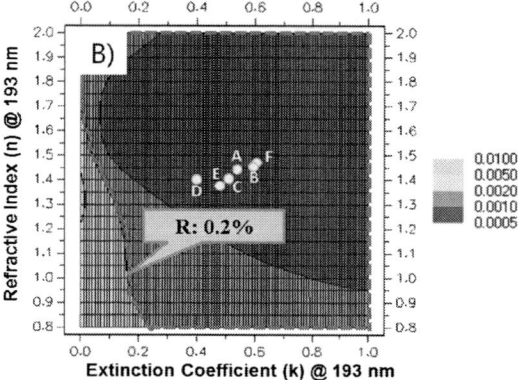

Figure 1: The n/k windows of C-SOH for A) tri-layer which has Si-ARC (51nm) and B) quad-layer system which has BARC (25nm) and SiON (32 nm) with reflectivity below 0.2% (red dashed line area)

During high temperature baking (300~400 ℃), C-SOH film gets harder (Figure 2) with excellent pattern fidelity after etch without wiggling. Seino *et al.* reported that chemical reaction by CFx plasma replaces H with F atoms on methylene linker position of polymer during etching C-SOH layer, and then relatively large F atoms cause wiggle patterns [6]. However, the baking process at high temperature substitutes H with O atoms on the methylene group, so that it prevents fluorine replacement which induces wiggling. Furthermore, the density of the film will increase in proportion to the heating temperature as shown in Figure 2.

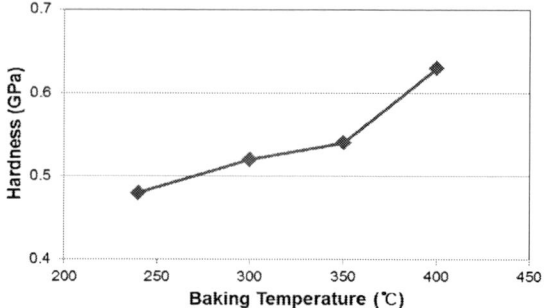

Figure 2: Film hardness is improved as bake temperature increases

C-SOH candidates need to be thermally stable at high temperature, because SiON layer is deposited onto the C-SOH layer at high temperature (300~400℃). The cross-sectional Scanning Electron Microscope (SEM) image shows a smooth, flat, and void-free interface of SiON and C-SOH (sample A) layers after 400℃ treatment as shown in Figure 3A. The weight losses of the C-SOH candidates were analyzed by Thermal Gravimetric Analysis (TGA) to confirm the reliability of the C-SOH films at a high temperature condition. For the sufficient durability at deposition step, the weight loss of film must be below 5% after additional heat treatment. All the six candidates shown in Figure 3B fulfilled the required level for quad-layer system.

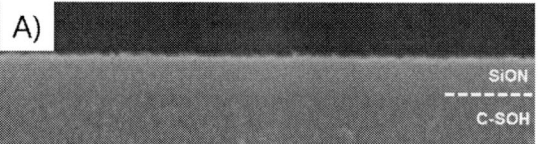

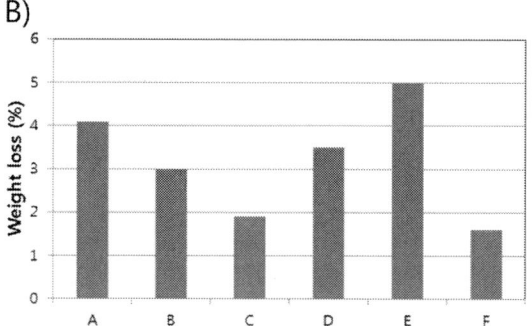

Figure 3: A) Cross-sectional SEM image of SiON and C-SOH(sample A) layers and B) the weight loss(%) of C-SOH candidates at 400 ℃ treatment

978-1-5386-5309-8/18 $31.00 © 2018 IEEE

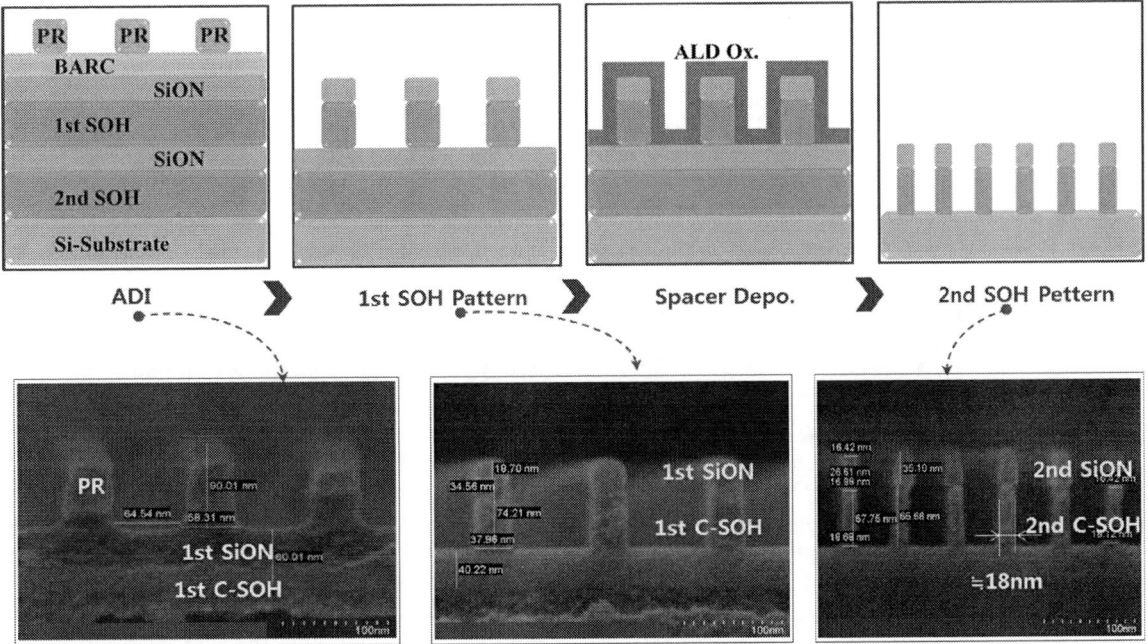

Figure 4: Scheme and SEM images for self-aligned double patterning with quad-layer system of sample A

Well-defined high resolution patterns are constructed in the quad-layer system as shown in Figure 4. Sample A is used for self-aligned double patterning as a representative example. C-SOH and SiON layers are coated and deposited twice, respectively, to make double stacks of alternating C-SOH/SiON layers. After patterning PR, first and second C-SOH layers are etched and examined with SEM. Through vertical SEM images, it exhibits well-defined high resolution patterns with bar CD of about 36 nm for the first C-SOH and about 18 nm for the second C-SOH patterns.

CONCLUSIONS

Compared to tri-layer system, quad-layer system requires an additional layer which needs high temperature treatment. However, it is an excellent stack strategy for the fine patterning with distinct advantages. It is efficient to control the reflectivity including pitch dependency in quad-layer system with a wider optical window for C-SOH candidates, whereas tri-layer system needs delicate reflection control with narrow n/k window of C-SOH. Furthermore, high temperature treatment allows excellent pattern fidelity after etch without wiggling phenomena. However, C-SOH candidates must be thermally stable at high temperature of the next SiON deposition, and its weight loss must be below 5% at 400℃. As a representative example of a selected candidate (Sample A), it was possible to obtain well- defined sub-20nm patterns in quad-layer system combined with multiple-patterning technology.

ACKNOWLEDGEMENTS

The authors gratefully acknowledge technical support for lithography and patterns etch from TEL Technology Center, America, LLC.

REFERENCES

[1] W. –H. Wu, E. Y. Chang, H. –S. Cheon, S. K. Kim, H. M. Cho, K. –H. Yoon, J. S. Kim, T. Chang, and S. Shin. *Proc. SPIE*, vol. 7140, 2008, pp. 71402Q.

[2] H. –S. Cheon, K. –H. Yoon, M. –S. Kim, S. B. Oh, J. Y. Song, N, Tokareva, J. S. Kim, and T. Chang. *Proc. SPIE*, vol. 7140, 2008, pp. 71402R.

[3] C, –I. Oh, J. –K. Lee, M. –S. Kim, K. –H. Yoon, H. –S. Cheon, N. Tokareva, J. –Y. Song, J. S. Kim, T. Chang. Proc. *SPIE*, vol. 6923, 2008, pp. 69232V.

[4] Y. Park, S. Shin, Y. Kim, S. Kim, J. Lim, S. H. Kim, H. Jung, C. Lee, M. Han, S, Lim, J. Y. Yu. Proc. SPIE, vol. 6923, 2008, pp. 69232V.

[5] Y. Wei, M. Glodde, H. Yusuff, M. Lawson, S. Y. Chang, K. S. Yoon, C. –H. Wu, and M. Kelling. Proc. SPIE, vol. 7972, 2011, pp. 79722L.

[6] Y. Seino, K. Kobayashi, K, Sho, H, Kato, S. Miyoshi, K. Kikutani, J. Abe, H. Hayashi, T. Ohiwa, Y. Oonishi, and S. Ito. *Proc. SPIE*, vol. 6923, 2008, pp. 69232O.

Comparative Analysis of Resist Model Stability in Negative Tone Development Process

Hua Cai, Hanmo Gong, Qian Ren

SMIC Advanced Technology Research & Development (Shanghai) Corporation
18 Zhangjiang Road, Shanghai, China
*Corresponding Author's Email: Hua_Cai@smics.com

Biography

Hua Cai obtained his doctoral degree in Optical Science and Engineering from Fudan University in June 2016, thereafter entered SMIC Advanced Technology Research & Development (Shanghai) Corporation. In SMIC, he focus on negative tone development (NTD) resist model buildup and Source Mask Co-Optimization (SMO) simulation.

Abstract

We discussed a method of checking the stability of a physical resist model and verifying the model accuracy, using simulation contour data in conjunction with scanning electron microscope (SEM) images. Firstly, we introduced the hybrid modeling flow, involving the optical parameters and the resist parameters. Whereafter, we compared the stability of two resist models with various resist terms. One model (Model 1) used several effective negative tone development (NTD) resist terms and the other one (Model 2) adopted more high order NTD resist terms were built up. Comparing the SEM image and simulation contour of complex 2D patterns, we found that the Model 1 did better in predicting the final wafer critical dimension (CD), and it could also successfully indicate the potential pinch risk. The Model 1 rendered reduced fitting CD errors for the 2D verification patterns without sacrificing the accuracy for 1D features, while the Model 2 lost its predication ability, because high order resist terms might induce mathematical over fitting into the model. In other words, the model terms should be chosen in accordance to their physical background: the NTD resist shrinkage, development loading effects and the various effective diffusion lengths of different types of patterns.

Keywords—Resist Model; Negative Tone Development (NTD); Model Stability; NTD Resist Terms

Introduction

Over the last decades, computational photolithography has been one of the key elements in extending ArF immersion lithography tools, processes and materials to meet challenges in advanced integrated circuits (IC) fabrication driving down the feature sizes. [1] Multiple patterning methods are being used to meet the design requirements at the 20 nm technology node and the beyond for the interconnect levels in the BEOL, and they are always coupled with negative tone development (NTD)

process due to its several advantages.[2] The negative tone process has been shown to improve the imaging contrast and mitigate the optical proximity effects. In the negative tone develop process, the trenches or contact holes are formed by dissolving away the unexposed resist using organic solvent, thus it is easy to achieve narrow features in the patterning. The NTD process also helps to have higher depth of focus compared to the conventional positive tone develop process, because the width of the illuminated region of the aerial image increases as the width of the to be patterned trench decreases, therefore, the image log slope of the aerial image formed by the negative tone imaging is higher than the positive tone image. [3] However, compared with PTD resist, three characteristic effects of the NTD resist, the physical resist shrinkage, resist development loading and sub-resolution assist feature (SRAF) induced effect, were demonstrated in our experiment data. Thus a NTD resist model is needed to simulate those complex effects.

Furthermore, as device feature sizes continue shrinking, it is increasingly difficult to achieve a highly accurate prediction model owing to the model complexity, since the correction amounts for edge displacements greatly vary in accordance with the surrounding environment of fragmented edges and optical proximity effects (OPEs), which contribute to one of the factors that increases difficulty of model training.[4] It results in the following two disadvantages: (1) Over fitting: Model learning with limited samples or non-representative samples causes over fitting or over generalization that lowers predictive performance of the regression model. (2) Limited applications: It is difficult to apply the linear regression method to complex nonlinear phenomena because the algorithm includes a linear assumption in which all model parameters are linearly correlated with input feature vectors. However, the over-fitting issue is still open since a sufficient amount of supervised data is not necessarily given in all cases. Furthermore, practical applications of the regression model are also restricted even with nonlinear algorithms because there are relatively few adjustable model parameters that can improve predictive performance.[5]

In this article, we introduced the hybrid modeling flow and then the estimation for model accuracy and stability verification. Finally, we compared two resist models with different resist terms to systematically study and go deep into the three main characteristic effects of NTD resist.

978-1-5386-5309-8/18 $31.00 © 2018 IEEE

Metrology

A Hitachi scanning electron microscopy (SEM) was used to measure pattern width and space. The threshold algorithm was employed for detecting the pattern edges of line in top-view images obtained at the magnification of 250000.

This study used an optical only immersion exposure system with an NA of 1.35, a pixelated source, and a wavelength of 193 nm. The source is sector polarized. The pixelated source is produced using the SMO process. The background of the photomask is bright field, and the main features have a phase shift of 180 degrees with an intensity transmission of 6%.

Model calibration was performed on a platform by minimizing the root mean square (RMS) error between simulation and actual data with selected resist parameters constrained to their physical values. After calibration, the quality of the match can be evaluated by inspection of the RMS error. Validation of the portability of the calibrated NTD resist model was done by collecting further experimental top-down SEM CD data and comparing it to the simulation contours. Comparative analysis of stability for resist model parameters was studied when the altered NTD resist terms are set to the two NTD resist models, TABLE I shows the comparison of resist parameters for Model 1 and Model 2. Again the metric of RMS error was used to evaluate the quality of the prediction.

Hybrid Modeling Flow

The hybrid modeling flow, which includes the optical and resist parameters tuning, has been shown to have significant value in OPC model calibration[6]. For the model setup, contour data is also very expensive in terms of metrology bandwidth extracting, storing, and transferring full contour information compared to a single measurement of a CD value.

Optical Parameters Tuning

For the optical parameters, most of them can be set to the certain value in the actual scanner, except that one of the most important pieces in modeling, the definition of the optical focus plane and simulated measurement plane. In a model calibration job, the first one is the aerial image location (AI_location) of the optical system, in the absence of a film stack. It is the distance between the imaging plane, where the optical image is taken for the simulation of resist image, and the interface of the free space and the top film. The second is called defocus or defocus start, which is the distance between the top film surface and the focal plane with no films. Fig1 schematically shows the aerial image location and defocus. These two parameters are not independent and generally correlate with each other. A mismatch in these parameters will lead to tilted Bossung plots, focus shifts, and poor model quality in general. Through process data is necessary for setting the focus parameters in the optical model. The OPC model must be accurate through focus and dose for building a photomask robust to process variation seen in a manufacturing environment.

Bossung curve

Rather, we can calibrate the optics using the FEM CD data, especially choosing focus sensitive structures and well printing structures with clean data using a simple constant threshold model. Fig. 1(a) schematically shows the relationship between AI_location and defocus, the important parameters in optical terms tuning. Theoretically, by adjusting the Bossung curve of certain features, the minimum RMS can be achieved, which used to determine the best optical operating settings for the model. The dense patterns show a flat and symmetrical Bossung curve, which is not focus sensitive because of its large depth of focus. The wafer CD of isolated feature is optically sensitive and shows out a tilt Bossung curve, as it can be seen in Fig. 1(b).

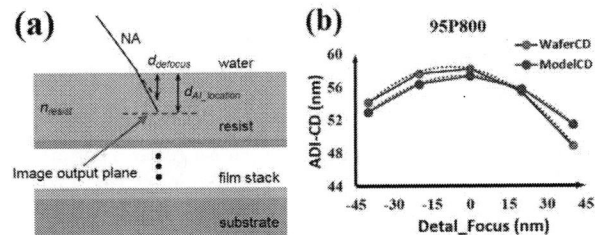

Fig.1 (a) the schematic relationship between AI_location and defocus. (b) Bossung curve of representative isolated trench feature.

For this data set, the value is 0.075 um for AI_location and - 0.060 um for defocus by minimizing the model RMS. These two values can then be used for the next phase, calibrating the empirical resist model using the nominal CDs and contours. In this experiment, the optical parameters were kept the same in Model 1 and Model 2, while resist parameters were selected and to be calibrated against resist wafer CDs.

Resist Parameters Tuning

Resist model which simulates development behavior is one of keys to decide the accuracy of the simulation, to realize a NTD resist model, the parameter tuning was divided into two part: the PTD resist terms and the NTD resist terms.

PTD resist terms

The positive tone developer (PTD) resist model form can be written as:

$$R_{(PTD)} = c_A \times A + c_{Ap} \times Ap*G_{Ap} + c_{Bp} \times Bp*G_{Bp} + c_{Am} \times Am*G_{Am} + c_{Bn} \times Bn*G_{Bn} + c_{AG1} \times A*G_{AG1} + c_{AG2} \times A*G_{AG2} + c_{Mav} \times Mav + c_{MG1} \times M*G_{MG1} + c_{MG2} \times M*G_{MG2} + c_{Slope} \times Slope + theta - thresh \tag{1-1}$$

Where R represents the resist contour, $c_{A...}c_{slope}$ and *theta* are weighting coefficients to be fitted, and thresh is the threshold of resist image. The * operator in this equation represents convolution. A is the aerial image intensity. The term $c_A \times A$ models the effects of aerial image exposure, the initial acid distribution in the resist on the wafer. The Gaussian filters (G_{Ap}, G_{Bp} ...) are modeling various effects in diffusion or 3D resist development. The two separate Gaussian terms $c_{AG1} \times A*G_{AG1}$ and $c_{AG2} \times A*G_{AG2}$ model effects of aerial image diffusion. With different sigma$_{AG1}$ and sigma$_{AG2}$ values, the double Gaussian terms give more flexibility in modeling more complex diffusion effects during exposure, baking and development. The Mav is mask image intensity and the two separate Gaussian terms $c_{MG1} \times M*G_{MG1}$ and $c_{MG2} \times M*G_{MG2}$ present the mask loading effect. The term thresh models the resist image threshold at which the resist contour is extracted.

NTD resist terms

The PTD resist model form can only simulate the existing acid and base diffusion and reaction, but cannot simulate the characteristic NTD effects such as physical resist shrinkage, organic developer, and SRAF induced developer redistribution. The NTD process, which includes resist ingredient modification, utilization of organic solvent developer, and other process modification, shows the greatly different behavior against the PTD process. The NTD terms describe the effects with proper image signal from aerial image and mask image to pattern dependent resist bias trend. By using the NTD terms, we can write the NTD resist model form written as:

$$R_{(NTD)} = R_{(PTD)} + A + B + C \qquad (1\text{-}2)$$

Where A, B and C represent the physical resist shrinkage, developer loading and SRAF induced develop consuming, respectively.

In this experiment, we introduced two models with different NTD terms to study the resist parameters systematically and to understand the real physical meaning behind them, Table I shows the comparison of resist parameters for Model 1 and Model 2. We mainly focus on the discussion of physical resist shrinkage and developer loading effects, so we escape the SRAF introduced effect, although it can introduce de-protection and developer distribution, and then impact dissolution rate and main feature CD.

Table I
Comparison of resist parameters for Model 1 and Model 2

NTD Resist Terms	Terms Order Chosen			
	Model 1		Model 2	
Resist Shrinkage	Basic	--	Basic	High
Developer loading	Basic	--	Basic	High
SRAF developer loss	Basic	--	Basic	--

Model Simulation Result

Simulations were run for each of the validation structures and the predictions were compared to the experimental measurements. This process yields a set of kinetic parameters describing the resist behavior. In this subsection, we present the concept of the proposed resist model regression, followed by the solution of the parameter estimation technique.

With the remaining features, a model calibration job can filter out gauges which are sensitive to resist model parameters and optical parameters. The first data set is measured in nominal condition only, while the optically sensitive features are measured through focus energy matrix (FEM) condition. The modeling was done using a mask three dimension model with 9 slices in incident angle.

A. RMS

We evaluated model CD prediction errors using RMS (Root Mean Square) error given by,

$$RMS = \sqrt{\frac{1}{N} \sum_{1}^{N} d_n} \qquad (2\text{-}1)$$

Where N is the gauge number, and d_n is the square of the difference between simulated model CD and measured wafer CD. The group RMS error at nominal condition was mainly considered for the first validation data set. Assuming that the

built resist model has enough coverage and accuracy, the wafer data should be a close match to modeled data. Below is a summary of the final Model 1, as shown in Fig. 2. It can be seen that the 1D line/space features are calibrated quite well. The RMS error of the model is almost approximate 1 nm and the range is on the order of 4 nm, as shown in Fig.2 (c). And for the 2D tip-2-tip, including head-to-head (HTH) and head-to-line (HTL) data, the RMS was 2.1 nm with model fitting error distribution of +/- 5.0 nm, as shown in Fig. 2(d). The table II shows the 1D and 2D group RMS and range under nominal and FEM condition.

We also noted the RMS errors at different pitches. The RMS errors observed for the F-E 1D line/space data were approximately 0.9 nm, 1.2 nm and 1.4 nm for the 140 nm, 180 nm and 800 nm pitches, respectively. The RMS error for the F-E 2D HTH/HTL data were 1.5 nm, 2.4 nm and 2.0 nm for these pitches, respectively.

Table II
1D and 2D group RMS and Range under nominal and FEM condition

Type	NC (nm)		FEM (nm)	
	RMS	Range	RMS	Range
1D	1.2	4.1	1.3	4.6
2D	2.1	9.4	2.5	10.4

Capturing the tip-2-tip behavior was a bit more problematic, a small systematic offset is observed but cannot be explained. This is still under investigation and can potentially be related to selected resist terms.

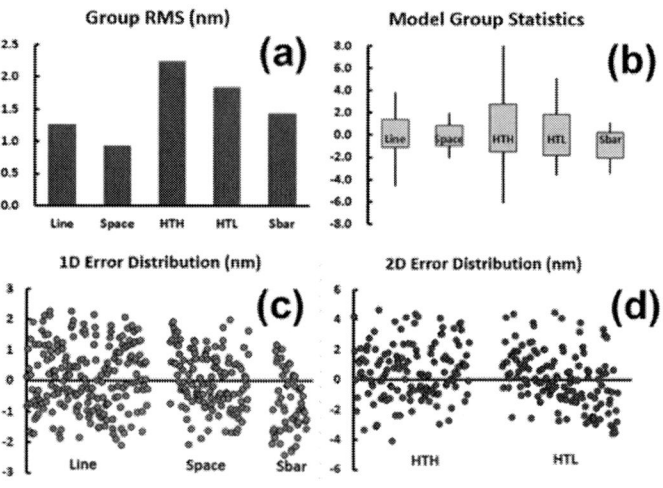

Fig.2 Summary of the final calibration results of the compact model of the 1D line/space and 2D through pitch model fitting error with distribution of +/- 2 nm

B. Trend

The model CD trend is a very import evaluating factor for NTD model resist terms, which is the difference between model CD and aerial image CD, since aerial image CD is very physical concept derived from optical aspect. The aerial image intensity distribution forms the latent image in the resist that can directly influence the wafer CD. Fig.3 (a) and (b) show the 1D and 2D model trend, respectively. The model trend describes the CD differences as a function of through pitches. In Fig.3 (a), it can be observed that the dense feature shows a positive CD difference value, and the isolated pattern presents a negative value. While at the forbidden pitch, it exibits a transition from positive to negative value. As it can be seen in detail, the CD difference is an oscillating value. So we choose three dense, semi-dense and isolated features as the representative pattern, as shown in the upright plot of Fig.3 (a). For the dense and semi-dense features, the value ascend linearity, since the small resist line is small amount of resist, and its CD is very sensitive to line size. While for big pitch, the CD difference is saturated, because the leaving base depleted and mechanical resistance increase.

As the trend line shown, small pitch and big pitch have a flatten slope, which looks like there is a constant value between wafer CD and aerial image CD, that is to say, the wafer CD could be simulated by a constant threshold model. While for the semi-dense pattern, it has a slightly sharp slope, and the CD difference value changes linearity, i. e., the wafer CD should be simulated by a variant threshold model. Physically speaking, the diffusion lengths of different features are not constant but variant. The compact resist model function is calibrated to wafer CD to minimize the difference between model CD and wafer CD. To achieve an accurate simulation, it is necessary to fix a complicated non-linear model function.

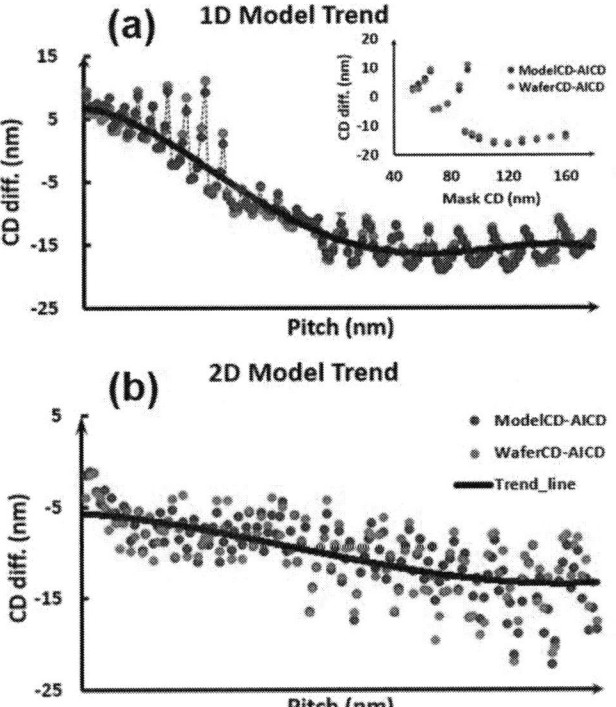

Fig.3 (a) and (b) illustrate the through pitch CD difference between wafer (model) CD and aerial image CD (AICD) of 1D and 2D group patterns, respectively. The upright plot in Fig. 3(a) shows the CD difference as a function of the mask CD for dense, semi-dense and isolated 1D features.

For 2D tip-2-tip patterns, the CD difference is almost the negative values, as shown in Fig.3 (b). The trend line lentamente change from nearly zero to big negative value, and finally to a constant value at big pitch. It means that the gap CD of tip-2-tip is not that sensitive to the pitch change.

C. Contour

In order to thoroughly study the resist model terms and go deep into their physical meanings. We built up two resist models, the Model 1 and Model 2. The Model 2 choose high order not only basic order terms like Model 1. Comparison of the SEM image with the simulated contour is straight method to evaluate the model stability and accuracy. Further, lithography simulation is an indispensable technique for device manufacturing and it requires accuracy which means small difference between the simulated value and experimental value on wafer. The contact holes and tip-2-tip were chosen as the validation features.

Fig.4 shows the overlap information of SEM image and simulated contour by Model 1 and Model 2. In Fig. 4 (a) and (c), it is shown that Model 1 catch the resist trend properly, and the contour matches the SEM image quite well for the contact holes and tip-2-tip validation pattern. The Model 2 traces the resist change trend well for contact hole and tip-2-tip features, while miss the model trend for reversed tip-2-tip feature, clearly shown in the right side of Fig.4 (d). The reason is the model over fitting which is caused by high order resist terms added in Model 2. The model with high order resist terms can reduce RMS while sacrifice the model stability. For the certain input data, the model fitting result is very well, because it excessively depends on the data value and ignores the common characteristic of the pattern in the physical background, thus lose the model prediction. So it is challenging to train a robust regression model owing to the over fitting problem, whereby a model has poor predictive performance. This indicates that it is expected to realize flexible modeling even of a complicated phenomenon having a large variation in input data or including unknown variables that cannot be measured.

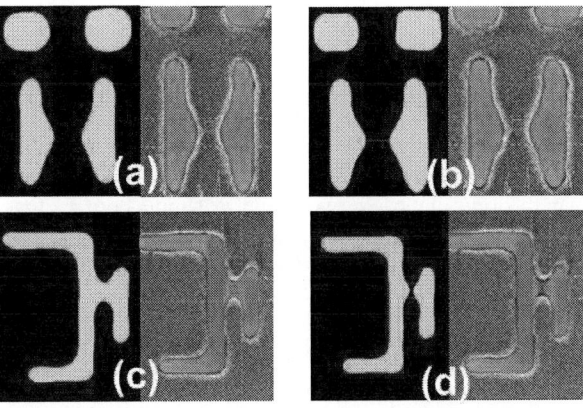

Fig.4. (a) and (c) show the overlay of SEM image and Model 1 simulated contour, (b) and (d) show the overlay of SEM image and Model 2 simulated contour for two validation clips.

Physical resist shrinkage term represents post exposure bake and development shrinkage effects, which can directly change resist thickness and side wall angle, then influence lateral CD by mechanical relaxation. High order resist term might introduce mathematical over fitting into the model thus compromise the model coverage. The development loading effect resist term presents the combination of resist chemical and physical processes, thus influence the developer dilution and diffusion. This term is very sensitive to pattern type. Since the NTD development is slow and low contrast, which brings into the issue of pattern dependent development difference. The high order terms could produce pattern dependent model error, thus deteriorate the model accuracy.

We put a cutline across the gap of reversed tip-2-tip pattern in Model 2, and then can obtain the model threshold and ILS (image log slope), as shown in Fig. 5. It is well known that the ILS is the gradient of the natural logarithm of the aerial images in lateral direction, and is a much important estimation factor that directly relates with energy latitude and the process window in lithography.

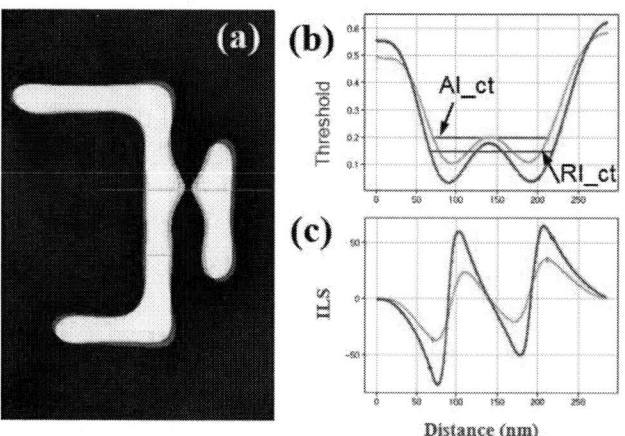

Fig. 5 shows the model threshold and image log slope (ILS) at the gap of the reversed tip-2-tip validation pattern in resist Model 2.

The Fig. 5 (b) shows the threshold of aerial image (AI_ct) and that of resist image (RI_ct). It can be seen that the RI_ct has an apparent truncation with resist, formed bridge resist line. The AI_ct threshold line cuts at the top of resist line, which is closer to wafer SEM image, when compared with the result of simulated resist image. This further demonstrates the high order resist terms lead the model to incorrect direction, resulting in model over fitting. It should be noted that ILS values only represent the optical and physical terms that predict relative process window for different features, the ILS of image plotted in the low-right of Fig. 5 (c), and we should confirm that the contrast of resist image is higher than that of aerial image in resist Model 2, although this model misses catching the resist trend.

Conclusion

There exists challenges of negative tone process at the 20 nm node and the beyond, including resist materials, resist model stability and reliability. In this experiment, we introduced the hybrid modeling flow, involving the optical parameters and the resist parameters (PTD resist terms and NTD resist terms tuning), and then used the RMS and model trend to evaluate the model. The 1D group RMS was 1.2 nm with model error distribution of 4.1 nm, and the 2D group possesses the RMS of 2.1 nm with model error distribution of 9.4 nm at nominal condition. The model trend (the difference between model CD and aerial image CD) is also an important factor for estimating the capability of resist model to capture the changing trend of wafer CDs. It can clearly show the diffusion properties of a NTD resist.

In order to systematically study resist model terms and go

deep into the three main NTD resist characteristics, the physical resist shrinkage, the development loading effect and SRAF introduced effect. We built up two resist models with different resist terms, namely, Model 1 and Model 2. Comparing their simulation contour, one can find that Model 2 with high order resist terms lost the model prediction, because high order terms might induce mathematical over fitting into the model, thus weakened the physical meanings in the model.

Besides the model stability and accuracy impacted by the chosen resist terms, the model still faces the over fitting issue, since the separate model requires a larger number of supervised data. A new efficient regression algorithm is further required to realize flexible modeling for complex phenomena based on a small number of data.

References

[1] G. Amblard, S. Purdy, R. Cooper, M. Hockaday, An improved method for characterizing photoresist lithographicand defectivity performance for sub-20nm node lithography, Proc. SPIE 9779, 97790U (2016).

[2] S. Mehta, L. Ganta, V. Chauhan, Y. Wu, S. Singh, C. Ann, L. Subramany, C. Higgins, B. Erenturk, R. Srivastava, P. Singh, H. Koh, D. Cho, Process Variation Challenges and Resolution in the Negative Tone Develop Double Patterning for 20 nm and Below Technology Node, 9425, 94250B (2015).

[3] Y. Miyamoto, J. Sagan, M. Padmanaban, G. Pawlowski, T. Nagahara, Chemical Shrink Materials and Process for Negative Tone Development (NTD) Resist, Proc. SPIE, 9051, 905123 (2014).

[4] A. Chen, Y. M. Foong, T. Thaler, U. Buttgereit, A. Chung, A. Burbine, J. Sturtevant, C. Clifford, K. Adam, P. DeBisschop, Effective use of Aerial Image Metrology for Calibration of OPC Models, Proc. SPIE 10147, 101470Y (2017).

[5] Y. Y. He, C. S. Chou, Y. P. Tang, W. C. Huang, R. G. Liu, T. S. Gau., Resist Profile Simulation with Fast Lithography Model, 9052, 90520Y (2014)

[6] K. Hitomi, S. Halle, M. Miller, I. Graur, N. Saulnier, D. Dunn, N. Okai, S. Hotta, A. Yamaguchi, H. Komuro, T. Ishimoto, S. Koshihara, and Y. Hojo, Hybrid OPC Modeling with SEM Contour Technique for 10 nm Node Process, Proc. SPIE 9052, 90520W (2014).

[7] M. Miller, K. Hitomi, S. Halle, I. Graur, T. Bailey, Application of SEM-Based Contours for OPC Model Weighting and Sample Plan Reduction, Proc. SPIE 9426, 94260Y (2015).

ADVANCED PHOTORESIST AND MATERIAL DEVELOPMENT IN CHINA

Ran Ruicheng, Mao Guoping, and Sun Yousong
Jiangsu Hantop Photo-Material Company,
Liaohe Road, Pizhou Economic Zone, Xuzhou, China, 221300
Email: sunys@htphotrom.com

ABSTRACT

The paper describes a brief situation of advanced photoresist development in China, which is either in mass production or in customer qualification loop, mainly for lithography process in 8-12 inch wafer fabrication of large scale integrated circuit. The resists and the materials introduced here are produced from different domestic factories. In this paper, their KrF and ArF types of photoresists and the materials are reported. The paper also codes the usage roadmap supplied by domestic resist makers in the future.

INTRODUCTION

Recent years more and more new IC Fabs have been established in different locations in China. Meanwhile the existing IC Fabs expand their capacity or build their second phase plant to increase manufacturing scale. In the other side, in equipments and materials used in the manufacturing start to be developed and produced by China domestic suppliers. So China market of semiconductor equipments and materials encountered full of competition between traditional brand and emerging producers.

Photoresist is one of the most important materials used in semiconductor industry. It is image transfer media to help creating patterns which, with other technologies, form electric circuit on Si wafer or other substrates. There are many types of photoresists can be used in different industries, like semiconductor, LED, MEMS, LCD, and nanofabrication. For the advanced microelectronic fabrication, circuit width becomes tiny to cope with circuit integrate density desire. So the resist patterning size has to be created with shorter wavelength lithography process. KrF and ArF resist is then developed for these purposes.

There are several resist makers in the world providing very good resist products for semiconductor industry and for other industries as well. As rapid developed IC fabrication in China, resist made from domestic supplier is also found in this market. Here only advanced resist supply from domestic manufacturers will be presented.

The major advanced resist supply chain from domestic source

As general, resist material include monomer, polymer, PAG, solvent and additives. As upstream supply of resist,

there are some suppliers in China provide monomers both of KrF and ArF. Table 1 shows the main supply on the chain.

Materials		Supplier and Capacity
KrF	Monomers	Xuzhou B&C Informatic Chemicals, facility capacity of 350ton/Yr.
	Polymers	Xuzhou B&C Informatic Chemicals, facility ready for capacity of 50ton/Yr.
	PAGs	● Shanghai B&C Chemicals, lab scale production, 150KG/Yr ● Changzhou Strongly new Materials, mass production for initiators. Partially for KrF resist
	Resist	● Beijing Kempur Microelectronics, small scale production, facility ready for 50 ton per year ● Jiangsu Hantop Photo-Materials, small scale production, capacity ready for 120 ton per year
ArF	Monomers	Xuzhou B&C Informatic Chemicals, capacity of 80ton/Yr.
	Polymers	Jiangsu Hantop Photo-Materials, small scale, facility ready for capacity 30 ton per year.
	PAGs	● Shanghai B&C Chemicals supply with small sample ● Some universities supplying with samples
	Resist	Some companies in R&D stage for Dry process

Table 1: resist material suppliers from China, 2016

Besides above resist and material, DSA material investigation is active in China by some universities and companies. E-Beam resists have also been developed by Jiangsu Hantop Photo-Material, who supplies now two types of resist for E-beam writing.

For BARC and Gap fill products, some raw materials in electronic grade are supplied from China to be abroad in the past a few years.

Electronic grade Solvents, as basic material, are

978-1-5386-5309-8/18 $31.00 © 2018 IEEE

already supplied by Jiangsu Dynamic Chemicals, Jianghua Microelectronics Materials, and others. Those materials support electronics material industry development in China.

Local Fabs call for Domestic Resist Supply as safer logistic concern

In the past, local fabs start to accept domestic resist supply as safer logistic supply chain, taking the advantages of no longer importation and custom clearance delay, urgent batch delivery and cash flow burden. This change purely caused by domestic supplier continue investing on the advanced resist R&D in the past ten years and now can deliver qualified resist products to the industry.

Fig 1 shows KrF resists are provided by domestic suppliers. 2017 is the key time for local resist makers to get their products into big fab. There are a lot of evaluation activities in the wafer fab manufacturing lines. As the result, more resist from domestic source will be adopted in 2018 by these IC fabs.

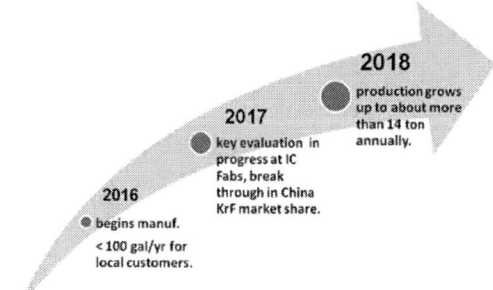

Fig. 1: local supplied KrF resist Volume Chart

Based on ICMTIA collected data, by 2030, the local sourced resist consumption in IC fab will be exceed 30% in average. It means KrF and ArF from local makers will be used in manufacturing lines and gradually increases into certain level.

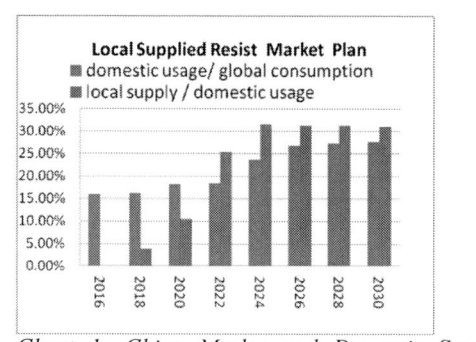

Chart 1: China Market and Domestic Suppliers Growing

Main Domestic Resist and Material Suppliers in China

A. ***Xuzhou B&C Informatic Material Co., Ltd*** starts to produce resist monomer in 2010. Till now it provides more than 50 kind of ArF monomer to Japan and Korea market. It also sold KrF monomer abroad as much as over 300 ton in last year.

Some of ArF monomers is open for selection and year round manufactured by Xuzhou B&C.

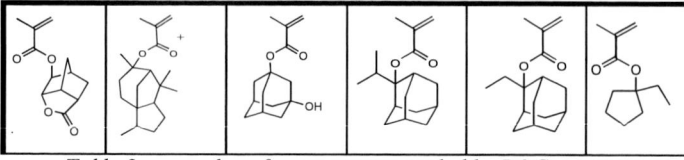

Table 2: examples of monomers provided by B&C

Except monomers, B&C supplies other electronics materials as lithography material, such as TPPA, 4HBP, 3HBP etc. These are in mass production as well and purchased by many customers globally.

B. ***Beijing Kempur Microelectronics Inc.*** is major advanced resist maker in China. It speeds up KrF resist development as early as in 2009. Now its KrF resist product is promoted into almost all big fabs in China. Some are already normal supplied and some are under evaluation by customers.

Product	Model	Performance
KMP DK	**DK1080**	**DUV resist, R 0.25um, THK: 0.55-1.0um**
	DK1086	**DUV resist, R 0.18um, THK: 0.55-1.0um**
KMP C	**C7000**	**I-Line PR, R 0.35um, THK 0.75-1.0um**
	C 7500	**I-Line PR, , R 0.45um, THK 1.0-2.5um**
Thk Film	**C6111**	**I-Line PR, R 2.0um, THK 3.0-6.0um**

Table 3: Some of Kempur advanced resist products

Kempur owns very powerful lithography verification facility including updated KrF exposure tools, CDSEM and coating track. It can test advanced KrF resist with CD down to 130nm features.

C. ***Jiangsu Hantop Photo-Material Co., Ltd.*** Located in Mid-East of China. It equipped with advanced photography test kit which can assess KrF resist and polymer as well as other materials. It starts involved in KrF resist in 2014 and now it provides products to customers already.

	P/N	Resist	Polymer	Feature	Supply
Thin Resist	P	HTK-100	ESCP T14	Film 800nm, R=0.18um	Production
	P	HTK-109	ESCPT14	Film 1000nm, R=0.25um	Production
	P	HTK-150	ESCPT14	Film 300nm, R=0.13um	R&D
	N	HTKN-601	PHS6015	Film 800nm, R=0.25um	Production
Thick Resist	P	HTK-510	PHST25	Film 3-5um, R=0.7um	Evaluation
	P	HTK-812	PHST29 Mixed	Film 8um, R=1-2um	Evaluation
	P	HTI 560	PHST25 Mixed	Film 3-5um, R=0.6um	Evaluation

Table 4: Hantop advanced resist product list

Hantop has developed its own polymers for resist makeup. It synthesizes 3-, 4-copolymers with different performance. The polymer types cover KrF, ArF and E-Beam applications.

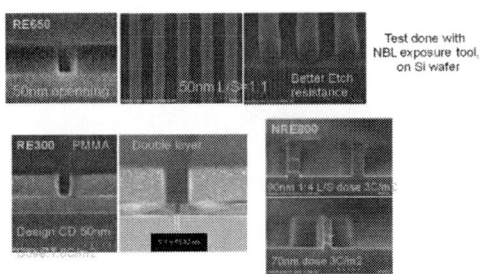

Figure 2: examples of E-Beam resist patterning
Re650 is high resolution positive resist with good etch resistance. RE300 is high resolution E-Beam resist, can be thicker. NRE800 is negative E-beam resist with good etch resistance.

In 2017, Hantop is promoting thick KrF resist to local customers. The evaluation is undergoing.

Below are some examples of KrF resist lithography test results. It proves the performance can meet customer's requirements.

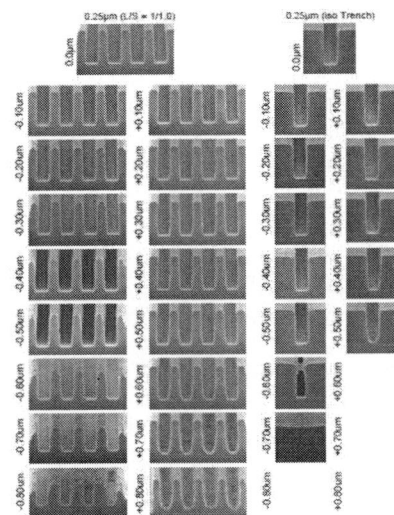

Figure 3: 250nm pattern capability of Hantop KrF resist. Wide DOF and EL% with straight profile.

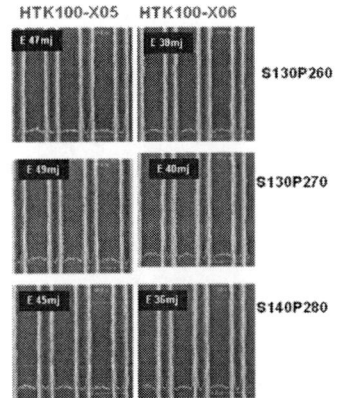

Figure 4: 130nm pattern capability of Hantop KrF resist. Acceptable LWR.

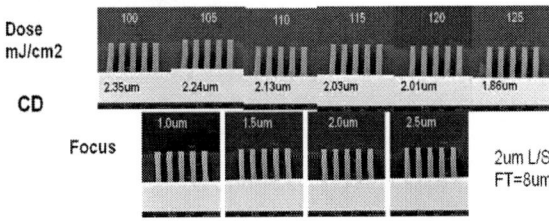

Figure 5: CAR I-Line resist. Thickness is 8um. 2um patterning with satisfied DOF and EL% process window.

Opportunities and Challenges in China

With the point of view that global IC manufacturing volume increase mainly contributed by China in near future, we will expect IC process materials including

978-1-5386-5309-8/18 $31.00 © 2018 IEEE   165

photoresist will increase greatly. Year 2020 will be a milestone for domestic tool and material makers to supply more products for IC manufacturing, and we estimate the figure will be over 30% in market share.

As resist maker, we hope we can work together with international resist suppliers to provide better services and products for China IC manufacturers.

ACKNOWLEDGEMENTS

Here we are appreciation to ICMTIA in the past year to have organized advanced resist road map forum and data report. Also we shall thank all participators who devote into China domestic resist manufacturing business.

REFERENCES

[1] 中国半导体支撑业发展状况报告, ICMTIA, 2017

A STUDY OF MULTI-FACTOR INFLUENCE ON RESIT LWR PERFORMANCE

Dongxu-Yang[1], Lei-Ye[1]
Technology R&D, Semiconductor Manufacturing International Corp.
Pudong New District, Shanghai, P. R. China 201203
+8621 38610000x15973, Don_Yang@smics.com

ABSTRACT

Reduction of line width roughness (LWR) has been one of the biggest challenges in advanced lithography and continuously being more crucial as technology node moving forward. In semiconductor industry, any approach to improve LWR performance has to be carefully evaluated considering possible impacts on other parameters. In this study, multi-factor influence on resist LWR performance has been evaluated. Illumination conditions with various source type, NA and sigma value, and polarization type were tested. Optimized illumination condition was selected based on LWR values as well as other factors including exposure latitude (EL), depth of focus (DOF), critical dimension uniformity (CDU) and mask error enhancement factor (MEEF). With fixed illumination condition, the effect on LWR by other process conditions such as resist film thickness, post-application bake (PAB) temperature and substrate type was also evaluated.

Keywords — LWR; resist; substrate; illumination condition

INTRODUCTION

LWR reduction has been considered as one of the biggest challenges in advanced lithography. As critical dimension (CD) continues to shrink, the value of typical LWR becomes no longer negligible. As a consequence of large LWR, the chance of considerable device performance variation or even failure increases dramatically. Although not fully understood, a number of factors have been observed to contribute to pattern LWR, including illumination shot noise, resist material uniformity, photo-acid diffusion, resist-substrate interaction, development impact, etc. [1] Vast number of research works, with both experimental and simulational approaches, have been carried out to investigate the underlying mechanism of LWR formation. And many approaches have demonstrated effective LWR reduction. [2] However, industry has been cautious about adopting any modification in manufacturing due to process complexity,

process window and stability considerations, which, unfortunately, are the cost that most LWR reduction approaches have to pay. Therefore, systematic evaluation on multi-factors including EL, DOF, CDU, MEEF, etc. is necessary for any LWR improvement approaches. In this study, LWR reduction through optimizing illumination condition as well as other process conditions (baking temperature, film thickness, substrate type, etc.) was evaluated using industrially standard tools, patterns and commercially available materials. Multi-factor influence was studied when applying various illumination conditions.

EXPERIMENTAL

Line-space patterns with pitch 100 nm in Y direction (related to shot grid on wafer) were used in all LWR evaluation tests. Resist patterns were exposed by an ASML 193 nm ArF immersion scanner with various illumination conditions (different source shape, NA, sigma and polarity). Pattern images (top-down) were obtained and analyzed by a Hitachi CG series CDSEM. A commercial 193 nm ArF immersion resist was used in all tests. All underlayer materials, including organic bottom anti-reflection coating (BARC), silicon-containing bottom anti-reflection coating (Si-BARC), spin-on-carbon (SoC), were standard commercial materials

RESULTS AND DISCUSSIONS

Illumination Conditions

Using standard resist and organic BARC substrate, a number of illumination conditions were selected to test the LWR variation as well as other factors including EL, DOF and MEEF. An ASML 193 nm ArF immersion scanner was used to expose the resist film. The detailed illumination conditions are listed in Table 1. Various source type, NA and sigma value, and polarization type were tested.

Test No.	Source mode	NA	Sigma	Polarity
1	Dipole A	small	Sigma A	XY
2	C-quard	large	Sigma B	XY
3	Dipole A	large	Sigma C	XY
4	Dipole A	large	Sigma C	Y
5	Dipole A	large	Sigma D	Y
6	Dipole B	large	Sigma C	Y

Table 1: Illumination conditions for LWR test. Dipole angle A > B.

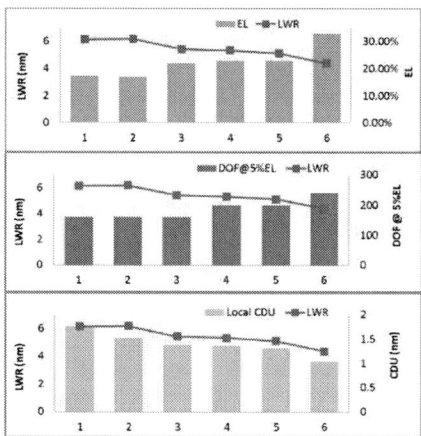

Figure 1: Measured LWR, EL, DOF (5% EL) and local CDU (which reflects MEEF) versus different illumination conditions. LWR is shown in every diagram to compare with the other three parameters.

The measured LWR, EL, DOF (5% EL) and local CDU of the six illumination conditions are shown in Figure 1. The modification that contributes most to LWR reduction was the conversion of source type from large dipole angle to small angle. Near 1 nm LWR improvement was achieved with other conditions fixed, which was benefited from the contrast enhancement in X direction by the reduction of dipole angle. As expected, increasing NA, pushing the sigma value closer to edge, and utilizing Y polarization also help improve LWR slightly. In terms of EL, DOF (5% EL) and local CDU, significant improvement was also observed when dipole angle reduced. Note that local CDU can be considered as indicator of MEEF due to the same photomask was used. In addition, the change of polarization from XY to Y direction resulted in obvious enhancement in DOF.

Illumination with more flexible dipole angle selections was then applied to give a complete comparison of the lithography performance with various dipole angle conditions (as shown in Figure 2). Significant improvement in all parameters was achieved when the dipole angle started to decrease from large value. As the dipole angle continued to go down, the improvements of LWR, EL and CDU tended to flatten, whilst DOF constantly increased. Based on these results, and also considering the disadvantage of small dipole angle (impact on patterns of another direction), dipole condition No. 2 was selected as the optimum illumination condition for all the following experiments.

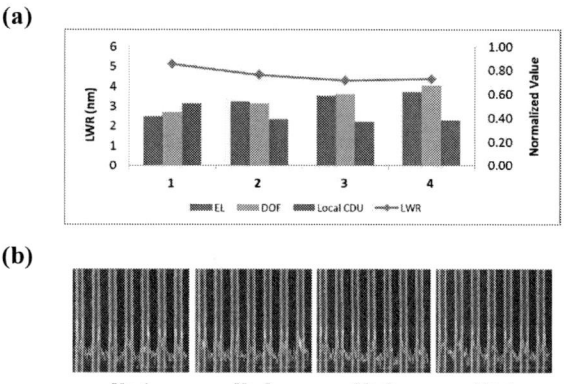

Figure 2: (a) Measured LWR, EL, DOF (5% EL) and local CDU (which reflects MEEF) versus different dipole angle settings; (b) SEM images of the L-S pattern with different dipole angles. From condition No. 1 to No. 4, dipole angle continuously decrease.

Other Process Conditions

The effects of other process conditions on LWR were evaluated. Variation on resist film thickness, post-application bake (PAB) temperature and underlayer type was tested. It is known that resist film thickness variation can have complex physical impacts on pattern performance, including change in DOF, reflection (i.e. aero-image), range of resist-substrate interface interaction etc., all of which can potentially change LWR. Variation in PAB temperature results in different amount of residual solvent in the resist film, impacting the photo-acid diffusion property. In terms of substrate type variation, resist-substrate interface interaction as well as film property itself is likely to influence LWR. Variation in post-exposure bake (PEB) was also tested in previous work and found to have little contribution to LWR. This is possibly due to the low activation energy of this specific resist, making it insensitive to PEB temperature. Therefore PEB variation was not tested in this study.

Figure 3 (a) shows LWR values as function of focus at reduced resist film thicknesses (enhancing light intensity at the bottom of the resist) and various PAB temperatures. Film thickness of 105 nm (reference), 90 nm, 80 nm; and PAB temperature of 90 °C (reference), 80 °C and 110 °C were tested. No obvious improvement was observed in any given conditions. Contrarily, most of the splits made LWR worse. Marginal decrease of ~0.1 nm in LWR was achieved with reduced PAB temperature of 80 °C. It might result from the increase amount of residual solvent that enhanced the acid diffusion, thus "smoothing" the pattern edges.

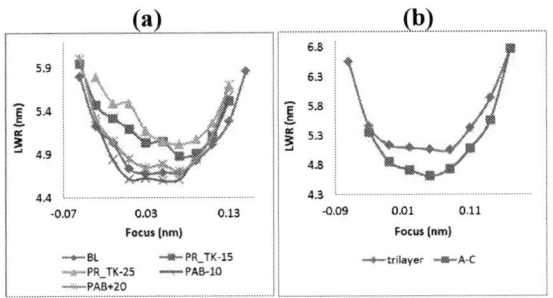

Figure 3: LWR values as function of focus with (a) various resist film thicknesses and PAB temperatures; and (b) two different substrate types.

In order to evaluate the effect of different substrates on LWR, two common film stack types were tested. One was "amorphous carbon (A-C) layer + dielectric anti-reflection coating (DARC) + organic BARC", and another was the tri-layer approach with "SOC + Si-BARC". The film thicknesses of both multi-layer stacks were adjusted according to simulation to give a negligible amount of reflection. Figure 3 (b) shows LWR values as function of focus with the two substrate types. Clearly, patterns on the organic BARC (A-C approach) had a lower LWR than those on the Si-ARC (tri-layer approach) with ~0.5 nm improvement. Profile images of the resist pattern were also taken to give a closer look at the difference of the two samples (Figure 4). Both samples had near vertical side wall without obvious footing. The pattern on Si-BARC had slightly higher side-wall roughness comparing to the organic BARC sample, but the difference was minor. It is also worth noticing that the Si-BARC used in this study has a higher contact angle than the organic BARC, which might result in lower compatibility with the organic photo-resist, causing a degraded profile and higher LWR.

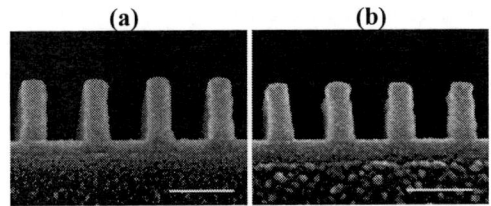

Figure 4: Profile SEM images of the resist patterns on (a) organic BARC, and (b) Si-BARC. The scale is 100 nm.

CONCLUSION

Multi-factor influence on resist LWR performance has been evaluated in this study. At the first stage, the effect of illumination condition on LWR, as well as other factors including EL, DOF and MEEF, was tested. Optimum illumination condition was selected based on the overall performance of all factors. At the second stage, LWR performance with various process conditions was evaluated using the optimized illumination condition. It was found that only marginal reduction of LWR was achieved with decreased PAB temperature, which might be a result of enhanced acid diffusion. Considerable difference in LWR performance with two substrate types (organic BARC vs. Si-BARC) was also observed. It is speculated that different contact angle of the two substrate materials might be a contributor to this effect.

ACKNOWLEDGEMENTS

The authors would like to thank the support from Technology R&D team.

REFERENCES

[1] A.P.G. Robinson，R. Lawson (Ed.), "Materials and Processes for Next Generation Lithography", Elsevier Press, (2016)

[2] K. Cho, S. Tarutani, N. Inoue, H. Tsubaki, M. Neisser. "LWR study on resist formulation parameters". Proc. of SPIE Vol. 8682, 868217 (2013)

ULTRA-FAST DIRECTLY SELF-ASSEMBLY MATERIALS FOR SUB-5 NM LITHOGRAPHIC PATTERNING

*Xuemiao Li, Chenxu Wang, and Hai Deng **

Department of Macromolecular Science, State Key Laboratory of Molecular Engineering of Polymers，Fudan University，Shanghai 200433, China
* Corresponding Author's Email: haideng@fudan.edu.cn

ABSTRACT

The directed self-assembly (DSA) of block copolymers has attracted a great deal of interest due to its potential applications in sub-10 nm or sub-5 nm lithography [1-3].

The conventional organic-organic DSA materials such as poly-(styrene-block-methyl methacrylate) (PS-b-PMMA) have been extensively studied [4,5], however, the low etch contrast and the difficulty to reduce L_0 limits its application. In this study, we designed and synthesized the novel DSA materials based on PS-b-PMMA. Through the modifying of acrylics part, segment−segment interaction parameter (χ) can be significantly increased, which leads to rapid self-assembly and higher resolution.

INTRODUCTION

In the semiconductor industry, the reduction in feature sizes on integrated circuitry is quite an essential part, which closely based on the patterning technology. Since 193 nm patterning technology is approaching its resolution limits, semiconductor industry urgently needs a new patterning technology to continue Moore's Law.

EUVL is getting more mature and widely considered as the next generation litho technology for 7 nm and 5 nm nodes. However, EUV optic system is high cost with low throughput that may limit its applications. Compared with the EUVL, the DSA technology has been investigated worldwide due to its low cost and high potential, [6-10] Some phase separating block copolymers are directed to self-assembled (DSA) into alternative lines, and then form litho patterns through selective etch.

Our research interest is to design and synthesize optimal DSA materials with sub-5 nm resolution, and shortest assembling time. (Figure 1)

EXPERIMENT AND DISCUSSION
Synthesis of block copolymers and characterization

In this study, some block copolymers with low PDI were synthesized. All the solvent and monomers were distilled to remove the moisture and impurities. After the polymerization the polymer was purified by dissolution and precipitation following with the nuclear magnetic resonance (NMR) and gel permeation chromatography (GPC) measurement.

^{1}H-NMR spectra were acquired on a 400 MHz AVANCE III instrument using $CDCl_3$ as solvent and TMS as internal standard. From the ^{1}H-NMR spectra we can identify the characteristic peak of each component.

GPC measurements were used to calculate the molecular weight and PDI. (Polydispersity index) Narrowly distributed polystyrene samples were used as calibration standards. All the block copolymers show a narrow distribution up to 1.15 during this synthesis process.

Field emission scanning electron microscope (FESEM) images were observed using a Zeiss Ultra 55 with In-lens detector at 3 KV.

Small-angle X-ray scattering (SAXS) and Grazing-incidence small-angle x-ray scattering (GISAXS) measurements were collected on a Nanostar SAXS (Bruker AXS GmbH, Germany), with pinhole collimation for point focus geometry.

The annealing process are shown in Figure 2. After 1 min thermal annealing, we got the lamella or hexagonal structures with sub-5 nm half-pitch. (Figure 3) Phase separation morphologies and micro domain spacings (D) of the block copolymers were characterized by small-angle X-ray scattering (SAXS).

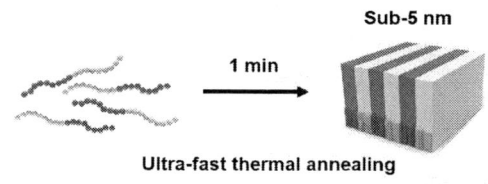

Figure 1: Schematic diagram of sub-5 nm lamellar packing structure in block copolymers after ultra-fast annealing process.

To make sure the exact time of the thermal annealing process is 1 min, we chose the icy plate (copper: 1 mm thickness, good thermal conductivity) to chill the samples after annealing at 80 °C (as shown in the following Figure 2a). After annealing, the morphologies and d-spacings of the samples were measured by SAXS. The temperature reflected the average of three samples during the cooling process. During the self-cooling process, about 200 s were needed for the samples to return to room temperature.

However, the cooling time were reduced to 12 s during the chilled process (Figure 2b).

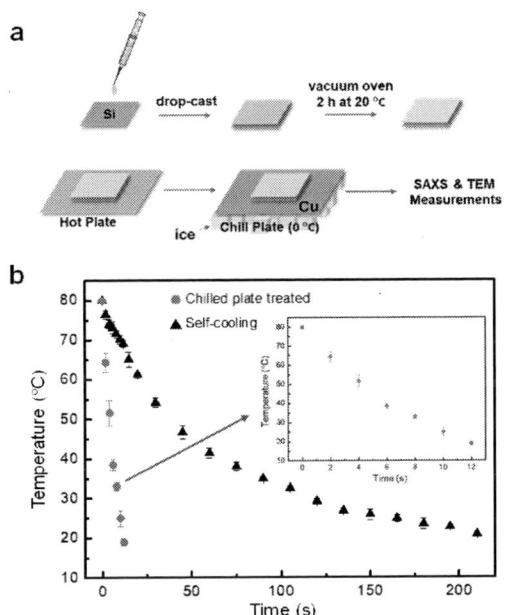

Figure 2: a) Schematic diagram of the samples preparation and thermal annealing process. b) The cooling curves of samples by self-cooling process and fast-chilled process.

By changing the elemental ratio or the block segment ratio in the copolymers, different lamella phase separation pitches were obtained. These block copolymers show the potential as DSA material with high intrinsic resolution for sub-5 nm and beyond nodes.

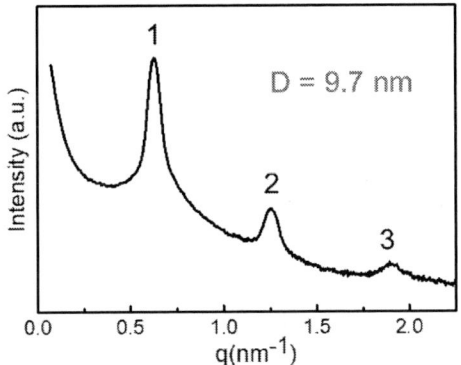

Figure 3: One-dimensional SAXS patterns for the bulk samples of novel DSA materials.

In order to examine the DSA patterning capability of block copolymers, we spin-coated and annealed our DSA materials in the prepared template (by 193 nm lithography) (Figure 4). In the concept of templated self-assembly, the topography of the templates also affects the assembly orientation. We found that the SEM image showed fingerprint pattern outside the template, while the line assembled parallel to the wall of the template by the topological guiding.

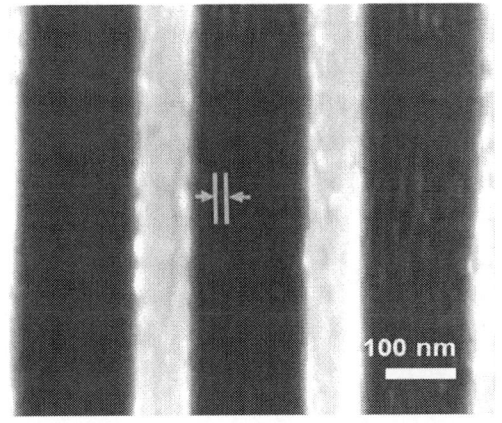

Figure 4: SEM images of BCP material directed assembled in bare silicon templates.

CONCLUSIONS

In this study, we designed and synthesized a novel BCP which show ultra-fast assembly process to achieve sub-5 nm micro phase-separated structures. Different pitches and morphology are obtained by precise control of the two blocks. It is worth noting that the finest half pitch of these BCPs is less than 5 nm, which meets the requirements of semiconductor patterning materials with small features. We further exhibit that ordered nanopatterns are assembled to appear straight lines in the silicon template.

ACKNOWLEDGEMENTS

Funding for this research was provided by the "1000 Talent plan" in China. The authors also acknowledge experimental support from State Key Laboratory of Molecular Engineering of Polymers in Fudan University.

REFERENCES
[1] Kim, H.-C.; Park, S.-M.; Hinsberg, W. D. *Chem. Rev.* 2010, 110, 146.
[2] Herr, D. J. C. *J. Mater. Res.* 2011, 26, 122
[3] Christophe S, Frank S. B, Marc A. Hillmyer *ACS Macro Lett.* 2015, 4, 1044−1050
[4] Benoit, H.; Wu, W.; Benmouna, M.; Moser, B.; Bauer, B.; Lapp, A. *Macromolecules* 1985,18, 986
[5] Bang, J.; Kim, S. H.; Drockenmuller, E.; Misner, M.

J.; Russell, T. P.; Hawker, C. J. *J. Am. Chem. Soc.* 2006, 128, 7622−7629

[6] Kennemur, J. G.; Yao, L.; Bates, F. S.; Hillmyer, M. A. *Macromolecules* 2014, 47, 1411.

[7] Cheng, J. Y.; Rettner, C. T.; Sanders, D. P.; Kim, H.; Hinsberg, W. D. *Adv. Mater.* 2008, 20, 3155.

[8] Tsai, H.; Pitera, J. W.; Miyazoe, H.; Bangsaruntip, S.; Engelmann, S. U.; Liu, C. C.; Cheng, J. Y.; Bucchignano, J. J.; Klaus, D. P.; Joseph, E. A.; Sanders, D. P.; Colburn, M. E.; Guillorn, M. A. *ACS Nano* 2014, 8, 5227.

[9] Wan, L.; Ruiz, R.; Gao, H.; Patel, K. C; Albrecht, T. R.; Yin, J.; Kim, J.; Cao, Y.; Lin, G. *ACS Nano* 2015, 9, 7506.

[10] Jung, Y. S.; Chang, J. B.; Verploegen, E.; Berggren, K. K.; Ross, C. A. *Nano. Lett.* 2010, 10, 1000.

STUDY OF UNIQUE PSEUDO BURIED LAYER IN 0.18UM SIGE BICMOS PROCESS

Donghua Liu[1,2]*, Xi Chen[2], David Wei Zhang[1]

[1]State Key Laboratory of ASIC and System, School of Microelectronics, Fudan University, Shanghai 200433, China

[2]HuaHong Grace Semiconductor Manufacturing Corporation, Shanghai 201206, China

*Corresponding Author's Email: donghua.liu@hhgrace.com

ABSTRACT

SiGe is an important semiconductor material after Si and GaAs. SiGe HBT has obvious advantages over Si HBT in terms of frequency characteristics, DC characteristics and noise characteristics, especially compatible with mature Si process. In our low cost and high performance 0.18um SiGe BiCMOS with collector epitaxial layer free, a unique pseudo buried layer is developed to replace the conventional n-type buried layer under collector epitaxial layer. This pseudo buried layer is important to keep the high performance of SiGe HBT due to the skipping of collector epitaxial layer and conventional n-type buried layer. The lithography is the most crucial step in pseudo buried layer process. In this paper, we report the study result of lithography process of the pseudo buried layer. The lithography conditions suitable for fabrication are achieved. A negative tone photoresist is selected with 7770A thickness and 80 ~ 100nm resolution. The mask bias is 100nm per side and STI side wall angle 82 degrees. The process window (CD + OVL) control is within 50nm and the recommended minimum window CD of the device is 300nm.

INTRODUCTION

SiGe is an important semiconductor material after Si and GaAs. Because SiGe has better properties than pure Si and is technically compatible with Si process, SiGe and its Si / SiGe heterojunction are used to fabricate devices and circuits with almost the same properties as GaAs and other semiconductor devices at circuits Level[1]. Even in many fields, it conveniently alternatives compound semiconductor devices.

SiGe material has the main advantage of high mobility, wide variable band gap and low cost compared with GaAs HBT. It also has obvious advantages over Si HBT in terms of frequency, DC and noise performances, as well as good compatibility with mature Si process. All these make it gradually becomes a new generation of microelectronic devices research hotspot[2]-[4].

Figure 1 is a cross-section diagram of a conventional SiGe HBT device. The important features of the device include the n-type buried layer (NBL), the collector epitaxial layer, and deep trench.

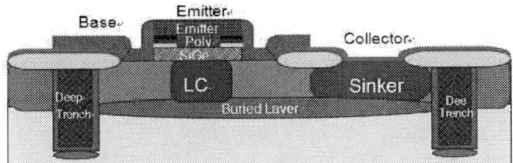

Figure 1: SiGe HBT cross section diagram

Huahong Grace Company conducts the research of SiGe BiCMOS technology for several years and develops a low-cost, high-performance 0.18um SiGe BiCMOS (BCS18) process[5]. Figure 2 is a cross-sectional view of the HBT device, the PMOSFET, and the NMOSFET in the BiCMOS process. MOSFET devices and their process are exactly the same as pure CMOS technology. But the HBT device has a huge difference from the traditional structure. It abandons the buried layer, collector epitaxial layer and deep trench process. In order to maintain the high performance of the device, a pseudo n-type buried layer (PNBL) and deep contact are developed.

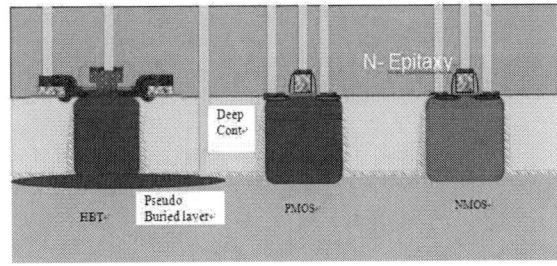

Figure 2: SiGe BiCMOS cross section diagram

This paper will focus on the research and development of lithography process for pseudo-buried layer of Huahong Grace BCS18 technology.

PNBL PROCESS AND EXPERIMENTS

The NBLs in BiCMOS are specifically designed for NPN and PNP transistor to reduce the parasitic effects, such as the gain of the longitudinal PNP, substrate current, and so on. A parasitic PNP will thus affect the performance of the tube. CMOS is working at low voltage levels, so the parasitic effect is negligible and this buried layer is not necessary.

978-1-5386-5309-8/18 $31.00 © 2018 IEEE

Traditional NBL is to put in the active area before growing an epitaxial layer and the devices are fabricated in the epitaxial layer. In our device design, in order to reduce costs, the NBL is in the bottom STI. The following high temperature anneal will drive the NBL diffuses to the active area.

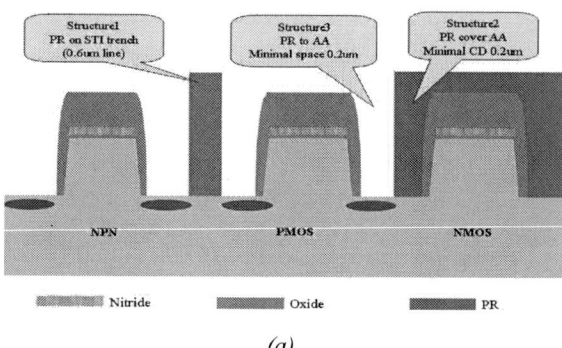

(a)

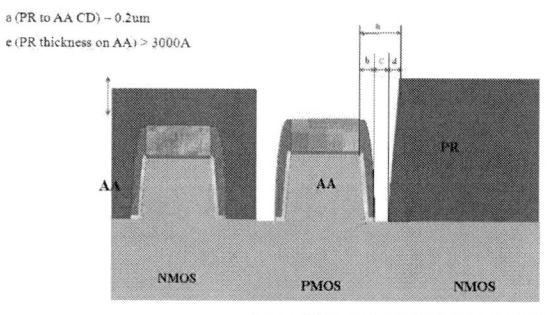

(b)

Figure 3: Key parameters of PNBL lithography

Figure3 (a) and (b) illustrates the rule parameters of a, b, c, d and e of PNBL lithography process. According to the steepness design of the device STI, the c-parameter depends not only on the photoresist resolubility, but also on alignment accuracy. The d-parameter depends on the characteristics of the photoresist. The e-parameter depends on the doping and implantation energy of the PNBL.

Lithography Process Challenge of PNBL

The purpose of lithography is to etch the same geometry on the wafer or on the film as the mask, thus achieving selectivity of different pattern. The important parameters of the lithography process are: printing accuracy and line width, while the process window depends on the total tolerance of line width and printing accuracy.

The process of the photoresist has a relatively high degree of resolution requirements: need to reach about

0.2um. At the same time the photoresist morphology has certain requirements: requires photoresist steepness greater than 87 degrees, cannot have adhesion. The entire process window is only 0.1umgiven the size of the device used in the project is 0.13um / 0.18um node. So the process of the entire lithography process window is tight.

Combined with the production capacity of Huahong Grace Company and the process requirements of the device, we have got the technical spec of the target lithography as summarized in table I.

TABLE I: Technical Spec of PNBL lithography

Item	Spec
PR TO AA CD	0.2um+/-0.02um
PR Cover AA	Thickness >0.3um
Minimal PR Space	0.08um~0.1um
PR Slope	PR Slope < 0.065um

Research Contents and Experiment

The two important parameters of the lithography process are line width and printing accuracy, and these parameters have a very strong correlation with the alignment of the lithography. Combined with SiGe device structure and manufacturing process characteristics, following issues are mainly studied:

1. The thickness of the photoresist required for the PNBL and the Sidewall Angle simulation of the device STI.

2. PNBL required lithography process parameter simulation.

3. PNBL process photoresist selection.

4. PNBL process optimal solution.

For photolithography, the choice of photoresist is very important. It not only represents the ability to resolve lithography, but also represents the basic morphology of photoresist. So based on different SiGe device structure and the characteristics of the process of manufacturing PNBL, the required photoresist thickness and the depth of STI as well as other lithography parameter are studied by simulations and experiments.

RESULTS AND DISCUSSIONS
Photoresist Thickness

We first simulated the thickness of the photoresist required for the NBL process and the side wall of the STI.

From the simulation results in figure 5, the photoresist has a thickness of at least 3000A on the active surface. It is a combination of the height of the device and the photoresist swing. We need a thickness of more than 6000A and have a resolution of 0.2um to meet the process requirements.

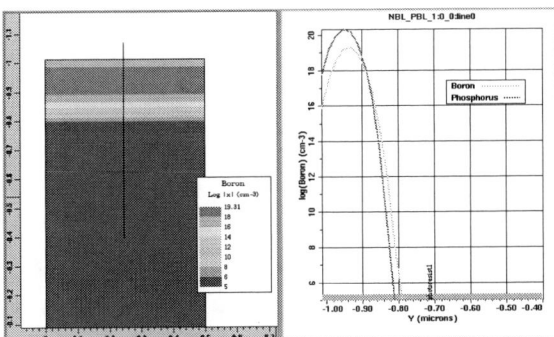

Figure 5:Thickness simulation of photoresist

Side Wall Angle of STI

From the results of simulation in figure 6: STI side wall angle and NBL overlay precision on the device has a very big impact upon the simulation data. The slope of STI side wall should be designed to 82 degrees or larger and NBL corresponding to the requirements of the accuracy of the overlay is 0.1um within.

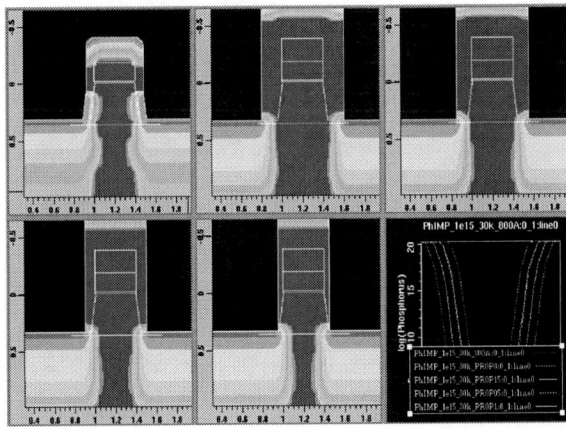

Figure 6: STI side wall angle simulation

Process Parameters of Lithography

In order to reduce the work loading, we use the simulation software to simulate the three typical structures of PNBL process because the photolithography process is complicated and many parameters will affect the final result.

From figure 7 we can see that the thickness of the photoresist of structure 1 is 0.6um and the line width requirement is 0.2um, so that the aspect ratio exceeds the ordinary photolithography requirement of 1: 2.8. The structure 2 and 3 require the distance from NMOS area to PMOS area to reach 0.2um. From the above analysis which is to see that 0.1um overlay accuracy is unrealistic, the entire process window is only 0.1umand photoresist morphology control is also a difficult point.

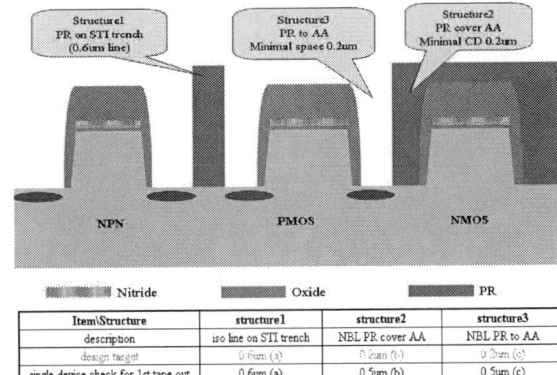

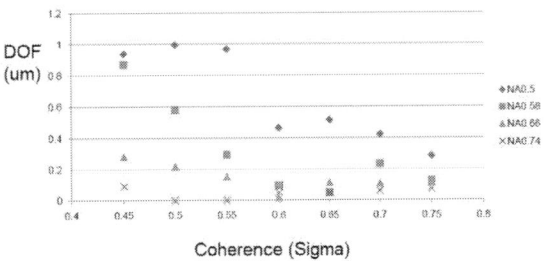

Item\Structure	structure1	structure2	structure3
description	iso line on STI trench	NBL PR cover AA	NBL PR to AA
design target	0.6um (a)	0.2um (b)	0.2um (c)
single device check for 1st tape out	0.6um (a)	0.5um (b)	0.5um (c)

Figure 7: PNBL process requirement and parameters

NA (Numerical Aperture) and Sigma

Based on analysis of the above data and the requirements of the device combined with three different structures, we carried out the study of lithography related parameters, such as NA / Sigma, mask bias, alignment error and other related parameters.

Figure 8: Simulation of DOF versus Sigma at different NA

From the results of the simulation in figure 8, the smaller the NA (numerical aperture), the larger the lithography window, which is the same as the basic understanding of lithography, i.e. the depth of focus (DOF) = optical wavelength / numerical aperture (equation 1),

$$DOF = \frac{\lambda}{2(NA)^2} \quad (1)$$

So we need to meet the premise requirement of the resolution and minimize the size of NA.

According to the theory of Sigma = Illumination lens NA / Project lens NA, the smaller the NA, the larger the photolithography window. But smaller NA causes different DOF and sigma combinations, so we need to choose a matching NA and sigma.

Mask Bias

Because of the photoresist and lithography machine matching ability is not the same, while the photoresist threshold is distributed in a certain range; usually photoresist has a best exposure energy which requires the mask plus a certain bias to achieve such exposure energy. At the same time, mask bias cannot be unlimited because the mask bias to a certain extent will affect the lithography line width. From the simulation results in figure 9, we can see that greater mask bias gives better photoresist profile, but the CD cannot be guaranteed. The mask bias is 0.1um to give a good balance, so 0.1um mask bias is our first choice.

Defocus -0.5 Dose 10

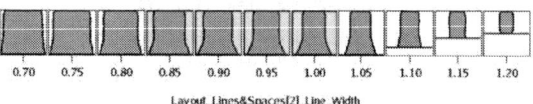

Figure 9: Simulation of mask bias (Mask CD=1.0um)

Alignment Error

From the above data analysis, we know that overlay accuracy is a crucial parameter and is the key to the success of our entire lithography process. From the results shown in figure 10, alignment error value will seriously affect the lithography process window. This means that we must make the line width more stable and reliable in the lithography and then we need tighten the specifications of inline control. A lithography machine with better overlay accuracy also will give more tolerance in the line width control.

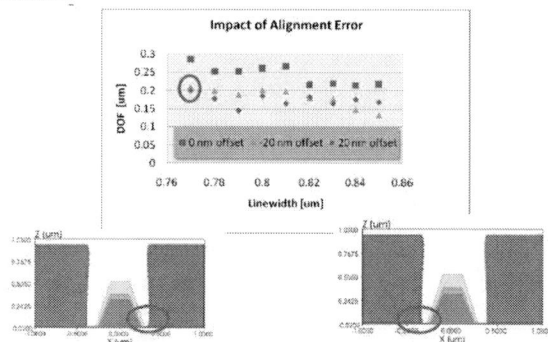

Figure 10: Simulation of alignment error

Selection of Photoresis of PNBL lithography Process

Several of the most important factors in determining photoresist are coating uniformity, sensitivity, resolution, process window size and defect problems. In photo lithography, the two factors that play a key role in image quality are resolution (R) as expressed in equation (2) and depth of focus (DOF).

$$R = \kappa \, \frac{\lambda}{NA} \qquad (2)$$

Where λ is the wavelength and NA is the numerical aperture of the optical system. It is obvious that reducing the wavelength of the exposure light source and increasing the NA of the projection lens can increase the resolution.

For the depth of focus (DOF), the lens can capture more optical details after the numerical aperture increases. So in the lithography to get a better resolution is important to form a key size graphics. But to maintain the appropriate depth of the shadow is a contradictory complex. Although the resolution is highly dependent on the exposure equipment, but the high-performance exposure tools with a suitable high-performance photoresist can really get high-resolution processing capacity.

The sensitivity can be reflected in the contrast curve of the photoresist. The contrast is defined as equation (3):

$$\gamma = \left[\log_{10} \left(\frac{D_{100}}{D_{10}} \right) \right]^{-1} \qquad (3)$$

Where D_{100} is the minimum energy dose required for all photoresists to be removed which is the sensitivity, also known as the exposure threshold. D_{10} is the lowest energy that photoresist begin to perform photochemical reactions. Contrast can be considered as a measure of the ability of the photoresist to distinguish between the bright and dark areas of the reticle and the intensity of the radiation varies smoothly near the edges of the photoresist lines and spaces. The greater the contrast of the photoresist is, the steeper the edge of the line. The typical photoresist contrast ratio is from 2 to 4. For an ideal photoresist, if the exposure dose is above the threshold, the photoresist is fully sensed. Otherwise, it is not sensitive at all. And the exposure threshold of the actual photoresist has a distribution, the narrower the distribution range, the better the performance of the photoresist. In addition to resolution and sensitivity, photoresist also need to have excellent plasma resistance, chemical resistance and corrosion resistance. The general performance of photoresist as above discussed needs to be evaluated when an engineer in choosing photoresist. And below items, Such as the existing product requirements, equipment capacity and photoresist costs also need taken into account.

Combined with the theory and the above studies, the negative KrF photoresist performance is proposed for this PNBL lithography process due to other types of photoresist have a certain limit.

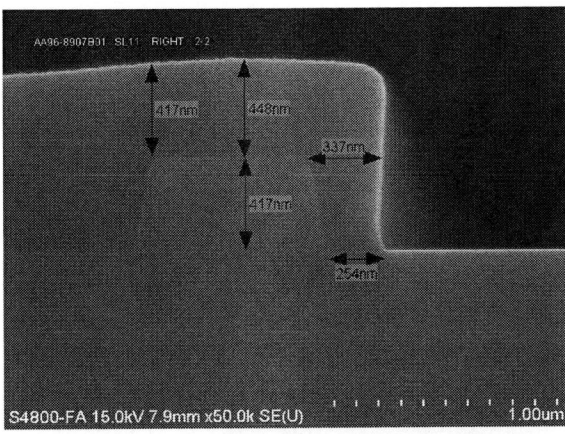

Figure 11: Photoresist thickness on active area

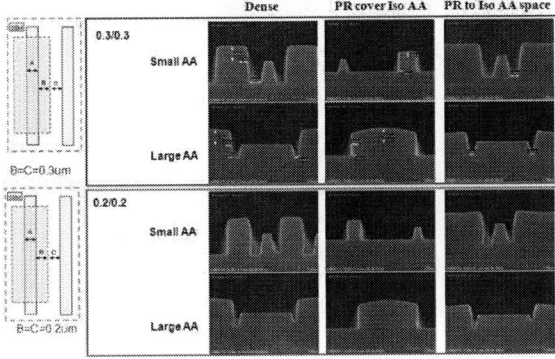

Figure 12: Photoresist cross section on ISO and Dense active area

CONCLUSIONS

After a few rounds of experimental data collection, theoretical and simulation analysis, combined with the experimental results in *figures 11 and 12*, the final solution is obtained. As the lithography process ensures that the photoresist has a certain degree of resolution and thickness with collapsing free, the final lithography conditions include: negative photoresist with 7770A thickness, 100nm mask bias per side, negative photoresist resolution 80 ~ 100nm, STI side wall angle 82 degrees, lithography process window (CD + OVL) control within 50nm and minimum window CD of the device is 300nm.

ACKNOWLEDGEMENTS

The authors would like to thank all members of device group, Lei Wang and SiGe group of HHgrace for their great support on this work.

REFERENCES

[1] H. Li, B. Jagannathan, J. Wang, T.-C. Su, S. Sweeney, J. Pekarik, Y. Shi, D. Greenberg, Z. Jin, R. Groves, L. Wagner, and S. Csutak. *IEEE Symposium on VLSI Technology*, June 2007, pp. 56–57.

[2] Z. E. Fleetwood, B. R. Wier, U. S. Raghunathan, N. E. Lourenco, M. A. Oakley, A. J. Joseph, and J. D. Cressler. *IEEE Bipolar/BiCMOS Circuits and Technology Meeting - BCTM*, Oct 2015, pp. 72–75.

[3] B. R. Wier, U. S. Raghunathan, Z. E. Fleetwood, M. A. Oakley, A. J. Joseph, V. Jain, and J. D. Cressler. *IEEE Bipolar/BiCMOS Circuits and Technology Meeting (BCTM)*, Sept 2016, pp. 21–24.

[4] A. K. Sutton, A. P. G. Prakash, B. Jun, E. Zhao, M. Bellini, J. Pellish, R. M. Diestelhorst, M. A. Carts, A. Phan, R. Ladbury, J. D. Cressler, P. W. Marshall, C. J. Marshall, R. A. Reed, R. D. Schrimpf, and D. M. Fleetwood. *IEEE Trans. Nucl. Sci.*, vol. 53, no. 6, Dec 2006, pp. 3166–3174.

[5] D.H. Liu, W.S. Qian, X.B. Chen. *ECS Transactions*, vol.34(1), Mar. 2011, pp. 173-181.

LITHOGRAPHY PROCESS OPTIMIZATION FOR EMITTER WINDOW IN SIGE-HBT DEVICE

Donghua Liu[1,2], Ziquan Fang[2], Zhaozhao Xu[2,3]*

[1]State Key Laboratory of ASIC and System, School of Microelectronics, Fudan University, Shanghai 200433, China

[2]HuaHong Grace Semiconductor Manufacturing Corporation, Shanghai 201206, China

[3]State Key Laboratory of Functional Materials for Informatics, Shanghai Institute of Microsystem and Information Technology, Chinese Academy of Sciences, Shanghai 200050, China

*Corresponding Author's Email: Ziquan.Fang@hhgrace.com

ABSTRACT

Emitter window (EW) of SiGe-HBT is the region which connects the Base and Emitter and forms the EB junction. The size of EW determines the device operation current. Larger size provides larger current and even 0.02um of EW size shift will cause device characteristic drift. Thus the size and shape of EW are critical in process and EW is a great challenge for lithography to form an ideal small rectangle. Therefore special lithography process solutions are needed to ensure the fidelity of EW. In this paper, the minimum EW size requirement is 0.2μm ×0.5μm and low cost KrF tool is used for the process. The reflectivity properties of SiGe, DARC, BARC and photoresist films impacts on lithography process are studied. The optical coefficient impacts to pattern resolution and production capability are investigated. An optimized OPC solution is also demonstrated. Finally, a high fidelity small size EW lithography solution is realized with features of low cost, high stability and easy manufacturing.

INTRODUCTION

Germanium (Ge) is the earliest semiconductor material being used and developed in semiconductor industry as it is easy to refine and fabricate into single crystal. But with the development of crystal pulling and manufacturing technology, Silicon (Si) receives more attention. Especially after planar silicon processing techniques have developed, Si almost completely replaced Ge in microelectronics area.

Ge has distinctive features, including smaller bandgap, higher carrier mobility, etc.. Now SiGe compound is used as a new semiconductor material which has both advantages of Ge and Si such as higher carrier mobility and mature processing techniques. Nowadays, SiGe is widely used in wire & wireless communication and satellite communication field, while SiGe-HBT is the most studied device[1]-[5].

The SiGe film thickness, Ge concentration, crystal quality and defect are critical control parameters during SiGe film grow and which decide the performance of SiGe device[6]. Besides the SiGe film growth technology, the corresponding process technology like lithography, etch, etc. also need to be developed for mass fabrication purpose.

Figure 1 shows the cross section of SiGe-HBT device studied in this paper. Emitter window (EW) of SiGe-HBT is the region which connects the Base and Emitter and form the EB junction. The EW is opened on SiGe film and the size is small.

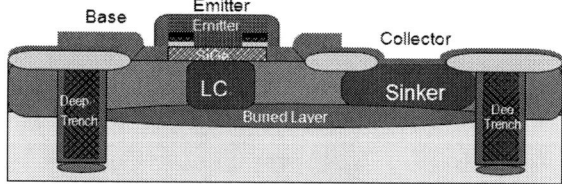

Figure 1: Common SiGe-HBT device cross section

In this paper, the minimum EW size is 0.2μm × 0.5μm to fulfill device performance requirements. As this size is close to the optical proximity effect, it is difficult to form such an ideal rectangle. A series of studies have been done on film stacks, film properties, OPC, etc.. A whole set of lithography process solution is confirmed successful for high fidelity small size EW open on SiGe film. In order to lower the process development and mass production cost, KrF tool is used instead of ArF tool which has better properties but higher cost.

FILM OPTICAL CHARACTERISTICS

Film reflectivity is a critical parameter which impacts the lithography pattern fidelity. Material with high reflectivity leads to poor lithography process stability. The reflective index of SiGe film is relatively high, thus in order to get a stable pattern, dielectric films are added above SiGe film to overcome the effect of reflection.

In this section, reflection properties of SiGe film, anti-dielectric film and photoresist are studied to get the minimum reflective index. All the studies are using pure KrF process for cost concern. Figure 2 shows the film structure studied in this section, the film from bottom to top is silicon→SiGe→dielectric film→photoresist. The

978-1-5386-5309-8/18 $31.00 © 2018 IEEE

dielectric film consists of one or more types of inorganic substance which can help reduce reflection. Moreover, organic anti-reflect film is also included in photoresist layer.

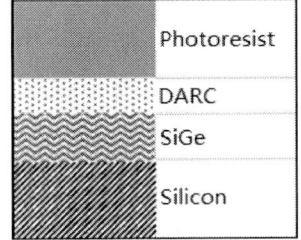

Figure 2: Film structure for lithography process on SiGe film

Figure 3a and 3b shows the EW open process discussed in this paper. Two dielectric anti-reflect films are added on top of SiGe film.

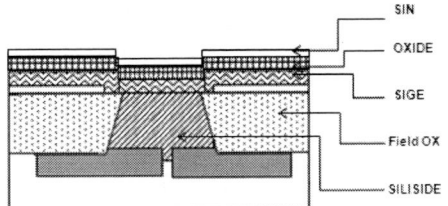

Figure 3a: SiGe film grow

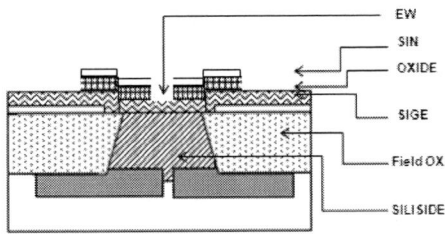

Figure 3b: EW window open

Lithography process simulation software is used in this paper for reflectivity study and all the simulations below are using KrF scanner. With film characteristic, film thickness and other parameters setting, the reflectivity can be simulated and the data is reliable through a large number of experimental verification.

SiGe Film Reflective Index

Figure 4 shows the simulation result of SiGe film reflectivity. When SiGe film is lower than $0.03\mu m$, the reflectivity variation is large along with SiGe film thickness change. However, the reflectivity value is almost fixed when SiGe film thickness is thicker than a certain value.

Normally, in SiGe device the SiGe film thickness

varies from $0.02\sim0.15\mu m$. In this range, the reflectivity is high (>49%) and this value cannot meet the requirement for small size lithography process. Adding a dielectric anti-reflect film on SiGe film is quite necessary.

The reflectivity has no correlation with Ge concentration in Si, SiGe with different Ge% has similar reflective index.

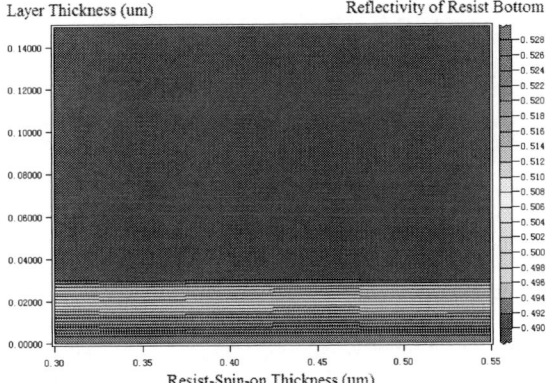

Figure 4: Reflective index of SiGe film

Dielectric Anti-reflect Coating (DARC) Reflective Index

As discussed above, a dielectric anti-reflect film will be deposit on top of SiGe film before EW window open due to process requirement. Normally the DARC film is composed of some inorganic substances, like SiO2, SiN, SiON, etc.. In order to minimize the impact to subsequent etch process, the DARC film is required to be as thin as possible and normally less than 50nm.

Two types of DARC film are discussed as below to seek for better anti-reflect function with thinner thickness. Figure 5 shows the simulation result of DARC film composed of SiN and SiON. The reflectivity is improved but the thickness need to be around 80nm, cannot meet the requirement of thickness less than 50nm. Figure 6 shows the simulation result of DARC film composed of SiO2 and SiN, the reflectivity are further improved and the thickness is only around 45nm. This film is selected to form the DARC film.

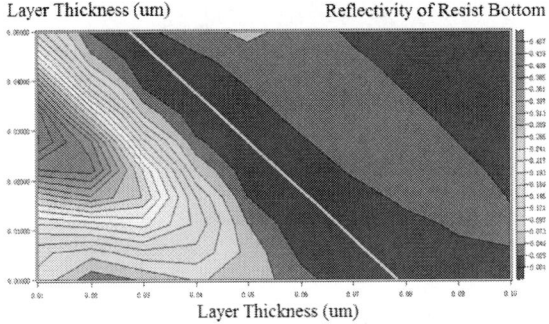

Figure 5: Reflective index of DARC composed of SiN and SiON

978-1-5386-5309-8/18 $31.00 © 2018 IEEE   179

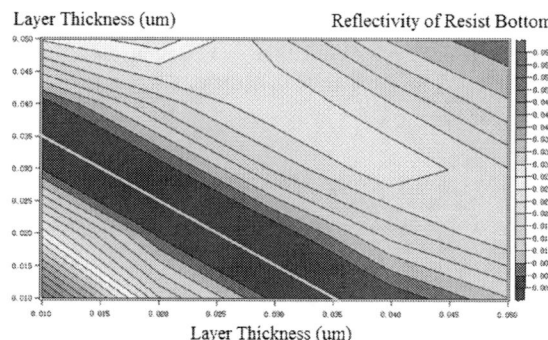

Figure 6: Reflective index of DARC composed of SiO2 and SiN

Bottom Anti-reflect Coating (BARC) Reflective Index

BARC is an organic film locating between DARC and photoresist. Similar to photoresist, BARC layer also uses spin coating method to grow. Both the type and thickness of BARC film will impact the reflectivity. Moreover, the photoresist type also need to take into consideration as the type of BARC film needs to match to certain type of photoresist for process.

Figure 7 shows the reflectivity of BARC type I. The reflective index reduces to about 21% when BARC thickness thicker than 0.06μm. Figure8 shows the reflectivity of BARC type II. The reflectivity greatly reduced to less than 5% when BARC thickness thicker than 0.07μm.

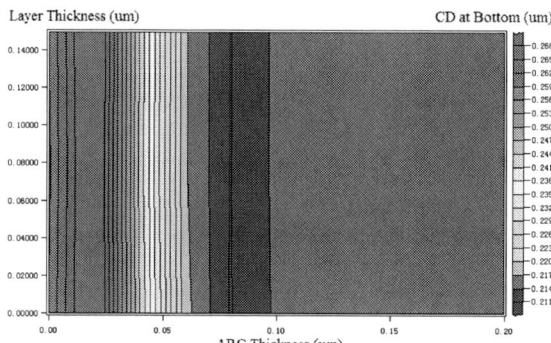

Figure 7: Reflective index with BARC type I

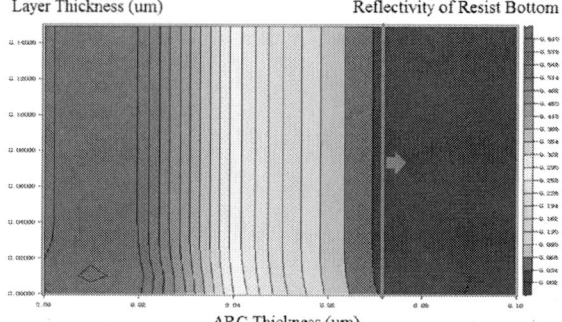

Figure 8: Reflective index with BARC type II

Photoresist Thickness Reflective Index

Photoresist is the film that displays the pattern after whole lithography process. The process tool, size requirement and etch condition need taking into consideration in selection of resist type and thickness.

Figure 8 and 9 show the simulation of photoresist reflectivity with different thickness. There is a region for resist thickness to keep the reflective index at a very low level when SiGe film thickness reaches a certain value.

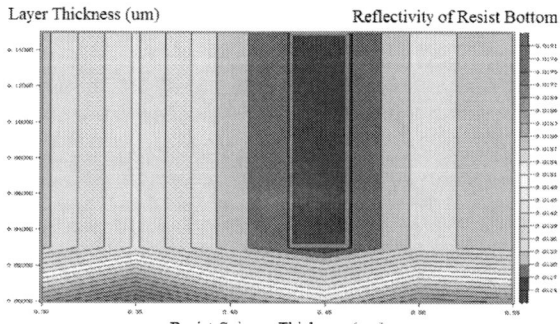

Figure 9: Photoresist reflective index

As show in Figure 10, photoresist thickness and pattern size has swing curve effect in lithography process. The pattern size will reach a critical value which is relatively stable with small variation of resist thickness when the resist thickness at specific values.

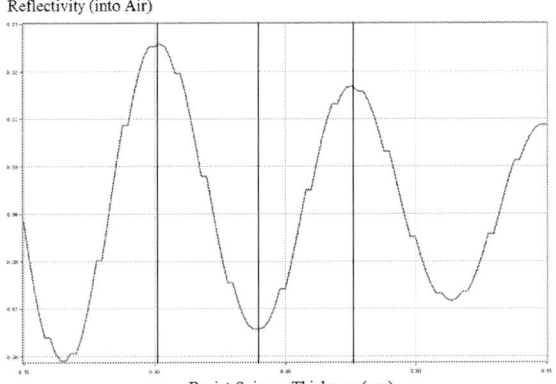

Figure 10: Photoresist swing curve

With the simulation results as discussed above, the film layers are confirmed and film thickness are optimized and a minimum reflective index is obtained for lithography process.

Figure 11 is the final film structure for lithography process on SiGe film in this paper. The film stack from bottom to top is silicon→SiGe→SiO2→SiN→BARC→photoresist.

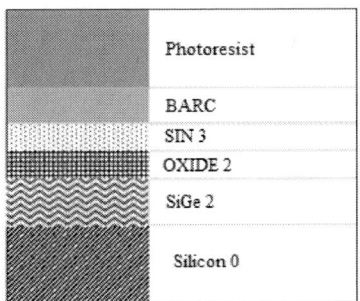

Figure 11: Final film structure for lithography process on SiGe film

LITHOGRAPHY OPTICAL COEFFICIENT
Numerical Aperture (NA)

In optical technology, larger NA value means higher resolution. But the process design capability will be lost. Simulation is carried out based on the film condition and process condition discussed in previous sections. Figure 12 shows the simulation result of process capability at different NA value. Proper NA value is selected to balance the trade-off of resolution and process capability.

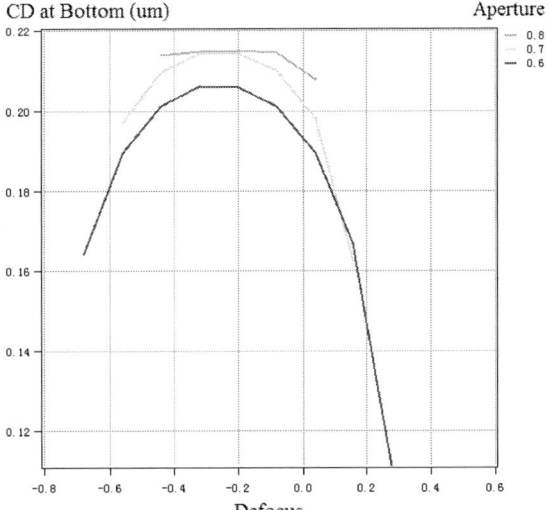

Figure 12: Process design capability at different NA value

Sigma Value

Normally sigma value will impact the pattern critical dimension (CD) at isolate and dense area. Figure 13 shows the CD difference between isolate and dense area at different sigma setting.

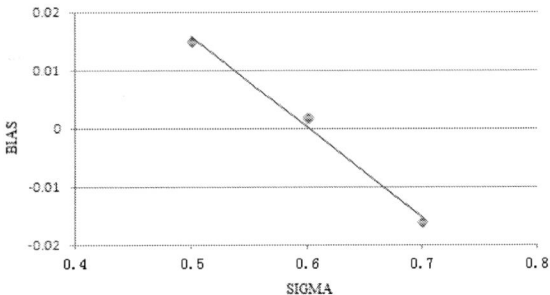

Figure 13: Sigma vs Iso/Dense CD bias correlation

OPC SOLUTIONS

Optical proximity correction (OPC) is used in lithography process for small size patterns or patterns with special requirement. Process tool, film stacks, photoresist and optical parameters setting should be taken into consideration for OPC solution.

OPC approaches are optimized step by step. KrF tool, SiGe film, photoresist, optical coefficient are set based on above simulation results. Figure 14a is the original pattern image without any OPC amendment, which has severe corner rounding effect. OPC model has been gradually optimized after OPC being fitted and tested. Figure14d shows the pattern image with excellent pattern fidelity which has been corrected by final optimized OPC approach.

978-1-5386-5309-8/18 $31.00 © 2018 IEEE

solutions

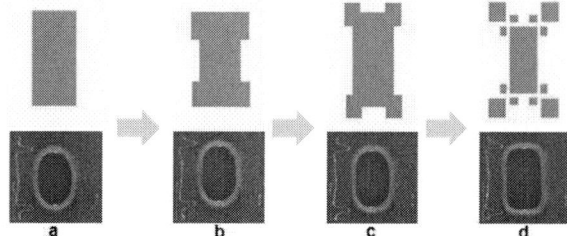

Figure 14: OPC approaches and pattern image

On top of the OPC solution, KrF tool process capability is another critical factor to determine the minimum pattern size. Figure15 shows four pattern sizes, 0.2μm×0.5μm, 0.2μm×1.0μm, 0.2μm×9.9μm and 0.2μm space line, the process capability for all the patterns are close to 0.4μm, thus mass production on KrF tool is proved to be capable.

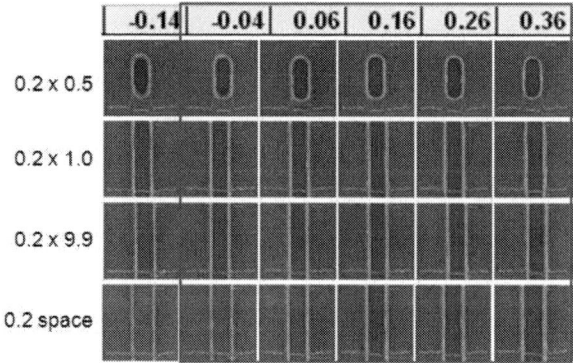

Figure 15: KrF lithography process capability

Figure 16 shows the pattern fidelity for small size patterns with different length and width. Four OPC approaches which show in Figure 15 are used for comparison. It is clear that the last OPC solution gives the best pattern fidelity for all sizes.

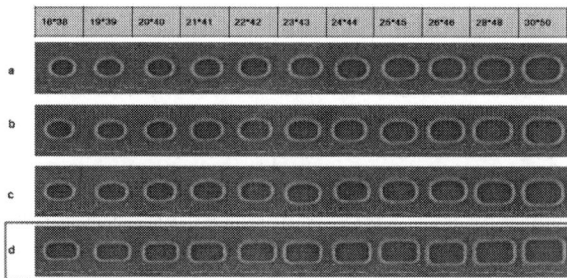

Figure 16: Pattern fidelity comparison among OPC

CONCLUSION

The lithography solutions for small EW window opening on SiGe film are studied based on KrF scanner tool. DARC, BARC, photoresist film properties and lithography optical coefficient are optimized. Optimized OPC model is also obtained. Finally the excellent pattern fidelity has been realized and the lithography process for EW is cost effective, high stability and capable for mass production.

ACKNOWLEDGEMENTS

The authors would like to thank all members of Device Design group, Lei Wang and Advanced Module group of HHgrace for their great support on this work.

REFERENCES

[1] G. Avenier, et al, "0.13um SiGe BiCMOS Technology Fully Dedicated to mm-Wave Applications," *IEEE J. Solid-State Circuits*, Vol. 44, No. 9, Sept. 2009, pp. 2312-2321

[2] P. Chevalier, et al, "A Conventional Double-Polysilicon FSA-SEG Si/SiGe:C HBT Reaching 400 GHz fMAX," *Froc. BCTM*, 2009, pp. 1-4

[3] J.I. Pekarik, et al, "A 90nm SiGe BiCMOS Technology for mm-wave and high-performance analog applications," *Proc. BCTM*, 2014, pp. 92-95

[4] B. Geynet, P. Chevalier, B. Vandelle, et al, "Sige hbts featuring ft >400ghz at room temperature," *Proc. BCTM*, Oct 2008, pp. 121–124

[5] Y. Yang, S. Cacina, and G. Rebeiz, "A SiGe BiCMOS W-Band LNA with 5.1 dB NF at 90 GHz," *IEEE Compound Semiconductor Integrated Circuit Symposium (CSICS)*, Oct 2013, pp. 1-4

[6] Donghua Liu, Wensheng Qian, Xiongbin Chen, et al. "0.18 micron BiCMOS process with novel structure SiGeC HBT", *ECS Transactions*, 2011, pp. 173-181

STUDY ON PLANARIZATION PERFORMANCE OF SPIN ON HARDMASK

Huichan Yun*, Jinhyung Kim, Yoona Kim, Seulgi Jeong, Sanghak Lim, Jeong Yun Yu

Development group, Electron Materials Division, Samsung SDI

130, Samsung-ro, Maetan-Dong, Yeongtong-Gu, Suwon-Si, Gyeonggi-Do, Korea 443-370

*Corresponding Author's Email: hcyoon.yoon@samsung.com

ABSTRACT

For multilayer process, the importance of spin on carbon (SOC) material that replaces amorphous carbon layer (ACL) is ever increasing. SOC is an organic polymer with high carbon content formulated in organic solvents for spin-coating application that is cured through baking. In comparison to ACL which is applied by CVD process, SOC material can offer several benefits: lower cost of ownership (CoO) and shorter process time as well as better gap-fill and planarization performances. Thus SOC material of high etch resistance, good gap-fill properties, and global planarization performances over various pattern topographies is desired to achieve fine patterning with high aspect ratio (A/R). In particular, good level of global planarization of spin coated layer over the underlying pattern topographies is important for self-aligned double patterning (SADP) process as it dictates the photolithographic margin. Herein, we report a copolymer SOC resin formulation that exhibits favorable film shrinkage profile and ideal etch resistance properties. By combining the favorable characteristics of each resin – one resin with good shrinkage property and the other with excellent etch resistance into the copolymer, it was possible to achieve desirable level of etch resistance and planarization performances across various pattern topographies of underlying substrate.

INTRODUCTION

As the feature size of electronic device continues to shrink, carbon based spin-on carbon (SOC) can offer benefits of good gap-fill and planarization performances as well as lower CoO and improved process throughput compared to traditional amorphous carbon layer (ACL).[1] Thus SOC material of high etch resistance, gap-fill, and planarization performances is becoming more important to achieve fine patterning with high A/R for advanced multiple patterning processes.[2] In particular, global planarization level of SOC over underlying pattern topographies has a close relationship with the photolithographic margin of multiple patterning processes. [3-5] While SOC materials of high carbon contents are desirable for etch resistance hardmask[1-2], there is a trade-off between etch resistance and gap fill performance. There are some reports of influencing the planarization of the SOC material by altering spin-coating process such as spin speed and solution viscosity.[6] However, such optimizations are not viable from a practical point of view in semiconductor manufacturing.

In this paper, we present SOC hardmask materials with good etch resistance as well as planarization performance across various pattern topographies of underlying substrate based on film shrinkage profile across different areas of patterned substrates (pad and open/cell dense areas).

EXPERIMENTAL

1. Materials

Based on the properties such as etch resistance, shrinkage, and flexibility of resins, four monomers were chosen to be synthesized as SOC resins. Those monomers could be categorized into core monomers (monomers A and C) and linker monomers (monomers B and D). We conducted polymerization with the monomers to obtain the core-linker structures shown as below (Figure 1). Molecular weight of the polymers was controlled to be the same as each other.

Figure 1: Synthesized core-linker structures

2. Evaluation methods

Performance evaluation was conducted with polymer thin films on Si bare wafers or Si patterned wafers. The samples were prepared by spin coating PGMEA solutions of the polymers at 1,500rpm (Mikasa MS-A200) and baking at 400 °C for 2 minutes (As One HTH-500N).

For evaluation of etch resistance, the polymer films with a thickness of ca. 5000 Å were prepared on 8 inch Si-wafers. Etching was performed with Exelan HPT from Lam Research Inc. The etch conditions were similar to those used for etching SiN : 100 mTorr / 600 W / CF_4 (42 sccm) / Ar (600 sccm) / O_2 (15 sccm) for 50 sec. The etch rate was calculated from the film thickness change arisen from etching measured by a reflectometer (KMAC ST4000).

Etch resistance = (etch rate of A-B copolymer/etch rate of another copolymer) × 100

Etch rate (Å/s) = (reduced amount of the thickness of the films after etching process/etching time) × 100

To measure reflow distances, shrinkage, and thickness bias, polymer thin films with thickness of 2000Å were used. Reflow distances of the polymers were evaluated by subtracting flow distances of as-coated films from flow distances of baked films on trench patterns. The pattern was 100 um wide and 100 nm deep. The flow distances were measured in the manner described in Figure 2.

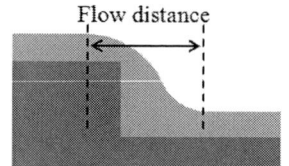

Figure 2: Definition of flow distance

Thermal shrinkage was evaluated by calculating the rate of reduced thicknesses during baking processes to the thickness of as-coated films with a reflectometer (KMAC ST4000). The formula could be described as below.

Thermal shrinkage (%) = (reduced amount of the thickness of the films/thickness of the as-coated films) × 100

The thickness bias was evaluated by using a patterned wafer with 180nm line-and-space patterns. Figure 3 shows the SEM (scanning electron microscope, Hitachi SU 8230) images of the patterns and the scheme depicting the thickness measurement points to evaluate the thickness bias.

Thickness bias (Å) = film thickness at block edge - film thickness at center

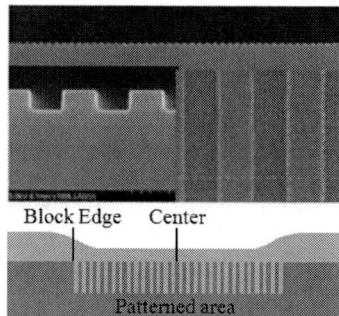

Figure 3: Pattern geometry of SDI's wafer and film thickness measuring points to evaluate thickness bias

Both flow distance and thickness bias were measured with an AFM (Atomic Force Microscope, Park systems XE-300).

RESULT & DISCUSSION

The evaluated properties of the samples are summarized in Table I.

Table I. Summary of planarization performances

Copolymer	Etch resistance	Reflow distance	Shrinkage	Thickness bias
A-B	1.00	3 um	23%	400Å
A-D	1.01	31 um	23%	180Å
C-D	1.04	28 um	15%	140Å

First of all, etch resistance against CF_4 is one of the important characteristics of hardmask materials and the etch resistance of the polymers were confirmed to be similar to each other from the etch rate data. We found that the reflow distances of A-D and C-D copolymers were much larger than that of A-B copolymer. That means the reflow distances of the polymers which consist of the linker monomer D are much larger than the reflow distance of the polymer made of the linker monomer B. Because the D linker has more flexibility in the molecular structure than the B linker, the reflow distance is considered to be influenced tremendously by flexibility of molecular structures of linker monomers. For the film shrinkage, C-D polymer showed smaller film shrinkage than A-B and A-D polymers and it is thought to be due to a better thermal stability of the monomer C than that of the monomer A. From the observation that A-B and A-D polymers showed the same film shrinkage, the structural flexibility of linker monomers did not affect thermal stability of the polymers. The thickness bias was measured from the profile of each sample as shown below (Figure 4). Compared to A-B copolymer, A-D and C-D copolymer showed improved thickness bias as the reflow distance increased. Between A-D and C-D copolymer, C-D copolymer showed smaller thickness bias than A-D copolymer although the reflow distance of C-D copolymer was a bit shorter than that of A-D copolymer. The reason would be smaller thickness loss of C-D copolymer than that of A-D copolymer.

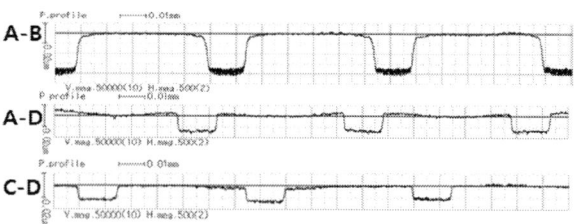

Figure 4: Film profile of copolymer on pattern topology

From the result above, we could conclude that a long reflow distance is the essential property to achieve a good thickness bias and thermal stability of copolymers should be good to maintain the planarization performance in addition.

CONCLUSION

To sum up, we designed three polymers out of four monomers and evaluated their planarization performances. From the result, we found out that reflow distance and shrinkage should be controlled to achieve the better planarization performance. Among the three polymers, C-D copolymer showed the best planarization performances, which is highly recommended to be applied to fabrication of organic planarization layers.

REFERENCES

[1] Hwan-Sung Cheon et al., *Proc. SPIE*, vol. 7140, 2008, 71402R.

[2] Takanori Kudoiro et al., *Proc. SPIE*, vol. 9051, 2014, 90511X-1.

[3] Ruei Hung Hsu et al., *Proc. SPIE*, vol. 5753, 2005, pp. 1158-1169.

[4] White, L. K., *J. Electrochem. Soc.*, vol. 130, 1983, pp. 1543-1548.

[5] Punit Chiniwalla et al., *IEEE*, vol. 24, 2001, pp. 41-53.

[6] Kazuhiko Komura, Yoshi Hishiro et al., *Proc. SPIE*, vol. 9051, 2014, 905115-6.

A Novel Lithographic Material Constructed by Polymerizable Liquid Crystal Molecule

Hui Cao, Hai Deng *

Department of Macromolecular Science, State Key Laboratory of Molecular Engineering of Polymers，Fudan University
Shanghai, China
Phone number: 18280304058, E-mail:16210440001@fudan.edu.cn
* Corresponding author: haideng@fudan.edu.cn

Biography

Hui Cao earned her B.S. degree in polymer science and engineering at the Sichuan University in 2016. She is currently pursuing a master under the guidance of Dr. Hai Deng in Fudan University. Her research mainly focuses on synthesis and directed self-assembly (DSA) of polymerizable liquid crystal and controlling the orientation of liquid crystal in thin films.

Abstract

This study is aimed at bringing up a novel approach to form higher-resolution patterns which can be used in the lithography field. The idea is operated through taking advantage of the arranging ability of liquid crystal molecules. A polymerizable liquid crystal molecule with specific structure is synthesized at first, then POM and SAXS are done to confirm the specific morphology of the molecule, and application in lithography will be accomplished in the near future. With the continuous stream of integrated circuits with significant size reduction, traditional photolithographic methodologies are thwarted to reduced feature sizes. Meanwhile, other existing methods or materials still have their own shortcomings. Therefore, a new method needs to be conceived to form long-order patterns with small feature sizes.

Keywords—polymerizable liquid crystal molecules; lithography; self-assembly; morphology

Introduction

The sustained shrinking of device components is an embodiment of human progress that has led to an explosion of micro- and nanotechnologies. In order to support the rapid development of microelectronics industry, manufacturing process has been developed to the size of 10nm node [1]. In an effort to overcome the size limitations facing top-down microelectronics fabrication , several methods including EUV (extreme ultraviolet)、DSA (Directed self-assembly) have been proposed in these years.

Extreme ultraviolet (EUV) lithography was first put forward in 1980's by Hawryluk and Seppala [2]. Gwyn suggested that these extensions are relied on incremental decreases in illumination wavelength and increases in optical numerical aperture (NA) for the system [3]. Extreme ultraviolet lithography (EUVL) based on the industrial optical experience uses 10–14 nm extreme ultraviolet light and is expected to support IC fabrication at 10 nm. Though EUVL has several advantages such as good depth of focus, this technique is still plagued by large capital investments, source power limitations, resist outgassing complications and its considerable cost [4].

Block copolymer (BCP)-based nanopatterning has paved a strong foundation in nanotechnology including nanoporous membranes, patterned magnetic media, and nanolithography. These materials have attracted much attention because of there excellent self-assembly ability. A previous research of Leibler showed that there were three important physical parameters dictating the self-organizing behavior of di-block copolymers: the Flory−Huggins interaction parameter, χ; the overall degree of polymerization, N; and the polymer composition, expressed as the volume fraction of block A (fA) or block B (fB) [5]. With the view of the Leibler's theory, many researchers are inclined to focus di-block copolymers that have high χ and low N. Unfortunately, an immiscibility-driven approach will not suffice when sub-5 nm periodicities are desired, preventing extendibility to smaller dimensions. In addition, although the assembly of block copolymers can be directed within a photolithographic prepattern, the formation of defect-free periodic structures remains a challenge. [6-8].

Lyotropic liquid crystal (LLC) mesogens are amphiphilic molecules containing a hydrophobic organic tail section and a hydrophilic headgroup [9-10]. Depending on the overall shape and the interfacial curvature energy of the system, monodisperse aqueous domains ranging in structure from lamellae to cylindrical channels with dimensions in the 1−10 nm size range can be formed [11]. As early as 1975, Ekwallp has put forward that the smectic phase formed by lyotropic liquid crystals is a lamellar structure with periodically repeated surfactant layers [12].liquid crystalline ordering suggests the possibility of long-ranged pattern formation with very small resolution and few defects, which can't get from di-block polymers [13].These special materials have the advantage that they can be easily aligned into defect free and sub-5 nm patterns [14]. Exciting work has been performed on the confinement of liquid crystals in structures such as microchannels,[15] nanogrooves,[16] or nanopores,[17] but the directed self-assembly of thin films with sub-5 nm periodicities has not been reported. Therefore, this study synthesized some kinds of polymerizable liquid crystal molecule to form the target lamellar template with photolithography value, which could also provide a new idea for the design and construction of lithography materials.

Experiment and discussion

In this study, we synthesized a kind of polymerizable LLC

molecule, the synthetic route and molecule structure are shown in Fig.1. The structure has been confirmed by NMR, ^{1}H NMR (400 MHz, d5-methanol): δ 7.27 (s, 2H), 5.82 (dd, 3H),5.03 (dd, 3H), 4.96 (dd, 3H), 4.08 (t, 6H), 3.95 (t, 6H), 1.52 (m, 56 H). Then we mixed the LLC molecule with appropriate amount of water to form ordered structure at the same time explore the impact of water. The effect of different water contents has been studied. When the LLC monomer content is 80%, the target lamellar phase can be obtained, as shown in Fig.2. When the LLC monomer content is lower, a mixed structure occurs, in which we guess it exists lamellar and hexagonal structure simultaneously. With the reduction of water content, lamellar is disappeared while hexagonal is appeared, as shown in Fig.3. Since the monomer has three double bonds, after crosslinking the original ordered structure can be maintained. The crosslinking process needs photoinitiator and a UV light source with a wavelength of 365nm. Then after some specific processing, the polymer free-standing film is able to have the potential value as a mask that can be applied in the lithography field. In the next step, we will try to introduce this LLC molecule into a prepared template made by EBM to explore its further application, which may be not only limited in the lithography field.

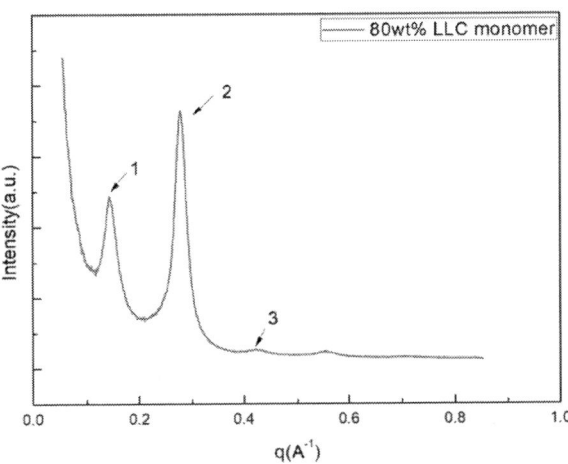

Fig. 2 (a) SAXS profile of 80% LLC monomer

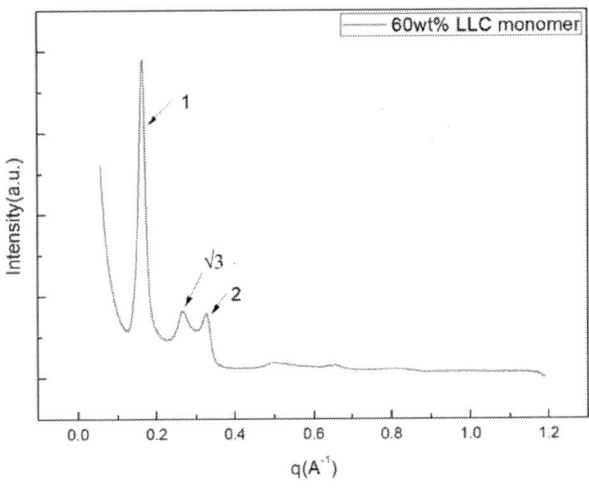

Fig. 2 (b) POM image of 80% LLC monomer

Fig. 2 When the LLC monomer content is 80%, the lamellar structure is obtained, as shown by the SAXS(a) and POM results(b).

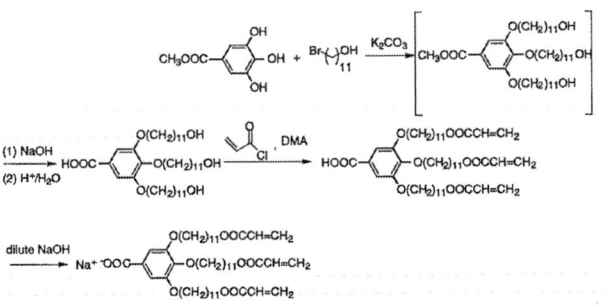

Fig. 1 (a) LLC Monomer synthetic route

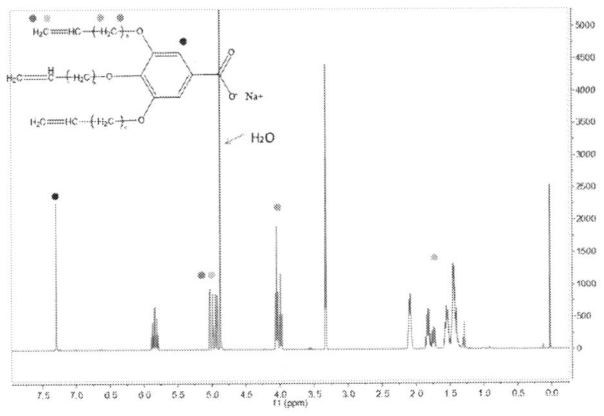

Fig. 1 (b) NMR result of the final product.

Fig. 1 LLC Monomer synthetic route (a) and NMR result of the final product (b).

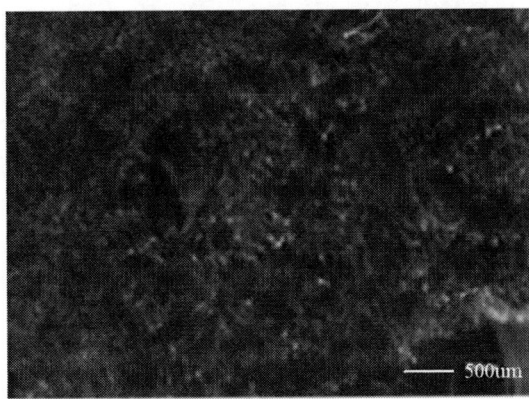

Fig. 3 When the LLC monomer content is 60%, the hexagonal structure is obtained, as shown by the SAXS result.

Conclusions

Through the multi-step synthesis, a liquid crystal molecule with polymerization ability is successfully obtained. After adjusting the concentration of deionized water and crosslinking, the polymer mask with lamellar structure which has value being used in lithography field is successfully obtained. The follow-up work is to perfect the lamellar phase structure, and achieving the ultimate goal that transfers the polymer mask pattern to the wafer substrate.

References

[1] Markoff, J. IBM Discloses Working Version of a Much Higher Capacity Chip. The New York Times; The New York Times Company: New York, July 9 2015; p B2.J. Clerk Maxwell, A Treatise on Electricity and Magnetism, 3rd ed., vol. 2. Oxford: Clarendon, 1892, pp.68–73.

[2] A. M. Hawryluk; L. G. Seppala. J. Vac. Sci. Technol. B, 1988, 6, 2162.

[3] Gwyn, C. W.; Stulen, R.; Sweeney, D.; Attwood, D. J. Vac. Sci. Technol. B, 1998, 16, 3142-3149.

[4] Wagner, C.; Harned, N. Nat. Photonics, 2010, 4, 24–26.

[5] Leibler, L. Macromolecules, 1980, 13, 1602.

[6] M. Luo, T. H. Epps, Macromolecules 2013, 46, 7567.

[7] C. Sinturel, F. S. Bates, M. A. Hillmyer, ACS Macro Lett. 2015, 4, 1044.

[8] H. Pathangi, B. T. Chan, H. Bayana, N. Vandenbroeck, D. Van den, Heuvel, L. Van Look, P. Rincon-Delgadillo, Y. Cao, J. Kim, G. Y. Lin, D. Parnell, K. Nafus, R. Harukawa, I. Chikashi, M. Polli, L. D'Urzo, R. Gronheid, P. Nealey, J. Micro/Nanolithogr. MEMS 2015, 14,031204.

[9] G. J. T. Tiddy, Phys. Rep. 1980, 57, 1.

[10] J. N. Israelachvili, Intermolecular and Surface Forces with Applications to Colloidal and Biological Systems, Academic Press, London, 1985, pp. 249–257.

[11] S. M. Gruner, J. Chem. Phys. 1989, 93, 7562.

[12] EKWALLP., in Advances in Liquid Crystal Physics, edited by G. M. BROWN (Academic, New York, N.Y.) 1975.

[13] Luckhurst, G. R.; Zannoni, C. Nature, 1977, 267, 412–414.

[14] K. Nickmans, J. N. Murphy, B. de Waal, P. Leclère, J. Doise, R. Gronheid, D. J. Broer, A. P. H. J. Schenning, Adv. Mater. 2016, 28, 10068.

[15] J. Cattle, P. Bao, J. P. Bramble, R. J. Bushby, S. D. Evans, J. E. Lydon, D. J. Tate, Adv. Funct. Mater. 2013, 23, 5997.

[16] P. O. Mouthuy, S. Melinte, Y. H. Geerts, A. M. Jonas, Nano Lett. 2007, 7, 2627.

[17] a) C. V. Cerclier, M. Ndao, R. Busselez, R. Lefort, E. Grelet, P. Huber, A. V. Kityk, L. Noirez, A. Schonhals, D. Morineau, J. Phys. Chem. C 2012, 116, 18990; b) R. B. Zhang, X. B. Zeng, B. Kim, R. J. Bushby, K. Shin, P. J. Baker, V. Percec, P. Leowanawat, G. Ungar, ACS Nano 2015, 9, 1759.

A STUDY OF PHOTORESIST RESIDUE DEFECT INDUCED BY SUBSTRATE SURFACE CONDITION

Zhou Fang[1]*, Chang Liu[1]

[1] Technology R&D, Semiconductor Manufacturing International Corp.
Pudong New Area, Shanghai 201203, P. R. China
*Corresponding Author's Email: Zhou_Fang@smics.com

ABSTRACT

In order to improve the semiconductor device performance, a special lithography process with high mask transmission rate is needed. But we recently found a kind of serious photoresist residue defect on substrate surface. This defect was like a circular-pattern about one hundred micron in diameter with thin photoresist residue, and some massive photoresist residue was also observed at the interface between Active Area (AA) and Shallow Trench Isolation area (STI). After performing some experiments by improving lithography recipe parameters, such photoresist residue defect count will be decreased but cannot be totally removed. So this residue defect generation mechanism was analyzed and the substrate surface condition was found contribute to the residue defect. Therefore, different surface treatment methods were tried, and finally an optimized process was applied to remove the photoresist residue defect totally.

Keywords—Photoresist Residue Defect; Surface Treatment

INTRODUCTION

As semiconductor industry being driven into more advanced node, the more complex design is usually used to meet device demand, which will bring the challenges of manufacture technology. As for lithography process, a larger process window and fewer defects are good for process integration. Defect becomes one key factor to wafer yield with continues development of semiconductor industry. And even a very small defect will cause the failure of device.

In our work, a special lithography process with high mask transmission rate (~90%) is used to adjust implantation (IMP) condition and then it will improve the semiconductor device performance. But a kind of serious photoresist residue defect was found after lithography process. The characteristic of this defect was clarified, and the generation mechanism of such defect was discussed and one effective method was adopted to avoid the generation of this residue defect successfully.

EXPERIMENT AND RESULTS

The typical semiconductor process consists of below steps, including the formation of AA and STI area, Well IMP, Gate formation, Source and Drain formation and Metal connection [1]. After the formation of AA and STI,

the Well IMP is used to define Well area and then adjusts device parameter. So IMP parameters, such as the area and depth, will directly determine the device performance. And usually the IMP area is defined by lithography process. Here, a special mask with high transmission rate (~90%) is used, but we recently found a kind of serious residue defect after lithography process.

As we can see in Figure 1, some residue defect can be found within wafer level, and the defect is random distributed. It's found that the defect is like a circular-pattern about one hundred micron in diameter with thin residue layer. And some massive residue is also observed within this area, most of it is located at the interface between AA and STI area, as shown in Figure 2. Furthermore, the chemical composition of this residue defect was measured by EDS micro-analysis technique, and Si, O, and C element are found in the defect area. So this residue defect must come from the residue of photoresist, which may be not totally removed after lithography develop process at dark area.

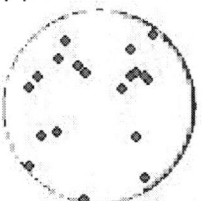

Figure 1: The photoresist residue defect map

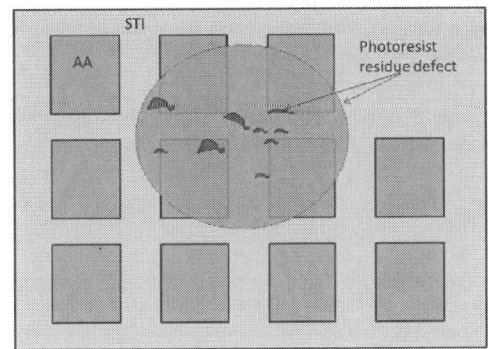

Figure 2: Schematic image of a typical residue defect

As the defect was defined as photoresist residue,

978-1-5386-5309-8/18 $31.00 © 2018 IEEE

the lithography process, especially develop process, was firstly thought to need to be optimized. So a lot of lithography process optimize methods were proposed to remove this residue defect, the method and residue defect ratio summary was shown in table 1. The post-exposure bake (PEB) temperature increase was thought to enhance the diffusing of photo-acid, and then the photoresist will be removed easily after light chemical reaction. But the result in table 1 shows that defect ratio of PEB split is higher than that of the baseline wafer. Hexamethyldisilazane (HMDS) is commonly used to increase the adherence between wafer and photoresist, which is the first step of lithography process. So HMDS split was also tried to decrease such adherence. But serious side effect was happened, isolated long line was easily collapsed under this condition. The PEB and HMDS split both didn't achieve the purpose of residue defect removal. So the develop optimization method was carried out. Usually the lithography develop process consists of three step, wafer surface pre-wet, develop and then water rinse, all these three steps are high-speed spin process. Develop chemical is used to dissolve photoresist which is illuminated by laser, then water rinse is used to clear residue develop chemical and photoresist. The spin speed, process time and the develop mode all carried out optimized splits, and the rinse process also did many splits. But we can see from Table.1, the defect ratio of some split will decrease due to the enhancement of develop ability, but the residue defect can't be totally removed with develop optimized method used. It seems that the rinse split got a better defect result. After so many lithography methods used, this residue defect still existed.

TABLE I
THE SUMMARY TABLE OF DEFECT RESULT

Split condition	Residue defect ratio
Baseline	1
PEB Split	1.6/1.3
HMDS split	NA
Develop split	0.8/0.7/1.1/1.5
Rinse split	1.2/0.4/0.8/0.7

From above results, we realized that this residue defect can't be totally removed if only photography process changed, and the generation mechanism of this residue defect need carefully analyzed. The two characteristics of this residue defect, ~100 micron in diameter with thin photoresist residue layer and massive residue was usually found at the interface between AA and STI area, attached our attention. We suspect that the wafer surface condition will contribute to the production of this residue defect, so two kinds of experiment were used to clarify. Firstly, Q-time (the time between former process step and photography) comparison experiments were

carried out. The results showed that the residue defect becomes worse with longer Q-time, but the residue defect also exists if we strictly control Q-time at very short time. On the other hand, the wafer with residue defect was send to photoresist rework. The rework method consists of two steps, including dry etch and wet chemical treatment. Afterwards, the second lithography was carried out and the defect condition was measured. We found from the result that the residue defect was obviously decreased and only 1~2 (sometimes none) residue defect was found.

Based on the above results, we can deduce that the residue defect should be originated from the surface condition effect. The longer stay time before lithography will result in the generation of more residue defect after lithography. And the number of defect was obviously decreased after lithography rework. According to the shape and size of the defect, a simple model is built to clarity the defect generation mechanism, as described in Figure 6. After AA and STI area formation, the wafer surface structure changed and surface condition variated at different area. During lithograph process, the remained develop chemical and residue photoresist will be removed. But in fact, the residue photoresist was found at the interface between AA and STI area. We suspect two factors may contribute to the defect generation. Firstly, the physics height is different for AA and STI due to design and process requirement. Secondly, the material of AA and STI is different, and the relatively hydrophobic and hydrophilic area may exist at wafer surface. After water droplet comes across this area, the water is stretched and separated on surface, then the larger size circular-pattern thin layer is produced, and the massive anomalistic residue defect will stay at the interface due to step height between AA and STI area.

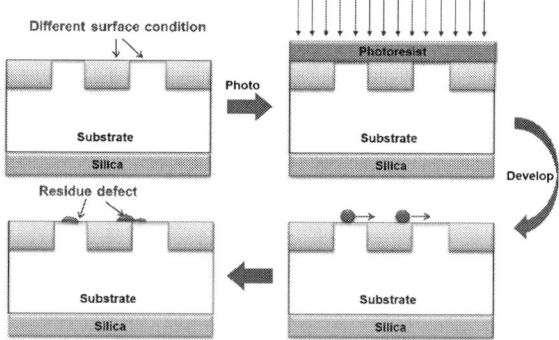

Figure 3: Schematic of the defect formation mechanism

Based on the residue defect generation mechanism, we suppose that the surface pre-treatment before lithography process can solve this residue defect problem. The former result showed that the residue defect was obviously decreased after lithography rework. But this method is not suitable for semiconductor manufacture process from both economic and operational. So we tried another surface treatment method before lithography process. Figure 4 (a) shows the defect result of wafer without surface treatment. We can see that the defect count is significantly decreased and almost no residue defect is found after we use optimized surface treatment method, as shown in Figure 4 (b,c). The defect result also in turn confirms the generation mechanism that we proposed.

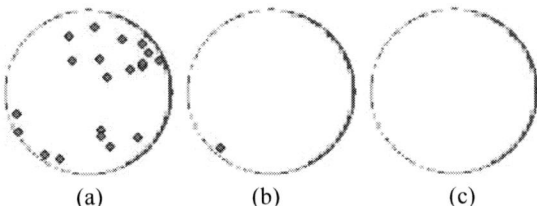

(a)   (b)   (c)

Figure 4: The defect map of wafer (a) without surface treatment, (b,c) with optimized surface treatment method

CONCLUSION

In this paper, we have investigated a photoresist residue defect on wafer surface if high transmission rate mask is used. This defect was like a circular-pattern about one hundred micron in diameter with thin photoresist residue. And some massive photoresist residue was also observed at the interface between AA and STI area. We found from the defect result that this residue defect can't be totally removed after using several lithography process optimized methods. After performing some experiments, a model was proposed to explain the generation mechanism of this residue defect, and wafer surface condition may contribute to the generation of such defect. An optimized surface treatment method was carried out to avoid the generation of this defect successfully. Furthermore, the generation mechanism needs to be studied carefully to find the connection between wafer, mask and photoresist.

ACKNOWLEDGEMENTS

The authors would like to express their gratitude to SMIC TD colleagues who have contributed to this work through discussion, experiment and tool support.

REFERENCES

[1] R.J. Zhang, Nanoscale Integrated Circuits – The Manufacturing Process. Tsinghua University, 2016.

CHALLENGES AND SOLUTIONS OF FEW-LINE PATTERNS WITH FREEFORM ILLUMINATION IN IMMERSION LITHOGRAPHY

Peipei Liang, Zhifeng Gan, Qiang Wu

Technology R&D, Semiconductor Manufacturing International Corp.
Pudong New Area, Shanghai, P. R. China 201203
+8621 38610000x17478, Peipei_Liang@smics.com

ABSTRACT

Lithography is facing great challenges from the continuous shrinkage of critical dimension in order to satisfy the development of the technology node. To extend ArF immersion lithography to sub 20 nm technology node, it is necessary to print smaller pitch size such as 80 nm for 1D pattern with enough process window. But it is hard to compromise 2D and other more complicated features. Source-mask optimization (SMO) is one of the key techniques for sub 20 nm technology node, which combines freeform light source with mask optimization to achieve a good process window for a variety of key features which optical proximity correction has limited solutions. In this paper, we use SMO technique for different kinds of pattern which is typical in sub 20 nm node, including 1D pattern, 2D regular pattern such as line-end to line-end, and 2D irregular pattern. It is shown that a proper source with a proper mask optimization strategy can be obtained, which gives enough common process window on most of the test pattern. However, few-line pattern like 3-line, 4-line or 5-line which have some special pitch and are common in real chip, are very hard to print with enough process window even with SMO technique. In this paper, we present several solutions such as pattern shift and broaden for few-line pattern. The process window can be improved with these solutions.

Keywords—Source-Mask-Optimization, process window, sub 20 nm node, few-line pattern

INTRODUCTION

Optical lithography is facing great challenges with the continuous shrinkage of the technology node. In order to extend ArF immersion lithography to smaller and smaller dimensions, it is necessary to print continued shrunk features with acceptable process window. In the past years, various resolution enhancement techniques (RETs) have been developed such as OPC, SRAF, shorter wavelength (EUV) [1-3]. Source-mask optimization (SMO) [4-6] is one of the necessary techniques for technology nodes below 20 nm, which is aimed at balancing process window for a variety of features through optimization of illumination pupil shape and mask dimensions of critical design pattern.

Generally, after iteratively updating the source image and the mask shape, a source with balanced performance on various pattern can be obtained. However, some special pattern still could not get enough process window even with the help of SMO. The process window can be solved with some pattern shift, while it can also be solved through the relaxation of the

design, such as head-to-head distance. Or the small process window can be solved through the tightening of the mask line-width uniformity and process control in exposure energy and overlay. In addition, it can also be improved by selecting a better photoresist material.

In this paper, we find that few-line patterns which have some special pitch, are very hard to print with enough process window. We present several effective solutions for these patterns such as pattern shift and broaden to greatly improve the process window.

EXPERIMENTAL AND RESULTS

In this study, we use ASML's Tachyon SMO software to do the simulation. We use SMO technique for different kinds of pattern which is typical in sub 20 nm node, including 1D through pitch (TP) pattern and few-line pattern (bi-line, BL and tri-line, TL), 2D regular pattern such as line-end to line-end (head to head, HTH), and 2D irregular pattern (Figure 1). Most of the pattern is in horizontal direction and the minimum pitch in this direction is 80 nm, which is called critical pitch (CP). While the vertical direction is called non-critical pitch (NCP) direction. In order to get enough process window in CP direction which has very small pitch and more complicated patterns, it is necessary for the NCP patterns to do some trade-off. Thus the minimum pitch of NCP direction must be twice of that of the CP direction.

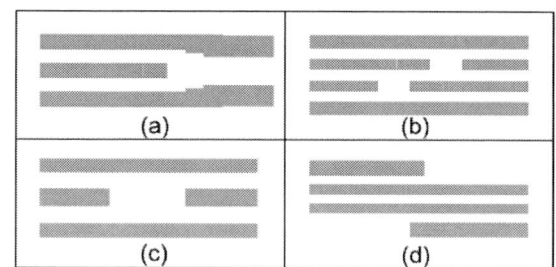

Figure 1: 2D irregular pattern

In the SMO simulation, several representative clips were selected through diffraction pattern analysis to determine unique pitch components under model conditions and were used to do the source and mask co-optimization. All the test patterns were used to verify the output source. In order to get the largest common process window, we have to do some balance among all these patterns. There were several splits of the input test pattern for source tuning which have different weight and number of 1D and 2D patterns. We did mask

978-1-5386-5309-8/18 $31.00 © 2018 IEEE

optimization of all the sources and got a freeform source which has the balanced process window for all the test pattern. The source shape is somewhat like the dipole illumination for most of the patterns are in horizontal direction.

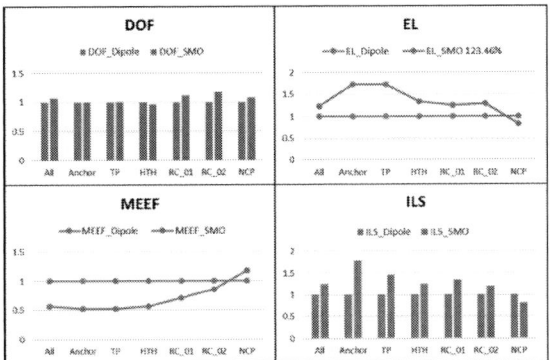

Figure 2: Comparison of depth of Focus (DOF), exposure latitude (EL), mask error enhancement factor (MEEF) and image log slop (ILS) of the dipole and freeform source (Normalized)

We compared the process window of freeform source with dipole source. The evaluation parameters of the process window include depth of focus (DOF), exposure latitude (EL), mask error enhancement factor (MEEF) and image log slop (ILS) and were simulated after mask optimization of the two sources, respectively. They are shown in Figure 2, which illustrates that the common process window of SMO source is higher than dipole source, shown as the first point in each chart. The DOF of all the pattern is equivalent for the two sources, while the EL, MEEF and ILS are all improved obviously except the pattern in NCP direction which has a lower ILS and EL. But the ILS and EL of NCP patterns are still higher than those of HTH pattern and some 2D irregular pattern.

Better yet, we find that few-line patterns which have some special pitch, are very hard to print with enough process window even the other pattern especially NCP and 2D irregular pattern do some trade-off. The few-line patterns are shown in Figure 3. If the space between the isolated line and the first line in the dense area (labeled as il and 1l in Figure 3) is about 100nm, the process window of these patterns are extremely low no matter how to tune the source.

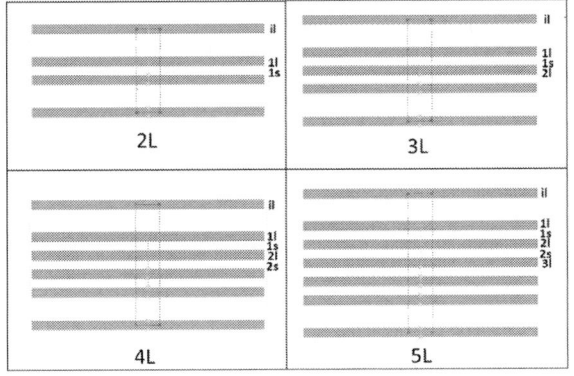

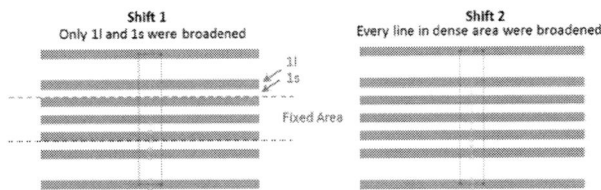

Figure 3: few-line patterns which have some special pitch

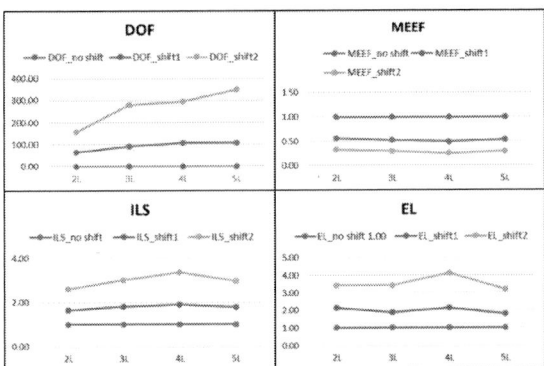

Figure 4: Shift strategies to solve few-line pattern

To solve this problem, we present several solutions for these features such as pattern shift and re-target. Figure 4 illustrates two shift strategies: shift 1 means only the first line and first space (labeled as 1l and 1s in Figure 4) are broadened out; shift 2 means that every line and space in dense area is broadened out.

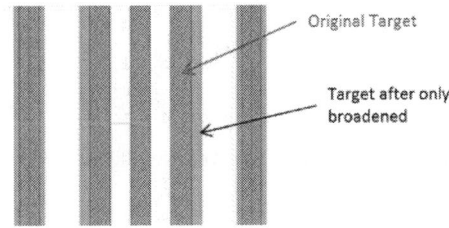

Figure 5: Process window comparison of different shift strategies (Normalized)

Figure 5 displays the process window of the two shift strategies and compares with that of the initial pattern. It is obvious that both the two shift strategy is conducive to the improvement of the process window, while the second shift strategy is more effective.

Figure 6: Broaden strategy to solve few-line pattern

However, sometimes the line patterns could not be shifted especially for metal layers, there may be contact hole in the metal lines. For this purpose, we have another strategy which only broadens out the outside edge of the first line and is shown in Figure 6. The EL improvement of the broaden strategy is displayed in Figure 7, which means that the broaden strategy is also an effective solution for the few-line patterns.

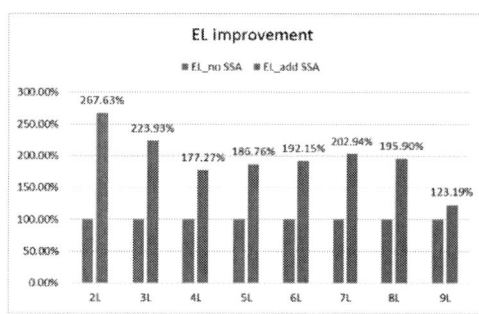

Figure 7: EL improvement of the broaden strategy

CONCLUSION

We have applied SMO technique to solving different kinds of pattern which is typical in sub 20 nm node in lithography and obtained a proper freeform source with a corresponding mask optimization strategy. The freeform source has balanced performance on both critical-pitch line/space and non-critical pitch line/space. It also has acceptable performance on line-end structure in densest head-to-head patterns and other more complicated 2D irregular patterns. However, few-line patterns which have some special pitch and are common in real chip, are very hard to print with enough process window even with SMO. To solve this problem we present several pattern shift and broaden strategies. With these strategies, the freeform source still have capability to solve the few-line patterns with enough process window.

REFERENCE

[1] A. K. Wong, Resolution Enhancement Techniques in Optical Lithography (SPIE, 2001).

[2] M. Rothschild, "A roadmap for optical lithography," Opt. Photon. News 21(6), 26–31 (2010).

[3] E. Y. Lam and A. K.Wong, "Computation lithography: virtual reality and virtual virtuality," Opt. Express 17(15), 12259–12268 (2009).

[4] L. Pang, P. Hu, D. Peng, D. Chen, T. Cecil, L. He, G. Xiao, V. Tolani, T. Dam, K.-H. Baik, and B. Gleason, "Source mask optimization (SMO) at full chip scale using inverse lithography technology (ILT) based on level set methods," in Lithography Asia 2009, vol. 7520 of Proc. SPIE, p. 75200X (2009).

[5] T.Mˉulders, V. Domnenko, B. Kˉuchler, T. Klimpel, H.-J. Stock, A. Poonawala, K. N. Taravade, andW. A. Stanton, "Simultaneous source-mask optimization: a numerical combining method," in Photomask Technology 2010, vol. 7823 of Proc. SPIE, p. 78233X (2010).

[6] X. Ma and G. R. Arce, "Pixel-based simultaneous source and mask optimization for resolution enhancement in optical lithography," Opt. Express 17(7), 5783–5793 (2009).

Research of Lithograph Process of Polyimide Photo Resist for Passivation Thick Films

Zhao Wang*, Ke Feng

R&D Center, CRRC Yongji Electric Co., Ltd, Xi'an 710018, China

*Corresponding Author's Email: wangz_crrc@163.com

ABSTRACT

Polyimide has been applied in semiconductor field for a long time. For its stable chemical characteristics, it's usually used as the passivation layer on the chip. In this paper, litho process was setup to research the influence of polyimide as photo resist. Several experiments were performed to solve the spin coating uniformity, polyimide resist residue and the edge bead removal (EBR). Optimization of the develop recipe were carried on step by step, ultimately, existing problems were solved by modifying several parameters, such as, soft baking time and temperature, spin speed, spin time, exposure time and develop time. Smooth and compact 43 μm thick polyimide films were prepared under the parameters below: soft baking under 100 °C for 90 s, spin time of 25 s, spin speed of 2500 rpm, 50 μm approximate-mode exposure, exposure time of 100 s, develop time of 80 s. The forming polyimide films were applied on the power (insulated gate bipolar transistor) IGBT chip as passivation layer.

INTRODUCTION

Polyimides (PIs) have many desirable properties, such as low dielectric constant, high breakdown voltage, chemical resistance, wear resistance, stability at elevated temperatures, low thermal expansion, as well as excellent mechanical and good optical properties. As a result, the polyimides have found a wide range of applications particularly in microelectronics and optoelectronics. Examples are flexible printed circuit boards, IC packagings, liquid crystal (LCD) alignment layers organic thin-film transistors [1]. In many applications of polyimide materials, the capability of patterning and fabricating PIs into micro-scaled or even nano-scaled structures plays a critical role and dominates the future developments of new polyimide-based devices. It is, however, a challenging issue still within academia and industry as well, owing to the mechanical and chemical stability of PIs and the requirements for small feature size, precise dimension control, process compatibility, low-cost and high throughput.

There are many different ways of patterning and fabricating polyimide structures at the micro- and nanoscales depending on whether the PIs are photosensitive or non-photosensitive. Traditionally, PIs are non-photosensitive and hence cannot be used as a photo-resist (PR) in photolithography. However, efforts have been made in the past few decades in synthesizing a variety of photosensitive polyimides (PSPIs) which have become popular now.

Schematic representatives of photolithographic processes are shown in Figure 1. For patterning non-photosensitive PIs (Figure 1(a) and (b)), also known as indirect photolithography, is to spin coat a PR layer on top of the PI layer and then pattern the PR with standard photo lithography. The process is then followed by wet or dry etching on PIs using the patterned PR as an etching mask. Quite often an inorganic layer such as SiO_2 in between the PR and PI layers is also needed as an intermittent etching mask since directly using a PR as the etching mask for PI can be very difficult [2]. On the other hand, by incorporation of photosensitivity into PIs (or PI precursor), PSPIs enable to simplify a patterning process owing to elimination of a PR layer (Figure 1(c) and (d)).

In this article, litho process was setup to research the influence of PSPIs as photo resist. Researches were carried out to solve problems existing in spin coating uniformity, polyimide resist residue and the edge bead removal (EBR). Serial experiments were executed step by step, in order to optimize the lithographic recipe.

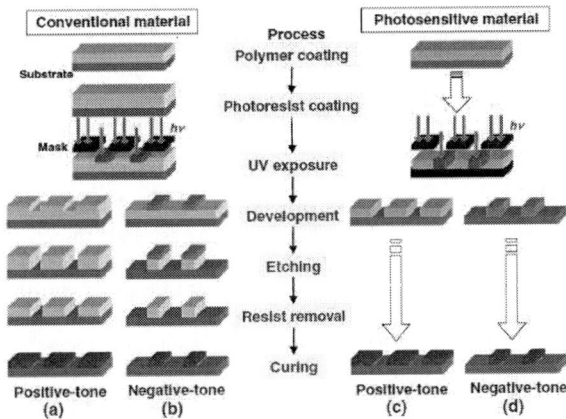

Figure 1: Photolithographic patterning processes in conventional methods using photoresists (a & b) and in advanced methods using photosensitive materials(c & d). Positive-tone can be formed if the exposed area is washed out at the development (a & c), whereas negative-tone can be patterned if theremoval of unexposed area is carried out (b & d) [3].

Ultimately, the cured PSPIs thick films for IGBT chip passivation layer were prepared by modifying several specific litho parameters.

EXPERIMENTAL DETAILS

The chemicals chosen in this study are commercially available polyimide (Fujifilm Durimide 7020), developer (Fujifilm HTR D-2) and rinse (Fujifilm RER600). 6-inch Si wafers were pre-cleaned by a washing machine (ADT WCS-977). Polyimide was spin-coated on the Si wafer to form a thin polyimide layer by a spin coater (MIDAS SPIN-4000A). Then, the polyimide layer and the silicon substrate were soft baked on a hot plate at a temperature of 100 °C for 90 s. The most important part for exposure and developing were carried out in the litho equipment (SUSS MA/BA 6) and the developing machine (MIDAS DEVELOP-4000A), respectively. The PI/Si sample was then moved into an oven (DTPN-612C) for thermal imidization for 2.5 h at 350 °C.

RESULTS AND DISCUSSION

The PSPIs photolithography process can be subdivided into three main operations: polyimide coating, alignment and exposure, and polyimide developing.

In PI coating process, a series of experiments were designed to observe how the film thickness changes with the spin speed. The result is shown in Figure 2. The film thickness decreases as the spin speed increases. As the requirement of a 40 um-thick PI film, spin speed around 2500 rpm was chosen for design of experiments (DOE).

In this study, we selected the spin speed and the spin time for controllable parameter in the PSPIs coating process. Accordingly, this study performed the experiment work using 3-level factors and two response variables. Table I lists two factors of spin speed (i.e. A/rpm), spin time (i.e. B/s), with three levels in this study. Namely, the experiment work in this study is based on the orthogonal array $L_9(3^2)$.

The mean thickness and thickness uniformity were measured using an optical probe film thickness analyzer of Filmtek 1000M. The mean thickness and thickness uniformity are measured as average and standard deviation values of 16 points on wafer surface from different directions, respectively. The mean thickness and the thickness uniformity are listed in Table II.

TABLE I.   EXPERIMENTAL FACTORS AND FACTORS LEVELS

Levels of experimental factor	Experimental factors	
	A(rpm)	B(s)
1	2100	15
2	2300	20
3	2500	25

TABLE II.   L9 OF THE EXPERIMENTAL RUNS AND RESULTS

L9	A	B	t_m (um)	t_u (um)
1	1	1	62.77	0.42
2	1	2	55.84	0.25
3	1	3	50.88	0.89
4	2	1	56.32	0.28
5	2	2	51.23	0.34
6	2	3	47.07	0.21
7	3	1	52.37	0.28
8	3	2	47.12	0.28
9	3	3	42.99	0.25

The Taguchi method could be applied for improve process performance. The different signal–noise ratios, corresponding to n experiments, are presented below:

$$\frac{S}{N} = -10\log(\frac{1}{n}\sum_{i=1}^{n} y_i^2) \qquad (1)$$

Where S denotes the standard deviation, y_i is the data from on-line experiments and n represents the times of experiments. Typically, lower S/N ratio of thickness and thickness uniformity during the PI coating process is preferred [4].

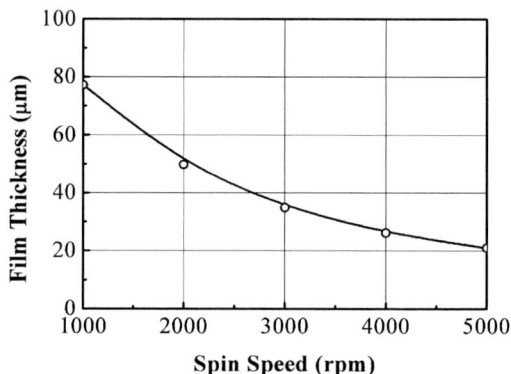

Figure 2: Polyimide film thickness vs. coating spin speed.

Spin speed of 2500 rpm and spin time of 25 s were chosen to be the optimal conditions to perform PI spin coating. Furthermore, in order to found out optimal develop patterns for PI thick film, exposure time of 40s, 60s, 80s and 100s were set up with 50 μm approximate-mode exposure. All films were developed for 80 s. Figure 3 shows the exposure results, which were taken by a digital microscope (Keyence VHX-5000). Clear chip edge and inner corner profile could be observed from the microscope. Accordingly, as the exposure time was below 80 s, PI residues were discovered in the pattern corners. Excellent pattern profile was obtained after the PIs were exposed for 100s.

Experiments show that PSPIs have taken over excellent chemistries developed by PR for microlithography. Two distinctive respects are mainly

978-1-5386-5309-8/18 $31.00 © 2018 IEEE

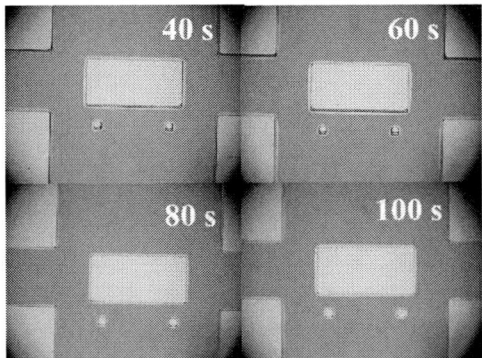

Figure 3: Polyimide development results vs. exposure time.

demanded in PSPIs: photolithographic characteristics to form patterns; quality and durability required in final products. PSPIs are consistent and satisfied with these factors simultaneously. Therefore, lithographic technique depends on photosensitivity introduced into a PSPI system as well as design of polymer structure which is also effective for the PI film properties.

In order to improve the IGBT performances in the insulation and partial discharge, polyimide films are usually used upon the chip surface, which could effectively block the influence of humidity, dust and other factors, meanwhile, weaken the phenomenon of partial discharge in IGBT modules. Polyimide thick films used for passivation application were prepared following the process flow shown in Figure 4. First, the wafer is coated with a thick layer of photosensitive polyimide. Then, it is exposed by ultraviolet (UV) light through a mask or reticle with the pattern of clear and dark areas generated by the plotter, based on IC design. The chemistry of the exposed polyimide is changed by the photochemical reactions under the clear areas through which the UV light can pass. After the photolithography process is completed, the final baking is carried out in the oven with nitrogen. This process takes 90 min to increase from room temperature to 350 °C at a speed of 3-5 °C per second. And keep baking at 350 °C for 150 minutes. After baking, it takes 160 min to cool it. In order to prevent the formation of the oxide film, the oven must be cooled to less than 100 °C to take out the wafer.

CONCLUSIONS

Litho process was setup to research the influence of PSPIs as photo resist. Researches were carried out to solve problems existing in spin coating uniformity, polyimide resist residue and the edge bead removal (EBR). Serial experiments were executed step by step, in order to optimize the lithographic recipe. Ultimately, the cured PSPIs thick films for IGBT chip passivation layer were prepared by modifying several specific litho parameters, such as, spin speed, spin time, soft baking time and temperature, exposure time and develop time. Smooth and compact 43 μm thick polyimide films were prepared using parameters below: soft baking under 100 °C for 90 s, spin time of 25 s, spin speed of 2500 rpm, 50 μm approximate-mode exposure, exposure time of 100s, develop time of 80 s.

ACKNOWLEDGEMENTS

The authors would like to thank for the experiment equipments support from Semiconductor Department, Xi'an CRRC Yongji electric Co., Ltd.

REFERENCES

[1] B. Cui, Y. Cortot and T. Veres. *Microelectronic Engineering*, vol. 83, 2006, pp. 906-909.
[2] C.-Y. Chiu and Y.-C. Lee. *Journal of Micromechanics & Microengineering*, vol. 19, 2009, pp. 105001.
[3] K. Fukukawa and M. Mitsuru. *Polymer Journal*, vol. 40, 2008, pp. 281-296..
[4] Y.-K. Yang and T.-C. Chang. *Microelectronic Journal*, vol. 37, 2006, pp. 746-751.

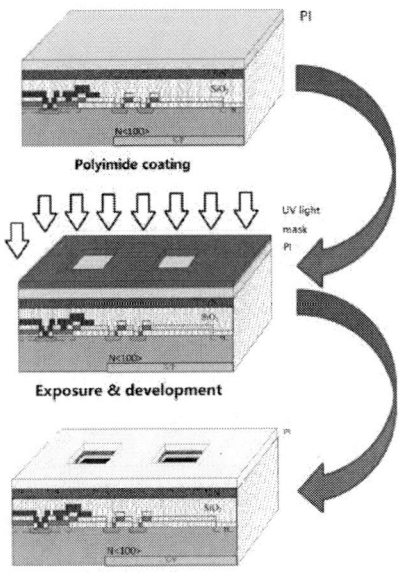

Figure 4: Polyimide passivation thick film preparation process among the IGBT chip process flow.

THE STUDY ON THE DIFFRACTION SPECTRUM OF ADDING S-BAR DURING THE SOURCE MASK OPTIMIZATION (SMO) PROCESS

Miao Xia*, Song Bai, Qin Wang

Technology R&D, SMIC Advanced Technology R&D (Shanghai) Corporation
Pudong New District, Shanghai, 201203, China
*Corresponding Author's Email: miao_xia@smics.com

ABSTRACT

To keep the practicability and applicability of source, source-mask optimization (SMO) only optimizes a group of representative test patterns. As for the selection of 1D patterns, dense pitches which could not be improved by adding scattering bar (S-BAR) are selected as test patterns. For the patterns with pitch equal to multiples of the minimum pitch, the method of adding S-BAR was involved to make the diffraction pupil spectra [1] close to the idealized one from the dense pitches. The position and size of S-BAR should also vary with different condition to ensure the ideal spectra. We calculated the diffraction spectra for varies pitches with S-BAR, and demonstrated what the appropriate location of S-BAR is during SMO process. By adjusting the position and width of S-BAR, we have also studied its importance on improving the process window for different source types. Finally, similar effect and spectrum of S-BAR for 2D patterns is also discussed.

Keywords—Source mask optimization, Scattering Bar, test pattern, process window, diffraction spectrum

INTRODUCTION

It is widely known that source-mask optimization (SMO) only optimizes a group of representative test patterns extracted from design rules to keep the practicability and applicability of source. The rule of pattern selection is also part of strategy of source optimization to enhance process window. For example, as for the selection of 1D patterns, dense pitches which could not be improved by adding scattering bar (S-BAR) are selected for source figuration. To utilize the source advantage for dense pitch, the patterns with pitch equal to multiples of the minimum pitch would be segmented by the S-BAR, to realize the strategy of making the diffraction pupil spectra [1] of isolated pattern close to the idealized spectra from the dense pitches. The position and width of S-BAR should also vary with the ADI target and mask bias to ensure the ideal spectra. However, the resulting spectrum of post OPC for SMO source is not always ideal and the process window may be limited. The lack of simplicity and ideality of pupil spectrum could give the explanation for the deficiency of process window such as exposure latitude (EL) and density of focus (DOF). SMO gives the assumption that the results of mask optimization (MO) could always provide the relatively ideal spectrum. But in fact, limited by a given ADI target, mask bias and rule-based scattering bar rule, post OPC results which even evolves from continuum transmission mask (CTM) could not always give a satisfactory performance.

In this study, with detailed calculation of the diffraction spectra for varies pitches with S-BAR, we present a study on the ways to evaluate what is the appropriate position to add S-BAR during SMO process. By adjusting the position and width of S-BAR, we have also studied its importance on improving the process window for different source types. Finally, similar effect and spectrum of S-BAR for 2D patterns is also discussed. Our result is helpful to understand the underlying mechanism of how to add S-BAR, and is inspiring to enhance the process window of photolithography.

ONE-DIMENSIONAL DIFFRACTION INTENSITY

For the one-dimensional condition, our calculation was performed on MathWorks MatLAB R2012b. The complex number k can be expressed as

$$k = e^{i\frac{2\pi}{\lambda}X(\sin\theta - \sin\theta_i)} \tag{1}$$

Where λ is the wavelength, X stands for the position on the mask, and θ and θ_i are the diffraction angle and the incident angle separately. The complex amplitude A is the accumulation of k

$$\widetilde{A} = \sum k \tag{2}$$

The intensity is expressed by

$$I = \left|\widetilde{A}\right|^2 \cdot \left\{\frac{\sin\left[\frac{N\pi a\cdot(\sin\theta - \sin\theta_i)}{\lambda}\right]}{\sin\left[\frac{\pi a\cdot(\sin\theta - \sin\theta_i)}{\lambda}\right]}\right\}^2 \tag{3}$$

where a stand for the pitch value and N represent the repeat times of the pitch.

For the dense pitch, take the pitch of about 100nm as an example, (here after denoted as P), the diffraction spectrum is quite simple. Assume the light incident angle is θ_i, and $\sin\theta_i = 0.7$. As the $\sin\theta$ varies from -1 to +1, the interference of light transmitted through mask should only form two peaks, and the separation of these peaks is larger. However, if the pitch was 3 times of the dense pitch, denoted as $3P$, besides the zeroth diffraction peak is on the $\sin\theta = 0.7$ position, several higher order peaks are in the left position, which will finally form interference of multi-beams on the wafer. The diffraction intensity vs $\sin\theta$ in the one-dimensional condition is shown in Fig. 1a. To optimize the light intensity propagates through the isolated mask patterns, two S-BARs are inserted between the main patterns. The resulting intensity is re-calculated as shown in Fig 1b for one pitch. If the pitch is repeatedly arranged several times, the diffraction intensity with S-BAR demonstrates only two main peaks (shown in Fig 1c), similar with that of dense pitch of P.

Our calculation demonstrates that adding S-BAR is an effective method to optimize isolated patterns. $3P$ is three times as minimal pitch P, so SMO adds two S-BARs to increase the third order diffraction and decrease the first& the second in order to form two- beam interference so that the density of focus (DOF) will be enhanced as for the source we have. All pitches multiple times as wide as P could be taken care.

However, for the semi-isolated pitch which is 5times of dense pitch, the optimized spectrum is like that shown in Fig. 2a. Two main peaks are similar to the above discussion, but the peak position is not exactly at -0.7, it locates at -0.5. The enhanced is the forth order peak, not the fifth. This is the best optimized S-BAR position. So the 5 times of dense pitch is another DOF limited pattern.

For the semi-isolated pitch which is 6 times of dense pitch, the optimized spectrum is like that shown in Fig. 2b. It can be found that, the diffraction forms double peak. It is worth noting that, not only two- beam interference could work, but also four beam interference could also work.

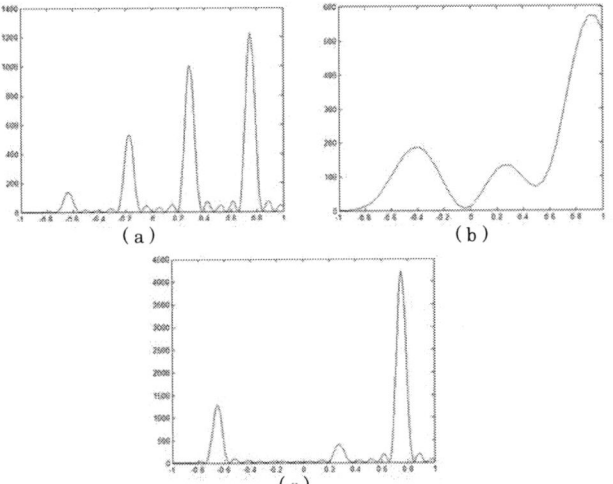

Fig. 1: (a) The diffraction intensity vs. sine of the incident angle in the one-dimensional condition. (b) The intensity calculated for one pitch with S-BAR. (c) The intensity of several pitch arrays with S-BAR, which is more like that of dense pitches.

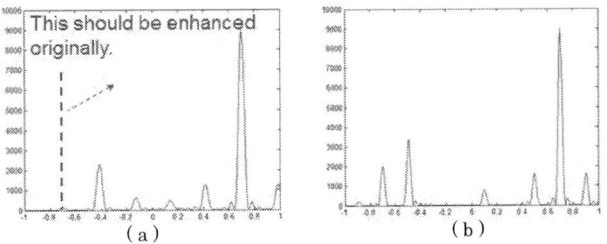

Fig. 2: (a) the intensity of 5 times dense pitch after S-BAR optimization. (b) The intensity of 6 times dense pitch after S-BAR optimization.

DIFFRACTION INTENSITY OF TWO-DIMENSIONAL SQUARE ARRAYS

In order to further study the effect of S-BAR in a more applicable condition, we calculated the diffraction spectrum of the two-dimensional mask. In 2D condition, more item is added into the expression. The complex number k is

$$k = e^{i\frac{2\pi}{\lambda}(X*sin\theta_x + Y*sin\theta_y)} \qquad (4)$$

The complex amplitude A is the accumulation of k

$$\widetilde{A} = \sum k \qquad (5)$$

The intensity is expressed by

$$I = \left|\widetilde{A}\right|^2 \cdot \left\{\frac{\sin\left[\frac{N\pi a \cdot (\sin\theta_x - \sin\theta_{xi})}{\lambda}\right]}{\sin\left[\frac{\pi a \cdot (\sin\theta_x - \sin\theta_{xi})}{\lambda}\right]}\right\}^2 \cdot \left\{\frac{\sin\left[\frac{N\pi b \cdot (\sin\theta_y - \sin\theta_{yi})}{\lambda}\right]}{\sin\left[\frac{\pi b \cdot (\sin\theta_y - \sin\theta_{yi})}{\lambda}\right]}\right\}^2 \qquad (6)$$

All the meanings of variables are similar to those in formula (1-3).

We use the array distributed squares with ADI target about 50nm, and the dense pitch is about 100nm (here after denoted as P), with the mask pattern shown in Fig 3a. Note that the mask pattern was obtained by MO with a typical source. The corresponding diffraction intensity is shown in Fig 3b. The horizontal axis is the sine of the incident angle in the direction of x, and the vertical axis is that of y direction. The intensity is represented by the color. The incidence angle was set to $\sin\theta_x = 0.5$ and $\sin\theta_y = 0.5$, which is also the brightest spot position. The other diffraction spot are in the lower left position. As the pitch becomes larger, if there were no S-BAR (Fig. 3c and 3d), several higher order diffraction spot would appear.

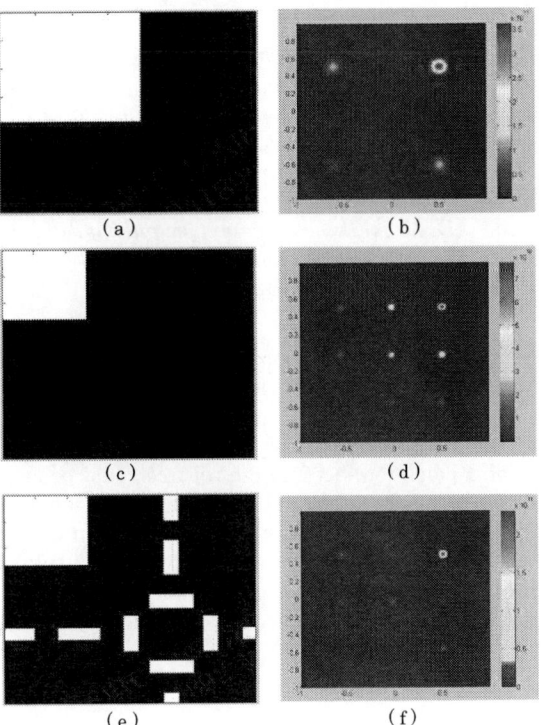

Fig. 3: (a) the single pitch in the mask of dense patterns

978-1-5386-5309-8/18 $31.00 © 2018 IEEE

with the pitch as P. (b) The diffraction spectrum of a. (c) the single pitch in the mask of semi-isolated patterns, the ADI target is the same and the pitch is 2P. (d) The diffraction spectrum of c. (e) the single pitch in the mask of semi-isolated patterns with S-BAR, the ADI target is the same and the pitch is 2P. (f) The diffraction spectrum of e.

The S-BAR of the pitch $2P$ on the mask is shown in Fig. 3e, and the resulting diffraction spectrum is shown in Fig. 3f. Obviously, some high order spot was suppressed, and only the third spot was enhanced. The zeroth and the third spot forms the diffraction spectrum very like that of dense pitch. More obvious phenomenon was also observed in the $5P$ and $10P$ pitch, illustrated in Fig 4. For the mask with no S-BAR, many higher order diffraction spots exist. After properly adding the S-BAR, the diffraction spectra are more like those of dense patterns.

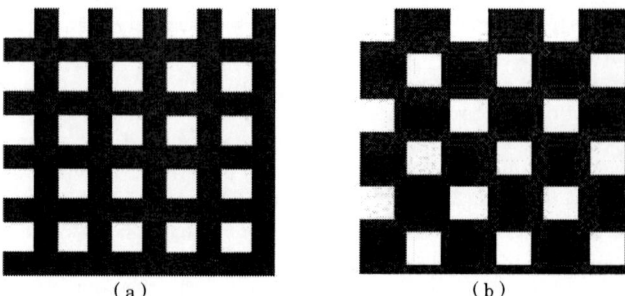

Fig.5: The schematic pattern arrangements in the mask of 2D squared (a) and staged (b) conditions.

The calculation results are summarized in Fig.6. The incident angle is set to x=0.4, and y=0.4, which is the zeroth order spot position. For the dense pitch, in contrast with previous 2D condition, the first order diffraction spot locates in the lower left position to the zeroth spot, which is a natural result of the staged squares. For semi-isolated and isolated pitches, the diffraction spectrums with/without S-BAR are separately shown in Fig. 6b-6e. Obviously, adding S-BAR is effective to suppress some order diffraction spots, and make the final spectrum more like that of dense pitches.

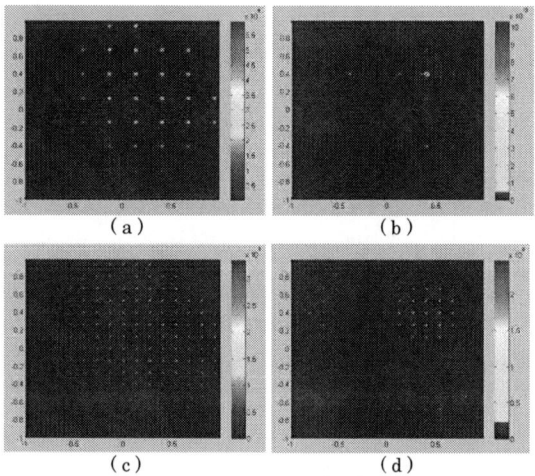

Fig.4: Comparison of diffraction spectrum with and without S-BAR. (a) The diffraction spectrum of semi-isolated patterns without S-BAR, the ADI target is the same and the pitch is 5P. (b) The diffraction spectrum of semi-isolated patterns with S-BAR, the ADI target is the same and the pitch is 5P. (c) The diffraction spectrum of isolated patterns without S-BAR, the ADI target is the same and the pitch is 10P. (d) The diffraction spectrum of isolated patterns with S-BAR, the ADI target is the same and the pitch is 10P.

DIFFRACTION INTENSITY OF TWO-DIMENSIONAL STAGED ARRAYS

Two-dimensional mask is more complicated than one dimensional condition, because the squares could be arranged in a different manner. These arrangements are compared in Fig.5. Above discussion is based on the condition of Fig. 5a. For the staged arrangements in Fig. 5b, we also calculated the spectrum.

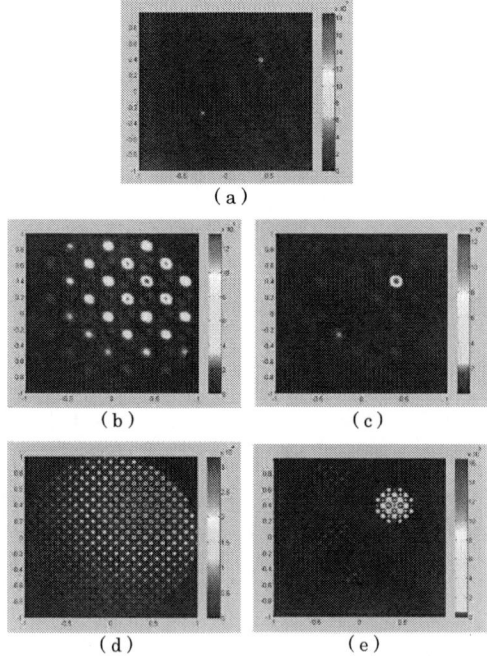

Fig.6: Comparison of diffraction spectrum with and without S-BAR. (a) The diffraction spectrum of dense patterns, the ADI target is the same and the pitch is P. (b) The diffraction spectrum of semi-isolated patterns without S-BAR, the ADI target is the same and the pitch is 3P. (c) The diffraction spectrum of semi-isolated patterns with S-BAR, the ADI target is the same and the pitch is 3P. (d) The diffraction spectrum of isolated patterns without S-BAR, the ADI target is the same and the pitch is 10P. (e) The diffraction spectrum of isolated patterns with S-BAR, the ADI target is the same and the pitch is 10P.

SUMMARY

To summarize, we calculated the diffraction spectrum from dense patterns to isolated patterns, and compared the results of isolated patterns with and without S-BAR. We found that after adding S-BAR, the diffraction spectrum of isolated patterns is similar to dense pitches. Our simulation results show the effect of S-BAR and help to understand the underlying mechanism of SMO.

REFERENCES

Alan E. Rosenbiuth, Scott Bukofsky, Michael Hibbs, Kafai Lai, Antoinette Molless, Rama N. Singh, and Alfred wong, "Optimum Mask and Source Patterns to Print a Given Shape", Proc. SPIE 4346, 486, 2001.

Nanometer Scale Mixed Patterns Formed by Self-aligned Spacer Image Transfer and Optical Lithography Technology

Zhaohao Zhang[1], Qingzhu Zhang [1,2], Junjie Li [1], Xiaobin He [1] and Huaxiang Yin[1,3]

[1] Key Laboratory of Microelectronics Devices & Integrated Technology, Institute of Microelectronics of Chinese Academy of Sciences (IMECAS), Beijing, 100029, China; [2] State Key Laboratory of Advanced Materials for Smart Sensing, General Research Institute for Nonferrous Metals, Beijing 100088, China, [3] University of Chinese Academy of Sciences, 100049 Beijing, China;
*Corresponding Author's Email: yinhuaxiang@ime.ac.cn; zhangqingzhu@ime.ac.cn

ABSTRACT

A mixed lithography with self-aligned spacer image transfer (SIT) technology and I-Line lithography is successfully demonstrated in this paper. The large patterns are defined in the resist film through photolithography whereas the smaller patterns are defined through self-aligned SIT. The formed nanometer hard mask (HM) patterns have a good compatibility with I-Line lithography. The image transfer process by etching Si is further verified in our experiment. The mixed patterns with well uniformity after Si etching illustrate the feasibility and provide a useful candidate for the real needs in the future.

Keywords: mixed lithography; SIT technology (SIT); optical lithography; hard mask (HM); reactive ion etch (RIE)

INTRODUCTION

Since 1950s, optical lithography has been the main field of semiconductor device manufacturing. Photolithography has gradually advanced to the point where it is capable of routinely pattern structures down to less than 20 nm, across an entire 300 mm silicon wafer [1]. Many separate advances have contributed to the highly developed photolithography state. It includes short wavelength deep ultraviolet (DUV) laser sources, projection optics [2], lens immersion technology [3], phase shifting masks [4, 5] and self-aligned double pattern (SADP) or self-aligned quadruple pattern (SAQP) using extra-deep ultraviolet ArF laser [6]. Although these methods have made photolithography capable of producing rarely small structures, the process for continue shrinking is very difficult and the EUV equipment is also very expensive (~100 million dollars).

The continuous demands for the integrated circuit (IC) chips in the market promote the continuous development of device scaling. The continuous scaling of patterns face great challenges to the present optical lithography technology. Therefore, it is urgent to develop a new lithography technology. At present, the fabrication methods of nanoscale structure are formed by many ways, e.g. extra-deep ultraviolet lithography

[2], electron beam direct writing [7], plasma ashing and SIT technology. The extra-deep ultraviolet lithography and electron beam direct writing have been considered as the most important candidates in the future. The equipment of electron beam direct writing is quite inexpensive compared with EUV and it can form mixed patterns without any lithography masks. Because of low efficiency and output rate, it is very difficult to be applied to industrial mass production, so electron beam direct writing technology is mainly used in laboratories for making nanostructure. Nowadays the SIT technology is the most popular technology for nanoscale patterns formation using SADP or SAQP. The nanoscale patterns could be formed after once lithography, therefore the efficiency of SIT is much higher than electron beam direct writing process and it makes the CD of formed patterns more convenient to adjust. However this method requests equipment with high quality and forms unchangeable HM on one wafer once. Therefore we can not achieve the mixed patterns using the SIT technology only. In the actual demand for patterns formation, mixed patterns with large and nanometer scale are demanded, however there is few proposed methods for mixed patterns formed.

In this paper, a high efficiency mixed lithography with self-aligned SIT technology and I-Line lithography was successfully integrated. The large patterns were defined through I-Line lithography, whereas the smaller patterns were defined through self-aligned SIT. The nanometer HM patterns formed by self-aligned SIT have a good compatibility with I-Line lithography. The further image transfer process was verified in our experiment by etching Si substrate. The mixed patterns with well uniformity after Si etching illustrate the feasibility of the proposed method and provide a useful candidate for the real needs in the future.

EXPERIMENTAL PROCEDURE

As is briefly shown in Fig.1, the mixed lithography was fabricated on 8-inch *p*-type Si (100) wafers. A sequential multi-layer SiO_2/Si (α-Si)/Si_3N_4 (25nm/100 nm/25nm) materials were deposited on Si wafers (see

978-1-5386-5309-8/18 $31.00 © 2018 IEEE

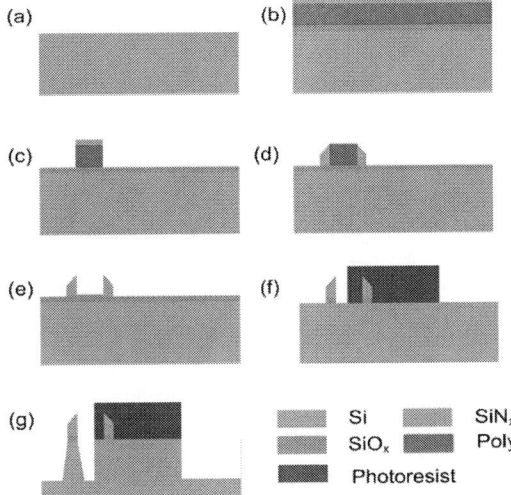

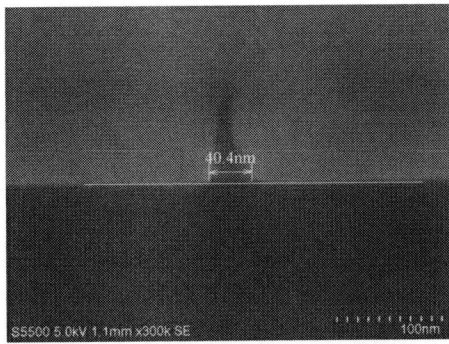

Fig. 2. SEM image of formed spacer HM by SIT. ~40nm width HMs are achieved.

Fig.1. Scheme and the flow of mixed pattern formation.(a) 200mm Si substrate with (100) crystal face, (b) 250 Å Oxide/1000 Åα-Si/ 250 Å Si3N4 deposition, (c) lithography and Si3N4/α-Si reactive ion etching, (d) Si3N4 spacers hard mask formation, (e) α-Si dummy gate remove (f) I-Line lithography for large area, (g) Si etching.

Fig.1(b)). The SiO2 and Si3N4 films were formed by plasma-enhanced chemical vapor deposition (PECVD). The 100 nm α-Si layer was thermally grown on the top of SiO2 layer using rapid thermal annealing processing (RTP). To form rectangular masks, a lithography process was carried out. Then, the process of etching Si_3N_4 and α-Si films was performed on 8-inch Lam Research Exelan HPT and Lam Research 9400 reactors, respectively. The schematic structure profile of nanometer Si_3N_4 is shown in Fig.1(c). The top Si_3N_4 HMs were removed by hot H_3PO_4 solution at 140°C for 15 minute. A 20 nm Si_3N_4 film was deposited by PECVD and the process of etching Si_3N_4 film was performed by Lam Research Exelan HPT tools, as is shown in Fig.1 (d). Two Si_3N_4 spacers with back-to-back wedge shape were formed on both sides of rectangular α-Si. The wafer was soaked in tetramethylammonium hydroxide (TMAH) at 70°C for 3 minutes on single-wafer cleaning Lam SP 203 equipment immediately for removing inner α-Si. Nanometer scale Si_3N_4 spacers HMs were left.

The further image transfer process was performed by etching Si. The process of etching 25 nm bottom SiO_2 was performed by using top Si_3N_4 spacers as the HMs and the larger patterns by I-line lithography were carried out subsequently.

The process of Si etching was performed at 8-inch Lam Research reactors. A one-step anisotropic etching process was carried out for Si etching. Firstly the CF_4 fluorocarbon-based plasma was performed for removing the native oxide. Secondly, anisotropic etching with HBr/Cl_2 and a little mount O_2 plasma was

carried out, as is shown in Fig.1 (g) [8]. The final transfer structure was formed after removing the top remain HMs. To characterize the morphological profiles of the structure, the cross-sectional views of the final structure were photographed in a Hitachi (Japan) scanning electron microscope (SEM) S-5500, and the top views were carried out in by Hitachi S-4800.

RESULTS AND DISCUSSION

The final structure after etching the bottom oxide by RIE is shown in Fig. 2. The bottom oxide layer was etched entirely and the etching process well stoped on Si, which means there are few damages on Si material. The width of the spacer is about 40 nm, which is used to form nanometer patterns in the flowing steps.

After forming the nanometer scale spacer HMs, the process of mixed large patterns was formed by I-Line lithography. As can be seen from the SEM images, the large images were formed across the HM array and the width of big patterns is about 10 times than the spacer HMs. The enlarged SEM image is inserted in the Fig. 3. The linked areas between big patterns and small HM are quite smooth after the optical exposure.

In order to conform the robust of the mixed patterns, a standard anisotropic dry etch was carried out on the whole wafer. The process of Si etching was performed at 8-inch Lam Research reactors. The wafer was placed on cathode with an electrostatic chuck (ESC) which was cooled to 60°C, in order to maintain a good thermal contact between the ESC and the wafer. The system is equipped with a high-frequency (HF) power source (27MHz) and a low-frequency (LF) power source (2MHz) that allows a separate control of plasma power and substrate bias. The process of Si etching was carried out in inductively coupled plasma etch tool Lam TCP 9400DFM with the same frequency for source power and bias power.

The related results of mixed profiles after Si etching are shown in Fig. 4. Fig. 4 (a) and (b) show the bird views of final structure. As can be seen from the SEM pictures, the formed silicon images were well transferred from the top mixed patterns and both the

978-1-5386-5309-8/18 $31.00 © 2018 IEEE

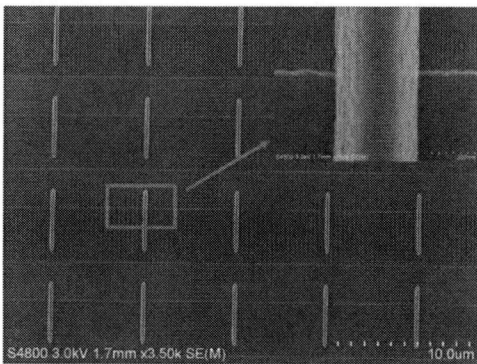

Fig. 3. SEM images of mixed patterns after I-Line lithography.

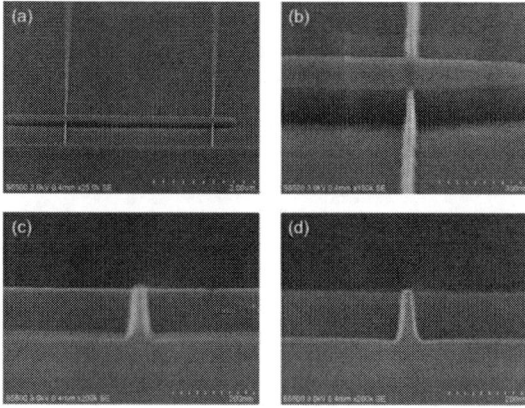

Fig. 4. SEM images of mixed patterns after RIE etching.

large and small images are achieved with good uniformity. On the basis of SEM results, the linked areas between big patterns and small HMs are very smooth after the anisotropic etching and there is no obvious difference between different locations for nanometer Si profiles, which further prove the effectively of the proposed method.

The cross section images of mixed patterns after etching are shown in Fig. 4 (c) and (d). The final structure of Si with ~110 nm height and ~30 nm width

was formed. The formed patterns with well uniformity after Si etching further illustrate the feasibility for the needs in the future.

SUMMARY AND CONCLUSION

We have demonstrated an effective method combining I-Line lithography and self-aligned SIT, which achieves large (300 μm) and small (20 nm) structures simultaneously on Si (100) wafers. Photoresist and nanometer scale HMs patterns, could be obtained using I-Line lithography and self-aligned SIT respectively. Both large and small patterns could be achieved at the same time with high efficiency. The patterns in photoresist and HMs can be transferred to the subsequent substrate. Such techniques allow relatively simple fabrication of large and small mixed structures simultaneously on one substrate.

REFERENCES

[1] Wallraff G M, and Hinsberg W D. "Lithographic imaging techniques for the formation of nanoscopic features" Chemical Reviews, 1999, 99(7):1801.

[2] Rothschild M 2005 Projection optical lithography Mater. Today 8 18

[3] Switkes M, and Kunz R R, "Extending optics to 50 nm and beyond with immersion lithography" Journal of vacuum science & technology. B, Microelectronics and nanometer structures: processing, measurement, and phenomena: an official journal of the American Vacuum Society, 2003, 21(6):2794 - 2799.

[4] Tritchkov A, and Kenyon C 2005 Lithography enabling for the 65 nm node gate layer pattern with alternating PSM Proc. SPIE 5754 215

[5] Hong S, and Jeong S. "Stochastic resist pattern simulation using attenuated PSM for EUV lithography" Proc. SPIE 8679 867928

[6] Babin S, and Glushenko G "A 2012 Application of double pattern technology to fabricate optical elements: process simulation, fabrication, and measurement" J. Vac. Sci. Technol. B 30 031605

[7] Katine J A, and Rooks M J. "E-beam writing: a next-generartion lithography approach for thin-film head critical feature". IEEE Transaction on Magnetics, 2002, 38 (1): 95-100.

[8] Q. Zhang, and H. Yin. "FOI FinFET with ultra-low parasitic resistance enabled by fully metallic source and drain formation on isolated bulk-fin", in Proc. IEEE IEDM, Dec. 2016, pp. 452-455, doi: 10.1109/IEDM.2016.783843

THE STUDY OF STI ETCHING MICRO-LOADING IN REACTIVE ION ETCH (RIE)

Yunhe Dong, Zhongwei Jiang, Yahui Huang

Beijing NAURA Microelectronics Equipment Co., Ltd, Beijing, 100176, P.R. China

*Email: dongyunhe@naura.com

ABSTRACT

As the semiconductor industry moves to the advanced technology nodes, the process micro-loading in reactive ion etch (RIE) becomes more and more severe. It is critical to control micro-loading to ensure precise profiles among different patterns. In this paper, the mechanism of profile micro-loading in STI etch is investigated from the view of a comprehensive effect of free reactive radicals concentration and direction, passivation polymer deposition of radical etching. The micro-loading can be optimized by adjusting various process parameters properly such as source power, bias power, pressure and different etching gas ratio etc. Based on these investigations, we demonstrate an improvement on the control of STI etching micro-loading in RIE.

Keywords: STI etch, micro-loading, sidewall angle

INTRODUCTION

With the continuous development of semiconductor technology, the density of transistors has been multiplied every 18 months following Moore's law. Especially as technology evolves into the sub 40nm node, requirements on all unit modules of chip fabrication become more stringent. Shallow trench isolation (STI), used to achieve device isolation, is one of the critical structures for transistor fabrication[1]. It is found that the profile micro-loading between narrow and wide trenches in STI etch become more severe as the dimension continuously shrinks. Thus there is a compelling need to control the process loading in RIE to ensure precise patterning and to avoid performance degradation from pattern to pattern[2].

In this paper, we focus on the optimization of STI etching micro-loading especially depth loading in RIE by investigation the effect of several critical process parameters such as source power, bias power, pressure and different etching gas ratio. During the study, the mechanisms responsible for the profile and depth micro-loading will be discussed together.

EXPERIMENTAL

Patterned test wafers with a standard STI film stack consisting of photoresist, anti-reflective coating, hard mask, pad oxide and silicon were used to demonstrate different patterns and for the study of micro-loading effect of STI etch. The experiments were carried out in a 300mm inductively coupled plasma (ICP) reactor, NAURA 612C Series. In this system, process gases, a mixture of Br⁻ based gas/Cl⁻ based gas/CF⁻ based gas/O_2, are injected on top of source through mass flow controllers.

After the etch, the characterization of profile depth and sidewall angle (SWA) on pattern wafers is performed on an optical critical dimension (OCD) measurement in a non-intrusive way.

RESULTS AND DISCUSSION

Mechanism of micro-loading in STI etch

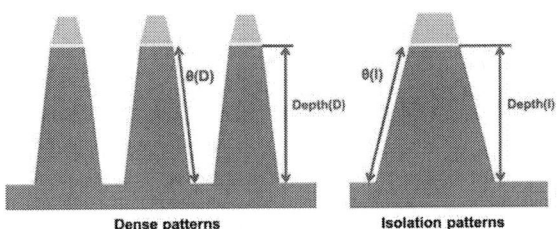

Fig.1. Dense and isolation patterns and parameters adopted for process loading evaluation

The profile micro-loading is usually expressed by the difference between dense and isolation areas. As illustrated in Fig. 1, depth loading and profile loading (sidewall angle loading) is demonstrated respectively by depth(I)-depth(D), θ(I)-θ(D). A smaller micro-loading has a beneficial effect on the device performance. Basically, the loading effect in RIE comes from its inherent local environment discrepancy on the contents of reactant and by-product around different patterns[2,3]. The ISO-dense micro-loading is attributed to a comprehensive effect of free reactive radicals and passivation polymer deposition:

a.　The difference of reactive radicals concentration among dense and isolation areas.

More reactive radicals and ions are incident to the isolation patterns because of its small pattern density and large space CD. More etching amount in isolation area will be induced than dense area, resulting in ISO-dense micro-loading.

b.　The difference of arriving angle distribution of positive and neutral radicals.

Due to the acceleration of negative sheath voltage, the positive ions have a smaller angle distribution than neutral radicals, thus causing more anisotropic etch. The micro-loading will be relatively reduced when the shielding effect is no longer obvious.

c. The difference of passivation polymer deposition among dense and isolation areas.

More polymer produced by plasma etching reaction deposited at the trench sidewall and give more protection from attacking laterally at the isolation area during the etch step, etching uniformity among different patterns would be improved, thus causing an improved ISO-dense micro-loading.

Effects of etching parameters on micro-loading
1. Effect of source power

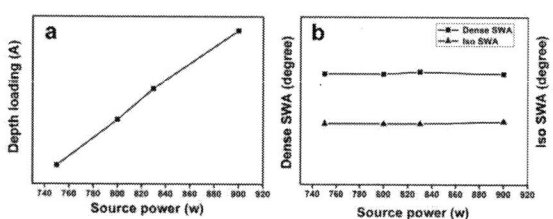

Fig.2. Effects of source power on the depth loading and profile sidewall angle

Keeping the other operating parameters (including bias power, pressure, gases and temperature) constant, we vary the depth control step source power with a broad range to study the effect of this parameter on the micro-loading. The results in Fig.2 show the source power is an important parameter for trench depth micro-loading optimization. The results also agree with the mechanism of reactive radicals concentration difference between dense and isolation patterns. As the source power increases, the depth loading is increased gradually. It is because more source power results in higher concentration of ions and neutral radicals into trenches in the isolation areas. And as a consequence it yields a faster etching rate. Thus it causes a more obvious ISO-dense depth loading. It is also worth noting that both the dense and isolation profile sidewall angles are not affected by the source power.

2. Effect of bias power

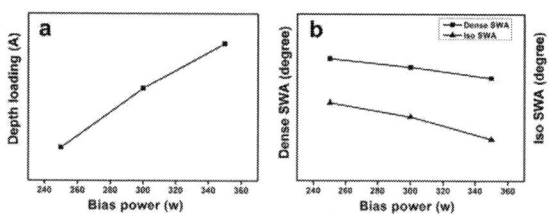

Fig.3. Effects of bias power on the depth loading and profile sidewall angle

Fig.3 shows that the bias power plays a significant role in affecting the depth loading and profile sidewall angle. It is observed that the depth loading is increased along with the increasing bias power. In the RIE process, the bias power contributes to the direction convergence of the ions, which causes more ion bombardment toward the substrate. The etch efficiency of isolation area will be more enhanced than dense area. As a result, the depth loading becomes more significant as bias power increases. In addition, the etch profile tends to be more tapered and the profile sidewall angle loading become more obvious with gradually increasing bias power, since the bombardment of heavier ions results in less isotropic etch.

3. Effect of etch pressure

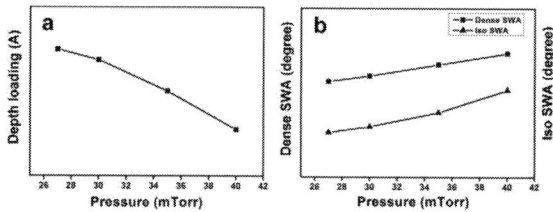

Fig.4. Effects of etch pressure on the depth loading and profile sidewall angle

The chamber pressure at the etch step is varied to investigate the effect on the profile of features. As revealed in the trend chart of Fig.4, the depth loading is reduced and dense and isolation profile sidewall angle become larger gradually as the etch pressure increases. At a higher pressure, the probability of reactive radicals collide with the ions is higher, isotropic etching becomes more prominent. The sidewall will be more pulled back continuously during etching, resulting in an enhanced lateral etch and a larger sidewall angle. At the same time, the higher pressure leads to more difficult polymer removal and more by-products deposited at the trench sidewall. It results in that the etch rate tends to slow down. Such change would be more imposed on the isolated sidewall areas, thus the ISO-dense micro-loading would be reduced.

4. Effect of Br^- based gas/Cl^- based gas ratio

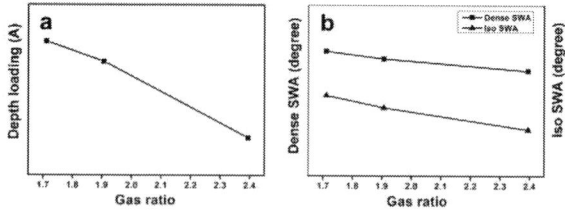

Fig.5. Effects of Br- based gas/Cl- based gas ratio on the depth loading and profile sidewall angle

Fig.5 displays the correlation between micro-loading and Br^- based gas/Cl^- based gas ratio. The increasing Br^- based gas/Cl^- based gas ratio is thought to provide more Br^- reactive radicals for the formation of sufficiently thick $SiBr_xO_y$ passivation, so that the mechanism of passivation polymer deposition will dominate the etch process. It results in the reduction of micro-loading. At the same time, the profile tends to be more tapered correspondently.

STI etching micro-loading optimization

Based on the above investigations, low source power, low bias power, high pressure and a little higher Br- based gas/Cl- based gas ratio can be adopted during STI etch process. It is demonstrated that the depth loading between dense and isolation patterns has been drastically improved.

CONCLUSIONS

In this paper, a systematic etch study has been presented in order to achieve low micro-loading among different patterns by varying source power, bias power, pressure and Br^- based gas/Cl^- based gas ratio during the STI etch process in RIE. It is found that the following process parameters have significant influences on micro-loading. As source power or bias power decreases, the depth micro-loading decreases gradually. With etch pressure increases, the depth micro-loading can be reduced. In addition, high Br^- based gas/Cl^- based gas ratio also has a beneficial effect on the improvement of micro-loading. And especially, It is observed that the sidewall angle is very sensitive to these etch process parameters except source power. Based on these investigations, an optimized etch condition could be selected to achieve an improved STI profile as well as appropriate sidewall angle among different patterns.

ACKNOWLEDGEMENTS

The authors would like to gratefully thank SMIC for sample characterization and the team members of the Etch department in NAURA for providing the technical discussion and guidance ideas during the experiments.

REFERENCES

[1] R. Zhang, et. al., Nanoscale Integrated Circuits – The Manufacturing Process, Beijing, 2014, pp. 209.

[2] D. Wang, Y. Zhang, Q. Han, Q. He, Y. Fu, M. Yang and H. Zhang. ECS Transactions., vol. 60(1), 2014, pp. 349-354.

[3] Q.Li, X. Chen, Y. Zhang. Paper.edu.cn, 2014.

PATTERNING TECHNOLOGY OPTIONS FOR FUTURE SCALING

Kenichi Oyama

TOKYO ELECTRON TECHNOLOGY SOLUTIONS LIMITED

650 Mitsuzawa, Hosaka-machi, Nirasaki City, Yamanashi, Japan

Email: Kenichi.Oyama@tel.com

ABSTRACT

Self-aligned multiple patterning technique has enabled the further down scaling through 193 immersion lithography extension [1-4]. In particular, focus on the logic device scaling, we have finished the verification of patterning technology of up to 10nm node [5-6], we will discuss about some patterning technologies that are required to 7nm node. For critical layers in FinFET devices that presume a 1D cell design, there is also a need not just for the scaling of grating patterns but also for pattern cutting process. In 7nm node, cutting number increase in metal or fin layer, and also pattern splitting of contact or via is complicated, so both cost reduction and process controllability including EPE are strongly required. This paper presents the possibility of immersion-based multiple patterning techniques and EUV lithography applicability for up to 7nm node.

INTRODUCTION

Lithographic scaling that conforms to the Rayleigh equations has reached the 40nm half-pitch (hp) by ArF immersion patterning, but advanced devices such as MPU logic products have already arrived sub-40nm hp in 14nm node. Development delay of EUV lithography has been accelerated the adoption of multiple patterning technique and process-friendly design layout by 193 immersion extension. However multiple patterning technology has a clear risk of cost growth and patterning fidelity accuracy degradation.

In this paper, we discuss the potential possibility of immersion-based multi-patterning extension for 7nm node FinFET. In particular we introduce about the line cutting performance for fin or gate layers, the various trench blocking processes for metal interconnect layer, and pitch shrink technique with self-aligned for via patterning.

Process key components for 1D patterning

Here are two basic types of patterning for 1-D cell design (Fig.1, Fig.2). One is line cutting for the Fin and Gate layers and the other is trench blocking for Mx. In this paper, we describe sub-10-nm hp LS process performance, 10-nm-order cut technology, and 10-nm-order block technology, and pitch shrinking, which is typically used for vias and cuts together with pattern shrink technology for cut masks and block masks. First, the finer grating patterns that are common to Fin, Gate, and Mx may be easily achieved with SAQP, which is used to extend the

existing mass production technology. Sub-10-nm hp is said to be achievable by starting with the 193 resolution limit, such as for SAQP case 2, and SAOP is a prominent candidate for 5-nm hp order patterns. (Fig.2)

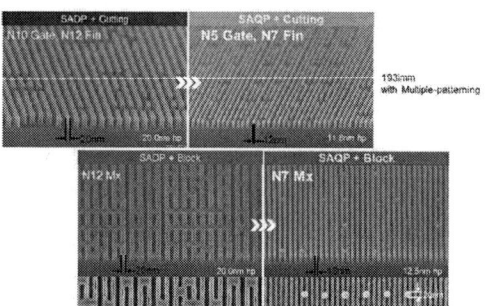

Figure 1: N7 1D layout images for Fin, Gate, Mx

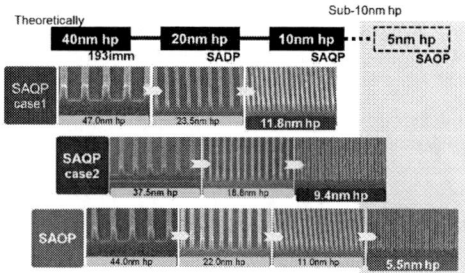

Figure 2: Promised scaling solution

N7 Via patterning assumption

We would like to take an overall view of the V0 scaling scenario (contact layer) for N7. As we can see from the trend in V0 scaling (Fig. 3), the number of splits in immersion exposure increases remarkably with critical design rules beyond N10 (N7 and N5). As the final N7 cell design, the 30-nm minimum pitch with a 12-nm hole size is required. For the five-color case shown in Fig. 3 the minimum half-pitch per split is around 50-nm hp, and the current immersion scanner using 1.35 NA provides sufficient resolution. However, the increase in cost and the overlay control are severe problems. For the four-color case, construction with the 40-nm half pitch level is possible, but that is in the limiting region of immersion resolution, so the process margin is lower and hole CD variation becomes dominant. Furthermore, for three exposures in the three-color case, we can expect an extreme reduction in the number of splits, but that requires patterning at 32nm hp, which is below the resolution limit. As we can surmise from actual resist imaging, it is extremely difficult to ensure this hole pattern separation. We introduce, here, technology that enables interpolation of patterns below this resolution limit.

At past SPIE, we touched on the feasibility of hole repairing as a motif N11 V0. [7] We use 'Eda-mame process' as the generic term for that hole repairing technology combined with hole shrink technique, shortened to simply 'Eda-mame'. Resist images for each layer after splitting in N11 are shown in the upper part of Fig. 4. The minimum hole pitch is 92 nm, and the k1 factor is 0.31. In this case, there is a design that can ensure sufficient process margin, and this can be taken as the first result that verifies Eda-mame for large surface areas. Although hole separation is not possible at the resist pattern stage in the region below the ultra-low k1 resolution limit, as described above, the Eda-mame process can shrink nearly all of the holes and at the same time ensure separation, as we see in the lower part of Fig. 13. The hole CD variation in the original pattern can also be improved by the loading effect during dry etching process. We can also see that there is almost no change in the positions of the hole centroids before and after lithography and etching, which deserves particular attention as a potentially major benefit for further scaling node.

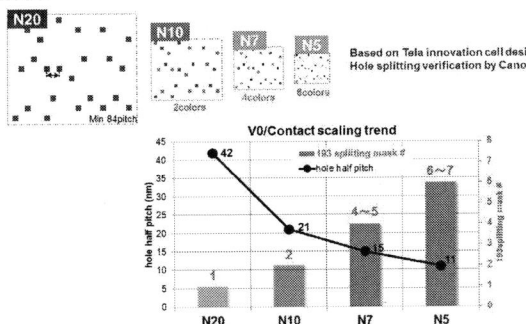

Figure 3: Via patterning trend in N7 and beyond

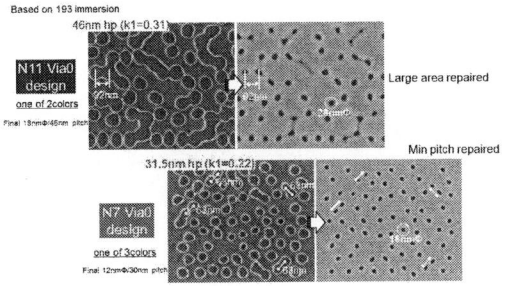

Figure 4: Hole repairing and CD healing opportunity

CONCLUSION

We have described here process technology for 7-nm node in terms of the possibility of existing immersion lithography extension. We have focused on the each critical layers of SRAM, for which fine design rules are considered to be particularly difficult, even among the general logic devices. Our conclusion is that the scaling technology for grating patterns and hole shrinking for cut masks provides sufficient process controllability for fin and gate layers. We also explored the possibility of an inversed block scheme for the Mx layer, and proposed a full inversion scheme that can ensure highly accurate transferring that is independent of the block pattern material and layout design. This is robust process technology for suppressing CD variation in the final copper wiring. Also, in application to Vx, the Eda-mame process is a very attractive technology for its ability to break through the resolution limit of immersion exposure. In addition to contributing to raising the theoretical resolution, it has the effects of expanding the process margin and reducing CD variation, and can easily be extended to the fin and gate layers and the Mx layer for the previously described full inversion approach. It may also serve as a complementary technique to EUV, the future mass production lithography technology.

ACKNOWLEDGEMENTS

The authors would like to extend their appreciation to the R&D staff of Tokyo Electron group for their support in carrying out the experiments described in this paper.

REFERENCES

[1]: H.Yaegashi, et al, "Novel approaches to implement the self-aligned spacer DP process toward 11-nm node and beyond", Proc. SPIE, 2011

[2]: H. Yaegashi, et al, "Overview: Continuous evolution on double-patterning process", Proc. SPIE, 2012

[3]: K. Oyama, et al, "Process variability of self-aligned multiple patterning", Proc. SPIE, 2013

[4]: H. Yaegashi, et al, "Sustainable scaling technique on Double-patterning process", Proc. SPIE, 2013

[5]: H. Yaegashi, et al, "Recent progress on multiple-patterning process", Proc. SPIE, 2014

[6]: K. Oyama, et al, "Robust complementary technique with multiple-patterning for sub-10nm nm node device", Proc. SPIE, 2014

[7]: K. Oyama, et al, "Sustainability and applicability of Spacer-related patterning towards 7nm node", Proc. SPIE, 2015

DUMMY GATE CRITICAL DIMENSION UNIFORMITY IMPROVEMENT FOR 14NM AND BEYOND

Jie Luo[a], Dongping Zhang[a], Haiyang_Zhang[a]*

Technology R&D, SMIC Advanced Technology R&D (Shanghai) Corporation

Pudong New Area, Shanghai, P. R. China 201203

+8621 38610000x15973, Jie_Luo@smics.com

*Corresponding Author's Email: *Steven_Z@smics.com

ABSTRACT

In advanced CMOS technology node, the dummy gate critical dimension (CD) and its line head to head (HTH) space CD will directly affect the device performance and its long term reliability. Macro- and micro-uniformity of dummy gate CD and its HTH CD are key parameters for wafer yield and product quality. In dummy gate process, both the thickness uniformity of film stack (silicon nitride, amorphous carbon and etc.) and the CD uniformity (CDU) of lithography line will influence final dummy gate CDU. Feed forward APC mode for dummy gate CDU improvement requests a long term work in semiconductor manufacturing which is not a focus point here. In this paper, an advanced temperature compensation function of a commercial inductively coupled plasma (ICP) etcher will be mentioned. At the same time, both dummy gate and HTH CDU tuning by this function with feedback mode will be introduced. Some possible enhanced solutions of CDU improvement will also be addressed in this paper.

INTRODUCTION

Nowadays, most CDU improvement tasks of dummy gate process are focused on lithography module with DoseMapper (DoMa). However, poor illumination condition might be demonstrated on the scribe line between different illuminated blocks. At the same time, though pitch etch bias (TPEB) variation within each block and other uncontrollable shift are also strongly limiting DoMa's effectiveness in 14 nm technology node and beyond.

For ICP etcher, traditional CDU controlling methods are limited radial tunable functions, which include source power distribution, gas flow distribution and multi-zone electrical static chuck (ESC) tuning, shown in Fig.1. In order to achieve a more precise CDU improvement with azimuthal tunable function, an advanced ESC system within more than one hundred heaters is introduced. In this paper, we will study dummy gate critical dimension uniformity improvement effect using this advanced function [1].

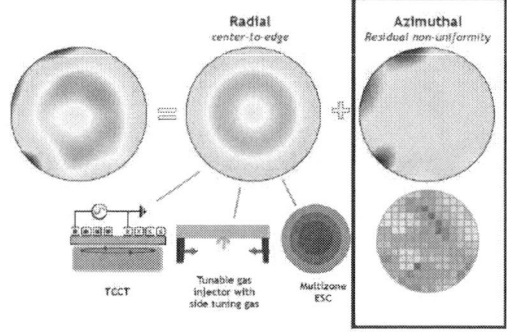

Fig. 1 An advanced CDU tuning system with an advanced temperature controlling function. For this kind of etcher, CDU tuning is not limited in radial distribution. Azimuthal direction's uniformity improvement also become possible in etching process.

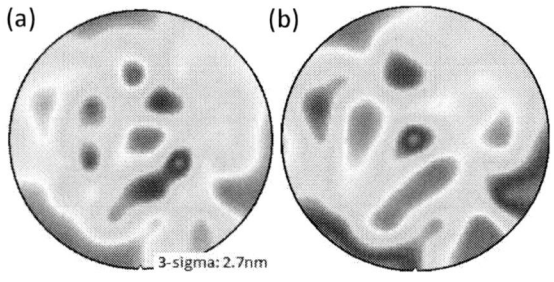

Fig.2 An importing map of poly CD initial data and its temperature compensation are demonstrated. (a) The uniformity of poly CD can be barely improved by traditional radial tuning functions. (b) The temperasure compensation map is sumilated by hydra software with range around 5 degrees Celsius.

By importing initial data, hydra will simulate a temperature compensation map, shown in Fig. 2(b). CDU map with unsmooth CD variation and varied etch bias within wafer could be easily overcome by this advanced CDU improvement system. At the same time, the short-comings of hydra are obvious as listed. 1) Temperature sensitivity has to be a key factor in etch tuning step. 2) The tunable temperature range is limited. 3) Wafer to wafer and lot to lot's long term process stability directly impact the feedback mode CDU improvement performance.

978-1-5386-5309-8/18 $31.00 © 2018 IEEE

EXPERIMENT

In this paper, CDU improving with feedback compensation mode is focused on. A compensation map after poly etching is requested using this method. Two related metrology methods, critical dimension SEM and optical critical dimension (OCD), are usually applied for poly CD measurement. By considering both the tool stability and measurement accuracy, metrology of OCD is used in this study. A dose of maps are collected to average a compensation map as the incoming.

The phase one poly CD compensation map is shown in Fig.3(a) which is long term average data of more than one hundred wafers. At the same time, a map which is sited only on center zone is researched as phase two, shown in Fig.3(b).

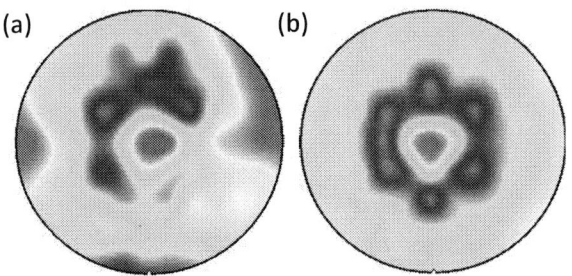

Fig.3 Poly CD maps which are provided by OCD tool. (a) The phase one poly CD compensation map is delivered by long term average measurement data. (b) The phase two poly CD compensation map is a partial map of phase one to minimize wafer to wafer variation.

The temperature sensitivity of poly CD is another key parameter for this advanced function. A trend chart, shown in Fig.4, is figured out by exploring the relationship between ESC temperature and poly line CD. A linear relationship with ESC temperature and OCD inline CD, which is suitable for our mentioned advanced temperature compensation method, can be easily demonstrated in Fig 4.

By combining the compensation maps and the temperature sensitivity of poly CD, this advanced system will automatically simulate a temperature compensation map to improve the poly CDU. For a 12 inch wafer, the locally temperature sensitivity would not perfectly meeting its global trench. As a result, sensitivity calibration is needed which is briefly introduced in following section.

Temperature Trend

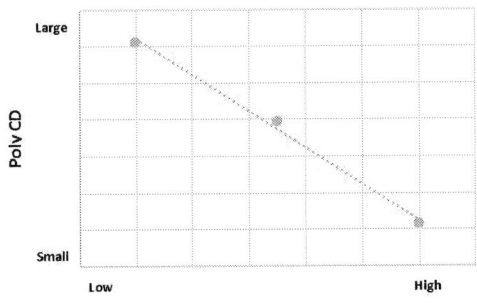

Fig.4 ESC temperature is adjusted +/-5 degrees Celsius to trace the variation of poly CD. From this trend chart. A linear

relationship with ESC temperature and OCD inline CD, which is suitable for our mentioned advanced temperature compensation method, can be easily demonstrated.

	Temp I/MI/MO/O	Poly CD
Split1	a/a/a/a	B
Split2	a+b/ a+b/ a+b/ a+b	B+C
Split3	a+2b/ a+2b/ a+2b/ a+2b	B+2C

Table 1 The ESC temperature contribution and splits are demonstrated in this table. Poly CD results of different ESC temperature are in the last row

RESULTS & ANALYSIS

After several rounds, the sensitivity is sited around 2nm per degrees Celsius which is far from what we simulate initially. The root cause of this case is micro-environment effect of poly etching process [2]. For the same temperature compensation map, different sensitivities perform totally different outcome which can be easily figured out in Fig.5.

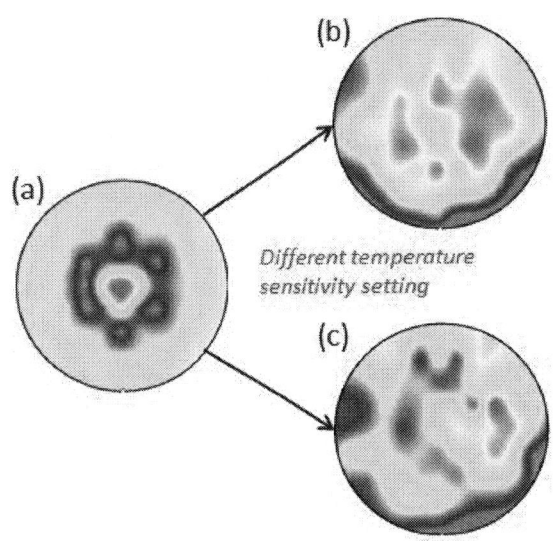

Fig.5 The poly CDU improvement results of same compensation map with different import temperature sensitivities. (a) A map of center zone temperature compensation. (b) The CDU improvement result with the sensitivity of p nm per degrees Celsius. (c) The CDU improvement result, 3-sigma is improved 30 persentage with the sensitivity of q nm per degrees Celsius.

According to the results in Fig. 5, the temperature sensitivity setting of poly CD is significantly impacting the compensation result. An overshooting can be easily observed as a consequence of temperature sensitivity mismatch. A nearly 2.0nm per degrees Celsius sensitivity difference between the temperature trend experiment and the CDU improvement test is obvious. The root cause for this phenomenon is a global and local temperature non-uniformity distribution effect. Due to this effect, the temperatures of isolation points on compensation map need to be much higher than the global ones. And it leads the temperature sensitivity mismatching in this case.

CONCLUSION

CDU improvement with temperature compensation by hydra function is an effective CDU improvement method for etching process. The temperature sensitivity and incoming compensation map are two key tuning nodes for this process. The incoming compensation map should be carefully chosen by long term stable inline data. The temperature sensitivity cannot be simply chosen by inline correlation curve for global and local temperature distribution effect. Different sensitivity value has great influence on final output CDU data. As a whole, this advanced temperature compensation system can overcome several shortages of DoMa compensation. From the other point of view, the poly CDU improvement is limited to 3-sigma around 1.5 times of DoMa compensation result based on current study. However, this advanced temperature compensation system still shows great potential to solve dummy gate CDU issue especially in future smaller CD node.

In the future, feedback mode with this advanced temperature compensation system which is used to overcome wafer to wafer variation will be our future focus point.

ACKNOWLEDGE

This work has been supported by Semiconductor Manufacturing International Corporation. Authors would like to acknowledge Yan Wang and Yuchen Li for their contributions to this project. I also like to deliver my sincerely thanks to Kefang Yuan for her kindly support.

REFERENCES

[1] Liu Jian Gang, *Lam Conductor Etch. Lam Research Corp.* September 2017.

[2] Samer Banna and Ankur Agarwal, *Pulsed high-density plasmas for advanced dry etching processes.* Applied Materials Inc.,974 E. Arques Avenue, M/S 81312, Sunnyvale, California 94085. May 2012.

Zero Lag Dispense to Increase the Etching Uniformity in a Single Wafer Wet Cleaner

Wei Liu, Xiaoyan Liu, and Yi Wu

Beijing NAURA Microelectronics Equipment Co., Ltd.
Beijing Tianzhu Export Processing Zone, ShunYi District, P.R.China
Phone: 010-69458732 and Email: liuwei@naura.com

Biography

Wei Liu, PhD obtained in Shanghai Institute of Ceramics, Chinese Academy of Science, major in material physics. Since August 2011, she has been working in Beijing Sevenstar Electronics Co., Ltd.(now is NAURA), IC research center as a senior process engineer, mainly working for 28nm FEOL wet clean process development on 300mm wafers.

Abstract

With the increase of the wafer size, it becomes more and more challenging to control the etching rate and etching non-uniformity precisely in the wet process of semiconductor manufacturing. In this paper, the effects of the DHF (diluted hydrofluoric acid) starting dispensing point, the dispensing time, and the wafer clean time on etching rate and non-uniformity of the silicon oxide wafer were investigated. The results indicated that the first reaction of DHF with silicon wafer surface plays an important role to the final etching rate and non-uniformity. Based on the analysis, a new method of zero lag dispense was proposed and tested to reduce the non-uniformity of etching. The method uses a special dispensing device to inject DHF on the surface instantaneously and uniformly before it is dispensed from the swinging nozzle. The non-uniformity within a wafer reduces to 1.8% under DHF process (1:220, 20℃, 60s).

Keywords—DHF; Zero lag dispense; Etching rate; Etching uniformity

Introduction

As semiconductor devices continue to increase integration complexity and reduce in design size, it requires a very clean and flat silicon surface in the process of manufacturing[1,2]. Particularly, for fabricating a silicon-on-insulator (SOI) wafer, the thickness variation of the silicon dioxide film must be well controlled[3]. Although the film thickness of silicon dioxide can be adjusted by various oxidation process conditions, the wafer surface still should be effectively improved after the oxidation. The single wafer wet etching method is a useful technique for this purpose[4]. Particularly the wet etching method on a single wafer has a function for choosing the region to be deeply etched on the wafer, by means of adjusting the motion of swinging nozzle which injects the etchant[5]. While with the increase of wafer size, it becomes more and more challenging to control etching uniformity precisely[6].

In this paper, the wet etching process of a 300 mm single

wafer cleaner was experimentally designed. The effects of the DHF starting dispensing point, the dispensing time, and the wafer clean time on etching rate and etching non-uniformity of the silicon oxide wafer were investigated. Based on the analysis, a new method of zero lag dispense was proposed and tested to reduce the etching non-uniformity.

Experiment

Materials and methods

Fig.1 shows the schematic diagram of zero lag dispense in a single-wafer wet etcher used in this study. The zero lag dispense was that, before DHF was dispensed from the swinging nozzle 1, 500mL DHF using a special device (nozzle 2) was instantaneously and uniformly injected on the wafer surface. DHF was then injected downward from the 4-mm diameter nozzle 1 to the 300-mm diameter wafer surface at the flow rate of 1 L/min.

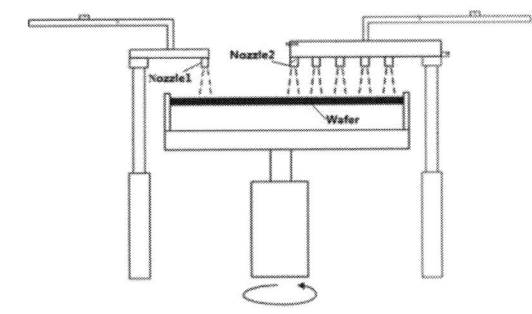

Fig.1.The sketch map of zero lag dispense in single-wafer wet etcher.

The wafer used in this study had a 100-nm thick silicon dioxide film, formed by thermal oxidation. HF (49%w/w in water) had a purity grade which was standard for the ULSI industry.

To decrease the experimental error, the wafers were pre-etched with HF in ultra-pure water (UPW) of 1:200 ratios for 60s to remove the contamination and native oxide on the surface. After pre-etching, the wafers were etched using 1:220 DHF at 20℃. In order to obtain the etching rate, the silicon dioxide film thickness was measured using SFX200 (KLA-Tencor, USA) before and after etching. The measured wafers were tested by a 49-point scan with 3 mm of edge exclusion. In order to obtain statistic results precisely, three wafers were evaluated at each condition.

The etching non-uniformity (ENU) within a wafer was

978-1-5386-5309-8/18 $31.00 © 2018 IEEE

calculated according to equation (1),

$$ENU = \frac{3\sigma}{ave} \times 100\% \qquad (1)$$

Here ave and σ are the average etching rate and standard deviation of etching rate, respectively.

Experimental design

The main etching process and parameters are shown in TABLEI. The etchant arm swung with a parabola motion on the surface. The velocity of DHF arm was controlled from 0 m/s to 15 m/s and finally to 0 m/s from edge to center and then to the other edge of the wafer in one cycle. Normally, the etchant was dispensed from the swinging nozzle and the wafers were cleaned only one time.

TABLE I.
THE MAIN ETCHING PROCESS AND PARAMETERS

Proc	Name	Dispense time [s]	Chuck speed [rpm]	Flow [L/min]
1	UPW	20	100	1.5
2	DHF	60	200	1
3	UPW	40	400	1.5
4	N_2	32	1700	10

Based on TABLE I, the effect of the DHF starting dispensing point, the dispensing time, and the wafer clean time on ER and ENU of a silicon oxide wafer were investigated. Based on the results, a new method of zero lag dispense, using a special device that DHF was instantaneously and uniformly injected on the surface before DHF dispensed from a swinging nozzle, was proposed and tested to reduce the etching non-uniformity. The parameters of the four designed experiments were shown in TABLE II. In experiment 3, one group of wafers were cleaned only one time that DHF was dispensed 20s, and the other group of wafers were cleaned 5 times that DHF was dispensed 4s each time.

TABLE II.
THE PARAMETERS OF THE DESIGNED FOUR EXPERIMENTS

Exp	DHF starting dispensing point	DHF Dispensing time [s]	Wafer clean time	DHF dispensing method
1	center/ edge	60	only one	normal
2	center	15/30/45/60/75/ 90/200/ 300	only one	normal
3	center	20	only one /5 times	normal
4	center	60	only one	zero lag dispense

Results and discussion

The effect of DHF starting dispensing point on ER and ENU

The results of the average ER and ENU with different DHF starting dispensing points were listed in TABLE III. The

distributions of etching rate on silicon surface with DHF starting dispensing points were shown in Fig. 2.

TABLE III.
THE ETCHING RATE AND ETCHING NON-UNIFORMITY WITH DIFFERENT DHF STARTING DISPENSING PIONTS

DHF starting dispensing point	ER [Å/min]	ENU [%]
center	10.28	2.72
edge	10.30	2.83

From TABLE III, it can be seen that ER and ENU in the different process cases in which DHF started dispensing from wafer center and from wafer edge, were almost the same. ER was about 10.30 Å/min and ENU was about 2.7-2.8%.

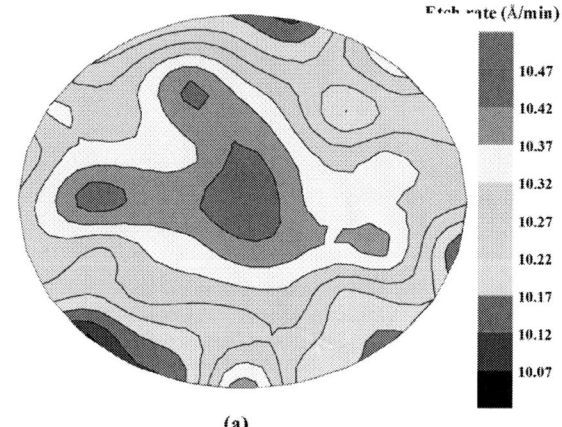

(a)

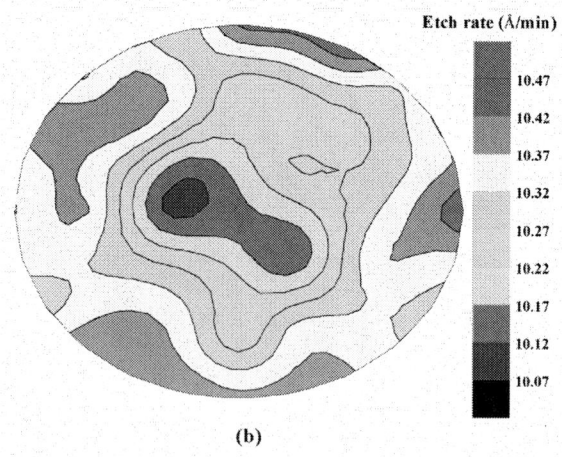

(b)

Fig.2 The distributions of etching rate on silicon surface with DHF start dispensing points from wafer center(a) and from wafer edge(b).

As exhibited in Fig.2(a), when DHF started dispensing from the wafer center, the etching rate of the wafer center (red region) was obviously higher than that of the wafer edge (blue region); while when DHF started dispensing from the wafer edge, as shown in Fig.2(b), the results were just opposite. The reason was maybe due to the change of the first reaction position of DHF on silicon wafer surface. DHF solution was transported along the rotating wafer surface by means of swinging nozzle. Since the diameter of DHF nozzle was quite small, only 4-mm, it needed at least 4s that DHF covered with the whole wafer surface (the first reaction time of DHF with silicon wafer). Before the wafer surface wholly covered with DHF solution, at the position that DHF started dispensing, the wafer surface first reacted with the freshest etchant, which resulted in the etching rate was the highest on the surface.

The effect of DHF dispensing time on ER and ENU

The results of the average ER and ENU with different DHF dispensing times were shown in Fig.3. As can be seen from Fig.3, with the increase of DHF dispensing time, the ER decreased from 11.42Å/min at 15s to 10.08Å/min at 300s, and the ENU also gradually degraded from 5.60% to 1.51%. That might because the effect of the first reaction phenomenon on the etching properties decreased. Before the whole wafer surface covered with DHF (about 4s), the surface reacted with the freshest DHF, so the etch rate was very high. While after the whole wafer surface covered with DHF, the DHF concentration on the silicon surface decreased than that of the first 4s due to the consumption by the silicon dioxide reaction, so the etching rate became lower than that of the first reaction. In the first reaction 4s, the etching rate at the wafer center where DHF started dispensing was higher than the position wasn't covered with DHF. The etching rate within wafer was quite different, so the etching uniformity was not good. While after the whole wafer surface covered with DHF, the DHF concentration on the surface was nearly the same, so the etching rating within wafer was quite uniform. That is to say, with the increase of DHF dispensing time, the ratio of the first reaction 4s to the total DHF dispensing time (from 15s to 300s) decreased, the effect of the first reaction phenomenon on the etching properties degraded, so the average etching rate decreased, the etching rate uniform within wafer became better.

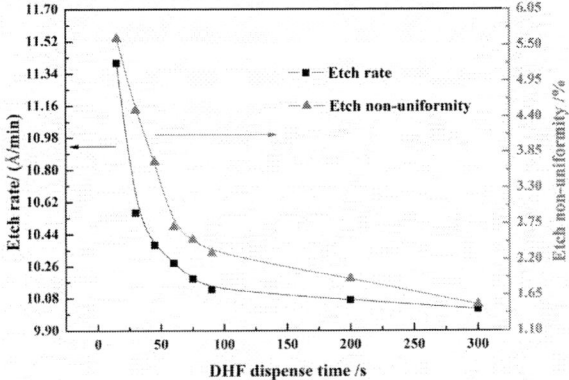

Fig.3 The etching rate and non-uniformity with different DHF dispensing times

Based on the above results, we deduced that the first reaction of DHF with silicon wafer surface played an important role to the final etching rate and non-uniformity. To confirm the conclusion, the additional experiments were carried out.

The effect of wafer clean time on ER and ENU

The results of etching rate and etching non-uniformity with different wafer clean times were list in TABLE IV. The distributions of etching rate on silicon surface were shown in Fig. 4.

TABLE IV.
THE AVERAGE ETCHING RATE AND ETCHING NON-UNIFORMITY WITH DIFFERENT WAFER CLEAN TIMES

Wafer clean time	Methods of DHF dispensing	ER [Å/min]	ENU [%]
only one	20s only one time	10.98	5.03
5 times	4s each time and repeated 5 times	11.87	10.99

From TABLE IV, it can be seen that the etching rate (11.87 Å/min) that cleaned 5 times was much higher than that (10.98 Å/min) cleaned only one time, and the etching non-uniformity greatly increased from 5.03% to 10.99%.

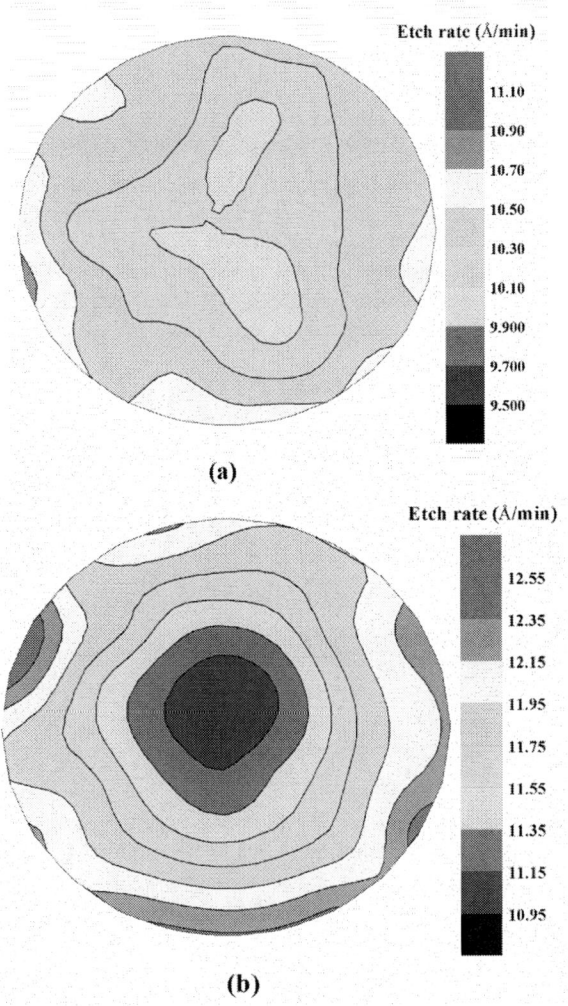

(a)

(b)

Fig.4 The distributions of etching rate on silicon surface with DHF dispensing 20s only one time(a) and DHF dispensing time 4s each time and repeated 5 times(b).

As exhibited in Fig. 4(b), when wafers were cleaned 5 times, there were many concentric circle contour lines and 8 kinds of colors (from red to blue) separated by the same etch rate interval on the wafer surface. The span of etching rate between the maximum and the minimum was quite large, about 1.6 Å/min, and the etching rate within a wafer was quite non-uniform. While when wafers were cleaned only one time, as shown in Fig. 4(a), the etching contour lines decreased, and there were only 5 kinds of colors. The span of etching rate between the maximum and the minimum became smaller, about 1 Å/min, and the etching non-uniformity within a wafer degraded. Although the total DHF dispensing time of the two methods was the same, the method that the wafers were cleaned 5 times was equivalent to enlarge 5 times of the first reaction time of DHF connected on silicon wafer surface. So the etching rate increased and the non-uniformity decreased. And then the experiment proved our conclusion that the first reaction of DHF with silicon wafer surface played an important role to the final etching rate and non-uniformity.

From the above three experiments, it can be inferred that there were two ways to reduce the etching non-uniformity in a single wafer cleaner, by decreasing DHF concentration or enlarging the etching time of DHF on the silicon wafer each time. But both methods will add the manufacturing cost due to the lower process efficiency. To solve this problem, a new method was proposed and tested as below.

The effect of zero lag dispense on ER and ENU

Before DHF was dispensed from the swinging nozzle, 500mL DHF using a special device was instantaneously and uniformly injected on the wafer surface.

The average, the maximum and the minimum of ER and ENU of each wafer were listed in TABLE V. The distribution of ER on silicon dioxide surfaces was shown in Fig. 5.

TABLE V.
THE AVERAGE ETCHING RATE AND ETCHING
NON-UNIFORMITY WITH ZERO LAG DISPENSE

N	Ave ER [Å/min]	Max ER [Å/min]	Min ER [Å/min]	ENU [%]
1	10.49	10.67	10.42	1.86
2	11.78	11.91	11.61	1.81
3	10.65	10.81	10.49	1.89

As can be seen from TABLE V, using zero lag dispense method, the etching rate of all the wafers increased than that normally cleaned, which might be due to the increase of DHF solution quantity in the DHF dispensed 60s. While the etching non-uniformity gradually decreased to 1.8%, which was comparable to the data of "Dainippon Screen SU3200" single wafer clean tool ever reported. From Fig.5, it was shown that there was no obvious area with very fast or very low etching rate, which indicated that this zero lag dispense technology was effective to improve the etching uniformity.

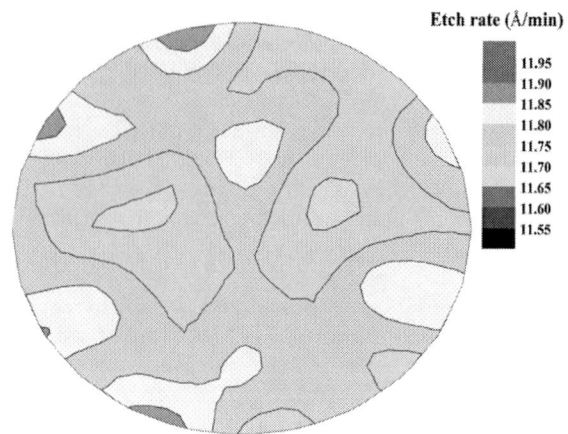

Fig.5.The distribution of etching rate on silicon surface with zero lag dispense.

Conclusions

The effects of the DHF starting dispensing point, the dispensing time, and the wafer clean time on ER and EUN of silicon oxide wafers were investigated. The results showed that 1) With the same dispensing time, where DHF started dispense, where the ER was higher. 2) With the increasing of DHF dispensing time, the ER and the ENU both decreased. 3) The ER of wafer cleaned 5 times was much higher than that of cleaned only one time, and the ENU greatly increased, which may be ascribed to the longer time that the first reaction of DHF with silicon dioxide wafer surfaces. Based on the analysis, a new method of zero lag dispense, was proposed and tested to reduce the ENU. Before DHF dispensed from the swinging nozzle, DHF was instantaneously and uniformly injected on the wafer surface. The ENU reduced to 1.8%, which was comparable to the ever reported.

References

[1] G. Choi, "Necessity of cleaning and its application in future memory devices,"Solid State Phenomenal, vol. 219, pp.3–10, 2015.
[2] Y. Ogawa, "Cleaning technology for advanced devices beyond 20 nm node,"Solid State Phenomenal,vol.195, pp. 7–12, 2013.
[3] W. Cheng, A. Teramoto, and T. Ohmi, "Very high performance CMOS on Si using radical oxidation technology and accumulation-mode SOI device structure," Journal of the Electrochemical Society, vol. 157,pp. 389–393, 2010.
[4] H. Lee, D. Ahn, S. Lim, and T. Kim, "Effect of DI-water dilution and etchant arm movement on spinning type wet etch,"Solid State Phenomenal,vol. 219, pp. 125–127,2015.
[5] B. Yu, S. Huang, M. Yeh, C.C. Chen, S.M. Jang, A. Ratkovich, D. Yang, and J. Lauerhaas,"Novel wet etching of silicon nitride in a single wafer spin processor,"Solid State Phenomenal,vol. 195, pp.46–49, 2013.
[6] B. Yu, F. Ku, C. Taft, A. Larrea, J. Lin, T. Hayashi, A. Morita, K. Arai, and H. Naohara, "Challenges and solutions for 450nm FEOL wet clean tool, ECS Transactions, vol. 58, pp. 87–92, 2013.

STUDY OF DOWNSTREAM CF4 CONTAINED PLASMA PROCESS IMPACT ON CHAMBER CONDITION

Yali Fu[1,2]*, Shawming Ma[3], Yi Wang[1], Ken Wang[4], Mingjie Jiao[2], Nancy Zhang[2], Vijay Vaniapura[3], Hai-Au Phan-Vu[3] and Bob Elliston[3]

[1]Peking University, School of software and microelectronics, Beijing, China
[2]Mattson Technology, China Field Process Group, Beijing, China
[3]Mattson Technology, Plasma Product Group, Fremont, California, USA
[4]SMNC Etch Department, Beijing, China
*Corresponding Author's Email:Yali.fu@mattson.com

ABSTRACT

CF$_4$ gas chemistry is widely used in plasma strip process as addition to other main chemistry such as oxygen (O$_2$) and nitrogen (N$_2$) because of its benefit of removing Si contained residue or some other inorganic crust layers. However, it is not widely accepted due to the concern for the chamber ash rate shift during mixing run conditions. In this article, low ratio CF$_4$ addition downstream plasma impact is studied on the chamber ash rate with substrate temperature <300 °C. It is found that CF$_4$ contained process impacted the chamber condition and cause ash rate shift of O$_2$/N$_2$ or O$_2$/4%H$_2$-N$_2$ Forming Gas (FG) ash process. The CF$_4$ gas ratio, running accumulation sequence and chamber post condition impact on the chamber ash rate are analyzed. The results from this study also provide one effective way to avoid ash rate drop post CF$_4$ contained process.

Keywords—strip, post etch treatment, CF$_4$, ash rate shift, downstream plasma

INTRODUCTION

Plasma strip used to be a less demanding process step with wide process window, but it has become more and more complex and critical since 28nm and beyond, not only in logic process but also in NAND or NOR process. Typical plasma strip process steps include post implant photoresist strip, post etch photoresist strip [1,2], polyimide strip, film surface treatment, etc. Fig. 1 shows a typical scanning electron microscope picture of photoresist structure after ion implantation before plasma strip. Fig.2 shows a typical post etch structure with inorganic crust or polymer on the sidewall.

Traditional O$_2$/N$_2$/H$_2$ process combined with soft WET clean can not remove all crust with Si, or inorganic contained polymer. Low ratio CF$_4$ addition is usually needed to get clean results. However, it is not widely accepted due to the concern for the chamber ash rate shift. Little study was done on the CF$_4$ impact of strip chamber condition, and no solution was provided in literature for the chamber maintenance of CF$_4$ contained plasma strip process.

In this work, CF$_4$ gas ratio and cycling accumulation sequence impact on chamber condition was studied. The chamber condition was checked by non-CF$_4$ process ash rate. Ash rate was found impacted by CF$_4$ addition process , and the change is feasible to be reversible. Post condition process is one good solution to keep the chamber to maintain stable ash rate.

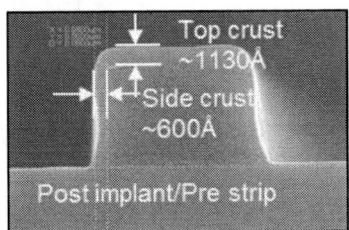

Fig.1 Typical crust formation on photoresist after high dose ion implantation and before plasma strip process step.

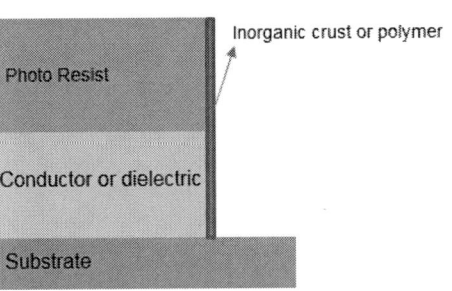

Fig.2 Typical post plasma etch structure with inorganic crust or polymer on the sidewall.

EXPERIMENTAL

This work was performed in a Suprema XP™ downstream plasma tool from Mattson Technology, as shown in Fig. 3. Downstream plasma reached the wafer with dominant neutral reactive species and most ionic species were screened out by the charge separation grid. With a Faraday shield, capacitive coupling and high ion energy peak can be minimized. The test sample sat on a heated pedestal, and the temperature was controlled by an imbedded heater.

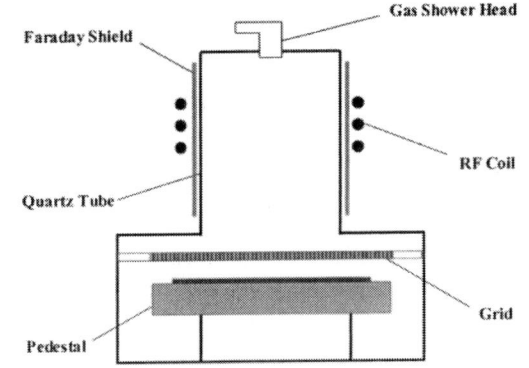

Fig.3 Schematic diagram of Suprema XP downstream plasma reactor.

978-1-5386-5309-8/18 $31.00 © 2018 IEEE

The O_2/N_2 strip ash rate was tested on two heads in one chamber with 300mm photoresist (PR) blank wafers to check the chamber condition.

CF_4 process running impact on the chamber wall was first studied in Test A by adding 1% total flow of CF_4 into O_2/N_2 mixed process. Collected O_2/N_2 ash rate Pre/post CF_4 contained process running to check the chamber condition impact, and repeated the test to check the variation.

Further CF_4 ratio and cycling accumulation effect was studied in Test B with the procedure as shown in Fig.4. Six different CF_4 ratio splits were done in random order, details of CF_4 ratio and recipe gas flow can be found in Table I. The test sequence was in the following order: Split 3 -> 1-> 2-> 5-> 6-> 4-> 1-> 5-> 2-> 3-> 4-> 6-> 5-> 2-> 4-> 6-> 1-> 3.

There was no post condition in the process of both Test A and Test B, which means no wafer-less dry run after process. The post condition was auto done by each 25pcs wafers' running in the Test C and Test D.

The post condition gas splits were tested in Test C to choose the most efficient one for chamber recovery. Then the marathon test was followed in test D to verify the effect after continuous CF_4 addition process running.

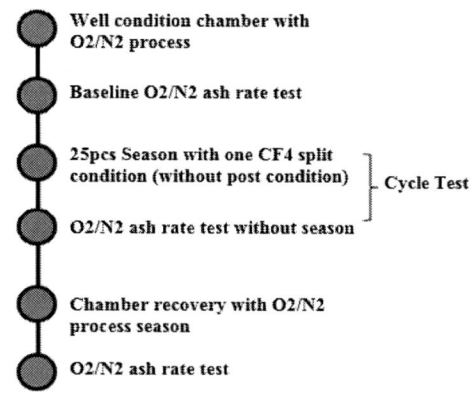

Fig.4 Schematic diagram of chamber ash rate test procedure

TABLE I
CF_4 SPLIT TEST CONDITION

Split ID	CF_4 gas ratio/%	Gas Flow/sccm		
		O_2	N_2	CF_4
1	0.5	3000	300	17
2	1	3000	300	33
3	1.5	3000	300	50
4	2	3000	300	66
5	2.5	3000	300	83
6	3	3000	300	99

RESULTS AND DISCUSSION

In test A, about ~5.5% ash rate drop was observed post the 25 pieces wafers' test with 1% CF_4 contained process, as showed in Fig. 5. And ~0.4% ash rate normal variation was found.

Post the continuous cycle testing with varied CF_4 ratio in

Test B, ash rate dropping was found at the beginning as expected, but no further ash rate drop was found after the CF_4 process cycling accumulation as showed in Fig. 6. And there was also no obvious ash rate uniformity change post cycling accumulation. The correlation between non-CF4 PR ash rate variation and CF_4 ratio was shown in Fig. 7. Only a bit ash rate trend down was found when CF4 ratio increased, and 0.5% CF4 ratio had less ash rate dropping impact than other higher ratio splits. During the O_2/N_2 seasoning recovery process, it was found that the ash rate was increased continuously with the accumulation of PR wafer seasoning. After ~75 pieces PR wafers' seasoning, the ash rate was increased to baseline level.

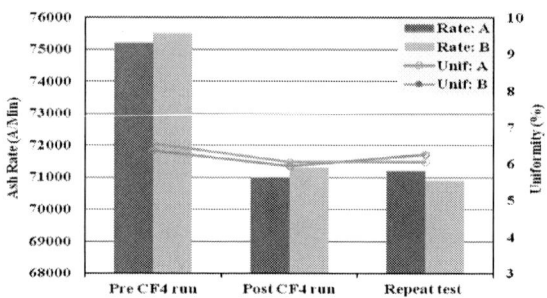

Fig.5 Ash rate test on two heads after 25pcs 1% CF_4 process test

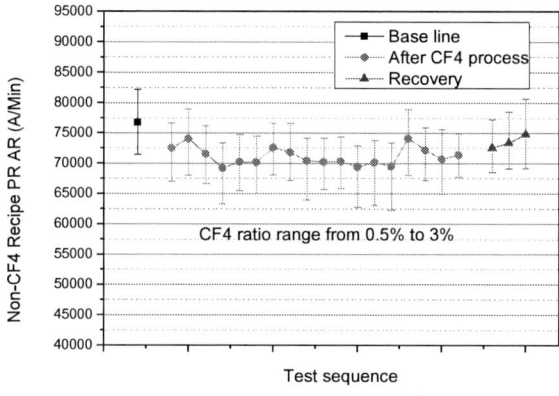

Fig.6 Average and max/min ash rate results after each 25pcs season of split condition.

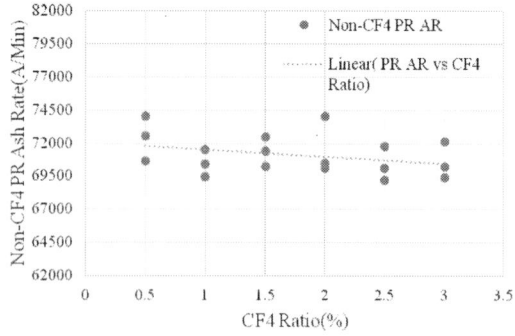

Fig.7 CF_4 ratio impact on Non-CF4 PR ash rate mean value

The CF_4 process impact on the chamber wall was suspected as the CF_4 removed the chamber wall surface passivation layers,

978-1-5386-5309-8/18 $31.00 © 2018 IEEE   218

as illustrated in Fig. 8. A well-conditioned chamber wall has polymer conditioned surface (a), after CF4 process, it will be removed and changed to state (b). Florine radical can further react with oxidized surface, to change the surface to with or without fluorinated, as shown in (c) and (d). The lower ash rate was due to increased re-combination of reactants as the chamber wall metal substrate exposed. Plasma–wall interactions play a crucial role in the evolution of the plasma properties, L. Stafford et al [3] estimated the recombination rates for Oxygen on various surfaces. The reactants loss will lead to the loss of ash rate.

The effective way to rebuild the chamber wall condition is to do chamber wall oxidation again right away. O_2/N_2 process running can recover the chamber wall condition. Seasoning with PR wafer can also help condition the chamber wall to state (a) quickly, by decreasing the effect of reactants re-combination with chamber wall. As shown in Fig. 6, continuous chamber O_2/N_2 PR seasoning can help recover the chamber condition quickly after the continuous running CF4 contained process without any post conditioning.

In order to recover the chamber condition after CF4 process, the best way is to implement by wafer post conditioning process immediately, i.e., to do wafer-less chamber auto clean. If change to other strip process after recovery, 3 to 6 pieces of PR wafers per process head can also help build up stable chamber polymer condition of (a).

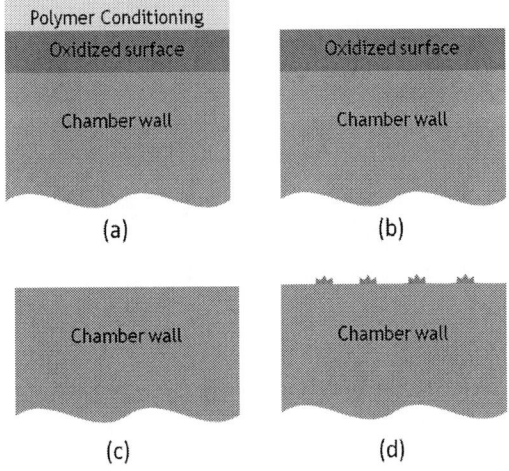

Fig. 8. Chamber condition states: (a) Chamber surface condition after PR wafer running with O_2/N_2 or O2/FG normal chemistry; (b) After polymer conditioning being removed; (c) After removal of oxidized surface film; (d) fluorinated surface film

The effect of post conditioning was also studied in Test C by testing O_2/N_2 base recipe PR ash rate after seasoning with 25 pieces 1% CF4 process and implemented post condition right away. Without any post condition, ~5% ash rate drop was found, while by lot post condition can help recover the chamber from the impact of one lot CF4 contained process running. There were several types of post condition processes, including O_2/N_2 only process, $4\%H_2$-N_2 Forming Gas only process, and the mixture of the two. All these different approaches can help keep ash rate variation <1% as shown in Fig. 9. O_2/N_2 base process and forming gas process can both help reducing

fluorinated surface film. If doing post condition with O2/FG followed by O_2/N_2, no ash rate dropping was found. O2/FG can help recover chamber condition from state (d) to (c), then oxidize the wall into state (b) showed in Fig.8.

Marathon test was done in Test D with by lot O2/FG followed by O_2/N_2 wafer-less dry run, as shown in Fig. 10. The ash rate was found kept stable for both heads.

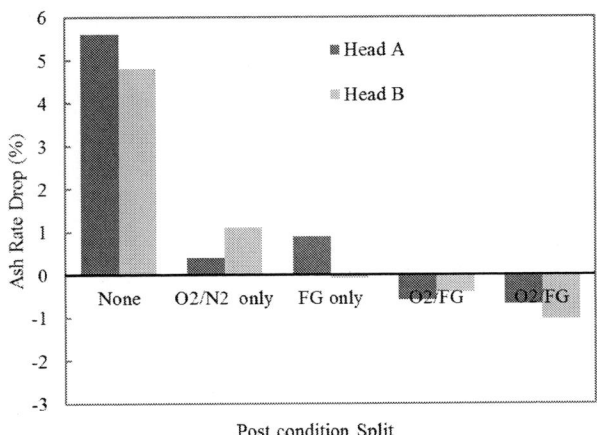

Fig.9. Post ash chamber conditioning effect comparison.

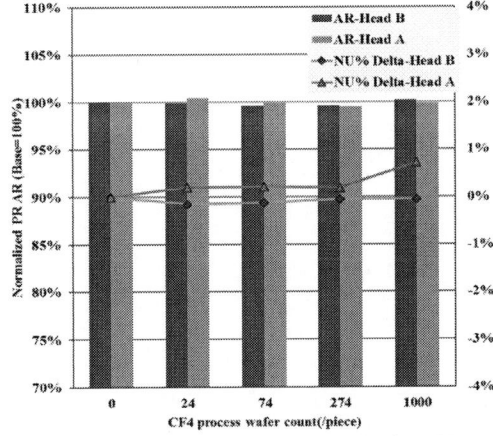

Fig. 10. Normalized O2/N2 PR ash rate after CF4 process on two heads with by lot post condition of O_2/FG followed with O_2/N_2

By wafer count post condition is also recommended as another way for strong CF4 process with higher ratio CF4 or longer process time.

CONCLUSION

CF4 contained plasma strip process impacted the chamber condition and caused ash rate shift of O_2/N_2 ash process but kept the uniformity stable. ~5% ash rate drop was found in this study. The CF4 ratio increase worsened the ash rate dropping a little when CF4 ratio >1%. Continuous running did not further cause more ash rate dropping, which was found kept stable at ~5% lower level than baseline. One effective way to avoid ash rate shift when CF4 process mix-running with other process is to add post condition by wafer or by lot for chamber recovery. O_2/N_2, $4\%H_2$-N_2 forming gas process or the mixture can all help keep the chamber in stable ash rate, the O_2/FG followed with O_2/N_2 post condition was the most effective. The chamber condition of O_2/N_2 or O_2/FG recipe mix-running with CF_4 process can be kept stable with the setting of post condition. CF_4 ratio is recommended to be limited as <1% to prevent chamber from irreversible damage.

REFERENCES

[1] Songlin Xu and Li Diao, "Study of Tungsten oxdation of in O2/H2/N2 drown stream plasma" J. Vac. Sci. Technol. A 26(3), May/Jun 2008.

[2] Li Diao and Songlin Xu "Study of film growth of Si in advanced high dose implant photo resist" ECS Transactions, 18 (1) 663-668 (2009).

[3] Luc Stafford, Joydeep Guha, etc. " Experimental and modeling study of O and Cl atoms surface recombination reactions in O2 and Cl2 plasmas" Pure Appl. Chem., Vol. 82, No. 6, pp. 1301–1315, 2010. DOI: 10.1351/PAC-CON-09-11-02

ADVANCED TECHNIQUES FOR ENHANCED ION BEAM ETCH UNIFORMITY

Peter Goglia[1], Hari Hegde[1], and Yannick Pilloux[2]*
[1]Plasma-Therm, LLC, Fremont, CA, USA
[2]Plasma-Therm, LLC, St. Petersburg, FL, USA
*Corresponding Author's Email: peter.goglia@plasmatherm.com

ABSTRACT

Ion Beam Etch (IBE) is an existing technology used for many years in hard drive recording head applications where accurate, nanometer etched scale is required. IBE has been adopted in selected areas of Semi-Conductor processing such as STT-MRAM memory devices and IR Sensors. The accuracy and low damage of the IBE process are ideal for sensitive structures such as magnetic sensors. Plasm-Therm Planetary IBE system introduces a sweeping wafer stage feature producing an order of magnitude improvement in uniformity, excellent center-to-edge feature symmetry and shaping the radial etch profile. The design architecture not only improves etch uniformity and symmetry for 200mm applications but enables scaling to larger wafer diameters while maintaining performance.

INTRODUCTION

Background

Formation of a contained, ionized plasma is fundamental to Ion Beam Etch technology. Ionized plasmas were demonstrated by Geissler in 1857 [1] in the form of a weakly ionized plasma in glass tubes. "Geissler tubes" are the basis for neon lights. An early application of ions accelerated through a grid were used as thrusters for the NASA program in the early 1960's [2]. The ions were forms from the collision of hollow cathode emission of electrons colliding with the propulsion gas resulting in positive ions. The ions were directed by DC magnets and accelerated by a two grid system.

Early research and industrial application of gridded broad ion beam systems began in the early 1970s [3, 4]. Modern IBE tools apply both wafer tilt relative to the collimated ion source and wafer rotation to optimize etch conditions and uniformity. Early application of these techniques was reported by Spencer and Schmidt in 1971 [5]. While system design has evolved to enhance reliability, increase the process area and improve throughput, there have been no significant changes in broad ion beam etch systems architecture in the intervening time.

The IBE technology discussed here is based on extracting a stream of ions, typically argon, from a remote plasma source and accelerating these ions toward a surface to be etched using a three grid structure. The collimated beam of ions are neutralized once they leave the source. In a sense, the ion beam etch process is a type of atomic scale, precision sand blasting. The ions are created using an RF coil that energizes free electrons in the working gas generating a plasma. The velocity of the ions is controlled by the beam grid plate voltage. The energy range is precisely controlled in the range from 75eV to greater than 1000eV. Figure 1 shows the details of a Plasma-Therm RF Ion source with 3 grid acceleration plates.

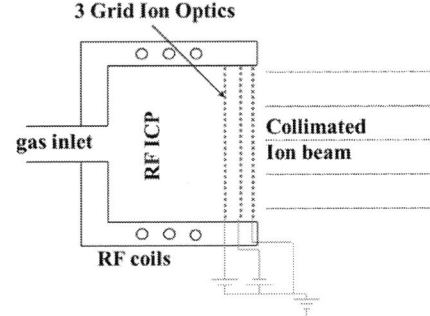

Figure 1: Three grid RF Ion Source

Application

Ion beam etching is widely used in the production of magnetic recording heads for hard disk drives. The read element is a tunneling giant magneto resistive sensor requiring precision patterning. Writing is performed with a complex thin film perpendicular recording element which requires precision patterning and shaping. IBE is used widely in the process of forming these elements. Dimensions width is on the order of 25 nm and film thickness etched range from 8Å to 2 µm. Etch uniformity over a 200mm wafer is < 0.5% σ and the beam imparts little damage to the sensitive reading element since the beam arrives neutralized and there is no plasma exposure. IBE is also commonly used in MEMS, STT-MRAM as well as FBAR, BAW and SAW filter production.

Plasm-Therm Pinnacle family of full wafer IBE tools have an advanced feature set allowing wafer rotation for improved uniformity, back-side helium cooling for high power applications utilizing photoresist and wafer tilt capability for process optimization.

Advances Discussed

Plasma-Therm is introducing an innovative broad beam IBE technique of in the latest generation of Pinnacle system that significantly improves uniformity and eliminates feature etch bias from center-to-edge. We will describe the first innovation in ion beam etch process technology since Spencer and Schmidt reported their results in 1971. The Plasma-Therm Planetary IBE tool introduces a sweeping stage feature that enables these improvements in uniformity, feature profile and advanced techniques for thickness profile shaping. Profile shaping flattens the thickness of incoming wafers with radially symmetric (doughnut shaped) non-uniformity. The Plasma-Therm Planetary sweeping stage IBE systems is a

978-1-5386-5309-8/18 $31.00 © 2018 IEEE

major step forward that evolved from years of experience with the reliable Pinnacle Ion Beam Etch platform. These systems can be configured for R&D applications with a single process module and simple wafer loader through high volume production systems with a wafer handler and cassette loader supplying material one or two process modules. All of the configurations have wafer rotation capability and can tilt the wafer relative to the ion beam. Systems are designed for reliability and repeatable performance. Run-to-run etch repeatability is controlled to an industry leading < 0.25% σ through the use of a proprietary charge integration technique. Source and grid technology is central to Plasma-Therm Planetary IBE system performance not only providing industry leading uniformity but also long grid life and stable performance from the time a grid is installed until it is replaced. Large ion beam sources are notoriously unstable so scaling to 300mm has historically been challenging. The sweeping source system solves this problem through inherent capability to correct for source variations.

PLANETARY IBE SYSTEM DESIGN

Limitations of Previous IBE System Designs

Ion beam etch systems have followed the architecture known since the early 1970's [5]. This architecture provides two geometric variables for process optimization for process; tilt and rotate. The inherent limitations of this architecture are 1) limited ability to optimize uniformity, 2) inherent symmetry bias of feature etching from wafer center to edge and 3) no ability to change radial etch thickness profile.

Best uniformity optimization is achieved through optimizing source conditions, wafer tilt and rotation. Further optimization for specific power levels can be achieved through grid design. The inherent 200mm wafer uniformity for a system with grids optimized for a specific power level and tilt is about 4% 5σ. Other power levels will be higher variation.

Patterned feature asymmetry results from etching with a collimated beam incident on a rotating wafer tilted relative to the ion source. Etching will occur faster when the patterned feature on the outer edge of the wafer is closer to the source (the "outboard" edge of the feature) and slower when the outer edge is farther away (the "inboard" edge of the feature). This is illustrated in figure 2.

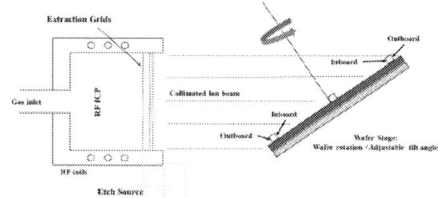

Figure 2: Inboard / Outboard definition

Feature asymmetry, as defined in Figure 3, is inherent to this system architecture.

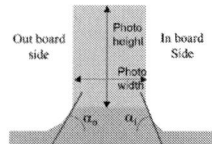

Figure 3: Feature Inboard/Outboard Asymmetry

After pattern etching, the outboard angle, α_i is typically less than α_o. The features at the edge of the wafer have the largest asymmetry while the center of the wafer yields symmetric features. Many times this leads to yield loss at the outer edge of the wafer.

Planetary IBE System Design

The Plasma-Therm Planetary IBE System (Figure 4) introduces innovative features eliminating patterned feature asymmetry and improving uniformity by an order of magnitude.

Figure 4: Plasma-Therm Planetary IBE Module

Wafer Sweep Innovation

Sweeping the wafer across the collimated ion beam while rotating the wafer eliminates feature asymmetry. The Plasma-Therm Planetary IBE tool accomplishes this by utilizing a robust swing arm to hold the wafer rotation fixture. The swing arm sweeps the wafer fixture in a back-and-forth semi-planetary motion across the path of the ion beam at the tilt angle selected. The sweep arm is given an additional degree of freedom to tilt with respect to the direction of the ion beam. This is illustrated in Figure 5. The tilt, rotate and swing capability is critical for correcting the inboard / outboard asymmetry. Notice that all parts of the beam etch features at each radius of the wafer equally. The patterned features at the edge of the wafer are etched the same as the features in the thus forming inherently symmetric etch profiles.

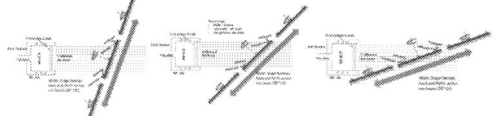

Figure 5a: 20°, 5b: 45°, 5c: 70° wafer tilt
Figure 5: Oscillating Wafer Sweep Operation.

Variable Sweep Speed Innovation for Uniformity

Varying the sweep speed of the wafer stage during the oscillation allows compensation for radial variation in ion source etch characteristics. Uniformity is optimized by first observing the uniformity of the source with the wafer rotating and the sweep are fixed with the wafer centered on the ion source. This reveals the inherent etch profile

978-1-5386-5309-8/18 $31.00 © 2018 IEEE   222

with of the source. Through a short set of experiments varying the sweep speed profile, the etch uniformity can be improved to < 1% 5σ over a 200mm wafer. The speed profile is controlled by a set of velocity/position pairs. The speed is varied in an arbitrary, step-wise fashion using a coordinate axis with the wafer centered on the ion beam as X=0.

Optimizing the velocity profile for radial etch profile uniformity results in dramatic improvements. Two examples are shown in figure 6, and Figure 7. In Figure 6, 200mm wafer etch radial profile is shown for conventional etch process for wafer tilt angles of 10 degree and 80 degree with respect to the ion beam. The non-uniformity sigma in these cases are 0.44% and 0.73% 1σ respectively. These uniformities are quite good but not spectacular. In conventional systems, uniformities must be optimized with specific grid systems for each range of operating conditions. For example, there may be an optimized grid design for a low beam energy range and a different optimized grid for a high beam energy range. However, the planetary tool has a universal grid design. Using the variable velocity sweep function, < 0.2%, 1σ uniformity is obtained for all tilt angles and beam energies. Typical radial profiles of such processes are shown in Figure 7. This is significant optimal uniformity is achieved independent of wafer tilt, and beam energy.

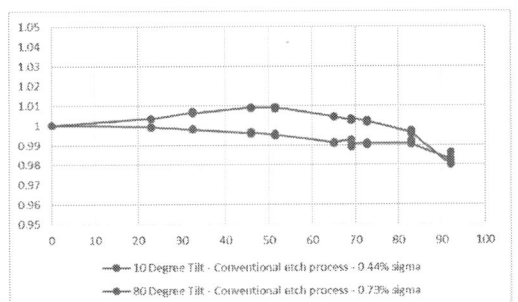

Figure 6: Conventional etch process with Plasma-Therm Pinnacle source

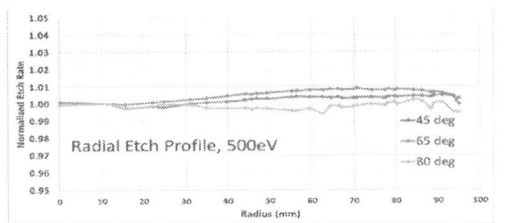

Figure 7: Optimized uniformity with velocity profiling for 500eV etching at wafer tilt of -45°, -65° and -80° wafer tilt.

Variable Sweep Speed Innovation for Etch Profiling

Varying the wafer stage sweep velocity can also be used to produce a radial profile of etch rates. This is particularly useful for wafers with non-uniform films containing radially symmetric non-uniform films. These non-uniformities may be any radial pattern including concave, convex and bulls-eye patterns. An example of radial etch is shown in figure 8.

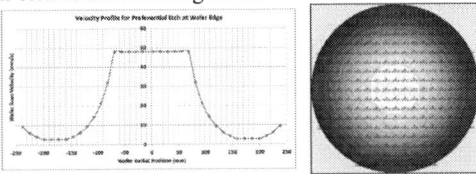

Figure 8a: Velocity Profile;   8b: Etch depth

Slowing the velocity near the OD results in a convex etch pattern with greater etching at the OD. Similar technique can be used to achieve a concave or bulls-eye etch pattern.

CONCLUSION

The innovation of sweeping the wafer fixture across the ion source with variable sweep speed function introduces the first major improvement in ion beam etch processing in more than 40 years. These features enable correction of ion beam etch long standing patterned feature asymmetry issue, significant improvement in uniformity and capability to etch radially symmetric patterns. Improved wafer yields are possible for existing designs. Design rules can be extended utilizing tightened tolerance for feature control and uniformity thus enabling improved device designs.

REFERENCES

[1] S. Reif-Acherman. *Proceedings of the IEEE | Vol. 103, No. 9*, September 2015, pp. 1672-1684.
[2] *NASA Facts*, FS-1999-03-008-GRC, 1999.
[3] D. T. Hawkins, *Journal of Vacuum Science and Technology*. 12, 1975, p. 1389.
[4] D. T. Hawkins, *Journal of Vacuum Science and Technology*. 16, 1979, p. 1051.
[5] E. G. Spencer and P. H. Schmidt, J. Vac. Sci, and Technol, 8, 52 (1971)

THE DEVELOPMENT OF NOVEL STI SPACER IN AN EPI FREE SIGE BICMOS PROCESS

Donghua Liu[1,2], Ziquan Fang[2], Wensheng Qian[2]*

[1]State Key Laboratory of ASIC and System, School of Microelectronics, Fudan University, Shanghai 200433, China

[2]HuaHong Grace Semiconductor Manufacturing Corporation, Shanghai 201206, China

*Corresponding Author's Email: donghua.liu@hhgrace.com

ABSTRACT

Silicon germanium (SiGe) BiCMOS process is now becoming the hot technology for radio-frequency (RF) applications. In a 0.18um SiGe BiCMOS process (BCS18), the device structure and process of SiGe HBT have significant difference from conventional one. These differences includes epitaxial layer (EPI) free and pseudo buried layer. This paper focuses on the study and development of the STI spacer process. Finally the STI spacer process in the BCS18 is successfully developed. STI spacer film thickness and profile all meet technical requirement. After a long period of operation, the process is in a stable state, the process window has a sufficient tolerance range, CP / CPK can be controlled within the target, the number of defects can be maintained at a very low level. All these data shows that the BCS18 process with a novel STI spacer is applicable to production.

INTRODUCTION

Silicon germanium (SiGe) BiCMOS process is now becoming the hot technology for radio-frequency (RF) applications [1]. This process offers high performance SiGe HBT, standard CMOS and good integration of SiGe HBT, CMOS and passive devices to realize RF, analog and digital functions on a single chip. The major application area of SiGe HBTs is for amplifiers including power amplifiers (PA) and low noise amplifiers (LNA). High voltage SiGe HBTs and high speed SiGe HBTs are convenient for fabrication in one process flow and easy to form PA and LNA circuits on one chip, and the low noise figure of SiGe HBT meets the requirements of LNA[2][3].

SIGE BICMOS FABRICATION FLOW

This paper reports a 0.18 micron manufacturable SiGe BiCMOS process (BCS18) which integrates new structure SiGe HBTs and 0.18 micron foundry compatible CMOS. The device structure and process of SiGe HBT have significant difference from conventional one [4]. These differences includes epitaxial layer free, pseudo buried layer and deep contact.

The basic structure of the new device concepts are depicted in Figure 1. After a standard STI etch is complete, a lithography step is used to define a region surrounding the bipolar active area. A heavy implant is then introduced to serve as pseudo buried layer. The heavy implant is blocked from the active region by the STI dielectric stack and sidewall (Figure 1a). After the STI oxide refilling and CMP, the bipolar active base area is opened for SiGeC epitaxial layer growth (Figure 1b). A SiGeC layer of 60nm with a base sheet resistance of about 2000 ohm/sq is grown. The base poly area is then lithographically defined and etched. An emitter region is defined and etched the oxide /nitride bilayer to expose the epitaxial SiGeC layer (Figure 1c) and an undoped polysilicon layer is deposited followed by a high dose emitter implant. The emitter poly is then defined and etched (Figure 1d). A high dose self aligned extrinsic base implant is then introduced. The dopant are activated by drive in and RTA. A self-aligned salicidation of emitter/base poly follows.

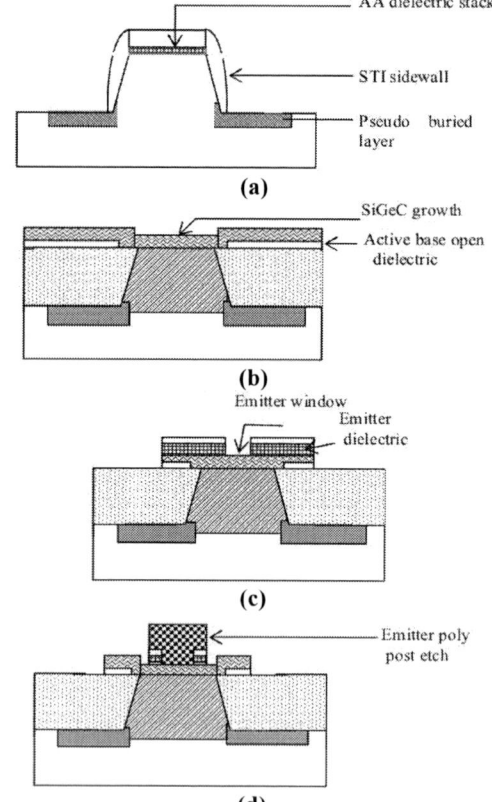

Figure 1: Outline of the new HBT Fabrication with pseudo buried layer

After IMD was deposited, a deep collector contact over the pseudo buried layer was defined and etched. A second contact was opened to expose emitter and extrinsic base region. Standard barrier layer and W plug was deposited into the collector, base and emitter contacts. After W CMP, Al metallization follows (Figure 2).

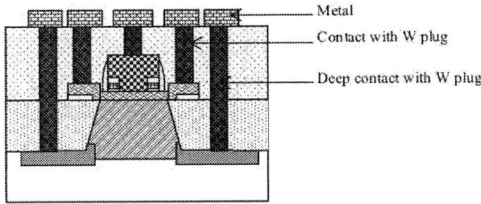

— Metal
— Contact with W plug
— Deep contact with W plug

Figure 2: HBT device Cross section

In this process, the skip of the epitaxial layer achieves a reduction in process cost. In order to ensure the performance of the device, pioneered the proposed buried process [1]. That is, implanting a high concentration of impurities at the bottom of the STI to form a pseudo-buried layer to achieve the low resistance connection of the collector. In the implantation of high concentration of pseudo-buried layer, it is necessary to avoid impurity into the sidewall of the active area (AA). Otherwise it will lead to doping concentration of HBT device collector area is too high and cause device failure. The process uses the STI sidewall protective layer process (STI Spacer) to protect the sidewall active area. The STI sidewall protection is a critical step in this SiGe BiCMOS process. Therefore, this paper focuses on the development of the STI side wall spacer.

EXPERIMENTS AND RESULTS

After the STI etching is completed, the liner oxidation is carried out and then an oxide layer is deposited. The relevant films after deposition are as follows:

(1) film stack in trench：700Ox+180 liner oxide/Sub

(2) Film stack on AA area：700Ox /1630SiN /125Ox /Sub

Using Dry Etching Process to etch away the oxide film, stopped on the silicon substrate at the trench position and on the nitride film at the AA position. Figure 3 shows the Diagram of STI sidewall spacer after film deposition and etching. The less damage to the underlying silicon substrate and the nitride film is better. Mainly using CxFx gas, which has high selection ratio of oxide to nitride and silicon during main etch step. Endpoint detection (EPD) technology is adopted and precisely control the oxide film etching time, reducing the nitride film and silicon Loss of substrate. The requirements of side wall width is greater than 550A and substrate silicon loss is less than 50A.

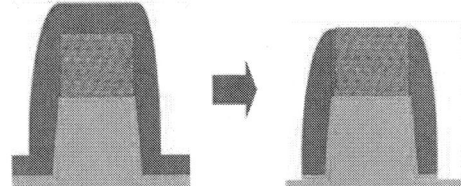

Figure 3: Diagram of STI sidewall spacer after film deposition and etching

STI Spacer Etching Process

The etching process are divided into three steps. The first step is to dry etch the large-scale oxide film. The EPD is used to make sure the large oxide film removed, the good morphology and good etching uniformity obtained. The second step is to remove the polymer after the main etching. The third step is based on the EPD, to do the oxide overetch (OE). This step is importan to to remove oxide film residue after the large pieces oxide etching, especially remove the oxide film of the graphics corner. After the three steps, the oxide film can be completely removed, the damage of the nitride film and the silicon substrate is controlled within a certain range, and the best morphology can be obtained.

Experiment of different etch time

(1) Etch time: EPD + OE 6s:

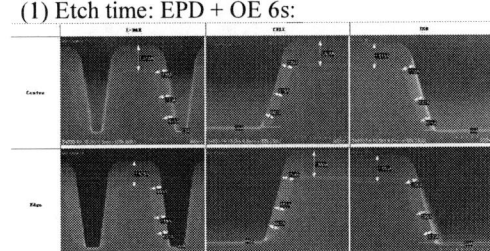

Figure 4: SEM of STI spacer with EPD + OE 6s

Based on the shape of sidewall etched with EPD + 6 seconds overetch shown in figure 4, the width of the side wall between 500A to 750A, the substrate loss of silicon is about 60A, AA area of the remaining of nitride film on AA is about 1500A .

(2) Etch time: EPD + OE 4s:

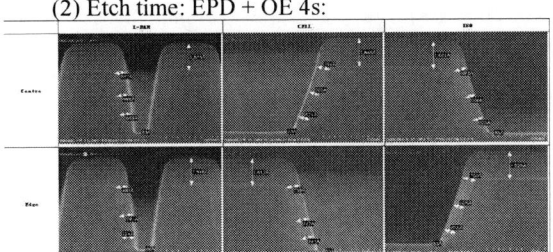

978-1-5386-5309-8/18 $31.00 © 2018 IEEE

Figure 5: SEM of STI spacer with EPD + OE 4s

Reducing the overetch time to 4 seconds and keep EPD, the width of the side wall between 500A to 750A, the substrate loss of silicon is about 60A, AA area of the remaining of nitride film on AA is about 1600A as shown in figure 5.

(3) Etch time: EPD+OE 8s:

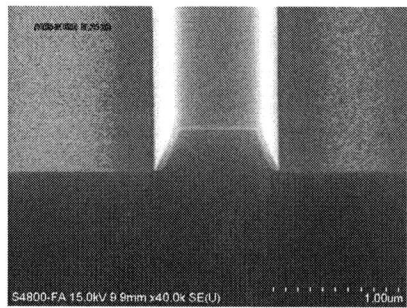

Figure 6: SEM of STI spacer with EPD + OE 8s

Increasing the overetch time to 8 seconds and keep EPD, the width of the side wall between 500A to 750A, the substrate loss of silicon is about 70A, AA area of the remaining of nitride film on AA is about 1480A as shown in figure 6.

(4) The etching conditions without polymer removal step：

Figure 7: SEM of STI spacer etch without polymer removal

When the etch time is 40 seconds, the end detection time is 27 seconds, 40 seconds is equivalent to 50% overetch. However, the oxide film remains on the silicon substrate as shown in figure 7.

So the best etch condition is EPD + 6 seconds overetch and with polymer removal . Figure 8 is the SiGe HBT device SEM cross section pitures with all process completed.

Etching Process Stability Monitoring

The film thickness, Cp/Cpk and particcal number have been monitored. A long period of time to maintain a relatively stable state and with spec. It provides a mature process for the HBT device fabrication with unicle pseudo buried layer.

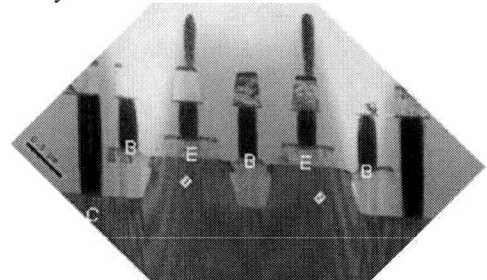

Figure8. SiGe HBT device SEM cross section pitures with all process completed

CONCLUSIONS

Through unremitting efforts, STI spacer etching process in the BCS18 project is successfully developped. STI spacer film thickness and profile all meet the technical requirement. After a long period of operation, the process is in a stable state, the process window has a sufficient tolerance range, CP / CPK can be controlled within the target, the number of defects can be maintained at a very low level, the etching process has matured in the BCS18 project.

ACKNOWLEDGEMENTS

The authors would like to thank all members of device group, RF group and advanced module of HHgrace for their great support on this work.

REFERENCES

[1] Marco Racanelli and Paul Kempf, *IEEE TRANSACTIONS ON ELECTRON DEVICES*, 1259 , VOL. 52, NO. 7, July 2005.

[2] Zhenqiang Ma and Ningyue Jiang, *IEEE TRANSACTIONS ON ELECTRON DEVICES*, 875 , VOL. 53, NO. 4, Apr. 2006.

[3] J. F. W. Schiz and Andrew C. Lamb, *IEEE TRANSACTIONS ON ELECTRON DEVICES*, 2492 , VOL. 48, NO. 11, Nov. 2001.

[4] D.H. Liu, W.S. Qian, X.B. Chen, et al., *ECS Transactions*, vol.34(1), pp. 173-181, Mar. 2011.

FABRICATION OF SILICON NANOPORES WITH TUNABLE SHAPES AND SIZES

Qi Chen, Yifan Wang, and Zewen Liu

Institute of Microelectronics, Tsinghua University, Beijing 100084, China

*Corresponding Author's Email: liuzw@tsinghua.edu.cn

ABSTRACT

Silicon nanopores were successfully fabricated by wet anisotropic etching method. The minimum feature sizes of square and rectangular nanopores were 33 nm and 6 nm, respectively. To reduce the sizes of larger pores prepared by wet etching method, e-beam induced deposition (EBID) of hydrocarbon compounds was used to shrink as well as modify the shape of the nanopores. Hydrocarbon compounds are common contaminants existing in the scanning electron microscope (SEM) chamber, and are easily deposited on the sample surface under e-beam irradiation. By using this strategy, nanopores with tunable shapes and sizes were controllably obtained with a visual feedback.

INTRODUCTION

Nanopore-based sensing has attracted growing attention in DNA/RNA sequencing [1], drug delivery [2], molecule separation [3], and so on. Common methods for solid-state nanopore fabrication are FIB/TEM drilling [4, 5], e-beam lithography based methods [6, 7], etc.

In this paper, wet anisotropic etching is used to fabricate nanopore arrays and individual nanopores in silicon substrates. It is generally known that anisotropic etching is widely applied in micromachining for micro-electromechanical systems (MEMS) [8, 9] and microfluidics [10, 11]. Herein, pores at nanoscale are also obtained by using this conventional wet etching method, with the help of a color-based feedback control. To improve the controllability of nanopore shapes and sizes, e-beam induced deposition (EBID) of hydrocarbon compounds was used to shrink and modify the shape of nanopores. Under localized e-beam irradiation in a scanning electron microscope (SEM), nanopores with tunable shapes and sizes were successfully obtained.

EXPERIMENTAL

The fabrication process of silicon nanopores is illustrated in Figure 1. The fabrication started with a 4-inch double polished (100) silicon wafer. A layer of chromium (Cr) with thickness of 300 nm was spurted on the front side of the wafer. Two layers of SiO_2 and Si_3N_4 with respective thickness of 150 nm and 300 nm were consecutively deposited on the back side. Standard double-side lithography was performed to prepare patterns on mask layers, followed by the three step single-side wet etching in 33 wt% KOH solution. Firstly,

inverted-pyramid structures were created on the front side by anisotropic etching at 70 °C. Secondly, fast etching was conducted from the back side at 80 °C, until the residual thickness of the back-side silicon was about 10 μm. After dicing of silicon wafer into 1.5 × 1.5 cm^2 chips, finally, 30 °C was adopted to open the tips of the inverted-pyramid structures. The pore opening event was monitored by a color feedback mechanism, where phenolphthalein acted as the indicator based on its color change property under PH variation [12]. After the pore-opening event, the chip with manufactured nanopores was rinsed by deionized water and cleaned by oxygen plasma, and then inserted into the SEM chamber for observation and modification of the pore morphology.

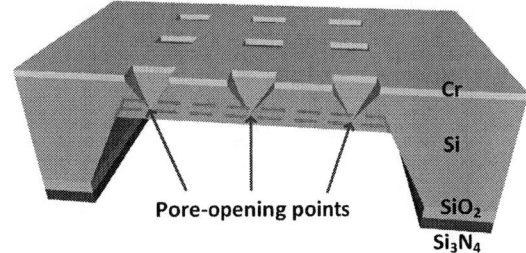

Figure 1. Fabrication process of silicon nanopores.

RESULTS AND DISCUSSION

Massive fabrication of ordered nanopore arrays was realized by wet anisotropic etching method, as shown in Figure 2(a).

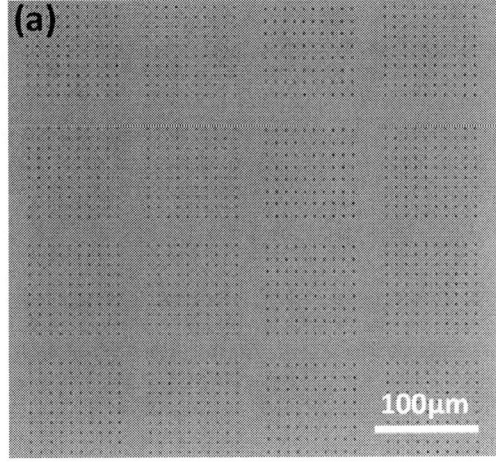

978-1-5386-5309-8/18 $31.00 © 2018 IEEE

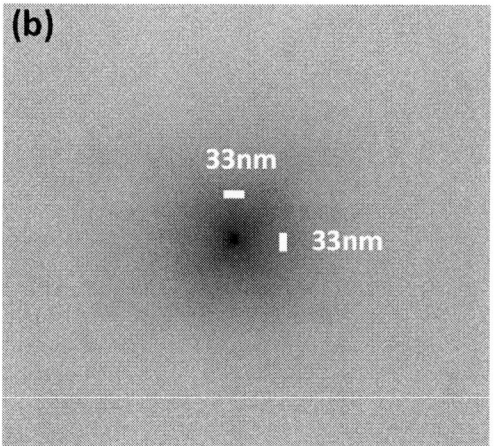

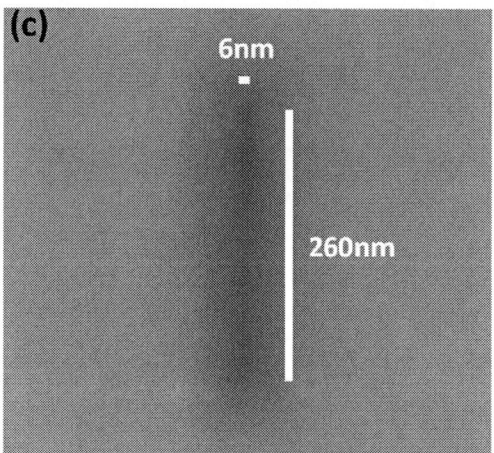

Figure 2. (a) Nanopore array and (b) individual nanopores with various shapes and sizes.

By careful designing masks on the front side of silicon wafer, nanopores with various shapes and sizes were successfully fabricated. As shown in Figures 2(b) and (c), square and rectangular nanopores with respective minimum feature sizes of 33 nm and 6 nm, were obtained.

Nanopores with larger feature sizes were also fabricated because of the size error of the pores by using wet etching method. However, the feature size can be reduced by nanopore shrinking methods, such as atomic layer deposition (ALD) [13] and thermal treatment [14]. Among the existing advanced shrinking methods, EBID approach provides a reproducible way and a visual feedback. Figures 3(a)-3(e) show a shrinking process of a nanopore with initial size of 82 nm × 125 nm. The pore was gradually shrunk to smaller sizes, and eventually closed (Figure 3(e)), as the e-beam irradiation process was prolonged. To increase the shrinking rate of the nanopore under e-beam irradiation, the e-beam was guided into a

scanned area through the SEM fine-tuning tool. The scanned area is easily distinguished by the contrast in Figures 3(b)-3(e), and only indicated by the yellow dashed box in Figure 3(e). Additionally, the shrinking rate of nanopores can be improved by increasing the e-beam current, decreasing the accelerating voltage of e-beam, decreasing the scanned area. Hydrocarbon compounds can be removed by oxygen plasma. As shown in Figure 3(f), the shrunk nanopore in Figure 3(e) recovered to its initial shape and size after oxygen plasma cleaning treatment. The structural recovery phenomenon indicates that EBID method offers a reproducible way to shrink the silicon nanopore.

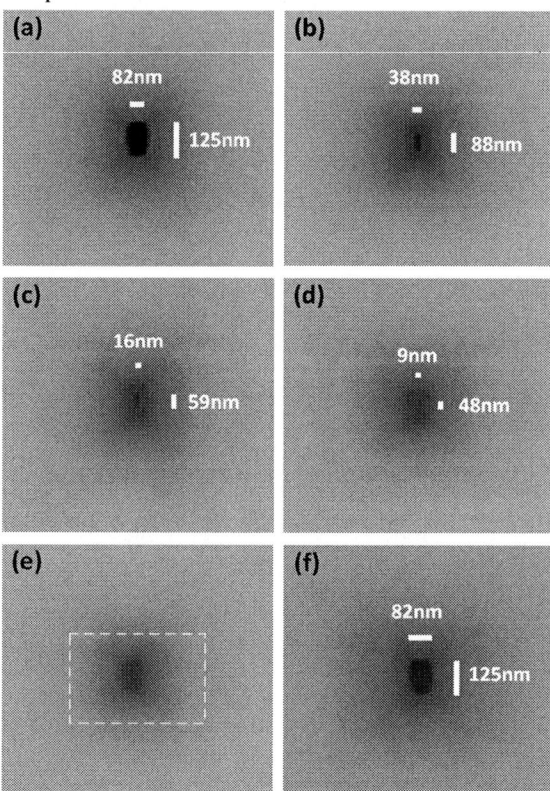

Figure 3. Shrinking process of a nanopore with initial size of 82 nm×125 nm. The pore was irradiated under e-beam for (a) 0 s, (b) 100 s, (c) 150 s, (d) 170 s, and (e) 200 s. (f) Nanopore recovery to its original size and shape after oxygen plasma cleaning. The accelerating voltage of SEM e-beam is 10 kV.

Inspired by localized e-beam irradiation via the SEM fine-tuning tool, shape modification of nanopores was realized by guiding e-beam into a scanned area residing partially in the nanopore. As shown in Figure 4(a), by guiding e-beam into the right part of a rectangular nanopore, the length-width ratio of the nanopore was

978-1-5386-5309-8/18 $31.00 © 2018 IEEE   228

altered. The right part of the pore shrank to closed, while the left part kept its initial morphology. Figure 4(b) shows the result of partially shrinking of a square nanopore to a triangular pore. By continually changing the location of the e-beam during the pore observation process in the SEM, the nanopore shape was modified gradually. The inset on the right top of Figure 4 (b) demonstrates the original morphology of the nanopore before shape modification.

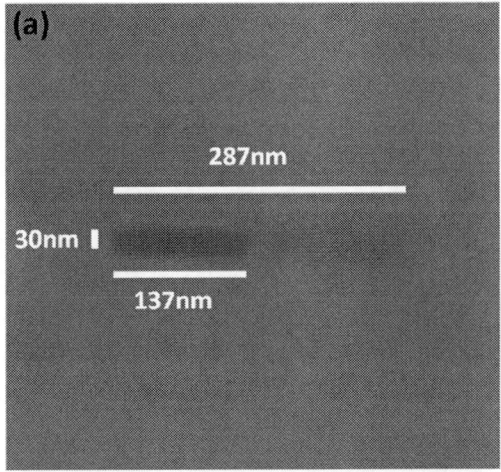

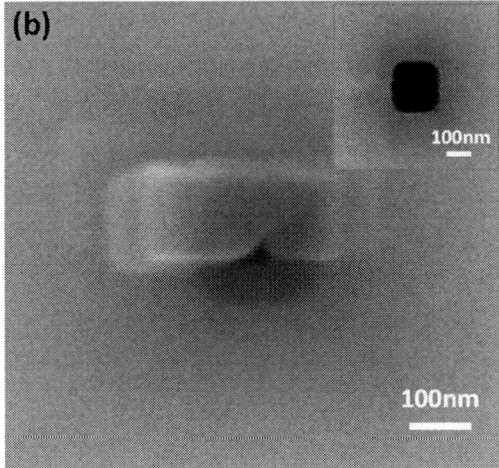

Figure 4. Shape modification of nanopores. (a) Change the length-width ratio of a rectangular nanopore. (b) A triangular nanopore obtained by shrinking partially of a square nanopore.

CONCLUSION

Nanopores with square and rectangular shapes were prepared by wet anisotropic etching method. The feature sizes of obtained nanopores range from 6 nm to several hundred of nanometers. EBID method was used to shrink larger nanopores by using hydrocarbon compounds which exists in the SEM chamber. The nanopore feature size was successfully reduced to sub-10 nm, as the e-beam irradiation process was prolonged. By oxygen plasma cleaning, the shrunk nanopore recovered to its initial size and shape, without any structural damage. Shape modification was achieved by using EBID method, by guiding e-beam into a certain location of the nanopore. Nanopores with tunable sizes and shapes were obtained.

ACKNOWLEDGEMENTS

This research was supported by the National Natural Science Foundation of China (NSFC, Grant Nos 61273061).

REFERENCES

[1] Jay Shendure and Hanlee Ji. Nat. Biotechnol., 26, pp.1135–1145 (2008).
[2] Ruixue Duan, Fan Xia, and Lei Jiang. ACS Nano, 7, pp. 8344–8349 (2013).
[3] Liang Huang, Miao Zhang, Chun Li, and Gaoquan Shi. J. Phys. Chem. Lett., 6, pp. 2806−2815 (2015).
[4] Zhu Chen et al. Nature Mater. 9, pp. 667–675 (2010).
[5] Christopher A. Merchant et al. Nano Lett., 10, pp. 2915–2921 (2010).
[6] Jingwei Bai et al. Nanoscale, 6, pp. 8900-8906 (2014).
[7] Amir G.Ahmadi, Sankar Nair. Microelectronic Engineering, 112, pp. 149–156 (2013).
[8] Nijdam A J. University of Twente, (2001).
[9] JackW Judy. Smart Mater. Struct. 10, pp. 1115–1134 (2001).
[10] Ciprian Iliescu, Hayden Taylor, Marioara Avram, Jianmin Miao, and Sami Franssila. Biomicrofluidics 6, 016505 (2012).
[11] E. Verpoorte, N. F. De Rooij. Proceedings of the IEEE, 91, pp. 930-953 (2003).
[12] T. Deng, J. Chen, C. N. Wu, and Z. W. Liu. ECS J. Solid State Sci. Technol., 2, pp. 419–422 (2013).
[13] Peng Chen, Toshiyuki Mitsui, Damon B. Farmer, Jene Golovchenko, Roy G. Gordon and Daniel Branton. Nano lett. 4, pp.1333−1337 (2004).
[14] Tao Deng et al. Nanotechnology 24, 505303, (2013).

STUDY OF PLUG ETCH BACK PROCESS MATCHING ON FARADAY SHIELDED ICP AND TRADITIONAL CCP PLASMA ETCH CHAMBERS

Yali Fu[1,2]*, Shawming Ma[3], Yi Wang[1], Jammy Xiao[4], MH Kim[3], Nancy Zhang[2], Mingjie Jiao[2], Frank Wang[3], Tammy Lee[3]

[1]Peking University, School of Software and Microelectronics, Beijing, China
[2]Mattson Technology, China field process Group, Beijing, China
[3]Mattson Technology, Plasma Product Group, Fremont, California, USA
[4]SMNC ETCH department, Beijing, China
*Corresponding Author's Email: 010-50959351; Yali.fu@mattson.com

ABSTRACT

It is well known that different plasma etch tools running exactly the same process step may get mis-matched results and impact other process steps, which can cause issue of the integration to incorporate different type of tools into the same process flow. Very little prior work studied the mechanism of etch process matching between inductive coupled plasma (ICP) and capacitive coupled plasma (CCP) tool. This paper studied BEOL Plug Etch Back(PEB) process using O_2 chemistry between Faraday shield ICP etch tool and traditional CCP etch tool. The process mis-match was found to cause Photo CD mis-matched at the following process step. The mechanism was studied through wafer surface analysis, it was found that film surface modification generated different surface morphology between the two types of tools with different ion energy. One potential solution was also provided to realize the mass production of the two different type tools both runing in line.

Keywords—Plug Etch Back, O_2 chemistry, Faraday shield, ICP plasma, CCP plasma, Ion energy control, surface modification

INTRODUCTION

There are two main types of plasma etcher widely used in semioconductor manufacturing: Capacitively Coupled Plasma (CCP) Etcher and Inductively Coupled Plasma (ICP) Etcher [1]. Under certain situation of capacity limitation, fab operation might need to transfer process between different type of tools. However, process mis-matching issue is usually one key road block to implement the same process step into different etch tools at the same time. In this study, PEB (Plug Etch Back, or BARC etch back) is one of the etch process steps in via first approach of the dual damascene structure integration [2].

This paper studied BEOL BARC etch back process on the two types of etcher, as shown in Fig. 1 (b). The process mismatching was found at following layer, for which photoresist CD was found with several nanometers difference,

as shown in fig. 1 (d). All the process parameter optimization can not obviously help improve the CD difference. It was suspected that the film surface was differently changed after etching in different types of tools. The tests were done to study the different surface impact.

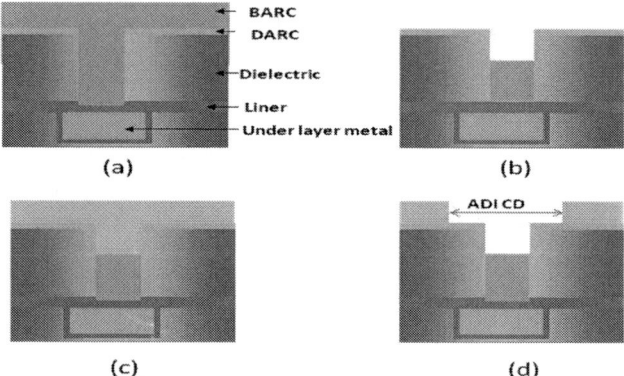

Fig. 1 Film stack of BEOL inter metal or top metal layer with via first dual damascene integration scheme. After via etch to stop on the bottom liner then (a) After BARC coating to fill the via; (b) After PEB etch; (c) After trench PR coating; (d) After inter-metal or top metal photo lithography step.

EXPERIMENTAL

This work was performed in one Faraday shield ICP tool and one traditional CCP tool. The ICP tool was Paradigm E plasma tool from Mattson Technology as shown in Fig. 2.

Five control wafers deposited with DARC film were prepared to study the different surface modification effect of ICP and CCP etcher. The split table was shown in Table I. The PEB process in this study was process using O_2 chemistry. Two wafers were run 3 cycles' PEB process, compared to one cycle process and reference control wafer without process. Scanning probe microscopy (SPM) and X-Ray Photoelectron Spectroscopy (XPS)/ Electron Spectroscopy for Chemical Analysis (ESCA) were done to study the surface roughness and atomic concentrations which might impact the photoresist development. The XPS detection spot size was about 0.3mmx0.3mm and the penetration depth was about 60Å-80Å.

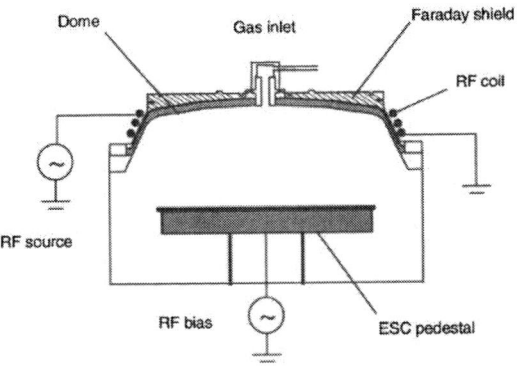

Figure 2. Schematic diagram of Paradigm E ICP plasma reactor with single 13.56MHz source on top and 13.56MHz bottom bias.

Table 1. Split table for control wafer process

Sample ID	Process
Slot 1	CCP 1X
Slot 2	ICP 1X
Slot 3	CCP 3X
Slot 4	ICP 3X
Slot 5	No process

RESULTS AND DISCUSSION

Scanning probe microscopy analysis was carried out to study the surface roughness difference between the processed wafers. Top view Atomic Force Microscope (AFM) images were given in Fig.3, and the surface roughness results was showed in Fig.4. From Fig.3 (e), it can be found that the original film surface was with highest roughness. Both PEB processes on CCP and ICP smoothened the DARC surface. The more processing, the smoother DARC surface was got. However, the relative ability to smoothen the DARC surface was comparable between the two etchers. And the photoresist CD at next process stage should be little impacted by the surface roughness.

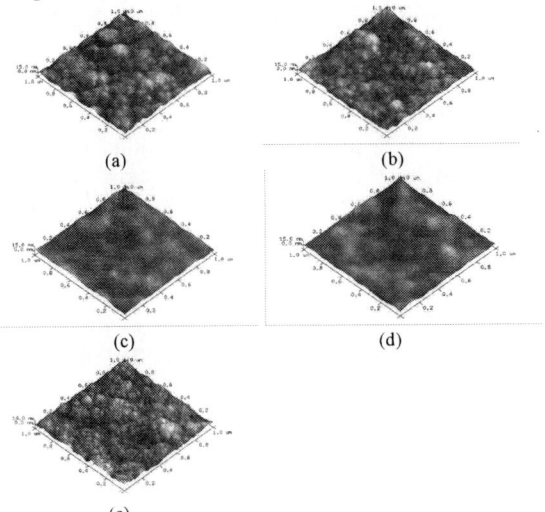

Fig. 3 Top view AFM images: (a) slot 1, (b) slot 2, (c) slot 3, (d) slot 4, (e) slot 5.

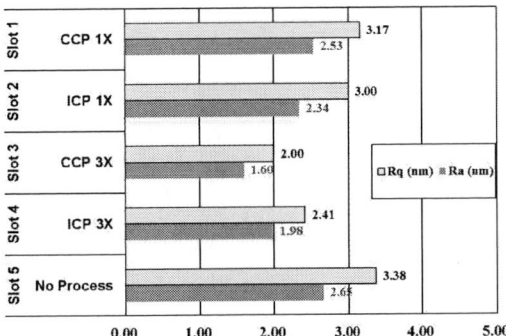

Fig. 4 Surface Roughness Test results, R_a is mean roughness, R_q is standard deviation of roughness

Surface film was surface checked by X-Ray Photoelectron Spectroscopy (XPS) to analyze the chemical components after the PEB process, as shown in Fig. 5. Nitrogen concentration after process was reduced from 3.3% to 0.1-0.5% while that of oxygen was increased from 61.6% to 65.5-66.8% of the original film, which suggested that there was possibly conversion of SiN to SiO_2.

The Si component of original film was 30.7% as shown in Fig.5, but almost the same after one cycle PEB process, and kept the same after 3 cylce's running. XPS silicon chemical states was further studied to check the difference, as shown in Table 2 and Fig.6, no elemental Si(<0.1%) existed after PEB process while there was 1.2% of elemental Si on the original DRAC surface, which suggested that PEB process turns elemental Si into SiO_2. There was 2.7% of SiN and 26.9% of SiO_2 on the original film, which was changed to <0.3% and 28.3% after PEB process of CCP tool on slot 1(Table 2), respectively. It also indicated that surface SiN was converted into SiO_2 by the PEB process.

Comparing the difference between 1X and 3X processes, it shows that the longer process time, the higher degree of conversion. The most surface Si and SiN were converted into SiO_2 after 3X process.

The CCP tool in this study was more capable of converting Si and SiN into SiO_2, as it was found that after one cycle PEB process, the wafer surface of paradigm E ICP tool had higher O concentration or higher SiO_2, as shown in both Fig. 5 and Fig.6. After 3 cycles testing, almost all the surface N has been converted to O, thus the two etchers showed comparable results.

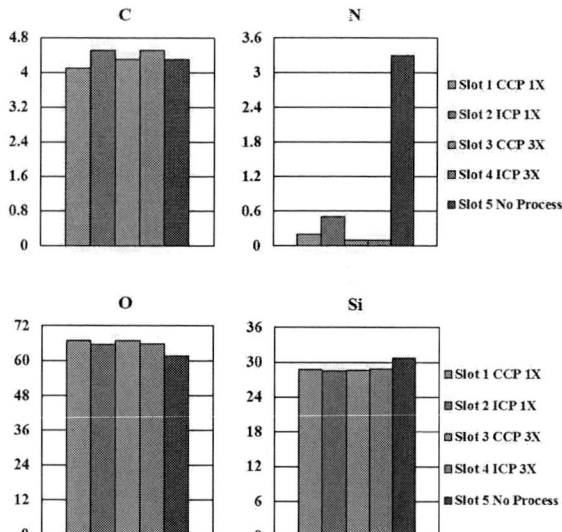

Fig. 5 XPS atomic concentrations (in atomic %). F and Ar were also tested, as most of the concentration of the two elements <0.1, which was not shown in the graph.

Table 2 XPS Silicon Chemical States (in atom%)

Sample ID	Process	Si	SiN	SiO_2
Slot1	POR 1X	<0.1	0.3	28.3
Slot2	ICP 1X	<0.1	0.4	27.9
Slot3	POR 3X	<0.2	<0.1	28.5
Slot4	ICP 3X	<0.1	<0.1	28.7
Slot5	No Process	1.2	2.7	26.9

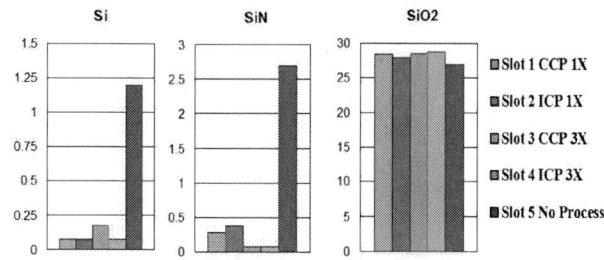

Fig. 6. Diagram of XPS Silicon Chemical States

Fig. 7. Film structure of the test wafers: (a) before PEB process, (b) after PEB process

Base on the AFM data and XPS data, the most apparent difference between CCP and ICP was the surface component change after only 1X PEB processing. The difference disappeared after 3X PEB processing. This behavior matched with the photo CD results. Anti-reflectvie coating layer's reflectivity will impact the Litho development process[3], and film surface material change may impact the anti-reflectvie efficiency, which might finally impact the CD after developing.

The ion energy was theoretical calculated to explain the different behavior of the two types tools. Base on the PEB process recipe setting in this study, the ion energy of ICP tool Paradigm E was calculated to be ~22 eV, while the that CCP tool was up to ~50 eV, which was much higher. The detailed calculation will be shown in another paper. The higher ion energy of plasma was suspected to cause higher surface chemical reaction, which matched with the surface convertion of SiN to SiO_2 from XPS test results.

To optimize the ion energy by fine tuning process parameters, there is limit to overcome the big ion energy gap, and the process loading performance or other etch output performance will also be impacted, thus it is trade-off. In this case, in mass production line, it is recommended to dedicate process on each types of tools, or control the mix running by Atuo Process Control (APC) system[4].

CONCLUSION

Plasma of Faraday Shield ICP etcher has less ion energy than that of traditional CCP tool, which will cause different behaviors on the surface film coversion. It is highly suspected that the degree of surface SiN conversion to SiO2 on the wafer surface impacted the photoresist development in next process stage, thus ADI CD was impacted. For plasma etching process, usually there is limit to match the same process between different types of etchers with different plasma system. To incorporate different type of tools into the same process flow for mass production, APC control system is highly recommended.

ACKNOWLEDGMENTS

The authors would like to thank their colleagues at etch department of Mattson Technology for technical assistance and useful discussions during the course of this work

REFERENCES

1. T.K. Kang and W. Y. Chou. "Avoiding Cu Hillocks during the Plasma Process" Journal of The Electrochemical Society, 151 (6) G391-G395 (2004).
2. K. Nojiri, Dry Etching Technology for Semiconductors, P57., Springer International Publishing Switzerland(2015)
3. B. Tang, M. Seamons, and etc. "PECVD bi-layer ARC for BARC-less Immersion Lithography". ECS Transactions, 19(2), 765-772 (2009).
4. M. James. "A blueprint for enterprise-wide deployment of advanced process control." Solid State Technology, July 2009, p. 35+. Academic OneFile, Accessed 23 Dec. 2017.

DE-WETTING MODEL STUDY ON SPINNING SUBSTRATE – CHALLENGES FOR LOW CHEMICAL CONSUMPTION

David Mui[1], Mark Kawaguchi[1], and Christian Haigermoser[2]*

[1] Lam Research Corp., Fremont, USA

[2] Lam Research Service (Shanghai) Co., Ltd., Shanghai, China

*Corresponding Author's E-mail: Christian.Haigermoser@lamresearch.com

ABSTRACT

A model is proposed to describe the wetting and de-wetting dynamics on a spinning wafer with a non-wetting surface. The model is based on the dynamic balance between the wetting and de-wetting velocities. The proposed model predicts a theoretical lower limit for the media flow rate to completely wet a wafer surface as a function of its contact angle and wafer rotational speed. The model predictions are found to agree reasonably well with experimental results. The model is also used to develop a low flow process for the removal of Titanium Nitride. A 66% flow rate reduction is achieved in comparison to the baseline process.

INTRODUCTION

Continuous device shrinkage beyond 3X and 2X nm increases the number of critical cleaning processes in semiconductor manufacturing. In order to avoid cross contamination from reclaimed chemistry, single-pass processes have become more commonly used. This migration to single-pass started in the front-end-of-line pre-gate cleaning process. In recent years, some back-end-of-line cleaning processes have also started using single-pass due to the risk of defect formation on wafer surfaces with reclaimed chemistries. The major drawback of single-pass is the cost of the chemistry. In an effort to contain this higher cost, processes with lower medium flow rates are rapidly becoming more popular. In this paper, the risks associated with lower flow processes are discussed.

Single-wafer spin clean tools have been widely used in semiconductor manufacturing. In processes where mass transport is not the rate limiting step, reducing chemical consumption can be achieved by decreasing the medium flow rate. This approach yields acceptable uniformity results when the process is operated near ambient temperature or heating is provided by a substrate heater. The other risk associated with low flow processes is wafer surface de-wetting which may result in non-uniform etching and/or reduces cleaning efficiency. A de-wetting model for single-wafer spin clean processes has been developed to predict the onset of de-wetting. The model is used to calculate the theoretical minimum media flow rate required to completely wet a surface as a function of contact angle and rotational speed. Good agreement between model predictions and experimental observations is obtained.

MODEL

The phenomenon of de-wetting on a stationary surface has been studied in the past [1]. However, there is very limited information in the published literature on this topic on a spinning surface. In this situation, a dynamical balance exists between the flowing liquid and the wafer surface. In this paper, this balance is characterized by the wetting and de-wetting velocities. The former is modeled as the liquid radial flow velocity (U_r) at the air/liquid interface and is approximated by the following expression:

$$U_r = \frac{\omega^2 r h^2}{2\nu} \tag{1}$$

where ω is the substrate rotational speed, r is the radial distance from the wafer center, h is the thickness of the liquid film, which is a function of the liquid volumetric flow rate, Q, and ν is the kinematic viscosity of the liquid. The force balance that exists on the de-wetting front is illustrated in Figure 1.

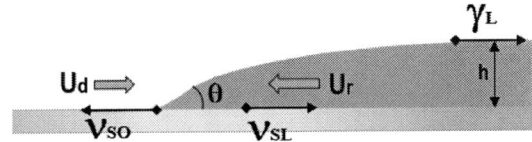

Figure 1: De-wetting contact line force balance illustration.

The de-wetting driving energy, E_D, is given by:

$$E_D = \gamma_L + \gamma_{SL} - \gamma_{SO} - \frac{1}{2}\rho g h^2 = \rho h U_d^2 \tag{2}$$

where γ_L is the liquid surface tension, γ_{SL} the solid/liquid interfacial energy, γ_{SO} the dry-solid surface energy, ρ the mass density of the liquid, g the gravitational acceleration, and U_d the de-wetting velocity. In the inertial de-wetting regime, as in the current case of the rotating wafer, it can be shown that U_d can be expressed as follows:

$$U_d = \sqrt{\frac{g(h^2 - h_c^2)}{2h}} \tag{3}$$

where the critical de-wetting thickness, h_c, is given by:

$$h_c = 2\sqrt{\frac{\gamma}{\rho g}}\sin\left(\frac{\theta}{2}\right) \tag{4}$$

978-1-5386-5309-8/18 $31.00 © 2018 IEEE

with γ the liquid surface tension and θ the equilibrium contact angle.

The empirically corrected film thickness expression obtained in [2] is used here and is repeated below:

$$h = 0.886 Q^{0.348} v^{0.328} \omega^{-0.676} r^{-0.70} \qquad (5)$$

The dynamical balance between wetting and de-wetting on the spinning surface is modeled by the ratio of the wetting velocity over the de-wetting velocity. By combining equations (1), (3), and (4), it can be shown that this ratio is given by:

$$R = \frac{U_r}{U_d} = \frac{0.739 \omega^{0.31} Q^{0.87}}{\sqrt{2g h_c} v^{0.18} r^{0.75}} \qquad (6)$$

When $R > 1$, wetting dominates over de-wetting and a wetted surface results. On the contrary, when $R < 1$, the surface de-wets. A dynamical balance occurs when $R = 1$, which also defines the minimum flow rate required to maintain a wetted surface at a given rotational speed and contact angle. The model predicted minimum wetting flow rate as a function of the contact angle calculated at various rotational speeds is shown in Figure 2.

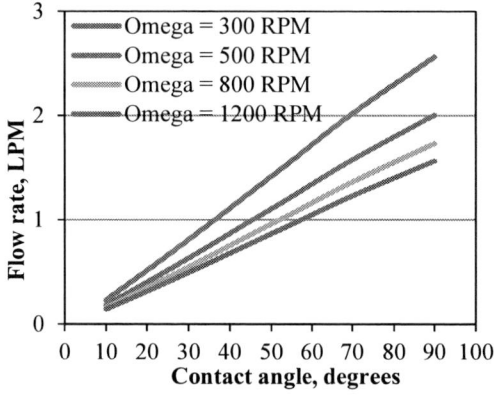

Figure 2: Minimum wetting flow rate model calculation as a function of contact angle and rotational speed (RPM = Revolutions Per Minute).

EXPERIMENTAL

The equilibrium contact angle is measured on a KRUSS DSA 100 contact angle measurement tool. This study focused on wafer-edge contact angles. Silicon (100) substrates are pre-cleaned by SC-1 (NH$_4$OH:H$_2$O$_2$:DIW = 1:2:50) at room temperature. In order to generate wafers with various contact angles, a dilute HF (HF:DIW=1:200, at room temperature) process with different exposure times was applied to the pre-cleaned substrates. For hydrophobic surfaces, HF (1:50) is used. Table I tabulates some of the surface preparation processes used. All the surface preparation and wafer coverage tests were performed on the EOS™ system designed and manufactured by Lam Research Corp.

TABLE I. Contact angle surface preparation

Contact Angle	Process
78°	HF 1:50 30s
60-64°	HF 1:50 30s, followed by DIW rinse 2 min
40-50°	HF 1:200 15s
30-34°	HF 1:200 10s
~10°	HF 1:200 5s

RESULTS AND DISCUSSION

<u>Model Justification:</u> In the first step to test the validity of the model, liquid surface coverage with DIW was visualized with a camera on wafers with different contact angles. Figures 3 show examples of a de-wetted and wetted surface. Figure 3a shows severe de-wetting from the middle to the edge of the wafer and Figure 3b shows complete coverage of the wafer surface by DIW. The latter result shows the benefit of a higher flow rate process in wetting a surface with high contact angle.

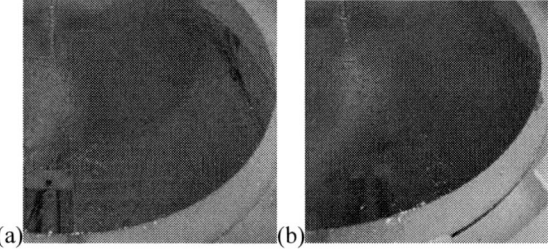

Figure 3: Dispensed DIW at 300 RPM on a 64° surface at a flow rate of (a) 1.0 LPM (Liter Per Minute) and (b) 2.0 LPM.

Figures 4 summarize the surface coverage results which are plotted with an overlay of the model results for the predicted minimum flow required to completely wet the wafer. Most of the data points represent good correlation with the model results. The figure indicates that process conditions below the model line are expected to result in de-wetted surfaces (open symbols) and conditions close to and above the model line complete wafer surface coverages (closed symbols). Based on this model, one can predict the minimum flow requirement at a given rotational speed just by knowing the equilibrium contact angle of the wafer surface.

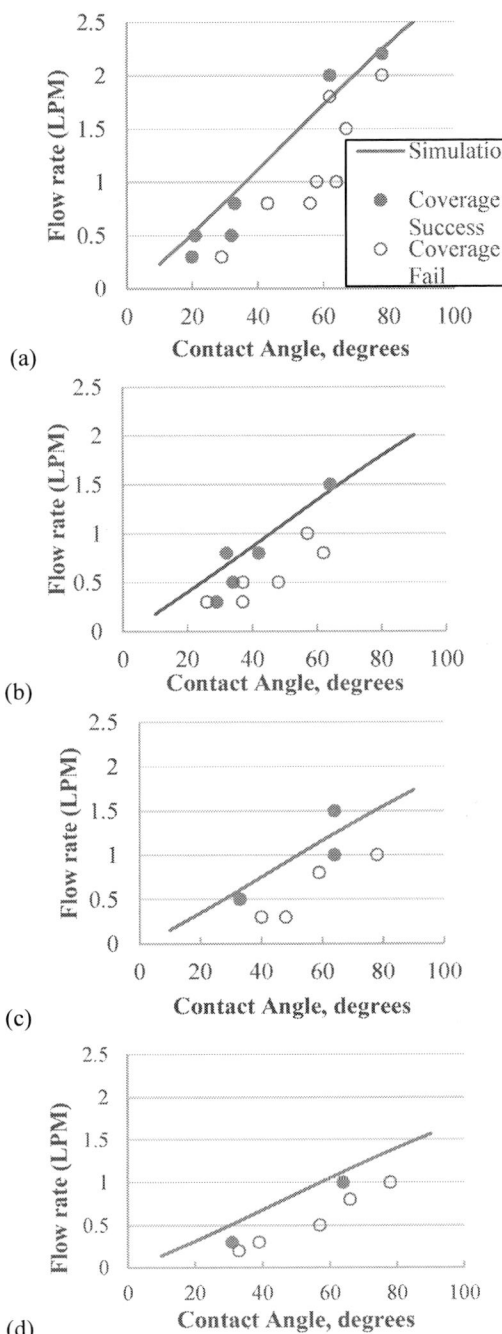

(a)

(b)

(c)

(d)

Figure 4: Liquid surface coverage results at (a) 300 RPM, (b) 500 RPM, (c) 800 RPM, and (d) 1200 RPM.

Low flow application: Reducing chemistry consumption is one of the key process requirements. Titanium nitride (TiN) hardmask removal process commonly uses an oxidizing chemistry which necessitates a single pass operation. Figures 5a, b, and c show the TiN etch rate dependency on the chemistry flow rate at 500 RPM. The surface contact angle was measured to be 22°. In order to reduce total chemistry consumption, the wafer backside uses no media dispense. As the flow rate decreases, the etch rate decreases by 32% due to a reduction in heating effect which is solely provided by the heated chemistry. Figure 5d shows the result of a challenged low flow rate process. According to Figure 1, a 22° contact angle at a flow rate of 0.5 LPM combination requires a rotational speed of 500 RPM for full wafer coverage. The resulting etch rate is found to be comparable to the 1.5 LPM process. The heat loss of low flow rate process is compensated by adding hot DIW of 1.5LPM at 70°C to the wafer backside. Overall, a 66 % chemistry reduction is achieved by this low flow process.

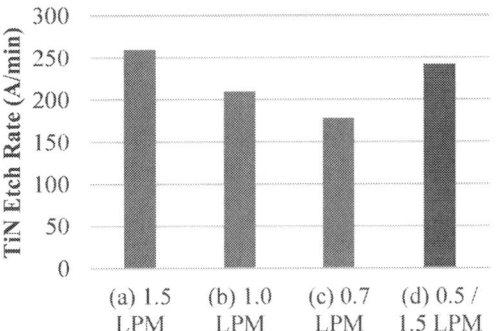

Figure 5: TiN etch rate variation of flow rate split (a) Chemistry 1.5 LPM, (b) 1.0 LPM, (c) 0.7 LPM with no backside dispense and (d) Front side 0.5 LPM Chemistry with backside 1.5 LPM Hot DIW 70°C, all processes were run at 500 RPM

SUMMARY

A model has been developed based on the dynamical balance between wetting and de-wetting velocities to describe the complex wetting/de-wetting phenomenon on a spinning wafer. The model also satisfactorily predicts the minimum flow rate required to completely wet a wafer surface at a given rotational speed and contact angle. The proposed model showed a good correlation with experimentally collected liquid coverage data. The TiN removal process is used as a challenge application in media consumption reduction. Using the model, a 66% chemistry reduction is achieved while maintaining baseline process performance.

REFERENCES

[1] M. Schneemilch, Langmuir, 14, P.7017 (1998). R. Smith, *Electrochem. Solid-State Lett.*, **10**, A1 (2007).
[2] I. Leshev *Chemical Engineering and Processing R*, 42 p. 925 (2003).

Advanced Material for Process Performance in Plasma Process

Xingjian Chen*, Tuqiang Ni, Shenjian Liu
Advanced Micro-Fabrication Equipment Inc.
*Corresponding Author's Email:
xingjianchen@amecnsh.com

For the IC chip production, the process stability is always the vital for the quality control in the plasma dry etching process. However, we found E/R on SiN always showed down-trend with the increasing of processed wafers in the aggressive process, such as HBr and Cl_2, it really brought a lot of trouble. The feedbacks from on-site productions (more than two customers) all are suffering from etch rate drifting issue. That need we are look into new advanced material to resolve the issue.

In both ICP (Inductive coupled plasma) and CCP (Capacitive coupled plasma) etch chambers, the chamber liner is designed to encapsulate the plasma inside the plasma reactor and its big surface is contacted with plasma during the plasma etching process. The variations of the composition and properties on the liner surface will sensitively affect the plasma etching process stability. Usually, the chamber liner is manufactured by Al alloy, while the liner surface is coated by plasma sprayed Y_2O_3 or other plasma resistant coating, to maintain the process stability and to extend the service life time.

The present work investigated the impacts of the plasma chemistry and process stability on SiN etch rate in plasma etching. It was found that the variations of the composition and properties on the liner surface will sensitively affect the plasma etching process stability.

So it suspects that if process recipe is aggressive enough and then process kit with high porosity and rough surface will be bad for process stability. Especially, as for PS coated kits, the feedback from other customers is suffering from etch rate drifting issue. AMEC study the productive issue and develop an advance coating material HD YO (high density and smooth surface coating) to improve the process stability.

Keywords—component; process stability; plasma etching; plasma etch chamber

1. Introduction

In the semiconductor manufacturing industry, one major challenge to process chips with high production yield is the creation of process stability chamber in the plasma dry-etching process. Especially, to some ICP processes, such as poly-Si etching and metal etching, the steps of the polymer deposition and de-deposition on the chamber surface, called as coating and de-coating, are required for the process to etch each wafer. The chamber coating means the in-situ pre-polymer deposition onto the chamber surface

or environment, so that the followed plasma etching on the wafer is conducted in a constant artificial environment. Since the plasma etching enables plasma to etch wafer and at the same time to erode the chamber surface, including liner surface, the former coated polymer on the chamber surface is damaged and cannot be used for the next wafer process. Therefore, the residual coating on the chamber surface or liner surface has to be removed, which is the called as the de-coating process that is always followed after each wafer etching.

Regarding to the coating and de-coating processes, the liner surface is very sensitive to have the influence on the plasma etching process stability. One possible issues is the over etching in the de-coating process which will erode the liner surface. Another issue is the roughness on the liner surface. As the rough surface makes more difficult to clean away the deposition through the de-coating, the remained residual coating will become thick along the plasma etching process and will late or soon to affect the plasma process stability. Again, another issue is the chemical composition stability on the liner surface. When the liner surface is coated with PS Y_2O_3, the Y_2O_3 surface is going to convert as YF3 or YOF (here YOF means that the material "YOF" contains the elements of Y, F and O, but the composition ratio of them varies according to the plasma chemistry) surface when the F contained plasma is used in the plasma etching. As the same thing happens to PS YF_3 coated liner, the YF_3 surface is converted as the YOF surface when the liner is used in the O contained plasma etching process. The unstable surface composition will definitely affect the process stability during the plasma etching.

With the rapid development on the ICP plasma etching processes, such as Logical Si Etch and Memory Poly Carbon Etch, the process stability requirement is very strict and critical [1]. Typically, it has been required that the limitation on the non-uniformity on the plasma etching rate (ER) is less than 1% and the limitation on the 3 sigma ER repeatability is less 1%. In addition, it has been limited that the process recovery time from PM to PM is less 8 hours and the chamber particle count through 10 pcs wafer process is less 1 at the size of 0.19nm. To meet all these or the similar technical specs, the plasma etch environment or the chamber surface, especially the chamber liner surface, has to be extremely and enduringly stable.

Therefore, it is necessary to improve the structure stability of the inside chamber surface, including the liner surface, to upgrade the plasma etching process performance and to meet the production technical spec in the advanced plasma etching process.

2. Experimental

The paper study the techniques and approaches to develop ICP chamber liners with the advanced coating and the coating surface is dense, smooth and stable so that the ICP chamber with the surface modified liner will maintain stable through the ICP etching processes. The paper also

978-1-5386-5309-8/18 $31.00 © 2018 IEEE

includes the development of the chamber liner with the dense and smooth surface that is formed by the plasma resistant coating. Again, the paper is the development of the chamber liner protected by the advanced coatings which has the top dense and smooth layer that is the plasma resistant coating and the bottom coating that has the stable structure and good adhesion on the liner surface, so that the application of the liner into the plasma etching process can improve the process stability and increase the production yield in the chip production.

The figure 1 shows the Standardized etch rate (E/R) monitor sequence is regularly performed to ensure the process stability. But Etch rate of OE on patterned nitride wafers gradually reduced with the increasing amount of processed wafers. The Question is: How to explain this phenomenon?

We did a lot work in order to try to find the real root cause behind and solve this issue. We found the E/R of OE on patterned SiN wafers differed a lot for different pre-conditions, but the ER drift always appeared. So, ER of OE on patterned SiN wafer is related to not only the pre-condition, but also the chamber process kits. We developed new process kits with high dense and smooth surface to solve this issue. As figure 1 show, we can create one process stability chamber with the new and advance material during the aggressive plasma process.

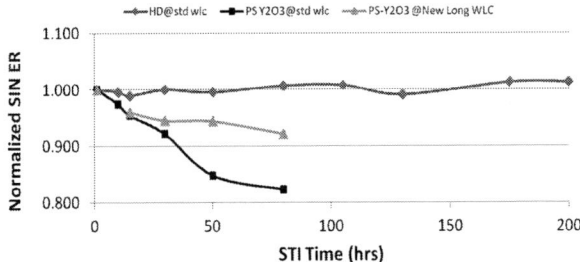

Figure 1. The SiN ER drifts during the aggressive process

3. Results and Discussion

Figure 2 shows the PS Y_2O_3 and PS YOF coupons placed near ESC in the SAB plasma etching process. Table 1 indicates the variation of the surface composition through 150 RF hours plasma etching process. It can be seen that the PS Y_2O_3 coating surface is changed as the YOF surface, and the F composition on the surface is increased from zero to 57%. However, for the PS YOF, the surface composition does not change so much, and all 3 elements of O, F and Y do not change more than 12% through 150 RF plasma etching process, which suggests that PS YOF can remain the stable surface composition even through the long service time in the coating and de-coating plasma etching process.

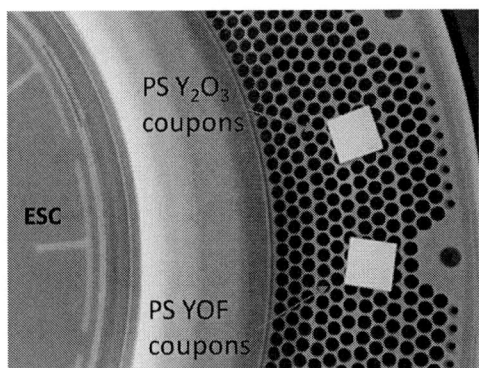

Figure 2. PS Y_2O_3 and PS YOF coupons placed near ESC for the surface composition stability test.

Table 1. Coating composition variation through 150 RF hours plasma etching process

Coating Element	PS Y_2O_3			PS YOF		
	0 RF hr	150 RF hrs	Chang	0 RF hr	150 RF hrs	Change
O % at	59.9	18.06	41.84↓	19.82	26.32	6.5↑
F % at	0.0	57.04	57.04↑	54.39	43.28	11.11↓
Y % at	40.1	24.90	15.20↓	25.79	30.40	4.61↑

The present paper disclose the formation of the chamber liner with the dense and smooth surface, and the liner surface is actually formed by PS Y_2O_3 or YF_3 coating with fine raw ceramic powders and is post-polished to reach a surface roughness (Ra) less 1.0um. Comparing with the conventional PS coating formed by the ceramic powder of 25-40um in size and with the porosity of 3-5 % and the surface Ra of 4.0 um or above, the PS coating formed by the fine ceramic powder of 10 um or less in size will have the reduced porosity less than 2% and the surface Ra less than 1.5um. In addition, the post polish can further reduce the surface Ra of fine powder formed coating to below 1.0um. The dense and smooth surface with the good plasma resistance could help to maintain the coating and de-coating without the change on the surface morphology and composition of chamber liner, and thus improve the stability of the plasma etching performance. The typical cross-sectional structure, the as-coated surface morphology and the polished surface from both the conventional PS coating and the fine powder PS coating are shown by the images in figure 3. It can be seen that the PS coating formed by the fine ceramic powder has the low body porosity or high volume density, and the smooth surface.

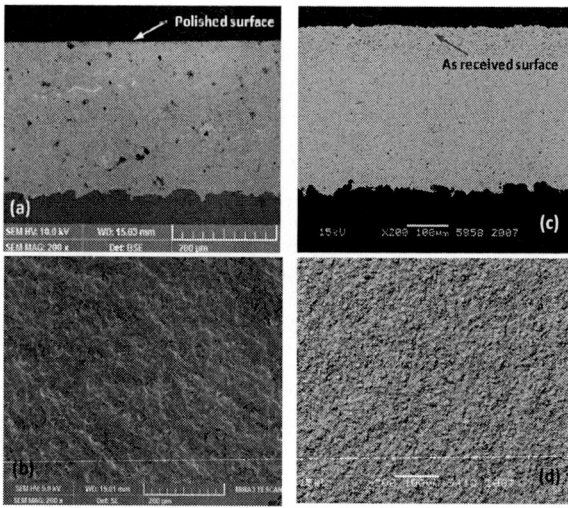

Figure 3. (a) The cross-sectional structure and (b) the as-coated surface of the conventional PS coating. (c) The cross-sectional structure and (d) the as-coated surface of the fine powder PS coating.

In one embodiment, the present paper disclose the formation of the chamber liner with the dense and smooth surface, and the liner surface is actually formed by plasma resistant coatings, such as Y_2O_3, YF_3, etc., by the advanced technology or deposition approaches so that the as-coated coating has the porous free dense structure and smooth surface with Ra less than 1.0 um, or less than 0.1um. The typical deposition approaches including the powder sprayed deposition but at the room temperature, such as cooled sprayed deposition (CSD) and/or aerosol deposition (AD) [2]. Other kinds of deposition approaches are the vacuum and plasma related deposition, such as the plasma enhanced physical vapor deposition (PEPVD), physical vapor deposition (PVD), chemical vapor deposition (CVD), and their combinations. For the CSD and AD, the ceramic powder is injected out off the spray gun and bombarded the material surface with the speed above the sound velocity, and the coating is formed through the impendence of the ceramic powders on the materials surface and the collision and micro plastic deformation of the ceramic powders. According to the process conditions, the CSD can be used to form the dense coating (without any porous defect) on any materials surface with the thickness of 100um or thicker and with the smooth surface (Ra<0.6um). In this invention, the CSD and AD Y_2O_3 coating is deposited over the chamber liner surface with the thickness in the range of 2um up to 200um and with the surface Ra below 1um or below 0.1um, based on the requirement from the plasma etching processes. Figures 4(a) to (d) demonstrates the top view and cross-sectional morphology of AD and CSD Y_2O_3 coating over Al substrates. It can be seed that very dense Y_2O_3 coatings with the thickness form 5um to 114um are prepared on Al_2O_3 and Al substrates. The polished surface has the low Ra below 0.1um. Since CSD and AD coating are formed at the ambient environment and the coating

usually has the very high residual stress, the post treatment has to be executed so that the coating could have a stable structure and good adhesion to the substrate.

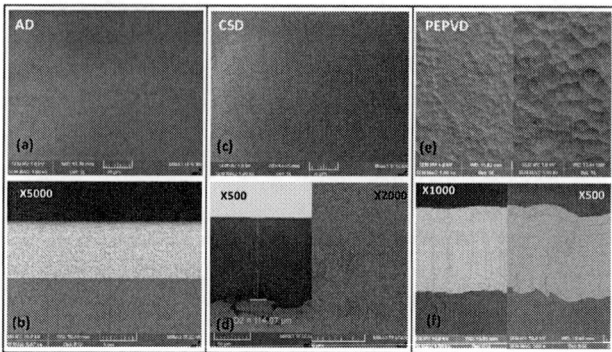

Figure 4. Advanced plasma resistant coatings prepared by AD in (a) and (b), CSD in (c) and (d), and PEPVD in (e) and (f).

Comparing to CSD approaches, PEPVD can be executed from the room temperature up to 200C according to the substrate materials and the service condition. In addition, PEPVD can prepare the coating with the multi-layered structure so that the coating stress can be released through the interfacial area. All these process conditions help to prepare the thick coating on the chamber liner for the long time service life. Figures 4 (e) and (f) show the top view and cross-sectional morphology and structure of PEPVD coating. It can be seen that PEPVD can synthesize the plasma resistant coating such as Y_2O_3 with different surface Ra and the different thickness through the multi-layered combinations

In one embodiment, the paper disclose the formation of the chamber liner with the dense and smooth surface, and the liner surface is actually formed by plasma resistant coatings that have the stable surface chemistry or chemical composition, such as YOF or YF_3, etc., through the plasma etching processes. One typical invention is to prepare YOF coating on the chamber liner that will be used in the plasma etching processes which contain F and O ionized plasma. Another application is to synthesize YF_3 coating on the chamber liner that will be used in the plasma etching processes which contains F-based plasma. This is because the conventional Y_2O_3 coated surface will be converted as YF_3 surface in the F-based plasma etching process, and will be reacted with F and O to form the YOF surface while the plasma etching chemistry consists of F and O contained plasma. However, the surface chemical change will take some process time during which the unstable surface chemical composition will continuous react with plasma and thus affect the plasma density and plasma distribution above the wafer surface. This phenomenon happens in both ICP and CCP plasma etching processes and makes the plasma etching rate changes over the wafer surface, which leads to the unstable production yields in the chip production. Especially, for to the coating and de-coating in

978-1-5386-5309-8/18 $31.00 © 2018 IEEE

ICP etching processes, the plasma etching in the de-coating process may more or less erode the Y_2O_3 coated surface on the chamber liner, the unstable surface in chemistry will always impact the coating formation in the next coating process and thus make the plasma etching process unstable in the etching performance.

This paper disclose the formation of the chamber liner with the dense and smooth surface, and the liner surface is actually formed by coatings that have the composite structure or multilayered structure in which the top layer is plasma resistant coating and the bottom layers have the composition and structure features so that the coatings over the liner surface have the desired functions, such as thermal conductivity, electrical insulation or conductance, structural stability, etc. One paper includes the deposition of the multilayered coating with the porous structure over the liner surface and the plasma resistant top coating is over the porous layers. In this invention, the top coating will not crack and keeps stable when the liner is heated to the relative high temperature, e.g. 80C, in some processes according to the plasma etching requirement. The typical application is to deposit monolayer or multilayered PS Y_2O_3 coating or Al_2O_3 coating or their combinations as the bottom layers on the liner surface and the porosities may be decreased from 10 % or higher as the layer contacted with liner surface to the 3-5% as the middle layer between the bottomed layer and the top plasma resistant layer. The relative high porosity helps to release the volume expansion or variation of the liner due to the temperature variation, and the top plasma resistant layer keeps the smooth and dense surface, which maintains the stable surface during the plasma etching process. In addition, the bottom layer(s) with enough thickness can provide the good electrical insulating performance. Another typical application is the deposition of thermal conductive layers as the bottom layers contacted with the liner surface, such as PS Yttrium Silicate, or PS Si, and then the plasma resistant coating, such as PEPVD Y_2O_3 or YOF is deposited over the bottom layers. With the adjustment on the layer thicknesses, the formed composite coatings on the liner surface will have the good thermal conductance and plasma resistance during the plasma etching process.

To ensure the yield in the actual chip production, the in-situ Etch rate (ER) of chamber is monitored at intervals, especial in original status and chamber PM. Based on material study in the plasma process; we have the followed hypothesis and conclusions:

1. The modifications on the chamber liner by the advanced PS Y_2O_3 or YF_3 coating with fine raw ceramic powders, so that the formed PS coating has the reduced porosity and improve d surface Ra. Plus the post-polished to reach a surface Ra less 1um, the formed chamber liner or parts can maintain the stable plasma etching performance in various plasma etching processes.

2. The modifications on the chamber liner by the advanced Y_2O_3, YF_3, and related coatings that is prepared on the chamber liner surface by the advanced deposition technologies, such as PEPVD, CSD or AD. The formed chamber liner or parts has the dense (or porous free) structure and smooth surface, and extremely stable resistance to plasma erosion, so that the formed chamber liner or parts can maintain the stable plasma etching performance in various plasma etching processes.

3. The modifications on the chamber liner with the dense and smooth surface, in which the chamber liner is formed by coatings that have the bottom PS or multi-layered PS coatings deposited on the liner surface and the top dense and smooth advanced plasma resistant coatings deposited on the PS Y_2O_3 surface. Therefore, the invented chamber liner can be modified with the bottom functional coatings to improve the thermal conductivity, the thermal expansion and the electrical properties and the top smooth, dense and plasma resistant coating to maintain the stable surface composition and geometric dimension. The implement of the invented chamber liner should improve and maintain the stable plasma etching performance in the various plasma etching processes.

4. Summary

It was found that chamber is sensitive to only the pre-condition, but also the chamber process kits surface. AMEC can create process stability chamber for the aggressive plasma process to meet the very strict and critical plasma process by developing advanced high dense material.

Reference

[1]95Developments_of_Plasma_Etching_Technology_for_Fabricating_Semiconductor_Devices-3. Jpn.J.Appl.Phys, Vol.47
[2] Aerosol Deposition of Ceramic Thick Films at Room Temperature: view issue TOC Volume 89, Issue 6 june 2006 Pages 1834-1839.

Germanium Compatible Photoresist Strippers and Residue Cleaners
For 5nm and Below Technology Nodes

Chien-Pin Sherman Hsu

Avantor

Avantor, No. 38-1, Tai-Yuan St., Chu-Bei, Hsinchu, Taiwan
sherman.hsu@avantorinc.com

Biography

Chien-Pin Sherman Hsu, Ph.D., serves as a Technical Fellow and Global R&D Director, Electronic Materials at Avantor. Current programs focus on products for advanced semiconductor technology nodes of 5nm and beyond, including high-performance implant photoresist strip, residue clean, selective etch, surface preparation and Ge/III-V integrations. With more than 25 years in research, Hsu has established a series of Cu and low-k compatible cleaning chemistries and novel techniques that include supercritical fluid cleans in microelectronics and nanoelectronics. His recent work on selective etchants and cleaners provide fine profile tuning, surface conditioning and integration of a wide range of materials in FEOL, MOL and BEOL applications. Hsu received a Ph.D. degree in Chemistry from Cornell University, Ithaca, NY, USA. He has authored more than 50 patents, publications and conference presentations.

Abstract

This study describes the development of both new acidic- and alkaline-based SiGe and/or Ge compatible clean chemistries for efficient removal of photoresist (PR) and post-etch residues in 5nm or below FEOL and BEOL applications.

Keywords— Ge Compatible Cleaner, Stripper, Residue Clean, SiGe, Ge, Germanium Oxide, SPM

Introduction

Advanced semiconductor technologies with shrinking geometries result in new challenges in wet cleans. New materials, such as SiGe, Ge, and III-V, demonstrate very different properties from silicon-based substrates. For example, germanium oxides show much lower stability than silicon oxides. SiGe with high Ge content or 100% Ge show high etch rates with many common wet clean products. As a result, processes that use these substrate materials require new clean chemistries to ensure compatibility. This study describes the development of both new acidic- and alkaline-based SiGe and/or Ge compatible clean chemistries for efficient removal of photoresist (PR) and post-etch residues in 5nm or below FEOL and BEOL applications.

Experimental

Wet clean experiments were conducted in beakers and the SEZ (LAM) SP304 Single Spin Tool. Film thicknesses were measured with X-ray Fluorescence (XRF), ellipsometry or 4-point probe. Removal of PR films or residues was examined with optical microscopy or High Resolution Scanning Electron Microscopy (HR-SEM).

Results and Discussion

Ge Compatibility and Cleaning Challenges

Due to the instability of germanium oxides, especially germanium dioxides, many common and effective cleaners cannot be used for cleans when Ge compatibility is required. Ozone-based cleaners and sulfuric acid-hydrogen peroxide mixtures (SPM, or Piranha Solution) cannot be used for wet cleans. [1]. In addition, alkaline-based cleaners can attack Ge and poly Si.

To overcome such challenges, various approaches to protect Ge surfaces have been proposed: alkanethiolate self-assembled monolayers (SAMs), sulfur-covered Ge surface or thin Si capping on Ge [2, 3, 4]. However, the effects of these added layers or their subsequent removal (if needed) will need to be carefully studied.

Thus, it is desirable to develop efficient Ge compatible cleaners without any added layers on Ge.

Ge Compatible Cleaner Development

Thin native oxide layers can be commonly found on Ge surfaces. They usually are a mixture of Ge dioxide (GeO_2) and Ge sub-oxides (GeO_x, $x < 2$). [1] The sub-oxides have better stability than GeO_2.

Oxidant types, content levels and processing conditions can generate different mixtures of germanium oxides, which exhibit tunable etch behaviors.

As previously reported, standard H_2O_2-based chemistries lead to excessive Ge etch. [6] In our study, both semi-aqueous alkaline chemistries with H_2O_2 (AK-1) and semi-aqueous alkaline chemistries without H_2O_2 (AK-2) have been demonstrated to result in high Ge etch rates (Table I). However, Ge film can be protected through the controlled growth of a protective layer by the use of a similar semi-aqueous chemistry with special additives (AK-3). A layer of approximately 10-20Å thickness, likely consisting of Ge sub-oxides, may be formed to act as passivation. However, such sub-oxide layers are preferably kept at a minimum thickness or totally avoided for device integration.

Acidic-based and Alkaline-based Ge Compatible Cleaners

Balancing between controlled growth and etch of oxides and sub-oxides for both silicon and germanium (e.g. SiO2, SiOx, GeO2, GeOx; x<2), various alkaline-based or acid-based clean chemistries have been successfully developed to provide excellent substrate compatibilities with sufficient cleaning capabilities. They demonstrate excellent compatibilities with a wide range of substrates, including Ge, poly Si, TEOS, SiN, Ti, W (Table II). They are proven to have effective PR and residue cleaning performance with typically $45 - 65$ C processing temperatures on single wafer processing tools. Clean options include 1-step and 2-step processes (Table III).

TABLE I

Etch Rates of Selected Alkaline-based Semi-Aqueous Chemistries (SAC)

Chemical Description	Poly Si Etch Rate (Å/min)	Ge Etch Rate (Å/min)	TEOS Etch Rate (Å/min)	Chemistry Types
40:1 (AK-1)-30% H2O2, 75C, 5m	<4	>80	2	SAC Alkaline + Oxidizer
AK-2, 70C, 5m	>100	55	2	SAC Alkaline
AK-3, 75C, 5m	>100	<1 (17Å increase of film thickness)	<1	SAC Alkaline + Additive 1A

TABLE II
COMPATIBILITY COMPARISONS BY ETCH RATE (Å/MIN) MEASUREMENTS

Chemical Description	Poly Si Etch Rate	TEOS Etch Rate	Ge Etch Rate	SiGe Etch Rate	W Etch Rate	TiN Etch Rate	Chem Types
AC-1, 60°C	<1	2	1	<1	<1	<1	Acidic
AC-2, 55°C	<1	<1	<1	<1	<1	<1	Acidic
AK-4, 60°C	<1	<1	<1	---	<1	<1	Alkaline
AK-5, 60°C	<1	<1	<1	<1	<25	<1	Alkaline

TABLE III
EVALUATION OF PR AND RESIDUE CLEAN EFFICIENCIES IN STRUCTURE WAFERS WITH GE FILM

STACKS

Process Description	PR and Residue Clean Efficiency	Chemistry Type
AK-4 based 2-Step Clean (AK-4, 60°C, 5min + DHF, 25°C, 0.5min)	100%	Alkaline
AC-1 based 2-Step Clean (AC-1, 60°C, 5min + DHF, 25°C, 0.5min)	100%	Acidic
AK-5 based 1-Step Clean (AK-5, 60°C, 5min)	100%	Alkaline
AC-2 based 1-Step Clean(AC-2, 55°C, 2min)	100%	Acidic

References

[1] B. Onsia et al., Proceeding of the Int. Symp., UCPSS, Belgium (2004)
[2] P. Ardalan et al., Langmuir, 25(4), p. 2013-2025 (2009)
[3] S.M. Han et al., J. Am. Chem. Soc., 123(10), p.2422-2425 (2001)
[4] P.W. Mertens et al., ECS Transactions, 41(5), p.3-13 (2011)
[5] J-I. Song, R. Novak, I Kashkoush and J. Molinaro, Semiconductor Fabtech, p.1-5 (1996)
[6] Virginie Loup et al., Solid State Phenomena Vol. 134 (2008) pp 37-40

MEGA-SONIC EFFECT ON SUB-RESOLUTION ASSIT FEATURES DURING PHOTOMASK CLEANING

Sophia Xue*, Irene Shi, Alan Li, Schafer Qin, Eric Tian, Max Lu, Eric Guo
Semiconductor Manufacturing International Corporation
18 Zhangjiang road, Pudong New Area, Shanghai, China 201203
*Corresponding Author's Email: Sophia_Xue2@smics.com

ABSTRACT

As the shrinkage of feature size and the increase of pattern density, specification requirements for photomask cleaning is getting more and more challenging. Mega-sonic energy is widely applied in photomask cleaning, while it may lead to the damage of Sub-Resolution Assist Features (SRAF) with small size. In this work, different magnitude of mega-sonic energy has been evaluated and its effects on SRAF are studied. High end phase shift mask (PSM) cleaning could be accomplished under the condition that the power of mega-sonic is suitable.

Keywords—mask clean, sub-resolution assist pattern, mega-sonic, phase shift mask, pattern damage.

INTRODUCTION

The task of mask clean is to remove resist, particle, glue and other contamination from the mask surface. Hydrogen peroxide based wet chemical cleaning method is invented by Werner Kern at RCA[1]. The usage of mega-sonic on chemical such as SC1 provided more opportunities to apply external forces on particles which is helpful for the removal of contaminant, especially for mask with high pattern density[2]. Whereas, the mega-sonic energy may cause small size pattern damage during mask clean[3, 4]. Therefore, it becomes increasingly difficult as smaller size pattern designed on optical masks is sensitive to mega-sonic vibration. On the other hand, tetramethylammonium hydroxide (TMAH) is an effective choice for its favorable pH and Zeta potential, which plays an important role in the process of particle removal[5]. Meanwhile, the big cation size of TMAH molecule ($N(CH_3)_4^+$) hinders the access of etching agents into the photomask materials[6]. In addition, there is no transient cavitation risk during the clean process using TMAH which may lead to less SRAF damage[5, 7].

In our work, we researched the clean performance and SRAF damage condition under 3MHz mega-sonic with different power use the chemical of SC1. And the result indicates that traditional wet clean is still a good choice for high end mask cleaning as long as the mega-sonic energy is controlled at an appropriate range. At the same time, the SRAF damage condition under 4MHz mega-sonic with different power use the chemical of TMAH is investigated as well.

EXPERIMENTAL

The damage conditions of SRAF features with different length, from short to medium and long, were investigated. The mega-sonic power was altered between 0 and 80% for the frequency of 3MHz assisted with SC1, and 4MHz assisted with TMAH. After each clean procedure, the test mask was investigated by inspection system at P72 via die to die setting. The detailed SRAF structures were analyzed with Scanning Electron Microscope (SEM) and Atomic Force Microscope (AFM) image.

RESULTS AND DISCUSSION

The test mask is designed with SRAF features of different length and main pattern. Figure 1 shows atomic force microscope (AFM) images of a patterned MoSi-quartz PSM photomask. The SRAF features and main pattern are well aligned. The distance between SRAF pattern to main pattern and one SRAF head to another one are controlled at a reasonable level by following the photomask design rule. Meanwhile the pattern density of this test mask is similar with online production.

The clean performance and SRAF damage condition of 3MHz mega-sonic with different power (from 80% to 0) using the chemical of SC1 is studied on the test mask. After each test, by comparing the inspection coordinates with base inspection results, the new added SRAF damage and particle count are calculated. It is obviously shown in figure 2 that the SRAF damage count is reduced dramatically with the decrease of mega-sonic power. Under the condition that the mega-sonic power is the highest, the damage of SRAF with short length is the most serious, and the medium length and long length SRAF are at the same level. When the mega-sonic power reduced to next grade, there is small account of SRAF damage, and continue to lower the mega-sonic power, the phenomenon of SRAF damage is disappeared. On the contrary, the particle count is the highest when no mega-sonic power is applied to the photomask. With the increase of mega-sonic power, the particle count could be controlled at an acceptable level. Therefore, a suitable clean recipe of no SRAF damage and reasonable clean performance could be obtained by tune the 3MHz mega-sonic power.

978-1-5386-5309-8/18 $31.00 © 2018 IEEE

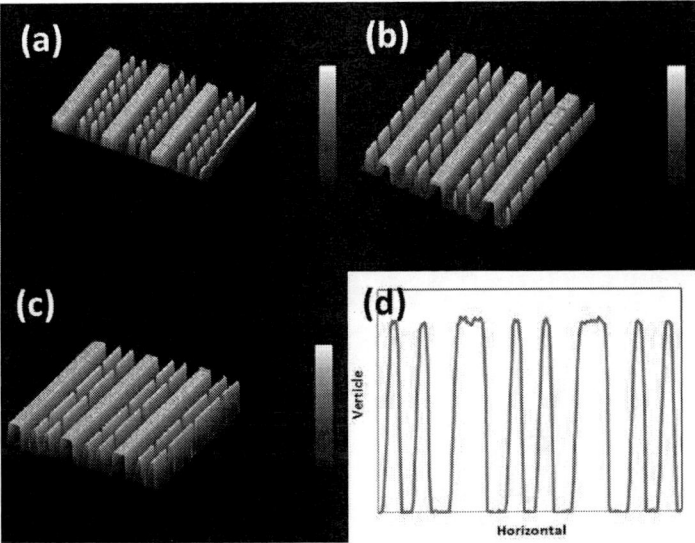

Figure1: AFM images of SRAF features with different length (a) short length SRAF, (b) medium length SRAF, (c) long length SRAF feature, (d) schematic diagram of SRAF's cross section.

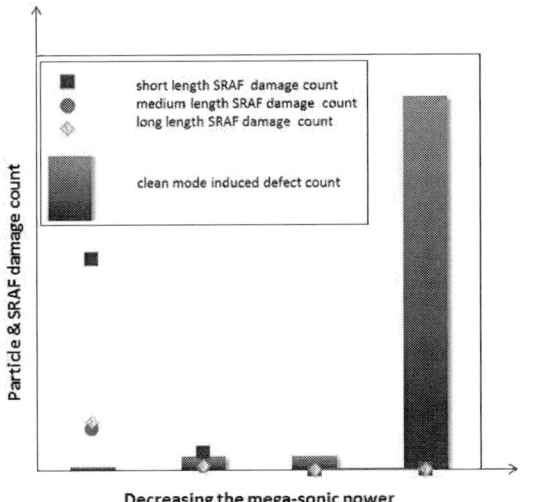

Figure2: The SRAF damage condition and particle count under 3MHz mega-sonic with different power use the chemical of SC1.

Figure 3 shows an example of damaged SRAF features. The adhesive force of short and medium length SRAF features are relative small compared with long length SRAF feature, and they tend to be totally removed under the action of mega-sonic. While, the impact surface of long length SRAF feature is larger, and it seems more likely broken at the middle of SRAF line.

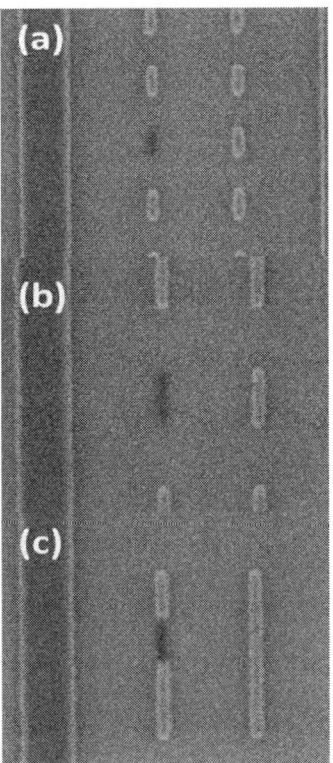

Figure 3: Example of damaged SRAF features of different length (a) short length SRAF, (b) medium length SRAF, (c) long length SRAF.

978-1-5386-5309-8/18 $31.00 © 2018 IEEE

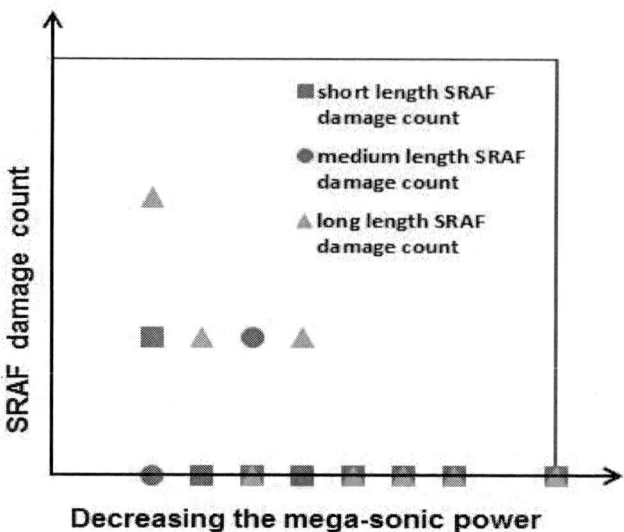

Figure4: The SRAF damage condition under 4MHz mega-sonic with different power use the chemical of TMAH.

In order to widen the adjustable range of mega-sonic power, the SRAF damage condition under 4MHz mega-sonic using the chemical of TMAH is studied as well. The preliminary results are shown in figure 4. It is clearly that the trend of SRAF damage count is similar with 3MHz, however, the SRAF damage count is less than 5(the detailed data is not shown here), even under the highest mega-sonic power. Therefore, the mega-sonic power could be tuned at a large range which is benefit for the adjustment of process.

CONCLUSIONS

The clean performance and SRAF damage condition under the action of 3MHz mega-sonic with different power using the chemical of SC1. With the decrease of mega-sonic power, the SRAF damage condition is improved, while the clean performance becomes terrible. And the reasonable mega-sonic power is found based on the balance of SRAF damage and clean performance. On the other hand, 4MHz mega-sonic with TMAH shows a large adjustable range of mega-sonic power for high end photomask technology node.

ACKNOWLEDGEMENTS

This work was supported by Shanghai City Industrial Innovation and development of the Internet (special application), the development and demonstration application of complete equipment on 14nm high-end photomask cleaning, Shanghai Municipal Commission of Economy and Information.

REFERENCES

1. Kern W., P., D. A., RCA Review, vol. 31, 1970.
2. H. Shende, S. Singh, J. Baugh, et al., *Proc. of SPIE*, vol. 8522 , 2012 , 852217.
3. R. Singh, *Photomask Technology and Management*, vol. 2322, 1994, pp 165-177.
4. S. Helbig, S. Urban, E. Klein, et al., *Proc. of SPIE*, vol. 7122, 2008, pp 712210.
5. D. Dattilo, U. Dietze, *Proc. of SPIE*, vol. 9256, 2014, pp 925604.
6. I. Zubel, I. Barycka, K. Kotowska, et al., *Sensors and Actuators A*, vol. 87, 2001, pp 163-171.
7. S. Singh, U. Dietze, P. Dress, *Proc. of SPIE*, vol. 8701, 2013, pp 870104.

978-1-5386-5309-8/18 $31.00 © 2018 IEEE

Fast model for Positive Tone Development (PTD) Source Mask Optimization (SMO) calibrated by dose tuning

Song Bai, Xuan Li, Manhua Shen

Technology R&D, SMIC Advanced Technology R&D (Shanghai) Corporation
Pudong New Area, Shanghai, P. R. China 201203
+8621 38610000x15957, Cypress_Bai@smics.com

ABSTRACT

The measurement of critical dimension (CD) using scanning electron microscopy (SEM) brings the blur of CD measured compared with the absolute CD from optical simulation. Although optical property is dominant in PTD process, the CD bias between measurement and simulation with optical model will disqualify the accuracy of CD prediction, EPE calculation and mask correction in SMO. In this paper, using multiple different 1D patterns with different exposure latitude (EL) could discovery the global bias between simulation with optical model and wafer data measured. Due to the property of photoresist shrinkage and the identical profile of PTD through pitches, this bias works well in 1D patterns and assists in anchor re-matching between wafer data and simulation, which could control dose shift between wafer situation and simulation under 1mJ. This bias is also not relevant to source type and will improve the accuracy of PTD SMO highly without resist model.

INTRODUCTION

It's widely known that SMO is the optimization process requiring iteration of building resist model, setting more effective process settings and tuning source. The difference of optimization settings between first and final version of SMO could be represented in different mask limit settings, scattering bar settings and After Development Inspection (ADI) target reformed, which is unforeseen at the first optimization. Although optical model considering photoacid diffusion could provide good trend description of CD for PTD, there is still CD prediction error as large as 10nm for model not calibrated by wafer data. It will influence the process window prediction and the balancing of source among multiple types of patterns. Moreover, the application of mask limit into source optimization and the prediction of scattering bar printing demands the accuracy of aerial image, whose CD prediction error is under 3nm, and it is normally achieved by resist model building. However, resist model building for every step in iteration means heavy data collection.

Data collection for optical proximity correction (OPC) is normally completed by SEM. However, CD value obtained by SEM is an artificial value partially determined by SEM beam condition and method applied in obtaining CD value from SEM signal. For example, a fixed pattern about 50nm, the difference among values obtained by multiple typical CDSEM settings could be larger than 6nm. But in simulation, the CD value is a physical value obtained by slicing aerial image calculated. The gap between the artificial and physical value will interfere the anchor pattern CD matching between wafer data and simulation without resist model, which could then interfere the dose matching between real wafer situation and simulation without resist model. Sometimes it's the primary reason of CD prediction error for simple optical model with image blur that the dose shift between real wafer situation and simulation could be as large as 20%. Not only photoresist (PR) but also mask suffer from the gap between physical CD in simulation and artificial CD in measurement, and the global bias of mask is under 8nm(4x). In this paper, the method to find the amount of dose shift is proposed, and it only needs limited wafer data. The fast new model calibrated by dose fine-tuning could give CD prediction accuracy under 3nm for the mass of patterns, which meet the initial requirement of source optimization for PTD, accurate prediction of mask limit and scattering bar printing.

THE ASSUMPTION OF MEASURMENT BIAS

There are two inducement that induces the blur of CD measurement in SEM. The one is the signal broadening during SEM scanning, which requires artificial rule to slice it to extract CD information, and the other one is the shrinkage effect of ADI. Due to the consistent profile through pitches for PTD process as figure 1 shows and the charging property of photoresist, the two inducement should lead to an almost fixed bias through pitches for 1D pattern.

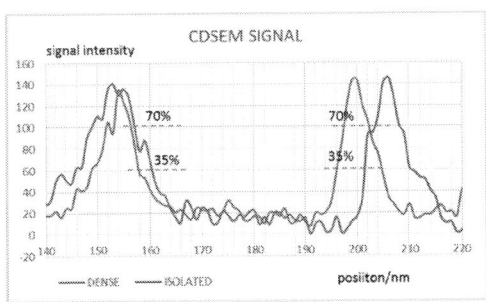

Figure 1: Dense pitch and isolated pitch SEM signal comparison. Threshold values are set to intercept CD values.

Mismatch between wafer data and simulation

Figure 2 shows the CD and slope mismatch between simulation with optical model and real wafer situation. Three patterns include a high EL pattern (anchor pattern), a medium EL pattern and a low EL pattern. We select the shot under the energy which makes the CD of anchor pattern match with simulation CD directly, but the other pattern such as medium EL pattern or low EL pattern will mismatch severely.

978-1-5386-5309-8/18 $31.00 © 2018 IEEE

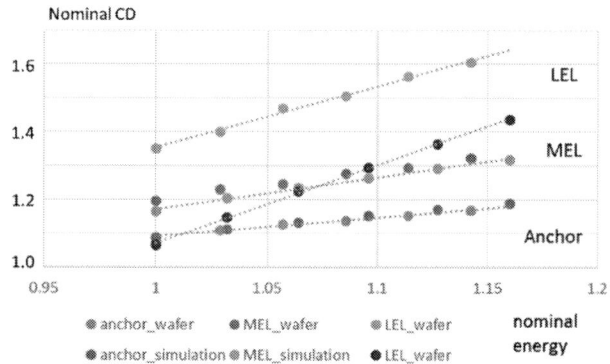

Figure 2: EM CD curve with nominal energy for three patterns in both wafer data and simulation with optical model.

Re-matching between wafer data and simulation

As for the effect of CD blur because of two inducement, it assumes a fixed CD bias between real wafer situation and simulation with optical model for 1D patterns. And then we set it as an independent variable to watch matching results for the low EL group of patterns and medium EL group of patterns. When the bias varies from 0 to 5-6nm, the root mean square (RMS) varies from 9nm to under 3nm for ten 1D patterns (dense and isolated pitched, forbidden pitch, none-uniform pitch). It means that trench CD 60nm in simulation means 57nm in wafer data and trench CD 45nm in simulation means 42nm in wafer data. Figure 3 shows the re-matching results of the three patterns. Figure 4 shows the EL prediction improvement after rematching.

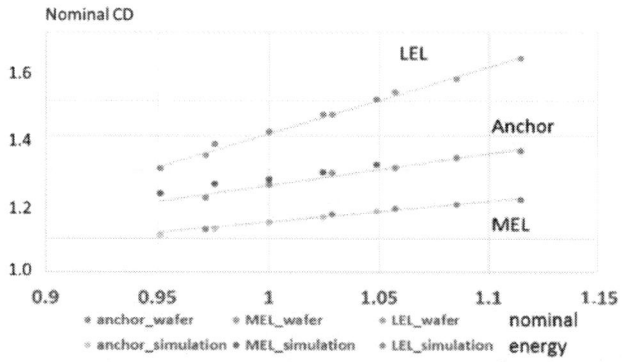

Figure 3: EM CD curve with nominal energy for three patterns in both wafer data and simulation with optical model after 5nm pre-bias between wafer data and simulation.

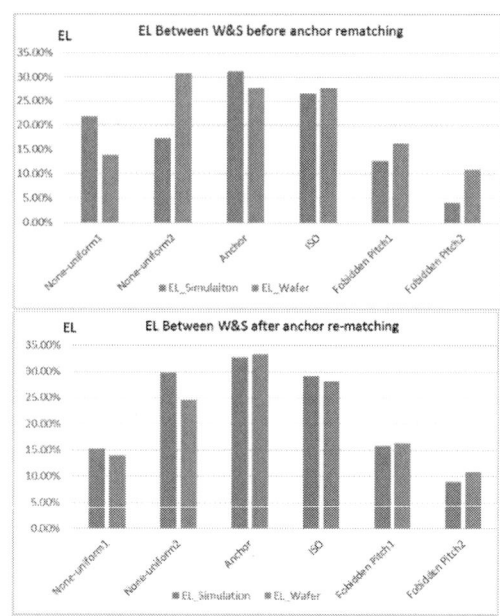

Figure 4: two figure shows the EL matching between wafer data and simulation before and after re-matching.

All the evidence above illustrate that the measurement bias exists as simple and effective resist terms which could fit wafer data well. Furthermore, we also found that the prediction of scattering bar printing energy is as accurate as 1-2mJ, while before re-matching it's about 6-7mJ. It indicates that the measurement bias is not only a mathematical fitting terms but also a concept with physical meaning which could calibrate the energy shift between simulation and wafer situation.

We also check the accuracy of prediction for 2D patterns. The contour comparison between a resist model and an optical model for the same mask shows that the line-end contour and CD prediction is good, while the "Z" shape pattern owns big discrepancy.

MEASUREMENT BIAS FOR PTD SMO

For initial source tuning for SMO, it's without any resist model or accurate process information such as scattering bar rule or mask bias. After the first tuning of source, the wafer data could be collected to build a resist model and then to fine tune the source. If measurement bias could keep constant among different source type, the optical model with measurement constant should provide much better performance for initial source tuning. Figure 5 shows the accuracy of prediction with the optical model of source B and the measurement bias from source A. Not only the error of CD prediction under nominal condition but also the CD under exposure energy highly off target are controlled under 3nm. It indicates the exact placement error calculation with exposure energy variation for the optical model of source B and measurement bias from source A. Even for the SMO source of dipole type, which covers mainly the patterns in one direction, the optical model with measurement bias is enough for PTD source tuning.

978-1-5386-5309-8/18 $31.00 © 2018 IEEE   246

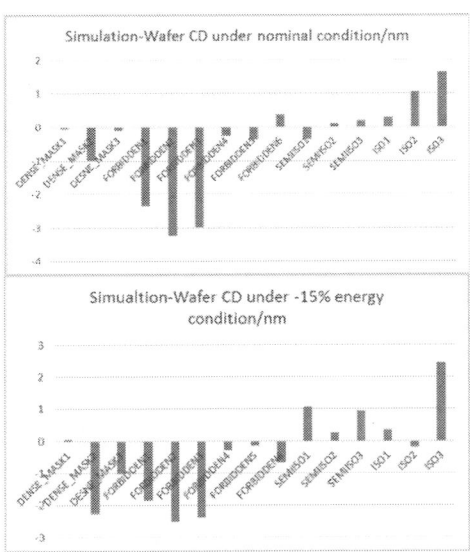

Figure 5: CD predict error of ten 1D patterns for both the nominal condition and the -15% energy condition under optical model with measurement bias obtained from the other source but the same film stack.

SIMILAR EFFECT FOR MASK AND NEGATIVE TONE DEVELOPMENT (NTD)

The measurement bias for mask

Mask measurement also uses CDSEM as tool, so the similar problem also exists for mask CD definition. The typical settings for mask measurement could lead to 5-6nm (4x) bias, i.e.1-2nm (1x), which could bring inherent error for OPC resist model fitting. The real mask CD, real ADI, exposure energy and resist model are four parameters entangled mutually and distinguished hardly. The simple optical model with photo-acid diffusion length could help in investigating the bias of mask and ADI with patterns sensitive to exposure energy and mask variation to alleviate the inherent error of OPC model.

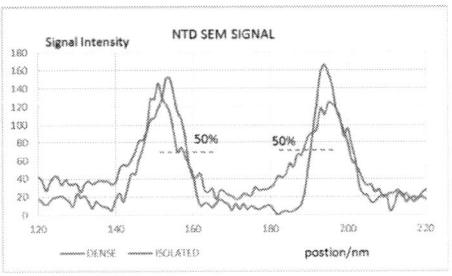

Figure 6: Dense pitch and isolated pitch SEM signal comparison for NTD. Threshold values are set to intercept CD values.

The similar effect for NTD

The chemical reaction happens in the exposed parts of NTD photoresist and will not be developed. For the same trench CD, the shrinkage effect by chemical reaction will become much more severe for larger pitches. It indicates that the SEM signal will vary largely from dense trench to isolated trench as figure 6 shows. Hence the measurement bias will not be a constant for 1D pattern.

SUMMARY

The assumption are proposed that the simulation CD with optical model owns a constant bias with wafer data of PTD process. By this assumption, the re-matching of anchor point and energy coordinate between simulation and wafer situation are completed. More accurate CD, placement error and scattering bar printing prediction appears after the re-matching. The measurement bias is also a constant for 1D pattern not relevant to source type, which could be applied into the initial source tuning for SMO effectively. Similar effect exists for mask or NTD ADI measurement, but could not be easily extracted and applied.

ANALYSIS OF REVERSED DEPTH MICRO-LOADING BASED ON PULSED PLASMA

Guodong Chen, Liming Yao, Zhongwei Jiang, Yahui Huang

Beijing NAURA Microelectronics Equipment CO.,Ltd

No. 8 Wenchang Road, Beijing, 100176, P.R. China

chenguodong@naura.com

ABSTRACT

Depth Micro-loading control is being generally concerned for the latest semiconductor manufacturing technology during plasma etching, e.g. 14nm CMOS and 3D NAND device etch process. Most current leakage between two adjacent devices can be optimized by changing depth micro-loading. In this paper, a silicon etching method is developed at NAURA etch tool by using CHF3 as the main etch gas under a pulsed plasma. It shows a widely controllable tuning window for depth micro-loading. Normal depth micro-loading can be eliminated or reversed. The trend is shown that the depth micro-loading decreases with the ratio of CHF3/HBr increase.

Keywords—Depth Micro-loading; CHF3; pulsed plasma

INTRODUCTION

With the semiconductor manufacturing technology getting to 14nm node and beyond, the critical dimension of device is rapidly reduced. The depth micro-loading effect has become one of the most important issues for device electrical performance control [1]. Nowadays, the common approaches for this problem are to adjust the process condition, or to optimize the hardware structure of etching tool [2]. However, these methods can only reduce the depth micro-loading effect to a certain level and hardly to eliminate it completely.

DISCUSSION

Halogen gas (Cl_2) 、 fluoride gas(SF_6,NF_3,CF_4) and halogen hydride gas(HBr) are usually applied in silicon etch process for their good etching capability. Most of these etching process will cause a normal depth micro-loading. Normal micro depth loading means the depth of open area with a larger trench critical dimension is deeper than the depth of dense area with a smaller trench critical dimension. **Fig1(a)** shows one typical STI profile etched by continuous plasma using Cl_2/HBr. The image shows that the open area has an obviously larger depth than the dense area. The normal depth micro-loading is 43nm. **Fig1(b)** shows another typical STI profile etched by continuous plasma using CHF_3/HBr. It displays a normal depth micro-loading of 30nm similar like in **Fig1(a)**. Usually, normal depth micro-loading is easily to find during silicon etch using continuous plasma regardless of different etch gas group.

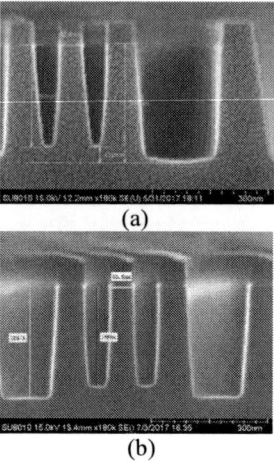

Fig.1 STI profile etched by continuous plasma using ((a)etching by Cl_2/HBr, (b)etching by CHF_3/HBr)

It has been reported that the pulsed plasma can be applied to effectively reducing the normal depth micro-loading [3,4]. But the max capability of reduce depth loading still not be reported. To finding the tuning window, based on Fig 1(b) etch gas condition, a bias pulsing plasma was added into silicon etch process by using CHF_3/HBr on NAURA etch tool (pulsing frequency=200Hz, pulsing duty=25%). **Fig2(a)** shows the optimized STI profile etched by bias pulsing plasma using CHF_3/HBr. With pulsed plasma, the depth micro-loading was effectively decreased from 30nm to 14nm. It means the pulsed plasma has a great positive impact on reducing depth micro-loading.

For further optimization of depth micro-loading, some other extending etching conditions using bias pulsing plasma were investigated based on **Fig.2(a)** etch condition. With the increase ratio of CHF_3/HBr, the normal depth micro-loading can decrease to a lower level, which can equal zero near. If the gas ratio increase further, the depth micro-loading can be reversed to a more negative value. The reversed depth micro-loading means the depth of open area with a larger trench critical dimension is shallower than the depth of dense area with a smaller trench critical dimension. **Fig.2(b)** gives a zero near(-2nm) depth micro-

978-1-5386-5309-8/18 $31.00 © 2018 IEEE   248

loading STI profile SEM image. **Fig.2(c) and Fig(d)** give two different level reversed depth micro-loading. The depth micro-loading is -13nm in **Fig.2(c)** and -20nm in **Fig.2(d)** respectively. This result is totally different from the traditional silicon etch process.

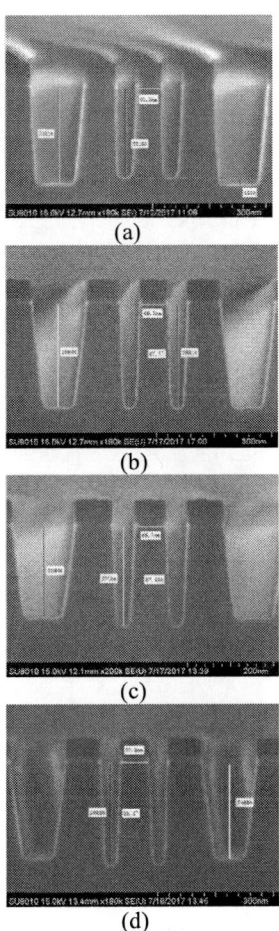

(a)

(b)

(c)

(d)

Fig.2. STI profile etched by optimized bias pulsing plasma using CHF_3/HBr(the gas ratio of CHF_3/HBr: (a)>(b)>(c)>(d))

Fig.3 shows the relation between the depth micro-loading and the etching gas ratio of CHF_3/HBr. With the bias pulsing plasma, the depth micro-loading continuously decreases to zero, and eventually gets into the reversed profile when the CHF_3/HBr gas ratio increases gradually. During the process, a better way to tune the gas ratio is by changing CHF_3 gas flow. The results also illustrate that adding more CHF_3 has a decreasing effect on depth micro-loading under bias pulsing plasma while more HBr has an increasing effect on depth micro-loading. Further investigation to control the depth micro-loading with

different gas were carried out. For other typical silicon etching gas like CF_4, NF_3, SF_6 and Cl_2, they all have increasing effect on depth micro-loading just similar like HBr.

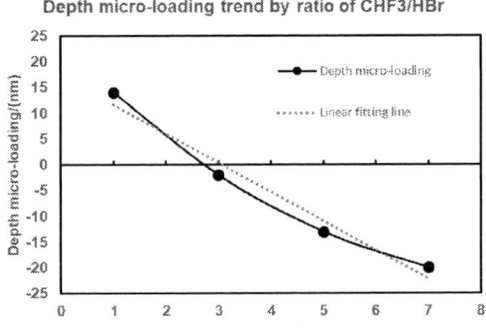

Fig.3. Depth micro-loading trend by ratio of CHF3/HBr after silicon etching using bias pulsing plasma

The relationship between other process parameters and reversed depth micro-loading is also investigated. **Fig.4** shows the relationship between the depth micro-loading and the etching gas ratio of bias pulsing duty. **Fig.5** shows the relationship between the depth micro-loading and pressure. **Fig.6** shows the relationship between the depth micro-loading and the ratio of SRF power/BRF power. From these trend curves, it is easily to find that depth micro-loading can be adjusted by tuning these process parameters under during bias pulsing plasma etch. Besides higher ratio of CHF_3/HBr mentioned above, lower bias pulsing duty, higher pressure and higher ratio of SRF power/BRF power can also help to reduce or reverse the depth micro-loading. However, a key element for all recipe conditions which can get zero near or reversed depth micro-loading is that CHF_3 must be used as the main etch gas during silicon etching process.

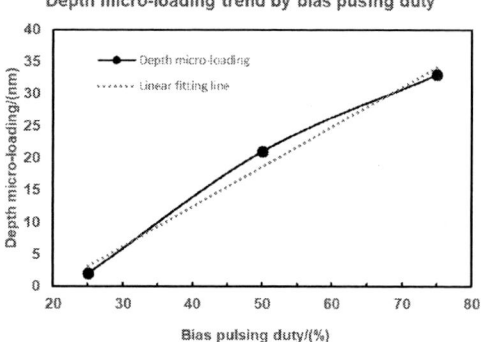

Fig.4. Depth micro-loading trend by bias pulsing duty after silicon etching using bias pulsing plasma

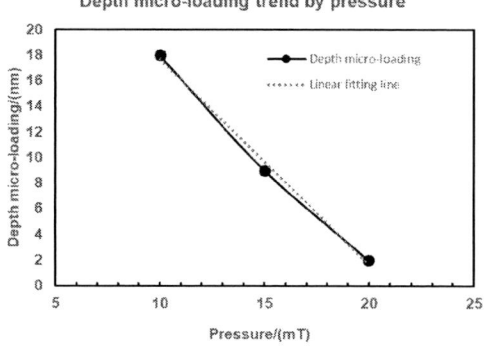

Fig.5. Depth micro-loading trend by pressure after silicon etching using bias pulsing plasma

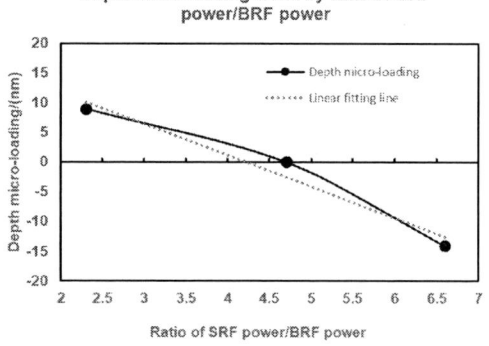

Fig.6. Depth micro-loading trend by ratio of SRF power/BRF power after silicon etching using bias pulsing plasma

As an etching gas, CHF_3 has very special characteristics in silicon etch process under the pulsed plasma. It shows a widely controllable tuning window for depth micro-loading. Normal depth micro-loading can be eliminated or reversed. An explanation for reversed depth micro-loading is that CHF_3 etch gas can have a deposition effect in open-area during the RF off time of bias pulsing plasma, which finally slows down the silicon etch rate of open area in total etch process. Usually, higher gas ratio of CHF_3, lower bias pulsing duty, higher pressure and higher ratio of SRF power/BRF power will help to enhance the deposition effect. So these parameter changes can also have a positive impact on reducing depth micro-loading. That can explain that why reversed depth micro-loading usually emerges under process conditions with higher gas ratio of CHF_3, lower bias pulsing duty, higher pressure and higher ratio of SRF power/BRF power during silicon etching using bias pulsing plasma.

CONCLUSION

Reversed depth micro-loading is hardly to find in silicon etch process. Using CHF_3 as main etch gas under bias pulsing plasma is an effective way to eliminate or reverse the depth micro-loading. It can further increase the process window of silicon etch process. As the consequence, the control window of device electrical performance can also be effectively expanded by using well reversed depth micro-loading.

ACKNOWLEDGEMENTS

Thank all supports and helps from colleagues in Beijing NAURA Microelectronics.

REFERENCES

[1] Y. Wang, et al, "Process Loading Reduction on SADP FinFET Etch", CSTIC, 2015 :1-3.

[2] Chang, Shih-Ming, et al, "Study of loading effect on dry etching process", SPIE, Volume 5130, p. 228-233 (2003)

[3] Samer Banna, et al, "Inductively Coupled Pulsed Plasmas in the Presence of Synchronous Pulsed Substrate Bias for Robust, Reliable, and Fine Conductor Etching", IEEE TRANSACTIONS ON PLASMA SCIENCE, VOL. 37, NO. 9, SEPTEMBER 2009.

[4] M. Schaepkens and G. S. Oehrlein, "Effects of radio frequency bias frequency and radio frequency bias pulsing on SiO_2 feature etching in inductively coupled fluorocarbon plasmas", J. Vac. Sci. Technol. B, Microelectron. Process. Phenom., vol. 18, no. 2, pp. 856–863, Mar. 2000. 10-15.

A STUDY OF NON-BOSCH DEEP TRENCH ETCH

Wang Jing, Nie Miao, Jiang Zhongwei, Huang Yahui
Beijing NAURA Microelectronics Equipment CO. Ltd
wangjing02@naura.com

ABSTRACT

A non-Bosch deep trench plasma etch process which is suitable for trench-gate IGBT has been developed on NMC612C Inductively coupled plasma etching(ICP) tool. An undercut-free 1um CD and 3.5um-Deep Trench with round bottom corners and around 89deg sidewall slope is obtained by carefully adjusting additive gases like HBr, O2 and SF6. After a systematic optimization of process parameters such as gas flow, process pressure and RF power, uniformity of <5% and an etch rate of 2000nm/min can be achieved. The trench profile is further improved by additional processing including multi step etching. Based on the above learning, a non-Bosch etch process is optimized and a good trench profile is obtained.

INTRODUCTION

Because of its advantage of high power, high frequency and low loss, IGBT has gradually replaced the traditional power devices, it is widely used in the key equipment of emerging industries such as rail transits, high voltage power transmission and transformation. Compared with planner IGBT, the channel gate IGBT significantly improves the balance between on state loss and switching loss, therefor it is more suitable for medium, low voltage and high frequency applications. Aiming at the challenge of trench gate technology, different of key technologies of trench gate IGBT are studied and developed, and a good trench etching profile is obtained.

EXPERIMENTAL

The experiments were performed in a 300 mm inductively coupled plasma reactor, NMC612C Series.

First, blanket tests were performed on Si wafers. The etch rate and etch uniformity of the substrate material were obtained using thickness meter. To minimize the etch variance, the process steps were optimized by tuning current ratio, plasma gradient, and gas composition. Best recipes were chosen to perform on the patterned wafer.

The study was performed to measure Trench width CD 1um. The patterning stack was composed of the following layers: PEOX/SiN/pad OX/Si. During the etching process of the patterned wafer test, Trench profile was studied through SEM analysis.

RESULTS AND DISCUSSION

The trench etching is usually based on SF6 based and plasma etching with O2[1-2], the section profile as shown in figure1. It often appears as: the top of the trench bowing,

the top corner obvious undercut effect and bottom corner rounding is not enough. In the deep trench etching process, the top morphology seems to have a correlation with the process gas. The bombardment of the hard mask and sidewall byproduct polymer provide the protection of the top corner. Compared with mask prepared by photoresist mask, hard masking can only provide limited protection. There is not enough protection at the top to reduce isotropy etching.

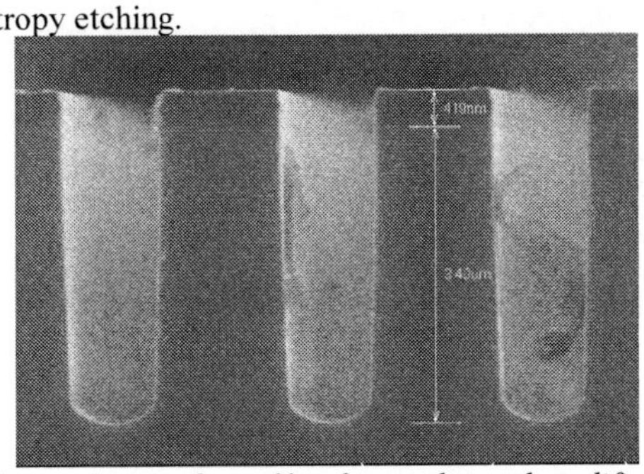

![Figure 1: trench profile after traditional modification]

Figure 1: trench profile after traditional modification

Changing the ratio of power and gas is the traditional way to change the top shape of the top. By reducing the high source power and increasing the bias power, adjusting the proportion of O2 and increasing the dilution gas, we can reduce the undercut of the top morphology. In the history of etching recipe adjustment, we found that increasing the number of polymers and increasing the bombardment are beneficial methods to improve the morphology of the top. By adjusting the plasma gas and conditions, we can reduce the top undercut and bowing to a very small degree. As shown in Figure 2, the undercut can be reduced from 100nm to 15nm by changing the etching recipe. But the morphologies of bowing may still exist.

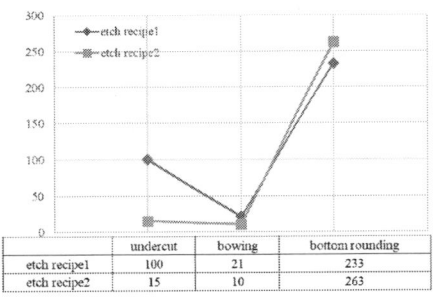

	undercut	bowing	bottom rounding
etch recipe1	100	21	233
etch recipe2	15	10	263

Figure 2: trench profile after traditional modification

978-1-5386-5309-8/18 $31.00 © 2018 IEEE

To better optimize the morphology of the top undercut, HBr was added as a process auxiliary gas. We used the same etching recipe but only change SF6 and HBr gas ratio to etch. Better top profile is obtained by increasing the HBr process gas. After etching, there are a lot of etching by-products at the top. The presence of by-products completely disappears the top undercut, and the bowing morphology is also completely free, but at the same time, the rounding at the bottom of the trench is worse，as shown in Figure 3. HBr gas produces many reaction byproducts to protect the side walls. A strong top deposit can cover the bowing problem which caused by the etch of the top. At the same time, there will be many byproducts at the bottom of the trench make the bottom rounding profile is poor.

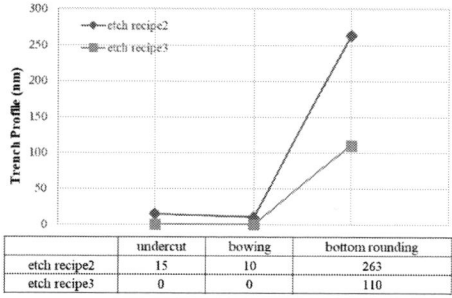

	undercut	bowing	bottom rounding
etch recipe2	15	10	263
etch recipe3	0	0	110

Figure 3: trench profile after added HBr

To better optimize the morphology of the bottom rounding, the deposition distribution of the polymer has been adjusted. In this experiment, we used two different HBr/O2 gas ratios to adjust polymer distribution. As shown in Figure4, recipe4 bottom rounding is significantly higher than the recipe3.

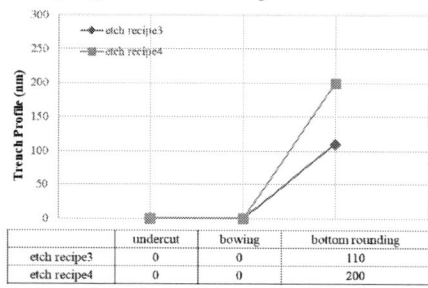

	undercut	bowing	bottom rounding
etch recipe3	0	0	110
etch recipe4	0	0	200

Figure 4: trench profile after adjust gas ratio

Multi step process has been tested. In this method, we can change two extra steps instead of simply changing the gas ratio and power. The additional steps give us more opportunities to optimize the plasma properties, such as etching components and plasma dissociation. We use a multipart process to increase better angle control and for the bottom rounding profile. As shown in Figure 5. After optimization, multi process of recipe5 is undoubtedly a very powerful plan to control bottom rounding profile. At the same time, the pulse also shows smoother sidewall, which will help the gate oxygen growth and polysilicon filling requirements.

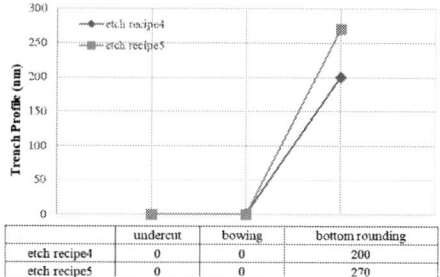

	undercut	bowing	bottom rounding
etch recipe4	0	0	200
etch recipe5	0	0	270

Figure 5: trench profile using multi step

CONCLUSIONS

An undercut-free 1um CD and 3.5um-Deep Trench with round bottom corners and around 89deg sidewall slope is obtained by carefully adjusting additive gases like HBr, O2 and SF6. The top profile was controlled by HBr gas. Bottom rounding with optimized top profile were obtained by controlling the plasma density and ions bombardment energy independently in multi steps. Based on the above learning, a non-Bosch etch process is optimized. A good trench profile is obtained.

ACKNOWLEDGEMENTS

The authors wish to express their gratitude to NAURA Etch PE group for provide me resource and some guidance ideas. Special thanks to Wu YunFei for SEM pictures.

REFERENCES

[1] S. Panda, R. Wise, and A. Mosden, et al. Microelectronic Engineering, 75(3), 2004, pp.275-284.

[2] J. M. Lane, K. H. A. Bogart, and F. P. Klemens et al. . J. Vac. Scl. Technol., A, 18(5), 2000, pp.2067-2079.

INVESTIGATION ON THE LEAKAGE OF TRIPLE SPLIT-GATE FLASH DEVICE AND ITS IMPROVE SOLUTION

Zigui Cao[1]*, Lingyue Zhang[1], Yan Sun[1], Buchun Su[1]

[1]Huahong Grace Semiconductor Manufacturing Corporation, 1399 Zuchongzhi Road, Zhangjiang Hi-Tech Park, Shanghai 201203, People's Republic of China

*Corresponding Author's Email: Steam.cao@hhgrace.com

ABSTRACT

In this paper, the leakage issue of triple split-gate flash device was studied and P out-gassing from high doping floating gate poly is found to be the root cause for high leakage issue. Base on the formula of P diffusion in silicon material, we get the ideal silicon oxide thickness to be 1.72A as the limit value for P out-gassing prevention from floating gate poly. By introducing of rapid temperature oxidation anneal(RTO) method for oxide forming, we found that oxide thickness above 2A can suppress P out-gassing effectively, which is similar as the ideal value calculated by formula.

Key words: P out-gassing, leakage, rapid temperature oxidation anneal, diffusion

INTRODUCTION

With high storage density and good data retention performance after power off, non-volatile memory application is rising rapidly in consumer electronics industry, especially for mobile electronics market. Among these, split-gate flash memory is a typical one which is widely used in either standalone NOR flash or e-flash field due to the characteristics of high program efficiency and over erase immunity [1-4]. As the increasing of storage density, it is necessary to shrink cell area more aggressively. For this purpose, triple split-gate flash device is invented and has gotten more and more attractions of investigation [5-9].

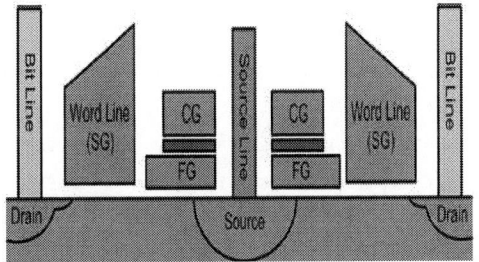

Fig.1 Triple split-gate flash device structure schematic

The triple split-gate flash device is fabricated by HHG 0.12μm self-aligned flash process, the key process steps can be described as below: similar as traditional stack floating gate (FG) device, the cell structure is fabricated with self-aligned process. The isolated FG is formed by following dry etch process with SiO2 on control gate as hard mask; Then, an anisotropic dry etch process is applied to form select gate structure by using of control gate and following isolation film as hard mask. The final triple split-gate flash structure is shown in **Fig.1**. Due to good electronic conductivity requirement, FG is formed with highly phosphorus doped at high temperature in furnace. In the following process of word line and tunneling oxide fabrication steps, FG is exposed outside and thus phosphorus will outgas in the furnace of tunneling oxide deposition, penetrate into silicon surface of device and word line area. It will modify the channel doping concentration and shift its threshold voltage, finally result in the product function failure.

In this paper, an investigation is carried on the phosphorus contamination issue during the fabrication of triple split-gate flash device. Then

978-1-5386-5309-8/18 $31.00 © 2018 IEEE

an optimal tunneling oxide forming process is put forward based on the theory of phosphorus diffusion into silicon, which can solve phosphorus contamination issue effectively.

MECHANISM STUDY

During the fabrication of triple split-gate flash device, the threshold voltage of peripheral device and flash device itself was always found shift in some level, which result in the big leakage issue of the chip. Fig.2 shows inline lots flash device leakage versus single batch lot count during tunneling oxide forming in furnace, while flash device channel leakage is measured under the condition of word-line and source line biased at 0 volt. From the correlation, we can see that the leakage current of flash device is much smaller with tighten distribution if 1lot/batch is processed; As the lot count is above 2 in one batch, the leakage will increase significantly, especial for the lot at position1 of 4 lots/batch condition, it will reach the peak among 4 lots.

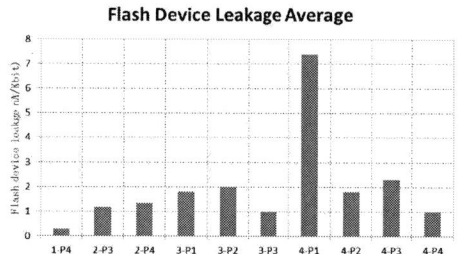

Fig.2 Inline lots flash device leakage versus single batch lot count, where 1st number is lot count/batch processed in furnace, P1~P4 is the position in furnace.

Through the investigation on the fabrication of triple split-gate flash device, we found out the root cause of high flash device leakage, as shown in Fig.3: during the forming of tunneling oxide, floating gate poly is exposed to outside. Since a waiting time is needed in furnace before the deposition of tunneling oxide, phosphorus in floating gate is accelerated for out-gassing by initial temperature of 500 degree in furnace. These gasified phosphorus elements

will condense and penetrate into the channel surface of peripheral device and flash device itself because the channel surface is wet cleaned. It finally causes the threshold voltage shift and leakage increasing issue on device. More lots/batch, higher gasified phosphorus concentration, heavier impact on the doping of channel. Because the lot is transferred into furnace in sequence of first in last out, the time waiting in furnace of 1st is the longest, thus a worse one for phosphorus contamination, which can explain the phenomenon observed in Fig.2.

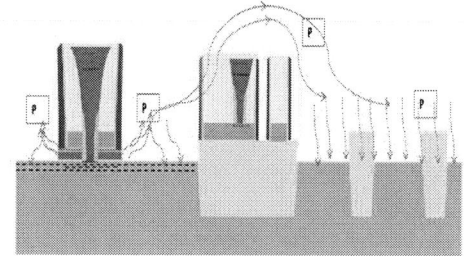

Fig.3 Schematic of phosphorus contamination in triple split-gate flash device

To solve this issue, we optimized the process of tunneling oxide forming and phosphorus precipitation is inhibited by adding one rapid thermal processing step at oxygen environment before the deposition of tunneling oxide. However, due to thicker oxide caused by rapid thermal processing, the erase performance will become worse and induce erase function failure issue, therefore, it is important to form a suitable oxide film which can block phosphorus precipitation and avoid flash erase performance degradation simultaneously.

OXIDE THICKNESS DEFINITION

In this paper, similar as the theory of phosphorus diffusion into silicon substrate, the oxide thickness forming by rapid thermal oxidation is calculated. Base on below formula of phosphorus diffusing into silicon:

$$Xj = 2\sqrt{Dt} * \text{erfc}\left[\frac{CB}{Cs}\right]$$

(1)

Where Xj: depth of diffusion; D: diffusion velocity, CM2/s; CB: initial concentration; CS: diffusion source concentration; t: diffusion time, s. The velocity of phosphorus diffusion into SiO_2 is related to atmosphere temperature, height of barrier and initial constant value of phosphorus in SiO_2, which is shown in formula (2) :

$$D = D_0 * e^{-Ea/KT}$$

(2)

where D0 is initial constant value of phosphorus in SiO_2, here we take the default value +0.19; Ea is the barrier height, 4.03eV; K is Boltzmann constant, 1.38e-23; T is the kelvin temperature unit.

Introduce formula (2) into (1), we can derive the depth of phosphorus diffusion into SiO_2:

$$Xj = 1.72A$$

EXPERIMENT RESULTS AND DISCUSSION

Based on the results calculated by section3, we planned experiments with different oxide thickness forming by rapid thermal oxidation process. In order to verify the value calculated by formula, we plan thinner and thicker oxide thickness around the ideal value on purpose. The details are shown in table1.

Table 1 Oxide thickness split of rapid thermal oxidation

Condition	#1	#2	#3	#4
RTO Oxide 1.5A	√			
RTO Oxide 2A		√		
RTO Oxide 4.7A			√	
No RTO(1lot/batch)				√

Fig.4 shows the results of flash device leakage under different oxide thickness grown by rapid thermal oxidation process. Compared to the process without additional protection oxide, the leakage of flash device is much smaller with

a tighten distribution in additional oxide process splits. Among the splits of additional protection oxide, the leakage decreased gradually with oxide thickness increased. As the oxide thickness approaches to 2A, the leakage caused by phosphorus contamination is similar as that of 4.7A split, in other words, it proves that 2A oxide covering on floating gate can inhibit the outgassing effectively, which is consist to the result calculated theoretically.

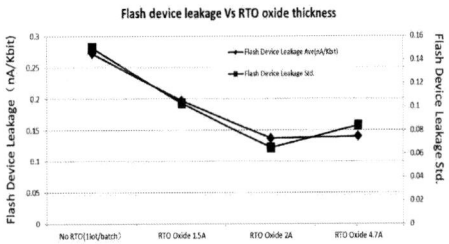

Fig.4 flash device leakage comparison under different oxide thickness grown by rapid thermal oxidation process.

CONCLUSION

In this paper, the leakage issue of triple split-gate flash device was studied and P out-gassing from high doping floating gate poly is found to be the root cause for high leakage issue. Base on the formula of P diffusion in silicon material, we get the ideal silicon oxide thickness of 1.72A as the limit value for P out-gassing prevention from floating gate poly. By introducing of rapid temperature oxidation anneal (RTO) method for oxide forming, it is found that oxide thickness above 2A can suppress P out-gassing effectively, which is consist to the ideal value calculated theoretically.

REFRENCE

[1] J. V. Houdt, G. Groeseneken, and H. E. Maes, "An analytical model for the optimization of source-side injection flash EEPROM devices," IEEE Trans. Electron Devices, vol. 42, pp. 1314–1320, 1995.

[2] Y. Ma, C. S. Pang, J. Pathak, S. C. Tsao, and C. F. Chang, "A novel high density contactless flash memory array using split-gate source-side

978-1-5386-5309-8/18 $31.00 © 2018 IEEE

injection cell for 5 V-only application," in 1994 Symp. VLSI Technology Dig. Tech. Papers, vol. 5A, pp. 49–50.

[3] K. Chang, W. Chen, C. Swift, J. M. Higman, W. M. Paulson, and K. Chang, "A new SONOS memory using source-side injection for programming," IEEE Electron Device Lett., vol. 19, pp. 253–255, 1998.

[4] A. T. Wu, T. Y. Chan, P. K. Ko, and C. Hu, "A novel high-speed, 5-volt programming EPROM structure with source-side injection," in IEDM Tech. Dig., 1986, pp. 584–587.

[5] C. Y.-S. Cho, M.-J. Chen, C.-F. Chen, et al. A novel self-aligned highly reliable sidewall split-gate flash memory. Electron Devices, IEEE Transactions on, 2006(3):465 – 473.

[6] S. Saha. Scaling considerations for sub-90 nm split-gate flash memory cells. In Solid-State and Integrated Circuit Technology, 2006. ICSICT '06. 8th International Conference on. 2006 pp. 705 – 708.

[7] S. Saha. Design Considerations for Sub-90-nm Split-Gate Flash-Memory Cells. Electron Devices, IEEE Transactions on, 2007(11):3049 – 3055.

[8] S. Saha. Scaling considerations for sub-90 nm split-gate flash. Circuits, Devices Systems, IET, 2008(1):144 – 150.

[9] 曹子贵. 先进三栅分栅闪存器件工作特性及耐久特性的研究. 博士学位论文, 中国科学院上海微系统与信息技术研究所, 2009.

HYBRID INTEGRATION OF FDSOI AND SI BULK FOR CMOS PROCESS

Yang Song*, Kang Ye, Changfeng Wang

Huali Microelectronics Corporation, Shanghai 201203, China

* Corresponding Author's Email: songyang@hlmc.cn

ABSTRACT

The FDSOI (Fully Depleted Silicon on Insulator) technology has proven to be a promising candidate for next CMOS generations, due to its very high speed, low power and moderate cost. In addition, recent reports on combining FDSOI with high k metal gate have raised considerable interest since it is able to achieve low power devices. But it should be noticed that a drawback of using this SOI structure is that the thickness of active Si and buried oxide (BOX) limit the implementation of ESD protections. Hybrid co-integration of SOI thin film with bulk will be a solution of this problem, and also makes it possible to better control Vt with use of back biasing. This paper will demonstrate experimentally about one approach of the hybrid co-integration scheme in FDSOI CMOS process. The influence of back biasing on threshold voltage and ESD protection will also be discussed.

INTRODUCTION

As integrated circuit scales down to 28nm and beyond, FDSOI devices become considerable attractive due to better control of short channel effect for low-power devices. The use of a thin silicon film combined with ultra-thin buried oxide layer makes it possible to improve the short channel effects (SCE) due to the reduction of lateral electrostatic coupling between the source and drain. [1] In addition, the introduction of high-k and metal gate technology for FDSOI devices makes it possible to adjust the NMOS and PMOS threshold voltage separately. However, a further slight modulation of threshold voltage is still necessary to decrease Ioff, moreover, low threshold voltage should be reached in applications such as analog devices and other LP multimedia platform. To achieve multiple Vt devices, the concept of hybrid co-integration is proposed.

Hybrid co-integration of FDSOI and Si bulk makes it possible to apply local back biasing separately on both NMOS and PMOS transistors, making it possible to modulate Vt for a low power platform, including forward backside bias (FBB) and reverse backside bias (RBB). Figure 1 shows a schematic cross-section view of FDSOI device with hybrid bulk region.

In addition, it has been reported that the implementation of back biasing can also help improve the transistor performance, such as improves DIBL (drain induced barrier lower) and the SS (subthreshold slope). [2] Besides, compared with conventional bulk devices, the thermal dissipation of BOX layer is not good and the

conductivity of ultra-thin Si film is lower, FDSOI devices are difficult to integrate ESD protections, but hybrid technology also makes it possible to solve this problem. [3]

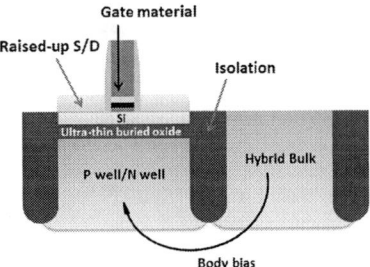

Figure 1: Cross-section view of FDSOI with hybrid bulk region

RESULTS AND ANALYSIS

Hybrid process

The FDSOI devices were processed on 300mm (110) SOI wafers with BOX thickness of 20nm. In this work, we propose one scheme to hybrid integrate FDSOI and bulk area, as shown in Figure 2. This process includes the removal of silicon and BOX layers, and epitaxial growth of Si in subsequent process.

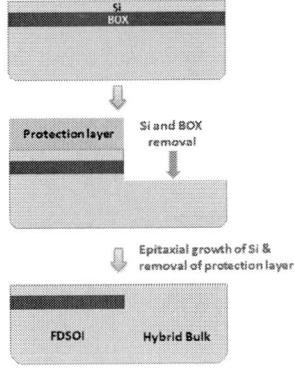

Figure 2: Hybrid process

The hybrid process exhibits many challenges, for instance, vertical sidewall profile needs to be achieved during hybrid etch, as the sidewall of epitaxial growth of Si may become facet if the sidewall profile is sloping. During epitaxial growth, the epitaxial layer thickness has to be precisely controlled as it is necessary to ensure the epitaxy surface to be flat with SOI surface, due to the

978-1-5386-5309-8/18 $31.00 © 2018 IEEE

existence of gate on the SOI region and to ensure uniform metal gate deposition during subsequent process.

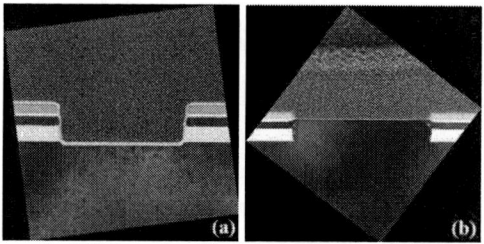

Figure 3: TEM result of hybrid integration of SOI and Si bulk: (a) hybrid etch; (b) hybrid epitaxy.

Figure 4 (left) shows an example of the subsequent FDSOI process. After formation of hybrid area, STI isolation is performed to isolate the SOI area and bulk Si area. High k metal gate material can be introduced for better control over Vt. Figure 4 (right) also shows a TEM example of the 45nm node 0.299µm² 6T SRAM bitcells. [4]

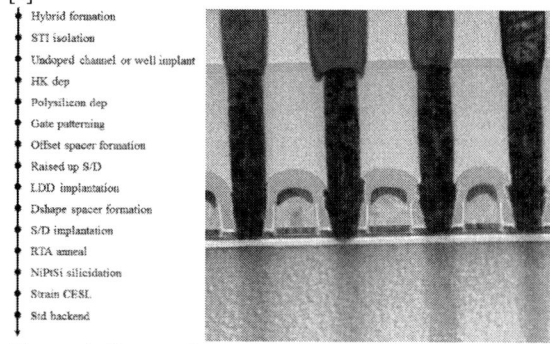

Figure 4: Process flow scheme of FDSOI (left) and TEM cross section of a FDSOI device (right) [4]

Threshold voltage reduction

Three approaches can be used to tune the threshold voltage. One is to work with work-function metal film stack and thickness; the second is implementation of well implant and back biasing; the third one is to play with halo implant. In previous study of NMOS transistor, it has been shown that by decreasing the TiN thickness from 10nm to 3nm, a 150mV Vt reduction can be obtained. [4]

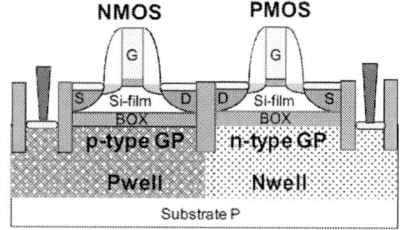

Figure 5: Schematic view of FDSOI transistor with back bias contact [2]

Well implant and back biasing also play roles on Vt tuning. Figure 5 illustrates an example of FDSOI transistor structure with back bias contact. For different BP (back plane) type, Vt variation on Vb (body bias voltage) can be easily seen. The LVT options can be achieved for both NMOS and PMOS respectively with n-type BP (N well) and p-type BP (P well) if the body bias is set to Vdd, whereas the HVT options be obtained on a p-type BP and n-type BP set to 0 for NMOS and PMOS, as is shown in Figure 6. Table 1 also gives a summary about the BP type and Vb influence on the NMOS and PMOS Vt. [5]

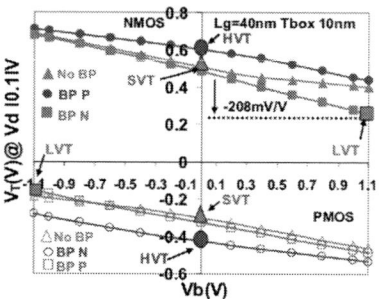

Figure 6: Vt vs Vb for each BP type[4]

TABLE I.   NMOS & PMOS Vt for Various BP Types & Vb [5]

Vb	NMOS		PMOS	
	OV	*Vdd*	*OV*	*Vdd*
n-BP	SVT (i)	LVT	HVT	SVT (ii)
p-BP	HVT	SVT (ii)	SVT(i)	LVT
w/o BP	SVT (iii)	SVT (iii)	SVT (iii)	SVT (iii)

Improved ESD protection

In FDSOI devices, due to the existence of ultra-thin buried oxide layer and thin Si layer, the integration of ESD protections are limited. Though the thin BOX layer can reduce the leakage current, it also makes it difficult for thermal dissipation.

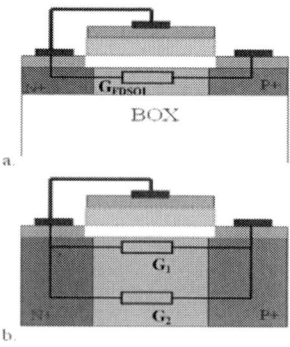

Figure 7: Schematic view of a gated diode: a. FDSOI; B. bulk [6]

978-1-5386-5309-8/18 $31.00 © 2018 IEEE   258

In 2010, T. Benoist *et al.* evaluated the ESD protection robustness in FDSOI technology. Compared with the bulk devices or PDSOI (partial depleted SOI), as shown in Figure 7, FDSOI devices shows lower conductivity (1/Ron) due to the ultra-thin silicon film and It2 (break down current) is around four times lower due to the poor thermal dissipation of the BOX. It is found that the co-integration of SOI/bulk devices exhibits much better results compared to conventional FDSOI and provides improved ESD protection---higher conductivity and higher break down current. [6]

CONCLUSIONS

FDSOI technology is proposed as a promising candidate for low power device applications. Conventional FDSOI devices exhibits many challenges such as ESD protection limitations, hybrid co-integration of FDSOI and Si bulk is shown to be a solution for this problem. In this study, hybrid co-integration scheme in FDSOI CMOS process is presented. Moreover, back biasing combined with well implant can be considered as a Vt tuning knob for FDSOI devices, and also improve the device performance as compared to bulk devices.

ACKNOWLEDGEMENTS

The authors wish to express their gratitude to Haizhou Wang, Yizheng Zhu and Jun Tan at HLMC for their kind support. Special thanks to Duanquan Liao for his expertise, feedback and advice during the whole work.

REFERENCES

[1] C. Fenouillet-Beranger *et al.*, *ESSDERC 2008 - 38th European Solid-State Device Research Conference*, Edinburgh, 2008, pp. 206-209.

[2] C. Fenouillet-Beranger et al., 2012 13th International Conference on Ultimate Integra*tion on Silicon (ULIS)*, Grenoble, 2012, pp. 165-168.

[3] A. Dray *et al., Electrical Overstress / Electrostatic Discharge Symposium Proceedings 2012*, Tucson, AZ, 2012, pp. 1-7.

[4] C. Fenouillet-Beranger *et al.,2011 IEEE International Conference on IC Design & Technology*, Kaohsiung, 2011, pp. 1-4.

[5] J. P. Noel *et al., IEEE Transactions on Electron Devices,* vol. 58, no. 8, pp. 2473-2482, Aug. 2011.

[6] T. Benoist *et al., Electrical Overstress/Electrostatic Discharge Symposium Proceedings 2010*, Reno, NV, 2010, pp. 1-6.

Patterning challenges in 193i-based tip to tip in N5 interconnects

Basoene Briggs, Janko Versluijs, Juergen Bömmels, Christopher J. Wilson, Zsolt Tőkei, Arindam Mallik, Job Soethoudt[#]

imec,
Kapeldreef 75, 3001 Heverlee, Belgium.
Telephone: +32 16 28 8511, basoene.briggs@imec.be
[#]KU Leuven (University of Leuven), Leuven, Belgium

ABSTRACT

CMOS area scaling to N5 dimensions will have interconnect metal pitch around 30nm. Patterning such small features, using 193 ArF immersion lithography (193i), is only possible with pitch multiplication techniques such as SADP, SAQP, SAOP, etc. An additional keep or block patterning process is often used to achieve line interruptions and turns essential to have functional electrical devices. In this paper, we review three block patterning approaches experimented on imec's 32nm metal pitch N5 test vehicles. We discuss the merits and challenges of each patterning option and describe, qualitatively, process interactions with lithography parameters such as alignment and overlay. Lastly, we show that EUV block approach is a simpler benchmark in terms of process complexity and cost.

Keywords: N5 interconnect, block, tip to tip, tone inversion

INTRODUCTION

Continued CMOS area scaling will derive from pitch reduction of the interconnect metal lines. At foundry N5 dimensions interconnect pitches will be in the sub-30nm regime, with typical critical dimension (CD) less than 20nm [1]. To pattern such small dimensions, the conventional 193 ArF immersion lithography single exposure and etch patterning scheme will not be sufficient; multi patterning techniques such as SADP, SAQP, SAOP, etc will be involved to achieve pitch scaling [2, 3]. Feasibility of some of these multi patterning approaches have been demonstrated to realize very small CD interconnect lines [4-7]. Very often, design and technology constraints limit multi patterning techniques to produce one-dimensional (1D) line-space patterns which define the interconnect lines and dielectric spacings.

Usually, additional keep or block processing is included in the damascene integration process flow to realize interruptions in the 1D tracks needed to produce functional electrical devices. The block size defines the tip to tip (t2t) distance between metal islands while the block spacing mostly dictates the minimum island rule. To obtain high density blocks at tight CD, multiple color block design is the main approach available in 193i process assumptions. This approach leads to increased mask count, film stack complexity, as well as overlay and edge placement error (EPE) challenges.

In this paper, we present a qualitative review of different 193i-based block integration flows. The block process flows discussed use 193i lithography LELELE (LE3) to define dense hole patterns which are then reversed to pillars using tone inversion processing techniques. These flows were experimented on imec's 32nm metal pitch test vehicles. As a benchmark, we show that EUV block patterning approach largely minimizes the patterning challenges associated with the 193i approaches, and could even provide cost advantages. In this review paper, where applicable, citations to more detailed descriptions of the process flows are provided for further reference.

'HOLE' VERSUS 'PILLAR' BLOCK APPROACHES

Before describing the block patterning schemes, it is useful to begin with some definitions and distinctions. Generally, there are two ways to define the block patterns at lithography. For simplicity, these can be defined as 'pillar' or 'hole' approach as shown in Fig. 1. In the pillar case, dark field mask is used to define the resist islands in a positive tone photoresist, while in the hole case, light field mask exposes via-like holes in a positive photoresist.

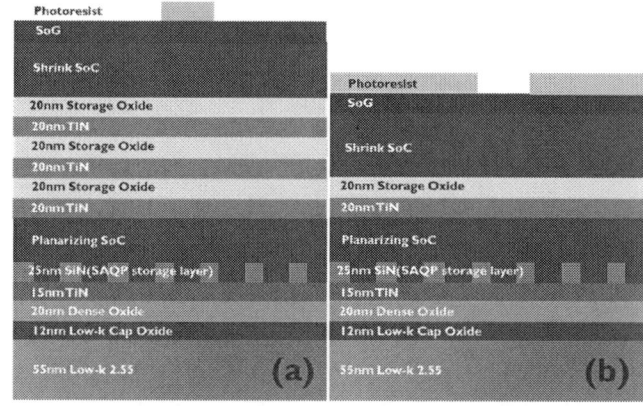

Fig.1. different stacks required for three-color block in (a) 'pillar' approach and (b) 'hole' approach. The pillar approach requires more complex stack, etching, and is susceptible to alignment challenges

Due to resolution constraints, the minimum pitch of holes or pillars achievable with 193i lithography is around 105nm at typical CD of 40nm; too large for N5 relevant blocks in sub 20nm lines. To resolve the CD issue, the block patterns are exposed larger at lithography followed by CD trim (for pillars) or shrink (for holes) during dry etch transfer. For the test vehicles described, CD of the trench is 16nm at 32nm pitch and a final block CD corresponding to metal t2t distance of 21nm.

978-1-5386-5309-8/18 $31.00 © 2018 IEEE

To obtain high density blocks using multiple color block masks, a minimum of two contrasting dielectric storage layers are used to temporarily memorize the contents of each mask during the multiple LE transfer. Fig.1 also shows by means of illustration that the pillar approach will incur more stack and patterning complexity when compared to the hole approach. The hole approach is therefore more practical as it does not require many consecutive storage layers as well as complex selective etch processes. However, the hole approach requires additional tone inversion steps to convert the holes into pillars. The hole approach was studied and is further discussed.

BLOCK PATTERNING OPTIONS

A. Spin-On Tone-Inversion

This tone-inversion approach employs spin-coated films such as spin-on carbon (SoC) at different steps of the process flow. SoC is used to planarize the topography created by the SAQP trenches before stack deposition and LE3 patterning of the holes into a storage layer of PECVD oxide. A layer of SoC is used to derive 40% CD shrink of the holes during transfer into the storage oxide. The holes created in oxide are then filled with SoC and the excess materials recessed by dry plasma etch thereby exposing the storage oxide which is selectively removed, leaving behind SoC block pillars illustrated in Fig. 2.

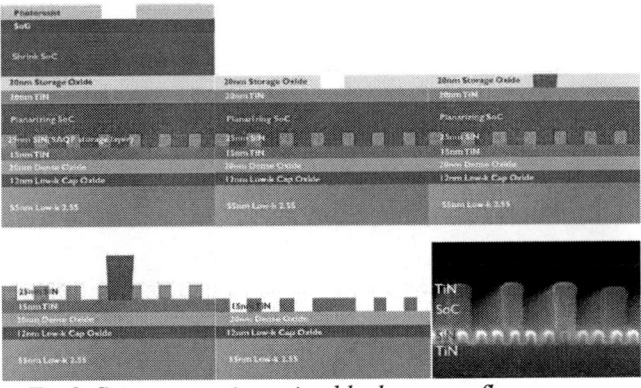

Fig.2. Spin-on tone inversion block process flow.

The material choices in this approach favors etch selectivity because of the use of contrasting carbon-based and oxide films. However, one major challenge is that most spin-on films are only able to planarize topographies of a few hundred

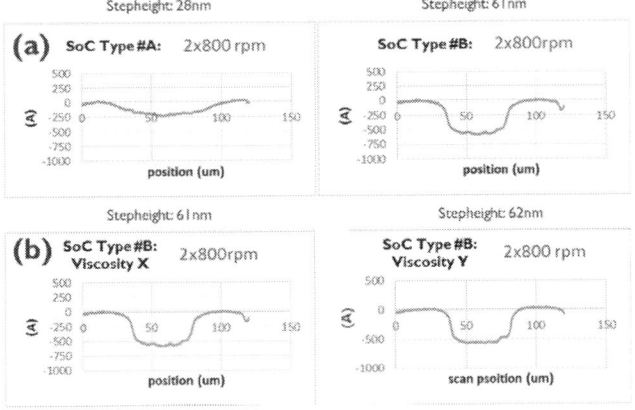

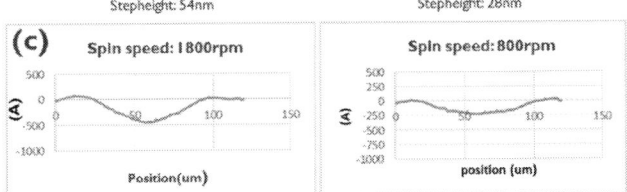

Fig. 3. stepheight measured with high resolution profilometry after different SoC spin coat on 42um wide, 60nm deep trench patterned in PECVD oxide. (a) shows that some SoC are inherently better planarizing than others (b) changing viscosity of a particular SoC does not improve its planarization (c) increased spin speed during coat of SoC worsens the planarization performance

nanometers width. Fig. 3 shows that some SoC films are inherently better planarizing than others, and that process enhancements such as spin speed reduction can improve the planarization length, but not lead to complete planarization.

The impact of structure aspect ratio and pattern density on the planarization behavior of several SoC films has been presented in "unpublished" [8], with similar conclusions. Hence, the spin-on approach will command design rule restrictions that presently do not support wide design features, for example unsegmented VSPM marks for alignment, and conventional ellipsometry thickness measurement pads.

B. Double CMP Tone-Inversion

The double CMP tone-inversion approach uses oxide (for SAQP pattern gap fill) and nitride (for hole fill) accompanied each time by a dielectric CMP planarization process. Detailed patterning description of the double CMP approach can be found elsewhere [9]. The planarization of large design features is resolved in this approach, with the CMP processes optimized to improve selectivity and minimize dishing and erosion. However, further challenges can arise; within-wafer and wafer-to-wafer non-uniformity, impairment of alignment and overlay markers, etc.

The signal strength obtained during block lithography in both the spin-on and double CMP approaches is generally poor on most traditional alignment marker designs as shown in Fig. 4. While the signal strength is close to the 0.1% threshold when aligning to narrow SAQP/SADP-defined spacer patterns, significantly larger signal strength is observed when alignment is done to larger patterns defined by a dedicated 'ZERO' mask. The micrometer large ZERO patterns are etched into silicon and back filled with oxide thereby producing better material contrast when addressed at higher metal layers. Poor alignment and large ROPI residuals can worsen block placements accuracy as shown in Fig. 5.

C. Selective Deposition Tone Inversion

The number of dielectric CMP steps needed in the previous approach can be reduced to one if after the hole etch into oxide, a selective growth technique is used to grow a blocking material at the bottom of the hole. Detailed patterning description of block tone-inversion using Ru area-selective deposition is submitted to "unpublished" [10]. Some of the process interactions with litho parameters in the double CMP approach will also apply in the selective deposition method.

PROCESS INTERACTIONS WITH LITHOGRAPHY

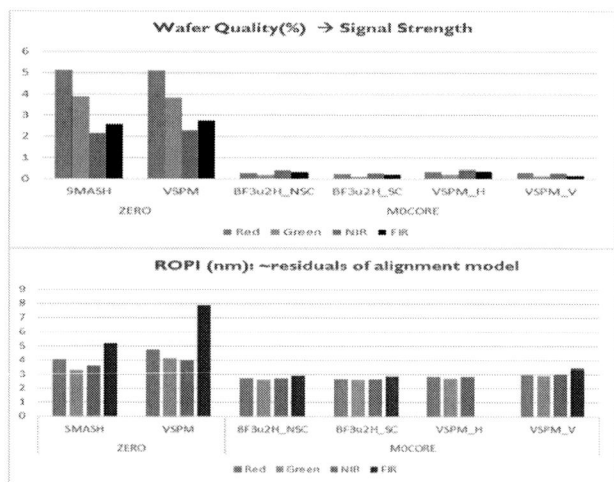

Fig.4. signal strength and ROPI values for various alignment markers at block lithography step where block mask is aligned to 21nm wide SADP-defined oxide spacers

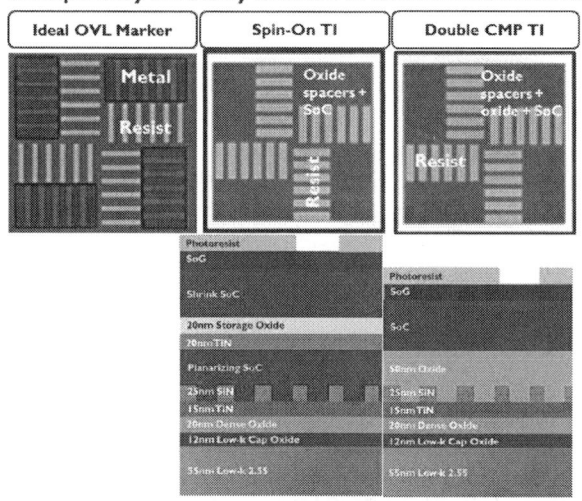

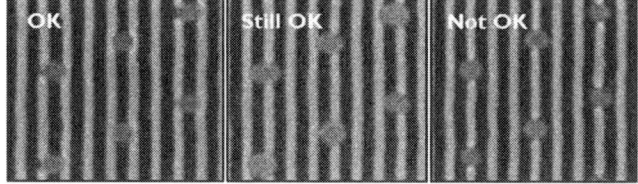

Fig.5. (top row): poor visibility of under-lying overlay structures. Main reason is because structure to align is defined by narrow 16nm SAQP-defined SiN spacers. (middle row): corresponding stack at block lithography for two tone inversion options. (bottom row): varying degrees of block placement error because of alignment difficulty

D. EUV Block

EUV single exposure (EUV SE) at 36nm pitch line-space has been demonstrated with level of maturity enabling bi-directional patterns as shown in Fig. 6(a) [11]. However, at 32nm pitch, processing defects such as line bridging, line interruption, and pattern merging at lithography contribute to, relatively, high stochastics. Another concern for the 32nm pitch EUV SE is high CD variabilities, especially t2t variability [12]. Table 1 shows reported LCDU 3s values for t2t obtained by two EUV patterning approaches at 32nm line-sapce pitch. Although different t2t CD were targeted, LCDU values reported in the hybrid SAQP with EUV block approach is around half the values reported for the single exposure option.

Table 1: LCDU values of t2t obtained by EUV patterning

Line-Space Pitch(nm)	Block/t2t CD (nm)	LCDU(nm) - 3s	Patterning
36	-	-	1D, 2D EUV SE
32	26.5	8.5 [11]	EUV SE
32	21	3.8 [13]	iSAQP + EUV Block
32	21	-	iSAQP + 193i Block (LE3 + TI)

The feasibility of 32nm pitch EUV SE, discussed in [12], suggests that with current resist materials and best processes, t2t as small as 25nm can be obtained. However, large t2t LCDU was observed for both line-style block of regular line-space patterns (Fig. 6 b), and non-regular logic patterns (Fig. 6 c).

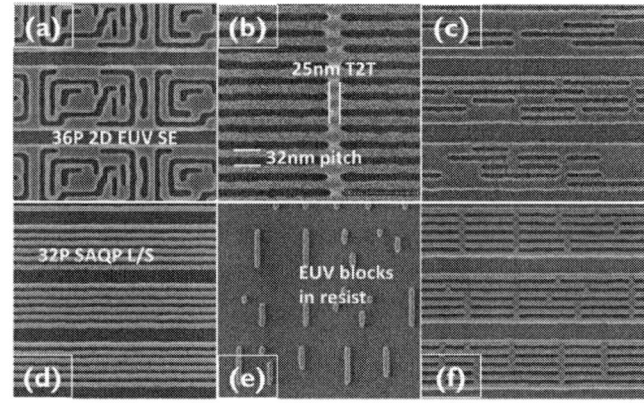

Fig. 6. Illustration of EUV patterning options. (a) 36P EUV 2D print (b) 32P EUV SE: line-space structure with regular t2t grid simultaneously blocking multiple lines. (c) 32P EUV SE: logic structure with non-regular t2t. t2t CD variability and pattern merging observed. (d, e, f) Hybrid patterning using SAQP-defined line-space and EUV block process. Improved pattern fidelity and t2t/block uniformity. Figures re-used from [12, 13]

It is important to add that EUV SE, when qualified for 32nm interconnect pitch will eliminate the need for extra block patterning, potentially reduce cost, and potentially improve device yield. But while research continues on the single exposure solutions, EUV block patterning onto SAQP-defined 16nm line-space patterns already show significant simplification of the patterning, improved block CD uniformity, and reduced EPE errors [14].

Fig. 7 illustrates the simplified stack needed for the EUV block option. This consists of a thin SoC film to planarize the SAQP topography and a thin metal-oxide negative tone photoresist (Inpria™) used in combination with dark field EUV mask [13, 15]. EUV mask allows tighter block pitch, thus does not need multi patterning, storage layers and tone-inversion

978-1-5386-5309-8/18 $31.00 © 2018 IEEE

processing. Patterning results obtained with this method is shown in Fig. 6 (d - f).

Fig. 8 shows 43% normalized cost savings for the EUV block approach compared to 193i LE3 flavors, with majority of the cost reduction attributed to the reduction of deposition, etch, clean, and metrology operations.

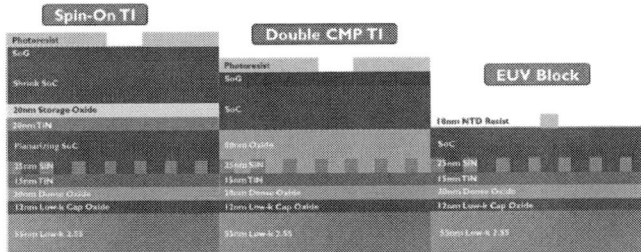

Fig. 7. Stack comparison of 193i versus EUV block patterning options. Inpria^TM negative tone photoresist used in EUV stack to print pillars

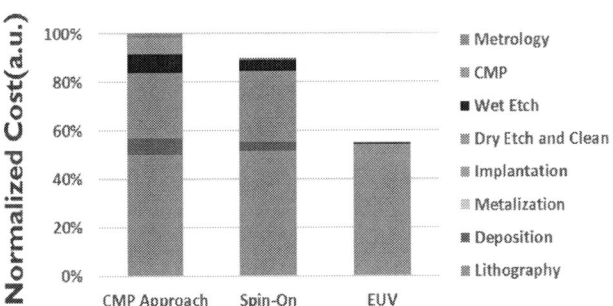

Fig. 8. Cost comparison of three block patterning options. The cost of each process module is normalized to show a relative benchmark. Relative to the double CMP approach, the EUV block option can potentially provide as much as 43% cost saving, using assumptions described in [16].

CONCLUSION

Common block processes required in 193 immersion-based N5 interconnect patterning was reviewed. Patterning challenges of multi-color 193i blocks is greatly reduced in the EUV block alternative with potential yield improvement and cost saving anticipated.

REFERENCES

[1] L. Liebmann, et al, "Overcoming Scaling Barriers through Design Technology CoOptimization", VLSI 2016.

[2] J. Ryckaert, et al, "DTCO at N7 and beyond: patterning and electrical compromises and opportunities", SPIE 2015.

[3] P. Raghavan, et al, "Metal stack optimization for low power and high density for N7-N5", SPIE 2016.

[4] J. S. Chawla, et al, "Demonstration of a 12 nm-Half-Pitch Copper Ultralow-k Interconnect Process", IITC 2012.

[5] M. van Veenhuizen, et al, "Demonstration of an electrically functional 34 nm metal pitch interconnect in

ultralow-k ILD using spacer-based pitch quartering", IITC 2012.

[6] T. Nogami, et al, "Through-Cobalt Self Forming Barrier (tCoSFB) for Cu/ULK BEOL: A Novel Concept for Advanced Technology Nodes", IEDM 2015.

[7] T. Standaert, et al, "BEOL Process Integration for the 7 nm Technology Node", IITC 2016.

[8] "unpublished" Decoster, S., Piao, X., Lazzarino, F., Feurprier, Y., "Tone reversal technology targeting below N5 technology node applications", AVN 64th International Symposium and Exhibition 2017.

[9] B. Briggs, et al, "N5 Technology Node Dual-Damascene Inter-connects Enabled Using Multi Patterning", IITC 2017.

[10] "unpublished" Briggs, B., Soethoudt, J., Delabie, A., et al., "A method to pattern tight tip-to-tip in 32nm-pitch N5 interconnect using Ru area selective deposition tone inversion process", submitted to Materials for Advanced Metallization Conference (MAM 2018).

[11] Peter De Bisschop, "Stochastic effects in EUV lithography: local CD variability, and printing failures", JM3, Journal of Micro/Nanotechnology, MEMS, and MOEMS, 16(4), 041013 (2017)

[12] Blanco Carballo, V., Bekaert, J., Mao, M., et al., "Single exposure EUV patterning of BEOL metal layers on the IMEC iN7 platform", Paper 10143-41 Proceedings of SPIE 2017.

[13] Bakaert, J., Di Lorenzo, P., Mao, M., et al., "SAQP and EUV block patterning of BEOL metal layers on IMEC's iN7 platform", Paper 10143-13 Proceedings of SPIE 2017.

[14] Halder, S. et al., "EPE analysis of sub-N10 BEOL structures with Coventor's SEMulator3D", Paper 10145-81 Proceedings of SPIE 2017.

[15] Lariviere, S., Briggs, B., Wilson, C. J., et al., "N5 BEOL process options patterning flow comparing 193immersion to hybrid EUV or full EUV", SSDM 2017

[16] Mallik, A., Ryckaert, J., Mercha, A., Verkest, D., Ronse, K., Thean, A., "Maintaining Moore's law: enabling cost-friendly dimensional scaling", Proceedings of SPIE 9422, Extreme Ultraviolet (EUV) Lithography VI, 94221N, April 2015

OVERVIEW OF ALD APPLICATIONS FOR ADVANCED CMOS TECHNOLOGY

X. Shi[*], C. Li, H. Ji, H. Qin, W. Zhang, W. Xia, P. Ding

Beijing NAURA Microelectronics Equipment Co. Ltd

No. 8 Wenchang Avenue, Beijing E-Town, Beijing 100176, China

*Corresponding Author's Email: shixiaoping@naura.com

ABSTRACT

Atomic Layer Deposition (ALD) is being used for many applications in advanced CMOS technologies, e.g. hi-k materials as gate dielectric, metal materials as metal gate electrode, dielectric spacers for self-aligned double patterning (SADP) or quadruple patterning (SAQP), and metal materials as Cu diffusion barrier layer etc.

Beijing NAURA Microelectronics is the largest Chinese semiconductor equipment vendor. Its products cover many IC manufacturing fields, e.g. Etch, PVD, CVD, Furnace, and Cleaning etc. With the growing Chinese semiconductor market, NAURA expends its business into ALD field and develops ALD tools and processes to meet its customers' requirements.

In this paper, we are going to review the ALD applications in advanced CMOS technologies, and discuss the ALD solutions which NAURA is capable to offer in advanced CMOS technologies.

INTRODUCTION

With CMOS device down-scaling from 45nm to 10nm, new materials and device structures are introduced. Hf-based hi-k materials are applied as gate dielectric to replace traditional SiON gate dielectric to reduce gate leakage [1~4]; High mobility channels are used to improve carrier mobility; FinFET 3D device structure is invented to replace planar device to improve device Ion/Ioff performance [5]. With these innovations, traditional thin film deposition techniques like PVD and CVD face more challenges. In contrast, ALD deposition process has many advantages over PVD and CVD processes, e.g. accurate film thickness control, excellent step coverage and low thermal budget etc. Therefore ALD is accepted for more and more applications in advanced CMOS technologies.

DISCUSSION

One ALD application in leading CMOS technologies is to deposit hi-k materials as gate dielectric for 45nm technology and beyond. In 2007, Intel announced that Hf-based hi-k dielectric grown by ALD was used as gate dielectric for its 45nm technology [3, 4]. Since then, ALD hi-k dielectric becomes the main stream of gate dielectric for leading CMOS. NAURA develops its own ALD clustered platform for hi-k/metal gate applications, Polaris A630 300mm ALD platform.

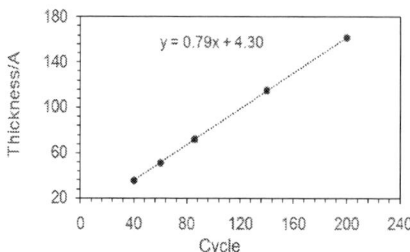

(a). Growth curve of ALD HfO$_2$.

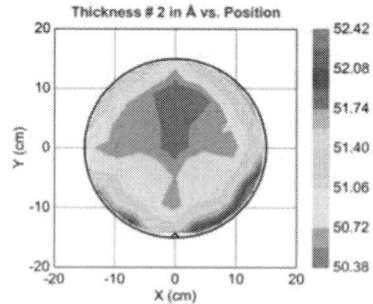

(b). Film contour map for 5nm HfO$_2$

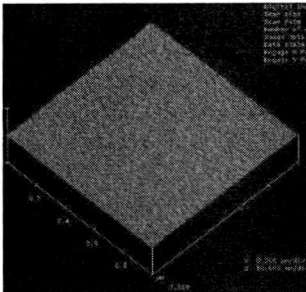

(c). AFM results of 5nm HfO₂.

Fig.1. Growth curve, film contour map, film roughness of ALD HfO₂ grown in NAURA Promi-300 single wafer ALD chamber. The contour map shows a film uniformity of ~ 1.1%. The AFM result shows a RMS of 0.19nm for 5nm HfO₂ film.

One Polaris A630 platform can hook-up maximum 6 process chambers. Thus the system is capable to perform more complex clustered process routines, and to yield a maximized tool throughput as well. NAURA 300mm thermal ALD process chamber, Promi-300，is developed for an optimum of ALD process cycles. The ALD HfO₂ process developed in Promi-300 ALD chamber exhibits typical ALD features, e.g. linear ALD growth curve, excellent film uniformity and superior step coverage performance etc. Some process results, like growth curve, film contour map, and film surface roughness etc., are shown in Figure 1.

Figure 2 shows some characterization results of HfO₂ grown in Promi-300 thermal ALD chamber. In Figure 2 (a), the CV measurement results of 4nm HfO₂ are shown. The CV measurement is done on MOS capacitor structure. The extracted k value is ~ 18 for the as-deposited ALD HfO₂ films. Figure 2 (b) shows a gate stack TEM picture of 4nm HfO₂ covered by a thin ALD TiN gate electrode. These results proof that the performance of hi-k materials grown in Polaris A630 ALD platform could be benchmarked with those from others.

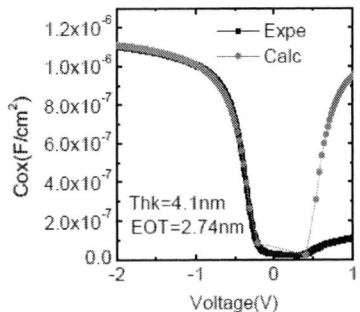

(a). CV curve for 4nm HfO₂ dielectric.

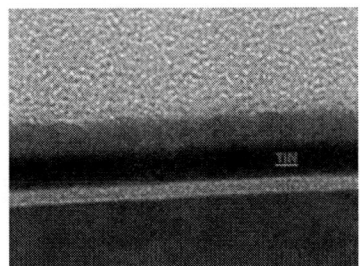

(b). TEM picture of a gate stack of HfO2/TiN.

Fig.2. Performance of 4nm ALD HfO₂ grown in NAURA Promi-300 single wafer ALD chamber. The CV measurement in (a) is done on MOS capacitor structure. The measured hi-k dielectric yields an EOT of 2.74nm. TEM picture in (b) exhibits excellent interfaces between substrate/hik layer, and hi-k/metal gate.

With CMOS down-scaling to 14nm and beyond, FinFET device structure is applied [5]. One major challenge is lithography patterning due to very critical features. SADP/SAQP is invented to meet the requirements. In SADP/SAQP process module, the control of spacer film thickness becomes very crucial. One of ALD applications is to deposit conformal spacer in SADP/SAQP process module. Figure 3 shows a typical SADP process flow. In the flow, an ALD dielectric layer, usually PEALD SiO₂ in this case, is used as spacer. Naura A630E PEALD system is developed for such applications. Similar to A630 thermal ALD platform, NAURA A630E platform can also hook-up maximum 6 process chambers. Thus it can provide flexible solutions for different customers, e.g. single process chamber set-up for R&D purpose, or a configuration with full 6-chambers to maximize throughput for mass production.

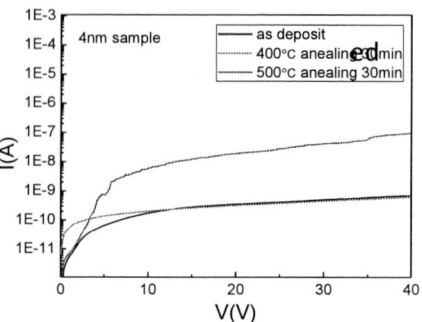

Fig.4. Performance of 4nm ALD TaN as Cu diffusion barrier layer grown in NAURA Promi-300 metal ALD chamber.

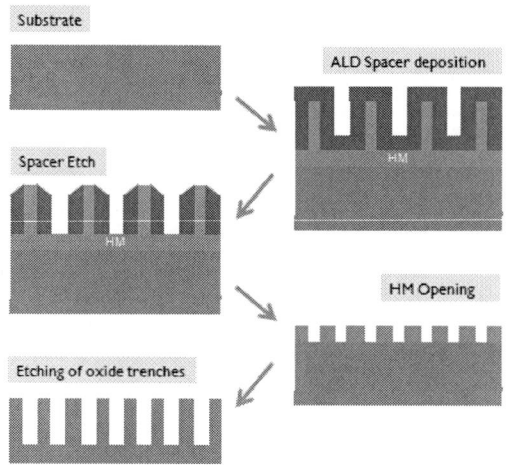

Fig.3. SADP process flow using PEALD SiO_2 for spacer deposition.

Cu diffusion barrier layer is another crucial application for ALD in advanced BEOL integration. With device down-scaling to 14nm node and beyond, metal line features become very critical. Thus traditional barrier deposition technique like PVD cannot meet the process requirements. ALD TaN process, with excellent step coverage, is adopted to replace PVD TaN process as Cu diffusion barrier layer in leading CMOS BEOL integration. NAURA eVictor C830 clustered PVD/ALD platform is developed for the applications. Figure 4 shows the barrier property characterization of ALD TaN deposited in NAURA ALD chamber. The results show that 4nm ALD TaN grown in NAURA ALD chamber yields an excellent performance as Cu diffusion barrier layer.

Metal-Insulator-Metal (MIM) capacitors have attracted great attention because of their high capacitance density, especially in high-performance mixed-signal and RF circuits built in CMOS. MIM capacitors can be integrated in BEOL modules. MIM capacitors consist of a bottom metal electrode, dielectric isolation layer, and a top metal electrode. The metal electrodes prefer to use metal materials with a low resistance. Usually metal materials deposited from PVD technique yield low resistance. Due to constrain of thermal budget and film quality, usually the dielectric layer is deposited by ALD. NAURA Polaris C630 PVD/ALD system is a clustered platform developed for MIM capacitor applications. The system is capable to cluster PVD chambers and ALD chambers together with Degas/Preclean chambers on one platform. Thus it can provide an integrated solution for MIM capacitor processes, no air-breaks are needed between bottom metal and dielectric, and dielectric and top metal processes. Consequently it can provide a better contamination control between these interfaces. The clustered system also yields a higher throughput and a shorter cycle-time.

Figure 5 shows a TEM picture of a MIM capacitor processed at NAURA Polaris C630 platform. Both top and bottom metal electrodes are processed by PVD. A sandwich metal structure TiN/Al/TiN, is applied for a

978-1-5386-5309-8/18 $31.00 © 2018 IEEE

better adhesion and conduction. The middle dielectric is ALD Al_2O_3 layer in this case.

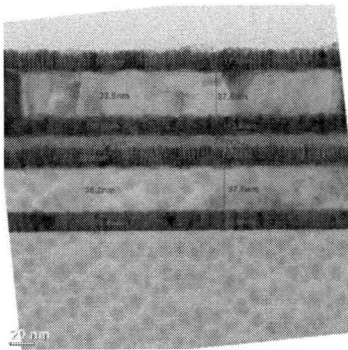

Fig.5. TEM picture of a MIM capacitor processed at NAURA Polaris C630 clustered PVD/ALD system.

SUMMARY

ALD is adopted for more and more applications in advanced CMOS technologies, due to its intrinsic process features e.g. precise film thickness control, excellent step coverage and low thermal budget etc. In the paper we review the ALD applications in advanced CMOS process modules such as gate stack, SAQP, interconnect barrier layer and MIM capacitor etc.

NAURA Microelectronics, the largest Chinese semiconductor equipment supplier, expands its process capability into ALD filed. With NAURA existing process technologies, NAURA could provide more clustered process solutions for its customers.

ACKNOWLEDGEMENTS

Authors appreciate the helps from Institute of Microelectronics of Chinese Academy of Sciences of assisting MOS capacitor measurements, and helps from School of Microelectronics of Fudan University of assisting Cu barrier characterization. Authors also would like to thank all supports and helps from colleagues in Beijing NAURA Microelectronics Equipment Co. Ltd.

REFERENCES

[1] X. Shi, H. Tielens, et al. *J. Electrochemi. Soc.*, 158 (1) H69 2011.

[2] X. Shi, M. Schaekers, L. Date, et al. *ECS Trans.*, 3(3) 417 2006.

[3] M. T. Bohr, R. S. Chau, T. Ghani, and K. Mistry, *IEEE Spectrum*, 44, 29 2007.

[4] K. Mistry, C. Allen, C. Auth, B. Beattie, D. Bergstrom, M. Bost, M. Brazier, M. Buehler, A. Cappellani, R. Chau, et al., *Tech. Dig. - Int. Electron Devices Meet.*, 2007, 247.

[5] D. Hisamoto, W. Lee, C. Hu et al., *Tech. Dig. - Int. Electron Devices Meet.*, 1998.

Conformal SiGe selective epitaxial growth for advanced CMOS technology

Yiqun Liu[*], Lan Jin, Kunshan Song, Qiong Wu, Youfeng He, Yonggen He

Technology Research and Development, SMIC

No. 18 Zhangjiang Rd., Shanghai, 201203, China

[*]Corresponding Author's Email: Yiqun_Liu@smics.com

ABSTRACT

For advanced CMOS technology, embedded Silicon Germanium (SiGe) alloys with high Ge% in source/drain is necessary to boost PMOSFET device performance through hole mobility enhancement. Both incoming Si surface quality and SiGe seed layer conformality are critical for the subsequent epitaxial growth of SiGe material with low defect density. In this paper, we reported an in-situ clean method to achieve a pristine Si surface as well as an optimized epitaxial growth process to form conformal SiGe seed layer. With these solutions, a high-quality embedded SiGe source/drain is achieved resulting in 97% reduction in defect density and 7% improvement in PMOS device performance.

INTRODUCTION

Embedded Silicon Germanium alloys (SiGe), selectively grown at the source/drain (S/D) areas, have been widely utilized in CMOS technology since 90nm technology node with the aim to boost PMOSFET device performance through the enhancement of hole mobility[1]-[3].

As CMOS technology scales down to 14nm and beyond, the transistor design has evolved from planar device structure to 3D FinFETs in order to improve device performance by exerting good electrostatic control and modifying short channel effect [4]-[5]. In order to boost the channel mobility for FinFET, the Germanium concentration of the S/D SiGe stressor has also been demanded to increase to higher than 50%. A good silicon interfacial bonding prior to epitaxial growth is critical to initiate the epitaxial growth of defect-free SiGe material [6]-[7]. A good-quality and conformal seed layer with intermedium Ge% is also necessary to serve as a strain buffer layer in order to form high quality and high Ge% bulk SiGe stressor in the subsequent step. However, the shrinking geometry of FinFET has resulted in a smaller dimension for both channel and S/D area. Abnormal epitaxial growth of SiGe in the dry-etch recessed area, such as underfill of bulk SiGe, can be usually found if there is polymer residue or lattice damage caused by reactive ion etch (RIE) process prior to SiGe process, as illustrated by schematics shown in Fig. 2a. The underfill of bulk SiGe may lead to high resistance of source/drain areas and failure of metal Silicide formation in the following steps. Therefore, it brings in challenges on how to achieve a clean and defect-free Si surface prior to SiGe process as well as how to grow conformal SiGe seed layer for bulk SiGe growth at FinFET source/drain areas.

We reported that a clean Si surface with no polymer residue and free of lattice damage can be achieved by an in-situ pre-clean method inside the SiGe epitaxial growth process chamber. Besides, the process parameters of SiGe seed layer epitaxial growth have also been optimized to minimize the difference in growth rates among different Si crystalline planes. Combining these efforts, a high-quality SiGe seed layer with conformality higher than 95% has been selectively grown in the recessed S/D areas. With these solutions, an embedded high-quality SiGe stressor with higher Ge concentration can be formed at S/D areas. Its underfill defect density is significantly reduced by 97% and PMOS Ion-Ioff performance is improved by 7%.

EXPERIMENT AND DISCUSSION

The experiments were carried out on 300mm patterned wafers following the advanced FinFET integration scheme. The process flow in the SiGe loop is briefly described in Fig. 2. After the dummy gate patterning process, a set of spacer films are deposited by ALD processes. Then the NMOS area is covered up by lithography, and the PMOS source/drain area is anisotropically etched to a U-shape trench by RIE. After photoresist strip and wet clean, the POR in-situ oxide removal and hydrogen bake steps are done before selective SiGe epitaxy process. The SiGe epitaxy process is tuned to selectively deposit SiGe epitaxial film on recessed PMOS source and drain areas, but not on the dielectric area. The SiGe stressor contains a SiGe seed layer with intermedium Ge concentration and a SiGe bulk layer with higher Ge concentration serving as the main stress layer.

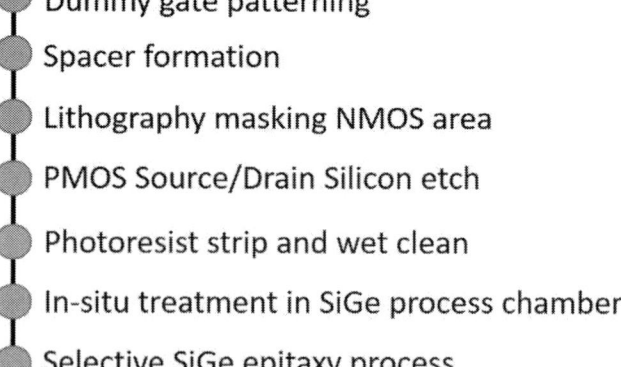

Dummy gate patterning

Spacer formation

Lithography masking NMOS area

PMOS Source/Drain Silicon etch

Photoresist strip and wet clean

In-situ treatment in SiGe process chamber

Selective SiGe epitaxy process

Fig. 1. Brief description of SiGe loop process flow in advanced FinFET integration scheme.

Poor step coverage of SiGe seed layer inside the trench as well as underfill defects of SiGe bulk layer are easily found through cross-sectional TEM, as illustrated in Fig. 2a. In order to solve the underfill defect, the conformality of SiGe seed process is first investigated by checking the deposition rates on blanket Silicon wafers with (100), (110) and (111) orientations, since trench sidewall is with (110) orientation and bottom (100) orientation. The seed layer is expected to be more conformal if the gap of growth rates on different Silicon crystalline planes is smaller. The deposition rate on (110) plane is lower than (100) plane for baseline seed process. Through Design of Experiments (DOE), several process parameters, such as temperature, pressure and gas flow are optimized in order to minimize the growth rate gap on different planes. As shown in Table 1, the growth rate ratio of (110) v.s. (100) is much improved from 0.78 to 1.09, comparing the optimized Process A and Baseline seed process. However, with Process A, the SiGe seed layer is not continuously grown on the sidewall and the underfill defect is not significantly reduced, as illustrated in Fig. 2b.

TABLE 1. The growth rates of SiGe seed epitaxial process on different Silicon crystalline planes.

SiGe seed epitaxial process	Growth rate ratio	
	(110)/(100)	(111)/(100)
Baseline Process	0.78	0.35
Process A	1.09	0.69

Two major root causes are suspected that may lead to the poor coverage of SiGe seed layer. One is the polymer residue generated by the anisotropic etch process. The aspect ratio of dummy gate top to trench bottom in FinFET technology is much higher than that in planar technology, which makes it more challenging to clean away polymer residue inside the trench. However, enhancing the wet clean process is not effective for this case.

Another possibility is the damage of the crystal lattice on the Silicon surface caused by plasma during the dry etch process. An in-situ pre-clean method is used in the SiGe epitaxy process chamber prior to the POR (Process of Record) Hydrogen bake step. The pre-clean step is to clean Silicon surface and etch away the damaged Silicon lattice on the surface. With this in-situ pre-clean process, Process A seed layer is grown inside the trench conformally and the underfill defect of SiGe bulk layer is resolved effectively, as illustrated in Fig. 2c. It is therefore demonstrated that a pristine Silicon surface with no lattice damage or polymer residue is critical for high-quality SiGe epitaxy and can be achieved by this in-situ pre-clean method in the SiGe epitaxy chamber.

The process window of the in-situ pre-clean process is also carefully checked by evaluating the etch rate of Si on blanket wafers with (100), (110) and (111) orientation at different process temperature, as shown in Table 2. The etch rate of Silicon is increased at higher process temperature. And the etch rate on (110) plane is faster than that on (100) plane, which is preferred for controlling the proximity between source/drain SiGe and channel and also good for preserving the trench depth during the in-situ pre-clean process.

TABLE 2. The etch rates of Silicon on Silicon blanket wafers with (100), (110) and (111) orientations using in-situ pre-clean method at different process temperatures

In-situ pre-clean temperature (°C)	Etch rate of Silicon (A/s)		
	(100)	(110)	(111)
T	0.091	0.118	0.049
T+20	0.134	0.155	0.099

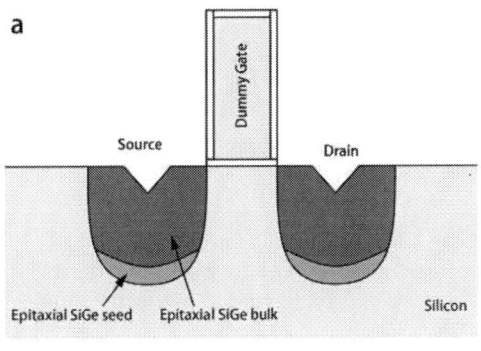

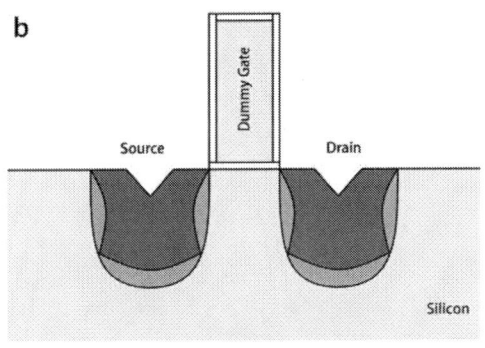

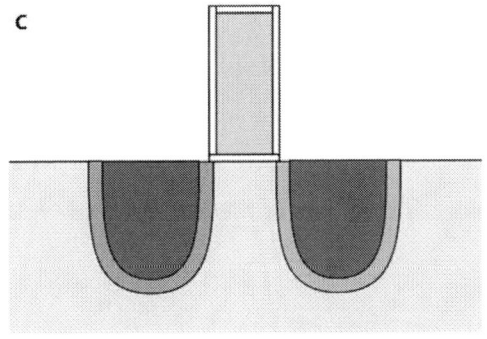

Fig. 2. Schematics of a Silicon MOSFET with dummy gate and recessed source/drain areas filled up with epitaxial SiGe stressor material. (a) Abnormal growth of bulk SiGe, known as underfill defect, caused by the poor coverage of SiGe seed layer; (b) With Process A SiGe layer, there is growth of seed layer on the sidewall but not continuous. The underfill defect is still not resolved; (c) Normal growth of bulk SiGe thanks to the pristine Si surface achieved by in-situ pre-clean method as well as the optimized conformal seed layer epitaxial process.

With the in-situ pre-clean method and the conformal Process A SiGe seed layer, an embedded high-quality SiGe

978-1-5386-5309-8/18 $31.00 © 2018 IEEE

stressor with higher Ge% concentration can be formed at S/D areas. Its underfill defect density is significantly reduced by 97% according to the defect scan result and PMOS Ion-Ioff performance is improved by 7%.

SUMMARY

We have achieved a conformal selective epitaxy SiGe seed process with a step coverage > 95% by DOE method. We also demonstrated an in-situ pre-clean method in the SiGe epitaxy process chamber that can effectively clean Silicon surface and remove the damaged lattice. By adopting the in-situ pre-clean method as well as the conformal SiGe seed layer process, an embedded high-quality SiGe source/drain is formed free of underfill defect. PMOS Ion-Ioff performance is improved by 7% due to the increased strain from SiGe bulk layer with improved epitaxy quality.

ACKNOWLEDGEMENTS

This project is sponsored by Shanghai Pujiang Program and also sponsored by Program of Shanghai Subject Chief Scientist (B type).

REFERENCES

[1] T. Ghani, M. Armstrong, C. Auth, M. Bost, P. Charvat, G. Glass, T. Hoffmann, K. Johnson, C. Kenyon, J. Klaus, B. McIntyre, K. Mistry, A. Murthy, J. Sandford, M. Silberstein, S. Sivakumar, P. Smith, K. Zawadzki, S. Thompson, and M. Bohr, "A 90 nm high volume manufacturing logic technology featuring novel 45 nm gate length strained silicon CMOS transistors," in IEDM Tech. Dig., 2003, pp. 978–980.

[2] H. Ohta, Y. Kim, Y. Shimamune, T. Sakuma, A. Hatada, A. Katakami, T. Soeda, K. Kawamura, H. Kokura, H. Morioka, T. Watanabe, J. Oh, Y. Hayami, J. Ogura, M. Tajima, T. Mori, N. Tamura, M. Kojima, and K. Hashimoto, "High performance 30 nm gate bulk CMOS for 45 nm node with Σ-shaped SiGe-SD," in IEDM Tech. Dig., 2005, pp. 10.6.1–10.6.4.

[3] N. Yasutake, A. Azuma, T. Ishida, N. Kusunoki, S. Mori, H. Itokawa, I. Mzushima, S. Okamoto, T. Morooka, N. Aoki, S. Kawanaka, S. Inaba, and Y. Toyoshima, "Record-high performance 32 nm node pMOSFET with advanced two-step recessed SiGe-S/D and stress liner technology," in VLSI Symp. Tech. Dig., 2007, pp. 48–49.

[4] T. Chiarella, L. Witters, A. Mercha, C. Kerner, R. Dittrich, M. Rakowski, C. Ortolland, L.Å. Ragnarsson, B. Parvais, A. De Keersgieter, Migrating from planar to FinFET for further CMOS scaling: SOI or bulk? Proceedings of the 35th European Solid-State Circuits Conference (ESSCIRC 2009) 2009, pp. 85–88.

[5] C. Auth, C. Allen, A. Blattner, D. Bergstrom, M. Brazier, M. Bost, M. Buehler, V. Chikarmane, T. Ghani, T. Glassman, A 22 nm high performance and low-power CMOS technology featuring fully-depleted tri-gate transistors, self-aligned contacts and high density MIM capacitors, 2012 IEEE Symposium on VLSI Technology 2012, pp. 131–132.

[6] J Ming Mao Chu and June-Hua Chou, "Physical yield improvement for SiGe Selective Epitaxial Growth fabrication process on nano scale pMOS strain engineering," 2009 IEEE Nanotechnology Materials and Devices Conference, Traverse City, MI, 2009, pp. 42-45.

[7] Tao Han, Man Gu, S. Grunow, Huang Liu, S. Sankaran and Jinping Liu, "Forming a more robust sidewall spacer with lower k (dielectric constant) value," 2017 China Semiconductor Technology International Conference (CSTIC), Shanghai, 2017, pp. 1-3.

EFFECT OF IDLE TIME ON THE PROPERTY OF ULTRA LOW-K DIELECTRIC FILM

Wenrong Hou[1]*, Jack Chen[1], Neil Li[1], Eric Li[1], Xiaorong Wang[1], Ming Feng[1], Xiaowei Shu[1], and Jianqiang Liu[1]

[1]Semiconductor Manufacturing North China (Beijing) Corporation (SMNC)
No.18 Wenchang Road, Beijing Economic-Technological Development Area, 100176 Beijing, China
*Corresponding Author's Email: Wendy_HWR@smnchina.com

ABSTRACT

The effect of idle time before process started on the ultra low-k (ULK) film's property (thickness, refractive index, stress) was studied in this paper. The results show that as idle time increase from 0s to 600s, the ULK film's thickness is decreased about 100~150A while the refractive index has no evident change. As idle time increase from 0s to 420s, the ULK film's stress is decreased about 4~5Mpa. The thickness decrease mainly comes from the main-deposition stage, slightly come from the initial stage, while the transition stage nearly has no influence on it. The root cause of thickness changes is related to the competitive relationship of film deposition between wafer and chamber body when chamber idle. Base on the experiment results, the automatic process control (APC) system can be introduced to compensate the film thickness loss caused by idle time automatically, thus realize the stability of ULK process.

Keywords—Ultra low-k; dielectric film; thickness; idle time

INTRODUCTION

Ultra low-k (ULK) dielectric film, with dielectric constant (k) lower than 2.5, is used as back-end copper interconnects dielectric layer to reduce RC delay for 45nm technology nodes and beyond. Because of higher carbon content and porosity, the ULK film is more easily to be damaged from etch plasma process and from exposure to the wet clean or strip chemicals [1]. Thus the introduction of ULK film brings many new challenges and attracts a large number of researches around the world. The challenges and improvement for the integration of ULK by film deposition, etch/ash and liner processes optimization was reported by Liu et al. [2]. The impact of ULK material deposition process and barrier layer deposition process on BEOL TDDB performance were investigated by Feng-lian Li et al. [3]. The reactive ion-etch (RIE) process for ULK film and its influence on electrical yield and dielectric breakdown is presented by B.R. Murthy et al. [4]. The effect of various barrier slurries on topography, defect, within-wafer uniformity and k-value of ultra low-k film was evaluated by Feng Zhao et al. [1].

As the introduction of ULK film narrowed process

and reliability margins, the film performance stability is very important to the back-end integration performance, such as thickness, with-in-wafer uniformity, refractive index, stress, k-value. However, the ULK film's property is found to be greatly affected by the idle time before process started, which has not been well studied. Thus the effect of idle time on the ULK film's thickness, refractive index and stress was studied in this paper.

EXPERIMENTS

Porous SiCOH ultra low-k films with thickness of about 2500A (250nm) are deposited on the Si substrate by plasmas enhanced chemical vapor deposition (PECVD) using APPLIED MATERIALS equipment. The thickness, refractive index and stress of ULK film are measured on KLA-Tencor metrology tools.

In order to study the effect of different idle time on the property of ULK film, eight split conditions are designed as shown in TABLE 1 and the experiments are processed on two same type tools, called as EQP1 and EQP2, for validation. Idle time 0s represents the normal ULK deposition process. Idle time 10s means there is 10s delay before wafer transfer into chamber and so on.

TABLE I.   IDLES TIME EFFECT SPLIT CONDITIONS

Wafer ID	Idle Time(s)	Thickness (A)		Reflectivity		Stress(MPa)	
		EQP1	EQP2	EQP1	EQP2	EQP1	EQP2
#1	0	2476.6	2491.0	1.4267	1.4232	40.1	41.8
#2	10	2472.1	2465.5	1.426	1.4243	42.3	40.9
#3	30	2457.4	2443.8	1.4262	1.4237	41.9	39.7
#4	60	2449.3	2443.8	1.4265	1.4238	41.4	39.9
#5	180	2431.8	2411.5	1.4256	1.4253	40.9	41.0
#6	300	2409.2	2394.2	1.4248	1.4245	40.0	40.9
#7	420	2391.8	2349.0	1.4258	1.4246	37.5	38.1
#8	600	2377.8	NA	1.4254	NA	NA	NA

RESULTS AND DISCUSSIONS

Figure 1 shows the influence of idle time on the thickness, refractive index and stress of ULK film. Figure 1 (a) is the effect of idle time on the ULK film's thickness. It shows that ULK film's thickness trend down linearly when idle time increases for both two tools. But the influence of idle time on ULK film's thickness for two tools is not exactly the same. For EQP1, when idle time increases from 0s to 600s, ULK film's thickness decreases about 100A. For EQP2, when idle time increases from 0s to 420s, ULK film's thickness decreases about 140A. This means that the different tools have different response to

978-1-5386-5309-8/18 $31.00 © 2018 IEEE

the idle time. EQP2 is more sensitive to the idle time than the EQP1. Figure 1 (b) is the effect of idle time on the ULK film's refractive index. When idle time increases from 0s to 600s, the refractive index of EQP1 decreases from 1.4267 to 1.4254. When idle time increases from 0s to 420s, the refractive index of EQP2 increases from 1.4232 to 1.4246. The results indicate that the refractive index has no evident change with idle time changing. Figure 1 (c) is the effect of idle time on the ULK film's stress. It shows that ULK film's stress decreases when idle time increases for both two tools. As idle time increases from 0s to 420s the ULK film's stress decreases about 4~5Mpa, which means the ULK film be less tensile.

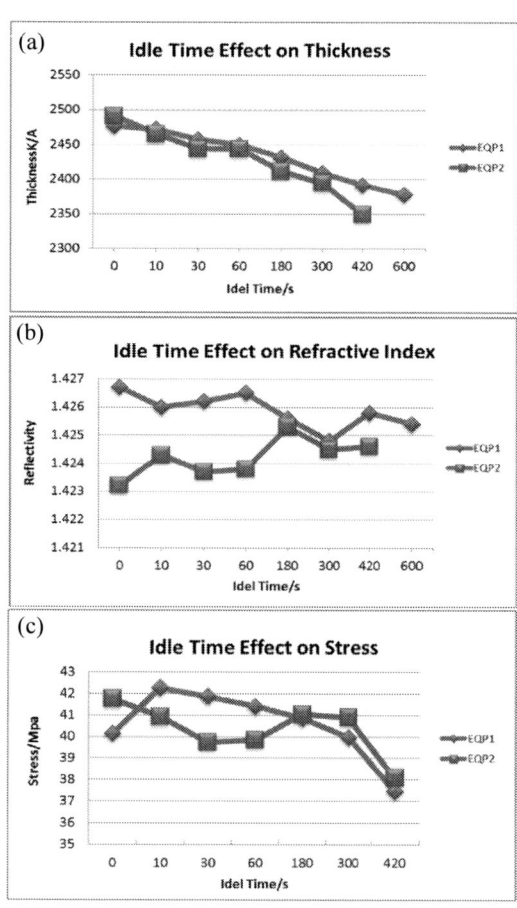

Figure 1: The influence of idle time on (a) thickness (b) refractive index (c) stress of ULK film.

The deposition process of ULK film is combination of three stages named initial, transition and main deposition. The major difference of three stages is oxygen content. The Initial deposition stage is oxygen rich, the main deposition stage is oxygen less, and the oxygen ratio

of transition deposition stage is in between of them. In order to find out which stage does the film's thickness changes mainly come from, we continue studied the idle time effect on ULK film's thickness during these three steps independently on EQP2. The results are shown in Figure 2.

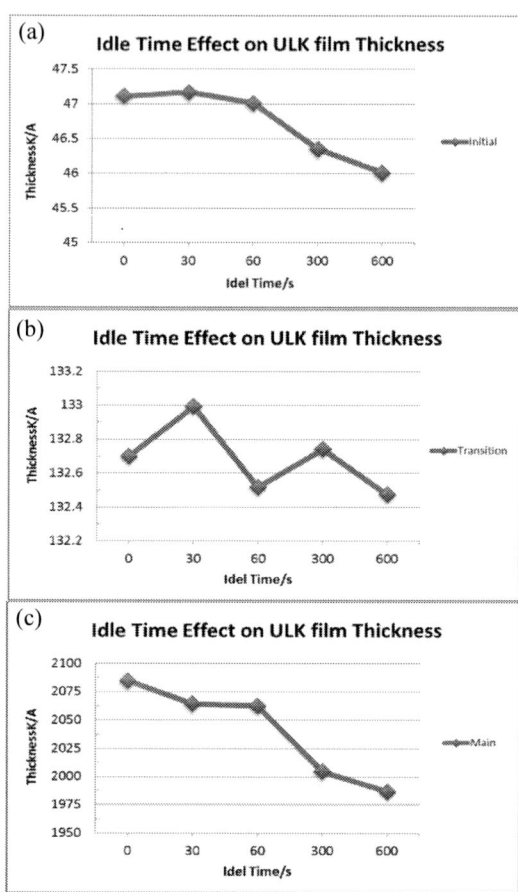

Figure 2: The influence of idle time on the thickness of ULK film during (a) initial (b) transition (c) main deposition stage.

It can be seen from Figure 2 that the influence of idle time on the thickness of ULK film during (a) initial (b) transition (c) main deposition stage is different. When idle time increase from 0 to 600s, the changes of ULK film's thickness is very small during the initial and transition deposition stages, which is about 1A (2%) at most. However, during the main deposition stage the ULK film's thickness trend down from 2084A to 1986A, with almost 100A's decreased. The results indicate that the effect of idle time on ULK film's thickness mainly comes from the main deposition stage, with about 83% nearly.

978-1-5386-5309-8/18 $31.00 © 2018 IEEE   272

The root cause of idle time effect on the thickness of ULK film is considered to be temperature changing in chamber body and environment when chamber idle. The wafer substrate temperature is very important to the deposition rate. Long Gu et al. [5] has studied the substrate temperature influence on properties of amorphous Silicon-Germanium thin films prepared by RF-PECVD. The results show that with the substrate temperature increasing, the deposition rate decreased. In our experiments, the wafer substrate temperature is equal to the heater temperature. Figure 3 is the heater's temperature with different idle time in main deposition stage. It shows that the wafer substrate temperature keeps no change and stable.

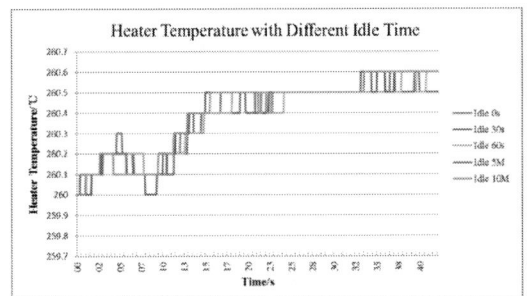

Figure 3: The heater temperature with different idle time during main deposition stage of ULK film.

It seems that the deposition rate should not change when substrate temperature keeps no change. However, as the film deposition on chamber body during PECVD process is unavoidable, there will be a competitive relationship of film deposition between wafer and chamber body. As there is always have a clean process before ULK film deposition process and the clean process is exothermic, thus the chamber body is hot before film deposition started. The longer idle time before film deposition started is, the more heat dissipated from chamber body. Thus the chamber body's temperature decreases when idle time long. As the deposition rate increased with the substrate temperature decreasing for ULK film, film deposited more on the chamber body but less on the wafer when chamber idle. Thus the thickness of ULK film deposited on the wafer decreased with idle time increasing although the deposition rate is not changed.

Marie Netrvalová et al. [6] has studied refractive index at different deposition temperature. The results show that the refractive index increases with higher temperature. As the refractive index is related to the atomic structure and the mass density of the material, it imply that film deposited at higher temperature have better atomic order and higher mass densities. In our experiments, the wafer substrate temperatures are not changed with different idle time, which can explain the results in Figure 1 (b) that the refractive index has no

evident change with idle time changing.

Base on the results in this paper, the automatic process control (APC) system can be introduced to compensate the film thickness loss caused by idle time automatically. By using APC system, different extra deposition time can be compensated based on different idle time, thus realize the thickness stability of ULK process.

CONCLUSIONS

The effect of idle time on the ULK film's thickness, refractive index and stress was studied in this paper. The results show that the thickness of ULK film decreases when idle time increase, and different tools have different response sensitivity to the idle time effect. The refractive index of ULK film has no evident change with idle time increases, while the ULK film's stress decreased about 4~5Mpa when idle time increases from 0s to 420s. The thickness changes of ULK film when chamber idle mainly come from the main deposition stage. The root cause of thickness changes is related to the competitive relationship of film deposition between wafer and chamber body when chamber idle.

ACKNOWLEDGEMENTS

This work is strongly supported by Jianqiang Liu, Ming Feng, and Xiaowei Shu as leaders in the group. Furthermore, I wish to express a lot of thanks to Jack Chen for many constructive discussions, and to Neil Li, Eric Li, Xiaorong Wang and other colleagues for the sincere helps. At last, the acknowledgement is extended to my company, the Semiconductor Manufacturing North China (Beijing) Corporation.

REFERENCES

[1] Feng Zhao, Laertis Economikos, Wei-tsu Tseng, et al. *International Conference on Planarization/CMP Technology*, 2007, pp. 258-260.

[2] H. Liu, J. Widodo, S. L. Liew, et al. *IEEE International Interconnect Technology Conference*, 2009, pp. 258-260.

[3] Feng lian Li, Jie Zhou, Li Fei Zhang, et al. *IEEE 24th International Symposium on the Physical and Failure Analysis of Integrated Circuits (IPFA)*, 2017, pp. 1-5.

[4] B.R. Murthy, M. Mukherjee-Roy, A. Krishnamoorthy, D.C. Frye. *IEEE Transactions on Semiconductor Manufacturing*, vol. 18, 2005, pp. 174-181.

[5] Long Gu, Hui dong Yang, Guo guan Wen, Yan min Li. *Symposium on Photonics and Optoelectronics (SOPO)*, 2011, pp. 1-3.

[6] Marie Netrvalová, Marinus Fischer, Jarmila Mullerovág, et al. *The Eighth International Conference on Advanced Semiconductor Devices and Microsystems*, 2010, pp. 329-332.

In-situ plasma monitoring of PECVD a-Si:H(i)/ a-Si:H (n) surface passivation for heterojunction solar cells application

Yu-Lin Hsieh[1], Li-Han Kau[2], Hung-Jui Huang[2], Chien-Chieh Lee[3], Yiin-Kuen Fuh[1*]
and Tomi T. Li[1]

[1]Department of Mechanical Engineering, National Central University, Taiwan;
[2]Institute of Opto-mechatronics Engineering, National Central University, Taiwan;
[3]Optical Science Center, National Central University, Taiwan;

Biography

Yu-Lin Hsieh Currently is a PhD candidate at Institute of Mechanical Engineering in National Central University, Taiwan. His main research focuses on a solar cell process development with the application of in-situ plasma diagnostics such as OES and QMS in PECVD.

Abstract

Silicon-based solar cell manufacturing via plasma enhanced chemical vapor deposition (PECVD) of both active/passive layers is investigated. In addition, in-situ plasma diagnostics of the deposition process can be monitored in real-time. Two types of complementary diagnostics, namely optical emission spectroscopy (OES) and quadruple mass spectrometry (QMS) are applied to an PECVD reactor. Furthermore, the impact of chamber wall conditioning on the solar cell performance is experimentally investigated based on symmetrical stacks structure (a-Si:H(i) / CZ wafer (n)/ a-Si:H(i)) and the n-type hydrogenated amorphous silicon (a-Si:H) growth process conditions were optimized. Silicon heterojunction (SHJ) solar cell back surface field (BSF) layer was prepared by conventional radio-frequency plasma enhanced chemical vapor deposition (RF-PECVD) and the processing conditions in terms of the phosphine flow (1-20sccm), hydrogen dilution ratio ($R = H_2/SiH_4$) as 1-5 and symmetrical stacks structure using a ($PH_3/SiH_4/H_2/Ar$) mixture were experimentally optimized. In addition, characterization of effective carrier lifetime (τ_{eff}), electrical and structure properties as well as correlation with the hydrogen dilution ratio were systematically discussed with the emphasis on the effectiveness of passivation layer. A high quality intrinsic/n-type a-Si:H layer stack BSF layer with measured effective carrier lifetime (τ_{eff}) of 1.8ms (which counts for implied V_{OC} 0.707 V), can be consistently obtained and this improved passivation layer can be primarily attributed to the synergy of chemical and field effect to significantly reduce the surface recombination

Keywords: PECVD; plasma diagnostics; a-Si:H; silicon heterojunction (SHJ); Back surface field (BSF); RF-PECVD; effective carrier lifetime (τ_{eff}); Passivation.

1. Introduction

The development of high efficiency in heterojunction silicon solar cell needs to have an excellent P or N-type doped amorphous silicon layer. Boron-doped amorphous silicon layer processed by RF-PECVD is used as a P-type passivation (emitter) layer stacked on silicon intrinsic layer as part of a silicon heterojunction solar cell. The high conductivity of the film is strongly dependent on the doping element [1]. Excellent electric field can form a field-effect passivation[2] on cell and limit to generate minority carrier for recombination. In addition, an excellent emitter layer can reduce the surface carrier recombination rate and improve the carrier collection rate at interface, thus enhance the short-circuit current (Jsc) [3]. Excellent of hydrogenated amorphous silicon oxide (a-SiOx:H) can yield a maximum effective lifetime of 400 μs [4]. Furthermore, boron doped nano crystalline silicon / amorphous silicon hybrid thin films deposited by PECVD was reported to have a high dark conductivity of 6.5×10^{-4} S/cm and minority carrier lifetime (τ_{eff}) of 1740 μs on Czochralski (Cz) Si wafers with the hybrid p-type Si films [5]. In this study, the processing conditions in terms of the phosphine flow (1-20sccm), hydrogen dilution ratio (0.5-4) and symmetrical stacks structure using a ($PH_3/SiH_4/H_2/Ar$) mixture were systematically optimized. The in-situ OES and QMS are applied to monitor the stabilization of an PECVD reactor in order to secure process conditions and thus the best quality of passivation layer can be received consistently. In addition, the field effect combines with chemical passivation is primarily attributed to the enhancement of effective lifetime. The spectroscopic ellipsometer and Hall measurements were used to study their correlations with microstructure and electrical property of the deposited film. Trends relating deposition conditions to relevant film characteristics, such as resistivity, effective carrier lifetime (τ_{eff}), TEM interface morphology, and the optimized the passivation quality as well as doping process of the samples were investigated and discussed. High quality of BSF on passivation with effective carrier lifetime up to 1.8ms can be obtained for heterojunction solar cells application. Furthermore, the growth of a-Si:H thin films by PECVD was monitored by means of two complementary plasma diagnostics: OES and QMS.

978-1-5386-5309-8/18 $31.00 © 2018 IEEE

2. Experimental

In this study, the intrinsic and n-type thin films were deposited onto CZ n-type c-Si wafer (<100>; 300μm; 1-10 Ω·cm; double sided polished) by a commercially available plasma enhanced chemical deposition (PECVD) equipment (Creating Nano-Technologies, Taiwan, PE-001) operating at the conventional radio-frequency (13.56 MHz) for the growth of passivation layers in the symmetrical stacks structure (a-Si:H(i) / CZ wafer (n)/ a-Si:H(i)) and the n-type hydrogenated amorphous silicon (a-Si:H).

The wafers were initially cleaned using $H_2O_2:H_2SO_4 = 1:2$ solution for 5 min, rinsed in DI water, then dried in N_2 atmosphere. Prior to film deposition, wafers were dipped in 2% HF for 1min to remove native oxide, and then the wafers were immediately transferred into vacuum chamber to prevent wafers from the native oxide layer growth.

The effective lifetime was determined using the lifetime tester WCT 120 from Sinton Consulting Inc. by the quasi-steady-state photoconductance (QSSPC) method with the transient mode and the generalized mode [6] at an injection level of $10^{-15} cm^{-3}$. The thickness and structure properties of the thin films on wafer were determined by spectroscopic ellipsometry (SE) with the Tauc-Lorentz (TL) and Effective Medium Approximation (EMA) model [7][8]. Hall measurement were used to obtain the electrical properties of the films. Transmission electron microscopy (TEM) micrographs were made to study the integrity of interface. The samples were prepared by standard ion milling method. Amorphous and defect-free structure can be vividly identified as shown in Fig. 1 of TEM picture for the a-Si:H(i) / a-Si:H(n+) interface. There is no discernable interface between doping layer and intrinsic layer, and a smooth interface between c-Si and a-Si:H can be obtained. As a result, it could effectively enhance the passivation quality and obtain a favorable field effect [9].

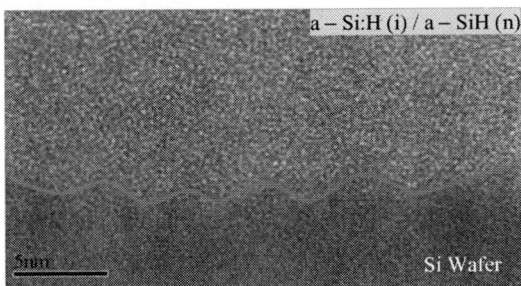

Fig. 1: TEM micrograph picture for the a-Si:H(i) / a-Si:H (n) interface, amorphous structure can be clearly seen.

3. Results and Discussion

3.1 Passivation properties as a function of PH₃ flow.

As shown in the Fig. 2(a), the measured effective carrier lifetime was experimentally determined to be highly dependent on film resistivity, i.e., the effective carrier lifetime has reached the maximum value ~ 587us at 3 sccm condition (coincides with the resistivity minimum value of ~ 7.3×10^{-4} ohm-cm.). On the other hand, Fig. 2(b) shows the deposition rate is approximately in the range of 0.5~0.65 (Å/s), achieving the minimum at 3 sccm. Fig. 2(c) indicates that the resistivity is approaching the saturation value ~ 3.1×10^{-3} ohm-cm, while the resistivity is reaching the minimum value of ~7.3×10^{-4} ohm-cm. The effective carrier lifetime deteriorated as the PH₃ flow rate is over 3 sccm, which can be primarily attributed to the notactivated dopant carriers, though the phosphine dopant concentration is proportionally increasing. Previous study also indicated that a reduction in conductivity was attributed to the more coordination number defects as compared with the electrically active dopants [9]. The impact on the lifetime can also be correlated to the saturated carrier activation rate, those inactivated dopant carriers form defects, generating a carrier recombination center to cause the lifetime to decrease [9]. An optimum flow of PH₃ was experimentally obtained in order to achieve the minimum resistivity and thus, maximum lifetime in films.

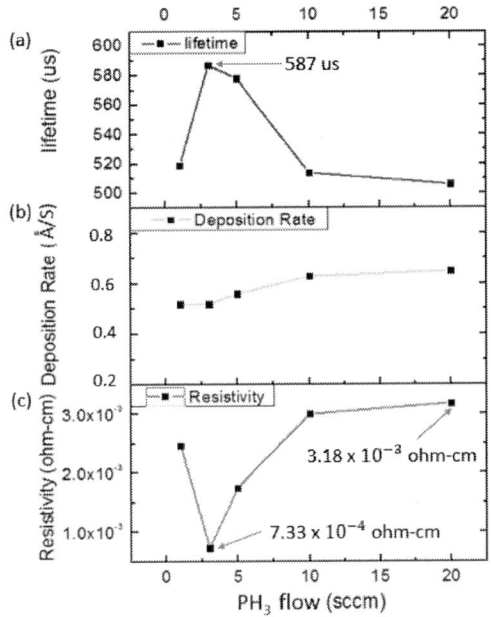

Fig. 2: Effective carrier lifetimes (τ_{eff}), resistivity and deposition rate on different PH₃ flow rate for the stacked structure.

3.2 Passivation properties as a function of intrinsic layer with varying dilution ratio R.

As previously reported that the intrinstic layer can provide more H atoms to fill surface dangling bonds, in particular, the hydrogen dilution ratio (R) in intrinsic buffer layer has a significant impact on the passivation properties during the deposition process. Fig. 3(a) shows the effective carrier lifetimes (τ_{eff}) are increasing almost linearly with increasing dilution ratios before the saturation of R approaches the value of 3, with the effective carrier lifetime is measured as ~800us, then the effective carrier lifetimes drops slightly as dilution ratio of R of 4 (lifetime is 700us). The pseudo dielectric function ε2 peak for the dilution ratio (R) was measured in the range 1 to 4, coincides at 3.7eV and experimentally confirming the deposited structures are in the amorphous phase as presented in Fig. 3(b).

3.3 Passivation properties as a function of BSF

In order to test the integrated passivation quality, the samples from the maximum effective carrier lifetime value (PH$_3$=3 sccm, 587μs) in stacks and the effects of dilution ratio R were compared and optimized. From Fig. 4(a), the maximum effective carrier lifetime value as measured up to 1865μs, which analytically calculated as the implied V_{OC} 0.707 V at R=2. It is contrast to the experimental results in Fig. 4(a) and the integrated passivation effect substantially increases more than doubled when R<2, but no obvious change when R over 3. Furthermore, Fig. 4(b) shows the SE measurement that the double layer stack structure has no crystal peak in the dilution ratio range of 0.5 to 4, which highly validates the introduced H atoms via the hydrogen dilution ratio will not result in adversely crystalline structures .

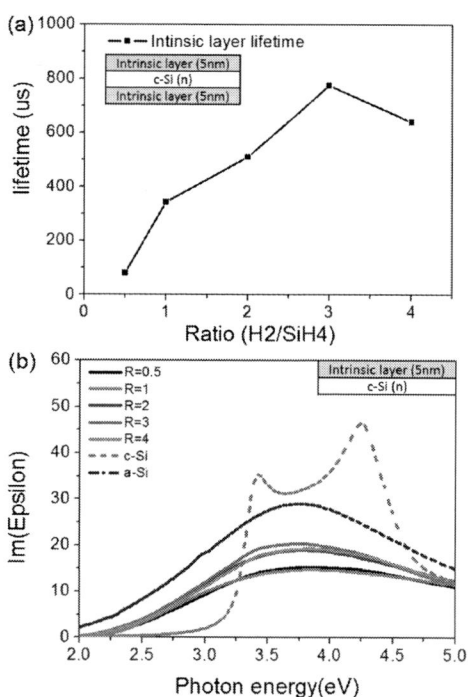

Fig. 3 (a) Effective carrier lifetimes and (b).the pseudo dielectric function ε2 for the intrinsic passivation layer as a function of the dilution ratio R for the stacked structure as shown in the inset of Fig. 3(a) 3(b). (Adapted from ref 10.)

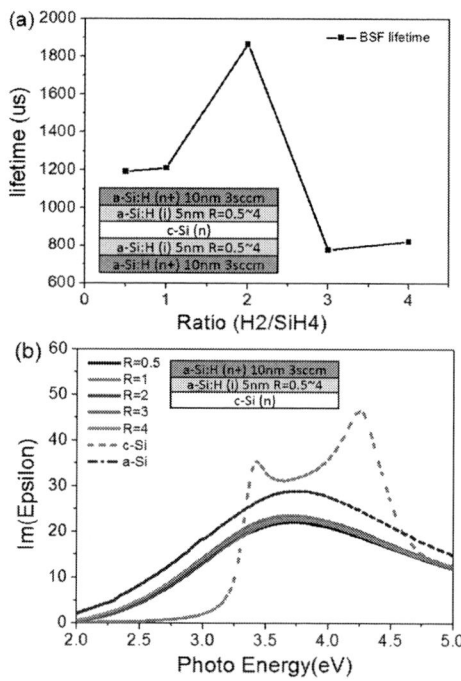

Fig. 4 (a) Effective carrier lifetimes and (b) the pseudo dielectric function ε2 for the intrinsic passivation layer as a function of the dilution ratio R. (Adapted from ref 10.)

3.4 Optical emission spectroscopy (OES)

Optical emission spectra was experimentally performed using an anti-coating device with honeycomb structure to enshroud the built-in OES be using in the chamber. The light is optically coupled via a fiber into equipment (Ocean Optical USB2000+). By fixing the total flow and hydrogen dilution ratio (R= SiH_4/H_2), Fig. 5(a) shows the most important measured peaks and bands (spectra) during a-Si:H deposition of the PECVD processes for various phosphine flow rates (1,3,5,10,20 sccm) respectively. Some prominent optical emission peaks and band structures are labelled such as Si* at 288.10 nm, SiH* at 414.30 nm, H_α at 656.30 nm, Ar at 750.30 nm. The curves of the crystallization rate index $H\alpha$ / SiH* (red line) and ratio of Si* / SiH* electron temperature (black line) for adjusting the PH_3 ratio are shown in Fig. 5(b). The tendency of the obtained crystallization rate index $H\alpha$ / SiH* is consistently proportional to that of crystallization rate index. This is primarily attributed to the ratio of electron temperature Si* / SiH* increases as the PH_3 flow increases.

3.5 Quadruple mass spectra (QMS)

A gas analyzer/ Quadruple mass spectrometer (Hiden PSM003P) was applied near the process chamber exhaust exit [11]. The position of the QMS orifice is about 12 cm away from the outer edge of the chamber. Residual Gas Analyzers (RGA) was used and consists of a quadrupole mass spectrometer and can detect masses up to 200 atomic mass unit (amu). An orifice is directly mounted in front of the QMS at measurements pressures up to 600 mTorr during deposition. A range of 0- 90 amu is performed to increase the time resolution(~30s sweeping time) during measurements. Some distinguishable fingerprints of atoms and molecules during process plasma of the SiH_x(x<4) signals can be identified. Fig. 6 shows the mass spectra measured during a-Si:H deposition. RGA results shows four distinishable peaks as H_x, H_2O, SiH_x, and Ar. Signal arises in form of SiH_x molecules measured in the range of m = 30 by mass spectrometry is indicating that the OES signals do not originate from SiH_4 but from SiH_x (x = 0 − 4) radicals. [11]

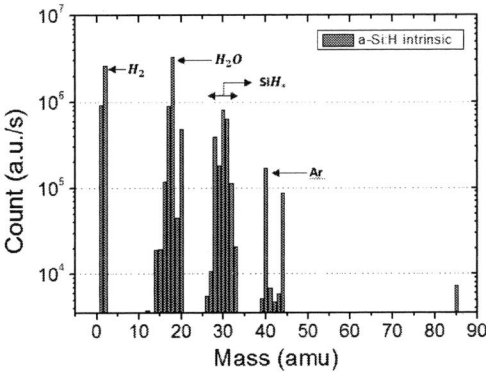

Fig. 6 Quadruple mass spectra measured during typical a-Si:H deposition.

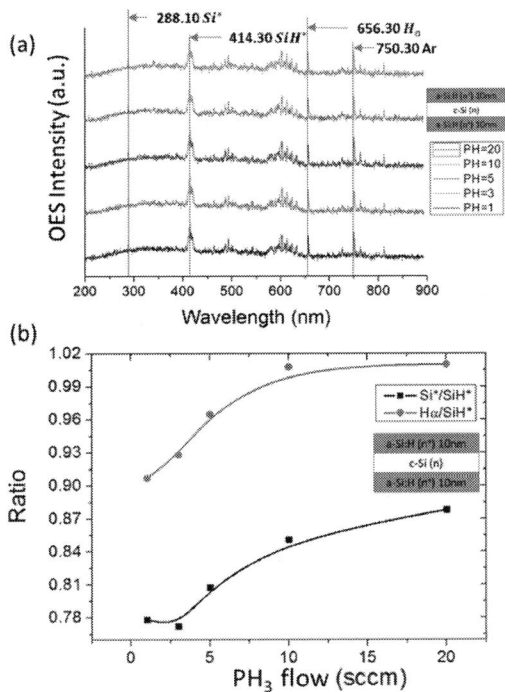

Fig. 5 (a) Optical emission spectra measured during the deposition of amorphous thin film. Some prominent optical emission peaks and band structures are labelled. (b) Ratio of crystallization rate index $H\alpha$ / SiH* (red line) and ratio of electron temperature Si* / SiH*(black line) calculated during the deposition of amorphous thin film as a function of varying PH_3 flow.

4. Conclusions

In this study, the processing conditions in terms of the phosphine flow (1-20sccm), hydrogen dilution ratio (0.5-4) and symmetrical stacks structure using a (PH$_3$ / SiH$_4$ / H$_2$ / Ar) mixture were systematically optimized. In addition, the field effect combines with chemical passivation is primarily attributed to the enhancement of effective lifetime. The spectroscopic ellipsometer and Hall measurements were used to study their correlations with microstructure and electrical property of the deposited film. Trends relating deposition conditions to relevant film characteristics, such as resistivity, effective carrier lifetime (τ_{eff}), TEM interface morphology, and the optimized the passivation quality as well as doping process of the samples were investigated and discussed. High quality of BSF on passivation with effective carrier lifetime up to 1.8ms can be obtained for SHJ solar cell application. In addition, the growth of a-Si:H thin films by PECVD can be quantitively monitored by means of OES, QMS omplementary plasma diagnostic. The process parameter of PH$_3$ gas flow rate and the influence of plasma chemistry as well as the PECVD deposited thin films can be closely correlated. Therefore, the in situ plasma diagnostic tools of OES and QMS can provide the valuable data for timely optimizing the properties of the a-Si:H films in the PECVD process developmenton.

5. References

[1] J.-W. aSchüttauf, K. H. M.van derWerf, I. M.Kielen, W. G. J. H. M.vanSark, J. K.Rath, andR. E. I.Schropp, "Excellent crystalline silicon surface passivation by amorphous silicon irrespective of the technique used for chemical vapor deposition," *Appl. Phys. Lett.*, vol. 98, no. 15, p. 153514, 2011.

[2] M.Taguchi *et al.*, "24 . 7 % Record Efficiency HIT Solar Cell on," vol. 4, no. 1, pp. 96–99, 2014.

[3] M.Taguchi, A.Terakawa, E.Maruyama, andM.Tanaka, "Obtaining a higher voc in HIT cells," *Prog. Photovoltaics Res. Appl.*, vol. 13, no. 6, pp. 481–488, 2005.

[4] H.Search, C.Journals, A.Contact, M.Iopscience, andI. P.Address, "Surface Passivation of Crystalline and Polycrystalline Silicon Using Hydrogenated Amorphous Silicon Oxide Film" vol. 3296.

[5] F.Wang *et al.*, "Boron doped nanocrystalline silicon/amorphous silicon hybrid emitter layers used to improve the performance of silicon heterojunction solar cells," *Sol. Energy*, vol. 108, pp. 308–314, 2014.

[6] H.Nagel, C.Berge, and A. G.Aberle, "Generalized analysis of quasi-steady-state and quasi-transient measurements of carrier lifetimes in semiconductors," J. Appl. Phys., vol. 86, no. 11, pp. 6218–6221, 1999.

[7] A. S.Ferlauto et al., "Analytical model for the optical functions of amorphous semiconductors from the near-infrared to ultraviolet: Applications in thin film photovoltaics," J. Appl. Phys., vol. 92, no. 5, pp. 2424–2436, 2002.

[8] T.Yuguchi, Y.Kanie, N.Matsuki, and H.Fujiwara, "Complete parameterization of the dielectric function of microcrystalline silicon fabricated by plasma-enhanced chemical vapor deposition," J. Appl. Phys., vol. 111, no. 8, 2012

[9] H.Fujiwara and M.Kondo, "Impact of epitaxial growth at the heterointerface of a-Si:H/c-Si solar cells," Appl. Phys. Lett., vol. 90, no. 1, 2007.

[10] Y.L.Hsieh, C.C.LEE, Y.K.FUH, J.Y.Chang, J.Y.Lee and T.T.Li, "Structural and electrical investigations of a-Si:H(i) and a-Si:H(n$^+$) stacked layers for improving the interfaceand passivation qualities," *J. Photon. Energy.* Vol. 7, Issue 3, 2017

[11] O.Gabriel, S.Kirner, M.Klick, B.Stannowski and R.Schlatmann, "Plasma monitoring and PECVD process control in thin film silicon-based solar cell manufacturing," EPJ.Photo., 5, 55202 (2014)

PVD SYSTEMS FOR ADVANCED PACKAGING APPLICATIONS

Jinguo Zhen[*], Kuanmao Wang, Wei Xia, Hougong Wang, Peijun Ding

Beijing NAURA Microelectronics Equipment Co. Ltd

No. 8 Wenchang Avenue, Beijing E-Town, Beijing 100176, China

*Corresponding Author's Email: zhengjinguo@naura.com

ABSTRACT

Physical vapor deposition (PVD) technology plays an important role in the advanced packaging (AP). It has a direct impact to the performance of final products. For AP application, e.g. bumping, fan-out and CIS , a clustered PVD approach is required to ensure a low contact resistance (Rc) and better film properties for better device performance. A clustered PVD platform consists of several modules including a transfer module, a Degas module, a Pre-clean module and PVD deposition modules. In this paper, we are going to report the developments of a clustered PVD system, the NAURA Polaris PVD for AP applications.

NAURA Polaris PVD system is widely used in many advanced packaging processes. There are significantly different type of wafer in AP fields due to different applications, e.g. SOG wafers, EMC wafers, glass wafers, thin wafers, taiko wafers. It is a big challenge to handle wafers with a large warpage. Thus it is required that the transfer module must have a strong capability to handle different wafer type and to detect failures. Wafers with PI/PBO/EMC (Polyimide/ Poly-p-phenylene benzobisoxazole/ Epoxy Molding Compound) materials require longer time degas process to ensure full outgasing.However, longer degas process results in lower throughout. Therefore it is necessary to develop a multi-wafer degas chamber to ensure competitive throughput. Proper temperature control(proper temperature and proper time) is critical for better Rc uniformity. An ICP Pre-clean module with stable plasma is developed to remove organic residues and native oxide on metal surface efficiently. To avoid re-oxide on metal surface, low process temperature is required. Furthermore, low process temperature is also capable to yield a better Rc performance. The PVD deposition module has wide process windows on film stress, uniformity, resistance and film crystalline phase .Thus it is easier to achieve a stable optimized process condition that improve device reliability.

Keywords—Advanced Packaging, PVD, Contact Resistance (Rc), Warpage, NAURA (key words)

INTRODUCTION

The role of advanced packaging in semiconductor industry is changed from the follower of the main down-scaling CMOS process technology to the enabler for future semiconductor applications. For many new applications, such as industry internet of everything (IOT), advanced automotive electronics, 5G communications, augmented reality (AR)/virtual reality (VR), and artificial intelligence (AI), higher performance, reliability and security are required. Thus, the modern advanced packaging needs to incorporate many innovations to meet the requirements. These innovations cover many fields, for instance from multi-chip heterogeneous integration to the latest high density interconnection technologies. In near future, the highly customized applications together with a well-controlled low cost will still be the main trend of the advanced packaging. Fan-out, Copper pillar, Gold bump and 2.5D / 3D TSV technologies will continue to be the main and the fastest growing advanced encapsulation platform. PVD process, as one of the most important technologies in the advanced packaging field, will continue to face more and more challenges.

DISCUSSION

There are many challenges in the advanced packaging field. NAURA Polaris PVD system is developed to meet these challenges.

The first challenge in advanced packaging filed is the wafer handling. There are too many different wafer types involved in AP, including standard/SOG/EMC/Glass/Thin wafers. The warpage for these wafers is so different. NAURA Polaris PVD platform develops a wafer transportation system using passive ATR, Aligner, VTR and AWC(active wafer centering), to be compatible with all wafer types. The transfer module in Polaris AP PVD system is equipped with full range sensors. Thus the system is capable to detect all failure modes and eliminate risk of wafer breakage during wafer transfer The maximum tolerant of wafer warpage is about 10mm.Figure 1 and 2 show the ATR, Aligner and AWC.

Fig.1. Passive ATR(with reflective sensor or through beam sensors) and passive VTR.

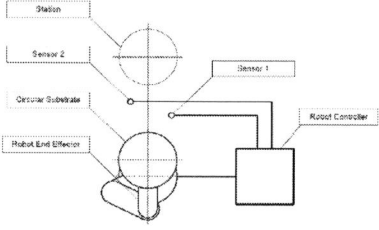

Fig.2. Schematic of AWC

The second challenge is about the optimization of Degas process module. the properties of materials involved in AP field such as PBO/EMC is sensitive to temperature, thus stringent process temperature control is required. Furthermore, to eliminate the moisture and other impurities in the layer to ensure a better contact resistance, a sufficient degassing process is a must. However, a sufficient degassing process often has a negative impact to the tool's productivity. NAURA Polaris PVD platform configures a multi-wafer Degas(MWD) module with

turbo or cryo pump to pump out gas rapidly. It is capable to process 50 wafers simultaneously. The temperature uniformity can be controlled within a variation of $\pm$ 3@150℃ during a long degas process.

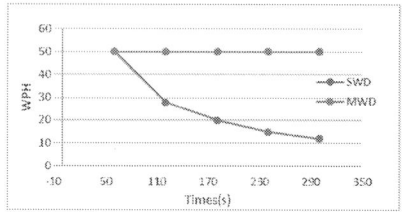

Fig.3. WPH of single wafer degas and multi-wafer degas

Another challenge for AP applications is the Etch module optimization. The Etch process is applied to remove the metal oxide to ensure a stable and low contact resistance. After etch process, the metal oxide needs to be eliminated completely without secondary contamination and plasma damage. Figure 4 shows contact resistance based on different temperature NAURA Polaris Etch module equipped with an auto RF match that guarantee precise ignition control and fast auto matching. This will ensure sufficient stable plasma to avoid PID damage (see Figure 5). A cooling pedestal with backside gas function and/or with water or chiller fluid cooling is configured to ensure well controlled process temperature. Figure 6 shows the Al_2O_3 etch amount of corresponding to SiO_2. A 300A SiO_2 etch recipe is able to etch away ~90A Al_2O_3 layer. Figure 7 shows TEM images. After processing with a 300A SiO_2 etch recipe, the native oxide (usually ~40A thick) Al_2O_3 on Al surface is completely removed. Figure 8 shows Rc performance at different temperature.

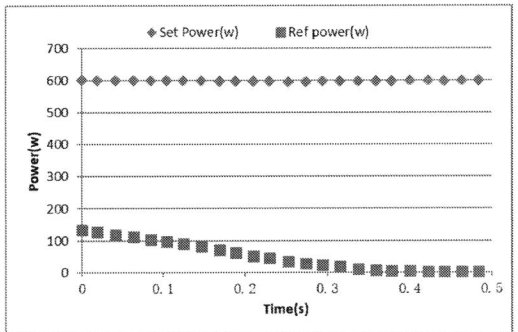

Fig.5. Precise ignition control and fast auto match:

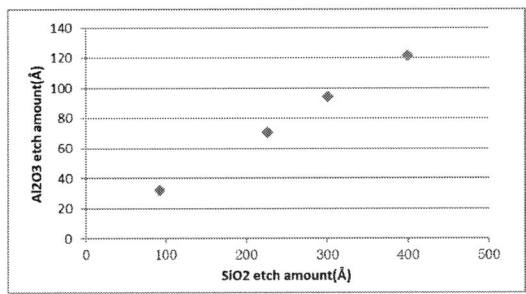

Fig.6. The results of etch removal correlation between Al$_2$O$_3$ etching and SiO$_2$ etching.

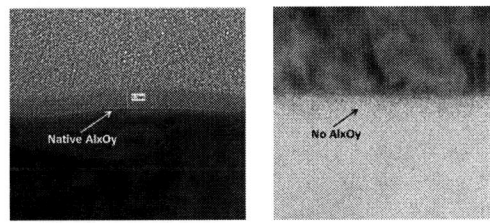

Fig.7. TEM images show the etch removal of native Al$_2$O$_3$.

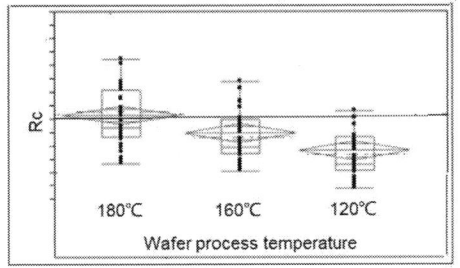

Fig.8. Rc performance at different process temperature

Stress control of metal film plays an important role in AP field. TiW is taken as an example. The native stress of TiW film (about 3000A) is between -1000MPa to -2000MPa. The stress level is too high and can seriously influence the ultimate reliability in practical applications. The unique design of magnetron is developed in NAURA Polaris PVD module to ensure a wide tunable process window on film stress, uniformity, resistance, film crystalline phase. Furthermore the system is designed to maximize the utilization of the target. The target utilization can be improved above 45%.

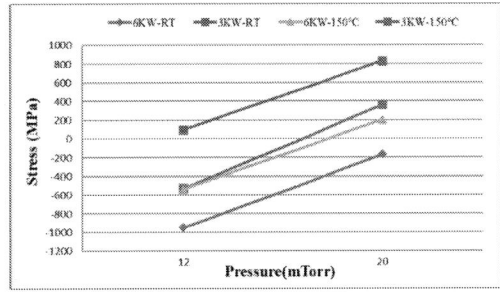

Fig.9. The results of TiW stress window

Particle control is always a crucial element for process equipment. NAURA Polaris PVD platform has unique structure design, cooling design, surface treatment of process kits, and an auto pasting function of etch module. As a result, it is able to improve the platform particle performance effectively. Figure 10 is the results of particle marathon.

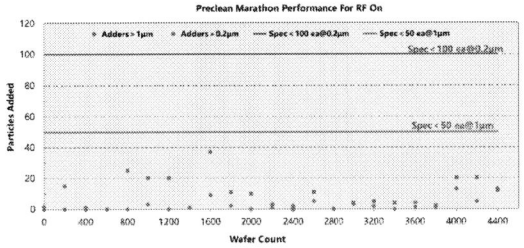

Fig.10. The results of particle marathon.

To improve the contact resistance and eliminate the risk of secondary contamination and cross contamination, the queue time between etch process and the following process is very crucial. Therefore the wafer transfer sequence and software interlock is very critical. For instance, to avoid the secondary contamination after etch process, the wafer needs to be taken out of etch chamber immediately and transferred to next PVD chamber. During the transfer, it needs to be guaranteed that in the transfer chamber there is no other wafer from Degas chamber or from Loadlock, to avoid the cross contamination. It is also important to ensure the vacuum base pressure and the pump speed of each module is optimum. It is always very challenging to fill TSV without any voids. TSV often is with a high aspect ratio. To improve the device reliability and to meet the requirements for the latest technology, the void-free filling of TSV becomes more and more important. It requires a high step coverage seed deposition to ensure void-free plating. The NAURA Polaris PVD Unique Magnetron design with high ionization rate and bias RF technology capable to achieve excellent Cu seed coverage in TSV with an aspect ratio up to 12:1

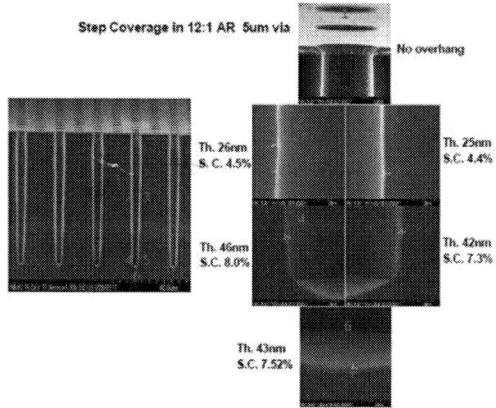

Fig.11. SEM pictures show the step coverage for TSV with AR=12:1.

SUMMARY

There are a lot of challenges in the advanced packaging field, for instance, wafer warpage, outgassing control, secondary contamination control and high aspect-ratio TSV. NAURA Microelectronics, the largest Chinese semiconductor equipment supplier, developed the Polaris PVD system that provide perfect solutions for all challenges in advanced packaging field.

ACKNOWLEDGEMENTS

Authors would like to thank all supports and helps from customers and colleagues in Beijing NAURA Microelectronics.

REFERENCES

[1] Yole Développement, edition of advanced packaging industry-2017.

The proposal and application of the hole Smoluchowski effect to explain the current-voltage characteristics of high-k MIM capacitors

W.S. Lau

Zhejiang University, Department of Information Science and Electronic Engineering
No. 38 Zheda Road, Hangzhou 310027, Peoples' Republic of China
E-mail: liuweicheng@zju.edu.cn

Biography

W.S. Lau has been a visiting Associate Professor in Zhejiang University with 30 years of experience working on high-k dielectric materials.

Abstract

This paper explains the theory of the simultaneous existence of the hole and electron Smoluchoski effects. The proposed theory is then used to explain the polarity asymmetry in the current-voltage (I-V) characteristics of high-k MIM capacitors.

Keywords—MIM capacitor; high-k dielectric; leakage current; asymmetry; Schottky emission; Poole-Frenkel mechanism; hole injection; electron Smoluchowski effect; hole Smoluchowski effect

Introduction

In earlier work [1] by the author, it was reported that the I-V characteristics of high-k MIM capacitors can be asymmetrical even though the MIM capacitor has an apparently symmetrical structure with the same top and bottom electrodes. The author attempted to explain this experimentally observed effect by proposing that the roughness of the top metal-insulator interface could be different from the bottom metal-insulator interface and the effective work function of the metal could be influenced by the roughness through the Smoluchowski effect [1]. Subsequently, the author noticed that the I-V characteristics of high-k MIM capacitors can be asymmetrical in the following strange way. The polarity asymmetry in the higher bias voltage region can be opposite to the lower bias voltage such that the polarity asymmetry can be reversed by applying a higher bias voltage. In this paper, the author will use the electron Smoluchowski effect to explain the polarity asymmetry in the lower bias voltage region and the hole Smoluchowsi effect to explain the polarity asymmetry in the higher bias voltage region.

Theory

For the electron Smoluchowski effect involving a high-k dielectric, the energy for an electron (work function) to jump from a metal to the conduction band of the high-k dielectric at a metal-high-k-dielectric interface is influenced by the morphology of the metal-vacuum interface. For example, the work function is smaller for a rougher metal-high-k-dielectric interface. For an MIM capacitor, the effective electron Schottky barrier height has a more direct impact on the leakage current than the work function of the metal. Let the effective electron Schottky barrier height be ϕ_{Bn}. In addition, there exists the effective hole Schottky barrier height. Let the effective hole Schottky barrier height be ϕ_{Bp}. For the hole Smoluchowski effect involving a high-k dielectric, the energy for an electron to jump from the valence band of the high-k dielectric to the metal at a metal-high-k-dielectric interface is influenced by the morphology of the metal-vacuum interface such that the energy involved is smaller for a rougher metal-high-k-dielectric interface. In other words, the energy for a hole to jump from the the metal to the valence band of the high-k-dielectric at a metal-high-k-dielectric interface is influenced by the morphology of the metal-high-k-dielectric interface such that the energy involved is smaller for a rougher metal-high-k-dielectric interface.

When the metal-high-k-dieletric interface is smooth, there is no Smoluchowski effect, the sum of the effective electron and hole barrier heights is just equal to the bandgap of the high-k dielectric

$$\phi_{Bn} + \phi_{Bp} = E_g \qquad (1a)$$

$$\text{or } \phi_{Bp} = E_g - \phi_{Bn} \qquad (1b)$$

Under the above situation, an increase in ϕ_{Bn} with lead to a decrease in ϕ_{Bp} or a decrease in ϕ_{Bn} with lead to an increase in ϕ_{Bp}. When the metal-high-k-dieletric interface is rough, both electron Smoluchowski effect and hole Smoluchowski effect exist such that both ϕ_{Bn} and ϕ_{Bp} can decrease. Then the sum can be smaller than the bandgap.

$$\phi_{Bn} + \phi_{Bp} < E_g \qquad (2)$$

The reason that hole injection tends to occur at higher voltage than electron injection is that, in a practical metal-high-k-dielectric interface, usually $\phi_{Bp} >> \phi_{Bn}$. For example, if $\phi_{Bn} = 0.8$ eV and $E_g = 3.2$ eV for TiO_2, then $\phi_{Bp} = 3.2 - 0.8 = 2.4$ eV. Usually, the range of ϕ_{Bn} is limited; for example ϕ_{Bn} is within the range of 0.5-0.8 eV. From the above equation, it can be predicted that hole injection is more likely observed in smaller bandgap high-k dielectric like TiO_2 or $SrTiO_3$, etc.

In 2012, the author [2] proposed his extended unified

Schottky-Poole-Frenkel theory for MIM capacitors with examples for various high-k dielectric materials. Subsequently in 2017, the author applied a combination of the Smoluchoski effect and Lau's extended unified Schottky-Poole-Frenkel theory to high-k MIM capacitors [1]. Modification of the previous theory [1],[2] is required in order to explain the hole Smoluchowski effect which involves hole injection. In Fig. 1(a), a high-k MIM capacitor involving only electrons for conduction is shown as a series combination of two diode and a nonlinear resistor. In Fig. 1(b), a high-k MIM capacitor involving only holes for conduction is shown as a series combination of two diode and a nonlinear resistor. In Fig. 1(c), a high-k MIM capacitor involving both electrons and holes for conduction is shown as a parallel combination of the two networks shown in Fig. 1(a) and Fig. 1(b). The diode represents the Schottky emission mechanism together with the Fowler-Nordheim tunneling mechanism. The nonlinear resistor represents the Poole-Frenkel mechanism together with simple Ohmic conduction.

Results and Discussion

An example of application is shown in Fig. 2; the I-V characteristics of a $RuO_2/TiO_2/RuO_2$ capacitor re-plotted from Frolich et al. [3] are obviously not the same for positive gate bias and negative gate bias. In Fig. 2, J_+ and J_- are the magnitude of the leakage current density for positive voltage applied to the top RuO_2 and bottom RuO_2, respectively. 16 nm TiO_2 was deposited by ALD at 350ºC while RuO_2 was deposited by MOCVD. The author suggested that the electron and hole Smoluchowski effects can be seen in the lower and upper bias regions, respectively. TiO_2 deposited by ALD tends to be amorphous below a critical thickness but polycrystalline above the critical thickness. It can be predicted that below the critical thickness ALD TiO_2 is amorphous. It is known that amorphous Al_2O_3 deposited by ALD has a surface smoothing effect [4]; similarly, amorphous TiO_2 deposited by ALD should also have a surface smoothing effect. MIM capacitor based on amorphous TiO_2 will have the top electrode interface with TiO_2 smoother than the bottom interface. It can be predicted that above the critical thickness ALD TiO_2 is polycrystalline. Polycrystalline TiO_2 will create a surface roughening effect. MIM capacitor based on polycrystalline TiO_2 will have the top electrode interface with TiO_2 rougher than the bottom interface. The existence of a critical thickness for crystallization during ALD can be explained by thermodynamics according to Nie et al. [5]. For the sample shown in Fig. 2, the thickness is above the critical thickness. (Note: According to Kim et al. [6], the critical thickness is about 6 nm for TiO_2 ALD at 250ºC. Now the ALD temperature is 350ºC. It can be roughly estimated that 20 nm is above the critical thickness.) Under such a situation, the top $RuO_2/TiO_2/RuO_2$ interface can be predicted to be rougher than the bottom interface. Experimental evidence that hole injection tends to occur at higher voltage than electron injection can be found from Chong et al. [7]. As shown in Fig. 3, the leakage current of $Al/TiO_2/n^+$-Si was significantly higher than that of $Al/TiO_2/p^+$-Si when the substrate bias was negative, according to Chong et al. [7]. In addition, as shown in Fig. 3, the leakage current of $Al/TiO_2/n^+$-Si was equal to that of $Al/TiO_2/p^+$-Si when the substrate bias was positive and < 2 V. The author interprets this leakage current as electron injection from the Al

contact. The leakage current of $Al/TiO_2/p^+$-Si became bigger than that of $Al/TiO_2/n^+$-Si when the substrate bias was positive and > 2 V. The author interprets this leakage current as hole injection from the p^+-Si contact in addition to electron injection from the Al contact. Thus hole current tends to dominate over electron current at higher bias voltage. The reason why hole injection tends to occur at higher voltage than electron injection has been discussed above.

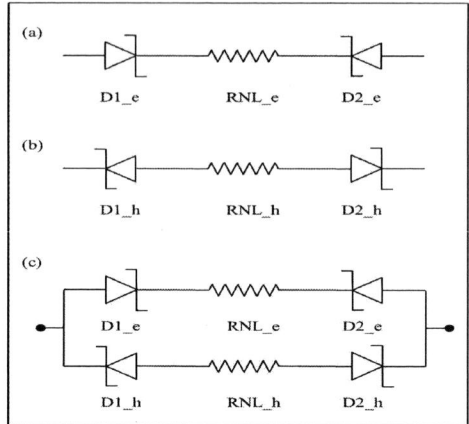

Fig. 1 (a) If electrical conduction is entirely due to electrons, an MIM capacitor structure involving a high-k dielectric can be thought as two back-to-back Schottky diodes D1_e and D2_e with a non-linear resistor RNL_e in between. (b) (a) If electrical conduction is entirely due to holes, an MIM capacitor structure involving a high-k dielectric can be thought as two back-to-back Schottky diodes D1_h and D2_h with a non-linear resistor RNL_h in between. (c) If electrical conduction is due to both electrons and holes, an MIM capacitor structure involving a high-k dielectric can be thought as the parallel combination of the two networks shown in (a) and (b).

Similarly, for $RuO_2/TiO_2/RuO_2$ capacitors, in the lower voltage range, electron injection dominates over hole injection; electron injection will be stronger from the top RuO_2 than from the bottom RuO_2 because of the electron Smoluchowski effect. In the higher voltage range, hole injection dominates over electron injection; hole injection will be stronger from the top RuO_2 than from the bottom RuO_2 because of the hole Smoluchowski effect. The reversal in polarity asymmetry at the transition from the lower to higher voltage range is due to the fact that holes are positively charged while electrons are negatively charged. More evidence for the above theory can be obtained by switching the top electrode material from RuO_2 to Au. As shown in Fig. 4, the leakage current became smaller for negative voltage applied to the top electrode whereas the leakage current became larger for positive voltage applied to the top electrode when the top electrode was changed from RuO_2 to Au. The data came from Hudec et al. (same research group as Frolich et al. [3]) [8]. This can be explained by an increase in work function and ϕ_{Bn} and a corresponding decrease in ϕ_{Bp} due to the change from RuO_2 to Au. Hudec et al. [8] pointed out that the use of Ni as top electrode caused strong leakage current. This is unreasonable because Ni is a high work function metal such that the adoption of Ni as electrode material usually leads to lower leakage current. However, if there is hole injection, Ni can cause a corresponding decrease in ϕ_{Bp}, resulting in a larger hole current.

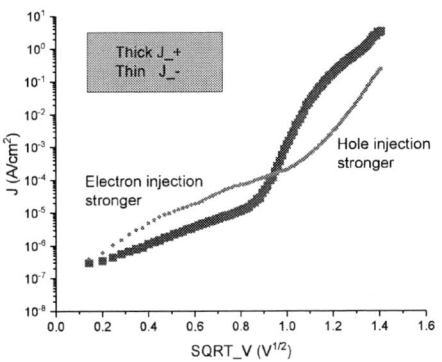

Fig. 2 I-V characteristics of a $RuO_2/TiO_2/RuO_2$ capacitor re-plotted from Frolich et al. [3].

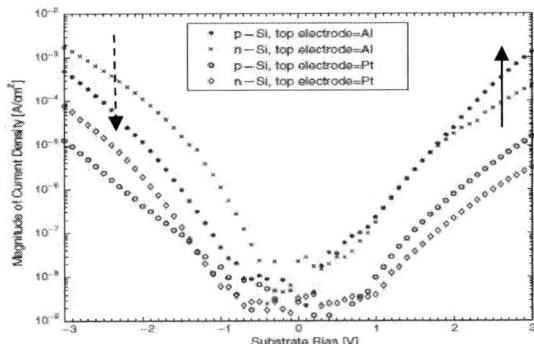

Fig. 3 I-V characteristics of $Al/TiO_2/n^+$-Si, $Al/TiO_2/p^+$-Si, $Pt/TiO_2/n^+$-Si and $Pt/TiO_2/p^+$-Si at room temperature according to Chong et al. [7]. Hole injection dominated over electron injection at a positive substrate voltage > 2 V as indicated by a solid arrow showing the change from n^+-Si to p^+-Si substrate. (Electron injection dominated over hole injection for negative substrate voltage as indicated by a dashed arrow showing the change from n^+-Si to p^+-Si substrate. The two arrows are in opposite directions because of eqn. (1); a larger ϕ_{Bn} for p^+-Si implies a smaller ϕ_{Bp} for p^+-Si.)

Conclusion

Experimental evidence for the existence of both electron and hole conduction in TiO_2 can be obtained by making use of eqn. (1), as shown in Figs. 3 & 4. In addition, the electron and hole Smoluchowski effects can be used to interpret the I-V characteristics of high-k MIM capacitors, as shown in Fig. 2. In this paper, the author has applied the hole Smoluchowski effect to explain the I-V characteristics of TiO_2 MIM capacitors. Besides TiO_2, the same theory can be applied to Ta_2O_5, $SrTiO_3$ and ZrO_2 MIM capacitors, etc.

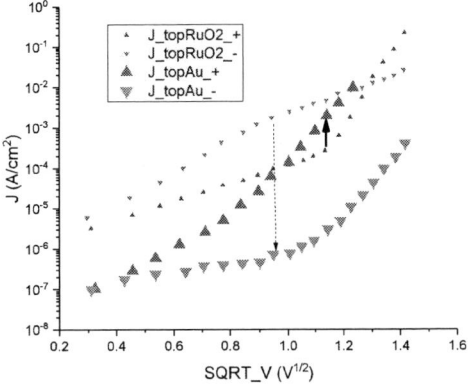

Fig. 4 The leakage current became smaller for negative voltage applied to the top electrode whereas the leakage current became larger for positive voltage applied to the top electrode when the top electrode was changed from RuO_2 to Au. The direction of the arrow shows the move from RuO_2 to Au. (The two arrows are in opposite directions because of eqn. (1); a larger ϕ_{Bn} for Au implies a smaller ϕ_{Bp} for Au.)

References

[1] W.S. Lau, "The application of the Smoluchowski effect to explain the current-voltage characteristics of high-k MIM capacitors," CSTIC 2017 (China Semiconductor Technology International Conference, Shanghai, 2017, IEEE), pp. 1-3, 2017.

[2] W.S. Lau, "An extended unified Schottky-Poole-Frenkel theory to explain the current-voltage characteristics of thin film metal-insulator-metal capacitors with examples for various high-k dielectric materials," ECS J. Solid State Sci. Technol., vol. 1, pp. N139-N148, 2012.

[3] K. Frolich, M. Tapajna, A. Rosova, E. Dobrocka, K. Husekova, J. Aarik and A. Aidla, "Growth of high-dielectric-constant TiO_2 films in capacitors with RuO_2 electrodes," Electrochemical and Solid-State Letters, vol. 11, pp. G19-G21, 2008.

[4] W.S. Lau, L. Du. D.Q. Yu, X. Wang, H. Wong and Y. Xu, "The application of a selective etch to conclusively show the surface smoothing effect of an amorphous thin film deposited by atomic layer deposition," ECS J. Solid State Sci. Technol., vol. 6, pp. N111-N116, 2017.

[5] X. Nie, F. Ma, D. Ma and K. Xu, "Thermodynamics and kinetic behaviors of thickness-dependent crystallization in high-k thin films deposited by atomic layer deposition," J. Vac. Sci. Technol. A, vol. 33, article number 01A140, 2015.

[6] W.D. Kim, G.W. Hwang, O.S. Hwang, S.K. Kim, M. Cho, D.S. Jeong, S.W. Lee, M.H. Seo, C.S. Hwang, Y.-S. Min and Y.J. Cho, "Growth characteristics of atomic layer deposited TiO_2 thin films on Ru and Si electrodes for memory capacitor applications," J. Electrochem. Soc., vol. 152, pp. C552-C559, 2005.

[7] L.H. Chong, K. Mallik, C.H. de Groot and R. Kerstling, "The structural and electrical properties of thermally grown TiO2 thin films," J. Phys.: Condens. Matter, vol. 18, pp. 645-657, 2006.

[8] B. Hudec, K. Husekova, E. Dobrocka, T. Lalinsky, J. Aarik, A. Aidla and K. Frolich, "High permittivity metal-insulator-metal capacitors with TiO_2 rutile dielectric and RuO_2 bottom electrode," IOP Conf. Series: Materials Science and Engineering, vol. 8, article number 012024, 2010.

Engineering implication of the correlation between the leakage current in high-k dielectric materials and the electronic defect states detected by zero-bias thermally stimulated current spectroscopy

W.S. Lau

Zhejiang University, Department of Information Science and Electronic Engineering
No. 38 Zheda Road, Hangzhou 310027, Peoples' Republic of China
E-mail: liuweicheng@zju.edu.cn

Biography

W.S. Lau has been a visiting Associate Professor in Zhejiang University with 30 years of experience working on high-k dielectric materials.

Abstract

It is not obvious to correlate the leakage current with the electronic defect states in high-k dielectric materials. In this paper, the author will show the correlation between them. The engineering implication of this success will also be discussed.

Keywords—MIM capacitor; high-k dielectric; leakage current; electronic defect states; thermally stimulated current; annealing in an oxidizing ambient; ultraviolet (UV) annealing; recombination enhanced diffusion

Introduction

The significance of high-k dielectric materials has been demonstrated for the microelectronics industry. The leakage current mechanism in this class of materials is not well understood. However, it is expected that the electronic defect states are probably responsible. It has been difficult to correlate the leakage current in high-k dielectric and the electronic defect states in high-k dielectric materials. The author has overcome the difficulties involved and has made a successful correlation. The engineering implication of this success will be discussed in this paper.

Experimental

For this work, tantalum oxide (Ta_2O_5) was chosen as the high-k dielectric material. Ta_2O_5 was deposited onto (100) n^+-Si wafers by low-pressure metal-organic chemical vapor deposition (LP-MOCVD) [1]. The physical thickness of the film was about 80 nm. As deposited Ta_2O_5 films tend to be very leaky and a post-deposition anneal is necessary to reduce the leakage current. A set of Ta_2O_5 samples were annealed in O_2 at 800°C for 30 s in a RTA (rapid thermal annealing) tool. Another set of Ta_2O_5 samples were UV annealed in O_2 at 400°C for 30 min and then further annealed in O_2 at 800°C for 30 s in a RTA tool. For the top metal electrode, electron-beam evaporation method was used to form 1-mm-diameter Al dots using a metal shadow mask. Zero-bias thermally stimulated current (ZBTSC) measurements were performed at a ramp rate of 0.5 K/s as discussed before in [2]. For conventional thermally stimulated current (TSC) measurement, a bias voltage is applied across the sample; this bias voltage causes a parasitic current, which is particularly serious for an ultrathin sample. The energy level of the electronic defect state was estimated using

$$E_T = 23kT_m \qquad (1)$$

where E_T is the energy level of the defect, T_m is the peak temperature and k is the Boltzmann constant. The initial rise method has also been used to determine the energy level; the validity of equation (1) is reasonably confirmed for a temperature ramp rate of 0.5 K/s. The I-V characteristics of $Al/Ta_2O_5/n^+$-Si capacitor structures were also measured at room temperature.

Results and Discussion

As shown in Fig. 1, the sample with 400°C UV O_2 anneal has an important electronic defect state (Defect C) suppressed compared with the sample without 400°C UV O_2 anneal according to ZBTSC. This defect state was probably related to carbon contamination. As shown in Fig. 2, the sample with 400°C UV O_2 anneal has less leakage current compared with the sample without 400°C UV O_2 anneal according to I-V curve measurement.

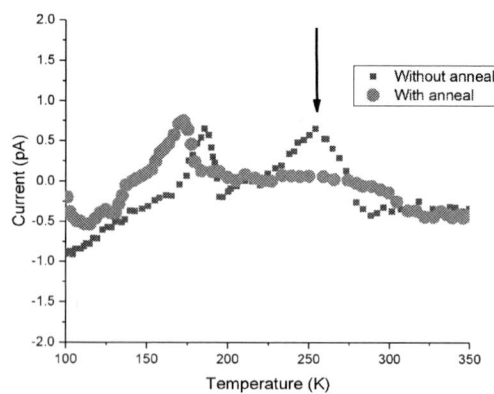

Fig. 1 ZBTSC spectra of $Al/Ta_2O_5/n^+$-Si samples with and without 30 min. 400°C UV O_2 anneal. 800°C RTA in O_2 for 30 s was done for both samples. Ta_2O_5 thickness = 80 nm. Defect C is highlighted by an arrow.

The electronic defect state highlighted by an arrow is found in samples with thick Ta_2O_5. When the thickness is reduced to < 10 nm, this electronic defect state can be easily suppressed by high temperature RTA in an oxidizing ambient. As shown in Fig. 3, Defect C was present in a sample with Ta_2O_5 thickness = 986 nm after 900°C RTA in O_2 for 30 s according to ZBTSC. As shown in Fig. 4, Defect C disappeared in a sample with Ta_2O_5 thickness = 8.2 nm after 800°C RTA in N_2O for 30 s according to ZBTSC. Thus UV O_2 anneal is more important for thick samples.

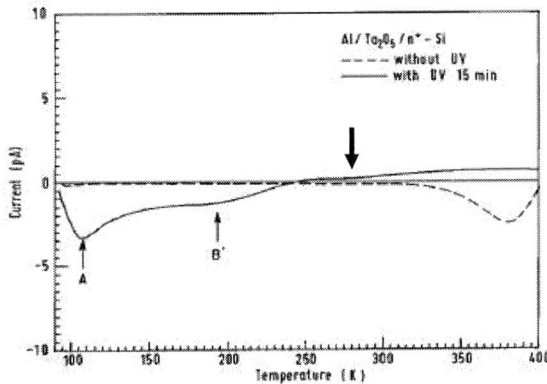

Fig. 4 ZBTSC spectra of $Al/Ta_2O_5/n+$-Si samples with and without UV excitation showing the absence of Defect C in ultrathin (8.2 nm) Ta_2O_5 with the help of an arrow. 800°C RTA in N_2O for 30 s was done.

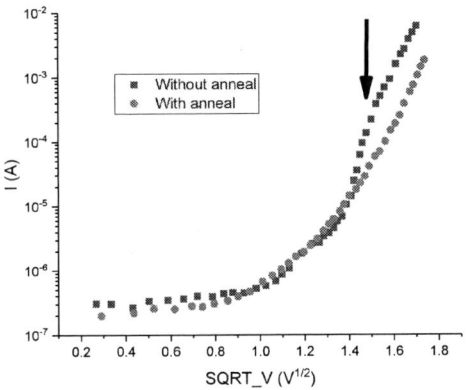

Fig. 2 I-V curves of $Al/Ta_2O_5/n+$-Si with and without 400°C UV O_2 anneal for 30 min. 800°C RTA in O_2 for 30 s was done for both samples. Ta_2O_5 thickness = 80 nm. Reduction of leakage current by UV O_2 anneal is highlighted by an arrow. Positive voltage was applied to the Al gate.

The UV light makes "recombination enhanced diffusion" possible [3]. (Electrons and holes can be generated in a high-k dielectric by UV light. The recombination of electrons and holes in a high-k dielectric is a highly energetic process which can enhance a lot of microscopic processes, for example, diffusion.) Impurities like carbon can be removed more efficiently by UV O_2 anneal than by high temperature annealing in an oxidizing ambient; without UV O2 anneal, only the carbon impurities near the surface can be removed by high temperature RTA in an oxidizing ambient for a thick sample. This can be seen in the SIMS (Secondary Ion Mass Spectrometry) depth profile shown in Fig. 5.

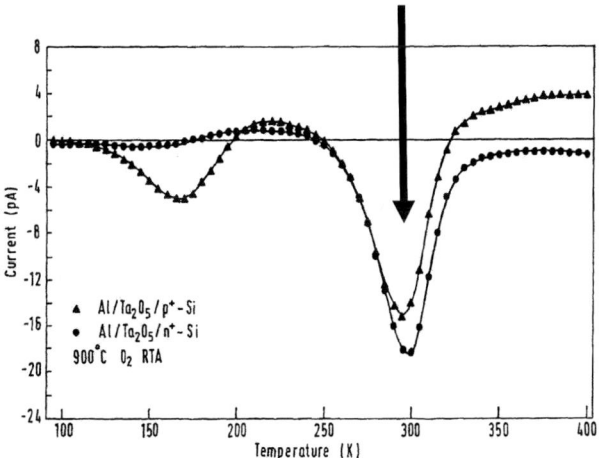

Fig. 3 ZBTSC spectra of $Al/Ta_2O_5/n+$-Si and $Al/Ta_2O_5/p+$-Si samples with UV excitation showing the presence of Defect C in 986 nm thick Ta_2O_5 with the help of an arrow. 900°C RTA in O_2 for 30 s was done.

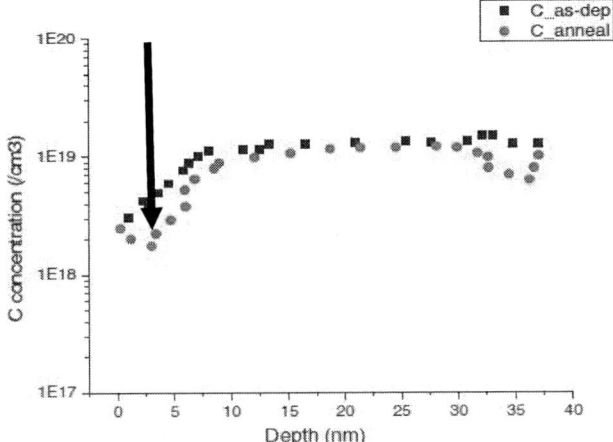

Fig. 5 SIMS depth profile of Ta_2O_5 before and after annealing in an oxidizing ambient at high temperature. The thickness of Ta_2O_5 was 40 nm. The arrow in this figure shows that annealing in an oxidizing ambient at high temperature can only remove carbon near the surface.

The engineering implication of this conclusion can be applied to other high-k dielectric materials. When the high-k dielectric material is < 10 nm thick, a short high temperature RTA in an oxidizing ambient without UV O_2 anneal can be enough to remove impurities (e.g. C, H, N) which can be oxidized into a gaseous compound.

Previously, Miki et al. measured I-V curves of Ta_2O_5 capacitor structures for various temperatures (I-V-T) and for various Ta_2O_5 thicknesses and pointed out that for Ta_2O_5 treated by the same RTA process in an oxidizing ambient, the thinner sample seemed to have less electronic defect states [4]. ZBTSC performed by the author confirmed the above observation as discussed above. Thus the thickness of the high-k dielectric is important. When the high-k dielectric material is thick, a short high temperature annealing in an oxidizing ambient may not be enough. It can be imagined that for thicker high-k dielectric, several cycles of deposition of an ultrathin high-k film plus a short high temperature annealing in an oxidizing ambient can be done. In fact, this approach has been demonstrated for HfO_2 as a DADA (D=deposit and A=anneal) process [5].

The readers should note that Fleming et al. pointed out that the electronic defect state density increases for thinner samples [6]. The conclusion by Fleming et al. and the conclusion by Miki et al. seem to be contradictory. Actually, Fleming et al. pointed out that there are much more interfacial electronic defect states compared to bulk electronic defect states. If the defect state characterization cannot distinguish interfacial electronic defect states from bulk electronic defect states. The measured electronic defect state density is approximately equal to the total amount of interfacial electronic defect states divided by the thickness of the high-k dielectric. If the total amount of interfacial electronic defect states does not change with the thickness of the high-k dielectric, the measured electronic defect state density will appear to increase when the high-k dielectric becomes thinner. The conclusion by Miki et al. was that for a high-k dielectric treated by the same RTA process in an oxidizing ambient, the thinner sample seemed to have less "bulk" electronic defect states [4]. Miki et al did not measure the defect states directly; the significance of this work is that the defect states can be measured directly by ZBTSC for both thick and ultrathin samples in a spectroscopic matter. The conclusion by Fleming et al. and the conclusion by Miki et al. and the author are not contradictory; in fact, they are complementary. However, the conclusion by Miki et al. and the author can in fact serve as the theoretical basis of the DADA process. Fleming et al. actually pointed out that there are much more interfacial electronic defect states compared to bulk electronic defect states in a very thin high-k dielectric film. A very large quantity of interfacial electronic defect states can decrease the effective Schottky barrier height, resulting in leakage current for some samples [7]. Thus, in addition to the reduction of bulk electronic states, the reduction of interfacial defect states can sometimes be important for the reduction of the leakage current. The usual method to decrease interfacial defect states is just the optimization of the post-deposition annealing (PDA) and post-metallization annealing (PMA) processes. PDA and PMA can affect both the interfacial and bulk defect states for a very thin high-k dielectric film

Conclusion

It has been difficult to correlate the leakage current in high-k

dielectric and the electronic defect states in high-k dielectric materials particularly for very thin samples. The author has overcome the difficulties involved and has made a successful correlation. The engineering implication of this success has also been discussed. The thickness of the high-k dielectric is important: thinner high-k dielectric film tends to have less defect states in the bulk after a short anneal in an oxidizing ambient at high temperature. Thus it is necessary to detect defect states in ultrathin high-k dielectric films because conclusions drawn from thick samples may not be valid for ultrathin samples. For an ultrathin sample, ZBTSC is better than conventional TSC; however, ZBTSC may still be insufficient. Zero-temperature-gradient zero-bias thermally stimulated current (ZTGZBTSC) is superior to ZBTSC [8].

References

[1] K.A. McKinley and N.P. Sandier, "Tantalum pentoxide for advanced DRAM applications," Thin Solid Films, vol. 290, pp. 440-446, 1996.

[2] W.S. Lau, K.K. Khaw, T. Han and N.P. Sandler, "Mechanism of leakage current reduction of tantalum oxide capacitors by postmetallization annealing," Appl. Phys. Lett., vol. 89, article number 262901, 2006.

[3] W.S. Lau, M.T.C. Perera and P.K. Chu, "Recombination enhanced diffusion of impurities at low temperature in high-k dielectric with tantalum oxide as an example," Appl. Phys. Lett., vol. 96, article number 083501, 2010.

[4] H. Miki, M. Kunitomo, R. Furukawa, T.Tamaru, H. Goto, S. Iijima, Y. Ohji, H. Yamamoto, J. Kuroda, T. Kisu and I. Asano, "Leakage-current mechanism of a tantalum-pentoxide capacitor on rugged Si with a CVD-TiN plate electrode for high-density DRAMs," Symp. VLSI Tech., pp. 99-100, 1999.

[5] H. Jagannathan, R.D. Clark, S. Consiglio, P. Jamison, B.P. Linder, M. Hopstaken, G.J. Leusink, V.K. Paruchuri and V. Narayanan, "Methodolgy of ALD HfO_2 high-k gate dielectric optimization by cyclic depositions and anneals," ECS Trans., vol. 33(3), pp. 157-164, 2010.

[6] R.M. Fleming, D.V. Lang, C.D.W. Jones, M.L. Steigerwald, D.W. Murphy, G.B. Alers, Y.-H. Wong, R.B. van Dover, J.R. Kwo and A.M. Sergent, "Defect dominated charge transport in amorphous Ta_2O_5 thin films," J. Appl. Phys., vol. 88, no. 2, pp. 850-862, 2000.

[7] W.S. Lau, "ULSI front-end engineering", World Scientific, Singapore, 2017, pp. 209-226.

[8] W.S. Lau, K.F. Wong, T. Han and N.P. Sandler, "Application of zero-temperature-gradient zero-bias thermally stimulated current spectroscopy to ultrathin high-dielectric-constant insulator film characterization," Appl. Phys. Lett., vol. 88, article number 172906, 2006.

Optimization of backside metal deposition in power IC process for suppression of wafer warpage and film peeling issue

Wei Zhang, Weijie Liu, Bingxin Ma,

AUTHORS

Wei Zhang, **Bingxin Ma**, **Weijie Liu** are Process Engineers of FAB7 Center/PVD department at Semiconductor Manufacturing International Corp(SMIC), Tianjin, China.
E-mail:Rick_Zhang@smics.com,Matthews_Ma@smics.com,Jerry_LWJ@smics.com

ABSTRACT

In power IC manufacture process, wafer backside metal deposition as contact layer is a critical step to improve whole chip contact resistance and thermal conduction. The conventional metal films applied in backside deposition are titanium (Ti), Nickel (Ni) and Silver (Ag) with physical vapor deposition (PVD) technique. In order to reduce contact resistance, higher metal film thickness is requested. This result in high wafer stress and warpage challenge. The adhesion between wafer backside silicon and Ti/Ni/Ag film is also critical index to ensure electrics stability. In general, — wafer substrate roughness, metal process temperature and out-gassing control are key points in process optimization. In this paper, we will introduce basic process of backside metal deposition and discuss the experiments / results on wafer warpage and film peeling suppression.

Keywords: Backside metal, warpage, film peeling.

INTRODUCTION

Power semiconductor is applied in almost all fields of national economy and national defense industry. Power device includes power IC and power discrete device such as MOSFET, high power transistor and IGBT. [1].

Some common power devices are the power diode, thyristor, power MOSFET and IGBT. A power diode or MOSFET operates on similar principles to its low-power counterpart, but is able to carry_higher current and typically to support a larger reverse-bias voltage at the off-state.

Structural changes can help to accommodate higher current density, higher power dissipation and/or higher reverse breakdown voltage. The vast majority of the discrete power devices a use a vertical structure, whereas small-signal devices employ a lateral structure. For the vertical structure, the current rating of the device is proportional to

978-1-5386-5309-8/18 $31.00 © 2018 IEEE

its area, and the voltage blocking capability is relative to the height of the die. With this structure, one of the connections of the device is located on the bottom of the die [2].

The backside metallization process is a necessary step to control RDS (on). As the wafer gets thinner and thinner, it become more and more difficult to make process efficient and stable. The following diagram shows schematically the backside metallization [3].

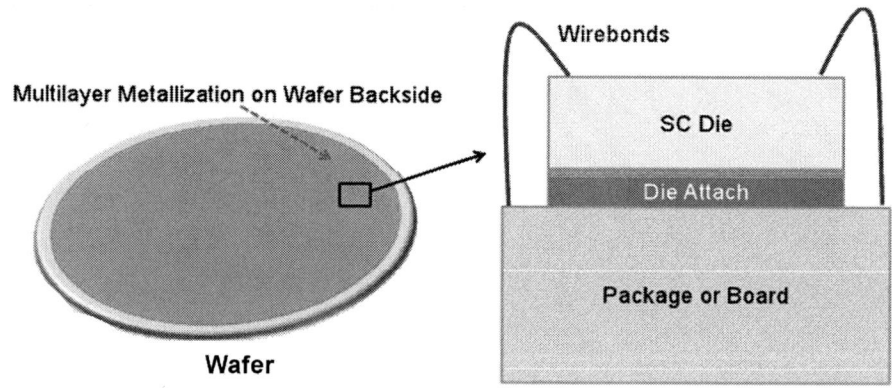

Fig1 Backside Metallization scheme

For vertical trench power device, backside metallization structure works under higher current than normal device to provide power supply for electronic terminal. So the value of RDS (on) takes the most important place in power device performance evaluation. Be different from most MOS device, power device internal current direction is perpendicular to wafer surface as shown in Fig2 [4].

$$RDS_{(on)}=R_{source}+R_{ch}+R_J+R_D+R_{su} \qquad [4]$$

In backside metallization process, we need deposit one or several metals on wafer backside in order. The Ti/Ni/Ag structure is usually used, as shown in Fig3.

978-1-5386-5309-8/18 $31.00 © 2018 IEEE

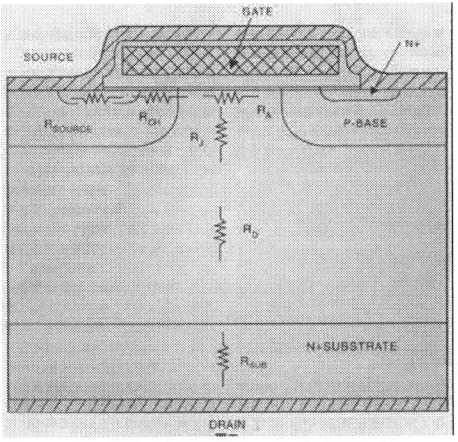

Fig2 Power MOSFET structure

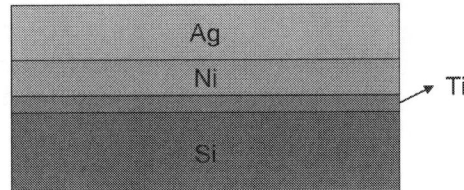

Fig3 Ti/Ni/Ag structure on wafer backside

One method of above deposition is vacuum metal evaporation. This process is usually carried out in a bell jar, in which a filament is used to generate Joule heat.

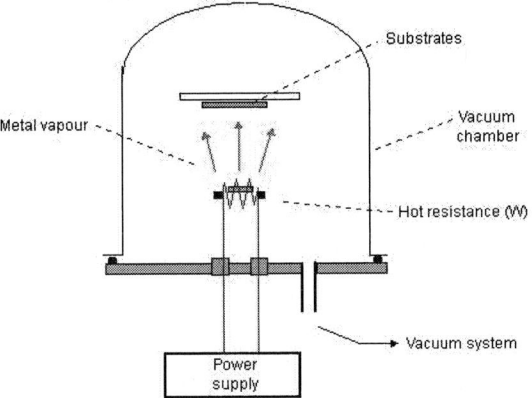

The other method is physical vapor deposition (PVD).

In mass production, backside metal PVD process suffers two issues:

1) <u>Wafer Warpage High</u>. As the wafer thickness become thinner and thinner, wafer warpage gets worse and worse, which lead to higher wafer broken rate and incorrect CP & final test. It can also affect device performance, reliability and line width or critical dimension (CD) control in various semiconductor processes [5]. Fig6 shows the warpage issue

978-1-5386-5309-8/18 $31.00 © 2018 IEEE   291

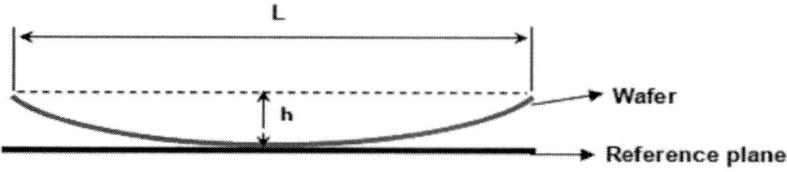

Fig6 wafer warpage

2) <u>Metal Peeling Issue</u>. When the adhesion performance between Si and Ti is not good enough, metal peeling takes place. It could   be deteriorated by cutting wafer after package and lead to electronic function failure. The Fig7 is wafer with metal peeling [4].

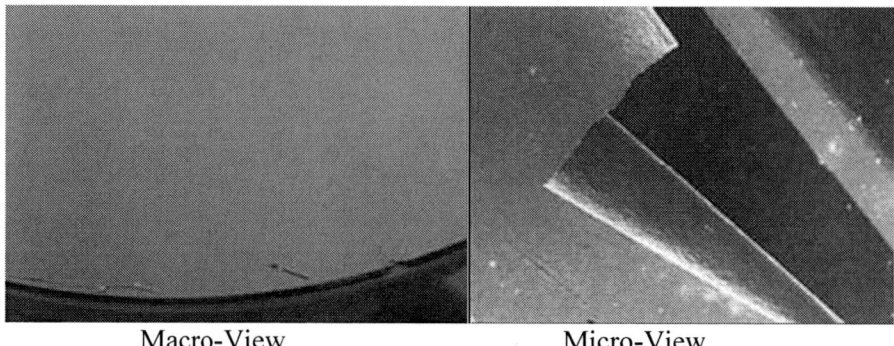

Macro-View   Micro-View

Fig7 Wafer Backside Metal Peeling

RESULT AND DISCUSSION

1. <u>Wafer Warpage Reduction</u>

   ➤ Factors That Impact Wafer Warpage:

      ✓ Wafer Thickness Reduction After CMP
         After CMP backside polishing, 8" wafer total thickness decreases to a large extent. The thickness reduction results in high wafer warpage. The Fig8 shows wafer warpage comparison among different thickness.

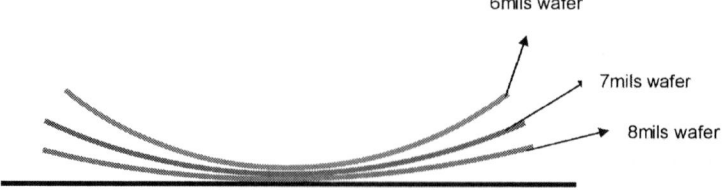

Fig8 Wafer Warpage with Different thickness scheme

978-1-5386-5309-8/18 $31.00 © 2018 IEEE   292

✓ Metal Deposition

During metal backside deposition process, wafer would get high temperature because of the plasma energy. This high temperature can make materials in every lay band to different direction because of the different coefficient of thermal expansion (CTE). The Tab1 includes some metals' CTE.

Metal	CTE $(25^{o}C)/10^{-6}K^{-1}$
Al	23.2
Cr	8.5
Cu	16.8
Au	14.2
Ni	12.7
Ag	19.3
Ti	8.5
Si	2.44

Tab1 Metal CTE

➢ Solutions to Reduce Warpage:

✓ Ag Recipe Optimization

For the power device process, the wafer thickness and metal thickness are fixed. We can not do any change on them. But we can still improve warpage during metal deposition.

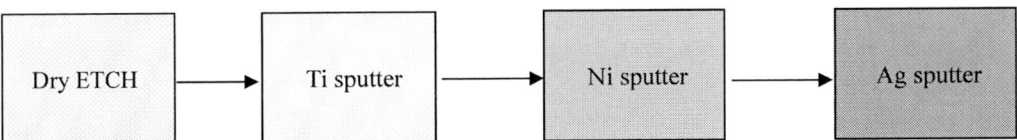

Fig9 Metal Deposition Process Flow

The above Fig9 is metal deposition process flow. In the proportion of Ti/Ni/Ag, Ag is the most. So Ag deposition should give out very high thermal energy. The scheme of temperature changing in metal deposition process is showed at below Fig10.

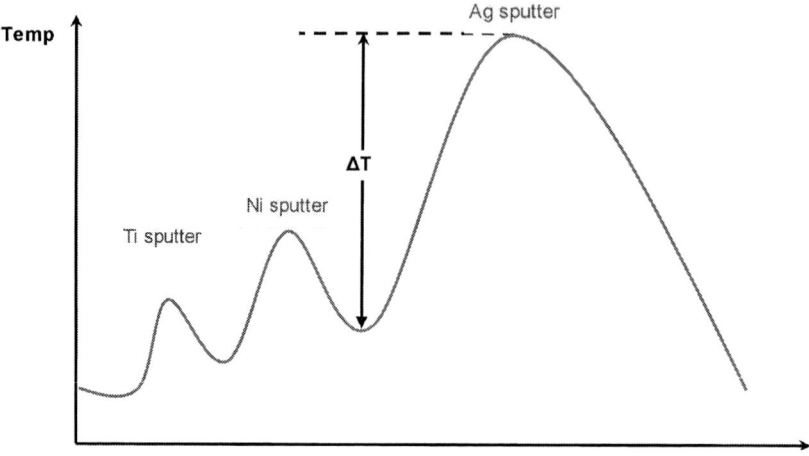

Fig10 Wafer temperature during metal sputtering

According to coefficient of thermal expansion formula:

$$\Delta L = L * \alpha * \Delta T \qquad [4]$$

ΔL is expansion length, α is expansion coefficient depending on material, L is material length and ΔT is temperature changing. In the deposition process of three metals, Ag requires longer time with high DC power in recipe setting. Through optimizing recipe parameter，we reduced sputter DC power from high to low and split one main sputtering step to many sub-steps. Then we added cooling function step between two sub-steps with cooling gas. It can interrupt thermal accumulation and avoid high temperature. The model is shown in Fig11.

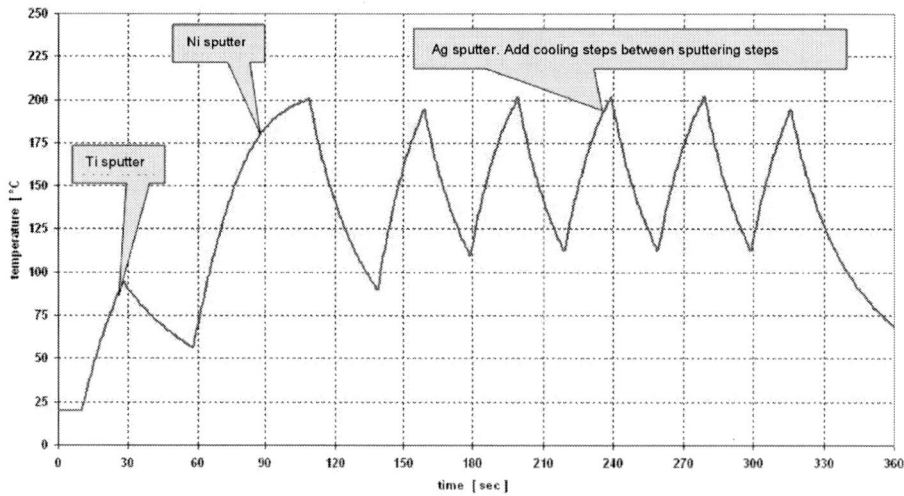

Fig11 Temp chart after Ag recipe optimization

We used 8" wafer to do experiment and compared original and new Ag deposition recipe. The result is showed in below Tab2. Slot1 is original 1 step recipe, slot2 is 2 steps, slot3 is 3 steps, slot4 is 4 steps and slot5 is 5 steps. We can see the warpage value trends down obviously (from 1663 to 1098).

Slot	Ag	Warpage	THK
1	Ag(1 step)	1663	205
2	Ag(2 steps)	1446	205
3	Ag(3 steps)	1285	205
4	Ag(4 steps)	1137	205
5	Ag(5 steps)	1098	205

Tab2 Wafer warpage data with different Ag recipe body

✓ Ni Recipe Optimization

As it is showed in Tab2, different metals have different CTE. Ni is a good material to be used as buffer layer between Ti and Ag because of its medium CTE value. But in the real manufacture process, the effect of Ni is not enough. To solve this issue, we introduce N_2 gas during Ni sputtering and change Ni film structure. The model is shown in Fig12; the N atom is embedded in Ni atoms which changes the characteristic of Ni and make it more flexible. So the effect as buffer layer is better than original Ni recipe.

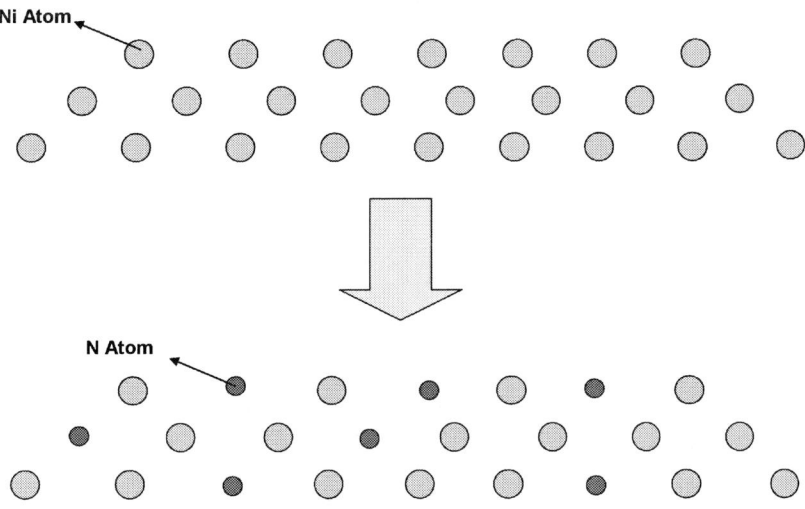

Fig12 New Ni structure with N atom

We compared baseline recipe and optimized Ni recipe in 8" wafer with different thickness and got expected result in Tab3. The warpage was reduced with huge extension(almost 50%).

Slot	wafer thickness	warpage(baseline)	warpage(new recipe)
1	208	669	415
2	185	1085	526
3	185	988	536
4	185	891	471
5	160	1380	650

Tab3 Wafer warpage data with different Ni recipe body

2. Adhesion Improvement

If adhesion performance between metal and Si layer is not good enough, metal peeling happens. The following Fig13 is adhesion test and peeling.

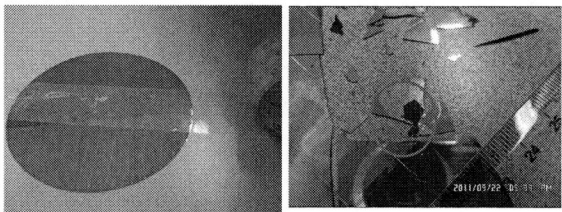

Fig13 Adhesion test and Metal peeling

➢ The Reasons of Metal Peeling:

✓ SiO2 Layer Forming

After CMP and WET ETCH, Si in wafer backside is exposed into the atmosphere. The interface layer SiO_2 will be formed between Ti and Si(Fig14). This layer's thickness is no more than 20A. But the adhesion performance of SiO_2-Ti is worse than Si-Ti which leads to metal peeling.

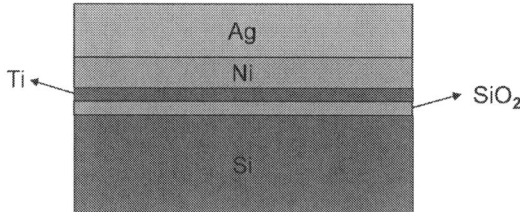

Fig14 SiO$_2$ between Ti and Si layer

✓ Out Gassing

Water residue and impurity is always included in wafer backside surface. During the sputtering, those water and impurity will be out gassing under thermal effect(Fig15). Then lead to adhesion between Ti and Si gets worse.

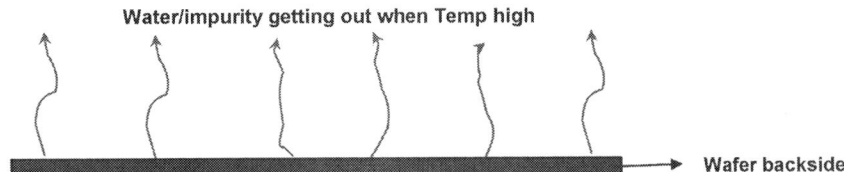

Fig15 wafer out-gassing

✓ Not Rough Enough on Backside Surface

Before metal sputtering process, dry etch is required as an important process to remove native SiO_2 and get better contact resistance. But it makes Si surface more smooth which is harmful to adhesion.

➢ Peeling issue solutions:

✓ Using N2 box and control waiting time.

To avoid Si oxidation as less as possible, we should control the time between WET ETCH and metal deposition within 1~3 hours. And we put the wafers that wait for metal deposition process in the N_2 box to protect Si from oxidation.

✓ Improving backside roughness with soft dry etch recipe.

In the dry etch process, the typical RF power is high. The energy of Ar^+ is high enough to damage original wafer backside surface structure into smooth one(Fig16). To avoid this issue, we reduced etch recipe RF power from high to low and increase etch time. The power is enough to remove SiO_2 but not enough to continue remove Si structure.

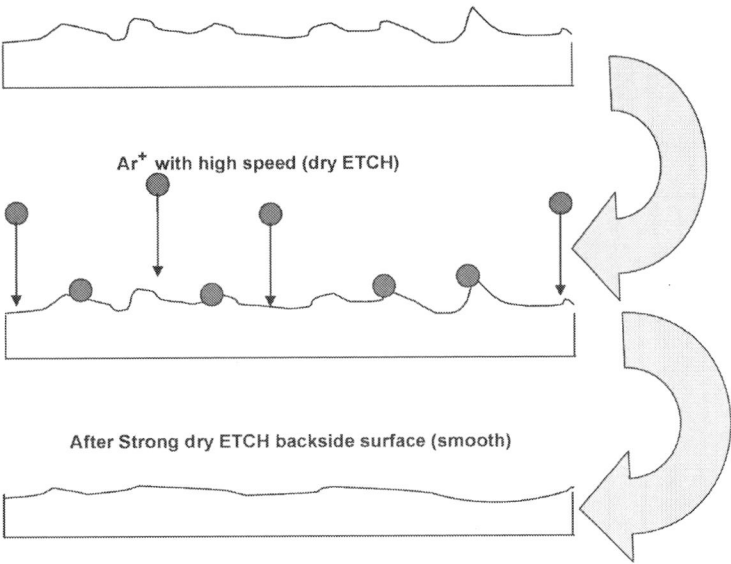

Fig16 Strong Dry etching Scheme

✓ Add Degassing Process at the Beginning.

Degassing process is heating the wafer with lamps and gets rid off impurity and water. Wafer backside surface become clean after degassing process and make improve the adhesion between Ti and Si. The typical temperature setting and degassing time depends on wafer environment and wafer quality(Fig17).

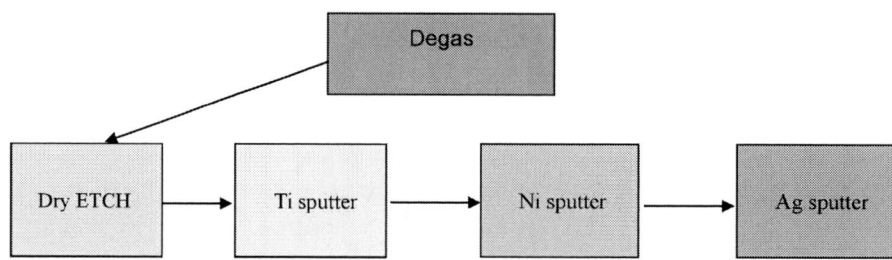

Fig17 New backside metal process flow with degassing

We did peeling test with baseline process flow and new flow with degassing and new etch recipe (Tab4). The result proved that degassing and soft etching are necessary and could improve adhesion indeed.

Wafer	baseline	use soft ETCH and degas
1	Peeling	OK
2	OK	OK
3	Peeling	OK
4	Peeling	OK
5	OK	OK
6	Peeling	OK
7	OK	OK
8	OK	OK
9	Peeling	OK
10	Peeling	OK

Tab4 Adhesion test comparison

CONCLUSION

In backside metal deposition process, two problems exist and impact device quality:

1. Wafer warpage high issue
2. Metal peeling issue

After analyzing the reasons of above issues, we design several solutions to improve them:

1. Reducing Wafer Warpage Solutions
    ➢ Optimizing Ag deposition recipe and separate the main sputtering steps.
    ➢ Introducing N_2 gas during Ni deposition to make Ni film more flexible as buffer layer.

2. Improving Adhesion Performance
    ➢ Controlling waiting time between WET ETCH and metal deposition to avoid oxidation.
    ➢ Changing dry etching recipe body to soft etching which can improve wafer surface roughness.
    ➢ Adding degassing process to get rid of water and impurity and make sure wafer surface clean before metal deposition.

REFERENCE

1. Xiao Gang. Power semiconductor discrete device[R]. ShenZhen: SHENZHEN CAPITAL GROUP CO., LTD, 2010: Page 6
2. Jacques Arnould, Pierre Merle Dispositifs de l'électronique de puissance, Éditions Hermès, ISBN 2-86601-306-9 (in French)
3. Harris, J. Semiconductor Wafer Backside Metallization: Die Attach for High Thermal Demand Applications[C]. US: CMC interconnect technologies, 2007: page 2
4. Liu Yu. 0.35μm Vertical Trench Power Device Backside Metallization Process Optimization and Improvement[C]. ShangHai: Shanghai Jiao Tong University, 2008: page 2,9,16, 50.
5. W.K. Ho, A. Tay, Y. Zhou and K. Yang. Fault Detection and Estimation of Wafer Warpage Profile During Thermal Processing in Microlithography [J]. IEEE: Transaction Semiconductor Manufacturing, 17(3), 2004.

A TECHNIQUE FOR IMPROVING CONTACT FILLING WITH ALUMINUM

Chunling Liu*, J.K. Shen, Xingjie Wang, Zhihui Ji

Shanghai Huahong Grace Semiconductor Manufacturing Corporation

Shanghai 201206, China

*Corresponding Author's Email: chunling.liu@hhgrace.com

ABSTRACT

Contact filling with aluminum (Al) is widely employed in Power discrete and Power IC. Good Al step coverage is a challenge for the high aspect ratio (A/R) contact process. In this article, we propose a Two-Step-Deposition (TSD) method for Al sputtering to improve contact filling. Firstly, a thinner layer warm Al with low power is deposited; then, a cooling step is followed, finally, we deposit the second layer warm Al with high power. With the suitable Al thickness ratio of low/high power, contact filling with void free is achieved, and the step coverage is also improved to more than 80%. The TSD process is accomplished in two chambers of one toolset, and the throughput of SPT tool can also be balanced extremely.

INTRODUCTION

Thick metal is generally needed in Power discrete and Power IC due to the big drive current requirement. Contact Filling with Aluminum Directly (CFAD) is widely employed due to the lower cost, lower resistance and simpler process compared with W-plug. However, CFAD has a drawback, worse step coverage and Al thinning, cracking, and electric migration are related with the poor step coverage. In general, hot Al can get better step coverage, but Al void is easy to happen since Al deposition rate is too fast at the contact top corner. Moreover, the cost of hot Al chamber is very high because of more expensive E-chuck and much shorter lifetime of the E-chuck. This article provides a Two-Step Deposition (TSD) technique to improve contact filling performance based on the current warm Al sputtering (SPT) tool. Good filling performance with void free contact has been achieved together with improved step coverage to more than 80%.

EXPERIMENT

Since contact profile will impact contact filling performance, all experiments based on bowl contact. The contact hole and filling process includes:

(1) Interlayer dielectric (ILD) deposition on wafer: SiO2 (TEL ALPHA-8SE-LP), BPSG (AMAT / Centura Canon), plasma enhanced TEOS (ASM / Eagle-10).

(2) Contact litho and etching process, to form the bowl contact.

(3) Wet pre-clean to remove the native oxide in contact.

(4) Contact barrier metal Ti/TiN sputtering, TiN can stop Al diffuse into Si to improve Al spiking issue.

(5) Al/Ti/TiN SPT (Endura 5500), the top TiN as anti-reflection layer of the litho process. Detail Al SPT process as below:

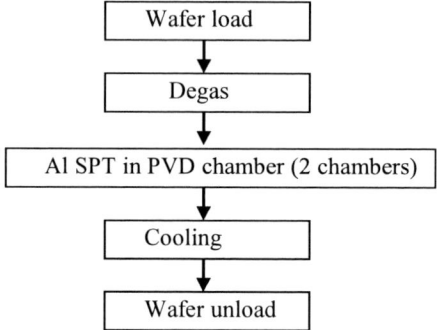

Normally, Al SPT is processed in one chamber: cold Al (50°C) nucleation with high power firstly, then the deposition of the remaining Al. Here we designed some experiments to improve contact filling, included splitting Al SPT temperature (250 °C, 380 °C), power, and SPT sequence. Al thickness is 2.2um in this experiment.

The contact filling profile was surveyed by SEM cross-section.

RESULTS AND DISCUSSION

A measure of how well a film maintains its nominal thickness is expressed by the ratio of the minimum thickness of a film as it crossed a step, t_s, to the nominal thickness of the film on flat regions, t_n (See Figure 1). This film property is referred to as the step coverage of the film, and is expressed as the percentage of the nominal thickness that occurs at the step.[1]

$$Step\ coverage\ (\%) = (t_s\ /t_n) * 100\ \% \qquad (1)$$

Step coverage depends on many factors include contact profile and SPT process. The contact profile has been fixed as bowl contact here. There are then several conditions of the SPT process can be adjusted to improve step coverage, including SPT temperature, power and sequence.

978-1-5386-5309-8/18 $31.00 © 2018 IEEE   300

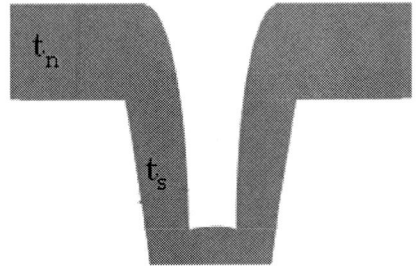

Figure1: schematic cross section for step coverage

The temperature experiment (1 step SPT: Al SPT in one chamber) results were summarized in TABLE I, and the contact filling profile showed in Figure2 and Figure3. The result showed that the step coverage was better with the higher temperature, but Al void occurred, which can be explained by the following mechanism:

During Al sputtering, Al atoms transportation mainly comes from surface diffusion; atom counts crossed surface at unit time and unit distance is expressed as surface flux (E_a: activation energy; T: temperature; $\frac{\partial K}{\partial x}$: curvature gradient).

$$Flux_s \cong -C \frac{\partial K}{\partial x} \exp\left(-\frac{E_a}{kT}\right) \qquad (2)$$

Temperature plays an important role in surface flux of atoms. The deposition rate is lower with lower temperature of Al SPT, and no Al void occurs. However, the surface flux of atoms decreases at lower temperature, and resulted in worse step coverage (see Figure2 and TABLE I).

TABLE I   THE RELATIONSHIP BETWEEN CONTACT FILLING AND SPT TEMPERATURE

Contact filling performance	SPT Temperature	
	250 ℃	380 ℃
void	No	void
step coverage	37%	96%

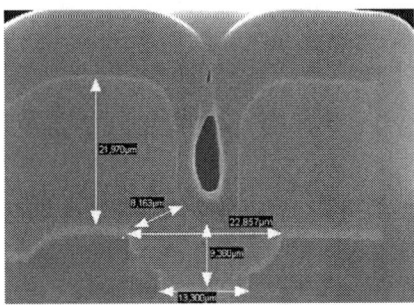

Figure2: contact filling profile (250 ℃ SPT) with 1 step deposition

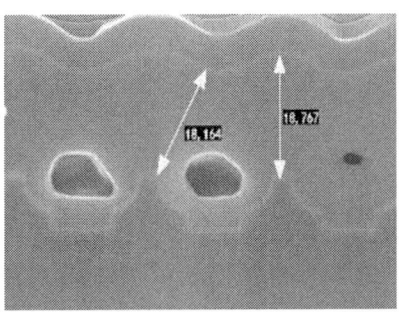

Figure3: contact filling profile (380 ℃ SPT) with 1 step deposition

Step coverage is better at higher temperature since surface flux of atoms increase with temperature, but Al void occurred (see Figure 3). This happens because the atoms are more likely to arrive at the upper corners of the contact from larger arrival angles than at locations within contacts [2]. That is, the deposition rate at contact upper corner is higher than at the sidewall and bottom of contact, this makes the opening of the contact easy to be closed off. As time goes on, the higher deposition rate further stops atoms diffuse into contact bottom, the atoms that have no chance to diffuse into the hole remain at the upper corner, finally the opening of contact is closed off and void is formed.

The above experiment indicates that void is easy to form at the-high temperature due to fast deposition; SPT power also plays an important role in deposition besides temperature. The lower the SPT power, the lower the deposition rate, the less chance of void formation. For bottom-up filling, we prefer to a deposition rate as low as possible. But too low deposition rate will depress the throughput. To balance the quality, performance and throughput, appropriate SPT power and SPT sequence should be found.

As we know, there are 2 Al SPT chambers for PVD tool Endura 5500. We split the thick Al into 2 steps in two different processing chambers, and Al deposition temperature is set at 380℃. At the 1st step, 100nm to 200nm Al nucleation layer is deposited with low temperature (about 50 ℃) and high power firstly. Then deposition of another Al layer is conducted with low power (3KW~6KW) at high temperature (380℃) for lower deposition rate. The low deposition rate can prolong deposition time. That is expected to be enough for Al to reflow and coalesce, and thus to improve the contact gap fill performance. After this, the wafer was sent to the cooling chamber to cool for certain time. After cooling, the wafer was sent to the other process chamber to process the 2nd step deposition with high power (10~14KW). As shown in Figure4, Al void can be avoided, and Al step

978-1-5386-5309-8/18 $31.00 © 2018 IEEE   301

coverage is also improved obviously compared with one step deposition (Figure2 and Figure3). Because surface status also plays an important role in surface flux of atoms according to formula (2).The surface status can be improved by TSD process, especially by the cooling between 2 steps. Cooling between the 2 steps can make Al grain size much smaller than one step deposition. Furthermore, in the 2nd step Al deposition process, Al nucleation layer is also deposited firstly at low temperature (50℃),Al nucleation is small and continuous, which further improve the surface status, the smaller the grain size, the smoother the surface, the faster the surface atom diffuse, then step coverage performance is improved .

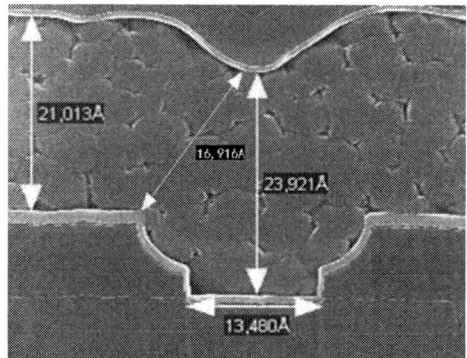

Figure 4: contact filling profile with 2 steps SPT (380℃)

The thickness ratio between 1st step and 2nd step of TSD is important for step coverage, the higher the thickness ratio, the better the step coverage (see Figure5); the step coverage can be improved to more than 80% if the thickness ratio is more than 0.8. To balance step coverage and tool's throughput, we do the process margin check for Al thickness ratio between 2 steps. 0.8 is reasonable

TSD process can improve the contact filling performance, moreover, Al grain will also be smaller than one-step deposition, Al can be etched easily, no etching residue is found from the KLA scan result (see Figure 6)

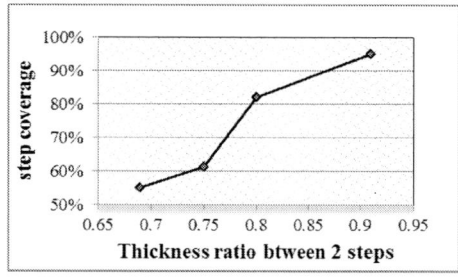

Figure 5: step coverage Vs thickness ratio between 2 steps in two-step deposition

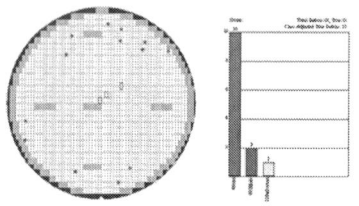

Figure 5: Defect map after metal etching

CONCLUSION

Two-step deposition in different two chambers can remarkably improve contact filling performance with warm Al (380℃). 1st step Al deposition with low power can decrease Al deposition rate, and makes Al reflow driven by the difference in surface energy, the following cooling step decreases Al grain size. After that, 2nd step deposits the remaining Al with high power in the other Al SPT chamber. With the suitable Al thickness ratio of low/high power, contact filling with void free is achieved, and the step coverage is also improved to more than 80%, and at the meantime, the tradeoff in throughput is still acceptable for mass production.

ACKNOWLEDGEMENTS

The authors would like to thank the metal SPT team and FA group in HHGRACE for their contributions to this paper.

REFERENCES

[1] S.WOLF, *Silicon processing for the VLSI Era Volume1*, 1993

[2] Liu Jian. *International IEEE conference on Asian Green Electronics 2004*, Hong Kong, Jan6-8, 2004, pp.92

EMITTER-BASE SHORT ISSUE STUDY AND IMPROVEMENT IN A LOW COST AND HIGH PERFORMANCE 0.18UM SIGE BICMOS PROCESS

Donghua Liu[1,2]*, Zhaozhao Xu[2,3], Wensheng Qian[2]

[1]State Key Laboratory of ASIC and System, School of Microelectronics, Fudan University, Shanghai 200433, China

[2]HuaHong Grace Semiconductor Manufacturing Corporation, Shanghai 201206, China

[3]State Key Laboratory of Functional Materials for Informatics, Shanghai Institute of Microsystem and Information Technology, Chinese Academy of Sciences, Shanghai 200050, China

*Corresponding Author's Email: donghua.liu@hhgrace.com

ABSTRACT

This paper studied the short issue of the emitter and the base, which is frequently found in the fabrication of the SiGe HBT devices. The physical failure analysis reveals that the main reason is the formation of cobalt silicide in the emitter poly-Si sidewall. Reducing the lower electrode voltage and source power of the main etch step can reduce the byproducts accumulation during etching and improve the emitter poly-Si profile. Adjusting the etch time ratio of main etch and soft landing makes the polysilicon angle improved from less 80° to 88° and eliminate the footing issue. - It is found that the PECVD film has poor coverage, and that by LPTEOS shows better coverage. And by increasing the oxide film thickness from 80nm to 160nm, the width of emitter poly (EP) sidewall spacer after etching is 105nm. Before cobalt silicide formation, the sidewall spacer width is 63nm, which can play a role of a good protection layer for the emitter polysilicon and isolate the emitter from the base of the SiGe HBT.

INTRODUCTION

With the development of information technology, the frequency performance of semiconductor devices is increasingcontinuously. Simple silicon devices is difficult to meet the application requirements. The SiGe HBT,which has frequency characteristics, gain characteristics and noise characteristics much better than the traditional silicon bipolar transistor, and can be made by mature CMOS compatible silicon processing technology , will have a very broad application prospects in future mobile communications [1-3].

The In this work, we conduct a continuous and comprehensive study of Silicon Germanium (SiGe) HBT BiCMOS process, and developed a low cost process to make high performance 0.18um SiGe BiCMOS (referred to as BCS18) [4-6]. Polysilicon emitter process is one of the critical steps. Figure 1 (a) and (b) shows the cross-sectional views before and after etching of the emitter polysilicon layer, respectively. If the topgraphy before etching is not smooth, and the etching process conditions can not be set properly, short between the emitter and the base (E-P) is frequently formed in the fabrication of the device. In this paper, the fabrication process of polysilicon emitter is studied and the optimized processing conditions are presented

EMITTER POLY ETCHING PROCESS OF SIGE HBT

In the BCS18 process, the etch area of SiGe HBT emitter is defined with photoresist. At the same time, in order to ensure the polysilicon on the gate of CMOS being completely removed, etching stop layers on silicon nitride / silicon oxide are deposited underneath. Dry etching process is used for stable etching rate, good uniformity and relatively high choose rate. Etching conditions are selected with cautionto avoid resulting in damage of the underlying nitride & gate poly film.

Film stack in the emitter poly etching process is composed of 930nm PR / 76nm BARC / 150nm poly // 20nmSiN2 / 20nmSiO2

Dry Etching ADI target is: 0.6 +/- 0.06um AEI target: 0.6 +/- 0.06um

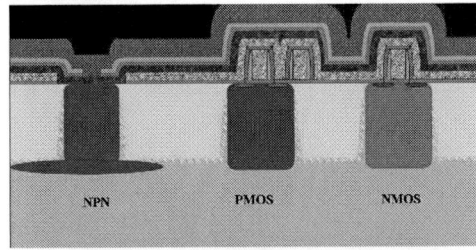

(a)

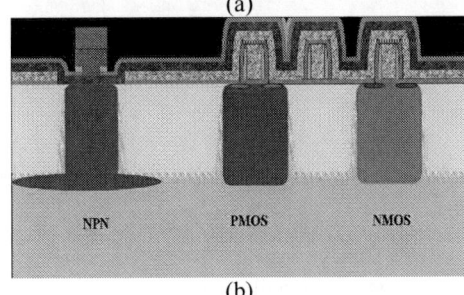

(b)

Figure 1: (a) Cross-section before Emitter Poly etch; (b) Cross-section after Emitter poly etch

RESULTS AND DISCUSSION

Mechanism Analysis of E-B short

BCS18 electrical performance test results reveal that E-B junction is easily short (BVEBO failure). SEM pictures in figure 2 (a ~ d)shows that the dielectric sapcer of emitter polysilicon has been removed. There is no isolation between emitter and base. EDX analysis result in Figure 3 shows clearly the presence of cobalt elements, indicating that cobalt silicide forms on the emitter sidewall because of no oxide spacer protection during cobalt silicide process. The metal silicide in the emitter poly side wall causes the short of emitter and base. This attributes to two reasons: 1) the emitter polysilicon tilt, which makes the sidewall difficult to form; 2) side wall oxide film coverage is poor and the width is not enough to sustain subsequent dry or wet etch step.

(a) SEM of Y direction   (b) SEM of Y dirction

(c) SEM of X direction   (d) SEM of X direction

Figure 2: SEM pictures of emitter-base short samples. (a) & (b) are Y dirction; (c) & (d) are X direction

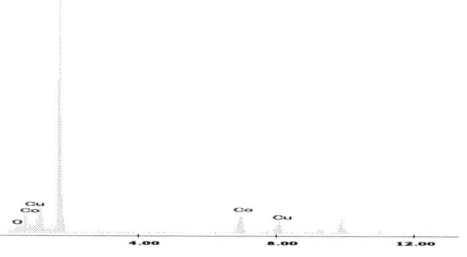

Figure 3: EDX analysis of emitter-base short sample

Technical Scheme for Optimizing EP Profile

In order to avoid the problem due to the tilt, it is necessary to form a relatively vertical profile. The dry etching needs to remove different layers from top to bottom: 1) the bottom of the anti-reflective layer (Barc) , 2) emitter polysilicon, 3) etching stop bilayerformed by oxide and nitride The etch process is thus divided into following four steps:

1) anti-reflective layer etching step
2) polysilicon main etching step
3) polysilicon soft landing etching step
4) over etching step

The polysilicon main etch step is generally used to define the emitter shape, while soft landing step using the etching endpoint detection (EPD) method to avoid plasma demage. The parameters used for the main etch step are as follows:

a) Pressure 5 ~ 20MT,
b) Upper electrode power 200 ~ 350ws,
c) Positive voltage -100 ~ -450V,
d) Gas: CF4 50 to 150 Sccm, Cl2: 50 to 150 Sccm HBr: 50 to 150 Sccm O2: 2 to 5 sccm

After forming the proferable polysilicon profile of the emitter, we can start to improve the sidewall oxide film quarlity. nThe original PECVD oxide film, which has normally poor coverage, is replaced by the low-pressure TEOS (LPTEOS) film with better coverage performance. Considering the width loss of the sidewall in the subsequent dry or wet steps, it is necessary to use a larger thickness of the oxide filmto conpensate the loss.

The etching process of the sidewall oxide film needs also to be modified to fit the new LPTEOS film. The etch process consists of one main etch step with etch endpoint detection and one over etch step. The main etch parameters are as follows:

a) pressure 300 ~ 800MT
b) electrode power 300 ~ 500w
d) Gas: CF4 5 to 20 Sccm, CHF3: 5 to 15 Sccm He: 150 to 350 sccm

Optimization of EP Surface Morphology

Figure 4 (a) is the SEM Cross-Section picture of original emitter polysilicon etched, showing emitter poly slightly tilted. Figure 4 (b) is the SEM image obtained using the optimized processby reducing the lower electrode voltage and source power during etching. One can see that the polysilicon profile has been significantly improved. It is believed that new conditions can reduce the accumulation of byproducts and yields a better topography.

With an profile angle about 80°, there is still a little bit of footing at the bottom, as shown in figure 4 (b). Based on the optimization of the first step, adjusting the etch time ratio of the main etch step and the soft landing step, the footing of emitter poly bottom can be further improved. After optimization of the ratio, the profile angle is improved from 80° to 88°, as figure 4 (c) shows.

978-1-5386-5309-8/18 $31.00 © 2018 IEEE   304

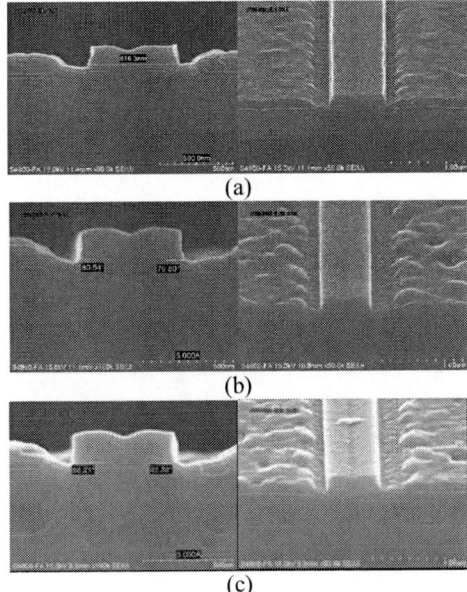

(a)

(b)

(c)

Figure 4: SEM pictures of emitter poly before and after optimization by etch process. (a) before optimization; (b) optimization result 1; (c) optimization result 2

Emitter Poly Spacer Optimization

Figure 5 (a) shows the SEM images of the original emitter side wall after spacer etching. The oxide thickness before etching is 80nm and it was deposited by PECVD. After two steps of dry and wet etch, the width of the side wall is reduced to 45nm. Figure 5 (b) is the SEM picture of side wall spacer with the oxide film deposited by LPTEOS and the thickness is increased to 110nm. After dry and wet etching, the width of the side wall spacer has increased from the original 45nm to 69nm.

Figure 5 (c) is the result of LPTEOS film of 160nm. After the etching, the sidewall spacer width increases to 105nm.

Figure 5 (d) is the SEM picture before formation of cobalt silicide film. The side wall width is 70nm, 35nm oxide loss caused by the middle process. Figure 5 (e) is the SEM picture after the formation of cobalt silicide film. The sidewall width is 63nm, which can be a good protection of the emitter polysilicon and isolate the emitter and base of SiGe HBT.

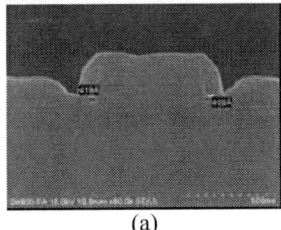

(a)

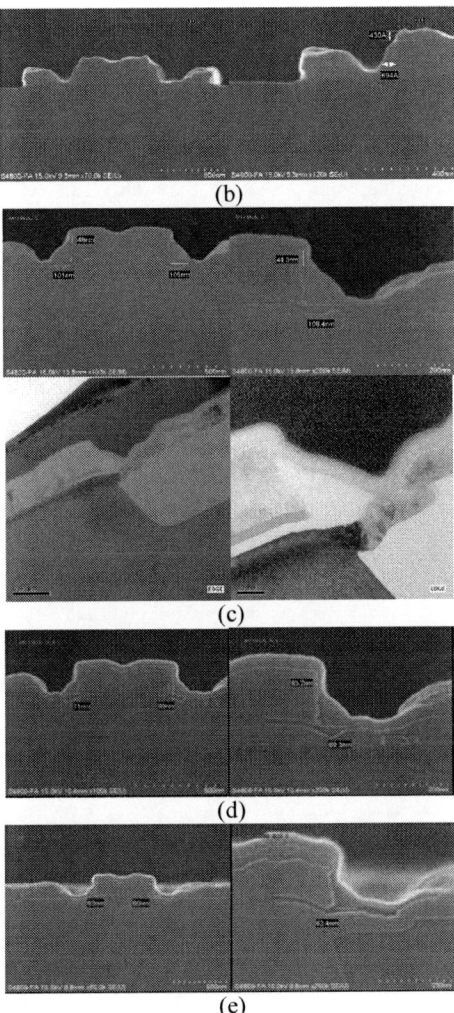

(b)

(c)

(d)

(e)

Figure 5: Spacer of emitter poly before and after optimization. (a) SEM picture of emitter poly side wall spacer after etch of original; (b) emitter poly side wall spacer after etch with LPTEOS 1100A; (c) emitter poly side wall spacer after etch with LPTEOS 1600A; (d) emitter poly side wall spacer before silicidation; (e) mitter poly side wall spacer after silicidation

CONCLUSIONS

The physical analysis of emitter-base short failure of SiGe HBT reveals that the main reason of the failure is the formation of cobalt silicide in the emitter poly sidewall, implying that the emitter poly spacer does not function well.

Reducing the lower electrode voltage and source power of the main etch step can reduce the byproducts during etching and improve the EP profile. Adjusting the

978-1-5386-5309-8/18 $31.00 © 2018 IEEE

etch time ratio of the main etch and the soft landing makes the polysilicon angle improved from less 80° to 88° and eliminate the footing issue. Optimizing the side wall oxide film can be realized by using LPTEOS film and increasing the oxide film thickness from 80nm to 160nm. The width of EP side wall spacer after etching is increasing to 105nm. Before cobalt silicide formation, the sidewall spacer width is 63nm, which is good enough to protect the emitter polysilicon from Co and isolate the emitter and base of SiGe HBT.

ACKNOWLEDGEMENTS

The authors would like to thank all the members of device group, RF group and advanced module of HHgrace for their great support on this work.

.

REFERENCES

[1] A. Keerti and A. Pham, "SiGe power devices for 802.11a wireless LANapplications at 5GHz," *Electron.Lett.*, vol. 39, pp. 1218–1220, 2003.

[2] A. J. Joseph, J. Dunn, G. Freeman, D. L. Harame, et al., "Product applicationsand technology directions with SiGeBiCMOS," *IEEE J. Solid-State Circuits*, vol. 38, no. 9, pp. 1471–1478, 2003.

[3] W. Bakalski, W. Simbürger, R. Thüringer, et al., "A fully integrated 5.3-GHz 2.4-V 0.3-W SiGe bipolar power amplifierwith 50- output," *IEEE J. Solid-State Circuits*, vol. 39, no. 7, pp. 1006–1014, 2004.

[4] D. H. Liu, W.S. Qian, X.B. Chen, et al., "0.18 micron BiCMOS process with novel structure SiGeC HBT," *ECS Transactions*, vol.34(1), pp. 173-181, 2011.

[5] D. H. Liu, J. Shi, J. Hu, et al., "PNP Transistor Design & Simulation in SiGe BiCMOS with Low Cost & High Performance," *ECS Transactions*, vol.52(1), pp. 23-31, 2013.

[6] D. H. Liu, J. Shi, et al., "Base–region optimization of SiGe heterojunction bipolar transistors (HBT)," *Proceeding of IC&SI*, vol.17, pp. 320-323, 2011.

ADVANCED MASS FLOW CONTROLLERS (MFC) WITH ETHERCAT COMMUNICATION PROTOCOL AND EMBEDDED SELF DIAGNOSTICS

Kevin Findleton*, Berwin Banares and Mohamed Saleem
Brooks Instrument, Hatfield, PA, USA
*Kevin.Findleton@BrooksInstrument.com

ABSTRACT

Mass Flow Controllers (MFCs) are the most critical components of a gas delivery system used for semiconductor manufacturing. As the technology transitions to advanced nodes below 10 nm, so have the requirements for the MFCs to have embedded self-diagnostics to support Fault Detection and Classification. A few examples of factors contributing to MFC performance degradation are sensor drift and control valve leak-by. Sensor drift can affect accuracy and repeatability of the device which in turn, can cause shifts in CD uniformity. Valve leak-by can cause 'first wafer effect' essentially impacting yield. Therefore, developing an in-situ method to detect sensor drift and valve leak-by is beneficial to process tool owners to minimize tool downtime and improve process yield. An algorithm based approach to detect sensor drift and valve leak-by has been developed without interruption to process flow steps or requiring any hardware changes. Utilizing advances in EtherCAT communication protocol as a backbone, data from the devices on these two critical parameters can be logged and monitored to ensure fault-free and consistent device performance. The algorithm and implementation relies on standard EtherCAT indices for commissioning and diagnostic routines.

INTRODUCTION

It has been well documented over the years that issues found in the field associated with Mass Flow Controllers (MFCs) have been very difficult to troubleshoot due to lack of information available from the devices. Issues that are random in occurrence and are specific to the field configurations are often the most difficult to reproduce. This results in longer troubleshooting times and often forces the tool owners, in the semiconductor fabs, to replace the MFCs without a clear understanding of the root cause of the problem.

With advanced nodes and complex processes in chip technology, the semiconductor companies (Integrated Device Manufacturers (IDMs)) require superior performance such as better repeatability, higher accuracy and low drift for these mass flow controllers. For certain applications and gases, these devices may be operating under non-ideal conditions and the IDMs would benefit from having a method of determining performance degradation before causing major failures in wafer production and thereby impacting yield.

A major contributor to performance degradation in MFCs is sensor drift caused by improper zeroing or sensor aging. If there was a method to determine true zero flow drift signatures before the device impacts wafer production it would be very beneficial to tool owners as they can prevent loss of wafers and prevent process tool downtime. Current wafer quality control is managed downstream of process, resulting in all wafers being processed before anomalies are detected. Furthermore, the process tool, which is based on high speed data acquisition, can only detect hard failures such as a MFC not reaching a given set point or not communicating with the process tool. In addition, detecting soft failure modes such as MFC calibration changes require the user to stop process runs and perform time consuming calibration checks. Therefore, automatically detecting soft failure modes directly supports yield enhancement programs and tool up time.

EXPERIMENTAL DETAILS

Process tools are now moving towards advanced communication protocols, such as EtherCAT for high speed data transfers and control. For example, the GF125-Enhanced series of Mass Flow Controllers takes advantage of this high-speed interface, which enables fast actuation of isolation valves and information transfer to the MFC using the same interface. This enables the MFC to run its leak and drift diagnostics algorithms in a standard gas panel configuration without hardware requirements or changes.

A diagnostics measurement is triggered when the tool transitions the MFC from FLOW to OFF state. During this transition, valve leak measurements are made by measuring the devices pressure transducer decay (rate-of-decay (ROD)), and translating that decay into a leak rate value. This leak rate, if there are no external leaks (leak check is a standard practice during tool installation), then becomes our diagnostics valve leak. During this measurement period, the device also analyzes its internal sensor output. Valve leak information is vital in determining the zero offset to be able to decouple leaks from true sensor zero.

During GF125-Enhanced installation, the device needs to perform a 'commissioning' step, in which it establishes a coefficient to convert the rate-of-decay measurement into volumetric flow rate (sccm). Due to variations in gas panels, it is not possible to convert ROD measurements to a flow rate via a fixed volume parameter.

978-1-5386-5309-8/18 $31.00 © 2018 IEEE

Therefore, to have good leak measurements, the device must calculate this conversion constant in the active setup with actual process gas. During commissioning, mathematical calculations performed by the device require certain variable values, which can only be obtained during the commissioning process. The most important variable is the volume representation of gas between the upstream isolation valve and the valve seat.

MFC Commissioning Steps

1. Inlet pressure check - Inlet pressure is checked by the on-board pressure transducer.

2. Checking Isolation Valve - Verifies the function of the isolation valve to stop inlet gas during measurement.

3. Leak check - Checks the amount of Leak-by when the valve is in the 0% setpoint (closed) position.

4. Starting Auto - Volume Commissioning - This is the process of determining the volume of gas between the inlet isolation valve and the device itself. Tool needs to provide process gas supply and respond to isolation valve actuation requests from the device.

5. Commissioning Completed - At this point, the Commission process routine is completed.

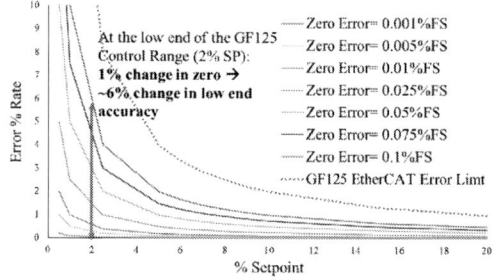

Figure 1: Zero offset induced flow error

Figure 1 shows how a MFC with varying change in zero offset directly impacts low end accuracy of gas delivery. For example, a 0.1% F.S. change in zero causes a 6% change in rate at a 2% F.S. setpoint.

RESULTS AND DISCUSSION

The definition of mass flow rate (sometimes called mass flux or mass current) is the derivative of mass with respect to time.

$$\dot{m} = \frac{\Delta m}{\Delta t} \qquad (1)$$

Flow rate can also be called the rate of change in number of molar mass in a given volume. Using the Ideal Gas Law equation

$$PV = nZRT \qquad (2)$$

Where: P is pressure
V is volume
n is number of moles
Z is compressibility
R is the universal gas constant
T is absolute temperature

Solving for molar mass n results in

$$n = \frac{PV}{ZRT} \qquad (3)$$

In this equation, pressure, Z and temperature changes over time. Grouping constant terms and terms that change over time results in

$$n = \left[\frac{P}{ZT}\right]\frac{V}{R} \qquad (4)$$

Since what we are after is mass flow rate, taking the derivative of the number of molar mass n with respect to time gives us

$$\frac{dn}{dt} = \frac{d(\frac{P}{ZT})}{dt}\frac{V}{R} \qquad (5)$$

The term $[P/ZT]$ in equation 5 is redefined as Dv.

Since Dv is P/ZT, the rate of change in Dv with respect to time (dDv/dt) multiplied by some constant, which will be defined as k, gives the mass flow rate Q_{ROD}.

$$Q_{ROD} = \frac{d(Dv)}{dt}k \qquad (6)$$

Valve leak measurement uses the signal from the MFC's internal pressure transducer in a form where temperature and compressibility (gas dependent) is taken into account. This new form of signal is called Dv (mass of gas in inventory volume). The MFC controls the upstream valve and deliberately shuts off the gas source to the MFC at zero flow. The device bleeds off the gas in inventory volume (gas between the upstream isolation valve and the MFC control valve). While in the process of depleting gas in inventory, the mass of gas in inventory Dv is measured over time. This rate of change of mass in inventory multiplied by the conversion constant, k (calculated during device commissioning) is the valve leak rate. This measurement routine takes place every time the device is given a zero setpoint for longer than 30 seconds.

Figure 2 illustrates the linear decay of Dv and the timing of the measurement routine. During a single transition from flow to off state the valve leak is measured during the first 10 seconds with a 10 second delay for sensor stability before true sensor zero offset is measured.

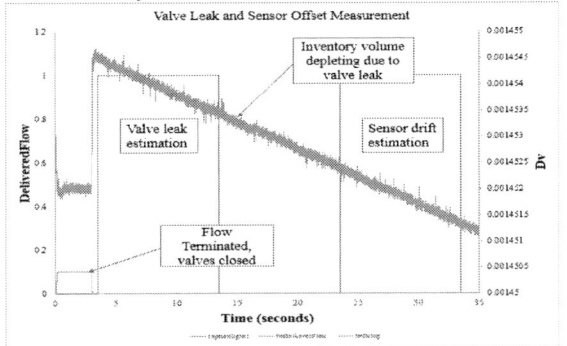

Figure 2: Valve leak and sensor drift measurement

Data Analysis
Table 1.

Run	Sample of Data collected over EtherCAT		
	Zero (%FS)	Leak (%FS)	Pressure (psia)
1	0.4996108	0.002792713	31.70451
2	0.4981688	0.001827049	31.78647
3	0.4996979	0.001891514	31.73146
4	0.5019056	0.001171776	31.71181
5	0.5006123	0.001976435	31.66978

Table 1 is an example of data collected over EtherCAT from a GF125-Enhanced running a simulated recipe. IDMs currently store information from MFCs like flow, set point, temperature, and pressure. With the GF125-Enhanced monitoring valve leak and zero drift is as easy as recoding any other EtherCAT index. The data collected can then be used for long term data mining and tool troubleshooting.

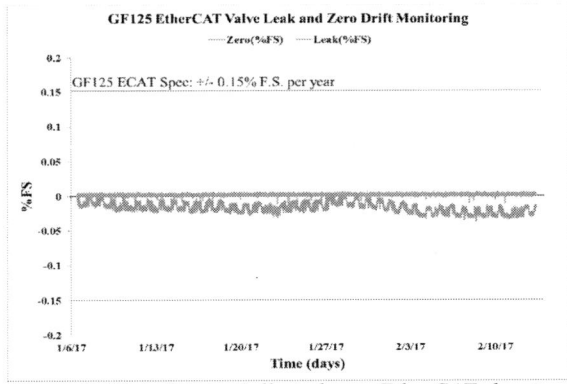

Figure 3: Plot of Data Collected over EtherCAT showing valve leak and zero drift

Figure 3 shows valve leak and zero drift measurements while running a simulated tool recipe collected over 1 month. Valve leak measurement resolution is 0.001%.

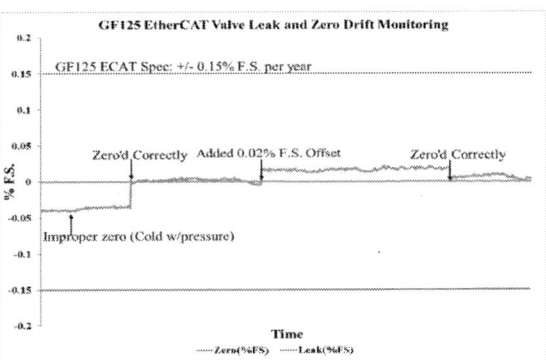

Figure 4: Plot of Data Collected over EtherCAT forcing zero offset to the device

Figure 4 shows how Self-Diagnostics can detect zero shift anomalies that were caused by an improper zeroing procedure and simulated sensor drift. Furthermore, a forced offset of 0.02% F.S. was induced to illustrate zero drift tracking abilities while running a simulated recipe. The GF125-Enhanced zero drift monitoring can track minuet changes over time. This information is important since zero drift directly impacts low end accuracy and can be used to set limits or alarms to schedule routine maintenance.

CONCLUSION

The use of Brooks' Self-Diagnostics technology in mass flow controllers provides excellent process gas accuracy and repeatability via superior predictive maintenance with the help of data generated by our advanced self-monitoring algorithms. This generates consistent and repeatable device performance enabling better chamber matching and decreased downtime due to maintenance, which increases wafer production throughput and tool productivity. Integration on tool is extremely easy, which allows configurations to be modified quickly and with little effort by leveraging high speed data control and acquisition. GF125-Enhanced enables the process tool owner to maximize value proposition in mass flow by combining the various benefits associated with Self-Diagnostics along with the quality and reliability that comes with the Brooks Instrument's GF125 series MFCs.

ACKNOWLEDGEMENTS

Authors would like to give special thanks to our co-workers, Anthony Kehoe, Jeffrey Whiteley, Andrew Staudt, Shaun Pewsey, and Alexandra Liso.

REFERENCES

[1] Fox and McDonald, *Fluid Mechanics, 2nd Edition, 1978*

[2] S. Pewsey, B. Valentine, *A New Class of MFCs with Embedded Flow Diagnostics*

The Study of TiN Residues Formation Mechanism and Removal Solution on Bondpad Surface in PI/PA Mask Combined Process

Xu Jie, Huang Chong, Li Zhiguo, Ding Tongguo

Shanghai Huahong Grace Semiconductor Manufacturing Corporation
Shanghai, China
Corresponding Author's Email: Gavin.Xu@hhgrace.com

Abstract

In PI/PA mask combined process, the problem of TiN residues on bondpad surface is often encountered. To obtain TiN residues free bondpad, the top metal sputtering recipe splits were done. The bondpad TiN residues formation mechanism was analyzed, finding that the alloy process location change produced $TiAl_3$ and then induced the TiN residues issue after PI/PA etching. The Ti/TiN/Al/TiN top metal film stack can get clear bondpad surface, but it would bring in crown defect after top metal etching. Optimized the top metal sputtering recipe to the film stack of Ti/TiN/Al/TiN/Ti/TiN, the $TiAl_3$ formation was prevented and the bondpad TiN residues can be fixed, meanwhile the crown defect issue resulted from TiN cracking was resolved.

Keywords—PI/PA mask combined; Bondpad TiN residue; Top metal film stack

Introduction

In semiconductor manufacturing, the passivation (PA) is to isolate chip from the environment, preventing iron contamination on device, and protecting it from mechanical and chemical damage. At present, SiO_2 and Si_3N_4 are widely used as the inorganic passivation film[1]. Polyimide (PI) can be covered with the inorganic material on the chip surface to form a composite passive film, which can effectively block the migration of electrons, prevent corrosion and increase the moisture resistance of components[2].

The graphics of the two layers' mask of PI and PA are the same, and the exposed area is called bondpad for subsequent bonding process. Therefore, PI can be used as photoresist for PA etching, in this way, the layer of PA lithography process can be eliminated and cost is much down in manufacturing, which is the so-called PI/PA mask combined process. This process change of PI as photoresist for PA etching, usually results in the TiN residues on bondpad surface, which will lead to subsequent bonding reliability issue.

In this paper, the formation mechanism of TiN residues on bondpad surface in PI/PA mask reduction process is studied, and a solution to optimize the sputtering process of top metal is proposed, and further improvement is studied.

Experimental

Fig. 1 is the comparison of the PI/PA two reticles process and the mask combined process. Besides the mask reduction, the biggest difference is that the alloy process need move to the step after passivation SIN deposition.

According to the process, five samples as shown in Tab. 1 are prepared.

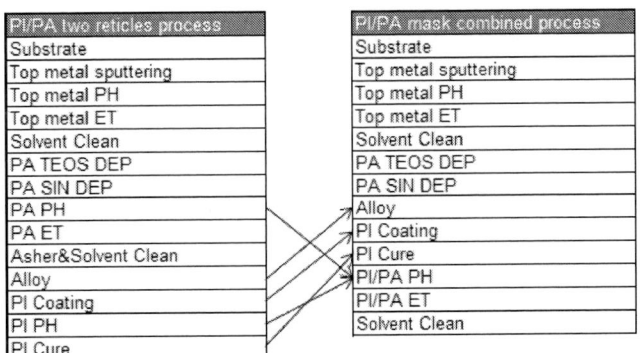

Fig. 1 Comparison of the PI/PA two reticles process and the mask combined process

Tab. 1 Samples with different process conditions

	PI/PA mask condition	Top metal sputtering condition
S1	Two masks baseline process	Ti/TiN/Al/90A Ti/TiN (N2 2s dealy)
S2	Mask combined process	Ti/TiN/Al/90A Ti/TiN (N2 2s dealy)
S3	Mask combined process	Ti/TiN/Al//TiN (N2 0s dealy)
S4	Mask combined process	Ti/TiN/Al/180A Ti/TiN (N2 4s dealy)
S5	Mask combined process	Ti/TiN/Al/TiN/Ti/TiN

Results and discussions

1. Formation mechanism of TiN residue on bondpad surface

Contrasted the condition of sample S1 and sample S2 in Tab.1, the top metal sputtering condition is the same, and the only difference between them is the mask condition. Fig. 2&3, respectively, are the bondpad image of baseline process sample S1 and PI/PA mask combined process sample S2 under optical microscope. It can be seen that there are a lot of tawny spots on the bondpad surface with mask combined process.

978-1-5386-5309-8/18 $31.00 © 2018 IEEE

Fig. 2 Bondpad surface of sample S1 with baseline process

Fig. 3 Bondpad surface of sample S2 with PI/PA mask combined process

Auger analysis of Tab. 2 shows that the tawny spots are TiN residues on the bondpad surface after PI/PA mask combined process modification. It is also found that TiN residues always appear at the Al grain boundaries.

Tab.2 Auger analysis comparison of sample S1 and sample S2

Sample	Detect element (At %)						
	C	O	F	Al	Ti	N	Si
S1 Center	11.82	45.12	6.02	34.51	-	2.53	-
S1 Edge	7.77	41.98	10.72	37.33	-	2.2	-
S2 Center	19.59	23.35	5.16	10.07	15.26	24.42	2.16
S2 Edge	24.87	21.12	7.75	23.74	7.92	14.6	-

In PI/PA mask combined process, the alloy process need change to the step after passivation SIN deposition. This relocation makes the Ti and Al of top metal film stay together for 30min at about 500 degree. From Fig.4 Ti-Al two element phase diagram, Ti and Al will undergo phase transition to generate TiAl$_3$[3]. Fig.5 illustrates the bondpad with TiN residues formation process flow of sample S2. TiAl$_3$ will grow at Al grain boundary due to the low activation energy. TiAl$_3$ is hard to remove by PI/PA etching, so in the end, the TiN residues at the Al grain boundary are left.

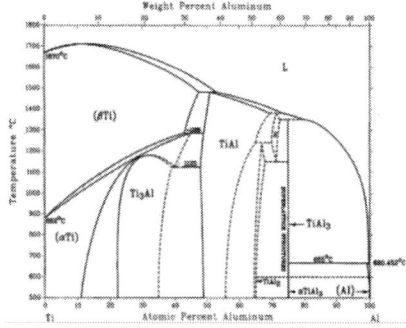

Fig.4 Ti-Al two element phase diagram

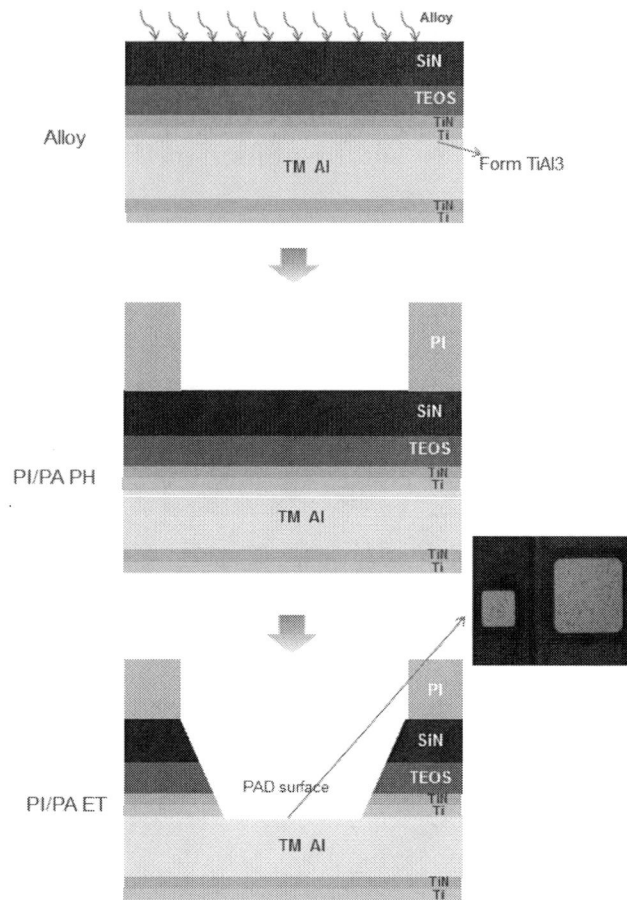

Fig. 5 Bondpad with TiN residues formation process flow of sample S2

To further validate this model, we do TiN N2 delay experiments split under the condition of PI/PA mask combined process, and obtain samples S3 and sample S4. As shown in Tab. 1, samples S3 is the split condition of TiN N2 0s delay, and get the stack structure of Ti/TiN/Al/TiN. There is no Ti on the top metal, only the TiN layer. The sample S4 is the worse condition and gets a thicker Ti layer on the top metal. Fig. 6 shows the bondpad surface observed by optical microscope. It can be seen that the thicker the Ti layer on the top metal, the more serious the TiN residues on the bondpad surface, which can be explained as that the thicker the Ti layer, the more TiAl$_3$ produced by Ti reaction with Al, and the more TiN remains after PI/PA etching.

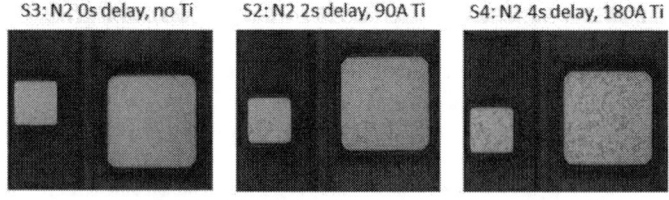

Fig. 6 Bondpad surface of different TiN N2 delay splits

2.   Study on removal method of TiN residue

From above discussion, we know that the Ti/TiN/Al/TiN stacking structure can solve the problem of TiN residue. But it brings in the crown defect(as shown Fig.7) after PA etching. Fig. 8 EDX analysis shows that the main components of the defect are Al and O. If this defect occurs in areas where the wire

978-1-5386-5309-8/18 $31.00 © 2018 IEEE   312

spacing is small, it will lead to metal bridging, and induce functional failure.

Both the TiN and Al have the same preferred crystalline orientation (111). In the Ti/TiN/Al/TiN film structure, the stress mismatch between Al and TiN results in the fracture of the TiN along the grain boundaries (Fig. 9). In the laminated structure of Ti/TiN/Al/Ti/TiN, the metal Ti (preferred crystal phase (002)) has a good ductility, and play a role in the adhesion between Al and TiN, so in this structure TiN fracture is not easy to happen.

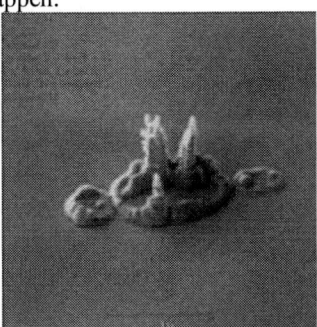

Fig.7 Crown defect after PA etching under Ti/TiN/Al/TiN film stack

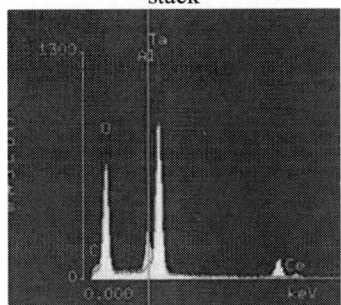

Fig. 8 EDX analysis of crown defect

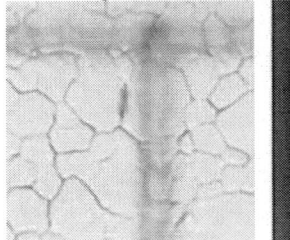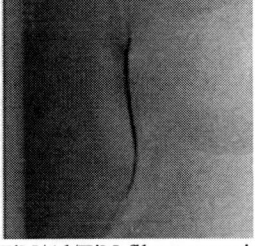

Fig. 9 TiN fracture after Ti/TiN/Al/TiN film sputtering

Fig. 10 is a schematic diagram of crown defect formation. In the development step of top metal lithography, alkaline developer flows to the Al surface along the crack and reacts with Al to form Al_2O_3. Al_2O_3 is very difficult to etch away in the top metal etching process, eventually leaving the defects like the crown in the metal layer open area[4].

In order to achieve the goal of no TiN residue on the bondpad surface and no crown defects after the top metal etching, we tune the TiN film on top metal to the TiN/Ti/TiN sandwich structure. That is the sample S5 film stack: Ti/TiN/Al/TiN/Ti/TiN (as Fig.11 shown). This structure separates the Al from the Ti, avoids the formation of $TiAl_3$ during alloying, and solves the crown defects caused by TiN cracking. The results of samples S2, S3 and S5 are summarized in Tab. 3, and the results shows that the conditions of the sample S5 can optimize the process.

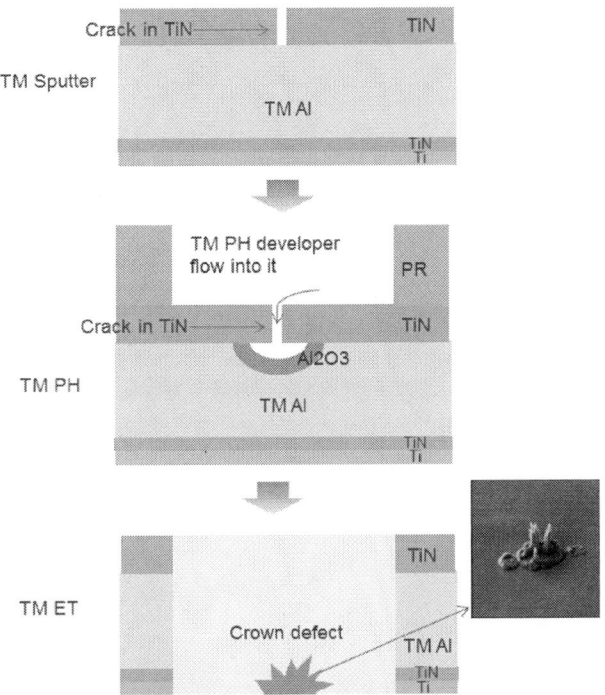

Fig.10 Schematic diagram of crown defect formation

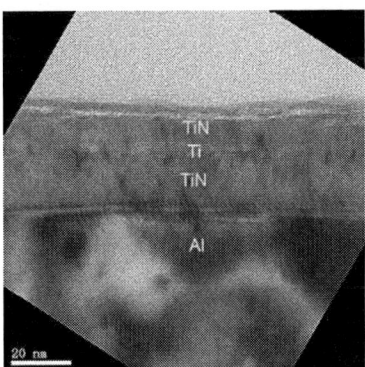

Fig. 11 TEM image of sample S5 film stack Ti/TiN/Al/TiN/Ti/ TiN

Tab.3 Crown defect and bondpad surface results under different top metal sputtering recipe

Smaple	Condition	TM AEI Defect	PAD surface
S2	Ti/TiN/Al/Ti/TiN	No crown defect	TiN residue
S3	Ti/TiN/Al/TiN	Crown defect	No TiN residue
S5	Ti/TiN/Al/TiN/Ti/TiN	No crown defect	No TiN residue

Conclusions

1. In PI/PA mask combined process, $TiAl_3$, which is produced by the alloy process location change, is hard to remove by PI/PA etching and this leads to pragmatic TiN.
2. The Ti/TiN/Al/TiN top metal film stack can get clear bondpad surface, however it will bring in crown defect after top metal etching.
3. The top metal film stack of Ti/TiN/Al/TiN/Ti/TiN can prevent $TiAl_3$ formation, resolve TiN residues on bondpad surface issue, and fix the crown defect resulted from TiN

cracking.

References

[1] Xuejian Liu. Study on surface passivation films for integrated
[2] circuits. "Journal of Ceramics" 2002 , 23 (2) :112-115.
[3] Jingang Liu. Polyimides for microelectronics packaging. "Semiconductor technology", 2003 , 28 (10) :37-41.
[4] Yanbo Sun. Phase transition process of Ti-Al intermetallic compounds prepared by additive sintering of elemental foils. "Proceedings of the National Symposium on solid phase transition, solidification and Applications", 2014
[5] Bingyuan He. Influence of base pressure and residual gases on metal film growth in integrated circuit fabrication. "Chinse Jouranal of vacuum science and technology", 2012, 4:318-323.

MANAGEMENT OF TEOS-CVD PROCESS TOOL EXHAUSTS IN 3D-NAND MANUFACTURING

Andrew A. Chambers

Edwards, Clevedon BS21 6TH, UK

Author's e-mail: andrew.chambers@edwardsvacuum.com

ABSTRACT

CVD film precursor flow rates continue to increase as device manufacturers seek to optimise film uniformity and deposition rate. Meanwhile, pressure to reduce total manufacturing costs will increase as additional 3D-NAND and DRAM capacity comes on-line.

Safe and cost-effective process exhaust management has now become a priority and this paper presents current risk control measures and emerging concepts which can enable safe process operation with reduced total cost.

INTRODUCTION

Due to the impracticality of further scaling in planar NAND structures, future bit cost reduction will be achieved by increasing the number of layers in 3D-NAND devices. This strategy will increase both process complexity and the number of CVD and etching steps, but the need to reduce manufacturing costs will also demand corresponding cost reductions per process step. In particular, the introduction of high-flow TEOS CVD processes into high volume manufacturing (HVM) of 3D-NAND memory devices demands a strong focus on both process tool exhaust safety and sub-fab equipment reliability, and these objectives must be achieved in a cost-effective manner.

TEOS-CVD CHALLENGES IN 3D-NAND

Requirements to optimise film thickness uniformity, stress control and high deposition rate tend to drive TEOS flow rates upwards in oxide CVD processes. It is common now to find CVD processes which use 15 g/min. of TEOS or more in oxide deposition steps, whilst in the case of very thick films TEOS flow rates as high as 60 g/min. have been encountered. Such high TEOS flow rates can have a serious impact on the sub-fab equipment which supports the process tool (dry-pumps, abatement systems, exhaust pipes). Typical issues include: high powder load on the downstream equipment which may cause reliability issues in dry-pumps and abatement devices; condensation of excess TEOS precursor or process byproducts in the exhaust system; a need for high dilution gas flow rate to control fire risks when high flows of TEOS are mixed with oxidizing process gases such as oxygen, and nitrous oxide; high gas abatement cost due to very high total gas flows passing through the exhaust pipes. Note that total cost includes not only the cost of utilities consumed by the abatement system, but also its capital cost, the cost of the sub-fab footprint it occupies and high maintenance costs.

Fortunately, current state-of-the-art dry-pumps and point-of-use (POU) abatement systems are now able to deal adequately with high powder loads provided they are specified and set-up according to the manufacturer's Best Known Method (BKM). Therefore, this paper will present ideas that can enable safe and cost-effective operation of the whole process gas exhaust management system.

THERMAL MANAGEMENT SYSTEMS

It has long been understood that thermal management of exhaust systems is critical to ensure that process precursors such as TEOS and process by-products like ammonium hexafluorosilicate do not condense in the pipe-work[1]. If condensation occurs, the exhaust pipe may become blocked and cause an unscheduled tool shut-down due to high exhaust pressure. If the condensed material contains partly-reacted Si-compounds, it may burn in the exhaust pipe when exposed to fluorine gas during the chamber cleaning process. Furthermore, when the exhaust pipe is removed for cleaning, condensed F-containing material may release HF gas – this can cause severe injury to sub-fab service technicians.

However, exhaust thermal management systems are only fully effective if they are designed, installed and set-up to suit each process. It is also essential that thermal management is applied both to the exhaust pipe and the abatement inlet pipes to avoid any possibility of blockage by condensed material. The temperature profile of the exhaust pipe and abatement inlet should be set-up according to the type of process – for example, LPCVD nitride tools require pipe temperatures greater than 200°C to avoid condensation of ammonium chloride, whereas TEOS CVD processes require only 150 – 180°C to avoid blockage. Furthermore, if the thermal insulation is ineffective or badly fitted, a cold spot in the exhaust pipe may occur which can result in localized condensation[1].

Even when exhaust thermal management systems are well fitted and cold spots in pipes eliminated, any additional gas injected into the exhaust pipe (for example, nitrogen to dilute flammable process gases) will cool the exhaust gas stream and promote condensation.

Therefore it has become necessary to heat all gases which are injected into the exhaust system, including fuel or oxygen injected into the abatement system inlets to improve the destruction efficiency of the process gases as well as nitrogen dilution gas.

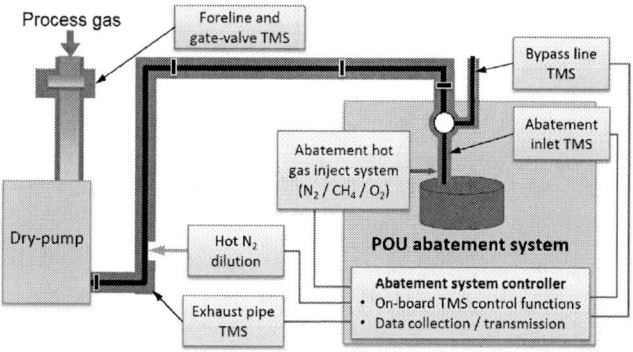

Fig. 1: Full thermal management system (TMS) applied to a process exhaust system

SMART EXHAUST DILUTION

Dilution of flammable gases is widely used as a way to avoid fires in exhaust pipes during CVD processes. The NFPA®69 standard[2] recommends that flammable gases are diluted to a concentration less than 25% of their Lower Flammable Limit (LFL) to avoid formation of flammable mixtures when mixed with oxidizing gases. Usually, the maximum capacity of the flammable gas mass flow controller (MFC) is used when calculating the flow rate of diluting gas, so when high flows of TEOS are used the resulting total nitrogen dilution flow rate can be huge. For example, consider a flow of 20 g/min. of TEOS:

– TEOS mass flow rate = 20 g/min.
– Relative vapour density of TEOS ≈ 7.19
– Density of air = 1.292 g/l (at 20°C, 1 atm.)
– TEOS flow = 20 / (7.19 x 1.292) slm = 2.15 slm
– TEOS LFL ≈ 1.3%
– N_2 flow rate ≈ 2.15 / (0.25 x 0.013) ≈ 660 slm

Such high total gas flows require corresponding high capacity abatement systems to destroy the waste process gas. These systems consume large amounts of fuel, water, PCW, and electrical power and significantly increase the total cost of ownership (TCOO) of the CVD process.

However, the application of "intelligence" to the flammable gas dilution process allows total cost to be reduced. For example, if the sub-fab equipment can access process gas flow information from the tool, the actual nitrogen dilution flow can be adjusted in real time to match the 25% LFL dilution rule. The on-going deployment of process tool control systems based on EtherCAT® (Beckhoff Automation, Verl, Germany; m.beckhoff.com) opens up opportunities to use such information to adapt sub-fab equipment operation to the actual process conditions during each process step.

Furthermore, the NFPA®69 standard does allow dilution rates of flammable gas to be relaxed to 60% of LFL if *"automatic instrumentation with safety interlocks*

is provided". Therefore, if the entire exhaust dilution system / abatement system / process tool interface is designed to be intrinsically safe and fitted with suitable safety interlocks, the introduction of "Smart" exhaust dilution systems becomes a real possibility. This approach will enable optimised nitrogen consumption and reduce the peak load on the fab nitrogen supply system. To be safe and effective, however, the exhaust dilution system needs to be integrated with the rest of the sub-fab equipment and supervised by a single system controller.

SMART POU ABATEMENT

Information from the process tool can also be used to set the POU abatement system into the most efficient operating mode to suit the process gases which flow in the exhaust during each process step.

For example, during process steps which contain low flows of easily-reacted flammable process gases with low or moderate dilution, the abatement system can be set to "low fire" mode in which fuel consumption, city water and process cooling water (PCW) consumption are all minimised. Alternatively, when flammable and oxidising gases flow together and high dilution flow must be used to mitigate fire risks, the POU abatement system can be set to "high fire" mode to ensure full destruction of all process chemicals including, for example, N_2O and NH_3.

During process steps where chamber cleaning gases which don't present a flammability risk (such as F_2, NF_3 or ClF_3) flow in the exhaust system, dilution can be reduced to a lower level or turned off completely: the reduced total flow into the POU abatement system allows these and other PFC gases to be abated with greater efficiency and better cost effectiveness than if they were heavily diluted.

Finally, when no process gas is flowing in the exhaust system, it makes complete sense to reduce the exhaust dilution flow as far as possible and set the POU abatement system into idle or "Green" mode where utilities consumption is minimised. However, the system's ability to recover rapidly to full abatement mode when the process restarts should also be taken into consideration.

Some commercial POU abatement systems already have the capability to change mode to match each process step, but as more information becomes available from the process tool and fully variable firing rate is introduced, further reductions in nitrogen, fuel gas, PCW, city water, and electrical power will become possible.

INTEGRATED SUB-FAB SYSTEMS

Many of the complex processes used in high volume 3D-NAND production contain very hazardous chemicals and sub-fab systems are now expected to operate with a high level of safety integrity. In the worst case, hazardous gas leaks from process exhaust pipes could lead to fires, equipment damage and harm to sub-fab personnel. More usually, local gas leak detectors will shut down the process to prevent further gas escape into the work environment. However, the resulting wafer scrap and loss of production capacity do result in a significant cost penalty.

In sub-fabs where individual pumps, POU abatement systems and other equipment are purchased from different suppliers and installed independently, it is frequently unclear who has "ownership" of (responsibility for) the process exhaust pipes, and this causes uncertainty about who has overall responsibility for ensuring leak integrity, correct thermal management and safe operation. This strategy can also lead to uncoordinated and inefficient equipment maintenance, which results in increased downtime and lack of continuous improvement programmes (CIPs) to improve sub-fab performance.

An attractive alternative to installation of standalone equipment from individual suppliers is to integrate dry-pumps, process exhaust pipes, point-of-use abatement units, thermal management systems, double-containment and "Smart" operation into a single unified package from a single supplier. Each integrated system is configured to support a specific process and contains all the supporting equipment to ensure safe, cost-effective and reliable operation in HVM. Integrated systems enable rapid installation and commissioning of the sub-fab equipment and allow faster fab start-up than installation of multiple stand-alone sub-fab systems. Furthermore, since the integrated system is provided by a single supplier, the responsibility for system set-up, safe operation, routine maintenance, optimisation of up-time and implementation of CIPs is completely clear to the fab operator.

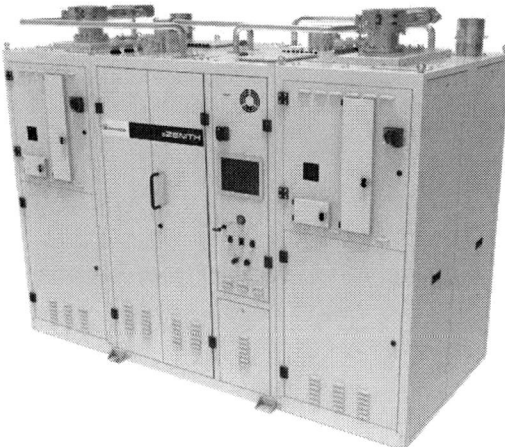

Fig.2: Commercial integrated sub-fab system including dry vacuum pumps, POU abatement and exhaust pipes

In future, integrated sub-fab systems will be built from process-proven modules, each designed with standard fixed interfaces, so configuration and manufacturing becomes highly efficient, and delivery times from the manufacturer can be made shorter to align with delivery of process tools.

CONCLUSIONS

Safe and cost-effective management of exhaust gas from CVD process tools has become an important issue, and safety especially must be maintained despite pressure to reduce memory manufacturing costs as currently planned additional manufacturing capacity comes on-line.

Whilst still at an early stage of development, and not yet widely available, introduction of "Smart" safety dilution control and "Smart" abatement system operation into HVM-optimised sub-fab systems will allow total cost of ownership to be reduced while maintaining safety at a high level. In part, this transition will be enabled by the ability to share data between process tools and sub-fab equipment as EtherCAT® becomes widely accepted.

Instead of trying to optimise the safety and performance of each piece of sub-fab equipment in isolation, it makes more sense to integrate all sub-fab equipment into a single coherent system for which safety and total cost of ownership (TCOO) are optimised. For best performance, each integrated system is configured and set-up according to the supplier's process BKM:

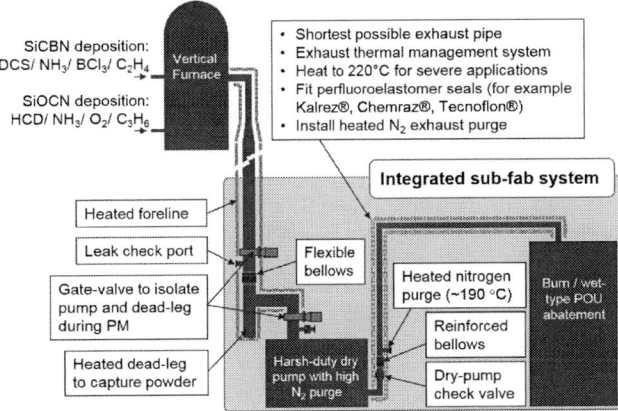

Fig.3: Example of an ALD process BKM applied to an integrated sub-fab vacuum and abatement system

One of the key benefits of integrated sub-fab systems is that they provide complete clarity of ownership and responsibility for all the sub-fab equipment. Furthermore they also ensure that sub-fab systems are designed and supported according to a well-defined and proven BKM, and can deliver safe operation, cost-effective performance and assured environmental stewardship[3].

REFERENCES

[1] C. O'Rourke, *Silicon Semiconductor,* vol.38, iss.2 2016 pp. 26-31

[2] National Fire Protection Association, *NFPA 69 Standard on Explosion Prevention Systems*

[3] A. Ifould, A. Chambers, *Solid State Technology*, July 2016, pp.32-37

HYDROGEN BASED HIGH STRAIN SI:P EPITAXY PROCESS DEVELOPMENT AND OPTIMIZATION

Jin Yang

Applied Materials China, Shanghai 201204, China

*Corresponding Author's Email: Jin_Yang@amat.com

ABSTRACT

High strain in-situ phosphorus doped silicon (called HS -Si:P) epitaxy in the source/drain (S/D) areas has been adopted for nMOS transistors for 14nm node and beyond, which can boost nMOS drive current by applying tensile strain on device channel and lower contact resistivity. N_2 carrier gas were widely used for strain benefit, however, some customers still prefer to hydrogen gas as carrier gas due to better thermal conductivity. In this paper, a H_2 based HS-Si:P process was developed in Applied Materials Centura® RP Epi chamber, various process tuning parameters including reaction temperature, process gas, carrier gas flow etc and their impacts on film thickness, strain and resistivity are discussed.

1. INTRODUCTION

As devices are shrunk aggressively, the scaling–down of transistor S/D contact area makes it more challenging to reduce the S/D parasitic resistance which becomes comparable to (or even higher than) the channel resistance itself [1] . A heavily phosphorous in-situ doped silicon epitaxial film on S/D area is crucial to reduce the parasitic resistance in nMOSFET transistors. Besides the low S/D contact resistance, 2D/3D nMOSFET transistors favor the tensile strain induced in the channel to enhance the electron channel mobility [2]. Therefore, N_2 based in-situ doped Si:P epitaxial process has been widely adopted by the leading foundry for n type source/drain epitaxy on 14nm and beyond node because of providing a higher phosphorus concentration and high tensile strain in the channel. Table. I listed the typical N_2 based HS-Si:P film characterization.

However, some customers observed lower resistivity of HS-Si:P film achieved more benefit than strain gain on their real FinFET devices on 14nm and beyond node, Hence, lower resistivity were preferred even some tensile stain loss. Hence, Hydrogen gas as carrier gas regained more attention due to better thermal conductivity. Various process parameters are explored in this paper to achieve in-situ phosphorus doped selective Epi with lower resistivity and comparable strain.

2. EXPERIMENTS

Those selective Si:P epitaxial process were performed using dichlorosilane (DCS), phosphine (PH3), and hydrochloride (HCl) gases in Applied Materials Centura® RP Epi system. Various process tuning parameters including reaction temperature, reaction gas, carrier gas flow and rotation speed were split on blanket wafers. The grown HS-Si:P epitaxial films were thus analyzed by high-resolution XRD (HR-XRD) for thickness and strain measurement, four-point Rs probe for resistivity result, and SIMS for phosphorus concentration calculation.

Table. I: Typical N_2 based HS-Si:P film characterization

Process	Strain (Eq.C_sub%)	Resistivity (mOhm-cm)	Total [P] by SIMS (at/cc)
N_2 based HS-Si:P	0.7~0.8	0.6~0.7	1.3~1.7E21

3. RESULT AND DISCUSSION

3.1 Process Temperature Effect

Fig. 1 plots out H_2 based HS-Si:P epitaxial process sensitivity to growth temperature (T-20C~T-10C, T mean process temperature used in N_2 based HS-Si:P process) in aspects of center growth rate, strain and resistivity. Although strain has been somewhat lost both growth rate and resistivity increased with increasing temperature, that mean lower resistivity could be achieved at low temperatures, but lower process temperature may have productivity concern due to lower growth rate. Hence, moderate temperature will be considered. Also compared with N_2 based HS-Si:P process, similar growth rate were achieved at 15~20 degree lower temperature on H_2 based HS-Si:P process due to hydrogen better thermal conductivity.

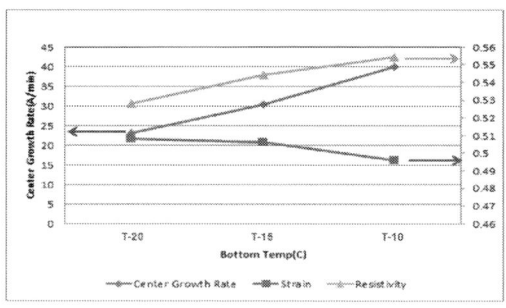

Fig. 1: H_2 based HS-Si:P process sensitivity to growth Temperature

3.2 HCl Gas Flow Effect

As known, HCl as etch gas provides the process selectivity window. Fig. 3 plots out hydrogen based HS-Si:P epitaxial process sensitivity to HCl gas flow (-10~+10 sccm split). Results show HCl gas flow has linearity relation with thickness and the tensile strain: Lower HCl gas flow, higher growth rate and better strain while similar resistivity. However, too low HCl gas will have the selectivity concern. Normally growth rate less than 40 A/min mean the selectivity acceptable.

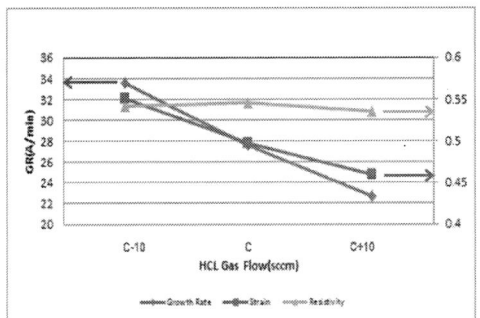

Fig. 2: H_2 based HS-Si:P process sensitivity to HCl gas flow

3.3 H_2 Gas Flow Effect

Besides reaction gas, carrier gases, either N_2 or H_2 also have great impact to epitaxy film. The tunable carrier gas flow includes main carrier gas and slit purge gas. The former is designed to carry reaction gases and purge the process chamber; the latter flows below the susceptor which helps to isolate the bottom of the chamber from the top of the chamber, and prevents silicon deposition from occurring on the bottom of the susceptor on the bottom dome.

Table.II describes the HS-Si:P film sensitivity to different main H_2 carrier gas and slit purge gas. We can see higher main H_2 gas and split purge gas flow degraded the tensile strain obviously because of lower process gas partial pressure. However, HS-Si:P film resistivity seems similar between different carrier gas flow.

Table. II: H_2 based HS-Si:P process sensitivity to H_2 gas flow

Main/Slit (slm)	Growth rate (A/min)	Strain (Eq.C_sub%)	Resistivity (mOhm-cm)
H_2=5/5	36.68	0.567	0.577
H_2=10/5	28.32	0.408	0.555
H_2=10/2.5	31.43	0.441	0.552

Fig. 3 indicates H_2 based HS-Si:P film normalized thickness data with different main carrier gas flow. Less main H_2 contributes to wafer edge thickness roll-up, which will be one of tuning knobs for thickness profile manipulation.

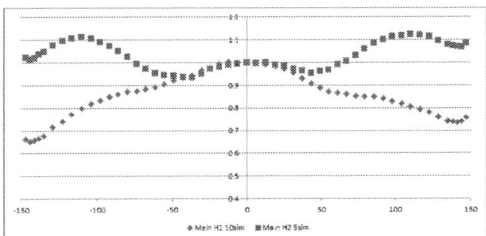

Fig. 3 HS-Si:P film normalized thickness data with different main carrier gas

3.4 Rotation Speed Effect

Rotation speed will impact gas distribution in Epi chamber under the process condition of higher pressure.

Table.III listed HS-Si:P film characterization with different rotation speed. As seen, more process gas introduced to wafer center under lower rotation speed which resulted in center higher growth rate. However, reaction gas may over-inject once rotation speed are too slower than one given value;

Table. III: H_2 based HS-Si:P process sensitivity to rotation speed

Rotation speed (rpm)	Growth rate (A/min)	Strain (Eq.C_sub%)	Resistivity (mOhm-cm)
32	30.36	0.535	0.560
24	40.62	0.613	0.551
16	36.68	0.567	0.577
8	30.53	0.512	0.538

3.5 Optimized Process Results on Blanket Wafers

After lots of process tuning iterations, we achieved H_2 based HS-Si:P with lower resistivity around 0.53 mOhm-cm. Fig. 5 describes H_2 based HS-Si:P film epitaxial characterization on (110) Si crystallographic plane substrate. The HR-XRD profile in Fig. 5(a) indicates well-ordered HS-Si:P epitaxial film growth with the tensile strain (equivalent to 0.6 at% Csub). Also SIMS profile indicates a total [P] of $\sim 1.2 \times 10^{21}$ at/cc in Fig. 5(b).

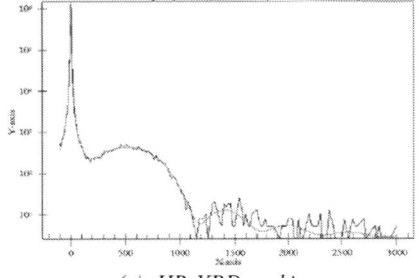

(a). HR-XRD rocking curve

978-1-5386-5309-8/18 $31.00 © 2018 IEEE

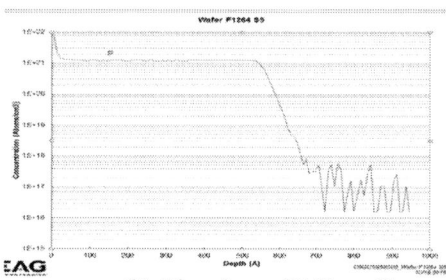

(b). Phosphorus SIMS profile
Fig. 5: H_2 based HS-Si:P film characterization

4. CONCLUSION

In this paper, a preliminary process tuning of H_2 based HS-Si:P epitaxy are accomplished on Applied Centura® RP Epi chamber. By optimizing the process parameters, we achieved the phosphorus doped silicon film with lower resistivity and similar strain compared with N_2 based HS-Si:P film. Such high quality selective HS-Si:P epitaxial films will grow on planar structure and fin structure for further optimization.

REFERENCES

[1] S. Senturia. *Proceedings of Transducers2003*, Boston, June 8-12, 2003, pp. 10-15.

[2] T. Tsuchiya, O. Tabata, J. Sakata and Y. Taga. *J. Microelectromech. Syst.*, vol. 7, 1998, pp. 106-113.

New CMP Processes Development and Challenges for 7nm and Beyond

Haigou Huang[a], Dinesh Koli[a], John H Zhang[b], Stan Tsai[b], Taifong Chao[a], Yuanfang Lu[a], Hong Jin Kim[a], Qiang Fang[a] Wenyin Lu[a]

[a] Advanced Technology Development, GLOBALFOUNDRIES, 400 Stone Break Drive, Malta, NY, USA 12020
[b] GLOBALFOUNDRIES, Albany Nanotechnology Center, 257 Fuller Road, Albany, NY 12203, USA
E-mail: huanghg@globalfoundries.com

Abstract

Either new material or new integration usually requires a new CMP process. CMP is still growing for CMOS applications. And many newer applications are now also being developed beyond "traditional" CMP. In each new application, CMP's stop-on capability, high selectivity, with-in-wafer and with-in-chip uniformity control are increasingly seen as essential elements in gauging device integrity, and thus are critical to the successful implementation of a technology. This presentation discusses oxide CMP stop on SiN and new Si-poly CMP. Poor SiN quality in oxide CMP stop on SiN causes center residues and with-in-wafer SiN uniformity issue. Center residues is strong related with incoming step height, which affects CMP slurry removal rate on pattern wafer. New slurry application is required with good stopability or new stop hardmask is implemented. In Si-poly slurry evaluation, we demonstrate that the silica-based slurry has better selectivity, dishing, and nitride loss performance than the ceria-based slurry, but the process is heavily dependent on the incoming profile and topography. In further process improvement and optimization work will continue.

Keyword: CMP, Uniformity, Selectivity, Interface

Introduction

In advanced node semiconductor manufacturing, Chemical Mechanical Polish (CMP) is a critical process step in formation of many modules, from fin formation, replacement metal gate (RMG), MOL contact formation, to interconnect. [1-2]. In each application, CMP's stop-on capability, high selectivity, with-in-wafer and with-in-chip uniformity are increasingly seen as essential elements in gauging device integrity, and thus are critical to the successful implementation of a technology [4-5]. In order to enable new integration schemes, CMP steps doubled from 28nm to 10/7nm as shown in Figure 1. Therefore, CMP is becoming complex process.

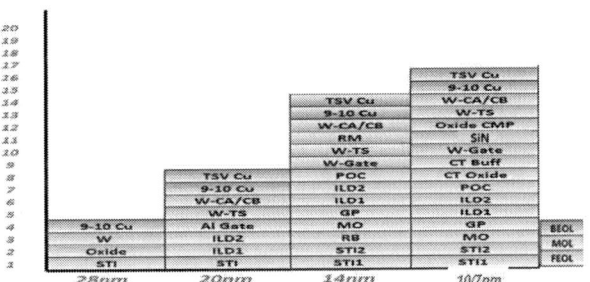

Figure 1 CMP is becoming complex.

Results and Discussion

Oxide CMP stop On SiN Development and Challenge

This process is similar with STI CMP, polishing oxide and stopping on the SiN. Figure 2 shows the simple cartoon stack. The main difference point from STI CMP is the SiN quality. It has the thermal budget limitation in this step due to the device performance concerns. The low quality SiN gives CMP significantly challenge on process control (such as SiN loss, with-in-wafer uniformity, with-in-die uniformity and process window).

Process window narrow is observed on oxide CMP stop on SiN. Figure 3 shows the post SiN thickness chart and OPI images. Center SiN thickness outfly is observed and OPI image show center underpolish residues issue. Further investigation found the oxide residues is strong correlation with head. Head 1 shows clean, without residues. And residues are dedicated on head2 and head3. Those heads are closed to life time and head 1 is in the middle of life time. The residues issue is possible related with head process margin.

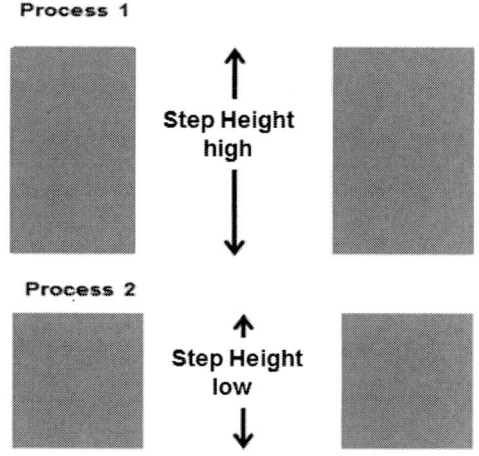

Figure 2 Simple stacks in Oxide CMP Stop on SiN.

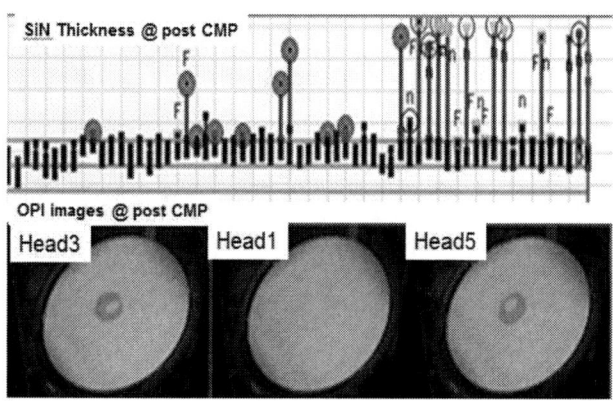

Figure 3 Inline thickness chart and OPI macro image @ post oxide CMP stop on SiN.

But there are two similar oxide CMP stop on SiN processes (as shown in Figure2) in the same tool. Only process 1 has the center residues issue on head2 and head3, and process 2 is not residues on all the heads. It indicates that there has other fact affected on the residues performance. Figure 2 also compared the step height on process 1 and process 2. They are with different step height. It seems step height is a key for the center residues issue. Further experiment splits are explored on different step height pattern wafer to check the removal rate of slurry A. The results shows on Figure 4, indicate the removal rate of slurry A is strong dependent on the step height. The step height as 0A is with maximum removal rate and when the step height reaches to 10X A, the removal rate is close to 0A. Based on the finding, we can explain why there is with the residues on process 1 rather than on process 2. Step height of process 1 is high, which is close to the limitation of slurry A, indicates less process window on it comparing with on process 2.

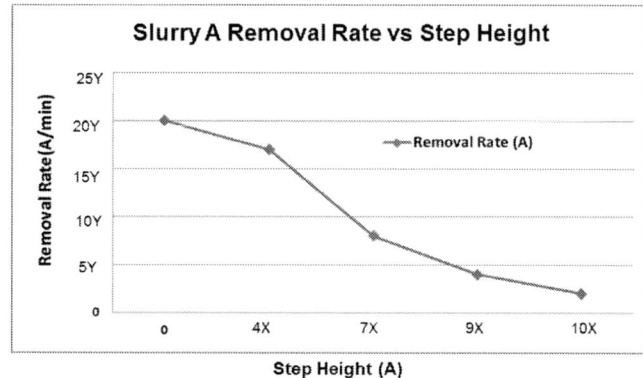

Figure4. Slurry A removal rate vs step height evolution.

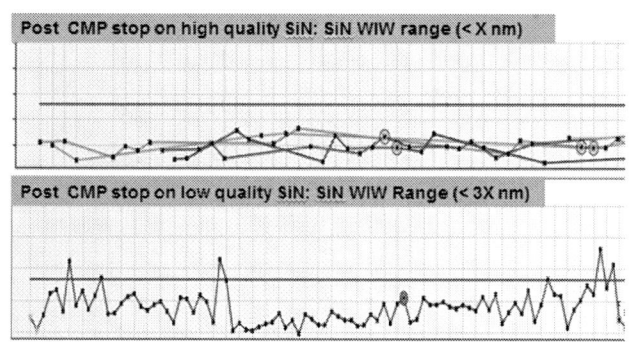

Figure 5 WIW SiN uniformity comparison between oxide CMP stop on SiN with high and low SiN quality.

Poor SiN quality in oxide CMP stop on SiN will also cause another issue. It will be challenged on CMP shopability, which will effect on the with-in-wafer and with-in-die uniformity. Figure 5 shows the comparison on the WIW uniformity between high and low quality oxide CMP. The results indicate there has huge different, low quality SiN is above 3 times worse than high quality SiN. Higher SiN loss is normal with worse sigma performance. It required new slurry application, which is good stopability or new stop hardmask implemented.

Poly CMP Development and Challenge

978-1-5386-5309-8/18 $31.00 © 2018 IEEE   322

Polysilicon is a highly critical process for tighter gate height control, at both wafer level and within-die level, with improved uniformity. Uniform gate height not only determines the transistor performance but also impacts downstream processes. As the nodes are shrinking, the contact space becomes narrower, where historical gapfill materials such as TEOS cannot be etched by conventional RIE methods without causing significant damage to contacts/junctions. The aim of this work is to understand if polysilicon can be used as a gapfill material to exploit superior etching capabilities of polysilicon in narrow structures compared to TEOS. To enable this structure, CMP plays critical role. Figure 6 shows the simple cartoon of new ploy CMP.

The two main requirements for testing CMP slurry are selectivity and topography. Slurries with good selectivity and tunable characteristics can be applied to different applications. This process requires the high poly selective to stop on SiN hardmask. This had been proved to be very challenge [6-7].

Batch of slurry are evaluated list in Table1 and hard and soft pad properties in Table2. The first research and development (R&D) ceria slurry with hard pad show high frictions at the interface. Wafer slipped out and caused the retain ring damage as shown in Figure7. The second R&D ceria slurry with hard pad show high macro to macro variation (about above 60% delta between dense structure and CMP house) and poor dishing performance as shown in Figure 8 and 9. The second R&D ceria slurry with soft pad is enable to improve within-die SiN uniformity significantly, but dishing performance is still worse. Two main facts will be affected on the dishing: incoming overburden and slurry properties with less protecting on recessing location. Overburden splits show that dishing is not obvious changed even three time overburden as shown in Figure 10. The third R&D ceria slurry with soft pad can achieve comparable within die SiN uniformity with good dish performance and without color variation as shown in Figure 11.

The silica-based slurry is also explored since the silica-based slurry has a better selectivity and capability to stop on nitride based on the blanket wafer results as shown in Table 3. Also, a soft pad and hard pad were tested to understand the dependency of the pad hardness on the selectivity of the silica-based slurry. Figure 12 shows that the SiN losses for the silica-based slurry CMP are quite similar for both pad types but the removal rate for the soft pad decreases 50% dramatically. The silica-based slurry with hard pad can meet Si-poly removal rate target. Patterned wafers with polysilicon gates were used to understand how the silica-based with hard pad will perform on real product wafers. Figure 13 shows the within-die SiN uniformity

cross dense and CMP house. It is better than Ceria-based slurry performance. Also, SiN loss is better than Ceria-based slurry, which is consistent with blanket wafer. Dishing performance of Silica-based with hard pad is also better than Ceria-based as shown in Figure 13. So far, the silica-based slurry was proven to have better selectivity, dishing, and nitride loss performance than the ceria-based slurry, but the process is heavily dependent on the incoming profile and topography.

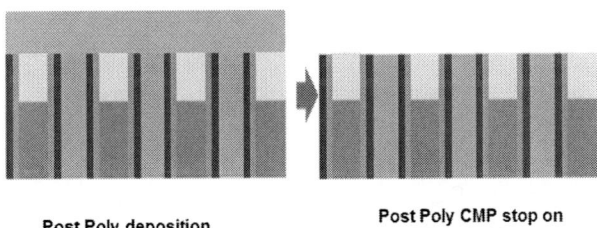

Figure 6 Pre and post of new Si-poly CMP.

Table1: The summary of the evaluation results.

Slurry Type and Pad	High level summary of the evaluation
Ceria-based -1 with hard pad	High friction at the interface and wafer slipped
Ceria-based-2 with hard pad	Hugh macro to macro variation and high dishing (6.5Y)
Ceria-based-2 with soft pad	Good with-in-die uniformity and high dishing (6.5Y)
Ceria-based-3 with soft pad	Good with-in-die uniformity and dishing significant improvement (1.5Y)
Silica-based-1 with soft pad	Low poly-Si removal rate
Silica-based-1 with hard pad	Good with-in-die uniformity and good dishing (Y)

Table2: Hard and Soft pad properties comparison.

	Hard pad	Soft pad
Specific Gravity (g/cm3)	0.8-1.0	0.82-1.0
Hardness	59-63	50-55
Material	Polyurethane	Polyurethane

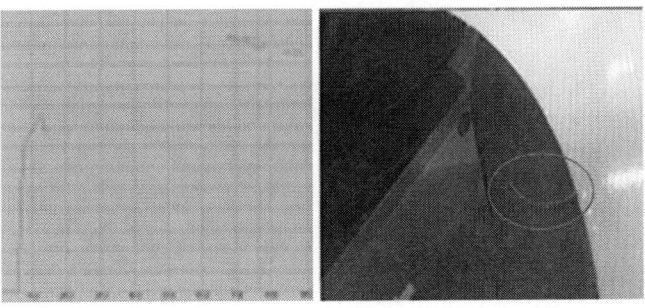

Figure 7 Ceria-based-1 slurry with hard pad test with damaged retain ring (RR).

978-1-5386-5309-8/18 $31.00 © 2018 IEEE

SiN Remaining Thickness Cross Macro

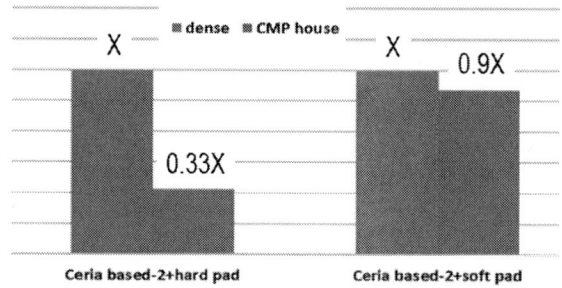

Figure 8 With-in-die SiN uniformity comparison on Ceria based-2 slurry with hard and soft pad.

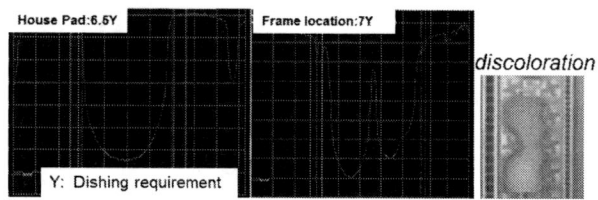

Figure 9 High dishing on CMP house pad and frame location with Ceria based-2 slurry.

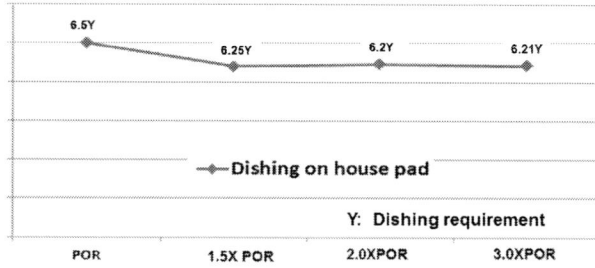

Figure 10 Dishing of CMP house pad with Si-poly overburden increasing on Ceria based-2 slurry with hard pad.

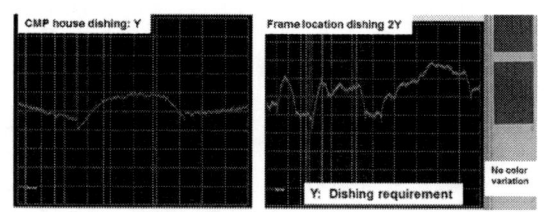

Figure 11 Dishing on CMP house pad and frame location on Ceria based-3 slurry with hard pad.

Table3: Removal rate of hard pad for ceria- and silica-based slurry.

	Poly silicon	SiN
Ceria-based slurry (30 s)	0.87 A	3.6 B
Silica-based slurry (60 s)	A	B

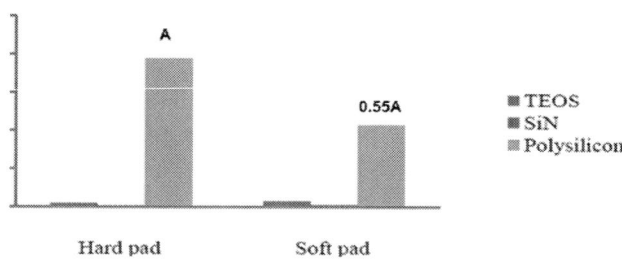

Figure 12 Removal rates of silica-based slurry polished by a hard pad and a soft pad.

Silica-based Slurry SiN uniformity cross Macro

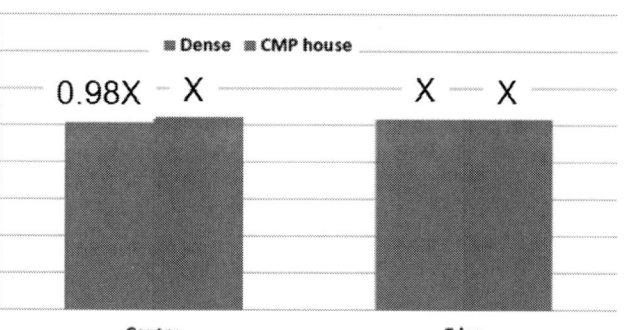

Figure 13 Silica-based slurry on hard pad SiN uniformity cross macro.

Silica-based slurry with hard pad: dishing vs overpolish

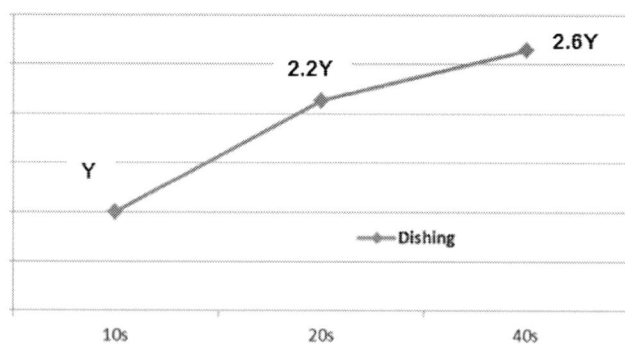

Figure 14 Silica-based slurry on hard pad: dishing vs overpolish.

Conclusion

This presentation discusses oxide CMP stop on SiN and new Si-poly CMP. Poor SiN quality in oxide CMP stop on SiN causes center residues and with-in-wafer SiN uniformity issue. Center residues is strong related with incoming step height, which affects CMP slurry removal rate on pattern wafer. New slurry application is required with good stopability or new stop hardmask is implemented. In Si-poly slurry evaluation, we demonstrate that the silica-based slurry has better selectivity, dishing, and nitride loss performance than the ceria-based slurry, but the process is heavily dependent on the incoming profile and topography. In further process improvement and optimization work will continue.

Acknowledgement

Thanks for helpful discussions from many colleagues of Advanced Module Engineering (AME) and Advanced Technology Development (ATD) teams in GLOBALFOUNDRIES.

References

[1] K. Ahmed and K. Schuegraf, "Transistor wars: Rival architectures face off in a bid to keep Moore's Law alive", IEEE Spectrum, Nov. p.50, 2011.
[2 K. Seo, B. Haran, D. Gupta, D. Guo, T. Standaert, R. Xie, H. Shang, E. Alptekin, D. Bae, G. Bae, "A 10 nm platform technology for low power and high performance application featuring FINFET devices with multi workfunction gate stack on bulk and SOI, Dig. Tech. Papers VLSIT (2014) 1–2.
[3] S. W. King, "Dielectric barrier, etch stop, and metal capping materials for state of the art and beyond metal interconnects", ECS J. Solid State Sci. Technol.4, p3029, 2015
[4] J.H. Han, et al, "A Study on Selectivity during SiN Chemical Mechanical Polishing for Sub-10nm Logic Device" International Conference on Planarization/CMP Technology (ICPT 2016), Beijing, China. Oct 17-19, 2016.
[5] S Tsai, H Amanapu, R Xie, J Zhang, K Chung, C Labelle, H Huang, et al. ," CMP: Consideration of Stop-on Selectivity in Advanced Node Semiconductor Manufacturing Technology". ECS Transactions 77 (4), 169-177, 2017.
[6] JH Zhang, H Huang, AM Greene, R Xie, SC Seo, P Montanini, WT Tseng, "CMP Challenges for Advanced Technology Nodes", MRS Advances, 1-12, 2017.

[7] TF Chao, D Penigalapati, JIC Yang, H Huang, DR Koli, "The Role of Polysilicon Slurry and Its Application in 7nm CMP" ECS Transactions 77 (5), 227-233.

Corresponding Author:

Haigou Huang
Tel: +1 518-305-6035
Fax: +1 518-305-6178
E-mail: huanghg@globalfoundries.com
Advanced Module Engineering, Globalfoundries
400 Stone Break Rd. Ext. Malta, NY 12020 USA

STI SCRATCH DEFECTS REDUCTION BY USING SOLID PAD IN 1X TECHNOLOGY NODE

Kuang-Wei Chen[1] *, Yen-Ting Chen[1] , Chun-Fu Chen[1], Tuung Luoh[1], Lin-Wuu Yang[1], Ta-Hone Yang[1], Kuang-Chao Chen[1], Yuchi Tay[2], Yi Ming Chen[2], Johnson Ton[2] and Yen Pin Lu[2]

[1]Macronix International Co., Ltd., Technology Development Center, Hsinchu, 30078, Taiwan ROC
[2]Thomas West Inc. 470 Mercury Dr.Sunnyvale, CA94085
* Email: conwaychen@mxic.com.tw

ABSTRACT

CMP scratch defects reduction plays an important role in yield improvement. In this paper, pad debris reduction and solid pad evaluation are applied to reduce scratch formation. Novel solid pad with high removal rate and less dressing amount demonstrate the good performance of scratch defects generation and get the better productivity. As compared to POR fully dressing on porous pad, scratch defect density greatly reduces 65% by using partial dressing on solid pad.

Keywords—scratch; solid pad; dressing; down force; pad debris

INTRODUCTION

Chemical mechanical planarization (CMP) process consists of chemical and mechanical interactions between the wafer and polishing pad with a slurry; the wafer surface is reacted chemically and softened with slurry and the wafer material can be removed by the mechanical action of abrasive particles in the slurry. Based on this mechanical interaction, the scratches formation is inevitable, and it affects device performance severely as device shrunk to 1x to 2x technology node [1]. Several scratches formation mechanisms are proposed and discussed; such as agglomeration of abrasive particle [2], pad conditioning [3], polishing pad debris, slurry byproducts/residue, conditioning methods [4], etc. Pad conditioning with a diamond conditioner is to keep the pad performance consistent through a wide range of CMP wafer to wafer runs. However, the pad conditioning process is the wearing of pads, and it may generate the sources of contamination and scratch formation [5]. Appropriately pad conditioner is required to maintain the steady CMP uniformity; however, it can also increase the risk of pad debris and induce scratch issue [6-7]. The aim of this paper is to demonstrate the pad debris induced scratches issue, and propose a solid pad polishing process, which have the benefit of scratch reduction with shorter dressing process.

Results and Discussion

Scratch Reduction in POR Porous Pad

Figure 1 shows the current POR STI CMP defect morphology. These kinds of scratches defect will induce Poly floating gate damage and further leads to WL short to Si substrate in subsequent WL formation process, then results in programing failure of 1x NAND products finally.

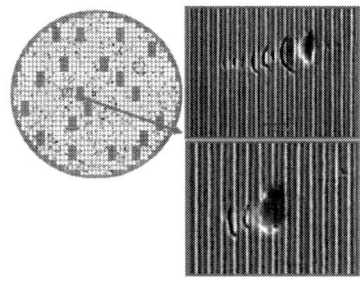

Figure 1: POR CMP scratch morphology

The scratch defects size and density are dependent on the level of friction stresses applied on interface between wafer and pad interface; hence, reducing mechanical polishing down force is feasible way for the scratch defects reduction. As shown in Figure 2, changing the down force of over-polishing step from POR LDF (low down force/2psi) to optimized- LLDF (low/low down force/1psi) can improve the scratch defect density 30% above effectively, as shown in Figure 2.

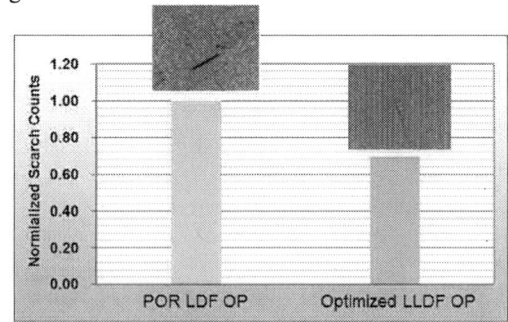

Figure 2: Normalized scratch count reduced with optimized LLDF OP process

Generally, pad conditioner will generate pad debris, and it plays an important role on scratch formation. In order to resolve the scratch issue enhanced by pad debris, the POR in-situ 100% fully dressing is reduced. The scratch defect density can further improve 40%~50%, and it indicates that the scratch defect reduces with decreasing the dressing time. Although the partial conditioner dressing can achieve better scratch defect performance, the removal rate of LLDF over-polishing drops around 50% obviously as shown in Figure 3. This unstable oxide removal behavior of requires longer polishing process time to cover oxide residue during STI CMP process.

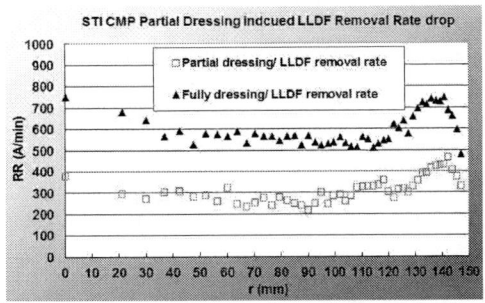

Figure 3: Removal rate drop in partial dressing over-polishing step

Scratch Defect Reduction by Implementing Solid Pad and Optimized PAD conditioner

As shown in Table I, novel solid pad is made by polyurethanes material which is similar to conventional porous one. Novel pore-free solid pad is not only designed for lower defect level but also got shorter conditioner dressing benefits; it can reduce the scratches sources of trapping polishing residues and pad debris dramatically. Furthermore, solid pad can trap the abrasives by short-time conditioner dressing, which can also reduce the generation of pad debris during longer conditioner dressing process.

Traditional pads trap slurry abrasives with pores, and conditioner dressing can keep pores fresh to achieve stable removal rate. Since the solid pad traps slurry abrasives with surface roughness, over-dressing may disturbs the performance of slurry abrasive trapping on the solid pad and then reduces the HDP removal rate.

Figure 4 summarizes the HDP removal rate on solid pad polished with CeO_2 slurry. Solid pad can achieve less in-situ dressing time but have higher HDP oxide removal rate. Hence, it has benefit of reducing debris during polishing and leads to less scratch defect. In addition, reducing the dressing time seems to result in larger WIW range on solid pad. We further optimize the removal rate profile with multi-zone pressure tuning, and the tunable performance of removal rate profile on solid pad is demonstrated as shown in Figure 5. Figure 6

and Figure 7 show the STI step-height monitor and STI profile comparison on 1x NAND flash. Novel solid pad can reveal comparable WIW range and it also achieves similar dishing performance as compared to POR porous pad.

Furthermore, in the partial dressing STI CMP process, the over-polishing step without dressing exhibits higher removal rate with solid pad and requires shorter over-polishing time, which also decreases the probability of scratch formation. It leads to 60% scratch improvement, as shown in Figure 8.

TABLE I. THE PROPERTIES COMPARISON BETWEEN NOVEL SOLID PAD AND POROUS PAD

	Solid Pad	Porous Pad
Material	Solid Polyurethane	Foamed Polyurethane
Top Pad Hardness	48	48
SG(g/cm²)	1.02	0.75
Top View		
Cross-section View		

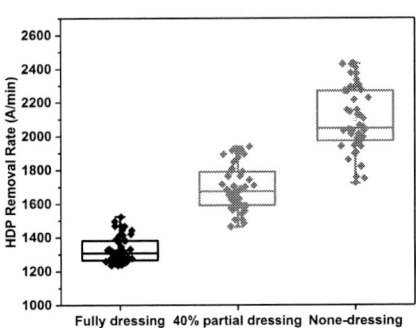

Figure 4: HDP oxide removal rate with different dressing splits.

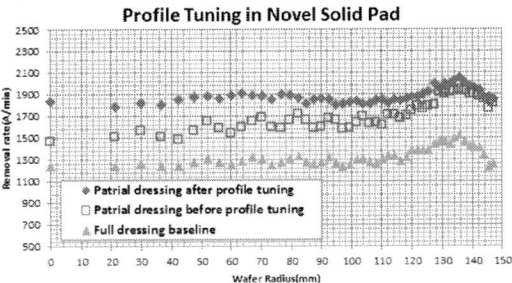

Figure 5: Profile tuning on Novel Solid Pad

978-1-5386-5309-8/18 $31.00 © 2018 IEEE

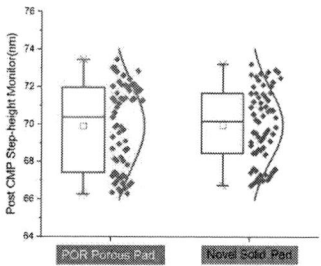

Figure 6: In-line step-height comparison between POR Porous pad and novel Solid pad

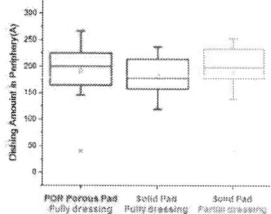

Figure 7: STI dishing amount comparison between POR Porous pad and novel Solid pad

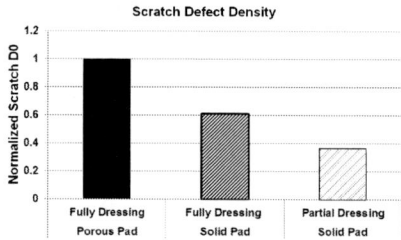

Figure 8: Comparison of scratch performance between POR Porous pad and Solid pad

Proposed Mechanism of Scratch Reduction

Novel pore-free solid pad provides flat surface with less pad debris stuffing, and then reduce the contamination and debris sources for scratch formation. Based on novel solid pad characteristics, partial dressing implemented on solid pad can achieve the comparable or higher oxide removal rate which shortens the CMP process time with less debris formation; thus the scratch defects generated by pad debris reduce significantly, as shown in Figure 9.

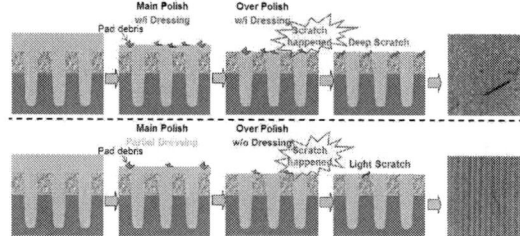

Figure 9: Partial dressing decreases scratch formation probability.

CONCLUSION

Traditional porous pad STI CMP process suffers serious scratch defects which induced by high down force polishing and full conditioner dressing. The longer the dressing time applied will exhibit higher pad debris stuffing inside the pores of pad, and lead to scratch formation. Down force reduction cannot resolve the scratch defect issue totally; furthermore, partial dressing will get the drawback of unstable oxide removal and longer process time.

Novel Solid pad with partial dressing can exhibit 65% scratch reduction than that on porous pad with fully dressing. The mechanism is reducing the pad debris number and probability for scratch formation. Higher HDP removal rate on solid pad with less in-situ dressing amount can improve productivity and scratch performance simultaneously.

REFERENCES

[1] Tae-Young Kwon, , "Scratch formation and its mechanism in chemical mechanical planarization (CMP)", *Friction*, vol. 1, issue 4, pp. 279–305, 2013.

[2] Hong Jin Kim, Tae Hoon Lee, Anne Ryan, "Material Removal Behavior of Nano-sized Cerium Hydroxide Abrasive Slurry for Chemical Mechanical Polishing" *ECS Transactions*, 72 (18) pp. 27-36, 2016.

[3] Tae-Young Kwon, Manivannan Ramachandran, Byoung-Jun Cho Ahmed A. Busnaina, Jin-Goo Park, "The impact of diamond conditioners on scratch formation during chemical mechanical planarization (CMP) of silicon dioxide" *Tribology International*, vol. 67, pp. 272–277, 2013.

[4] In-Ha Sung, Hong Jin Kim and Chang Dong Yeo, "First Observation on the feasibility of scratch formation by pad-particle mixture in CMP process" *Applied Surface Science*, vol. 258, pp. 8298–8306, 2012.

[5] Yang, Ji, "Experimental Evaluation of the Effect of Pad Debris Size on Microscratches during CMP Process", *Journal of Electronic Materials*, vol. 42 issue 1, pp. 97-102., 2013.

[6] Y. Nagendra Prasad, "Generation of Pad Debris during Oxide CMP Process and Its Role in Scratch Formation" *J. Electrochem. Soc.*, vol. 158, issue 4, H394-H400, 2011.

[7] Ji Chul Yang, "Effects of diamond size of CMP conditioner on wafer removal rates and defects for solid (non-porous) CMP pad with micro-holes", *International Journal of Machine Tools & Manufacture* vol. 50, pp. 860–868, 2010.

LOW SCRATCH HIGH THROUGHPUT AUTO STOP SLURRY

Jinfeng Wang[*], Juyeon (Jay) Chang, and Jeongdeog (Jeidy) Koh
Cabot Microelectronics Corporation, Aurora, IL 60504, USA
*Corresponding Author's Email: Jinfeng_wang@cabotcmp.com

ABSTRACT

CMP (chemical mechanical planarization) applications on both DRAM and 3D NAND have gained a lot of attentions. As one of the most important processes for high step height pattern, CMP has challenges to meet all the requirements. For high step height pattern CMP, high removal rate is required when high step height is necessary while low removal rate is required when step height is low. Meanwhile the defect is also critical. In other words, the CMP process is required to have high removal rate and good planarization efficiency (PE) with good trench protection and good defectivity performance. Here we will introduce a slurry with great self-stop features necessary for high step height DRAM and 3D NAND applications. The slurry is proven to meet high step height reduction rate, good trench protection and great throughput.

INTRODUCTION

As chip size continues to shirk, new chip design and chip manufacture process are necessary. For the memory market, 3D NAND and DRAM are designed to overcome physical constraints to enable larger memory space. New chip designs introduce some challenges for CMP processes at the same time.

For high aspect ratio of DRAM and 3D NAND designed patterns, the initial step height is over 1um. For 3D NAND design, the initial step height is even higher [1]. To have great final planarization and cost of ownership, high step height reduction rate is necessary. At the same time, high standard trench protection is required after the CMP process. Low scratch count is another necessary consideration for high volume mass production.

Historically, auto stop slurry used acrylic-acid based chemicals, consequently exhibiting low removal rates due to the electrostatic repulsion between the particle and oxide surface. Thus, to achieve high removal rate, the particle solid and down force for polish need to be high. However, high defect count (especially scratch count) follows as a necessity from these conditions.

Here we will introduce a wet based ceria slurry with great auto stop capability. Two packs of the slurry are used for auto stop applications. The A pack has wet processed ceria particles and the B pack has self-stop control agent and topo control agent. When step height is high, the local high down force takes place to remove self-stop agent from ceria particles, resulting in oxide removal at the same rate as high rate ceria slurries. But, as the step height becomes low, the rate drops with the protection of the self-stop agent and topo-control agent. With a two pack system, we can tune the slurry with different mixing ratios to have desired removal rate and planarization efficiency (PE) performance.

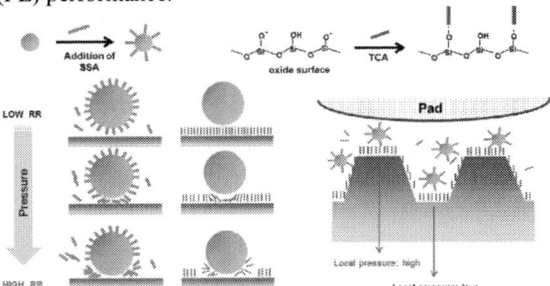

Figure 1. Auto stop slurry polishing mechanism

REMOVAL RATE AND THROUGHPUT

In our study, if the initial step height is about 1um, TEOS removal rate can be as high as traditional high rate ceria slurry. But comparing to traditional high rate ceria slurries, auto stop slurry needs to have great trench protection. With the ceria based slurry, the wafer also has defect reduction benefits, especially scratch reduction.

Table 1. CMC iDIEL[TM] D7400 (A pack 1:6 dilution with DI water) + D7422B-02 (B pack) performance comparison (Reflexion, Saesol DS8051, IC1010, 3psi & 93/87rpm, 250ml/min)

Slurry mixing ratio	Blanket rate (Å/min)	Pattern rate (Å/min)	Polish time for 1um step height reduction	Trench loss (Å)
1:6:0	9000	11400	53	>1000
1:6:2	670	11400	53	250
1:6:3	500	9000	67	<50

Comparing high ratio ceria slurry (A pack only) and auto stop slurries, similar pattern removal rate is observed while the blanket removal rate is significantly different. Auto stop slurry has a much lower removal rate on blanket wafers. On pattern wafers, the removal rate is very close between traditional high rate ceria slurry and auto stop slurry. Although exact reason is not clear, we assume that such a high removal rate of auto stop slurry on pattern is attributed to its relatively higher pH compared to traditional high rate ceria slurry. Oxide polishing generally requires OH⁻ ions, and thus leads to higher oxide polish rates with increasing pH [5]. This could be

978-1-5386-5309-8/18 $31.00 © 2018 IEEE

especially enhanced under higher local pressure due to topography, since the oxide surface needs certain amount of pressure to be hydrated by the diffusion of OH⁻ ion [5]. On the other hand, the auto stop slurry has a low polish rate on a blanket wafer due to lower local pressure compared to a pattern wafer. A low removal rate on a blanket wafer helps on the trench protection when step height is low. The removal rate behavior helps PE performance of a pattern wafer.

For high step height pattern planarization, two steps polishing is recommended. The first step with high concentration of A pack to have high removal step height reduction. And the second step with high concentration of B pack to have good trench protection. With this process, the CMP process satisfied both throughput and trench protection purposes.

AUTO STOP PERFORMANCE

For auto stop slurry, another factor is the auto stop behavior when step height is low. The final topo and within chip uniformity are very important. For traditional slurries, the removal rate is consistent during polishing. The removal does not have slow down even when at low step height features. For our auto stop slurry, we notice that the removal rate is low when step height is low.

Figure 2. Auto stop slurry removal active oxide and trench loss comparison at different polishing time. (Reflexion, E6088-15CF, DS8051, 3psi, 93/87rpm, 250ml/min)

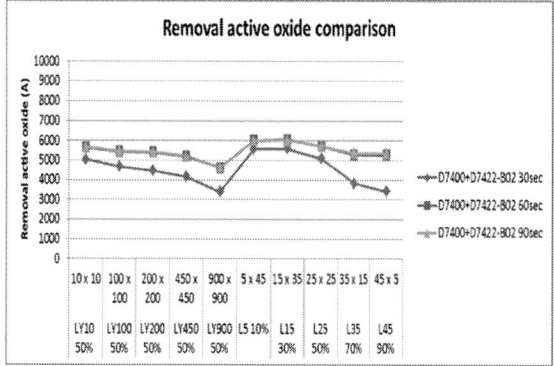

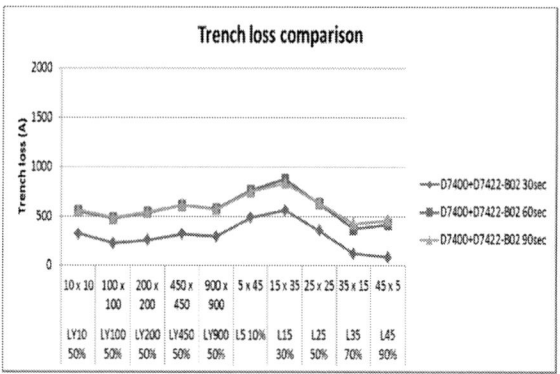

Not even the removal rate is slow, trench loss is observed to be much lower than traditional high rate ceria slurries.

DEFECT

For defect performance, since auto stop slurry iDIEL™ D7400 +D7422B-02 is using wet process ceria particles, the defect and scratch performance are much better than traditional fumed silica and calcined ceria slurries.

Figure 3. Defect and scratch comparison between fumed silica slurry, calcined ceria slurry and wet processed ceria slurry.

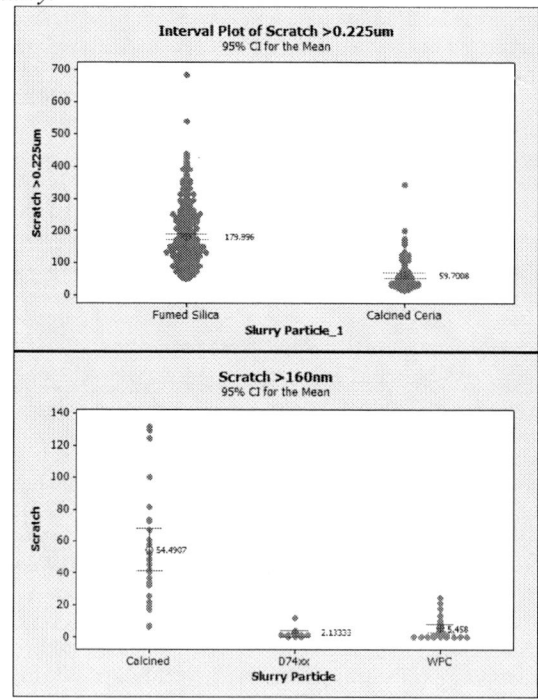

With low solid percent of wet process ceria based slurry, scratch performance is much better than high solid percent of self-stop slurries, calcined ceria slurries and other fumed silica based slurries. With ceria particle inherited surface charge property at high PH level, high down force is not needed to achieve high step height reduction.

CONCLUSIONS

CMC iDIEL[TM] D7400 (A pack) and D7422B-02 (B pack) slurry exhibits high step height reduction, great trench protection and excellent defect performance. The slurry established great performance on high aspect ratio DRAM and 3D NAND device applications.

REFERENCES

[1] K. Parat, C. Dennison, *IEEE International Electron Devices Meeting (IDEDM)*, 2015, pp. 48-51.

[2] G. Hong, D. Lee, N. Kim, K, Kang, *US 8314028 B2*, 2012.

[3] T. Parj, I. Lee, B. Choi, *US 20080121839 A1*, 2008.

[4] J. Ahn, J. Park, S. Kim, E. Jeong, D. Han, H. Park, K. Baek, T. Lee, *US 20110045741 A1*, 2011.

[5] M. Krishnan, J. W. Nalaskowski, L. M. Cook, *Chem. Rev.* 2010, **110**, pp. 178–204.

Characterization of Lanthanide Elements Doped Ceria Nanoparticles and Its Performance in Chemical Mechanical Polishing as Novel Abrasive Particles

Jie Cheng [1], Yang Li [2], Xinchun Lu [1]*

[1] State Key Lab of Tribology, Tsinghua University
Tianjin Hwatsing Technology Company Limited
[2] Department of Chemistry, Tsinghua University
* xclu@tsinghua.edu.cn

Biography

Jie Cheng received a B.S. degree in mechanical engineering from Shandong University (in 2011) and a Ph.D. degree in tribology from Tsinghua University (in 2016) where she got the Outstanding Doctoral Thesis Prize of Tsinghua University. The Ph.D. project was chemical mechanical polishing (CMP) mechanism of Cu interconnect for the BEOL application. Jie went to KTH Royal Institute of Technology as a visiting scholar (in 2015) to conduct relevant research on bimetallic micro-galvanic corrosion. Jie works as research associate in the State Key Lab of Tribology, Tsinghua University since Sep. 2016. Now Jie is in charge of a National Natural Science Foundation of China (No. 51705278).

Jie has gained international awards such as Young Researcher Award on International Conference on Planarization/CMP Technology (ICPT) in 2014, Best Student Paper Award and Young Engineer Award on China Semiconductor Technology International Conference (CSTIC) in 2016 and 2018, respectively. Also, Jie is the reviewer of international journals such as Applied Surface Science and ECS Journal of Solid State Science and Technology. In 2016, Jie was recognized as the Outstanding Reviewer of Applied Surface Science. Till now Jie has published 14 scientific articles. Her book titled *Research on Chemical Mechanical Polishing Mechanism of Novel Diffusion Barrier Ru for Cu Interconnect* has been published as the Recognizing Outstanding Ph.D. Research of Springer Thesis.

Abstract

The conventional modification method is mainly to tune the mechanical wear of CeO_2 abrasive particles in CMP, but the chemical reaction of CeO_2 is unique, which is rarely reported. In this paper, cerium nanoparticles doped with lanthanide (La, Nd and Yb) were modified by incipient-wetness impregnation method. In this way, the percentage of Ce^{3+} increased in CeO_2 nanoparticles, which improved the chemical reaction activity of CeO_2, resulting in higher removal rate of dielectric material in alkaline slurry and better surface quality.

Keywords—CMP; ceria; doping; chemical reaction reactivity; ytterbium; lanthanide; neodymium

Introduction

Chemical mechanical polishing (CMP) is the critical method to planarize the structures to remove topography and create a planar surface for the next level metallization. As one of the most widely used abrasive particles in CMP, ceria (CeO_2) has been used in traditional integrated circuit (IC) media polishing process, such as inter-level dielectric (ILD) and shallow trench isolation CMP [1]. Ceria -based slurry is cable of achieving a high ratio of oxide material removal rate to nitride, which is critical to minimize the loss of nitride in the active region in STI. Therefore, in the process of preparing advanced IC chip, the application of cerium based paste in the process of medium polishing is emphasized. Consequently, there is a continuous emphasis on the investigation of ceria-based slurry during dielectric polishing process in the fabrication of advanced IC chips.

It is generally assumed that CeO_2 has good affinity for silicon oxides, which helps to break the chemical bonds on the SiO_2 surface. The SiO_2 surface is terminated with Si-OH, while the surface of CeO_2 is Ce-OH. CeO_2 will react with SiO_2 to form the Ce-O-Si bound. The Ce-O-Si bound is stronger than the Si-O-Si bound, and the surface layer of SiO_2 will be removed under the mechanical abrasion function. In order to improve the performance of CMP, some methods can realize the surface modification of CeO_2 particles, such as the surface modification by organic functional group and synthesizing the CeO_2 core-shell particles. Ce^{3+} is the reactive ion that functions during the material removal of SiO_2, the improvement of which could help to increase the material removal rate of dielectric material [2].

In this paper, three lanthanide elements (La, Nd, Yb) were selected in the surface doping of CeO_2. The method is an ideal method to realize surface doping and does not change the intrinsic crystal structure of CeO_2. The results of this paper will provide a new method for improving the performance of CeO_2 abrasive particles in CMP applications.

Experimental

Ceria nanoparticles (99.95%, 20-50 nm) were purchased from Aladdin. $La(NO_3)_3 \cdot 6H_2O$ (99.99%) and $Yb(NO_3)_3 \cdot 5H_2O$ (99.9%) were purchased from Aladdin and $Nd(NO_3)_3 \cdot 6H_2O$ (99.9%) was from MACKLIN. The load of dopant was set to 1%-10%. The doped ceria nanoparticles were prepared by modified incipient-wetness impregnation method. The calcination temperature was from 300 °C to 500 °C. The heating rate was 5 °C/min-10°C/min.

The particle morphology was observed using transmission

electron microscopy (TEM, JEOL) and scanning electron microscopy (SEM, S-5500, Hitachi). The lattice structure was analyzed by X-Ray diffraction (XRD, Bruker). CMP experiments were conducted by polishing a 2 inch SiO_2 wafer on a polishing machine (Universal-150, Hwatsing). The polishing pad used was politex from Dow Chemical. The material removal rate (MRR) of SiO_2 was calculated by film thickness gauge (Resmap, Creative Design Engineering). The topography of polished surface was measured by white light surface profilometer (Talysurf PGI, Taylor-hobson).

Results and discussions

According to the pyrolysis experimental results reported previously, the decomposition temperature of nitrates is ca. 400 °C at which $La(NO_3)_3 \cdot 6H_2O$, $Nd(NO_3)_3 \cdot 6H_2O$ and $Yb(NO_3)_3 \cdot 5H_2O$ will fully decomposes. The XRD patterns of the pure ceria nanoparticles calcined at 350 °C, 400 °C and 500 °C are shown in Fig. 1. The crystal structures of the M-doped ceria (M= La, Nd and Yb) are shown in Fig. 1. It is obvious that the calcination temperature ranging from 350-500 °C will not change the crystalline structure of CeO_2. Figure 2 shows the (111) lattice plane of the doped CeO_2 at different temperature and different dopant load. The M-doped ceria nano-particles maintain the typical fluorite structure. When the calcination temperature is over 400 °C, the diffraction peak of (111) lattice plane weakens and shifts to lower diffraction angular direction after doping. However, when the calcination temperature is 350 °C, there is no regular shift of the diffraction peak, indicating that doping at this temperature could be ineffective. Based on the above analysis, the calcination temperature is set to be 400 °C. Higher temperature might cause further sintering of CeO_2, which should be avoided. Figure 3 shows the XPS results of the 1% M-doped CeO_2 nano-particle. In this figure, the M-doped CeO_2 nanoparticles were compared with the pure CeO_2. It is obvious that the M elements can be detected in the doped CeO_2 particles. This phenomenon indicates that the doping elements exist in the CeO_2 nano-particles. But further validation could be provided by the TEM images. The energy dispersive system (EDS) mapping by TEM is provided in Fig. 4. Here we take the 5%La doped CeO_2 as an example. It is obvious that the La element uniformly distributes on the CeO_2 nano-particles.

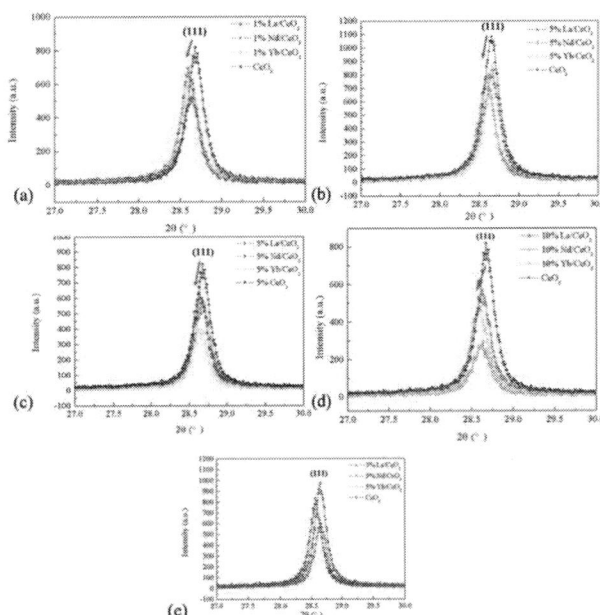

Fig.2. （111）lattice plane of the XRD patterns of the M-doped ceria nanoparticles at different load and temperature: (a) 1% load calcined at 500 °C; (b) 5% load calcined at 400 °C; (c) 5% load calcined at 500 °C; (d) 10% load calcined at 500 °C; (e) 5% load calcined at 350 °C. M refers to La, Nd and Yb.

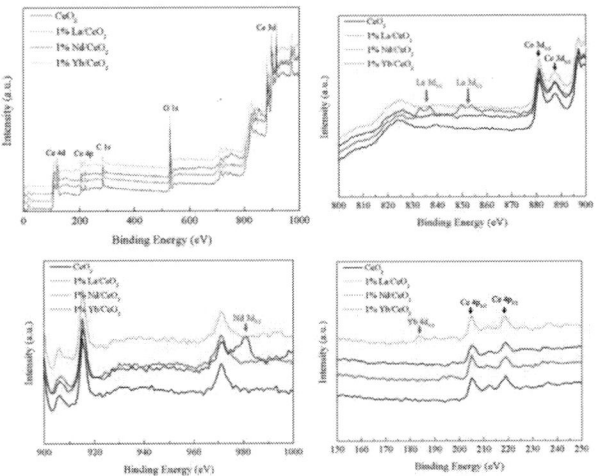

Fig.3. XPS results of the 1%M-doped CeO_2 particles. M refers to La, Nd and Yb.

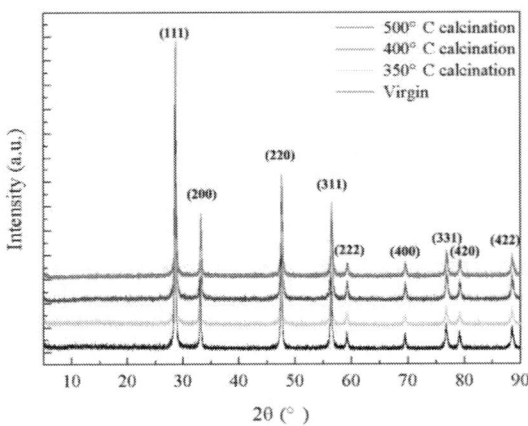

Fig.1. XRD pattern of the pure ceria nanoparticles at 350 °C, 400 °C and 500 °C, respectively.

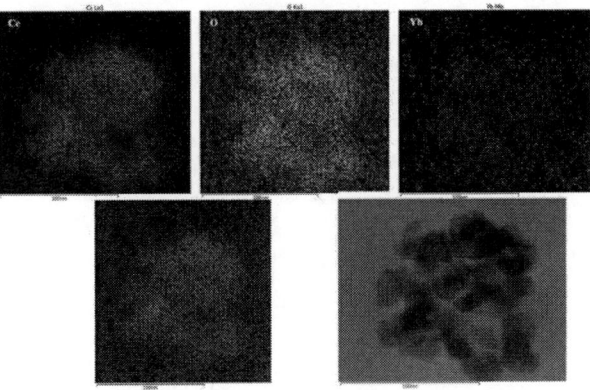

Fig.4. Energy dispersive system (EDS) mapping of the 5%La doped CeO_2.

TEM images show that the M-doped CeO_2 nanoparticles maintain good morphology after doping process, as is shown in Fig. 5 with 5% La doped CeO_2 as an illustration. The shape of the particle is near-spherical without introduction of any secondary phase. The average size of the M-doped CeO_2 could be calculated by Debye-Scherrer equation. The average size of three main crystal planes (220), (311) and (111) was calculated. The results are shown in Fig. 6. The average size of the M-doped CeO_2 remains at ca. 33-35 nm, which is quite uniform.

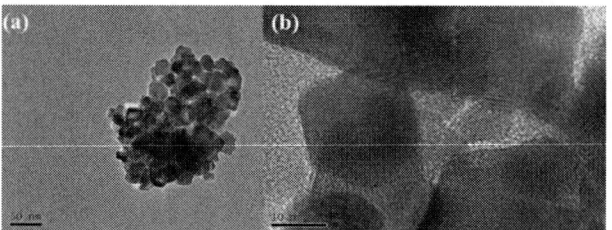

Fig.5. TEM image (a) and high resolution (HRTEM) image (b) of the 5% La doped CeO_2.

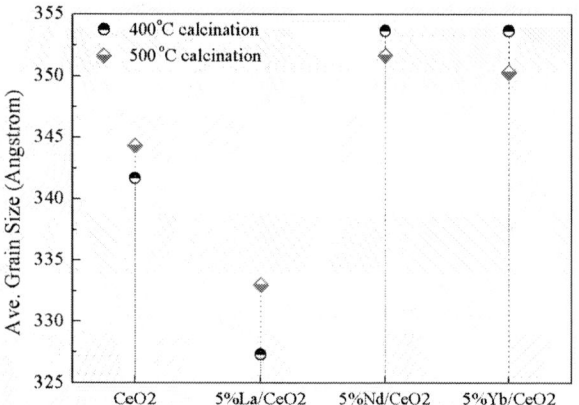

Fig.6. Average size of the CeO_2 particles calculated based on XRD results

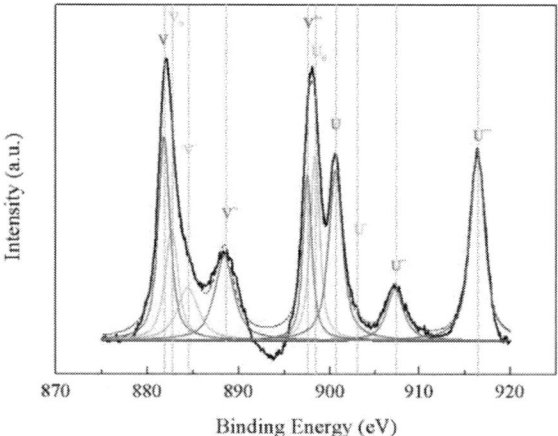

Fig.7. The Ce $3d_{3/2, 5/2}$ XPS spectrum and the peak fit of 1% Nd doped CeO_2.

The semi-quantitative calculation of Ce^{3+} content could be conducted by XPS analysis according to Fig.7. Different peaks refer to different valences of Ce. The calculation results are shown in Table I. It could be seen that doping could effectively increase the Ce^{3+} content in the 1% M doped CeO_2.

TABLE I
CONCENTRATION OF CE^{3+} IN CEO_2 BY XPS

Particle	CeO_2	5%La/CeO_2	5%Nd/CeO_2	5%Yb/CeO_2
Con. of Ce^{3+}	0.298	0.305	0.310	0.321

CMP experiments were carried out both by using commercial CeO_2 and the doped CeO_2 as the abrasive particles. The pH values of the polishing slurry is adjusted and the abrasive content is 0.25 wt%. Figure 8 shows the material removal rate (MRR) of SiO_2 using pure CeO_2 abrasive particles at different pH values. The MRR of SiO_2 reaches the maximum at a pH value of ca. 8 using the commercial CeO_2 nanoparticles No.2815. Figure 9 shows the MRR of SiO_2 using the M doped CeO_2 as abrasive particles. The MRR of SiO_2 could be significantly increased by using the doped CeO_2 abrasives in alkaline slurries, however decreases in acidic slurries. Therefore, the M doped CeO_2 could enhance the CMP performance in alkaline slurries.

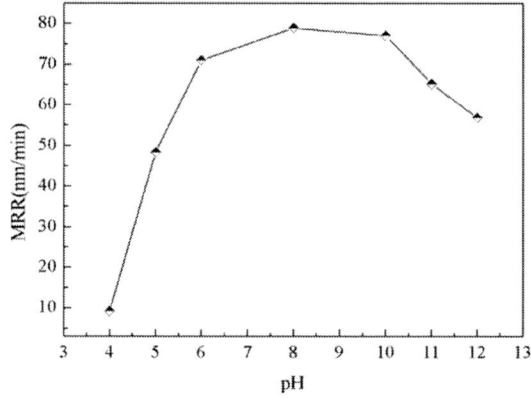

Fig.8. The material removal rate (MRR) of SiO_2 using pure CeO_2 as abrasive particles.

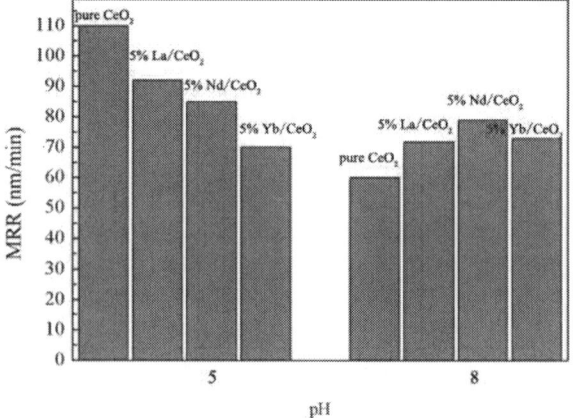

Fig.9. The material removal rate (MRR) of SiO_2 using doped CeO_2 as abrasive particles.

Conclusions

In the paper, lanthanide metals (La, Nd and Yb) doped ceria nanoparticles were synthesized by modified incipient-wetness impregnation method. By using this method, the doped CeO_2 nanoparticles could achieve near-surface doping without changing the bulk mechanical properties as a CMP abrasive particle. Results show that the doped CeO_2 nanoparticles maintain good morphology, and the trivalent Ce content significantly increases, which could be one of the decisive factors for the improved material removal rate of SiO_2 in alkaline slurries.

References

[1] R. Srinivasan, P.V. Dandu, S.V. Babu, Shallow Trench Isolation Chemical Mechanical Planarization: A Review, ECS J SOLID STATE SC, 4 (2015) P5029-P5039.
[2] H. Doi, M. Suzuki, K. Kinuta, Effects of Ce^{3+} on removal rate of ceria slurries in chemical mechanical polishing for SiO_2, International Confenrence on Planarization and CMP Technology, IEEE, 2014.

INVESTIGATION OF EFFECTS OF PATTERN STRUCTURES ARRANGEMENT ON CHEMICAL MECHANICAL POLISHING PROCESS

Lixiao Wu [1], Sookap Hahn [2], and Changfeng Yan [1]*

[1] School of Mechanical & Electronical Engineering, Lanzhou University of Technology, Lanzhou, China, 730050

[2] SKW Associates, Santa Clara, CA 95054, USA

*Corresponding Author's Email: lixiao_wu@163.com

ABSTRACT

Different pattern arrangement will have different wafer profile after CMP. In order to get a planar wafer, the influence of the pattern structure arrangement on the polishing process and on the final wafer profile needs to be investigated. In this paper, CMP_29 of SKW Associates is taken as an example. The initial profile of CMP_29 is divided into three signals and three areas. By using wavelet analysis, the influence of the initial profile on the contact pressure distribution of the different areas is analyzed. The interaction between the splitting signals and the areas during one-material polishing is given.

INTRODUCTION

Pattern structures can be arranged in different way on wafer surface. It is found that different pattern arrangement will have different polishing process during CMP and final wafer profile after CMP. As shown in Fig. 1, the pattern structure with line width 0.14μm and pattern density 50% has large edge over erosion in CMP_28 and little edge over erosion in CMP_29 after CMP. It seems that the pattern structures around pattern 0.14μm affect its polishing process greatly. In order to solve the problem, H. Cai [1] established a multi-level model in which envelope and step-height are defined. Superposition method is used to get the pressure distribution in the model. L. Wu et al. [2] gave an EOE model based on only single pattern structure.

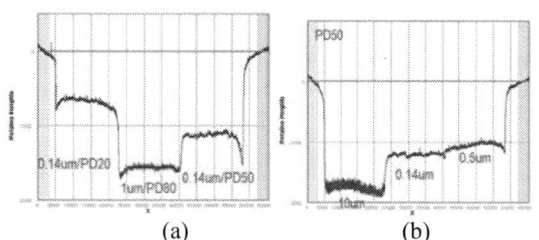

(a)   (b)

Figure 1: Experimental data of effects of pattern structures arrangement: (a) CMP_28; (b) CMP_29.

In this paper, CMP_29 of SKW Associates is taken as an example and is analyzed. By using wavelet analysis, the influence of the pattern structure arrangement on the contact pressure distribution of different areas is investigated.

WAVELET ANALYSIS OF INITIAL PROFILE OF CMP_29 AND ITS SPLITTING SIGNALS

The initial profile of CMP_29 is shown in Fig. 2(a) and is denoted as profile F. As shown in Fig. 2, the profile F is divided into three signals of profile $f1$, profile $f2$ and profile $f3$ according to different topography along x axis. It can be expressed by Eq. 1.

$$F = f1 + f2 + f3 \qquad (1)$$

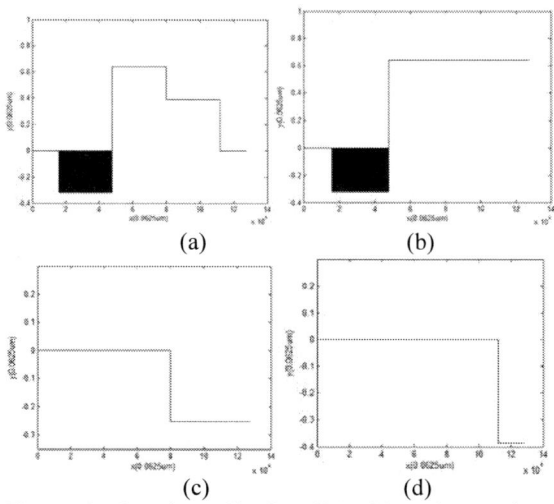

(a)   (b)

(c)   (d)

Figure 2: Initial profile for CMP_29 and its splitting signals. They are: (a) initial profile F; (b) profile $f1$; (c) profile $f2$; (d) profile $f3$.

By using wavelet analysis, the frequency components in the initial profile F and their position along x axis are obtained. The wavelet analysis is expressed by Eq. 2, Eq. 3 and Eq. 4.

$$F(x) = F_{j0} + \sum_j F_j(x) \qquad (2)$$

$$F_{j0}(x) = \sum_k c(k)\varphi(x - k) \qquad (3)$$

$$F_j(x) = \sum_k d_j(k)2^{j/2}\psi(2^j x - k) \qquad (4)$$

where $\psi(2^j x - k)$ is the wavelet function and here is taken as 'db8'; $\varphi(x - k)$ is the scaling function.

In Fig. 3, the amplitudes of some of the frequency components in the initial profile F and their position along x axis are shown in blue color. The same frequency components and their position of the splitting profile $f1$, $f2$ and $f3$ can also be obtained based on wavelet analysis method and are shown in red color. Comparing the frequency components of profile F with those of profile $f1$, $f2$ and $f3$, the initial profile F is divided into three areas and is shown in Fig. 4.

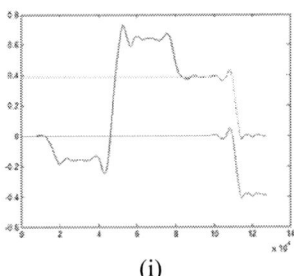

(i)

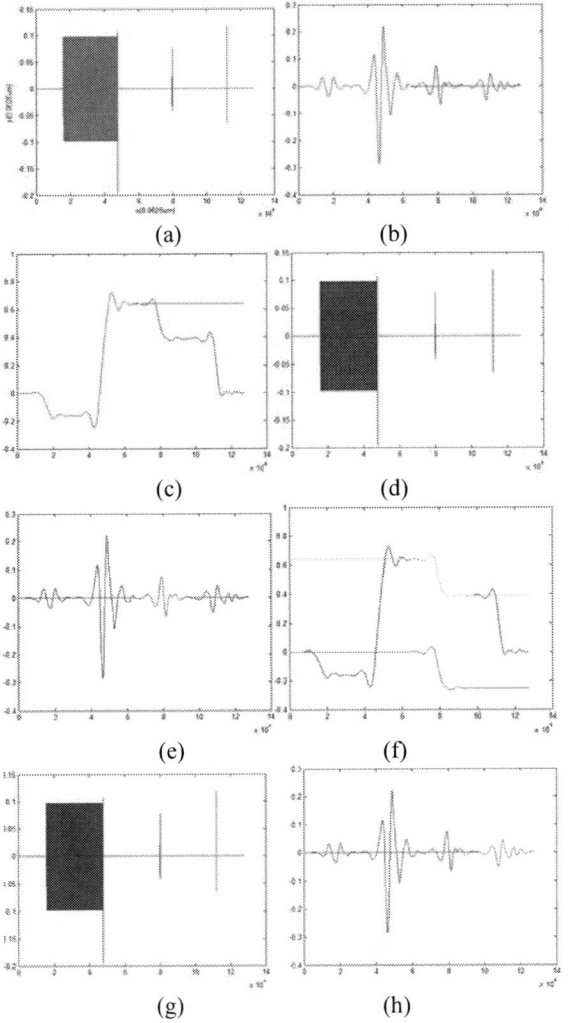

(a)　　　　　　(b)

(c)　　　　　　(d)

(e)　　　　　　(f)

(g)　　　　　　(h)

Figure 3: Frequency components of the profile $f1$, $f2$ and $f3$ in red color and of profile F in blue: (a), (b) and (c) frequency components of profile F and profile $f1$; (d), (e) and (f) frequency components of profile F and profile $f2$; (g), (h) and (i) frequency components of profile F and profile $f3$.(x: 0.0625μm; y:0.0625μm)

In Fig. 3 it is found that the frequency components in area A in the initial profile of CMP-29 are mainly determined by profile $f1$ and are almost not affected by profile $f2$ and profile $f3$. In the same way, the frequency components in area B are mainly influenced by profile $f2$ and the frequency components in area C are determined by profile $f3$. Because of this, the contact pressure distribution and thus the removal rate distribution during one-material polishing in area A is mostly defined by profile $f1$ [3]. Meanwhile, the contact pressure distribution in area B and area C are mainly affected by profile $f2$ and profile $f3$, separately.

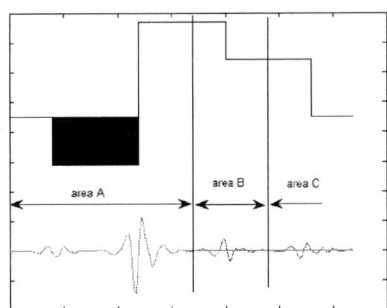

Figure 4: Different areas in the initial profile F .

SIMULATION RESULTS OF EVOLUTION PROCESS OF CMP_29 AND ITS SPLITTING SIGNALS

This conclusion can also be drawn in another way. From the results of the wavelet analysis, it is known that Eq. 5 is true.

$$F(t) = f1(t) + f2(t) + f3(t) \qquad (5)$$

978-1-5386-5309-8/18 $31.00 © 2018 IEEE

In Eq. 5, it shows the evolution process of the profile F equals to the sum of evolution process of the profile $f1$, $f2$ and $f3$. For instance, the wafer profile F at polishing time 6s is almost same as the sum of profile $f1$, $f2$ and $f3$ at polishing time 6s shown in Fig. 5. It should be noted that Eq. 5 is not applicable when the interaction occurs in high frequencies and the amplitudes of the interacting frequency components are large.

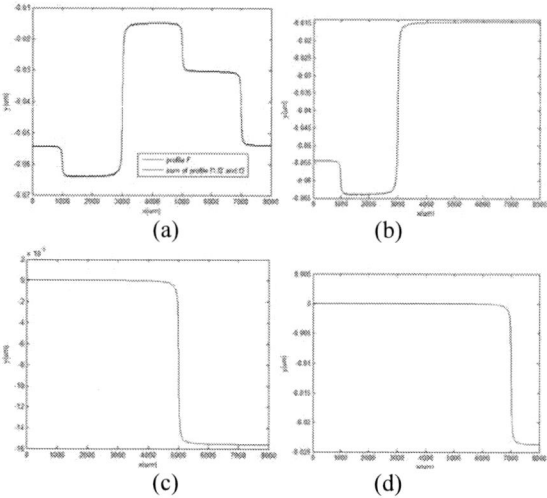

(a)　　　　　　　　(b)

(c)　　　　　　　　(d)

Figure 5: Evolution profiles at polishing time 6s: (a) profile F and sum of profile $f1$, $f2$ and $f3$ at 6s; (b) profile $f1$ at 6s; (c) profile $f2$ at 6s; (d) profile $f3$ at 6s.

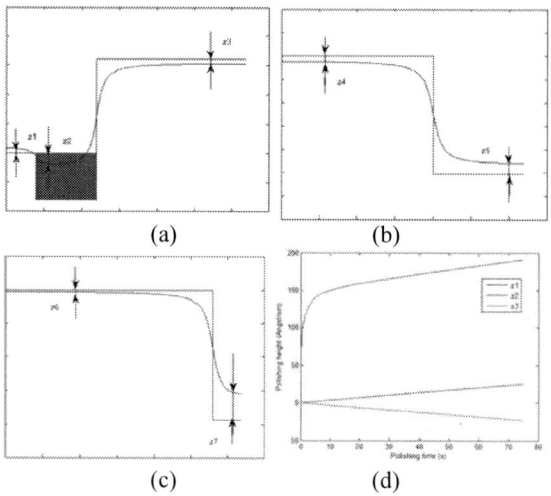

(a)　　　　　　　　(b)

(c)　　　　　　　　(d)

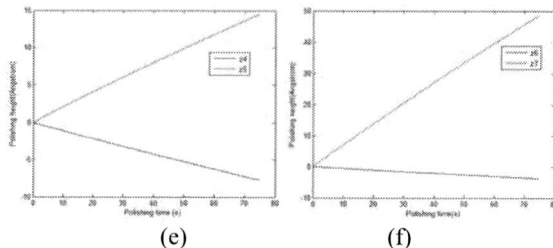

(e)　　　　　　　　(f)

Figure 6: Denotation of changes of profiles and their evolution processes during polishing: (a) changes of profile $f1$; (b) changes of profile $f2$; (c) changes of profile $f3$; (d) evolution process of profile $f1$; (e) evolution process of profile $f2$; (f) evolution process of profile $f3$.

The changes of the profile $f1$, $f2$ and $f3$ during polishing are denoted by z1, z2, z3, z4, z5, z6 and z7 as shown in Fig. 6(a), (b) and (c). The evolution processes of these values are shown in Fig. 6(d), (e) and (f). it is seen that the evolution process of area A during one-material CMP will be determined by profile $f1$ and approximately the constant values from profile $f2$ and $f3$. In the same way, the evolution process of area B during one-material CMP will be affected profile $f2$ and the constant values from profile $f1$ and $f3$. The evolution process of area C will be affected profile $f3$ and the constant values from profile $f1$ and $f2$. From Fig. 6 it is also seen the interaction constants are very small and are changed at a very low speed during polishing.

CONCLUSIONS

Wavelet method is used to analyze the initial profile of CMP-29 and the splitting signals obtaining the frequency components and their positions along x axis. Based on the results of wavelet analysis, the contact pressure and thus the removal rate distribution at different areas in CMP-29 is affected by its relating splitting signal. Little interaction occurs among different splitting signals.

ACKNOWLEDGEMENTS

This work was supported by National Nature Science Foundation of China under Grant 61564006.

REFERENCES

[1] H. Cai. Ph.D. dissertation, Dept. Electrical Engineering and Computer Science, Massachusetts Institute of Technology, Cambridge, MA 2007.

[2] L. X. Wu, S. Hahn and C. F. Yan. IEEE Trans. Semicond. Manuf., vol. 30, no. 1, 2017, pp. 69-77.

[3] L. X. Wu. ECS Journal of Solide State Science and Technology, vol. 1, no. 3, 2012, pp. 133-139.

COLLOIDAL SILICA: CHEMISTRY, PROPERTIES AND ADAPTATIONS FOR ELECTRONIC POLISHING APPLICATIONS

Francois Batllo, Yulia Tataurova, Michael Kamrath

Nalco Water: An Ecolab Company, 1601 West Diehl Road, Naperville, IL 60563

*Corresponding Author's Email: francois.batllo@ecolab.com

ABSTRACT

Colloidal silica products are stable dispersions of non-agglomerated, amorphous, nanometer-size, and spherical particles of silica. The good stability, adjustable particle size distribution and mechanical properties have made colloidal silica a preferred abrasive for many CMP applications. Recently, research and analytical efforts have focused on the development of colloidal products with tunable physical and chemical properties to open up new opportunities in the CMP industry segment.

INTRODUCTION

Optical lithography processing used for the laying of ever smaller interconnects during IC manufacturing requires that the planarization step be achieved at the nanolevel while minimizing defectivity. CMP is an essential planarization process step to meet this requirement. It consists of a polishing process combining both chemical and mechanical forces using slurries containing different chemical reagents and abrasives[1].

The chemical function of the CMP slurries is typically achieved by the chemicals added to the dispersions. The type of chemicals depends on the film being polished. For example Cu CMP slurries contain an oxidizer to facilitate the film removal, a complexing agent to help evacuate the Cu debris, a corrosion inhibitor to prevent over oxidation of the film, a surfactant to keep the suspension stable, a pH controlling agent, etc., and abrasives. Standard abrasives include: silica, alumina, and ceria. The choice of abrasive also depends on the film being polished. Silica is always used to polish prime silicon wafers, TSV (Through silicon Vias) back side. Silica and alumina particles are commonly used for polishing copper and tungsten, silica and ceria are used to polish SiO_2, nitride and low-*k* films. Materials used for optoelectronic devices including sapphire, SiC, GaN are usually planarized using silica.

Among all the abrasives utilized in CMP, colloidal silica present several benefits. Colloidal silica is a stable suspension of amorphous silica particles with shelf life up to 2 years or more. The particles remain suspended and do not require excessive agitation prior to use. The particle size can be accurately adjusted from 4 to 200nm. Colloid purity depends on the manufacturing process and ranges from few ppb's to 100's ppm's total contaminant concentration. One drawback of colloidal silicas is their tendency to form gel under certain conditions of pH and/or conductivity.

There has been a fair amount of development work dedicated to modifying colloidal silica both physically and chemically to alter the abrasive mechanical and chemical properties. The results of this work have led to colloids with tunable morphology, mechanical properties and surface chemistry that have contributed to new design and improved performance of CMP slurries.

MANUFACTURING

There are 2 main routes for manufacturing colloidal silica. Both processes start from the same raw materials including sand or quartz precursors. One route consists of dissolving the raw material in alkali (usually caustic) to produce a liquid alkali silicate that is then transformed into a silicic acid monomer by ion exchange and polymerized into silica in a basic environment[2].

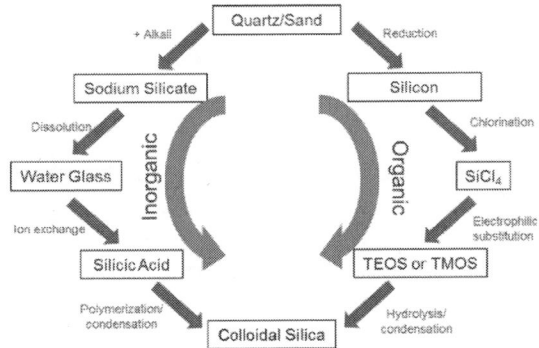

Figure 1: Preparation methods of colloidal silica

The second route involves a series of steps aimed at producing an alkoxyde of silicon that is hydrolyzed into colloidal silica. The resulting alcohol based product is then solvent exchanged to an aqueous silica suspension. Advantages and drawbacks from manufacturing routes are summarized in Table I.

The main advantage of the colloids prepared from organic method is the high purity although recent developments in the inorganic method have enables colloids with acceptable purity for most CMP applications. An additional benefit from the inorganic route is the ability to functionalize the particles to alter their chemical and physical properties.

Table I: Comparison of colloids prepared from inorganic and organic methods

	Inorganic	**Organic**
Raw Material	Na silicate	Si alkoxyde
Reaction synthesis	Ion exchange and polymerization	Hydrolysis and solvent exchange
Purity	Moderate	High
Cost	Moderate	High
Stability	Excellent	Good
Functionalization	Easy	Difficult

PROPERTIES

Morphology

Colloids manufactured by both inorganic and organic routes are in general dense and display spherical or ovoid morphology. As both routes involve growing particles from a solution either by polymerization/condensation or hydrolysis/condensation they enable the addition of elements capable of altering the properties of the colloids. The kinetics of reaction can also be tuned by changing the conditions of the polymerization. The inorganic route presents an advantage over the organic route as the element or dopant can be introduced via several ways during the colloid nucleation or growth and the kinetics of polymerization can be controlled by a simple shift in pH[3]. Figure 2 shows a few examples of colloid morphologies achieved by modification of the polymerization process.

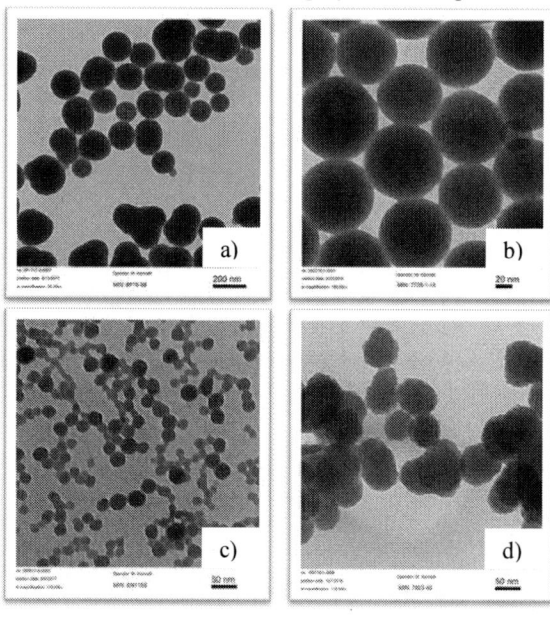

Figure 2: TEM of modified colloidal silica showing particles a) mono, di and tri dispersed, b) monodispersed with smooth surface, c) fused chain like and d) monodispersed with jagged surface

It is evident that the particles presented in Figure 2 will have very different tribological properties.

Surface chemistry

Concentrated silica colloids are stabilized by either 1- electronic repulsion from an ionic charge on the surface of the particles at high pH or 2- Steric stabilization from an adsorbed layer of inert material on the particles at low pH. Between the 2 pH regions, collisions between particles result in siloxane bonds and eventually irreversible gelling. Certain dopants like aluminum added in the silica network or at the surface of the particles[4] act as a stabilizer to the suspension and enable colloid stability over a wide range of pH.

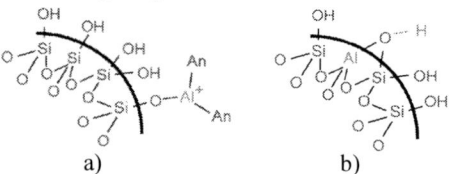

Figure 3: Aluminum modified silica via a) surface coating and b) bulk doping

Addition of an aluminum salt molecule at the surface of the particle in place of a silanol increases the cationic charge of the particle and enhances its ionic stability at medium to low pH. The insertion of a trivalent aluminum in substitution of silica generates a brønsted acid in the colloid acting as a permanent negative charge in and at the surface of the particle and henceforth enhancing its stability across a wide pH range. The stabilizing effect of doping can also be illustrated by the zeta potential curves of colloidal silica unmodified and modified shown in Figure 4.

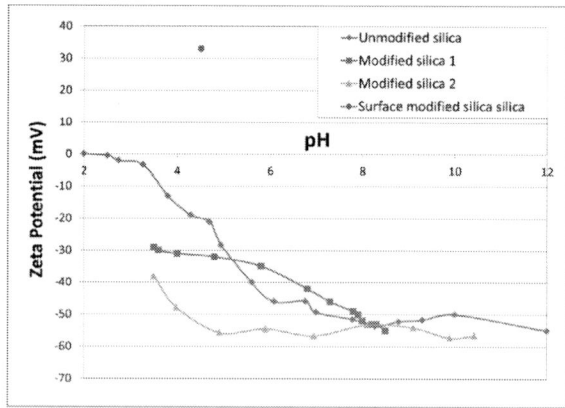

Figure 4: Zeta potential vs. pH for unmodified and modified colloidal silica

The ability to tune the surface chemistry of the

colloids and regulate their stability expands the possibility of formulating CMP slurries in a wide range of pH and conductivity without affecting the properties of the abrasives.

Mechanical properties

To assess the impact of doping technology on the mechanical properties of the particles, we performed nanomechanical testing using a Hysitron Picoindenter/TEM analysis. The analysis consists in placing a minute amount of sample onto a silicon wedge and compressing a single particle between the wedge and a picoindenter transducer tip. The set-up is placed under a TEM and the compression is followed via a camera. Figure 5 show the TEM images of the particle before and after compression.

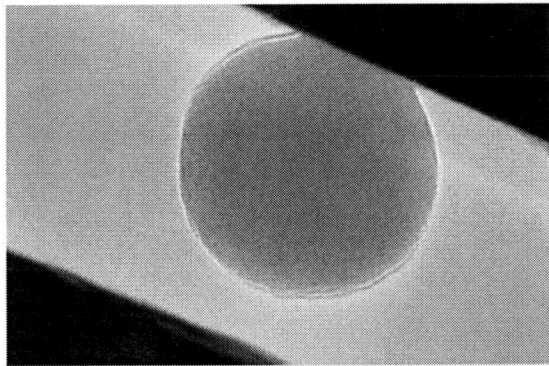

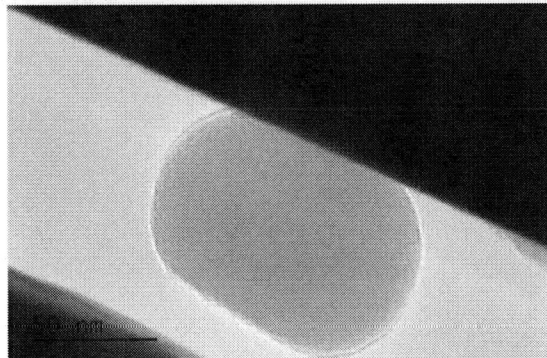

Figure 5: TEM images of the particle before and after compression

From the load-displacement graphs collected we can calculate the stiffness of the particles from the elastic portion of the compression signal. Figure 7 shows the stiffness for high and standard purity as well as unmodified and modified particles made from the inorganic route.

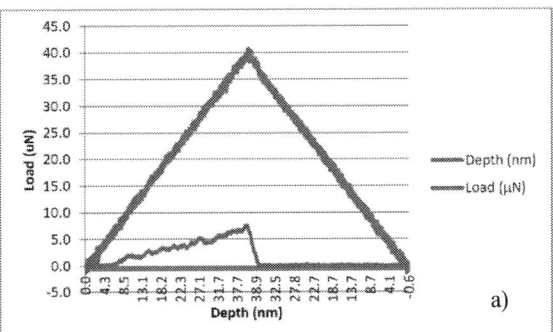

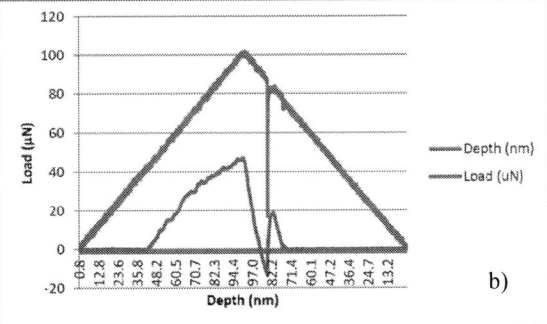

Figure 6: Transducer load vs indenter travel for a) unmodified and b) modified particles

The results show a low stiffness from the particles manufactured from organic precursors compared to inorganic precursors. There is slight decrease in stiffness with the level of purity in the particles made by inorganic route. Modified particles display double stiffness value of the unmodified particles. Work is underway to determine the variation in stiffness by changing the manufacturing conditions and the extent of modifications of the particles.

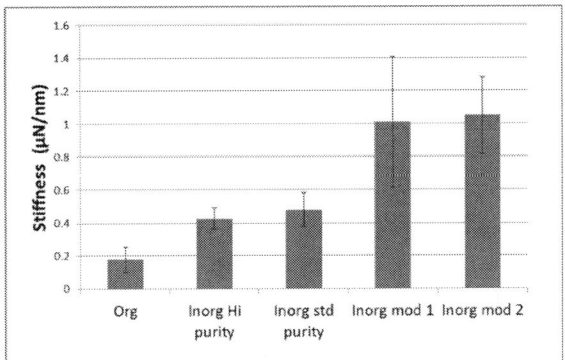

Figure 7: Stiffness of particles manufactured from: Org=Organic Route, Inorg = Inorganic Route, Mod 1 and Mod 2 = Modified 1 and 2. Error bars = standard deviation

CONCLUSION

Colloidal silica is a unique component entering the formulation of CMP slurry. The chemical modification of the particles from the colloids enables control of the morphology, stability and mechanical properties of the abrasives. The ability to tune the abrasive properties brings an additional tool to design CMP slurry for many planarization applications.

ACKNOWLEDGEMENTS

The author wished to thank Changqiang Chen, Ph.D., Research Scientist, Frederick Seitz Materials Research Lab, Univ. of Ill. at Urbana Champaign for providing the Hysitron picoindenter measurements.

REFERENCES

[1] Parshuram B. Zantye, Ashok Kumar, A.K. Sikder; Materials Science and Engineering R 45 (2004) 89–220

[2] Hyung Mi Lim, Jinho Lee, Jeong-Hwan Jeong, Seong-Geun Oh, Seung-Ho Lee; Engineering, 2010, 2, 998-1005

[3] Iler R. K.; The chemistry of silica, John Wiley & Sons, 1979

[4] Ming-Shyong Tsai, Wen-Chang Wu; Materials Letters Volume 58, Issues 12–13, May 2004, Pages 1881-1884

NOVEL CMP PADS BY SPECIAL STRUCTURAL DESIGN

Yijie Luo, Min Liu*, Weitung Wang and Jun Yan

Hubei Dinghui Microelectronics Materials Co., Ltd, Wuhan 430057, China

liumin@dl-kg.com

ABSTRACT

We provide a serial of chemical mechanical polishing （CMP）pads，which are produced by pre-polymers and multi-components of curing agents. The curing agents containing aromatic diamines and secondary aliphatic polyetheramines are mixed with the pre-polymers to create new CMP pads which has obturator structure by microspheres. Average diameter of the microspheres with smaller size is less than 30 micron, which bring higher remover rate and less defects to the polished wafer. A serial of CMP pads with different hardness are manufactured by the above method for polishing copper processes. By carefully study of the effects of hardness to the polishing performance, we choose appropriate pads for advanced semiconductor manufacturing processes.

INTRODUCTION

As the feature size of the integrated circuits developing towards deep-nanometer scale, the manufacturing process puts forward higher requirements for the dielectric materials [1]. In order to suppress the interconnection delay caused by the increase of the crosstalk between the metal lines, more and more low-κ, even ultra-low-κ dielectric materials are applied to it. However, low-κ and ultra-low-κ dielectric materials are typically obtained in such a way as to increase the porosity of the material [2]. Thus, low κ and ultra-low κ dielectrics tend to have lower mechanical strength and poorer adhesion than conventional dielectrics, making the planarization process more difficult [2]. In addition, defects such as scratch caused by chemical mechanical polishing (CMP) processes become more and more problematic in advanced process as feature sizes are reduced, and therefore, feature sizes of 14 nm and below technology requires lower defects, fewer scratches, and better removal rate of low-κ and ultra-low-κ materials in the CMP process. The development of new polishing pads for advanced processes is one of the important solutions. In fact, many products have been developing specifically for it. Herein, we report our works in the developing of new polishing pads for the advanced process.

Conventional polishing pads are usually prepared using a single pre-polymer and a single curing agent, which has the advantages of simple composition, easy mixing during preparation, and good uniformity of the product [3]. However, the defect of single curing agent system is usually high, which is difficult to be applied in the advanced process, so a more complex multi-component pre-polymer and multi-component curing agent system is developed subsequently [4]. Its significant advantage is that the ratio of hard segment and soft segment of the pre-polymer and curing agent can be adjusted separately to improve the physical and chemical parameters of the polishing pad to improve the polishing performance. However, in the multi-curing agent system, aromatic polyamines and aliphatic polyols are basically used in combination, which represent the hard segment and the soft segment of the curing agent, respectively. In this kind of curing agent composition, there is an inevitable birth defect, which is the different reactivity between aromatic polyamine and aliphatic polyol with the free isocyanate. The reaction activity of the former is usually tens or even hundreds of times of the latter, which leads to the first reaction of aromatic polyamine in the reaction process, and makes the reaction system gel, which in turn leads to the phenomenon of aliphatic polyols that are more difficult to participate in the reaction and the reaction degree is not equal, which ultimately influences the polishing performance.

Based on the above reasons, we consider whether a new curing agent system can be found, which makes the aliphatic curing agent and aromatic curing agent very close to the reaction activity, avoiding the uneven-difference caused by the difference of reaction activity from the root cause, thus obtaining the CMP polishing pad with improved polishing performance. In the selection of aromatic curing agents, we consider using the most widely used of 4, 4'-Methylene bis(2-chloroaniline)(MOCA). Although MOCA contains chlorine elements that are potentially harmful to the environment, and have a lot of work on the basis of MOCA alternative research in recent years, but they have been unsuccessful. The main reason is the unique chemical structure of MOCA, which guarantees the suitable operation time in the preparation of the polishing pad, and the obtained polishing pad has good polishing performance. For the selection of aliphatic curing agents, because of the low and unimprovable activity of the polyalcohols, we choose aliphatic polyamine. But the aliphatic polyamine activity is much higher than that of the aromatic polyamine, how to reduce its reactivity by structural modification is the primary problem that needs to be faced. In theory, to reduce the activity of a certain group, it can usually be carried out from two aspects, steric effect and electronic effect. Specific to the aliphatic polyamine curing agent, an effective method is to increase

978-1-5386-5309-8/18 $31.00 © 2018 IEEE

the amino steric hindrance and reduce the amino electron cloud density. After carefully screening, we found that the secondary polyetheramine can meet our requirements, and then apply it to the polishing pad preparation. After preparing of a series polishing pad of different hardness, we evaluated their performance.

EXPERMENTAL SECTION

100 parts by mass of the isocyanate-terminated pre-polymer (containing 8.75% to 9.05% by mass of unreacted NCO groups) obtained by the reaction of tolylene diisocyanate (TDI) and polytetrahydrofuran was heated to 80 °C, after degassing in a vacuum, followed mixing with 1.50 parts by mass of hollow microspheres (having an average particle diameter of 20 μm), the mixture were degassed again and the temperature was lowered to 50 °C. A liquid curing agent combination of 23.5 parts by mass of MOCA and 10 parts by mass of a terminal secondary polyetheramine at a temperature of 115 °C was added to the above mixture, then mixed with high-speed, casted into lumps, and then heated to 100 °C for curing 16 hours. After the process is completed, the composite polishing pad is obtained after skiving, grooving, laminating and punching process.

Fixing the total amount of pre-polymer, just adjusting the ratio of MOCA to the terminal secondary polyetheramine can prepare polishing pads with different hardness. The physical and chemical parameters as shown in Table 1. As the reactivity of terminal secondary polyetheramine is very colse to MOCA, we can easily adjust its ratio, which in a large range to adjust the hardness of the polishing pad to obtain polishing pads with different polishing performance.

Table 1: Composition and physical parameters of the polishing pad

	MOCA	secondary polyetheramine	Thickness (mm)	Specific Gravity (g/cm3)	Hardness (Shore D)	Compressibility (%)
Pad-1	23.5	10	2.01	1.004	68.4	0.42
Pad-2	21.5	25	2.01	0.989	61.2	0.66
Pad-3	19.5	40	2.01	0.994	54.7	0.72
Ref pad	-	-	2.01	1.003	66.7	1.61

ref pad is a polishing pad which is widely used in copper polishing in the advanced process.

RESULT AND DISSCUSION

For the three polishing pads prepared above and the ref pad (commercial CMP Pad with similar hardness and specific gravity), we conducted a comprehensive evaluation of the polishing performance through on-line testing. The testing conditions are as follows: testing machine is AMAT Refelxion (Modify 5 Zone); polishing slurry is ANJI 3060 (1: 9 Dilution , H_2O_2 % = 1 %), flow rate is 250 mL/min; conditioner is Saesol Disk 6045C4, P/C downforce is 5 lbf, rotation rate of head and platen is 93 and 87 rpm, respectively; wafer is pattern wafer of Semitech 754, Cu Blanket Wafer Pre Thickness 10 KA.

First is the removal rate comparison. Figure 1 shows

the removal rate tested for pad -1 to 3 and the ref pad at 2.5 psi. Pad-1 and pad-2 show higher removal rates than the reference polishing pad, the removal rate of pad-3 is lower, probably due to lower hardness. In addition, all three polishing pads exhibited similar rate uniformity to the reference pad, with no large rate fluctuations from the center of the wafer to the edges with NU values less than 5%, which is very advantageous for improving the yield of the wafer.

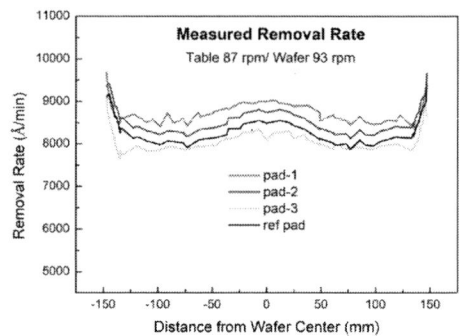

Figure 1: Removal rate comparison.

Table 2: Average Removal Rate and none uniformity comparison

	Pad-1	Pad-2	Pad-3	Ref pad
Average Removal Rate (Å/min)	8746	8614	8098	8363
none uniformity (%)	3.22	4.18	3.62	4.31

In addition to the removal rate, the other two important evaluation criteria for the polishing performance are the defect and the planarization effect. In the advanced process of 14 nm and below, extremely high demands are placed on the defect and the planarization effect, and the required defect as well as the dishing and erosion values characterizing the planarization effect, usually the smaller the better. Figure 2 shows a comparison of the defect between pad 1-3 and the reference polishing pad. As we can see, the hardness of the polishing pad is reduced due to the incorporation of the soft segment of polyetheramine, leading the defect shows a significant decrease. Pad-1 also showed a lower defect rate than the reference pad at similar hardness.

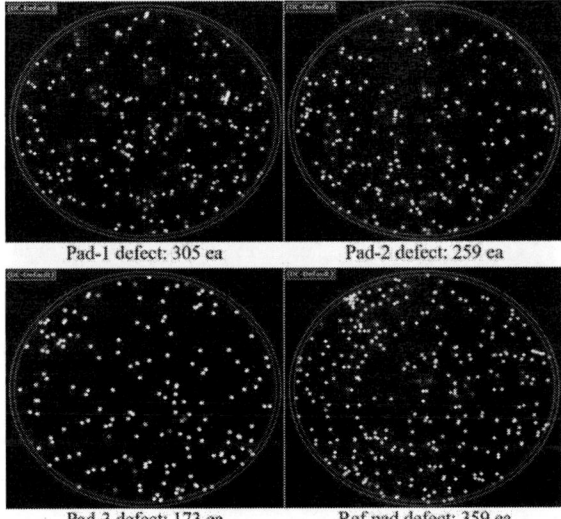

Pad-1 defect: 305 ea   Pad-2 defect: 259 ea

Pad-3 defect: 173 ea   Ref pad defect: 359 ea

Figure 2: Defect comparison

Through atomic force microscopy (AFM), we also observed the degree of planarization of the wafer after polishing. We can see that pad-1 ~ 3, with the decreasing of hardness, dishing and erosion show an increasing trend, while the values of pad-1 and 2 are still lower than the reference pad, only pad-3 shows worse flatness than the reference polishing pad.

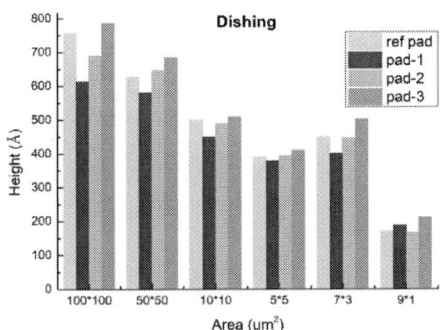

Figure 3: Dishing comparison

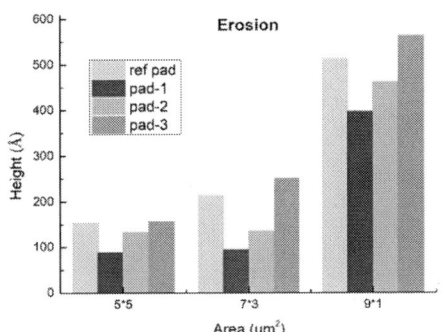

Figure 4: Erosion comparison

CONCLUSION

Through the introduction of the terminal secondary polyetheramine, we have prepared three novel polishing pads. By adjusting the ratio of MOCA to the secondery polyetheramine, the physical parameters of the polishing pad are effectively adjusted. In-line evaluation results show that reducing the hardness of the polishing pad can reduce the defects effectively, but the side effects are reducing of planarization effect. Comprehensively considering the removal rate, defect and planarization effect, we think pad-2 is a high performance polishing pad, which has the potential to replace the reference polishing pad in advanced process.

ACKNOWLEDGEMENTS

Financial support from Wuhan Huanghe Talent Plan, Wuhan Talent Innovation Project, High-tech Office of Hubei Provincial Science and Technology Department, Talent Office of Hubei Provincial Party Committee Organization Department and High-tech office of Hubei Development and Reform Commission is gratefully acknowledged.

REFERENCES

[1] Ryan J. G.; Geffken R. M.; Poulin N. R.; Paraszczak J. R. *Ibm. J. Res. Dev.* **1995**, *39*, 371–381.

[2] Volksen, W.; Miller, R. D.; Dubois, G. *Chem. Rev.* **2010**, *110*, 56–110.

[3] Reinhardt H. F.; Roberts J. V. H.; Mcclain H. G.; Budinger W. D.; Jensen E. W. US5578362.

[4] Qian B.; James D. B; Murnane J.; Yeh F.; Degroot M. W.; US9144880.

Achieving Highly Accurate Profile Control by Applying a Multi-Zone Polish Head for 200mm Thin-Film SOI CMP

Mike Liu[1], Jamie Leighton[2], Mike Rosa[2].*
[1]Applied Global Services, Applied Materials, Xi'an, Shaanxi, China
[2]Applied Global Services, Applied Materials, Santa Clara, California, USA
*Corresponding Author's Email: mike_liu@amat.com

ABSTRACT

The specification for the total thickness variation (TTV) of the top Si layer of silicon on insulator (SOI) wafers is getting tightened for the future application [1] [2]. Therefore, the current Chemical Mechanical Pol ishing (CMP) process cannot meet the flatness requirement, especially at wafer edge, which could cause up to 5% yield loss.

This paper describes how an Applied Materials Mirra® CMP configured with a multi-zone polish head was introduced to develop a new thin film SOI polishing process. The resulting TTV improvements demonstrated by the CMP step generates a competitive and effective method for achieving the best possible profile control.

INTRODUCTION

Various factors, such as consumables, process parameters and polisher hardware design, affect CMP processing. The polishing head is one of the most important factors because it directly affects the performance of the wafer-polish (e.g., across-wafer uniformity and with removal rate stability).

Titan™ Head – Conventional 200mm CMP Polish Head

The smart cut process (Fig.1) is the most popular SOI manufacturing method. The mainstream thin film SOI wafer polishing process within the smart cut process was developed with Applied Materials, Inc. (on the Mirra platform, configured with the Titan polish head).

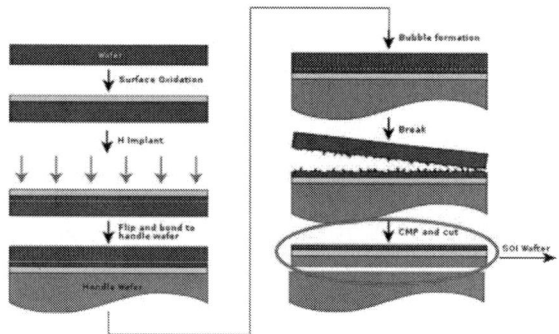

Fig.1. CMP step in the smart cut process

The design of the one pressure zone polish-head constrains polish-removal profile tuning and control. The inherent non-uniformity leads to high TTV issues associated with removal profile control during the thin film SOI polish process. Regions near the wafer's extreme edge (radius 94–97mm) are over polished during the 200mm Si CMP step (Fig.2). The over polishing region results in significant wafer yield loss.

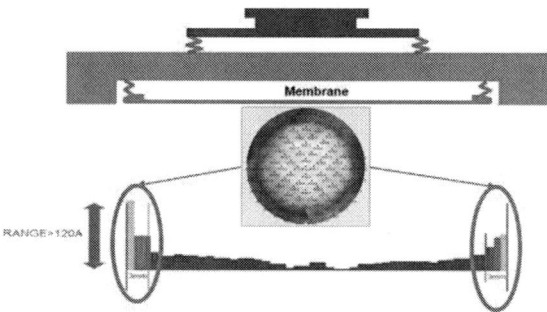

Fig.2. Titan head process result—100% range of post CMP thin Si film in the last 3mm

Contour™ Head – Multi-zone Polish Head

To solve the known TTV issues, Applied Materials provides a 200mm multi-zone (Contour) polish head design [3][4]. The Contour head configuration provides access to a total of six (6) independent pneumatic zones (including the retaining ring) (Fig.3).

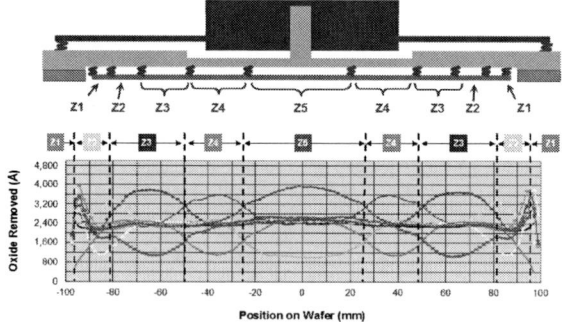

Fig.3. Contour polish head zone configuration showing active polish region by zone

Figure 3 illustrates the general zone regions in response to Contour head polishing on oxide film with a

978-1-5386-5309-8/18 $31.00 © 2018 IEEE   346

specific consumable set (IC1010/Klebosol 30N50/3M). It indicates the Contour head has the over-and under-polish capability on edge.

For the top Si polishing of thin SOI wafers with different consumable sets, the work is evaluated and performed by edge zone response DOE, profile tuning and a repeatability check.

EXPERIMENT

A high pressure (5 psi) process geared toward the thin SOI polishing application was developed on the Mirra Mesa system using the 200mm Titan Contour polish head. Thin SOI monitor wafers were used for DOE, profile tuning and a repeatability check. The wafer run was conducted using Nanopure slurry, SPM3100 pads and a groove smoothed peek ring. A brush disk was employed in-situ to wafer polishing. The Nova 2040 tool measured the film thickness and profile.

RESULTS

Edge Zone Response DOE

A DOE on zone 1 and 2 pressure was completed on thin SOI wafers to determine the wafer edge response of each zone.

To demonstrate performance and facilitate understanding, Table 1 below identifies and defines the wafer regions associated with the greatest zone response. Transitional areas between zone 1 and zone 2 are not included.

TABLE I.   Titan Contour Edge Zone Control Regions

Region 2	84mm-92mm
Region 1	95mm-97mm

In the DOE, zone 1 and zone 2 were over- and under-pressurized by 10%. For zone 2, the change in removal from the baseline was ~+/-5% (Fig.4). For zone 1, the change was ~+/-20% (Fig.5).

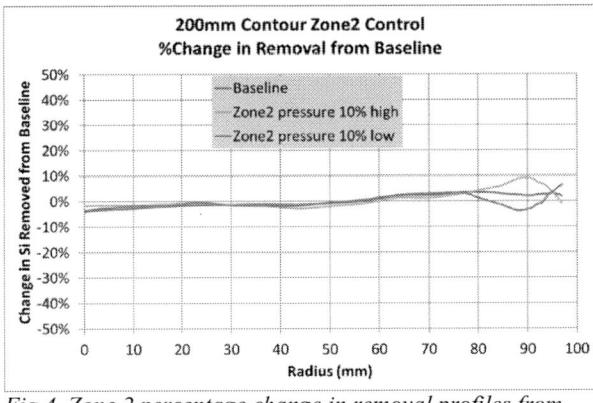

Fig.4. Zone 2 percentage change in removal profiles from baseline

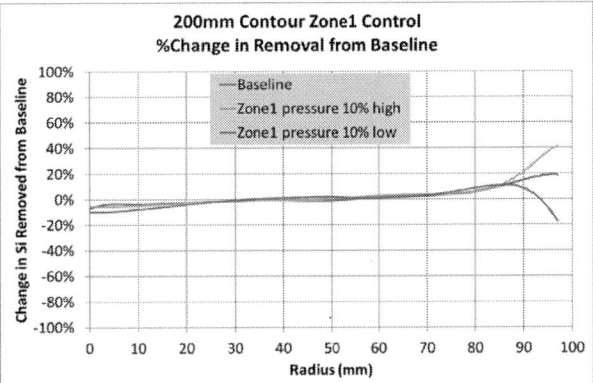

Fig.5. Zone 1 percentage change in removal profiles from baseline

The removal rate for zone 2 pressure tuning changes was between 75mm and 95mm from wafer center, with the maximum effect exhibited around 87mm. Due to the membrane design, zone 2 tuning results in a significant change in the zone 1 region. To neutralize the impact at the very edge, zone 1 must be changed accordingly.

The response of zone 1 is highly edge focused, starting at ~90mm from the wafer center and increasing rapidly towards the edge. The removal at the very edge (95–97mm) is very sensitive to zone 1.

Removal Profile Tuning

The top Si removal profile tuning was conducted by zone 1 and zone 2. A significant improvement in TTV was observed on top Si film removal thickness (Fig.6). The TTV result was dramatically decreased from over 100Å to under 50 Å. Meanwhile, the impact of edge zone tuning on the 0–70mm region was almost zero.

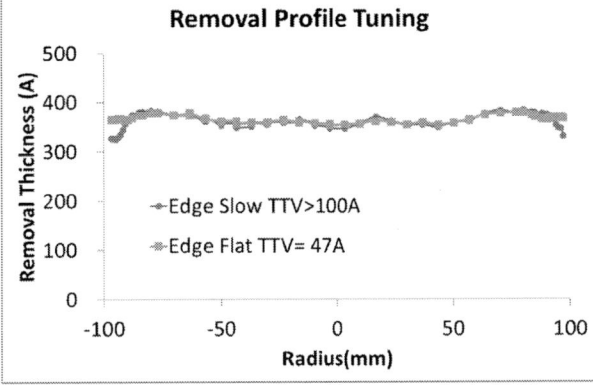

Fig.6. Contour head removal profile tuning of thin Si film

Applying the Contour polish head enabled profile tuning of the top thin Si film and resulted in a significant flat removal thickness TTV.

Removal Amount DOE

Another DOE on removal amount was completed to determine the impact to TTV (Fig.7). Each removal amount is conducted in a different polishing time by using an edge flat polishing recipe, which is used in removal profile tuning. It is obvious that an upward tendency of the TTV exists, which is increased gradually from a ~38Å to 350Å removal amount to a ~85Å to 1200Å removal amount. The lower removal amount helps reduce TTV as shown.

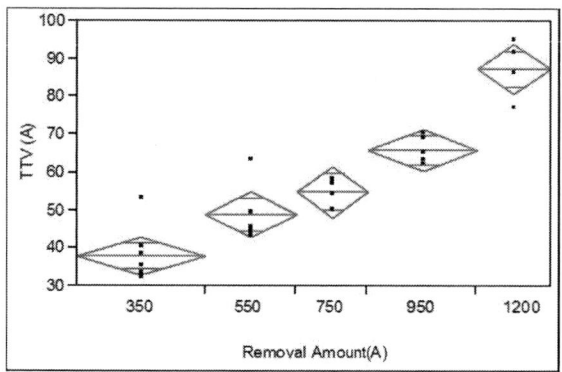

Fig.7. Analysis of TTV by removed amount

Repeatability Check

A repeatability check on post CMP Si thickness is performed with a 350Å removal amount edge flat recipe (Fig.8). Note that by applying a Contour polish head, it is possible to keep post CMP thickness and TTV stable by processing 14 wafer runs.

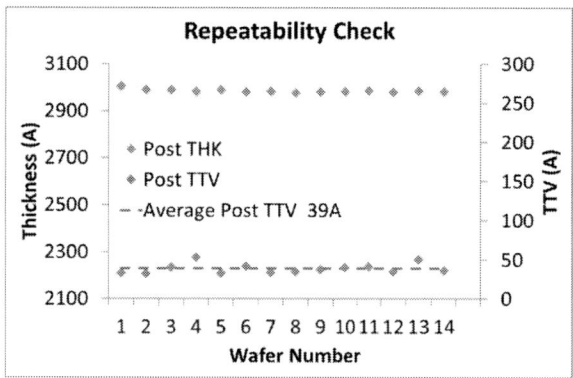

Fig. 8. Overall performance of post thin Si Film, 93% post range≤50A

CONCLUSION

This paper highlights the capability of a Titan Contour polish head to correlate edge zones with pressure tuning. We demonstrated that zone 1 and zone 2 over- and under-polishing by 10% can result in a **5%** removal change from 84mm to 92mm and **20%** from 95mm to 97mm. Our repeatability test also demonstrated process stability. Average TTV narrowed to 39 Å, and 93% post TTV was controlled within 50 Å. In summary, the Titan Contour polish head solved the TTV issue, which resulted from the inability of the traditional Titan polish head to control edge area.

ACKNOWLEDGEMENTS

The authors would like to thank E. Liu and M. Xue of Applied Materials Inc. for their support and willingness to review the content included in this paper.

REFERENCES

[1] G. Celler and M. Wolf. Unpublished.
[2] L. P. Allen, W. Skinner and A. Cate. *IEEE International SOI Conference. Proceedings, 2001,* Durango, Oct 1-4, 2001, pp. 5–7.
[3] B. Milmore, S. Wills, Y.M. Xing and M. Liu, *International Conference on Planarization/CMP Technology (ICPT), 2015,* Chandler, Sept 30 – Oct 2, 2015, pp. 1–3.
[4] Y.H. He, M. Lube and J. Leighton, *China Semiconductor Technology International Conference (CSTIC), 2017,* Shanghai, March 12–13, 2017, pp.1–4.

CMP PAD SURFACE UNIFORMITY OPTIMIZATION AFTER POLISH

Ying Emily Lu*, Wei William Guo, Jenny Wu

Corporate Q&R Center, Semiconductor Manufacturing International Corporation,
Shenzhen, Guangdong, China

*Corresponding Author's Email: Emily_LY@smics.com

ABSTRACT

Aiming at CMP (Chemical Mechanical Polisher) pad thickness uniformity optimization and pad lifetime extension, we performed optimal design to do experiments finding the optimal process condition. We obtained much longer using time of pad and better pad uniformity on CMP tool from wafer stay time distribution on pad by zone. Base on the results of DOE (Design of Experiment), we get the best setting of stay time ratio by zone for optimal pad thickness uniformity is 18%/ 20% (zone1/ zone10). Also we estimated a mechanism of pad thickness uniformity maintenance after polish in CMP process.

INTRODUCTION

Both cost down and good quality are two development directions in semiconductor manufacturing. In order to control the cost, one of the effective methods is to extend parts' or materials' using time. Currently CMP application is widely used in interlayer dielectric planarization to metal CMP and silicon oxide trench planarization [1]. As the devices of ULSI scale down, the CMP step requires high remove rates and a planarized surface. It is essential to expedite the remove rates to raise the efficiency of manufacturing process and avoid the surface damage at the same time [2]. Pad which used in CMP is a huge consumable item and need to be replaced with a new one in very short time. The high frequency of replacement induces the cost raising and if don't change the pad, the wafers processing will get broken easily with the bad uniformity of pad. Many papers had analyzed some aspects of kinematics in CMP [3, 4], and the pad image with wafer processing is shown in Fig. 1 and the zone distribution is shown in Fig. 2. During process, wafer move between zone1 and zone10, while pad and wafer go round and round.

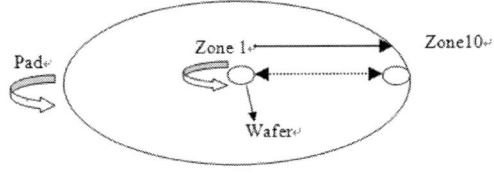

Figure 1: CMP Process Schematic – Pad & Wafer Movement

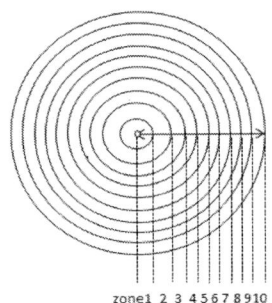

Figure 2: CMP Process Zone Distribution

It is indispensable to find a reasonable process time distribution by zone to optimize pad uniformity and prolong pad using time. The polish process time by zone is critical to the performance of pad surface except the condition of pad surface texture in chemical mechanical planarization processes [5]. In order to ensure CMP process stable, the total polish time should be fixed and unchangeable, but wafer stay time by zone on pad could be tunable. The baseline condition for wafer stay time ratio by zone is shown in TABLE I, and the baseline pad remain thickness by zone after polish is uneven and the pad must be discarded and replaced every day. The baseline performance is shown in Fig. 3.

TABLE I
Baseline Condition 1 & 2 Wafer Stay Time by Zone

Zone	1-Stay Time Ratio	2-Stay Time Ratio
1	22%	15%
2	8%	9%
3	6%	9%
4	8%	9%
5	10%	7%
6	9%	7%
7	6%	8%
8	6%	8%
9	5%	12%
10	21%	16%

From the wafer stay time ratio in TABLE 1 and the baseline pad remaining thickness by zone in Fig. 3, we can see the wafer stay time is longest in zone1 and zone 10, also the pad thickness in zone 1 and zone 10 is different from other middle zones. The pad surface uniformity

978-1-5386-5309-8/18 $31.00 © 2018 IEEE   349

optimization is equal to the adjustment of zone 1 and zone 10 thickness standard deviation with other central zones. In view of the cost control, the optimal design is a good experimental method to save runs. Considering the pad is expensive and tool process time limitation, just takes the major effect into account in following experiments.

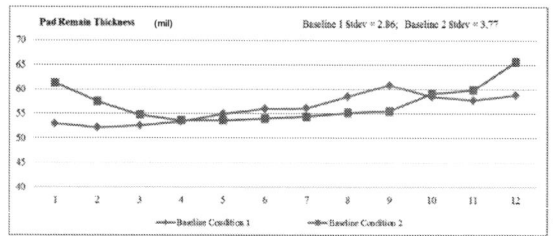

Figure 3: Baseline Pad Remain Thickness by zone

EXPERIMENT

Before implementing DOE, the factors and levels, response, and DOE experiment conditions should be considered firstly.

Many factors such as pad conditioning and staying time setting which are more critical for wafer WIW and WTW uniformity benefit, and had been studied in experiments and research before. This time we focus on stay time ratio by zone to study the impact of time distribution by zone on pad uniformity. Zone 1 stay time ratio can be chosen from 16% to 20%, while zone10 stay time ratio can be chosen from 20% to 24%. The stay time ratio on zone9 is fixed in 11%, and the remaining stay time ratio is allocated from zone2 to zone8 almost averagely. Except the stay time ratio by zone changing, all process conditions keep the same, using new pad and same wafers in the experiment. Customer design with the factors of zone1 and zone10 stay time ratio is selected in JMP software to achieve the experimental table.

Factors and Response

Base on previous experiences, not only the stay time in zone1 and zone10 is the most key parameter, but also the zone10 stay time must be greater than or equal to zone1, since the removal rate in center area is higher than edge area.

In our experiment, we select zone 1 and zone 10 stay time ratio as two factors, the pad remaining thickness standard deviation as response. Both zone 1 and zone 2 have 2 levels, zone 1 tune from 16% to 20%, zone 10 tune from 20% to 24%, the sum of other zones' stay time ratio is equal to (1-zone1-zone10). The response of DOE is pad thickness uniformity which should target to minimize. That is to say, pad thickness standard deviation (Std Dev.) should be minimize.

Experiment Table and Results

For pad count and process time limitation, we just consider major effect (zone1: 16%, 20%, zone10: 20%, 24%) and add 2 center points (zone1: 18%, zone10: 22%). There have 5 runs, which is the sum of intercept, two major effects (zone1 and zone10) and two center points. In the experiments, the total stay time in zones for each run keep the same value (zone1+ zone2+ zone3 + … + zone10=1), as showed in TABLE II. For the central zones' stay time have little influence on pad uniformity, we just lightly tune the middle zones' stay time after the harmonization of zone1 and zone10. As the edge thickness of pad on zone1 and zone10 have a big fluctuating, so we stress on these two zones' stay time tuning.

TABLE II
Experiment Design Table (stay time ratio by zone)

Zone	Run #1	Run #2	Run #3	Run #4	Run #5
1	18%	16%	18%	20%	20%
2	8%	8%	8%	8%	7%
3	8%	8%	8%	8%	6%
4	7%	7%	7%	7%	6%
5	6%	6%	6%	6%	6%
6	6%	6%	6%	6%	6%
7	7%	7%	7%	7%	7%
8	7%	7%	7%	7%	7%
9	11%	11%	11%	11%	11%
10	22%	24%	22%	20%	24%

After the same process time close to tolerance limit on the same tool, we collect pad remain thickness by zone in TABLE III.

TABLE III
Experiment Result Summary (pad remain thickness/mil)

Zone/ thickness	Run #1	Run #2	Run #3	Run #4	Run #5
0 (center)	60.09	60.83	60.06	57.77	58.63
1	58.37	61.11	58.58	59.5	59.7
2	56.46	61.11	56.29	58.94	59.82
3	55.3	60.05	55.22	56.82	57.84
4	54.79	57.71	54.83	57.38	58.29
5	54.91	57.71	54.81	56.99	57.9
6	53.57	56.15	53.77	57.38	56.49
7	54.07	56.15	53.85	56.99	55.59
8	53.4	54.98	53.36	55.82	54.22
9	55.24	52.07	55.42	53.87	51.46
10	56.02	49.95	56.12	54.26	50.41
10-1 (edge)	56.41	50.34	56.55	55.82	50.47
Ave.	55.72	56.51	55.74	56.80	55.90
Range	6.69	11.16	6.70	5.63	9.41
Std Dev.	1.95	4.04	1.98	1.67	3.48

978-1-5386-5309-8/18 $31.00 © 2018 IEEE

Data Analysis

From above TABLE III of experiment results, Run #1, #3, #4 pad thickness uniformity less than 2 and meet the requirement. We can see that the sum stay time of zone1 and zone2 not more than 40% and the difference between zone1 and zone2 not more than 4%, which ensure the pad uniformity lower than 2. Since the RSquare is 0.999956 and more than 0.75, the p-value of ANOVA is less than 0.05, which means that the linearity model of fit is good and successful, as shown in Fig. 4.

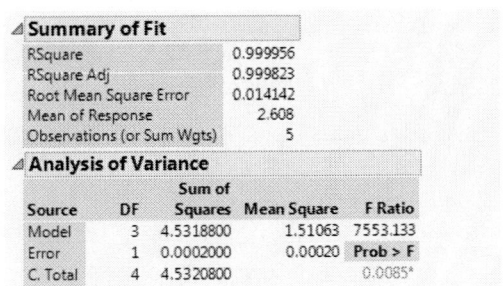

⊿ **Summary of Fit**	
RSquare	0.999956
RSquare Adj	0.999823
Root Mean Square Error	0.014142
Mean of Response	2.608
Observations (or Sum Wgts)	5

⊿ **Analysis of Variance**

Source	DF	Sum of Squares	Mean Square	F Ratio
Model	3	4.5318800	1.51063	7553.133
Error	1	0.0002000	0.00020	Prob > F
C. Total	4	4.5320800		0.0085*

Figure 4: DOE Simulation and AVOVA Check

For the parameter estimates analysis shown in Fig. 5, the p-value is less than 0.05, which means the significant effect on the response. The sorted parameter estimates shows the parameters which have significant effect from big to small influence, and all factors are significant in the model. Zone10 stay time has strong effects on the response, zone1 stay time have quadratic model and also has big effect too.

⊿ **Sorted Parameter Estimates**

Term	Estimate	Std Error	t Ratio		Prob>|t|
zone10(20,24)	0.89	0.01	89.00		0.0072*
zone1*zone1	0.915	0.014142	64.70		0.0098*
zone1(16,20)	-0.295	0.01	-29.50		0.0216*

Figure 5: Key Factors Checking

According to the model and set desirability for the responses, the best process condition 18%/ 20% (Zone1 stay time ratio / Zone10 stay time ratio) can be determined with prediction profiler by maximize desirability, as shown in Fig. 6. The desirability is 0.781582, and when zone1/ zone10 are equal to 18%/ 20%, the pad remaining thickness standard deviation is close to 1.

As the baseline pad thickness uniformity showed above is near 3, which bring in high risk to wafer processing and the pad must be replaced with a new one no more than one day. With the optimized result of DOE, we can keep low pad uniformity then prolong pad using time to 2 or 3 days. It is a great breakthrough for pad thickness uniformity after polish with tolerance limit. Also it is a big cost down item for pad expense in IC manufacturing.

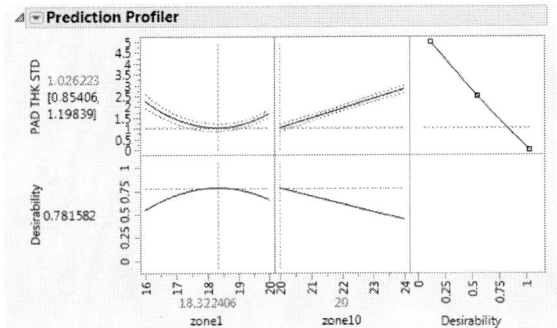

Figure 6: DOE Best Condition Analysis

From the contour profiler charts in Fig. 7, we set the constraint for zone1 and zone10 with the experience summary. We can see that, when selecting the best condition of 18%/ 20% (Zone1 stay time ratio / Zone10 stay time ratio), we can get the appropriate response value 1.05, and the white area in the picture shows the process window and adjustable tuning space.

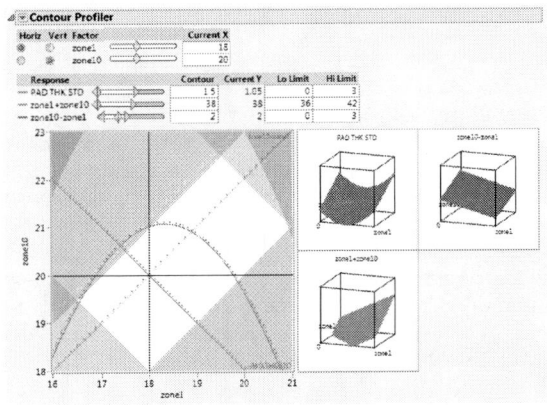

Figure 7: Process Window Analysis

DISCUSSION

According to the DOE analysis results above, wafer stay time in zone10 is the most critical parameter for pad thickness uniformity improvement. Thus, the shape of deformation across the pad radius will get worse as the stay time in zone10 become too high. This phenomenon can be observed from TABLE III, in which we get the result that when zone10 ratio change to 24%, the pad remain thickness STDEV increase to over 3. Data were collected to verify response under the best process condition (18% in zone1/ 20% in zone10). The charts in Fig. 8 show that the data under DOE condition meet the requirement and is stable in two verified conditions. And the results are proofed by the pads' longer using time in mass production. Since the sum of all the zone stay time must keep the same, zone1 and zone10 are the most sensitive area on the pad, suggest following DOE result to find the suitable distribution ratio for all zones.

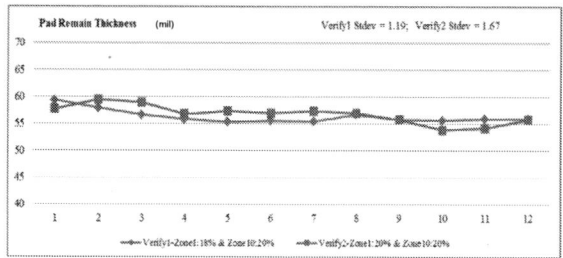

Figure 8: Pad remaining thickness verify under DOE condition

CONCLUSION

In this paper, we study the non-uniformity optimization of polishing pad from kinematic point of view. An important observation for polishing pad is that, the thickness in zone1 and zone10 are hard to control and also the sensitive to the pad uniformity. It is shown by the experiments that the pad's uniformity is determined by two parameters, wafer polish stay time in zone1 and zone10. With the adjustable ratio from zone1 to zone10, the pad remain thickness distribution is a dynamic changing trend. In general, the primary factors affecting the pad uniformity are mechanical (down force on the wafer, platen speed, pad's structure, etc.) and chemical (slurry chemistry). In our paper, we fix all the above factors and select wafer stay time by zone to improve the throughput. Our experiments prove that control wafer stay time in zone1 and zone10 can fix up pad thickness uniformity issue. Then by zone to distribute the wafer stay time on polishing pad is proposed for improving the non-uniformity of the pad. We have used DOE to determine the final process condition of 18% / 20% (Zone1 stay time ratio / Zone10 stay time ratio) as the optimal CMP polish condition in mass production. In the future, other mandatory factors including pressure distribution, interaction with slurry, etc. will be taken into account step by step. In addition, using DOE method to find the optimal process condition is a good solution in semiconductor problem solving.

ACKNOWLEDGEMENTS

Authors want to specially thank CMP team in SMIC SZ site for data acquisition, technical discussion and process discussion.

REFERENCES

[1] R. DeJule. CMP Challenges below a quarter micron, Nov. 1997, pp. 54-60.
[2] Michael Quirk, Julian Serda. Semiconductor Manufacturing Technology [M], 2007, pp. 50-53.
[3] Feng Tyan. American Control Conference, Portland, June 8-10, 2005, vol. 3, pp. 2046-2051.
[4] H. Hocheng, H. Y. Tsai, and M. S. Tsai. International Journal of Machine Tools & Manufacture, 2000, vol. 40 (11), pp. 1651-1669.
[5] G. P. Muldowney and D. B. James. Proc. of 2004 MRS Spring Meeting, 2004, pp. 816.

METHODS FOR TESTING SIGNALS CONSIDERING JITTER TRANSFER FUNCTION[*]

Kenichi Nagatani[*]

Advantest (China) Co., Ltd. Shanghai, China

*Email: kenichi.nagatani@advantest.com

ABSTRACT

A signal with a jitter in excess of one unit interval will result in a failure of functional testing because the signal and its expected data are misaligned. Additionally, jitter measurements performed without considering the correct jitter transfer function (JTF) will yield incorrect results. Here, two methods are presented: one for functional testing while countering jitter and the other that considers the JTF for measuring jitter. The new methods use a time measurement unit (TMU) in automatic test equipment (ATE) and signal processing.

INTRODUCTION

High-speed IO is an increasingly popular technology that has continually increased data rates. Although high-speed IO was originally developed for network and computer applications, it is now incorporated into consumer devices that are equipped with several types of high-speed IO such as USB, SATA, PCI-Express, and HDMI. These interfaces comprise clock recovery (CR), which extracts clock signals from serial data or multiplies the reference clock; and a series-to-parallel circuit, which samples incoming data with the clock signals from the CR to convert to parallel data. The architecture of the clocking and transfer functions of HDMI2.0 is shown in Figure 1 [1].

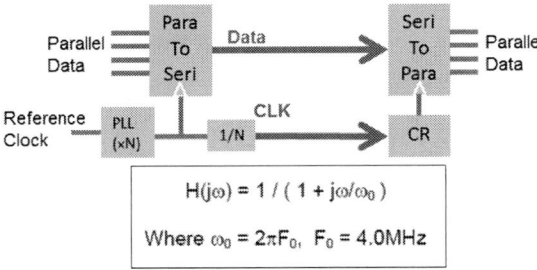

$$H(j\omega) = 1 / (1 + j\omega/\omega_0)$$

Where $\omega_0 = 2\pi F_0$, $F_0 = 4.0\text{MHz}$

Figure 1: HDMI2.0 clocking architecture and JTFs of an ideal CR unit for recovery clock definition

To evaluate the high-speed IO on the bench, oscilloscopes equipped with a software CR are typically used. The signal quality of the transmitter can be evaluated using the oscilloscopes. However, a functional testing is required to verify the logical states of the transmitter. In contrast, ATEs are designed to test DUT functionality as their main purpose, thereby allowing high throughput at a low cost. As a result, ATE channels for high-speed IO comprise comparators, which are equivalent to 1-bit ADC instead of multi-bit ADC. The ATE channel is equipped with a hardware-based CR. Therefore, the jitter transfer function (JTF) of the CR cannot be controlled flexibly. Additionally, the CR cannot operate at a high data rate. We observe that certain video-related interfaces, such as HDMI, have large-amplitude jitter (>1 UI). Figure 2 shows an example of mismatching between DUT output and the expected data owing to the large-amplitude jitter. In this example, the signal is delayed (shifted right) by the jitter, resulting in a failed test. To obtain an accurate test result, this jitter must be counteracted using the CR as in actual applications.

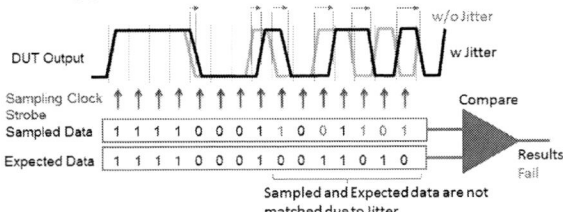

Figure 2: Example of DUT output that is mismatched with the expected data because of jitter

METHODS

Functional testing method while countering Jitter

The following explanation using Figures 3–9 shows the process of the new functional test that we developed.

First, Figure 3 shows the TMU stamps taken at the time when the signal crosses a threshold level (Vt); it then calculates the time interval between one edge and the next edge. From this interval, we calculate the number of consecutive same-level bits (NCSLB) using the equation

$$\text{NCSLB} = \text{integer } \{(\text{Time } [n + 1] - \text{Time}[n]) / \text{Ideal\_UI}\} \tag{1}$$

where integer() represents a function that returns an integer obtained by nearest rounding of its argument and Ideal_UI is equivalent to 1 UI without jitter. The NCSLB is compared with the expected data to obtain the test results.

978-1-5386-5309-8/18 $31.00 © 2018 IEEE

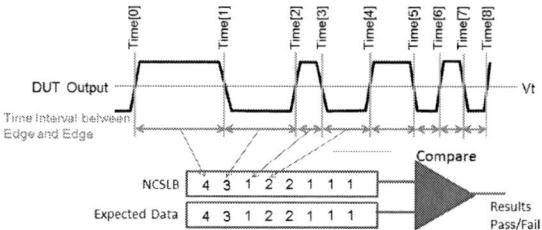

Figure 3: Functional test based on NCSLB

Figure 4 illustrates the process employed by this method to cancel the effects of jitter. In this example, the edge of the data at cycle 19 (19.0) is shifted to cycle 20 (20.63) because of accumulated jitter, and the data is therefore mismatched. However, in the short period from cycle 0 to cycle 4, the bits between the leading edge and the following edge becomes to 4.40, and its NCSLB is therefore 4. The elimination of this fractional part can function as a jitter counteraction. The acceptable range for edge-to-edge variation from the ideal signal is ±0.5 UI.

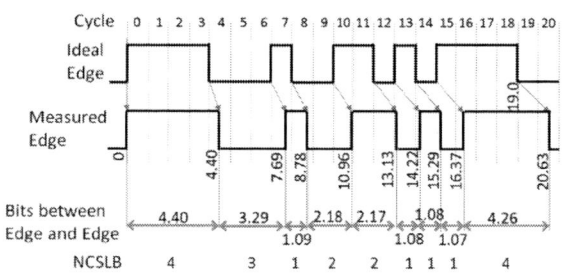

Figure 4: Illustration of jitter counteraction

Jitter measurement method considering Jitter Transfer Function

Jitter is defined as variations in the timing of significant instants in a signal; there are several definitions such as cycle-to-cycle jitter, period jitter, and time interval error (TIE). For high speed IO evaluation, the TIE is commonly used and is also known as accumulated jitter or phase jitter. The TIE is the difference between a measured edge and its ideal edge over all periods. It can be calculated by subtracting the time of the measured edge from the time of the ideal edge. Figure 5 shows an example of TIE causing jitter in a signal.

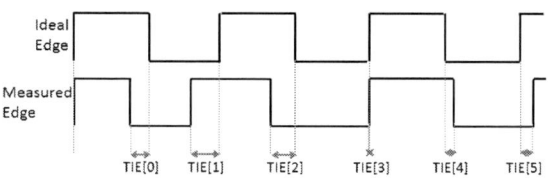

Figure 5: Example illustration of TIE

The JTF should be considered while measuring the jitter. The JTF is a function of frequency, which expresses the ratio of output jitter to input jitter in a given signal. The clock signals from CR is affected by the JTF because there is a PLL (or circuit which has a feedback system) inside the CR. The PLL has a closed loop response, which shows the characteristics of a low-pass filter. Jitter components below the cut-off frequency passes through and those above it were attenuated. Figure 6 shows a jitter transfer characteristic for HDMI2.0 as an example. The gain becomes 0, −3, and −10 dB at 100 KHz, 4 MHz, and 36 MHz, respectively. The JTF must be implemented in our measurement system to accurately measure jitter.

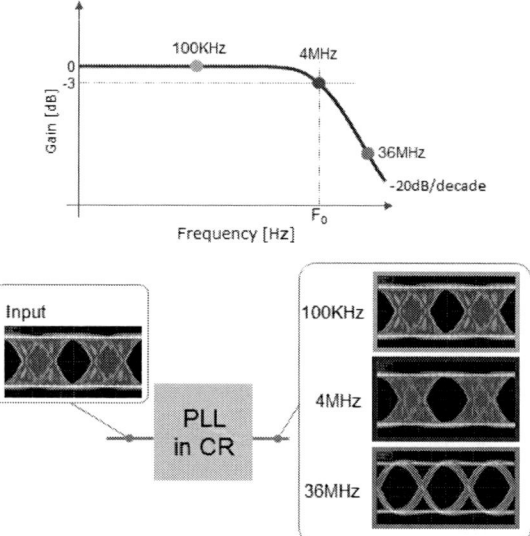

Figure 6: Jitter transfer characteristics for HDMI2.0

The following explanation with Figures 7–10 shows a jitter measurement process. In this example, DUT outputs clock signal, which is modulated with a periodic jitter (PJ). First, the TMU captures both rising and falling edges of the signal and calculates TIE over all edges to obtain a TIE trend, as shown in Figure 7.

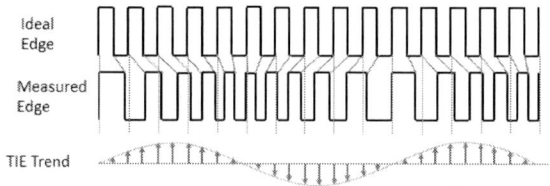

Figure 7: Plots of the TIE trend

Next, a FFT of the TIE trend produces a jitter spectrum. As shown in Figure 8, this spectrum reveals the PJ.

978-1-5386-5309-8/18 $31.00 © 2018 IEEE   354

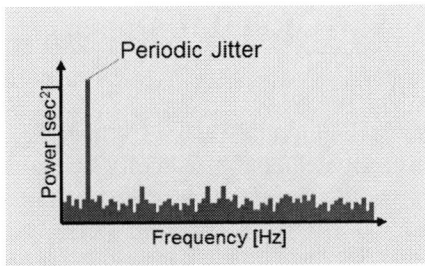

Figure 8: Jitter spectrum produced by FFT

A low-pass filter is applied in order to implement the JTF. Figure 9 illustrates this process.

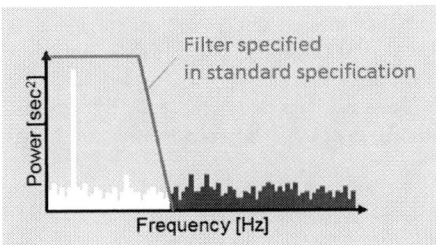

Figure 9: Filtering process of lower frequencies

Figure 10 demonstrates the process of applying an IFFT to the filtered TIE spectrum to obtain a TIE trend. Finally, jitter peak-to-peak values are calculated from this TIE trend.

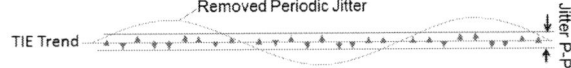

Figure 10: TIE trend after filtering low-frequency jitter

The FFT requires continuous time domain data; however, the TIE of a data stream, such as a PRBS, is non-continuous. To generate a complete continuous time series dataset, the missing edges must be interpolated. In Figure 11 the interpolated edges are colored in red. From the numerous possible interpolation methods, a linear proportional interpolation was chosen herein, as it is sufficient for this task and easy to implement [2].

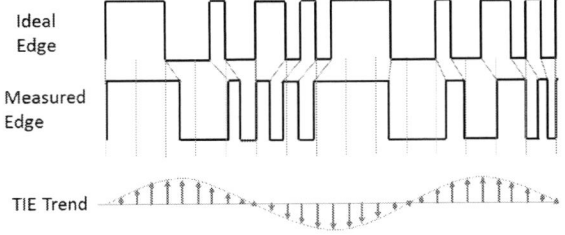

Figure 11: Interpolated TIE trend

RESULTS

To verify our methods, we have tested HDMI2.1, which contains a jitter that is observed to be greater than 1UI, using a PinScale SL digital card on a V93000 SoC tester and a PRBS7 pattern at 12 Gbps. We executed the functional testing method and obtained pass results while countering jitter. Additionally, we were able to apply the JTF based on the HDMI specification to the jitter measurement method and obtain the correct jitter value. The results are shown in Figure 12. When a 8MH filter was not applied, jitter peak-to-peak was127 ps (168 ps/UI); however, when the 8MHz filter was applied, it became 9.4 ps.

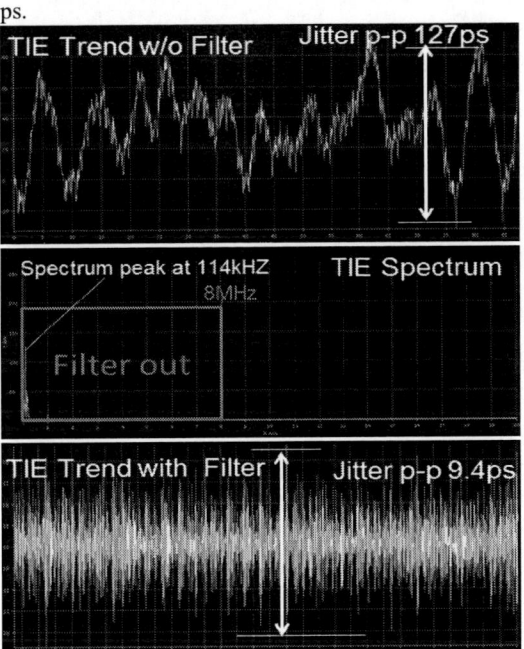

Figure 12: TIE Trend, Spectrum, and Jitter p-p of an actual HDMI2.1 device

CONCLUSION

We have presented unique test methods, which consider the JTF on ATE. Our methods have been successfully applied to evaluate some devices. These methods can be further used for mass production and improve the quality of various devices.

ACKNOWLEDGEMENTS

The author would like to acknowledge Takashi Iino, Qiang Zhou, and Haizhou Xu who assisted with this study.

REFERENCES

[1] High-Definition Multimedia Interface Specification 2.0b HDMI Licensing LLC, March 3, 2016, pp. 31

[2] V93000 Technical Documentation Center DSP-Based Testing - Fundamentals 33– Interpolation, Topic 125436, Advantest Corporation 2011

THE MONITORING EVOLUTIONARY ALGORITHM
IN DEVICE PRODUCTION TESTING

Kun Xu[1]*, Songmiao Wang[1], Wei Xu[1], and Qin Wang[2]

[1] System Engineering Department, Advantest (China) Co., Ltd, Shanghai 201203, China
[2] Department of Micro/Nano Electronics, Shanghai Jiaotong University, Shanghai 200240, China
*Corresponding Author's Email: kun.xu@advantest.com

ABSTRACT

There is Test Data Robustness (TDR) judgment issue in device production testing process. Part Average Testing guideline (PAT) is proposed to fix this issue through tight limits establishment in automotive device testing.

In this paper, a new Monitoring Algorithm (MA) is proposed based on the frame of Evolutionary Strategy (ES) and set Complex Process Capability Index (CPK) as constraint condition to be suited for consumer integrated circuits (IC). The issue test data simulation result displays the ratio of detection to abnormal data is 97.4% while the dynamic PAT is invalid on same data.

INTRODUCTION

The Test Data Robustness (TDR) will influence the device stability and quality in production process. To judge TDR, statistics methods should be built that test results outlier will be transferred from preset limits to statistics results [1,2]. At the same time, Part Average Test (PAT) is proposed to fix this issue in automotive device testing by Automotive Electronics Council (AEC) [3]. In this guideline, AEC suggests all of automotive electronics productions should follow PAT before ship out.

Base on standard PAT method, serial new adaptive test methods are developed [4-6]. These methods import evolutionary algorithms (EAs) into device testing, such as support vector machine, decision tree algorithm, heuristic algorithm and so on. But all of them are suited for six sigma tolerance and zero target of defect parts per million productions (DPPM) in automotive device testing. But the DPPM keeps away from zero for consumer devices. Therefore, these traditional PAT methods are not appropriate for consumer ICs.

A new Monitoring Algorithm (MA) is proposed to monitor consumer devices' TDR under production in this paper. MA is based on dynamic PAT method, involved Evolutionary Strategy (ES) and set Complex Process Capability Index (CPK) as constraint condition. MA is developed and realized under V93000 automatic test equipment (ATE) of Advantest. Now, it is compiled to shared library and can be called in test program.

THEORY AND METHODOLOGY
Part Average Test Method and Disadvantage

PAT method that introduced in guideline includes two manners to establish new test limits. They are static PAT and dynamic PAT. All of them use same formulas to set new test limits. The new test limits calculation process is based on quartile theory in statistics. In this theory, the relationship of three points $\{Q_1, Q_2, Q_3\}$ will reflect distribution status of population. In PAT methods, test data mean value is robust mean that is equal to Q_2, the middle point of the ordered data. And sigma value is robust sigma gets from Q_1 and Q_3 that are the points of 1/4 and 3/4 position.

Then, PAT sets six sigma tolerance to tighten test limits as equation 1 than upper/lower specification limits (USL/LSL).

$$PAT\ Limits = Q_2 \pm 6 \times \left| \frac{Q_3 - Q_1}{1.35} \right| \qquad (1)$$

PAT methods use quartile theory because they are insensitive to outlier than statistics' mean value and sigma. But PAT limits may be invalidated if the test data type is integer and the distribution is very concentrated. Equation 2 is one of examples. In this case, robust sigma is zero because both Q1 and Q3 are 56.

$$Data = \{55,56,56,56,56,56,56,56,56,56,57\} \qquad (2)$$

MA needs to consider it because these similar data distributions will appear in some test items, such as low accuracy analog-to-digital convertor (ADC) regulation and so on. MA avoids failure by recombination operation in $(\mu+\lambda)$-ES that will be introduced on next sub-section.

Algorithm Frame based on Evolutionary Strategy

ES, one type of EAs, is an optimization algorithm based on the model of evolution. Whatever variations of ES, they include reproduction, mutation, recombination and selection always. ES is developed step by step, such as $(1+1)$-ES, $(1+\lambda)$-ES, $(\mu+\lambda)$-ES and so on [7]. MA selects $(\mu+\lambda)$-ES as basic algorithm frame because most of device production testing data distributions are one dimensional normal distribution to match with advantage of $(\mu+\lambda)$-ES.

Standard $(\mu+\lambda)$-ES flow is an algorithm to solve equations in artificial intelligence, and it will return a group of candidate solutions as result. But TDR judgment issue is not same. It should analyze the TDR of device under testing (DUT) depends on historical data and passed data. MA optimized algorithm frame as Figure 1.

978-1-5386-5309-8/18 $31.00 © 2018 IEEE

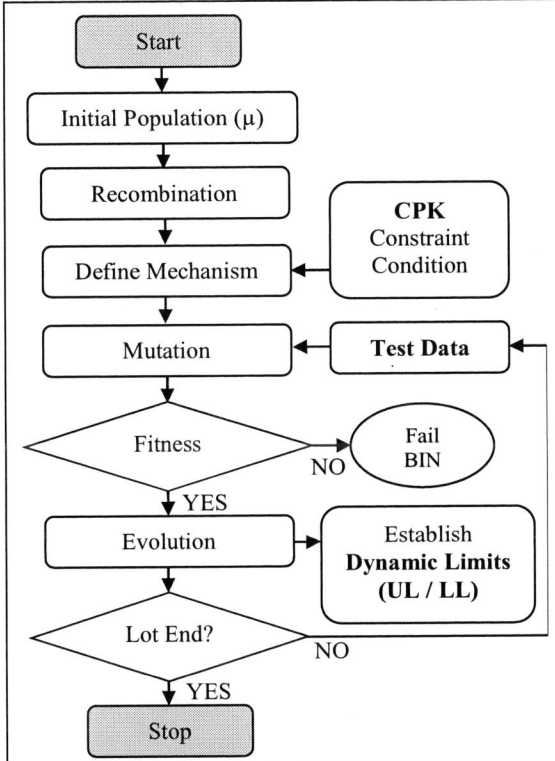

Figure 1: Optimized Evolutionary Strategy Frame in MA

At the first step, MA will build initial population from historical test data, such as STDF files. Secondly, initial population will be recombined considered equation 2 case and other situation. After CPK constraint condition setting finished, the first tighten limits is established.

Then, automated testing starts in mass production. Each DUT will be tested and generates test data. Through additional special library API calling, new focused test data can be pushed into MA and TDR can be checked by fitness function. MA will establish new limits if fitness result is accepted, or fail bin information if it is rejected.

When DUT test data is accepted, population evolution is another task of MA. MA will put test data into original population as mutation operation. Population is evolved and new limits will be recalculated and re-established based on the new population. That means the limits are dynamic and their values are related with the statistics of the lot under testing. In other words, MA is adaptive as same as other evolutionary methods, such as [4-6].

Fitness function is one of core functions in MA based on ES frame. The design of fitness function is on the basis of practical problems. In MA, it is designed for check the relationship between population statistics and CPK target as constraint condition, that will be introduced on next sub-section.

Process Capability Index as Constraint Condition

MA has better performance than other test methods because CPK is imported as constraint condition in fitness function. CPK is a statistical measure of the ability of a process to produce output within specification limits. CPK reflects the relationship between the test data distribution and specification limits.

In general understanding, CPK should be the bigger the better. But bigger CPK means test data distribution is far away with specification limits. As mentioned as PAT guideline, it means the specification limits is too wide and some parts with abnormal characteristics will be passed because their test results are in specification limits.

PAT used six-sigma tolerance as the only acceptable terms. It is too inflexible to be fit for most consumer devices. MA adds CPK control level setting to adapt consumer devices, that can be set by user depends on the device preference. New dynamic test limits (UL/LL) will be more tight through iterative computing population CPK value to match CPK control level. Then, dynamic test limits are established after each device testing finished until the lot is end. As a result, the dynamic limits can be established more scientifically from Figure 2 to Figure 3.

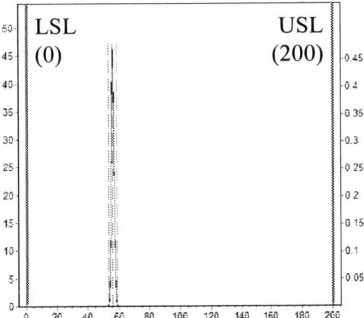

Figure 2: Specification Limits vs. Test Data Distribution

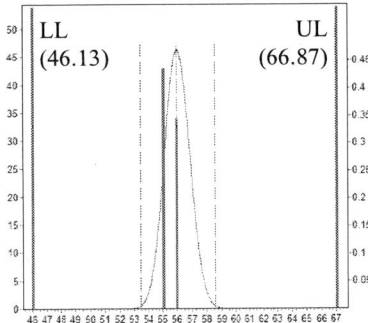

Figure 3: Tightened Limits by CPK and Data Distribution

SIMULATION RESULT

MA is used in devices mass production testing process. It can monitor device test data on time, analyze

TDR and detect issue test data devices.

To verify the performance of MA, one production lot that has issue test data is used to generate simulation input. In this test item, USL is 200 and LSL is 0. There are some abnormal test data from distribution. Because USL is far away from mean value, these abnormal devices cannot be detected by traditional test methods. Figure 4 displays detection process by MA, and Figure 5 displays dynamic PAT detection process under the same input.

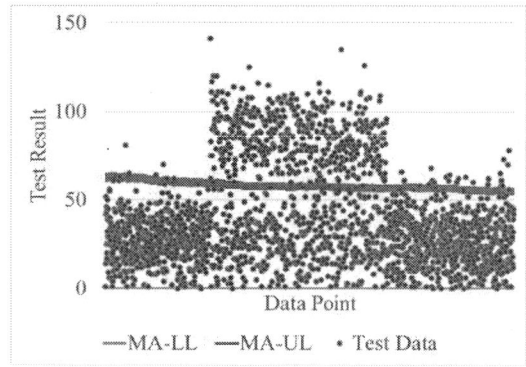

Figure 4: Abnormal Devices Screen out by MA

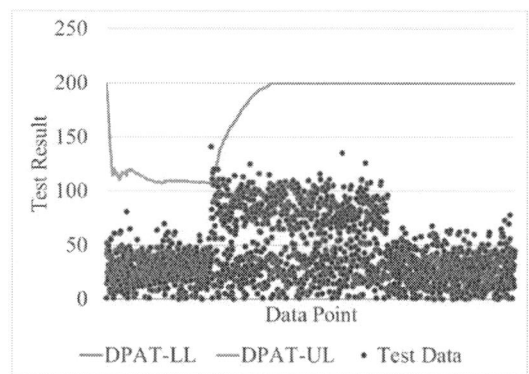

Figure 5: Abnormal Devices Screen out by Dynamic PAT

MA's dynamic UL trend is much more stable than dynamic PAT compared with the two figures. Because MA uses evolutionary population and CPK to calculate limits for each DUT. When this population size is big enough, these continuous marginal-pass test data will influence statistics of population finitely.

Table 1 displays the simulation result for this issue lot. After analyze the issue lot statistic distribution, the data that is more than 60 (Mean value plus three sigma) will be identified as abnormal. Under MA's real-time monitoring process, total 97.4% issue units are detected. And DPPM is reduced by one order of magnitude at least. While traditional dynamic PAT only detects 4 devices on the beginning, less than 1% detection rate.

TABLE I. THE SIMULATION RESULT OF MA

Lot Size	Issue Units	Org DPPM	Detect Count	Detect Rate	New DPPM
11440	498	43531	485	97.4%	1187

CONCLUSION

In this paper, a new real-time monitoring algorithm, MA, is proposed for TDR judgement issue under device production testing. MA is based on PAT and is expanded to consumer device testing after ES is imported as algorithm frame and CPK as constraint condition.

Simulation result shows MA has better performance than standard PAT method. The detection rate of abnormal devices is 97.4% compared with the less than 1% of dynamic PAT.

ACKNOWLEDGEMENTS

This work is initiated and supported by Advantest (China) Co., Ltd. Dr. Qin Wang, from Shanghai Jiaotong University, provides the academic support for MA design.

The authors would like to thanks Y. X. Tian from Spreadtrum for algorithm discussion and data analysis.

REFERENCES

[1] W. R. Daasch, G. Shirley, and A. Nahar. "Statistics in Semiconductor Test: Going beyond Yield", *IEEE Design & Test of Computers,* vol. 26, no. 5, 2009, pp. 64-73.

[2] A. Nahar, K. M. Butler, J. M. Carulli and C. Weinberger, "Quality improvement and cost reduction using statistical outlier methods", *2009 IEEE International Conference on Computer Design*, Lake Tahoe, CA, 2009, pp. 64-69.

[3] Guidelines for Part Average Testing, Dec. 2011, [online] Available: http://www.aecouncil.com/.

[4] H. C. M. Bossers, J. L. Hurink and G. J. M. Smit, "Online Univariate Outlier Detection in Final Test: A Robust Rolling Horizon Approach", *2011 Sixteenth IEEE European Test Symposium*, Trondheim, 2011, pp. 201-201.

[5] D. Drmanac, N. Sumikawa, L. Winemberg, L. C. Wang and M. S. Abadir, "Multidimensional parametric test set optimization of wafer probe data for predicting in field failures and setting tighter test limits", *2011 Design, Automation & Test in Europe*, Grenoble, 2011, pp. 1-6.

[6] C. Streitwieser, "Real-time test data acquisition and data processing enabling closed loop control systems for adaptive test", *2016 IEEE International Symposium on Circuits and Systems (ISCAS)*, Montreal, QC, 2016, pp. 902-905.

[7] H. G. Beyer, and H. P. Schwefel, "Evolution strategies – A comprehensive introduction". *Natural Computing*, vol. 1, no. 1, 2002, pp. 3-52.

978-1-5386-5309-8/18 $31.00 © 2018 IEEE

ACHIEVING MASTERY IN CHIP MANUFACTURING

Cyril-Patrick Fernandez and Jim Jensen

Fuping Semi Yield Masters, Fuping, China, Advanced Fab Technology, Dallas, Texas, U.S.A

Author's Email: patrick@advancedfabtechnology.com

ABSTRACT

Losses in wafer fabs can be $2-10 millions/day and result in lost market share, characterized by delays in product qualification, yield targets, output and cost targets. These performance problems are rooted in high defects, poor non-uniformity and high equipment downtime caused by out-of-control charts, parts failure, recovery from maintenance, process marginalities, etc. Oftentimes, the problem is not observed until final test putting considerable wafers at risk.

We have a unique and proven methodology, categorized by 17 fundamentals that address all causes. The methodology is based on i) experienced knowledge (as different from academic knowledge) based on scientific principles, ii) software to process big data and iii) the right team that executes the methodology in a discipled and systematic manner. Like wheels of a car, all wheels must function perfectly for a car to win the race.

INTRODUCTION

Fabs exist to grow revenues and make profits. However, manufacturing is so complex that commonplace unobserved events hinder profitability and competitiveness. Why? Because of the hectic nature of the fab, the nature of experts, big data and large number of problems exceed the fab's bandwidth to observe the right information to be acted on to avoid problems. Also, Fabs spend 80% of time on relearning old solutions to problems that could have been avoided (Table 1). This takes away the time to learn new technology or innovate.

TABLE 1
EQUIPMENT-PROCESS FAILS RECURRENCE

Bottom's up- inline fails	Percent
Equipment- FDC-PM	65
Process- seasoning, dry clean	18
Fab Infrastructure	6
Design-Process Integration	10
Manufacturing	1
Total	100

The cornerstone of the methodology is the useful knowledge of inline causes for final test failures. Knowing this, one can avoid >80% of fab problems. Proper setup, proper controls and proper ramp cuts learning cycles into half, thus proving mastery in chip manufacturing (figure 1)

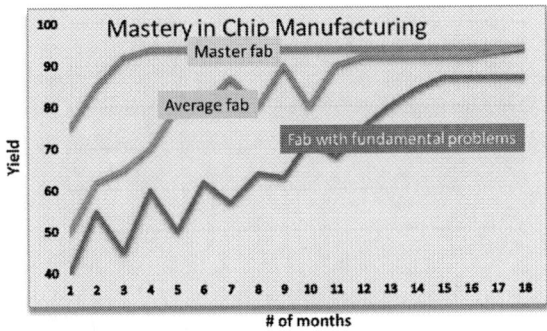

Fig1: Mastery in Chip Manufacturing

CONCEPTS and PRINCIPLES

Given the varied backgrounds of fabs, an implicit understanding of fab principles is necessary to guide fab decisions. (much like the law of gravity, stepping out on the ground floor is fine, stepping out of the 2nd or 50th floor will hurt or kill you). The main principles are:

1. Knowledge gained from experience, based on science and engineering principles and proper application drives mastery of chip manufacturing
2. Even though knowledge experts exist in each area, it is the integration of knowledge of 17 functions that is needed for mastery in chip manufacturing. (Table 2)
3. 80% of SPC 3-sigma controls and big data analysis produces actions that do not impact results.
4. Constants do not stay constant, they vary.

TABLE 2
17 FUNDAMENTALS: % Occurrence

S#	Function	Example	%
1	Design	DFM: Metal Overlap	10
2	Proc.Integ	Interactions between steps	
3	Equipment	Parts, PMs: MFC overshoot	83
4	Process	Marginal recipe- gas nucleation	
5	Facilities	Ammonia causing metal opens	6
6	Manufact.	Misprocessing	1
7	Materials	Slurry filtration, ECP chemicals	1
8	Product E.	Incorrect testing	1
9	Quality C.	Variables and change control	
... 17	Training, IP, Info. Systems, Transfer, KB, Management, Metrology, Yield Enhancement		

PHASES OF A FAB AND 17 FUNDAMENTALS

The three phases of a fab are Setup, Daily Operations and Ramp.

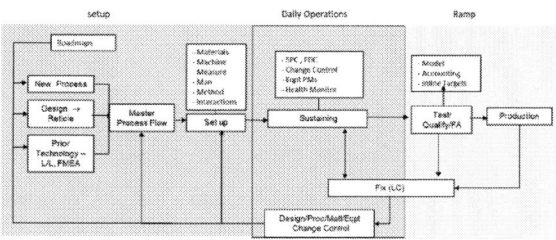

Figure 2: Model of 3 Phases of a Fab

Phase 1: Setup

Setup of the fab for a new product or technology is the foundation for mastery in chip yield; to avoid needless problems later. Like laying a foundation for a high-rise building if 10% of the foundation is weak for a 50-story building, then, the building is at risk. This is what happens in fabs. For example, an equipment setup problem of a 4 second time-to-base pressure difference between two fabs caused a 9-month qualification delay, caused by the lack of understanding of cryopump issue that led to maintenance specifications being different to save cost.

As new products or technology are added to the fab, problems will increase on a weak foundation. Fabs, too often rely on vendors and not themselves, which is a problem as vendors are not responsible for end of line yield only for uptime of their equipment. An example of this is the liberal use of equipment constants by vendors to cover up a component problem, which leads to chamber matching of device performance (iDrive) at final test because a critical real gas flow being different between chambers even though the readout monitored by FDC (fault detection and classification) was correct. This resulted in both tool shut downs (that created a bottleneck), scrap and loss of shipments. Likewise, a process recipe marginality because of an incorrect nitrogen gas flow caused chronic defect excursions that resulted in metal opens. Usually fabs and equipment vendors clean the chamber for a particle excursion, which makes the problem chronic. A study of the last 5 nodes show that 80% of problems in a new technology were previously solved and thus can be defined as a setup problem.

The solution is to use the right knowledge and methodology rigorously at setup. The elements of setup are described in Figure 3,

Phase 2: Daily Operations

The ability to observe a yield problem inline is the cornerstone of steady state manufacturing, as assumed constants do not stay constant. The elements of steady state manufacturing are i) Control (SPC and FDC) for variation and stability, ii) Maintenance and iii) Change Control. Too often too many charts are put under tight 3-sigma control resulting in needless action. It is imperative to know what to prioritize and what control limits to set. For example, 75% of charts do not need 6 sigma limits

and are thus overcontrolled. The root cause of this is a lack of understanding the science and engineering experience of the control team. For example, nitrogen flow is controlled to 3-sigma limits, when it does not need to be and a Silane flow overshoot is not controlled at all (which caused a reliability failure, where the ramifications were severe.

There are similar problems with maintenance, preventive maintenance and change control, discerning between what is critical and what is nice to have but not necessary is key to gaining mastery. It is imperative that the engineer understand the underlying science and engineering causes and interactions. Changes can occur out of ignorance to solve one problem that creates a worse problem later. The solution to this is training and periodic reinforcement of training. The consequence between a good setup and a poor one is variation of yield and output and excursions, both minor and severe.

Phase 3: Ramp

This refers to both yield ramp and production ramp. A perfect setup does not require a yield ramp, except in instances of a new tool and technology. This is minimized by understanding the capability of the new tool or technology, which is done as a step in the setup methodology at equipment and process release.

However, when yield is not perfect, a methodology is to be followed. In most cases a Technology FMEA does not foresee a tightening of a specification is required for the tighter geometry.

Yield Methodology: Methodology is a precise step-by-step execution of all elements that affect yield. The methodology combines the three pillars of success; i) experienced-knowledge based on scientific principles, ii) software to process big data and iii) the right team that executes the methodology in a discipled and systematic manner. Like wheels of a car, all must function perfectly for a car to win the race.

The steps in the methodology are;

1.ESTABLISH PRIORITY-1 process steps using a top's down-bottoms up analysis of yield fail pareto and inline parameters with the knowledge base of failures seen by all fabs. This focuses the right resources and priority to the right areas, (Figure 3), comprehending interactions from upstream and downstream process steps, edge yield and chamber matching. This step establishes defect targets for median and percent out of control at each priority process steps

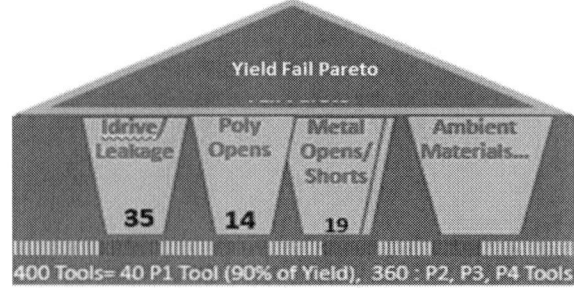

Figure 3: Tops Down-Bottoms Up Priority1 Steps

2) Assess and fix Tool-process at priority process steps
3) Model electrical results=f(inline parameters) for new sensitivities learning
4) Adjust control limits and specifications required for yield goals that has been learned from modelling.
5) Update the fab setup specification covered in Phase 1. Most often this is in the PRD (process release document).

For production ramp, the correct setup specification (including the right wafer-per-hour throughput) and release of the second-of-a-kind tools is critical to avoid chamber matching problems. Most fabs go through the motions of these practices, but do not go deeply and rigorously enough. They are dependent on the vendors, which is a mistake, as vendors have little yield knowledge. Best in class fabs do not depend on vendors.

Finally, the continuous improvement system which innovates and optimizes manufacturing capability for cost reduction, increased capacity and a smoother introduction of new products are routine for best in class fab. Their engineers have time to focus on this as they have less firefighting.

EXECUTION OF METHODOLOGY

The methodology is described in the previous section. The execution of the methodology is fully dependent on the team discipline and rigorousness. When right methodology and execution is done, then the fab has gained mastery in chip manufacturing and fabs make money and are best in the world. Conversely, if fabs are running at a loss or do not qualify products on time, then something is unknown to the fab or execution is lacking because of experience. It is best that fab management get independent feedback (a second opinion) and get benchmark data. The importance of the methodology at each phase of the fab is shown below (Table 3)

TABLE 3
METHODOLOGY Vs. PHASES

	Knowledge	Software	Team
Setup	●	●	●
Operations	●	●	●
Ramp	●	●	●

An interaction and technique example are showing in Figure 4

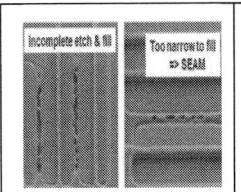

Figure 4: Wafer Edge Gate shorts. All processes were in control, yet the problem occurred. The issue was interactions. A special technique was used to detect the problem

Some tools and techniques and techniques are:
1. Process marginality due to interactions
2. Rapid Learning Cycle
3. Stress test

4. Golden line
5. FDC methodology
6. Sensitivity of process window.

RESULTS

Figure 54 shows the improvement made in a mature fab by correcting the SETUP Phase. This is the cause that affects final test yield result

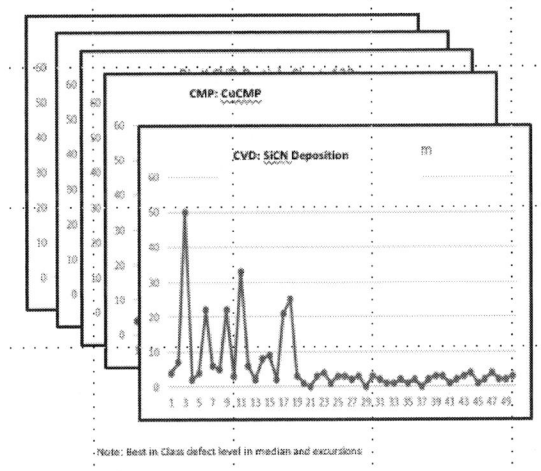

Figure 5: Setup issue produced erratic particle counts.

Fab performance results in improved yield from 10 to 70%, (which improve revenues by up to $2M/day) output by 10%, reduced cost in the millions of dollars and shortened new technology qualification by 12 months. (Figure 6). Note: today the weight on yield is >10X, Output is 3X and Cost is 1X. Yield is crucial.

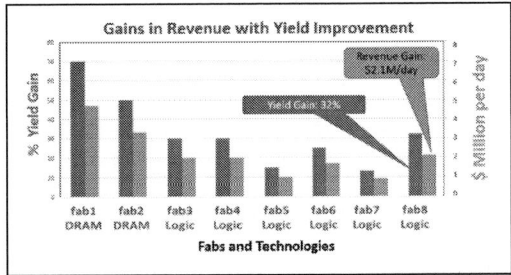

Figure 6: Revenues from Yield Gain with practicing Methodology rigorously

In summary, this unique strategy of using holistic knowledge, software and the right team allows fabs to gain mastery in chip manufacturing

ACKNOWLEDGEMENTS

The authors wish to thank Howard Tigelaar and Sarma Gunturi for editing and contributing key insights to clarify this paper.

REFERENCES

[1] S. Parikh, Fernandez, *Nanochip2013*, p34-38.

APPLICATION OF AN ENHANCED FAILURE BITMAP METHOD FOR MEMORY FAULT ISOLATION

Zhimin Zeng, Yun Xu, Qingwen Zhang, Mingming Li, Yutian Zhang, and Xiaojun Xu, Xiao Zou
Product Engineering Division, Shanghai Huahong Grace Semiconductor Manufacture Corporation
No.1188 Chuanqiao Road, Pudong New Area, Shanghai, China
Corresponding Author's Email: zhimin.zeng@hhgrace.com

ABSTRACT

This paper is to describe an enhanced failure bitmap analysis method which can significantly improve the hit rate on memory cell fault isolation than conventional failure bitmap method. In this method, during bitmap test analysis, a critical parameter is enhanced by corresponding setting, to make the difference between faulty memory cell and non-faulty memory cell large enough to be recognized and then be represented by failure bitmap. This paper presents the methodology and several typical application cases.

Key words: failure bitmap; fault isolation; memory

BACKGROUND

In semiconductor industry, memory is widely used as standalone or embedded products. Usually, critical design rule is employed in memory cell array design, therefore, memory yield shows high sensitivity on process window and defect, the root cause identification of memory failure is very important for process optimization and defect reduction. In memory failure analysis procedure, failure bitmap analysis is the most effective EFA (Electrical Failure Analysis) method for fault isolation before PFA (Physical Failure Analysis) work. Depending on the defect size and location, the failure bitmap can show as SB(Single Bit),TB(Twin Bits),BL(Bit Line),WL(Word Line),Cross(BL-WL cross),Cluster or Full-Array failure mode. Since the SB, TB, Cross and Cluster failure modes provide accurate fault address information, it can help PFA with high hit rate to identify the root cause. Meanwhile, other modes could not provide accurate fault address information, which brings big challenge for PFA work.

In this paper, a method called "Enhanced failure bitmap analysis" will be introduced, which is to improve the conventional failure bitmap method and try to provide more accurate fault location for the failure modes of BL,WL and Full-Array by conventional method.

METHODOLOGIES

Fig.1 shows a simplified structure of memory cell array, in which each cell locates at the cross point of word line and bit line. This kind of structure allows to operate a cell independently by selecting a word line and a bit line. In some situation, when a cell is faulty, it could only lead a single bit failure of itself, without causing other cells' failure. This feature is very different from pure logic circuit. And this is why we can capture the failure bitmap, such as SB or TB mode, to help failure analysis work.

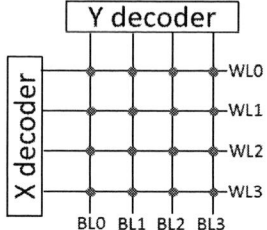

Figure 1: Simplified structure of memory array

Meanwhile, every word line and bit line are shared by multi bits, therefore, if a line suffers a bridge fault, it could cause the signal distortion of the line or of full array and then cause failure as BL, WL or Full-Array bitmap mode. For example, we can hypothesize that the voltage signal parameter of the cells close to the fault should have little difference with those far away from the fault because of the voltage drop due to the parasitic series resistance along the transfer path. If the voltage signal parameter is critical for the memory cell operation and make it pass/fail the bitmap test by a criteria value, we name this parameter as a critical parameter and this criteria value as Vpass. Fig.2a shows a distribution of critical parameter by cell address under conventional test conditions. We can see that the critical parameters of all cells are below Vpass and they should fail the bitmap test due to the impact by a fault at a cell address.

From the distribution image we also see that there is a difference between the cells close to faulty address and that far away from faulty address. Considering the parasitic series resistance model, we can enhance the difference by promoting the source voltage level of the critical parameter, and then make some non-faulty cells' critical parameters over Vpass, while that of most close to the real fault address still below Vpass (Fig.2b). Under this condition the fault address could be isolated in failure bitmap testing, we name this method as "enhanced failure bitmap method".

978-1-5386-5309-8/18 $31.00 © 2018 IEEE

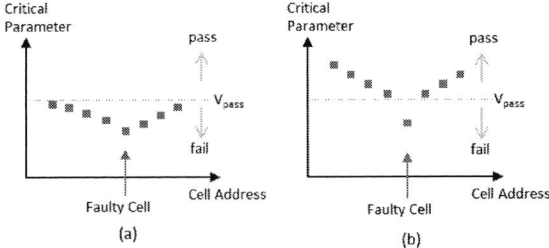

Figure 2:(a)With conventional method, the critical parameters of faulty cell and non-faulty cells below Vpass (b) With enhanced method, the difference of critical parameter between faulty cell and non-faulty cells are enlarged, and two groups separately stand on both sides of Vpass

To help understand this method, we have a discussion on an example of memory Word Line failure. Fig.3 shows a simplified structure of 4x4 memory cell array, we define that the selected WL voltage level is Vcc and the un-selected WL voltage level is 0V at normal "read" operation. We assume that a WL has a bridging to ground (GND) fault (Stuck-at 0) at one cell address (bit2 on WL1), it will cause WL failure mode at bitmap test due to WL voltage drop.

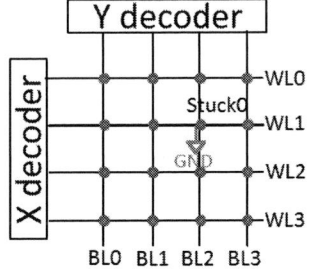

Figure 3: WL with stuck-at 0 fault

Considering the parasitic resistance model of WL with Stuck0 fault, when the WL is selected, the voltage distribution of each bit on this WL is shown in Fig.4a. The voltage of the specified position can be calculated, equation (1) shows an example for V_{bit1} calculation:

$$V_{bit1}=V_{cc}*\ (R_2+R_{short})\ /\ (R_o+R_0+R_1+R_2+R_{short}) \qquad (1)$$

Here, R_o is the total output resistance of tester power supply and WL decoder, R_0 to R_3 are the WL parasitic resistances corresponding to bit0 to bit3, and R_{short} is corresponding to the bridge. V_{bit0} to V_{bit3} are the localization WL voltage level corresponding to Bit0 to Bit3, and they are the critical parameters in this case. So Vpass is the WL voltage value at which the bit can be selected and pass the read function test.

In general, R_o is higher than others, therefore, V_o has larger drop from Vcc, which causes V_{bit0}, V_{bit1}, V_{bit2} and V_{bit3} all falling below Vpass, and makes WL fail in conventional bitmap test (Fig.4a). When we promote Vcc value, based on (1), all the WL localization voltages will rise up. At the Vcc enhanced condition as Fig.4b, V_{bit0} and V_{bit1} are both higher than Vpass, therefore, bit0 and bit1 turn to pass at bitmap test. Comparing to the conventional bitmap test, the failure address is narrowed down from WL to two bits, which is a significant improvement for the fault isolation.

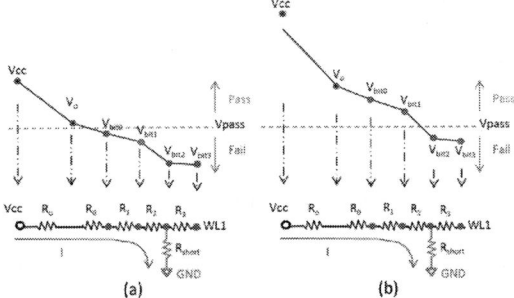

Figure 4:(a) At conventional condition, resistance model of WL with bridge fault, V_{bit0}~ V_{bit3} below Vpass, cause WL failure
(b)With Vcc enhanced condition, V_{bit0} and V_{bit1} rise up, cause partial row failure

APPLICATION CASES

In this section, several real application cases will be discussed.

Case I:

Fig.5a shows a low yield CP bin map of a 130nm embedded SONOS (silicon-Oxide-Nitride-Oxide-Silicon) flash product, random Bin9 is the major failure bin. Refer to the CP test flow, we knew Bin9 failure associates with VPOS, a charge pump voltage for flash cell Erase/Program operation. From the test data log, we knew the normal value is about 7V, and the abnormal value is about 2V~4V. Based on our experience, we have two suspects, one is that the charge pump's function is abnormal, and the other one is that VPOS normal value is dropped down by large leakage loading from flash cell array's bridge fault. To clarify this issue, we tried to do failure bitmap analysis on the fail dies, but only captured a Full-Array failure bitmap after erase operation (Fig.5b).

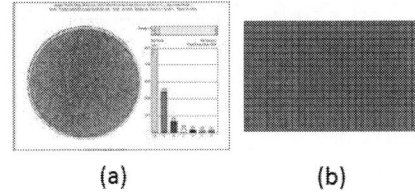

978-1-5386-5309-8/18 $31.00 © 2018 IEEE   363

Figure 5: (a) Wafer CP bin map with random bin9 failure (b) Full-Array failure bitmap by conventional bitmap method

At this situation, the enhanced failure bitmap method was applied. Firstly, a VPOS measure/force path was setup by a test mode (Fig.6), and then a normal or higher VPOS voltage was forced through the VPOS pin, at the same time the flash erase operation was launched. After that a read test was performed and a failure bitmap was captured by the tester. We successfully got the valuable failure bitmaps of Cross mode on most of the samples (Fig.7a). Based on the fault address information, PFA was implemented and a poly bridge defect was identified (Fig .7b).

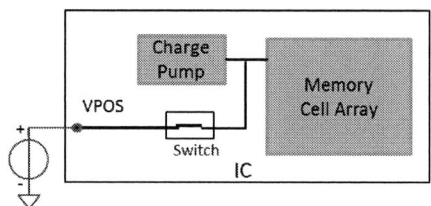

Figure 6: A VPOS measure/force path setup by a test mode

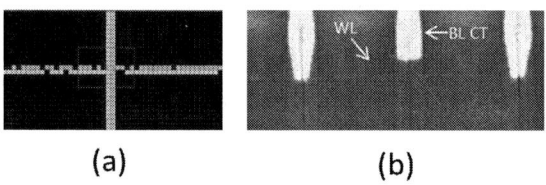

Figure 7: (a) WL-BL Cross failure bitmap (b)WL poly-BL contact bridge on Cross location was shown at FIB cross-section picture

In this case, the critical parameter is VPOS value for erase operation of the flash cell. We can imagine that the hard bridge fault induced a big leakage that the charge pump driving capability cannot afford, which causes the root VPOS value lower than the necessary for any cell's erase operation. After we force the VPOS directly by the tester, with strong driving capability, the internal VPOS value recovers and most of the cells can pass erase operation except the WL and BL corresponding to the bridge fault.

Case II:

Another real application case is on a 90nm 512K bits SRAM IP test chip, which suffered a baseline low yield with function failure. Based on conventional bitmap analysis, the major failure mode is BL failure. Several rounds of PFA were tried according to the BL address, but

no abnormality was observed. A big challenge for PFA is that the BL is too long to examine and the defect could be missed. To provide more accurate fault address, EFA engineer had a try of enhanced failure bitmap method with Vcc as the critical parameter. At first, promoting the Vcc value from 1.5V (typical value) to 1.8V (up spec value), no change was observed on the failure bitmap. It seemed that the enhanced method could not work in this case. But when some extremely high Vcc value (2.0V/2.1V/2.2V) was tried, he found the fail bits count decrease with Vcc rising up and the fail bits finally concentrate at the boundary of the block (Fig8). This phenomenon was confirmed on more samples. After that, product engineer implemented the VC (voltage Contrast) analysis, Nano Probing test, and TEM analysis on the failure address, finally identified the root cause: source/drain leakage due to dislocation related with STI etch stress.

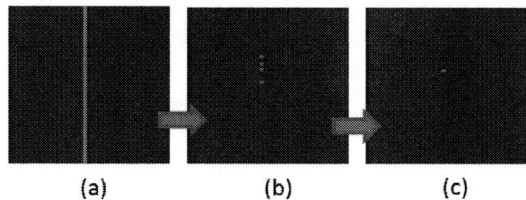

Figure 8: fail bits count decrease with Vcc rising up and the fail bits finally concentrate at the boundary of the block (a)at Vcc=1.5V (b)at Vcc=2.0V (c)at Vcc=2.2V

Case III:

Previous two cases show a way to reduce the fail bits count and then isolate the fault address by using enhanced failure bitmap method, the case below shows another way to increase the fail bits count by using enhanced failure bitmap method.

The sample is an 130nm SONOS embedded flash product failed in endurance reliability qualification. By conventional bitmap method, a partial WL failure bitmap was captured. Since it is a reliability failure, we suspect the physical defect is very tiny and hard to be observed according to the fail WL address, we need to isolate the fault address with more accuracy. In this case we used the cell VTP scan test mode to enhance the failure bitmap. The VTP scan test mode is a method to measure the cell's VTP (threshold voltage after program) that a critical parameter for NVM (Non-Volatile Memory) cell. At VTP mode, a VTP voltage is forced on the gate of cell from 0V, once the voltage is higher than the threshold voltage, the cell should be read as "0" from "1". Fig.9 shows the curve of fail bit count vs. VTP scan, the VTP value of point A is the value employed by normal read mode, at which the endurance fail sample shows WL failure bitmap. When we increased the VTP scan value, we found the fail bit count increased step by step, and at point B the failure bitmap turned to Cross failure mode. That is what we wanted because we

978-1-5386-5309-8/18 $31.00 © 2018 IEEE   364

suspect the cross point is the real fault location and it is accurate enough for PFA. Later, PFA confirmed our suspicion and found a weak bridge between WL poly and BL contact at the poly shoulder due to contact overlay shift and spacer over etch issue (Fig.10).

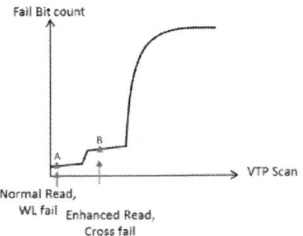

Figure 9: Curve of fail bit count vs. VTP on endurance fail sample

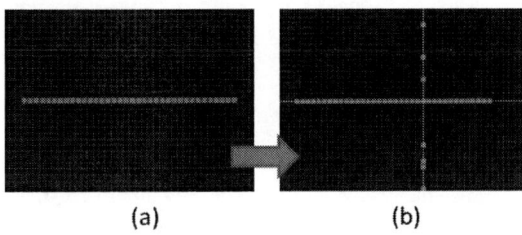

Figure 10: (a) Partial WL failure bitmap at normal read condition
(b) Cross failure bitmap at enhanced VTP read condition

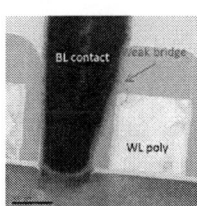

Figur11: TEM picture shows a weak bridge between WL poly and BL contact at the poly shoulder

CONCLUSION

The enhanced failure bitmap method has been proved to be an effective method for fault isolation for memory product by lots of real application cases. It is an improvement of conventional failure bitmap method, it could make breakthrough in some situation that conventional method cannot work. The key point of this method is to choose the proper critical parameter in each real case, so we hope it can be used with flexibility and it would be helpful for your failure analysis work.

ACKNOWLEDGEMENTS

We would like to thank FA team for their great support.

REFERENCES

[1] Chih-Wea Wang et al., "Fault Pattern Oriented Defect Diagnosis for Memories", *Test Conference 2003. Proceedings. ITC 2003*, vol. 1, pp. 29-38, Sept. 30–Oct. 2, 2003.

[2] J. Segal, A. Jee, D. Lepejian, and B. Chu, "Using electrical bitmap results from embedded memory to enhance yield", IEEE Design & Test of Computers, vol. 15, no. 3, pp. 28-39, May 2001

Advanced Process Equipment Matching Methodology in Semiconductor Manufacturing

Ziqian Javaer Liu, Hongtao HT Qian, Mengyang Elaine Liu,

Corporate Quality & Reliability Center, Semiconductor Manufacturing International
Corporation, Shanghai, 201203, China

(Javaer_Liu@smics.com)

ABSTRACT

The typical features in semiconductor manufacturing are complex processes, varieties of tools and tremendous data. In a mega FAB, hundreds of products are processing on thousands of tools, which request the consistency between manufacturing tools (no matter same type tools or different type tools) on the same process stage, called process equipment matching. This paper tries to explain how to set up a high efficiency and accuracy evaluation methodology on process equipment matching with mechanism and real practice. Meanwhile, the systematical application of this methodology is presented on this paper.

Keywords – Semiconductor manufacturing; Equipment matching; Comparison methods; Systematical application

INTRODUCTION

As we all know, semiconductor manufacturing is a complex process with hundreds of products on thousands of manufacturing tools with huge database come out. In semiconductor foundry, the same product with the same layer is impossible to run just in one machine due to the production capacity. However, the production capability of different models or even the same type of machine can inevitably have many differences. Normally, the industry will use the CP/ CPK to reflect the stability of the process in the actual production performance. (CP: Process Capability, used to represent the degree of discrete process distribution, the greater value mean the smaller fluctuation of the process; CPK: Process Performance, used to exhibit the process control capability, the greater the value means better process control capability). The engineers can through the warning rules detect the warning automatically in advance, which are setting in the chart and do action. However, during the control process, if CPK improvement is only by optimizing the performance of actual measurement process, the effect will be limited. It will be very difficult to do the corresponding action for more improvement if the performance already reaching certain value. At this point, the differences' elimination of the different tools are introduced to further enhance the CPK approach, which is the reason why the tool matching control mode came up.

Tool match is a management method according to the SPC (Statistics Process Control) control mode. It is based on the inline (data collection on real products)/ offline (data collection on dummy wafers) monitor data to determine whether the EQP's actual process capability is able to achieve the golden level of the machine group which it belonging to, the golden level is setting by its belonging machine group's performance. Through this method, the abnormal can detect and the action can done in time. Moreover, CPK can improved by avoiding the adverse condition caused by the mismatching between the tools.

Nowadays, some semiconductor manufacturing factories began tool matching review work step by step and achieved some results. For example, One foundry judges whether one machine in the same group went abnormal by offline monitor data analysis[1,2], but it confined to the offline control. For most of the fab, a nature system is missing for the control ways. At the same time, due to the lower frequency of offline data collection than inline, it was very easy to cause the delay of inline performance abnormal review. By far, most corporations with tool matching review fulfilled their control by manual. If so, review will be delayed inevitably, which are unable to fulfill the prescription requirement. It will be very difficult to detect and easy to miss some problems. Moreover, most companies are hard to achieve control by chamber, thus, owner cannot do action targeted that will waste lots of time to find worse chamber. Meanwhile, tremendous data was a huge loading for human.

Therefore, the authors try to create an advanced control method and management system in semiconductor manufacturing basing on statistical concept which is presented in this paper. This paper presents the complete and precise system of tool matching reviews from data collection, analysis and application and introduces a special tool auto system for matching review. This method and auto system have already applied in an actual mature and leading automotive semiconductor manufacturing.

METHODOLOGY

This paper presents a new control method which has applied in a mature semiconductor line and be performed through two aspects of inline/ offline. Until now, the method has already reached chamber performance matching [4]. The specific practices are as follow [3]:

978-1-5386-5309-8/18 $31.00 © 2018 IEEE

Golden performance definition:
1. Golden mean is SPC chart target.
2. Golden sigma is tool/ chamber minimum sigma from 3-month data.
3. Mean shift: MR1 = Abs (Tool mean-golden mean)/golden sigma.
   Sigma shift: MR2 = Tool sigma/golden sigma.

Remark:

MR: Matching Rule，MR is used for the comparison of two samples. Define one sample as baseline, and based on the samples' mean and standard calculate the deviation. Then, determine the comparison results based on the accepted degree of engineering. MR-1 > 0.5 means the obvious deviation between the compared sample with the mean baseline of the standard. If not, it means no obvious difference. MR-2 > 1.2 means the obvious deviation between the compared sample with the standard deviation baseline of the standard. If not, it means no obvious difference (specific reason for the criteria was shown in Figure 1). For few samples, it is normally to use the difference between MR-1 > 1 and MR-2 > 1.5 as the comparison standard.

As depicted in the figure 1, the criteria set up by the demoing out the comparison standard of HC-2. HC-2 is a tool of engineering comparison for baseline. HC-2 ≥ 99% is suitable to compare the key (risk high) items (5% false). HC-2 ≥ 95% is suitable to compare the non-key (low risk) items (1% false). If item fails HC criteria, it can refer MR to judge the failure caused by mean shift or variation shift. Both HC and MR tool are suitable for normal distribution only. Recommend data count ≥ 30 (The statics is meaningful only if the samples were higher than 30). The corresponded criteria: HC-2 < 99% refer to MR1 > 0.5 MR2 > 1.2 (Fig. 1a); HC-2 < 95% refer to MR1 > 1 MR2 > 1.5 (Fig. 1d).

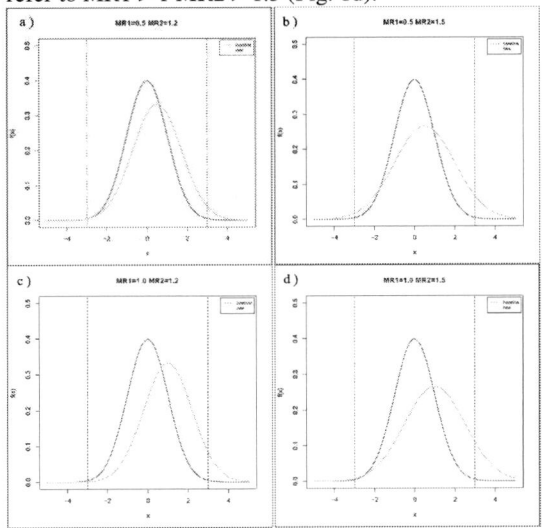

Figure 1: HC-2 compare between different MR

Review mechanism:

Base on the theoretical basis of SPC, this paper present one complete and precise review mechanism as follows [5,6]:

Inline Review: By product, we are maintaining inline Key + Sub-key charts in the system. System will base on the inline charts' SPC target and long-term performance set common golden mean and golden sigma respectively for one EQP group, thus, all EQPs in the same EQP group have same criteria. Judging if one machine matches with other machines or not based on the inline monitor data. The specific criteria are:

(1) Measured lot count ≥ 30 points: MR1 < 0.5, MR2 <1.2;
(2) 15 points≤ measured lot count < 30 points: MR1 < 1, MR2 < 1.5.
(3) Measured lot count <15 points: Not review due to data is not statistically significant.

Offline non-particle review: Put all the offline charts which fitting the normal distribution into the calculation system. Golden mean and golden sigma settings are similar with inline review mechanism. The specific criteria are:

(1) Monitor count ≥ 15 times: MR1 < 1, MR2 < 1.5.
(2) Monitor count < 15 times: Not review due to statistically significant.

Offline particle review: Put all the offline particle charts which fitting the review criteria into the review system. The system will set up a golden value according to the long term performance, and then according to the real performance to judge whether the chamber conditions is OK or not, and judge if the chamber conditions is trend worse or not. The specific criteria are:

(1) Monitor count ≥ 15 times: P90 < Golden.
(2) Monitor count < 15 times: Not review due to data is not statistically significant.

Routine review flow:

The handle procedure of tool matching review on actual automotive manufacture is as follows:

978-1-5386-5309-8/18 $31.00 © 2018 IEEE

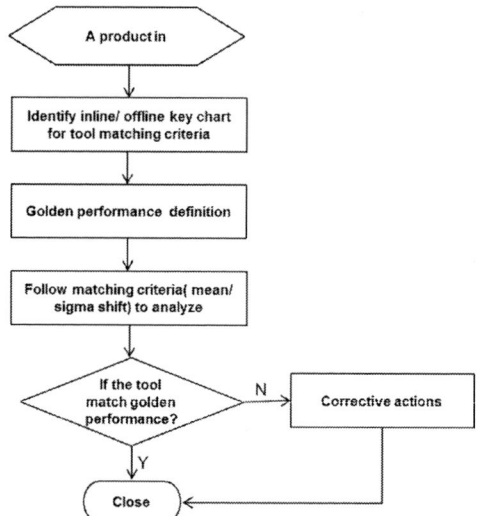

Figure 2: The handle procedure of tool matching review on actual automotive manufacture

Application:

The calculated result of one chart by chamber's performance is shown as figure 3. Under the premise of the monitor data count that meets the minimum sample size requirement, we can judge whether the chamber's performance match with the golden or not clearly. For example in Table I, all the chambers are match golden except chamber 8, which sample size is not enough. Meanwhile, Based on the box chart which shown in figure B, we can judge whether the chamber's performance is within SPEC (the two dashed lines) or not more intuitively. The SPEC setting is based on golden mean and golden STD.

TABLE I: A tool matching calculate result sample on actual automotive manufacture

Process	Golden Parameter	Chamber1	Chamber2	Chamber3	Chamber4	Chamber5	Chamber6	Chamber7	Chamber8	Chamber9
Sample Count	150	195	230	265	160	190	195	175	140	215
Golden Mean	0.18	0.1806	0.181	0.1802	0.1805	0.1804	0.181	0.181	0.1805	0.1789
Golden Sigma	0.003027	0.0028	0.0028	0.0032	0.0031	0.0032	0.0031	0.003	0.0033	0.0027
Mean Shift	0.5	0.2084	0.3334	0.0788	0.1524	0.1217	0.3468	0.3187	0.1602	0.3775
Sigma Ratio	1.2	0.9286	0.9352	1.0583	1.0278	1.0595	1.0086	0.977	1.083	0.8908
Min Sample Size		Enough	Enough	Enough	Enough	Enough	Enough	Enough	Not Enough	Enough
Result		Match	Match	Match	Match	Match	Match	Match	~~Match~~	Match

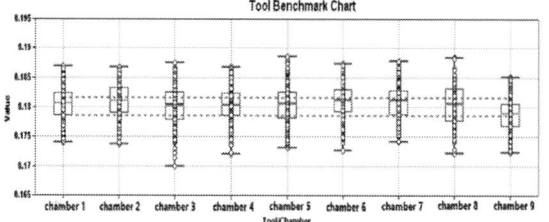

Figure 3: A box chart of tool matching calculate result sample on actual automotive manufacture

Linking to SPC:

Tool matching also has a great effect on inline SPC performance improvement. For example, before the tool matching review mechanism set up in the foundry, the CPK of chart A is 0.81 and after a series of process improvement CPK raised to 1.04, which still does not reach the pass criteria 1.33. More seriously, it is difficult to find a breakthrough to improve. However, after the tool matching review started in fab, the value rose to 1.35 gradually. After SOV analysis, the reason for this chart CPK worse mainly from the variation of lot to lot when CPK = 0.81. When the CPK reaches 1.04, the most important problem is found in the variation between different tools (The value has reached 60.7%) and the impact by variation of lot to lot have reduced to 20.3%. When the CPK reached 1.35, the variation between different tools dropped to 25.6% (Fig.4). Namely, if it did not carry out this control mode, the CPK improvement just only reaches a limited degree due to lack of one effective direction

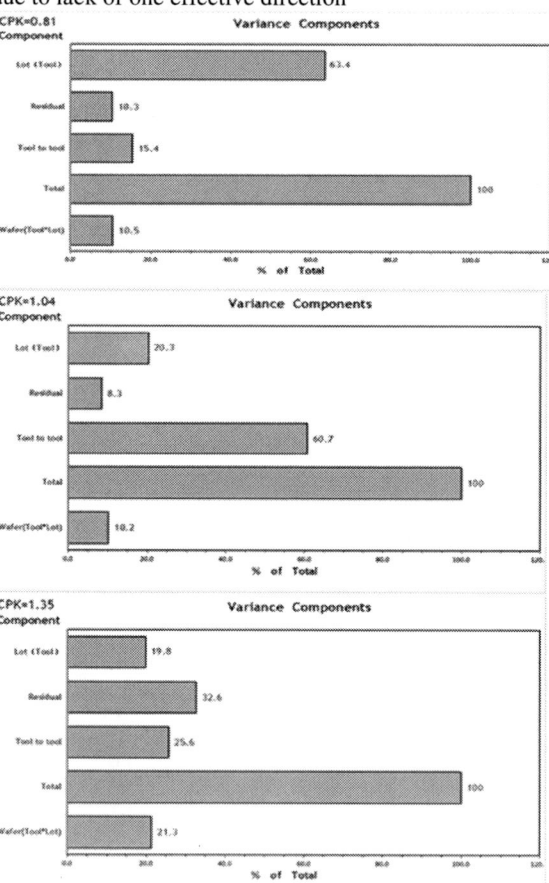

Figure 4: SOV analysis of one inline chart

SYSTEMATIC

As mentioned above, the production process of the

semiconductor manufacturing is uninterrupted and the data count is huge. It is difficult to execute the method efficient if only it relies on manual work. This paper presents an auto system (tool matching system) which has applied in a mature semiconductor line, the system including several aspects as follows:

(1) The golden mean is the target which sets in EDC (Engineering Data Collection). Tool matching system will link MES (Manufacturing Execution System), if the chart through MES does retarget OK, it's golden mean setting in tool matching system that will auto changed.

(2) Golden sigma for each chart calculated through the auto formula in the system. The system will be based on long-term data auto refinement of the golden sigma tri-monthly, then triggering the chart leads to tighten for owner sign off, and ensuring the golden setting real-time and rationality. If golden sigma needs loosen, an internal review and sign off procedure will be triggered.

(3) System has auto inform function, charts' owner will be informed if the charts have abnormal. Thus, fail reason can be analyzed in time, action can be done more targeted, efficiency can be improved and owner's loading can be reduced.

(4) Tool matching system also links iEMS (Equipment Monitoring System) system[7] and e-SPC system. If one chart fails the down tool criteria which sets in system, iEMS will alarm and inform owner to detail check. If one chart CPK<Bar & K shift fails criteria in e-SPC, tool matching system can help detect which chamber's abnormal induces the fail.

Specific process of tool matching system is as follows:

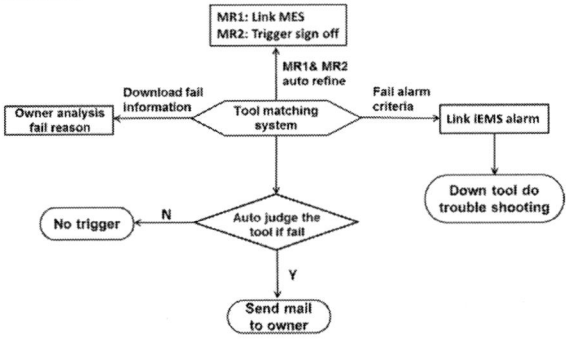

Figure 5: The basic function of tool matching system

Remark:
1. MES (Manufacturing Execution System): Records of all equipment combinations and equipment abnormal logs during the whole automotive wafer manufacturing.
2. iEMS (Equipment Monitoring System): Records of equipment real time monitor results and loadings (lifetime).
3. SPC (Statistics Process Control System): Records of all measurements results on actual products and monitor records on process equipment.

4. EDC (Engineering Data Collection): Records of all information of chart setting.

ACHIEVEMENTS

After the implementation of this control method and systematical control methodology on a mature semiconductor production line, more than five years data show obvious benefits which means current optimized control method for common equipment group has received great achievements. As shown in the figure 6，With the improvement of tool matching pass ratio，the ratio of   CPK>Bar is continuous promotion。

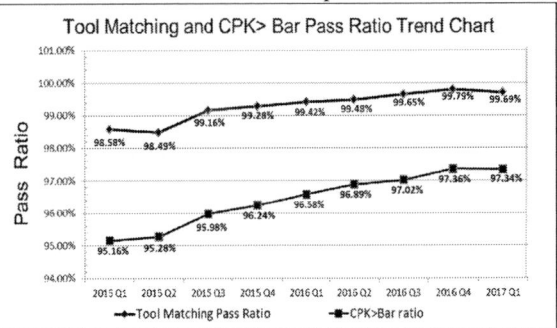

Figure 6: Tool Matching and CPK>Bar Pass Ratio Trend Chart

REFERENCES

[1] P.F. Tsai; C.T. Ho; R.L. Chen; J.H. Huang; W.P. Liu; J.M. Wu, J.F. Wang; J.I.Mou, Sensitivity Integrity for tool monitoring, tuning and matching, *eMDC*, 1-2, 2014.

[2] Guilherme S. Rosa; Maiquel S. Canabarro; José R. Bergmann; Fernando L. Teixeira, A comparison of numerical mode-matching techniques for the analysis of well-logging tools, *ACES*, 1 – 2, 2017.

[3] Sarah Huang, Tool Matching Statistical Methodology & Application, *eMDC & ISSM*, 1 – 8, 2011

[4] Richard G. Cosway; Steven R. Burch; Andrew D. Rosser; Phillip T. Lazok, A novel technique for epitaxy tool-to-tool and chamber matching and optimization: ER: Equipment reliability and productivity enhancements, *ASMC*, 429 – 434, 2016.

[5] B. Funsten; A. Ogawa; D. Lui, In-line defect detection metrology tool matching, *ISSM*, B59 - B62, 1997.

[6] J. D. Boyd; M. Banan, Run by run (generalized SPC) control of semiconductor processes on the production floor, *IEMT*, 60 – 64, 1995.

[7] Ziqian Javaer Liu; Hongtao H. T. Qian; Yuhong Betsy Xu, Equipment assessment methodology and automatic management system in automotive semiconductor manufacturing, *IEEM*, 1584 – 1587, 2016.

978-1-5386-5309-8/18 $31.00 © 2018 IEEE

APPLICATIONS OF ADVANCED TECHNIQUES OF TRANSMISSION ELECTRON MICROSCOPE IN CHARACTERIZATION OF SEMICONDUCTOR DEVICES

Jinghong Li

Globalfoundries

2070 Route 52, Hopewell Junction, NY 12533, USA

jinghong.li@globalfoundries.com

ABSTRACT

In this paper, practical methods of quantification analysis of electron energy loss spectrum (EELS) and X-ray energy-dispersive spectrometry (EDX) will be introduced. Quantification analysis of SiGe by EELS and EDX will be compared. And it will also be presented how to use PCA to improve analysis of EELS. Dark-filed holography (DFH) has been used to measure channel strain with high spatial resolution at sub-nanometer scale, it will be shown first how to set-up holography fringe's spacing down to a few Angstrom which is related to spatial resolution of strain map by DFH. Then high spatial resolution DFH strain measurement in FinFet 3D devices will be demonstrated.

INTRODUCTION

Transmission electron microscopy (TEM) is the most powerful tool to analyze chemically and physically semiconductor devices. As continuing shrinking devices of semiconductor devices into sub-nanometer scale, it becomes more and more important to know quantification chemistry information such as chemical compositions of device components of P-fet, N-fet, channel and gate stacks. Electron energy loss spectrum (EELS) and X-ray energy-dispersive spectrometry (EDX) are most widely used in characterizing chemically devices [1]. However, sometimes, the signal/background ratio is so low or their spectrums are so noisy, it is very difficult to obtain quantification information. Principle Component Analysis (PCA) is widely used in data analysis [2]. One of its advantages is to reduce noisy level. So it is perfectly suitable for EELS and EDX analysis when experimental signals are weak. In this paper, it will be first to compare quantification analysis of EELS and EDX, then it will also be shown that how to use PCA to improve the ratio of signal to intensity and obtain high quality chemical maps.

It is also well known that strain within channel or RX region increases mobility of devices [3]. In the last 10 years, there is a lot of research works in the development of metrology for strain measurement for semiconductor devices, including nano-beam diffraction (NBD), precession electron diffraction (PED), and DFH. Due to its high spatial resolution, DFH is more suitable for strain measurements [4], [5]. In this paper, it will be shown first how to set-up holography fringe's spacing down to a few

Angstrom which is related to spatial resolution of strain map by DFH. Then high spatial resolution DFH strain measurement in 3D devices will be demonstrated.

EXPERIMENTS

All devices used in the paper are made from 14nm technology in the Globalfoundries. TEM analysiscarried out in either FEI Osiris or Titan microscopes, equipped with super X-gun and four Esprit EDX detectors and Gatan EELS detector. For EELS quantification analysis, Digital Microscope (DM) was used and for EDX quantification analysis, the FEI/Bruker Esprit system was used. A practical PCA software developed by the TEM analytical group in Globalfoundries has been used.

RESULTS AND DISCUSSIONS

Quantification analysis of EELS vs EDX

EELS and EDX have been used to quantitatively analyze devices' chemical compositions in order to control process and to achieve the best device performance. Quantification analysis of chemical composition by EELS and EDX has been well established [1]. For EELS quantification analysis for binary system, one element composition could be determined by:

$$\frac{NA}{NB} \approx \frac{IA(\beta,\Delta)\sigma B(\beta,\Delta)}{IB(\beta,\Delta)\sigma A(\beta,\Delta)} \qquad (1)$$

where N_A and N_B are atomic percentage of each element and $N_A+N_B=100\%$

$I_A(\beta,\Delta)$ *and* $I_B(\beta,\Delta)$ *are* intensity integrated up to energy loss Δ *for element A and B respectively;*

$\sigma_A(\beta,\Delta)$ and $\sigma_B(\beta,\Delta)$ are *the partial cross-section for element A and B respectively.*

For EDX quantification analysis for a binary system, one element composition could be determined by:

$$\frac{NA}{NB} = KAB\left(\frac{IA}{IB}\right) \qquad (2)$$

where N_A and N_B are atomic percentage of each element and $N_A+N_B=100\%$

K_{AB} *is the Cliff-Lorimer factor;*

I_A *and* I_B *are X-ray intensity for element A and B, respectively.*

978-1-5386-5309-8/18 $31.00 © 2018 IEEE

The whole calculations and related physical parameters for EELS and EDX quantification analysis have been fully integrated into the Gatan Digital Microscope for EELS/EDX, and the FEI/Bruker Esprit of FEI microscope operation system for EDX. But, some precautions are needed to be taken during such processes, such as background fit models for EELS. Here EELS analysis is involved in low-loss spectrum for elements such as Si, B, N, and O and high-loss spectrum for Ge and Hf etc., both spectrums are spliced together. For line scan profiles, a line with 30 pixel are used to average EELS signals. Figure 1(a) shows an EELS and EDX analyzed area for SiGe/FIN; and Figure 1 (b) and (c) for Si, O and Ge line profiles from EELS and EDX, respectively. The epitaxy SiGe was grown on Si FIN, and it was capped with amorphous Si. It is shown that Ge% and Si% concentrations within SiGe EPI are ~50% by EELS for each of them from Figure 1(b) and there are around 47% for Ge% and 45% for Si% and the rest is for O% from Figure (c). Comparing line scan profiles from EELS and EDX, it can be seen that the profile from EELS is smoother and less various and EDX is a little noisier. The Ge% concentration from EELS and EDX is close to each other.

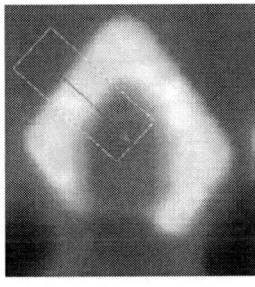

Figure 1. (a)

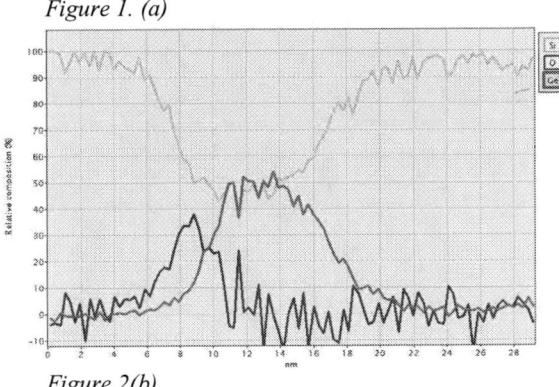

Figure 2(b)

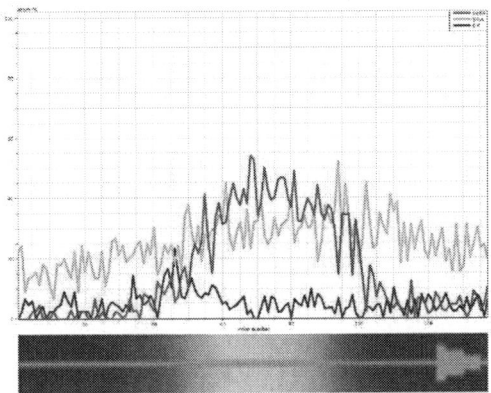

Figure 1(c)

Figure1: (a) STEM image of SiGe/FIN with EELS line scanning direction and location; (b) EELS quantification line profiles of Ge, Si and O; (c) EDX quantification line profiles of Ge in red, Si in green and O in blue.

Application of CPA in EELS analysis

PCA is a well-established mathematical procedure to reduce the dimensionality of date by creating a new, artificial variables called principle components while representing the dominant patterns of original experimental data. So it is very useful to reduce noisy level and reconstruct experimental data. Boron is widely used in doping P-fet. But the cross-section of boron is very small and produces very weak EELS intensity, producing high noisy EELS map. PCA has been used in this work to process low ratio of signal to noise of EELS spectrum, such as B and N. The PCA software used is developed by our TEM group. Figure 2 shows (a) dark-field STEM image of SiGe/FIN; (b) EELS analyzed area; (c) B EELS map without PCA and (d) EELS map with PCA. The target B concentration is 2.5E21 by in-situ doping during epitaxy growth. From Figure 2 (a) and (b), it can be seen that there are some B precipitates. Comparing Figure 2(c) with Figure 2(d), without PCA, the B map from Figure 2(c) is noisy and did not show B is well and uniformly distributed within SiGe, while with PCA, the B map from Figure 2(d) not noisy and B is uniformly distributed within SiGe even B precipitates show much higher intensity. So application of PCA in EELS analysis can greatly reveal distribution of B dopant which is very difficult to show-up by without processing with PCA, because of large deduction of noise.

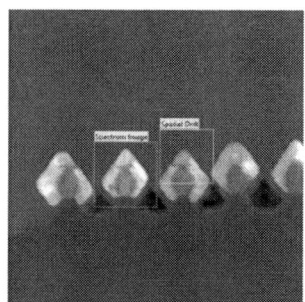

Figure2 (a)

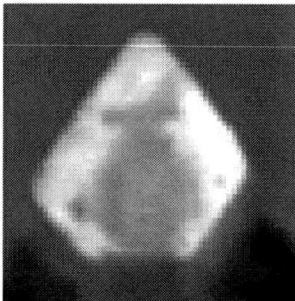

Figure2 (b)

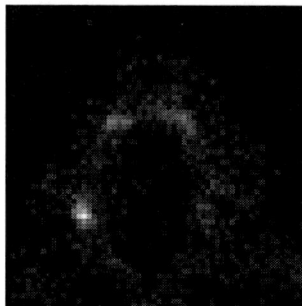

Figure2 (c)

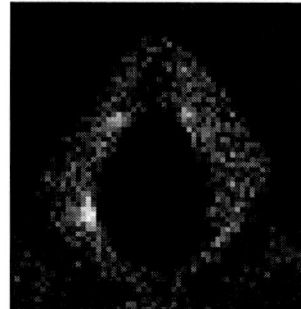

Figure2 (d)

Figure2: (a) STEM image of SiGe/Fin; (b) EELS analyzed area; (c) B EELS map without PCA and (d) B EELS map with PCA.

Application of high resolution of DFH in measurement of channel strain

Strain within channel or RX region increases mobility of devices. DFH has been used in measurements of channel strain. Dual lens dark-field holography has been established and used to measure strain within channel and reaction region. The spatial resolution depends on fringe spacing of dark-field holography [6]. The smooth and eventual strain map also depends on the contrast of the holography. Figure 3(a) shows profiles of fringe spacing, and Figure 3(b) shows strain map of SiGe/FIN; and Figure 3(c) shows strain across Line1 in Figure 3(b); Figure 3(d) shows strain profiles across SiGe epitaxy and Si Fin. The Ge% concentration of SiGe grown on the Si Fin is ~35% based on EDX and X-ray measurement. It can be measured and calculated from Figure 3(a) that the fringe spacing is ~0.3nm and contrast is ~21%, which would produce a strain map with spatial resolution of ~6A. Finer fringe spacing can be achieved by adjust dual lens settings such as the voltage applied to bi-prism and strength of objective lens. The voltage applied to the bi-prism is ~80ev and the strength of objective lens is 60%. The in-plane strain along <220> direction within epitaxy SiGe (35%Ge) over Si FIN is ~1.73% and it is tensile strain. Si Fin is also in tensile strain of ~0.2%. Tensile strains are due to lattice expansion of SiGe which pulls lattice of Si Fin in slight tensile as well.

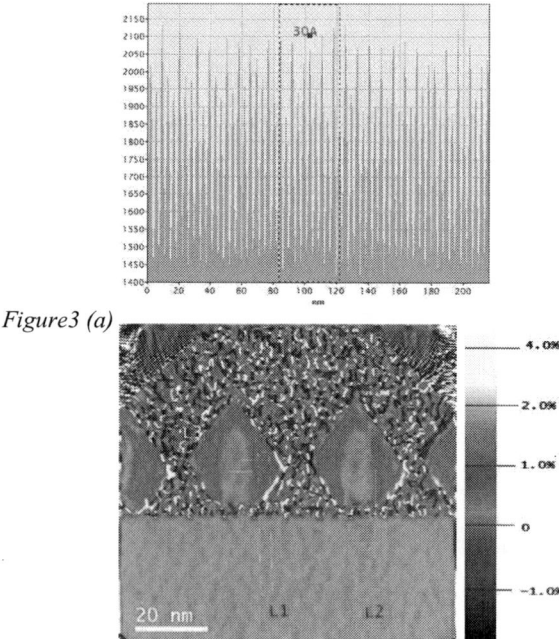

Figure3 (a)

Figure3 (b)

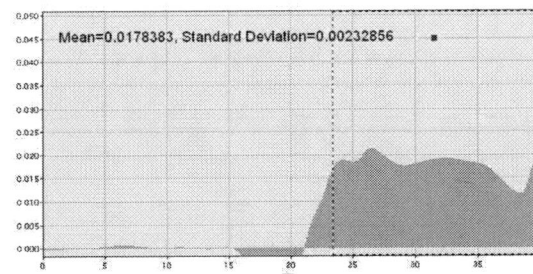

Figure 3(c)

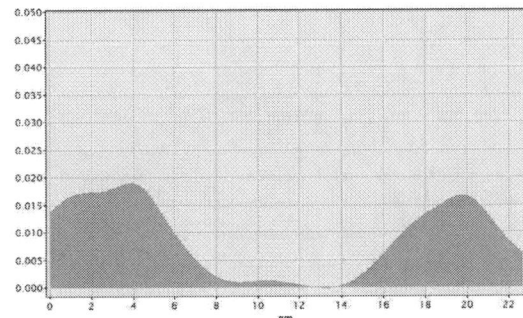

Figure 3 (d)

Figure3 (a) DFH fringe spacing; (b) DFH strain map; (c) strain profile within SiGe and (d) strain within SiGe and Fin

SUMMARY

Advanced TEM techniques such as quantified EELS and EDX, PCA analysis of EELS spectrum and high spatial resolution DFH have been used in characterization of semiconductor devices. Quantification analysis of EELS and EDX of SiGe grown over Si FIN shows that Ge% concentration is very close to each other (~50%). High spatial resolution strain map with a spatial resolution of ~6A can be achieved by adjusting parameters of dual lens dark-field holography, such as the voltage applied to bi-prism and the strength of objective lens. The strain measured by DFH is ~1.73% of tensile strain within SiGe grown over Si Fin and ~0.2% of tensile strain within Si Fin as well.

ACKNOWLEDGEMENTS

The author would like to thank the Globalfoundries, USA for support the research and development work.

REFERENCES

[1] D. Williams and C. B. Carter, <Transmission Electron Microscopy>, Plenum, pp.599-600 and pp. 673-674, 1996.

[2] H. Abdi, L. J. Williams, "Principle Component Analysis", Wiley Interdisciplinary Review: *Computational Statistics*. 2 (4): pp. 433-459, 2010

[3] S. E. Thompson, G. Sun, K. Wu, J. Lim and T. Nishida, Technical Digest, IEDM (2004), pp. 221-224

[4] J. Li, A. Lamberti, A. Domenicucci, L. Gignac, H. Utomo, Z. Luo, N. Rovedo, S. Fang, H.Ng, J. Holt, A. Madan, C. W. Lai, J. H. Ku, D. Schepis, J.-P. Han, M. Lagus, "Channel Strain Characterization in Embedded SiGe by Nano-beam Diffraction", *ECS Transactions*, v 16, n 10, pp. 545-549, 2008

[5] D. Cooper, T. Denneulin etc, *Micro 80* (20016), pp. 145-165.

[6] Y. Y. Wang, J. Li, A. Domenicucci, J. Bruley, *Ultramicroscopy*, Vol.124, pp. 117-129, (2013)

THE DETECTION AND INVESTIGATION OF POLY LINE LEAKAGE BY ELECTRON-BEAM INSPECTION

Rongwei Fan*, Hunglin Chen, Kai Wang, Kun Cai, Yin Long, Qiliang Ni, Xiaofang Gu
Shanghai Huali Microelectronics Corporation, 201210, Shanghai, China
*Corresponding Author's Email: fanrongwei@hlmc.cn

ABSTRACT

The study aims at the inline detection of invisible defects of poly line leakage. The defect was induced by extreme tiny residue at the bottom of the poly line of the SiGe loop process. A novel electron beam (E-beam) scan method was developed. And with voltage contrast (VC) inspection of E-beam, poly line leakage with bright VC (BVC) was detected just post SiGe deposition (DEP) nitride (SIN) hard mask (HM) remove (RM). The failure model of tiny SiGe residue was setup and established. The impacted factors of the defect were also discussed, including spacer SIN and SiGe HM SIN thickness, the step height of the AA and STI, the OX loss induced by SiGe ETCH clean/SiGe DEP pre-clean even the wet strip of low dose drain (LDD) implantation (IMP) loop. The corresponding improvement actions including controlling OX loss, increasing the thickness of SiGe SIN HM and reducing the step height of AA and STI were executed. The poly line BVC defects obviously trended low accordingly.

POLY LINE LEAKAGE DETECTION BY E-BEAM SCAN

An e-beam scan with VC images comparison is an effective inspection method and a good alternative to bright and dark field ones as tolerance of defects in the semiconductor process decreases [1]. With the SEM (scanning electron microscope) image observation, VC inspection is able to detect tiny defects nonvisual to optical inspection [2], furthermore, the under layer defect is possible to be found by the surface charging [3]. Some applications even use dedicated test structures [4] for VC inspection as a routine monitoring vehicle to sweep the line defects. Usual cases of those applications are mostly aimed at detecting physical defects or short open defect post CTW CMP at MEOL of the SRAM. The study herein further extends VC inspection to detect the junction leakage of poly gate at the FEOL steps.

As the feature size continuously shrinks, embedded SiGe process was used more and more, especially under 28nm technology. The defect induced by the SiGe loop process also was introduced. Most of the defect types could be detected by the optical inspection system except extreme tiny residue at the bottom of the poly line.

In order to setup the index of this type tiny residue, E-beam inspection method was carried out. To achieve the signal of the tiny residue defect, physical mode is not the right choice, but the voltage contrast mode needs to be used. Various combinations of electron gun settings, wafer charging modes and scan sequences were tested, a novel scan method was developed targeted to those extreme tiny SiGe residue at the bottom of the poly line. And with voltage contrast inspection of E-beam, poly line leakage with bright VC (BVC) was detected just post SiGe deposition(DEP) nitride (SIN) hard mask (HM) remove(RM), shown in Fig. 1 (a), (b) and (c).

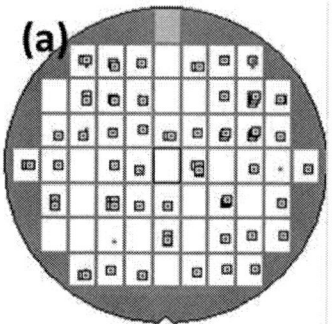

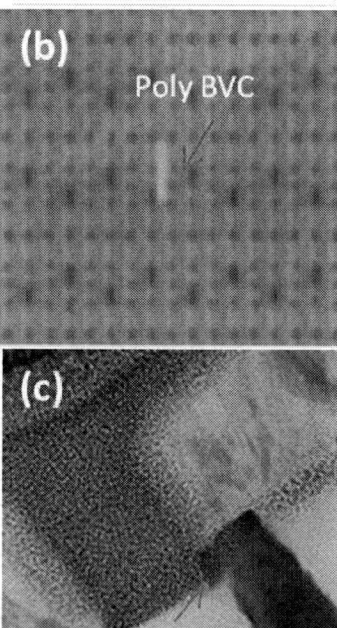

Fig.1. (a) Poly line leakage defect distribution map. (b) poly line leakage defect image of E-beam review and (c) poly line leakage defect TEM image.

The poly line BVC was induced by tiny SiGe residue at the bottom poly line. And the schematic diagram of poly line leakage with BVC induced by tiny SiGe residue was explained by Fig. 2(a) and (b).

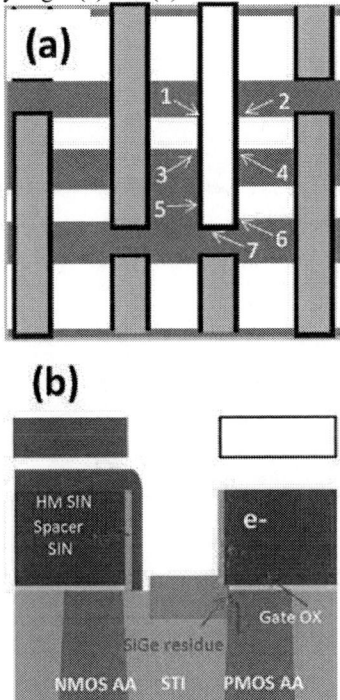

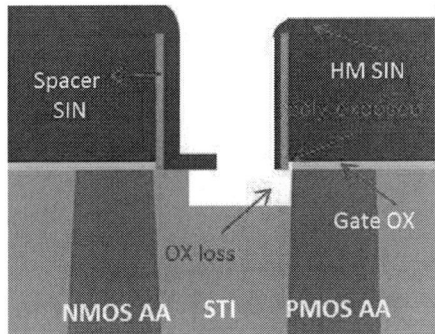

Fig.3. the schematic diagram of film stack just before SiGe deposition

Fig.2. the schematic diagram of (a) the top view of poly line leakage with BVC and (b) the schematic diagram of cross section of poly line leakage

It was shown from Fig.2 (a), every OX loss at the corner of STI and poly line facets would induce poly line BVC. If there were too much OX loss at any of the position, poly bottom corner would exposed before the process of SiGe deposition. So the poly would connect with PMOS AA by embedded SiGe then the poly line leakage generated.

EXPERIMENTS AND RESULTS

There were major 2 factors would impact final OX loss at the corner of the poly line and STI just before SiGe deposition. The first one was the SiN thickness beside the poly line which was determined by the thickness of deposition and the SiN loss induced by etch/wet process. The second one was the OX thickness beside the poly line and the accumulated OX loss at the bottom of the poly line by wet process from the process of LDD implantation to just before the SiGe deposition. It was showed by Fig.3 that SiN/OX thickness and their loss would decide if the bottom of the poly line would be exposed.

Therefore, the first action to do defect reduction was that to optimize the process of SiN HM deposition and the correlative process of etch. With these optimized processes, poly BVC defect could reduce from more than 20thousands to less than 10thousands. Furthermore, OX loss was controlled by optimizing LDD loop wet strip processes and pre-clean process just before the SiGe deposition. Then, the defect count of poly line BVC decreased obviously from ~10thousands to be close to zero. Fig.4 showed the poly line BVC defect trend chart with optimized action.

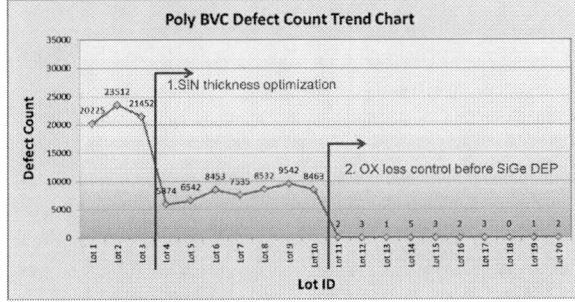

Fig.4. Poly BVC defect obviously trend low by inline process optimization

CONCLUSIONS

The study built up an inline index for poly line leakage defects induced by extreme tiny SiGe residue at the bottom of poly line. With a series of experiments, the corresponding improvement actions including LDD loop wet strip process optimization and controlling OX loss by SiGe etch/pre-clean process optimization just before DEP; increasing the thickness of SiGe SIN HM and reducing its loss by optimizing etch process were executed. The poly line BVC defects obviously trended low accordingly. Furthermore, the proposed voltage contrast inspection method not only targeted to the detection of 28nm poly line BVC defects but also potentially extends to the smaller line width of next technology nodes.

ACKNOWLEDGMENTS

The authors would like to thank Dr. Fang and Runling

978-1-5386-5309-8/18 $31.00 © 2018 IEEE

Li of Huali for coordinating the split experiments and process direction. Fiona Pan, Lucien Yeh, Family Wu and Tommy Lin of Hermes-Epitek are appreciated for the support of performing the e-beam inspection.

REFERENCE

[1] H.Chen,R.Fan,H.Lou,M.Kuo,Y.Huang, Mechanism and application of NMOS leakage with intra-well isolation breakdown by voltage contrast detection, J. Semicond. Technol.Sci.13(4)(2013)402–409.

[2] H.Chen,R.Fan,H.Lou,Y.Huang, Alternative voltage contrast inspection for pMOS leakage due to adjacent nMOS contact-to-poly misalignment, Mater.Sci. Semicond. Process.16(6)(2013)1873–1878.

[3] Mingsheng Kuo, HunglinChen, RongweiFan, DavidWeiZhang，Mechanism and detection of poly gate leakage with nonvisual defects by voltage contrast inspection Materials Science in Semiconductor Processing 20 (2014) 17–22

[4] T.R. Cass, D. Hendricks, J. Jau, H.J. Dohse, A.D. Brodie, W.D. Meisburger, Application of the SEMSpec electron-beam inspection system to in-process defect detection on semiconductor wafers, Microelectronic Engineering, Volume 30, Issues 1–4, January 1996, Pages 567-570, ISSN 0167-9317, 10.1016/0167-9317(95)00311-8.

LEGACY PROFILER CAPACITY/UTILIZATION IMPROVEMENT WITH AUTOMATION FOR HIGH VOLUME BAW PRODUCTION

Yanghua He[*], Dario Nappa, Hector Nevarez, and Michael Lube

[1]Process Engineering, Qorvo, Inc., Richardson, Texas, USA

*Corresponding Author's Email: Yanghua.He@Qorvo.com

ABSTRACT

Measurement will become a bottleneck when the metrology capacity is insufficient for high volume BAW production. Profilers are critical in providing step height data in key areas including CMP. Legacy profilers do not support multitasking functions while measuring wafers. This paper describes an innovative methodology to integrate profiler automation with a data analysis system. The profiler automation significantly improves metrology capacity and utilization, transforming legacy profilers into state-of-the-art tools.

INTRODUCTION

Profilers used in high volume Bulk Acoustic Wave (BAW) wafer fabrication are critical to the production line in order to define/monitor critical processes. For example, the post-metal etch step height (SH) data is fed-forward to determine the oxide deposition depth. The Chemical Mechanical Planarization (CMP) SH data is used to determine polish time and monitor post-polish performance. Some profilers used for BAW production are legacy models [1], which do not support multitasking functions such as review of a measured lot while measuring another lot. The profiler is virtually idle during the data review/transfer steps, which limits the capacity of the tool and drives additional tools to service high-volume ramps. An innovative methodology is thus needed for the capacity and utilization improvement of legacy profilers.

Major shortcoming of legacy profilers

Legacy profilers may be cost-effective, but they lack the efficiency required by modern semiconductor fabs. The major shortcoming of legacy profilers is that they were not designed to support multitasking functions such that the measurement data could be reviewed on them while they continue measuring wafers. Typical usage of single-tasking legacy profilers is shown in Fig.1.

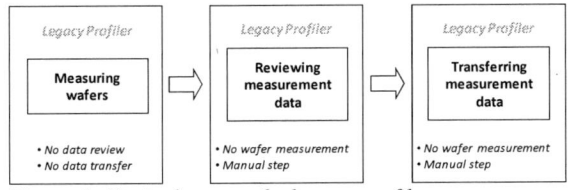

Figure 1: Typical usage of a legacy profiler

The data review/transfer performed on a legacy profiler prevents the metrology tool from continuing measuring wafers. This is considered the biggest contributing factor affecting the effective utilization of a legacy profiler.

CONVENTIONAL WAYS OF PROFILER CAPACITY IMPROVEMENT

The legacy profilers used in our fab would not have sufficient capacity to meet the challenges of the high volume BAW production ramp. To improve the capacity and utilization of the legacy profilers and to prevent the measurement step from becoming a bottleneck in the production line, a couple of conventional approaches have been practiced and implemented, as described below.

One option is to reduce the actual measurement time by measuring fewer sites per wafer with the fastest possible scan speed. The number of measurement sites has been reduced to bare minimum with only a few sites covering wafer center and edge areas. Various scan speeds have been tested with the highest possible scan speed adopted and the scan length has also been shortened with little redundancy. All these were done without any compromise of the SH data integrity.

Another option is to measure fewer wafers per lot, which went from 100% to 25% with the across-lot wafer sampling plan. More than half of the production routes that require SH measurement have further reduced the wafer sampling from 25% to 12.5% per lot. The profiler recipe and wafer sampling plan optimization were a big leap forward for the profiler capacity improvement.

INNOVATIVE SOLUTION OF PROFILER CAPACITY AND UTILIZATION IMPROVEMENT

For BAW production, the aforementioned common line practices alone were still insufficient to prevent the measurement step from becoming a bottleneck. The resort to increase the number of profiler tools was also subject to the equipment budget and clean room space limitation. Therefore, an innovative solution was required, which became critical to BAW production.

Transfer measurement data from profiler to network

The measurement data review/transfer on a legacy profiler is detrimental to the metrology tool's capacity and utilization because the step prevents the tool from performing continuous measurements. Thus, it is logical

978-1-5386-5309-8/18 $31.00 © 2018 IEEE

to get the as-is measurement data offloaded to an outside system so that the data review is no longer needed to be performed directly on the tool. This step is made possible with the automation system, which retrieves the measurement site trace data from the tool and uploads it to a database. The data is then made available via a web application for further review and analysis. Just the implementation of this step immediately increased the capacity and utilization of a legacy profiler. It allowed the continuous measurement of production lots without any interruption and interference from the data review/transfer performed from a network computer. Fig.2 (left) shows the as-is pre-CMP SH profile of a measured site, which is captured directly from a profiler. Fig.2 (right) is the pre-CMP SH profile of the same measured wafer and site, which is regenerated from the automatically transferred measurement data by the metrology automation. Except the heights of the two peaks are different, which will be explained later, the actual numeric SH data between the central portion and side banks from the two profiles are almost identical.

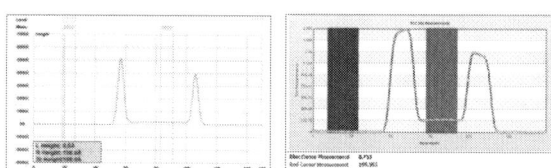

Figure 2: Typical as-is pre-CMP SH profile (left); Regenerated pre-CMP SH profile (right)

Establish an analytical model for measurement data analysis

The measurement data review/correction by a person is known to be manual, which, by nature, is slow and time-consuming. It can also be inevitably subjective and inconsistent between different reviewers. In order to improve the data review/correction efficiency and data integrity, an innovative analytical model with a built-in mathematical algorithm has been developed for automatic data review and analyses. This model is mathematically sophisticated and practically powerful in that it can automatically analyze different types of SH profiles.

The scan profile analyses start with leveling a scan profile first, then placing two virtual measurement cursors in appropriate locations, and finally calculate the SH value objectively. Fig.3 illustrated the model's leveling of a pre- and post-CMP SH profiles. For the pre-CMP SH profile like that shown in Fig.2, the SH target is around 200A, but the actual SH could occasionally be even lower than 50A. In this scenario, the profiles would all visually appear to have near zero SH values. That is the reason why the analytical model mathematically truncates the two high peaks of the profiles and switches to a much smaller graph scale. The central portion of the profile between the two peaks hence gets magnified, which not only enhances the

visual review of the profile, but also enhances the profile analyses to yield more accurate SH data.

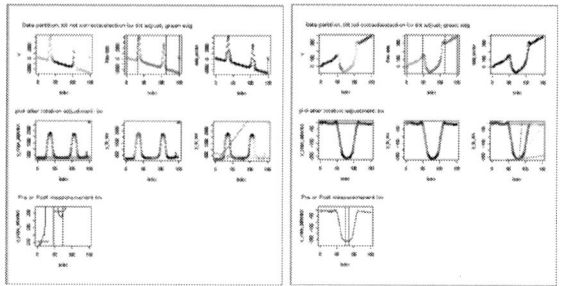

Figure 3: Pre-CMP (left) and Post-CMP (right) SH profiles evaluated by the analytical model .

Fig.4 (left) illustrates the model's capability of placing the two virtual measurement cursors for the SH calculation even when the profile was actually shifted to one side. This kind of profile shift would likely cause the actual measurement cursors to be misplaced. The analytical model can also handle different types of the SH profiles encountered in our BAW production. Shown in Fig.4 (right) is an example of another SH profile type at the backend of the line.

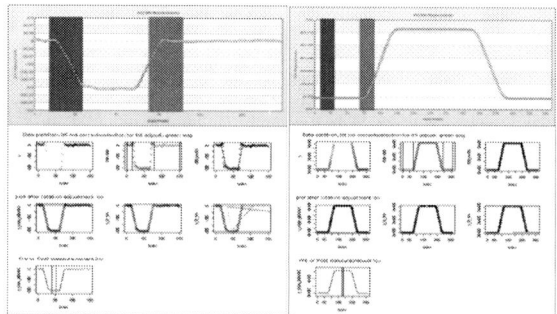

Figure 4: A shifted scan profile (left); A different type of scan profile (right)

The automatic data review and analyses by the analytical model not only improves the data review efficiency quite dramatically, but also ensures the SH data accuracy and integrity. Before being implemented in production, the model was first applied to a total of 3140 past measurement data values for the verification and justification of the analytical model's validity. This passive production data was collected the old way from profilers with operator's review and correction, and it covered almost all of the SH profile types in BAW production. Each of the individual SH profiles included in the passive data sample were carefully re-reviewed for the verification of their real values. This was for the fair comparison and/or judgment of the model with respect to the conventional method. Pre-CMP SH profile is one of

978-1-5386-5309-8/18 $31.00 © 2018 IEEE

the most critical profile types. Fig.5 shows a solid correlation of the model-analyzed data with the operator data for the pre-CMP SH case. To verify the model's reliability and robustness, over 3100 various SH values and profiles were analyzed and the comparison results are summarized in TABLE I. It is evident that model provides more reliable and accurate SH data than the human proved data. The measured values with operator review and correction have 5X higher error than that analyzed by the model even though both percentages are quite small. Therefore, the used-to-be manual data review step has since been updated and carried out automatically by the analytical model, virtually eliminating the manual step. As a result, all profilers in the BAW production line measure product wafers non-stop while the SH data is analyzed real-time in the background via the built-in model. This not only improved the SH data integrity but also significantly reduced the measurement cycle time.

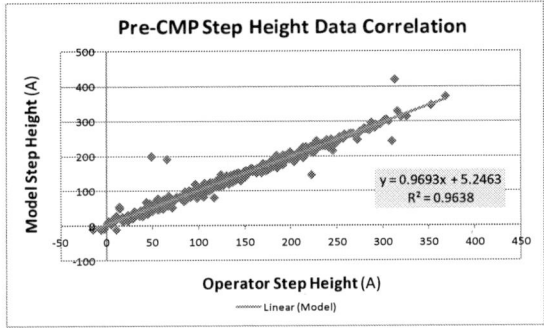

Figure 5: Model vs. Operator pre-CMP SH data

TABLE I: PASSIVE DATA ANALYSES BY THE ANALYTICAL MODEL

Model Integrity Verification	Counts	Percentage
Total number of measurements	3140	100.0%
Model is wrong	10	0.3%
Uncertain (not sure or no good signal)	10	0.3%
Measured value is wrong	47	1.5%
Difference > 37A	67	2.1%

Perform final data review/correction through network connections

The analyzed SH data at each of the measurement sites by the analytical model is summarized in color-coded green or red numeric values with clear visibility, which can expedite the final data review on any system-connected computers. This is because the model has built-in criteria for the self-assessment of its confidence in the SH data analyses. The green-red color-code has a distinctive indication about the data analysis confidence: green means high confidence when the built-in criteria are satisfied; and red means low confidence when the built-in criteria are dissatisfied. That is, green indicates high data integrity per the model while red implies potential abnormalities and/or uncertainties in the data or data analysis. Hence, the red-colored data does require final data review/correction with actions to correct, accept or even delete the data values.

Furthermore, the analyzed SH data is summarized in a table format for all measured sites of wafers in a production lot. Since the data table lists the SH values of individual sites with color codes, it is very clear for operators to spot and focus on any existing "red" sites at the final data review step. To facilitate the final data review/correction, each individual value in the SH data table (no matter it is "green" or "red") is actually linked to the SH profile. When a SH data value is clicked, the associated scan profile will be displayed under the summary SH data table, as shown in Fig.6, where measurement site 10 of wafer 13 is clicked for verification purposes. One can easily identify any abnormality, if any, about the scan profile (e.g., form, leveling, and the placement of the two virtual cursors). Whenever necessary, one can move the blue and red "bar-like" cursors to appropriate locations in the displayed profile to either correct or verify the model-calculated SH value. The blue "bar-like" cursor is for the SH of either left bank (as default) or right bank (as alternate). The red "bar-like" cursor is for the SH in between the two high peaks. The calculated SH is then defined as the SH difference between the red and blue "bar-like" cursors.

Two other colors associated with the SH data are blue (e.g., site 4 of wafer 18) and purple (e.g., site 14 of wafer 1 and site 4 of wafer 3), as shown in Fig.6, which are examples of "corrected" and "accepted" data values, respectively. These three sites required the final data review because they were all marked red by the analytical model. The profile at site 4 of wafer 18 was reviewed and confirmed with proper leveling. But the blue and red cursors were relocated to proper locations of the profile and the site SH was then re-calculated and submitted, which turned the original red color to blue. The profiles at the sites of the other two wafers mentioned above were reviewed with nothing wrong and were all accepted with no changes, which turned the original red color to purple. In summary, the SH data analyses performed by the built-in analytical model have virtually replaced the conventional "data review" that used to be done manually site-by-site for all sites. The final data review step required over the red-flagged data, if any, is greatly simplified and expedited due to the SH data color coding and the automation setup, which usually is a very quick process through a network computer.

978-1-5386-5309-8/18 $31.00 © 2018 IEEE   379

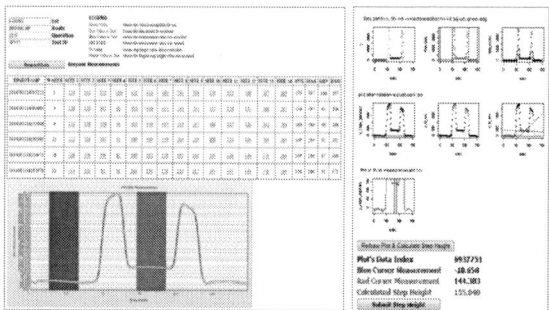

Figure 6: SH data review, correction and verification

Implement profiler automation for automatic data transfer to production line

After the profiler automation was implemented in BAW production line, the SH data analyzed by the model was automatically transferred into a chart for inline process monitoring. Illustrated in Fig.7 are the key elements for legacy profiler capacity and utilization improvement. With profiler automation integration, all legacy profilers will keep measuring wafers non-stop. All of the measurement data gets analyzed by the analytical model real time in the background and only small percentage of "red" data gets reviewed and/or corrected in the network.

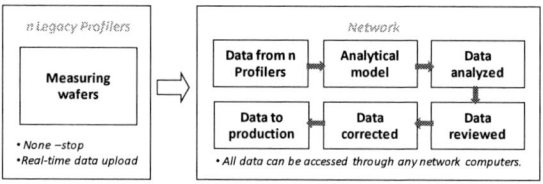

Figure 7: Legacy profilers with automation integration

SUMMARY AND CONCLUSIONS

Profiler capacity/utilization has been dramatically improved by the innovative integration of legacy profilers with data analysis/collection systems. First, the measurement data review, correction and transfer no longer needed to be performed directly on and from any profilers performing the measurement. This allowed the profilers to continuously measure production lots. Second, the data review and correction were automatically done by an analytical model, which mostly eliminated the manual data review/correction step. Finally, the corrected SH data after the final data review/correction was automatically transferred into online charts and database for process performance monitoring. All of these steps have been integrated into the profiler automation. Consequently, any legacy profilers can measure lots non-stop while the measurement data is reviewed and analyzed by the built-in analytical model. The model analyzed SH data that is summarized in color-coded numeric values with links to the associated scan profiles can be reviewed on any network computers in the Fab. Green-colored data has high data integrity and it generally does not need any review. In contrast, red-colored data implies possible abnormalities in either the data itself or data analysis, and it does require final data review. It has been confirmed in BAW production that the vast majority of the measurement data analyzed by the model is classified "green" and does not need any correction at all. The overall percentage of the "red" data that would need to be reviewed or verified is much less than 5%. The profiler automation is an innovative integration of legacy profilers with the built-in data analysis and collection systems. It significantly improved the profiler capacity/utilization, saving ~54 minutes per lot when coupled with a 6-wafer sampling plan and up to a few hours per lot for 100% measurement as demanded by special work requests.

ADVANCEMENT IN METROLOGY APPLICATIONS

The profiler automation with integrated data analysis/process model has been proven to be able to utilize lower-cost legacy profilers as if they were the leading-edge state-of-the-art metrology tools. This in turn has provided maximum profiler capacity and measurement cycle time improvement for high volume production. It resulted in great cost reduction with fewer profilers and operators required in the fab. The methodology described in this paper can be easily applied to improve other similar legacy metrology tools in the semiconductor industry.

REFERENCES

[1] KLA-Tencor Corporation. *Tencor P-22 Automated Surface Profiler Reference Manual*, Milpitas, California, 1998.

MONITORING CRITICAL PROCESS STEPS IN 3D NAND AND ADVANCED RF USING PICOSECOND ULTRASONIC METROLOGY

Johnny Dai[1], Robin Mair[1], Kwansoon Park[2], Xinglin Zeng[1], Priya Mukundhan[1],*
Cheolkyu Kim[2], and Tim Kryman[1]

[1]Rudolph Technologies, 550 Clark Drive, Budd Lake, NJ 07828
[2] Rudolph Technologies, 16-6, Sunae-dong, Bundang-gu, Sungnam-si, Gyunggi-do, 3595 Korea
*Corresponding Author's Email: priya.mukundhan@rudolphtech.com

ABSTRACT

Picosecond Ultrasonics (PULSE™) is a rapid, non-contact, non-destructive first principles acoustic metrology technique for in-line metal film thickness measurements. We demonstrate measurement capability for transparent and semi-transparent films, by analyzing the oscillatory component of the signal described as Brillouin oscillations. Longitudinal sound velocity, thus calculated, is used in determining film thickness. In this paper, we discuss the application of the technology in 3D NAND and advanced RF filters. The additional parameter not only helps characterize the material and improve the accuracy of the reported thickness but is also correlated to critical process parameters; etch rate in 3D NAND and resonant frequencies in RF filter.

INTRODUCTION

Picosecond ultrasonic technology (PULSE™), as implemented in MetaPULSE® G system, is a pump-probe laser acoustic technique for the measurement of metal film thickness. A 0.1ps laser pulse (pump) is focused to a small ($\sim 5 \times 7 \mu m^2$) spot onto a wafer surface to create a sharp acoustic wave. The acoustic wave travels away from the surface through the film at the speed of sound. At the interface with another material, a portion of the acoustic wave gets reflected and comes back to the surface while the rest is transmitted. The probe pulse detects this reflected acoustic wave as it reaches the wafer surface. One can detect the change of optical reflectivity that is caused by the strain of acoustic wave or alternatively detect the deflection of reflected probe beam that is caused by the deformation of surface due to the acoustic wave using a position sensitive detector (PSD). Both of these modes, reflectivity and PSD are used in characterizing metal films (Figure 1). Knowing the speed of sound in the material, and the arrival time of the echoes, thickness is readily extracted using first principles technique. Information on film density and surface roughness, depending on the application, can also be obtained by fitting the damping rate of the echoes and the width of the echoes, respectively.

In the case of transparent or semi-transparent films on silicon or other metal transducers, the opaque substrate absorbs energy from the pump pulse, launching a sound wave that travels up through the transparent film at the speed of sound. The strain causes a local change in the index of refraction of the film. The partial reflection of the probe beam from the moving sound wave, combined with the partial reflection from the film surface, leads to destructive and constructive interference at the detector. This effect can also been described as a Brillouin oscillation given that it is an interaction of photons and acoustic phonons [1]. As a result of this time dependent interference, the measured signal oscillates with a period, τ, from which the sound velocity (V) in the material can be determined by (1)

$$V= \frac{\lambda}{2n\tau\cos\varphi} \qquad (1)$$

where n is the refractive index and φ is the angle of optical probe beam within the film.

Applications in 3D NAND

3D NAND, driven by data intensive applications, changes the paradigm for manufacturing by providing an opportunity for vertical scaling using a highly repetitive and precise deposition and etch process. Hardmasks are used for etching deep, high aspect ratio features that conventional photoresists cannot withstand. Amorphous carbon (a-C)-based hard masks provide superior etch selectivity, chemical inertness and mechanically strong [2]. Monitoring a-C thickness is critical to the 3D NAND process as it goes through an iterative etch process. Film thickness and repeatability affects the active area of cell and consistency of the litho/etch performance. This is especially relevant as the films get thicker in advanced NAND stacks where high aspect ratio structures (64X) are critical and the a-C films are ~2μm thick. The films are opaque and optical metrology solutions are not robust for process monitoring

Shown in Figure 2 is the raw data of change in reflectivity versus delay time from PULSE measurements. Oscillations seen at earlier times (<100ps) are used in determining the longitudinal sound velocity. As the sound wave travels through the film and reflects at the interface

978-1-5386-5309-8/18 $31.00 © 2018 IEEE

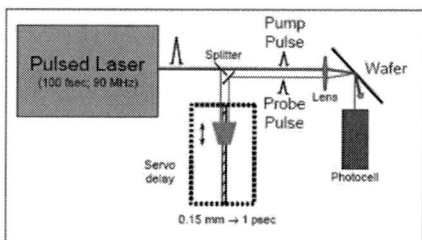 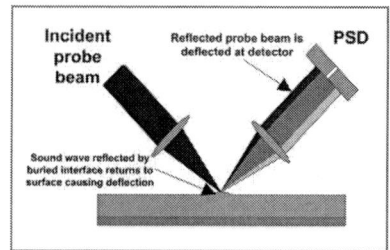

Figure1 (a): Schematic representation of picosecond ultrasonic technology. (b) In position sensitive detection (PSD), reflected sound wave causes a displacement of the wafer surface and thus the deflection of the reflected probe beam on the detector.

between film and underlying layer, it generates acoustic echoes (~700ps). Thickness of the film is calculated using the velocity determined from the early signal and the echo arrival position. The variation in the echo position from ~620ps to ~770ps represent real variation in thickness. Velocity values provided by the technique served a two-fold purpose. First, when used in the calculation of the thickness, it provided a more accurate representation of wafer level variation and second, the velocity values have helped provide feedback to the etch process optimization.

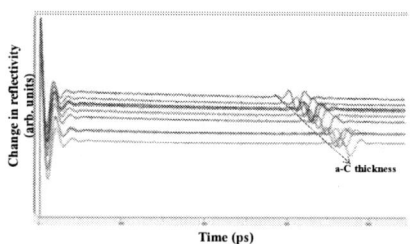

Figure 2: PULSE signal from a-C hardmask. Oscillations seen at earlier time is used for calculating the velocity and echoes at ~700ps for calculating thickness.

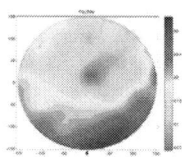 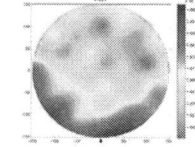

Figure 3(a): Within wafer uniformity profile of velocity.

Figure 3(b): Within wafer uniformity profile of thickness.

The small spot technique allows for measurements very close to wafer edge. Within wafer uniformity profiles of

velocity and thickness, Figure 3, as well as high resolution line scans (0.5mm EE) are useful during process development and optimization (Figure 4). 3σ repeatability performance for thickness and velocity are < 0.5%.

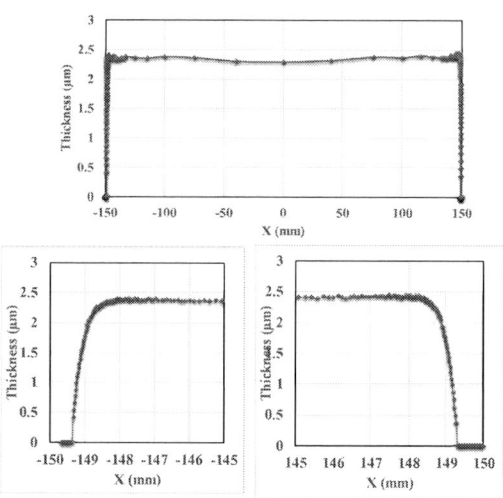

Figure 4: High resolution line scans across the wafer. Zoomed-in view of left and right edge profiles.

Applications in SAW and BAW

Surface acoustic wave (SAW) devices are used in telecommunications, and other areas where precise filtering is need. As smartphones integrate more SAW filters and duplexers for multiple band on a single chip, requirements for process control are tighter. The drawback for SAW filters is their thermal stability. A SAW filter's response may shift downward by as much as 4 MHz as temperature increases. Most devices make use of silicon dioxide's (SiO_2), positive temperature coefficients of elasticity (TCE) to compensate other materials' negative TCE and CTE (coefficient of thermal

978-1-5386-5309-8/18 $31.00 © 2018 IEEE

expansion). Typical within wafer uniformity of frequency is < 2MHz for such samples and the thickness uniformity is ~ 0.25%. At > 2.5GHz SAW filter performance is limited and bulk acoustic wave (BAW) filters is the primary technology that is employed. Film bulk acoustic wave resonators (FBAR) are the dominant BAW filter and performance is determined by the thickness and acoustic property of the piezoelectric aluminum nitride (AlN) layer either with or without metal electrodes, such as Molybdenum (Mo) [3]. Accurate measurement of the film thickness and sound velocity is critical to the design of high performance RF filters as the resonant frequencies are directly related to these values.

PULSE measurements are made in-line and on patterned structures. In Figure 5, raw data from the temperature compensation (TC) oxide layer is shown. Raw data is characterized by high signal to noise ratios (SNR) from which both velocity and thickness values can be reliably extracted using signal processing techniques. The high SNR contributes to the excellent repeatability for both velocity (Vel.) and thickness (T), 3 sigma <0.05% (Table 1). We were able to verify that the reported sound velocity correlates to filter performance (Figure 6), thus eliminating the need for send ahead wafers and/or monitor wafer measurements.

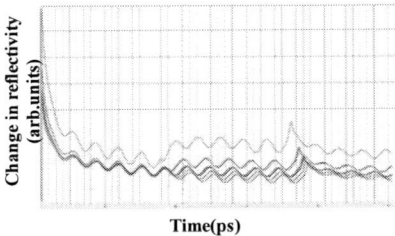

Figure 5: Raw data of TC oxide layer.

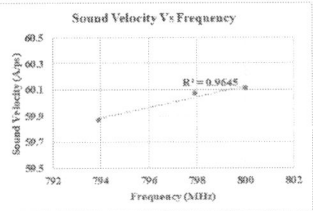

Figure 6: Correlation between sound velocity and RF filter frequency.

Similarly, during BAW process development, radial line scan profiles for both velocity and thickness are used in process development and optimization. Shown in Figure 7 is an example from a 5G development process. Piezoelectric aluminum nitride film is characterized in a Mo/AlN/substrate stack. In this example, AlN thickness varies between 1.7-2.5% in the radial scan depending on

the quadrant. Sensitivity of the method to track process changes have been validated by designing a two dimensional skew of wafers of varying AlN (5000Å-1.5μm) and Mo thickness (100-140Å). The technique had excellent accuracy as verified by using cross-sectional scanning electron microscopy (SEM) with correlation coefficient R^2> 0.95. Optical constants used in the calculation of thickness was obtained using the S2000™ focused beam ellipsometer system.

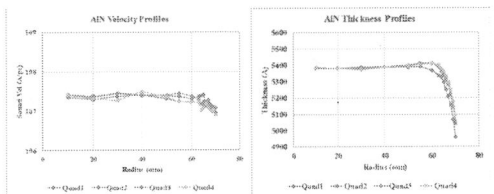

Figure 7: Longitudinal velocity and thickness profiles in different quadrants of the AlN wafer.

Table 1: Summary of velocity and thickness repeatability on oxide samples.

#	Vel. (Å/ps)	Vel. 3σ (%)	T (Å)	T 3σ (%)
1	59.87	0.015	11712.9	0.015
2	60.07	0.012	11945.1	0.009
3	60.11	0.012	11799.0	0.015

CONCLUSIONS

In summary, we have successfully demonstrated the measurement of longitudinal sound velocity and thickness of critical films in both 3D NAND and RF filter processes. Measurements are highly repeatable and exceed the stringent requirements needed for process monitoring and control. Additionally, the non-destructive nature of the technology enables in-line measurements on product wafers thus providing direct feedback for process control and eliminating the need for test or send ahead wafers.

REFERENCES

[1] J. L. Arlein, S. E. M. Palaich, B. C. Daly, P. Subramonium, and G. A. Antonelli, J. of Appl. Physics, vol 104, 2008, pp 033508 1-6.

[2] H. Singh, Solid State Technology, July 2017, pp 18-21.

[3] MEMS Industry: RF is changing the game, Status of the MEMS Industry report (2017).

THE TOTAL INSPECTION SOLUTIONS OF EXTREME TINY DEFECT IN BEOL ADVANCED SEMICONDUCTOR PROCESS

Xingdi Zhang[1], Hunglin Chen[1]*, Yin Long[1], Qiliang Ni[1], Rongwei Fan[1] and Kai Wang[1]

[1]Shanghai Huali Microelectronics Corporation, Shanghai 201210, China

*Corresponding Author's Email: chenhunglin@hlmc.cn

ABSTRACT

In this paper, extreme tiny defects are detected and monitored by using brightfiled inspection system in 28nm back-end-of-the-line (BEOL) processes. Two main defects, including tiny bridge and tiny bump defects, were studied by a novel combination of dedicated scan settings such as spectrum mode, directional electrical field (DEF), focus offset and so on. The purpose of these studies is to capture defects more efficiently. After defect detecting, failure models of these defects were established and the effective defect reduction actions were carried out. Combining effective defect monitor and defect reduction actions can make BEOL yield improve rapidly.

INTRODUCTION

The tolerance of tiny physical defects in the integrated circuit (IC) manufacturing becomes smaller and smaller, since advanced semiconductor process continually drive the line width shrinking. Tiny defects that are gentle in 55nm technology processes would however become fatal killers in 28nm node. The study proposed effective methods allowed inline detection of extreme tiny defects in back-end-of-the-line (BEOL) processes. The tiny defects in the BEOL contenting multi stacks of repeated metal lines and inter metal dielectric (IMD) would lead to end-of-line (EOL) yield failures. However, the transparent IMD and rough copper grains make the detection difficultly. A novel combination of dedicated scan settings on the equipment and designed layout patterns on the wafer is proposed to achieve the inline monitor [1]-[4].

After-development inspection (ADI) is a particularly important process monitoring method to identify tiny defects which can affect yield. It is a challenge to capture tiny defects at ADI because of very low contrast for defects and transparent photoresist which can introduce interference of pre-layer defects [5]-[6]. Using after-etch inspection (AEI) instead of ADI is a common method to monitor defects, but this method lengthens the feedback loop. If lithography defects are detected at ADI, they can be removed by rework, but it is impossible at AEI. Furthermore, it is difficult to identify the defects source according to images captured at AEI, so it is necessary to monitor lithography defects at ADI.

Defects in film are difficulty to identify because there are several layers of film. In order to confirm the defects source, defect scan and TEM analysis should effectively combine. Scan is for defect capturing and TEM analysis is to confirm that the defect comes from which layer of film.

In this paper, two main defects were effectively detected by series of studies. After defect detecting, failure models of these defects were established and the effective defect reduction actions were carried out. Combining effective defect monitor and defect reduction actions make average D0 of 28 nm BEOL improve from 1.2 to 0.2.

TINY BRIDGE DEFECT DETECTION AND IMPROVEMENT

Tiny Bridge Defect

Bridge defects were found at AEI. As shown in Fig.1a, bridge defects were very tiny and less than 40 nm. Fig.1b shows that bridge defects can lead to the metal open which affects yield and the killer ratio was 100%. According to image at AEI, it is difficult to identify the defect source, both etch and lithography are possible. The suspicion of lithography is the largest, so it is necessary to detect them at ADI.

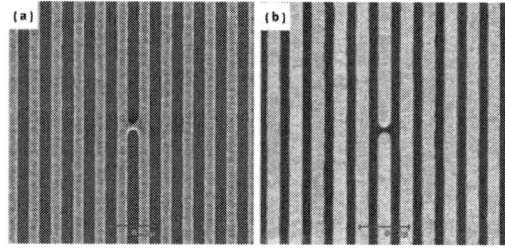

Fig. 1 (a) SEM image of bridge defect at AEI, (b) SEM image of bridge transferred to CMP layer

Bridge defect detection and monitoring methodology

Firstly, the bridge should be detected at ADI. Hot scan was used and bridge defects were found after a large number of reviews. As shown in Fig.2a and b, about 40nm

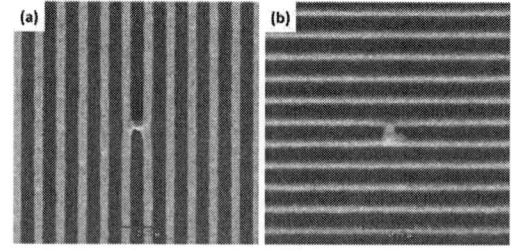

Fig. 2 SEM image of bridge defects at ADI

size defects were found and they are comparable with defects captured at AEI.

Then, the study of defect performed for effective detection. The spectrum model and offset were chosen as items of defect study. As shown in Fig 3a, shortwave has good performance in detection of bridge defect and blueband shows the highest signal to noise ratio. Fig 3b shows that negative focus offset value has good performance and -0.15 um shows the highest signal to noise ratio. After determining the relevant parameters, the bridge defects can be effectively detected at ADI. Compare with AEI result, the capture rate of bridge defects at ADI is more than 90%.

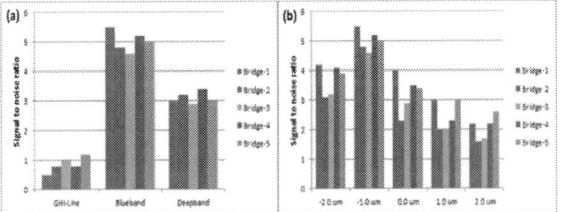

Fig.3 (a) Signal to noise ratio of bridge defect in different spectrum mode, (b) signal to noise ratio of bridge defect in different offset

Bridge defect failure model and reduction

Suspected pipeline pollution is the main reason of introducing the bridge defects. Firstly, old pipeline was purged to improve bridge defects, but the bridge defects were still detected at ADI. Then, the old pipeline was purged several times but it was still invalid. Changing a new pipeline was seen as a possibly valid strategy. Some bridge defects were still detected, but the quantity was lower than before. After purge of new pipeline, the bridge defects were not found at ADI which also had been confirmed by AEI results. The tiny bridge defects trend chart of Fig.4 shows that the bridge defects were effectively detected with the developed defect scan recipe, and the defect was trend to 0 post the changing and purge of pipelines.

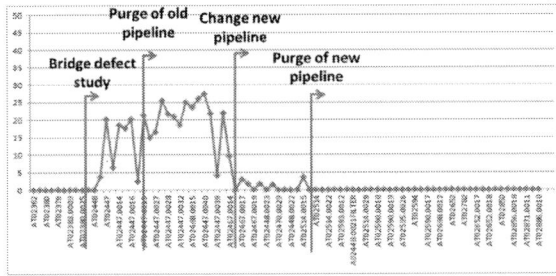

Fig.4 The tiny bridge defects trend chart

TINY BUMP DEFECT DETECTION AND IMPROVEMENT

Tiny Bump Defect

Tiny bump defects were found at AEI. As shown in Fig.5a, bridge defects were very tiny and less than 40 nm. According to image at AEI, they could be confirmed in the film, but there are 6 layers of film (Fig.5b), so the location was not still clear. In order to confirm the actual location, TEM analysis was used as an effective strategy. In Fig.5c, it can be seen that the tiny bump defect is above the NDCHM layer which is the defect source layer.

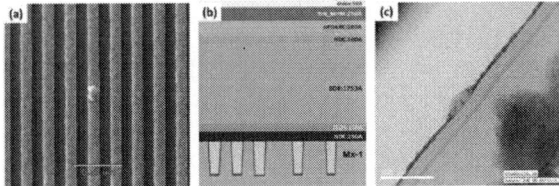

Fig.5 (a) SEM image of tiny bump defect, (b) diagram of film structure, (c) TEM image of tiny bump defect

Tiny bump defect failure model and reduction

According to TEM analysis, tiny bump defects was above NDCHM film, but there was scrubber process after NDCHM film deposition, so it is difficultly to confirm that defect came from NDCHM film or scrubber. Planning a split test was a valid method to find the defect source. Fig 6a shows that tiny bump defect was not found in every wafer of group skipped scrubber, but still existed in every wafer of group kept scrubber. The result shows that the scrubber was the defect source. There are four scrubber tool, and another split shows that tool A and B is worse than others. The tiny bump defects trend chart of Fig.7 showed the defect was trend to 0 after inhibiting the tool A and B.

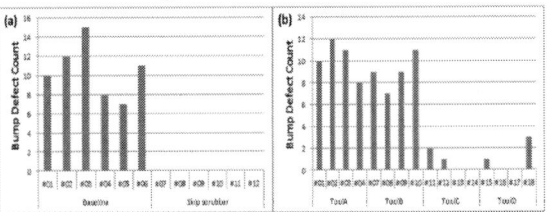

Fig.6 (a) Skip scrubber split result and (b) different scrubber tool split result

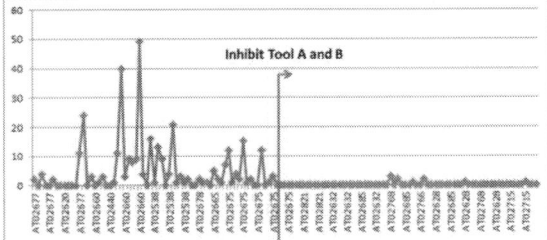

Fig.7 The tiny bump defects trend chart

978-1-5386-5309-8/18 $31.00 © 2018 IEEE

CONCLUSION

In this paper, two main defects were effectively detected by series of studies. After defect detecting, failure models of these defects were established and the effective defect reduction actions were carried out. Combining effective defect monitor and defect reduction actions make average D0 of 28 nm BEOL improve from 1.2 to 0.2.

ACKNOWLEDGEMENTS

Thank you very much for the leadership and colleagues for their support of my work

REFERENCES

[1] H.Chen, R.Fan, H.Lou, M.Kuo, Y.Huang, Mechanism and application of NMOS leakage with intra-well isolation breakdown by voltage contrast detection, J. Semicond. Technol.Sci. 13(2013)402–409.

[2] H.Chen, R.Fan, H.Lou, Y.Huang, Alternative voltage contrast inspection for pMOS leakage due to adjacent nMOS contact-to-poly misalignment, Mater.Sci. Semicond. Process.16(2013)1873–1878.

[3] M.Kuo, H.Chen, R.Fan, D.Zhang，Mechanism and detection of poly gate leakage with nonvisual defects by voltage contrast inspection, Mater.Sci. Semicond. Process. 56(2016) 362-367.

[4] H.Chen, R.Fan, H.Lou, Y.Huang, Electrochemical mechanism of layout-dependent corrosion of tungsten in contact plugs, Mater.Sci. Semicond. Process. 20(2014)17-22.

[5] I. Malik and B. Pinto, "Immersion Changes Litho Cluster Qualification," Semiconductor International, September 2006.

[6] L. Peters, "Defectivity Issues Drive Immersion Lithography," Semiconductor International, April 2006.

THE DETECTION AND INVESTIGATION OF TUNGSTEN-PLUG VOIDS BY ELECTRON-BEAM INSPECTION

Rongwei Fan*, Hunglin Chen, Kai Wang, Zuoyue Liu, Yin Long, Qiliang Ni, Xiaofang Gu

Shanghai Huali Microelectronics Corporation, 201210, Shanghai, China

*Corresponding Author's Email: fanrongwei@hlmc.cn

ABSTRACT

The study aims at the inline detection of invisible defects of tungsten (W) plug voids. Those voids inside contact holes would directly lead to chip yield fail or, even worse, reliability issues. A dedicated electron-beam (E-beam) inspection (EBI) method with "penetration mode" was performed to detect W plug voids. Then experiments showed the failure model involved poor CT hole profiles, the accumulating polymers in the CT hole, incompatible ILB/W DEP processes; furthermore, the quantified impacted factors of void defects were also discussed, including the aspect ratio (AR) and shape of CT holes, CT etch polymer status, the thermal effect of ILB/W DEP processes, and even the FOUP materials. The corresponding improvement actions including changing new material FOUP, reducing CT hole AR by optimization of the ILD loop and CT loop process and optimizing the ILB/W DEP process were executed. The W plug void defects obviously trended low accordingly.

W PLUG VOIDS DETECTION BY VOLTAGE CONTRAST INSPECTION

W with superior step coverage capability is a traditional material used for filling contact holes; however, the contact dimension continually shrinks, and incomplete filling of contact holes were found in 28 nm technology-node wafers. The defect was worse at the wafer edge dice, showed in Fig.1 (a). The defect detected by electron beam inspection was showed as dark voltage contrast (DVC), showed in Fig.1 (b), and the defect was just only the void inside the W plug which could not be detected by optical inspection system, showed in Fig.1 (c).

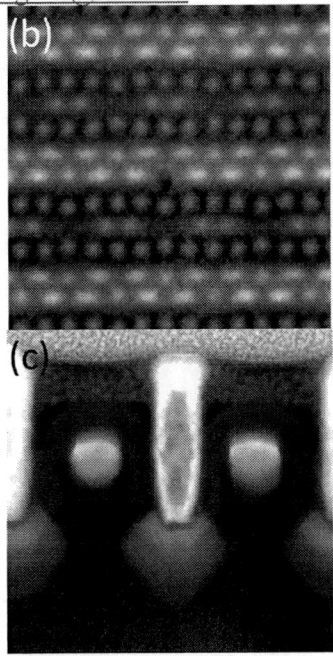

Fig.1. (a) W plug void defect distribution map, (b) W plug void defect image of E-beam review and (c) W plug void defect TEM image.

An e-beam scan with VC images comparison is an effective inspection method and a good alternative to bright and dark field ones as tolerance of defects in the semiconductor process decreases [1]–[3]. To achieve the voltage contrast signal of the void defect is a challenge for EBI since those partial filled contact holes do not really make contact open. Various combinations of electron gun settings, wafer charging modes and scan sequences were tested; a novel scan method with "penetration mode" was developed targeted to those partial filled contacts.

Under the normal inspection mode, shown in Fig.2 (a), the primary beam of EBI tool could not touch the W plug voids, and there would be no VC signal difference between the normal CT plug and the CT plug with voids. However, with the novel "penetration mode" of EBI tool, shown in Fig. 2(b), there would be VC signal difference between the CT plug with voids and the normal CT plug, since the secondary electron amount stimulated by void plug and normal plug would be different.

978-1-5386-5309-8/18 $31.00 © 2018 IEEE

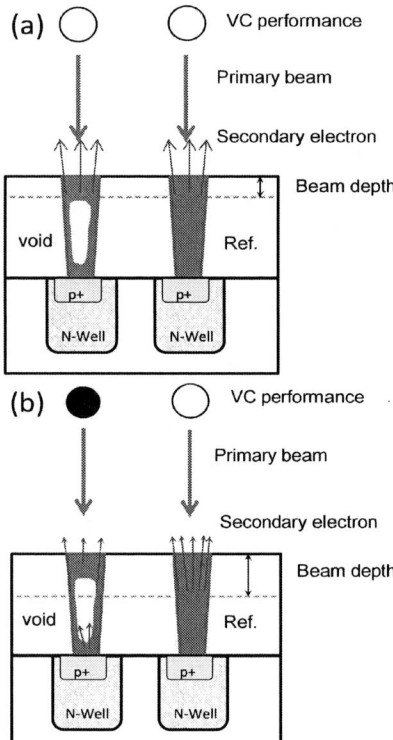

Fig.2. the schematic diagram of (a) the traditional E-beam scan method and (b) novel E-beam scan method

Combining other inspection conditions, inline index for W plug void defects was built up, which would speed up the development progress of 28nm technology node. Furthermore, the proposed "penetration mode" not only targeted to the detection of 28nm W plug void defects but also potentially extends to the smaller line width of next technology nodes.

EXPERIMENTS AND RESULTS

As the contact dimension continually shrinks, the filling of the contact holes were more and more difficulty. There are many factors could impact the W filling process, including CT hole aspect ratio (AR), by product polymer performance post CT etch, the minienvironment of the FOUP and the CT glue layer/CT W DEP. With the inline index built up by the proposed novel detection method, a series of inline process experiments were then carried out to fix void defects.

From some of the experiment results, high correlation between W plug voids count and CT hold aspect ratio (AR) was found, shown in Fig.3. It means that, the higher AR ratio, the more defect count.

Furthermore, the AR full map signature was very similar with the defect distribution, shown in Fig.4. The AR at the wafer edge was obviously larger than the center dice. So the correlative processes, such as inter layer

dielectric (ILD) chemical and mechanical polish (CMP) which would impact thickness and its uniformity, CT photo and etch which would impact CT hole CD and its uniformity .etc, need be optimized firstly.

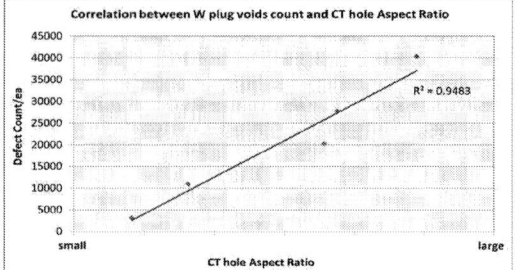

Fig.3. there exist high correlation between W plug voids defect count and CT hole aspect ratio

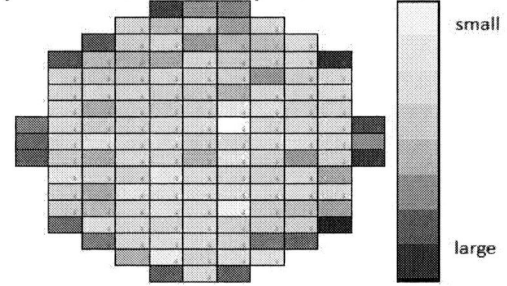

Fig.4. AR full map distribution which was similar with the defect distribution signature

Post the process optimization, the uniformity of the CT hole AR ratio changed better, then the wafer edge signature of W plug void defects was eliminated.

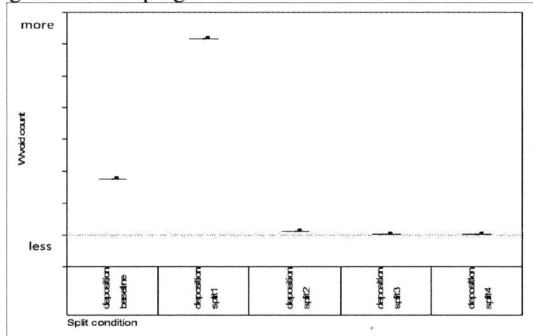

Fig.5. Split results of the CT glue layer amorphous state

However, the W plug void defects of random distributed in the wafer were still worse. Then a serious of process tuning was carried out in order to optimize deposition efficiency, to optimize the polymer post CT etch, even to optimize the microenvironment of FOUP.

The split results of improving deposition efficiency indicated that, shown in Fig.5, the defect count reduced obviously from thousands to nearly 0 by deposition process optimization. However, the deposition process change would induce the higher Rs of the CT W plug. The

978-1-5386-5309-8/18 $31.00 © 2018 IEEE

tradeoff need be evaluated between the defect status and the Rs.

Therefore, to avoid defect suffer completely, further split need be carried out. W plug void defects also impacted by the polymer of CT etch, moisture and gas from the environment and even the microenvironment of the FOUP. To avoid polymer reacted with the moisture in the environment, new type FOUP with barrier material was used from just before the step of CT etch to the step of W DEP. Furthermore, FOUP purge with inert gas splits also was carried out.

Fig.6 showed the split results of the new type FOUP and inert gas purge time results. It shown that, new type FOUP only could reduce the W void defect obviously. Furthermore, with the extra inert gas purge, the defect could be reduced to a lower level, but it could not eliminate the defect absolutely even with the condition of longest inert gas purge.

Finally, with the optimized processes that could reduce the CT hole AR and improve its uniformity, new type FOUP, inert gas purge combining deposition optimization could reduce the W void defects nearly to zero. At the same time, the Rs of CT W plug was acceptable.

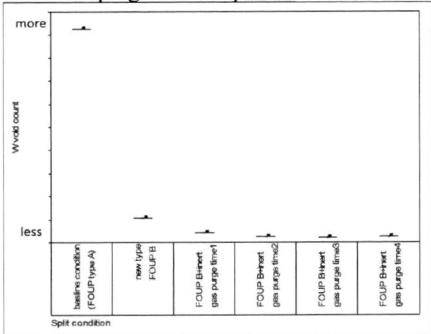

Fig.6. new type FOUP and inert gas purge time split results

CONCLUSIONS

The study builds up an inline index for W plug void defects and speed up the development progress of 28nm technology node. A serious experiments were carried out and showed the defect was obviously impacted by CT hole AR. The failure model involved the accumulating polymers in the CT hole, incompatible ILB/W DEP processes even the microenvironment of the FOUP was also discussed. Furthermore, the corresponding improvement actions including reducing CT hole AR, by optimizing the processes of ILD DEP and CMP, tuning the deposition process, changing the new type FOUP and adding inert gas purge were executed. The W plug void defects were trend low accordingly.

ACKNOWLEDGMENT

The authors would like to thank Dr. Fang and Kang Ye of Huali for coordinating the split experiments and process

direction. Fiona Pan, Lucien Yeh, Family Wu and Tommy Lin of Hermes-Epitek are appreciated for the support of performing the e-beam inspection.

REFERENCES

[1] H.Chen,R.Fan,H.Lou,M.Kuo,Y.Huang, Mechanism and application of NMOS leakage with intra-well isolation breakdown by voltage contrast detection, J. Semicond. Technol.Sci.13(4)(2013)402–409.

[2] H.Chen,R.Fan,H.Lou,Y.Huang, Alternative voltage contrast inspection for pMOS leakage due to adjacent nMOS contact-to-poly misalignment, Mater.Sci. Semicond. Process.16(6)(2013)1873–1878.

[3] Mingsheng Kuo, HunglinChen, RongweiFan, DavidWeiZhang，Mechanism and detection of poly gate leakage with nonvisual defects by voltage contrast inspection Materials Science in Semiconductor Processing 20 (2014) 17–22

PROBING AND MANIPULATING THE INTERFACIAL DEFECTS OF INGAAS DUAL-LAYER METAL OXIDES AT THE ATOMIC SCALE

Xing Wu,[1,4], Chen Luo[1], Peng Hao[2], Tao Sun[2], Runsheng Wang[2], Chaolun Wang[1], Zhigao Hu[1], Yawei Li[1], Jian Zhang[1], Gennadi Bersuker[3], Litao Sun[4], and Kinleong Pey[5]*

[1] Department of Electrical Engineering, East China Normal University, Shanghai 200241, China
[2] Institute of Microelectronics, Peking University, Beijing100871, China
[3] The Aerospace Corporation, Los Angeles, CA 90009, USA
[4] SEU-FEI Nano-Pico Center Key Laboratory of MEMS of Ministry of Education School of Electronic Science and Engineering, Southeast University, Nanjing 210096, China
[5]Singapore University of Technology and Design, 487372, Singapore
*Corresponding Author's Email: xwu@ee.ecnu.edu.cn

ABSTRACT

The interface between III–V and metal-oxide-semiconductor materials plays a central role in the operation of high-speed electronic devices, such as transistors and light-emitting diodes. The design of high-performance devices requires a detailed understanding of the electronic structure at the interface. However, the relation between the interface state charges to the electrical failure, such as breakdown of the oxide in the transistor remains unknown. Herein, the defect-driven interfacial electron structure of the Ti/ZrO$_2$/Al$_2$O$_3$/InGaAs system are probed and manipulated using a specifically designed in situ transmission electron microscopy experimental method. The interfacial defects induced by oxygen-atom missing is found the main reason for the device failure. This study unearths the fundamental defect-driven interfacial electric structure of III–V semiconductor materials and paves the way to future high-speed and high-reliability devices.

Keywords—interfacial defects; III-V semiconductors, breakdown; oxygen vacancies; in situ transmission electron microscope

INTRODUCTION

Scaling the high-κ/Si system for future technology nodes requires a reduction in the thickness of the interfacial layer oxide [1]. However, such reduction results in a decrease in the carrier mobility of the metal-oxide-semiconductor feld-effect transistor inversion layer, which has been attributed to the presence of charges in the high-κ gate stack and the possible contribution of remote phonon scattering from the high-κ layer. One potential solution is the heterogeneous integration of the silicon platform on high-mobility semiconductors, such as III–V materials [2]. InGaAs is one of the best candidates for the III–V substrate because of its high electron mobility and is thus of great interest for *n*-channel metal-oxide-semiconductor feld-effect transistors (MOSFETs) [3,4]. One of the principal challenges associated with the formation of insulating gate InGaAs MOSFETs is the characterization and passivation

of electrically active defects at the interface between the InGaAs semiconductor and the high-dielectric-constant metal oxide [5,6]. Interface states affect the device operation in several ways. Interface states below the edge of the conduction band increase the sub-threshold swing, while those inside the conduction band trap electrons. Both effects reduce I_{ON} for a given I_{OFF}. Interface states can also shift the threshold voltage, degrade the channel mobility, and be a source of instability [7,8]. Therefore, the understanding and control of electrically active defect evolution at the high-κ/InGaAs interface is essential for the successful implementation of high-mobility-channel materials [9]. However, the relationship between the interface state charges and the electronic structure, for example, the breakdown of the insulating oxide material, remains unknown. The inadequate understanding of the physical mechanism is due to the diffculties involved in determining and analyzing the type of microscopic breakdown that forms nanoscale conductive paths within oxide materials. In this work, we present a specifcally designed in situ transmission electron microscopy (TEM) system equipped with electron energy loss spectroscopy (EELS) that can probe and manipulate the breakdown process of metal oxides in real time at the atomic scale under an external electrical feld, during which the deformation behavior of dual-layer oxides and contrast propagation can be observed directly. Breakdown occurs at the ZrO$_2$/Al$_2$O$_3$ interface, and then, structural deformation extends vertically and horizontally. The transformation from monoclinic ZrO$_2$ to cubic ZrO can be seen from a series of high-resolution TEM (HRTEM) images. Obtaining these pieces of information could help improve the performance and reliability of future III-V devices composed of thinner metal oxide layers [10].

EXPERIMENTAL SECTION

To understand the physical mechanism of the breakdown process, a specifically designed *in situ* TEM experiment was carried out. First, a 1 µm thick Pt layer is deposited on the sample to protect the device surface. Second, two rectangular cross-section patterns are used to

978-1-5386-5309-8/18 $31.00 © 2018 IEEE

cut "U" shaped with an ion beam, and the distance between the edge of the Pt layer and the rectangular pattern is typical ~ 0.5 to 1 μm. Then, the thickness of the TEM lamella is reduced to ~ 1.5 to 2 μm by ion milling. Third, a Pt layer is used to connect the TEM lamella to the nano-tip. Then, the nano-tip is moved to lift out the TEM lamella. Fourth, after the lift-out process, the lamella is mounted on top of a TEM grid by depositing Pt to attach the lamella to the Cu grid. Finally, the thickness of the TEM lamella is reduced to ~ 50 nm by FIB cutting. By using this method, one TEM specimen can contain multiple independent, regular-shaped nano-devices with widths less than 20 nm. Hence, the experimental efficiency is improved. Moreover, a HRTEM image can include the whole device without losing sight of the other positions. Figure 1 shows the manipulation and evolution of interfacial defects during *in situ* electrical stressing. A two-step defect-generation method is used in this study. First, a positive sweep voltage from 0 to 3 V is applied to the stack. The trends in current evolution show that the stress-induced leakage current (SILC)occurs at 2.6 V. Second, a stressing voltage of 1.9 V was chosen as the constant voltage stress (CVS) in the experiment, which can generate defects with controllable growth. At the initial stage, the interface between the TiN gate electrode and ZrO_2, ZrO_2 and Al_2O_3 is smooth, and no additional contrast was observed, as shown in Figure 1c. At ~ 0.5 s, breakdown occurs with an abrupt current increase. After breakdown, a black dot appears in the ZrO_2 layer, as shown in Figure 1d.

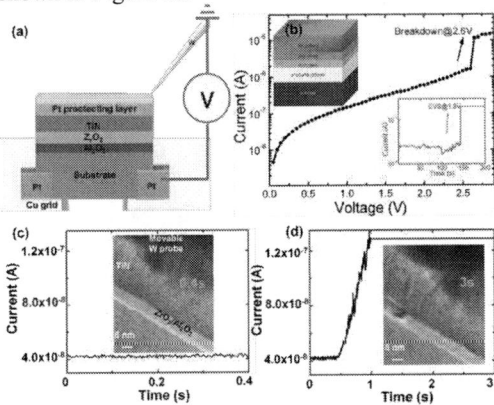

Figure 1: Microstructural changes during in situ electrical stressing. (a) Schematic sketch of the in suit TEM experimental set-up. (b) Schematics of the InGaAs based III-V gate stack and corresponding electrical data during stressing. (c,d) A series of TEM images during the real-time stressing.

RESULTS AND DISCUSSION

Figure 2a-c shows a series of low-magnification TEM images during real-time stressing. No additional contrast is observed in any layer in Figure 2a. At ~ 60 s, SILC occurs with an abrupt current increase, and the area highlighted by a green dashed ellipse shows the vertical

and horizontal expansion of the defects, and a heavier contrast was observed during the whole process. After breakdown (Figure 2d), grain boundary twists, not present in the fresh sample, are observed in the ZrO_2 layer. Together with the grain boundary twists, a hillock-like shape appeared in the InGaAs substrate. These findings are in agreement with the previous study of dielectric breakdown-induced epitaxy (DBIE) in a SiO_2/Si gate stack [11]. Compared with the non-breakdown region, the ZrO_2 layer is a mixed amorphous/crystalline film, exhibiting that transitions mainly occur in the dual Al_2O_3/ZrO_2 layer. As shown in Figure 2g, the fast fourier transform (FFT) patterns show that the area after breakdown is cubic ZrO, while the initial ZrO_2 phase is monoclinic before breakdown. The *in situ* results show many dislocations in the ZrO_2 layer. We suggest that the crystallographic similarity between the cubic TiN metal and cubic ZrO structures is the reason that this phase forms, similar to the hexagonal Zr metal and hexagonal ZrO structures reported by Nicholls *et al* [12]. To analyze the chemical nature of interface state defects, EELS analysis was performed at the breakdown location. The background-corrected O K edge spectra measured from the breakdown and non-breakdown metal oxides is given in Figure 2e. The first absorption peak at 541 eV corresponds to the non-breakdown metal oxide. The peak position shifts from 541 to 539.6 eV upon breakdown. Such shift is believed to be a result of conduction-band *p*-DOS redistribution within the percolation path. The breakdown metal oxide shows significantly lower O K core-loss signals than the non-breakdown metal oxide. The decrease in the intensity arises from the missing O atoms at the breakdown site. The deficiency of O atoms within the breakdown dual ZrO_2/Al_2O_3 layers can be calculated using the core-loss signal intensities. Compared with the non-breakdown signal intensities, the deficiency of O atoms is ~50%, which is the same as the result obtained by the FFT patterns. The EELS mapping results of the breakdown region also show the missing O atoms at the breakdown site.

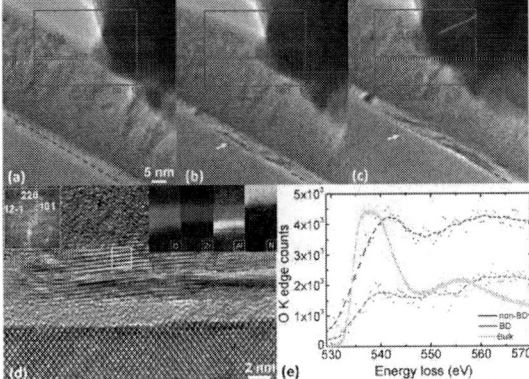

Figure 2: In situ TEM images of the gate stack under a

constant stressing voltage of 1.9 V and STEM/EELS spectra of the breakdown and nonbreakdown locations. (a–c) Low-magnifcation TEM images of in situ electrical stressing. The inset shows the time evolution of the leakage current.(d) High-magnification TEM images after breakdown. The left inset is the corresponding fast Fourier transform (FFT) pattern of the area highlighted by the yellow rectangle. The right insert shows EELS mapping results of the breakdown region. (e) Background-corrected O K edge spectra of the breakdown (red line) and nonbreakdown (blue line) metal oxides.

Figure 3 shows a set of simulated images of the whole breakdown process. Simulation combined with experiment is performed to examine the real situation of a dual-layer metal oxide stressed over its the whole lifetime. In our experiment, the pre-breakdown process is used to generate trap defects, and the sudden increase in current can be attributed to a lack of trap generation. Figure 3a-f shows the corresponding simulated heat distribution profile. At the beginning (Figure 3a-d), heat accumulates at the middle of ZrO_2; the heat is mild (close to room temperature) and cannot lead to breakdown. Then, there is a sudden increase in heat at the ZrO_2/Al_2O_3 interface (marked by white dashed line in Figure 3e), and the breakdown process occurs, extending vertically and horizontally (Figure 3f). The movement from the middle to the interface may be induced by the low thermal conductivity (0.055 $Wcm^{-1}K^{-1}$) of InGa compared with that of TiN (0.688 $Wcm^{-1}K^{-1}$). Compared with the simulated breakdown process, the length of the breakdown process in the experiment is short, and there is not enough time for heat dissipation. The lack of time for heat dissipation produces more breakdown locations compared with the simulated result, and expansion is more significant. Both the simulated and experimental results show the breakdown process to occur at the interface between ZrO_2 and Al_2O_3 and then expand, indicating the important role the interface plays in the breakdown process.

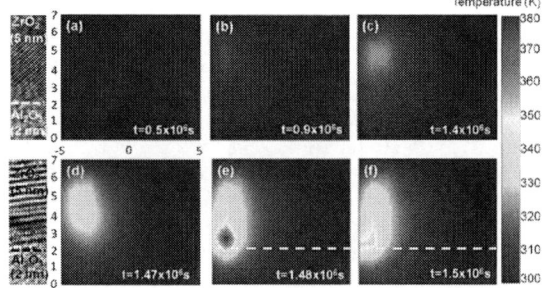

Figure 3: Temperature distribution at a cross section of TiN-ZrO_2-Al_2O_3-InGaAs MOS structure during CVS-TDDB stress.

CONCLUSIONS

In summary, a specifically designed *in situ* TEM experiment was developed, and the site-specific structure and elemental composition, as well as the chemical environment of dual-layer metal oxides along the breakdown percolation path, were analyzed using HRTEM and EELS mapping and STEM-EELS. We unveiled the breakdown mechanism of the metal oxide under an external electrical field. The results show that O atoms missing from the breakdown site are the critical cause of atomic-scale structural changes along the breakdown path. Grain boundaries are observed at the breakdown region, which may be caused by the missing O atoms. Similar to other high   dioxide materials, the local energy gap at the breakdown path is believed to narrow, indicating that these localized states in the band gap are mainly contributed by unbonded Zr d-orbitals, which have the potential to greatly contribute to the breakdown leakage current with the help of grain boundary twisting, temperature *etc*. Such understanding is critical to probing the breakdown mechanism of the metal oxide and advantageous for the development of future device applications.

REFERENCES

[1] M. Lundstrom. *Science*, vol. 299, 2003, pp. 210-211.

[2] J. A. del Alamo. *Nature*, vol. 479, 2011, pp. 317-323.

[3] R. Engel-Herbert, Y. Hwang, S. Stemmer. *J. Appl. Phys.*, vol. 108, 2010, pp. 124101-124115.

[4] M. Benedicto, B. Galiana, J. M. Molina-Aldareguia, S. Monaghan, P. K. Hurley, K. Cherkaoui, L. Vazquez, P. Tejedor. *Nanoscale Res. Lett.*, vol. 6, 2011, pp. 400-405.

[5] W. Spicer, P. Chye, P. Skeath, C. Y. Su, I. Lindau. *J. Vac. Sci. Technol.*, vol. 16, 1979, pp. 1422-1433.

[6] B. Shin, J. R. Weber, R. D. Long, P. K. Hurley, C. G. Van de walle, P. C. McIntyre. *Appl. Phys. Lett.*, vol. 96, 2010, pp. 152903-152906.

[7] W. S. Cho, M. Luisier, D. Mohata, S. Datta, D. Pawlik, S. L. Rommel, G. Klimeck. *Appl. Phys. Lett.*, vol. 100, 2012, pp. 063504-063506.

[8] M. Luisier, G. Klimeck. *IEEE Electron Device Lett.*, vol. 30, 2009, pp. 602-604.

[9] J. Ballingall, P. Martin, J. Mazurowski, P. Ho, P. Chao, P. Smith, K. Duh. *Thin Solid Films*, vol. 231, 1993, pp. 95-106.

[10] Q. Liu, J. Sun, H. Lv, S. Long, K. Yin, N. Wan, Y. Li, L. Sun, M. Liu. *Adv. Mater.*, vol. 24, 2012, pp. 1844-1849.

[11] K. L. Pey, N. Raghavan, X. Wu, W. Liu, X. Li, M. Bosman, K. Shubhakar, Z. Z. Lwin, Y. Chen, H. Qin. *Microelectron. Eng.*, vol. 88, 2011, pp. 1365-1372.

[12] R. J. Nicholls, N. Ni, S. Lozano-Perez, A. London, D. W. McComb, P. D. Nellist, C. R. Grovenor, C. J. Pickard, J. R. Yates, *Adv. Eng. Mater.*, vol. 17, 2015, pp. 211-215.

978-1-5386-5309-8/18 $31.00 © 2018 IEEE

RELIABILITY ENGINEERING: HELP ENABLE TECHNOLOGY SCALING

S. Pae, HC. Sagong, M. Jin, T. Jeong, J. Kim, I. Baick, H. Shim, J. Park, H. Kim, Y.C. Choi, and S. Shin
Foundry Business, Samsung Electronics, Gi-Heung City, Korea, sangwoo.pae@samsung.com

ABSTRACT

The paper discusses many aspects of reliability engineering, modeling, and characterization methods that could be used to improve reliability of the technology. As results, higher transistor Vmax or reduced dppm, more tolerable aging at the circuit and/or product level can be utilized by accurate wafer level reliability characterization and modeling. On the BEOL and far-BEOL, EM and BEOL metal-via spacing requirements will be discussed with areas where improvements can be made with help of simulation and low voltage stress data. CPI stress results will also be covered. Product level reliability stack up is assessed to provide more accurate dppms projections. Extra reliability margins could be applied for further dimension scaling, or upfront commit for technology development, increasing over-drive to achieve higher performance where needed and also to support different grade levels for automotive applications.

INTRODUCTION

The paper discusses FEOL & BEOL/far-BEOL (including CPI) reliability followed by Product stress data. In addition to the leading edge process technology development, reliability engineering based on physics and realistic condition provide more accurate reliability assessments which greatly aids the process development. Any uncovered reliability margin could be leveraged for product benefit and accurate modeling may provide needed gap. The study is based on 10nm and 14nm technology nodes where the 3D FinFET transistors with high-k/metal-gate (HK/MG) and ultra low-k (ULK) BEOL dielectrics were used. Many million parts of products have been shipped to many different market segments and robust intrinsic reliability as well as low dppms have been demonstrated.

EXPERIMENTS
FEOL Reliability (TDDB, BTI)

With transition of Poly/SiO_2 (including oxynitride) to MG/HK process technology, conventional 0.7X dielectric scaling (in terms of T_{inv}) no longer applies. HK scaling in terms of ALD cycles can be done but this doesn't offer signicant T_{inv} scaling benefit at the cost of exponential leakage increase. Thus, voltage (Vdd) scaling seemed slowed down with HK/MG transistor technology nodes. While higher Vdd is still wanted for boosting up performance in certain applications on SOC chips, such as over-drive, achieving lower Vmin continues to provide unparalled product benefits but can be challenging. Note that with introduction of FinFETs, Vmin can be further lowered due to reduced Vt and its variations. Between 20nm to 10nm process nodes, Vmax based on gate TDDB breakdown hasn't essentially changed much (again, small HK cycles of scaling does not provide large gains in oxide TDDB, especially on NFETs which limits the Vmax [1]). **Fig.1** shows an example of gate TDDB distributions for NFETs in inversion and PFETs in accumulation (both with positive gate bias stress). Note that the lifetime distribution isn't just a Weibull but curves off at both lower and higher percentiles. TDDB rolling-off at lower percentiles can lead to lifetime gain if carefully modeled. The dual interface layer (IL) and HK layer at reduced bias has been discussed to improve this behavior [2].

N/PFETs fabricated with HK/MG shows some interesting polarity dependent behavior. All 4 models (N/P, inv/acc.) are needed to accurately assess product Vmax for dppms (shown in **Fig.2**) and compared to product HTOL results.

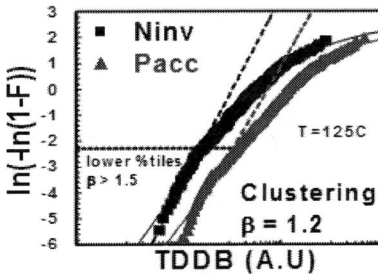

Fig.1: Weibull slope on Pinv/Nacc TDDB is improved (PFET: beta:0.8→1.2, NFET: beta>1.5) at lower percentiles by applying the clustering model [1].

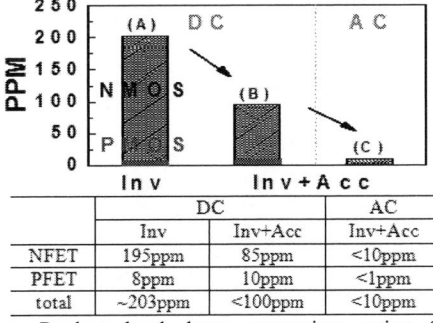

	DC		AC
	Inv	Inv+Acc	Inv+Acc
NFET	195ppm	85ppm	<10ppm
PFET	8ppm	10ppm	<1ppm
total	~203ppm	<100ppm	<10ppm

Fig.2: Product level dppm comparison using flat DC (worst-case) bias. With addition of correct usage modeling, dppm can be projected with better accuracy (>2X reduction in example) at 10yrs of lifetime [1].

978-1-5386-5309-8/18 $31.00 © 2018 IEEE   393

Different schemes of assisted-circuitry have been used to achieve lower Vmin. It's been shown that PFET NBTI limits the transistor aging and also for product reliability ΔVmin guardband. Reliability stress must ensure die-level ΔVmin is within guardband design limits. More details will be discussed at the conference.

Self-Heating and HCI, RO Aging

Fin transistor self-heating can be directly or indirectly measured with special test structures and have been modeled vs. power [3]. Despite the higher DC level self-heating of these transistors, the actual self-heat during AC operation is much less and should be accounted for in design and also during reliability stresses. An odd stage Ring Oscillator stress showed similar level of degradation, dominated by BTI, rather than self-heating and HCI [4]. This is because the duty to induce excess self-heat is low due to fast switching near 1/2Vdd. HCI aging with de-coupling self-heat effects showed significantly reduced PMOS HCI than in NMOS HCI due to higher Ea sensitivity with PMOS [3, 5]. **Fig.3** shows NFET/PFET HCI stress with and without self-heat decoupling. **Fig.4** shows flow chart of circuit simulations with self-heat model to help more accurate aging simulations.

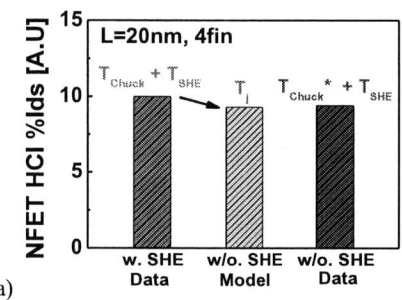

(a)

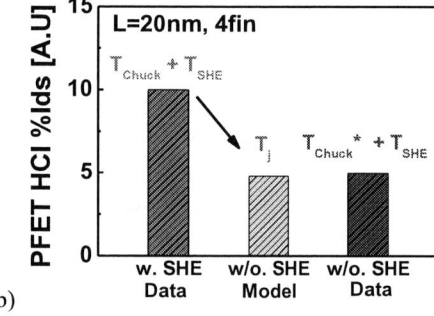

(b)

Fig.3: HCI before SHE correction ($T_{chuck} + T_{SHE}$) vs. HCI after SHE correction model (Tj), further confirmed by HCI test by lowering temperature ($T_{chuck}^{*} + T_{SHE}$) (a) NFET HCI : SHE is small (b) PFET HCI : SHE correction lead significantly reduced HCI in Tj due to higher Ea in PFET HCI [3].

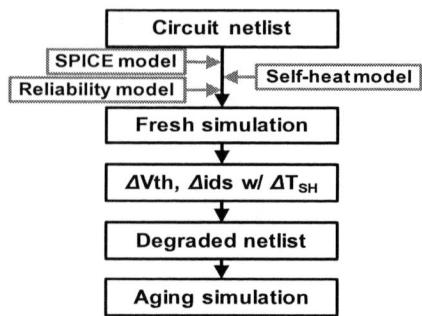

Fig.4: Self-heating model is embedded into SPICE simulation along with aging runs. The aging is estimated based on self-heating. ($Tj + T_{SHE}$) [3].

Fig.5 shows actual duty of IP blocks and corresponding transistor types. SG refers to thin/core oxide, and EG refers to I/O or thick gate oxides. Self-heating effect is reduced with lower duty in AC operations.

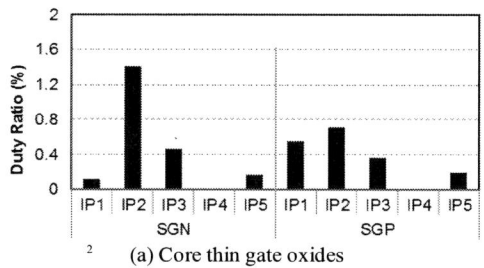

(a) Core thin gate oxides

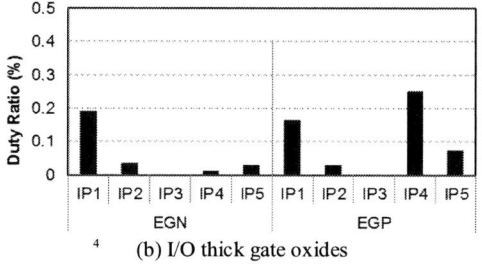

(b) I/O thick gate oxides

Fig.5: The HCI duty ratio on the typical IP blocks by aging simulation is less than 2%. Thick oxide (termed as EGN, EGP) showed even less than thin gate (termed as SGN, SGP). Duty in this case was normalized to the HCI aging at the Vg=1/2Vd condition [4].

BEOL Reliability (EM, IMD TDDB)

With metal line width and spaces shrink, the spacing between metal lines are becoming more critical (physical spacing can be reduced to below 10nm as scaling progresses), especially for those at the minimum 1x DR, despite using double patterning techniques and even with self-aligned via processes. Our experiments showed that the voltage and field acceleration parameters for BEOL IMD TDDB increased more than the model predicted below certain physical space (~18nm) [6]. Power-law

(V^{-n}) model predicted better than the square-root E-model ($\sqrt{E}$) with more than 8 months of constant voltage stress (CVS) done in package level (and also through comparison of 1000hrs of product HTOL data). Very high stress voltage conducted on product HTOL results scaled back to test chip area assuming the worst case spacing (within the process variation) also better supported the power-law model line. The product HTOL resulted in 1000hrs of passing at 125C. Based on the sample size tested for HTOL, t63.2% point was added using the IMD TDDB distribution (from the IMD TDDB beta, **Fig.6**). Through this simple analysis, we can conclude that the long-term results of test-chip level are also in good agreement with the product level reliability results to a first order. With demonstration of the power-law IMD TDDB model in both 10nm and 14nm technology nodes, further opportunities for process technology scaling is viable. The long term test structure stress and product level stress data allows us to gain further insight into the lifetime projection more accurately [6-7].

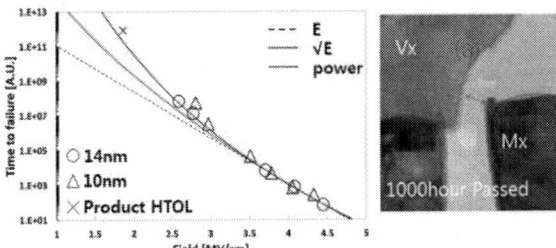

Fig.6: t$_{63.2\%}$ vs. stress field trends of 14 and 10nm technology nodes from long-term (>8month) TDDB and HTOL results [6].

Operating current density increases with chip size scaling (**Fig.7**) and moreover heat removal from the chip becomes increasingly difficult as low-k or ultra-low-k film has been adapted in advanced technology nodes which have much lower thermal conductivity. The heat dissipation will be a continuous threat especially when air-gap process is used in BEOL processes. The remaining heat induced by the high current density increases the temperature of metal interconnect lines and can hurt EM/SM (Electro-/ Stress-Migration) reliability performance. 3D-FEM (Finite Element Modeling) was performed to investigate metal length effect and empirical model was proposed based on the FEM result. Temperature profile of short interconnects can be modeled more accurately and 25% upside in EM Irms performance has been quantified through both electrical-thermal 3D-FEM simulations and verification of silicon on short (<0.5um) interconnect lines on advanced BEOL interconnect technology [8]. This current relaxation in short interconnect is beneficial especially for circuit designs where local routing and high current is needed (**Fig.8**).

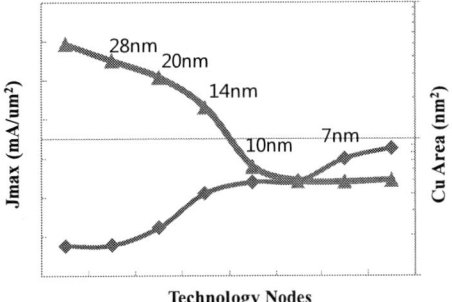

Fig.7: Electromigration (EM) current density increase vs. technology scaling. Innovations are needed to further continue scaling and improve EM (Jmax) performance.

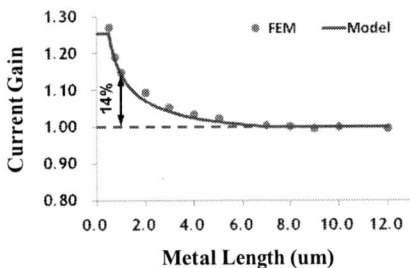

Fig.8: Comparison between 3D-FEM simulations and empirical model. Note that for practical reasons, line lengths below 0.5um has been clamped [8].

CPI, EMI, and MIMCAPs

Increasing hand-held products require robust board level reliability (BLR) in addition to IC level stresses. Most of flip-chip packages use either C4 or Cu pillar bumps. Any crack propagation between the bump and the PCB is critical and must be thoroughly characterized through reliability stress. Due to material thermal expansion (CTE) differences, the stress is distributed differently among the bumps, PCBs, and to silicon die. EMC (elastic molded compound) is an underfill material beteen the bumps and the Die and PCBs to mitigate the CTE and thereby relieving stress (**Fig.9**). Temp cycling tests along with HTS and HAST stresses show power or signal line fails and upon X-ray and PFA, to detect any joint cracks or other CPI issues. Many factors including EMC properties, ball size, ball content can be adjusted to meet and improve reliability requirements.

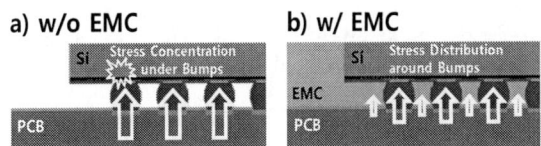

Fig.9: CPI (a) stress concentration under bumps w/o EMC and (b) stress distribution around bumps w/ EMC. The stress is induced by CTE mismatch between Si and PCB [9].

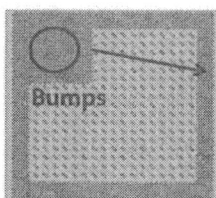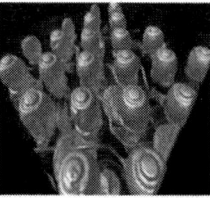

Fig.10: Reliable MUF process results from the optimization of intrinsic process parameters [10].

Fig.10 shows reliable underfill process without any bump shorts (through 3D X-ray) [10].

MIMCAPs are used as decoupling capacitors. It helps reduce power fluctuations as well as enable benefit in reduction of EMI (Electromagnetic Interference). Noise spectrum vs. frequency for MIM vs. package embedded CAP showed 11dB reduction with having MIM in noise characteristics [11-12]. MIM process defects includes punchthrough defects which can be fixed and monitored using MIM Vramp characteristics. MIM and CPI reliability results using both conventional POP, Intersposers, and FO-PLP packages are summarized along with drop tests in Table 1.

TABLE I. CPI RELIABILITY TEST RESULTS

Reliability Items	Stress Condition	Stress Time/Cycle	Result
HTS	150C	1000hrs	Pass
Precond.	30C/60% RH	192hr	Pass
TC	-55~125C	1000cy	Pass
Biased-HAST	130C/85%RH, Bias	168hr	Pass
Drop Test			Pass

Product Reliability (HTOL, BLR, Drop Tests, Set)

In addition to conventional IC level reliability stresses like HTOL and environmental (HAST, TC) stresses, set level reliability stresses have been devised and leveraged (in **Fig.11**) to identify system fails and marginality issues that are not detectable through DFT and screen vectors have been implemented to ATE/ASB to help screen and significantly reduce dppms prior to shipping the parts to customer. In ASH (At Speed HTOL) [13], set level stress tests running firmware with user-based scenarios are conducted to accelerate more realistic usage case. Different scenario tests to accelerate larger area CPUs and GPUs could be done to highlight their failures. Any fail modes detected from ASH creates screening vectors then reflected on ASB (At Speed Board) test to screen similar fails from happening. Design block pareto of failures in ASH stress, mostly coming from the big CPU block. Thermal characterization of AP under the worst case scenario tests can be verified. More details can be found in [13].

CONCLUSION

Future technology scaling renders little room for reliability and often times, reliability can be perceived as technology limiter and barrier. With accurate reliability modeling and clever ways to conduct stresses combined with real usage scenarios can help greatly uncover margins and continue to help and contribute to the technology scaling. Here, the reliability does not become limitation any more but as part of more keen technology development itself.

ACKNOWLEDGEMENTS

Authors would like to thank Technology Development PA, PIE, MTC teams for fabricating the materials, PTE and WLR/PLR labs for readout testing and stress characterizations.

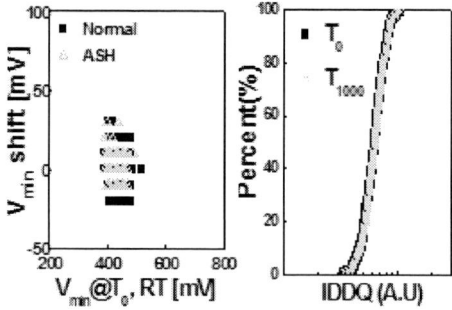

Fig.11: (a) The amount of Vmin-shift is below 50mV after 1000hrs of product stress, accelerated at much higher than the technology Vmax, still doesn't exhibit Gate oxide breakdown. (b) There is no IDDQ shift [1].

REFERENCES

[1] H. Kim et al., "A Systematic Study of Gate Dielectric TDDB in FinFET Technology", *Proc. of IEEE IRPS 2018 in press*.

[2] S. Pae et al., "Gate dielectric TDDB characterizations of advanced High-K and metal-gate CMOS logic transistor technology", in *Proceedings of IEEE International Reliability Physics Symposium (IRPS)*, 2012, pp.5C.1.1-5C.1.5.

[3] M. Jin et al., "Hot Carrier Reliability Characterization in Consideration of Self-Heating in FinFET Technology", in *IEEE IRPS*, 2016, pp.2A.2.1-2.4.

[4] Y. Kim et al., "Investigation of HCI effects in FinFET based Ring Oscillator Circuits and IP blocks', in *IEEE IRPS*, 2017, pp.4C.2.1-2.4.

[5] M. Jin et al., "Reliability Characterization of 10nm FinFET Technology with multi-V_T gate stack for Low Power and High Performance", in *IEEE IEDM*, 2016, pp.380-383.

[6] T. Jeong et al., "Opportunities for further BEOL Technology Scaling using Power-Law IMD TDDB model on 10/14nm BEOL Process Technologies and beyond", in *IEEE Proceedings of International Interconnect*

978-1-5386-5309-8/18 $31.00 © 2018 IEEE

Technology Conference (IITC), 2017, pp.1-3.

[7] T. Jeong et al., "Low Voltage IMD-TDDB lifetime model for Advanced Future Logic Technology Nodes", in *IEEE IITC*, 2015, pp.299-301.

[8] J. Kim et al., "EM performance upside of short BEOL interconnects in advanced process technologies: Electrical-Thermal Finite Element Simulations and Silicon Verifications", in *IEEE IITC*, 2014, pp.227-229.

[9] I. Baick et al., "Effect of EMC Properties on the Chip to Package Interaction (CPI) Reliability of Flip Chip Package", in *IEEE International Integrated Reliability Workshop (IIRW)*, 2016.

[10] K. Choi et al., "Two-Phase Flow Simulation of Molded Underfill Process to Mitigate Void and Assure Reliability at Product Level", in *Proceedings of IEEE Electronic Components and Technology Conference (ECTC)*, 2016, pp.660-665.

[11] H. Chun et al., "CPI Reliability and EMI Benefit for MIM CAP Embedded C4 Package", in *Proceedings of IEEE International Reliability Physics Symposium (IRPS)*, 2015, pp.5C.5.1-5C.5.4.

[12] S. Kim et al., "Simulation-based Analysis on EMI Effect in LPDDR Interface for Mitigating RFI in a Mobile Environment", in *Proceedings of IEEE Signal and Power Integrity (SPI)*, 2016, pp.

[13] J. Park et al., "Scenario-based Set-level HTOL Test (ASH III) for Product Quallity and Reliability Qualifications on High-Speed APs', in *IEEE IRPS*, 2016, pp.7C.3.1-3.4.

978-1-5386-5309-8/18 $31.00 © 2018 IEEE

CHANGEABLE ELECTROMIGRATION FAILURE MODE IN WIDE CU INTERCONNECTS

Hui Zheng[*], Binfeng Yin, Leigang Chen, Ke Zhou, and Chinte Kuo
Department of Reliability Engineering, Shanghai Huali Microelectronics Corporation,
Shanghai, 201203, P. R. China
*Corresponding Author's Email: zhenghui@hlmc.cn

ABSTRACT

The electromigration (EM) failure mode of wide via-line Cu interconnects, together with the current density exponent, is shown to be changeable with current density at common package-level acceleration conditions. As the stress current density increases, the resistance of upstream interconnect is more likely to exhibit open jump at EM failure. Failure analysis shows Ta-based liner at the Via is broken at high currents, and isolated Via voiding overwhelms the line voiding and becomes responsible solely for the EM failure. Finite element simulation suggests the inhomogeneous current density in the Via and line may dominate the EM kinetics at high currents.

INTRODUCTION

JH (joule heating) effect is widely utilized in wafer-level electromigration (EM) test, e.g., IsoEM [1], whereas it needs to be avoided in package-level EM test. However, it is often inevitable that wide Cu interconnects, which may be used as power lines in the back end of line of integrated circuits, may suffer JH effect during conventional package-level acceleration test. This is because, in comparison with narrow Cu interconnects, wide wires have much enhanced JH power but less increased heat-dissipating area. The JH power per unit length can be expressed as:

$$P = I^2 R = (jS)^2 \rho \frac{1}{S} = j^2 \rho S$$

where j, ρ and S are the stress current density, Cu resistivity and cross-sectional area, respectively. Obviously, the JH power is in proportion to S. On the other hand, the area for thermal dissipation would be roughly proportional to $S^{0.5}$, which implies JH effect would be nonnegligible even at conventional package-level acceleration test as long as the Cu interconnects are sufficiently wide.

It is well known JH may cause the current-density exponent increasing and deviating from its true value during acceleration test of electromigration reliability [2]. It is believed this is attributed to underestimation of the effective temperature of the whole line [2,3] or local temperature [4] as a result of current-temperature coupling. Some researchers pointed out dislocation rearrangement at high current may be responsible for the large current-density exponent [5], which may further complicate the interpretation of JH effect during EM. In

this paper, it will be shown that with the enhanced JH effect as current density increases, the electromigration failure mode of wide via-line Cu interconnects will be changed from line voiding to via voiding although the nominal current densities of the metal line are in the range of operation and conventional package-level acceleration conditions.

EXPERIMENTAL

An upstream test structure was investigated in this paper, in which the Via has a nominal dimension of 0.324×0.324 μm, and the line is 0.36 μm wide, 0.78 μm thick and 600 μm long. The interconnect lines were fabricated with Cu damascene process, in which Ta-based film was used as the liner and SiN-based film as the cap layer. The inter-metal dielectric is SiO_2. For package-level EM test, nominal current density of 0.65~3.12 MA/cm^2 was applied on the metal line at 300 °C with Quali-Tau Mira Electromigration Test System. Additional EM tests under 275 and 325 °C were done at 3.12 MA/cm^2 for extracting the EM activation energy, which will be used to correct the *TTF* (Time to Failure) measured at different current densities with the integrated software in Mira. The initial resistance jump was taken as the failure criterion. Failure analysis was implemented with FIB-SEM.

Furthermore, preliminary finite element analysis (FEA) on the distribution of current density and temperature was conducted with Ansys 15. A simplified 3D geometrical graphic was produced to model the actual test structure, in which two pads were used with their full area being close to the actual four Pads, the lead structure was trimmed for convenience of meshing and solving, and the dielectric encompassing the structure was generated to be homogeneous with no dummy metals or other dielectrics inside. Two boundary conditions were used:

a) The temperature at the bottom surface of the sample block and surfaces of the two Pads was set to be the oven temperature, because theses faces are bonded to the package socket with highly-thermally-conductive materials like Al, Si, and Au.

b) At the rest five surfaces of the sample block, heat transfer to the ambient was set to be by natural convection with a constant convection coefficient.

RESULTS AND DISCUSSION

Fig. 1 shows the typical variation of resistance with

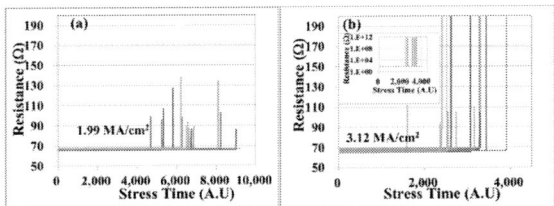

Fig. 1 R-t curves of the wide interconnects measured at (a) 1.99 and (b) 3.12 MA/cm².

time during EM test of the test structures. It shows the resistance varies little before a jump appears which indicates EM failure of Cu. Such R jumps are all below or around 200% at j=1.99 MA/cm², whereas many open jumps occurs at j=3.12 MA/cm². Fig. 2 summarizes the

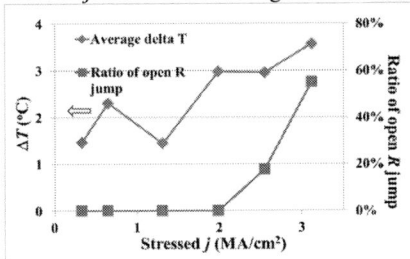

Fig. 2 Variation of JH induced initial temperature rise (ΔT) and the ratio of open R jump at EM failure of the wide interconnects as a function of current density.

ratio of the samples with open R jump at EM failure. Obviously, there is no open R jumps when $j \leq 1.99$ MA/cm², whereas they increase with j when j>1.99 MA/cm².

Fig. 3 gives the typical FIB-SEM images of the test structures failed at three current densities. For j=0.65 MA/cm², the EM induced voiding that's responsible for the observed EM failure is located at Cu line end and upper part of the Via near the cathode. For j=1.99 and 3.12

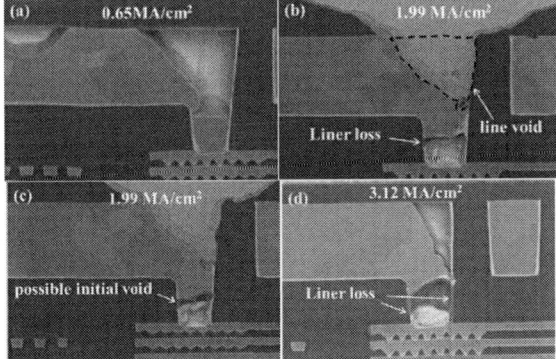

Fig. 3 FIB-SEM (BSE) images of the cathode of the as-failed wide interconnects. The sample of (d) was stressed at 285°C, and the others were at 300 °C. Note the R jump ratio shows no considerable temperature dependence.

MA/cm², although voiding at the line corner also takes place, the responsible voiding changes to be formed at the under part of the Via. It can be clearly seen from the deeper profile in Fig. 3(c) that such voiding is isolated from that at the line corner, indicating independent nucleation of the two voids. Moreover, the ratio of the size of Via void to line void appears to increase with increasing the current density, which is accompanied by reinforced destruction of the Ta-based liner at the Via. It demonstrates there exists competition of Via failure and line failure for EM of such wide Cu interconnects, and the former may overwhelm the latter as the current density increases.

Obviously, the open R jump observed at high current densities can be attributed to the complete interruption of the liner at the failure location where voiding occurred. It implies the JH effect played a role during EM of such wide test structure, although the stress currents are of the same order of magnitude with the operation current (~1 MA/cm²). In Fig. 2 also summarizes the JH-induced rise in the initial effective temperature of the whole line (ΔT) which was calculated from the *TCR* (temperature coefficient of resistance). ΔT tends to increase from 1.5 to 3.6 °C as j increases from 0.33 to 3.12 MA/cm². Table I summarizes these results, together with the activation

TABLE I. SUMMARIZATION OF THE PARAMETERS EXTRACTED IN THE EXPERIMENTS.

j (MA/cm²)	0.33	0.65	1.31	1.99	2.55	3.12
Ratio of open jump at fail	No fail in 2256 hr	0.00%	0.00%	0.00%	17.65%	55.17%
average ΔT (°C)	1.47	2.31	1.45	2.98	2.96	3.57
E_a (eV), before correction	—					0.994
E_a (eV), after correction	—					0.946
n, before correction	1.489			1.813		
n, after correction	1.472			1.773		

energy (E_a) and current-density exponent (n) extracted with Black's equation separately from the *MTF*s (mean time to failure) at different temperatures and different current densities. For calculating E_a and n, firstly the ΔT obtained from *TCR* was used to correct the test temperatures, then the E_a obtained after such correction was utilized to normalize the *TTF*s measured at different j, which have different ΔT, to the same temperature, and finally n was extracted from the normalized *TTF*s. This procedure was implemented by use of the integrated software of Mira and is similar to that used in Ref. 3. It is clear the n value extracted from the low j is remarkably lower than that from the high j, which further confirms the nonnegligible JH effect. However, after correction with ΔT the n value only decreases a little both at high and low j. Considering ΔT herein is too small to melt the Ta-based liner, it suggests the JH-induced initial effective temperature rise in the test structure alone can't be able to interpret the JH effect in this case.

As a matter of fact, during EM test, local high temperature that can melt or damage the refractory Ta-based liner can be produced only by the local Joule heating. Particularly, when a void forms in the Via, the temperature rise of the liner nearby the void could be much higher than the initial effective temperature, ΔT, due to much larger resistivity and much smaller cross-sectional area of Ta-based liner than Cu, which may lead to local temperature running away at high j. Fig. 4 shows the temperature distribution in the Ta-based liner encompassing a 0.4 μm long void inside the cathode Via simulated with Ansys 15. It is observed from Fig. 4 that the liner temperature can locally exceed 3000 °C, owing to giant local JH power density which appears to be ~5 orders of magnitude higher than that in the metal line. Note such high local temperature would result from positive self-feedback of local void growth and local temperature rise, rather than from initial JH.

Fig. 4 Simulated temperature distribution in the liner encompassing a 0.4 μm long void in the Via. The nominal current density is 3.12 MA/cm^2 and the ambient temperature is set to be 300 °C during simulation.

Fig. 5 shows the simulated initial temperature distribution at the initial state when there's no void in the structure. It shows JH-induced temperature rise of 0~4.4 °C along the test structure at j=3.12 MA/cm^2, which is comparable with that obtained from the TCR. The maximum temperature at j=3.12 MA/cm^2 is located at line center due to relatively poor heat dissipation in comparison with the line ends. However, the Via temperature is only a little lower than the line center, although the current density at the via is 3~5 folds larger (Fig. 6). This demonstrates Cu, as a good heat conductive material, obviously is very efficient for temperature homogenization. It implies the initial local temperature or temperature gradient could not be responsible for the observed nucleation at line end and via. According to Fig.

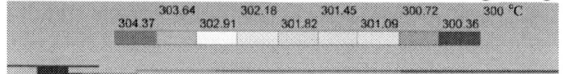

Fig. 5 Simulated variation of temperature along the interconnect. The initial conditions are the same as Fig. 4.

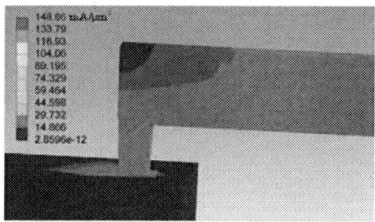

Fig. 6 Simulated variation of current density nearby the Via of the interconnect. The initial conditions are the same as Fig. 4.

6, it is suggested: a) the line corner may act as vacancy "sink" during EM due to its much smaller j compared to the vicinity region [6], which may be responsible for the void nucleation at line corner as observed; b) at via/lead interface, the atomic flux divergence would be much larger than inside the line, then the much higher j and j gradient in the Via in comparison with the line may trigger independent nucleation inside the Via, through coupling with some local weakness in the Via, such as defects and stress concentration region which however can't be clearly determined so far. As j decreases, the latter becomes difficult to occur; then, only the voiding at line corner and Via upper part can be observed when j is around the operation condition.

CONCLUSION

The Via voiding and accompanying catastrophic failure at common package-level acceleration current density implies the inhomogeneity in current density and temperature at via-line interconnects, which are made with varying dimension and different materials, should be carefully accounted during reliability design and assessment of wide Cu interconnects.

ACKNOWLEDGEMENTS

Hui Zheng would like to thank Lingqiao Cheng for her supports on this study.

REFERENCES

[1] J. –C. A. Huang, W. –T. K. Chien, and C. H. -Jia Huang, *IEEE Trans. Semi. Manu.*, vol. 14, 2001, pp. 387-394.

[2] Y. Wang and Y. Yao, *J. Appl. Phys.*, vol. 121, 2017, pp. 065701-1-7.

[3] A. Royand and C. M. Tan, *J. Appl. Phys.*, vol. 103, 2008, pp. 093707-1-7.

[4] C. Pennetta, E. Alfinito, L. Reggiani, F. Fantini, I. DeMunari, and A. Scorzoni, *Phys. Rev. B*, vol. 70, 2004, pp. 174305.

[5] A. S. Budiman et al., *J. Electron. Mater.*, vol. 38, 2009, pp. 379-391.

[6] C.-K. Hu et al., *Appl. Phys. Lett.*, vol. 78, 2001, pp. 7-9.

IMPACT OF TID RADIATION ON HOT-CARRIER EFFECTS IN 65NM BULK SI NMOSFETS

Zhexuan Ren[1], Xia An[1], Jianing Wang[2], Xing Zhang[1], and Ru Huang[1*]*

[1] Institute of Microelectronics, Peking University, Beijing 100871, China

[2] Faculty of Information Technology, Beijing University of Technology, Beijing 100124, China

*xia.an@pku.edu.cn, ruhuang@pku.edu.cn

ABSTRACT

The impact of γ-ray exposure on Hot-Carrier effects of 65nm bulk Si NMOSFETs and the dependence on device geometry is experimentally investigated. Experimental results show that stress-induced degradation on irradiated devices is severer than un-irradiated devices. Besides, the geometry dependence of degradation is illustrated: the narrower the device the larger the stress-induced degradation, which is attributed to the radiation-induced trapped charges in STI. The results indicate that the reliability issue of MOS devices is more challenging in harsh environment due to the synergetic effect of TID and HCI.

INTRODUCTION

With the development of space technology, the demands for the reliability issue of microelectronic devices working in harsh environment increase. Radiation damage and reliability damage exist simultaneously in these devices. Among the radiation effects and reliability issues, the total ionizing dose (TID) and hot carrier injection (HCI) play important roles and have been separately investigated for decades [1-9]. It's quite necessary to focus on the combined effects of TID and HCI to evaluate the reliability of devices in radiation environment. The hot carrier degradations of 130nm NMOS after 136Mrad(SiO$_2$) X-ray exposure is severer than un-irradiated devices [5]. Similar phenomena are observed in 0.35μm NMOS [6] and 65nm NMOS [7][8]. However, the research on the correlation of TID and HCI for nano-scale bulk Si MOS devices is quite limited.

In this paper, the synergetic effect between TID and HCI of 65nm NMOS is experimentally investigated. And the dependence on device size is demonstrated.

DEVICES AND EXPERIMENT DETAILS

The devices used in this work are fabricated by commercial 65nm bulk Si CMOS process. The channel length (L) is 0.06μm and width (W) are 0.5μm and 5μm, and nominal operating voltage is 1.2V. The irradiated samples were encapsulated in ceramic packages.

The TID irradiation was performed using cobalt-60 source at room temperature. The irradiation bias condition was ON state (where $V_{gs}=1.2V$ and $V_{ds}=0V$), which was regarded as the worst irradiation bias condition [9]. The accumulative total dose was 300 krad(Si). The CHC stress condition was $V_g=V_d=V_{stress}=2V$, with other terminals grounded. The stress duration was 1000s and was interrupted to measure I_d-V_g curves.

All I-V curves of the devices were obtained using Agilent B1500A semiconductor parameter analyzer.

RESULTS AND DISCUSSION

Fig.1 shows the I_{ds}-V_{gs} curves and g_m-V_{gs} curves during CHC test for irradiated and un-irradiated devices with W/L of 0.5μm/0.06μm. It can be seen from Fig.1(a), the threshold voltage (V_{th}) of the irradiated narrow device (W=0.5μm) decreases slightly (~5mV) after 300krad(Si) irradiation. While for the irradiated wide device (W=5μm), V_{th} shift is not observed (not shown here for brevity). The Hot-Carrier degradation behavior of irradiated device is similar to un-irradiated one, as shown in Fig.1. The threshold voltage increases about 100mV after 1000s stress time. And the maximum transconductance decreases about 50% after CHC test, as shown in the insets of Fig.1.

Fig.2 gives the stress-induced threshold voltage shift (ΔV_{th}) as a function of stress time (T_{stress}) for un-irradiated and irradiated devices. The threshold voltage of irradiated and un-irradiated narrow devices increases about 101.3mV and 71.6mV after 1000s stress, respectively. While for the irradiated and un-irradiated wide devices, V_{th} shifts are about 70.1mV and 60.2mV after 1000s stress, respectively. The stress-induced degradation on irradiated devices is larger than un-irradiated devices, especially for the narrow devices.

The V_{th} shift and stress time follow a power-law relationship:

$$\Delta V_{th} = \alpha (T_{stress})^n \qquad (1)$$

where α and n are fitting parameters. The n value of irradiated narrow device increases about 30.2% compared to un-irradiated narrow device, which indicates much faster degradation and shorter predicted lifetime. While for the wide devices, n is almost the same. From the results, it can be seen that the degradation is geometry dependent: the narrower the device the larger the stress-induced degradation.

978-1-5386-5309-8/18 $31.00 © 2018 IEEE

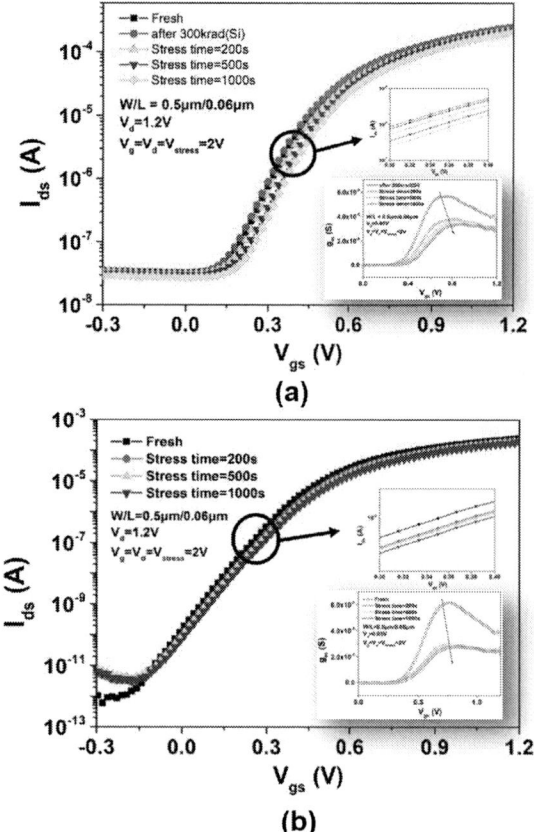

(a)

(b)

Figure 1: The I_{ds}-V_{gs} curves of devices before and after CHC stress with W/L of 0.5μm/0.06μm. (a)after 300krad(Si) irradiation and (b)un-irradiated. The insets in (a) and (b) are the degradation of the transconductance of irradiated and un-irradiated devices, respectively.

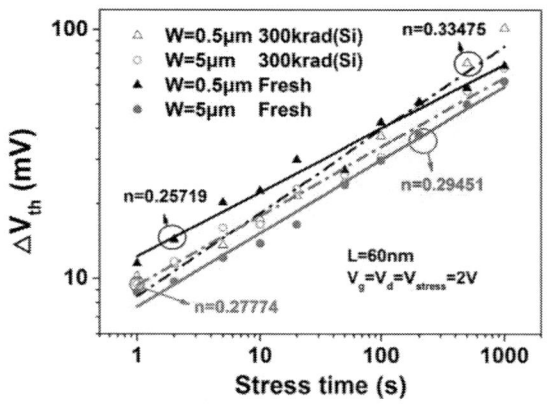

Figure 2: V_{th} shift as a function of stress time for irradiated and un-irradiated devices.

Furthermore, the synergetic effect between TID and HCI of 65nm NMOS is demonstrated, as shown in Fig.3. ΔV_{th_post} is the stress-induced V_{th} shift of irradiated device, which is discussed above. And ΔV_{th_total} is the total V_{th} shift induced by TID and HCI. ΔV_{th_post} and ΔV_{th_total} are calculated by equation (2) and (3), respectively.

$$\Delta V_{th\_post} = V_{th\_stress} - V_{th\_irradiated} \quad (2)$$

$$\Delta V_{th\_total} = V_{th\_stress} - V_{th\_fresh} \quad (3)$$

The n value of ΔV_{th_total} is 65.4% larger than that of un-irradiated device. The results indicate that the reliability of MOS device in radiation environment becomes worse due to the synergetic effect of TID and HCI, which should gain more attention.

The measured average bulk and drain currents during CHC test are listed in TABLE I. The normalized average bulk and drain current ratio (I_b/I_d) of irradiated narrow device is 45% higher than that of un-irradiated narrow device, which indicates the enhanced impact ionization in the irradiated device, attributed to the radiation-induced trapped charges in STI [5].

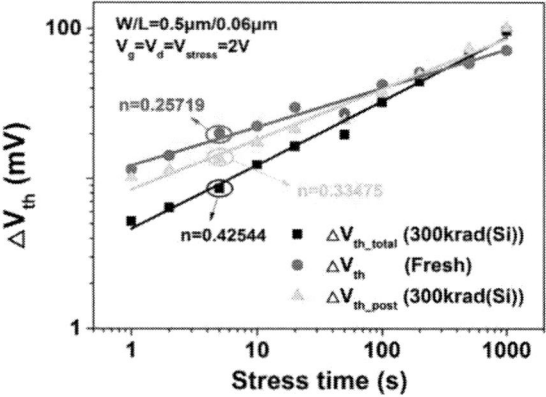

Figure 3: V_{th} shift as a function of stress time for irradiated and un-irradiated narrow devices.

TABLE I.   MEASURED AVERAGE BULK AND DRAIN CURRENTS DURING CHC TEST

Narrow Devices	Normalized Bulk Current	Normalized Drain Current	I_b/I_d
Un-irradiated	1	1	1
Irradiated	1.51	1.04	1.45

To further understand the synergetic effect between TID and HCI, 3-D TCAD simulation was carried out. A 65nm bulk Si NMOSFET with W/L of 0.5μm/0.06μm was developed in Sentaurus. Fig.4 shows the stress-induced

978-1-5386-5309-8/18 $31.00 © 2018 IEEE   402

electric field in the channel along the channel/STI interface with or without trapped charges in STI. Different trap densities in STI are simulated. It can be seen from Fig.4(a), the stress-induced electric field in channel is enhanced by radiation-induced STI trapped charges, resulting in an enhanced impact ionization in the irradiated device.

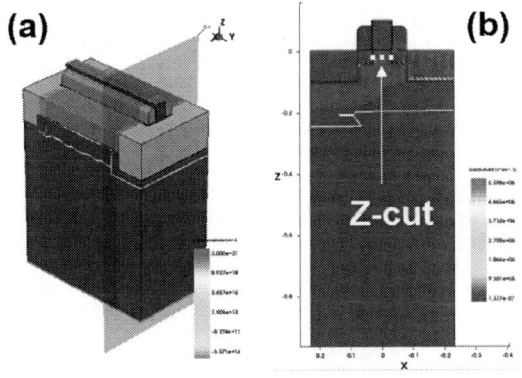

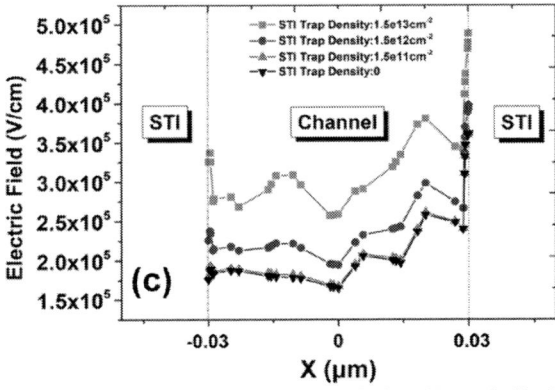

Figure 4: (a), (b) Illustration of the 65nm bulk Si NMOSFET showing the cuts in the device. (c) Electric fields along the Z-cut for different trap densities in STI.

SUMMARY

The synergetic effect between TID and HCI of 65nm bulk Si NMOS is experimentally demonstrated. The results show that irradiated devices degrade much severer than un-irradiated ones, especially for the narrow devices, which are attributed to radiation-induced charges in STI, resulting in enhanced impact ionization of irradiated devices. As a result, the interplay between TID and HCI may bring reliability challenges of MOS devices in radiation environment.

ACKNOWLEDGEMENTS

This work is supported in part by the National Natural Science Foundation of China (No.61421005, 61434007 and 60806033).

REFERENCES

[1] Acovic. A, LaRosa. G, and Sun. YC, *Micro. Relia.*, vol. 36, no. 7-8, pp. 845-869, July 1996.

[2] C. Claeys and E. Simoen, *Radiation effects in advanced semiconductor materials and devices*, New York: Springer-Verlag, 2002.

[3] B He, Z Wang, et al. *Journal of Semiconductors.*, Vol. 37, No. 12, pp. 124003, December 2016.

[4] Y Xu, J Bi, et al. *SCIENCE CHINA Information Sciences.*, Vol. 60, No. 12, pp. 120401, December 2017.

[5] M. Silvestri, S. Gerardin, et al, *IEEE Trans. Nucl. Sci.*, Vol. 55, No. 4, pp. 1960-1967, August 2008.

[6] Y He, and X Zhang, *IEEE International Conference on Electron Devices and Solid-State Circuits*, Chengdu, June 18-20, 2014.

[7] Q Zheng, J Cui, et al. *Chin. Phys. Lett.*, Vol. 33, No. 7, pp. 076102, 2016.

[8] D Yu, Q Zheng, et al. *Journal of Semiconductors.*, Vol. 37, No. 6, pp. 064016, 2016.

[9] Y Chen, X An, et al. *IEEE International Conference on Solid-State and Integrated Circuit Technology*, Guilin, October 28-31, 2014, pp.1458-1460.

RELIABILITY COMPARISON BETWEEN AL PROCESS AND CU PROCESS BASED ON 0.11UM TECHNOLOGY NODE

XiangFu Zhao[*]

Semiconductor Manufacturing International Corporation, Shanghai 201203, China

*Corresponding Author's Email: XiangFu_Zhao@smics.com

ABSTRACT

Reliability of both Front-end-of-line (FEOL) and Back-end-of-line (BEOL) was evaluated on 0.11um Al process and Cu process. Time dependence dielectric breakdown (TDDB) of Gate oxide integrity (GOI) and time-to-failure (TTF) of hot carrier injection (HCI) degraded for Cu process due to the effect of fluorinated silicate glass (FSG). TTF lognormal distribution of isothermal electromigration (IsoEM) of Cu metal lines showed better uniformity. Inter metal dielectric (IMD) voltage ramp (Vramp) of Al bottom metal displayed better TTF distribution caused by receiving more F ions from undoped silicate glass (USG) etch process.

INTRODUCTION

To pursuit higher profit in semiconductor manufacturing, device is shrunk into nanoscale for higher density with lower unit cost. Nanoscale devices require smaller interconnect critical dimension which brings higher resistance [1]. Thus, copper metallization and low k dielectric materials are comprehensively applied to reduce the RC (resistance-capacitance) delay [2]. For advanced tech nodes, it will bring challenges in the material selection and the integration [3].

However, developing a new technology node involves a huge amount of money and sometimes fails. For mature technology nodes, fab engineering change proposal is an effective method to continually reduce the production cost without apparent degradation on its reliability. In this paper, we compared FEOL and BEOL reliability of both Cu and Al processes based on 0.11um tech node. We found that Al process can replace Cu process without apparent degraded reliability, which greatly enhance the fab production capability if a little larger RC delay is acceptable.

EXPERIMENTAL

FEOL process including the contact is exactly the same for both Cu and Al processes. From M1, Cu process is the standard dual-damascene technology and the dielectric is FSG. Metal line in Al process is the traditional sandwich structure (Ti/Al/Ti) and the dielectric is USG. Three metal layers were used in our reliability tests. The bottom metal (Mx) is narrower than the middle metal (Mx+1) and the top metal (Mx+2). Mx+1 and Mx+2 have the similar width. IsoEM (the fast wafer level Isothermal EM) and IMD Vramp (comb to comb structure) were conducted to evaluate the uniformity of a single metal line and the dielectric between two parallel metal lines in the same layer, respectively. GOI TDDB was to characterize the integrity of the gate oxide and HCI was to describe the performance of transistors. All reliability tests were measured using Agilent with TEL prober.

RESULTS AND DISCUSSION

Metal IsoEM TTF log-normal distributions of both Cu and Al processes are showed in Fig. 1 (a) and (b), respectively. It can be found from Fig. 1(a) that three metals display similar sigma value for Cu process, indicating good uniformity of each metal. For Al process, sigma values showed in Fig. 1(b) of three metals are OK. However, the value of each layer is a little larger than that of Cu process, especially for Mx. This is because the bottom metal received more F ions from upper USG etch process [4], which etched the side wall a bit and made the interface rougher. This can be revealed by its larger sigma value due to the degraded uniformity.

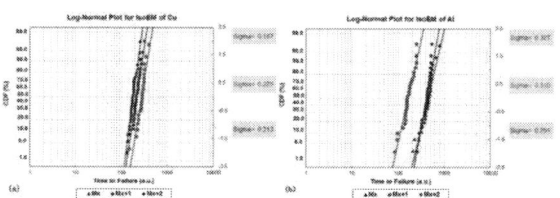

Figure 1: Metal IsoEM TTF log-normal distributions: (a) Cu, (b) Al

If the metal space and the dielectric are the same, IMD Vramp should show the similar performance. Figures 2(a), 2(b) and 2(c) display IMD Vramp Weibull plots of Mx, Mx+1 and Mx+2 for both Cu and Al processes, respectively. It is observed in Fig. 2(a) that IMD Vramp of Mx for Al process shows better performance. This is because that Mx received more F ions from upper USG etch process, which etched the side wall a bit and thus widened the space between two metal lines. This leads to a longer TTF distribution of Mx IMD Vramp for Al process. However, it can be found that IMD Vramp performance of Mx+1 became similar in Fig. 2(b) and even coincided with each other for Mx+2 in Fig. 2(c). This is because the upper metal received less F ions than the bottom metal and thus got less affected.

978-1-5386-5309-8/18 $31.00 © 2018 IEEE

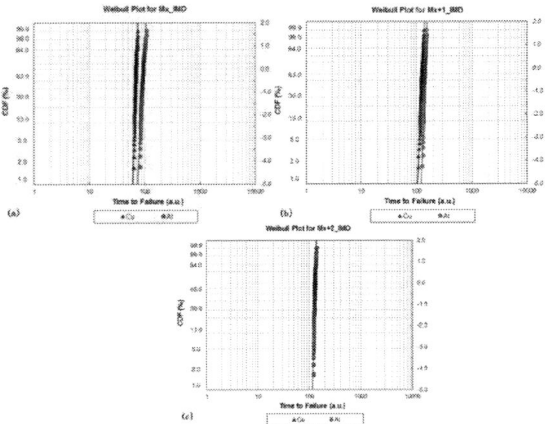

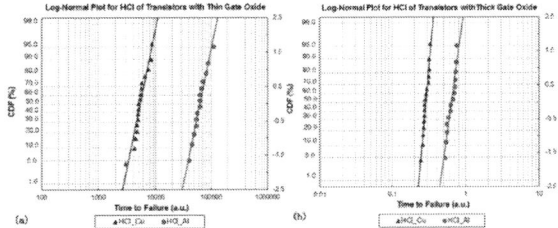

can cause more serious damage to gate oxide than USG of Al process, and thus produce more trap states in the interface of gate oxide, which degrades HCI performance of thin gate oxide MOS more obviously.

Figure 2: IMD Vramp Weibull plots for Cu and Al process of different metal layers: (a) Mx; (b) Mx+1; (c) Mx+2

TDDB log-normal plots of thin gate oxide and thick gate oxide were showed in Figures 3(a) and 3(b), respectively. Solid triangle dots stand for data collected from Cu process and solid circle dots represent data measured from Al process. It can be seen from Fig. 3(a) that thin gate oxide integrity from Al process behaved better uniformity (smaller sigma value) than that from Cu process. However, thick gate oxide integrity displayed similar performance in Fig. 3(b). It was explained that plasma-induced damage to 0.18um transistors from high density plasma (HDP) FSG is greater than that from HDP USG [5]. Thus, FSG of Cu process can cause more serious damage than USG of Al process, which can be apparently revealed through TDDB performance of thin gate oxide.

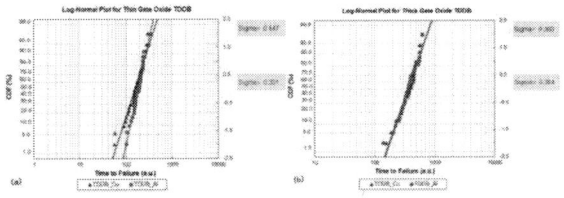

Figure 3: Log-normal plots of GOI TDDB from Cu and Al process: (a) Thin gate oxide, (b) Thick gate oxide

Similar results were obtained from the comparison of HCI performance in Figures 4(a) and 4(b), which shows HCI TTF distributions for transistors with thin gate oxide and thick gate oxide from Cu and Al processes, respectively. For transistors with thin gate oxide, HCI TTF of Cu process shows around 10 times difference from that of Al process in Fig. 4(a). Even for transistors with thick gate oxide, there is about 2 times difference shown in Fig. 4(b). It is well-known that HCI performance is strongly related with the interface trap density. FSG of Cu process

Figure 4: Log-normal plots of HCI from Cu and Al process: (a) Thin gate oxide, (b) Thick gate oxide

CONCLUSIONS

For 0.11um tech node, FEOL process was kept as unchanged while Cu BEOL process was replaced with Al BEOL process. It was found that Al metal lines showed a little worse uniformity than Cu metal lines due to the damage from USG etch process. The dielectric between the bottom metal lines of Al process showed better IMD performance than that of Cu process. This is because the bottom metal received more F ions during USG etch process and etched the side walls a lit, which widened the space between two Al metal lines and thus improved IMD performance. FSG of Cu process can cause a little worse damage to gate oxide than USG of Al process, which degrades TDDB performance of gate oxide a bit. Degraded HCI performance of thin gate oxide transistors from Cu process also confirmed it. Our results clearly prove that Al BEOL can replace Cu process without apparent degraded process reliability if a little RC delay is acceptable. This can greatly enlarge fab production capability.

REFERENCES

[1] X. F. Zhao, and W. T. K. Chien, *Microelectronics Reliability,* vol. 72, 2017, pp. 1–4.

[2] N. Suzumura, S. Yamamoto, D. Kodama, H. Miyazaki, M. Ogasawara, J. Komori, and E.Murakami, *IEEE IRPS* , 2008, pp. 138–143.

[3] B. Taylor, X. Lin, X. Zhang, H. Kim, M. He, and V. Ryan, *IEEE IITC*, 2012, pp.1-1.

[4] Y. Liu, and F. Li, *9th International Conference on Solid-State and Integrated-Circuit Technology*, 2008, pp. 1242-1244.

[5] A. T. Cheong, L. S. Yee, O. P. Ing, T. M. Sua, J. Sudijono, *5th International Symposium on Plasma Process-Induced Damage*, 2000, pp. 69-72.

ASSESSING THE ACCURACY OF STATISTICAL PROPERTIES EXTRACTED FROM A LIMITED NUMBER OF DEVICE UNDER TEST FOR TIME DEPENDENT VARIATIONS

J. F. Zhang, M. Duan, Z. Ji, and W. D. Zhang

Department of Electronics and Electrical Engineering, Liverpool John Moores University,
Byrom Street, Liverpool L3 3AF, UK
E-mail: j.f.zhang@ljmu.ac.uk

ABSTRACT

As device sizes scale down, device variations scale up. There are two types of device-to-device variations (DDV): as-fabricated or time-zero DDV and the time dependent variations (TDV). Even if two nano-scaled devices were identical at time-zero, they would be different after stresses and result in TDV, since the defect generation and charging-discharging are stochastic. To characterize TDV, statistical properties, such as the mean value and standard deviation, are extracted from tests. Their accuracy improves as the number of device under tests (DUTs) increases. Ageing is time consuming and the typical DUTs used are in the range of tens to hundreds. There is little information on the accuracy of the statistical properties extracted from such a limited DUTs and the objective of this paper is to propose a methodology to assess it. Based on the defect-centric model, the accuracy with a specific confidence level is evaluated for a given number of DUTs and a stress level.

INTRODUCTION

As CMOS technologies enter the nano-meter range, device-to-device variations (DDV) become a challenge for circuit design and optimization [1-12]. There are two types of DDV: the as-fabricated DDV at time zero and the time dependent DDV. The as-fabricated DDV has a number of origins, such as random discrete dopant, line edge roughness, gate work function variation, and FIN size variation etc [1]. They have been extensively investigated. This work focuses on the time dependent DDV (TDV). Under electrical stresses, transistors age through charging/discharging either as-grown traps or generated traps in gate dielectric [13-20]. Even though two devices can be identical at time zero, they become different after electrical stresses, because the defect generation and the charge-discharge of traps in the gate dielectric are stochastic [1-3]. One example is given in Fig. 1, where the ageing of two devices is clearly different under the same stress biases [9].

To characterize DDV, the common practice is to repeat the same test on multiple devices and extract the statistical properties, such as the mean value and standard deviation [1-12]. For as-fabricated DDV, the measurement is fast and a large number of devices can be used in a practical test time. For TDVs, however, device ageing can be a time consuming process and the number of Device Under Tests (DUTs) used is limited. While the DUTs can reach the order of ~ 10^5 [4] for relative short time such as 1000 sec, they are often only in the range of tens to hundreds for longer stress time [2-12]. As the accuracy of the statistical properties always improve with number of DUTs, the question is how accurate they are when extracted from a limited number of DUTs. In this work, we will develop a methodology to assess this accuracy, based on the defect centric model [2-4].

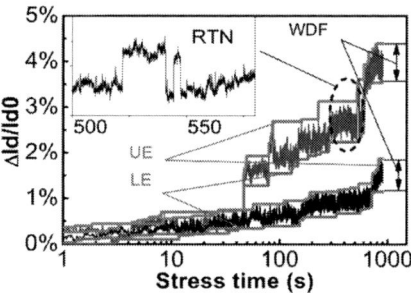

Fig. 1 An example of time dependent device-to-device variation (TDV). Two devices of the same size were subjected to the same electrical stress. Their ageing is clearly different [9].

DEFECT CENTRIC MODEL

The defect centric model [2,3] is based on two assumptions. One of them is that the number of traps in gate dielectric per device is random and follows the Poisson distribution. The other is that the impact of a trap on a device in term of parameter shift, such as threshold voltage shift ΔVth, follows an exponential distribution. This leads to a combined distribution function of [2,3,9],

978-1-5386-5309-8/18 $31.00 © 2018 IEEE   406

$$H_{\eta,Nt}(\Delta Vth) = \sum_{k=0}^{\infty} \frac{e^{-Nt}Nt^k}{k!}\left[1 - \frac{k}{k!}\Gamma(k,\frac{\Delta Vth}{\eta})\right], \quad (1)$$

where H is the cumulative distribution, Γ is the Gamma function, η is the average impact of one trap on the device and Nt is the average number of traps per device. They are related to the mean, μ, and standard deviation, σ, by:

$$Nt = \frac{2\mu^2}{\sigma^2}, \quad (2)$$

and

$$\eta = \frac{\sigma^2}{2\mu}. \quad (3)$$

Once the μ and σ is obtained from the test data, one can evaluate Nt and η by eqs. (2) and (3) and in turn the statistical distribution by eq. (1).

THE MODEL VERSUS TEST DATA

For the TDVs induced by the bias temperature instabilities, the defect centric model has been verified based on extensive amount of tests: 92,000 DUTs from 4000 process lots [4].

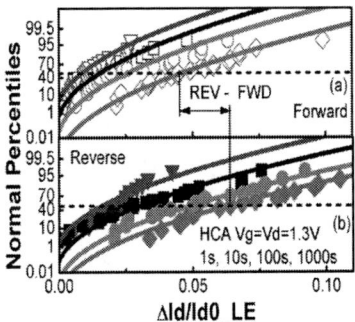

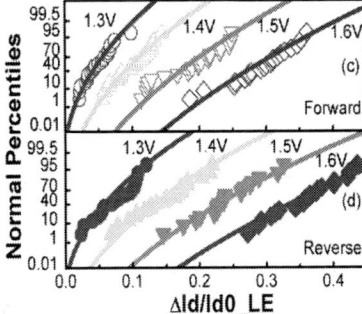

Fig. 2 An example of good agreement between the test data (symbols) and the defect-centric model (lines). 'Forward' means the source and drain for stress and measurement are the same, while 'Reverse' means that they were swapped after stresses [9].

For the TDVs induced by hot carrier ageing (HCA), there is also a good agreement between the model and the test data, which were taken after different stress times in Figs. 2(a) and 2(b) and different stress biases in Figs. 2(c) and 2(d).

The model predicts that standard deviation is related to the mean by a power law,

$$\sigma = \sqrt{2\eta\mu}. \quad (4)$$

This again agrees with the test data in Fig. 3 [9].

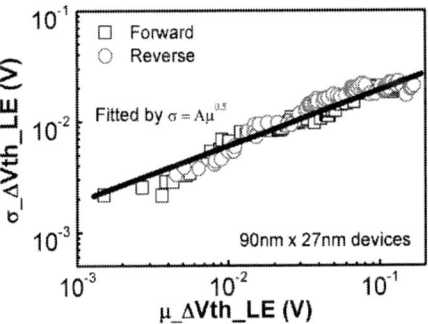

Fig. 3 The relation between standard deviation and the mean follows the prediction by eq. (4) well [9].

METHODOLOGY

For a given Nt and η, one can construct hypothetic devices with the number of traps in each device, nt, determined by the Poisson distribution and the threshold voltage shift induced by a trap, $\Delta Vth,i$, obtained by the exponential distribution. The total ΔVth of this device is the sum of each-trap induced shift,

$$\Delta Vth = \sum_{i=1}^{nt} \Delta Vth, i. \quad (5)$$

These hypothetic devices allow one to simulate the experiments that use a limited number of DUTs for extracting the μ and σ. For example, one test engineer used X DUTs can be simulated by generating X hypothetic devices for the 'Test 1' in Fig. 4. The X ΔVth for these devices is then used to evaluate μ and σ, which corresponds to one point in Figs. 5(a) and (b), respectively.

Now we assume that a different test engineer is doing the same test by using X DUTs again. These X DUTs of course can be different from those used by the previous engineer and we imitate this by randomly generating the second set of X DUTs, labeled as 'Test 2' in Fig. 4. This will produce another point for μ and σ in 5(a) and (b), respectively.

978-1-5386-5309-8/18 $31.00 © 2018 IEEE

Repeating the same simulation for a sufficiently larger number of sets, i.e. the M in Fig. 4, one can obtain the distribution of extracted μ and σ for a given X DUTs, as shown in Fig. 5. They can then be compared with the true μ and σ that is evaluated by eqs. (2) and (3) for the given Nt and η used to generate the hypothetic devices. At a given level of confidence, the accuracy of μ and σ for a specific DUTs can be evaluated, as shown in Fig. 7 and Fig. 8.

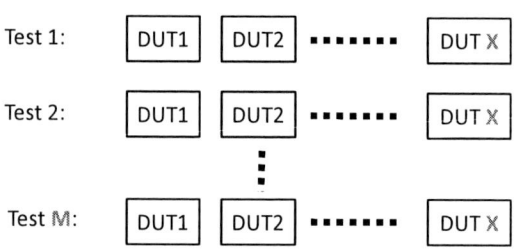

Fig. 4 An illustration of statistical tests: In a hypothetic Test 1, engineer 1 used X DUTs for extracting the μ and σ of TDV. In test 2, engineer 2 also used X DUTs, but will obtain different μ and σ, because a different set of devices were used [10].

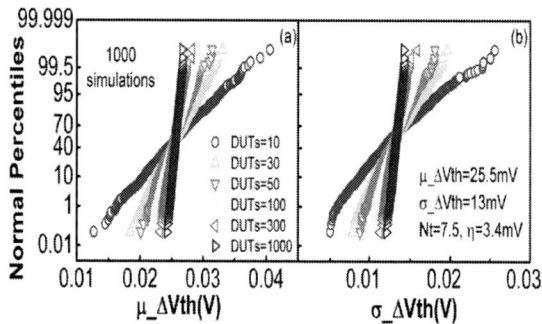

Fig. 5 The μ (a) and σ (b) extracted for different number of DUTs (X in Fig. 4). For a given X, the tests were repeated 1000 times (M=1000 in Fig. 4) [10].

APPLICATION EXAMPLES

Device lifetime is typically defined as the time for ΔVth reaching 25 ~ 50 mV. The average ΔVth induced by one trap, η, is ~ 3.4 mV, evaluated from the slope of the fitted line in Fig. 3. This gives an average number of traps per device, Nt=ΔVth/η, of 7 ~ 15.

Fig. 5 shows the distribution for Nt=7.5, when the same set of tests were repeated 1000 times, i.e. M=1000 in Fig. 4. As expected, the statistical spread is larger for smaller DUTs in both μ and σ.

The number of DUTs is not the only parameter

affecting the spread. Fig. 6 shows that Nt also has an impact. The spread reduces for larger Nt, since a larger Nt averages out the individual effect of a trap on a device.

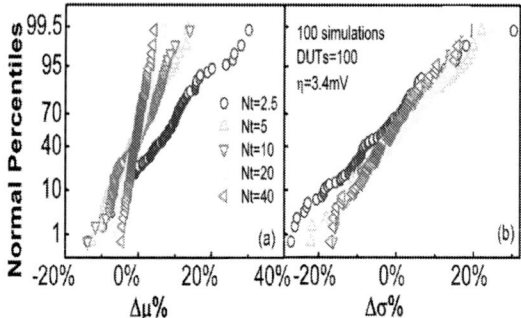

Fig. 6 The impact of the average number of traps, Nt, per DUT on the μ (a) and σ (b) extracted for DUTs=100 when the tests were repeated 1000 times (M=1000 in Fig. 4) [10].

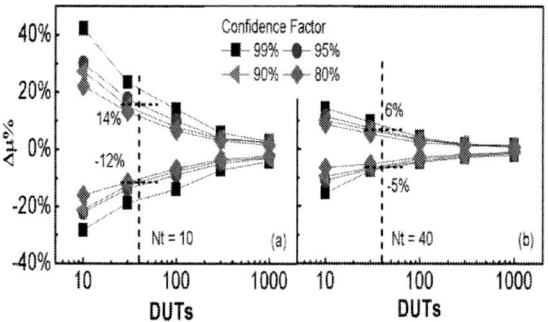

Fig. 7 The dependence of the accuracy of mean value, μ, on the number of DUTs used in a test for Nt=10 (a) and Nt=40 (b). The accuracy with a 95% confidence is marked out for 40 devices [10].

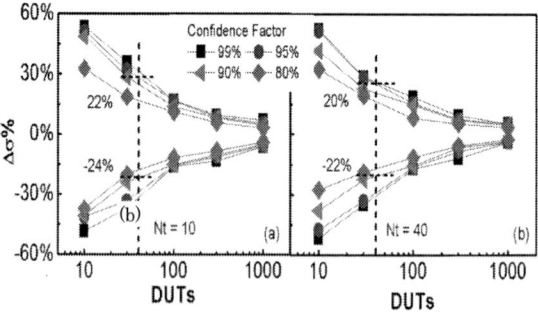

Fig. 8 The dependence of the accuracy of standard deviation, σ, on the number of DUTs used in a test for Nt=10 (a) and Nt=40 (b). The accuracy with a 95% confidence is marked out for 40 devices [10].

978-1-5386-5309-8/18 $31.00 © 2018 IEEE   408

Figs. 7(a) and (b) shows the accuracy of extracted μ for Nt=10 and 40, respectively. For Nt=10, μ has an accuracy within ±14% with a 95% confidence when X=40. It is improved to ±6% when Nt=40. If 1000 DUTs were used, the accuracy will be improved to ±2.6% for N=10 and ±1.3% for Nt=40.

For the same DUTs and Nt, the accuracy in σ is less. Figs. 8(a) and (b) shows the accuracy of extracted σ for Nt=10 and 40, respectively. For Nt=10, σ only has an accuracy within ±24% with a 95% confidence when X=40. It is improved only to ±22% when Nt=40. If 1000 DUTs were used, the accuracy will be improved to ±5% for Nt=10.

CONCLUSIONS

Defect generation and charge-discharge of traps in gate dielectric are stochastic processes. This induces a time dependent device-to-device variations (TDVs). In this work, we propose a methodology for assessing the accuracy of the statistical properties of TDVs extracted from a limited number of DUTs, based on the defect-centric model. An increase of either the number of DUTs or the number of average traps per device improves the accuracy of mean and standard deviation. When the average number of traps per device (~ 10) corresponds to typical definition for device lifetime, an accuracy for 40 DUTs is around 14% and 24% for the mean and standard deviation, respectively. They are improved to 2.6% and 5%, when 1000 DUTs are used.

ACKNOWLEDGEMENTS

The authors thank D. Vigar of Qualcomm for supplying the test samples and B. Kaczer of IMEC, Belgium, and A. Asenov of Glasgow University for discussions. This work is supported by the EPSRC of UK under the grant no. EP/L010607/1.

REFERENCES

[1] A. Asenov, B. Cheng, D. Dideban, U. Kovac, N. Moezi, C. Millar, G. Roy, A. R. Brown, and S. Roy, *Proc. IEEE Custom Integr. Circuits Conf.*, 2010, pp. 1–8.
[2] B. Kaczer, T. Grasser, P. J. Roussel, J. Franco, R. Degraeve, L. A. Ragnarsson, E. Simoen, G. Groeseneken, and H. Reisinger, *Proc. Int. Rel. Phys. Symp.*, 2010, pp. 26–32.
[3] L. M. Procel, F. Crupi, J. Franco, L. Trojman, and B. Kaczer, *IEEE Elec. Dev. Lett.*, vol. 35, 2014, pp. 1167–1169.
[4] C. Prasad, M. Agostinelli, J. Hicks, S. Ramey, C. Auth, K. Mistry, S. Natarajan, P. Packan, I. Post, S. Bodapati, M. Giles, S. Gupta, S. Mudanai, and K. Kuhn, *Proc. Int. Rel. Phys. Symp.*, 2014, pp. 6A.5.1-6A.5.7.

[5] M. Duan, J. F. Zhang, Z. Ji, W. Zhang, B. Kaczer, T. Schram, R. Ritzenthaler, G. Groeseneken, and A. Asenov, *IEEE Trans. Elec. Dev.,* vol. 60, 2013, pp. 2505-2511.
[6] M. Duan, J. F. Zhang, Z. Ji, W. D. Zhang, B. Kaczer, T. Schram, R. Ritzenthaler, G. Groeseneken, and A. Asenov, *IEEE Trans. Elec. Dev.,* vol. 61, 2014, pp. 3081-3089.
[7] M. Duan, J. F. Zhang, Z. Ji, J. G. Ma, W. Zhang, B. Kaczer, T. Schram, R. Ritzenthaler, G. Groeseneken, and A. Asenov, *Proc. IEDM*, 2013, pp. 774-777.
[8] M. Duan, J. F. Zhang, Z. Ji, W. Zhang, B. Kaczer, T. Schram, R. Ritzenthaler, A. Thean, G. Groeseneken, and A. Asenov, *Proc of IEEE VLSI Tech. Symp.,* 2014, pp. 74-75.
[9] M. Duan, J. F. Zhang, A. Manut, Z. Ji, W. Zhang, A. Asenov, L. Gerrer, D. Reid, H. Razaidi, D. Vigar, V. Chandra, R. Aitken, B. Kaczer, and G. Groeseneken, *Proc. IEDM*, 2015, pp. 547-550.
[10] M. Duan, J. F. Zhang, Z. Ji, W. Zhang, B. Kaczer, and A. Asenov, *IEEE Trans. Elec. Dev.,* vol. 64, 2017, pp. 2478-2484.
[11] R. Gao, Z. Ji, S. M. Hatta, J. F. Zhang, J. Franco, B. Kaczer, W. Zhang, M. Duan, S. De Gendt, D. Linten, G. Groeseneken, J. Bi and M. Liu, *Proc. IEDM*, 2016, pp. 778-781.
[12] R. Gao, Z. Ji, A. B. Manut, J. F. Zhang, J. Franco, S. W. M. Hatta, W. D. Zhang, B. Kaczer, D. Linten, and G. Groeseneken, *IEEE Trans. Elec. Dev.,* vol. 64, 2017, pp. 4011-4017.
[13] M. H. Chang, J. F. Zhang, and W. D. Zhang, *IEEE Trans. Elec. Dev.,* vol. 53, 2006, pp. 1347-1354.
[14] J. F. Zhang, I. S. Al-kofahi, and G. Groeseneken, *J. Appl. Phys.,* vol. 83, 1998, pp.843-850.
[15] J. F. Zhang, C. Z. Zhao, G. Groeseneken, and R. Degraeve, *J. Appl. Phys.,* vol. 93, 2003, pp.6107-6116.
[16] J. F. Zhang, M. H. Chang, Z. Ji, L. Lin, I. Ferain, G. Groeseneken, L. Pantisano, S. De Gendt, and M. M. Heyns, *IEEE Elec. Dev. Lett.,* vol. 29, 2008, pp.1360-1363.
[17] C. Z. Zhao and J. F. Zhang, *J. Appl. Phys.,* vol. 97, art no. 073703.
[18] M. Duan, J. F. Zhang, Z. Ji, W. Zhang, B. Kaczer, S. De Gendt, and G. Groeseneken, *IEEE Elec. Dev. Lett.,* vol. 33, 2012, pp. 480–482.
[19] J. F. Zhang, C. Z. Zhao, G. Groeseneken, R. Degraeve, J. N. Ellis, and C. D. Beech, *Solid-State Electronics*, vol.46, 2002, pp.1839-1847.
[20] Z. Ji, S. F. W. M. Hatta, J. F. Zhang, J. G. Ma, W. Zhang, N. Soin, B. Kaczer, S. De Gendt, and G. Groeseneken, *Proc. IEDM*, 2013, pp. 413-416.

ADVANCED N-CHANNEL LDMOS WITH ULTRALOW SPECIFIC ON-RESISTANCE BY 0.18 μm EPITAXIAL BCD TECHNOLOGY

Yao Yao[1], Linhui Hu[2], Gangning Wang[2], Shanon Pu[2]*,
Min-Zhi Lin[1], Zhiyuan Ye[1], and Peng-Fei Wang[1]

[1]School of Microelectronics, Fudan University, Shanghai 200433, China
[2]Semiconductor Manufacturing International Corporation, Shanghai 201203, China
*Corresponding Author's Email: Shannon_Pu@smics.com

ABSTRACT

An advanced n-channel LDMOS (nLDMOS) by 0.18 μm epitaxial BCD Technology is proposed with the best-in-class performance. Thin drift region oxidation by independent LOCOS process is adopted and optimized to improve the trade-off between breakdown voltage (BV) and specific ON-resistance ($R_{on,sp}$). Both P-Body and N-Drift region vertical doping optimization is developed accordingly to improve the reliability of the device. Both 20 V and 30 V nLDMOS devices are designed, and experimental results show that ultralow $R_{on,sp}$ has been demonstrated (i.e., $R_{on,sp}$ = 6.5 mΩ·mm² for BV = 27.6 V, $R_{on,sp}$ = 9.6 mΩ·mm² for BV = 37.3 V, respectively). Moreover, the electrical safe operating area (SOA) and hot-carrier injection (HCI) are also improved.

INTRODUCTION

The quick charging technology has been rapidly developed in recent years, as the result of the fast growing market of battery-powered mobile devices. With their ease of integration and compatibility with the control circuitry, laterally double diffused metal-oxide-semiconductor (LDMOS) devices are widely used as power devices in BCD technology [1-2]. Power consumption is one of the major problems of LDMOS devices as the key device trade-off between breakdown voltage (BV) and specific ON-resistance ($R_{on,sp}$). Ultralow $R_{on,sp}$ is required for quick charging devices to achieve high current capability. In this paper, an advanced n-channel LDMOS (nLDMOS) with the best-in-class performance by 0.18 μm epitaxial (EPI) BCD technology is demonstrated. Thin drift region oxidation by independent LOCOS process is adopted to obtain an optimal BV-$R_{on,sp}$ trade-off alongside with the reduced surface field (RESURF) technique. Accordingly,

a Medium-Voltage P-Well (MVPW) is introduced, and both P-Body and N-Drift region vertical doping optimization is developed to improve the reliability of the device. Both 20 V and 30 V nLDMOS devices are designed, and experimental results show that each of the devices has ultralow $R_{on,sp}$ as expected. In addition to $R_{on,sp}$, the electrical safe operating area (SOA) and hot-carrier injection (HCI) are improved.

DEVICE STRUCTURE & OPTIMIZATION

Figure 1 shows the schematic cross-sectional view of proposed advanced nLDMOS device. In order to obtain a sufficient BV, the structure of conventional nLDMOS devices contains an oxide layer in the top of drift region, mostly by the same process of STI for low voltage LDMOS by BCD technology [3-6]. The BV is improved as the surface electric field is reduced and Kirk Effect is suppressed. However, the path of ON-state current is also extended, which can contribute to the rise of $R_{on,sp}$. Thus, the thickness of the drift region oxidation should be considered while optimizing drift region doping. In this case, independent LOCOS process is applied to forming thin drift region oxidation, i.e., RESURF-oxidation (ROX), as shown in Figure 1. The relationship between the thickness of drift region oxidation and $R_{on,sp}$ is investigated

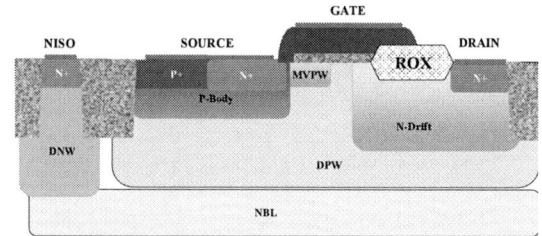

Figure 1: Schematic cross-sectional view of proposed nLDMOS device.

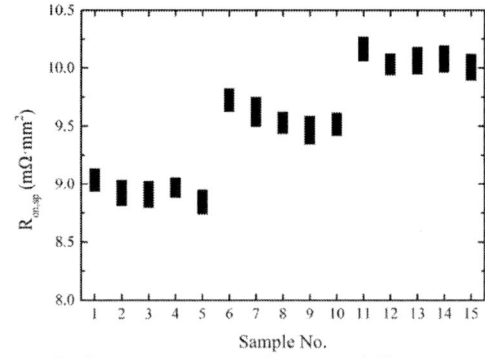

Figure 2: $R_{on,sp}$ comparison between different samples. The ROX thicknesses of Sample 1 to 5, 6 to 10, and 11 to 15 are ×1, ×1.25, and ×1.5, respectively. Each group of samples (1, 6, 11; 2, 7, 12; 3, 8, 13; 4, 9, 14; 5, 10, 15) shares same doping rule.

978-1-5386-5309-8/18 $31.00 © 2018 IEEE

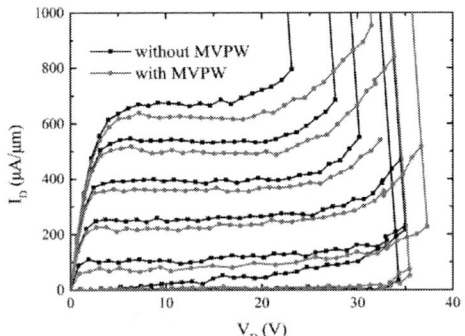

Figure 3: TLP measurement results of ID-VD characteristics of 30 V nLDMOS device with or without MVPW, respectively. Pulse width = 100 ns, V_G = 0 to 6 V.

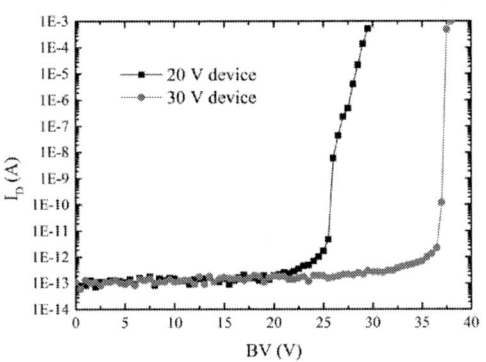

Figure 4: Breakdown voltage of 20 V device (BV = 27.6 V) and 30 V device (BV = 37.3 V), respectively.

and shown in Figure 2. The ROX thicknesses of each group of samples are ×1, ×1.25, and ×1.5, respectively.

To overcome the BV-$R_{on,sp}$ trade-off by the RESURF technique, the amount of charge in the N-drift region is optimized and breakdown occurs at the junction between N-drift and deep P-well (i.e., DPW in Figure 1) [8]. However, by typical RESURF drift doping profile, simulation results show that impact ionization at BV concentrates at the gate-oxide connected to bird's beak of LOCOS, which lead to significant reliability problems. For this reason, reduced surface electric field at the gate-side bird's beak of LOCOS extends device lifetime. Both P-body and N-drift vertical doping optimization under the gate should take into consideration accordingly as the limitation of gate-length, and a Medium-Voltage P-Well (i.e., MVPW in Figure 1) region is added in the surface of body region to support depletion of the surface of N-drift region. Moreover, the gate extension above ROX works as field plate to help reducing surface electric field.

RESULTS AND DISCUSSION

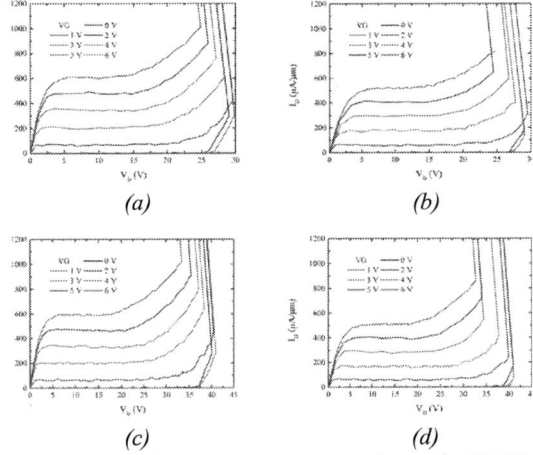

Figure 5: TLP measurement results of ID-VD characteristics of 20 V device at (a) 300 K; (b) 400 K; a...
1...

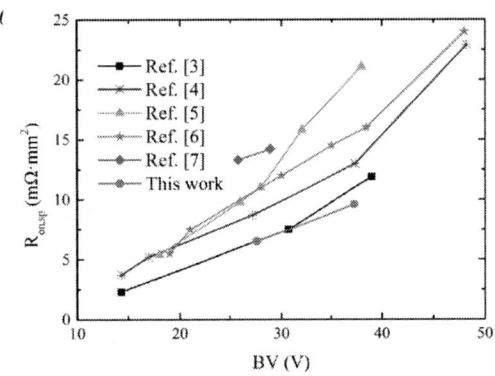

Figure 6: Performance of proposed advanced nLDMOS devices compared with some of the best reported devices by other latest BCD technologies [3-7].

As power device area shrinking, the enhancement of SOA has gained wide attention. The transmission line pulse (TLP) methods are widely used for LDMOS measurements [9]. The major solution to improving SOA without sacrificing BV or $R_{on,sp}$ is by introducing a buried body implant. However, this improvement only occurs above 30 V due to Kirk Effect [10-11]. Body region doping is vertical optimized to satisfy the BV-$R_{on,sp}$-SOA trade-off. Figure 3 shows TLP measurement results of ID-VD characteristics of 30 V nLDMOS device with or without MVPW, respectively. Increasing SOA is achieved by introducing MVPW with slightly decreasing of drain current, as a 0.16 V increment of the threshold voltage.

Figure 4 and Figure 5 show the BV and ID-VD characteristics of 20 V and 30 V devices, respectively. The 20 V device achieves BV of 27.6 V with SOA which

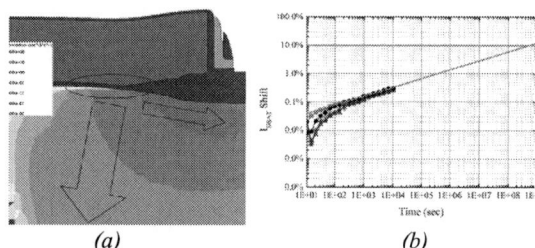

(a)   *(b)*

Figure 7: (a) Simulation result of the impact ionization distribution at BV under the gate-side bird's beak of LOCOS. Impact ionization at surface is reduced (circle) and shifted to the drift region (arrow). (b) Percentage of 20 V nLDMOS device I_{DSAT} shift versus HCI stress time. Measured and stressed at $V_D = 20$ V, $V_G = 2$ V.

covers $V_D = 0$ to 24 V at $V_G = 0$ to 6 V, and the 30 V device achieves BV of 37.3 V with SOA which covers $V_D = 0$ to 33 V at $V_G = 0$ to 6 V. Experimental results also show that the devices have ultralow $R_{on,sp}$ as 6.5 m$\Omega \cdot$mm^2 for 20 V device and 9.6 m$\Omega \cdot$mm^2 for 30 V device. TLP measurement results demonstrate the improvement of SOA by vertical doping optimization under the gate. Both of the resulting devices present the best-in-class performance compared with some of the best reported devices by other latest 0.18 μm BCD technologies [3-7], as shown in Figure 6.

In addition to SOA, the optimization of P-body, N-drift, and P-well doping contributes to improved device lifetime. Figure 7a shows the simulation result of the impact ionization distribution at BV after vertical doping optimization. Impact ionization at surface is reduced and shifted to the N-drift region under the gate-side bird's beak of LOCOS. As a result, the HCI lifetime of the device is significantly improved (10 years), as shown in Figure 7b. With MVPW applied, 20 V nLDMOS device I_{DSAT} shift is within 0.3% after 1×10^4 s under stress voltage ($V_D = 20$ V, $V_G = 2$ V), and 30 V device is within 0.2% ($V_D = 30$ V, $V_G = 2$ V).

CONCLUSION

In this paper, both 20 V and 30 V nLDMOS devices with the best-in-class performance by 0.18 μm EPI BCD Technology are demonstrated. Experimental results show that each of the devices has ultralow $R_{on,sp}$ as 6.5 m$\Omega \cdot$mm^2 for BV = 27.6 V and 9.6 m$\Omega \cdot$mm^2 for BV = 37.3 V. Meanwhile, improved reliability characteristics of SOA and HCI are achieved.

REFERENCES

[1] D. Disney, and Z. J. Shen, "Review of silicon power semiconductor technologies for power supply on chip and power supply in package applications," IEEE Transactions on Power Electronics, vol. 28, no. 9, pp. 4168-4181, Sep. 2013.

[2] A. Ludikhuize, "A review of RESURF technology," in The 12th International Symposium on Power Semiconductor Devices and ICs (ISPSD), 2000, pp. 11.

[3] H. L. Chou, P. C. Su, J. C. W. Ng, P. L. Wang, H. T. Lu, C. J. Lee, W. J. Syue, S. Y. Yang, Y. C. Tseng, C. C. Cheng, C. W. Yao, R. S. Liou, Y. C. Jong, J. L. Tsai, J. Cai, H. C. Tuan, C. F. Huang, and J. Gong, "0.18 μm BCD technology platform with best-in-class 6 V to 70 V power MOSFETs," in The 24th International Symposium on Power Semiconductor Devices and ICs (ISPSD), 2012, pp. 401-404.

[4] T. Y. Huang, W. Y. Liao, C. Y. Yang, C. H. Huang, W. C. V. Yeh, C. F. Huang, K. H. Lo, C. W. Chiu, T. C. Kao, H. D. Su, and K. C. Chang, "0.18um BCD technology with best-in-class LDMOS from 6 V to 45 V," in The 26th International Symposium on Power Semiconductor Devices and ICs (ISPSD), 2014, pp. 179-181.

[5] M. Li, J. M. Koo, and R. V. Purakh, "0.18μm BCD technology platform with performance and cost optimized fully isolated LDMOS," in IEEE International Conference on Electron Devices and Solid-State Circuits (EDSSC), 2015, pp. 820-822.

[6] H. Cha, K. Lee, J. Lee, and T. Lee, "0.18μm 100V-rated BCD with large area power LDMOS with ultra-low effective specific resistance." in The 28th International Symposium on Power Semiconductor Devices and ICs (ISPSD), 2016, pp. 423-426.

[7] M. H. Han, H. B. Chen, C. J. Chang, C. C. Tsai, and C. Y. Chang, "Improving Breakdown Voltage of LDMOS Using a Novel Cost Effective Design," IEEE Transactions on Semiconductor Manufacturing, vol. 26, no. 2, pp. 248-252, May 2013.

[8] T. Efland, P. Mei, D. Mosher, B. Todd, "Self-aligned RESURF to LOCOS region LDMOS characterization shows excellent Rsp vs BV performance," in The 8th International Symposium on Power Semiconductor Devices and ICs (ISPSD), 1996, pp. 147-150.

[9] P. L. Hower, "Safe operating area - a new frontier in Ldmos design," in The 14th International Symposium on Power Semiconductor Devices and ICs (ISPSD), 2002, pp. 1-8.

[10] P. Hower, J. Lin, S. Pendharkar, B. Hu, J. Arch, J. Smith, and T. Efland, "A Rugged LDMOS for LBC5 Technology," in The 17th International Symposium on Power Semiconductor Devices and ICs (ISPSD), 2005, pp. 327-330.

[11] S. Reggiani, G. Baccarani, E. Gnani, A. Gnudi, M. Denison, S. Pendharkar, R. Wise, and S. Seetharaman, "Explanation of the Rugged LDMOS Behavior by Means of Numerical Analysis," IEEE Transactions on Electron Devices, vol. 56, no. 11, pp. 2811-2818, Sep. 2009.

978-1-5386-5309-8/18 $31.00 © 2018 IEEE

RELIABILITY STUDY OF NOVEL GATE OXIDE PROCESS CONTROL METHOD

Guangpeng Zeng[1]*, Weidong Geng[1], Yi Shih Lin[2]

[1]Key Laboratory of Photo-Electronics Thin Film Devices and Technique of Tianjin, Key Laboratory Opto-electronic Information Science and Technology, Institute of Photo-electronics Thin Film Devices and Technique, Nankai University, Tianjin, China

[2]Technology R&D, Semiconductor Manufacturing International (Shanghai) Corporation, Pudong New Area, Shanghai, China

*Corresponding Author's Email: zengguangpeng@mail.nankai.edu.cn

ABSTRACT

As a key loop, gate oxide process loop is very necessary in logic schematic process flow. Especially in advanced semiconductor technology, intrinsic material properties of metal gate oxide have a major effect on device performance. So more accurate and precise inline metrology solutions for tightly gate oxide process controlling is very critical to monitor device performance.

In the paper, the current metrology solution for gate oxide process controlling is investigated. It is found that the current *WAT* (wafer acceptance test) method solution is not convenient enough to control gate oxide process loop tightly, which needs long measure period and complex analysis model. Then a novel solution is proposed to measure bandgap during gate oxide manufacturing process which can represent electrical and intrinsic material properties of gate oxide. What's more it has following advantages: shorter measure period, simpler measure model and less measure parameters. More importantly the correlation between bandgap and the sub-threshold leakage current was investigated in the paper. It has been shown that bandgap has strong correlation with the sub-threshold leakage current. So bandgap measurement can be regarded as an effective inline metrology method for tightly gate oxide process controlling. And it is found that the novel solution is more accurate and precise to measure bandgap during gate oxide manufacturing process, which is significant for tightly gate oxide process controlling.

INTRODUCTION

As MOS device scale decreasing leads to the observable short channel effects, MOS device will deviate from the ideal MOS model.

The equation (1) and (2) respectively represent the saturated current of the ideal MOS device under long channel effect and real model under short channel effect [1][2].

$$I_{dsat} = u_n C_{ox} (W/2L_{eff})(V_G - V_T)^2 \qquad (1)$$

$$I_{dsat} = \frac{W u_n C_{ox}}{L_{eff} + (V_{dsat} u_n / v_{sat})} (V_G - V_T - 1/2 V_{dsat}) V_{dsat} \qquad (2)$$

u_n: mobility of channel carriers; W: effective width of channel; L_{eff}: effective length of channel; C_{ox}: capacitance of inversion layer; V_G: gate voltage; V_T: threshold voltage; V_{dsat}: saturated drain voltage; v_{sat}: high field saturated drift velocity of channel carriers.

Comparing the ideal MOS device, the saturated current equation of the real model under short channel effect is more complicated and relating to more parameters.

So it's important for the MOS device to be free of short channel effect. However, it can be achieved only if the gate length is more than 6 times longer than the natural length (λ). The natural length is a model parameter based on the Poisson equation and represents the extension of the electric field lines from the source and the drain to the channel region, which is showed in Figure1. Therefore, as long as MOS device scales, the natural length must be scaled as well.

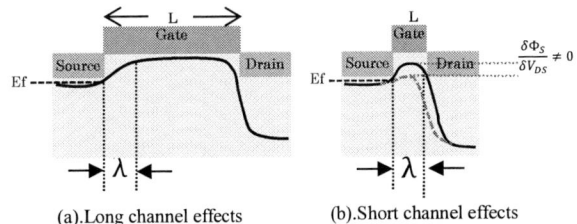

(a).Long channel effects   (b).Short channel effects

Figure1. MOS device in long and short channel effects

$$\lambda = \sqrt{\frac{\varepsilon_{si}}{4\varepsilon_{ox}} t_{si} t_{ox} + \frac{t_{si}^2}{16}} \qquad (3)$$

According to equation (3), the natural length (λ) depends on t_{ox} (the thickness of gate oxide), t_{si} (the thickness of the silicon layer), ε_{si} (the dielectric constant of silicon), ε_{ox} (the dielectric constant of gate oxide) [3][4][5][6]. So there are two solutions to avoid short channel effects: (1) decreasing the thickness of gate oxide; (2) increasing the dielectric constant of gate oxide.

Because the thickness of gate oxide decreases, tunneling phenomenon will happen to electrons existing in the gate oxide, which leads to higher leakage current. And according to the equation (4), the tunneling probability P of electrons is influenced by the thickness of gate oxide t_{ox} and the barrier potential V_b [7]. What's more, the thinner the thickness of gate oxide is, the higher the tunneling probability P of electrons is, as well as the higher the leakage current is.

$$P \propto \exp\left(-\sqrt{\frac{2mqV_b}{\hbar^2}} t_{ox}\right) \qquad (4)$$

978-1-5386-5309-8/18 $31.00 © 2018 IEEE

According to second given solution to avoid short channel effect, the dielectric constant of gate oxide can be increased by nitride doping. Along with the MOS device continuously scales, especially below the 45nm technology node, the leakage current can't meet the requirement any more if just by nitride doping. Instead of doping nitride into SiO_2, high dielectric constant material has been proposed to minimize the short channel effects.

In addition, below the 45nm technology node, it becomes very critical to control gate oxide process because the high dielectric constant material is used. The inline measurement method for gate oxide process controlling is critical to capture the property of gate oxide as well. However, the current WAT metrology solution is not good enough to control gate oxide process tightly, because this solution needs long measure period and complex analysis model. So a new metrology solution is required to control gate oxide process more tightly. In this paper, a novel metrology solution for gate oxide process controlling is introduced to measure bandgap energy. In order to study the reliability of bandgap energy measure method, the Hurkx tunneling current model is introduced in this paper.

The Hurkx model is a recombination model for the tunneling current, which includes trap-assisted tunneling current and band-to-band tunneling current [8] [9]. And the Hurkx model for tunneling current is explained as follow.

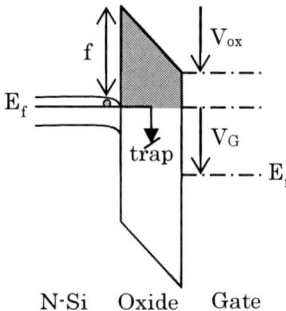

Figure 2: The trap-assisted tunneling mechanism.

$$J_{tat} = -q\sqrt{3\pi}c_{SRH}W\frac{F_\Gamma}{|F_m|}\cdot\left[exp\left(\frac{F_m}{F_\Gamma}\right)^2 - exp\left(\frac{F_m W_0}{F_\Gamma W}\right)^2\right] \quad (6)$$

$$F_\Gamma = \frac{\sqrt{24m^*(KT)^3}}{q\hbar} \quad (7)$$

J_{tat}: trap - assisted tunneling current density; W: depletion width; T : temperature.; q: trap charge density.

The trap-assisted tunneling mechanism is showed in Figure2 and the dependence of trap-assisted tunneling current is explained by equation (6). According to equation (6), the trap-assisted tunneling current depends on depletion width, temperature and trap charge density. The trap charge density can be measured directly by the existing charge pumping measure method. So the trap-assisted tunneling current can be primary controlled in the gate oxide process loop.

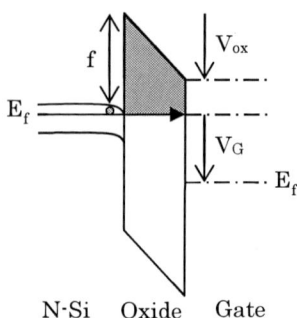

Figure 3: The band-to-band tunneling mechanism

$$J_{bbt} = c_{bbt}V_j(F_m/F_0)^{3/2}e^{-F_0/F_m} \quad (8)$$

$$F_0 \propto E_g^{3/2} \quad (9)$$

J_{bbt}: band-to-band tunneling current density; V_j: junction voltage; c_{bbt}: temperature - independent factor; E_g: bandgap energy; F_m: maximum electric field in junction.

However, the band-to-band tunneling current can be controlled only by the WAT, which is not convenient enough to control gate oxide process tightly. In this paper, a novel metrology solution for the band-to-band tunneling current controlling in gate oxide process is introduced to measure bandgap energy. According to the equation (8), the band-to-band tunneling current has strong correlation with the bandgap energy that represents the intrinsic material property of gate oxide. So it is effective for gate oxide process controlling to measure bandgap energy. In addition, comparing with the WAT, the bandgap energy measure method is more convenient and simple to control gate oxide process. So the novel metrology solution is necessary for gate oxide process controlling to measure bandgap energy.

Therefore, according to the Hurkx model for the tunneling current, the bandgap energy measurement can be an effective and necessary inline metrology method for gate oxide process controlling.

METROLOGY

WAT

The WAT test is a very necessary method that is used to measure the device performance. Usually, the gate oxide process loop can be control by the WAT test. However, it takes too long time about 1~2 month from front-end tool to WAT test [10]. In addition, the analysis model is too complicated to control the gate oxide process loop tightly by the WAT test. So The WAT test is not convenient enough to control gate oxide process loop tightly.

Bandgap energy measurement

Typically bandgap energy is measured by spectroscopic ellipsometry, but this measurement method is highly impacted by noise. So the LSPS (laser-sustained plasma source) is considered to decrease the influence of noise and increase the SNR (signal to noise ratio), which is showed in Figure 4. In this measurement method, the bandgap energy of a material is estimated by the tangent to the ε_2 [11].

More importantly, bandgap energy can be measured just after the gate oxide material deposition. And less measurement parameters need be considered to control gate oxide process by bandgap energy measurement method. Therefore, the bandgap measurement method is more convenient to control gate oxide process tightly.

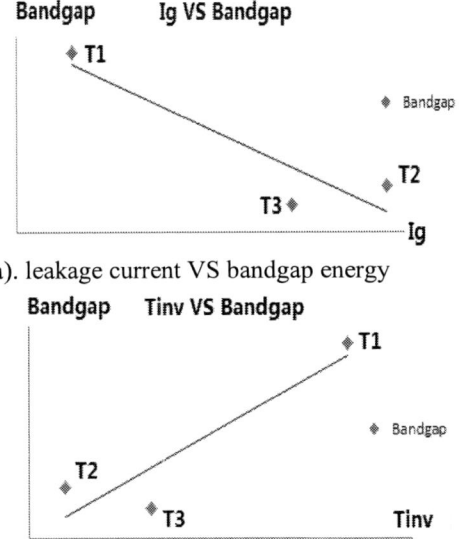

Figure 4: Bandgap energy measurement [11]

ILLUSTRATIONS

In order to verify the reliability of bandgap measurement method, an experiment is designed with different gate oxide process conditions, including T1, T2 and T3. In the experiment, the correlation between bandgap measurement results and WAT results is focused. According to the measurement results from Figure 5, bandgap energy has strong correlation with some WAT measurement results, such as leakage current, thickness of inversion layer and threshold voltage. So the bandgap energy measure method is an effective metrology solution to control gate oxide process tightly.

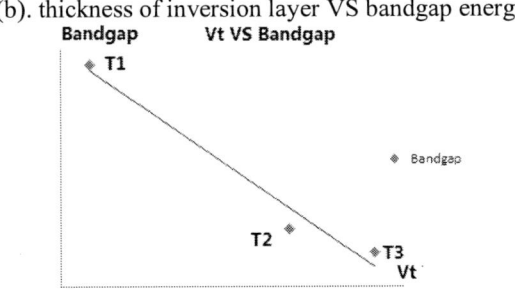

(a). leakage current VS bandgap energy

(b). thickness of inversion layer VS bandgap energy

(c). threshold voltage VS bandgap energy

Figure 5: Correlation between bandgap measurement and WAT results

CONCLUSION

As the MOS device scales, more tight process control method is required for gate oxide process loop. But the traditional WAT test need cost too long time, and the analysis model is too complex as well. So it is not convenient enough to control the gate oxide process tightly. In the paper, a novel metrology solution for tight gate oxide process controlling is proposed to measure bandgap energy. And it has been proved that the bandgap energy measure result has strong correlation with the WAT test results. In addition, it is more convenient for gate oxide process controlling to measure bandgap energy. Therefore, the bandgap measurement can be regarded as an effective and necessary inline metrology solution for tight gate oxide process controlling.

REFERENCES

[1] Arora N. "MOSFET Models for VLSI Circuit Simulation". Computational Microelectronics, 1993, 26(2-3):170-177.

[2] Nishi Y, Doering R, Wooldrige T. "Handbook of semiconductor manufacturing technology". 2008.

[3] Yan R H, Ourmazd A, Lee K F. "Scaling the Si MOSFET: from bulk to SOI to bulk". IEEE Transactions on Electron Devices, 1992, 39(7):1704-1710.

[4] Suzuki K, Tanaka T, Tosaka Y, et al. "Scaling theory for double-gate SOI MOSFET's". IEEE Transactions on Electron Devices, 1993, 40(12):2326-2329.

[5] Tosaka Y, Suzuki K, Sugii T. "Scaling-parameter-dependent model for subthreshold swing S in double-gate SOI MOSFET's". IEEE Electron Device Letters, 2002, 15(11):466-468.

[6] Wong H S P, Frank D J, Solomon P M. "Device Design Considerations For Double-Gate, Ground-Plane, And Single-Gated Ultra-Thin Soi Mosfet's At The 25 Nm Channel Length Generation". Electron Devices Meeting. iedm Technical Digest.international, 1998:407 - 410.

[7] Kane E O. "Theory of Tunneling". Journal of Applied Physics, 1961, 32(1):83-91.

[8] Hurkx G A M, Klaassen D B M, Knuvers M P G. "A new recombination model for device simulation including tunneling". IEEE Transactions on Electron Devices, 2002, 39(2):331-338.

[9] Hurkx G A M, De Graaff H C, Kloosterman W J, et al. "A new analytical diode model including tunneling and avalanche breakdown". Electron Devices IEEE Transactions on, 1992, 39(9):2090-2098.

[10] Chung F, Shih L H, Lin J, et al. "One methods to real-time control tool PM effect and predict WAT output". E-Manufacturing & Design Collaboration Symposium. IEEE, 2012:1-2.

[11] Mahendrakar S, Vaid A, Venkataraman K, et al. "Optical metrology solutions for 10nm films process control challenges". SPIE Advanced Lithography. International Society for Optics and Photonics, 2016:97780Z.

RESEARCH ON ULTRA-LOW R$_{DSON}$ WAFER MEASUREMENT

Lei Wang [1]*

[1]Shanghai Huahong Grace Semiconductor Manufacturing Corporation, Shanghai, China
*Corresponding Author's Email: Ray.Wang@hhgrace.com

ABSTRACT

With the rapid development of the automotive electronics and industrial application, power device technology attracts more and more researchers' eyesight [1][2]. Meanwhile, the smaller the chip size becomes, the lower the key parameter which is named as R$_{DSON}$ will be. Lower R$_{DSON}$ affects not only the system stability but also the power consumption. Therefore, it is crucial that the high measurement accuracy of this key parameter should be captured. Several test approaches which use floating GND (Ground) instrument are analyzed and compared on R$_{DSON}$ measurement in this paper. One test approach is proposed that the ultra-low R$_{DSON}$ could be previously detected without complex connection.

INTRODUCTION

From the view on global semiconductor industry, automotive electronics and industrial application have become the key drive force of the growth [3]. Meanwhile, more and more new opportunities are constantly proliferating, such as IOT (Internet of Things), MEMS (Micro Electro Mechanical System). It has brought new energy for semiconductor industry revival. What's more, the proportion of electronic devices in the car is increasing from car entertainment systems, car-assisted driving and driverless cars. The rapid development of power device semiconductor technology will play an important role to the challenges posed by these changes [4]. Therefore, the measurement approach on the key parameter named R$_{DSON}$ becomes more and more important.

THEORY ANALYSIS

The smaller the power device chip size becomes, the smaller the gate PAD is. It makes that the traditional test approach which calls Kelvin four wires connection couldn't be covered because there is not enough area to place one more probes. Fig.1 is shown as below to describe the detail connection of Kelvin test. R1, R2, R3, R4 both include the cable resistor, probe resistor and contact resistor while doing on-wafer test [5][6].

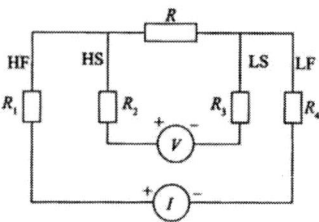

Fig.1. Kelvin Connection Theory

On Kelvin test, it could bypass the resistors on the cables and probes. But it needs more probing area for probe position while we use floating GND instruments.

Another popular test approach called adjacent dice test is introduced as Fig.2 shows. However, the approach needs two die bundled together. If one die is failed, the test approach will affect wafer yield.

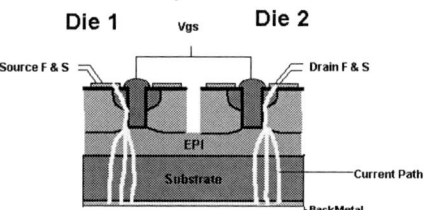

Fig.2. Adjacent Dice Test

Moreover, adjacent dice test approach could only be suitable to VDMOS (Vertical Double-diffused Metal Oxide Semiconductor). It is not suitable to LDMOS (Laterally Diffused Metal Oxide Semiconductor) because the drain is not connected to wafers substrate and probe chuck together with two dices. Therefore, a new test approach is analyzed in following paper to increase the measurement accuracy without complex connections.

MODEL ESTABLISHMENT

In this paper, two Keithley SMUs (Source Measurement Unit) with floating GND are utilized to do the experiments for detail analysis. It is shown in Fig.3 that four connection models are selected. DF/SF stands for drain force and source force. DS/SS stands for drain sense and source sense. Model 1 is the simplest setup that the second instrument's GND is isolated. Model 2 is shown that the second instrument low force point and low sense point are shorted and the short point is connected to the first instrument SF. Model 3 connects point 6 and 8 which

978-1-5386-5309-8/18 $31.00 © 2018 IEEE   416

connect to device probing source PAD directly. Model 4 shorts the second instrument's low force point and low sense point, then connects to the first instrument SS.

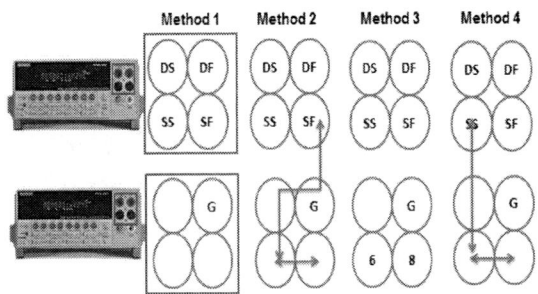

Fig.3. Connection of Different Models

As shown in Fig.4, when drain current is lower than 3A, the four connections don't have obvious differences. Model 1 would give a different trend when gate voltage is not high enough.

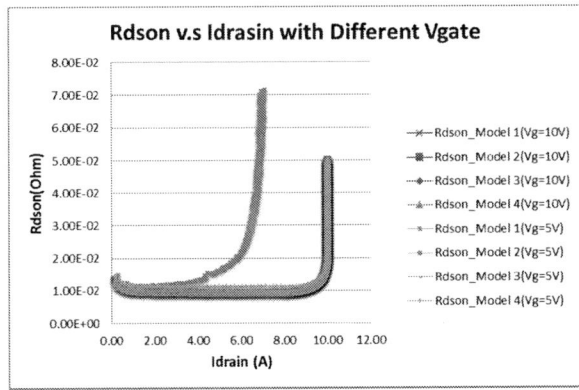

Fig.4. R_{DSON} with different Vg

DATA ANALYSIS

Before the correlation study, one ATE (Auto Test Equipment) with common GND and one selected golden die are selected. And the golden sample is tested as base line data. In the meantime, several test conditions are split. The test results are collected with these split.

Based on table 1 results, we could get conclusion that the actual gate voltage on the gate PAD is smaller than the voltage we expected to force on the gate PAD. There exist high current through the SF that leads voltage drop. It's possible that when gate = 6V, Id = 10A, DUT works closer nonlinear area. The R_{DSON} has obvious difference even if we only change it to closer working point.

TABLE I. R_{DSON} RESULTS IN VG=6V

	Vg=6V		
	Id=2A	*Id=5A*	*Id=10A*
Baseline (B)	7.67E-03	7.89E-03	7.86E-03
Method2 (M2)	8.10E-03	8.46E-03	9.20E-03
△(M2-B)/B	5.61%	7.22%	17.05%
Method3 (M3)	7.84E-03	7.81E-03	7.97E-03
△(M3-B)/B	2.22%	-1.01%	1.40%
Method4 (M4)	7.71E-03	7.86E-03	7.88E-03
△(M4-B)/B	0.52%	-0.38%	0.25%

TABLE II. R_{DSON} RESULTS IN VG=10V

	V_g=10V		
	Id=2A	*Id=5A*	*Id=10A*
Baseline(B)	6.88E-03	7.07E-03	7.01E-03
Method2(M2)	7.11E-03	7.36E-03	7.47E-03
△(M2-B)/B	3.34%	4.10%	6.56%
Method3(M3)	7.05E-03	6.96E-03	7.07E-03
△(M3-B)/B	2.47%	-1.56%	0.86%
Method4(M4)	6.91E-03	7.06E-03	7.05E-03
△(M4-B)/B	0.44%	-0.14%	0.57%

Finally, we choose to connect the low sense connection of gate to sense port of device source pad. Then it shows good match with baseline R_{DSON} result both on low Vg and high Vg as table 1, 2 shows.

CONCLUSION

Last but not least, ultra-low R_{DSON} could be measured with simple connect of floating GND instruments through the above experiment. And the optimized approach could not only control the ultra-low R_{DSON} accuracy around 0.5% in all working area, but also multi-site test could be involved and extended for actual industrial engineering.

ACKNOWLEDGEMENTS

First of all, I would like to express my gratitude to HHGrace that provide me the experiment environment and advanced instruments.

And I gratefully acknowledge all those who helped me during the writing of this paper and especially acknowledge the help of my supervisor Frank, who has offered me valuable suggestions in the academic studies.

I also owe the special debt of gratitude to Mia and Shangyi. In the preparation of the paper, they has spent much time to do the experiment and provided me with inspiring advice. Without their patience and discussion, the completion of this paper would not have been possible.

I should finally like to express my gratitude to my beloved parents who have always been helping me out of difficulties and supporting without a word of complaint.

REFERENCES

[1]. Emmanuel Marcault, Abdelhakim Bourennane, Patrick Tounsi, Marie Breil, Jean-Marie Dorkel, *VDMOS electrical parameters potentially usable as mechanical state indicators for power VDMOS assemblies*[J], IET circuits, devices & systems, 2014, 8(3)

[2]. WANG Lei, ZHANG Jinyi, *Definition and Optimization of VDMOS FBSOA[J]*, *Microelectronics*, 2011, 41(5)

[3]. *Essentials of Electronic Testing for Digital, memory and Mixed-signal VLSI CIRCUITS*[A]. New York, Boston 2002

[4]. Escalona-Cruz, Pedro J. Jimenez-Cedeno, Manuel A. Palomera-Garcia, Rogelio, *Automated RDSon characterization for power MOSFETS*[C], 2015 IEEE 6th Latin American Symposium on Circuits & Systems

[5]. HAN Xinfeng, GU Weimin, *Studies of MOS Turn-on Resistance Measurement Methods*[J], Electronics and Packaging, 2016, 16(11)

[6]. GU Hanyu, WU Qianwen, *A Skillful Technology for Low MOSFET RDS(on) Testing*[J], Electronics and Packaging,2014, (9)

A METHOD OF CONTACT LEVEL PVC ENHANCEMENT

Xiaojun Xu, Jinjin Xie, Yutian Zhang, Zhimin Zeng, Yun Xu

Product Engineering Division, Shanghai Huahong Grace Semiconductor Manufacture Corporation

No.1188 Chuanqiao Road, Pudong New Area, Shanghai, China

Corresponding Author's Email: alvin.xu@hhgrace.com

ABSTRACT

PVC (Passive Voltage Contrast), as a basic localization method by using SEM, has been widely used in IC failure analysis process. Because of the difference of under layer structure, the secondary emission induced charge accumulation and the surface potential may also be different. Generally, we use it to judge the structure under CT is good or not. But PVC method does not always work well for all cases. We need to consider about the PVC method used for fault isolation case by case, due to the difference of device structure. In this paper, we will provide some typical cases, such as in ROM, by using a particular modified PVC method enhancing PVC result. We will discuss the limitation of normal PVC in memory cell array. And an optimal solution will be given.

Keywords—Failure analysis; localization; PVC; memory

INTRODUCTION

Passive voltage contrast is a common defect localization method in IC failure analysis technology. Because of the difference of under layer structure, the secondary emission induced charge accumulation and the surface potential may also be different. By comparing the brightness of CT which connected to a similar structure in the VC image, we can easily localize the defect and judge the failure type. But PVC method does not always work well for all cases. In this paper, based on the principle of PVC, we will tease out the reason of PVC limitation in embedded memory failure analysis. A method of contact level PVC enhancement will also be given in the end.

PRINCIPLE AND PHENOMENON DISCUSSION

Case I: a PVC case in ROM

Figure1 shows a contact level PVC picture at fault bit address. In this picture, the brightness of fault bit CT is lower than the other. A preliminary estimate is that the structure under the fault bit CT suffered residue induced high resistance. With a further physical failure analysis progress (cross section TEM, Figure 2), we found the root cause was the thickness of fault bits' Gox. A marked thickening of Gox is visible in the fault bit's TEM image.

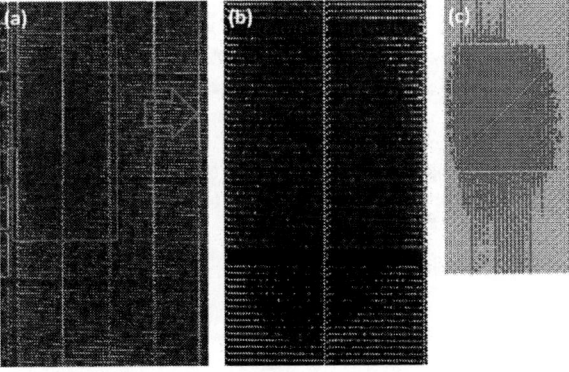

Figure 1: (a) contact level PVC image in ROM array, (b) fault bit address zoomed in, (c) bitmap of fault bits

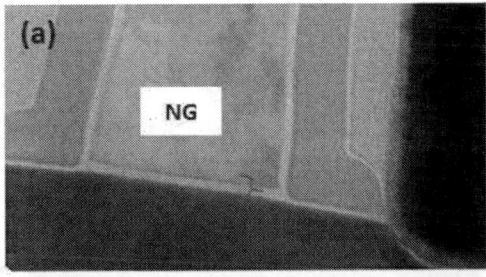

Figure 2: (a) the fault bit cross section TEM image, (b) the normal bit cross section TEM image

PVC principle and phenomenon discussion

In contact level PVC analysis, the sample was delayered to CT layer. By using a scanning electron microscopy, we can observe chip sample at the nanoscale. The scanning electron microscopy (SEM) excites and collects secondary electrons (SEs) of the sample surface

978-1-5386-5309-8/18 $31.00 © 2018 IEEE

by using focused electron beams bombardment the sample surface. Since the amount of SEs generated is affected by the incident electron beam energy, the secondary electron emission yield (gross secondary electrons / gross primary electrons) can be adjusted by changing the incident electron beam energy.

Under the specific incident electron energy conditions, the generation of secondary electrons is greater than the primary electrons means the secondary electron yield greater than 1 is called passive voltage contrast in the positive bias state. When sample at the contact layer, the atoms in the vicinity of the surface of the tungsten metal lose their electrons due to the incident electron excitation, and the total number of electrons decreases in the positive-bias mode, resulting in a large number of holes. When under layer of the CT is a floating structure, the surface holes accumulate, and the rapid rise of the positive potential prevents the further excitation of the secondary electrons.

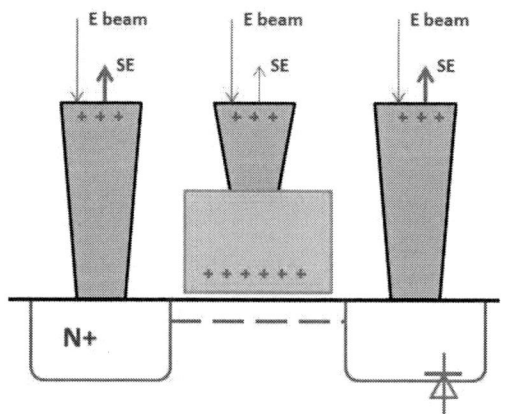

Figure 3: the schematic diagram of an NMOS surface charge in positive mode PVC

In the integrated circuit, the N / P type source / drain regions and gates all belong to the floating structure. However, due to the different structures of the contact holes and the different total atoms, the total amount of secondary electrons that can be excited is different, as Figure 3 shows. Therefore, under the same excitation conditions, different structures' CT has different secondary electron production. Scanning electron microscopy will collect and calculate the secondary electron of these CTs, then display them in different brightness. When the under-layer of the CT was connecting to the substrate directly, an endless stream of electrons would replenish to the surface of the CT. So there is no positive charge accumulated. Under normal circumstances, the same shape, size, material of CT, with the same incident electron excitation, the secondary

electron yield order: the CT of the substrate > P-type junction (junction positive bias, with N-type well electron supplement) > N-Type Junction > floating Gate. Figure 4 is one of examples in SRAM cell array.

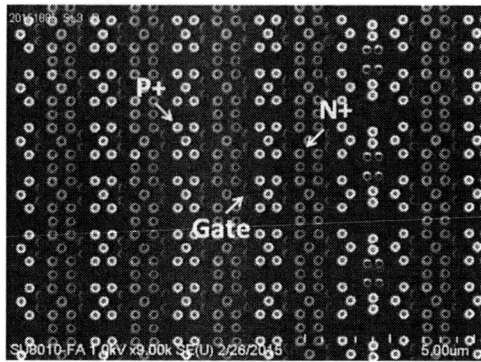

Figure 4: the PVC image of SRAM cell array

However, for memory circuits of high-density, same-type (N-type) transistors such as ROMs and NVMs, memory cells in the same row often have extremely long common word line (gate), which contact hole level VC analysis due to electron beam bombardment will have charging phenomenon. The gate is charged with a potential that affects the switching state of the corresponding device, thereby affecting the result of the VC on the source and drain contacts. As in Case I, the NMOS ROM stores the entire column of drain terminal contact holes which are more or less bright in the VC plot because the gate is positively charged by hole accumulation under this scanning condition, the NMOS is turn-on, the drain terminal and the source terminal is turned on, and the electrons are replenished by the source, so partial light. In the faulty region, due to the abnormal thickening of the gate oxide layer, the NMOS VT rises and the electron replenishment amount is lower than that of the normal region, which appears darker in the contrast image.

This phenomenon, besides being able to identify the faulty areas in some specific cases, may also induce interference in some analysis processes, confusing the real location of failure points. For example, when the floating gate charges, the NMOS enters the on state and the reverse-biased source / drain junction is connected as a larger charge storage area. Compared to a single reverse bias junction, it can accumulate more charges. The greater the SEs yield the brighter the contact is. In this state, it is difficult to judge whether there is a short circuit or a leakage defect under the contact hole.

OPTIMIZATION PROGRAMS AND CASE
CASE II: FIB Circuit Editing Assisted PVC Inspection

In another failure analysis case, it is confirmed that

978-1-5386-5309-8/18 $31.00 © 2018 IEEE

some of the columns in ROM suffered read "1" fault according to the testing result. It seems that there is an abnormal leakage path on the bit line of fault column which leads to reading error. So we delayer check the sample at fault columns. M1, as bit line metal layer, no abnormal was found. After delayering to CT layer, we try to use the PVC to check under layer structure due to which is connecting to BL and also has a possibility of defects.

According to the principle of PVC, the CT of source or drain junction in NMOS ROM cell will accumulate positive charge at the surface during electron beam scanning. The reverse junction prevents the electron supplement. The secondary emission induced a positive surface potential recapture the secondary electrons, which would decrease the secondary electron emission yield. However, due to the existence of leakage path, the CT connecting to the defect can get more electron replenishment. Compared with the CT of a normal N+ junction in bit cell, the fault bit's CT is brighter.

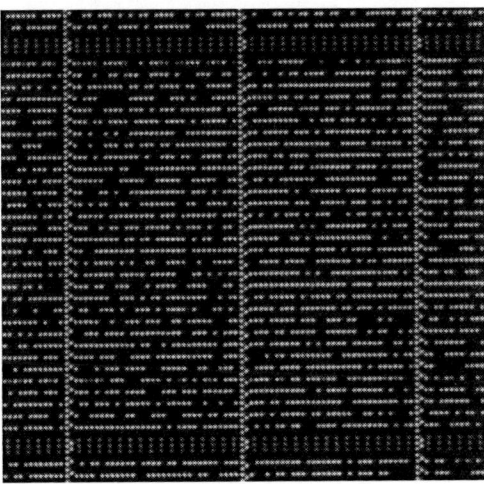

Figure 5: CT level PVC in ROM array.

However, the actual PVC diagram shows the CT of memory cell in array are all bright, which means a high yield of secondary emission. The specific location of the fault bit cannot be identified as expected. Based on the discussion of CASE I in the previous case, this phenomenon stems from the positive charge accumulation in the NMOS gate. Therefore, the corresponding optimization solution is to avoid the gate charge. In this failure analysis case, we shorted the gate to the substrate by using the FIB circuit repair method at the far end, as shown in the Figure 6. The brightness of PVC in line repair area was significantly lower than that in the area without FIB repairing. At this moment, the CT of each memory cell corresponds to only the isolated N + junctions below. As mentioned in the previous analysis,

the defect was highlighted as expected. Cross section found STI loss and poly residue induced WL-BL bridging.

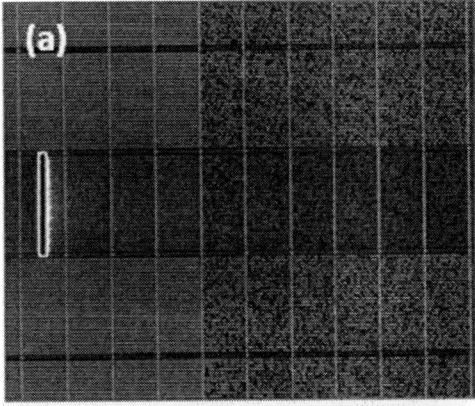

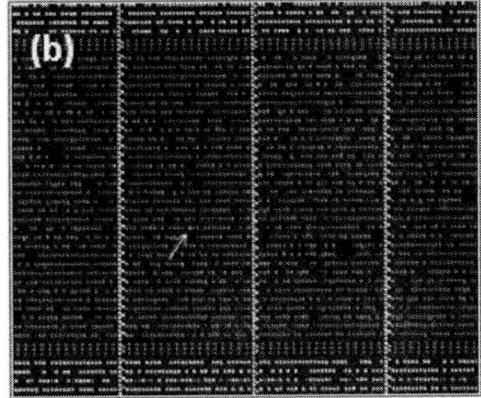

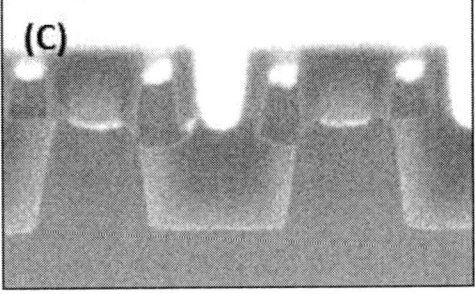

Figure 6: (a) top view of the circuit editing, (b) abnormal bit zoomed in at fault column address, (c) FIB image of cross section check abnormal bit

CONCLUSION

In this paper, we started with a special case. In the case, the defect (thickening Gox) did not connect to abnormal CT directly in PVC image. As we discussed earlier, what the defect affected is the ramification of NMOS in PVC. And this phenomenon, sometimes, will induce interference to make the judgment accurately. So, what we do to preventing gate charge is connecting it to a fixed potential power reference point. Sub is a easily accession choice, by using FIB circuit editing. CASE II is a good example.

ACKNOWLEDGMENT

The authors would like to thank the failure analysis team in Huahong Grace Semiconductor Manufacture Corporation for their work, help and advice.

REFERENCES

[1] Lai L.-L., Huimin G., Hong X., "Surface effect on SEM voltage contrast and dopant contrast", Conference Proceedings from the International Symposium for Testing and Failure Analysis, 2009, pp. 202-207.

[2] Rongwei F., Hunglin C., Yin L., "The detection and investigation of SRAM data retention soft failures by voltage contrast inspection", China Semiconductor Technology International Conference, 2015, pp. 1-3.

STUDY ON OPTIMIZATION METHOD OF FLASH CELL ANALOG MEASUREMENT

Wu Yuan, Zhang Qingwen, Li Mingming

Product Engineering Division, Shanghai Huahong Grace Semiconductor Manufacture Corporation
No.1188 Chuanqiao Road, Pudong New Area, Shanghai, China
* Email: Yuan.wu@hhgrace.com

ABSTRACT

With the rapid development of VSLI and deep submicron technology, the transistor count increases fast, fault chip often shows the soft failure, such failure mode even relying on traditional bitmap address is difficult to capture the clear physical failure phenomenon, so the analog parameter analysis of the fault zone and the surrounding environment is particularly needed. This paper discusses the method of flash cell performance detection through BIST（Build-in-self-test）circuit, At the end of this paper, an improved scheme of a BIST circuit is suggested, which can gain flash analog data automatically. It will bring great time saving to flash product's performance research and development and debug.

Key words：BIST, test vector, Icell, analog measurement

INTRODUCTION

The traditional method of flash failure analysis is to generate some high fault coverage test vectors through the BIST circuit, access the internal memory, test it and read out the output results, and compare them with the expected value, so that bitmap can be generated and it can help us do defect location analysis.

With the rapid development of VSLI and deep submicron technology, the transistor count increases fast, fault chip often shows the soft failure, such failure mode even relying on traditional bitmap address is difficult to capture the clear physical failure phenomenon, so the analog parameter analysis of the fault zone and the surrounding environment is particularly needed. For example, Icell (Flash cell current), Vep (programming voltage), Vee (erasing voltage), etc. are the parameters that are often concerned.

BIST (built-in self-test) application on flash memory

First, let us understand the basic structure and working principle of the BIST circuit. BIST (built in self-test) is a DFT design technology which can verify the circuit function or detects the processing faults effectively. BIST technique is widely applied in SOC design. The BIST circuit can send out the test signal on its own and check its output by itself. It works by combination of signal generator, output analyzer and test controller on the tested circuit. In order to ensure the completion of the test, some other necessary auxiliary circuit is also embedded. A framework diagram of the BIST circuit of a memory chip, as shown in Figure 1

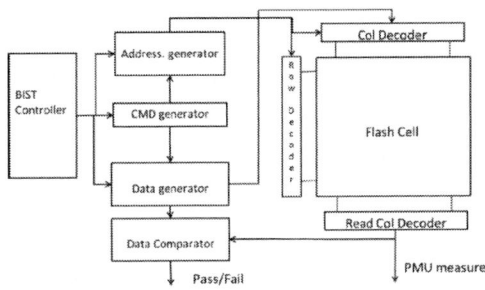

Figure 1: framework diagram of the BIST

To execute read and write operations on Flash cells, we usually need to send test vectors to the chip following BIST protocol. For example as Figure 2 shown below:

Figure 2: Current Schematic of the BIST command

The number and order of the byte of each test vector are fixed as protocol defined, and the content usually contains three parts – command / address / data. Each test vector is usually triggered by a strobe signal. After received test vectors, BIST circuit will act according to the CMD specified to enable/disable signals PIN, to enter the specified test mode (several high fault coverage test mode are designed), the serial input address information through the decoding circuit, positioning to flash unit operation, the data in the test vector information can be written data, or expected output value (read operation).

The Flash output is produced by the output response analyzer, Output response analyzer consists of an output responder, a register and a data comparator, while output responder receives the output results of CUT response, and performs the corresponding processing (compression or feature extraction), The register stores target value by different test sequence. The data comparator compares the output with target value and judge pass/fail, thus fault diagnosis and localization can be done, it can also generate a bitmap for analysis by extract failure address. [1]

Evaluation method of Flash cell performance

Usually we test and judge the Flash cell function by program and erase operations and read the result(compare with target value), read is the process of comparing the Icell(cell current) with Iref (reference current) and output "0" or "1", and then compared with the expected value, judge pass or fail, then generate the corresponding bitmap.

In order to evaluate the reliability performance of flash chip application, a special design of Test Mode is embedded to stress flash cell, and check the cell anti-disturb ability. For those cases, the defective Flash cells often exhibit soft failure, even if the failure can be located by bitmap, It is also difficult to capture clear physical failure phenomenon, so the analog parameters measurement of the fault unit and the surrounding environment is needed. For new product evaluation and failure debug, such as program/erase depth after stress mode is what we often concern.

There are two kinds of methods to know that, one called Margin Read, it is by changing the reference current (Iref) level, to read fail/pass result from the bitmap to know the distribution and change of the global situation for cell performance. (as shown in the Figure 3)

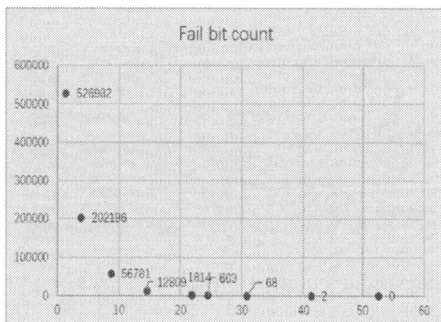

Figure 3: Margin read check fail bit counts

Another method called Sweep VT method, Iref does not change, but sweep the Word line voltage (Vwl), read the Cell results, FBC (Fail bit counts) will change according to Vwl, the definition of the first failure bit voltage is VT(Vtp for programed cell and Vte for erased cell), We usually read Vte and Vtp as characterization of flash depth.

Figure 4 shows Vte change for a flash product. Vte is -1V after erase but declined to -0.3V after pgm00 disturb. By this method, we can know the performance for one product and also can compare with others.

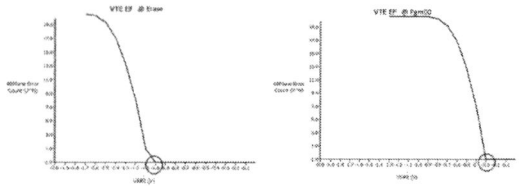

Figure 4: Vte changes after disturb

These two methods are used in different product design now, both are by changing the external conditions lead to FBC (fail bit count) change, let us grasp the depth of flash cells, it is indirect but fast, so it is effective methods for cell performance evaluation and soft failure debug.

FURTHER STUDY
About Analog measurement

Although the above two methods can examine the cell erase/program depth, but there may still be some reasons listed below, which may block the analysis, sometimes, may even go to wrong judgment. For

1. The number of steps in Iref setting is limited, so that the cell performance monitor is not accurate enough.
2. Iref itself may also be unstable with temperature or other abnormal interference.
3. FBC (fail bit count) is only macro information. It is not easy to observe the changes of specific bit under different conditions.

An actual case for example, a product evaluate M2M endurance, 3/50 failure at R.T temperature but 50/50 failure at high temperature (105c). By intuitive understanding, Flash cell declined worse at high temperature than R.T. but when we study the characteristics of individual cell, measure cell current and will find there is no obvious difference! To clarify this issue, we need to measure a large number of Icell data at different temperature and cycling to draw a conclusion.

Existing test mode does not support continuous analog measurement, We completed the task by editing vectors externally and import to the tester (will be discussed later), finally we got the following decline curve Figure 5 and Figure 6.

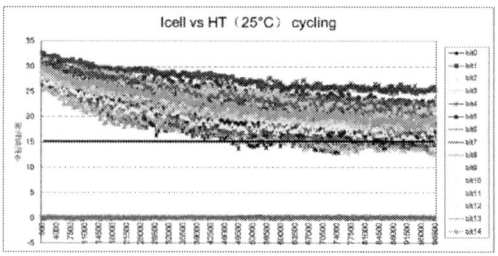

978-1-5386-5309-8/18 $31.00 © 2018 IEEE   424

Figure 5: Icell decline by 100K cyclying @25c

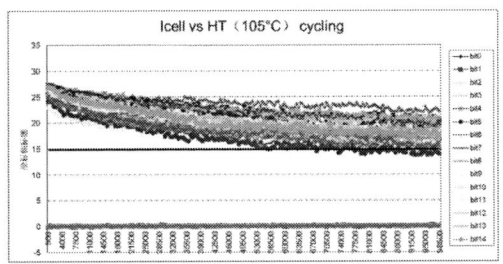

Figure 6: Icell decline by 100K cyclying @85c

From the large amount of data, we can draw conclusion that the Flash cell decline at high temperature is no worse than that at R.T temperature. But why is it fail more at endurance testing? By measuring Iref current, it is found that Iref is about 0.5uA higher at 105C temperature, so it is easily to judge that failure is due to the rise of the reference current. [2]

This example shows a necessary application for large amounts of analog measurements. The performance of some other monitor items (such as operating voltages in various test mode) is the same.

Optimization method of flash cell analog measurement

The following describes the detail operation we usually do for cell analog measurement. According to the test protocol, that is through the BIST circuit to send test vector, activate internal Test Mode, transfer analog signal (Icell/Vep/Vee etc.) to PMU port of tester, by holding a sufficiently sense time, output and print out data.

Existing design circuit which sends a section of test vector (including several bits of CMD command, ADD address, bits, DATA bits), can only operate on one cell with designated addresses. When a large amount of data acquisition is needed, the test vector must be edited (change address only) and the format instruction should be sent one by one. This kind of testing is no doubt very time-consuming. So, how to improve this test method and quickly and automatically measure the analog in cell?

One kind of improvement mentioned in the previous case, is to edit the repeat and similar commands by external tools according to a certain format, then import the tester to realize. But sometimes it may not work because of the vector length limit. On the other hand, a large number of test vectors   take much test time.

An optimization method discussed next is considered from the BIST design point of view. Can a brief test instruction be implemented to control the analog measurement one by one? After careful consideration, we think two things are needed, One is address adder (by row or by column), another is jump trigger signal. Since the address adder module in used in normal operation, and the

measurement jump trigger signal can be achieved by external CLK counter, it seems available if we add and multiplex some logic function as frame diagram shown below. After finishing the first unit measurement, the CLK counters and triggers the address adder to next unit.

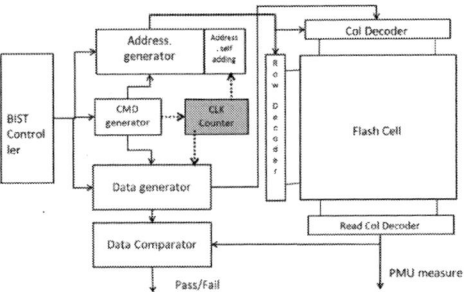

Figure 7: Improved framework diagram of the BIST

In combination with this design, a schematic change can be made in the BIST command as Figure 8   : Improved Schematic of the BIST command, only two sections of test vector are sent for continuous measurement. Address A1 in 1st section defines the starting address, Address A2 in 2nd section defines the end address, Add. control defines the addressing mode (row or column order), CMD generator activate CLK counter, so in the end the test unit after the starting address, CLK counter triggers address self adding (row or column), and go to measure the next unit, thus complete measurement until A1=A2 (address comparator).

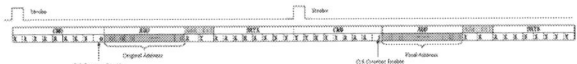

Figure 8: Improved Schematic of the BIST command

CONCLUSIONS

In this paper, an optimized BIST design is considered. Adding and multiplexing some hardware logic functions to realize the multi-point detection of cell analog changes conveniently and efficiently, which brings great time saving to Flash product's performance research and development and Debug.

ACKNOWLEDGEMENTS

Thanks to team leader XuYun, Zhang Yuxiang, Zeng zhimin's encouragement and thanks to Zhang Qingwen, for his experiment data support and technical discussion.

REFERENCES

[1] *Wang Ying, Xin Xiaoning, Wang Hong, Ma Jihu, BIST Techniques and its applications in Memory.2003,8*

[2] *Qingwen Zhang, 90nm G5XXX CTM Cycling Fail EFA Analysis Report. Unpublished. 2017.*

THICKNESS AND CONCENTRATION CONTROL FOR SILICON-GERMANIUM (SIGE) FILMS

Liying Wu[*], Xiaofeng Wang, Wei Li
Technology R&D, Semiconductor Manufacturing International (Shanghai) Corporation
Pudong New Area, Shanghai, P. R. China 201203
Phone: 86-21-20815350, *Email: Lynn_Wu@smics.com

ABSTRACT

This paper reports the study of channel SiGe metrology by using Broadband Spectroscopic Ellipsometry (BBSE) and High Resolution X-Ray Diffraction (HRXRD) simultaneous. The thickness and germanium concentration for each layer of SiGe graded stack have been identified by BBSE with regression model recipe, which is carefully built by using HRXRD as the reference of germanium concentration. The measurement is precise and linear in a wide range. Dynamic precision and repeatability of the thickness and germanium concentration measurements are discussed, which meet the inline monitor requirement. And the most advantage for optical is its high throughput, but because it is the model based measurement, the model will become invalid when process largely changes. This paper illuminates the applicable limits for the optical model measurement, and proposes a hybrid metrology method for inline epitaxy process control. We also studied the silicon channel strain by HRXRD, which is a direct method to monitor the SiGe strain and relaxation. With the function of RSM (reciprocal space map), the Si strain in OCD (optical critical dimension) structure pad has been well demonstrated by using (113) asymmetric measurement.

INTRODUCTION

Silicon-Germanium (SiGe) technology has played an important role in pFET performance boost and Vt tunability. PMOS channel is using a graded SiGe stack grown epitaxially in the source/drain region. It can significantly enhance the channel carriers' mobility due to germanium induced compressive strain [1]. In a production environment, the concentration and thickness of SiGe layers are key parameters for accurately controlling the carrier mobility enhancement. Capability of an accurate and effective in-line metrology, which is directly on product wafers, is highly desirable in high-volume manufacturing.

Broadband Spectroscopic Ellipsometry (BBSE) technology is widely used in thin film thickness measurement. Its advantage is the high throughput and non-contact. But BBSE also has several drawbacks for SiGe measurement in the process technology develop (TD) phase: it cannot monitor epilayer crystalline quality and cannot define the epilayer defect (relaxation etc.). Moreover, SE is not a direct metrology and needs timely calibrated for the build-in model, and the result is related to the reference tool (TEM etc.). Accordingly, we use the reference data to qualify an SE model. If the process change is beyond a calibrated range for film thickness and film reflective index, the qualified model may become invalid and would require recalibration. Thus, in TD phase, a hybrid metrology method for inline epitaxy process control should be implemented.

METROLOGY TOOL AND SAMPLE PREPARATION

A. BBSE

In our study, we use the BBSE to take the optical experiment and the optical model is built based on the spectrum collected with the wavelength 150nm to 800nm. The light is linearly polarized and focused on the wafer surface at an oblique angle 71°. After passing through the film layer stack on the wafer, the linear light will become elliptically polarized, which carries the information of polarization intensity ratio (tan Ψ) and phase shift (cos Δ) between P and S polarized direction. tan Ψ and cos Δ are sensitive to the film stack and they are the function of the thickness, reflective index and wavelength. The tool detector receives light signal, and extracts the value of the tan Ψ and cos Δ, it is the so called measured spectra. The model can be purely mathematical. Based on the model, a set of theoretical tan Ψ and cos Δ spectra can be calculated. If the model parameters are selected to floated, then the theoretical spectra will fit the measurement spectrum over the entire wavelength spectrum. Once the best fit is obtained, T, n and k can be derived from the model and report out the measurement value.

For SiGe composition and thickness measurement, BBSE is a relative measurement. It means a reference is needed in order to set up the model and determine the measurement value. In our experiment, we choose the most common concentration measurement tool XRD as the reference. Based on the reference data, we can build a regression lookup table model. This type of film model consists of several component n/k dispersions. Each of the dispersion corresponds to a certain germanium concentration. The algorithm floats the floating parameters in the model to interpolate between the two component dispersions that best capture properties of the measurement site. The type we choose for BBSE provides the best SNR (signal vs. noise ratio) optical design, which can obtain the detailed signal for the interfaces of the SiGe tri-layers stack.

B. HRXRD

HRXRD is used to study epitaxial thin films in our work. It can monitor the single crystal SiGe thickness and the

Germanium composition. HRXRD measurement principle is the Bragg's Law. Ge% concentration is calculated by measuring the lattice parameter change based on the Vegard's low:

$$a(Si_{1-x}Ge_x) = x*a(Ge)+(1-x)*a(Si) \qquad (1)$$

And SiGe is tetragonally distorted, the Ge composition of a fully strained Si1-xGex layer can then be determined from the given lattice parameter of unstrained layer with the Dismuke's Law through:

$$a(Si_{1-x}Ge_x) = 5.43105 + 0.1988x + 0.028x^2 \qquad (2)$$

where a is the lattice parameter (Å), x is the Ge composition expressed as an atomic fraction [2]. The lattice parameter of SiGe layer increases with Ge composition. For peudomorphic (fully strained) growth, the SiGe layer exhibits a compressive in-plane strain with elongated lattice parameter perpendicular to the surface. Symmetrical (around the Si 004 substrate peak) scan will determine the out-of-plain lattice parameter $a(Si_{1-x}Ge_x)$, and asymmetrical (around the Si 115 subtract peak) scan will allow us to determine the in-plain lattice parameter $a(Si_{1-x}Ge_x)$ hence solve for the composition and relaxation of the epilayer

The layer thickness is determined by the intensity fringes. As shown in Figure 1, in addition to the diffraction peak from the layer, the fringes are clearly seen in the intensity on either side of the peak. These fringes indicate a high degree of crystalline quality (low defect density) in the epitaxial layer and result from the interference of the x-ray wave fields within the sample. For a symmetric reflection, the angular period of the fringes is inversely proportional to the thickness of t of the layer according to below equation:

$$\Delta \omega = \frac{\lambda}{2t cos\theta} \qquad (3)$$

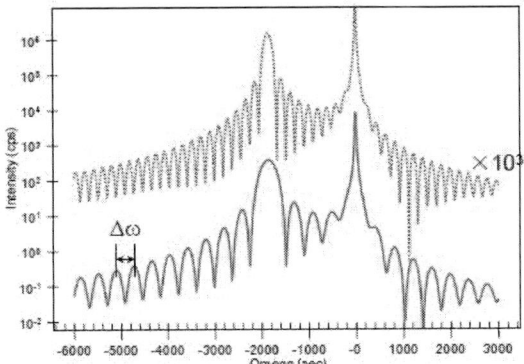

Figure 1. Simulated 004 reflection from Si1-xGex // Si with x = 0.20 and thickness values of 500 Å (solid) and 1000 Å (dots)

C. Test Material

Epitaxial growth of SiGe in the channel is dependent on process parameters. The germanium and Silane flow rates determine the stoichiometry while the temperature and deposition times determine the thickness of the film. A set of design of experiment (DOE) wafers as shown in Table 1, with the predictable target of the film thickness and composition around the baseline (BL) process specification, was generated using these process variables. This DOE set is designed for BBSE accurately modeling and calibration. Single layer and bi-layer wafers with the split Ge% concentration are designed to build up the Ge% concentration lookup model. All wafers relaxation is less than 3% (tested with HRXRD), thus can be considered as a fully strained sample. It is known that the relaxation might impact HRXRD and BBSE metrology model validity.

Table I. DOE wafers with designed variations for BBSE modeling and calibration. Thickness and Ge% concentration are normalized to DOE target. L1/L2/L3 indicate SiGe film stack individual layer from underneath to top.

DOE wafer	T1	L1 Ge%	T2	L2 Ge%	T3
#1	BL	BL			
#2	BL	BL+2.5%			
#3	BL	BL-2.5%			
#4	BL	BL	BL	BL	
#5	BL	BL	BL	BL+2.5%	
#6	BL	BL	BL	BL-2.5%	
#7	BL	BL	BL	BL	BL
#8	BL+50	BL+2.5%	BL+50	BL+2.5%	BL+50
#9	BL-50	BL-2.5%	BL-50	BL-2.5%	BL-50

RESULT AND DISCUSSION

a. BBSE Modeling

A model is derived using predefined DOE thickness and Ge concentration. The optical dispersions (T, n, k) for each of the distinct materials in the measured region are compared to the measured spectrum. A regression algorithm arrives at the solution for thickness and refractive index of the film layers using chi squared fitting across the model and measured spectrum.

Figure 2 shows the N and K dispersion curves for different Ge% concentration from table 1. The germanium dose and thickness both can affect the spectrum, and thickness effect is more apparent. For multiple SiGe layers, it is challenging to de-convolve the effects of thickness of cSiGe and superficial silicon and the Germanium content in the measured spectrum. A multi-pass regression is implemented that allows feeding forward the sub-layer thicknesses between passes to simplify the measurement of the layer of interest. To build this model, we use this feeding forward manner, but once the model is built, the tri-layers SiGe can be measured directly and the thickness and Ge concentration for every layer can be accurately derived from the model. We calibrated the model using the SEM and HRXRD as a reference.

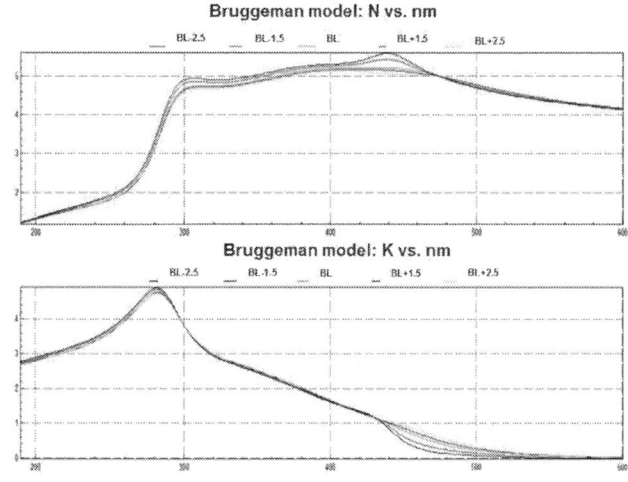

Figure 2. N and K dispersion curves for different germanium concentration

b.   BBSE Accuracy

The accuracy is tested by comparing the BBSE obtained parameters with the reference characterization techniques of TEM and HRXRD. The SiGe film stack individual layer thickness and germanium concentration derived from the above model have been compared with reference technology. We found all the parameters show a very good matching between BBSE and reference tool, indicating that the BBSE technique can obtain accurate measurement with an appropriate model. As an example, Figure 1 (a) (b) shows the L2 thickness and L2 Ge% concentration data, considering 2 points per wafer at center and edge. The correlation R square is larger than 0.9 and the slope is very close to 1. It is noted the model is only valid in the calibrated range. If the process change is beyond this calibrated range for thickness and film reflective index, the qualified model may become invalid and would require recalibration. This calibrated range is defined between the upper and lower limits of the designed thickness and of the germanium concentration, that is $BL-50\text{Å}<T<BL+50\text{Å}$, $BL-2.5\%<Ge\%<BL+2.5\%$.

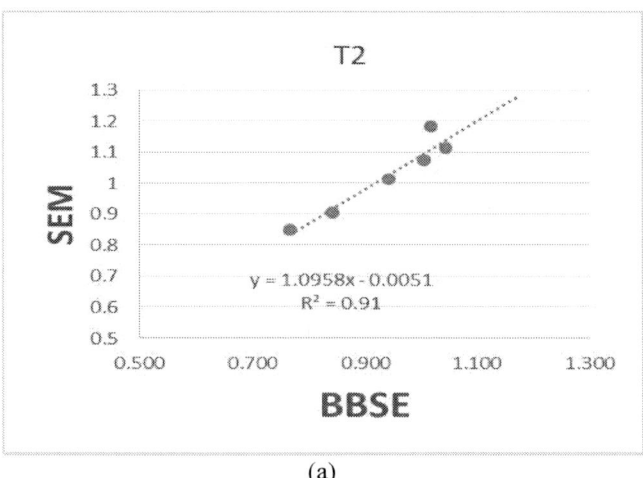

(a)

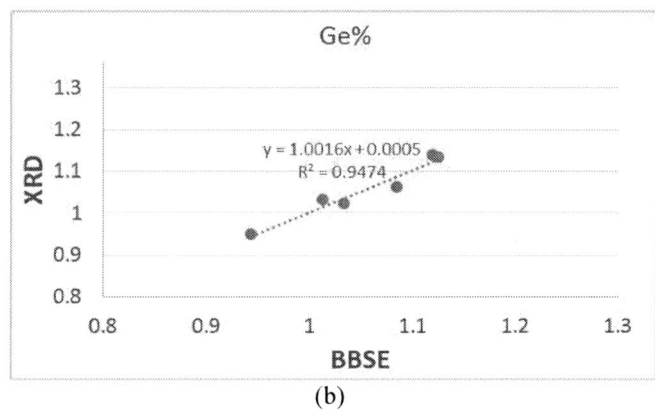

(b)

Figure 3 (a) BBSE L2 thickness vs. TEM (b) BBSE L2 germanium concentration correlation vs. HRXRD

BBSE and HRXRD map data have been collected for a baseline process inline wafer as Figure 4 shown. The calculated R square of correlation (HRXRD vs. BBSE) is above 0.9. The unusual outliers have been carefully checked, they are impacted by the SiGe film property. Either the relaxation or the mosaic is worse than the baseline sites for those outliers. Thus for inline SiGe monitoring, a hybrid metrology method needs to be implemented combining HRXRD and BBSE, especially for process develop phase. And it is proven that setting up a less HRXRD inline sampling size and a reasonable BBSE sampling size for throughput concern, is very effective.

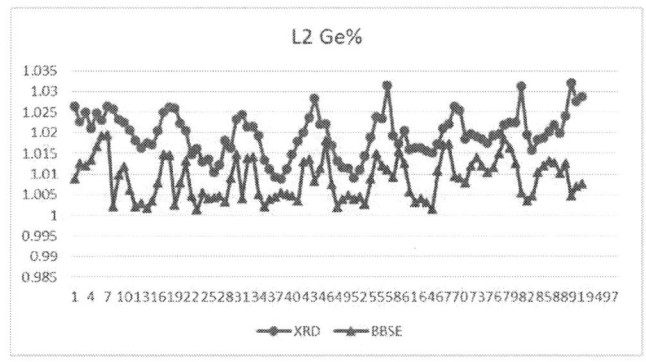

Figure 4 BBSE and HRXRD data comparison for L2 germanium for the baseline inline wafer

c.   BBSE stability

Gauge Repeatability and Reproducibility (GRR) is a key criterion for metrology tools. It is used to prove the metrology stability. In our work, GRR test is performed through a 10 times dynamic repetitive measurements on 13 points of an inline baseline wafer, all parameters has been collected and calculated. The result shows very good repeatability measured by BBSE (3σ are less than 1%). L2 Germanium concentration is shown in Figure 5 for example. Overall, it can be seen that the results from BBSE can be qualified as trustable and stable.

978-1-5386-5309-8/18 $31.00 © 2018 IEEE   428

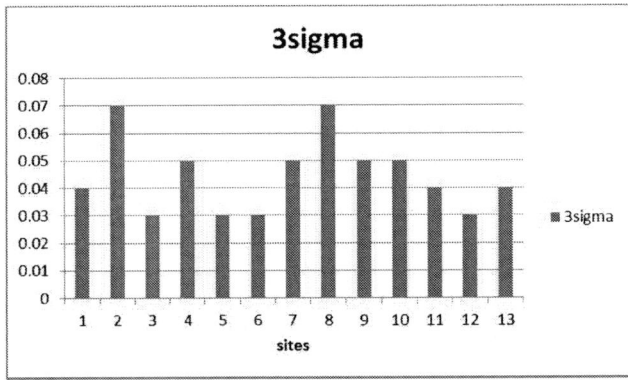

Figure 5. GRR test for L2 germanium concentration

d. Si strain on OCD pad by HRXRD

Si strain measurement is of particular interest as it reflects directly the channel strain status. All the above studies of Ge% concentration (indirectly reflect SiGe strain) are performed on SiGe solid pad, but OCD pad is the most representative structure of the in-device strain. We studied the silicon channel strain by HRXRD on OCD pad, which is a direct method to monitor the SiGe strain and relaxation. With the function of RSM (reciprocal space map), the Si strain in OCD (optical critical dimension) structure pad has been well demonstrated by using (113) asymmetric measurement. We could observe the Si-Strain increases with the increasing Ge% concentration as shown in the Figure 6. The error bar for out-of-plain stain is coming from the 3 measuring sites.

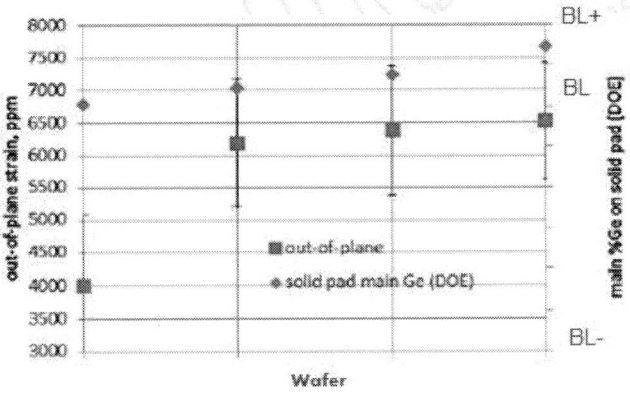

Figure 6. Si strain(out of plane) on OCD pad by HRXRD with varied germanium concentration

CONCLUSION

The thickness and germanium concentration for each layer of SiGe graded stack have been effectively identified by BBSE with a regression model which we built with a feeding forward method. The model accuracy and stability has been proven to be able to monitor the inline SiGe thickness and germanium concentration. This paper illuminates the applicable limits for the optical model measurement, and proposed a hybrid metrology method for inline epitaxy

process control. Silicon channel strain is also been successfully demonstrated by HRXRD with (113) asymmetric measurement.

[1]T. Ghani, M. Armstrong, C. Auth, M. Bost, P. Charvat, G. Glass, T. Hoffmann, K. Johnson, C. Kenyon, J. Klaus, B. McIntyre, K. Mistry, A. Murthy, J. Sandford, M. Silberstein, S. Sivakumar, P. Smith, K. Zawadzki, S. Thompson, and M. Bohr, "A 90nm High Volume Manufacturing Logic Technology Featuring Novel 45nm Gate Length Strained Silicon CMOS Transistors," IEDM Tech Dig., 2003.

[2] J.P. Dismukes, L. Ekstrom and R.J. Paff "Lattice Parameter and Density in Germanium-Silicon alloys", *Journal. Phyical Chemistry*, 1964, *68* (10), pp 3021–3027

ULK OPTIMIZATION FOR CU/ULK(K=2.5) BEOL INTERCONNECT TDDB AND EM IMPROVEMENT AT 28/14NM TECHNOLOGY NODE AND BEYOND

Xiaodong Zou, Tao Dou, Zheyuan Tong, Fenglian Li, Duohui Bei

Semiconductor Manufacturing International(Shanghai) Corporation

18 Zhangjiang Road, Pudong New Area, Shanghai, China

Johnston_Zou@smics.com

ABSTRACT

As CMOS is being scaled down to 28nm node and beyond, it is necessary to reduce k value in order to improve RC delay. ULK (ultra-low k) (k≤2.55) has been implemented into production from 45/40nm node. Compared with low-K film, ultra-low K film is more porous, softer and hydrophilic intrinsically. It is easy to be damaged by plasma, stress, thermal or aqueous environments. It brings more challenges for reliability improvement. This paper focuses mainly on ULK process optimization through simple key parameter optimization to improve voltage break down (VBD) and time-dependent dielectric breakdown (TDDB) and electro migration (EM) performance without scarifying RC delay.

Keywords—ULK; BEOL; VBD; time-dependent dielectric Breakdown (TDDB); electro migration (EM)

INTRODUCTION

In order to reduce chip RC (resistance capacitance) delay, porous SiOCH (p-SiOCH) ultra-low-k (ULK) dielectric material was adopted by 45/40nm and newer technology nodes. However, because of its porosity, ultra-low-k material generally suffers weak mechanical strength and lower intrinsic breakdown strengths. This fragile film is easily damaged by various fabrication process such as plasma etch/ash, barrier metal pre-clean, CMP, Wet clean, etc. [1]. Its C loss on exposed film surface leads to moisture absorption from air. And the Si-OH bonds absorbing H2O are considered as a major root cause for the reliability failure, as well as the capacitance increase [2]. Accordingly, the long-term reliability, especially time dependent dielectric breakdown (TDDB) of ULK and Electro-migration (EM) are becoming the most critical challenges for technology qualification. This problem is further exacerbated due to continuously shrinkage of the interconnect pitches.

To address these electrical and reliability challenges, it is necessary to optimize ULK deposition process. This study mainly focuses on ULK deposition key parameter investigations, such as precursors flow rate and plasma RF power. The effects on the properties of ULK films are summarized, including dielectric constant, elastic modulus, hardness, as well as RC and reliability performance. The purpose is to find the best recipe to get better TDDB and EM performance without mechanical properties and RC delay degradation.

EXPERIMENTAL

pSiCOH films are deposited using a commercial 300mm plasma enhanced chemical vapor deposition (PECVD) chamber. For the Film 1~3 series, two carbosilane precursors (P1 and P2) were admixed into the plasma during deposition. P1 is a carboxysilane precursor with Si-[CH2] x-Si, P2 is a pore-building CxHy precursor that acts as a porogen. P1, P2 flow and Plasma RF power can be adjusted to meet different carbon composition, porosity and film property requirements.

The porogen was removed by UV irradiation in a 300mm system around 400C to create porosity and lower k. UV curing also further crosslinks the SiOC skeleton to enhance the modulus. RF power and temperature also change film properties [3].

Film electrical and mechanical properties were measured by Hg probe, MTS nano-indentor and stress gauge on 3000A thick ULK films. Optical ellipsometry was used for film thickness measurement. C-V/I-V testing of MIS structures with aluminum dots was conducted to determine k and breakdown strength.

Low damage Cu/ULK dual damascene process scheme was employed including the trench first metal hard-mask (TFMH) scheme and remote plasma hydrogen (H) pre-clean for barrier metal with 90nm minimum metal pitch. Via face to face TDDB measurement was carried out on a comb-serpentine test-key and its TTF (Time to Failure) was evaluated based on the square root model. Electro-migration was measured for both up and down stream using the typical EM structure.

RESULTS AND DISCUSSION

A. Film Properties of ULK

Table 1 lists a series normalized parameters of ULK film properties. P1 flow rate is proportional to ULK mechanical properties such as modulus, hardness and k value, as figure 1a/1b/1c shows, while P2 shows inversed proportion to modulus and hardness. Be noted that P2 amount has no strong correlation with k value, as shown by figure 2a/2b/2c. It's understood that P2 is just a porogen, more porogen, higher porosity within film, it would

TABLE I

Blanket film properties of ULK materials (normalized)

Film	P1	P2	RF Power	Modulus	Hard-ness	K value
Ref. ULK	1.00	1.59	1.00	1.00	1.00	1.0000
ULK 1a	1.00	1.52	1.00	1.03	1.00	0.9945
ULK 1b	1.00	1.45	1.00	1.09	1.07	0.9974
ULK 1c	1.05	1.59	1.00	1.09	1.07	1.0079
ULK 1d	1.09	1.59	1.00	1.12	1.10	1.0129
ULK 1e	1.09	1.45	1.00	1.13	1.13	1.0154
ULK 2a	1.00	1.59	0.91	0.91	0.93	1.0030
ULK 2b	1.05	1.52	0.91	1.10	1.12	1.0151
ULK 2c	1.09	1.45	0.91	1.21	1.23	1.0296
ULK 3a	1.00	1.59	0.82	1.03	1.04	1.0201
ULK 3b	1.00	1.45	0.82	0.99	1.01	1.0204
ULK 3c	1.05	1.45	0.82	0.97	0.98	1.0275
ULK 3d	1.09	1.59	0.82	1.07	1.09	1.0371
ULK 3e	1.09	1.52	0.82	1.03	1.05	1.0334
ULK 3f	1.09	1.45	0.82	1.10	1.03	1.0444

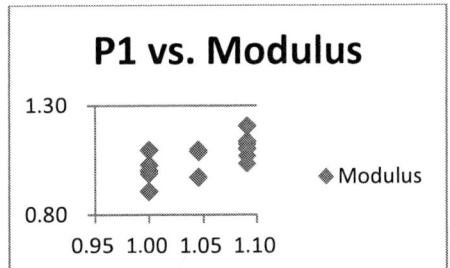

Figure 1a: Precursor 1 flow rate (X-axes) vs. Modulus (normalized) (Y-axes)

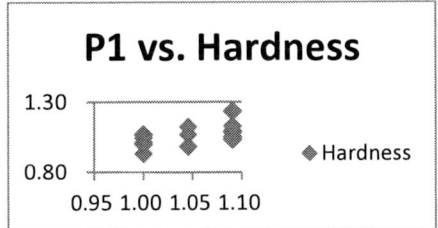

Figure 1b: Precursor 1 flow rate (X-axes) vs. Hardness (normalized) (Y-axes)

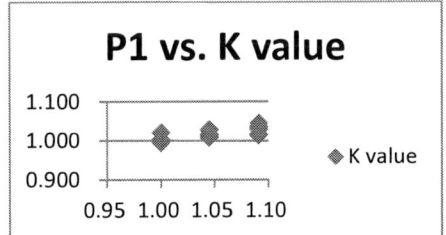

Figure 1c: Precursor 1 flow rate (X-axes) vs. k value (normalized) (Y-axes)

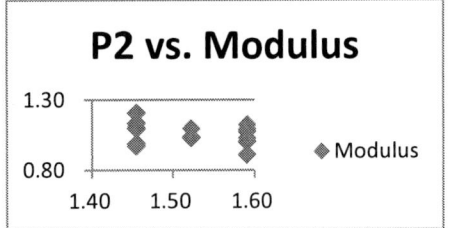

Figure 2a: Precursor 2 flow rate (X-axes) vs. Modulus (normalized) (Y-axes)

degrade mechanical strength while little impact on k value. Though higher porosity theoretically decreases k value, no strong correlation here was observed, possibility due to porosity saturation. Figure 3a/3b/3c shows RF power does not have much impact on ULK modulus and hardness, but has strong influence on k value, k increases dramatically with lower RF power.

B. RC performance of ULK

The RC for ULK 1 series is slightly higher (~0.5%) than and the RC for ULK 2/3 series is 2~3% higher than that of Ref ULK as shown in Figure 4a. Figure 4b shows RC curves of ULK1 series and Ref ULK. ULK 1b is 1.5% higher though its initial K value is comparable with Ref ULK. It means ULK 1b is damaged by various fabrication process and k value increased. As to other ULK 1 series, RC curves are comparable with Ref ULK though ULK 1a and 1c initial k are comparable and 1d and 1e are slightly higher (1.2~1.5%) than Ref ULK. It indicates that the resistance for ULK damage of 1d and

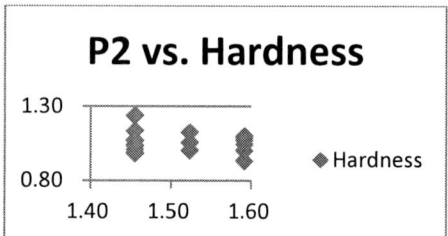

Figure 2b: Precursor 2 flow rate (X-axes) vs. hardness (normalized) (Y-axes)

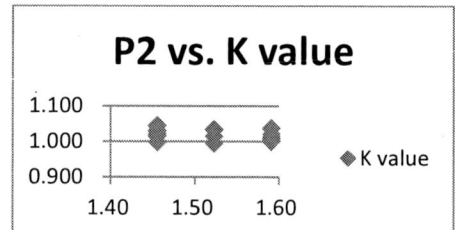

Figure 2c: Precursor 2 flow rate (X-axes) vs. K value (normalized) (Y-axes)

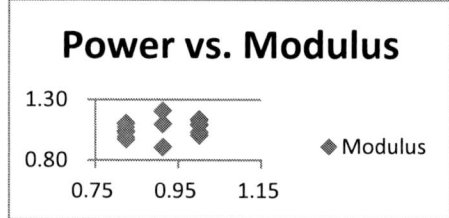

Figure 3a: RF Power (X-axes) vs. Modulus (normalized)(Y-axes)

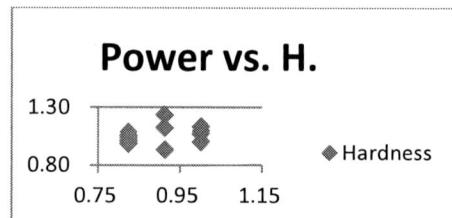

Figure 3b. RF Power (X-axes) vs. Hardness (normalized)(Y-axes)

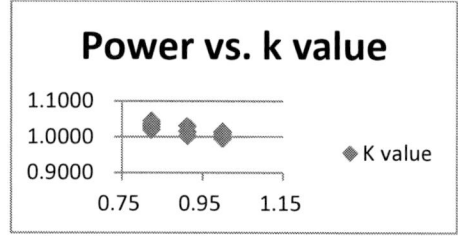

Figure 3c. RF Power (X-axes) vs. k value (normalized) (Y-axes)

1e is stronger than 1a and 1c, k value does not increase after integration process.

C. VBD performance of ULK

Both metal and via VBD performance of ULK 1a, 1b, 1c are comparable with Ref ULK, while 1d and 1e are obviously better as shown in figure 5a and 5b. VBD performance of ULK 2 and 3 series are better than ULK 1 series and Ref ULK but RC also higher. Therefore, ULK 1d and 1e are chosen to do further TDDB and EM test considering their lower RC performance.

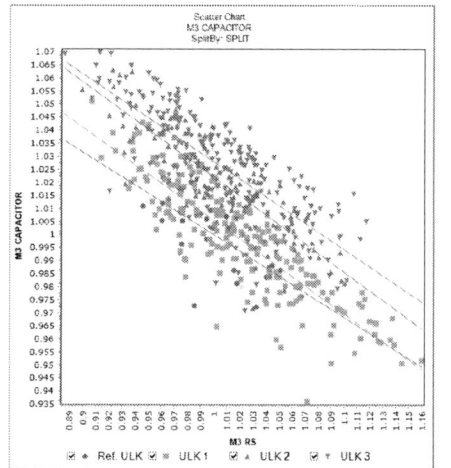

Figure 4a: RC Curves of ULK1/2/3 series and Ref ULK (normalized)

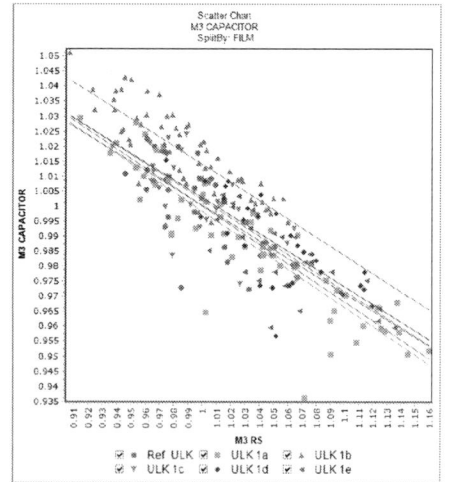

Figure 4b: RC Curves of for ULK1 series and Ref ULK (normalized)

D. TDDB performance of ULK

Figure 6 shows TDDB results with three types of ULK film with same integration processes. ULK 1d

978-1-5386-5309-8/18 $31.00 © 2018 IEEE   432

achieves 5 times and ULK 1e gets 7 times longer TTF (time to failure) than ref. ULK. The TTF uniformity is also improved 25% and 47% for ULK 1d and 1e respectively. Their longer life time is due to higher carbon content from higher P1 flow rate and stronger plasma resistance to ULK damage [4, 5].Si-OH bond of damaged ULK is well-known to attract moisture, which can contribute to the increase of leakage current by providing mobile H+ ions as well as trap sites [2]. Therefore, less ULK damage for ULK 1d/1e film is the key factor for TDDB improvement due to reduced leakage source. ULK 1e is slightly better than 1d because of less porosity, resulting in slightly higher k value and better VBD/TDDB performance [3].

for the EM failure [2].Therefore, these results are closely associated with 13% higher elastic modulus of ULK 1e.

CONCLUSIONS

ULK Process was optimized by key parameter and precursor flow optimizations. The ULK carbon content and porosity change accordingly. Stronger elastic modulus, hardness and plasma resistance can be achieved without compromising the final dielectric constant. It was confirmed that reliability performance including VBD, TDDB and EM were much improved without RC delay sacrifice.

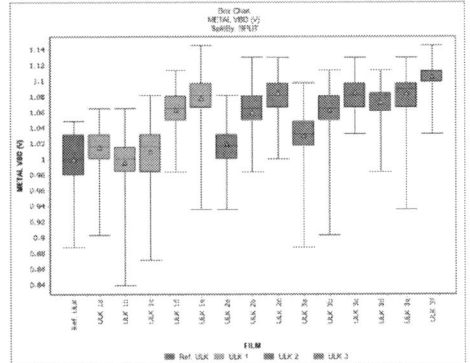

Figure 5a: Metal VBD of ULK1/2/3 series and Ref ULK (normalized)

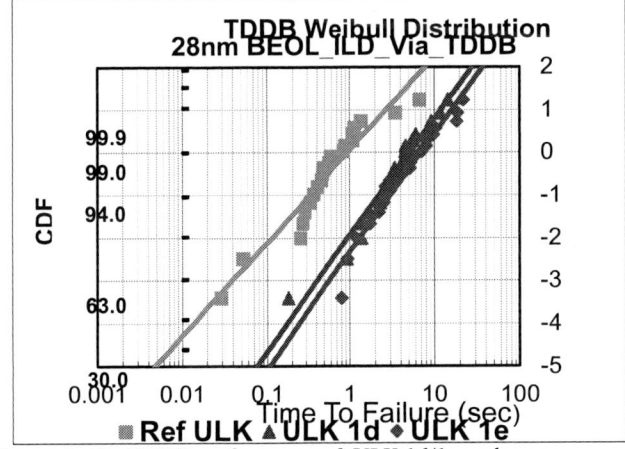

Figure 6: Via TDDB performance of ULK 1d/1e and Ref ULK(normalized)(Stress conditions:27V/125 ℃)

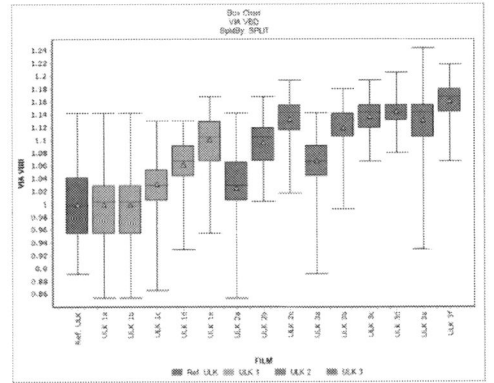

Figure 5b: Via VBD of ULK1/2/3 series and Ref ULK (normalized)

E. EM performance of ULK

Figure 7 shows electro migration (EM) results of two ULK materials. The life time of EM is improved by 25% for down-stream EM of ULK 1e while up-stream EM comparable with Ref ULK. Generally, it is known that higher elastic modulus of IMD materials increases the critical stress which is necessary for void nucleation

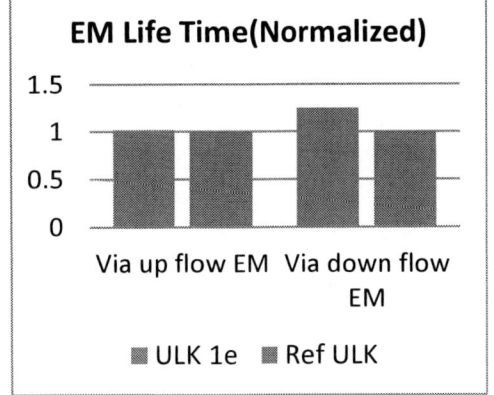

Figure 7. Up and down stream EM (normalized). (Stress conditions: 0.1mA/300C)

REFERENCES

[1] Chia-Lin Hsu1,2, Kuan-Ting Lu2, Wen-Chin Lin2, Jeh-Chieh Lin2, Chih-Hsien Chen2, Teng-Chun

Tsai2, et al. THE TDDB FAILURE MODE AND ITS ENGINEERING STUDY FOR 45NM AND BEYOND IN POROUS LOW K DIELECTRICS DIRECT POLISH SCHEME, IRPS10-918-921, 978-1-4244-5431-0/10/$26.00©2010 IEEE.

[2] Yunlong Li, Ivan Ciofi, Laureen Carbonell, Nancy Heylen, Joke Van Aelst, Mikhail R. Baklanov, Guido Groeseneken, Karen Maex, and Zsolt Tokei, "Influence of absorbed water components on SiOCH low-k reliability" J. Appl. Phys. 104, 034113 (2008)

[3] E.Todd Ryana, D. Priyadarshinib, S.M. Gatesc, H. Shobhab, J. Chen,b K. Virwanid, et al., Optimizing ULK Film Properties to Enable BEOL Integration with TDDB Reliability. 978-1-4673-7356-2/15/$31.00 ©2015 IEEE.

[4] E. T. Ryan, S. M. Gates, A. Grill, S. Molis, P. Flaitz, J.Arnold, M. Sankarapandian, S. A. Cohen, Y. Ostrovski, and C. Dimitrakopoulos, J. Appl. Phys. 104, 094109 (2008).

[5] E.T Ryan, S. Gates, A. Madan, D. Kioussis, E. Liniger, G.Bonilla, E. Zin, C.-U. Kim, U. Mayer, A. Peeva, and A. Grill,Adv. Metal. Conf. 2011.

FAN-OUT WAFER-LEVEL PACKAGING FOR 3D IC HETEROGENEOUS INTEGRATION

John H Lau

ASM Pacific Technology, Hong Kong, China

john.lau@asmpt.com

ABSTRACT

Two 3D IC heterogeneous integrations by fan-out wafer-level packaging (FOWLP) technology are investigated in this study. The emphasis of the first such method is on the design, and of the other method, the emphasis is on the manufacturing process.

INTRODUCTION

Moore's law [1] has been driving the system-on-chip (SoC) platform. Especially in the past 10 years, SoCs have been very popular for smartphones, tablets, and the like. SoCs integrate different-function ICs into a single chip for a system or subsystem. Two typical SoC examples are shown in Figure 1. The application processor (AP) A10 is

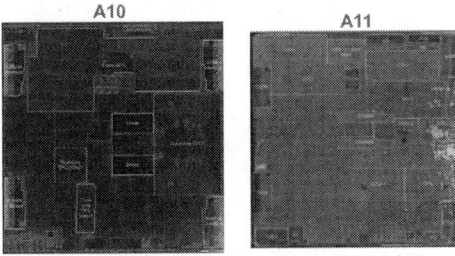

Figure 1: SoC platforms for the A10 and A11 APs

designed by Apple and manufactured by TSMC using its 16nm process technology. It consists of a 6-core graphics processor unit (GPU), two dual-core central processing unit (CPUs), 2 blocks of static random access memories (SRAMs), etc. The chip area is 125mm^2. The application processor A11 is also designed by Apple and manufactured using TSMC's 10nm process technology. The A11 consists of more functions, including a tri-core Apple designed GPU, etc. However, the chip area is about 30% smaller than that of the A10 because of the Moore's law, i.e., the feature size is from 16nm down to 10nm.

Why is heterogeneous integration of such great interest? One of the key reasons is because the end of Moore's law is fast approaching and it is more and more difficult and costly to reduce the feature size (to do the scaling) to make SoCs. Some of the early researches in heterogeneous integration have been provided by Georgia Institute of Technology [2, 3, 4] where they reported a differential Si CMOS (complementary metal-oxide semiconductor) receiver IC (operating at 1 Gbps) integrated with a large-area thin film InGaAs/InP I-MSM (metal-semiconductor-metal) photodetector (Figure 2).

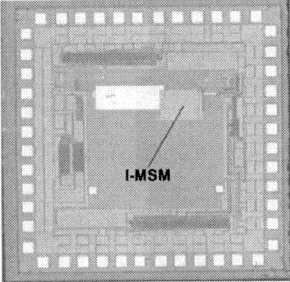

Figure 2: InGaAs/InP I-MSM integrated onto differential Si CMOS receiver IC

Heterogeneous integration contrasts with SoCs as follows. Heterogeneous integration uses packaging technology to integrate dissimilar chips with different functions from different foundries, wafer sizes, and feature sizes (as shown in Figure 3) into a system or subsystem, rather than integrating most of the functions into a single chip and going for a finer feature size. For the next few years, we will see more of a higher level of heterogeneous integration, whether it is for time-to-market, performance, form factor, power consumption, signal integrity, or cost. Heterogeneous integration is going to take some of the market share away from SoCs on high-end applications such as high-end smartphones, tablets, wearables, networkings, telecommunications, and computing devices. How should these dissimilar chips talk to each other, however? The answer is: redistribution layers (RDLs) [5]! How should those RDLs be made? In this study, we use FOWLP technology.

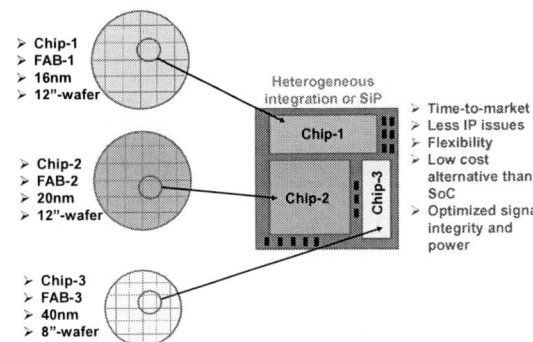

Figure 3: Heterogeneous Integration or SiP

978-1-5386-5309-8/18 $31.00 © 2018 IEEE

3D IC HETEROGENEOUS INTEGRATION BY FOWLP

The A10 and A11 application processors are packaged using TSMC's InFO (integrated fan-out) wafer-level packaging method [6-9]. The mobile dynamic

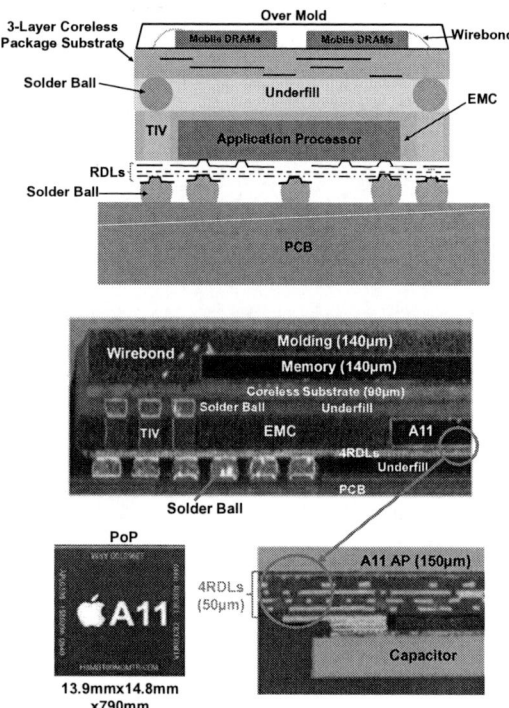

Figure 4: PoP for packaging the application processor and mobile memory. Top: Schematic. Bottom: SEM

random access memories (DRAMs) are wire bonded on a 3-layer core-less package substrate and the substrate is area-array solder balled on top of the application processor package - a package-on-package (PoP) format as shown in Figure 4. The interconnections between the AP and the

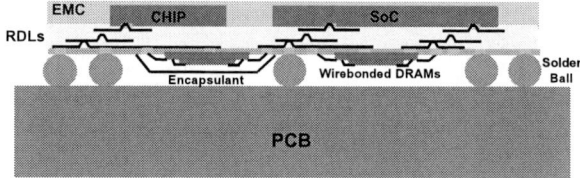

Figure 5: 3D IC heterogeneous integration by FOWLP

memory are mainly through the RDLs, through-InFO vias (TIVs), solder balls, and core-less substrate.

A new 3D IC heterogeneous integration by FOWLP, as shown in Figure 5, is proposed in this study. It consists of the SoC, chips, and the mobile DRAMs. Their interconnections are mainly through the RDLs, which can

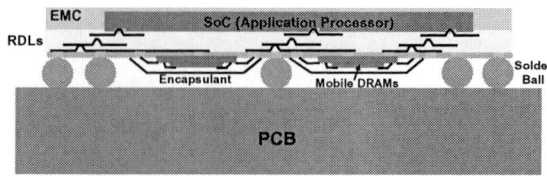

Figure 6: 3D IC heterogeneous integration to package the application processor chipset

be fabricated by the FOWLP method. Depending on the number of layer of the RDLs, usually the total thickness of a 3-layer RDL is about 40μm. The DRAMs (≤ 50μm -thick) are cross-stacked with wire bonds and then encapsulated. The diameter of the solder ball is 200μm.

Figure 6 shows a special case of Figure 5 (when there is no other chip and the SoC is the application processor). Comparing the new design (Figure 6) with that of Figure 4 (the 3D IC heterogeneous integration vs. the PoP), it is obvious that: 1) the new design leads to a lower package

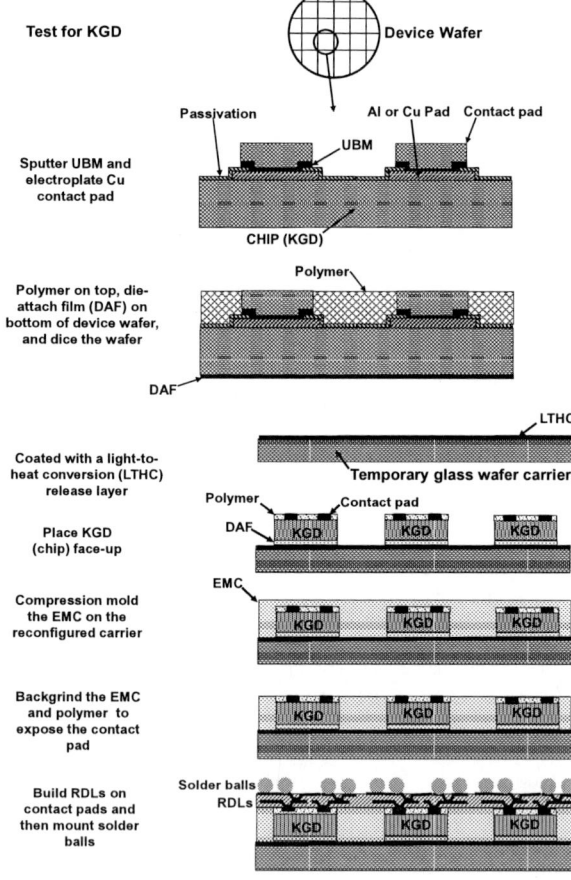

Figure 7: Manufacturing process for packaging the application processor chipset

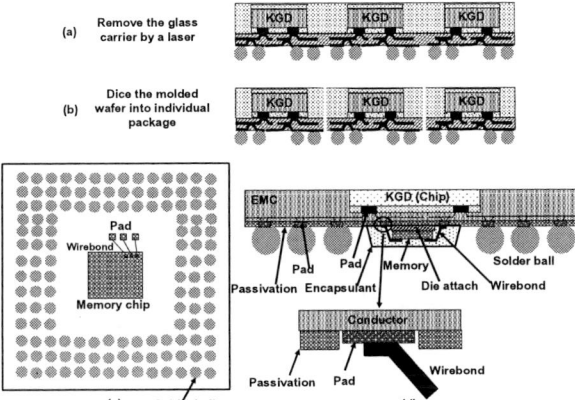

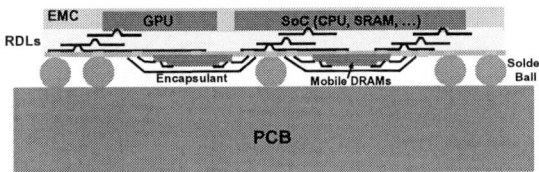

Figure 8: Wire bonding memory chip at the bottom of the individual application processor package

profile; 2) the new design has less interconnects; 3) the new design is more reliable because of less interconnects; 4) the new design has better electrical performance; and 5) the new design leads to lower cost.

The manufacturing process of the proposed 3D IC heterogeneous integration is very simple. First, the device wafer has to be modified by sputtering an under bump metallurgy (UBM) and electroplating a Cu contact pad (for building the RDLs later), as shown in Figure 7. This step is followed by spin coating a polymer on top of the device wafer and laminating a die-attach film (DAF) at the bottom of the device wafer. Meanwhile, a light-to-heat conversion (LTHC) layer is spin-coated onto a temporary glass carrier wafer. Then the individual known-good die (KGD) (chip) from the device wafer is placed face-up on the LTHC carrier. This step is followed by epoxy molding compound (EMC) dispensing, compression molding, and finally, post mold cure (PMC).

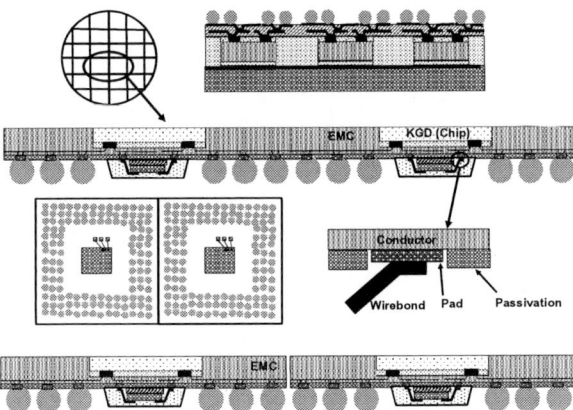

Figure 9: Wire bonding memory chip at the bottom of the application processor package on a wafer

These steps are followed by backgrinding the EMC and polymer to expose the Cu-contact pad for making the RDLs and for mounting the solder balls, as shown in Figure 7. This is the conventional FOWLP method to package the application processor [6-16].

There are two methods to attach the mobile DRAMs to the bottom of the AP fan-out wafer-level package. The first method comprises the following steps: 1) removing the glass carrier by a laser (Figure 8a); 2) dicing the reconstituted wafer carrier into individual package (Figure 8b); 3) wire bonding the memory chips to the bottom side of the individual package (Figures 8c and 8d); 4) and then glob topping the wires and memory chips with an encapsulant, Figures 8c and 8d.

The second method to attach the mobile DRAMs to the bottom of the AP package comprises the following steps: 1) wire bonding the memory chips to the bottom side of every package on the reconstituted wafer carrier; 2) glob topping the wires and memory chips with an encapsulant; and 3) then dicing the reconstituted wafer carrier into individual packages (Figure 9).

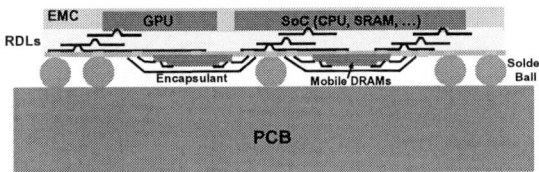

Figure 10: 3D IC heterogeneous integration to package the application processor chipset

Figure 10 is a special case of Figure 5. This is when it is difficult and costly to reduce the feature size to make the SoC. Therefore, some of the function (for example, the GPU) is not integrated into the SoC and the GPU chip is placed side-by-side with the SoC.

In [17] we asked the question: "What if there is no PoP for the application processor chipset?" We proposed

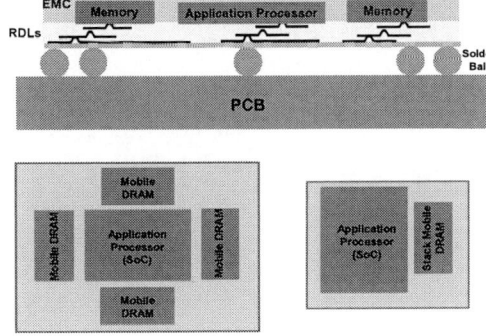

Figure 11: 2D IC heterogeneous integration to package the application processor chipset

978-1-5386-5309-8/18 $31.00 © 2018 IEEE

to place the AP and the DRAMs side-by-side on a build-up package substrate. The memory chips can be either cross-stacked or individually placed by wire bonding. Also, the memory chips can be placed individually by solder bumped flip chips. The memory chips can even be stacked and have TSVs. In this study, because we used the FOWLP method to construe the RDLs for the interconnections between the SoC and memory as shown in Figure 11, the package substrate was eliminated.

3D IC HIGH-PERFORMANCE HETEROGENEOUS INTEGRATION BY FOWLP

Figure 12 schematically shows a 3D IC high-performance heterogeneous integration by FOWLP technology. It can be seen that it consists of a GPU, a FPGA (field-programmable grid array), a CPU, or a high-performance application-specific integrated circuit (ASIC), and is surrounded by high-bandwidth memory (HBM) cubes. Each HBM cube consists of four DRAMs and a logic base with through-silicon vias (TSVs) [18-20]

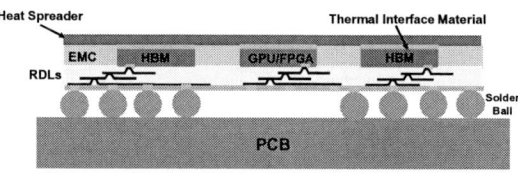

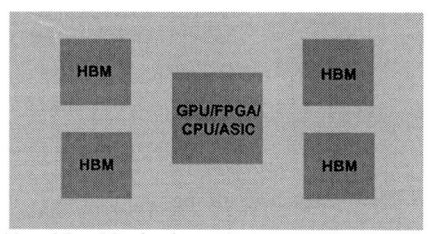

Figure 12: 3D IC high-performance heterogeneous integration by FOWLP

straight through them. Each DRAM chip has >500 TSVs. The interconnections between the GPU/FPGA/CPU/ASIC and HBMs are through the RDLs. The major heat path of this structure is from the backside of the GPU/FPGA/CPU/ASIC to the heat spreader. A heat sink can be added on top of the heat spreader if it is necessary.

In this case, the emphasis is placed on the manufacturing method (process) of this structure. This method comprises these steps: 1) testing for KGD of device wafers; 2) sputtering UBM; 3) electroplating the Cu-contact pad; 4) spin coating a polymer on top of the device wafers; and 5) painting a thermal interface material (TIM) on the bottom (backside) of the device wafers (Figure 13). The last step is different from the conventional method (the first case) which is laminating a DAF on the bottom of the device wafers.

After the steps outlined above are completed, the following are done: 1) the individual KGDs are picked and placed face-up on a metal such as copper, aluminum, steel, and an alloy 42 (with thermal expansion coefficient = 8 to $10 \times 10^{-6}/^{\circ}C$) carrier about 1mm-thick; 2) molding the EMC on the reconstituted wafer is accomplished by using the compression method and then post-mold curing (PMC) of the EMC; 3) backgrinding the EMC and polymer to expose the Cu-contact pad; 4) building up the RDLs; and

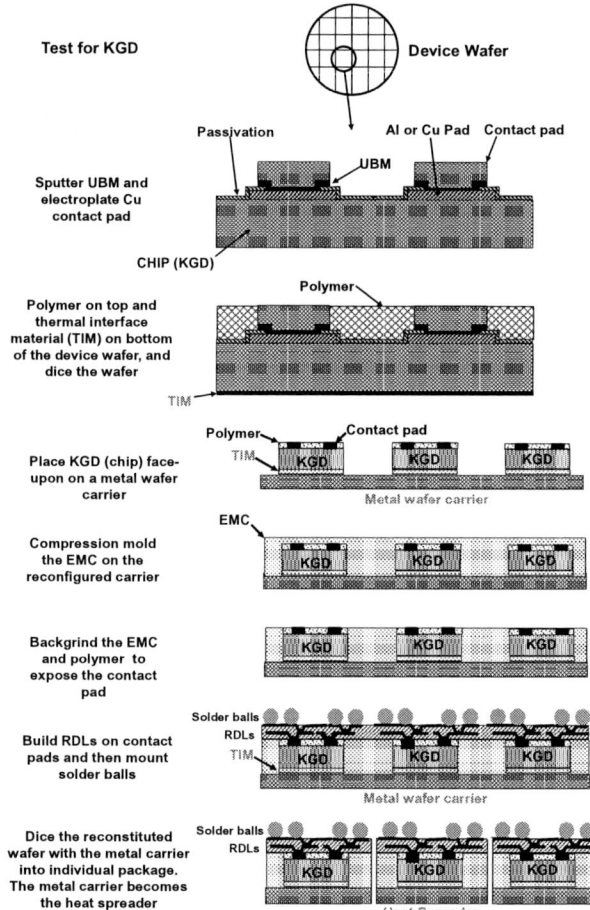

Figure 13: Manufacturing method for 3D IC high-performance heterogeneous integration

5) mounting the solder balls. Then, the reconstituted wafer is diced into individual packages (Figure 13). (Note: this process is different from the conventional method, which used a glass carrier and was coated with an LTHC layer).

It should be emphasized that unlike the conventional method, there is no debonding of the carrier. The metal carrier becomes the heat spreader of the individual high-performance heterogeneous integration package.

978-1-5386-5309-8/18 $31.00 © 2018 IEEE   438

This new method of manufacturing high-performance chips and memory cubes in a heterogeneous integration scheme with the FOWLP technology results in fewer assembly steps, lower cost, faster time-to-market, and higher assembly yield. Also, because of the metal carrier, the warpage is reduced during all the process steps. Furthermore, because of the metal carrier, the individual package size can be larger.

WAFER vs. PANEL

All pervious mentioned fan-out packaging are using the round 200- or 300-mm wafers as the reconstituted carriers for making the molds, RDLs, etc. (This is because of the existing equipment for fabricating the device wafers.) In order to increase the throughput, fan-out panel-level packaging (FOPLP) has been proposed by, e.g., OSAT and PCB/substrate manufacturers. At ECTC2014, SPIL published two papers on FOPLP, one is to develop and characterize a 370mm × 470mm FOPLP [21] and the other is to measure their warpage [22]. During ECTC2017, Unimicron presented the simulation and fabrication of FOPLP [23]. However, there are many issues in FOPLP and some are listed in the followings:

[1] Most OSATs and foundries already have the necessary equipment for FOWLP. For FOPLP, new capital will have to be expended on newly developed equipment.
[2] Inspection of wafers is a well-known process. FOPLP inspection must be developed.
[3] The yield of FOWLP is higher than that of FOPLP. (Assuming the size of panel is larger than that of wafer.)
[4] The cost advantages of panel over wafer need to be carefully determined. (Yes, the throughput is higher, but the pick & place and EMC dispensing time is longer, and the yield is lower.)
[5] A fully loaded high yield wafer line might be cheaper than a partially loaded low yield panel line.
[6] The panel equipment takes longer to clean than wafer equipment.
[7] Unlike FOWLP, the FOPLP is for medium chip size and line width and spacing.
[8] If indeed, the panel processing is developed and is high yield for fine line width and spacing, there is a chance to produce a major oversupply of capacity.
[9] IP, materials background, equipment automation, and manage the dimensional stability and yield of the panel in a large format are needed.
[10] Lack of panel standard for FOPLP, thus the equipment suppliers cannot make the equipment.

One of the bottlenecks for FOPLP is the availability of panel equipment such as the spin coating, physical vapor deposition (PVD), electrochemical deposition, etching, backgrinding, ball mounting, and dicing, for making the RDLs, molds, ball mounting, package, etc., due to

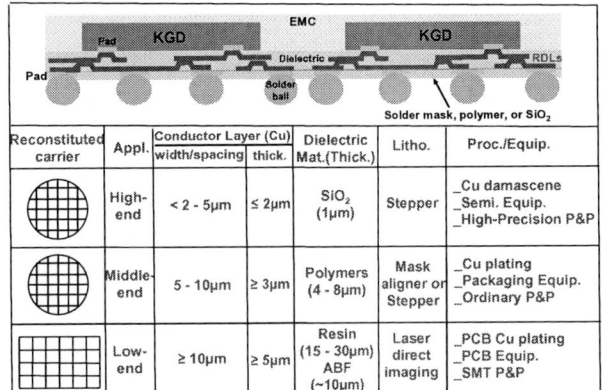

Reconstituted carrier	Appl.	Conductor Layer (Cu) width/spacing	thick.	Dielectric Mat.(Thick.)	Litho.	Proc./Equip.
(round wafer)	High-end	< 2 - 5μm	≤ 2μm	SiO₂ (1μm)	Stepper	_Cu damascene _Semi. Equip. _High-Precision P&P
(round wafer)	Middle-end	5 - 10μm	≥ 3μm	Polymers (4 - 8μm)	Mask aligner or Stepper	_Cu plating _Packaging Equip. _Ordinary P&P
(square panel)	Low-end	≥ 10μm	≥ 5μm	Resin (15 - 30μm) ABF (~10μm)	Laser direct imaging	_PCB Cu plating _PCB Equip. _SMT P&P

Figure 14: The geometry, material, process, equipment, and application of fan-out wafer/panel-level packaging

the lack of the standard of panel sizes. Also, FOPLP is for low-end applications with large line width/spacing of the RDLs as shown in Figure 14. It will be hard-pressed for HVM with FOPLP for small line width/spacing RDLs, except for very niche applications.

SUMMARY

Two 3D IC heterogeneous integrations by FOWLP technology have been presented. The first 3D IC heterogeneous integration is emphasized on the design and the other 3D IC high-performance heterogeneous integration is on the manufacturing method. Some important results and recommendations are as follows:

➢ A 3D IC heterogeneous integration of the application processor chipset has been proposed. The interconnections between the application processor and mobile DRAMs are through the RDLs, which are fabricated using the FOWLP method. The manufacturing processes for making the 3D IC heterogeneous integration have also been presented.

➢ When it is difficult and costly to reduce the feature size to make the SoC, one way is not to integrate some of the functions (for example, the GPU) into the SoC and instead place the GPU chip side-by-side with the SoC.

➢ The simplest heterogeneous integration of the application processor chipset is to place the application processor and the mobile DRAMs side-by-side on RDLs. One consideration is that the package size could be too large to be reliable. One of the alternatives is to stack up the mobile DRAMs by wire bonding (for lower cost) or TSV (for wider bandwidth.)

➢ A 3D IC high-performance heterogeneous integration of GPU/FPGA/CPU/ASIC and HBM/HBM2 by FOWLP technology has been proposed. Emphasis is placed on a simple and effective manufacturing method to fabricate the structure. Unlike the conventional method, there is no debonding of the temporary carrier. The metal carrier becomes the heat spreader of the individual high-performance heterogeneous integration package.

> The advantages of heterogeneous integration are time-to-market, performance, form factor, power consumption, signal integrity, and cost.
> In order to lower the package profile and enhance the electrical and thermal performance of the application processor chipset for mobile applications such as smartphones and tablets, the current PoP format should be eliminated.

REFERENCES

1. G. Moore, "Cramming more components onto integrated circuits," *Electronics*, Vol. 38, No. 8, April 19, 1965.
2. N. M. Jokerst, "Hybrid integrated optoelectronics: Thin film devices bonded to host substrates," *Int. J. High-Speed Electron. Syst.*, Vol. 8, No. 2, pp. 325–356, 1997.
3. M. Vrazel, J. Chang, I. Song, K. Chung, M. Brooke, N. Jokerst, A. Brown, D. Wills, "Highly Alignment Tolerant InGaAs Inverted MSM Photodetector Heterogeneously Integrated on a Differential Si CMOS Receiver Operating at 1 Gbps", *IEEE/ECTC Proceedings*, May 2001, pp. 1-6.
4. N. M. Jokerst, M. A. Brooke, S. Cho, and S. Wilkinson, M. Vrazel, S. Fike, J. Tabler, Y. Joo, S. Seo, D. Wills, and A. Brown, "The Heterogeneous Integration of Optical Interconnections Into Integrated Microsystems", *IEEE J. of Selected Topic in Quantum Electronics*, Vol. 9, No. 2, March 2003, pp. 350-360.
5. J. H. Lau, P. Tzeng, C. Lee, C. Zhan, M. Li, J. Cline, et al., "Redistribution Layers (RDLs) for 2.5D/3D IC Integration", *IMAPS Proc.*, October 2013, pp. 434-441. Also, *IMAPS Trans. Jour. of Microelectronics and Electronic Packaging*, Vol. 11, No. 1, 2014, pp. 16-24.
6. J. H. Lau, "Patent issues of fan-out wafer/panel-level packaging," *Chip Scale Review*, Vol. 19, Nov/Dec 2015, pp. 42-46.
7. J. H. Lau, N. Fan, M. Li, "Design, material, process, and equipment of embedded fan-out wafer/panel-level packaging," *Chip Scale Review*, Vol. 20, May/June 2016, pp. 38-44.
8. C. Tseng, C. Liu, C. Wu, D. Yu, "InFO (wafer-level integrated fan-out) technology," *IEEE/ECTC Proceedings*, June 2016, pp. 1- 6.
9. C. Hsieh, C. Wu, D. Yu, "Analysis and comparison of thermal performance of advanced packaging technologies for state-of-the-art mobile applications," *IEEE/ECTC Proc.*, June 2016, pp. 1430-1438.

10. J. H. Lau, M. Li, D. Tian, N. Fan, E. Kuah, K. Wu, et al., "Warpage and thermal characterization of fan-out wafer-level packaging," *IEEE/ECTC Proc.*, 2017, pp. 595-602. Also, *IEEE Trans. on CPMT*, Vol. 7, No. 10, Oct. 2017, pp. 1729-1938.
11. J. H. Lau, M. Li, N. Fan, E. Kuah, Z. Li, K. Tan, T. Chen, et al., "Fan-out wafer-level packaging (FOWLP) of large chip with multiple redistribution-layers (RDLs)," *IMAPS Proc.*, Oct. 2017, pp. 576-583. Also, *IMAPS Trans. Jour. of Microelectronics and Electronic Packaging*. Oct. 2017, pp. 123-131.
12. M. Li, Q. Li, J. H. Lau, N. Fan, E. Kuah, K. Wu, et al., "Characterizations of fan-out wafer-level packaging," *IMAPS Proc.*, Oct. 2017, pp. 557-562.
13. E. Kuah, J. Hao, C. Chan, K. Wu, N. Fan, M. Li, et al., "Challenges of large format packaging and some of its assembly solutions," *IMAPS Proc.*, Oct. 2017, pp. 747-753.
14. S. Lim, Y. Liu, J. H. Lau, M. Li, "Challenges of ball-attach process using Flux for fan-out wafer/panel level (FOWL/PLP) packaging," *Proc. IWLPC,* Oct. 2017, pp. S10_P3_1-7.
15. E. Kuah, W. Chan, J. Hao, N. Fan, M. Li, J. H. Lau, et al., "Dispensing challenges of large format packaging and some of its possible solutions," *IEEE/EPTC Proc.*, Dec. 2017, pp. S27_1-6.
16. X. Hua, H. Xu, Z. Li Zhang, D. Chen, K. Tan, J. H. Lau, et al., "Development of chip-first and die-up fan-out wafer-level packaging," *IEEE/EPTC Proc.*, Dec. 2017, pp. S23_1-6.
17. J. H. Lau, "The future of interposer for semiconductor IC packaging," *Chip Scale Review*, Vol. 18, No. 1, Jan-Feb, 2014, pp. 32-36.
18. J. H. Lau, *Reliability of RoHS compliant 2D and 3D IC Interconnects*, McGraw-Hill, New York, 2011.
19. J. H. Lau, *Through-Silicon Via (TSV) for 3D Integration*, McGraw-Hill, New York, 2013.
20. J. H. Lau, *3D IC Integration and Packaging*, McGraw-Hill, New York, 2016.
21. Chang, H., D. Chang, K. Liu, H. Hsu, R. Tai, H. Hunag, et al., "Development and Characterization of New Generation Panel Fan-Out (PFO) Packaging Technology", *IEEE/ECTC Proceedings*, 2014, pp. 947-951.
22. Liu, H., Y. Liu, J. Ji, J. Liao, A. Chen, Y. Chen, N. et al., "Warpage Characterization of Panel Fab-out (P-FO) Package", *IEEE/ECTC Proceedings*, 2014, pp. 1750-1754.
23. Lin, P., C.-T. Ko, W. Ho, C. Kuo, K. Chen, Y. Chen, and T. Tseng, "A Comprehensive Study on Stress and Warpage by Design, Simulation and Fabrication of RDL-First Panel Level Fan-Out Technology for Advanced Package", *IEEE/ECTC Proceedings*, 2017, pp. 1413-1418.

IMPROVEMENT OF PACKAGE WARPAGE THROUGH SUBSTRATE AND EMC OPTIMIZATION

Ken Lee[1]*, Min Sung Kim[1], Jaesung Kim[1], Sangkyun Kim[2], Donghwan Lee[2], Kyunghag Jung[2]
[1]Simmtech, Co., Ltd. 73, Sandan-ro, Heungdeok-gu, Cheongju, Chungcheongbuk-do, Korea
[2]Samsung SDI, 150-20 Gongse-ro, Giheung-gu, Yongin-si, Gyeonggi-do, Korea
*Corresponding Author's Email: ken.lee@simmtech.co.kr

ABSTRACT

Package warpage significantly affects the yield and reliability of package assemblies such as POP and board level surface mount. As the size of silicon die is fixed and its thickness is usually thinned down to the minimum, the only remaining variables to control the package warpage are the substrate and EMC. Therefore, the co-optimization of the substrate and EMC is critical to improve the package warpage behavior throughout the assembly temperature profile for the best yield and reliability.

Using simple bilayer structured specimens composed of various substrates and EMC's, we measure the warpage which is dependent on their CTE, modulus, and shrinkage. Then, we verify the mechanism of package warpage by comparing measured data with simulation results. Finally, we predict the warpage of real package structure using the verified warpage mechanism of simple cases. We explain the importance of shrinkage in solving room temperature package warpage. This systematic analysis and simulation approach improves the package warpage with minimal experimental iterations.

Keywords— Package Warpage; Shrinkage; SQBC; Simulation; CTE; Modulus; EMC; Substrate

INTRODUCTION

As the thickness of semiconductor packages decreases, the package warpage becomes more critical [1,2]. Systematic approaches are necessary to understand and control the warpage because it directly impacts the reliability and yield of the assembly [3,4]. In order to resolve warpage problems within extremely short development cycle time, we use various measurement equipment such as TSM (Thermal Shadow Moiré), DIC (Digital Image Correlation)[5], Micro Tensile Tester, 3D Scanner and software simulation tools, e.g. internally developed SQBC(Simmtech Quick Balance Checker) and commercially available ABAQUS.

In order to build up the warpage analysis process, progressively complex structures are used as shown in Table 1. As the first step, simple substrate structure is used to extract the key factors for warpage behavior through measurement and simulation. The second step is to analyze the simple package structure represented by EMC and Cu bilayer structure. Based on these analyses, we finally analyze and control the warpage behavior of the actual package structure which consists of complex substrate and EMC.

TABLE 1
SAMPLES OF SUBSTRATE AND PACKAGE

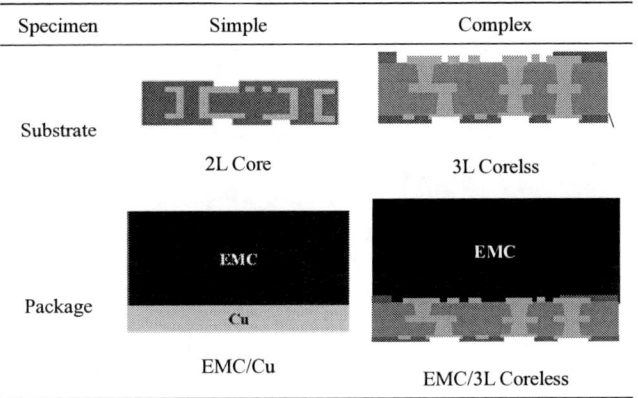

WARPAGE ANALYSIS PROCESS

The warpage analysis is carried out in the sequence of measurement, simulation, and verification as shown in Fig. 1.

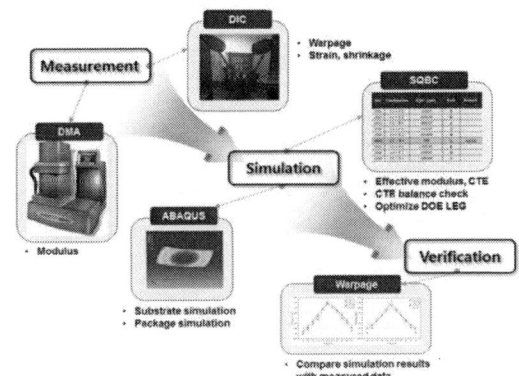

Fig 1. Process of Warpage analysis

In the measurement step, deformation of the sample and CTE (Coefficient of Thermal Expansion) of each material used in the sample are measured. The deformation consists of in-plane and out-of-plane part and they are measured in in-plane strain and warpage, respectively. CTE and in-plane strain are measured using DIC, a non-contact equipment, and the warpage is measured by TSM.

In the simulation step, warpage can be calculated by SQBC or ABAQUS based on the measured CTE and other material characteristics provided by the material manufacturers. Finally,

978-1-5386-5309-8/18 $31.00 © 2018 IEEE

we verify the key factors and mechanism of warpage by comparing the measurements and simulation results.

The verified simulation framework can be used to prioritize the DOE (Design of Experiment) legs to speed up the development of substrates. However, the simulation by ABAQUS takes undesirably long time for required fast development. SQBC can calculate the warpage based on simple input parameters much faster than ABAQUS, but still gives reasonably accurate results. For example, DOE legs are described in Table 2. ABAQUS and SQBC compute the warpage of package structure composed of 2L, EMC, Die, and 3L substrate. The results in Table 3 show that SQBC produces almost identical warpage ranking as ABAQUS.

TABLE 2
INFORMATION OF DOE LEG FOR SUBSTRATE

Type	Leg	Cu	Core	SR
2L Sub.	A 1	16/16	40	14/14
	A 2	20/20	40	10/10
	A 3	10/10	40	10/10
3L Sub.	B 1	18/15/15	30	15/15
	B 2	18/15/15	35	15/12
	B 3	15/12/12	30	15/15

Unit: um

TABLE 3
COMPARISON BETWEEN SQBC AND ABAQUS

Leg	Combination	ABAQUS Warpage (mm)	Warp Rank	SQBC (ppm)	SQBC Rank
1	A 1 + B 1	0.072	6	-1322.6	5
2	A 1 + B 2	0.059	5	-1343.7	6
3	A 1 + B 3	0.052	3	-1061.6	2
4	A 2 + B 1	0.058	4	-1111.6	3
5	A 2 + B 2	0.046	2	-1135.0	4
6	A 2 + B 3	0.037	1	-848.0	1
7	A 3 + B 1	0.168	9	-1925.9	8
8	A 3 + B 2	0.150	7	-1945.0	9
9	A 3 + B 3	0.155	8	-1681.5	7

EXPERIMENTS

Tables 4-6 show the dimensions and material properties of the samples shown in Table 1. We measured warpage and CTE of simple substrate and the curvature of EMC/Cu. 3L Coreless is selected as complex substrate to measure warpage, CTE and shrinkage of it. For three types of EMC/3L Coreless, warpage is measured respectively.

A. Simple Substrate and EMC/Cu specimen

TABLE 4
INFORMATION OF 2L CORED SIMPLE SUBSTRATE A

Layer	Thickness (um)	Portion (%)	CTE (ppm/K)	Tg (°C)	Modulus (Gpa)
SR-top	20	88	60/130		4.1/0.18
L1	15	69			
Core	60		11/16*	210	29
L2	15	72			
SR-bottom	20	92	60/130		4.1/0.18
Total(um)			130		

* measured by DIC

TABLE 5
MATERIAL PROPERTIES OF EMC AND COPPER

Type	Tg (°C)	E (Gpa)† 25°C	260°C	CTE (ppm/°C)* α1	α2
EMC-A	125	14.6	0.7	18.0	41.2
EMC-B	130	12.8	1.1	20.1	37.2
EMC-C	120	13.5	0.5	22.4	62.6
Cu		117	110	17	17

† measured by DMA, * measured by DIC

B. Real Substrate and Complex Package specimen

TABLE 6
INFORMATION OF 3L CORELESS REAL SUBSTRATE B

Layer	Thickness (um)	Portion (%)	CTE (ppm/K)	Tg (°C)	Modulus (Gpa)
SR-top	20	94	60/130		4.1/0.18
L1	18	77			
PPG	42		19/24*	210	17/9
L2	15	71			
PPG	47		19/24*	210	17/9
L3	18	68			
SR-bottom	20	93	60/130		4.1/0.18
Total (um)			180		

* measured by DIC

ANALYSIS AND VERIFICATION

A. Simple Substrate and EMC/Cu specimen

As shown in Table 7 and figures 2-3, SQBC successfully agrees with measured warpage and CTE of simple substrate.

TABLE 7
CTE BALANCE CHECK FOR 2L SIMPLE SUBSTRATE BY SQBC TOOL

CTE [ppm/K]	$T < T_{g\,SR}$	$T_{g\,SR} < T < T_{g\,PPG}$	$T_{g\,PPG} < T$
Upper	15.2	13.4	14.7
Lower	15.3	13.6	14.8
Warpage Change	Smile	Smile	Smile

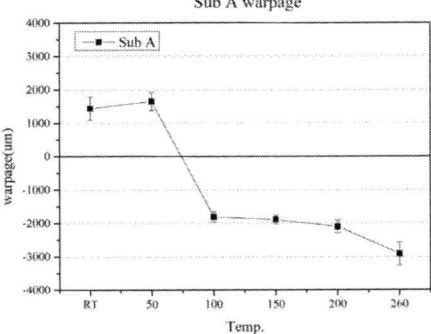

Fig. 2 Thermal warpage behavior of 2L Simple Substrate A (Positive warpage is cry and negative warpage is smile)

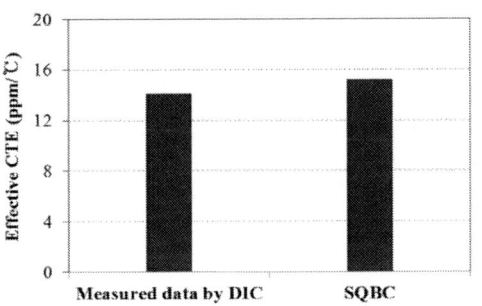

Fig. 3 Comparison of effective CTE of 2L Simple Substrate between masured data and SQBC calculation

The cure strain of EMC can be extracted from the curvature data of the EMC/Cu bilayer structure on Fig. 4 by subtracting the thermal expansion component [6] from the total curvature. As shown in Table 8, EMC has generally expansion cure strain and EMC-A and B have curing strain larger than EMC-C type.

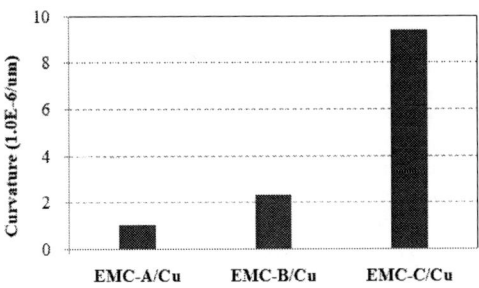

Fig. 4 Total curvature of Simple package EMC/Cu dependent on EMC type A, B and C

TABLE 8
TOTAL STRAIN, CTE MISMATCH STRAIN AND CURE STRAIN OF EMC TYPE A, B AND C ON COPPER RESPECTIVELY

Unit (ppm)	Total strain of EMC/Cu	CTE mismatch strain of EMC/Cu	EMC Cure Strain
EMC-A	263	1275	-1012
EMC-B	637	1220	-583
EMC-C	2268	2690	-422

B. Real Substrate and Complex Package specimen

Even though real substrate B has more complex structure than simple substrate sample, the warpage in room temperature is smile and consistently smile during temperature change in Fig. 5 because CTE is well-balanced between upper and lower part of real sample in Table 9.

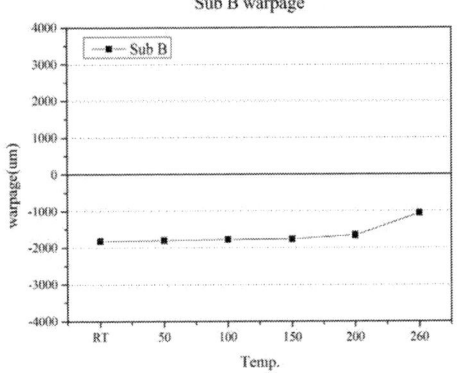

Fig. 5 Thermal warpage behavior of 3L Coreless Real Substrate B

TABLE 9
CTE BALANCE CHECK FOR 3L CORELESS REAL SUBSTRATE BY SQBC TOOL

CTE [ppm/K]	$T < T_{g\,SR}$	$T_{g\,SR} < T < T_{g\,PPG}$	$T_{g\,PPG} < T$
Upper	17.8	16.7	17.3
Lower	17.9	16.6	17.4
Difference	-0.07	0.06	0.04

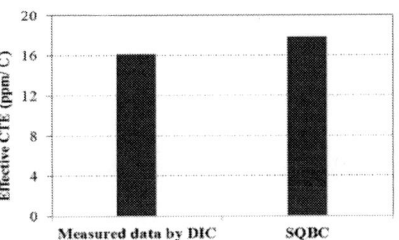

Fig. 6 Comparison of effective CTE of 3L Coreless real Substrate between masured data and SQBC calculation

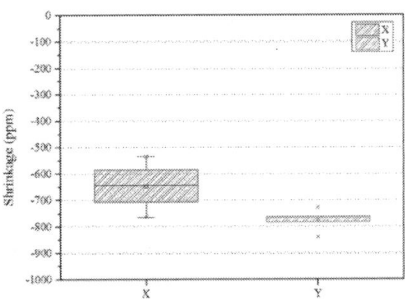

Fig. 7 Shrinkage of 3L Coreless real Substrate B

Room temperature warpage of EMC-A and B/3L Substrate sample are cry but don't agree with CTE mismatch between

EMC and 3L Substrate in Table 10. Taking shrinkage of substrate in Fig. 7 and curing strain of EMC in Table 8 into consideration, strain balances computed by SQBC are in good agreement with measured warpage in room temperature in Table 10.

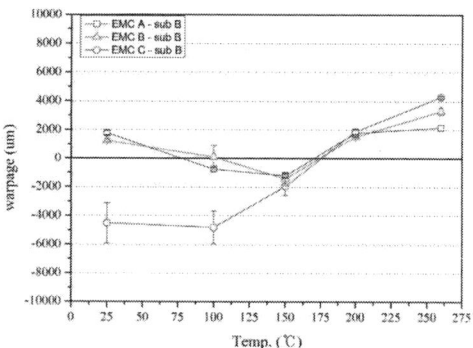

Fig. 8 Warpage of Complex package EMC/3L Coreless Subtrate B dependent on EMC type A, B and C

TABLE 10
STRAIN BALANCE CHECK FOR COMPLEX PACKAGE OF EMC/3L CORELESS REAL SUBSTRATE BY SQBC TOOL

Structure	EMC-A/Sub B		EMC-B/Sub B		EMC-C/Sub B	
Temp	α1	α2	α1	α2	α1	α2
EMC CTE	18	41.2	20.1	37.2	22.4	62.6
Sub CTE	17.8	16.6	17.8	16.6	17.8	16.6
EMC Cure	-1012		-583		-422	
Sub Shrink	760		760		760	
Upper	2858		3287		3448	
Lower	3351		3368		3372	
Difference	-493		-81		76	
Measured RT warpage	Cry		Cry		Smile	

TABLE 11
CTE COMPARISON FOR COMPLEX PACKAGE OF EMC/3L CORELESS REAL SUBSTRATE

Structure	unit	EMC-A/Sub B		EMC-B/Sub B		EMC-C/Sub B	
Temp	℃	α1	α2	α1	α2	α1	α2
EMC		18.0	41.2	21.6	37.2	22.4	62.6
Sub		16.1	19.1	16.1	19.1	16.1	19.1
Upper	ppm/℃	18.0	18.5	21.6	19.2	22.4	19.2
Lower		17.8	16.6	17.9	16.6	17.9	16.6
Difference		0.2	1.9	3.7	2.6	4.5	2.6
Warpage Change		Cry		Cry		Cry	

Table 11 shows that CTE balance by SQBC matches well with intuitive calculation comparing CTE between EMC and substrate to estimate high temperature warpage. The room temperature and high temperature warpage behaviors match well with SQBC calculations for the simple and complex samples.

The 3L coreless substrate has cry warpage in high temperature as well as low temperature (Fig 6) unlike 2L simple cored substrate (Fig. 2) because the maximum CTE mismatch is only 0.07 ppm/K compared with 0.16 ppm/K of 2L cored substrate.

As shown in Fig. 8, both of EMC-A/Sub B and EMC-B/Sub B sample have cry warpage in room temperature while EMC-C/Sub B, has smile warpage. It is because EMC-A and B has large curing strain as measured through simple EMC/Cu structure (Fig. 5) and CTE mismatch between EMC and substrate makes them smile but strain between them makes them cry larger than smile. In short, the curing strain is a warpage factor as dominant as CTE in room temperature.

CONCLUSION

In order to analyze and control the warpage of substrates and packages, basic contributing factors of warpage are extracted through simple test structures and these factors are applied to the real package structure to validate the factors and simulation models. The key contributing factor to the warpage in high temperature is CTE mismatch between substrate and EMC. However, in the room temperature curing strain of EMC and substrate are also a major factor depending on the type of EMC. In fast pace development environment, SQBC makes it possible to simulate warpage behavior of substrates and packages in tens of minutes based on the key contributing factors without using a time-consuming numerical simulation tool such as ABAQUS.

REFERENCES

[1] J.N. Burghartz, W. Appel, C. Harendt, H. Rempp, H. Richter, M. Zimmermann, "Ultra-thin chip technology and applications, a new paradigm in silicon technology," Solid-State Electron, vol. 54, pp. 818-829, 2012

[2] S. Miyajima, S. Nagamatsu, S.S. Pandey, S. Hayase, K. Kaneto, W. Takashima, "Electrophoretic deposition onto an insulator for thin film preparation toward electronic device fabrication", Appl. Phys. Lett., vol. 101, 193305, 2012

[3] S. Chung, S. Oh, T. Lee, M. Park, "Thermo-mechanical analyses of printed board assembly during reflow process for warpage prediction," Thermal, Mechanical and Multi-Physics Simulation and Experiments in Microelectronics and Microsystems (EuroSimE), 2014 15th International Conference on, IEEE, New York, NY, USA, 2014, pp. 1-5.

[4] S.Y. Yang, W.-S. Kwon, S.-B. Lee, "Chip warpage model for reliability prediction of delamination failures," Microelectron, Reliab. 2012 vol. 52, pp. 718-724.

[5] T. Lee, M.S. Kim, T.-S. Kim, "Contact-free thermal expansion measurement of very soft elastomers using digital image correlation," Polymer Testing, 2016, vol. 51, pp. 181-189.

[6] C. Kim, T. Lee, M.S. Kim, T.-S. Kim, "Warpage Analysis of Electroplated Cu films on Fiber-Reinforced Polymer Packaging Substrates," Polymers, 2015, vol. 7, pp. 985-1004.

Emerging Fine Line Panel Level Fan Out Technology

David Fang[*], Michael Hsu, CC Chang, KW Chung, Alex Liu, Irving Lin, and Daniel Fann
Powertech Technology Inc., Hsinchu, 30352, Taiwan
*Corresponding Author's Email: davidf@pti.com.tw

ABSTRACT

Advanced packaging technologies, such as 3D/2.5D IC and fan out, etc. has been changing the game rule of semiconductor chain, and blurring the boundaries between foundries and OSATs. Wafer level fan out package has been getting lots of attention since tsmc's InFO technology adopted for the flagship smartphones in 2016. Industry is working further on converting wafer level fan out to panel level fan out to achieve higher productivity and realize fine line interconnection and heterogeneous integration. Powertech Technology Inc. was the first company to setup the fine line trace and large panel level fan out solution in the world. This paper introduced the benefits, challenges and visions of this emerging package technology.

INTRODUCTION

In the past years, IC design engineers always tried to integrate more functions into one single chip and smaller dimension, the concept is commonly known as System on Chip (SoC). It is a good solution to achieve superior performance for speed-demanding applications, like CPU or GPU. However, the SOC design needs huge engineer resource, long development, and high cost. The situation has become more serious after 28nm node. Only a few devices request excellent performance can afford the extremely high cost of IC design, IP verification, and advanced wafer fab equipment, etc. Recently, many companies are revisiting the System in Package (SiP) solution to achieve a balance of performance, cost, and time to market.

In reality, not every functions required the state-of-the-art process node like 10 nm or other advanced node. Many functions can work just fine using matured wafer process generation, such as those fix-frequency devices like modem, audio, video, image processor, and WiFi, etc. Therefore, it is proposed that to split the SoC into multiple smaller discrete chips with different functions, each function can be built with the optimized wafer node and cost. The advanced packaging technologies would integrate these chips into an virtual-SoC device by reconstituting the chips back into one package. Such is the concept of System in Package.

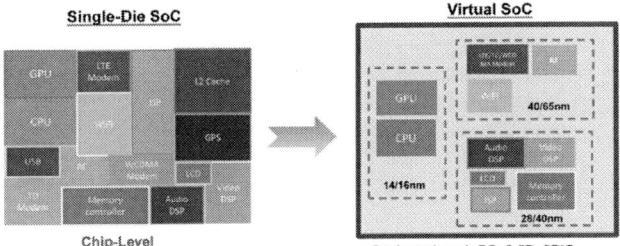

Figure 1: SiP realized virtual-SoC

BGA SiP

Conventional SiP used wire bond or FC technologies to mount the chips onto the substrate and connecting the signals of chips with long traces on substrate. The long traces will lead to large package form factors, high transmission resistance, and excessive power consumption. The performance of the chip may be deteriorated, especially for the high performance device. That is the reason the industry tried to develop other package solutions to support the high performance devices.

3D/2.5DIC

3DIC and 2.5DIC use TSVs to interconnect the signals between chips, which has been developed for years. The TSV designs on every stacked chips are required which increases the difficulties on chip design, and is very difficult for package with different chips. Besides, the 3DIC usually arranges high speed chip at bottom and its heat may cause data lost for upper thermal sensitive chips. It restricts the application of 3DIC on heterogeneous integration SiP.

The 2.5DIC uses silicon interposer as the integration bridge for different chips, and convert narrow trace pitch of chips to micro-meter level to connect with PCB substrate. It would be a good solution for heterogeneous integration but the high cost of the use of extra silicon spacer and the testing to verify silicon spacer are needed, which is a big concern to promote this technology.

Fan out on wafer level

Fan out packaging eliminated the need of substrate. The redistribution layer (RDL) replaces the substrate and it also provides better electrical performance, smaller form factor, and lower power consumption capabilities. Currently, the line/space of fan out is about 10 / 10um. Industry is working hard on developing the finer line/space solution. The fine line solution will allow chips

978-1-5386-5309-8/18 $31.00 © 2018 IEEE

to be placed closer to each other, and achieve better performance, lower power consumption, and higher IO density.

Fan out on panel level

Panel level fan out can process more chips at one time because the working panel is larger than the 300mm diameter wafer. The production efficiency gets improvement and package cost will be reduced. From our panel design, the throughput per panel is around 3 to 5 times higher than that of 300mm wafer, depending on the package size. Larger package enjoy more benefit from switching to panel. Usually the form factor of SiP is larger which shows more benefits when using this panel level fan out solution.

TABLE I. PACKAGE QUANTITY RATIO (PER PANEL)

Package size (mm²)	Wafer level	Panel level
15x15	1	3.6
20x20	1	3.9
30x30	1	4.0
40x40	1	4.5

CHALLENGES OF FINE LINE FAN OUT ON PANEL LEVEL

PTI is the first company to setup the fine line panel level fan out package production line in the world. Inevitably, it has faced some tough challenges in the development stages. Here are some to address when transiting fan out package from wafer to panel form.

Lack of Infrastructure

SEMI has defined the worldwide standards for wafer industry including wafer fabs and OSAT to follow. However, there is no standards for this emerging panel level fan out technology. It faces big challenges to decide relative equipments and materials specs and to convince equipment and material vendors to co-work on the new technology development.

We investigated tools of wafer foundries, LCD, and PCB substrate industries to benchmark the suitable tools for individual process. For example, the lithography process of the LCD industry from coating to development were adopted because of their existing experience with large size panel and fine line capability. The deposition would be utilizing both LCD industry's PVD tool and the expanded version of WLP & bumping tool for ECD. Since this technology is still at its infancy, the equipment suppliers were hesitant to jump into this market. But eventually we successfully built up the very first fully automated fine line fan out on panel level with the help of our partners. Some of the tools are our special designs.

Panel warpage control

Warpage is a result of thermal expansion coefficient mismatch between materials. It become more significant when handling the large dimension working panel because larger panel induces larger mis-match and warpage. It takes time to optimize the package structure, material property, and process parameters to reduce the warpage for each device. Computer simulation to predict the warpage for different structures and materials combination is also recommended to shorten development time.

Process tools may also help to tolerate panel warpage at processes. For instances, it can install guide rollers on conveyor for smooth transportation, and the clamper and vacuum chucking help to make the panel flat on stage so the following processes, such as lithography, can be handled without impact.

Chip shift and alignment

Chip-shift is a phenomenon frequently experienced during the molding and mold curing process. The poorly attached die may shift away from its original location due to the unmatched expansion property of each materials. Such shifting will subsequently cause the misalignment issue at patterning process and causes yield loss.

There are some possible ways to address the chip shift issue. For example, materials selection to add adhesive to prevent chip shift or adopting CTE matched materials. The process conditions optimization is also possible to improve the issue. One of the idea is to compensate the chip shift during process. There are several compensation methods: (a) intentional off-set of the chip placement, (b) real time adaptive mask alignment of stepper tool, and (c) laser direct imaging technique.

The intentional off-set method would require computer simulation to predict the trend and the amount of shifting. By off-setting the chip placement with a value that is inverse to the predicted shift, the chip will then shift back to the correct location during the molding and curing process.

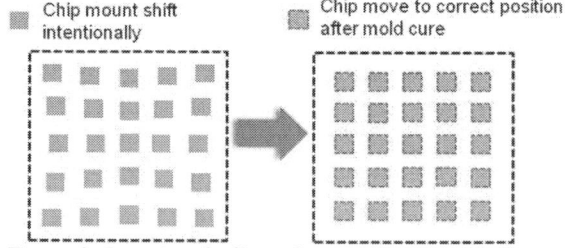

Figure 2: Intentional off-set chips at placement (left), chips will move to correct locations after mold cure (right).

The adaptive mask alignment detects chip shift tendency of certain area by AOI before lithography and feedback to stepper automatically. Stepper will adjust chuck table to make panel best fit to mask with minimum chip shift.

Another lithographic method is LDI (Laser Direct Imaging). LDI compensates individual chip shift one by one with good accuracy, however the throughput is still low for fine line application and takes time to be commercialized.

APPLICATIONS

Panel level fan out is expected to be a major package solutions for the upcoming "More than Moore" era. There are varied fan out package technologies developed to support different market applications. We can categorize the technology into three different areas: chip first, chip last, and chip middle. In the following section we will discuss four of our fan out packaging solutions: CHIEFS®, CLIP®, PiFO®, and BF²O®.

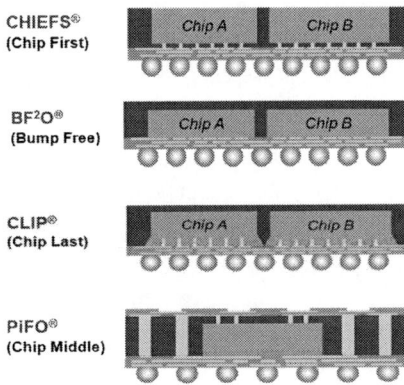

Figure 3: Varied fan out package solution

CHIEFS® (Chip Integration Embedded Fan out Solution) is a chip first solution which attaches the Cu post bumped die to the carrier and builds RDL after encapsulation. The BF²O® (Bump Free Fan Out) is also a chip first solution, but does not require pre-formed Cu-post on die. CHIEFS® is suitable for application processor, baseband processor, ASIC, and memory applications, while BF²O® is good for PMIC, audio, and RF transceiver applications.

Chip first package typically involves pick and place the chips on the carrier wafer or panel, followed by molding process. The chip may either face up or down. If the chip is facing down then the next step would be detaching the carrier and then do the RDL process on the revealed pads or Cu bumps, the backside of the chip is usually covered by molding compound as a result. If the chip is facing up, then the typical process would be mold grinding until the pads or Cu bumps reveals and do the RDL process and at last detach the carrier, the backside of the chip is usually not covered by the molding compound.

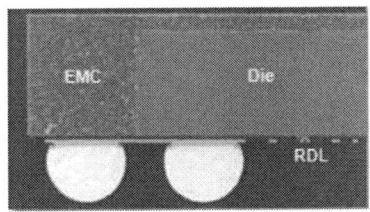

Figure 4: Fan out of chip first with face down structure

CLIP® (Chip Last Integration Package) builds RDL on panel before Cu post die flip bonding. Then followed by molding and RDL patterning process. It provides the advantage of known good RDL, die shift and misalignment issue are minimal, and finer L/S is achievable. CLIP® is ideal for devices with demand on high performance such as CPU, GPU, FPGA, or thermal sensitive devices.

Chip last, unlike chip first, will have less issue of the loss of known good die since both RDL and die can be tested before assembly. It is more possible to build finer traces because the RDL is built on a flat panel first, which is different from chip last that builds traces after molding.

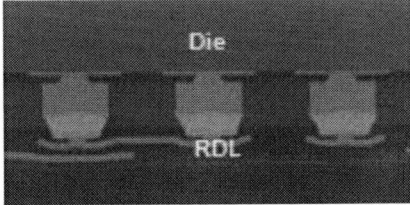

Figure 5: Fan out of chip last structure

In addition, the PiFO® (Pillars in Fan Out) is a chip middle solution. It builds pillars to connect package's top and bottom RDLs. It can provide fine electrical interconnection between chip stacking or package stacking with Fan-Out structures, which is very suitable for system level integration. The PiFO® is a recommended solution for RF module, sensor, mobile AP of PoP bottom, SiP and 3D stacking. Being chip middle process, both the chip and RDL can be tested before assembly, with an additional RDL built on top of the package, it can be either applied with an embedded antenna for RF applications, or to be stacked with another package such as memory if the application is the APU where fast access to memory is required. The Cu pillars also provide better electrical performance and reliability than conventional solder joints found in the packages with Through Mold Via (TMV).

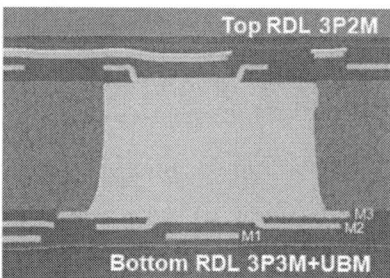

Figure 6: PiFO® uses Cu pillar to connect top and bottom RDLs

Panel level fan out is aimed to provide good solution for advanced SiP. The advanced SiP involves packaging for identical chips (homogeneous integration), and packaging for different kind of chips (heterogeneous integration) Figure 7~8 demonstrate some of the real cases of this technology.

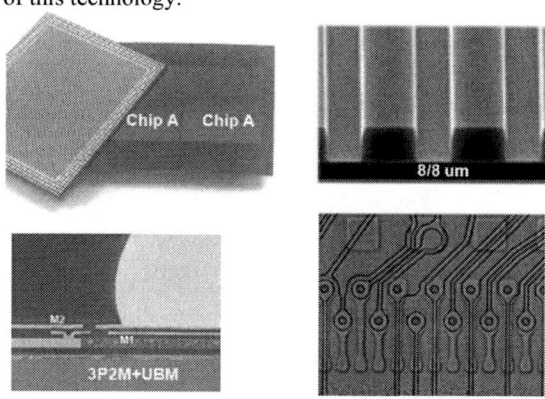

Figure 7: Homogeneous integration fan out package

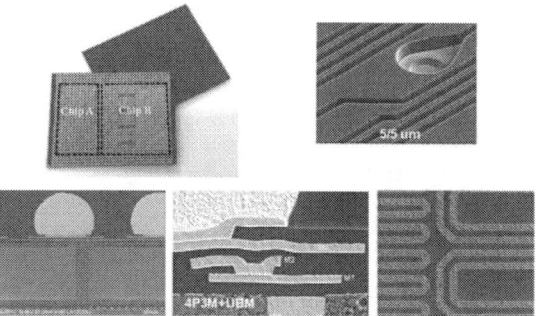

Figure 8: Heterogeneous integration fan out package

In summary, four different panel level fan out packaging architectures have been proposed in this paper, there are still many different forms that we can look into, such as multi-chip stack in a fan out package, fan out SiP with passive components, and embedded active die in the fan out RDL, compartmental EMI shielding, and antenna in package, etc. The panel level fan out solutions provide more package design freedom, smaller form factor and footprint by removing the need for a laminate substrate, and better electrical performance by utilizing the fine line RDL, and enhanced thermal performance due to lower power consumption, and much higher I/O density with high integration capability.

CONCLUSIONS

The advantage and the challenges of panel level fan out packaging has been thoroughly investigated in this paper. It is visible that diversified applications will be developed to achieve high performance, integration, and lower cost package based on this emerging technology. The panel level form enjoys the benefit of further high production efficiency. The industry's first production line has been setup and achieves good verification results. Production will be started in 2018. We expect more applications will adopt fine line fan out package for "More Than Moore" in coming years..

REFERENCES

[1] D. Yu "WLSI Extends Si Processing and Supports Moore's Law," SiP Summit, Semicon Taiwan, 2016.

[2] J. Tzou "Wafer Level System Integration," IEEE CMPT Heterogeneous Integration Technology Roadmap Taiwan Workshop, 2017.

[3] T. Strothmann "Optimizing Equipment Selection for Diverse Fan-Out Process Flows," Semicon Taiwan, 2016.

[4] B. Rogers, C. Scanlan, and T. Olson "Implementation of a Fully Molded Fan-Out Packaging Technology," IWLPC, 2013.

[5] S. Kumar and A. Pizzagalli "Status of Panel Level Packaging Manufacturing," Yole Development Report, 2016.

[6] C. Bishop, B. Rogers, C. Scanlan, and T. Olson "Adaptive Patterning Design Methodologies," IEEE ECTC, 2016.

[7] Y. Han, M.Z. Ding, B. Lin, and C.S. Choon. "Comprehensive Investigation of Die Shift in Compression Molding Process for 12 Inch Fan-Out Wafer Level Packing," IEEE ECTC, 2016.

[8] T. Nishio "Electronic Packaging Gears Up For 5G Mobile Race," ICEP, 2017.

[9] E. Jan Vardaman "FO-WLP Market and Technology Trends," ICEP, 2017.

[10] C.F. Tseng, C.S. Liu, C.H. Wu, and D. Yu "InFO (Wafer Level Integrated Fan-Out) Technology," IEEE ECTC, 2016.

PROCESS DEVELOPMENT OF FIVE- AND SIX-SIDE MOLDED WLCSP

Shuying Ma, Teng Wang, Zhiyi Xiao, Daquan Yu*

Huantian Technology (Kunshan) Electronics Co., Ltd.

Kunshan, Jiangsu Province, China

*Corresponding Author's Email: daquan.yu_ks@ht-tech.com

ABSTRACT

This paper reports the progress of 5- and 6-side (5s/6s) molded WLCSP development. A 6.3 mm × 6.7 mm image sensor test vehicle was used to develop the 5s molding process and a 0.63 mm × 0.33 mm test vehicle was used to develop the 6s molding process. Various molding compound materials and processes, including die attach, compression molding, de-flash, laser dicing, were investigated in depth. Experiments targeting at warpage reduction, chip crack prevention, and residue cleaning on solder balls were carried out. Reliability tests based on JEDEC standards were performed to characterize the package reliability and no failures have been found in the test samples using 5s and 6s molded package structures.

INTRODUCTION

Wafer Level Chip Scale Packaging (WLCSP) is widely used in consumer mobile electronics, such as image sensors, finger print sensors, RF and power IC, due to its small form factor, low cost, and good performance [1]. The conventional WLCSP structure has an active surface with polymers as protection layers, and bare silicon exposed on the side walls and backside. Such structure limits its long-term reliability performance, therefore is not widely used in the applications requiring high and long-term reliability, such as in automotive electronics. Inspired by Fan-out wafer level packaging (FOWLP) using epoxy molding compound [2-4], we used molding compound materials to provide additional protection of exposed silicon surfaces, to achieve a number of advantages. The molding cap can not only improve the strength of the chip but also protect it from external environment. Because the solder balls are partially encapsulated by molding compound, the board level reliability is also improved. In this paper, five- and six- side molded WLCSP technology is reported [5], including a detailed discussion on the development of wafer level molding process, de-flash process and backside molding process. The evaluation of suitable molding compounds (with a focus on CTE and Tg) is also presented.

5s molding process development

Fig. 1 shows the structure of a molded CSP [6]. Wafer level packaging of the image sensor using TSV technology is performed and then the package is further encapsulated by epoxy molding compound except the glass cover. The process of 3D wafer level compression molding can be divided into four steps, i.e., die-to-carrier attach, compression molding, de-flash, and laser saw, which is illustrated in Fig. 2. Firstly, the coming CIS wafer after WLP is diced into single packages. Secondly, thermal release film is laminated on 8 inch glass carrier. And the known good packages are attached to the film. Thirdly, the mounted packages are encapsulated by compression molding and the solder balls are exposed. Fourthly, the residue on the solder balls is removed by a CF_4 plasma. Fifthly, the molded wafer is released from the carrier plate. The tape is then pealed from the molded wafer. Finally, the molded wafer is singulated to individual molded packages.

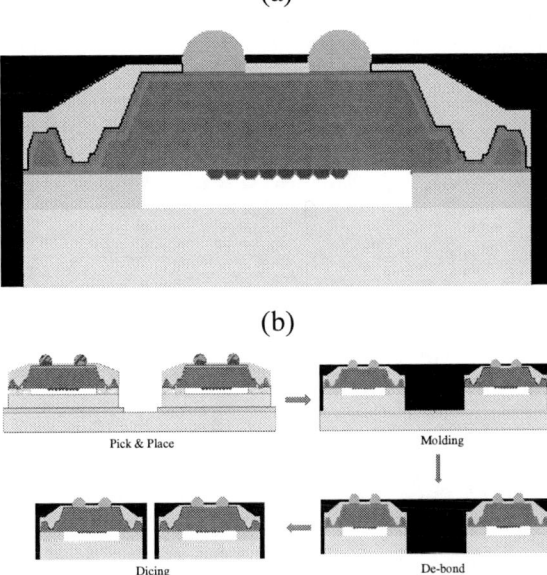

Fig.1 (a) Structure of molded package; (b) Process flow of 5s wafer level compression molding

Assembly on Carrier

A pick place tool is used for package placement on the glass carrier. An aluminum layer was deposited and patterned on the glass carrier to make the local fiducial marks. Packages were directly picked from the diced wafer and placed facing down onto the thermal release film laminated on the carrier. According to measurement results, a maximum package shift below 10μm was achieved. The shift can be further reduced by using an optimized nozzle in the future. Fig. 2 shows the image of the pick-and-place equipment and outlook of the reconstructed wafer.

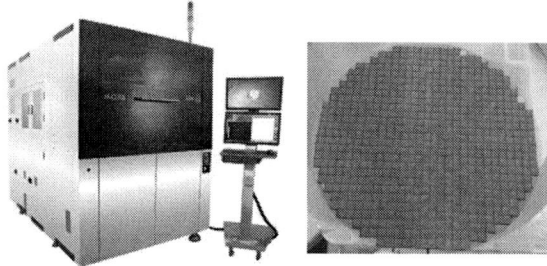

Fig. 2 Pick-and-place equipment and outlook of the reconstructed wafer

Compression Molding

There is a variety of compression molding compounds for wafer level molding from different suppliers on the market. For embedding technology, molding materials should have low curing shrinkage, low curing temperature, suitable thermo-mechanical properties, minimum die shift after molding, and good flow capability. Different material formats have been developed by suppliers, such as liquid, granular and sheet compounds. For present study, liquid molding compounds from different suppliers have been selected for wafer level compression molding process, with their key properties summarized in Table 1.

TABLE I. PROPERTIES OF MOLDING COMPOUNDS

Properties	A	B	C	D
State	Liquid	Liquid	Liquid	Liquid
Filler Loading	83%	81%	84%	76%
Filler Cut(μm)	10	5	25	10
CTE1	6	12	8	17
CTE2	8	45	12	68
Tg/℃	182	165	161	170
flexural modulus @ RT (GPa)	16	14	16	11
Viscosity (Pa s)	279	200-500	410	76
Mold condition	5min/125℃	7min/130℃	10min/130℃	10min/125℃
PMC condition	1h@150℃			

All of the molding compounds have a low molding temperature. Lower temperature molding is helpful for reducing the warpage of the molded wafer [7]. Because it has great influence on the molding cap thickness, the dosage of molding compound must be calculated

accurately. Post molding annealing condition is 150℃/1h. No mold ability problems such as void, flow mark and unfilling issues were observed after compression molding. Fig.3 is the overview after compression molding.

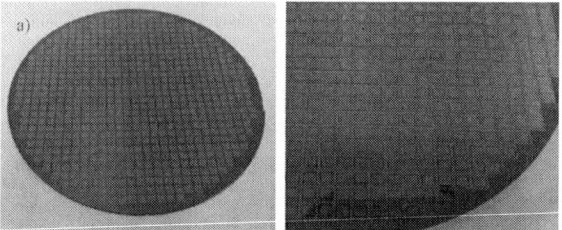

Fig. 3 Wafer after compression molding: (a) top view, (b) back view

Fig. 4 shows the warpage comparison and the results are listed in Table II. We can see that EMC B has the best performance, which has the smallest filler cut, lower CTE and flexural modulus.

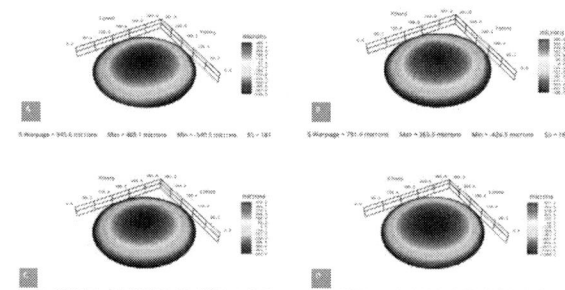

Fig. 4 warpage comparison

TABLE II. WARPAGE OF MOLDING COMPOUND

EMC	Warpage(μm)
A	0.9
B	0.8
C	1.1
D	2.5

De-flash

After molding process, we can easily find residue of epoxy and filler on solder ball, which is demonstrated by SEM. We try two cleaning methods, CO2 blast and plasma cleaning. A number of experiments have been carried out to investigate the optimum parameters. Results shows that both dry ice blast and CF_4 plasma are good candidates for epoxy cleaning on the solder ball. As shown in Fig 5, residue on the solder ball was successfully removed by the two cleaning methods.

(a)

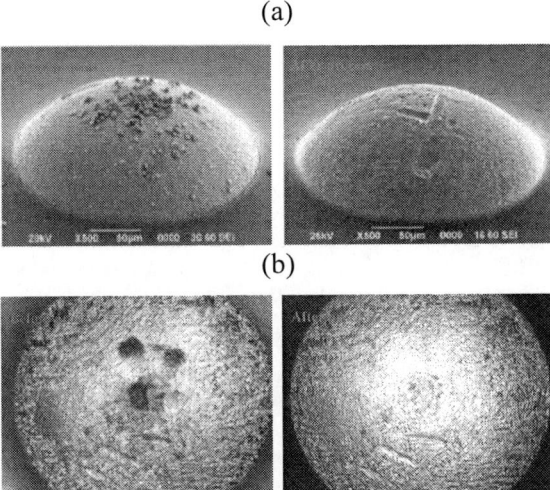

(b)

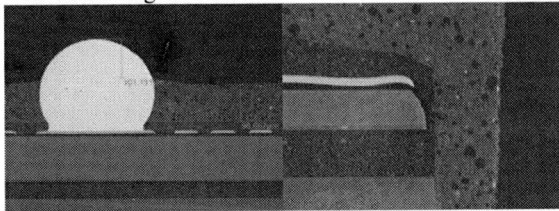

Fig. 5 (a) Dry ice blast cleaning effect; (b) Plasma cleaning effect

The cross-section of a molded package is shown in Fig. 6. No voids or cracks can be observed and solder balls are protected at the bottom and successfully exposed at the top. The thickness of molding cap and side wall coverage are well within targets.

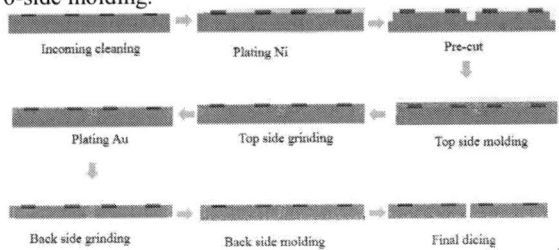

Fig. 6 Cross-section microstructure of the molded sample

6s molding process development

Based on 5s molding technology, we developed 6s molding technology. Fig. 7 shows the process flow of 6-side molding.

Fig. 7 Process flow of 6s molding

First, nickel bumps were plated on the incoming wafer after a cleaning step. Second, trenches were formed using mechanical dicing. Third, after pre-cut, the wafer was encapsulated using wafer level compression molding technology. The selection of molding compound plays a

critical role in 6s molding process development, which is related to the dicing width and depth. Fourth, the nickel bumps were exposed by top side grinding process. After that gold was plated to protect the nickel bumps. Sixth, backside was grinded to the targeted silicon thickness. Seventh, backside molding was performed to finish the six side protection. Finally, the molded wafer was singulated to individual packages. The cross-section of a 6s mold WLCSP component is showed in Fig.8.

Fig. 8 Top and cross-section images of the 6s molded sample

ACKNOWLEDGEMENTS

The authors appreciate the support by ASM Pacific Technology and E&R engineering Corp for process development.

REFERENCES

[1] J. M. Ryu, et al, US 2007/0080418 A1.

[2] John H. Lau, "patent issues of fan-out wafer/panel-level packaging", Chip Scale Review, Nov/Dec 2015, pp. 42-46

[3] CS Lee, EK Choi, UB Kang, MO Na, HC Kim, HJ Song, JS Le, MS Yoon, JH Hwang, TJ Cho, SY Kang, "A Study on Wafer Level Molding for Realizing 3-D Integration", Proc. of ECTC, 2011.

[4] Yoshimi Takahashi, Rajiv Dunne, "over molding process development for a stacked wafer level chip scale package with through silicon vias", Transactions of the Japan Institute of Electronics Packaging, Vol.5, No.1, 2012.

[5] Tom Tang, Albert Lan, Jason Wu, Joe Huang, Jensen Tsai, Jerry Li, "Challenges of Ultra-thin 5 Sides Molded WLCSP", Proc. of ECTC, 2016.

[6] Shuying Ma, et al, "3D Wafer Level Compression Molding Process Development For Image Sensor Package", Proc. of ICEPT, 2017.

[7] ToKu Hasegawa, Hidenori Abe and Takatoshi Ikeuchi; Wafer Level Compression Molding Compounds; Proc. of ECTC 2012.

THE SIGNAL INTEGRITY SIMULATION OF SIP PACKAGE USING TSV INTERPOSER TECHNOLOGY

Ge Sima*, Jian Xu,Peng Sun

The National Center for Advanced Packaging (NCAP China), WuXi 214135, China

*Corresponding Author's Email: gesima@ncap-cn.com

ABSTRACT

In this pater, we demonstrated the SI(Signal Integrity) Simulation of the SIP package for four chips using the TSV Interposer technology, which is mounted on the silicon interposer using TCB, and the interposer is then mounted on the package with Flip chip. The SI Analysis contained three parts, first is the 3D modeling and optimization analysis of TSV(Through-Silicon-Via) on the board; second is the simulation and comparison of eight different trace topologies, and the better topology has been gotten; third is the analysis of transmission and reflection characteristics by S-parameter extraction in the 0 ~ 7.5GHz bandwidth of the actual signal networks. Right choose of parameters and topologies can make the whole transmission link with better electrical performance.

INTRODUCTION

Moore's law has driven the IC industry to a billion transistor chip. One alternative path to this challenge seems to be stacked 3D ICs. In the 3D stacked chip packages, TSV and silicon interposer are main enablers that can take full advantages of 3D ICs.[1][2][4] However, a silicon substrate cannot be utilized as a good medium for signal transmission since interconnects on the silicon substrate suffer from substrate losses caused by the penetration of the electric and magnetic field. The losses and couplings may cause SI(Signal Integrity) issues. [3][5]

In this paper, the SI Simulation of the SIP package for four chips using the TSV Interposer technology as shown in Figure 1 has demonstrated.

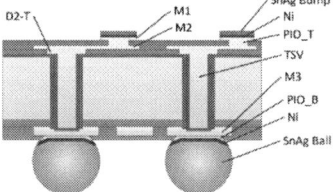

Figure 1: The stack up of the interposer

SIGNAL INTEGRITY FOR INTERPOSER

The SI simulation for interposer contains three parts. First is the 3D modeling and optimization of TSV which is the main source of impedance discontinuities of the interposer structure. Second is the trace topology optimization of some key signals (High-speed, high-frequency signals).And the last is the final transmission characteristics and reflection characteristics by S-parameter extraction of the actual signal networks.

TSV Simulation

The TSV is a core structure that provides a vertical interconnection to resolve the mismatching of interconnect density between advanced ICs and the coarse pitch of organic substrate.

However, the TSVs suffer from substrate losses caused by the penetration of the electric and magnetic field. The environment that thousands of interconnections are compactly packed with fine pitch on a lossy silicon substrate may cause crosstalk. It's necessary to simulate the electric characterization during the SI analyses of interposers.

We used the copper-filled type in this paper for the processing line we used. Figure 2 shows the 3-D model of the TSV. The characteristic impedance of the transmission line in the model is 50

Figure2: The 3D-model of TSV

By changing the radius of the inner-copper and the thickness of silicon-oxide, we can get different IL as figure 3 and figure 4 shown.

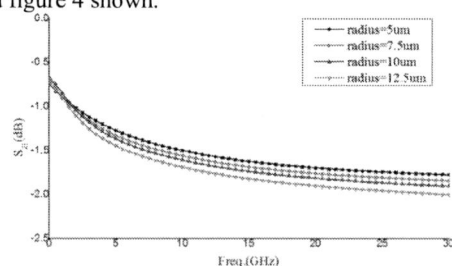

Figure3: The IL of TSV with different Cu-radius

978-1-5386-5309-8/18 $31.00 © 2018 IEEE   452

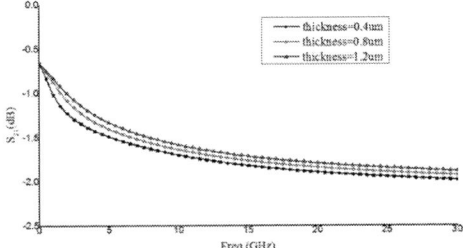

Figure4: The IL of TSV with different SiO₂-thickness

From figure 3, we can see that the IL decreases while radius of TSV decreases. In normal the characteristic impedance of the TSV is below 50Ω, the decrease of radius will increase the characteristic impedance of TSV which is better for high-speed or high frequency signal.

From figure 4, we can see that the IL decreases while the thickness of silicon-dioxide increases. Because of the semiconductor silicon itself, so there is a certain transmission loss, increase the thickness of silicon-oxide can effectively increase the isolation between signal transmission channel and the silicon, thus reduce the transmission loss in silicon and get better transmission performance.

TRACE Optimization

In high-speed, high-frequency situations, the signal performance of the different trace topologies will be different. In this part, we present 8 different topologies as figure 5 shown.

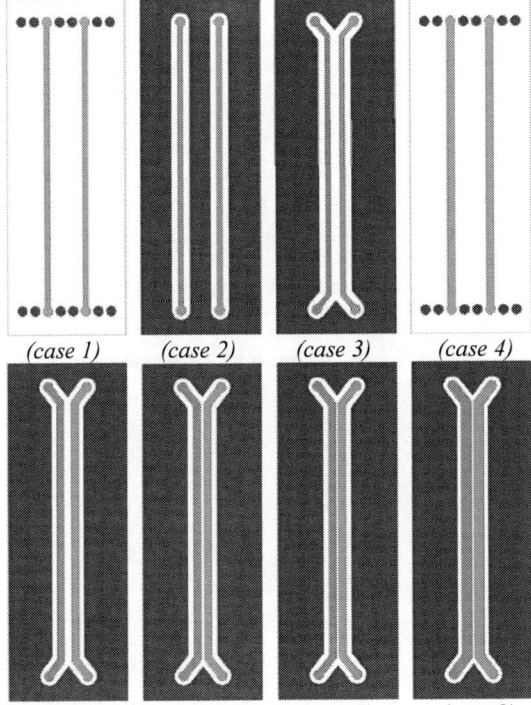

(case 1)   (case 2)   (case 3)   (case 4)

(case 5)   (case 6)   (case 7)   (case 8)

Figure 5: The eight different trace topologies

Case1 is the original trace, case2 ~ 7 make some topological changes based on case1, including line width, line spacing and ground plane, as shown inTable1

TABLE I.   THE DIFFERENCE BETWEEN CASE 1 TO 8

case	Trace width	Trace space	Ground
1	20um	/	No
2	20um	/	Yes
3	20um	40um	Yes
4	30um	/	No
5	30um	30um	Yes
6	30um	15um	Yes
7	30um	10um	Yes
8	40um	10um	Yes

Case 1 is the original trace topology；case2

TABLE II.   THE SELF-L、MUTUAL-L AND LOOP-L OF THE 8 DIFFERENT TRACE TOPOLOGIES

case	Self-L	Mutual-L	Loop-L
1	1.0249	0.3652	0.6597
2	0.9211	0.2877	0.6334
3	1.0252	0.4625	0.5628
4	0.9106	0.3726	0.5380
5	0.8394	0.3432	0.4962
6	0.8259	0.3550	0.4709
7	0.8415	0.3881	0.4534
8	0.8258	0.3552	0.4706
Net1：BT_TRX_P；Net2：BT_TRX_N			

The loop inductance is the difference value between the Self-inductance and the Mutual-inductance because of the opposite current magnetic field of the differential pair.

From table Ⅱ,we can get some conclusions:

a) Adding Ground around the signal line can reduce the loop area around the signal and the return path, thus reducing the self-inductance of the signal line.

b) Adding Ground around the signal linecan increase the coupling between signal and the ground, thus reducing the mutual-inductance between the differential pair.

c) Shortening the distance between the differential pair can significantly increase the mutual inductance, and increases the trace length slightly due to the decrease of the pitch which will increase the self-inductance; normally, the loop-inductance will decrease because the effect of the space on mutual inductance is more obvious.

d) Increasing the line width will increase the current distribution of the cross-section, reducing the number of magnetic lines, resulting in decreased signal self-inductance.

For self-inductance: case3 > case1 > case2 > case4 > case7 > case5 > case6 > case8.

For mutual-inductance: case1 > case2 > case3 > case4 > case5 > case7 > case6 > case8.

For both self-inductance and mutual-inductance, the topology of case 8 is the best choice, but it will take up more board space. We should choose the right topology based on the actual restrictions on self-inductance and mutual inductance.

S-parameter Extraction

S-parameter is the network parameter based on the incident wave, reflected wave which is suitable for high-speed, high-frequency analysis. Impedance discontinuity is the main cause of SI problems. Usually in the design of the package substrate, we will finally extract the S-parameters to understand the circuit performance of the entire transmission link.

Figure 6 shows the structure of the package including TSVs and RDL which discussed in the previous sections.

Figure 6: The structure of the package

Here we take the signal "FM_LANT_IPD" as an example(the red trace in figure 6), and the RL(S11,Return Loss) and IL(S21,Insertion Loss) of differential pairs are as figure7 shown.

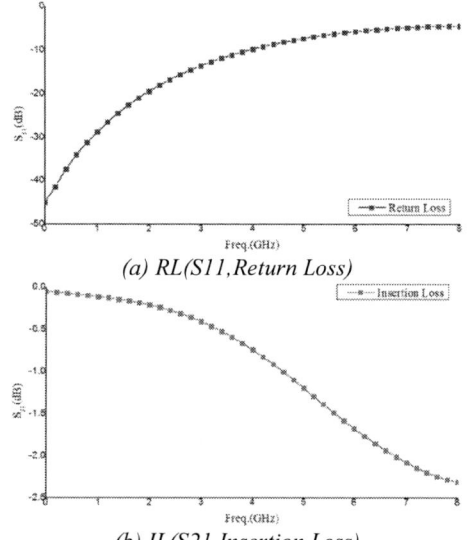

(a) RL(S11,Return Loss)

(b) IL(S21,Insertion Loss)

Figure 7: The S-parameter of the package

We can see that in the frequency range of interest(0~3GHz), S11 of the signal path is below -10 dB and S21 is within -0.5dB. Due to the introduction of TSV, bump, etc., the RL and IL performance of the link have declined(Normally without TSV,bumps,etc.,the IL will be within -0.5dB between 0~8GHz). However, the insertion loss is still less than 3dB in the simulated frequency range to meet the requirements.

CONCLUSION

In the study, we mainly carried out the SI simulation of interposer including the model and simulation of TSV, the optimization of trace topologies, and the extraction of the final S-parameter. Right choose of parameters and topologies can make the whole transmission link with better electrical performance. And while the radius is smaller and the silicon dioxide is thicker, the performance of the TSV is better for the better characteristic impedance continuity and isolation. Finally, the electrical characteristic of the transmission link meets the requirements with the RL is below 10dB and the IL is within -3dB between 0~3GHz.

ACKNOWLEDGEMENTS

The authors acknowledge the support of the National Science and Technology Major Project (Project No. 2014ZX02501-002). This work is financially supported by National Center for Advanced Packaging (NCAP China). The authors would also like to thank the help of their workmates.

REFERENCES

[1] Kihyun Yoon,Gawon Kim,Woojin Lee,Taigon Song,Junho Lee,Hyungdong Lee,Kunwoo Park, and Joungho Kim, "Modeling and Analysis of Coupling between TSVs,Metal,and RDL interconnects in TSV-based 3D IC with Silicon Interposer," Electronics Packaging Technology Conference, pp. 702–706, 2009.

[2] Joohee Kim,Jun So Pak, Jonghyup Cho,Eakhwan Song,Jeonghyeon Cho,Heegon Kim,Taigon Song,Junho Lee,Hyungdong Lee,Kunwoo Park,Seungtaek Yang,Min-Suk Suh,Kwang-Yoo Byun, and Joungho Kim, "High-Frequency Scalable Electrical Model and Analysis of a Through Silicon Via(TSV)," IEEE Transaction on components,packaging,and manufacturing technology , vol. 1, pp. 181–195, February 2011.

[3] Carmen Bachiller,Hector Esteban,Vicente Enrique Boria Esbert,Angel Belenguer Martinez and Jose Vicente Morro, "Efficient Technique for the Cascade Connection of Multiple Two-port Scattering Matrices," IEEE Trans. On Microwave Theory and Techniques, vol.55,pp.1880-1886, September 2007.

[4] Tseshih Sung,Kevin Chiang,Daniel Lee and Mike Ma, "Electrical Analyses of TSV-RDL-Bump of Interposers for High Speed 3D IC Integration,"IEEE,pp865-870,2012.

[5] Ren Xiaoli,Pang Cheng,Qin Zheng,Ping Ye,Jiang Feng,Xue Kai,Liu Haiyan,and Yu Daquan "Design,analysis and test of high-frequency interconnections in 2.5D package with silicon interposer," Journal of Semiconductors, vol.37, No.4,2016.

EFFECTIVE EMI SHIELDING FOR SEMICONDUCTOR PACKAGES THROUGH NOVEL CONFORMAL COATING

Xinpei Cao, Andrew Sun, Dan Maslyk, Junbo Gao, Qizhuo Zhuo, Jinu Choi

Henkel Electronic Materials, LLC.
Irvine, California USA
xinpei.cao@henkel.com

ABSTRACT

Enhanced wireless, memory, and data processing components enable innovative and optimal user experiences largely influenced by the growth of Internet of Things (IoT) applications. However, these radio-frequency (RF) emitting devices require effective isolation to limit the propagation of their interference to neighboring components to protect the end device from performance degradation. Consequently, advancement in electromagnetic interference (EMI) shielding technology is a critical factor as end products and components move toward miniaturization with higher density designs and increased function at higher speeds.

So far, the industry norm for EMI shielding has been to use custom-designed metallic cans. However, since cans consume large, valuable space and require complex and inflexible board layouts, alternative solutions, including physical vapor deposition (PVD) and spraying, are gaining interest.

To address these challenges, a unique material with rheological properties that provides reliable performance in stressful electronic conditions has been developed. The material can be applied at room temperature in a coating as thin as 3-5μm and provides high shielding effectiveness, uniform conformal coating coverage, and excellent adhesion to untreated organic surfaces. Furthermore, this EMI coating material technology allows easy production scalability and design flexibility with minimal cost of ownership.

Keywords: Package-level EMI shielding, novel conformal shielding, ultra-thin spray-coated metal ink, shielding effectiveness

INTRODUCTION

Electronic devices continue to push the limits of miniaturization, increased integrated functions, higher speeds, and lighter weights. With operating signals and harmonics of high-density RF modules and other various sensitive components placed close to each other, unwanted cross-talk among components causes electrical performance degradation and failures. Therefore, electromagnetic compatibility (EMC) is addressed [1-5] and is enabled by minimizing noise at the component-to-component level.

Since the conventional metal enclosure method introduces limiting factors for high-density designs such as space, height, and weight, package-level solutions such as PVD have gained greater traction. However, this metal deposition method is restricted by high process and operational complexities such as surface treatment, restricted material selection, strict maintenance requirements, low throughput, and high implementation costs.

This paper will discuss the development of a EMI adhesive coating, sprayable technology designed for use with cost-effective spray-coating equipment and which provides extremely high conductivity in a simplified process. The material properties, coating behavior, and reliability performance were characterized. Near-field radiation measurement was used to evaluate material shielding performance at the component level.

EXPERIMENT DETAILS

Spray-Coated Material

A Henkel product, LOCTITE® ABLESTIK® EMI 8880S was used for the coating study. The material is a silver-based ink designed to achieve a metal-like coating to address the high electrical conductivity requirements of package-level EMI shielding. The physical properties of the novel material are shown in Table 1. The volume resistivity of the coating material is 7.9×10^{-6} Ω·cm, which is in the same order as pure copper.

Spray-Coating Application

The material can be easily spray-coated onto epoxy molding compound surfaces to form a uniform thin coating layer on the top and four sidewalls. While the material is compatible with any type of spray technology, this evaluation was conducted using Ultrasonic Systems, Inc.'s (USI's) nozzleless ultrasonic spray coating technology, which can tightly control coating thickness with good uniformity and minimal material waste due to its precision digital dispensing mechanism.

This spray technology is currently used for fuel cell, solar, display, and other applications requiring fine coating layers with precision dispensing. The machine is operated in room temperature, in air, with adjustable parameters, and without moving internal parts in the spray head for stable liquid delivery. It uses frequency-dependent ultrasonic energy to atomize the material into fine uniform droplets, and shapes it with low pressure air flow for a flat "sheet-like" spray pattern.

The packages were placed on a carrier, as shown in Fig. 1, with a pitch corresponding to a 1:1 ratio of the package height, maximizing the number of components within the spray area to achieve high units per hour (UPH) for optimized production throughput. The spray head was programmed to maintain a tilted angle to enable maximum sidewall coverage on all four sides using four passes as shown in Fig. 2. Material flow rate and spray head speed were adjusted to achieve top coating thicknesses variations.

Fig. 1. Overall spray-coating application process

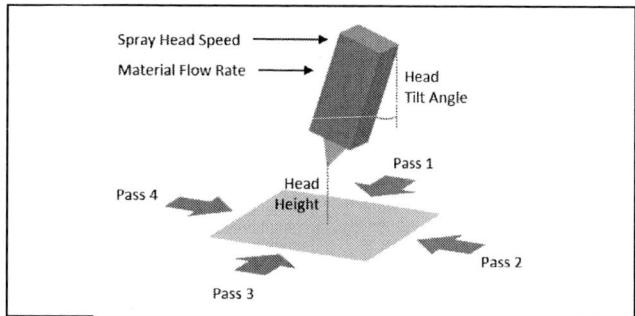

Fig. 2. Spray head and passes for sidewall coverage

Material Adhesion Measurement

Material adhesion measurements were conducted using the ASTM D3359 standard applying cross hatch cuts, then peeling with pressure-sensitive tape.

Material Shielding Effectiveness Measurement

To analyze comparative bulk material shielding performance, far-field measurements were used following the ASTM D5568-14 and ASTM D4935-10 standards.

For representative application-level performance measurements, a near-field testing method was used which aligns with the IEEE 299 standard. This method involves measuring the magnetic field of custom-designed radiating sources at select frequencies.

RESULTS

Material Properties

The LOCTITE ABLESTIK EMI 8880S is nearly organic-free after cure, and therefore provides very high conductivity, >1 order higher than conventional conductive inks and in the same order as pure copper. The material also utilizes unique fillers to increase magnetic permeability which enhances magnetic attenuation performance.

As shown in Fig. 3, far-field measurements showed the novel material can achieve up to 80 percent higher performance compared to traditional organic-based conductive inks at a similar thickness of 3-5μm enabling better laser mark visibility, thickness uniformity, and less material usage.

Furthermore, this EMI adhesive coating material is designed to have limited flow capability in its liquid state to prevent backside contamination at the bottom of the package during the spray process, confining the coating only up to the edge of the bottom sidewall surface.

Compared to metal deposition methods such as PVD, the new ink formulation approach enables greater flexibility through material design, allowing various desired functions that can be adapted to specific process and application needs.

Table 1. Properties of novel material

Material Name	LOCTITE ABLESTIK EMI 8880S
Technology	Electrically Conductive
Application Method	Spray
Volume Resistivity (Ω·cm)	7.9×10^{-6}
Coating Thickness Range (μm)	3 ~ >40
Viscosity, 5rpm (cps)	550
Thixotropic Index	1.3
Curing Temperature and Time	175°C, 1 hour, in air
Far-field Shielding Effectiveness (dB)	90 with 3-5μm thickness
Adhesion on EMC, Cu (ASTM D3359)	5B (0% peel)

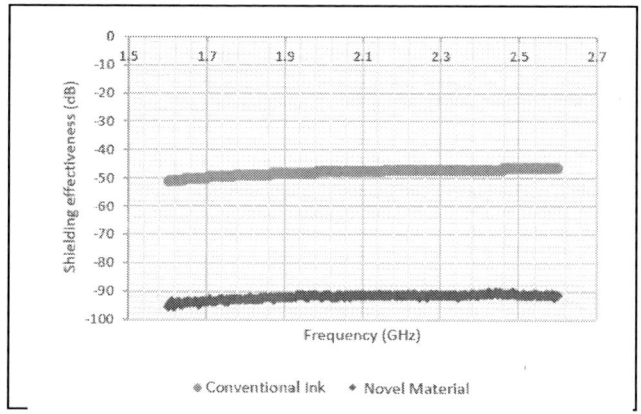

Fig. 3. Far-field shielding effectiveness of novel material vs. conventional ink at 1.6~2.6GHz

Coating Quality

Full uniform coating coverage was observed on the top and sidewall surfaces of each package, including the bottom of sidewalls and corners, which is required to achieve optimal shielding performance. (Fig. 4) The top to sidewall thickness ratio of 1:0.5~0.6 was achieved with a pitch-to-height ratio of 1:1. Further increasing the pitch between packages enabled a ratio of 1:0.7~0.8.

The fine coating surface of the material at minimal thicknesses allows for good recognition of laser markings as shown in Fig. 5. Furthermore, coating thickness was controlled by adjusting the spray head movement speed and material flow rate, yielding a thickness range of 3μm and above while achieving a deviation within +/- 1μm. (Fig. 6)

978-1-5386-5309-8/18 $31.00 © 2018 IEEE

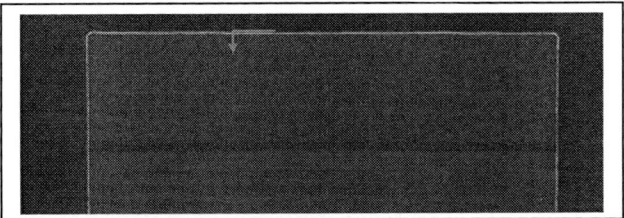

Fig. 4. Cross-section of coated package

Fig. 5. Laser marking after 5μm coating

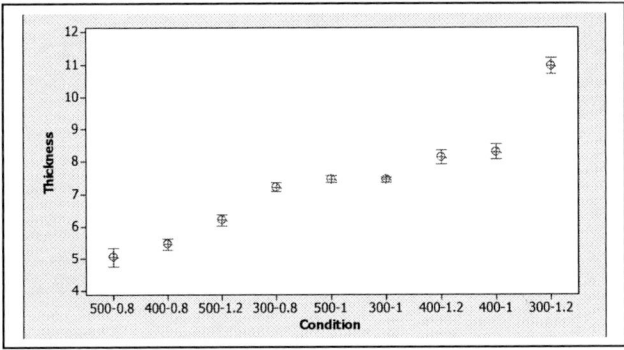

Fig. 6. Coating thickness (μm) vs. ultrasonic spray conditions (coating speed - mm/sec, flow rate – mL/min)

Adhesion Performance

Adhesion performance of the material was tested using the ASTM D3359 standard with applying cross hatch cuts on the surface of the coated substrate then peeling the surface with pressure-sensitive tape, Elcometer 99, over the cuts to analyze the remaining material to classify between the range of 0B (>65% peel) to 5B (0% peel) as shown in Fig. 7.

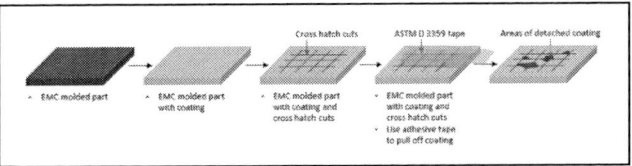

Fig. 7. Adhesion test procedure using ASTM D3359

The test results of the coated material showed good adhesion on both epoxy mold compound and copper surfaces without delamination. (Fig. 8 and Fig. 9).

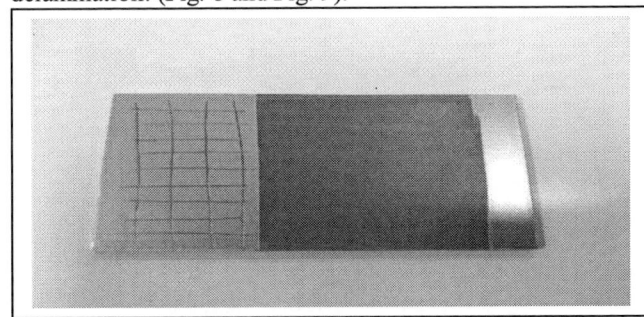

Fig. 8. Material adhesion on copper

Reliability Performance

Critical reliability tests were conducted to assess semiconductor packages, coated with EMI adhesive coatings ability to withstand the severe conditions experienced during the assembly process of semiconductor packaging, by examining adhesion level following the ASTM D3359 standard.

The results showed a 5B adhesion rating at 5μm and 10μm after Pressure Cooker Test (PCT). After High Temperature Storage (HTS) testing, the material also passed 5B level adhesion after 1000 hours at 150°C. In addition, semiconductor packages coated with EMI adhesives were tested for Moisture Sensitivity Level 3 (MSL3) followed by temperature cycling (TCT) using the JEDEC JESD22-A104 standard with results also passing 5B level for up to 1000 cycles at -40°C to 150°C as shown in Fig. 9.

978-1-5386-5309-8/18 $31.00 © 2018 IEEE

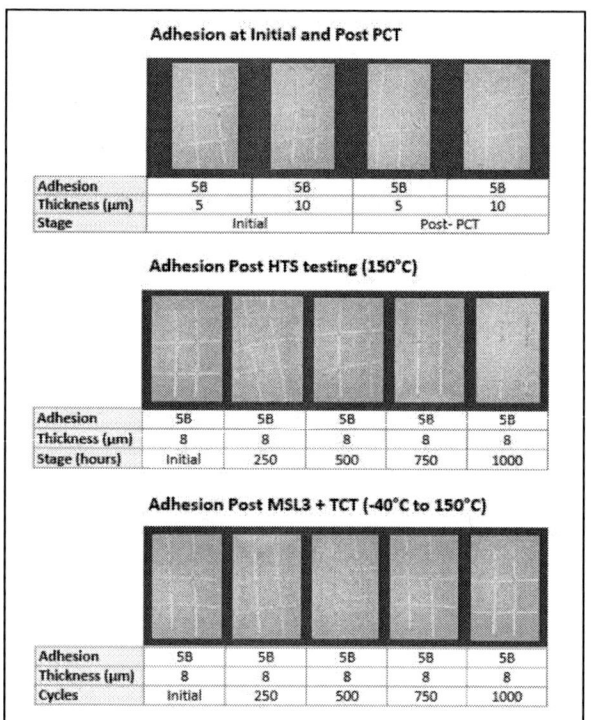

Fig. 9. Material adhesion post reliability testing

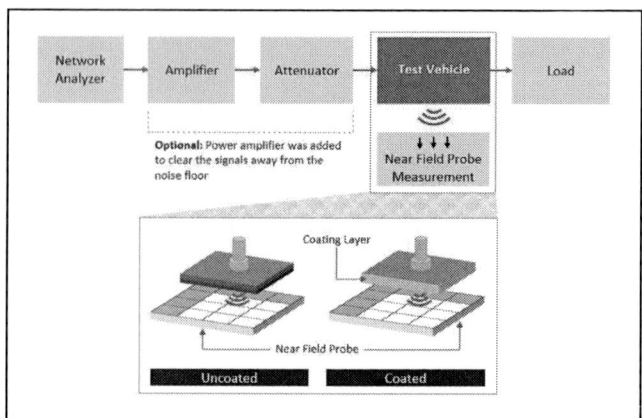

Fig. 10. Near-field shielding effectiveness (SE) test method

Far-Field Shielding Effectiveness

ASTM D55618-14 is a standard test method for measuring relative permittivity and magnetic permeability of solid materials at microwave frequencies using waveguides. The ASTM D4935-10 method provides a procedure for measuring shielding effectiveness of planar material for a plane using far-field waves.

LOCTITE ABLESTIK EMI 8880S demonstrated higher bulk shielding effectiveness compared to traditional organic-based conductive inks at a similar thickness as shown in Fig. 3. Far-field EMI shielding effectiveness measurements show the material can achieve up to 90dB at 1.6GHz to 2.6GHz with a cured coating thickness of 3-5µm.

Near-Field Shielding Effectiveness

While far-field radiation measurement was used to analyze comparative bulk material performance, near-field testing methods were developed to capture the representative shielding performance of the sprayed coating at the component level. This method, which aligns with the IEEE 299 standard, uses a custom radiating source and a scanner comprised of an array of loop antennae for measuring the magnetic field at a fixed distance for various frequencies as shown in Fig. 10.

Enabled by the material's high conductivity and optimized composition, sprayed top coating thickness as low as 5µm was able to achieve comparable or higher shielding effectiveness as PVD at a similar thickness as shown in Table 2. A thicker coating, controlled with head speed and flow rate, provided increased shielding performance at 1GHz and 5GHz.

Table 2. Near-field SE of spray by thickness vs. PVD

Method	Head Speed (mm/sec)	Flow Rate (mL/min)	Top Thickness (µm)	Side Thickness (µm)	SE at 5GHz (dB)	SE at 1GHz (dB)
Spray	400	0.8	5	3	48.2	45.9
	300	0.8	7	3.5	48.2	45.2
	400	1.2	8	4	51.3	46.6
	300	1.2	10	5	52.7	47.8
PVD	N/A	N/A	3	1	43.0	41.2

Furthermore, spatial radiation measurements of non-shielded, PVD-shielded (3µm), and spray-shielded (4µm) components showed that spray-coating provided better shielding. (Fig. 11)

Operational Benefits of Spraying

The cost of ownership required to implement and operate a typical PVD line for package-level shielding is high due to its equipment cost, which can range up to several million US dollars, and large physical footprint that can consume up to 35m². In contrast, a typical spray process requires only a fraction of the cost at less than half of a million US dollars and utilizing significantly less space -- only up to five m² -- allowing package-level shielding technology to become a much more scalable and affordable solution.

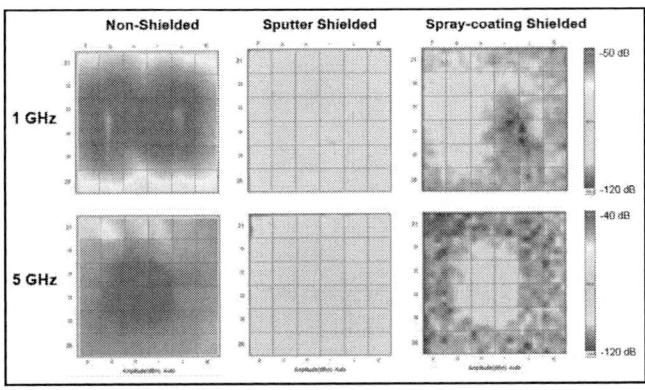

Fig 11. Near-field shielding effectiveness measurement comparison at 1GHz and 5GHz

CONCLUSION

The evaluation of the new spray-coated metal ink technology showed that it meets the stringent requirements for semiconductor package-level shielding. The material resulted in up to 80 percent higher shielding effectiveness than conventional conductive ink with optimal rheology for fine atomization and thin coating. Using nozzleless ultrasonic spray, parameters were characterized to control the coating thickness as thin as 3μm with good sidewall coverage, good uniformity, and good laser marking visibility without backside contamination.

Moreover, the novel material showed robust adhesion on epoxy mold compound and copper, passing the 5B level for time zero, post-PCT, post-HTS, and post-MSL3-TCT testing. In addition, the coating, as thin as 4-5μm, showed near-field shielding effectiveness comparable or better to PVD. With lower cost of ownership and higher throughput compared to PVD, spray-coating is a highly scalable and practical process for conformal shielding of semiconductor packages.

References

[1] J.V.Hoang, R.Darveaux, T.Lobianco, Y. Liu, and W.Nguyen, "Breakthrough Packaging Level Shielding Techniques and EMI Effectiveness Modeling and Characterization," IEEE proceeding, Electronic Components and Technology, pp1290-1296. Vol. 66, 2016

[2] E. Diaz-Alvarez ; J.P. Krusius, "Package and chip-level EMI/EMC structure design, modeling and simulation," IEEE Proceeding, Electronic Components and Technology, pp873-878, Vol. 49, 1999

[3] T. Sudo ; H. Sasaki ; N. Masuda ; J.L. Drewniak, "Electromagnetic interference (EMI) of system-on-package (SOP), IEEE Transactions on Advanced Packaging , Vol.27, Issue 2, pp 304-314, 2004.

[4] N. Karim, , "Electromagnetic Shielding for RF and Microwave Packages," RF and Microwave Microelectronics Packaging II pp 43-62, Springer, 2017.

[5] Namhoon Kim ; Leo Hongyu Li ; Sam Karikalan ; Reza Sharifi ; Henry Kim , "Package-level electromagnetic interference analysis," IEEE proceeding, Electronic Components and Technology, pp 2119-2123. Vol. 64, 2014

LATEST DEVELOPMENTS OF MOLDING COMPOUND MATERIAL FOR POWER SEMICONDUCTORS

Yusuke Tanaka*, Itaru Watanabe

Sumitomo Bakelite (Suzhou) Co., Ltd., Suzhou, China

*Corresponding Author's Email: yusuke_tanaka@sumibe.co.jp

ABSTRACT

Toward realization of "Low-Carbon Society", power semiconductors are becoming extremely important to use energy effectively. In recent years, application of SiC (Silicone Carbide) and GaN (Gallium Nitride) have been actively studied as a next generation semiconductor element. High heat resistance of epoxy mold compound (EMC) is required for SiC or GaN power devices. In this time, we found the new type resin structure based on the rigid biphenylene molecular structure which can achieve both high Tg and high heat resistance. EMC coefficient of thermal expansion (CTE) is also a key factor for reliability and warpage performance in a power module package.

INTRODUCTION

As the trend of energy saving in recent years, interest of power devices is increasing. So far power device is mainly used in home appliances such as air conditioners with relatively low power consumption and servers, but now it extends to the automotive industry and railway fields that require high power. By replacing the current Si devices with the next generation power semiconductors such as SiC or GaN, power loss is reduced, energy saving is improved efficiently and operation at a high temperature of 200 degC or more become possible. In order to apply those next generation power element, high heat-resistance materials are required.

In this study, we investigated factors of epoxy resin molding compound (EMC) affecting power devices. EMC is originally used for protecting semiconductor elements from mechanical external forces such as shock and pressure, humidity, heat and ultraviolet rays. Now we are working on high heat resistance of EMC for semiconductor encapsulation targeting operation at high temperature such as 175 degC and 200 degC (Table I). In order to operate a power device on which SiC or GaN is used in a high temperature environment, it is required that the characteristic change of EMC for semiconductor encapsulation is also small at high temperature.

TABLE I. Road map for power semiconductors

	2015		2020	2025
Devices	Si		SiC, GaN	
Power Density	5-20 kW/L	20-30 kW/L	30-80 kW/L	100 kW/L
T_j (max)	150degC	175degC	200degC	250degC

RESULTS AND DISCUSSION

Approach to high heat resistance of epoxy resin / hardener for EMC

Generally, EMC with high Tg is high bonding reliability at high temperature. If the EMC's Tg is higher than junction temperature (T_j), the molecular motion of the polymer chain in EMC is suppressed, and the ions contained in EMC are difficult to move.

On the other hand, as a disadvantage EMC with high Tg is high thermal-decomposability. To examine the more suitable resin structure for high heat resistance, we checked weight loss of EMC depending on various resin combinations after long-term storage up to 5,000 hrs at 200 degC in air.

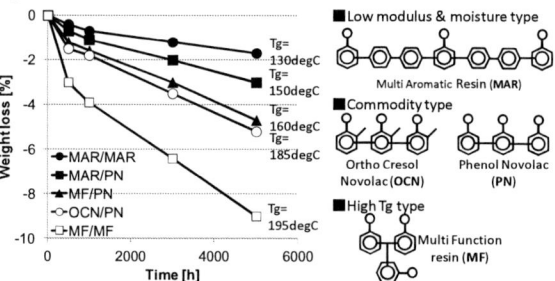

Figure 1: The time curse of EMC weight loss at 200 degC

Figure.1 depicts the time course of weight loss of EMC molded sample. Small weight loss was observed in the combination of Multi Aromatic Resin (MAR) type epoxy resin / hardener with low Tg. Its weight loss is within 2% after 5,000h. This is because the rigid biphenylene molecular structure is difficult to thermally decompose. As of Multi Function (MF) type epoxy resin / hardener with the high Tg shows large weight loss after long-term high temperature. We reasoned that the combination of MF type epoxy resin / hardener has many cross-linkers and those are easily to be decomposed by heat.

As shown in Figure 2, we observed the cross section of the molded samples after heat treatment at 200 degC for 1,000 hrs. A deteriorated layer of about 100 μm was seen from the surface layer in the combination of MF type epoxy resin / hardener with high Tg. In the degraded layer, voids were caused by filler-drop-off during sample rupture due to thermal decomposition of the resin matrix.

Table II shows the results of elemental analysis by EDS. A large composition change is not observed in the

978-1-5386-5309-8/18 $31.00 © 2018 IEEE

inside portion, but compositional change of C and O was observed in the surface layer. It is suggested that the resin layer oxidized and deteriorated in the surface layer. On the other hand, deterioration is hardly observed in the combination of MAR type epoxy resin / hardener with low Tg. This is also probably because the rigid biphenylene molecular structure is difficult to thermally decompose.

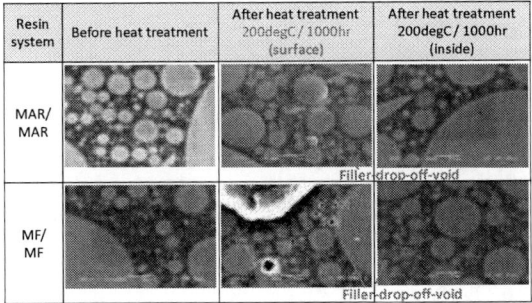

Figure 2: SEM observation of molded EMC sample after heat treatment

TABLE II. Element analysis of molded sample of MF resin by EDS

Element/Location	Before treatment		After treatment	
	Inside	Surface	Inside	Surface
C	46.6	47.7	47.4	40.0
O	11.5	11.2	11.9	12.6
Si	41.9	41.1	40.7	47,4

In addition, we checked the dielectric breakdown strength in Figure 3. The combination of MF type epoxy resin / hardener with high Tg shows the highest initial value of dielectrics strength among the five resin systems, but after long-term high temperature treatment, the decrease of dielectric strength tends to be large. The combination of MAR type epoxy resin / harder with low Tg shows that the lowest initial value of dielectrics strength, but the decrease of dielectrics strength is small.

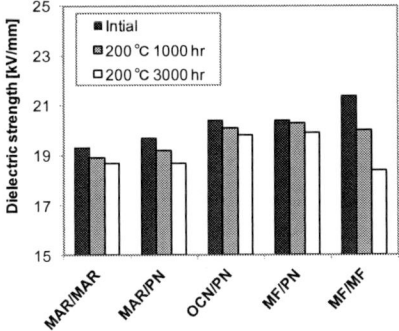

Figure 3: Dielectric breakdown strength depending on high-temperature environment

From those results, it can be concluded that, under high temperature environment of 200 degC, the EMC with high Tg has a good bonding reliability, but high cross-linking density cause high deterioration. In other words, it is necessary to make EMC which is both high Tg and high thermal decomposition resistance.

Study of heat cycle and warpage control

Conventional power module packages were sealed by silicone gel material, but the latest package design is to switch from silicone gel material to molding compound due to downsizing, weight saving, high heat resistance and high reliability (Fig. 4). In this trend, high heat cycle resistance, small warpage, high insulation property and high adhesion are necessary as EMC requirements in addition to high heat resistance.

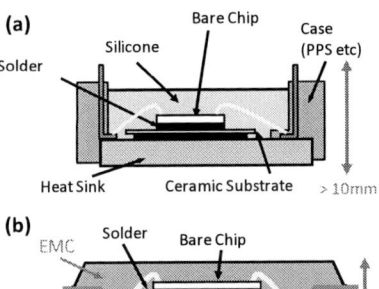

Figure 4: Power module package comparison ((a) conventional silicone gel type, (b) EMC molded type)

Power module packages are larger size than conventional discrete packages. The larger package size is, the larger failure risk becomes by temperature cycle (TC). Here, we conducted temperature cycle evaluation using two kinds of EMCs (high stress material and low stress material) which are different in coefficient of thermal expansion (CTE 1) and elastic modulus (Fig. 5).

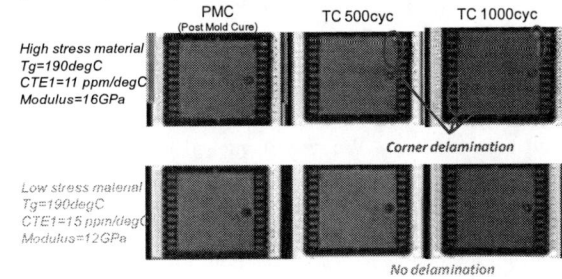

Figure 5: SAT (Scanning Acoustic Tomography) result after TC (PKG size: 65 x 75 x 6.5mm, Cu heat sink: 55 x 55 x 2.0mm, temperature cycle: -40 to 125 degC, 1000 cycles)

SAT result shows that delamination occurred at the corners of the heat sink after 500 cycles in the high stress material. On the other hand, delamination was not observed until the temperature cycle of 1,000 cycles in the low stress material. This result implies that CTE match between EMC and Cu of lead frame and low elasticity are key factor for the temperature cycle.

In the next, we carried out warpage evaluation. In case single-side molded packages or the mounted components are not be well balanced on substrate, we also need to consider CTE match between EMC and substrate. We selected Cu lead frame type with heat sink and ceramic substrate as model substrates. Figure 6 shows that the warpage level of package varies depending on EMC CTE, and the substrate type. This result suggests that we should prepare EMC matching the CTE of each substrate.

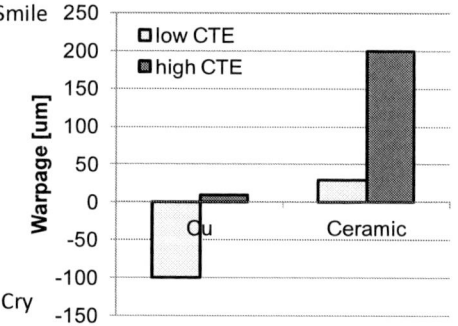

Figure 6: Package warpage results depending on substrate CTE (Cu PKG size: 65 x 75 x 6.5mm, Cu heat sink: 55 x 55 x 2.0 mm, Ceramic PKG size: 52 x 52 x 8 mm, Ceramic substrate size: 50 x 50 x 0.8 mm)

Toward further high heat resistance and high pressure resistance of epoxy resin / hardener for EMC

Currently we are working with some resin suppliers to develop the epoxy resin / hardener which can achieve both high Tg and high thermal decomposition resistance.

As results, we found the new structure based on the rigid biphenylene molecular structure which can achieve high Tg and high thermal decomposition resistance. Table III shows the EMC property using new high heat resistant resin. Its property shows high Tg (190 degC), excellent in thermal decomposition resistance, low water absorption and high adhesion. We have launched new EMC grades which contains high heat resistant epoxy resin / hardener from 2016 for power semiconductors.

In addition, we are also working on the development of the further higher heat resistant resin for power semiconductors which can handle high temperature operation more than 200 degC, and plan to launch a new EMC for power semiconductors that can handle junction temperature 225 degC. We are also developing high heat resistant resin for power semiconductor encapsulation

which can also cope with the junction temperature range of 250 degC.

TABLE III. The EMC property of new high heat resistant rein (Filler content is 85wt %).

Epoxy resin	MAR	MF	High heat resist
Hardener	MAR	MF	High heat resist
Tg [degC]	130	195	190
Weight loss [%]*1	0.12	0.58	0.22
Water absorption [%]	0.14	0.40	0.28
Ni Adhesion [-]*2	1.75	1.00	1.55

*1: After 200 degC for 1000h,
*2: Relative value (MF/MF is defined as 1.00)

CONCLUSIONS

We developed of epoxy resin molding compound (EMC) for power semiconductors and found some factors. Firstly, the rigid biphenylene molecular structure of epoxy resin / hardener is difficult to thermally decompose and it has an advantage for high heat resistance. Secondly, EMC CTE matching with substrate is key point for temperature cycle and small warpage. Thirdly, we found the new type resin structure based on the rigid biphenylene molecular structure which can achieve both high Tg and high heat resistance. Furthermore, EMC is expected to contribute to worldwide environmental preservation measures and future development.

ACKNOLEDGMENTS

We are grateful to raw material makers and equipment makers for cooperating us. The present work was supported mainly by Information & Telecommunication Materials Research Laboratory of Sumitomo Bakelite Co., Ltd.

REFERENCES

[1] BK. Bose, *Power Electronics and Motor Drives: Advances and Trends*, Academic press, 2010.

[2] S. Komori, Y. Sakamoto, *Materials for Advanced Packaging*, Springer Science, 2009, pp. 339-363.

[3] J. Millan, P. Godignon, X. Perpina, A. Perez-Tomas, J. Rebollo, *IEEE Transactions on Power Electronics*, vol. 29, 2014, pp. 2155-2163.

[4] F. Xu, TJ. Han, D. Jiang, LM. Tolbert, F. Wang, J. Nagashima, SJ. Kim, S. Kulkarni, F. Barlow, IEEE Transactions on Power Electronics, vol. 28, 2013, pp. 1464-1478.

[5] Y. Takahashi, H. Nogawa, A. Morozumi, Y. Nishimura, *CPMT Symposium Japan (ICSJ)*, 2016.

THERMAL AND MECHANICAL PERFORMANCE FOR DIFFERENT PACKAGE DESIGN OF ULTRA THIN 8 DIE STACKED FLASH PACKAGES

Hao Liu*, Qian Wang, Tianhan Wu, Xiaolin Bai, Fangfang Guo

Engineering Department, Ramaxel Technology Co., LTD, Dong Guan 523000, China

*liuhao@ramaxel.com

ABSTRACT

Based on the relationship between die size and package size, there were 3 possible package structures of 8 dies stacked memory flash as shown in figure 1, structure A1, A2 and A3 were shown in figure 1(a),1(b) and 1(c) respectively. All dies using in the 3 structures were numbered to U1~U8 from package bottom to top. Different package structures would induce different assembly process flows, which affected assembly process complexity and final cost. This paper explored thermal and mechanical performance of the 3 structures with finite element modeling (FEM). And the final finished packages with different stacked structures were also evaluated and compared. The warpage of different package structures were also studied and its possible impact to SMT process was also considered.

INTRODUCTION

To increased storage density of flash packages, multi-chip packages with stacked dies have become the main trend. However, traditional two resistors thermal analysis model is unapplicable[1] and liner superposition (LSP) theory is rarely used for flash packages with stacked dies[2]. Moreover, to stack 8, or even 16 dies, there are many different internal structures which will result in different warpage and stress. Meanwhile, thinning dies to ultra thin thickness are inevitable. Die crack might be induced during the assembly process. And severe warpage would also cause SMT process issue[3].

In this paper, flash packages with 8 dies stacked were studied. Package structure *A1, A2* and *A3* were shown in *Figure 1(a), 1(b)* and *1(c)* respectively. The thickness of ultra thin die used in this paper was 60µm. All dies using in the 3 structures were numbered to U1~U8 from package bottom to top. Detail dimensions of packages were shown in *Table I*. Firstly, thermal performance of different package structures was studied. Warpage and die stress were also studied and its possible impact to SMT process was also considered. In order to analyze the thermal and mechanical performance of the packages, thermal conductivity (λ), coefficient of thermal expansion (CTE), Young's modulus (E) and Poisson's ratio (v) were needed. Corresponding material properties were shown in *Table II*.

Different package structures would use different process flows, which affected assembly process complexity and final cost. So, engineering samples were built to evaluate key process units per hour (UPH) and cost.

And reliability test of the 3 structures were also conducted to further evaluate the performances.

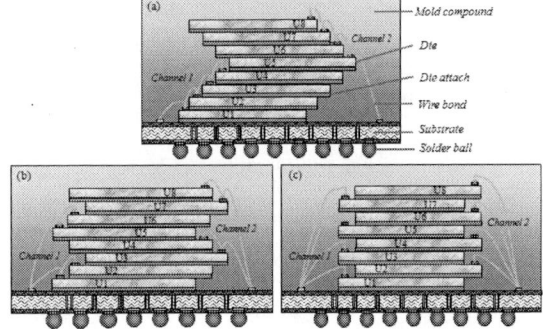

Figure 1: Structures of 8 die stacked two channel flash packages, (a): cross section diagram of structure A1; (b): cross section diagram of structure A2; (c): cross section diagram of structure A3

TABLE I. SIZE OF PACKAGES

Items	Size(mm)
Package	12 x 18 x 1.2
Die	6.36x12.25 x 0.6
Die attach thickness	0.02
Mold compound thickness	0.80
Substrate core thickness	0.136
Solder mask thickness	0.027
Ball Diameter/ Height	0.4/0.22

TABLE II. PROPERTIES OF MATERIALS[4]

Material	λ (W·m^{-1}·k^{-1})	CTE (C^{-1})	E (GPa)	ν
Mold compound	0.94	9	24	0.25
Silicon	180	2.6	131	0.28
Die attach	0.3	50	4	0.35
Substrate	0.8	15(xy) 58(z)	24	0.2
Solder mask	0.23	50	4	0.467
Solder ball	60	-	-	-

THERMAL PERFORMANCE

For IC packages, heat is generated when die is functioning. There are two thermal dissipating path as shown in *Figure 2*, one is through the top case of package to the ambient environment. The other is through the substrate and solder ball to print circuit board, and then to the ambient environment. The ambient environment temperature used in this paper was 70 degree celsius. Dissipating power of each functioning die was 0.1 watt.

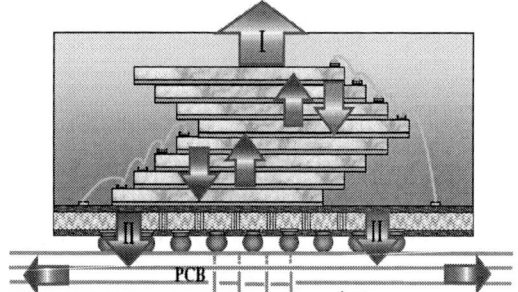

Figure 2: Thermal dissipating path schematic diagram

Firstly, the maximum temperature of structure *A1* with single die functioning was studied. Results were shown in *Figure 3*. As the functioning die changing from U1 to U8, the maximum temperature would raise gradually. So it was obvious that, the thermal dissipating path through substrate and solder ball was more significant than the one through package top case.

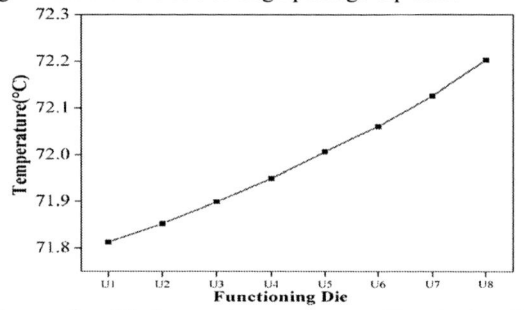

Figure 3: Maximum temperature with single die functioning

Two signal channel can support two die which belongs to different channel running at the same time. The corresponding chip enable signals and chips under control were shown in *Table III*. So there would be 16 functioning combinations for the 3 structures respectively. And the simulation results were shown in *Figure 4*.

TABLE III.   CHIP CHANNEL DETAILS

Channel Item	CH_1#	CH_2#
A1	U1, U2, U3, U4	U5, U6, U7, U8
A2	U1, U2, U5, U6	U3, U4, U7, U8
A3	U1, U3, U5, U7	U2, U4, U6, U8

It was found that, the further the selected functioning dies away from the substrate, the higher the maximum temperature would be for each structures. The maximum

temperatures of structures *A1*, *A2* and *A3* were 74.30°C, 74.24°C and 74.09°C recpectively. The maximum temperature differences among these 3 structures were negligible. And structure *A3* had a maximum temperature gap among selected functioning die groups

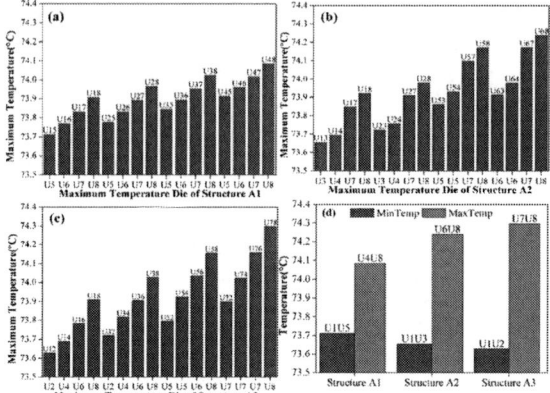

Figure 4: Maximum temperature of 3 structures, (a) maximum temperature of structure A1, (b) maximum temperature of structure A2, (c) maximum temperature of structure A3, (d) temperature comparison of 3 structures

MECHANICAL PERFORMANCE

Stress of dies were also simulated and compared. It was assumed that packages were stress-free and undeformed at 175°C[5]. Stress at room temperature (22°C) and reflow peak temperature (260°C) were simulated. Package warpage was supposed to be controlled within ±0.13mm[6]. And die stress less than 200MPa was acceptable[7].

Package Warpage Comparison

Warpage results at room temperature and reflow peak temperature were shown in Figure 5. Simulation results showed that the warpage was convex at room temperature and concave at reflow peak temperature. And when temperature changed from room temperature to reflow peak temperature, the maximum warpage change of structures *A1*, *A2* and *A3* were 81.79μm, 78.68μm, and 76.83μm respectively, which might be great challenge for SMT process.

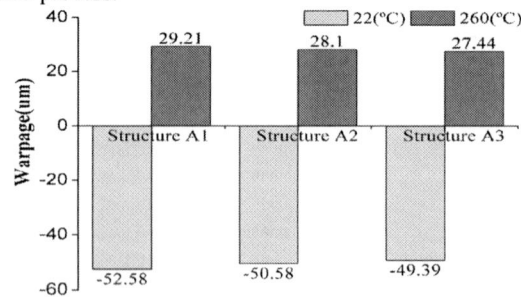

Figure 5: Warpage comparison of 3 structures

Samples were manufactured and warpage at room temperature were measured. Results of simulation and measured warpage comparison were shown in *Figure 6*. The simulation results were in agreement with the measured results. And the difference between simulation results and measured results were within 10% which was acceptable. The warpage performance of structure *A1* was the worst, while structure *A2* was better and structure *A3* showed the best result.

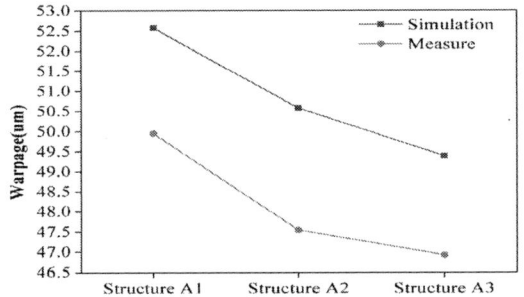

Figure 6: Simulation and measured warpage comparison

Mechanical Stress Comparison

Package warpage would induce die stress inside. Corresponding maximum die stress of 3 structures were shown in *Figure 7*. According to results, structure *A1* suffered the largest die stress at room temperature compared with other two structures, which was up to 93.69MPa. However, die stress of structure *A2* was the largest at reflow peak temperature which was up to 156.82MPa. And structure *A2* would have a maximum die stress difference when temperature changed from room temperature to reflow peak temperature. However, the maximum die stress of the 3 structures was all less than 200MPa which indicated risk of die crack was under control.

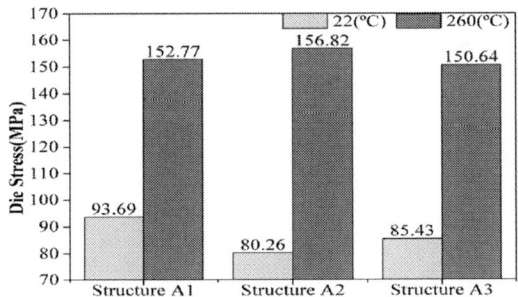

Figure 7: Maximum die stress comparison of 3 structures

PROCESS COMPARISON

Difference assembly process flows would be used for the 3 structures. The main difference among the 3 structures flows is die attaching (DA) and wire bonding (WB) process cycles (*Figure 8*). Structure *A1* would need two cycles of DA and WB process while structure *A2* and

structure *A3* would need four cycles of DA and WB process accordingly.

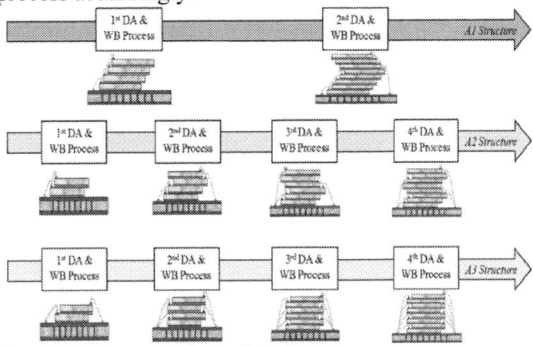

Figure 8: Process flow of the 3 structures

Moreover, the most challenging thing would be to bond all these dies with gold wires under limited overhang space (as was shown in *Figure 9*). For stacked die under overhang area here, there were only 60um or even less space for wire loop. To bond these die with wires, ultra low loop had to be adopted. On the other hand, if the size of package is certain, structure *A3* will be more advantageous because it can accommodate a larger die size.

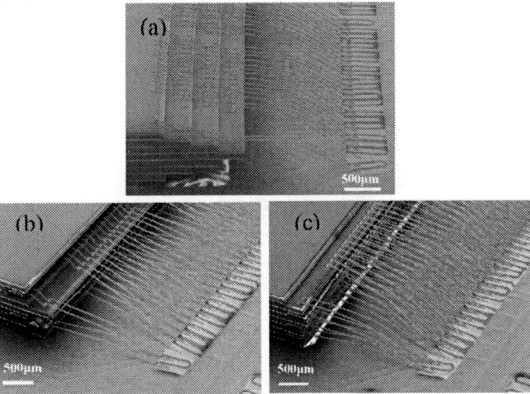

Figure 9: SEM of the 3 structures, (a): SEM of structure A1; (b): SEM of structure A2; (c): SEM of structure A3

UPH and Wire Consumption Comparison

To further evaluate the process difference between the 3 structures, UPH and wire consumption were also considered. The UPH difference of these three structures mainly existed in DA and WB process as was shown in *Table IV*.

Based on above Table *IV*, the structure *A1* needed only two process cycles of DA and WB process, which was only half of that structrue *A2* and *A3*. For the DA process, the UPH of first process cycle was lower than subsequent process cycles among the 3 structures. And for WB process, the UPH was continuously decreasing with process cycles. Moreover, the DA and WB process UPH of structure *A2* was higher than that of structure A3.

TABLE IV. UPH OF THE THREE STRUCTURES

Items	Process	1st (unit)	2nd (unit)	3rd (unit)	4th (unit)
A1	DA	367	375	-	-
	WB	100	70	-	-
A2	DA	600	622	622	622
	WB	220	155	150	145
A3	DA	581	597	597	597
	WB	170	150	145	140

TABLE V. WIRE CONSUMPTION COMPARISON

Structure	Wire consumption (mm / unit)
A1	353.64
A2	631.76
A3	1176.71

And in terms of the amount of gold wire was consumed, there were great difference among these 3 structures (*Table V*). Structure *A1* had consumed the least length of wire. And the amount of wire consumed by structure *A2* was 631.76mm, which was 1.78 times more of that of structure *A1*. Moreover, the wire consumption of structure *A3* was 1176.71mm, which was 3.33 times more of that of structure *A1* and 1.86 times more of that of structure *A2*.

PACKAGE LEVEL RELIABILITY

Reliability test of 3 structures were conducted. And they all passed the tests. The table below was the summary of criteria and results.

Test Type	Test Condition	Sample Size	Relialiably Failure		
			A1	A2	A3
Precond-itioning	MLS3(30C/60%RH 3xReflow @ 260°C Peak) 192 hrs	66	0/66	0/66	0/66
UHAST	130°C,85%RH 96hrs	22	0/22	0/22	0/22
TMCL	-65°C --150°C 500cycles	22	0/22	0/22	0/22
HTSL	150°C, 1000hrs	22	0/22	0/22	0/22

CONCLUSION

This paper presented the different structures of flash package with 8 die stacked, with the various aspects of thermal performance, mechanical performance, assembly process and reliability tests being addressed.

The thermal performance showed the thermal dissipating path through substrate and solder ball was more significant than package top case. And structure *A1*

had the best thermal performance among the three structures while structure *A3* had a maximum temperature gap among selected functioning die groups.

The results of mechanical simulation showed that the warpage changes of structure *A1*, *A2* and *A3* were 81.79μm, 78.68μm, and 76.83μm respectively during reflow process. So, the structure A1 would have a greatest challenge for SMT process. However, die stress of these 3 structures was all less than 200MPa, so the risk of die crack was under control.

For assembly process, structure A1 had the simplest process cycle flow and least wire consumption, so the structure A1 was preferred. On the other hand, for package with definite size, structure *A3* will be more advantageous because it can accommodate a larger die. At last, all 3 structures had passed package level reliability tests.

ACKNOWLEDGEMENTS

This paper was supported by DongGuan Major Special Project (No.2017215102009). The authors would like to thank all members of Ramaxel Design Center and Cost Accounting Engineering Department, Research and Development Engineering Department, QA laboratory and others help on this paper.

REFERENCES

[1] S. Krishnamoorthi, W. H. Zhu and C. K. Wang. *Electronics Packaging Technology Conference*, 2007, Eptc 2007, IEEE Xplore, 2008, pp. 278-284.

[2] P. Szabo, A. Poppe, G. Farkas, et al. *Thermal and Thermomechanical Phenomena in Electronics Systems2006*, IEEE, 2005, pp. 251-257.

[3] M. Kuczynska. *The influence of PBGAs post-manufacturing warpage simulation, viscoelastic material properties and evaluation methodology on accuracy of solder joints damage prediction*, international conference on thermal mechanical and multi physics simulation and experiments in microelectronics and microsystems , 2015.

[4] J. H. Yin, Bing. Du, S. Q. Wang, et al. *Thermal Stress Analysis and Optimization in Stacked Die Package Under Power Load*. Semiconductor Technology, 2007.

[5] C.Y. He, Z.Y. Liu and H.X. Wang, et.al. *Thermal-mechanical Simulation and Optimization Analysis for Warpage-induced PBGA Solder Joint Failures*.

[6] JEITA ED-7306E, *Measurement Methods of Package Warpage at Elevated Temperature and the Maximum Permissible Warpage*.

[7] A. T. Bradley, R. C. Jaeger, J. C. Suhling and K. J. Oconnor, *Piezoresistive characteristics of short-channel MOSFETs on (100) silicon. IEEE Transactions on Electron Devices*, 48(9), 2009-2015.

WAFER WARPAGE CONTROL BY EPOXY MOLDING COMPOUNDS FOR WAFER LEVEL PACKAGE

Kihyeok Kwon[1]*, Joo Young Chung[1], Donghwan Lee[1] and Sang Kyun Kim[1]

[1]Development 1 Group, Development 1 Team, Electronic Material Division, Samsung SDI,
130, Samsung-ro, Suwon-si, Gyeonggi-do, 16678, Korea
*Corresponding Author's Email: kihyeok.kwon@samsung.com

ABSTRACT

The recent interest of Fan-out wafer level packaging technology (FOWLP) comes from such benefits, thin package, board fan-out capability, high I/O, good thermal resistance, and electrical performance. However, its application is limited due to the difficulty in the warpage control of FOWLP. In order to control this difficulty, modulating the epoxy molding compound's properties is necessary. In this paper, we introduce our recent progresses in an evaluation tool and its use for tuning EMC properties.

INTRODUCTION

Semiconductor packaging technologies have been growing along with the development of various applications such as entertainment, sensor, communication, authentication, security, and environment systems for mobile, portable devices, and automotive. These highly functionalized electronic devices caused to evolve new packaging trends such as 3D integrated circuit (IC) packaging with through silicon-via (TSV), 3D and 2.5D system in package (SiP), 2.5D with TSV interposer (TSI), package on package (PoP), wafer level chip scale packaging (WLCSP), fan-out wafer level packaging (FOWLP), and so on [1]. Especially, FOWLP technology brought a surging attention in the packaging industries because of its advantages, such as high integration capability, small form factor, high I/O density, high performance, and cost effectiveness over a wide range of applications [2]. Additionally, a success on mass production of Application Processor (AP) in 2016 showed a huge market potential. However, there are certain challenges to be overcome.

Challenges that FOWLP packaging technology faced include wafer warpage, die shift/protrusion, and board level reliability [3]. Among those, wafer warpage is the most important characteristic to be resolved for mass production. There are many efforts to resolve this issue in FOWLP, but few effective tools have been reported up to date [4]. Among various tools, there are two distinctive wafer level molding materials in FOWLP, liquid and powder ones. The properties of each material must be understood in order to optimize FOWLP. Liquid materials have been successfully used in manufacturing since 2016, but their disadvantages such as low storage stability, high material cost, and complicated dispensing patterns have led to a new demand for powder type materials. These powder type materials show relatively better storage stability, lower cost, and simple dispensing patterns, which are more desirable for FLOWP.

RESULTS AND DISCUSSION

Seven commercially available epoxy molding compounds produced at Samsung SDI were compared according to the following properties: modulus, CTE1, CTE2, and T_g (Table 1). Then, a statistic simulation tool was designed to describe those properties by finite element modeling analysis (FEA) using ABAQUS 6.14. The calculation came out as a smile shape warpage, which indicates higher shrinkage of a compression molding material over that of the silicon wafer.

TABLE I. MATERIAL PROPERTIES AND CALCULATED WARPAGE

	Modulus (GPa)	CTE1 (ppm/℃)	CTE2 (ppm/℃)	T_g (℃)	Warpage (mm)
EMC A	21.6	9.5	43.1	155.9	4.47
EMC B	20.8	11.7	42.3	154.9	6.15
EMC C	17.8	6.3	26.3	164.7	1.61
EMCD	12.0	9.5	49.1	168.5	2.12
EMC E	12.7	16.4	51.7	164.4	5.00
EMC F	13.9	16.6	46.9	137.5	5.18
EMCG	14.0	16.1	52.4	146.3	4.18

Various parameters influencing on wafer warpage were screened by the simulation calculator. Among all these parameters, modulus, CTE, and T_g have a significant effect on the controlling warpage. Generally, wafer warpage become smaller, as the modulus and CTE gets lower and T_g gets bigger. The low CTE of EMC C was a main contributing factor that lowered wafer warpage by minimizing the CTE mismatch between the molding compound and silicon wafer. The low modulus of EMC D was found to reduce wafer warpage by lowering stress relaxation. In general, wafer warpage can be decreased by increasing the Tg of the molding compound because the

CTE of epoxy molding compound decreases with increasing T_g.

On the top of these results, the molding compound materials were further modified in terms of lower modulus and CTE and higher T_g to reduce wafer warpage after the post molding cure.

Screening epoxy resin systems

After screening the formulation of epoxy molding compounds to select a standard recipe, a series of alternative epoxy resin systems were surveyed for lowering modulus. Similarly, T_g was studied. The result of modulus and T_g values for each epoxy formulation are shown in Fig. 1. Those samples were prepared in the same condition under 135°C for 600 sec. The combination of Resin8 + Resin10 and Resin9 + Resin10 resulted in the lowest modulus. Especially, Resin10 gave the largest drop in modulus and CTE, as well as the largest elevation of T_g.

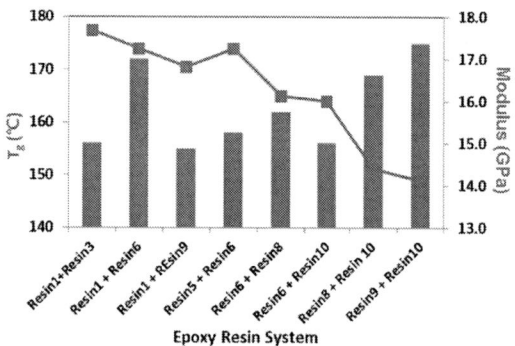

Figure 1: Screening epoxy resin system's properties

Screening low modulus/low CTE additives

Next, we investigated the effective additives, which can control Young's modulus and CTE for compression molding compounds. A total of 9 different additives were screened with various polymer functional groups and chain lengths. Results of each experimental group are shown in Fig. 2. Note the labeling nomenclature:

- A, B, and C represent polymer functional group.
- 1, 2, and 3 represent different chain length. (1) Long Chain, (2) Medium Chain, (3) Short Chain.

As shown in Fig. 2, additive A with medium and short chain length (A2 and A3) and additive B with medium and short chain (B2 and B3) were found to be the most highly efficient additives to control modulus. Further material property investigation also revealed that these additive and chain length combinations also provided the most optimum CTE 1, as shown in Fig. 8. In addition, these two functional additives with medium and short chain lengths showed a very good balance of reducing both modulus and

CTE while maintaining suitable flow-ability.

From these results, it was envisioned that different functional groups and chain lengths play an important role in the epoxy molding compound's properties after the epoxy resin-phenol hardener curing reaction, and these additives should be the key raw materials modulated to tune final warpage of compression molding compounds.

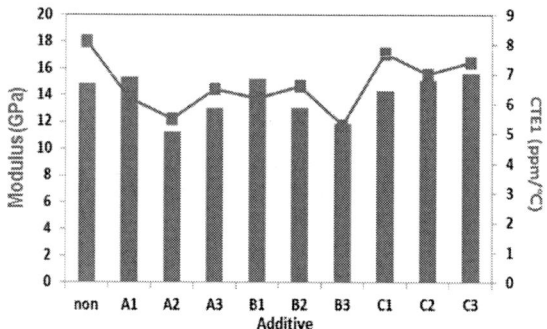

Figure 2: Screening additives with different functional group that have different chain length

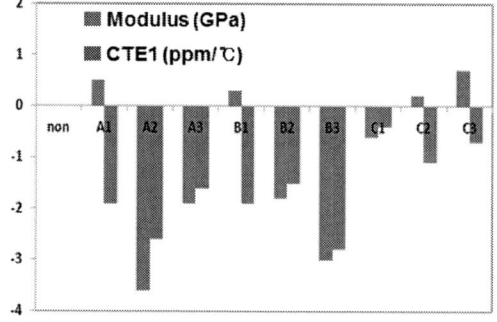

Figure 3: Modulus and CTE1 variation depending on additive

Evaluation of wafer warpage performance with newly developed EMC

We proposed a bilayer standard evaluation specimen which consists of a silicon wafer and epoxy molding compound. The thickness of silicon wafer and EMC molded layer are 700 and 300 μm respectively. For consistency, the molding condition and a post mold cure (PMC) were set to 135°C x 600 sec and 150°C x 2 hrs. Further, the warpages of mold compounds were directly compared.

Based on this standard procedure, we prepared several specimens having different modulus, CTE, and T_g properties. As shown by the data in Table 2, low modulus, low CTE, and high T_g resulted in the lowest wafer warpage. We set the experiment by using two different

978-1-5386-5309-8/18 $31.00 © 2018 IEEE

resin systems (Resin2 + Resin3 and Resin3 + Resin4) and two different additives (A and B) to screen the EMC for FOWLP, with warpage being the primary metric. The EMC behaviors after PMC were characterized by shadow möire, and the results were summarized in Table 2 and Fig. 2. Among all the options, the best result in terms of wafer warpage was the following condition, Resin3 + Resin4 and A additive, that showed the lowest Young's modulus, CTE, and highest T_g values. These experimental results were well matched with the simulation data.

TABLE II. PROPERTIES OF EPOXY MOLDING COMPOUNDS WITH LOW MODULUS/CTE MATERIALS

	Ref.	EMC1	EMC2	EMC3	EMC4
Epoxy resin system	Resin 1	Resin2 + Resin3	Resin2 + Resin3	Resin3 + Resin4	Resin3 + Resin4
Additive	-	A	B	A	B
Modulus (GPa)	17.7	10.8	11.8	10	11.4
CTE1 (ppm/℃)	6.5	5.7	5.9	5.5	5.7
T_g (℃)	162	156	153	164	159
Warpage (μm)	1779	826	1101	785	1089

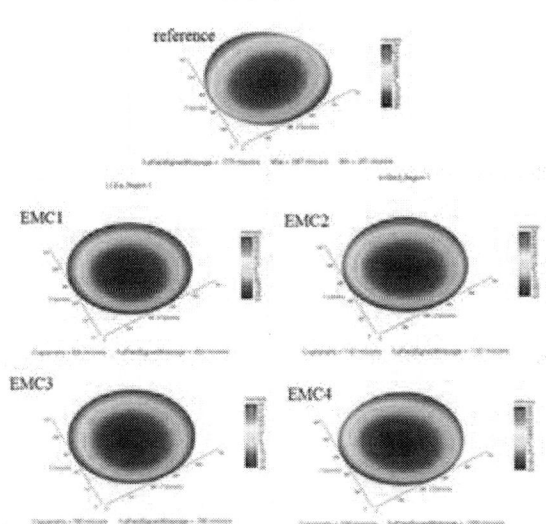

Figure 4: Wafer warpage for each EMC detected by Shadow möire

Conclusion

In this study, we first developed compression epoxy

molding compounds that wafer warpage met the minimum standard for mass production. Further, modulus, CTE, and T_g of screened molding compounds were modified. Finally, we developed a new compression molding compound of which modulus, CTE, and Tg are under control. The application of these novel compression molding compounds are suitable for a wide range of applications and mass production for successful fan-out wafer level package implementation.

ACKNOWLEDGEMENTS

The authors would like to thank Joongkeun Kwag, Hankyu Park, and Hoyun Kim for analysis of wafer warpage, sample preparation, and experiment set-up.

REFERENCES

[1] C. F. Tseng, C. S. Liu and C. H. Wu, "InFO (Wafer Level Integrated Fan-Out) Technology," 66th Electronic Components and Technology Conference (ECTC), 2016, pp. 1-6.

[2] T. Meyer, G. Ofner and S. Bradl "Embedded Wafer Level Ball Grid Array (eWLB)," 33th International Electronics Manufacturing Technology Conference (IEMT), 2008, pp. 9-12.

[3] H. W. Lin, Y. W. Liu, J. Ji, J. Liao, A. Chen, Y. H. Chen, N. Kao and Y. C. Lai "Warpage Characterization of Panel Fan-out (P-FO) Package," 65th Electronic Components and Technology Conference (ECTC), 2015, pp. 1750–1754.

[4] S. S. Deng, S. J. Hwang and H. H. Lee "Warpage Prediction and Experiments of Fan-Out Wafer level Package During Encapsulation Process," IEEE Transactions on Components, Packaging and Manufacturing Technology, 2013, pp. 452-458.

Advanced Panel Level Die Attach

Dr. Hugo Pristauz

Besi Austria GmbH
Radfeld, Tyrol, Austria
hugo.pristauz@besi.com

Biography

Dr. Hugo Pristauz has 18 years of experience in die attach equipment industry with focus on emerging packaging technologies, starting in 1999 at Datacon with major assignment to product management and development of the 8800 flip chip platform. After acquisition of Datacon by Besi in 2005 Hugo Pristauz was running emerging die attach business in VP position covering RFID, flip-chip, thermo-compression and fan-out bonder business. Since begin of 2016 Hugo Pristauz is focusing on technology scouting and networking to support Besi's roadmap definition process.

Abstract

A new generation of advanced die attach platform is in development, characterized by exceptional throughput on large panel area, supporting global placement accuracy (based on global alignment marks) down to 2μ@3sigma and the flexibility for both face-up and face-down die attach with optional thermo-compression capability, to be implemented either in traditional one-step approach or high throughput two-step approach utilizing collective thermo-compression bonding technique.

Keywords—die attach; advanced packaging; flip-chip; fan-out packaging; thermo-compression bonding, panel level packaging

Introduction

With the driving forces for smaller and cheaper advanced semiconductor packages, supporting higher performance and less power consumption, advanced packaging technologies like flip-chip, thermo-compression and fan-out packaging have to move from wafer to panel level processing.

A main driving force is related to the cost pressure on wafer level fan-out (WL-FO) packages. A major cost factor of WL-FO is RDL based, which is batch process related, and can be decreased by increasing the panel size (Fig. 1), while another major cost factor is the pick & place bottleneck, which has to be fixed by increasing pick & place throughput for panel level fan-out applications [3][4].

Another driving force has its origin in silicon bridge embedding applications, as it has been proposed for the EMIB technology [1]. Since such silicon bridges are used for high density data bus interconnection between two adjacent dies the placement of such bridges require high accuracy

(2-5μ@3sigma) [1]. As a further option high-accuracy/high-yield chips-last placement like thermo-compression (TC) might be considered to attach die at large panel level. This raises the need for a panel level thermo-compression bonder.

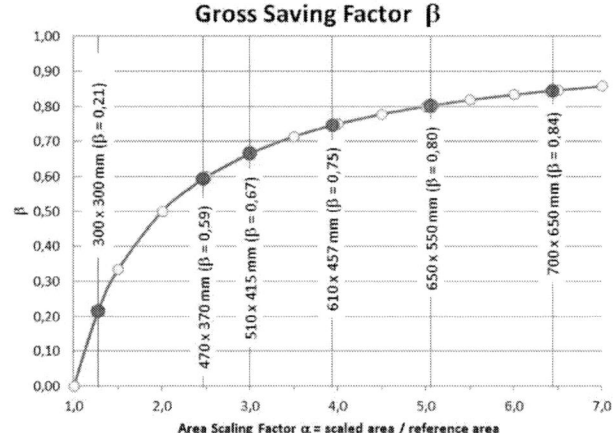

Fig.1. Gross saving factor β as a function of area scaling factor α for circular φ300mm reference area. β = 1-1/α would tell us the savings due to α-scaling if γ=100% of the costs would be batch related. For e.g. γ=35% batch related costs the total savings are γ·β = 35%·β [4].

Modular 8800 Platform Concept

To support emerging advanced die attach needs for flip-chip, fan-out and thermo-compression applications a modular 8800 platform concept has been developed with a common machine base and configurable sub-systems implementing the core capabilities (Fig. 2, Fig. 3) [2]:

- component (die, passive) feed (wafer, tape & reel)
- substrate handling (strips, carriers, wafers, panels)
- die flip (face-up, face-down)
- fluxing (in-situ dipping or up-stream)
- constant heat (top heat, bottom heat)
- ramping heat (top heat, bottom heat)
- accuracy (15μ, 10μ, 5μ, 3μ, 2μ)
- clean class (ISO6, ISO5)

This successful building block concept can be re-used when moving from wafer-level die attach to faster panel level die attach with the following additional improvements
1) the current universal 8800 gantry system needs to be forked into a low cost gantry system for moderate

accuracy (5-15µ@3sigma global) and a premium cost advanced gantry system supporting high accuracy (2-3µ@3sigma global) with less thermal drift (Fig. 3)

2) For higher throughput targeting 15.000-20.000 UPH the single bond head concept per gantry system needs to be replaced by a multi bond head concept

3) To propel cost-down of the slow and expensive thermo-compression process it is suggested to utilize a non-conductive film based TC process (TC-NCF) which can be implemented as a two-step die attach process [5] comprising fast panel level tacking with subsequent panel level collective bonding.

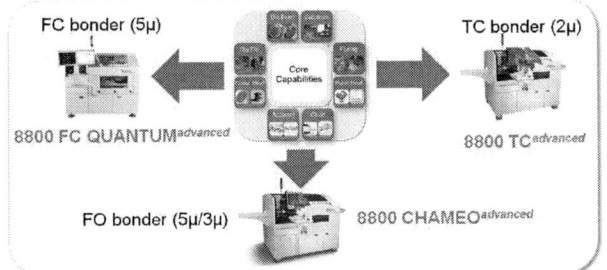

Fig.2. Modular advanced die attach platform (8800 platform) with configurable core capabilities to support flip-chip, fan-out and thermo-compression applications.

Fig.3. 8800 CHAMEO *advanced* dual gantry fan-out bonder with common wafer feeder and two pick and place systems.

Advanced Gantry System

For the existing 8800 wafer level platform the gantry system is based on an *indirect metrology*, which measures the position with encoders located close to the linear bearings. Due to thermal drift effects mainly caused by the heat flow induced by the motors in combination with bi-metal effects (aluminum structures, steel bearings) the tool center point "sees" thermal drifts in a range up to 3-4µm. For high accuracy wafer level applications (e.g. 2µ@3sigma) such drift effects can be successfully compensated by proper algorithms.

To support 2-3µ@3sigma based on pure global alignment on panel level, however, we introduced a water cooled advanced gantry system based on a *direct metrology* which is capable to measure the x/y-position of the tool center point without impacts caused by thermal gantry deformations. Fig. 4 compares the behavior of the two gantry system types.

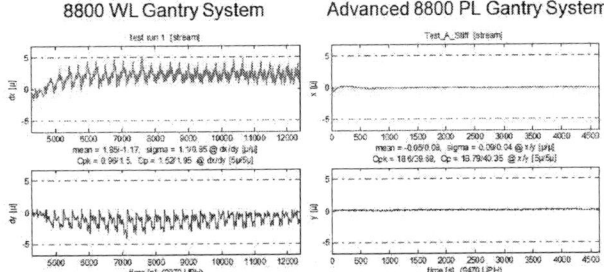

Fig.4. Comparison of x/y-drift at tool center point between conventional 8800 wafer level gantry system (left, 3-4µm drift) and advanced panel level gantry system (right, 0.5µ drift).

Multi Nozzle Bond Heads

To boost throughput from currently 7000 UPH of a wafer level fan-out bonder into the region of 15.000-20.000 UPH of a large panel level fan-out bonder the two single-nozzle bond heads of the 8800 platform are replaced by multi-nozzle bond heads with common theta-axis (Fig.5).

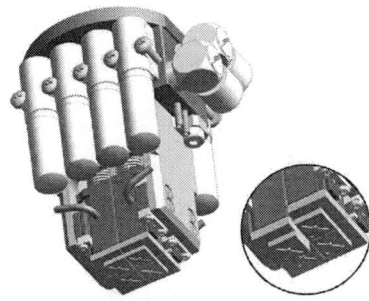

Fig.5. Multi-nozzle bond head concept with common theta axis and individual mini z-stroke (mm range).

The concept is in the first step implemented for a face-up die attach process with quattro-nozzle heads, where in the standard case 4 dies are picked concurrently from the wafer, are measured in parallel via an upward looking camera, and are bonded sequentially, one after each other. To enable sequential bonding each nozzle can perform an individual z-stroke in mm-range, while the whole bond head is mounted on a common theta axis which is carried by a z-axis with bigger z-stroke (100-150 mm range) to cover the height difference between substrate and wafer. To utilize wafer maps and to manage the boundary area of the wafer the multi-nozzle bond head needs its counterpart on the ejection side where a multi-die ejector needs to be implemented with 4 needle-packets which can be individually enabled for an ejection stroke.

A repeatability study for concurrent die ejection is shown in figure 6. The study was performed with a single nozzle and a single ejector. We expect the repeatability of the die ejection process to be within in a +/-50µ tolerance, which could be easily achieved in the feasibility study. Based on our experience an upward camera looking at the die edges should allow a placement accuracy of 10µ@3sigma with reasonable efforts. This makes us optimistic to build 600x600 mm panel level fan-out bonders with 10µ placement accuracy at a throughput level of 15.000 UPH based on a dual gantry concept where each gantry is equipped with a quattro-nozzle bond head.

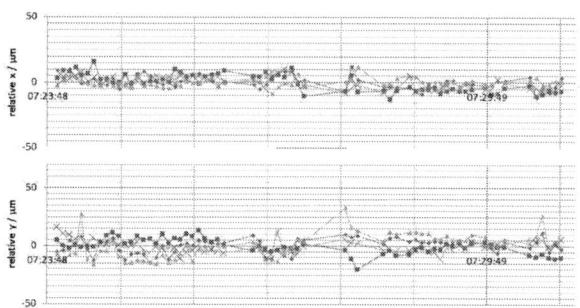

Fig.6. x/y-deviations of a multi die ejection process (4 concurrent die ejections) over time.

Collective TC-Bonding

We have developed a collective TC-bonder for 720x650 mm working area which can be utilized for TC-NCF processes (Fig. 7). Collective TC-bonding [5] splits up a one-step TC-bonding process into fast TC-tacking and collective TC-bonding. In a first step we target 4000 UPH for a panel level TC-tacker which supports hybrid force/position control for TC-tacking. The collective TC-bonder can be equipped with up to 12 TC bond heads which can be arranged in a matrix setup with configurable pitch. This matrix of bond heads is indexed over the panel, performing up to 12 parallel TC-bonds per cycle. During each bonding cycle each bond head can be switched either into active or passive mode in order to deal with x-out locations and out-of-boundary locations.

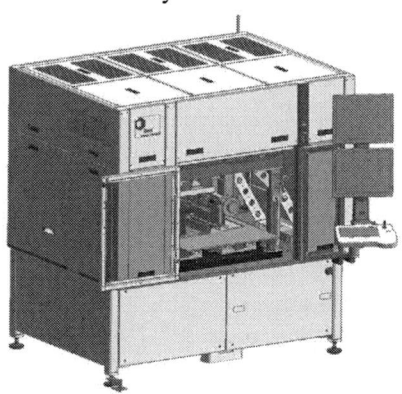

Fig.7. Collective TC-bonder with 12 parallel acting thermo compression bond heads, supporting a panel area of 720x650 mm, designed for 3000-5000 UPH, depending on NCF materials.

Conclusion

Advanced die attach methods like flip-chip, fan-out and thermo-compression are required also for panel level applications. The versatile field proven 8800 platform concept is extended to support panel level applications. For this reason three crucial extensions have been introduced: 1) water cooled advanced gantry system with less thermal drifts, 2) multi nozzle bond heads for 15.000-20.000 UPH, 3) collective bonding with panel level TC tacking and collective bonding in a separate equipment.

References

[1] R. Mahajan et all, "Embedded multi-die interconnect bridge (EMIB) – a high density, high bandwidth packaging interconnect", 2016 IEEE 66th Electronic Components and Technology Conference, pp. 557–565

[2] H. Pristauz, "Versatile platform for advanced die attach", NCAP & Yole Dèveloppement Symposium on Advanced Packaging & System Integration Technology, Wuxi/China 2017, pp.8–15.

[3] C. Palesko, A. Lujan, "Cost comparison of fan-out wafer-level packaging to fan-out panel-based packaging", 2016 International Microelectronics Assembly and Packaging Society - Vol. 2016, No. 1, pp. 180-184

[4] H. Pristauz, "Cost scaling of a panel level RDL process", LinkedIn blog, Dec. 2016, https://goo.gl/zgPXaC

[5] A. Attard, "Productivity improvements on thermo-compression bonding," European 3D Summit, Grenoble 2015, pp.: 13-16.

LASER RELEASE TECHNOLOGY FOR WAFER LEVEL PACKAGING

Dongshun Bai, Xiao Liu, Hong Zhang, Qi Wu, Ram Trichur, Rama Puligadda, and Tony Flaim

Brewer Science, 2401 Brewer Drive, Rolla, MO 65401, USA

*Corresponding Author's Email: dbai@brewerscience.com

ABSTRACT

Laser release technology provides advantages including high throughput and low stress during the release process, effective thin-substrate handling, and ease of application, even with large panels. It has drawn a large amount of attention and has become a very popular technology in advanced packaging applications. In this paper, we introduce our advanced materials for laser release technology that can be utilized in different applications in packaging areas such as temporary bonding, fan-out wafer-level packaging, lamination, and 2.5D/3D integration using through-silicon vias (TSVs). We also demonstrate what properties these laser release materials should possess to enable a successful processing. In addition, we will introduce a new laser release material, which can be used as both a bonding and laser release material and has broad applications for system-in-packaging (SiP), package-on-package (PoP), and other heterogeneous integration infrastructures.

INTRODUCTION

The laser release method provides advantages with regards to high throughput, lower stress release, and force-free separation from a carrier at room temperature [1,2]. It has shown the potential for high-volume production in various advanced packaging platforms. Currently, the three most commonly used laser systems focus on the UV wavelengths 308 nm, 343 nm, and 355 nm. Compared with other laser wavelength like 532 nm, or even IR laser, the UV laser source has advantages such as less generation of heat, less carbonized residue, and lower possibility of device wafer damage due to its higher photo energy density. A UV laser will induce photodecomposition of laser release material, instead of thermodecomposition [3,4].

UV laser systems can be classified as either excimer laser or solid-state laser [4,5]. The comparison of the three popular laser wavelengths are listed in Table 1. Technically, all UV laser systems show the advantages of low heat generation and less carbonized residue after the debonding process. However, in practice, the three tools will perform differently depending on the individual tool design. For example, both 308-nm and 343-nm laser tools possess better laser beam uniformity than the 355 nm tool because of the specially designed optical path of their laser beams. As a result, the extra laser energy of 355 nm from the uneven beam may cause more carbonized residue and damage device wafer due to the non-linear feature of the laser debonding process. On the contrary, it is well

acknowledged that the 308-nm laser tool is more expensive than the other two because of its laser generation mechanism and tool maintenance. Therefore, several factors need to be considered with respect to tool selection including the performance of laser release material(s) on each tool.

Table 1. Comparison of UV-laser systems

Laser Source	Excimer	Solid State	
	308 nm	343 nm	355 nm
Heat generated	Low	Low	Low
Frequency	Low	High	High
Laser size	Large	Small	Small
Beam uniformity	High	High	Low

LASER RELEASE FOR TEMPORARY BONDING/DEBONDING (TB/DB)

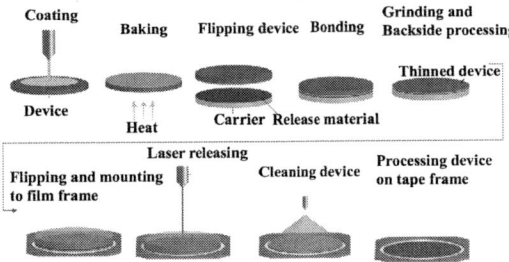

Figure 1: Typical TB/DB process flow

Figure 1 shows a typical temporary bonding and debonding process with laser release. In this application, the device will be bonded to a carrier substrate with temporary bonding and laser release materials. The laser release material will be coated on a transparent carrier (normally glass carrier) and the bonding materials can be applied on the device or coated on top of the laser release material. The laser release material will be coated to a few hundred nanometers while the temporary bonding material will be used to cover the topography on the device. After bonding, the device wafers will be ground down to target thickness and then go through the necessary backside processes depending on the applications. Then the bonded pair will be put on a tape frame with the carrier on top. The laser will be applied across the glass carrier and the laser release material will respond to the laser beam allowing for separation of the carrier from the device. Finally, the glass carrier will be removed using a vacuum chuck and the device wafer will progress to the

978-1-5386-5309-8/18 $31.00 © 2018 IEEE

cleaning step.

Figure 2 shows an example of the laser release for temporary bonding and debonding application. Commercially available BrewerBOND® 701 material was used as the laser release layer in this application. It has very high thermal stability to enable high temperature backside processing up to 400°C, and it can work with a wide variety of temporary bonding materials to meet the different requirement for downstream processing. In this example, 150 nm of BrewerBOND® 701 laser release material was coated on an 8" Eagle XG glass carrier. 50 μm of BrewerBOND® 305 material was coated on an 8" Si wafer. The wafer pairs were bonded at 200°C to get a good bond line. After that, the wafer pair was heat treated on a hot plate at 260° C for 30 min. The laser debonding was performed at both 308 nm and 355 nm. The results showed that the wafer pair can be successfully debonded at both wavelengths and the laser energy used on both tools are comparable. Figure 2 shows the wafers after laser release at 308 nm. The dose used for separation was 220 mJ/cm². Since BrewerBOND® 305 material also showed good thermal stability, this material set can be used for applications up to 300°C.

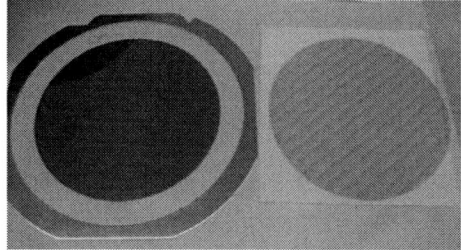

Figure 2: Laser release results of BrewerBOND® 305 and BrewerBOND® 701 materials at 308 nm.

LASER RELEASE IN FOWLP

Fan-out wafer-level packaging (FOWLP) is a cost-effective way to achieve high interconnect density and to manage larger I/O counts within an affordable package. It enables smaller footprints, higher interconnect density, better routing and thinner packages than current technologies [6]. FOWLP process flows fall into two categories: chip-first and chip-last, referring to the point in the process when chips are placed onto the substrate. Laser release technology can also be used in both applications.

In the chip-first process, wafer-handling challenges can occur after molding, because the high-temperature processing used during redistribution layer (RDL) processes to crosslink dielectric materials can introduce stress that leads to warpage due to the coefficient of thermal expansion (CTE) mismatch between Si and the mold compound. A temporary bonding and debonding process, including laser release technology, can be used to handle mold compound wafers to enable backside

processing. Here, we would like to focus on the chip-last (RDL-first) application.

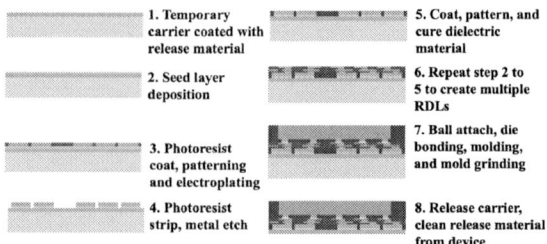

Figure 3: Fan-out application, RDL-first process flow

Figure 3 shows a typical process flow for RDL-first application. A laser release material will be applied on the glass substrate and then the RDL layers will be built on top of it. After all the buildup process, the glass will be separated with laser release. In this application, the laser release materials need to possess very good thermal, mechanical, and chemical stability to withstand the RDL-first process flow. Also, the warpage after coating needs to be well controlled to enable downstream processing.

Brewer Science developed a few novel laser releasable materials for this application, and these materials can be applied at both the wafer and panel level. Here, we focus on Material A developed for this application. Material A has thermal stability (T_d), above 400°C, and can be applied by spin or slot die coating to obtain film thickness ranging from a few hundred nanometers to a few microns. As shown in Table 2, it also exhibited very good chemical resistance against the common solvents that encounter in RDL application [7].

Table 2. Chemical Resistance of Material A

Chemicals	Results
NMP at 90 °C for 25 min	Pass
2% Ammonia for 10 min	Pass
2.5% TMAH for 10 min	Pass
PGMEA for 10 min	Pass
1.5% H_3PO_3 + 0.4% H_2O_2 for 2 min	Pass
0.05% NH_4OH + 0.1% H_2O_2 for 10 min	Pass

Material A can be laser released at both 308 and 355 nm. Figure 4 shows the glass carrier wafer after laser release at 355 nm. In this example, 500 nm of Material A film was coated on an 8" glass carrier and then it was heat treated at 400°C for 1 hour in a N_2 oven. Next, the glass carrier was bonded to a Si wafer with another temporary bonding (TB) material. The wafer pair was then laser released at 355 nm. These results indicate Material A can be used for high-temperature applications and still remain laser releasable.

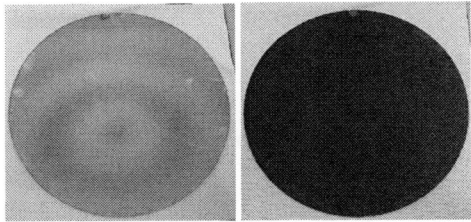

Figure 4: Glass (left) and Si wafer (right) after laser release at 355 nm.

LASER-RELEASEABLE TB MATERIAL

Currently, most of the temporary bonding/debonding platforms commercially available focus on multiple-layer structures, mainly on dual-layer systems consisting of a temporary bonding layer and a releasing layer. As a result, multiple steps of coating and baking of each layer lead to an increased cost of ownership as well as decreased throughput for the whole process. In contrast, a significant reduction of cost and improvement of throughput can be achieved by using a newly synthesized thermoplastic that combines both temporary bonding and laser releasing functions in one material. The concept is illustrated as in Figure 5.

Two layers with two functions

One layer with two functions

Figure 5: Illustration of laser releasable temporary bonding material.

Inspired by this new concept, a novel thermoplastic, Material B, was synthesized by Brewer Science with these targeted properties. The newly synthesized temporary adhesive has demonstrated a good bond line on an 8" blank Si wafer bonded to a glass wafer with commonly used thermal compression conditions, 230°C at 2000 N for 3 min. No defects or voids were detected after bonding.

Material B provided good mechanical support for the device wafer during the grind process. The wafer pairs bonded with Material B were successfully ground down to 60 µm. In addition, the thermal simulation on the thinned wafer pairs was conducted in nitrogen oven at 250°C for 30 minutes without severe observable defects for all the wafer pairs.

As one of the critical criteria for the success of the new laser debondable temporary adhesive, the material should demonstrate its laser debondability using commercial laser releasing tools, mainly at 308 nm, 343 nm, and 355 nm wavelengths. The bonded wafer pairs can be successfully debonded by using these three major UV-laser wavelengths available in the market. The laser debonding parameters for these wavelengths are listed in Table 3. Material B can be easily debonded at 308 nm using relatively low fluency. The image of debonded device wafer is shown in Figure 6. In contrast, higher fluency or Watts are needed to debond at either 343 nm and 355 nm, probably due to its lower absorbance at these two wavelengths compared to 308 nm. Any residue from Material A remaining on the device wafer can be cleaned effectively by using common organic solvents. After cleaning, the wafer is visually clean as verified under green light with no residue.

Table 3. Laser debonding parameter of Material B

308 nm	343 nm	355 nm
220 mJ/cm^2	1000 mJ/cm^2, 40 μm	6 W

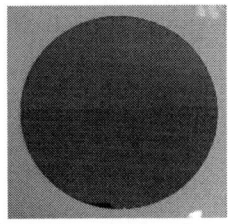

Figure 6: Blank device wafer after debonding using 308 nm laser.

ACKNOWLEDGEMENTS

We would like to thank Suss, EVG, and Kingyoup for their support for this work.

REFERENCES

[1] R. Trichur, T. Flaim, *Chip Scale Review*, November-December 2015, pp38-41.
[2] R. Trichur, T. Flaim, *Chip Scale Review*, September-October 2016, pp12-18.
[3] A. Phommahaxay, G. Potoms, et al., *Electronic Components and Technology Conference, 2016 IEEE 66th*, Las Vegas, May 31-June 3, 2016.
[4] K. Ohkita, Y. Maruyama, et al., *Chip Scale Review*, March-April 2017, pp 31-34.
[5] G. Oakes, *International Wafer-Level Packaging Conference 2017*, San Jose, Oct 23-26, 2017.
[6] Q. Wu, X. Liu, K. Han, et al., *Electronic Components and Technology Conference, 2017 IEEE 67th*, Orlando, May 30-June 2, 2017, pp 890-895.
[7] A. Lee, J. Su, B. Huang, et al., *IMAPS 2016 - 49th International Symposium on Microelectronics*, Pasadena, October 10-13, 2016, pp 190-195.

978-1-5386-5309-8/18 $31.00 © 2018 IEEE

CAPILLARY WEDGE BONDING TECHNOLOGY FOR STACKED DIE PACKAGES

Hao Liu[1]*, Maopeng Zhou[1], Xiansheng Duan[1], Jiantao Lin[1], Fangfang Guo[1] and Naoki Sekine[2]
[1]Engineering department, Ramaxel Technology Ltd, Dongguan, China
[2]Global Engineering Support Management Dept, Shinkawa Ltd, Tokyo, Japan
*Corresponding Author's Email: liuhao@ramaxel.com

ABSTRACT

BSOB is currently widely used in flash memory BGA packages, which use fine gold wire to connect each layer of stacked flash chips, that can be 2, 4, 8 or even 16 layers, and finally connect them to BGA substrate. Due to its special process characters, dummy stud balls are needed for continuous wire bonding process to link different layers of flash chips.

This special process occupies quite big capacity of wire bonding equipment. In this paper, a new wire bonding method, called Capillary Wedge Bonding (CWB), is introduced to perform interconnection for different layers of flash chips without or reducing the dummy stud balls. UPH is one of the key factors to be studied and compared, also ball shear , wire pull IMC and cratering test are included. The final packages reliability level with different wire bonding methods are also evaluated and compared.

INTRODUCTION

With the development of Multi chip packages (MCPs) with stacked dies, higher requirements are needed for the Wire bonding (WB) process. The normal WB process for MCPs with stacked dies is Ball Stick on Ball (BSOB). As figure 1 shown, the BSOB process, which will bond dummy stud balls first and then complete the normal bonding, has the lower Units per Hour (UPH) and shortens the lifetime of the capillary because of the extra three times contact on each pad of the middle dies.

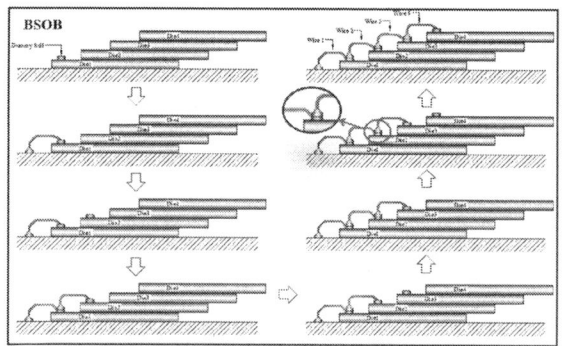

Figure 1: Schematic of BSOB

A novel wire bonding process called Capillary Wedge Bond (CWB) dies was developed for MCPs with stacked

chips. In this process, the second bond of the last wire is also the first bond of the next wire, which will save considerable time for the bonding motion. And CWBB is a special CWB process with the dummy stud balls on each layer of the flash chips except the top die as shown in Figure 2.

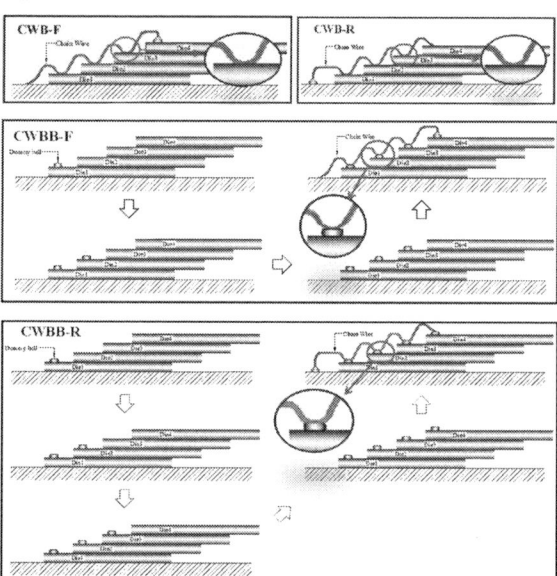

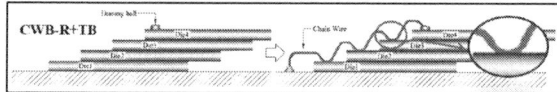

Figure 2: Schematic of CWB-based process

Higher UPH and lifetime of capillary are desired to be obtained by using CWB method. However, some problems like less contact surface are still existed, which may be a concern of inter-connective reliability.

In this study, different bonding methods were adopted to find out the suitable one for the MCPs with eight stacked dies, including BSOB, CWBB-F, CWBB-R, CWB-F, CWB-R and CWB-R+TB. UPH, wire pull, IMC and cratering test were studied and compared to seek trade-off point between efficiency and quality.

978-1-5386-5309-8/18 $31.00 © 2018 IEEE

TABLE I.   BONDING METHODS IN THIS STUDY

Code	Bonding TYPE	With/without dummy stud balls	Bonding direction	Refer
BSOB	BSOB	With	Reverse direction	Figure 1
CWBB-F	CWBB	With	Forward direction	Figure 2
CWBB-R	CWBB	With	Reverse direction	Figure 2
CWB-F	CWB	Without	Forward direction	Figure 2
CWB-R	CWB	Without	Reverse direction	Figure 2
CWB-R+TB	CWB+TB	Only Top die with dummy stud balls	Reverse direction	Figure 2

EXPERIMENTAL
Equipment and materials
The experiments were carried out on the wire bonding machine (UTC–5000 special version for CWB, Shinkawa). A high purity (99.99%) Au alloy wire with diameter of 18μm was used as the bonding material.

Test and Characterization
Scanning Electron Microscope (SEM) was adopted to characterize the structure of the bonded wire. In order to evaluate the bonding strength of the bonded wire, wire pull were tested by Bond Test machine (MFM1200, Clack). Inter-Metallic Compound (IMC) and Cratering were also tested to confirm status of the bond pad. The reliability tests were carried out for the final product after molding as shown in TABLE II.

TABLE II.   DESCRIPTION OF THE RELIABILITY TESTS

Test Type	Test Condition	Duration	Sample size	Reference Standard
Preconditioning	MSL3 (30°C / 60% RH, Reflow Peak Temperature : 260°C)	192 hours	66ea	J-STD-020E
UHAST	130°C, 85% RH	96hours	22ea	JESD22-A110D
TMCL	-65°C--150°C	1000cycles	22ea	JESD22-A104E
HTSL	150°C	1000hrs	22ea	JESD22-A103D

RESULT AND DISCUSSION
UPH and capillary cost
UPH is a key factor for the mass production. The MCPs with eight stacked dies were adopted to test the UPH of different bonding methods. As TABLE IV shown, the UPH of all the methods based on CWB is increased by at least 30%. The best one on the UPH performance is the CWB-F, which is increased by 142% compared with the normal BSOB.

TABLE III.   UPH OF DIFFERENT BONDING METHODS

Code	Capillary touch-down times per units	Capillary consumption rate (vs BSOB)	UPH (unit/hrs)	UPH rate (vs BSOB)
BSOB	1081	1	48.1	1
CWBB-F	752	0.70	66.9	1.39
CWBB-R	799	0.74	63.8	1.33
CWB-F	423	0.39	116.2	2.42
CWB-R	423	0.39	111.8	2.32
CWB-R+TB	470	0.43	92.4	1.92

Influence of Bonding direction
Even though the UPH of CWB-R is a little lower than CWB-F, wire bonding on the reverse direction is still necessary in some case like the ultra low loop. However, indentations were found on the passivations of the top dies after CWB-R as figure 3 shown. The reason is due to the 2nd bond on top die being formed by the capillary edge which will touch on the passivation.

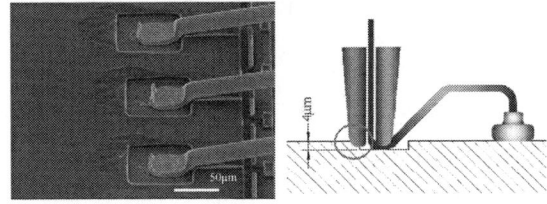

Figure 3: Capillary edge touched the passivation

In order to prevent this issue, the CWB-R+TB, which is the CWB-R with dummy stud balls on the top die, is designed to protect the passivation on the top dies while maintaining the high UPH as much as possible. With the extra dummy stud balls, the passivation is untouched because the stitches are formed on the dummy stud balls whose thickness is about 10μm to keep the capillary away from the passivation.

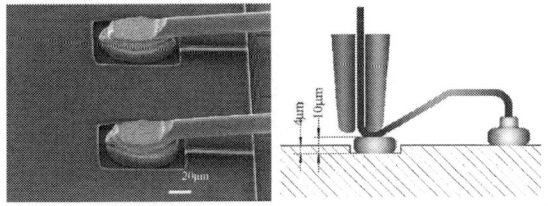

Figure 4: Extra dummy stud balls for the top dies

Influence of dummy stud balls
In order to seek balance point between efficiency and quality, the dummy stud balls during the CWB process were also discussed. As figure 5 shown, the dummy stud balls can increase the contact surface between the bond

pad and wire, which can increase IMC area and enhance the bonding strength. Moreover, the second bond level can be also increased with the dummy stud balls, which can keep the capillary away from the passivations as mentioned before.

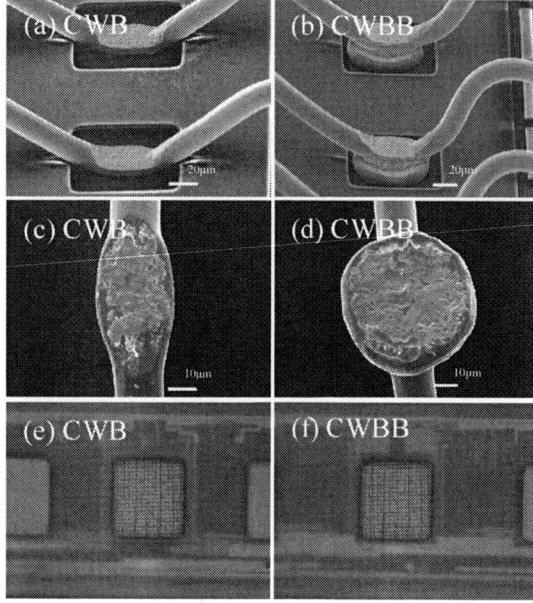

Figure 5: Bonding status of CWB and CWBB

The experimental result shows that all bond responses of the samples in this study can meet the production target. The IMC area can also meet the target, which is at least 60% of the contact surface. No crack was found on bond pad during the Cratering test.

Hence, the dummy stud balls are not necessary for the sample used in this study except the one on the top die of CWB-R. The extra dummy stud balls increased the wire-pull slightly, whereas they also decreased the UPH significantly.

TABLE IV.   BOND RESPONSE OF DIFFERENT BONDING METHODS

Code	Wire pull(g)				Ball shear(g)	
	On middle dies		On top dies		On top dies	
	AVG	STD	AVG	STD	AVG	STD
BSOB	5.49	0.18	5.35	0.18	23.83	0.88
CWBB-F	4.45	0.19	4.37	0.12	27.83	0.81
CWBB-R	5.37	0.15	5.44	0.20	31.18	0.77
CWB-F	3.56	0.11	3.60	0.19	28.25	0.72
CWB-R+TB	4.02	0.13	4.96	0.19	31.61	1.02
Target	>3	NA	>3	NA	>18	NA

Reliability test

After the processes study, the final products eMMC with 4 stacked flash dies and 1 controller fabricated via CWB-based methods are shown in Figure 6. No damage was occurred during the processes. The reliability tests including MSL3, UHAST, TMCL and HTSL were adopted to simulate the extreme environment that the product may be applied in.

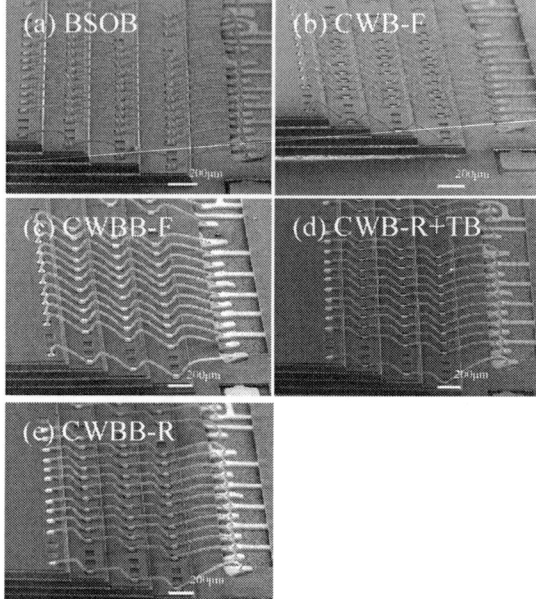

Figure 6: eMMC4+1 fabricated via CWB based methods

The final result shows that reliability tests of all the samples have been passed as TABLE VI shown, which means the devices fabricated via CWB-based process can work even in the extreme environment.

TABLE V.   RESULTS OF THE RELIABILITY TESTS

Code	Preconditioning	UHAST	TMCL	HTSL
BSOB	pass	pass	pass	pass
CWBB-F	pass	pass	pass	pass
CWBB-R	pass	pass	pass	pass
CWBB-F	pass	pass	pass	pass
CWBB-R+TB	pass	pass	pass	pass

CONCLUSION

UPH of all the CWB-based methods have been increased by at least 30%. CWB-F is the best method to replace the normal BSOB because of the excellent performance on UPH.

CWB-R+TB is suitable for the ultra low loop even with a little loss of UPH because the dummy stud balls on the top die can keep the capillary away from touching passivation.

Dummy stud balls on the middle dies can be removed because the extra balls decrease the UPH significantly. And all the bonding responses and reliability tests can meet the production target even without dummy stud balls on the middle dies.

Some problems are still existed for the CWB-based methods, especially the non-stick detection for the middle dies, which will be discussed in the further research.

ACKNOWLEDGEMENTS

This study was supported by DongGuan Major Special Project (No.2017215102009). The authors would like to thank the engineers of Shinkawa for their technical support on the WB machine.

REFERENCES

[1] H. Liu, B. K. Lim, Y. R. C. Desmond, W. Palei and H. B. Tan. *Proceedings of EMPC 2005*, Brugge, June 12-15, 2005, pp. 10-15.

[2] S. R. Martin and E. Christian. *Microelectronics Reliability.* vol. 63, 2016, pp. 336-341.

[3] R. Zhang, H Liu, B Li, T Sugiya. Semiconductor Technology International Conference, IEEE, 2016.

STUDY ON THE HIGH RELIABILITY PERFORMANCE AND HIGH THERMAL CONDUCTIVITY EPOXY MOLDING COMPOUND

Xingming Cheng[1], Wei Tan[1], Zhen Wang[2], Hongjie Liu[1], Yang Yang Duan[1]*

[1]Jiangsu HHCK Advanced Materials Co., Ltd

[2] Lianyungang Branch of Traditional Chinese Medicine, Jiangsu Union Technical Institute, Lianyungang, Jiangsu,2220007,China

wei.tan@hhck-em.com

ABSTRACT

High thermal conductivity Epoxy Molding compound (EMC) was formulated by using fully spherical aluminum oxide. With 88% of filler content, it is able to achieve thermal conductivity of 3.0 W/(m.K) with high reliability. The effect of crystalline silica: spherical aluminum oxide ratio on thermal conductivity was also investigated. Types of resin which affected the adhesion, flexural strength and water absorption has a great impact on the reliability performance of semiconductor devices.

Keywords—thermal conductivity, spherical aluminum oxide , reliability , epoxy molding compound

INTRODUCTION

More and more of power devices are demanded in semiconductor in recent years, especially with the introduction of SiC die. The power density has reached 100kW/L where the junction temperature (Tj) could reach $150^{o}C - 200^{o}C$. It needs a high thermal conductivity EMC to dissipate the heat. Conventional high thermal EMC with low filler loading has a challenge to meet this reliability requirement.

TABLE I. THE FILLERS USED IN THE EMC

	Alumina	Fused Silica	Crystal Silica	Silica Nitride	Aluminum Nitride
Chemical Composition	Al_2O_3	SiO_2	SiO_2	Si_3N_4	AlN
Specific Gravity(g/cm³)	3.9	2.2	2.7	3.2	3.3
Thermal Expansion Coefficient (X10⁻⁶)	8	0.5	14	3.5	5.6
Thermal Conductivity(W/m.K)	30	1	10	26	320

Table 1 show the common fillers which are used in EMC. Aluminum Nitride and Silica Nitride have high thermal conductivity but they are not widely used due to their high price. In addition, higher loading of such fillers would affect the flow of EMC.

EXPERIMENTAL

The spherical aluminum oxide, DAB-85FC was from DENKA Japan. All other ingredients were formulated into a (EMC) epoxy molding compound system. The ratio of Epoxy value / hydroxyl value was 1. The filler content was about 88% and the couple agent content was 0.3%.

The compound was prepared through dry blending, followed by melt-mixing, extrusion into sheet form, It was then fine-ground into powder, and then pelletized into performs.

Standard transfer molding techniques were used to fabricate the test specimens. All test specimens were prepared with in mold cure time of 120sec at175°C followed by a six-hour post mold cure at the same temperature.

The Shore D hot hardness was measured on an epoxy molding compound specimen at 175℃ which was cured for 90 second at the same temperature.

Tg and modulus were measured by DMA, TA instrument model, from 25℃ to 280℃ at 3℃ per minute. Water absorption was measured after subjecting the test specimens in a pressure cooker at 121C,100%RH for 24h. The reliability tests were done on SOP8 packages.

Water absorption measurement:

Specimens for water absorption with the size of Φ 50mm×3mm were prepared by Standard transfer molding techniques with the in mold cure time of 120sec at 175°C followed by a six-hour post mold cure at the same temperature. Water absorption was carried out by the Shanghai Sanshen PCT machine under 121°C, 100% relative humidity for 24 hours.

Calculation of water absorption by using the following equation:

$$W_a = \frac{(W_1 - W_0)}{W_0} \times 100 \qquad (1)$$

Where W_a is the water absorption, W_1 is the weight after 24h in pressure cooker test chamber, and W_0 is the initial weight of the specimen.

Adhesion measurement:

978-1-5386-5309-8/18 $31.00 © 2018 IEEE

A Tab pull sample can be taken as two pieces of lead frames connected by EMC as shown in Figure 1. The enclosed area is encapsulated by EMC and the shadowed area on the leadframe surface is attached by EMC. After transfer-molding, the tap pull samples were post cured in the oven at 175 °C for 6 hours. The tensile test was done with a Universal testing machine (AGS-5kNA,Shimadzu Corporation). The tensile force was reported as the indication of adhesion force between epoxy molding compound and leadframe. The leadframe surface are plated by silver or PPF.

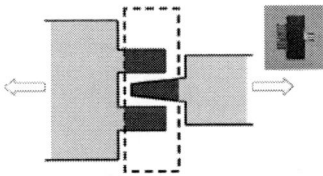

Figure 1: The schematic diagram of a tap pull sample for the tensile test. Two pieces of leadframes are connected by EMC. The enclosed area is encapsulated by EMC and the shadowed area on the leadframe surface is attached by EMC. The right graph is a tap pull sample photo taken by camera.

RESULTS AND DISCUSSION.

Figure 2 shows that the surface morphology of spherical alumina and spherical silica. The surface of spherical alumina is much smoother than spherical silica.

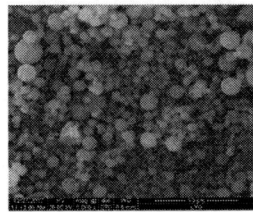 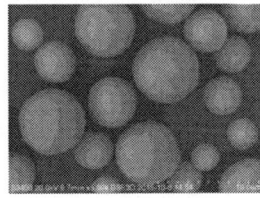

(1)Spherical Silica   (2)Spherical alumina

Figure2: the morphology of Spherical silica and spherical alumina

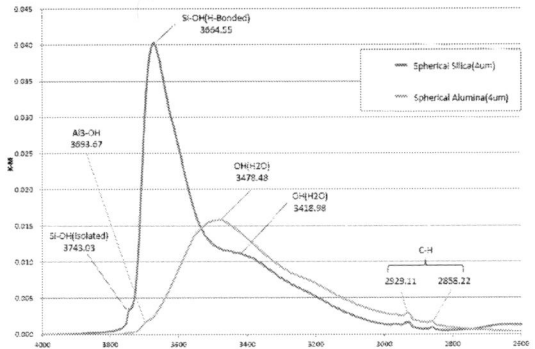

Figure3:the FTIR spectrum of Alumina and Silica[1]

Figure 3 is a FTIR spectrum of filler surfaces. It showed

the spherical alumina surface contained very small amounts of aluminum hydroxyl group. On the other hand, the spherical silica filler surface contained a lot of silica hydroxyl group which would make it difficult to attach to the aluminum oxide surface during the curing process of EMC, thus reducing key properties such as the flexural strength, hot hardness etc.

Table II showed the effect of different ratio of crystal silica and spherical aluminum oxide on the basic properties of EMC. The type of resin used in the formulation is MAR. Higher % of alumina content would increase the thermal conductivity, reduce the hot hardness, flexural strength, Tg. However it has no significant influence on the water absorption. The decrease in these properties was due to the lack of aluminum hydroxyl on the surface which was mentioned above.

The smoothening of the alumina surface and the lack of aluminum hydroxyl would eventually gave longer spiral flow. This was due to the smooth surface and a small amount of functional groups on spherical surface area of the alumina filler.

Table II also showed that the hot hardness decreased quickly with the increase of alumina % when MAR resin is used in the formulation.

TABLE II.   THE EFFECT OF DIFFERENT RATIO OF CRYSTAL SILICA/SPHERICAL ALUMINA OXIDE ON THE BASIC PROPERTIES OF EMC

The ratio of Crystal silica /Spherical alumina oxide		85/0	75/10	10/78	0/88
Resin Type		MAR	MAR	MAR	MAR
Thermal conductivity	W/m.K.	2.4	2.5	2.6	3.0
Gel Time	(s)	30	30	27	31
Spiral flow	cm	65	80	109	140
Hot Hardness	/	85	82	76	72
Flexural strength	MPa	152	130	112	98
Flexural Strength	MPa	28356	27325	26328	23127
Tg	℃	100	105	95	90
α1	PPM	19	18	13.4	14.9
α2	PPM	45.3	47	53.6	58.4
Water absorption (PCT，24h)	%	0.22	0.23	0.2	0.2

Table II showed when MFN resin was used, the hot

hardness would increase significantly from 65 to 88. According to the literature [2], the hot hardness was the key properties to balance the confliction between reliability and long terms moldability[2].when the hot hardness was larger than 85, the wax content can be decreased significantly in order to maintain long term moldability at the same level. Less wax content used in EMC would improve the reliability.

Table III showed the comparison of basic properties when different resin types were used. All formulations were kept in 88% percent spherical alumina. In formulation A and B, the hot hardness, Tg and water absorption were the highest and the adhesion was the lowest when using MFN resin. The results were in the opposite when MAR resins were used.

Formula C was the best among the formulations with balance of properties, acceptable hot hardness, Tg, water absorption and adhesion. It gave good reliability and moldability performance.

TABLE III.   THE BASIC PROPERTIES COMPARISON OF DIFFERENT FORMULATIONS

Formulation		A	B	C
Resin Type	/	MFN	MAR	MAR/MFN
Spherical Alumina	%	88	88	88
Thermal conductivity	W/K.℃	3.0	3.0	3.0
Gel Time	s	30	27	31
Spiral flow	cm	150	100	140
Hot Hardness	/	88	65	86
Flexural Strength	MPa	90	100	116
Flexural Modulus	MPa	25430	21576	27256
Tg	℃	175	100	116
α1	PPM	17	16	13.6
α2	PPM	45	66	56.8
Water absorption (PCT, 24h)	%	0.52	0.2	0.3
Adhesion After MSL3	Cu	N	400	520	550
	Ag	N	150	420	350
	PPF	N	50	130	110

This paper used the BGA package of fingerprint identification module to verify the reliability of formula C. Figure 4 showed the delamination performance after MSL3. It showed that formulation C could pass MSL3 without any delamination.

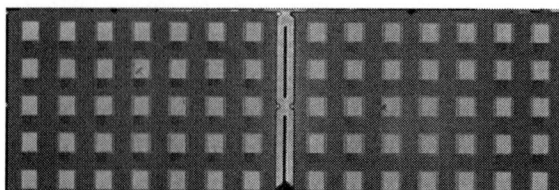

Figure 4: the CSAM imagine of formulation C on LGA package after MSL3

CONCLUSION

1. The spherical alumina could decrease the hot hardness property significantly compared to the silica due to the lack of aluminum hydroxyl on the surface.
2. The thermal conductivity was 3.0W/K.C when 88% percent of spherical alumina was used in EMC.
3. Use of spherical aluminum oxide would decrease the hot hardness, flexural strength, Tg significantly
4. Using MFN type epoxy resin could significantly increase the hot hardness.
5. Mixture of MFN and MAR resin could achieve an EMC with high thermal conductivity and good reliability.

REFERENCES

[1] Denka Co., Ltd. Denka Spherical Alumina.
[2] Wei Tan *et al.*, "The study on the moldability and reliability of epoxy molding compound," *2017 China Semiconductor Technology International Conference (CSTIC)*, Shanghai, 2017, pp. 1-3. DOI: 10.1109/CSTIC.2017.7919871

HIGHLY FUNCTIONALIZED EMC PACKAGE MATERIALS FOR FINGERPRINT SENSORS

Junghwa Kim[1], Junwoo Lee[1], Yongyeop Park[1], Jiah Yoon [1], Jiyoon Cho[1], Kyoungchul Bae[1], Donghwan Lee[1], and Sangkyun Kim[1]

[1]Development 1 Group, Development 1 Team, Electronic Material Division, Samsung SDI, 130, Samsung-ro, Suwon-si, Gyeonggi-do, 16678, Korea

*Corresponding Author's Email: junghwa0.kim@samsung.com

ABSTRACT

The current trends of chip packaging technologies are moving towards providing the functional solutions for modules, where chips cannot fully perform their roles without the support of functionalized packaging materials. Functionalized EMC specialized for future capacitive fingerprint sensors and further wearable, Fintech, and IoT technologies requires following functions or properties: high dielectric constant (Dk: >21 or28), flexibility (Modulus at 25°C: < 10Mpa), flow-ability (spiral flow: > 60 inch) and uniformity of appearance (Defection rate by black stains: <0.02%). Here, we present the state of the art achievements on those functions and properties of functionalized EMC.

INTRODUCTION

The world is on the verge of the fourth industrial revolution. The key of this revolution is the Internet of things (IoT) where machines are able to communicate with other machines with minimal human intervention. It results in improving economic benefits and the quality of life for population around the world exponentially [1-5].

As more IoT devices spread out, the security and privacy of IoT ecosystem have become a primary concern among consumers and businesses. These electronic applications closely linked with personal information employ biometric identification technologies for the security and privacy concern [2-9].

Fingerprint authentication is the most common and reliable method among other biometric authentications including face, palm, iris, voice, and retina [7-9]. The current fingerprint authentication sensors are classified in two main types in terms of the mass production and commercialization; optical and capacitive. The optical type fingerprint sensor provides a higher level of security than the other type, but its high power consumption and large size prevent the adoption of miniaturized electronic devices [6,10,11-12]. The optical fingerprint sensor is widely used in homeland security. On the other hand, the fingerprint sensors based on capacitive technology provide a lower level of security than an optical type fingerprint sensor, but its outstandingly low energy consumption and small size make it the most promising sensor technology, especially for miniaturized electronic devices [11].

While initially using capacitive fingerprint sensors for convenience, technologies such as IoT and Fintech, emerging new application that requires an especially high level of personal protection, are now using capacitive fingerprint sensors as a secure replacement for password authentication. There have been a lot of works to improve the security levels of capacitive fingerprint sensors by increasing fixers or building up 3D sensors [11-13]. To keep this endeavor alive, the packaging plastic of capacitive fingerprint sensor chips must be looked at further. The role of traditional packaging materials is the protection of Si chips from physical, chemical, and thermal environments. As electronic devices become miniaturized, integrated, and functionalized, the role of 2nd generation packaging materials has been shifting to the functionalized packaging technology solutions. In other words, chip makers request new functions on encapsulant which can either help the performance of electronic devices or prevent their malfunction in the aspect of high dielectric constant, flexibility, heat dissipation, electromagnetic interference, and heat resistance [11-13]. Here, we introduce two major EMC functions necessary for fingerprint sensors of four most highlighted technologies, IoTs, Fintech, wearable devices, and autonomous transportations.

1. CAPACITIVE SENSOR

A capacitive fingerprint sensor is composed of three main components, printed circuit board (PCB), Chip, and epoxy molding compounds (EMC). The distance from the valley and ridge parts of finger to electrodes induces different electric charges on electrodes. Electric charges influence fingerprint sensor voltage, which enables the storing of the fingerprint information that will be used for fingerprint authentication [11]. Higher capacitance leads to higher sensitivity of fingerprint sensors. In turn, this capacitance is tuned by three factors according to equation (1). Area of Sensors (s) is predetermined by the given sensor size in equation (1). Thus, the sensitivity of the sensors is determined by the remaining two factors, the distance between sensors and fingers and dielectric constant. To gain the better sensitivity, the sensors need thin package (d) and high Dk EMC (ε).

Table 2. Experimental Data

		Ref 1	Ref 2	EMC1	EMC2	EMC3	EMC4	EMC5*	EMC6*
	High Dk filler A	0	0	65	60	60	60	60	60
	High Dk filler B	0	0	0	5	10	10	0	10
Fillers	Alumina	0	95	35	35	30	30	34	22
	Silica	100	5		0	0	0	0	
	High Dk filler C	0	0	0	0	0	0	4	8
	Spiral Flow (inch)	81	71	69	66	62	51	69	63
	Dielectric constant	4	9	15	18	21	27	28	38
	Filler Contents (%)	89	89	89	87	87	89	89	89

* Flowability enhancer (0.3%) was added

Unfortunately, as the trend of capacitive fingerprint sensor being used in mobile devices will shift from button and cover types to under-materials and under-display type, the distance between a finger and an electrode becomes bigger. Thus, high Dk EMC is necessary to compensate for capacitive loss.

$$C = \varepsilon \times s / d$$

C : capacitance
ε : Relative Permittivity
s : Area of Sensor
d : Distance between Finger and Sensor

2. EMC FOR FINGERPRINT SENSORS

A. High Dk

The EMC packaging process of current fingerprint sensors is mainly conducted by a compression molding for the thinner mold clearance of FPS packaging. Initially, fingerprint sensors packaging has the mold clearance of between 100-150μm. These packaged sensors undergo a further grinding process to lower their mold clearance to 50-80μm. Lowering the packaging mold clearance below 50μm damages the reliabilities of sensors package by causing warpage, shrinkage, delamination, ununiformed appearance, and die mark. As packaging trends move to under-materials and under-display, the required mold clearance of FPS packaging increases from 50μm to 150 or 200μm. Therefore, depending on its mold clearance, EMC capacitance has to be improved between two and four times to compensate for this loss. Thus, high Dk fillers-based packaging materials (Dk=21 or 28) higher than current alumina EMC (Dk=7) have to be devised ahead of time. For that reason, various nonconductive high Dk fillers were screened in Table 1.

Among all of the candidates, Zirconia, Titanium Dioxide, and high Dk fillers were suitable for nonconductive and high dielectric constant EMC packaging composites.

However, the main barrier in applying these materials was flowability, mostly dominated by particle shape, size, and distribution. In term of morphology and rheology of commercially available fillers, Zirconia, Titanium dioxide, and crystal high Dk filler were not suitable for utmost flowabilty. On the other hand, alumina and silica fillers are relatively well established in terms of morphology and rheology. Zirconia, Titanium dioxide, and high Dk fillers A, B, C were added to these two base fillers to obtain high dielectric constant as well as high flowability.

On the basis of 87% filler content, the dielectric constants of Zirconia and Titanium dioxide were not enough to achieve the minimum specification (Dk >21 or 28 and spiral flow > 60 inch).

Table 1. Dielectric Constant of Fillers

Material	Dielectric constant (1MHz 20˚C)
Silica	3.9
Alumina	8.9
Zirconia	30
Titanium dioxide	90
High DK A, B, C*	500~2000

*DK value of high DK fillers: A< B<C

A series of experiments with different filler contents from 80 to 90% and various combinations of silica, alumina, and spherical high fillers A, B, and C were conducted. The content of high Dk fillers ramped up to 65% (EMC1) did not reach the minimum Dk value of 21 on basis of 89% filler content. On the other hand, high DK A and B filler content went up higher than 70% caused the moldability problems such as an incomplete mold and abnormal molding surface. Similarly, the mixture of high Dk fillers A and B with alumina fillers on the basis of 87% filler content (EMC2 and EMC3) did not reach Dk value more than 21. However, the increment of filler content from 87 to 89% gave dramatic Dk ramp up from 21 to 27 in the case of EMC3 and 4. Filler contents higher than 89% caused incomplete molding and abnormal molding surface. To achieve higher Dk value of more than 27, high Dk filler C was implemented. Additionally, 0.3% flow ability enhancer was added for uniformity of appearance. Two high Dk EMC (EMC5, 6) were prepared according to Dk values >28 and >38. All of these high Dk EMC were satisfied with the minimum specification of spiral flow (> 60inch.) and uniformity of appearance.

B. Color Control

The color of the compounds becomes an issue. Uniformity of EMC appearance has been considered to be an important property in fingerprint sensors [4]. Unfortunately high Dk fillers have been used as a white pigment. This pigment fades out dark color of EMC which makes any possible cosmetic defections visible such as black stains, die marks, and colored gels. To avoid these issues, organic and inorganic pigments were surveyed, especially in terms of their thermal stability. As a result, various colors of EMCs were developed and lined up by the order of darkness in Table. 3. EMC 3 is selected as a reference due to its mass production in fingerprint sensors, and its feasibility was mostly accepted and proved by the end customers. A newly surveyed inorganic pigment, additives, reduced the color difference between EMC 3 and EMC 4. In addition, a series of coloring organic compounds, which are thermally stable, were discovered and implemented.

Table 3. Colors for EMC with various fillers

	EMC 1	EMC 2	EMC 3	EMC 4
Color difference (ΔE)	18.3	12.6	0.0	1.1
High Dk Filler	O	O	X	O
Color				

Organic pigments showed higher shrinkage of EMC than inorganic pigments. A wide range of coloring EMCs can be prepared depending upon shrinkage, modulus, and die marks.

CONCLUSION

The capacitive fingerprint sensor is one of few commercially available features in miniaturized electronic devices. Due to low energy consumption and its size, its commercial preference will be maintained in the market. In the future, the application of capacitive fingerprint sensors is expected to be extended towards under-material and under-glass types of mobile devices, wearable devices,

and smart cards. In response to the future trends of fingerprint sensors, we have developed functional epoxy molding compounds with high dielectric constant (>21, >28, and >38), high flowability (>60 inch), and uniformity of EMC appearance.

REFERENCES

[1] Schwab, K. The fourth industrial revolution (Crown Business 2017)

[2] Gubbi, J. Buyya, R. Marusic, S. Palaniswami, M. Internet of Things (IoT): A vision, architectural elements, and future directions. Future generation computer systems, 29(7), 1645-1660 (2013)

[3] Saniee, I. Kamat, S. Prakash, S. Weldon, M. Will productivity growth return in the new digital era? An analysis of the potential impact on productivity of the fourth industrial revolution. Bell Labs Technical Journal (2017).

[4] Chahid, Y. Benabdellah, M. Azizi, A. Internet of things security. 2017 International Conference on Wireless Technologies, Embedded and Intelligent Systems (WITS), 1-6 (2017)

[5] Bughin, J. Chui, M. The Internet of Things: Assessing Its Potential and Identifying the Enablers Needed to Capture the Opportunity. The Internet of Things in the Modern Business Environment, 111-125 (2017)

[6] Sadeghi, A. R. Wachsmann, C. Waidner, M. Security and privacy challenges in industrial internet of things. Design Automation Conference (DAC), 52nd ACM/EDAC/IEEE, 1-6 (2015)

[7] Jain, A. Hong, L. Pankanti, S. Biometric Identification, Communications of the ACM 43, 91-98 (2000)

[8] Juniper, W. H. Securing public faith in biometrics. Biometric Technology Today 9, 7-9 (2016)

[9] Bhattacharyya, D., Ranjan, R., Alisherov, F., & Choi, M. Biometric authentication: A review. International Journal of u-and e-Service, Science and Technology 2, 13-28 (2009)

[10] Lu, D. & Wong. C. P. Materials for Advanced Packaging (Springer US, 2016)

[11] Tabei, J., Sasajima, H., & Mori, T. Epoxy molding compound for fingerprint sensor. In Electronics Packaging (ICEP), 553-559 (2016)

[12] Joshua, J. et. al. Universal 3D Wearable Fingerprint Targets: Advancing Fingerprint Reader Evaluations. arXiv preprint arXiv 1705.07972 (2017)

[13] Sheu, M.L. Tsao, L.J. A sub-fF capacitive fingerprint sensor with neighbor pixel difference sensing. 2016 5th International Symposium on Next-Generation Electronics (ISNE), 1-2 (2016)

STRUCTURING REDISTRIBUTION LAYERS DOWN TO 2 MICRON LINE SPACING WITH LASER DIRECT ABLATION

Dirk Müller[1]*, Jan Brune[2], and Rainer Pätzel[2]

[1]Coherent Inc., Santa Clara, CA 95054, USA

[2] Coherent Laser Systems GmbH & Co. KG, Hans-Böckler-Str. 12, 37079 Göttingen, Germany

*Corresponding Author's Email: dirk.mueller@coherent.com

ABSTRACT

We show how laser direct ablation can be used to yield μ-vias below 10μm diameter and line spacing feature down to 2μm. The laser direct ablation technique is applicable to photosensitive materials as well as non-photosensitive materials since the underlying principle of the laser ablation is common to all polymers. We show how by controlling the laser fluence (energy per unit area) and number of laser pulses, we can respectively control the wall angle as well as the feature depth. Eliminating costly photolithography chemistry makes laser direct ablation a cost effective approach to structure redistribution layers (RDL).

INTRODUCTION

Just as printed circuit boards allow the interconnection of different electrical subsystems, redistribution layers allow for the interconnection of IC dies within a package or to the circuit board. As integration densities increase the need for more I/O channels increases. Space becomes a premium and shrinking structures for electrical lines as well as vias become inevitable.

There are currently three methods that allow packaging houses to structure redistribution layers used for interconnecting dies and fanning out I/O lines: the incumbent photolithography, laser direct imaging (LDI, a variant on photolithography) and laser direct ablation (LDA). LDA aims to ameliorate some of the shortcomings of conventional photolithography and LDI, both of which rely on the use of photosensitive polymers. Photosensitive polymers are less mechanically stable, which makes them less ideal for RDL use. A polymer with much better thermal expansion coefficient matching is preferred. In addition, photolithography requires additional process steps that allow for yield loss and generate a higher production cost. On the other hand, it should be acknowledged that photolithography equipment is readily available and deployed tools are often fully depreciated. This leaves predominantly the high cost of consumables and the exponentially increasing cost of photosensitive materials as the features go down to 1μm.

LDI tools are generally significantly cheaper than steppers and allow the user to dynamically adjust the pattern by changing the bitmap that is being imaged. However, LDI tools struggle to reach the resolution of stepper based photolithography tools and generally have significantly lower throughput.

LDA combines photolithography's advantage of high resolution with lower cost and ability to use a larger range of polymers for the RDL.

TOOL SETUP

At the heart of LDA is a high power UV laser capable of delivering high pulse energies at repetition rates that allow for high throughput. Polymers generally have very strong UV absorption, which makes UV an ideal wavelength for RDL structuring. The strong absorption also prevents the laser radiation from unintended damage to the underlying circuitry. Excimer lasers offer the most cost effective means for high UV average power and pulse energy. Other lasers such as Diode Pumped Solid State Lasers could in principle also be used, but have much lower pulse energies and therefore can't expose a large area in a single shot.

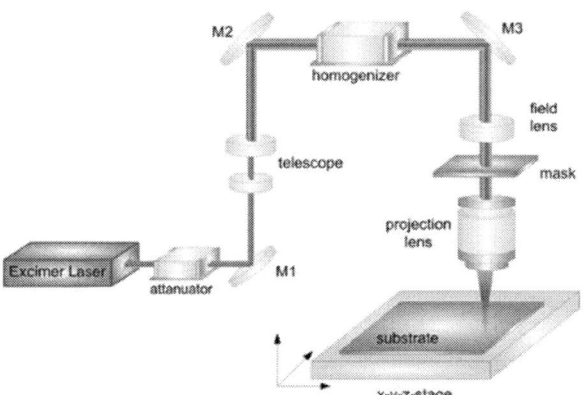

Figure 1: Schematic of Excimer laser ablation for RDL structuring. The Excimer laser's radiation at 308nm goes through beam shaping optics (telescope, homogenizer, field lens) before illuminating a mask. A projection lens de-magnifies the mask structure onto the substrate to ablate the desired circuitry with a few laser pulses..

The Excimer laser beam is conditioned by some beam shaping lenses and a homogenizer to form a rectangular beam that is very uniform over the entire area (Fig. 1). The laser beam is projected onto a mask that consists of a

978-1-5386-5309-8/18 $31.00 © 2018 IEEE

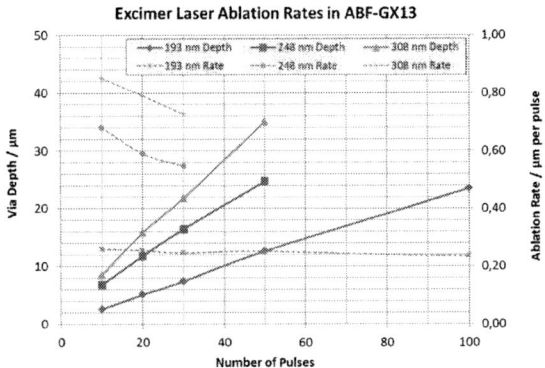

Figure 2: Graph showing the ablation depth with respect to the number of laser pulses and UV wavelength. All data taken with 1J/cm² fluence.

transparent blank with aluminum features that block out the light where ablation is not desired. The mask pattern is projected onto the substrate with a predetermined de-magnification ratio of 2 x 1, 2.5 x 1 or 5 x 1. The mask image illuminates an area of 10s to 100s of square millimeter on the substrate plane thereby covering the full RDL layer for a specific die or interconnect.

EXPERIMENTAL RESULTS

In order to ascertain the most suitable UV wavelength, we did ablation experiments on ABF-GX13 material with various UV wavelengths generated by the Excimer laser. We exposed the material to a varying

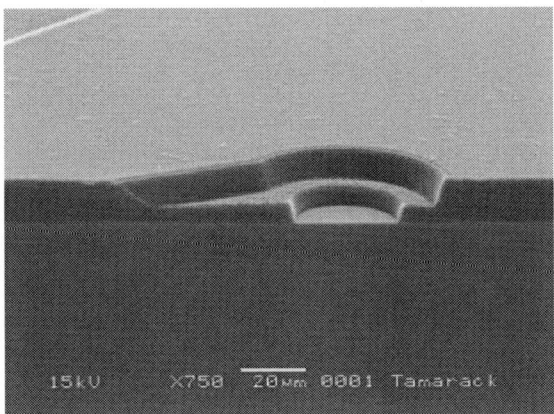

Figure 3: SEM image of via demonstrating depth control. Controlling the number of laser pulses at each location allows for precise depth control. Each pulse ablates about 0.3μm material thickness. (Courtesy of SÜSS MicroTec)

number of laser pulses and measured the depth of the material ablation (Fig. 2). As expected, 308nm shows the

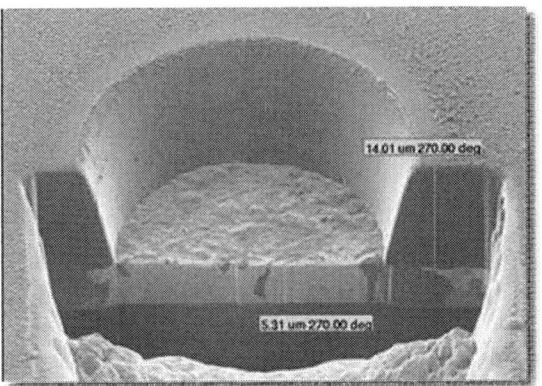

Figure 4: SEM image of laser drilled hole with metal layer underneath. The laser ablation automatically terminates at the >1μm thick metal layer.

largest ablation depth per pulse. The ablation depth is related to the absorption: the deeper the wavelength penetrates into the material, the larger the volume being ablated. This argument would speak for using an even longer wavelength than 308nm. However, as the wavelength gets even longer the polymer heats up and can char. Even more importantly, when the penetration depth gets too large, there is risk of damaging the underlying structure during the laser direct RDL structuring. A 308nm wavelength seems to be an ideal wavelength to maximize ablation rate without risk of damage. Each pulse ablates about 0.7μm in this specific material. Other materials were tested and the results showed similar ablation rates.

Further implied by figure 2 is the very precise depth control for forming vias and conductor lines that can be achieved by counting the laser pulses (Fig. 3). Alternatively, a >1μm thick metal layer can be used to terminate the ablation process. The metal layer reflects the UV radiation and automatically terminates the ablation at low fluences (Fig 4). Using a metal layer as a termination point also allows for a very steep angle between the polymer and the metal pad. In addition, we found that metal layers much thinner than 1μm can act as a seed layer. If the metal layer is much thinner than 1μm it can later be ablated by the LDA process, because such thin layers don't reflect as well. This means tailoring the metal layer thickness allows the user to either use it as a stop layer or a seed layer that is later removed.

Another knob that can be turned with LDA is the fluence level. In general one wants to operate at a fluence that provides sufficient ablation speed, but does not allow access energy to heat the material. A fluence level around 1J/cm² yields best results. Varying the fluence allows fine control on the wall angle of the ablated features. Figure 5 shows the different wall angles for three different fluence levels. At a laser fluence of less than 0.4 J/cm² we obtain a wall angle of 65 degrees. At 0.8 J/cm2 the wall angle

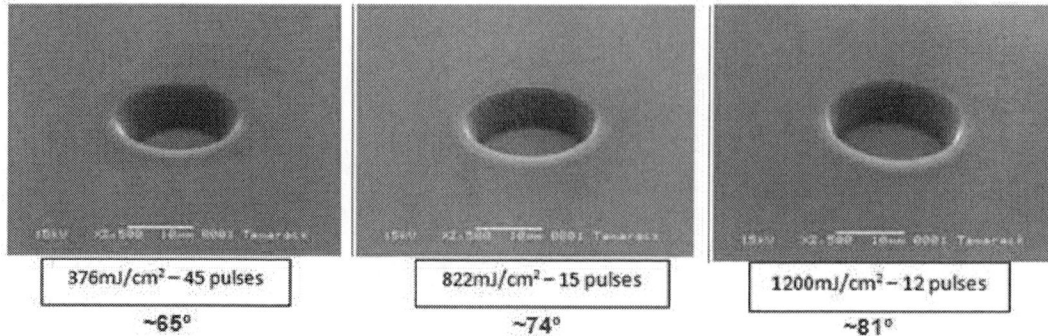

376mJ/cm² – 45 pulses	822mJ/cm² – 15 pulses	1200mJ/cm² – 12 pulses
~65°	~74°	~81°

Figure 5: SEM image showing the impact of laser fluence on wall angle. The wall angle can be adjusted by changing the laser fluence and number of pulses. As the laser fluence decreases (energy/area) the wall angle become more shallow and less material is ablated with each pulse. The reduction of material removal per pulse is compensated by using more pulses for each area.

increases to 74 degrees and at a fluence of 1.2 J/cm2 we obtain a wall angle of more than 80 degrees. It should be noted that increasing the fluence also increases the ablation thickness per pulse. So when the fluence is changed the number of pulses should be adjusted to yield the desired ablation depth.

Some advanced Fan-Out Wafer Level Packages demand line spaces down to 2µm. A similar mask pattern was used to create lines with 2µm, 3µm and 4µm L/S in ABF GY50 with small filler (Fig 6). The target depth of 4.5µm has been achieved and lines down to 2µm can be resolved with an aspect ratio of 1:2 width versus depth.

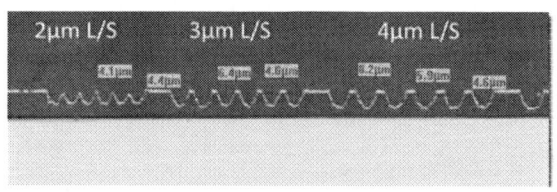

Figure 6: Microscope image showing structured line spacing down to 2µm as achieved by laser direct ablation.

Similar results have been achieved in Fujifilm polyimide films, where L/S of 2.4µm has been demonstrated.

One concern had been the particles that are being generated by direct laser ablation. This is a legitimate concern, but is easily addressable. The UV ablation is based on high energy photons breaking atomic bonds. This results in very fine particles that can be evacuated during ablation with an air flow across the substrate. Any remaining particles can be cleaned by a plasma. Alternatively one can apply a water soluble film on the substrate before the ablation process. This film can be rinsed off using deionized water. Rinsing off the protective film also carries away any remaining particles.

Another common concern is the cost of operating Excimer lasers. Excimer lasers are used excessively in the flat panel display industry for annealing of silicon films as well as for laser lift off during the manufacturing of flexible displays. As a consequence of the increased use of Excimer lasers much effort has been given to lowering the cost of operation through innovative engineering advances. These improvements have significantly increased the lifetime of Excimer lasers and lowered their operating cost. In addition, more recent power and pulse energy scaling have made laser direct ablation much more feasible as the mask image is large enough to cover the vast majority of RDL sizes used today.

CONCLUSION

We have demonstrated direct laser ablation for structuring redistribution layers used in advanced packaging architectures. Excimer-based laser direct ablation ameliorates some of the shortcomings of conventional photolithography or laser direct imaging. Laser direct ablation has been demonstrated to work on photosensitive as well as non-photosensitive materials allowing users access to a larger span of materials. By controlling the fluence and number of pulses users have control over the wall angle and depth of the features created. Metal layers of >1µm can serve as a termination layer. Thinner metal layers can be used as seed layers and removed with laser direct ablation after metal plating. Laser direct ablation tools are already operating in high volume manufacturing and growing interest points toward further proliferation in the coming years.

ACKNOWLEDGEMENTS

The authors would like to acknowledge contributions by SÜSS MicroTec Photonic Systems Inc. and the Georgia Tech Packaging Research Center.

978-1-5386-5309-8/18 $31.00 © 2018 IEEE

NEW APPLICATION METHOD FOR
PACKAGE LEVEL EMI SHIELD COATING

Stuart Erickson

Ultrasonic Systems, Inc.; Haverhill, MA 01835 USA

serickson@ultraspray.com

ABSTRACT

Package-level EMI (electromagnetic interference) shielding is increasingly important as highly sensitive components become more tightly packed in circuit assemblies. The use of sputtering and plating to apply a thin, conductive metal layer, less than 10μm, to individual packages has proven to be effective in providing EMI shielding, albeit at a substantial process cost.

New "sprayable" coatings have been developed that provide effective EMI shielding performance when applied in an ultra-thin layer. T-CAT (ultra-Thin Coating Application Technology), a novel application method has been developed to apply these new coatings in a uniform, ultra-thin layer on individual packages while reducing the process cost by up to 60%. [1]

INTRODUCTION

The need for EMI shielding is increasingly important as highly sensitive components become more tightly packed in circuit assemblies. To accommodate the move toward miniaturization and to reduce weight and thickness, each individual package requires EMI shielding [2]. Sputtering and plating are commonly used to apply an ultra-thin metal layer to individual components. A layer thickness between 3 and 6 μm has proven effective in providing the required EMI shielding. However, sputtering and plating require a substantial process cost.

New EMI shield coating materials have recently been developed that can be applied with a much simpler and cost-effective spray coating process. These new materials provide equivalent EMI shielding effectiveness as sputtering as long as they are applied in a uniform and conformal layer on the component surfaces.

This paper describes a novel direct-spray method called T-CAT to apply these new EMI shield materials in a uniform and conformal layer onto the top and side surfaces of individual components. T-CAT utilizes "nozzle-less" ultrasonic spray technology combined with a precision coating system platform.

T-CAT

T-CAT consists of a coating system platform that contains a nozzle-less ultrasonic spray head, a precision metering pump liquid delivery system, a spray head motion and positioning system, as well as a transport system for the substrates to be coated.

Nozzle-Less Ultrasonic Spray Head

With this technique, spray is produced without the use of any type of nozzle. Instead, liquid is broken up into small drops by ultrasonic energy, then accelerated and expanded by an adjustable velocity air stream.

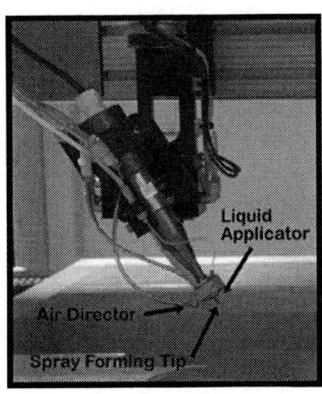

Fig. 1: Nozzle-less Ultrasonic Spray Head Assembly

As shown in Figure 1 the nozzle-less ultrasonic spray head assembly consists of an ultrasonic transducer with a spray forming-tip, an integrated liquid applicator, and air director.

The liquid applied directly to the spray forming tip is broken into small drops by ultrasonic energy. The ultrasonically produced spray is then expanded and accelerated by an adjustable velocity air stream to produce a uniform, rectilinear coating pattern.

This technology is capable of spraying a wide variety of materials from pure solutions to suspensions and slurries while producing a uniform coating layer on the substrate. [3]

The new EMI shield materials are high-density slurries with proprietary formulations that produce a continuous conductive layer on the substrate after coating.

Liquid Delivery System

A special liquid delivery system has been developed that incorporates a positive displacement metering pump as shown in Figure 2. The liquid flow rate is controlled by micro-stepping drive ensure that the coating material is delivered to the ultrasonic spray head at a precisely controlled flow rate.

Since some coating liquids used for EMI shields have suspended particles that can settle so this liquid delivery system also has the capability to keep the coating mixed while maintaining an accurate and repeatable flow rate of the coating to the spray head.

A dual pump version is also available that allows continuous operation, without the need to wait for the single pump to refill. With this configuration, as one pump is being used to feed the spray head, the other pump is being filled from the holding reservoir. The dual pump can also be configured to recirculate and stir the coating liquid, thus ensuring that the suspended particles do not settle out in the liquid feed lines or holding reservoir; see Figure 3.

978-1-5386-5309-8/18 $31.00 © 2018 IEEE

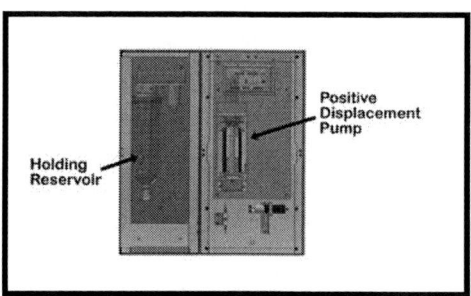

Fig. 2: Liquid Delivery System

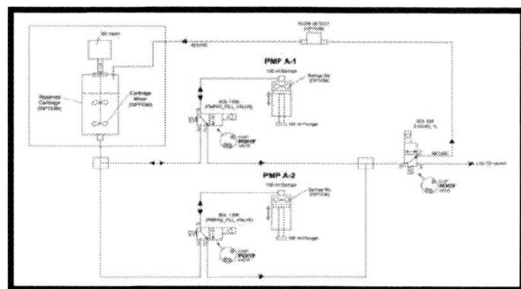

Fig 3: Dual Pump Liquid Delivery System

Coating System Platform

The coating system platform, as shown in Figure 4, consists of an X-Y-Z-Θ-Φ gantry system for the motion and positioning of the nozzle-less ultrasonic spray head, the liquid delivery system and a transport mechanism for the substrates. All critical process parameters are managed by the platform software and control system. Additionally, a log of the critical process data is recorded and stored in a process log file.

Fig. 4: Coating System Platform

In general, the uniformity and thickness of the applied coating layer are directly related to the following factors:

1) Uniformity of the spray pattern produced by the ultrasonic spray head

2) The stability of the coating flow rate delivered to the ultrasonic spray head

3) The consistency of the speed of the ultrasonic spray head relative to the substrate

4) The programmed coating recipe – number of applied layers, the application path, and the coating flow rate

The uniformity of the coating layer for EMI shield materials can be determined by cross-sectioning the components and measuring the thickness of the coating layer on the surface.

The speed and motion path of the ultrasonic spray head is controlled with a closed-loop servo drive system. The servo drive is tuned for uniform head speed so the variation of the head speed is negligible.

The liquid flow rate is determined by the rate of displacement of a piston in the pump. The motion of the piston is controlled by a micro-stepping drive system so the variation of liquid flow rate is also negligible.

For a given coating recipe, the variation in the total amount of liquid applied to the substrate is less than ±0.1%. This can easily be verified since the system software logs the total amount of liquid applied to each substrate that is coated. Any variations in the head speed and the liquid flow rate are accounted for in the process log.

EMI SHIELD COATING APPLICATION PROCESS

This direct spray application process for the new EMI shield materials is simple and easy. It consists of five (5) steps as shown in Figure 4: 1) place components on carrier; 2) load carrier into coating system; 3) activate spray coating process; 4) cure coated substrate with components in oven and 5) remove components from carrier after curing. There are no special pretreatment processes required for the components other than normal cleaning.

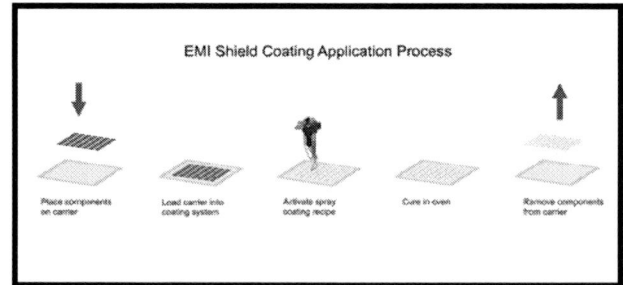

Fig.5: EMI Shield Coating Application Process

Spray Coating Application Sequence

When the carrier with the components enters the coating system the coating is applied in a defined sequence to ensure that all component surfaces are uniformly coated. In order to coat the top and side surfaces of the components on the carrier, four (4) coating layers are applied with the approach angle and direction of the spray head changed for each layer as illustrated in figure 5. The ultrasonic spray head is mounted on a 5-position rotate and tilt mechanism so that all component surfaces can be coated. The tilt angle is adjustable for process optimization.

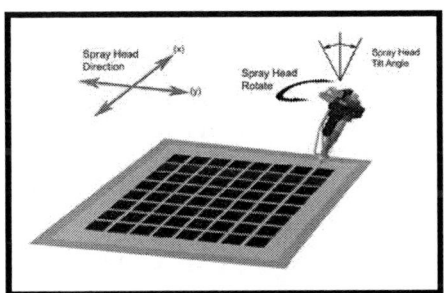

Fig 6: Coating Application Sequence

Coating Uniformity Data

The EMI shield coating thickness on the top and side surfaces of the components can be measured after the component is cross-sectioned. The average side surface thickness is approximately 50 to 60% of the top surface thickness as shown in Figure 6. The process is also very repeatable as shown in Figure 7. [4]

- Top surface coating uniformity is 6 μm ±0.5 μm.
- Side surface coating uniformity is 3 μm ±0.5 μm.

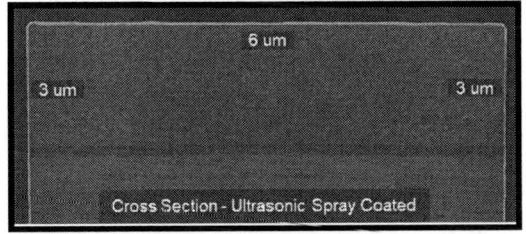

Fig 7: Typical Coating Uniformity on Component Surface

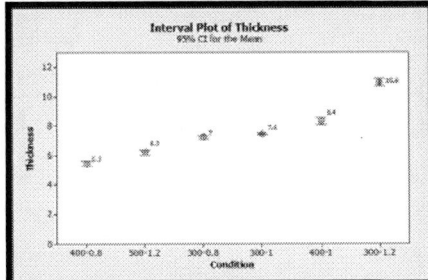

Fig. 8: Process Repeatability

Process Cost

Table 1 is a summary of the process cost between typical sputtering and T-CAT spray to apply an EMI shield layer on individual packages. The sputtering process involves a complex, multi-step process for the application of EMI shield coating and requires costly equipment that occupies a large amount of floor space. Alternatively, T-CAT is a simple process that requires much less capital investment and occupies much less floor space.

The capital investment for equipment for T-CAT vs. sputtering is up to 90% less and the throughput is up to 4 times higher. [4]

TABLE 1- Process Cost Comparison

Item	Performance Data	
	Typical Sputtering	**T-CAT Spray**
Carrier Shapes	Ring, square, or rectangle	Ring, square, or rectangle
Carrier Size*	350 mm x 350 mm	350 mm x 350 mm
Package Dimensions*	7 mm x 7 mm x 0.8 mm	7 mm x 7 mm x 0.8 mm
Carriers / hour Processed	~10	~40
Packages / hour Processed	~14,000	~72,000
Floor Space Required	12.5 to 35 m^2	2.5 to 4.5 m^2
Ave Equipment Cost	$3M to $8M**	$.2M to .5M**

*used for cost calculations
**depending upon configuration and target throughput

CONCLUSION

The increasing need for EMI shielding on individual components has prompted the development of new coatings and application methods. The combination of nozzle-free ultrasonic spray technology, a precision liquid handling and delivery system as well as an advanced coating system platform is a significant step towards applying these new EMI shield coatings reliably and consistently in a cost-effective manufacturing process.

ACKNOWLEDGEMENTS

I would like to thank my colleagues in applications engineering at USI for the countless hours of trials required to refine and collect data for this process.

REFERENCES

[1] Henkel Electronics, "EMI shield spray coating process cost", unpublished.

[2] Tatsuya Kawamura, "Package trends for mobile devices", February 2015

[3] Tao Wang and David Lidzey, Et al., "Fabricating high performance, donor–acceptor copolymer solar cells by spray-coating in air", Advanced Energy Materials, 2013

[4] Loctite Ablestik EMI 8880S Data Package, "Conformal EMI shielding material", August 2017

THIN DIE STACKING TECHNOLOGIES FOR 3D MEMORY PACKAGES

Jie Wu[1], Oranna Yauw[1], Andrew Tan[1], Horst Clauberg[2] and Daniel Buergi[2]
[1] Kulicke & Soffa Pte Ltd, 23A Serangoon North Avenue 5, Singapore 554369
[2] Kulicke & Soffa Industries Inc., 1005 Virginia Drive, Fort Washington, PA 19034, USA
*Corresponding Author's Email: jwu@kns.com

ABSTRACT

To meet the need for more compact memory packages, the thickness of memory die and die-attach film (DAF) is pushing to below 25um and 10um, respectively. Packages with 16 layers and 32 layers have become common in memory applications. In this paper, thin die tooling design utilizing FEA modeling is introduced. Thin die stacking up to 16 layers with both 50um and 15um thickness of dice will be studied. DOE of pickup process parameters are included to understand the critical factors most contributing to a fast and stable die pickup. Temperature gradient data and their effect on stacked dice over the entire 16 layers is also considered.

INTRODUCTION

To achieve superior performance with more compact package sizes, the memory industry is pushing the limits of the number of stacked die layers and total package thickness. With the help of new technologies and materials such as UV release foils and thermal release foils with low adhesion force [1][2], processing thin die having a thickness below 25um has become possible. 16-layer and even 32-layer of thin die stacking have been commonly running in high volume production. By reducing the die thicknesses, challenges arise in areas such as die pickup, and die warpage (i.e. the "potato chip effect"). By increasing the number of stacked layers, maintaining stable site temperature during die bonding becomes more difficult as the bond stage temperature is not able to reach the top die layers fast enough. These areas of concern in thin die stacked bonding will be the focus of this paper and will be discussed with solutions proposed and validated.

EXPERIMENT

To optimize the design of K&S patented needless thin die ejector, a finite element analysis (FEA) model was first built to simulate the deformation and stress distribution on the dice as well as the back tape during the die pick up process. In the FEA model, factors such as pick force, ambient pressure, vacuum pressure, deformation, and material properties have been all considered [3].

To determine the dominant factors that affect the thin die pickup process, a full 2-level factorial DOE of die pick based on 15um-thick die was conducted in this study. The details of the thin die used for the pickup DOE is listed in Table I.

TABLE I   CONFIGURATION OF DIE STACKING

	Pick & Place	Temperature Gradient
Die	9.7 × 17.1 × 0.015mm	10.1 × 10.6 × 0.050mm
DAF	10um thick	10um thick
Stacks	16-layer	16-layer
Bond Temp	110°C	130°C
Bond Force	20N	20N

Using the same configuration, the study of die warpage during die stacking was also conducted [4]. Thin die stacking was performed with 16-layer die stacks in Figure 1(a).

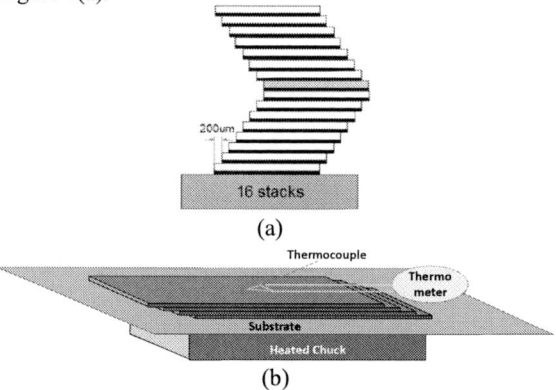

(a)

(b)

Figure 1: (a) Schematic of 16-layer die stacks. (b) Schematic of on-site temperature measurement with thermocouple on 4-layer die stack.

Lastly, the study of temperature gradient of stacked dice was conducted based on 50um-thick dice, as listed in Table I. A thermocouple was adopted to measure the onsite temperature. Noting that the wire diameter of the thermocouple is 50um, it could cause die crack if the thermocouple was embedded between die stacks. In this study, a thermocouple was attached to the top die of the stacked dice, as shown in Figure 1(b). The ambient temperature is about 25°C and the chuck temperature is set to 130°C. To understand the correlation between the die stacked number and the actual die temperature, the measurement was conducted on various numbers of layers of stacked dice, i.e., 4-layer, 8-layer, 12-layer and 16-layer of die stacks.

All of the thin die wafers were prepared with blade dicing and the operation of die bonding was conducted on a K&S APAMA® DA die bonder.

RESULTS AND DISCUSSION
Tooling Design with FEA

With FEA, the status of die peeling, as well as stress distribution in the center and neighboring dice on different designs of piston die ejector, was analyzed. An example of the FEA result for the stage of initial peeling is shown in Figure 2. The green color along the edges of the center die in Figure 2(a) indicates the peeled areas. The red spots in Figure 2(b) represent the max principle stress in the neighboring dice.

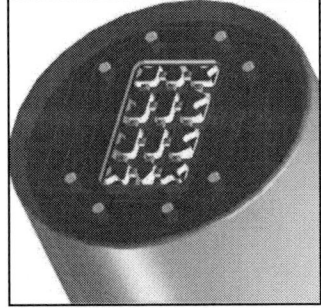

(a)   (b)

Figure 2: FEA example for the stage of initial peeling. (a) peeled/unpeeled areas, (b) stress distribution in die.

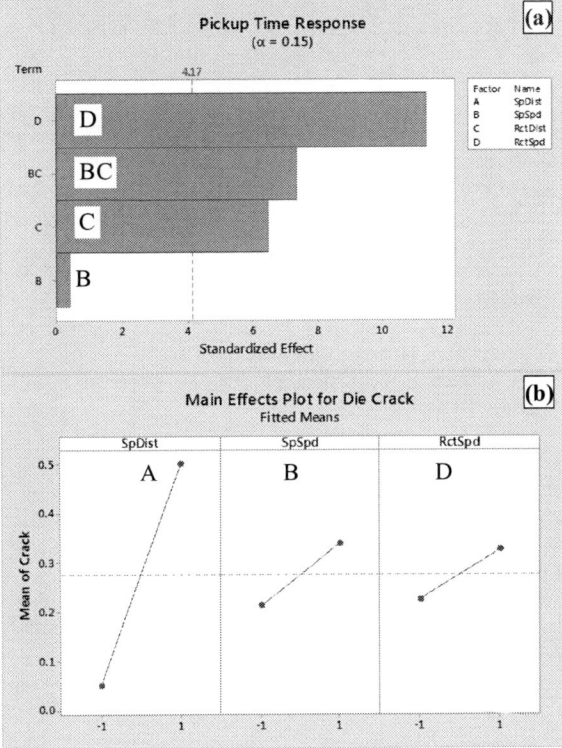

Figure 3: Design of piston die ejector

The optimum design of the piston die ejector is shown in Figure 3. The special piston corners can minimize the die stress but maximize the peel stress along the die edges at both the center and neighboring dice. The non-movable small pyramids on the top of the chuck of the die ejector can support the die while the die is peeling from the wafer tape.

DOE of Pickup Process

With a piston die ejector, picking up a die is accomplished by two stages, initial peeling and final peeling, as shown in Figure 4. During the initial peeling stage, the piston is ejected to push the target die up. Subsequently, the piston is retracted to complete the final peeling of the tape from the target die. The moving distance and speed of the piston during the two stages of

peeling are key parameters for thin die pickup.

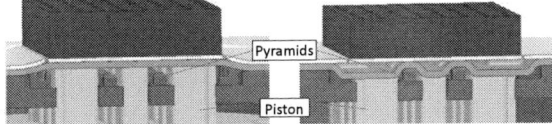

Figure 4: Thin die pickup process with a piston ejector: (a) initial peeling of tape at die periphery, (b) ejector retraction and final peeling.

Four factors were included in the DOE: separation distance (factor A); separation speed (factor B); retraction distance (factor C); and retraction speed (factor D). The responses to be investigated include pickup success rate (based on 10 pickup), die crack rate (based on 10 pickup) and pickup time. The response of pickup time with die crack rate as the weight is shown in Figure 5.

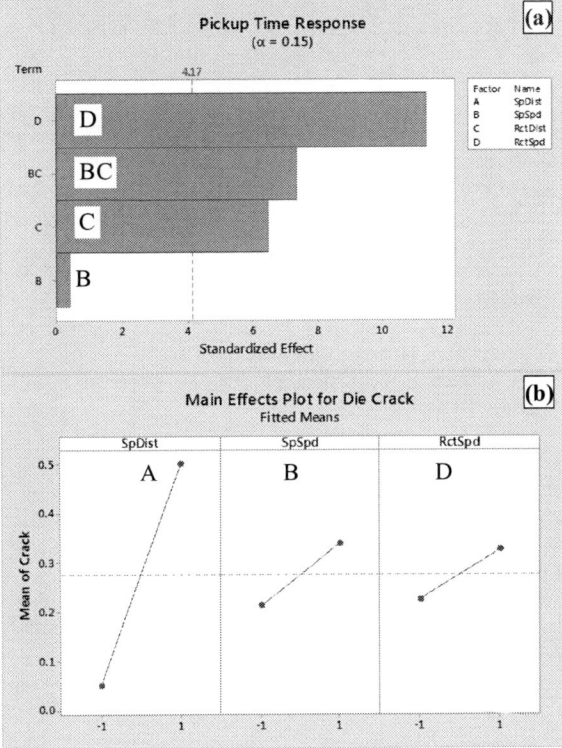

Figure 5: Analysis with the Factorial DOE. (a) Response of pickup time. (b) Main effects for Die Crack.

With the DOE analysis, some trends can be learned. The retraction distance (factor C) is not considered critical to die cracking. To reduce the cycle time of the die pickup process without increasing the risk of die cracking, the retraction speed (factor D) is the main factor to adjust. It is also noted that the interaction between the separation speed and the retraction distance also has a strong impact on the die pickup speed.

Thin Die Warpage

For dice with less than 20um die thickness, die warpage was occasionally observed. In this study, preliminary testing of die warpage based on bare silicon die with DAF was carried out to understand the correlation between stacked dice and die warpage.

With 16-layer of die stacks, no die warpage was observed. The measured overall die warpage of the stacked dice is less than 25um. The SEM picture of one bonded 16-layer die stack is shown in Figure 6.

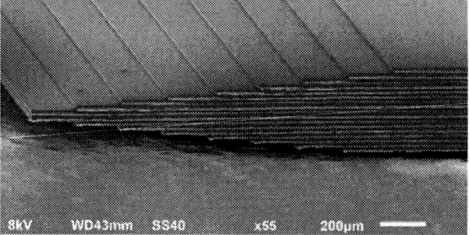

Figure 6: SEM image of a 16-layer of thin die stack.

In this study, no direct correlation between die thickness or number of die stacks and die warpage was observed. The cause of the die warpage is more due to the inherent deformation of dice before picking up and placing [4]. More specific tests will be carried out to determine the root cause of die warpage and to understand the correlation between process parameters and final die warpage.

Temperature Gradient on Stacked Dice

Conductive heat transfer can be expressed with "Fourier's Law" as

$$q = (k/s)AdT \qquad (1)$$

where q is heat transfer rate, k is thermal conductivity of material, s is material thickness, A is heat transfer area and dT is temperature difference. Equation (1) reveals that with the same values of change in temperature, the thicker material has a lower transferring rate.

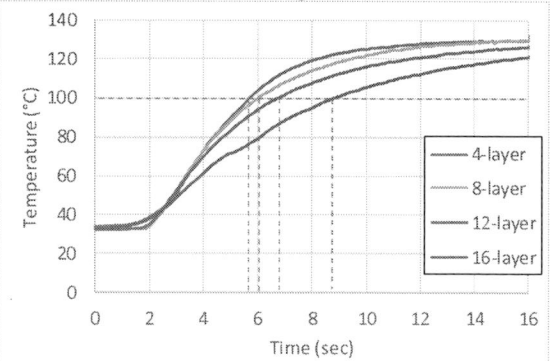

Figure 7: Temperature gradient on stacked dice.

As shown in Figure , the measured temperature gradient matched the theory well. The substrate with the bonded stacked dice physically contacted the hot chuck after approximately 2 seconds. The temperature ramping rate reduced with an increasing number of die stacks. From 80°C to 100°C, 4-layer stacks need less than 2 seconds while 16-layer stacks need about 4 seconds. After 16 seconds, the temperature of the 4-layer and 8-layer stacks saturated at 130°C while the temperatures of the 12-layer and the 16-layer of dice were still ramping up slowly.

Take note that the thermocouple was placed on the die surface and was exposed to air. The measured temperature is likely to be lower than the actual temperature at the DAF interface between dies, since the top die surface has a higher rate of heat loss to the ambient. However, the dramatic difference of the temperature ramping speed between 4-layer and 16-layer of dice is worth highlighting as it indicates that stacked dice with high stacks (more than 16 layers) is prone to bonding issues caused by insufficient bonding temperature.

K&S APAMA® DA provides a solution with a heated place tool to take care of this temperature gradient effect. With a heated place tool to provide heat directly to the top die surface, the temperature on the die to be bonded will be more controllable and the quality of the bonded DAF will be more consistent. Applying heat directly from the place tool to the bonded die also speeds up the heat rate, allowing shorter place times and a faster process cycle.

CONCLUSIONS

With the guide from the FEA model on the design of the piston die ejector, the geometry of the piston can be optimized for thin die picking up. The pickup DOE indicates how to optimize the pick process parameters to balance the speed and success rate. The DOE confirmed that the initial peeling of the die from the wafer tape is the key to achieve a reliable thin die pickup. This study also illustrated that the surface temperature of the dice with higher stacks is not consistent. A feasible solution using K&S APAMA® DA heated place tool has been proposed and further study will be conducted in the near future. All the findings in thin die stacking in this study are expected to benefit memory applications in volume production.

REFERENCES

[1] A. Marte, *IMAPS Conference*, 2005, pp. 386-396
[2] N. Ye, *Proceedings of IEEE 67th Electronics Components and Technology Conference (ECTC)*, 2016, pp 1840-1846.
[3] I. Qin, *Proceedings of IEEE 67th Electronics Components and Technology Conference (ECTC)*, 2017, pp. 1309-1315.
[4] M. Hassan, *Proceedings of Electronic System-Integration Technology Conference (ESTC)*, 2010, pp. 1-5.

ELECTROPLATED NANOTWIN COPPER FOR FINE LINE RDL

Stream Chung*, Yao-Tsong Chen, and Zong-Cyuan Chen
Chemleader Corporation, Hsinchu, Taiwan
*Corresponding Author's Email: stream_chung@chemleader.com.tw

ABSTRACT

Electroplated unidirectional (111) nanotwin copper has shown a high thermal stability and tensile strength. It is suitable for multi-layer fine line RDL in fan out wafer level packaging. With the line width keeps shrinking to 1μm, nanotwin copper with submicron grain size and low transition layer thickness from substrate is required. In this paper, we demonstrate a nanotwin copper of submicron grain plated using DC and proprietary additives as well as its mechanical properties and thermal stability.

INTRODUCTION

Fan out wafer level packaging has shown a cost efficient scheme for multi-chip packaging. In order to enhance the electric bandwidth between chips, multi-layer fine line RDL is needed. Multi-layer RDL built-up requires a high temperature annealing for dielectric polymer curing after copper line plating in each layer. The thermal process causes copper recrystallization, and results in copper RDL strength reduction. In the meantime, the RDL lines cross the interface between silicon and molding compound, and the coefficient of thermal expansion difference between these two materials induces significant stress on the copper line. Cracks are found in the RDL line when line width shrinks. A copper with high tensile strength and high thermal stability is then needed for this application.

Electroplated nanotwin copper is first introduced by Lu et al in 2004 [1]. Its superior mechanical strength and electrical conductivity are recognized as a suitable material for electric application. Potential on wafer level packaging RDL has been explored by Lee et al [2], and wafer warpage reduction was proposed over traditional RDL from high thermal stability and yield strength of nanotwin copper. Despite the advantage of nanotwin copper, it was not actually used in wafer mass production due to the process complexity of pulse plating and slow deposition rate. Hsiao et al [3] later developed a method using DC plating with reasonable deposition rate for unidirectional (111) nanotwin copper with excellent thermal stability [4]. In this method, a transition layer between copper seed layer and ordered nanotwin grains is formed during plating, and its thickness is found sensitive to the copper seed crystalline orientation [5]. The transition layer thickness could grow to several micrometers in some cases, and occupy most of the deposit volume in fine line RDL application [6]. Strong agitation and pulse current were found effective in reducing transition layer thickness. However, in order to be implemented in the industry, an evolved plating chemistry compatible with conventional plating apparatus is needed.

In this paper, we demonstrate a new copper plating chemistry for unidirectional (111) nanotwin copper. It generates (111) -oriented columnar grains of submicron grain diameter, and very thin transition layer on wafer substrate, suitable for fine line RDL in fan out wafer level packaging.

MATERIALS AND METHODS

Plating Chemicals and Cell

The electroplating bath (VMS) consists of 50g/L copper from copper sulfate, 50g/L sulfuric acid, and 50mg/L chloride from hydrochloric acid. All chemicals are reagent grade. Commercial additives, 1st generation and 2nd generation, from Chemleader are added to VMS. Plating is conducted in a 15 cm by 15 cm square cell with 2L plating solution. Cathode and anode are placed on parallel cell side walls. Phosphorous copper plate is used as anode, and sputtered Cu substrate is used as cathode. Solution agitation is fixed at 300 rpm using mechanical lab mixer with a 3cm diameter propeller in the center of the cell. A potentiostat (AUTOLAB 302N) is used to provide constant current and measure electrochemical potential during plating.

Tensile Test

Electroplated copper film was peeled off from blanket Si wafer, and punched to copper strip for tensile test. The test strip dimension is 18 μm thicknesss, 100 mm length and 12.7 mm width. Tensile test was conducted on INSTRON 4465 at strain rate 0.1 mm/mm per minute with gauge length 50 mm. For post anneal test coupon, electroplated Cu film was annealed in a nitrogen oven before peel-off from Si substrate.

Deposit Stress Test

Bent strip method following ASTM Standard B975 was used to study the internal stress of Cu film. Cu films of 50μm thickness from different plating chemistries were plated on test strip PN 1194 Cu-Fe, and leg distance is recorded without conversion to stress.

RESULTS AND DISCUSSIONS

Microstructure and Appearance

Figure 1 shows the cross section image and surface condition of deposits from 1st and 2nd generation nanotwin copper chemistries. Copper with nanotwins is in

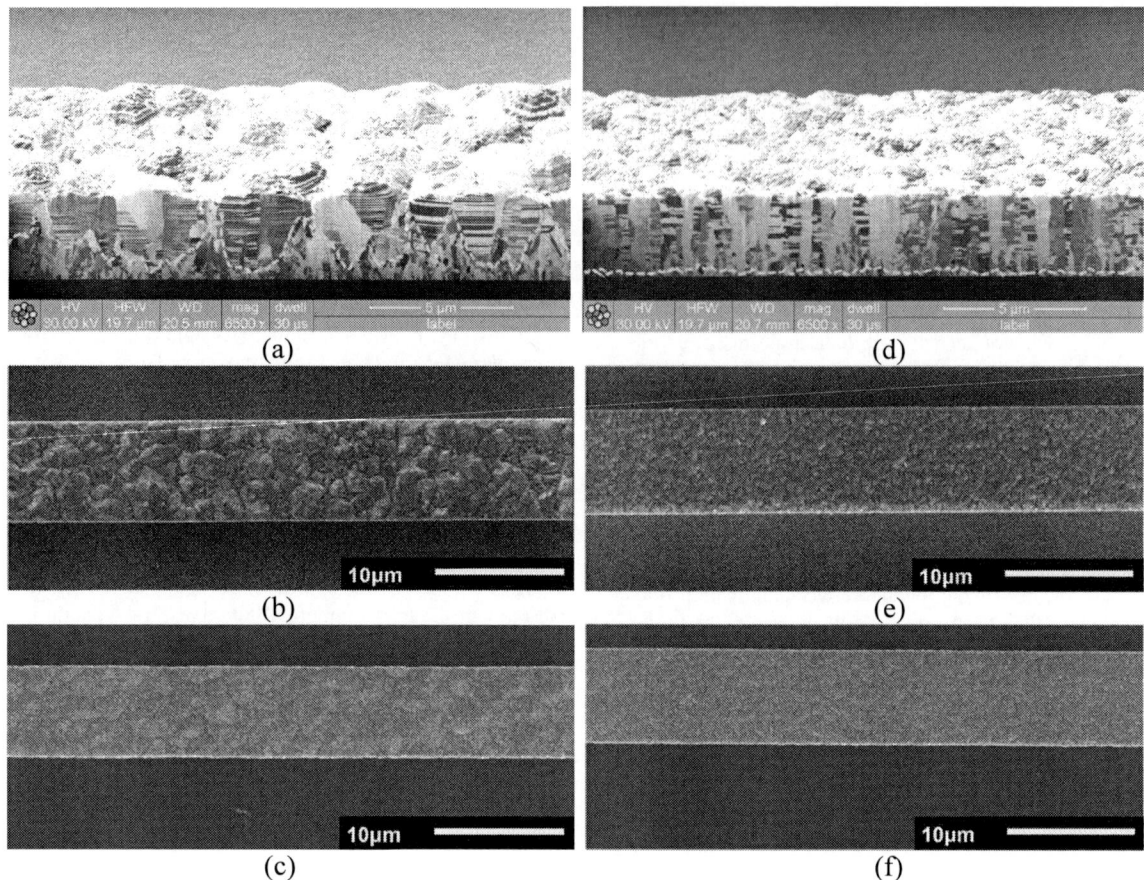

(a)   (d)

(b)   (e)

(c)   (f)

Figure 1. RDL with (111) nanotwin copper plated at 4ASD DC, 4μm thickness with 1st generation nanotwin copper chemistry (a) cross section FIB image, (b)as plated, (c) post UBM etched, and with 2nd generation nanotwin copper chemistry: (d) cross section FIB image, (e)as plated, (f) post UBM etched. The green dash line marks the boundary between nanotwin copper and transition layer to substrate. The 2nd generation nanotwin copper shows significant thinner transition layer and much finer grain.

columnar grain structure along (111) plane direction. A transition layer between copper seed and columnar grain is identified, and the frontline of transition layer indicated by the dash-green line has many peaks and valleys, as shown in Figure 1(a). This feature implies that columnar grains start from few nucleation sites, and later swallows the non-(111) grains during grain growth regulated by plating chemistry. Grains within transition layer do not have dense twins, and will recrystallize into larger grain during thermal anneal. With transition layer occupies such a big portion in the bulk, it weakens the line strength, and results in crack failure under external stress [6]. By increasing nucleation sites of (111) nanotwin grain with modified additive in the 2nd generation plating chemistry, as shown in Figure 1(d)(e)(f), the non-(111) grain growth diminish in earlier stage, and results in thin transition layer, finer columnar grains, and also smoother surface, which are critical for line width below 2μm. Surface roughness of

the RDL are shown in Table I.

Mechanical Properties

Tensile strength of nanotwin copper from 2nd generation plating chemistry is shown in Figure 2. The as plated samples were tested within 24 hours after plating. The post annealed samples were tested after 180°C 1hour annealing. Nanotwin copper shows much higher strength compared to randomly oriented copper, and does not lose much strength after annealing. Tensile strength of nanotwin copper increases with plating current density. Twin boundaries are known to block dislocation motion

TABLE I. Surface roughness of RDL with nanotwin Cu

	1st gen chemistry		2nd gen chemistry	
	as plated	*post etched*	*as plated*	*post etched*
Ra	0.159	0.150	0.033	0.030

(μm)

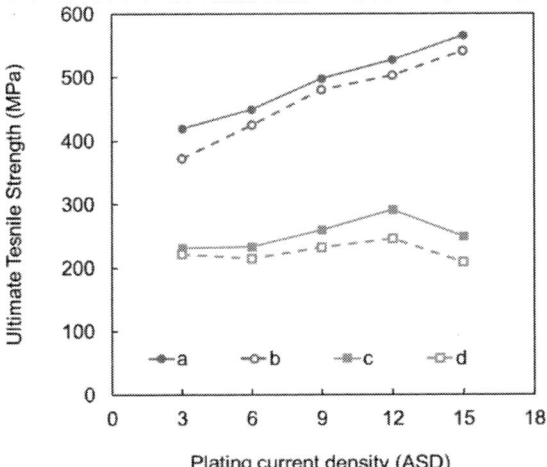

Figure 2. Tensile strength variation with plating current density (a) nanotwin copper as plated, (b) nanotwin copper post annealed, (c) general randomly oriented copper as plated, and (d) general randomly oriented copper post annealed.

for material strengthening, and also responsible for grain boundary thermal stability [7]. Without recrystallization, internal stress in the deposit is found stable during thermal process, and decreases with plating temperature, as shown in Figure 3.

Recent study from Pan et al reveals that highly oriented (111) nanotwin copper exhibits an unusual history-independent response under cyclic loading

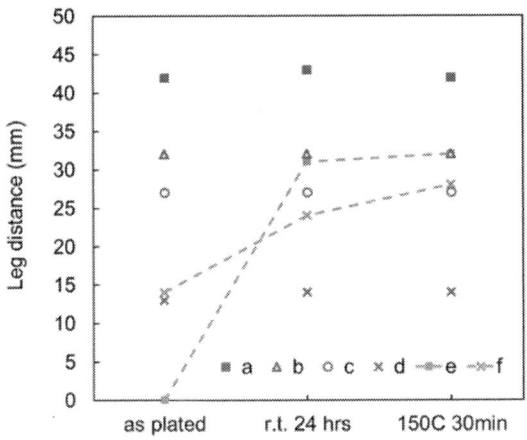

Figure 3. Deposit stress of copper as plated, room temperature storage for 24 hours, and 150 ℃ 30 min annealed. Nanotwin copper plated (a) 25 ℃ (b) 30 ℃ (c) 35 ℃ (d) 40 ℃. Commercial bright copper chemistry plated at (e) 25 ℃ and (f) 40 ℃

different from the general metal fatigue behavior [8]. A new type of dislocation, correlated necklace dislocation, identified moving back and forth along the twin boundary preserves the coherency and stability of twin boundary. It carries plastic strain in neighboring twin lamellae without strain localization and dislocation entanglements, and greatly reduces the fatigue failure. This finding provides an additional support of (111) nanotwin copper in the interconnect application where cyclic stress from temperature change during operation is the main reliability concern.

CONCLUSIONS

Nanotwin copper with thin transition layer and submicron fine grain from higher (111) nucleation density has been developed from a new chemistry compatible with existing industrial plating infrastructure. Superior tensile strength and thermal stability over general copper suggest an appropriate solution for fine line RDL in fan out wafer level packaging.

ACKNOWLEDGEMENTS

The authors would like to thank Prof. Chih Chen, Department of Materials Science and Engineering, National Chiao Tung University, Taiwan for FIB support, and Dr. Chiuyen Chiou and Dr. Jhen-Rong Chen, Material and Chemical Research Laboratories, Industrial Technology Research Institute, Taiwan for tensile strength measurement.

REFERENCES

[1] L. Lu, Y. Shen, X. Chen, L. Qian, and K. Lu, *Science*, Vol. 304, pp. 422-426, 2004

[2] H. Lee, W. Ning, C. Zhu, G. Xu, and L. Luo, *IEEE 64th Electronic Components and Technology Conference (ECTC)*, Orlando, FL, US, Sep. 15, 2014

[3] H.Y. Hsiao, C.M. Liu, H.W. Lin, T.C. Liu, C.L. Lu, Y.S. Huang, C. Chen, and K.N. Tu, *Science*, Vol. 336, pp. 1007-1010, 2012

[4] Y.S. Huang, C.M. Liu, W.L. Chiu, and C. Chen, *Scripta Materialia*, Vol. 89, pp. 5-8, 2014

[5] C.M. Liu, H.W. Lin, C.L. Lu, and C. Chen, *Scientific Reports*, 4:6123, DOI:10.1038/srep06123, 2014

[6] S. Chung, Z.C. Chen, Y.Z. Chen, Y.C. Chu, K.J. Chen, C.H. Tseng, and C. Chen, *13th International Wafer Level Packaging Conference*, San Jose, CA, US, Oct. 18, 2016

[7] R. Niu, K. Han, Y. Su, T. Besara, T. M. Siegrist, and X. Zuo, *Scientific Reports*, 6:31410, DOI: 10.1038 /srep31410, 2016

[8] Q. Pan, H. Zhou, Q. Lu, H. Gao, and L. Lu, *Nature*, Vol. 551, pp. 214-217, 2017

METHOD TO REDUCE METAL FILAMENT ON DIE WITH BACKSIDE METAL

Bo Peng, Deguang Zheng, Yue Yu

Wafer Test Department, NXP semiconductor (China) Ltd., Tianjin, China

*Corresponding Author's Email: deane.zheng@nxp.com

ABSTRACT

Wafer saw is one of the critical elements of the IC assembly process where improvements can make a major contribution to yield. Metal filament on die with backside metal has been identified by semiconductor manufacturers as a relevant area for improvement. Because the metal filament has chance to drop on the die surface which will result to application fail or reliability issues at final customers' side. [1]

This paper presents an investigation on the effect and optimization of machining parameters for the metal filament improvement in saw operation on die with backside metal. The results showed that the current method can effectively reduce metal filament. The experimental studies were conducted under varying table speed, spindle rotation and cut depth was examined. The setting of machining parameters was determined by using design of experiment (DOE) techniques.

This paper also presents a solution to pick up big size die with UV tape, which will resolve more product's pick-up issues in semiconductor assembly site.

INTRODUCTION

Traditional semiconductor back end process includes die sale process and package process. For die sale process it includes wafer mount, wafer saw, inspection and packing process. There are two packing process for die sale part. One is wafer level packing and another is die level packing. After packing it will ship to final customer. For package process it includes wafer mount, wafer saw, UV lumination, die bond, wire bond, molding, test, laser, inspection and packing process. Figure. 1 shows the flow chart for Die sale and package part back end process.

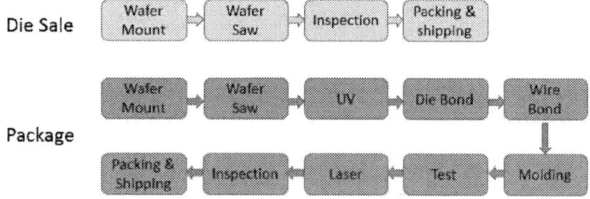

Figure.1. Die sale and package part back end process

From Figure.1 showed that die sale part back end process doesn't have UV stage. But the package part has UV stage. That is to say non-UV tape is used on die sale

part and UV tape is used on package part. UV tape means that there must be a UV lumination stage to reduce tape's adhesive to the backside of the semiconductor wafer and make it can be picked up. The non-UV tape's adhesive is much lower than UV tape. So it doesn't need take a UV lumination stage before die pick-up. The die sale part will directly ship to customer with tape. So if UV tape was used on die sale part and do UV lumination, the dice on the tape would have risk to drop during the way ship to customer. If UV lumination stage was not ran, the tape would abnormal adhesion after some days. And the final customer may hard to pick them up. So non-UV tape was selected on die sale product. And UV tape was selected on package product.

Die with backside metal was a big challenge on saw process. There would be more metal filament on the die edge. And there would be much more possibility to drop on the top of the die after die pick-up process, which would result to application fail to final customer. In this paper, the issue will be introduced in detail and will be provided detail method on how to resolve it.

METAL FILAMENT ISSUE RESOLVE

Figure.2 showed that the metal filament was found in the pocket of the reel. And it was just on the surface of the die. It made more electric yield drop after wire bond process. The filament was traced step by step and found that it came from saw process. Figure.3 showed the die image under microscope and SEM (scanning electron microscope) after saw process. It showed that metal residual was on the edge of the die after saw stage. But it was not stable. During the later process, the metal residual would have chance to drop on the surface of the die. And it would impact final test result.

Figure.2. Metal filament on the die surface

Figure.3. Metal filament under microscope and SEM

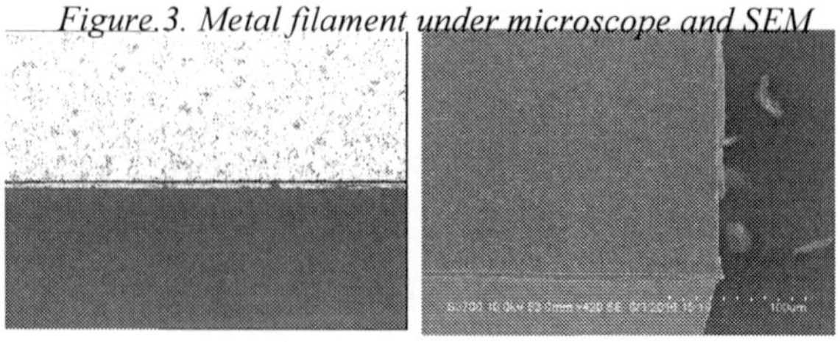

The metal filament might come from the improper depth of the blade cutting. The blade round zone might cut the die's metal layer and result to the metal filament. See from Figure.4. The cutting depth's setting should keep the blade round zone complete below the metal layer to minimize the metal filament created during saw process. But deeper blade cutting depth would get blade cloggy easier by foil and risk of broken foil during expansion in tape & reel process.

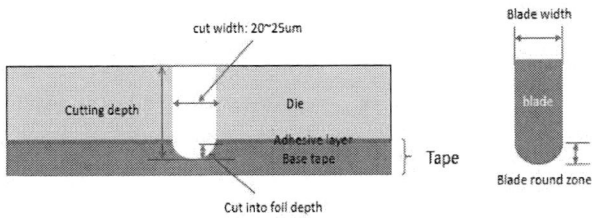

Figure.4. Blade cutting depth

Also the tape might not be thicker and harder enough. Currently non-UV tape was used on this product. Its thickness was 75 um. The die's backside metal was silver. As it was known that silver had good malleable of all metals. So UV tape was consider to replace the non-UV tape. Because UV tape was thicker and harder than non-UV tape. Table 2 showed the difference between these two kinds of tapes.

UV tape meant that after saw stage the tape must take UV lumination to reduce the adhesive of the tape. The non-UV tape meant that it was no need to take UV lumination after saw stage.

TABLE 1
UV and Non-UV tape comparison

Tape Type	Adhesion	Total thickness without release film	Material		
			Base film	Adhesive layer	Release film
UV	16.7 N/25mm (Before UV) 0.64 N/25mm (after UV)	110 um	PO (80um)	Acrylic (30um)	Polyester
Non-UV	1.2 N/20mm	75 um	PVC (65um)	Acrylic (10um)	PET

Note: PO: Polyethylene, Polypropylene
PVC: Poly Vinyl Chloride
PET: Polyethylene Glycol Terephthalate

So a DOE (design of experiment) was done to find an optimal parameter and proper tape during saw process to try to reduce the metal filament. Twelve wafers were used to mount on UV and non-UV tape separately. And the wafers with UV tape would run saw and UV lumination process. The wafers with non-UV tape would

just run saw process. Then a DOE was done to find the optimization parameter on saw stage. See table 3 which showed the experiment result. Table2 showed that all the result has a significant improvement with UV tape. And Cell9 was much better. So UV tape was selected as our production tape and choose Cell9's parameter as our production's parameter. But another issue was popped up. Most of the dice which mount with UV tape couldn't be picked up during tape & reel stage.

TABLE 2
DOE for tape and parameter selection

Tape type	Cell No.	Cut depth in tape	Spindle speed	Feed speed	% of unit with metal Filament	Metal filament qty per unit	% of die side with metal filament	Backside chipping
non-UV	Cell1	2.5mil	49k	1 inch/s	100%	11.4	100%	Yes
	Cell2	2mil	49k	1 inch/s	100%	9.5	100%	No
	Cell3	1.5mil	49k	1 inch/s	100%	7.9	100%	No
	Cell4	1.5mil	49k	0.5 inch/s	100%	12.8	100%	No
	Cell5	1.5mil	35k	1 inch/s	100%	24.8	100%	No
	Cell6	1.5mil	35k	0.5 inch/s	100%	15.4	100%	No
UV	Cell7	2.5mil	49k	1 inch/s	60%	1.8	20%	Yes
	Cell8	2mil	49k	1 inch/s	40%	0.5	10%	No
	Cell9	1.5mil	49k	1 inch/s	20%	0.4	5%	No
	Cell10	1.5mil	49k	0.5 inch/s	23%	0.6	7%	No
	Cell11	1.5mil	35k	1 inch/s	100%	4.6	60%	No
	Cell12	1.5mil	35k	0.5 inch/s	80%	2.2	40%	No

DIE PICKUP ISSUE RESOLVE

So UV tape was used on production wafers as experiment sample. And cell9's parameter was used on saw parameter and ran saw and UV lumination process. Then another DOE was done to find an optimization parameter on tape & reel machine. Push pin height was changed from 1mm to 1.5mm. Pickup head compression was changed from 80 um to 150 um. And tape tension was changed from 5mm to 10 mm . But none of those change was worked.

Table1 showed that the UV tape's thickness (especially the adhesive layer) was higher than non-UV tape. It was known that higher tape stickiness induced higher die tensile stress during die picks up process. [2] And when increase in die area also leads to an increase in the pull force. [3] So different die size wafers were selected to do an experiment to figure out if different die size would result to different die pick up performance. The result was in table 3. It showed that for the die with backside metal, if the die size smaller than 4mm*4mm, all the dice could be picked up. But if the die size larger than 6mm*6mm, some of the dice couldn't be picked up. And for the die without backside metal, no matter what was the size of the die, it could be picked up much easier.

So both the backside metal and the die size were the key reason to impact the pick-up performance. The issue production's die size was larger than 6*6mm and had backside metal. Nine this kind of wafers (issue production

978-1-5386-5309-8/18 $31.00 © 2018 IEEE

wafers) were selected to do a third DOE (see table 4) to find an optimization parameter on tape & reel machine. But no matter how to change the parameter, there were always some dice which couldn't be picked up. Vacuum of the pepper pot was also increased from -85Mpa to -100Mpa and changed the push pin's array from rectangle shape to diamond shape. None of the change was worked. Figure.4 showed the pepper pot and push pin position.

TABLE 3
DOE for different die size picking up

with backside metal	die size	Normal pick up parameter	If all the die can be picked up
Yes	2x2	Yes	Yes
	4x4	Yes	Yes
	6x6	Yes	No
	8x8	Yes	No
No	2x2	Yes	Yes
	4x4	Yes	Yes
	6x6	Yes	Yes
	8x8	Yes	Yes

TABLE 4
DOE for 8mm*8mm die picking up

push pin height (mm)	pickup head compression (um)	tape tension (mm)	If all the die can be picked up
1	80	5	No
1	110	5	No
1	150	10	No
1.2	80	5	No
1.2	110	10	No
1.2	150	10	No
1.5	80	5	No
1.5	110	5	No
1.5	150	10	No

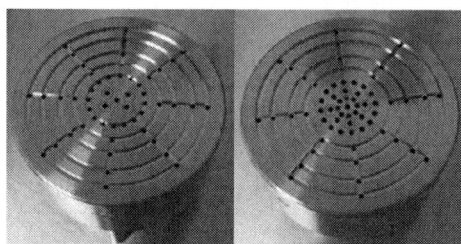

Figure.5. Pepper pot and push pin position
(left is rectangle shape and right is diamond shape)

After team's brain storming a novel method was thought out that UV lumination sequence was changed from post-saw stage to pre-saw stage. That was different with the packing's traditional process flow. The Figure.5 showed the difference between packing's traditional process flow and new process flow. But after evaluation there would be a risk that during the saw stage die flying issue would be happened. 10 wafers were selected as sample to proceed the new process flow. Set push pin

height as 1.2mm, set pickup head compression as 110um and set tape tension as 10 mm. Also diamond shape pepper pot was used and kept its vacuum pressure as -85Mpa.

The result was that all the dice could be picked up on tape & reel machine. And there were no die flying issue during saw process. Backside chipping and metal filament were also checked. And all of them were in our spec.

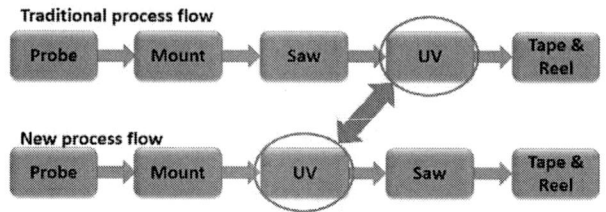

Figure.6. Difference between traditional and current flow

CONCLUSION

In this paper, UV tape was used to replace non-UV tape in production to reduce metal filament issue. And a DOE method was used to find out optimal saw parameter. After that the metal filament issue almost disappeared in production which make sure that there was no quality risk for this product. Die pick-up issue was resolved by change UV lamination sequence from post-saw stage to pre-saw stage. Both the method can give a good reference to assembly site in semiconductor area.

ACKNOWLEDGEMENTS

At the point of finishing this paper, I'd like to express my sincere thanks to all those who have lent me a hand on writing this paper. My sincere appreciation also goes to my colleague, who participated this study with great cooperation.

REFERENCES

[1] *Bo Peng, Deguang Zheng, Yue Yu, "Method to remove wafer surface particles", Journal of Semiconductors, Sep. 2017, 38 (9), pp. 096004-4*

[2] *Richard Qian, Yang Liu, "Thin and Large Die Assembly Pick Up Process Optimization by Dynamic Modeling", 17th Electronic Packaging Technology (ICEPT), Aug. 2016, pp. 147–152.*

[3] *Y. Sing Chan, Julie Chew, Chu Hua Goh, Siang Kuan Chua and Alfred Yeo, "Characterization of Dicing Tape Adhesion for Ultra-thin Die Pick-up Process", 16th Electronics Packaging Technology Conference (EPTC), Aug. 2014, pp. 554 – 557.*

Silane Oligomer in Epoxy Molding Compound

Hongjie Liu[1], Zhen Wang[2], Wei Tan[1], Lanxia Li[1], Xingming Cheng[1], Xiaojuan Jiang[1], Liang Cui[1]

[1]JiangSu HHCK Advanced materials Co., Ltd
[2]Lianyungang Branch of Traditional Chinese Medicine，Jiangsu Union Technical Institute
Lianyungang, Jiangsu, China
Hongjie.liu@hhck-em.com

ABSTRACT

With the rapid development of semiconductor packages, the reliability requirement of the epoxy molding compound as the packaging material becomes more and more rigorous. Introducing adhesion prompter is one of the key methods to increase the reliability of epoxy molding compound, and silane coupling agent with different functional group was one of the most common choice of researchers. Compared to common silane, silane oligomer was partially hydrolyzed and contain more condensed functional groups, it might be a good promising candidate of adhesion promoter. Here，we tried silane oligomer as adhesion promoter to replace the silane in epoxy molding compound, and the properties of resulting epoxy molding compound with silane oligomers in have been tested. The results showed that compared to normal silane, silane oligomer accelerated the reaction speed of epoxy molding compound and the gel time (GT) and spiral flow (SF) all were shorten with the silane oligomer in. Epoxy molding compound with mercapto group silane oligomer A and amino group silane oligomer B all disclosed similar water absorption compared to that with normal silane in, while epoxy molding compound with epoxy group silane oligomer C in exhibited about 25% higher than that with normal silane in. Adhesion test revealed that epoxy molding compound with silane oligomer in all possessed higher adhesion on NiPaAu (PPF) leadframes with the value above 100 N after MSL3 test.

Keywords—epoxy molding compound; silane; silane oligomer; water absorption; adhesion

INTRODUCTION

Silane coupling agents, as we all known, were generally dispersed on the filler surface to improve the mechanical strength of the composite materials [1-3]. In the field of epoxy molding compound, they also worked as adhesion promoters increasing the adhesion between epoxy molding compound and different metal lead frames. Furthermore, researcher also investigated the performance of physical properties of EMC according to the change of different coupling agent treatment process [4-5]. However, with the development of the semiconductor field, the performance improvement of epoxy molding compound is far from enough caused by the limitation of the types and treatment technology of coupling agent. Silane oligomer, partially hydrolyzed and contain more condensed functional groups, have been tried to increase the interfacial reinforcement between inorganic materials and polymers, especially when resin bonding is difficult to achieve using conventional method. O. Massahiro, etc., have applied silane coupling oligomer as the primer for zirconia[6], and found that the shear bond strength of increased after applying a non-activated ethanol solution of the silane coupling oligomer compared with that achieved when applying monomer. Also, I. Nobuo, etc., have tried polycondensates of silane coupling agent treating the surface of inorganic materials such as glass fiber and inorganic filler including silicates, and the results showed that it can increase the interfacial reinforcement between inorganic material and polymer [7]. Here, we investigated three different types of silane oligomer：mercapto group silane oligomer A, amino group silane oligomer B and epoxy group silane oligomer C in epoxy molding compound, and the influence on the performance of epoxy molding compound have been tested also.

EXPERIMENTAL

Materials

The epoxy resin used in this study was multi-aromatic ring type from Nippon Kayaku Co., Ltd，Hardener was xylokphenol novalac resin from Zhong He Techonology Co.Ltd. Triphenylphosphine as accelerator was got from HOKKO's Fine Chemicals. Filler was chosen as the spherical fused silica from Novoray Corporation with the mean diameter as 23 μ m and maximum filler size as 75 μ m.

Silane and silane oligomer were form shin-Etsu Chemical Co. Ltd., mold releasing agent was carnauba wax obtained from Foncepi Comercial Exportadora Ltd. Detailed descriptions of the chemical structures and grade names are summarized in Table I

TABLE I.　　RAW MATERIAL USED

Type	Raw Materials	
	Chemical structure/Name	*Maker*
Epoxy		Nippon Kayaku Co., Ltd
Hardener		Zhong He Technology Co.Ltd.
Accelerator	Triphenylphosphine	HOKKO's Fine Chemicals.
Filler	Fused silica	Novoray

978-1-5386-5309-8/18 $31.00 © 2018 IEEE

Type	Raw Materials	
	Chemical structure/Name	*Maker*
		Corporation
Coupling agent	KBM-403 Silane oligomer A Silane oligomer B Silane oligomer C	shin-Etsu Chemical Co. Ltd.
Wax	Carnauba wax	Foncepi Comercial Exportadora Ltd.

Sample preparation

All the ingredients were weighted up according to the formulation and mixed in the high speed mixing machine, after this, the mixture was feeding to double screw extruder under the heating condition for extrusion, and then the discharged material go through sheet formation, cooling, pre-breaker, granulation, post blend, and storage at 5 ℃.

Measurement

Tests were carried out on EMC samples prepared by the above raw materials to evaluate the performance. Gel time (GT) and Spiral flow (SF) as well as the Hot Hardener (HH) for 90s molding were tested. Meanwhile, following properties of EMC were tested also.

Flexural Modulus and Flexural strength: The specimens of 127mm×13mm×6.35mm were prepared through transfer molding method and post mold cure (PMC) at 175℃ for 6 hours. And then they were measured universal testing machine at room temperature.

Dynamic Mechanical Analyzer (DMA) modulus and Tg: The specimens of 50mm×13mm×3mm were prepared through transfer molding method and post mold cure (PMC) at 175℃ for 6 hours. The Dynamic Mechanical Analysis (DMA) of the cured specimens was performed in the three point bending mode under 1Hz sinusoidal strain loading on the Dynamic Mechanical Analyzer (DMA, TA Instruments, Q800) with the temperature ramping from room temperature to 265℃ at the rate of 5 ℃ /min. The storage modulus (SM) at room temperature, 175℃ and 260℃, peak of loss modulus (LM) and tan δ were recorded.

Moisture absorption: The specimens with the diameter of 50mm and thickness of 3mms were prepared through transfer molding method and post mold cure (PMC) at 175℃ for 6 hours. And then the samples were weighted and then placed in an 85℃/85%RH chamber for 168 hours. The samples were taken out of the chamber and weighted after the time run out. Weight gain of the EMC during aging at Pressure Cook Test (PCT) at100% relative humidity and 2 standard atmosphere for 24h was defined as moisture absorption.

Adhesion test: EMC samples was molded on the adhesion test leadframes plated by Ag and PPF separately which were custom tailored by SanJia electronic Co. Ltd. for HHCK.only. After PMC at 175℃ for 6 hours, the packages were placed in the 60℃/60%RH chamber for 40 hours, and then reflowed in the 260 reflow oven for three times according to JEDEC

standard. And the then adhesion test was done by the Universal Test Machine under the tensile manner after aged.

RESULTS AND DISCUSSION

The GT and SF and DMA test results of the epoxy molding compound with the silane coupling oligomer were summarized in Table II.

TABLE II.   GT SF AND DMA TEST RESULTS OF EMC

Test item	Coupling agent type			
	KBM-403	A	B	C
GT/s	35	20	24	26
SF/cm	111	74	76	103
SM @RT /Mpa	30600	27700	28700	27900
SM@175 /MPa	890	1010	1106	798
SM@260 /MPa	649	789	704	600
Peak of LM /°C	111	113	110	104
Peak of tanδ/°C	118	119	117	111
Flexural Strength /Mpa	143	154	146	149
Flexural Modulus/ Mpa	28000	27500	28000	27600

Compare to the normal silane coupling agent, silane oligomer all disclosed shorter GT and SF, and a little higher flexural strength with the same dosage in EMC formulation. Further, we tried to use differential scanning calorimetry (DSC) to get the reaction curves. From Figure 1, we can see that all the EMC with silane oligomer in with a lower onset point and lower reaction peak than that with normal silane coupling agent. Perhaps, this higher reaction rate was the reason which caused the shorter GT and SF of EMC with silane oligomer in.

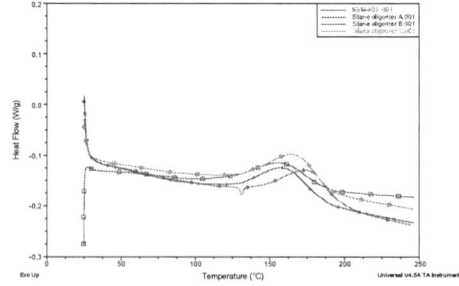

Figure 1: DSC overlay curves of EMC with silane oligomer

Water absorption of epoxy molding compound with silane oligomer in were also tested, and the results were revealed in Figure 2，From Figure 2，we can see that compared to EMC with normal coupling agent, EMC with silane oligomer C in possessed higher water absorption, while the water absorption of EMC with silane oligomer A and silane oligomer B in were at the same level as that with KBM-403 in.

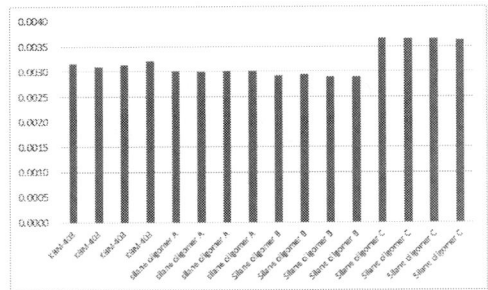

Figure 2 : PCT test of EMC with different coupling agent

The adhesion of epoxy molding compound on different leadframes after MSL3 with different coupling agent in have also been tested, and the results were disclosed in Table III.

TABLE III.   ADHESION OF EMC WITH DIFFERENT COUPLING AGENT

Lead frame	Coupling agent type			
	KBM-403	A	B	C
Cu/N	475	498	503	510
Ag/N	408	394	416	423
PPF/N	38	120	153	141

From the above test results, we can see that, silane oligomer A, B and C all can increase the adhesion on PPF after MSL3 compared to normal coupling agent although they show the same adhesion level on Cu and Ag.

CONCLUSION

In this study, silane oligomer A, B C have been tried to improve the performance epoxy molding compound. The results showed that compared to normal silane, silane oligomer accelerated the reaction speed of epoxy molding compound and the gel time (GT) and spiral flow (SF) all were shorten with the silane oligomer in. Epoxy molding compound with mercapto group silane oligomer A and amino group silane oligomer B all disclosed similar water absorption compared to that with normal silane in, while epoxy molding compound with epoxy group silane oligomer C in exhibited about 25% higher than that with normal silane in. Adhesion test revealed that epoxy molding compound with silane oligomer in all possessed higher adhesion on NiPaAu (PPF) leadframes with the value above 100 N after MSL3 test. Although the adhesion on PPF have been improved by introducing silane oligomer in, more studies will be carried out later such as the dosage and adding process of the silane oligomer to further enhance EMC performance.

REFERENCES

[1] W. J. Kim, J. H. Ryu, Physical Properties of Epoxy Molding Compound for Semiconductor Encapsulation According to the Coupling Treatment Process Change of Silica, John Wiley &

Sons, Inc., pp. 1975-1982, 1997.

[2] J. H. Lee, H. C. Yoo, Properties of Epoxy Molding Compound According to the Pretreatment Method of an Amino-Silane Coupling Agent on Epoxy or Phenol Resin, Journal of Applied Polymer Science, Vol. 100, pp. 2171-2179, 2006

[3] K. X. Pan, X. R. Zeng, H. Q. Li, X. J. Lai, Synthesis of Silane oligomers containing vinyl and epoxy group for improving the Adhesion of Addition-cure Silicone Encapsulant, Journal of Adhesion Science and Technology, Vol. 30(10), pp. 1131-1142, 2016

[4] N. Zheng, J. M. He, D. Zhao, Y. D. Huang, J. F. Gao, Y. W. Mai, Improvement of Atomic Oxygen Erosion Resistance of Carbon Fiber and Carbon Fiber/Epoxy Composite Interface with a Silane Coupling Agent, Materials and Design, Vol. 109, p 171-178, November 5, 2016

[5] H. Y. Sun, Z. Yang, G. X. Sun, X. F. Xu, The research of Silane Oligomer with Lower Valatilization Rate for Protecting Concrete, Applied Mechanics and Materials, Vol. 357-360, pp. 1318-1322, 2013

[6] O. Masahiro, I. Kazusa, I. Masao, T. Hiroaki, T. Yasuhiro, M. Takuya, Resin Adhesion Strengths to Zirconia Ceramics after Primer Treatment with Silane Coupling Monomer or Oligomer, Dental Materials Journal, Vol. 36(5), pp.600-605, 2017

[7] I. Nobuo, E. Ryuta, F. Seiji, T. Tomoya, A. Naoki, F. Fumiyasu, C. Satoru, K.Toyoki, Application of Polycondensates of Silane Coupling Agents to Adhesion and Hardening. Polymer Preprints, Vol. 54, (1), pp. 874, 2005,

Nanostructures for smart systems

Joerg Martin[1], Ray Saupe[1]*, Joern Langenickel[2], Martin Moebius[2], Kathleen Heinrich[1], Alexander Weiß[1] and Thomas Otto[1, 2]

[1]Fraunhofer Institute for Electronic Nano Systems, Chemnitz, Germany
[2]Center for Microtechnologies, University of Technology, Chemnitz, Germany
*Corresponding Author's Email: ray.saupe@enas.fraunhofer.de

ABSTRACT

Nanostructures offer a large potential for the realization of innovative microsystems with new functionalities. Various nanomaterials are relevant for microsystem technology, including dielectric, conducting and semiconducting nanoparticles. Here, we focus on the use of semiconducting nanoparticles in thin-film systems for light emitting, load indication and energy harvesting applications. The setup and basic functionalities of the respective systems as well as some characteristics are given.

INTRODUCTION

Nanotechnology is often mentioned as one of the key enabling technologies for the current century. It allows the development of new products in nearly all fields of our daily life, e.g. electronics, medical technology or automotive. Nanotechnology is connected to applications like smart home or smart health, and hence will support our well-being, while saving resources and protecting environment.

Due to their manifold properties and appearing effects nanomaterials are a very productive source for development of innovative microsystems. Depending on their properties, nanoparticles, nanorods, nanowires or tubes are worldwide incorporated in various smart systems to enable sensing of gases, humidity, forces/strain, absorption and emission of radiation (light) or conversion of energy.

Innovative nanocomposites and nanostructures provide completely new chances for the realization of sensor and actuator systems in microsystem technology. Processing from liquid dispersions via cost-saving technologies, like printing, allow for new sensor or actuator designs, even on flexible and curved surfaces. Consequently, sensor arrays can be manufactured easily on foils, which in turn can act as semi-finished products for further processing, e.g. integration in fiber-reinforced plastic materials.

A prominent type of nanostructures are semiconductor nanocrystals, so-called quantum dots (QDs). Due to their exceptional properties, like tunability of emission wavelength by changing particle size and their sensitivity to charges and electrical fields, QDs are applicable in actuators (in form of LEDs) as well as sensors.

In the following chapters an overview of our recent work with respect to the integration of quantum dots in various microsystems is presented. The application scenarios include quantum dot LEDs, indication of mechanical loads and energy harvesting with quantum dot solar cells.

QUANTUM DOT LEDS

Thin-film optoelectronic devices have attracted great interest over the past two decades because of their potential for low-cost fabrication on a freely chosen area and the possibility to apply the devices by coating on nearly any substrate [1, 2]. Especially light-emitting diodes based on colloidal quantum dots (QD-LEDs) show promising advances compared with their all organic counterparts, like size-tunable wavelength, high quantum yield and the chemical stability [3].

A substantial amount of attention has been dedicated by various groups to improve the efficiency of electron-hole recombination inside quantum dots. This is accompanied with an efficient charge transport across the respective thin-films and easy charge injection into the nanoparticles. Losses, e.g. via charge trapping or recombination already outside the quantum dots, have to be avoided. Also a balanced injection of electrons and holes without accumulation of charges on interfaces is beneficial for the efficiency.

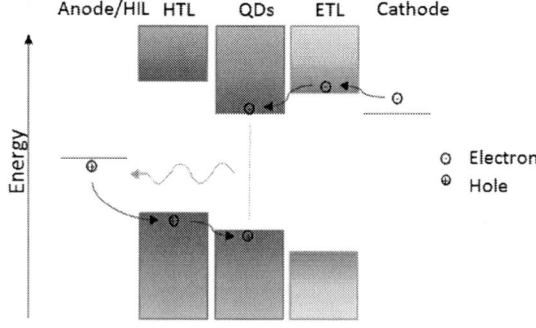

Figure 1: Band structure and scheme of charge injection in QD-LEDs

978-1-5386-5309-8/18 $31.00 © 2018 IEEE

A general band scheme of a thin-film QD-LED is sketched in Figure 1. While in a classical bulk semiconductor LED charge transport and recombination is driven by a pn-junction, which is created via doping profiles, in thin-film LEDs charge transport materials with appropriate band structure / HOMO-LUMO-levels and p- or n-mobilities, respectively, are used. According to Figure 1, holes are transported from the anode to the QDs in the valance band of a hole transport layer (HTL), while electrons are transported in the conduction band of an electron transport layer (ETL) from the cathode to ensure injection in the quantum dots at proper energy levels.

In recent years, four distinct generations of device architectures were proposed, namely: type I QD-LEDs with polymer charge transport layers (CTLs), type II QD-LEDs with organic small molecule CTLs, type III QD-LEDs with inorganic CTLs, and type IV QD-LEDs with hybrid organic–inorganic CTLs. The highest external quantum efficiencies can be found in type IV LEDs [4], hence we prefer this architecture for our own devices.

Figure 2 shows a photograph of a 15 x 15 mm² thin-film quantum dot LED. This QD-LED was prepared with an organic HTL and inorganic ETL, hence it represents a type 4 architecture. Cadmium selenide / zinc sulfide (CdSe / ZnS) core / shell quantum dots (Suzhou Xingshuo Nanotech Co., Ltd.) have been incorporated as emitter. LEDs of this type start to emit light at voltages under 3 volts and show external quantum efficiencies over 2 percent.

Figure 2: QD-LED with an active area of 15x15 mm²

Although thin-film QD-LEDs are still subject to research, they seem to be ideal light sources for a wide field of applications, not only for smart systems, but also for ambient lighting and communication, to mention just a few.

OVERLOAD INDICATION WITH QUANTUM DOTS

Electrical charges are injected into quantum dots for excitation of electroluminescence in case of quantum dot LEDs. At the same time the photoluminescence (PL) of these nanocrystals is very sensitive to electrical fields and charges. For example, electrical fields can lead to spectral shifts of the emission band (Stark effect) and reduction of photoluminescence intensity [5, 6]. Additionally, ionization of quantum dots, more precisely, a residual non-localized charge inside a nanocrystal is able to quench the photoluminescence of the respective particle completely [7]. We use this quenching mechanism for optical indication of mechanical (over-) loads, particularly on lightweight structures [8]. For that, we integrate piezoelectric elements together with thin-film quantum dot layer stacks in fiber-reinforced plastic materials.

If a mechanical load is applied on the sensor system, electrical charges are generated within the piezoelectric layer. These charges are subsequently transported to and injected into quantum dots, resulting in photoluminescence quenching of the respective part of the nanocrystal film. Consequently, the mechanical load is translated in a local optical contrast between charged and neutral areas of the quantum dot indicator layer, which can be read out even after a certain time e.g. via a common UV-LED lamp. An illustration of the sensor film with and without mechanical load is shown in Figure 3.

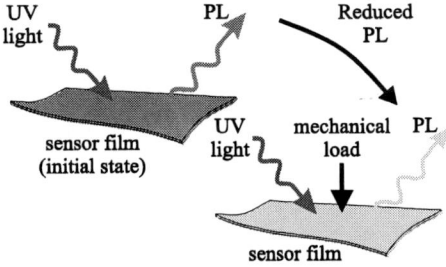

Figure 3: Initial state (upper left side) and reduced PL intensity of sensor film by mechanical load (lower right side)

Requirements on the sensor system are, among others, a high efficiency (low charge consumption) for nicely visible contrast and a storage time of at least some weeks. To meet the mentioned demands, we have carried out extensive experiments on the design of the layer stack and the used materials, especially with respect to charge transport materials and types of quantum dots. Thereby, photoluminescence characteristics of the quantum dot sensor stacks has been recorded with a home-built micro-spectroscopy setup. An example of a recorded series of PL spectra is shown in Figure 4. Here cadmium selenide / cadmium sulfide (CdSe/CdS) core/shell quantum dots (CANdot Series A plus) and N,N,N′,N′-Tetrakis (3 - methylphenyl) - 3,3′ - dimethylbenzidine (HMTPD) as charge transport material have been used.

978-1-5386-5309-8/18 $31.00 © 2018 IEEE   505

After an observation time of 5 seconds a voltage of 10 V was applied on the sample, leading to a quenching of the PL. With this sample configuration, a reduction in PL intensity to less than 70 % of the initial value could be achieved, which is visible by the naked eye. At a time of 30 seconds, the voltage was removed, resulting in a smooth increase of photoluminescence. We attribute this to a steady leakage of charges out of the particles [9].

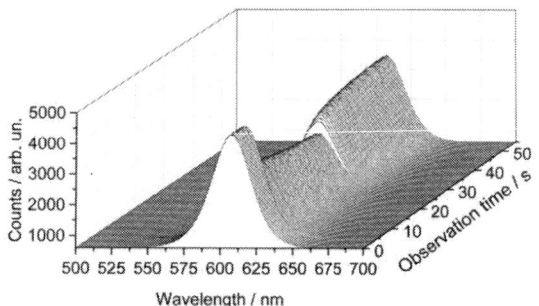

Figure 4. Series of PL spectra of a sample stack with CdSe/CdS quantum dots and HMTPD charge transport layer

With a different sensor layer design we have already measured storage times of more than 60 hours. Additionally, recently we demonstrated the generation of visible optical contrasts by combination of quantum dot layer stacks with commercial piezo elements. By integration of both elements in glass-fiber reinforced plastic components we were able to proof the general suitability of our sensor concept for structural health monitoring of lightweight structures.
Monitoring of the mechanical state of lightweight structures is of great importance, since overloads can lead to hidden damage and sudden failure of the material. With this respect, our approach can help to make these materials smarter and to promote the use of lightweight structures in automotive, aircraft engineering and other fields.

QUANTUM DOT SOLAR CELLS

In the field of energy conversion wind power, tidal power and solar cells have been established. Non-classical solar cells can be subdivided into dye-sensitized, perowskite, organic, hybrid and quantum dot sensitized ones. With regard to quantum dot sensitized solar cells (QDSSC) advantages like a tunable absorption region due to quantum dot size and multi exciton generation are used. So far efficiencies of more than 9.5 % have been reached [10]. In most cases the published solar cells are fabricated on solid substrates and completed with a liquid polyelectrolyte. But this is not favored, due to leakage reasons, when integrated onto highly-flexible substrates, e.g. polymer films.

Though, flexible substrates are crucial for several

modern smart systems, because of the demands according to integration and wearability. Consequently, we have prepared QDSSCs on different bendable substrates without using a liquid polyelectrolyte. The influence of parameters like surface activation, layer thickness, electrode materials as well as size and type of quantum dots have been investigated. Among other configurations, hybrid solar cells containing a poly(3-hexylthiophene) (P3HT) / quantum dot composite as absorbing layer have been realized on polyethylene terephthalate (PET) films. Instead of cadmium selenide quantum dots [11], we have used lead sulfide (PbS) nanocrystals (CANdot series C) to broaden the absorption band in the visible region to the near infrared range.

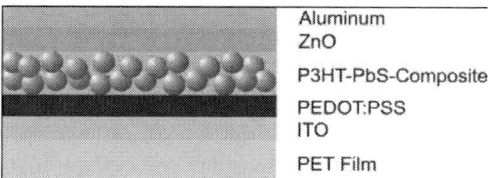

Figure 5: Illustration of layer stack of a hybrid polymer quantum dot solar cell

Figure 5 shows a sketch of a prepared hybrid polymer quantum dot solar cell composed of a layer stack of indium tin oxide (ITO) / poly(3,4 - ethylenedioxy - thiophene) - poly(styrenesulfonate) (PEDOT:PSS) / P3HT-PbS-composite / zinc oxide / aluminum on PET film. In this configuration P3HT and PbS quantum dots form a heterojunction network, which allows for the charge separation of generated electron-hole pairs.

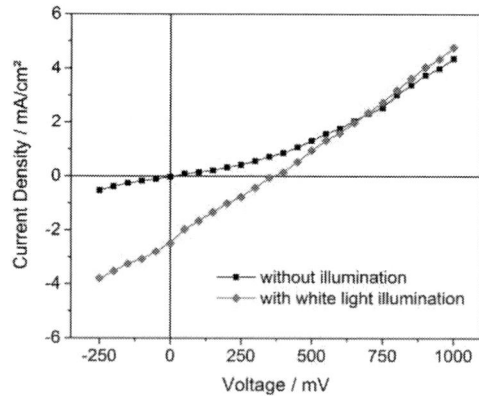

Figure 6: Current density-voltage characteristics of a hybrid quantum dot polymer solar cell test sample with an active area of 4 mm² on flexible substrate (black curve: without illumination; red curve: with 86 mW/cm² white light illumination).

978-1-5386-5309-8/18 $31.00 © 2018 IEEE   506

Additionally, fast electron transfer (and thus charge separation) is promoted by the ZnO nanoparticle layer because of energy level alignment of ZnO and the composite components [11].

Current density-voltage characteristics of a hybrid solar cell with an active area of 4 mm² corresponding to the described configuration are depicted in Figure 6. For recording of the data the samples were kept in darkness or illuminated with white light of 86 mW/cm² intensity of a halogen lamp with glass fiber output, respectively. From the characteristics of the sample under illumination a short-circuit current of 2.5 mA/cm² and an open-circuit voltage of 0.365 V can be extracted. At the maximum power point the sample delivered an electrical power of 0.2 mW/cm².

Although a moderate power output from the samples has been determined so far, the results are promising anyway. Modern smart systems often have a power consumption in the range of some milliwatts for data recording, storage and transmission. Hence, powering of such systems with appropriate energy harvesting and buffering regimes is already possible with the described solar cells at reasonable areas. Additionally, the recorded current-voltage characteristics exhibit a non-ideal curvature, which points to a considerable influence of series and shunt resistances of the layer stack, reflecting the challenging nature of flexible substrates. Nevertheless, with improved layer structure, being capable to withstand bends and kinks without drastically altering the performance, it will be possible to apply quantum dot sensitized solar cells on various technical substrates to realize power supplies for e.g. sensor nodes, on-board electronics and autonomous control of devices.

CONCLUSION

The presented three use cases for semiconductor nanoparticles are intended to give a first impression of the potential of nanostructures for smart system integration. Although materials and preparation technologies are similar, they have to be adapted in detail to the respective application, e.g. with respect to film thickness or stacking sequence of the layers. Still challenging in some cases is a specific connection of nanostructures to the macro world. Nevertheless, in the next years nanostructures will gain increasing importance for smart systems due to their large variety of possible applications.

ACKNOWLEDGEMENTS

Work according to quantum dot LEDs and load indication was performed within the Federal Cluster of Excellence EXC 1075 "MERGE Technologies for Multifunctional Lightweight Structures" and supported by the German Research Foundation (DFG). Work according to quantum dot solar cells was done in the project futureTEX, which is funded in the framework of the program "Zwanzig20 - Partnership for Innovation" by the German Federal Ministry of Education and Research. Financial support is gratefully acknowledged.

REFERENCES

[1] A. Ekimov, Al. L. Efros, and A. A. Onushchenko, *Solid State Comm.*, vol. 56 (11), 1985, pp.921-924.

[2] A. Alivisatos, *J. Phys. Chem.*, vol. 3654 (95), 1996, pp. 13226–13239.

[3] S. Reineke, *Nature Materials*, vol. 14 (5), 2015, pp. 459–462.

[4] Y. Shirasaki, G. Supran, M. G. Bawendi, and V. Bulović, *Nature Photonics*, vol. 7 (1), 2012, pp. 13-23.

[5] S. A. Empedocles and M. G. Bawendi, *Science*, vol. 278, 1997, pp. 2114-2117.

[6] J. Martin, F. Cichos, C von Borczyskowski, *J. Luminesc.*, vol. 132, 2012, pp. 2161-2165.

[7] Al. L. Efros and M. Rosen, *Phys. Rev. Lett.* vol. 78, 1997, p. 1110.

[8] M. Moebius, J. Martin, M. Hartwig, R. R. Baumann and T. Otto, *Proceedings of Smart Systems Integration 2017*, Cork, Ireland, March 8-9, 2017, pp. 423-426.

[9] M. Moebius, J. Martin, M. Hartwig, R. R. Baumann, T. Otto and T. Gessner, *AIP Adv.*, vol. 6, 2016, p. 085309.

[10] D. Sharma, R. Jha et al., *Sol. Energ. Mat. Sol. Cells.*, vol. 155, 2016, pp. 294-322.

[11] L. Quian, J. Yang et al., *J. Mater. Chem.* vol. 21, 2011, pp. 3814-3817.

RF MEMS RESONANT DEVICES FOR WIRELESS COMMUNICATION

Q. Yuan[1,2], X. Kan[1,2,3], Z.J. Chen[1,2,3], J. L. Yang[1,2,3]*
J.C. Zhao[4], L. Sun[4] and F. H. Yang[1,2]

[1]Institute of Semiconductors, Chinese Academy of Sciences, Beijing 100083, P. R. China
[2]University of Chinese Academy of Science, Beijing 100049, P. R. China
[3]State Key Laboratory of Transducer Technology, Shanghai 200050, P. R. China
[4]Jiangsu Key Laboratory of ASIC Design, Nantong University, Nantong 226019, P. R. China
*Corresponding Author's Email: jlyang@semi.ac.cn

ABSTRACT

Future multi-band, multi-mode wireless communication system requires high performance integrated MEMS resonant devices with low power consumption, such as resonator with high quality factor (Q) and high frequency, oscillator with high stability, and filter with narrow pass-band. MEMS resonators with $Q > 10^4$ are attractive candidates to build other RF components. Utilizing high Q resonators as frequency selective components, MEMS oscillators can achieve high frequency stability comparable with quartz-based oscillators. Consisting of coupled resonator array, MEMS filters can realize narrow pass-band frequency response with good shape factor and excellent stop-band. In this paper, a MEMS disk resonator with $Q>10^4$ is presented, of which vacuum packaging and frequency stability in different environments are well studied. Based on the resonator, the MEMS oscillator with low noise feedback circuit and the tunable MEMS filter of small percent bandwidth are achieved.

INTRODUCTION

High performance silicon-based MEMS resonators have become attractive devices for future wireless communication due to their high quality factors (Q), high resonance frequency (f), reduced size, low power consumption, and easy integration with IC chips [1-3]. MEMS resonators can realize high performance oscillators and narrow pass-band filters, which have potential application in future communication systems [4-6]. It is challenging to achieve higher frequencies above GHz with the disk resonator. Different routines have been studied, such as scaling down the dimension to sub-micro scale and utilizing new material like sapphire or diamonds with high acoustic velocity to realize high frequency.

MEMS oscillators have high performance such as low noise performance, low power consumption and high integration. With high performance resonator and low noise sustaining circuits, the MEMS oscillator can obtain high frequency stability comparable with conventional quartz-based oscillator [7]. However, the phase noise of the MEMS oscillator still needs to be improved for communication applications.

MEMS filters with excellent performance are realized with electrostatic or mechanical coupling of high Q resonators [8]. The center frequency is determined by its constituent resonators, while mechanical coupling enables desired narrow bandwidth with the coupling beam. However, for tunable MEMS filter, coupled resonator array is required to realize tunable bandwidth.

This paper summarizes the recent work on MEMS resonant devices, including the MEMS disk resonator, oscillator, and tunable filter. Section I introduces high performance disk resonators with $Q > 10^4$ and resonance frequency up to 150 MHz, their vacuum packaging methods and frequency stability in different environment. For higher frequency application, a novel multi-electrode disk resonator with high-order whispering gallery modes (WGM) excited by selecting electrode combination scheme has been developed as digitally tunable resonator. Section II describes MEMS oscillators built up with the high Q disk resonator and low noise feedback circuits. Section III focuses on MEMS tunable filter.

HIGH Q AND HIGH FREQUENCY RESONATOR

Device and fabrication

A contour mode disk resonator of 18 μm-radius with frequency up to 150 MHz is demonstrated. The supporting loss is effectively decreased with a $Q>10^4$ as the stem is localized at the center of the disk, which is the node of the mode shape. The resonator was microfabricated with three polysilicon layers as reported in previous paper [9].

To attain a high-performance resonator of excellent stability and reliability, the vacuum encapsulation of the disk resonator is essential. A low-cost vacuum encapsulation technique with less parasitic effect for the disk resonator is presented. A Pyrex 7740 glass cap is fabricated and bonded with the resonator using the method reported in [10]. Sn-rich Au-Sn solder bonding is utilized as it provides a low-cost vacuum encapsulation with high shear strength, simplifying the feedthrough structures of the device. Additionally, with the

978-1-5386-5309-8/18 $31.00 © 2018 IEEE

optimization of the sealing ring and the transmission lines, the parasitic effect is well suppressed, increasing the Q of the resonator from 10000 to 11000. Fig. 1 shows the fabricated disk resonator and measurement results.

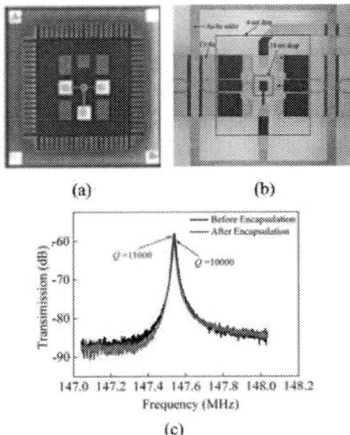

Figure 1: Illustration of resonator and packaging cap (a), SEM for the resonator (b), packaged resonator (c), and S_{21} parameter

Digitally tunable disk resonator

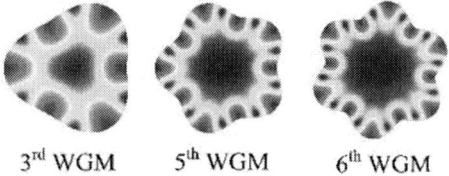

3rd WGM   5th WGM   6th WGM

Fig. 2: Simulated mode shapes of 2nd, 3rd, and 6th WGM's of the disk resonator.

For a resonator, maintaining high Q while vibrating in a high-frequency mode is a challenging issue as anchor loss tends to be more significant in higher frequencies. In addition, switchable vibrating modes are desired in digital systems as it consumes less power. A novel disk resonator with high-order whispering gallery modes with multiple electrodes is fabricated [11]. For the resonator vibrating in the WGM, the motion is distributed along the circumference of the disk, and the anchor loss can be reduced, leading to high Q at high frequency. The frequency of each mode is determined by the expressions as followed:

$$\left[(\zeta/\xi)\frac{J_{n-1}(\zeta/\xi)}{J_n(\zeta/\xi)}-n-q\right]\left[\zeta\frac{J_{n-1}(\zeta)}{J_n(\zeta)}-n-q\right]=(nq-n)^2 \quad (1)$$

$$q=\frac{\zeta^2}{(2n^2-2)} \quad, \zeta=2\pi fR\sqrt{\frac{\rho(2v+2)}{E}} \quad, \xi=\sqrt{\frac{2}{1-v}} \quad (2)$$

where ρ, v, and E are the density, Poisson's ratio, and Young's modulus, respectively. R is the disk radius, n is

the mode number and J is the Bessel function of the first kind.

The simulated mode shapes of the resonator are shown in Fig. 2. To achieve multi-mode excitation, the number of electrodes should be a common multiple of the number of the lobes for each mode. Furthermore, the electrodes should be aligned along the displacement maxima for a maximum electromechanical coupling factor. With the selectively connections of electrodes, the corresponding mode is attained while the others suppressed. Increasing the number of electrodes can excite higher-order WGM and get higher frequency. The resonance responses of the tunable disk resonator with different electrode connections are shown in Fig. 3. A high frequency of 275.7 MHz with Q up to 8500 is obtained.

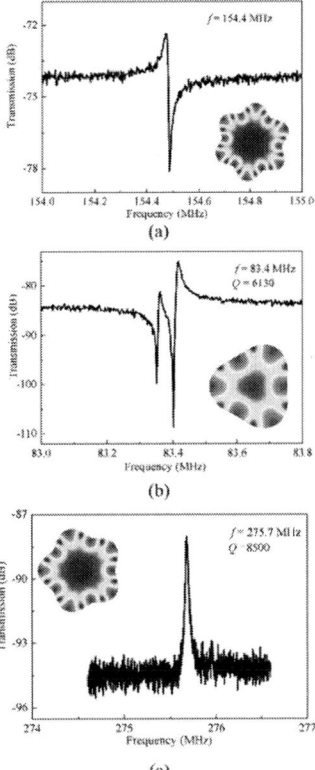

Fig. 3: Measured resonance responses of the tunable disk resonator with different electrode connections, (a) 6th mode of 37-μm disk (b) 3th mode of 37-μm disk (c) 5th mode of 18-μm disk

OSCILLATOR WITH HIGH Q MEMS RESONATOR

An oscillator based on the 18-μm disk resonator is built up to study the effect of temperature of stability and phase noise (PN) in this paper. The oscillator contains a low noise pre-amplifier (LNA) as the first stage and an automatic gain control (AGC) circuit [12] functioning as the second stage as shown in Fig. 4.

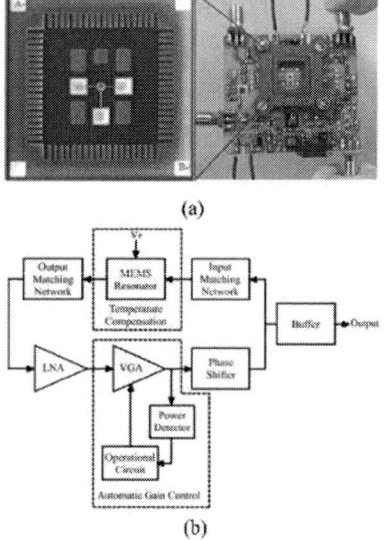

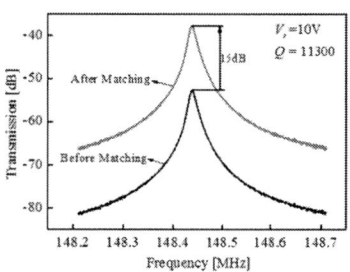

Fig. 4: (a). SEM photograph of a disk resonator with micro-oven. A dc current is applied to the heater through contacts A and B.

To match the motional resistance of the oscillator to 50 Ω, matching network is designed, which reduces the insertion loss of the disk resonator by 15 dB, as shown in Fig. 5

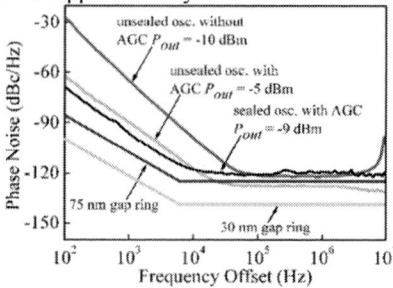

Fig. 5: The measured frequency response before matching and after matching of the disk resonator

The performances for three types of oscillators with different configurations have been measured. As shown in Fig. 6, the oscillator with sealed MEMS resonator and AGC module has better PN performance than the non-encapsulated one. Specifically, the output of a sealed oscillator with AGC circuit is -9 dBm, and the phase noise at 1 kHz offset is -96 dBc/Hz, while noise floor at far-from-carrier is approximately -120 dBc/Hz. In contrast, the signal output power of the unsealed oscillator with AGC circuit is -5 dBm, with its phase

noise performance of -90 dBc/Hz at 1 kHz offset, and noise floor of approximately -128 dBc/Hz.

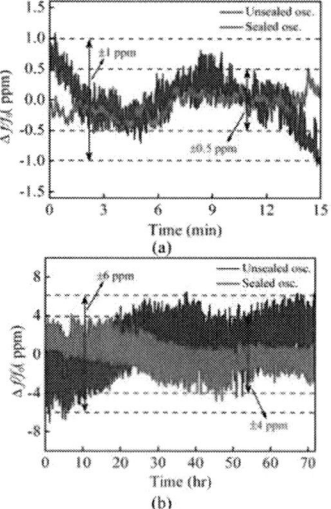

Fig. 6: Measured phase noise power spectral density-to-carrier power ratio versus frequency offset from the carrier for three types of oscillators

Frequency stability of the oscillator during a period was also studied, at room temperature (25 °C) including short-term (15 minutes) and middle-term (72 hours) stability measurements, as shown in Fig. 7. For non-encapsulated oscillator, the measured short term frequency stability (STFS) is ± 1 ppm; while the medium term frequency stability (MTFS) is ± 6 ppm. For the encapsulated oscillator, the measured STFS and MTFS are ±0.5 ppm and ±4 ppm, respectively. Frequency stability is improved by vacuum encapsulation, as the sealed resonator achieves higher Q value and is well protected. The results indicate that the oscillators show excellent frequency stability at room temperature.

Fig. 7: (a) Short term frequency stability with 0.1 second average time for 15 minutes, (b) middle term frequency stability with 1 second average time for 3 days.

DIGITALLY-TUNABLE NARROW PASS-BAND MEMS FILTER

Frequency selection modules in RF transceiver can be realized with frequency tunable MEMS filters. Tunable filters can greatly simplify the multiband selection system in the front end. With high Q MEMS filters, the system also benefits from loss reduction and low power consumption. Tunable MEMS filter provides desired characteristics, such as narrow bandpass width, sharp roll-off, large stop-band rejection and low insert loss which enable channel selection of multiple signals and eliminate sideband interferers.

Fig. 8 SEM image of the MEMS tunable filter (a), disk resonator with beams (b), coupling beam (c) and the center stem (d).

The coupled resonator array is used to construct multi-pole tunable filter as shown in Fig. 8. The MEMS filter comprises disk resonators and coupling beams fabricated with polysilicon. In the filter, four 18 μm contour-mode disk resonators are linked by longitudinal mode coupling beam and the beam length is 3λ/4 (λ is the wave length). RF signals are processed with micro fabricated coupling system and the the signal energy is transferred between electrical and mechanical domains in the coupling resonators. The coupled resonator systems generate different modes of vibration which define the pass-band and center frequency of the filter response.

The disk resonator is driven with electrodes surrounding it, which is important to enhance radial-contour mode of the resonators and suppress other potential vibration modes.

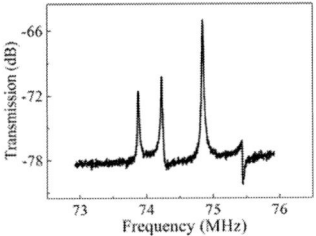

Fig. 9: The frequency response of 4-disk filter array.

The length of the coupling beam in the filter is equal to three multiple of a quarter-wavelength, which can cancel the mass and stiffness change of the resonators caused by the coupling beam. The resonators in the filter are identical to realize a symmetric pass-band to generate a symmetric pass-band and the low-velocity coupling is utilized to obtain a narrow bandwidth.

The measured result is shown in Fig. 9. As the Q values of the resonator are extremely high, the four resonance peaks are relatively discrete. Further optimization to enhance couplings between each modes and decrease ripples in the bandwidth is going on.

CONCLUSIONS

The high performance resonant devices based on the high Q and high frequency disk resonator are developed. The vacuum-encapsulated MEMS resonator has favorable performance. The tunable MEMS resonator is attained with high order vibration modes. High stable oscillator and tunable narrow pass-band MEMS filter are realized which can be used as frequency synthesizer and frequency selection module in RF front end. These resonant MEMS devices have potential applications in future wireless communication system.

ACKNOWLEDGEMENTS

This work was supported by NSFC projects (61234007 and 61404136) and the project from MOST of china (2013YQ16055103)

REFERENCES

[1] J. Xie, Y. F. Liu, M. L. Zhang, J. L. Yang, and F. H. Yang, *Technical Digest, 24th Int. IEEE MEMS,* Cancun, Mexico, Jan. 23–27, 2011, pp. 205–208.

[2] J. L. Lopez, J. Verd, J. Teva, G. Murillo, J. Giner, F. Torres, A. Uranga, G. Abadal and N. Barniol, *J. Micromech. Microeng.*, vol. 19, No. 1, 2009.

[3] J. T. M. van Beek, and R. Puers, *J. Micromech. Microeng.*, vol. 22, no. 1, pp. 13001, 2012.

[4] C.T.-C. Nguyen, *VLSI-TSA'08*, Hsinchu, Taiwan, April 21-23, 2008, pp. 3-4.

[5] J. R. Clark, M. A. Abdelmoneum, C. T.-C. Nguyen, *Frequency Control Symposium*, 2003, pp. 803-809

[6] Y. Yang, et al, *TRANSDUCERS & EUROSENSORS XXVII*, Barcelona, Spain, June 2013, pp.502 – 505.

[7] J. C. Salvia, R. Melamud, et al. *J. Microelectromech. Syst.*, vol. 19, no. 1, pp. 192-201, Feb. 2010.

[8] M. M. Shalaby, et al, *IEEE Trans. Ind. Electron.*, vol. 56, pp. 1022–1030, Apr. 2009.

[9] Y. F. Liu, J. Xie, et al. *J. Micromech. Microeng.*, vol. 22, no. 3, pp. 035003, 2012.

[10] Zhao J, Yuan Q, Luo W, et al. *J. Sensors & Actuators A Physical*, 2016, 252.

[11] Luo W, Zhao H, Yuan Q, et al. *J. Microsystem Technologies*, 2016:1-5.

[12] Yuan Q, Luo W, Zhao H, et al. *J. IEEE Transactions on Electron Devices*, 2015, 62(5):1603-1608.

Study of the Mesa Etched Tri-Gate InAs HEMTs with Extremely Low SS for Low-Power Logic Applications

Yueh-Chin Lin[1]*, Jing-Neng Yao[2], Hisang-Hua Hsu[3], Ying-Chieh Wong[1], Chi-Yi Huang[4], Heng Tung Hsu[5], Hiroshi Iwai[5], Life Fellow, IEEE, and, Edward Yi Chang[1, 2, 5]

[1]Department of Materials Science and Engineering, National Chiao-Tung University (NCTU), Hsinchu 30010, Taiwan.

[2] Department of Electronics Engineering, NCTU, Hsinchu 30010, Taiwan.

[3]Institute of Imaging and Biomedical Photonics, NCTU, Tainan 71150, Taiwan.

[4]Institute of Lighting and Energy Photonics, NCTU, Tainan 71150, Taiwan.

[5]International College of Semiconductor Technology, NCTU, Hsinchu 30010, Taiwan

*Corresponding Author's Email: nctulin@yahoo.com.tw

ABSTRACT

InAs HEMTs with a mesa etch structure that connects the Schottky gate through mesa sidewall with InAlAs layers were fabricated. The gate metal connection to the InAlAs layers increases the positive potential of the channel region through the gate bias, resulting in a steep SS due to a positive potential feedback. The mesa etch InAs HEMT shows an excellent performance with an extremely low minimum SS value of 46 mV/decade with the high $G_{m,max}$/SS of 33 and a high I_{ON}/I_{OFF} ratio of 6.6×10^4 at $V_{DS} = 1V$.

INTRODUCTION

Recently, $In_XGa_{1-X}As$ high electron mobility transistors (HEMTs) have exhibited superior DC and RF performances including higher transconductance (G_m), higher current density, and higher operating frequency [1, 2]. InAs HEMTs in particular possesses high electron mobility with low threshold voltage, and demonstrate high potential for high-speed and low power application and have been widely investigated for post CMOS applications [2]. However, as the transistor scales down, many problems such as V_{TH} roll-off and off-state leakage current issues still need to be solved. Therefore, searching an appreciate fabrication process to improve the device characteristics is a very important issue.

In order to further improve the subthreshold slope (SS) and I_{ON}/I_{OFF} values of the InAs HEMT, we introduced a gate-to-substrate contact so that the drain current can be increased by the positive back-gate bias from the contact. Connection to the channels at the mesa sidewall were connected to Schottky contact through the gate metal at the sidewall of the trench between the 2 mesas as shown in Fig.1 (a). The results show that the 3-D device structure results in better logic performance of the device, including lower SS, lower off-state leakage current and larger I_{on}/I_{off} ratio for low power logic applications(V_{DS}=0.5V). When V_{DS} was increased to 1V, the subthreshold characteristics and I_{on}/I_{off} ratio were further improved.

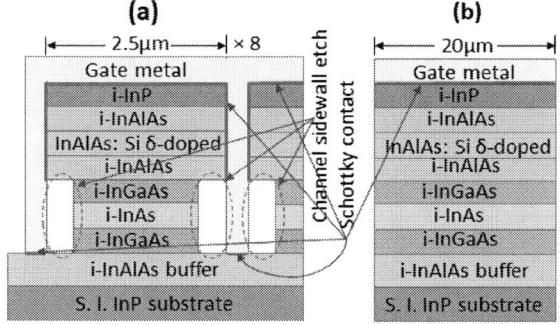

Fig. 1. Cross-section in the plane parallel to the gate electrode :(a) the mesa InAs HEMT, and (b) conventional InAs HEMT

DEVICE FABRICATION

The process of the mesa InAs HEMTs in this study is as following: First, the device isolation process was performed by etching the top InGaAs layer by H_3PO_4+H_2O solution at first, then, the InP layer was etched by HCl solution, and finally, the InAlAs/InGaAs/InAs/InGaAs/InAlAs layers by H_3PO_4+ H_2O solution. The etching depth was around 200 nm. Then, Ti/Pt/Au was deposited as Source/Drain ohmic metal with a 3 μm SD spacing. After that, the channel area was etched into 8 areas using mesa etch (2.5 μm width each) by electron beam lithography and Cl_2 ICP etching. An 80 nm SiN_x film was deposited by PECVD. Then, the electron beam lithography and CF_4 ICP etching were used to etch the SiN_x film, and $In_XGa_{1-X}As$ layer was etched by succinic acid and H_2O_2 solution through the SiN_x slits to form the gate recess areas (Fig.2). In this etching process, mesa-sidewall etching was also achieved at the same time, and $In_XGa_{1-X}As$/InAs/$In_XGa_{1-X}As$ layers were selectively side-etched as shown in Fig.1(a) to prevent the direct contact between the gate metal and InAs layer for low gate leakage current. Then, the Pt/Ti/Pt/Au film was evaporated and gate electrode was formed by lift off process. The gate electrode length was 500 nm at the top, but it was 170 nm at the bottom (Fig.2). Finally, 50 nm SiN_x film was deposited for device passivation. The SEM and TEM images of the plan view and cross section of the devices are shown in Figs. 3(a) and (b), respectively.

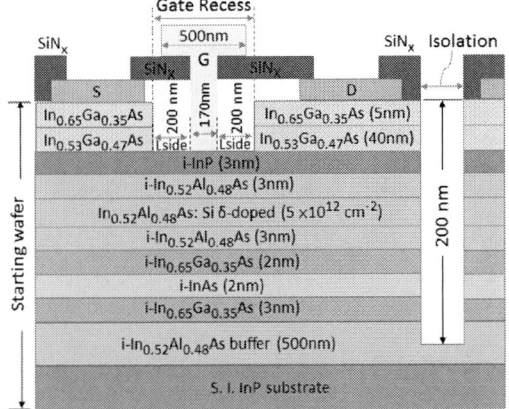

Fig. 2. Device structure of the mesa InAs HEMT

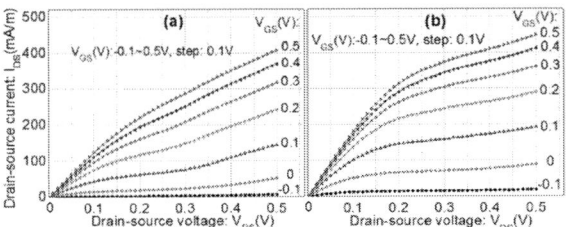

Fig. 3 (a) SEM and (b) TEM images of the mesa InAs HEMT.

RESULT AND DISCUSSION

Fig. 4 shows the DC characteristics of the 0.17 μm× 20 μm InAs HEMT. The maximum drain current ($I_{DS,max}$) and maximum transconductance ($G_{m,max}$) for InAs HEMT with mesa etch were 335 μA/μm and 998 μS/μm, respectively, when the device drain voltage (V_{DS}) was biased at 0.5 V.

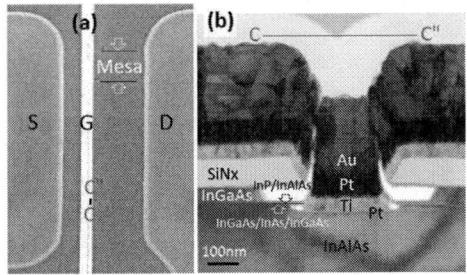

Fig. 4 I_{DS}-V_{DS} curves: (a) the mesa InAs HEMT, and (b) conventional InAs HEMT

Table I summarizes the device characteristics of the mesa InAs HEMT and a conventional InAs HEMT. The device with mesa etched structure has higher I_{ON}/I_{OFF} current ratio due to low OFF-state current, and lower SS values than the conventional HEMT. Fig. 5 (a) shows the I_{DS} vs. V_{GS} characteristics for the mesa HEMT with V_{DS} bias at 1V. Extremely small SS value of 46 mV/decade, high ON-state current of 654 μA/μm and high I_{on}/I_{off} ratio of 6.6×10^4 were achieved. Fig. 5(b) shows the I_{DS} vs. V_{GS} characteristics with V_{DS}. It noteworth that I_{DS} increased significantly in the higher V_{GS} region with the increase in V_{DS}, resulting in steeper I_{DS} vs. V_{GS} slope, and hence, extremely low SS values at high V_{DS}, presumably due to the 2D field effect from the drain.

Table I. Device characteristics comparison of the mesa InAs HEMT and conventional InAs HEMT

	Mesa HEMT	Conventional HEMT
V_T (V)	-0.17	-0.19
$I_{DS,max}$(μA/μm)	407	440
$G_{M,max}$(μS/μm)	998	1016
SS (mV/dec.)	64	70
Q ($G_{M,max}$/SS)	15.6	15.1
I_{ON}/I_{OFF}	2.9×10^4	1.9×10^4
OFF-state I_{GS} (A/μm)	$< 6.8\times10^{-10}$	$< 4.7\times10^{-9}$
DIBL (mV/V)	50	35

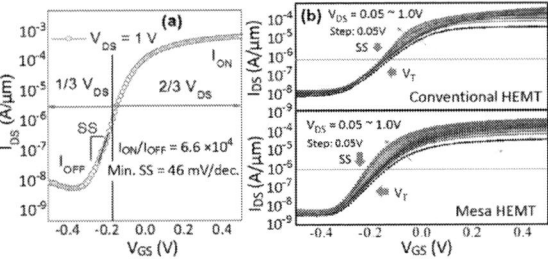

Fig. 5 (a) Sub-threshold characteristics of the mesa InAs HEMT (V_{DS} = 1.0 V), (b) The influence of V_{DS} increase on SS improvement for conventional HEMT and mesa HEMT

Fig. 6 shows the schematic band diagram of conduction band for the conventional InAs HEMT (Fig. 6(a)) and the mesa etched InAs HEMT (Fig. 6(b)). Because the Shottky contact from the gate electrode was connected to both the upper InAlAs and bottom InAlAs-buffer layers, the potential of the 2 InAlAs layers was shifted to positive direction when the gate voltage increases. This results in the $In_xGa_{1-X}As$/InAs/$In_xGa_{1-X}As$ layer potential shifts to the positive direction, resulting in the higher electron concentration at the channel. This is the reason for low SS and high I_{ON}/I_{OFF} ratio for the mesa HEMTs.

Fig. 7 (a) shows SS of the device as a function of the gate length for the mesa InAs etched HEMT and other devices from literature. The mesa etched InAs HEMT exhibits the best SS values. Fig. 7 (b) shows the maximum transconductance ($G_{m,max}$) as a function of SS, Q = $G_{m,max}$/SS which involves carrier transport properties and electrostatic integrity. Compared with other reports in the literature, the mesa etched InAs HEMT in this work exhibits high Q value of 33 when the device V_{DS} was biased at 1V. Figs. 7 (c) and (d) show the I_{ON}/I_{OFF} ratio as a function of SS and I_{ON} as a function of I_{OFF}, respectively. The mesa etched InAs HEMT in this work demonstrated a high I_{ON}/I_{OFF} ratio with extra low SS value. Furthermore, the high I_{ON}/I_{OFF} ratio over 4 order was obtained owing to high ON-state and low OFF-state currents for the mesa etched InAs HEMT.

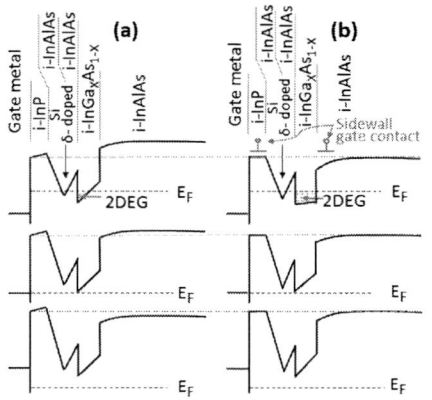

Fig. 6 Schematic of conduction band diagram of (a) conventional HEMT, and (b) the mesa HEMT.

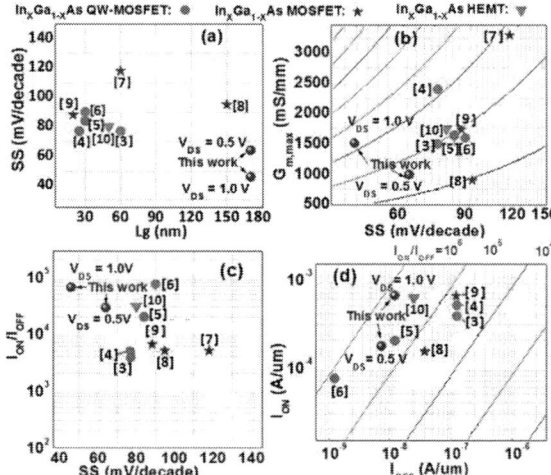

Fig. 7. (a) SS as a function of Lg, (b) Maximum transconductance ($G_{m,max}$) as a function of SS, (c) I_{ON}/I_{OFF} ratio as a function of SS, (d) I_{ON} as a function of I_{OFF} for the mesa InAs HEMT and other devices reported in the literature

CONCLUSION

Mesa etched InAs HEMTs with the sidewall-gate-metal connected to InAlAs layers using Shottky contact metal

were fabricated. Excellent logic performance including extra low SS value of 46 mV/decade, highest Q value of 33, high ON-state current, low OFF-state current and high I_{ON}/I_{OFF} ratio of 6.6×10^4 at $V_{DS} = 1V$ are demonstrated for the mesa etched HEMT device with the gate length of 170 nm. The excellent performance of the mesa etched InAs HEMT is attributed to the positive feedback from the gate electrode to the potential of the InAlAs layers.

ACKNOWLEDGEMENTS

This work was sponsored by National Chung-Shan Institute of Science & Technology, Taiwan, under Grant No. NCSIST-103-V312 (106) and TSMC, NCTU-UCB I-RiCE program, and Ministry of Science and Technology, Taiwan, under No. MOST106-2911-I-009-301.

REFERENCE

[1] S.-J. Yeon, M.-H. Park, J. Choi, and K.-S. Seo, in *IEDM Tech. Dig.*, 2007, pp. 613–616.

[2] D. H. Kim, J. A. del Alamo, J. H. Lee, and K. S. Seo, in *IEDM Tech. Dig.*, 2005, pp. 767–770.

[3] T.-W. Kim, D.-H. Kim, D. H. Koh, H. M. Kwon, R. H. Baek, D. Veksler, C. Huffman, K. Matthews, S. Oktyabrsky, A. Greene, Y. Ohsawa, A. Ko, H. Nakajima, M. Takahashi, H. Nishizuka, H. Ohtake, S. K. Banerjee, S. H. Shin, D.-H. Ko, C. Kang, D. Gilmer, R .J. W. Hill, W. Maszara, C. Hobbs and P.D. Kirsch., in *Proc. IEEE Int. Electron Devices Meeting (IEDM)*, Dec. 2013, pp. 16.3.1–16.3.4.

[4] S. Lee, V. Chobpattana, C.-Y. Huang, B. J. Thibeault, W. Mitchell, in Proceedings of the Symposium on VLSI Technology, (2014) T54.

[5] S. H. Kim, M. Yokoyama, R. Nakane, O. Ichikawa, T. Osada, M. Hata, M. Takenaka, and S. Takagi, in *IEDM Tech. Dig.*, Dec., (2013), pp. 429.

[6] C. Y. Huang, S. Lee1, V. Chobpattana, S. Stemmer, A. C. Gossard, B. Thibeault, W. Mitchell1, M. Rodwell, in *IEDM Tech. Dig.*, Dec., (2014) 586.

[7] C. B. Zota, L. E. Wernersson and E. Lind, in *IEDM Tech. Dig.*, Dec. (2015) 811.

[8] M. L. Huang, S. W. Chang, M. K. Chen, C. H. Fan, H. T. Lin, C. H. Lin, R. L. Chu, K. Y. Lee, M. A. Khaderbad, Z. C. Chen, C. H. Lin, C. H. Chen, L. T. Lin, H. J. Lin, H. C. Chang, C. L. Yang, Y. K. Leung, Y.-C. Yeo, S. M. Jang, H. Y. Hwang, and Carlos H. Diaz, in Proceedings of the Symposium on VLSI Technology, (2015) T56.

[9] J. J. Gu, X. W. Wang, H. Wu, J. Shao, A. T. Neal, M. J. Manfra, R. G. Gordon, and P. D. Ye, in *IEDM Tech. Dig.*, Dec. (2012) 633.

[10] D.-H. Kim, J. A. del Alamo, P. Chen, W. Ha, M. Urteaga, and B. Brar, in *Int. Electron Devices Meeting Tech. Dig.*, 2010, pp. 692–695 (30.6).

NEXT GENERATION OF WLCSP CONTACTING TECHNOLOGIES FOR 250 MICRON PITCH AND BELOW

Bert Brost and Valts Treibergs

Xcerra Corporation

St. Paul, Minnesota, United States of America

*Corresponding Author's Email: bert.brost@xcerra.com, valts.treibergs@xcerra.com

ABSTRACT

Spring probes are becoming a viable alternative to traditional probe card technologies for contacting Wafer Level Chip Scale Package (WLCSP) devices. Over the past decade, probe head designers have been challenged by several trends in semiconductor packaging. The most notably trend is a narrowing in the WLCSP "pitch" between balls, pumps pillars, and pads along with an increase in operating frequencies and a reduction of operating voltage. This paper explores a new probe architecture that is commercially available. The focus of this paper is a new and unique spring probe technology for WLCSP test.

INTRODUCTION

The focus subject of this manuscript is a new hybrid cantilever spring probe technology developed for contacting WLCSP devices with a pitch of 250 microns or less. The new technology is a more reliable, more rugged, and easier to maintain alternative to the traditional WLCSP probing technology

Spring probes are becoming the technology of choice for a variety of WLCSP test applications. The new WLCSP spring probes are an interconnect technology hybrid. As a hybrid, the new spring probe is a two-piece design: 1) a barrel and 2) a spring as shown in figure 1.

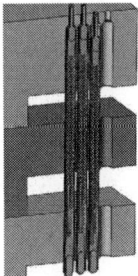

Figure 1: 250 micron WLCSP probe cross-section

Existing WLCSP Spring Probe Technology

Current state-of-the-art spring probes consist of 3 or 4 mechanically assembled components: a barrel, spring, and one or two plungers. The barrel can be formed by a mechanical deep-drawing process, screw-machined method, or by electroforming. The springs are typically wound with extremely small wire on a spring winding apparatus, and the plunger components are also screw-machined. Because of the fact that the spring resides inside the barrel and the assembly needs to be maintained, the probe tends to be very long in order to provide adequate force to make reliable contact. Because of the multiple-piece assembly, the contact resistance of the probe varies due to the sliding nature of the moving parts and overall lack of a lateral bias force that constantly ensures intimate contact between components. (1)

The Next-Generation WLCSP Probe Spring

The spring is a one-piece accordian fold cantilever design. The accordian fold is a simple zigzag cantilever pattern with three panels and two parallel folds that go in opposite directions, also known as Z-folds. Accordian folds are similar to the pleats on the musical instrument known as an accordion. Note the different spelling as the two are different yet the same. The accordian fold cantilever design compresses and expands at the bends as these are the flex points. This expansion and contraction provides the spring probe force biasing. The super-sharp tip of the spring functions as a plunger. There are shoulders at the base of the super-sharp tip that mate with hard stop surfaces on the probe head body (see figure 1). The mating of the probe shoulder features to the probe head body prevents the spring from coming out of the barrel. The location of the shoulder to the body determines the degree to which the plunger extends from the barrel in the probe's normal preload condition.

The mating surface of the spring to the test interface board compresses the accordian fold cantilever spring to some degree when the probe head is mounted to a test interface board or probe card. The amount of compression is referred to as the preload. This initial tension is referred to as the "preload force" of the spring. When the spring tip/plunger is pushed down while the barrel is held motionless, force builds as the folds of the cantilever spring compress. The shoulder and neck of the plunger tip move into the barrel when the probe is contacting the WLCSP device-under-test. The amount of motion/accordian fold cantilever spring compression is controlled by prober travel limits.

978-1-5386-5309-8/18 $31.00 © 2018 IEEE

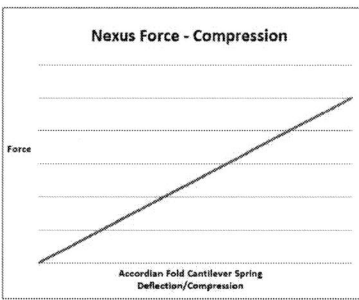

Figure 2: Spring force is linear

Compression force is primarily determined by the force required to compress the accordian fold cantilever spring. Compression forces are linear. This means that force at any degree of compression can be predicted with some amount of accuracy. The example of linear force versus deflection is shown in figure 2. In the figure 2 example, the force of the probe is estimated based on the amount of deflection/compression. (2)

The WLCSP Probe Barrel

The spring probe barrel is a hollow conductive housing for the spring contact probe. The barrel is open at both ends with the accordian fold cantilever spring protruding at both ends of the barrel for contact with the device-under-test and the test interface board. The barrel is concentric and the surfaces are smooth. WLCSP probe barrels typically have exceptionally thin walls with tight tolerances. Plating of the barrel is less of an issue since the barrel is open at both ends. The plating process for advanced WLCSP probes is typically a proprietary process. All processes are designed to provide consistent precious metal coverage of the surface of the barrel.

Probe Force, Compression, and Resistance

The Nexus WLCSP accordian fold cantilever spring probe is a two-point contact shunted with a barrel. The accordion fold cantilever spring is a single piece of conductive material with a contact point at each end of the material. A two point spring probe improves the current carrying efficiency of the probe by eliminating the internal IR drop variability of a multi-point contact probes. The reduction and/or elimination of variability improves the repeatable of contact resistance for improved DC measurement accuracy. The conductive probe barrel works as a shunt to increase the conductivity of the two piece accordian fold cantilever spring probe.

Current flows through the probe's accordian fold cantilever spring and barrel. Probe resistance in test is dependent on two variables and one constant. Force and deflection/compliance are the two variables and the static conductivity of the probe material is the constant. The contact resistance between the probe plunger tip and the contact surface at the device-under-test varies with the amount of penetration of the plunger tip into the contact surface. The amount of penetration will vary with the sharpness of the tip.

A Deflection/Overdrive (FReD) test fixture is used to evaluate the durability/reliability of a probe over a number of make and break contact connections. The FReD test fixture is a system that includes precision measurement instrumentation. The instrumentation is used to measure the resistance of the probe relative to force and deflection (compression or overdrive). The measures are collected and recorded over millions of make and break contact connections (touches).

The graph shown in figure 3 is a FReD plots and it illustrates the results of a 250-micron pitch Nexus probe at four million cycles. The deflection value is shown along the horizontal axis of the plot, as the probes are individually compressed and released. As you can see, force and resistance are inversely proportional. The A-frame roof top shape group of lines shows the insertion and withdrawal force required at each deflection increment. The force values are shown on the right side of the graph in grams. Each line is zero until the probe is touched, then jumps to a value due to the preload, then increments linearly (spring linearity was discussed earlier in this manuscript) as the probe is deflected, and decrements linearly as it is released, until contact is lost and it drops back to zero.

The U-shaped group of lines shows the contact resistance of the probes. The resistance value is shown on the left side of each graph, in milliohms. Each probe starts at infinity and then drops to a reasonable value when contact is first made to the probe. The resistance value drops slightly as the probe is compressed, and increases as it is released, finally jumping back to infinity as contact is lost. The resistance remains low and steady (no dramatic changes) throughout the operating range of the probe.

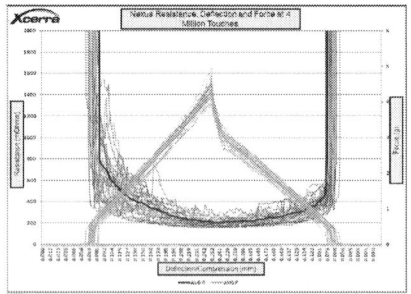

Figure 3: Force, Resistance, Deflection

These tests are primarily intended to show that the

probes are mechanically robust with little to no resistance deviation probe-to-probe. Since this testing is offline cycling and contact occurs between gold surfaces on both sides, these tests do to some degree show resistance to wear, but do not show resistance to solder build-up.

Contact resistance between the plunger tip and the contact surface at the device-under test must be considered as part of the total interconnect to the device under test interconnect resistive path. This type of resistance varies with the amount of penetration the probe/plunger tip has into the contact surface, through whatever nonconductive materials such as oxides and/or debris. A larger and cleaner surface area of contact will result in less interconnect resistance at the device-under-test.

The new WLCSP probes have super-sharp tips for lower interconnect resistance and longer uninterrupted run times between contactor cleaning with an increase in the number of devices tested and increased good part yielding between probe cleaning.

The super sharp tips have a smaller surface area at the tip of the probe (see figure 1) when compared to traditional probes. When a traditional probe tip and a super-sharp probe tip are pushed with the same force, the super-sharp probe tip will exert more pressure on a smaller area than a traditional probe.

It is the pressure that cuts through or pierces the surface oxides on the ball, pillar, or pad of the device-under-test (see figure 1). Like a sharp knife, the super-sharp probes make a cleaner cut than traditional probes. A cleaner piercing of surface oxides improves the interconnect quality of the probe to the device under test. Most notable is the improved repeatability of contact resistance over longer uninterrupted run times for increased cycle counts and yield between cleaning. Ratio of force to area over which it is applied is expressed in the following equation:

$$P=F/A \qquad (1)$$

Considering that pressure is the ratio of force to area over which it is applied. The super-sharp probe requires less force to effectively pierces the surface oxides on the pillar, ball, or pad of the device-under-test. As the area of the tip gets smaller (sharper) and force is a spring constant, the pressure increases for better and more controlled penetration.

To help increase runtimes as measured in the number of parts tested between cleaning, super-sharp WLCSP probes tips are manufactured with a smoother surface. Surface roughness plays an important role in the mean-cycle-count-between probe cleaning. The micro

surface texture or roughness of the probe tip abrades the solder ball, or the plating on the bumped pillar or pad of the device-under-test. The abrading action is like the abrading action of a file but at a micro level. The surface peaks of a file work to remove material and the valleys fill with debris: This is called loading. The same is true with the surface of probes but at a micro level. The smooth surface of a super-sharp probe does not fill with solder as fast as the not so smooth surfaces of traditional probes. The result is increased runtimes as measured in the number of parts tested between cleaning.

Ease of Maintenance

The probe is held in the probe head body by surface friction. As shown in figure 3, the probe head body narrows at the top (dark blue) and is open at the bottom (green). The probe protrudes slightly from the bottom of the barrel. This allows for easier removal of worn and/or damaged accordian fold cantilever spring probes.

Factors affecting WLCSP probe electrical performance

Probe spacing or pitch, and operating frequencies are two of the greatest challenges of WLCSP device testing. As probes are forced closer together by smaller pitch with increasing operating frequencies and as edge rates increase, a number of factors begin to have a significant impact on testing accuracy.

Inductance of the probe becomes critical. For most high performance testing applications, probe loop inductance can be no greater than 3 nano henries (nH}-a level usually exceeded by traditional wafer level probe technology contact designs.

Capacitance increases as contacts are forced into close proximity. The greater the contact surface area and the finer the pitch, the greater the capacitance and the greater the influence on test performance. Capacitance of two parallel plates can be found by the formula:

$$C=KA/4\text{-}rrd \qquad (2)$$

Where K is dielectric constant, A is surface area of plates, and d is distance. Moreover, test systems and device outputs may not be able to drive the higher capacitance.

Signal Interference may occur in the form of crosstalk. Crosstalk is caused by the hidden schematic on the unintended circuit created by the parasitic inductive and capacitive coupling. Crosstalk effects are minimized by reducing parasitic inductance and capacitance by reducing common length and length of individual probes.

The key to high performance, high reliability electrical contacting is the combination of mechanical and electrical performance

Conclusion

The novel design of the Nexus interconnect probe for WLCSP and wafer probe testing effectively addresses the challenges of testing fine pitch device operation at higher frequencies with lower operating voltage levels.

The main advantage of this new WLCSP probe technology is that it it is completely scalable to smaller wafer probe pitches without significantly increasing cost.

While spring probe technologies deliver additional benefits that are easy to see, the primary value of probe technology described in this paper is the maintained, high performance test signal fidelity over longer, uninterrupted run times for higher yields and increased test cell throughput.

References

(1) "Next Generation Contact Technology For Semiconductor Test" Burn-in & Test Strategies Workshop 2007, Mesa AZ, USA. Valts Treibergs, Jason Mroczkowski - Everett Charles Technologies

(2) "A Cost-Effective Approach for Wafer Level Chip Scale Package Testing" IEEE SW Test Workshop , Norman J. Armandariz, PhD Texas Instruments, Inc.

Photovoltaic Properties of Lateral Ultra-Thin Si p-i-n Structure

Suguru Tatsunokuchi[1]*, I. Muneta[1], T. Hoshii[1], H. Wakabayashi[1],
K. Tsutsui[2], H. Iwai[2], K. Kakushima[1]

[1] School of Engineering, Tokyo Institute of Technology,

4259 Nagatsuta, Midori-ku, Yokohama 226-8502, Japan

[2] Institute of Innovative Research, Tokyo Institute of Technology

*Corresponding Author's Email: tatsunokuchi.s.aa@m.titech.ac.jp

ABSTRACT

The effect of i-region length (L_i) in lateral ultra-thin Si p-i-n diodes on the photovoltaic (PV) properties has been investigated. The short-circuit current (I_{SC}) has increased along with L_i and has shown a saturation characteristic over L_i of 5 μm. On the other hand, it has been found that the open-circuit voltage (V_{OC}) becomes constant above L_i = 2 μm. The I_{SC} behavior can be modeled by the recombination in the i-region in combination with reduced lateral electric field.

INTRODUCTION

Although progresses in improving the conversion efficiency of Si-based solar cells are under way, power generation efficiency remains at the level of 25% [1]. In order to overcome this situation, various types of tandem solar cells have been proposed. Among them, Si nanowall (SiNW) solar cells have been expected to be used as the top cells for high-efficiency all-Si tandem solar cells [2]. Nanowall structures can introduce light trapping enhancement by high-aspect-ratio structure. Moreover, bandgap enlargement by quantum confinement effect is expected to increase the I_{SC} and V_{OC}. However, the PV characteristics of SiNW solar cells are degraded by surface effects due to its large surface-to-volume ratio [3]. In this work, lateral ultra-thin p-i-n Si solar cells were fabricated and the effect of i-region length dependence on PV characteristics are examined.

DEVICE FABRICATION

Fabrication process of a lateral SiNW cell and its cross-sectional image are shown in figures 1 and 2, respectively. The starting silicon-on-insulator (SOI) substrate of 88-nm-thick Si layer (p-type) was thinned down to 44 nm by the first oxidation process. After removal of the SiO$_2$ layer in the i-region, mesa patterns were formed by chemical-dry etching (CDE). Then, the second oxidation was performed to reduce the Si thickness of i-region (T_i) and the oxidation time was controlled to design the remaining Si thickness. With this 2-step oxidation process [4], T_i of 4.9 nm, and the anode and cathode regions as thick as 30 nm have been fabricated. The spacing between the anode and cathode regions, L_i, was set from 0.2 to 15 μm. The width of the solar cells is

80 μm. The backside Ohmic contact was used as an electrode to apply a back bias (V_b) to change the surface potential of the SiNW [5] through the 145 nm thick buried oxide layer (BOX). PV characteristics were measured under AM1.5G illumination with different V_b.

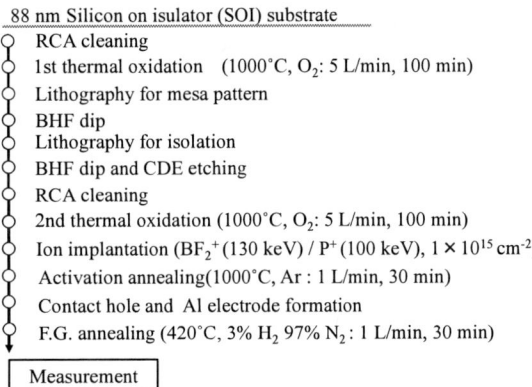

88 nm Silicon on isulator (SOI) substrate
- RCA cleaning
- 1st thermal oxidation (1000°C, O$_2$: 5 L/min, 100 min)
- Lithography for mesa pattern
- BHF dip
- Lithography for isolation
- BHF dip and CDE etching
- RCA cleaning
- 2nd thermal oxidation (1000°C, O$_2$: 5 L/min, 100 min)
- Ion implantation (BF$_2^+$ (130 keV) / P$^+$ (100 keV), 1×10^{15} cm^{-2})
- Activation annealing(1000°C, Ar : 1 L/min, 30 min)
- Contact hole and Al electrode formation
- F.G. annealing (420°C, 3% H$_2$ 97% N$_2$: 1 L/min, 30 min)

Measurement

Figure 1: Fabrication process of lateral SiNW solar cell.

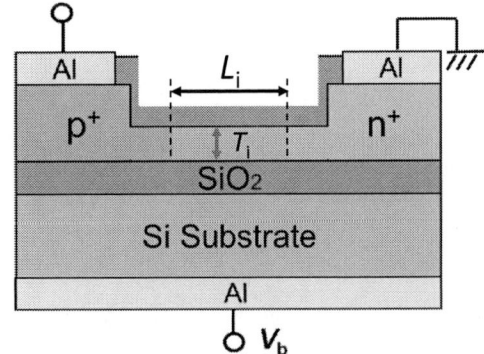

Figure 2: Cross section of lateral SiNW solar cell.

PV CHARACTERISTICS OF A SiNW SOLAR CELL UNDER V_b

Figure 3 shows PV characteristics lateral solar cells with different L_i. These PV characteristics were measured under each V_b where the maximum I_{SC} can be obtained. For example, the highest I_{SC} was obtained at $V_b = 0.6$ V for the device with L_i of 1.7 µm. As the vertical electric field in the SiNWs can be controlled by V_b, surface recombination velocities of the generated carriers at the surface and the interface to the BOX can be tuned. The maximum I_{SC} can be obtained by applying appropriate V_b to vanish the vertical electric field, thus suppressing the surface recombination velocity in the vertical direction. From figure 3, it can be seen that I_{SC} increases by increasing L_i, suggesting that increasing the L_i is one way of improving conversion efficiency.

Figures 4 (a) and (b) summarize the relationship between I_{SC} and V_{OC} on L_i, respectively. I_{SC} increases with L_i up to 5 µm and shows a saturation characteristic beyond this length. On the other hand, V_{OC} increases with L_i and shows a saturation characteristic at a length beyond 2 µm with a V_{OC} of 0.59 V. From this point, L_i of 2 µm or more is necessary to obtain a large V_{OC}.

Figure 4 (c) shows a proposed model of the L_i dependence on I_{SC}. In model A, which corresponds to the region A in figure 4 (a), I_{SC} is proportional to the area of the i-layer, namely L_i, thus the recombination at surface and interface can be ignored due to high lateral electric field. In model B with points in region B in figure 4(a), most of the photogenerated carriers are collected. However, due to weakened lateral electric field recombinations at surface and interface start to appear. In model C with points in region C in figure 4(a), recombination at surface and interface becomes dominant due to the lifetime of the carriers in the i-region. Moreover, as the electric field in the lateral direction decreases, the probability of recombination increases for the carriers generated near the anode and cathode regions before reaching the electrodes. Therefore, I_{SC} may decrease with longer L_i.

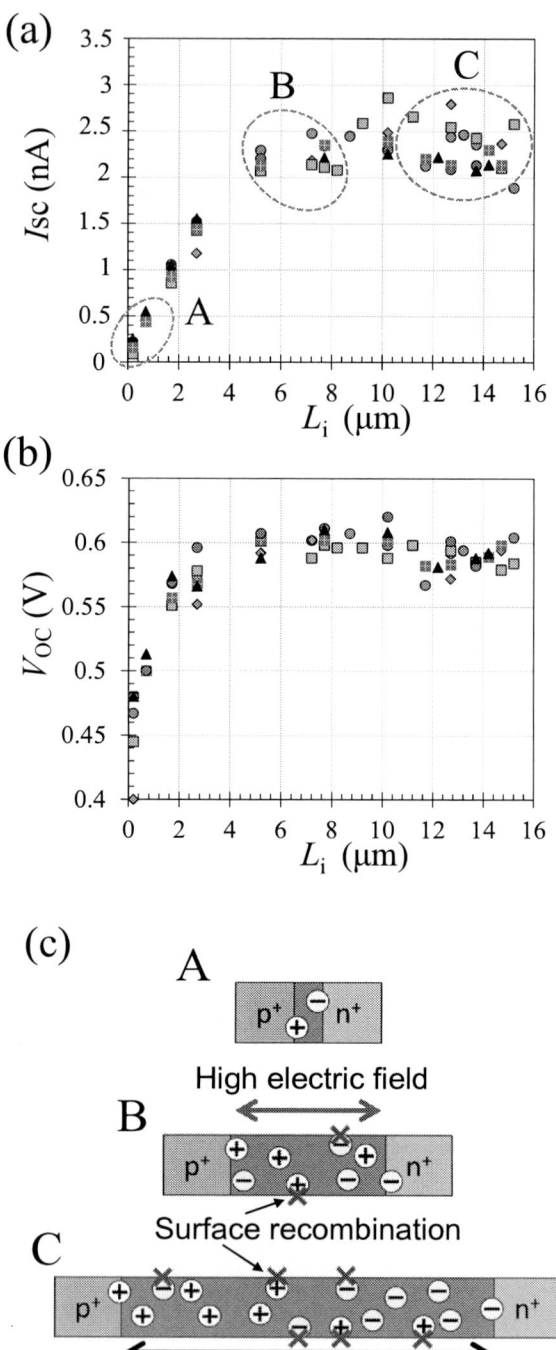

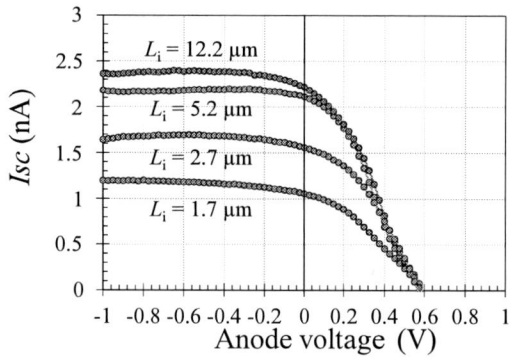

Figure 3: PV characteristics of SiNW solar cells under applying V_b.

Figure 4: L_i dependence of (a) I_{SC} (b) V_{OC}. (c) Model of the relation between L_i and I_{SC}.

CONCLUSION

The L_i dependence of PV characteristics of lateral ultra-thin Si solar cells was investigated. By comparing the PV characteristics of each sample while applying V_b that gives the maximum I_{SC}, we observe how I_{SC} increases with L_i. From this experiment it was found that I_{SC} saturates when L_i exceeds 5 μm. This phenomenon was explained from the model of the relationship between lateral electric field and recombination.

References

[1] M. A. Green, The path to 25% silicon solar cell efficiency: history of silicon cell evolution, Prog. Photovolt.: Res. Appl., 17, 183-189 (2009).

[2] Ichikawa, et al., Proc. SPIE, vol. 9178, p. 91780O (2014).

[3] E. Garnett and P. Yang, Nano Lett., vol. 10, pp. 1082 (2010).

[4] S. Tatsunokuchi, et al., IWDTF, pp. 84 (2017).

[5] T. Shoji, et al., 29th EUPVSEC Proceedings, pp. 844 (2015).

STUDY ON SENSITIVITY OF THE BILATERAL ULTRA-THICK SILICON SENSORS AS NEUTRON DOSIMETER

Y. H. Huang[1], M. Yu*[1], R. Y. Yin[1], Z. Zhang[1], X. Y. Zhao[1], Z. Y. Zhu[1], J. L. Liu[1], M.J. Wang[1], J. Y. Wang[1], Y. F. Jin[1]

[1] National Key Laboratory of Nano/Micro Fabrication Technology, Key Laboratory of Microelectronic Devices and Circuits (MOE), Institute of Microelectronics, Peking University, Beijing 100871, China

*Corresponding Author's E-mail: yum@pku.edu.cn

ABSTRACT

The radiation dose dependent sensitivity characteristics of bilateral silicon sensors are studied to improve the radiation hardness under neutron irradiation. Sensors with different effective intrinsic base lengths using 1.5mm ultra-thick silicon substrate are fabricated and the effect of structure parameters on neutron detection sensitivity is experimentally investigated. The simulations of devices with different structures are used to study the influence of potential distribution on radiation hardness characteristics. It is found that, increasing the effective intrinsic base length can improve the sensitivity. The effect of radiation does and working current are investigated experimentally.

INTRODUCTION

Neutron dose sensors are applied widely in nuclear research and radiation protection [1-3]. It is highly desirable to have high sensitivity for neutron monitoring due to the difficulty of neutron detection. The monitoring of the neutron dose by silicon sensors is based on the effect of damage caused by the flux of neutrons. Radiation-induced damage in silicon causes the decrease of the excess carrier lifetime and the increase of the output voltage, from which the sensitivity of the sensors can be obtained [4-5].

A new structure of neutron dosimeter has been proposed by our group [6]. In this paper, the structure is studied further. The influence of the geometric structure on the sensitivity of silicon sensors is investigated and the sensor response to different radiation dose is illustrated.

DIODE STRUCTURE AND EXPERIMENT

The structure of the bilateral sensor and its plane projection are illustrated in Figure 1. The rectangular doping regions are designed on the front and back sides of the silicon wafer. The length and width of the doping contact regions are represented by l and w. The interval between the electrodes is denoted by d. The thickness of silicon substrate is expressed as t. The w of p+ regions is set to 500 μm. High purity n-silicon wafers with the resistivity of 20 kΩ·cm and the thickness of 1500 μm are used for sensor fabrication. The high resistivity ensures

that the material has large minor carrier lifetime. The structure takes advantage of both thick wafer and lateral spacing to realize large effective intrinsic base length that is defined as the straight distance between p+ and n+ region. In this way the sensitivity can be improved.

Surface of the sensor is passivated by thermally growing 0.5μm SiO_2. The p+ and n+ contact regions are formed by ion implantation with Boron and Phosphorus respectively followed by high temperature annealing. The single sensor chip is achieved by careful dicing.

The diodes are exposed to neutron radiation for sensitivity testing. The radiation using Am-Be fast neutron radiation source, which dose equivalent average energy is 4.4 MeV. The V-I data are tested before and after irradiation with semiconductor parameter analyzer HP 4156B.

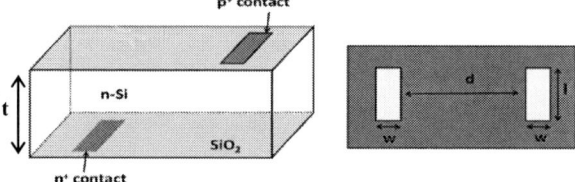

Figure 1: The structure of the sensor and the top view

RESULTS AND ANALYSIS

The V-I curve of the sensors are measured before and after irradiation. After neutron irradiation, the output voltage of the sensor rises. The sensitivity of the sensor, S, is derived from the formula with constant sensor current,

$$S = \frac{V_{after} - V_{before}}{D_{radiation}}$$

Where V_{before} is the initial output voltage before irradiation; V_{after} is the output voltage after irradiation; $D_{radiation}$ is the total irradiation dose.

The impact of the effective intrinsic base length on neutron detection sensitivity is experimentally investigated by performing different dose neutron radiation. Different effective intrinsic base lengths are achieved by different intervals between the electrodes. The experimental results of the bilateral sensor after irradiation are shown in Figure 2.

978-1-5386-5309-8/18 $31.00 © 2018 IEEE

The results show that, as the irradiation dose accumulates, the sensitivities increase, which is caused by the change of output voltage. The analysis of the sensitivities with different effective intrinsic base lengths shows that increasing the effective intrinsic base length significantly improves the sensitivity.

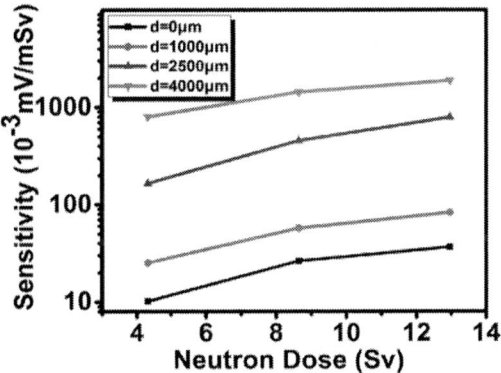

Figure 2: The sensitivity data of a bilateral sensor after irradiation, where intervals between the electrodes d= 0 μm, 1000 μm, 2500 μm, 4000 μm, respectively and all lengths of the doping contact are 2400 μm.

A simulation study on the bilateral diode is implemented to investigate the influence of structural parameters on electric potential distribution of the sensor by using TCAD Sentaurus [7]. The potential distribution between p+ contact and n+ contact is simulated to analyze the influence of electric field on radiation hardness and sensitivity. In the simulation, the current of the sensors is 1 mA.

As is shown in Figure 3(a-b), when the interval between the electrodes changes from 2500μm to 4000μm, the potential difference between p+ contact and n+ contact increases from 1.3V to 3V. The increase of potential difference may cause the increase of leakage current at the interface between silicon and silicon dioxide will increase. Detailed analysis on the simulation results indicates that the electric field strength is much larger in case of large electrode intervals. This can decrease the reliability and radiation hardness of the sensor. As shown in Figure 3(c-d), when the interval between the electrodes is changed from 1000μm to 0μm, the voltage drop changed from 0.6V to 0.5V. The effective intrinsic base length decreases from 1800 μm to 1500 μm. Although the sensitive region decrease leads to smaller potential difference, the radiation sensitivity decrease as well which is shown above.

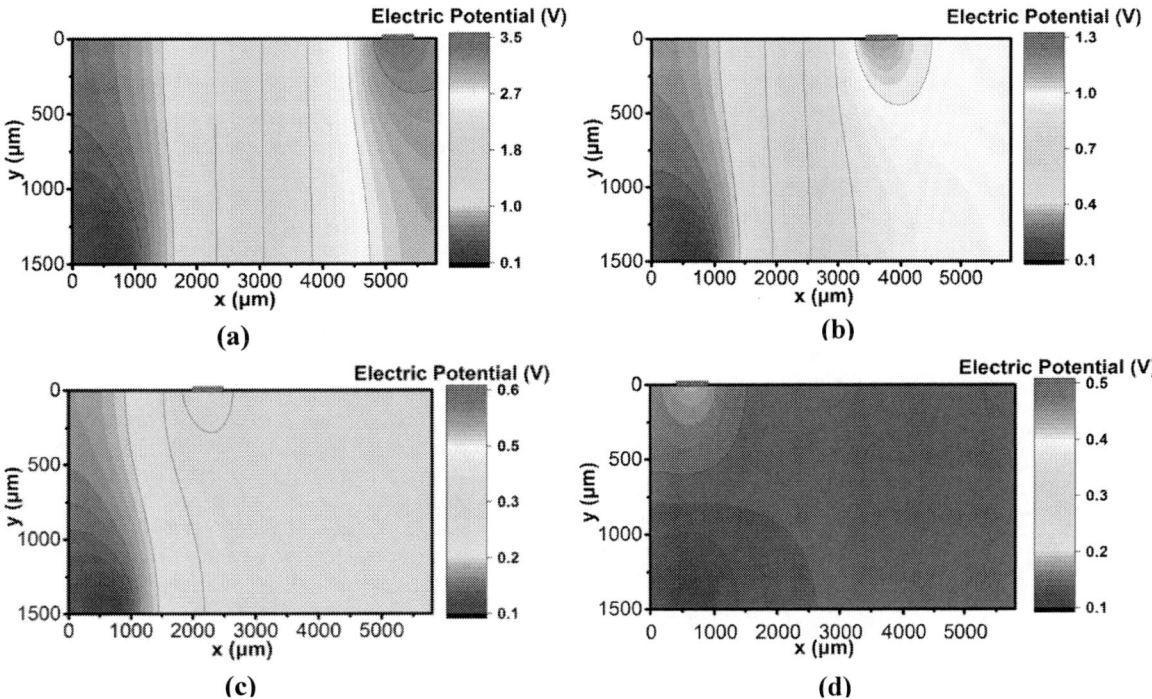

Figure 3: Simulation results of potential distributions at 1mA with different intervals between the electrodes, d, (a) d= 4000 μm, (b) d=2500 μm, (c) d= 1000 μm , (d) d=0 μm.

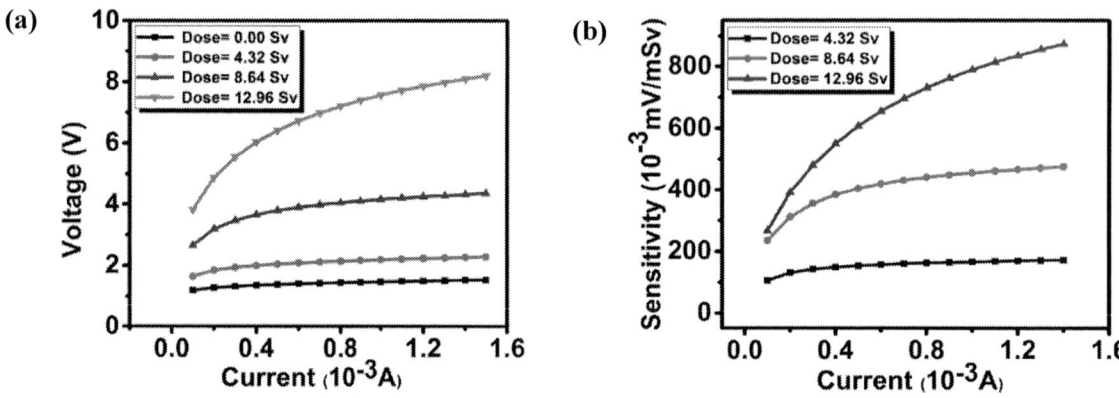

Figure 4: (a) The V-I data of the sensor before and after irradiation with length of the doping contact, l=2400 μm , (b) Sensitivity change against current for this diode. Different radiation doses are applied and indicated in the figures.

In order to further improve the sensitivity of the detector, the effect of current on sensitivity is studied based on the structure with the length of the doping contact of 2400 μm, the width of the doping contact 500 μm and the intervals between the electrodes of 2500 μm. The relationship between the sensitivity and the current under different doses of neutron irradiation is shown in Figure 4 (a) and Figure 4 (b).

As is shown in Figure 4 (a), when the irradiation dose changes from 0.00 Sv to 12.96 Sv, the potential difference between p+ contact and n+ contact increases. The increase of the potential difference is mainly due to neutron radiation damage in the silicon lattice. As the neutron irradiation dose increases, the damage in the silicon lattice gradually accumulates. In addition, the change in resistivity of silicon under irradiation cannot be neglected in the case of intermediate level injection. When the current is small, not only carrier lifetime decreases but also change in resistivity of silicon under irradiation has to be taken into account.

The relationship between sensitivity and current is shown in Figure 4 (b). When the irradiation dose is 4.32 Sv or 8.64 Sv, the sensitivity increases with current, and tends to saturate at higher current level. When the irradiation doses is 12.96 Sv, the sensitivity increases with current with no obvious saturation observed. It is found that under high dose neutron irradiation, high level injection needs higher current. Considering the sensor should work in the case of high level injection, 1mA is chosen as the working current in this work.

CONCLUSION

In this paper, structural parameters are investigated to improve sensitivity and radiation hardness in the bilateral neutron dose sensor. The effect of the effective intrinsic base length and the current are studied with considering the influence of the potential distribution and electric field distribution. By tuning the interval between the electrodes, which determines the effective intrinsic base length, and optimizing forward current, sensitivity and great radiation hardness and reliability can be optimized. Larger intervals induce higher sensitivities with larger electric potential differences. Current optimization is also necessary for the sensors.

REFERENCES

[1] Cheng L J, Lori J. Characteristics of Neutron Damage in Silicon[J]. Physical Review, 1968, 171(3):856-862.

[2] Lemeilleur F, Glaser M, Heijne E H M, et al. Neutron-induced radiation damage in silicon detectors[J]. Nuclear Science IEEE Transactions on, 1992, 39(4):551-557.

[3] Žontar D, Cindro V, Kramberger G, et al. Time development and flux dependence of neutron-irradiation induced defects in silicon pad detectors[J]. Nuclear Instruments & Methods in Physics Research, 1999, 426(1):51-55.

[4] Llacer J. Geometric Control of Surface Leakage Current and Noise in Lithium Drifted Silicon Radiation Detectors[J]. IEEE Transactions on Nuclear Science, 1964, 13(1):93-103.

[5] Braibant S, Demaria N, Feld L, et al. Investigation of design parameters for radiation hard silicon microstrip detectors[J]. Nuclear Inst & Methods in Physics Research A, 2002, 485(3):343-361.

[6] Hu, A.Q., et al., Bilateral PIN Diode for Fast Neutron Dose Measurement. IEEE Transactions on Nuclear Science, 2014. 61(3): p. 1311-1315.

[7] TCAD Sentaurus, Synopsys Inc. Mountain View, CA.

A novel three observation-windows measurement scheme for SPAD Fluorescence Lifetime Imaging Detector

Ding Li[1], Zhong Wu[1], Yue Xu[1,2]*, Tingchen Zhao[1]

[1] College of electronic and optical engineering & College of microelectronics in Nanjing University of Posts and Telecommunications, Nanjing, Jiangsu Province, China, 210023

[2] National and Local Joint Engineering Laboratory of RF integration & Micro-Assembly Technology, Nanjing, Jiangsu Province, China, 210023

*Corresponding Author's Email: yuex@njupt.edu.cn

ABSTRACT

This paper presents a novel three observation-windows scheme for fluorescence lifetime imaging (FLIM). The repeated integration is operated in three overlapped or discrete observation windows with nanosecond-range duration to obtain photon counting data, which can effectively reduce relative standard deviation in fluorescence lifetime evaluation. Further, based on this three observation-windows detection method, a single photon avalanche diodes (SPAD) pixel circuit including two analog counters is proposed to acquire two precise photon counting data in one fluorescence decay period. Accordingly, imaging speed and imaging quality can be improved.

Keywords—Fluorescence lifetime imaging (FLIM); single photon avalanche diodes (SPAD); three observation-windows; image sensor.

INTRODUCTION

Single photon avalanche diode (SPAD) has been widely applied in Raman spectroscopy, fluorescence lifetime imaging (FLIM), optical communication, and 3D imaging thanks to its outstanding advantages, such as large avalanche gain, high photon detection efficiency and quick time response [1]. Presently, research on SPAD FLIM sensor is very active in the application of chemical analysis and biophysical diagnosis. Most of SPAD FLIM sensors developed in the past years have been specifically designed with the time-correlated single photon counting technique (TCSPC), which allows measuring the time of arrival of the first photon in one fluorescence decay process [2]. However, the photon pile up effect of TCSPC will degenerate the fluorescence lifetime image quality. In recent years, a two observation-windows method is proposed to mitigate this effect [3]. But the traditional technology of FLIM by means of the observation-windows method is not mature enough, and the speed and quality of image need further promotion. This is due to the fact that the observation window of traditional FLIM focus on the early phase of fluorescence decay, therefore the detected data is not representative of the entire decay process,

leading to lager systematic error and higher sensitivity to very small uncorrected baseline interference. Moreover, only one data is detected in one fluorescence excitation for each pixel, which will cause the reduced speed of imaging.

This paper proposed a novel three observation-windows fluorescence lifetime detection scheme. Compared with the traditional FLIM technique, the new method is capable of detecting the entire decay of fluorescence. Therefore, the precision of fluorescence lifetime evaluation can be improved. Meanwhile, based on this technique, a pixel suitable for high-density array imager is designed. Two observation data can be obtained simultaneously in single fluorescence decay, thus the imaging speed and efficiency could be improved significantly. Due to the high accuracy and the high imaging speed, the novel detection method is very suitable for the high-density SPAD array level FLIM detector.

DESIGN of MEASUREMENT SCHEME

The operation principle of three observation-windows detection with fluorescence intensity decay is shown in Figure 1. The FLIM measurements utilize three predetermined timing observation windows to obtain photon counting data. Since the number of detected photons is proportional to the intensity of fluorescence, the gathering information including photon number, position and time distribution by three overlapped or discrete observation-windows timing can be used to predict the lifetime of fluorescence. As shown in Fig.1, the three observation windows will successively open after the excitation of a laser pulse. Note that the first two observation windows T_{w1} and T_{w2} are overlapped by half their width and the third observation window T_{w3} has a gap between them. The width of the observation windows is generally about the order of a few nanoseconds [4].

Based on the above proposed three observation windows scheme, a block diagram of pixel is designed, as shown in Fig.2. The pixel consists of a SPAD leading end, a gating circuit which can select and shape the avalanche current by signal Win, and two analog counters aim at counting the photon number.

978-1-5386-5309-8/18 $31.00 © 2018 IEEE

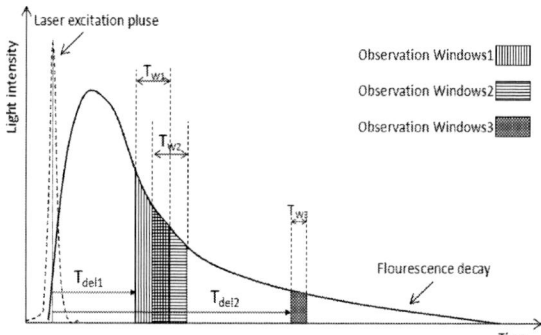

Fig. 1. Operation principle of three observation windows scheme.

The operation process of the pixel is described in the follows: Before the laser excites the sample, the two analog counters are reset by the clear signal CL (CL=high level). After that, a short signal E_x controls the laser to excite the sample, and subsequently entering the observation operation. At the same time, the control signals V_e and V_b enable SPAD working after a programmable delay T_{del1} from the laser excitation. Then, the SPAD avalanche process is controlled by the gate select signal Win in three timing windows (T_{w1}/T_{w2} and T_{w3}). The third widow occurs with T_{del2} delay from the laser excitation. After the avalanche pulses are shaped, they are transmitted to two analog counters (counter_1 and counter_2) in sequence by controlling of two signals SW1 and SW2 and the photon number related to a single fluorescence excitation is recorded. It is worth noting that the counter_1 operates in T_{w1} or T_{w2} windows, however, the counter_2 only records photon data in T_{w3} window.

After the repeated operation with hundreds of photons exciting and counting, we can obtain the total number of detected photons. Based on Newton-Rhapson formulas, we can finally acquire the average fluorescence lifetime τ corresponding to a pixel:

$$\tau = -T_W[\ln(N_2/N_1)^2] \qquad (1)$$

$$\frac{N_1}{N_3} = \frac{\exp(-s(x+y))-\exp(-ys)}{\exp(-s)-1} \qquad (2)$$

$$s = T_w/\tau \qquad (3)$$

where, N_i (i=1, 2, 3) denotes the counting number in the T_{wi} window. $x=T_{w3}/T_{w1}$ and $y=(T_{de2}-T_{del1})/T_{del1}$.

SIMULATION RESULTS

Figure.3 shows a transistor level schematic diagram suitable for three observation-windows FLIM technique. Transistor MN1 acts as a sensing only or as a passive quenching circuit element controlled by the signals V_e and

V_b. A clamping function of SPAD signals is realized with an isolation transistor MN2 following by transistor MN1. In this way, the SPAD quench circuit is in high excess bias voltage state and protects the next stage circuit. An NAND gate controlled by signal Win and an inverter, comprising of MP1/MP2/MP3 and MN3/MN4/MP5, respectively, are used to select and shape the avalanche signals. Transistors MN6 and MN7 select the avalanche signals from different observation windows and transmit them to two different analog counters. Transistors MN8~13 form two analog counters, which use the principle of charge sharing counting. Transistors MN14 and MN15 play the role of source follower. Transistors MN16 and MN17 act as select switches, delivering the counting data to row bus.

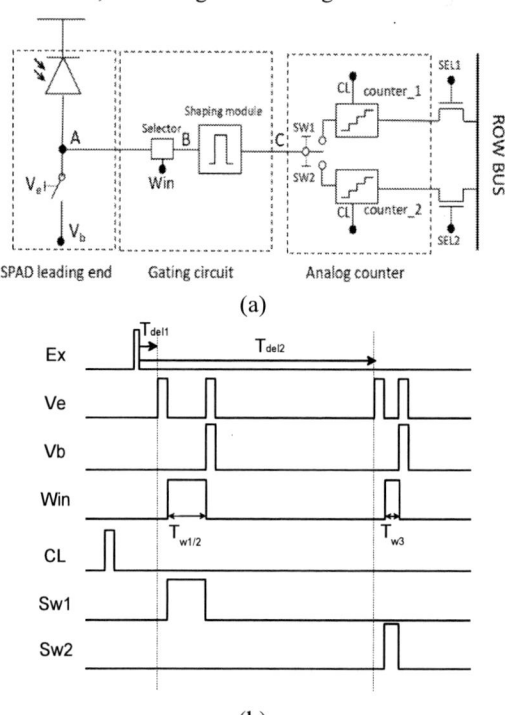

Fig. 2 (a) pixel block diagram; (b) timing signals.

The presented pixel circuit was designed by SMIC standard 0.18μm CMOS technology. To better account for the pixel circuit operation in three observation windows mode, Fig.4 gives the transient simulation waveform of the pixel circuit. In this simulation, a short pulse is assumed as the arrival of photon signal. During the observation window is active, SPAD is working in Geiger mode and the potential of node A is low. While a photon arrives, the potential of node A increases rapidly. Accordingly, the potential at node B quickly descends when a photon is detected in an observation window. As a result, a positive pulse is created at node C and a photon is recorded in analog counter. But it can be seen that there has no re-

978-1-5386-5309-8/18 $31.00 © 2018 IEEE

sponse to the photon if the SPAD is not working in Geiger mode (V_A=high level).

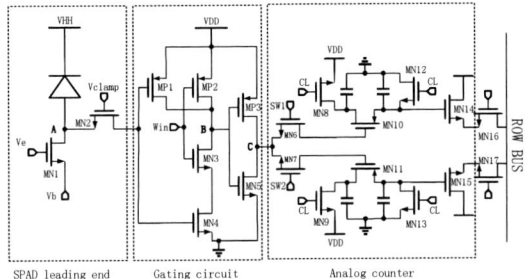

Fig. 3. Pixel transistor level schematic diagram.

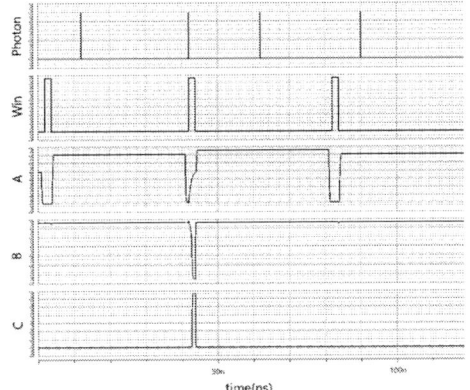

Fig. 4. Voltage signals in pixel circuit.

Fig. 5 shows the simulated countering results of pixel circuit versus different incidence photons frequency. Two pulses with different frequency are used to emulate the different light intensity in single fluorescence decay. An excellent linearity between the readout voltage and the counting number is observed, indicating that the output result of pixel circuit can precisely represent the number of detected fluorescence photons. Further, according to this countering result, we can accurately predict the fluorescence lifetime of each pixel.

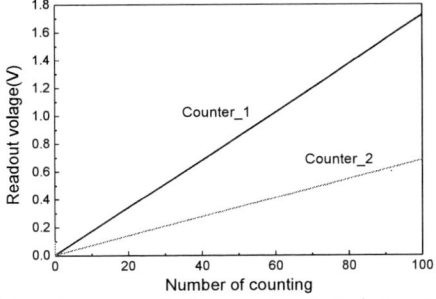

Fig. 5. Simulated countering results of pixel circuit versus different incidence photons frequency.

CONCLUSIONS

This paper demonstrates a novel three observation-windows fluorescence lifetime imaging method to detect and record photons. This method can effectively reduce systematic errors in fluorescence lifetime evaluation for the improved precision of photon counting. Further, a SPAD pixel circuit integrating two analog counters is designed, which can expand the size of collected data per unit time. The proposed pixel circuit based on this novel imaging method can achieve a lower systematic error and a higher imaging speed, making it suitable for the high-density SPAD array level FLIM detector.

ACKNOWLEDGEMENTS

This work is sponsored by the National Natural Science Foundation of China (No. 61571235) and Qing-Lan Project.

REFERENCES

[1] David S, Edoardo C, Kenneth L, et al.. "A Single-Photon Avalanche Diode Array for Fluorescence Lifetime Imaging Microscopy," IEEE Journal of Solid-State Circuits, vol. 43, pp. 2546-2557, November 2008.

[2] Ryan M, Simeon R, David S, et al. "A 100 fps, Time-Correlated Single-Photon-Counting-Based Fluorescence-Lifetime Imager in 130 nm CMOS," IEEE Journal of Solid-State Circuits, vol. 49, pp. 867-880, April 2014.

[3] Matteo P, Nicola M, David S, et al. "A 160×120 Pixel Analog-Counting Single-Photon Imager With Time-Gating and Self-Referenced Column-Parallel A/D Conversion for Fluorescence Lifetime Imaging," IEEE Journal of Solid-State Circuits, vol.51, pp. 155-167, November 2016.

[4] Sing P, Z.J. F, J.N.D, et al. "Optimized Gating Scheme for Rapid Lifetime Determinations of Single-Exponential Luminescence Lifetimes," Analytical Chemistry, vol. 73, pp. 4486-4490.September 2001.

AN IMPROVED BEHAVIORAL SIMULATION MODEL FOR AFTERPULSING PHENOMENON

Tingchen Zhao[1], Feiyang Sun[1], Yue Xu[1,2]*, Bin Li[1], Ding Li[1]

[1]College of Electronic and Optical Engineering & College of Microelectronics, Nanjing University of Posts and Telecommunications, Nanjing, Jiangsu Province, China, 210023

[2]National and Local Joint Engineering Laboratory of RF Integration and Micro-assembly Technology, Nanjing, Jiangsu Province, China, 210023

*Corresponding Author's Email: yuex@njupt.edu.cn

ABSTRACT

An improved behavioral model for simulating afterpulsing characteristic of single-photon avalanche diodes (SPADs) is introduced. The derived model fully considers the generation and trapping mechanism of carriers, and its important parameters are derived from TCAD simulation instead of empirical formula. The developed SPAD model is implemented in Verilog-A behavioral description language and has been successfully operated on Cadence Spectre simulator. The simulation model can successfully predict the features of afterpulsing phenomenon.

Keywords—Single photon avalanche diodes; behavioral simulation model; afterpulsing; Verilog-A

INTRODUCTION

Recently, single-photon avalanche diode (SPAD) solid-state detectors have become attractive in many fields, for instance high-resolution 3D imaging[1], fluorescence lifetime imaging[2] and spectroscopy, time-of-flight positron emission tomography[3] and so on. However, SPADs working at the Geiger mode suffers from severe noises of dark counts and afterpulsing. Different from generation mechanism of the dark counts, the afterpulsing is due to the trapping and detrapping of carriers relating to an initial avalanche pulse. In order to predict this statistical phenomenon, a behavioral simulation model to describe this phenomenon is required. The early model does not fully take into account the process of carrier generation and capture. Moreover, the avalanche trigger probability comes from empirical value. Those factors may reduce simulation precision of SPAD model to certain extent.

This paper presents an improved Verilog-A behavioral circuit simulation model to emulate afterpulsing of SPADs. Importantly, the vital model parameter of avalanche triggering probability is extracted from TCAD simulation not from empirical formula. As a result, the simulation model can accurately emulate the generation of afterpulsing.

MODELING OF AFTERPULSING

The paper takes a typical P+/Pwell/deep Nwell SPAD device structure as an example to illustrate modeling of afterpulsing. Fig.1 shows the SPAD device structure. An avalanche region of Pwell/deep Nwell junction is formed in a deep Nwell region. The Deep Nwell can solve the problem of latching up and noise coupling. SPAD in Geiger mode is working at excess bias voltage where the reverse bias voltage of SPAD is larger than breakdown voltage. When an incident photon reaches device surface, it can be directly absorbed in the depletion region, and then generates a pair of electron-hole. The electron-hole pair moves in two opposite directions in the high electric field, finally causing an avalanche in the depletion region.

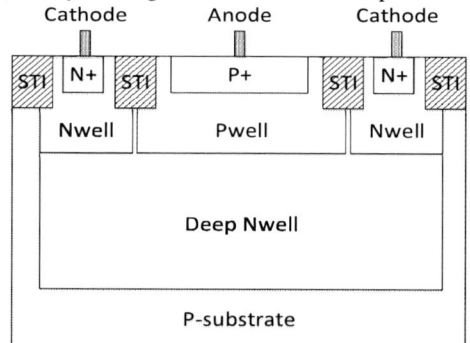

Fig. 1: Cross section of P+/Pwell/Deep Nwell SPAD

During avalanche period, if the carriers are captured by trap centers and the concentration of the ith level electron traps isn't saturate, we can get the electron capture rate for the ith level electron traps as following[4]:

$$U_{capt,ni} = N_{tni} \cdot \left(1 - f_{tni}\right) \cdot \frac{j_e(t)}{q} \cdot \sigma_n. \qquad (1)$$

Where N_{tni} is the density of ith level traps in the forbidden band, j_e is the electron-current density in the depletion region, and σ_n is the electron capture cross section. $(1 - f_{tni})$ is the fraction of unoccupied ith level electron trap at the beginning of the avalanche pulse.

At the end of avalanche pulse, the total number of captured electrons in the ith-level traps can be calculated by

978-1-5386-5309-8/18 $31.00 © 2018 IEEE

$$FT_i = \int_0^t U_{capt,ni} \cdot A_D \cdot W_D dt - k_{cri}. \qquad (2)$$

In which k_{cri} is the carriers number released from the ith-level traps during the avalanche duration time.

As for the ith-level hole traps, we can also obtain the similar equations to calculate the total number of captured holes during the avalanche process.

After avalanche pulse is quenched, the trapped carriers begin to release from the deep-level traps. The time intervals $\Delta \tau_{cri}$ between two adjacent carrier release events follow an exponential distribution function. The mean value of all these time intervals is expected to the lifetime of ith level traps, which can be expressed by [5-6]

$$\tau_{cri} = \tau_{0i} \cdot e^{\frac{E_{ai}}{k_B T}}. \qquad (3)$$

Where E_{ai} and τ_{0i} are the activation energy of ith-level traps and its pre-exponential factor, respectively.

Considering all energy levels of electron and hole traps, the total afterpulsing probability with the time decay can be estimated by

$$P_a(t) = \sum_i A_i \cdot \tau_i \cdot \exp\left(\frac{-t}{\tau_i}\right). \qquad (4)$$

In which τ_i and A_i are the lifetime and exponential pre-factor of ith-level traps.

In the Verilog-A modeling of afterpulsing, two deep-level traps (i=1, 2) are considered. Firstly, if $FT_i \geq 1$ and the returned random number (between 0 and 1) is less than avalanche trigger probability (P_{tr}), an afterpulsing event following an initial avalanche pulse may be generated. Then, taking τ_{cri} as mean value, an exponential distribution function is used to return a random number to mark the moment of after-pulsing phenomenon. This behavioral model can simulate the transient process of afterpulsing event.

SIMULATION RESULTS

Thanks to Geiger mode simulation, we can get a precise avalanche triggering probability of an electron or hole in the multiplication region. Fig. 2 shows the extracted triggering probability distributions of electron and hole at 3V excess bias voltage by TCAD simulation. It can be seen that the electron avalanche trigger probability achieves a maximum value of 0.63 at the upper boundary of PN junction, and it decreases to zero gradually at the bottom boundary of PN junction. On the opposite side, the hole avalanche trigger probability is zero at the upper boundary of PN junction, and it increases to a maximum value of 0.27 at the bottom boundary of PN junction. The triggering probability of carriers in the avalanche region

can be extracted from Fig. 2, and can be used for Verilog-A model.

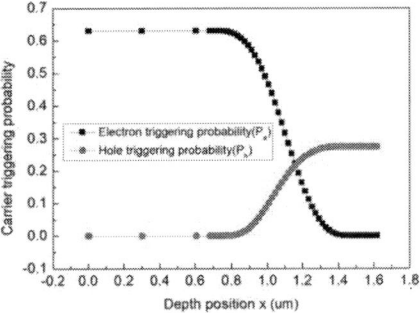

Fig. 2: Extracted distributions of electron and hole avalanche triggering probabilities at 3V excess bias.

An active quenching circuit is used to test and verify the function of the Verilog-A SPAD model, as shown in Fig. 3. In the transient simulation process, the simulation time is set to 1ms and the pulse width and pulse period are 1ps and 1μs, respectively. The transient simulation waveform is shown in Fig. 4. We can clearly find that there is an afterpulsing occurrence at 56 μs, following an initial avalanche pulse induced by incident photon. It is obvious that the simulated transient result is in accordance with experimental observation.

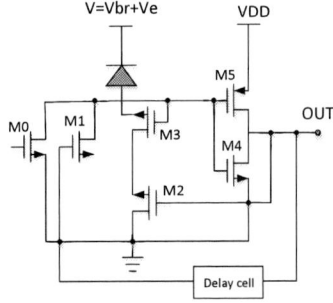

Fig. 3: An active quenching circuit.

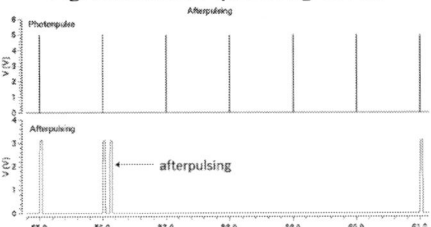

Fig. 4: Transient simulation waveform of afterpulsing event.

Fig. 5 shows the simulated afterpulsing probabilities of active quenching circuit at T=27°C and V_{excess}=3V using different hold-off times. When hold-off time equals 30 ns, afterpulsing probability is 1.35%, then the afterpulsing probability decreases with exponential distribution. (Activation energy of ith-level trap E_{a1}=0.1eV, E_{a2}=0.25eV, Pre-exponential factor of ith-level trap τ_{01}=2.2×10^{-11}s, τ_{02}=7.2×10^{-10}s). The attenuation trend of afterpulsing probability is consistent with the actual experimental results.

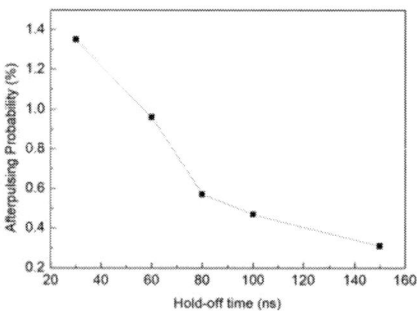

Fig. 5: Afterpulsing probability versus hold-off time for active quenching circuit (V_{excess}=3V, T=27°C).

The afterpulsing probability versus excess voltage is also simulated with a hold-off time of 100 ns, as shown in Fig.6. When V_{excess}=0.5V, the afterpulsing probability is 0.12%. The afterpulsing probability increases with excess voltage. When the excess voltage is 3.5V, the afterpulsing probability is increased to 0.54%. The increase trend of afterpulsing probability is also consistent with the actual experimental results.

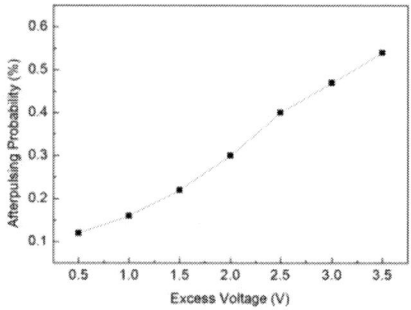

Fig. 6: Afterpulsing probability versus excess voltage for active quenching circuit ($t_{hold-off}$ =100ns, T=27°C).

CONCLUSIONS

An improved behavioral simulation model is proposed to accurately simulate the afterpulsing phenomenon of SPADs. The temporal dependence of afterpulsing model fully takes into account the generation and trapping mechanism of carriers. Importantly, the key model parameter of avalanche triggering probability in the depletion region is obtained from Geiger mode TCAD simulation instead of theoretical computation. The simulation model is realized by Verilog-A behavioral description language and can be used to predict the afterpulsing phenomenon.

ACKNOWLEDGEMENTS

This work is supported by National Science Foundation of China (No. 61571235) and QingLan Project of Jiangsu Province and Graduate Research and Innovation Projects of China Jiangsu Province (No. SJLX16_0324).

REFERENCES

[1] Villa F, Lussana R, Tisa S, et al. "CMOS Imager with 1024 SPADs and TDCs for Single-Photon Timing and 3-D Time-of-Flight," IEEE Journal of selected Topics in Quantum Electronics, vol. 20, pp. 364-373, September 2014.

[2] Perenzoni M, Massari N, Perenzoni D, et al. "A 160×120 Pixel Analog-Counting Single-Photon Imager with Time-Gating and Self-Referenced Column-Parallel A/D Conversion for Fluorescence Lifetime Imaging," IEEE Journal of Solid-State Circuits, vol. 51, pp.155-167, October 2015.

[3] L. H. C. Braga et al. "A fully digital 8 ×16 SiPM array for PET applications with per-pixel TDCs and real-time energy output," IEEE Journal of Solid-State Circuits, vol. 49, pp.301-314, October 2013.

[4] Xu Y, Xing P, Xie X and Huang Y. "A New Modeling and Simulation Method for Important Statistical Performance Prediction of Single Photon Avalanche Diode Detectors," Semiconductor Science and Technology, vol. 31, 065024, 2016.

[5] Giustolisi G, Mita R, and palumbo G. "Behavioral Modeling of Statistical Phenomena of Single-Photon Avalanche Diodes," International Journal of circuit theory and applications, vol. 40, pp.661-679, August 2012.

[6] Cheng Z, Zheng X, Palubiak D, Deen M J,and Peng H. A. "Comprehensive and Accurate Analytical SPAD Model for Circuit Simulation," IEEE Trans. Electron Devices, vol. 63, pp.1940-1948, November 2016.

[7] Giustolisi, G, R. Mita, and G. Palumbo. "Verilog-A modeling of SPAD statistical phenomena," IEEE International Symposium on Circuits and Systems, vol.19, pp.773-776, July 2011.

CELLULAR NEURAL NETWORK (CENN) FPGA IMPLEMENTATION USING MULTI-LEVEL OPTIMIZATION

Zhongyang Liu[1], Shaoheng Luo[1], Xiaowei Xu[2] and Cheng Zhuo[1]*

[1] Zhejiang University, Hangzhou 310058, China
[2] University of Notre Dame, Notre Dame, Indiana, 46655, USA
*Corresponding Author's Email: czhuo@zju.edu.cn

ABSTRACT

Cellular Neural Network (CeNN) has been widely adopted in image processing tasks, which is considered as a powerful paradigm for embedded devices. Recently, digital implementations of CeNNs on FPGA have attracted researchers from both academia and industry due to its high flexibility and short time-to-market. However, most existing implementations are not well optimized to fully utilize the advantages of FPGA platform with unnecessary design and computational redundancy that prevents speedup. We propose a multi-level optimization framework for energy efficient CeNN implementations on FPGAs. In particular, the optimization framework is featured with three level optimizations: system-, module-, and design-space-level, with focus on computational redundancy and attainable performance, respectively. Experimental results show that our framework can achieve an energy efficiency improvement of 3.54× and up to 3.88× speedup compared with existing implementations.

Keywords—FPGA; Cellular neural network; acceleration

INTRODUCTION

Cellular Neural Network (CeNN) is a nonlinear cellular processor array inspired by the functionality of neurons through modeling the working principles of human brain sensing. It has widely applied to various image processing areas such as noise cancellation, edge detection, path planning and segmentation [3]. Recently, CeNNs receive increasing interests from both academia and industry [5] for efficient hardware implementations.

CeNN accelerations on digital platforms such as ASICs [4], GPUs [6] and FPGAs [1][7] have been explored. Among them, FPGA has its unique potential and appears to be the most popular option due to its high flexibility and low time-to-market. With all these works sharing the same architecture and various acceleration techniques, there still lack a comprehensive study and design exploration for efficient CeNN implementation, which covers the following remaining issues:

· First, many existing works only employ a single processing element to compute one iteration, and hence do not exploit the full potential of parallelism.
· Second, many works are unaware of the repeated parameters in the template and incur unnecessary I/O overheads. Since I/O access is much slower than computation, such unawareness effectively leads to

computational redundancy.
· The last but not the least, many design space exploration do not well study the utilization of available resources and bandwidth of FPGAs to achieve the best performance.

Thus, in this paper, in order to address those remaining concerns, we propose a multi-level optimization framework for CeNN computation for scalability and efficiency. The framework is featured with three-level optimizations:

· System level optimization (SLO): A parallel and tilling processing (PTP) and a data-reuse optimization are proposed to enable the parallelism of CeNNs.
· Module level optimization (MLO): For processing elements, a parameter quantization scheme is adopted. By eliminating the processing of repeated parameters, we can further simplify the computation process.
· Design space level optimization (DSLO): Based on available resources and I/O bandwidth, an optimal model about attainable performance estimations to explore the design space is proposed.

FRAMEWORK OVERVIEW

As shown in Figure 1(a), the optimization framework is composed of external memory, memory interface controller, on-chip input and output buffers, a computation acceleration unit (CAU) and AXI4 bus. Due to on-chip resource limitation, data are stored in external memory and cached in on-chip buffers before processed in CAU. The workload is divided in temporal domain with a fully pipelined structure, where different iterations are processed simultaneously in different iteration units (IUs). The architecture is designed to load image stream from external memory, process it with CeNN and convert the result back. CAU constitutes the most important component, which is described in detail as follows:

· Iteration Unit A (IUA): The processor chain in CAU begins with IUA, which calculates the sharing parameters for all Euler iterations [7] in a time-invariant model. Thus only one IUA is implemented.
· Iteration Unit B (IUB): IUB indicates a number of identical modules in a fully-pipelined architecture. Pipeline depth is equal to the total number of Euler iterations desired.
· Data Transfer Controller: Data transfer controller manages the data transmission among the input buffer

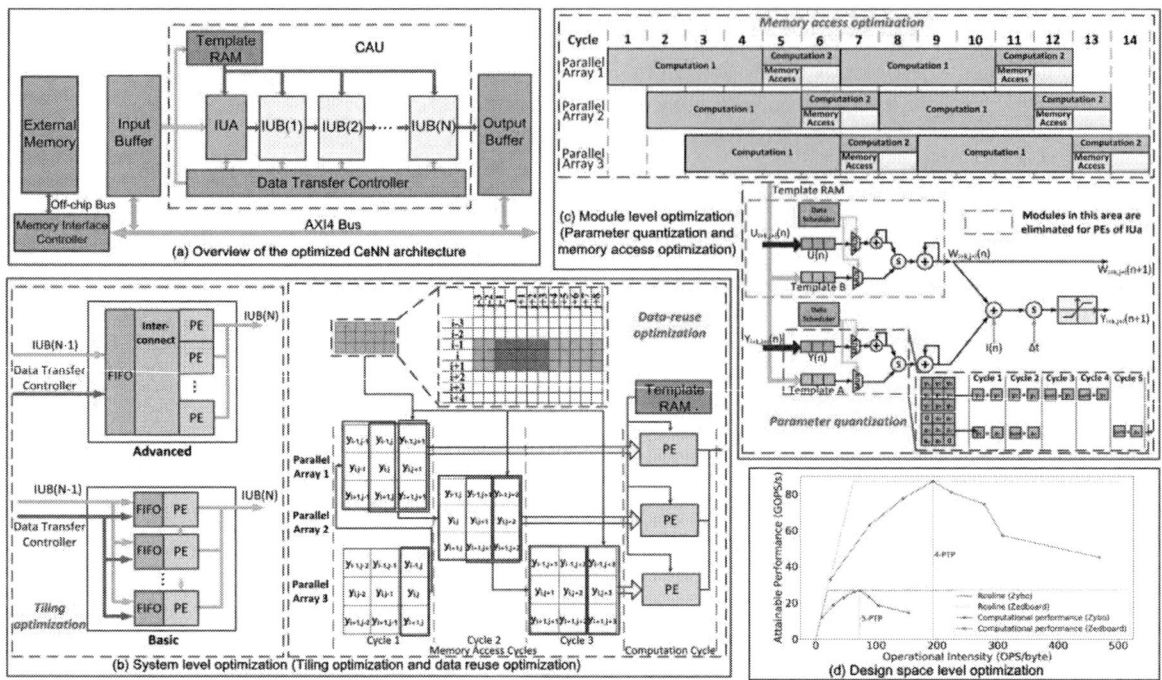

Figure 1: (a) The optimized architecture and associated multiple-level optimization framework: (b) system level optimization, (c) module level optimization, and (d) design space level optimization.

and the output buffer, template RAM and FIFOs in pipelined IUs. It also guarantees the synchronization of data transmission and computation in case of variant bandwidth features of the off-chip bus.

· Template RAM: The weights in CeNNs are space-invariant. Therefore, Template RAM needs only a few hardware resources for storage and memory access.

MULTI-LEVEL OPTIMIZATION

In system level optimization (SLO), tiling optimization and data-reuse optimization are proposed in Figure 1(b). N-parallel and tiling processing (N-PTP) divides the workload into N parallel arrays for parallel operations and improves the throughput. PTP can be implemented in two strategies and allocate IUs in different structures resulting in different internal memory access operations. Input images are split into separate data chunks for advanced strategy. Data in each data chunk are arranged spatially with continuity and operated in one cycle, which reduces DRAM ports and hardware resources consumption compared with basic strategy. Data-reuse optimization utilizes repeated values in adjacent pixels, which eliminates excessive read and write operations of IUs and the input buffer. When calculations are executed in IUs, data-reuse based on data sharing relationships results in 2/3 memory access reduction.

In module level optimization (MLO), a larger number

of multiplications are required using the limited embedded multipliers on FPGAs. Parameter quantization is proposed to reduce multiplier consumption by transforming multiplications into logic shifts, which enables repetition-induced and sparsity-induced optimization. It is found to be efficient for reducing multiplier consumption in most CeNN applications. Moreover, it is worthy to reorganize and manipulate the on-chip memory access and computation scheme to reduce processing latency. By investigating the memory access optimization in advanced PTP strategy as depicted in Figure 1(c), we overlap parallel phases, and arrange memory access cycles in a pipelined strategy. When no offloading operation is required in *Computation 2*, registers are prepared to load data for the next phase in memory access cycle.

Design space level optimization (DSLO) adopts the roofline model to recognize the peak performance for specific FPGA devices [2]. On one hand, a range of computational performance can be obtained by adopting optimization techniques. On the other hand, available resources and I/O bandwidth are two main limiting factors to CeNN. As shown in Figure 1(d), we explore the design space for the target platform and enumerate a set of implementations with the highest attainable performance. In most cases, we could find more than one design within the set, and the optimal solution can be selected for the given platform with desired performance and operational intensity.

TABLE I.   RESOURCE UTILIZATION AND PERFORMANCE COMPARISON

	Basic	SLO	MLO	SLO+MLO
LUTs	15336	10232	13632	7792
FFs	10824	6944	10656	6432
Computational performance (GOPS/s)	5.75	5.34	9.06	10.35
Computational density (GOPS/LUT)	3.75× E-4	5.22× E-4	6.65× E-4	13.28× E-4
Improvement	1×	1.39×	1.78×	3.54×

RESULTS AND CONTRIBUTIONS

Table 1 summarizes the performance comparison of the basic CeNN approach (Basic) [7] and the implementation of the proposed framework using different optimization schemes. Giga operations per second (GOPS/s) is used as the metric of computational performance. For a fair comparison, the concept of computational density is introduced as a measure of energy efficiency, which is defined as the average giga operations per area unit (GOPS/LUT). The combined optimization technique SLO+MLO achieves a speedup of 1.8× and an energy efficiency improvement of 3.54× with its resource utilization decreasing to 50.8% compared with the basic approach, which is much larger than utilizing MLO or SLO alone.

DSLO is also discussed in Figure 2, where resource utilization is constraint and pipeline depth is variable in the same FPGA platform. The optimal model based on SLO+MLO+DSLO has the highest computational performance and the least pipeline depth. The highest computational performance achieves a 3.88x speedup.

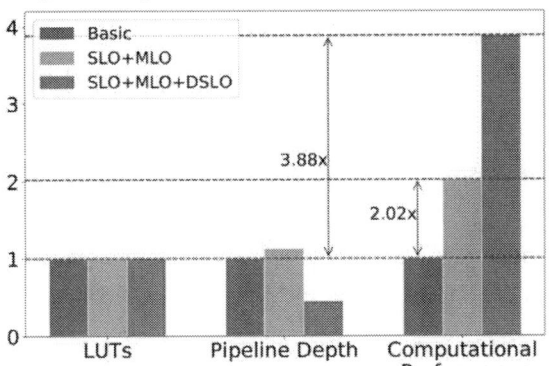

Figure 2: Pipeline depth and performance comparisons among Basic, SLO+MLO and SLO+MLO+DSLO.

CONCLUSIONS

We propose an efficient multi-level optimization framework for the FPGA CeNN implementations. In system level, parallel and tilling processing (PTP) and data-reuse optimization are applied. The computation flow is improved by separating the workload in spatial domain. In module level optimization, we propose parameter quantization to reduce multiplier overheads and memory access optimization for the data transfer mechanism. In design space level, we apply different aforementioned optimized designs and obtain roofline models accordingly of attainable performance and operational intensity. Then the optimal CeNN solution can be selected with desired performance and operational intensity. Finally, evaluations of the proposed framework are presented on different FPGA platforms, and a speed up of 3.88× is achieved compared with existing implementations.

ACKNOWLEDGEMENTS

The authors would like to thank the support from NSFC Fund (Grant No. 61601406).

REFERENCES

[1] H. Chen, *et. al.* "Image-processing algorithms realized by discrete-time cellular neural networks and their circuit implementations." *Chaos, Solitons & Fractals*, 29(5), pp. 1100-1108, 2006.

[2] C. Zhang, *et. al.* "Optimizing fpga-based accelerator design for deep convolutional neural networks." *Proc. of ACM/SIGDA International Symposium on Field-Programmable Gate Arrays*, pp. 161-170, 2015.

[3] D. Manatunga, *et. al.* "Cellular neural network based medical image segmentation using artificial bee colony algorithm." *Proc. of International Conference on Green Computing Communication and Electrical Engineering*, pp. 1-6, 2014.

[4] S. Lee, *et. al.* "24-GOPS 4.5-mm² Digital Cellular Neural Network for Rapid Visual Attention in an Object-Recognition SoC." *IEEE Transactions on Neural Networks*, 22(1), pp. 64-73, 2011.

[5] D. Manatunga, H. Kim, and S. Mukhopadhyay. "SP-CNN: A Scalable and Programmable CNN-Based Accelerator." *IEEE Micro*, 35(5), pp. 42–50, 2015.

[6] S. Potluri, *et. al.* "CNN based high performance computing for real time image processing on GPU." *Proc. of International Symposium on Theoretical Electrical Engineering*, pp. 1-7, 2011.

[7] N. Yildiz, *et. al.* "Architecture of a fully pipelined real-time cellular neural network emulator." *IEEE Transactions on Circuits and Systems I: Regular Papers*, 62(1), pp. 130-138, 2015.

Quadrature Amplitude Modulated Backscatter for 2.4GHz Self-Powered Chips

Luyao Zhang, Hao Zhang, Zhengkun Shen, Mingxiao He, Xiucheng Hao, Enbin Gong, Le Ye* and Huailin Liao

Laboratory of Microelectronic Devices and Circuits (MOE), Institute of Microelectronics, Peking University
Beijing, China
Phone: 86-10-62757761, E-mail: yele@pku.edu.cn

Biography

Luyao Zhang was born in Changchun, China in 1992. She received her B.S. degree in Microelectronics from Peking University, Beijing, China in 2015. Currently, she is pursuing M.S. degree at Institute of Microelectronics, Peking University, Beijing, China. Her research interests include wireless power harvesting system design and wearable bioelectronics.

Abstract

In this paper, we propose a 16 quadrature amplitude modulated (QAM) backscatter for self-powered chips with a symbol rate of 1 MHz and a 40 Mb/s bit rate in 2.4 GHz Internet of Things (IoT) system. A four-resistor array is implemented to shift amplitude and achieve quadrature amplitude shift keying (QASK). Transmission lines of four different lengths are applied to change backscattering phase and accomplish quadrature phase shift keying (QPSK). Jointly, by applying transmission lines of twelve lengths and resistors of three values, a 16 QAM backscatter is realized, which is compatible with extensive mainstream communication protocols. The whole system is off-chip, completely passive and low-cost, leading to an easy, economic and efficient communication within wireless sensor networks (WSNs).

Keywords—component; backscatter; quadrature amplitude modulation (QAM); self-powered; IoT

Introduction

Coming with the wide application and intensive nodes of Internet of Things (IoT) is the problem of power management. Traditionally, lower power and higher integrations are two attempts to achieve longer battery life in IoT nodes. However, electrical consumption and maintenance expenditure are inevitable and thus prevent IoT from large scale applications if battery exists.

As a result, completely passive system such as radio frequency identification (RFID) system has been applicated to minimize cost. However, there are two problems for RFID systems. The first problem is limited base station for UHF band where traditional RFID works, which confines the wide spread of IoT nodes and makes those nodes working merely locally without interacting into a net. Moreover, building new base stations would greatly expand the cost of IoT system. The second problem is the communication with IoT nodes. Traditional tag could carry only On-Off Keying (OOK) modulation, which is inefficient. Achieving more complicated modulation methods has been grabbing serious attention in recent years. Modulating complex-valued scatter field has been considered to achieve different modulations such as amplitude shift keying (ASK), phase shift keying (PSK) and quadrature amplitude modulated (QAM) [1]. A vector modulator based on a p-i-n diode which is controlled by bias current of two digital-to-analog converters (DACs) has been implemented to shift backscatter phase and thus achieve PSK in ultra-high frequency (UHF) [2]. A 32 QAM transmitter employing backscatter technique has been accomplished in 5.8 GHz [3]. Related work has also presented binary phase shift keying (BPSK) for X-band microwaves [4]. These works have provided effective solutions to the second problem, but still cannot solve the first problem.

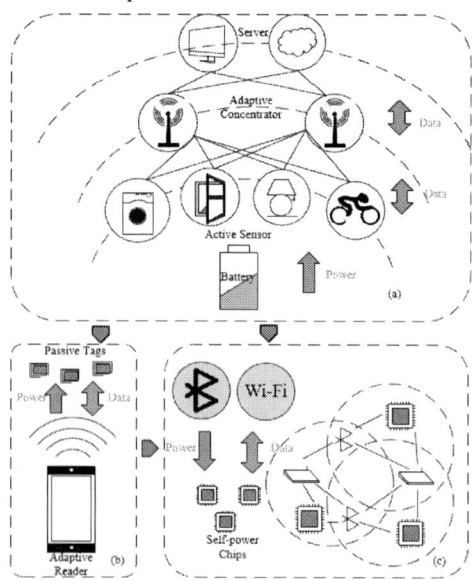

Fig. 1. (a) Traditional concept of IoT of LoRa/ZigBee. (b) UHF RFID system. (c) Wireless sensor networks (WSNs) of 2.4GHz self-powered chips for IoT.

978-1-5386-5309-8/18 $31.00 © 2018 IEEE

To solve the cost problem and make it possible to build tens of thousands of scattering IoT nodes, this paper comes up with a QAM backscatter 2.4 GHz IoT system. 2.4GHz is the frequency of widely scattering Wi-Fi and Bluetooth devices, eliminating base station construction costs. Backscatter enables communication in passive system, and QAM is compatible with OFDM in routers and Bluetooth devices. In this paper, a resistor series is implemented to change the magnitude of S parameters of antenna and the amplitude of backscattered signal. Transmission lines of different lengths are used to shift the phase of backscattered signal. The amplitude and phase modulation are totally independent, allowing ASK, PSK alone and QAM together, making the transmitter compatible with any kind of mainstream communication protocols. A symbol rate of 1MHz and a bit rate of 40Mb/s is achieved.

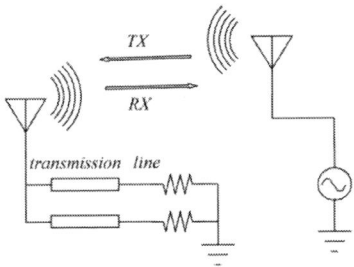

Fig. 2 Architecture of QAM backscatter achieved by resistors and transmission lines in this paper.

This 2.4 GHz backscatter system enables powering of trillions of wide spread IoT nodes and communication within them, making an authentic 'Internet'.

In this paper, backscattering modulation methods to achieve QASK, QPSK and 16 QAM are introduced separately, and simulation results on the system is analyzed.

Quadrature Amplitude Shift Keying

Resistors would change the real part of Z parameter of term and thus change the amplitude of backscattering signal. To make bit decision more accurate, we separate S11 parameter from 0 to 1 with four points, 0.125, 0.375, 0.625 and 0.875. Resistor values could be calculated with Smith chart tool in electromagnetic field simulation software.

Demodulated pattern is shown in Figure 3, and the constellation of QASK is shown in Figure 4.

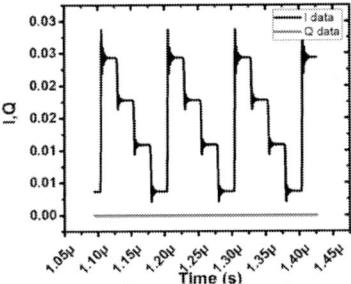

Fig. 3 Simulated demodulation pattern of QASK in time domain.

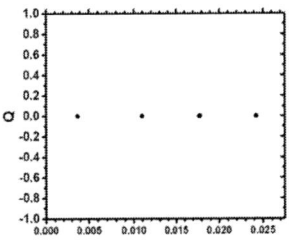

Fig. 4 Simulated QASK constellation

Quadrature Phase Shift Keying

To realize different phases, image part of Z parameter should be shifted. Transmission lines have a characteristic impedance of 50 Ohm regardless of length. However, they would affect the inductance and the phase of backscattering signal. As a result, changing the lengths of transmission lines is a good method to modulate phase without shifting amplitude. In this paper, a QPSK with four phases of ±45 degree and ±135 degree is presented. The transmission line could be manufactured on printed circuit board (PCB). The constellation is simulated with a 0.1mW power source (thermal noise at 16.85 Celsius) and a receiver of 8 dB noise figure.

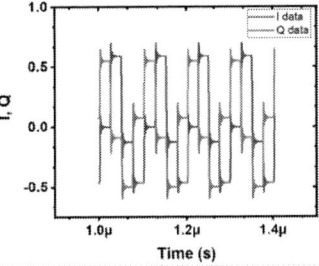

Fig. 5 Simulated demodulation pattern of QPSK in time domain.

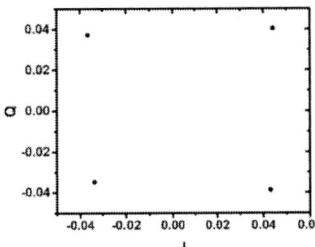

Fig. 6 Simulated QPSK constellation.

Quadrature Amplitude Modulation

Transmission lines of twelve lengths and resistors of three values could be combined to achieve 16 QAM backscattering. The architecture of simulation testbench is shown below.

978-1-5386-5309-8/18 $31.00 © 2018 IEEE

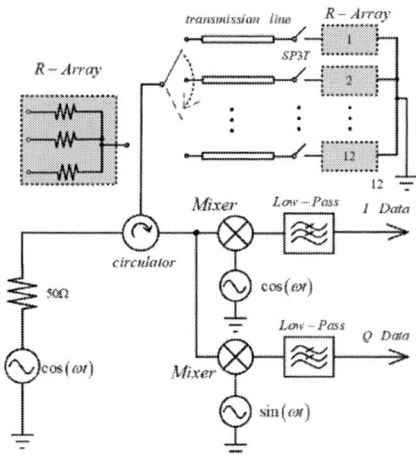

Figure. 7 Cabled test setup implemented in simulation tool.

The 1 tone power source generates a Tx of 0.1 mW, coupled by a circulator into the backscattering module including four transmission lines and four resistor arrays attached to each transmission line. The backscattered signal Rx and reflected Tx signal are coupled to the mixer and low-pass filter, generating I data and Q data. The constellation is shown below.

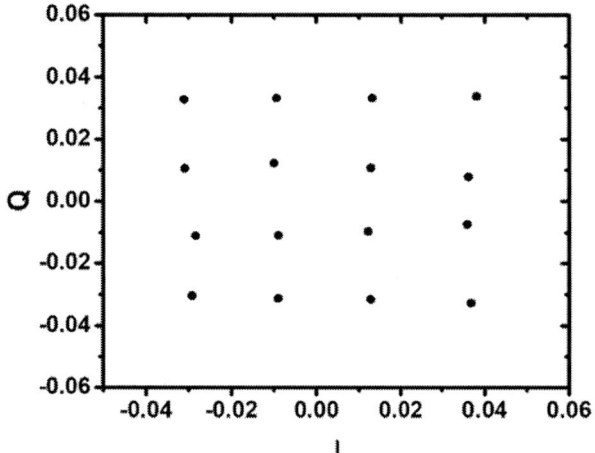

Fig. 8 Simulated 16 QAM constellation

Conclusions

This paper comes up with a 16 QAM backscatter achieved by resistor array and transmission lines for 2.4GHz self-powered chips in IoT systems, which is a solution to energy harvesting and communication problems in WSNs. With the method presented, IoT nodes could harvest energy from ubiquitous Wi-Fi and Bluetooth devices, with a near-zero cost and accomplish communications via these devices.

References

[1] Thomas, Stewart J., et al. "Quadrature amplitude modulated backscatter in passive and semipassive UHF RFID systems." IEEE Transactions on Microwave Theory and Techniques 60.4 (2012): 1175-1182.

[2] Winkler, Michael, et al. "An UHF RFID tag emulator for precise emulation of the physical layer." Microwave Conference (EuMC), 2010 European. IEEE, 2010.

[3] Shirane A, Tan H, Fang Y, et al. 13.8 A 5.8 GHz RF-powered transceiver with a 113μW 32-QAM transmitter employing the IF-based quadrature backscattering technique[C]//Solid-State Circuits Conference-(ISSCC), 2015 IEEE International. IEEE, 2015: 1-3.

[4] Fu X, Sharma A, Kampianakis E, et al. A low cost 10.0-11.1 GHz X-band microwave backscatter communication testbed with integrated planar wideband antennas[C]//RFID (RFID), 2016 IEEE International Conference on. IEEE, 2016: 1-4.

978-1-5386-5309-8/18 $31.00 © 2018 IEEE

A MEASUREMENT METHOD FOR THE CHIP-TO-BOARD TRANSMISSION OF RADIO FREQUENCY WAVES IN THE POWER DISTRIBUTION NETWORKS WITH SIMPLE EMBEDDED CIRCUITS

Lan Luo, Dihu Chen, and Tao Su[*]

School of Electronics and Information Technology, Sun Yat-sen University,
Guangzhou 510006, China
*Corresponding Author's Email: sutao@mail.sysu.edu.cn

ABSTRACT

The chip-to-board model is important for predicting the electromagnetic emission (EME) of an electronic system and checking whether it fulfills the emission standards at early design cycle. This paper proposes to build the chip-to-board model through directly measuring the chip-to-board transmission ratio with embedded circuit. The embedded circuit does not cause the modification the on-chip power distribution network. Only one signal pin is added to the chip. It preserves the original physical structure of the chip and package in a maximal way. The measurement method is verified through transistor level simulations and is demonstrated with a test chip. Using the measured transmission ratio, one can build a chip-to-board model with improved accuracy, especially in the high frequency range, for the EME simulation of electronic system

INTRODUCTION

It is compulsory that EME of an electronic system does not exceed the limits defined in the emission standards. The switching noise in the power distribution network of the digital integrated circuits (ICs) is a main source of the EME in an electronic system. The system designer need a EME model of the digital ICs so that the EME of the system can be predicted during the design phase of the system. The EME model of a digital IC contains two parts: a noise source and a chip-to-board path. This paper focuses on the chip-to-board model.

There are two approaches to build the chip-to-board model [1-3]: 1) simulation; 2) measurement approach. In current measurement approach, S-parameters of the package IC are measured. The chip-to-board model are obtained by fitting the measured S-parameters. The ports of the S-parameters are actually the external pins of the packaged IC. Parameter S21 is the transmission from one external pin to another external pin. A path from one pin to another contains two cascaded chip-to-board paths. Various chip-to-board models can satisfy the same S-parameters. Since the transmission of the chip-to-board path is directly measured, the current approach can not determine a unique model for the chip-to-board path.

The paper presents a method to directly measure the transmission of the chip-to-board with simple embedded circuits. In contrast to many reported on-chip techniques, our method does not introduce any new power domain into the chip. The original power distribution network (PDN) are entirely preserved. Section two describes the embedded circuit and the method for measuring the chip-to-board path transmission. The functionality of the method is checked through transistor-level simulation in section three. Section four uses a dedicated test chip to demonstrate the method.

PRINCIPLE OF THE METHOD

The embedded circuit is a ring oscillator (RO), built with a chain of inverters, as shown in Fig. 1.

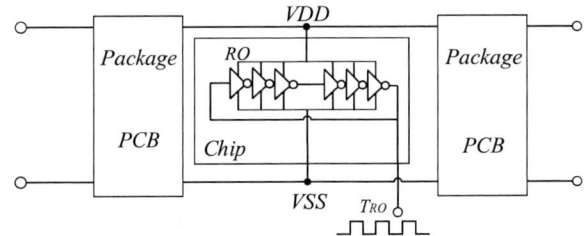

Figure 1: Embedded measurement circuit

There is a unique relationship between the operation frequency of RO and the amplitude of sinusoidal variation in the power supply [4]. The relationship is independent of the frequency of sinusoidal variation. Such an effect can be exploited to measure the transmission ratio between the board supply pin and the on-chip PDN. A radio frequency interference (RFI) source is connected to the VDD. Sweeping the frequency of RFI source, the transmission ratio of the chip-to-board over a wide frequency range can be obtained. The chip-to-board model can be extracted from the frequency response of the transmission ratio.

VERIFICATION WITH SIMULATIONS

A transistor-level simulation is implemented in HSPICE to verify the previous proposed method. The simulation setup is shown in Fig.2. The transistor model is from the standard library of SMIC 130nm technology.

978-1-5386-5309-8/18 $31.00 © 2018 IEEE

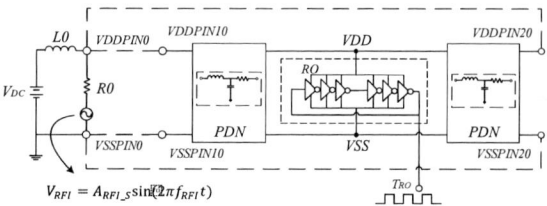

$$V_{RFI} = A_{RFI\_S}\sin(2\pi f_{RFI}t)$$

Figure 2: Simulation setup

The netlist is composed of three parts: 1) chip-to-board PDN; 2) power supply and RFI source; 3) embedded circuit. The chip-to-board PDN includes the PCB trace and package lead, package bonding wire, on chip interconnection, and on chip PDN, all described by lumped-element model. A 50 Ω RFI source is used to generate the sinusoidal variation in the PDN. The embedded circuit is a 101-stage RO.

To calculate the transmission ratio (TR) under interference, RO average period in the absence of RFI, T_{RO_0}, and in the presence of RFI, T_{RO_RFI}, for varying frequency of RFI, f_{RFI}, should be simulated first. T_{RO_0} is approximately 4.4ns. The RFI source amplitude A_{RFI_S} sets 0.4V. With (1), the so called normalized frequency shift (NFS) of the RO is calculated.

$$NFS = (f_{RO\_RFI} - f_{RO\_0}) / f_{RO\_0} \qquad (1)$$

$$TR = A_{RFI\_RO} / A_{RFI\_PIN} \qquad (2)$$

The NFS can be transformed into RFI amplitude of the local supply of the RO, A_{RFI_RO}, using method in [4]. The RFI amplitude between the supply pin VDDPIN0 and the ground pin VSSPIN0 is labeled A_{RFI_PIN}. Inserting A_{RFI_RO} and A_{RFI_PIN} into (2) yields the chip-to-board transmission ratio drawn with dot in Fig. 3.

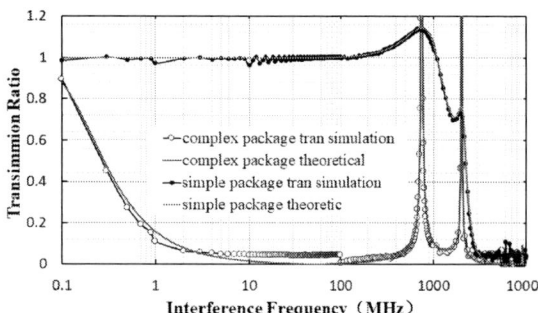

Figure 3: Simulation results

AC simulation is performed. The AC simulation gives the theoretical results of the transmission ratio. The result is shown as the solid line in Fig. 3. The simulation examines two different types of package, one without decoupling capacitor and the other with decoupling capacitor. The red line and black solid points are simple package while the blue line and black hollow points are complex package with decoupling capacitance. Both simulated results correspond well with the theoretical results in all aspects, including the local maximum, that's meaning the chip-to-board transmission ratios can be accurately measured by this method.

PRACTICAL APPLICATION

The measurement method is applied on a FPGA system. The device under test is the EP4CE55F23I7N FPGA from Altera. The inverter is formed with LUT in the FPGA, therefore and the RO is a LUT-based RO.

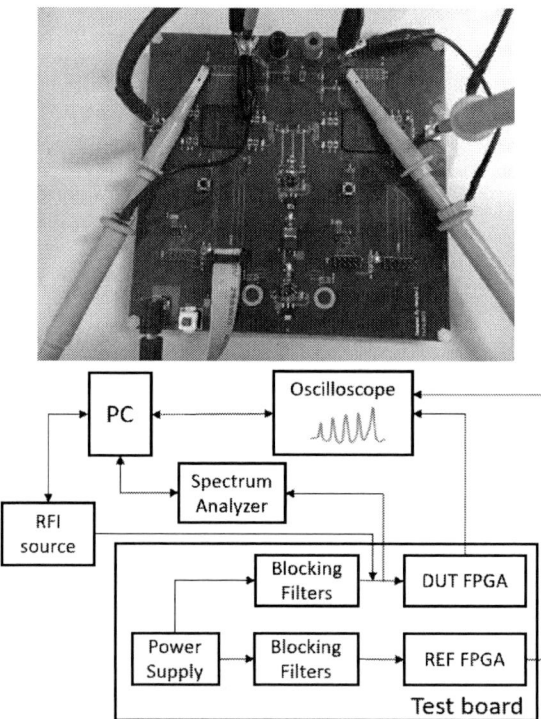

Figure 4: Test setup

Refer to Fig. 4, there are two FPGAs on the test board. The blocking filter isolates the RFI from the power supply. The RF interference is injected to only one FPGA, called DUT FPGA. The other FPGA, REF FPGA, works as monitor for the output voltage of the power supply. That is to make sure that any change of the signal from the DUT FPGA is not due to the output offset of the power supply. The power supply can be switched between a static

978-1-5386-5309-8/18 $31.00 © 2018 IEEE   538

on-board voltage regulator and a variable external supplier. The external supplier was used to measure the static response of the LUT-based inverter delay, as shown in Fig. 5a. With the static delay, the mapping curve from the normalized frequency shift of the RO to the A_{RFI_RO} is obtained, shown in Fig. 5b.

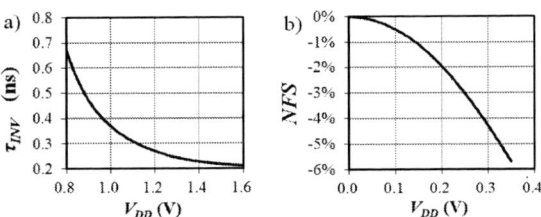

Figure 5: Static Response of the RO

During the measurement of the transmission ratio, the FPGA is supplied with the on-board voltage regulator. The RFI source is switched on. The interference frequency is swept from 1 MHz to 1 GHz. Both the RF voltage on the supply pin of the FPGA and the operation frequency of the RO are recorded. The results as shown in Fig. 6.

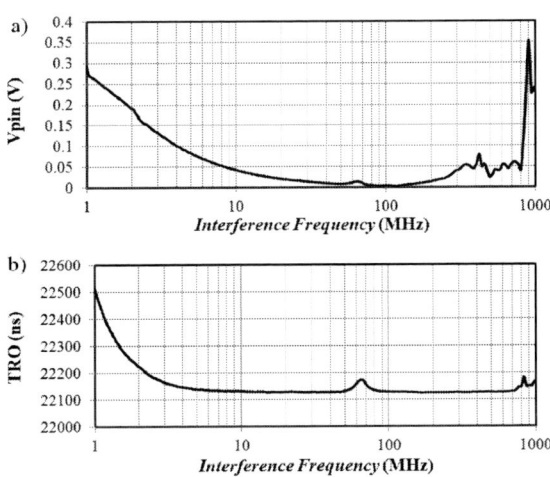

Figure 6: Measured pin voltage and RO frequency

Using the mapping curve in Fig. 5b, The RO frequency is converted into A_{RFI_RO}. The transmission ratio is thereby calculated with (2). The resulting frequency response of transmission ratio are depicted in Fig. 7. At the low frequency the TR is close unit. As the interference approaches 100 MHz, the inductance and capacitance of the board and package becomes active. The TR in that range varies strongly with the frequency. The peak at 100 MHz corresponds to the resonance of the package (including the board trace) inductance and chip (including

the package) capacitance. At that frequency, the pin voltage is almost zero while the pad voltage is large. The measured response is quite reasonable.

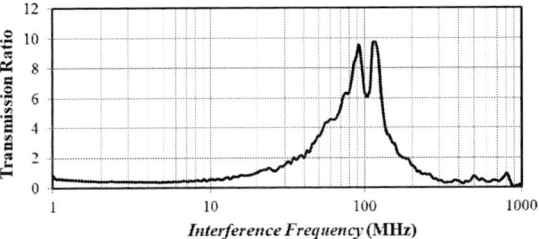

Figure 7: Measured the transmission ratio

CONCLUSION

This paper demonstrated a chip-to-board model accurate for modeling the complete PDN behavior with a simple embedded circuit. Good agreement between simulated and theoretical values of the chip-to-board transmission has been founded. The method has been demonstrated by a test chip. This model can be used for representing the PDN of chip, package, PCB in EME measurement. With a suitable current source model, it allows to predict the electronic system EME result so that the PCB power network can be tuned before fabrication.

ACKNOWLEDGEMENTS

Project 61471402 supported by NSFC. This work is also supported by the Project Science and Technology of Guangdong Province of China (2015B090909002, 2015B090912001, 2016B010123005, and 2017B090909005).

REFERENCES

[1] Labussière-Dorgan C, Bendhia S, Sicard E, et al. Modeling the Electromagnetic Emission of a Microcontroller Using a Single Model[J]. IEEE Transactions on Electromagnetic Compatibility, 2008, 50(1):22-34.

[2] Leca J P, Froidevaux N, Dupré P, et al. EMI Measurements, Modeling, and Reduction of 32-Bit High-Performance Microcontrollers[J]. IEEE Transactions on Electromagnetic Compatibility, 2014, 56(5):1035-1044.

[3] Ren L, Li T, Chandra S, et al. Prediction of Power Supply Noise From Switching Activity in an FPGA[J]. IEEE Transactions on Electromagnetic Compatibility, 2014, 56(3):699-706.

[4] Su T, Li F, Lian Z, et al. Frequency Shift of Ring Oscillators Due to Radio Frequency Interference on the Supply[J]. IEEE Transactions on Electromagnetic Compatibility, 2015, 57(6):1365-1373.

NOVEL APPROACHES TO CIRCUIT TIMING

Ulf Schlichtmann
Chair of Electronic Design Automation
Department of Electrical and Computer Engineering, Technical University of Munich
Arcisstr. 21, D-80333 Munich, Germany
ulf.schlichtmann@tum.de

ABSTRACT

In conventional sequential digital design, flip-flops are used to separate combinational logic gates. Signal propagation across logic gates ends at flip-flops. Accordingly, the minimum clock period is determined by the maximum of combinational delays between flip-flops. This partitioning of combinational logic into stages reduces design complexity significantly. However, in modern high-performance designs where clock frequency is usually pushed to the limit, this strict logic separation by flip-flops sacrifices timing performance. In addition, the simple assumption that all combinational paths work within one clock period makes the task to prevent counterfeiting very challenging, because a netlist extracted from reverse engineering represents all the functional information and can be processed using a standard IC design flow and used to produce chips in different foundries illegally. In this paper, we demonstrate two techniques that loosen the conventional strict separation of logic gates with flip-flops to enhance circuit performance and to reinforce netlist security.

INTRODUCTION

Traditionally, digital circuits are constructed using logic gates and sequential components such as flip-flops. The function of a circuit is essentially implemented by the logic gates, which process input data and produce corresponding outputs. Sequential gates are used to partition logic gates into consecutive stages, and logic signals are not allowed to pass through flip-flops freely. This approach can reduce design complexity significantly. This strict separation requires that a signal traveling along the longest combinational path should arrive at the flip-flop at the end of the path some time (the setup time) before the next rising clock edge. Therefore, the timing performance of the circuit is constrained by the largest delay of all the combinational paths in the circuit, so that designers need to focus only on these paths for timing optimization.

In the simplified model above, it is assumed that a flip-flop always works correctly with a given delay if the input data becomes valid setup time (t_{su}) before the active clock edge and stays unchanged hold time (t_h) after the clock edge, and the valid data appears at the output of the flip-flop after the clock-to-q delay (t_{cq}), as illustrated in Figure 1.

This simplified conventional timing model, however, does not accurately represent the real behavior of a circuit

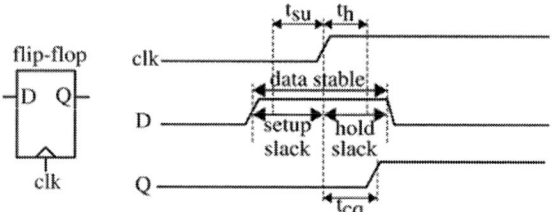

Figure 1: Conventional flip-flop model with setup time, hold time and clock-to-q delay in static timing analysis.

with flip-flops. For example, in reality a flip-flop does not have a clear open/close switching behavior. If a signal across a long combinational path reaches a flip-flop slightly inside the setup-hold window in Figure 1, the flip-flop can still deliver correct data at its output, although the delay of the flip-flop may increase. In addition, the assumption that the combinational logic between any pair of flip-flops works within one clock period implies that the circuit netlist represents the complete design information. If the logic gates and flip-flops as well as their connections can be recognized with reverse engineering, attackers can easily reproduce chips with the same function.

We will therefore discuss two techniques that break the conventional timing definition to improve circuit performance and to reinforce netlist security against counterfeiting.

SETUP AND HOLD TIME INTERDEPENDENCY

In reality a flip-flop can still work even if the window defined by setup time and hold time is not respected strictly. The relation between clock-to-q delay, the arrival time of the input data and the stable time of the data after the clock edge of a 45nm flip-flop is illustrated in Figure 2 [1]. According to this example, the flip-flop actually allows a smaller setup time and still maintains its function, if hold time is enlarged correspondingly and the increased clock-to-q delay can be tolerated by the timing constraints of the next flip-flop stage. Consequently, a special timing transparency across flip-flop stages is enabled to balance the delays between consecutive flip-flop stages.

When the setup time of a flip-flop is characterized in traditional static timing analysis, the latest data change is moved gradually toward the corresponding clock edge, until the delay of the flip-flop increases by a given amount, e.g.,

978-1-5386-5309-8/18 $31.00 © 2018 IEEE

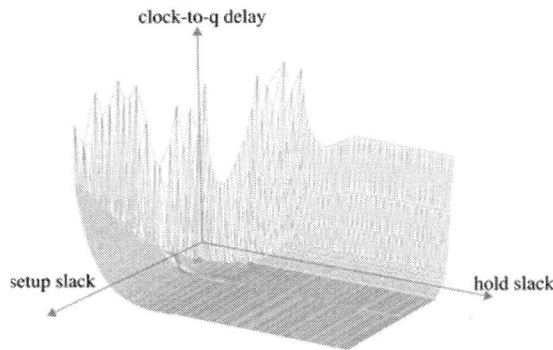

Figure 2: Simulated three dimensional clock-to-q delay surface of a 45nm flip-flop.

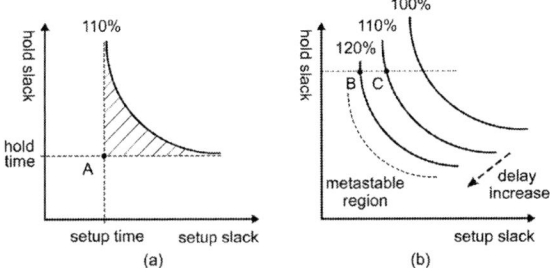

Figure 3: Delay curves of a flip-flop.

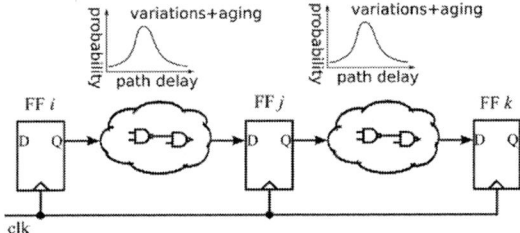

Figure 4: Circuit example with three flip-flops and two combinational circuits.

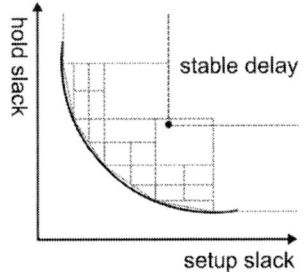

Figure 5: Polygon construction, split and merge of the delay surface of a flip-flop [1].

10%. The hold time is characterized similarly. Consequently, the setup and hold time point used for static timing analysis is shown as point A in Figure 3(a). The conventional setup time and hold time model, however, excludes the flexibility of the delay curves beneath the 110% curve in timing analysis, with which the clock period can be reduced [2]. For example, both B and C in Figure 3(b) are valid working points of the flip-flop, although the clock-to-q delay of the flip-flop may increase.

The setup and hold time interdependency is a very useful feature in satisfying timing specifications in modern IC design, where two special factors affect the design flow significantly: variations in the PVT space (Process, Voltage, Temperature) and reliability issues due to aging. In the nanometer era, the continuous increase in manufacturing variations have made the traditional worst- and best-case design methods unpractical by their large ratios to the nominal design parameters and their relatively low correlation caused by within-die variations [3, 4]. Recent studies show that FinFETs also suffer significant process variations [5, 6, 7, 8]. To deal with these challenges, analysis methods considering process variations for timing and clock tree have been explored extensively [4, 9, 10, 11, 12, 13, 14]. In addition, active methods such as the "risky" Razor [15] method and post-silicon tuning [16, 17, 18, 19, 20, 21, 22] have been deployed.

Unlike PVT variations, aging effects such as Hot Carrier Injection (HCI), Negative Bias Temperature Instability (NBTI), Electromigration (EM) and Time-Dependent Dielectric Breakdown (TDDB) are difficult to analyze, since they increase over time due to stress applied onto the circuit [23, 24, 25, 26]. Among these, the AgeGate model [27, 28, 29] is a leading method in gate-level aging analysis today. Combination with graph pruning can provide another four orders of magnitude of timing path reduction [30].

The methods above deal with PVT variations, unfortunately, with additional overhead in circuit area, design and analysis complexity. This overhead can actually be reduced by considering the interdependency of setup time and hold time of a flip-flop. For two consecutive combinational paths connected by a flip-flop in a sequential circuit, as shown in Figure 4, variations and aging may make their delays vary in different directions and with different magnitudes. If the first path becomes critical by entering the setup time window of the flip-flop in the middle, the delay of the flip-flop becomes larger. This delay is in fact a part of the delay of the second combinational path. If the second path is not critical so that it can absorb the increased clock-to-q delay of the flip-flop in the middle, the circuit still works properly.

To take advantage of the interdependency between setup and hold times, several methods have been introduced previously. The method in [31] uses a quadratic programming model to calculate the optimal clock period directly. To simplify the three-dimensional model, the method in [32] approximates it using an analytic function and calculates the minimum clock period by iterations. The method in [33] approximates the three-dimensional delay surface using linear planes, but the minimum clock period is verified with setup time and hold time only separately. Furthermore, the method in [34] proposes a very efficient algorithm to capture timing violations, but only considering the relation between clock-to-q delay and setup time. Finally, the method in [1] models the delay surface in Figure 2 with a piecewise model as shown in Figure 5 and solves the timing verification problem considering setup time and hold time and their interdependency together.

978-1-5386-5309-8/18 $31.00 © 2018 IEEE   541

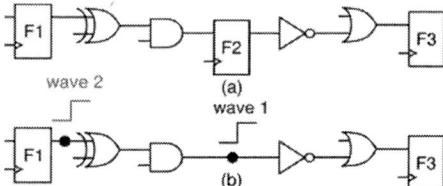

Figure 6: Conventional timing and wave-pipelining: (a) Single period clocking; (b) Pipelining with two data waves.

TIMING CAMOUFLAGE

In the conventional timing paradigm using the definitions of setup time and hold time shown in Figure 1, the netlist of a circuit represents all its design information at the logic level. If attackers can extract the netlist shown in Figure 6(a), they can then process this netlist using a standard EDA toolchain and map it to their own library. The design can thus be manufactured by a foundry to produce illegal chips. This poses a big threat to IC vendors since the cost of developing a modern IC is extremely high today. In addition, the business model of the IP (intellectual property) industry is also affected by this counterfeiting threat.

To thwart the attack attempt on a netlist, the conventional timing model can be invalidated in some parts of the circuit to form a circuit shown in Figure 6(b) [35]. If the data at the input of F3 is stable with respect to the clock edge two clock periods later, the data can still be latched into F3 correctly and the function of this circuit is not affected. This technique is called wave-pipelining. It has been explored for circuit design [36].

With wave-pipelining paths, attackers still face the challenge to determine the numbers of data waves on combinational paths even after the netlist is recognized through reverse engineering. If only one wave is assumed and the netlist is processed accordingly, the circuit loses data synchronization after manufacturing. To assume two waves, tremendous effort is still needed to identify where the pipelined waves are. Consequently, timing information provides another layer to strengthen the security of netlists.

To incorporate wave-pipelining paths to camouflage a netlist with timing information, the original function of the circuit needs to be maintained. More concretely, the second wave, as shown in Figure 6(b), should never catch the first wave on a wave-pipelining path constructed in the circuit. In real circuits, however, many short paths exist. If a flip-flop is removed directly, the short paths usually violate the wave-pipelining requirement. Therefore, wave-pipelining paths can only be constructed at the locations where short paths can be amended with a limited number of delay buffers or interconnect tuning. These delay changes should not be recognized in reverse engineering easily; otherwise, the locations of wave-pipelining paths are exposed.

To alleviate the challenge of constructing wave-pipelining paths, the original circuit can be partially duplicated, as shown in Figure 7, where Figure 7(a) shows the original circuit and Figure 7(b) shows the circuit with wave-pipelining paths. The flip-flop at which the wave-pipelining paths terminate is called a WP flip-flop. The

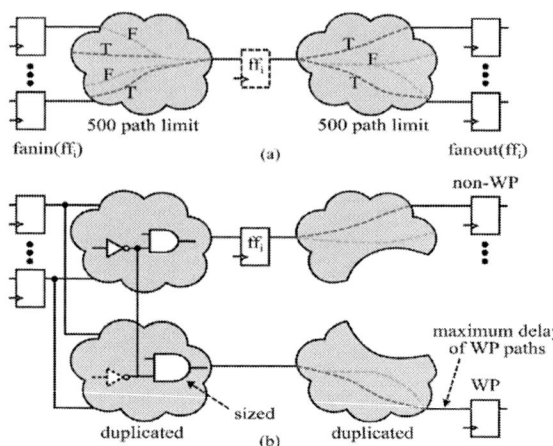

Figure 7: Circuit duplication in constructing wave-pipelining paths.

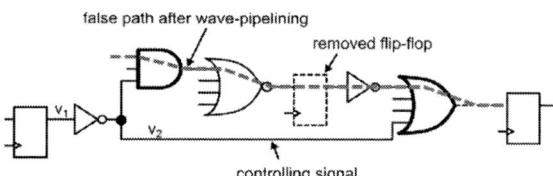

Figure 8: Two true paths form a wave-pipelining false path.

other paths without wave-pipelining terminate at a non-WP flip-flop and the delays along these paths are not modified. Since only the wave-pipelining paths in Figure 7(b) need to be padded, it is much easier to meet the timing requirements with only a few additional delay units inserted in the circuit. In Figure 7(b), logic gates unrelated to the normal paths to the non-WP flip-flop and unrelated to the wave-pipelining paths to the WP flip-flop are removed to save area, leading to the reduced "cloud" blocks.

Another challenge of this netlist security technique is that attackers might measure all the path delays exhaustively, no matter how much time and cost it requires. To counter this attack, wave-pipelining false paths can also be constructed in the circuit, as shown in Figure 8. After removing the flip-flop in the middle, the original two combinational paths are connected and a false path is formed. If this new path is considered as working within a single clock period, a signal switching at the beginning of the dashed path never reaches the final flip-flop, because the dashed path cannot be triggered no matter which value the signal v_2 has. Consequently, attackers cannot differentiate it from the other false paths in the original circuit using at-speed scan test, while the latter may contribute up to 75% of all the combinational paths in real circuits [37].

CONCLUSION

In this paper, we present two techniques to break the barriers currently taken for granted, which the conventional sequential design model imposes. As a result, the circuit performance can be improved further or the security of netlists can be strengthened against counterfeiting. With

978-1-5386-5309-8/18 $31.00 © 2018 IEEE   542

these new techniques, the flexibility of digital computing might potentially be rejuvenated in the era of high-performance design and highly collaborative manufacturing in today's semiconductor industry.

REFERENCES

[1] G. L. Zhang, B. Li and U. Schlichtmann, "PieceTimer: a holistic timing analysis framework considering setup/hold time interdependency using a piecewise model," in *IEEE/ACM International Conference on Computer-Aided Design (ICCAD)*, 2016.

[2] E. Salman, A. Dasdan, F. Taraporevala, K. Kucukcakar and E. G. Friedman, "Exploiting setup-hold-time interdependence in static timing analysis," *IEEE Transactions on Computer-Aided Design of Integrated Circuits and Systems*, vol: 26, no. 6, p. 1114–1125, 2007.

[3] L. Scheffer, "The count of Monte Carlo," in *TAU Int. Workshop*, 2004.

[4] D. Blaauw, K. Chopra, A. Srivastava and L. Scheffer, "Statistical Timing Analysis: From Basic Principles to State of the Art," *IEEE Transactions on Computer-Aided Design of Integrated Circuits and Systems*, vol. 27, no. 4, pp. 589-607, 2008.

[5] X. Wang, A. R. Brown, B. Cheng and A. Asenov, "Statistical Variability and Reliability in Nanoscale FinFETs," in *IEEE International Electron Devices Meeting (IEDM)*, 2011.

[6] V. B. Kleeberger, H. Graeb and U. Schlichtmann, "Predicting future product performance: Modeling and evaluation of standard cells in FinFET technologies," in *ACM/IEEE Design Automation Conference (DAC)*, 2013.

[7] S. Sinha, G. Yeric, V. Chandra, B. Cline and Y. Cao, "Exploring sub-20nm FinFET design with Predictive Technology Models," in *ACM/IEEE Design Automation Conference (DAC)*, 2012.

[8] S. Karapetyan, V. Kleeberger and U. Schlichtmann, "FinFET-based product performance: Modeling and evaluation of standard cells in FinFET technologies," *Microelectronics Reliability*, vol. 61, pp. 61: 30-34, 2016.

[9] H. Chang and S. Sapatnekar, "Statistical timing analysis considering spatial correlations using a single PERT-like traversal," in *IEEE International Conference on Computer-Aided Design (ICCAD)*, 2003.

[10] C. Visweswariah, K. Ravindran, K. Kalafala, S. G. Walker and S. Narayan, "First-order incremental block-based statistical timing analysis," in *ACM/IEEE Design Automation Conference (DAC)*, 2004.

[11] B. Li, N. Chen, M. Schmidt, W. Schneider and U. Schlichtmann, "On hierarchical statistical static timing analysis," in *Design, Automation & Test in Europe (DATE)*, 2009.

[12] B. Li, N. Chen, Y. Xu and U. Schlichtmann, "On Timing Model Extraction and Hierarchical Statistical Timing Analysis," *IEEE Transactions on Computer-Aided Design of Integrated Circuits and Systems*, vol. 32, no. 3, pp. 367-380, 2013 .

[13] C. Liu, J. Su and Y. Shi, "Temperature-aware clock tree synthesis considering spatiotemporal hot spot correlations," in *IEEE International Conference on Computer Design*, 2008.

[14] C.-L. Lung, Y.-S. Su, H.-H. Huang, Y. Shi and S.-C. Chang, "Through-Silicon Via Fault-Tolerant Clock Networks for 3-D ICs," *IEEE Transactions on Computer-Aided Design of Integrated Circuits and Systems*, vol. 32, no. 7, pp. 1100-1109, 2013.

[15] D. Ernst, N. S. Kim, S. Das, S. Pant, R. Rao and T. Pham, "Razor: A Low-Power Pipeline Based on Circuit-Level Timing Speculation," in *International Symposium on Microarchitecture (MICRO)*, 2003.

[16] D. Tadesse, J. Grodstein and R. I. Bahar, "AutoRex: An automated post-silicon clock tuning tool," in *International Test Conference (ITC)*, 2009.

[17] B. Li, N. Chen and U. Schlichtmann, "Fast Statistical Timing Analysis for Circuits with Post-Silicon Tunable Clock Buffers," in *IEEE International Conference on Computer-Aided Design (ICCAD)*, 2011.

[18] B. Li and U. Schlichtmann, "Statistical timing analysis and criticality computation for circuits with post-silicon clock tuning elements," *IEEE Transactions on Computer-Aided Design of Integrated Circuits and Systems*, vol. 34, no. 11, p. 1784–1797, 2015.

[19] G. L. Zhang, B. Li and U. Schlichtmann, "Sampling-based buffer insertion for post-silicon yield improvement under process variability," in *Design, Automation & Test in Europe (DATE)*, 2016.

[20] G. L. Zhang, B. Li, J. Liu, Y. Shi and U. Schlichtmann, "Design-phase buffer allocation for post-silicon clock binning by iterative learning," *IEEE Transactions on Computer-Aided Design of Integrated Circuits and Systems*, 2018.

[21] G. L. Zhang, B. Li and U. Schlichtmann, "EffiTest: Efficient delay test and statistical prediction for configuring post-silicon tunable buffers," in *ACM/IEEE Design Automation Conference (DAC)*, 2016.

[22] R. Kumar, B. Li, Y. Shen, U. Schlichtmann and J. Hu, "Timing verification for adaptive integrated circuits," in *Design, Automation & Test in Europe (DATE)*, 2015.

[23] A. H. Baba and S. Mitra, "Testing for Transistor Aging," in *IEEE VLSI Test Symposium*, 2009.

[24] S. V. Kumar, C. H. Kim and S. S. Sapatnekar, "NBTI-Aware Synthesis of Digital Circuits," in *ACM/IEEE Design Automation Conference (DAC)*, 2007.

[25] V. B. Kleeberger, M. Barke, C. Werner, D. Schmitt-Landsiedel and U. Schlichtmann, "A compact model for NBTI degradation and recovery under use-profile variations and its application to aging analysis of digital integrated circuits," *Microelectronics Reliability*, vol. 54, no. 6-7, pp. 1083-1089, 2014.

[26] H. Amrouch, B. Khaleghi, A. Gerstlauer and J. Henkel, "Reliability-aware design to suppress aging," in *ACM/IEEE Design Automation Conference (DAC)*, 2016.

[27] D. Lorenz, G. Georgakos and U. Schlichtmann, "Aging analysis of circuit timing considering NBTI and HCI," in *IEEE International On-Line Testing Symposium (IOLTS)*, 2009.

[28] D. Lorenz, M. Barke and U. Schlichtmann, "Aging analysis at gate and macro cell level," in *IEEE/ACM International Conference on Computer-Aided Design (ICCAD)*, 2010.

[29] D. Lorenz, M. Barke and U. Schlichtmann, "Efficiently analyzing the impact of aging effects on large integrated circuits," *Microelectronics Reliability*, vol. 52, no. 8, pp. 1546-1552, 2012.

[30] D. Lorenz, M. Barke and U. Schlichtmann, "Monitoring of aging in integrated circuits by identifying possible critical paths," *Microelectronics Reliability*, vol. 54, no. 6-7, pp. 1075-1082, 2014.

[31] A. Jain and D. Blaauw, "Slack borrowing in flip-flop based sequential circuits," in *Great Lakes Symposium on VLSI* , 2005.

[32] N. Chen, B. Li and U. Schlichtmann, "Iterative timing analysis based on nonlinear and interdependent flipflop modelling," *IET Circuits, Devices & Systems*, vol. 6, no. 5, p. 330–337, 2012.

[33] A. B. Kahng and H. Lee, "Timing margin recovery with flexible flip-flop timing model," in *International Symposium on Quality Electronic Design (ISQED)*, 2014.

[34] Y. Yang, K. H. Tam and I. H. Jiang, "Criticality-dependency-aware timing characterization and analysis," in *ACM/IEEE Design Automation Conference (DAC)*, 2015.

[35] G. L. Zhang, B. Li, B. Yu, D. Z. Pan and U. Schlichtmann, "TimingCamouflage: Improving Circuit Security against Counterfeiting by Unconventional Timing," *Design, Automation & Test in Europe (DATE)*, 2018.

[36] W. P. Burleson, M. Ciesielski, F. Klass and W. Liu, "Wave-pipelining: A tutorial and research survey," *IEEE Transactions on Very Large Scale Integration (VLSI) Systems*, vol. 6, no. 3, p. 464–474, 1998.

[37] K. Heragu, J. H. Patel and V. D. Agrawal, "Fast identification of untestable delay faults using," in *IEEE/ACM International Conference on Computer-Aided Design (ICCAD)*, 1997.

Electrical and Thermal Characterization of SiC Power MOSFET

Takashi Sato[1], Kazuki Oishi[1], Masayuki Hiromoto[1], and Michihiro Shintani[2]

[1] Graduate School of Informatics, Kyoto University, Kyoto 606-8501, Japan

[2] Graduate School of Information Science, Nara Institute of Science and Technology, Nara 630-0192, Japan

Abstract—**Design of efficient power converters that fully exploit the potential of SiC power MOSFET is supported by circuit simulations, wherein accuracy of power device models is extremely important. In this paper, a surface-potential based device model for SiC power MOSFETs is briefly presented. Then, measurement techniques suitable for extracting the model parameters are reviewed. Finally, the characterization of self-heating effect and input capacitance measurement based on switching capacitance are explained.**

I. INTRODUCTION

Circuit simulations are becoming increasingly important in designing highly efficient power converters. Among the components in the simulation deck, MOSFET models are critically important for obtaining highest accuracy. In the case of power MOSFETs, the models should correctly capture non-ideal switching behaviors.

Fitting-based models and threshold-voltage-based models have been widely adopted for simulating SiC power MOS-FETs [1]–[3]. Though these models are relatively simple to construct, accuracy of their current and capacitance characteristics may be often insufficient. Recently, the MOSFET models, which are originally developed for the simulation of integrated circuits, are applied for simulating power devices [4]–[9]. These models not just better reproduce measurement results, correct responses are obtained in the situations where the previous models become inaccurate, e.g., at the bias voltages beyond the fitted range. The model responses are also correct when there are variations of device parameters.

Physics-based compact models have been proposed for vertical SiC power MOSFETs [10], [11]. The model equations for the channel current and capacitances are fundamentally the same as those for Si planer MOSFETs, but the enhancements are necessary to reflect the differences in device structure. In addition, consideration of extremely high voltage and large current operations are required.

In this paper, characterization of physics-based device model for SiC MOSFETs is discussed. Being vertical double diffused MOSFET (DMOSFET) in Fig. 1(a) as an example, the enhancements necessary for modeling the power devices are described: implementation of self-heating effect (SHE), and model parameter characterization over a wide range of voltage and current.

II. PHYSICS-BASED MODEL FOR SiC VDMOSFET

Fig. 1(b) shows an equivalent model for SiC power MOS-FET. In the surface potential models, by solving a nonlinear

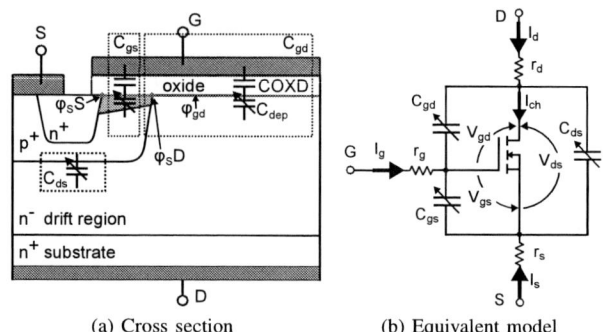

(a) Cross section   (b) Equivalent model

Fig. 1. Structure of a vertical DMOSFET.

Poisson equation about the surface charge, surface potential ϕ_s is obtained. Both channel current and capacitance are obtained through the potential — integrating the charge from drain to source gives the channel current I_{ch}, and differentiating the charge yields channel capacitance [6], [9], [12].

The parasitic capacitance of a MOSFET can be modeled as follows [10]. C_{gs} is the series connection of MOS capacitance at the channel surface and gate oxide capacitance. The dependence of MOS capacitance on ϕ_s is often ignored [10] as the bias voltage dependence of C_{gs} is much smaller than that of C_{ds} and C_{gd}. C_{ds} is modeled as a junction capacitance formed at the body and drift regions, which is bias dependent. C_{gd} is modeled as series connection of oxide capacitance C_{oxd} and depletion capacitance C_{dep} formulated at the MOS structure above the JFET region. Considering the substrate effect [12], C_{dep} is calculated by the surface potential ϕ_{gd}, obtained similarly to the calculation of channel surface potential.

As shown in Fig. 1(a), VDMOSFET is asymmetric in terms of source and drain structure. Whereas source resistance r_s is small and often ignorable, drain resistance r_d tends to become large due to the existence of drift region that realizes high breakdown voltage. Typically, r_d is represented as the series connection of accumulation resistance at the epitaxial layer below gate electrode, JFET resistance, and drift region resistance.

III. PULSE-BASED DRAIN CURRENT CHARACTERIZATION

A. Short-pulse I-V Measurement

Interdependence between junction temperature (T_j) and drain current has to be carefully considered in the device current characterization. The junction temperature must be

978-1-5386-5309-8/18 $31.00 © 2018 IEEE   544

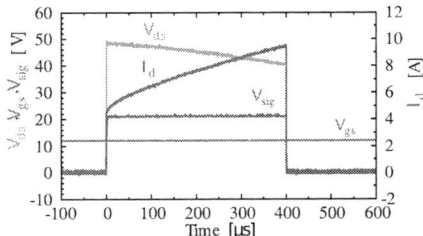

Fig. 2. Drain current transient in pulse-based measurement. Drain current roughly doubled in $400\,\mu s$ due to SHE.

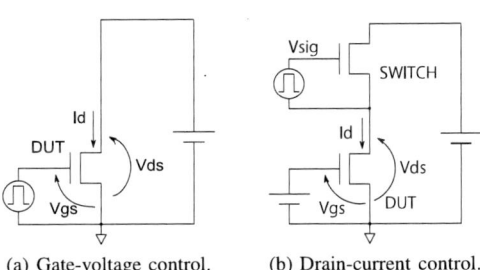

(a) Gate-voltage control. (b) Drain-current control.

Fig. 3. Two types of short-pulse measurement circuit.

kept constant to attain reliable measurement results. Tightly screwing the device to a thermal plate does not warrant a constant channel temperature. Fig. 2 shows the waveforms in a pulse measurement, which is widely adopted in commercial curve tracers to suppress SHE in the characterization of power devices. However, due to the rapid temperature rise at the channel, drain current increased from $4\,A$ to $10\,A$. In such bias conditions, a further shorter pulse than that realizable with versatile curve tracers [13], [14] has to be used.

Fig. 3 shows typical structures of the dedicated short-pulse measurement circuits. In (a), gate voltage of the device under test (DUT) is controlled, while in (b), drain current is limited by a switch MOSFET at the drain of the DUT. In [15], current pulse width is controlled in the order of μs to improve accuracy of I-V measurements beyond $1\,kW$ range.

B. MOSFET Model Considering Self-heating Effect

In order to reflect the change of device temperature, the model parameters are represented as the function of temperature. In addition, a thermal circuit model is incorporated with the electrical model to simultaneously consider SHE of the power MOSFETs in circuit simulations.

The channel current model considering SHE [16] is shown in Fig. 4. The model consists of the current model f, for example the surface potential model in the previous section, and thermal circuit model $G(s)$, where s is a complex number frequency parameter. Device current I_d is calculated being bias voltages and device (junction) temperature as inputs. Power consumption in the device, $P = I_d \cdot V_{ds}$, is the input of the thermal circuit $G(s)$, and the change of device temperature ΔT is the output. The updated device temperature $T + \Delta T$ again becomes the input of f.

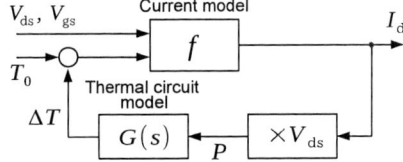

Fig. 4. MOSFET model considering self heating [16].

C. Identification of Thermal Circuit

$G(s)$, or its time-domain representation $g(t)$, can be obtained by knowing $P(t)$ and ΔT, assuming that $G(s)$ is a linear transfer function. The instantaneous power $P(t)$ at time t is calculated by the product of drain current $I_d(t)$ and drain voltage $V_{ds}(t)$. At the same time, channel temperature $T_j(t)$ is measured to find $\Delta T(t)$.

1) Channel Temperature Measurement: The channel temperature has to be estimated because device channel in an operating device is inaccessible. In JEDEC standard [17], the forward voltage of body diode is used to estimate the channel temperature, under an assumption that the temperature of pn junction is sufficiently close to that of the channel. In [16], channel temperature is estimated using the precharacterized temperature dependence of the drain current of the MOSFET. Once the channel current is represented as $I_d = h(T_j)$, the junction temperature can be reversely looked up as $T_j = h^{-1}(I_d)$. In order to obtain function h, device currents are measured at different temperatures in advance. The short-pulse circuits, such as the ones in Fig. 3(b), should be used to suppress SHE during the characterization.

2) Thermal Circuit Estimation: Let the impulse response of $G(s)$ be $g(t)$, the impulse response $y(t)$ for the input $x(t)$ is

$$y(t) = \int_0^t g(t-\tau)x(\tau)\mathrm{d}\tau. \tag{1}$$

While some known input power x is applied, channel temperature y is measured. Then, through deconvolution, the thermal transfer function is obtained. It may be approximated to an RC network [18]. In the case of Foster form, the approximation becomes

$$\Delta T_j = P \sum_{i=1}^{n} R_i \left[1 - \exp\left(-\frac{t}{R_i C_i}\right) \right]. \tag{2}$$

Since $\Delta T_j(s) = G(s)P(s)$, the thermal resistances R_i and capacitances C_i can be easily determined as

$$G(s) = \sum_{i=1}^{n} \frac{R_i}{1 + sR_i C_i}. \tag{3}$$

3) Thermal Circuit Identification: In the constant-voltage (CV) method, channel temperature is estimated through the MOSFET current [16]. The switch MOSFET is turned on to apply a constant drain voltage for the DUT. As seen in Fig. 2, the drain current changes due to SHE although the bias voltages are constant. As a result, input power for $G(s)$

978-1-5386-5309-8/18 $31.00 © 2018 IEEE

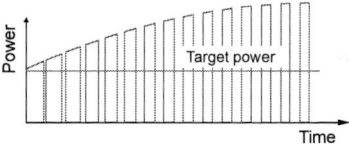

Fig. 5. Concept of CP method that equivalently achieves a step power input.

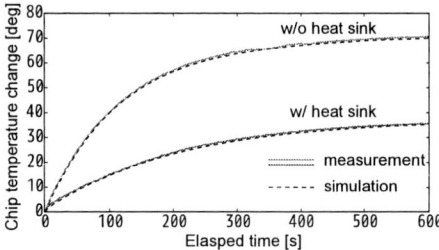

Fig. 6. The change of chip temperature for a 2 W step power input.

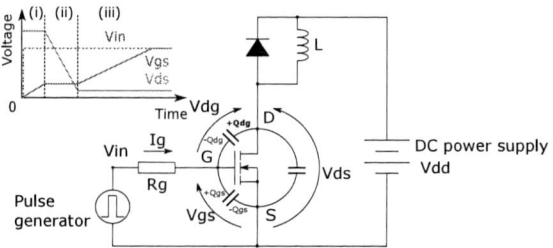

Fig. 7. Double-pulse circuit and its turn-on waveform.

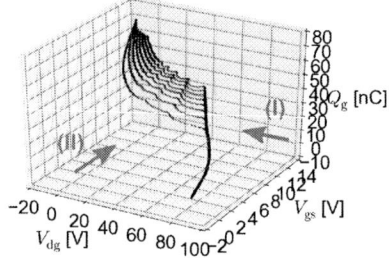

Fig. 8. Gate-charge trajectories in switching operation [23].

varies as opposed to the existing work [17]. By measuring $\Delta T(t)$ and $P(t)$, $g(t)$ is obtained by deconvolution.

In the constant-power (CP) method, pulse width of the switch MOSFET is modulated so that the constant-power input for $G(s)$ is equivalently achieved [16]. In Fig. 3(b), power consumption at the DUT is always $I_d V_{ds}$. When considering the duty ratio D of the switch MOSFET, it becomes $D I_d V_{ds}$. Now, upon the observation of input power, the duty ratio can be controlled in a pulse width modulation (PWM) manner to achieve desired constant input. Since the change of channel temperature can be obtained using MOSFET current as in the CV method, calculation of the thermal transfer function is significantly simplified.

D. Modeling Example

The CP method is applied to an SiC power MOSFET using 1 kHz pulse signal. Duty ratio is updated for every 1 s to maintain constant input power. SiC MOSFET is used as a switch in the circuit of Fig. 3(b). Operation point of the DUT is $V_{gs} = 8$ V and $V_{ds} = 10$ V. A step power of 2 W is applied for the DUT with and without a heat sink. The simulated predictions including SHE matched well with the measurement results as shown in Fig. 6.

IV. SWITCHING-BASED CAPACITANCE CHARACTERIZATION

A. Switching Capacitance

Conventionally, capacitance is measured through small signals, by applying a sinusoidal waveform to find terminal impedance at a fixed bias voltage [19], [20]. However, in actual switching operations, bias point of the MOSFETs changes from rail to rail. The conditions of the characterization and measurement are significantly different. In some literatures, it is reported that switching waveform cannot be reproduced well using small signal capacitance [21], [22]. In [23], a novel method for the measurement of the input capacitance of the

MOSFETs is proposed, which utilizes switching waveform of the double pulse circuit in Fig. 7.

In Fig. 7, gate charge Q_g is equal to $Q_{gs} - Q_{dg}$, and the gate charge is represented by using gate current I_g:

$$Q_g = \int I_g dt = \int \frac{V_{in} - V_{gs}}{R_g} dt. \tag{4}$$

Here, R_g is the external gate resistance, and V_{in} is the voltage given by the pulse generator. The gate charge Q_g can be calculated by measuring two terminals of the gate resistance. Let us assume $\partial Q_{gs}/\partial V_{dg} = \partial Q_{dg}/\partial V_{gs} = 0$ to separate Q_g into the contributions of C_{gs} and C_{dg} [23]. C_{gs} and C_{dg} can then be represented as

$$C_{gs} = \frac{\partial Q_{gs}}{\partial V_{gs}} = \frac{\partial Q_g}{\partial V_{gs}} = \frac{\partial(Q_{gs} - Q_{dg})}{\partial V_{gs}}, \tag{5}$$

$$C_{dg} = \frac{\partial Q_{dg}}{\partial V_{dg}} = -\frac{\partial Q_g}{\partial V_{dg}} = -\frac{\partial(Q_{gs} - Q_{dg})}{\partial V_{dg}}. \tag{6}$$

The turn-on waveform of the double pulse tester generally takes the form as shown in the inset of Fig. 7. According to (4), Q_g is equivalent to the area in between V_{in} and V_{gs}. Fig. 8 shows the measured switching trajectory in $V_{gs} V_{dg} Q_g$ space. Changing load current draws a different trajectory. According to (5) and (6), the slope of the trajectory from direction (I) is C_{gs}-V_{gs}, while that from direction (II) is C_{dg}-V_{dg}. These capacitances are called switching capacitance [23].

B. Modeling Examples

Fig. 9 shows measured C_{gs}-V_{ds} and C_{dg}-V_{dg} characteristics of a power MOSFET. C_{gs} obtained by switching is larger than small signal capacitance while small signal C_{dg} in negative V_{dg} is larger than that of switching capacitance. Transient waveform of a buck converter circuit is shown in Fig. 10. Two simulated waveforms are obtained by the same SPICE

978-1-5386-5309-8/18 $31.00 © 2018 IEEE

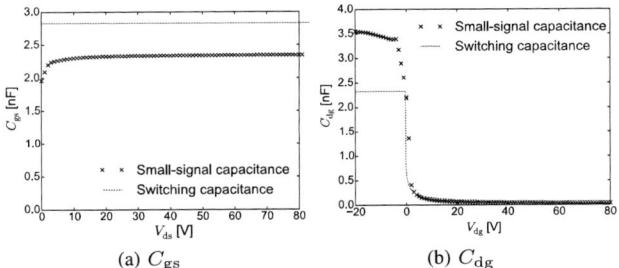

Fig. 9. Comparison of small-signal and switching capacitances [23].

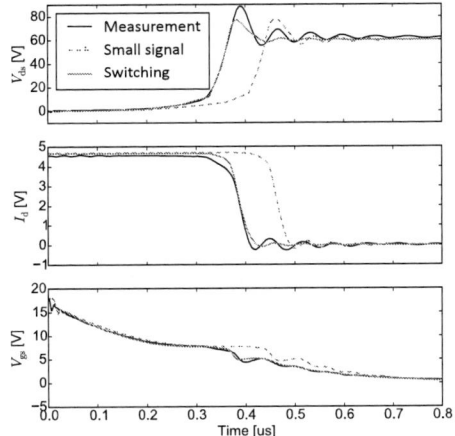

Fig. 10. Measured and simulated turn-off waveforms.

simulation decks except for the input capacitances: small signal and switching capacitances. The result of switching capacitance better matches with the measurement than small signal capacitance because the plateau duration of the turn off waveform is determined by C_{dg} for negative V_{dg} [24].

V. Summary

Modeling and characterization of SiC power MOSFET model are discussed in this paper. The nonideality associated with power devices is particularly emphasized including SHE and thermal circuit characterization, and input capacitance characterization through switching-based measurement.

Acknowledgment

This work was partially supported by JST Super Cluster Program and NEDO Cross-ministerial Strategic Innovation Promotion Program.

References

[1] I. Angelov, H. Zirath, and N. Rorsman, "A new empirical nonlinear model for HEMT and MESFET devices," *IEEE Trans. Microw. Theory Techn.*, vol. 40, no. 12, pp. 2258–2266, 1992.

[2] T. R. McNutt, A. R. Hefner, H. A. Mantooth, D. Berning, and S. H. Ryu, "Silicon carbide power MOSFET model and parameter extraction sequence," *IEEE Trans. Power Electron.*, vol. 22, no. 2, pp. 353–363, March 2007.

[3] N. Phankong, T. Yanagi, and T. Hikihara, "Evaluation of inherent elements in a SiC power MOSFET by its equivalent circuit," in *Proc. European Conf. Power Electronics and Applications*, 2011.

[4] M. Miura-Mattausch, U. Feldmann, A. Rahm, M. Bollu, and D. Savignac, "Unified complete MOSFET model for analysis of digital and analog circuits," *IEEE Trans. CAD*, vol. 15, no. 1, pp. 1–7, Jan 1996.

[5] Y. S. Chauhan, C. Anghel, F. Krummenacher, A. M. Ionescu, M. Declercq, R. Gillon, S. Frere, and B. Desoete, "A highly scalable high voltage MOSFET model," in *Proc. European Solid-State Device Research Conference*, Sept 2006, pp. 270–273.

[6] M. Miura-Mattausch, N. Sadachika, D. Navarro *et al.*, "HiSIM2: Advanced MOSFET model valid for RF circuit simulation," *IEEE Trans. Electron Devices*, vol. 53, no. 9, pp. 1994–2007, August 2006.

[7] G. Gildenblat, H. Wang, T.-L. Chen, J. Gu, and X. Cai, "SP: An advanced surface-potential-based compact MOSFET model," *IEEE Trans. Electron Devices*, vol. 39, no. 9, pp. 1394–1406, 2004.

[8] G. Gildenblat *et al.*, "PSP: An advanced surface-potential-based MOSFET model for circuit simulation," *IEEE Trans. Electron Devices*, vol. 53, no. 9, pp. 1979–1993, August 2006.

[9] H. J. Mattausch, M. Miyake, T. Iizuka, H. Kikuchihara, and M. Miura-Mattausch, "The second-generation of HiSIM_HV compact models for high-voltage MOSFETs," *IEEE Trans. Electron Devices*, vol. 60, no. 2, pp. 653–661, November 2013.

[10] Y. Nakamura, M. Shintani, K. Oishi, T. Sato, and T. Hikihara, "A simulation model for SiC power MOSFET based on surface potential," in *Proc. Int. Conf. Simulation of Semiconductor Processes and Devices*, September 2016, pp. 121–124.

[11] M. Shintani, Y. Nakamura, M. Hiromoto, T. Hikihara, and T. Sato, "Measurement and modeling of gatedrain capacitance of silicon carbide vertical double-diffused MOSFET," *Jpn. J. Appl. Phys.*, vol. 56, no. 04CR07, pp. 04CR07-1–04CR07-5, March 2017.

[12] Y. Tsividis and C. McAndrew, *Operation and Modeling of the MOS Transistor, 3rd ed.* Oxford University Press, 2011.

[13] *B1505A Power Device Analyzer/Curve Tracer*, Keysight Technologies, Inc., July 2015.

[14] M. Paggi, P. H. Williams, and J. M. Borrego, "Nonlinear GaAs MESFET modeling using pulsed gate measurements," *IEEE Trans. Microw. Theory Techn.*, vol. 36, no. 12, pp. 1593–1597, Dec 1988.

[15] Y. Nakamura, M. Shintani, T. Sato, and T. Hikihara, "A high power curve tracer for characterizing full operation range of SiC power transistors," in *Proc. Int. Conf. Microelectronic Test Structures*, March 2016, pp. 90–94.

[16] K. Oishi, M. Shintani, M. Hiromoto, and T. Sato, "Identifications of thermal equivalent circuit for power MOSFETs through in-situ channel temperature estimation," in *Proc. Workshop on Wide Bandgap Power Devices and Applications*, Nov. 2016, pp. 308–313.

[17] JEDEC, "Transient dual interface test method for the measurement of the thermal resistance junction-to-case of semiconductor devices with heat flow through a single path," JESD51-14. [Online]. Available: http://www.jedec.org/standards-documents/results/JESD51-14

[18] A. A. Merrikh and A. J. McNamara, "Parametric evaluation of Foster RC-network for predicting transient evolution of natural convection and radiation around a flat plate," in *Proc. Intersociety Conf. Thermal and Thermomechanical Phenomena in Electronic Systems*, May 2014, pp. 1011–1018.

[19] R. Elfrich, T. Lopez, and N. Koper, "Accurate behavioural modelling of power MOSFETs based on device measurements and FE-simulations," in *Proc. European Conf. Power Electronics and Applications*, 2005.

[20] T. Funaki, N. Phankong, T. Kimoto, and T. Hikihara, "Measuring terminal capacitance and its voltage dependency for high-voltage power devices," *IEEE Trans. Power Electron.*, vol. 24, no. 6, pp. 1486–1493, 2009.

[21] V. Höch, J. Petzoldt, A. Schlogl, H. Jacabs, and G. Deboy, "Dynamic characterization of high voltage power MOSFETs for behavior simulation models," in *Proc. European Conf. Power Electronics and Applications*, 2009.

[22] V. Höch, J. Petzoldt, H. Jacabs, A. Schológl, and G. Deboy, "Determination of transient transistor capacitances of high voltage mosfets from dynamic measurements," in *Proc. Int. Symp. Power Semiconductor Devices and ICs*, June 2009, pp. 148–151.

[23] K. Oishi, M. Shintani, M. Hiromoto, and T. Sato, "Input capacitance determination of power MOSFETs from switching trajectories," in *Proc. Int. Conf. Microelectronic Test Structures*, March 2017, pp. 1–6.

[24] B. J. Baliga, *Fundamentals of Power Semiconductor Devices.* Springer, 2008.

978-1-5386-5309-8/18 $31.00 © 2018 IEEE

MTTF-AWARE DESIGN METHODOLOGY FOR ADAPTIVE VOLTAGE SCALING

Masanori Hashimoto[1]*, Yutaka Masuda[1]

[1] Dept. Information Systems Engineering, Osaka University, Suita, Japan

*Corresponding Author's Email: hasimoto@osaka-u.ac.jp

ABSTRACT

Adaptive voltage scaling (AVS) is a promising approach to overcome manufacturing variability, dynamic environmental fluctuation, and aging. This paper focuses on design of AVS circuits. For pursuing efficient AVS, this work presents a design methodology that optimizes both voltage-scaled circuit and voltage control logic achieving practical long mean time to failure (MTTF), e.g., years. Evaluation results show the proposed AVS achieves 20.8% voltage reduction while satisfying target MTTF.

INTRODUCTION

Aggressive device miniaturization due to technology scaling has been improving the device performance. On the other hand, circuits have become sensitive to static manufacturing variability and dynamic environmental fluctuation. These static and temporal variations directly lead to circuit reliability degradation. The most effective tuning knob for post-silicon compensation is supply voltage control, and adaptive voltage scaling (AVS) is intensively studied [1][2]. AVS is expected to minimize process, voltage, temperature, and aging (PVTA) margin of each chip and allocate only a small margin for the entire lifetime as shown in Figure 1. The excessive conventional PVTA margins existing in most of the chips can be exploited as the source for power reduction. Conventional works [1][2] focus on where to insert sensors and how to control supply voltage and discuss the design methodology of voltage control system including sensing circuit.

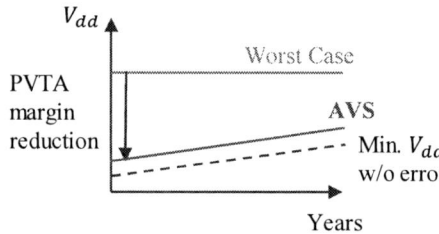

Figure 1 : Supply voltage of AVS and conventional worst case (WC) in device lifetime.

On the other hand, for implementing AVS systems that fully exploit run-time adaptation and eliminate the redundant margin, we have found that we should pay attention to the main logic circuit under AVS in addition to the sensing circuit. In the conventional VLSI design flow,

there are many critical paths since the timing slack is exploited for power and area reduction. However, we observe that inherent critical paths whose path delays cannot be reduced at all are limited. This observation suggests that adaptive slack assignment (ASA) to the main logic circuit under AVS, which allocates larger slack to highly active paths, could improve the efficacy of the AVS and enable further supply voltage reduction.

This work focuses on the error prediction based AVS (EP-AVS) and presents a design methodology for EP-AVS circuits [3]. The proposed methodology optimizes both the main logic under AVS and sensing circuit. In the main logic design, we perform mean time to failure (MTTF) aware ASA that utilizes critical path isolation (CPI) [4] and estimates MTTF of AVS circuits with a stochastic framework [5]. As for the sensing circuit design, we propose a novel sensor insertion method that minimizes the sum of gate-wise timing failure probabilities, where the timing failure probability is the joint probability of activation and timing violation. By exploiting the information on the paths with higher timing failure probability, the proposed sensor insertion makes EP-AVS efficiently monitor the timing-critical and highly-active FFs.

The rest of this paper is organized as follows. First, this paper proposes a design methodology that optimizes both the main logic under AVS and sensing circuit. Then, we demonstrate the supply voltage reduction and speed-up thanks to the proposed EP-AVS. Lastly, concluding remarks are given.

PROPOSED DESIGN METHODOLOGY

This section first explains the assumed EP-AVS and the overview of the proposed methodology. Then, The ASA and the sensor insertion are presented in series.

Assumed EP-AVS

Figure 2 illustrates an EP-AVS circuit assumed in this paper. The EP-AVS circuit is composed of the main circuit, timing error predictive flip-flop (TEP-FF) [6] and voltage control unit. The TEP-FF consists of a flip-flop, delay buffers, and a comparator (XOR gate), and works with the main FF. When the timing margin is gradually decreasing, a timing error occurs at the TEP-FF before the main FF captures a wrong value due to the delay buffer, which enables us to know that the timing margin of the main FF is not large enough. An error prediction signal is generated to predict the timing errors, and this signal is monitored during a specified period. Note that timing errors are predicted, not detected, which is a distinct difference from Razor [1]. Once an error prediction signal is observed, the higher supply voltage is given to reduce circuit delay. Note that clock frequency is fixed throughout this paper. If no error prediction signals are observed during the monitoring period, the circuit is slowed down for power reduction. This proactive AVS is expected to overcome the variation of the timing margin which is different in every chip and varies depending on operating condition and aging.

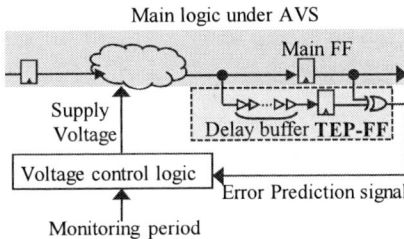

Figure 2 : Assumed EP-AVS.

Overview and Problem Definition

The proposed design methodology for EP-AVS consists of the ASA for the main logic under AVS followed by the insertion of TEP-FFs. Figure 3 illustrates the concept of ASA. In conventional design flow, cell instances that are included in non-critical paths are replaced with smaller cells and higher-Vth cells for reducing power dissipation and area. Consequently, this replacement decreases timing margin of many paths and may deteriorate MTTF. On the other hand, the ASA increases timing slacks of non-intrinsic critical paths as shown in Figure 3. Meanwhile, the path-based slack assignment is not efficient since the number of paths in a circuit is huge. Therefore, this work utilizes FF-based CPI proposed in [4] to adjust setup slacks of FFs. For each FF, we assign an individual target slack value. After the CPI, the paths ending at the FFs whose slack values increased are less likely to fail even when the gate delays in the paths vary, which contributes to MTTF extension.

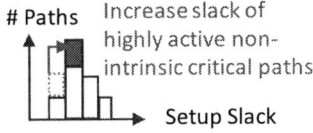

Figure 3 : ASA for main logic optimization.

Here, it should be noted that CPI increases area and power since the intentional increase in slack was originally exploited for area and power reduction in conventional design. In this sense, we need to identify FFs that have less impact on area and power yet contribute to remarkable

MTTF extension. Based on the discussion above, we formulate the design optimization of EP-AVS including CPI-based ASA and TEP-FF insertion.

Objective : To minimize V_{dd}
Variables : B_{TEP_i} $(1 \le i \le N_{FF})$, B_{CPI_i} $(1 \le i \le N_{FF})$
Constraint #1 : $MTTF \le MTTF_{const}$
Constraint #2 : N_{TEP} $(= \sum_{i=1}^{N_{FF}} B_{TEP_i}) = N_{TEP}^{max}$
Constraint #3 : N_{CPI} $(= \sum_{i=1}^{N_{FF}} B_{CPI_i}) = N_{CPI}^{max}$

The objective of this problem is to minimize V_{dd} aiming at power minimization. The variables for optimization are B_{TEP_i} and B_{CPI_i}. B_{TEP_i} is a binary variable, and it becomes 1 when i-th FF is replaced to TEP-FF. B_{CPI_i} is also a binary variable, and it is 1 when CPI is applied to i-th FF. The primary constraint is MTTF, and the lower bound of MTTF ($MTTF_{const}$) is given as a constraint. The second constraint gives the upper bound of the number of TEP-FFs (N_{TEP}^{max}), and this limits the area increase due to TEP-FF insertion. Similarly, the upper bound of the number of FFs to which CPI is applied (N_{CPI}^{max}) is also given as a constraint to limit the area increase originating from CPI. The proposed design methodology solves this problem with a two-stage procedure. The first stage designs the main logic under AVS using CPI [4], i.e., determines B_{CPI_i}, and the second stage performs TEP-FF insertion, i.e., determine B_{TEP_i}. The following subsections explain these two stages.

First Stage: CPI-based ASA in Main Logic

The CPI is performed referring to [4]. The CPI method focuses on gate-wise failure probability which is a metric that expresses the contribution to the timing failure probabilities at the downstream FFs. The detailed computation will be explained in the next subsection. Then, for maximally reducing the sum of gate-wise failure probabilities, this method selects target FFs by solving a covering problem of instances weighted with the failure probability as an integer linear programming (ILP) problem and thus determines B_{CPI_i}. Once B_{CPI_i} is determined, the FF-based CPI proceeds to the following two steps; (1) increase setup time of the i-th FF by $\Delta Setup_i$ artificially and re-synthesize the design as an engineering change order (ECO) process, and (2) restore the setup time for the successive analysis process. With this CPI, we enforce the paths ending at the target FF to have the slack of more than $\Delta Setup_i$. Referring to [4], $\Delta Setup_i$ is set to the upper bound value that can satisfy the setup constraint after ECO for simplicity.

Second Stage: Sensing Circuit Insertion

For making EP-AVS work well, TEP-FFs need to output the error prediction signals frequently to nicely adjust the supply voltage, and hence it is desirable that inserted TEP-FFs are highly activated. Also, FFs with small slacks need fewer delay buffers in TEP-FFs. Here, FFs

978-1-5386-5309-8/18 $31.00 © 2018 IEEE   549

having higher timing failure probability satisfies both the desirable properties above. Therefore, we propose a novel TEP-FF insertion method that minimizes the sum of gate-wise timing failure probabilities. Our insertion method consists of the following two steps; (1) calculate timing failure probabilities, and (2) find out a set of FFs that maximally reduce the sum of gate-wise failure probabilities by solving the instance covering problem as an ILP problem.

In the first step, the proposed method calculates timing failure probability of FFs, $P_{FF_i\_fail}$. Remind that timing failure probability is the joint probability of timing violation and activation. In this work, we calculate the timing violation probability by statistical static timing analysis (SSTA) and derive the activation probability of each path by associating the signal transition time in logic simulation with the path delay in STA. Then, we obtain $P_{FF_i\_fail}$ by multiplying the timing violation probability and the activation probability. Next, we compute the gate-wise failure probabilities, i.e., $P_{insk_k\_fail}$, as follows.

$$P_{insk_k\_fail} = \max\left\{\frac{P_{FF_i\_fail}}{\sum_{k=1}^{N_{inst}} B_{FF_i\_inst_k}}\right\} (1 \leq i \leq N_{FF}). \quad (1)$$

In Equation (1), N_{inst} is the number of instances in the circuit. $B_{FF\_inst_k}$ is a binary valuable which is determined by the circuit topology, and it becomes 1 when k-th instance is included in the paths ending at i-th FF. $\sum_{i=1}^{N_{inst}} B_{FF_i\_inst_k}$ is the total number of instances included in the fan-in cone of i-th FF. The above equation assumes that each instance included in the fan-in cone of i-th FF has the same contribution to the timing error at the FF, and hence the $P_{FF_i\_fail}$ is divided by $\sum_{k=1}^{N_{inst}} B_{FF_i\_inst_k}$. An instance can be included in the fan-in cones of multiple FFs. For coping with this, the max operation is performed.

In the second step, we select a set of FFs that maximize the sum of gate-wise failure probabilities. We formulate this FF selection problem as an ILP problem to derive the exact solution. Our ILP formulation is as follows:

Objective : To maximize $\sum_{k=1}^{N_{inst}}(P_{inst_k\_fail} \times B_{inst_k})$
Variables : B_{TEP_i} $(1 \leq i \leq N_{FF})$
Constraint #1 : $0 \leq B_{inst_k} \leq 1$ $(1 \leq k \leq N_{inst})$
Constraint #2 : $0 \leq B_{TEP_i} \leq 1$ $(1 \leq i \leq N_{FF})$
Constraint #3 : $\sum_{i=1}^{N_{FF}} B_{TEP_i} \leq N_{TEP}$
Constraint #4 : $B_{inst_k} \leq \sum_{i=1}^{N_{FF}}(B_{TEP_i} \times B_{FF_i\_inst_k})$

The objective of this ILP problem is to maximize the sum of $(P_{inst_k\_fail} \times B_{inst_k})$, where $P_{inst_k\_fail}$ is the gate-wise failure probability. B_{inst_k} is a binary variable, and it becomes 1 when k-th instance is included in paths ending at the target FFs for TEP-FF insertion. Therefore, the sum of $P_{inst_k\_fail} \times B_{inst_k}$ represents the gate-wise failure probability reduction. In this problem, we assign binary

variables B_{TEP_i}, where B_{TEP_i} becomes 1 when i-th FF is selected as target FFs for TEP-FF insertion.

The first and second constraints are given to restrict B_{inst_k} and B_{TEP_i} to binary numbers. The third constraint means that the number of target FFs for TEP-FF insertion should be equal or less than N_{TEP}. The fourth constraint is a key constraint that defines the relation between B_{inst_k} and B_{TEP_i}. B_{inst_k} becomes 0 only when the product of B_{TEP_i} and $B_{FF_i\_inst_k}$ is 0 for all the FFs. On the other hand, if k-th instance is included in the paths ending at the target FFs, at least one of the products of B_{TEP_i} and $B_{FF_i\_inst_k}$ become 1. In this case, B_{inst_k} can be 1. In our ILP formulation, we are maximizing the sum of $(P_{inst_k\_fail} \times B_{inst_k})$ and hence B_{inst_k} is necessarily assigned to be 1.

EXPERIMENTAL EVALUATION

This section experimentally evaluates the performance improvement thanks to the proposed EP-AVS.

Experimental Setup

In this work, we used the advanced encryption standard (AES) circuit and OR1200 OpenRISC processor, which is a 32-bit RISC microprocessor with five pipeline stages, as target circuits. These two circuits were designed by a commercial logic synthesizer with a 45nm Nangate standard cell library, where the synthesized OpenRISC has 2500 FFs and AES has 530 FFs. Also, standard cell memories [7] were used as SRAMs in OpenRISC processor. We used Gurobi Optimizer 7.0 to solve the ILP problem defined in the previous section. The solver was executed on a 2.4 GHz Xeon CPU machine under the Red Hat Enterprise Linux 6 operating system with 1024 GB memory. For calculating meaningful MTTF, practical delay variations should be considered. Our evaluation took into account the following variations; (1) Dynamic supply noise, which is assumed to temporally fluctuate between -50 mV and 50 mV by 10mV. (2) Intra-die random variation and inter-die variation, which consist of NMOS and PMOS threshold voltage variation of $\sigma = 30$ mV and gate length variation of $\sigma = 1$ nm. (3) NBTI aging, whose model was obtained by fitting the measured data in [8] to the trapping/de-trapping model [9]. Six degradation states of 0 mV, 0.5 mV, 1 mV, 5 mV, 10 mV and 15 mV are prepared.

As for workload in OpenRISC, we selected three benchmark programs (CRC32, SHA1, and Dijkstra) from MIBenchmark [9]. In AES, 1,000 random test patterns were used. We set MTTF of 1.0×10^{17} cycles, i.e., 3.3 years in Open-RISC and 1.6 years in AES, as $MTTF_{const}$. With this setup, we performed CPI-based ASA to both AES and OpenRISC. The constraint of ASA area overhead is set to 6.0% for AES and 1.0% for OpenRISC. Next, we inserted several TEP-FFs to the voltage-scaled circuits. The constraint of area overhead for TEP-FF is set to 1.0%

978-1-5386-5309-8/18 $31.00 © 2018 IEEE   550

for both AES and OpenRISC, i.e., N_{TEP}^{max} is set to 18 in AES and 13 in OpenRISC, respectively. When inserting TEP-FF, we need to determine the number of delay buffers for each TEP-FF. In this work, we inserted the delay buffers whose delay were comparable to the delay variation caused by 100 mV supply noise. This determination of the number of delay buffers includes room for improvement.

MTTF and average supply voltage under PVTA variation are evaluated by the stochastic MTTF estimation framework [5]. In our experiment, the monitor period for EP-AVS was set to 10^6 cycles. We prepared nine supply voltages from 1.20 V to 0.80 V with 50 mV interval.

Evaluation Results

Figure 4 shows the trade-off curves between the minimum average supply voltage and the clock cycle under the MTTF constraint of 10^{17} cycles, where (a) in OpenRISC and (b) in AES, respectively. The black square plots represent the conventional WC design with guard-banding for PVTA variation. The yellow circular and blue triangular plots correspond to the conventional EP-AVS which optimizes only the sensing circuit, and the proposed EP-AVS which optimizes both the main logic under AVS and sensing circuit, respectively. First, we compare the black square and blue triangular plots for clarifying the overall performance improvement thanks to the proposed EP-AVS. Figure 4 shows that the proposed EP-AVS reduces average supply voltage and clock cycle time while keeping

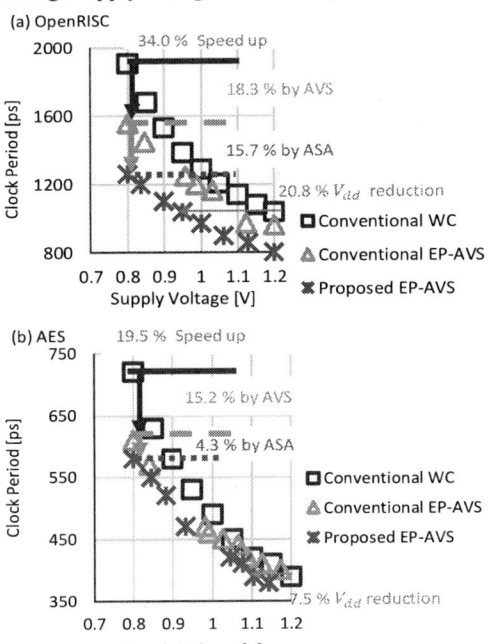

the target MTTF. For example, in Figure 4(a), at a clock period of 1040 ps, the proposed EP-AVS achieved the target MTTF with an average supply voltage of 0.95 V, whereas the conventional WC design required 1.20 V operation. In other words, EP-AVS achieved 20.8% V_{dd} reduction from 1.20 V to 0.95 V. Similarly, in Figure 4(b), at a clock period of 390 ps, the proposed EP-AVS achieved 7.5% V_{dd} reduction from 1.20 V to 1.11 V. As for clock period reduction, the proposed EP-AVS achieved 34.0% speed-up at 0.80 V in OpenRISC (Figure 4(a)) and 19.5% speed-up at 0.80 V in AES (Figure 4(b)), respectively. We experimentally confirmed that the proposed EP-AVS made the significant performance improvement at the cost of 7.0% area increase in. AES and 1.4% in OpenRISC.

Next, we compared the conventional EP-AVS and proposed EP-AVS. Figure 4 shows that the proposed EP-AVS further improves performance from the conventional EP-AVS. For example, at 0.80 V, the proposed EP-AVS achieved 15.7% speed-up from 1560 ps to 1260 ps in OpenRISC and 4.3% speed-up from 610 ps to 580 ps in AES. This reveals that the ASA for the main logic works well regarding speed-up and V_{dd} reduction and the simultaneous optimization of the main logic under AVS and the sensing circuit enhances the efficacy of EP-AVS.

CONCLUSION

This paper focused on EP-AVS and proposed a design methodology for EP-AVS circuits. The proposed methodology optimizes both the main logic under AVS and sensing circuits. The quantitative MTTF and supply voltage evaluation results showed that the proposed EP-AVS achieved 20.8% voltage reduction while satisfying target MTTF. One of our future works include power-oriented design optimization of EP-AVS.

ACKNOWLEDGEMENTS

This work is partly supported by STARC, Socionext, and ICOM Foundation, Japan.

REFERENCES

[1] S. Das, *IEEE Journal Solid-State Circuits*, vol. 41, no. 4, pp. 792–804, 2006.
[2] K. A. Bowman, *IEEE Journal Solid-State Circuits*, vol.46, no. 1, pp. 194–208, 2011.
[3] Y. Masuda, in *Proc. ASP-DAC*, 2018.
[4] Y. Masuda, in *Proc. ICCAD*, 2016.
[5] S. Iizuka, in *Proc. ITC*, 2015.
[6] H. Fuketa, *IEEE Trans. VLSI Systems*, 2012.
[7] J. Shiomi, in *Proc. PATMOS*, pp. 44–49, 2016.
[8] H. Awano, in *Proc. ESSDERC*, pp. 218–221, 2014.
[9] B. J. Velamala, *IEEE Trans. Electron Devices*, vol. 60, no. 11, pp. 3645–3654, 2013.
[10] M. R. Guthaus, in *Proc. IEEE Workshop on Work load Characterization*, pp. 3–14, 2001.

Figure 4 : Trade-off curves between supply voltage and clock period (a) OpenRISC, (b) AES.

Analysis of Affecting Factors in Neutron Interactions with Gate Oxide in CMOS Transistors

C.-Z. Chen[a*] and David Y. Hu[b]

[a]Qualchip Technologies, Inc., Wuxi, Jiangsu, China 214072
[*]Mobile: (+86) 13910199301, Email: czchen126@126.com
[b]MetroSilicon Microsystems, Kunshan, Jiangsu, China 215300
Mobile : (+86) 13764219094, Email: david@metroSilicon.com

ABSTRACT

High reliability of radiation resistant is required in CMOS based SoC designs for applications in today's commercial, automotive electronics such as in advanced driver-assistance systems (ADAS), in addition to well-studied avionics. Ionizing radiation particles can cause various types of single event effects (SEE), such as single event upset (SEU). For neutrons, the situation is much more complex than other charged particles (*ex.* protons, electrons and alphas), as neutrons are hard to detect and they may undergo several types of interactions with matter. Therefore, several factors can affect the neutron interaction probability, or "the effective target area", with target materials. This study calculates and analyses the interacting cross section (σ, in units of *barn*) of neutrons with CMOS transistor materials (gate oxide, or silicon dioxide, SiO_2) and temperatures of the target nuclei, under various energies (MeV) and *flux* ($n/cm^2 \cdot s$) and/or *fluence* (n/cm^2). Present study shows that neutron *energy*, *fluence* and *process feature size* have major impacts on CMOS damages than others. SiO_2 in transistors is known vulnerable under ESD (electrostatic discharge) stress typically known as CDM (charged device model) event, the combined study to be continued.

Keywords: single event effects (SEE), single event upset (SEU), single event rate (SER), advanced driver-assistance systems (ADAS), neutron interaction cross section, flux, fluence, silicon dioxide (SiO₂), electrostatic discharge (ESD), ionizing radiation.

INTRODUCTION

The issues of ionizing radiation generated single event effect (SEE), especially single event upset (SEU) in DRAM cells were studied in the late 1970s. Due to process shrinkage and widely used SRAM (DDRs) in advanced IC designs, which have much smaller capacitance in the RAM cells to render them more vulnerable to radiation damage. Some effort on the SEE detection of CMOS transistors by neutrons have been reported in recent years by either continuous measurement on SEU, FIT (failure in time, defined as the number of failures per 10^9 device hour, defined by JEDEC), and/or SER (single event rate, FIT/Mbit). These tests have been done on the ground level and above sea level (1,524m or 5,000ft), commercial 9,144m (30k ft) and military 18,288m (60k ft) aviation levels [1-3].

In view of reliability concerns, from commercial, automobile ECU (Electronic Control Unit) in ADAS (Advanced Driver-Assistance Systems) to commercial aviation and military needs, however, due to its physical nature, studies of the affecting factors by neutrons impacting on the CMOS gate oxide SO_2 have become increasingly important, as far as ESD protection is concerned. These factors may include neutron energy and range or mean free path, neutron flux (*neutron/cm²·s*) or neutron fluence (*neutron/cm²*), interaction types and their respective interaction cross sections (as typical nuclear radii (r) are of the order 10^{-12} cm, assuming spherical shape, the *physical cross-sectional area* (πr^2) of a heavy nucleus is to be of the order of 10^{-24} cm², which is defined to be 1 *barn*, or *b*). In radiation hardening (*RadHard*) study on CMOS transistors, the single event effects (SEE), mainly SEU are the evaluating measurements [4]. *RadHard* for CMOS also include Error Correction Code (ECC) on memories (especially double data rate SDRAM, DDR), and parity bits or checksums to detect whether the accessed data is correct. If a transmission is found bad, the system is allowed to request a retransmit or identify a failure and make the appropriate correction.

In this work, we first time calculated values of total cross section and mean free path at a given neutron energy of 10 MeV in several transistor gate materials (SiO_2 and GaAs, SiC, Si as a comparison); similarly for SiO_2 and Si at various incident neutron energies (1keV-100MeV). We compiled and analyzed available data the effect of SER (SEU) in several process nodes (40, 65, 90, 130nm) by neutron *fluence* values, which are used as *RadHard* references by JEDEC.

METHODOLOGY

Neutron sources and types of interaction

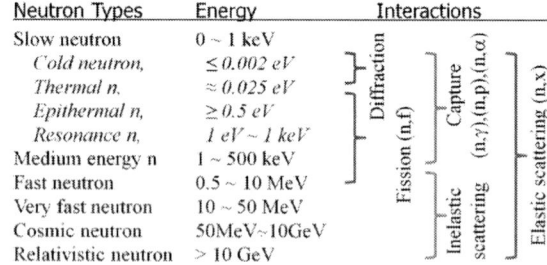

Figure 1: Neutrons and Their Energy Range

Neutrons come from various sources such as cosmic rays background, natural radioactivity such as synthesized fission elements (*e.g.* ^{252}Cf), spallation fission products from nuclear weapon and reactors, or from an irradiation facility, for example, through ^7Li(p, n)^7Be reaction to generate 'soft' neutrons (<100keV) [5] or very fast neutrons with sufficient flux [6]. According to their energies, neutrons can be termed as slow, medium energy and fast neutrons *etc.*, their interactions with matter are complex. Several types of interactions may take place at a given energy, see *Figure 1*.

978-1-5386-5309-8/18 $31.00 © 2018 IEEE

Neutrons can go through the target material directly via *transmission*; while most commonly neutron propagation is deviated with respect to the original direction via *scattering* or neutrons are captured by the target nucleus via *absorption*. Each absorption event may undergo one or more nuclear reaction(s), see *Fig. 2* (the letters in the parentheses show the incoming and outgoing particles) [7]. (*Note:* In that study, capture was also called electromagnetic interaction; neutral was called emission of neutrons; both charged interaction and neutral interaction were called transfer by others.)

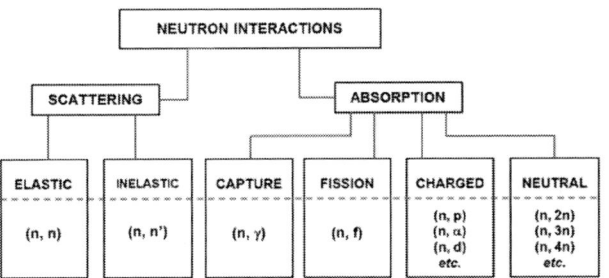

*Figure 2: Neutron can go through material without any interaction (*transmission*), or with interactions of* scattering *and* absorption [7] *(after P. Rinard, 1991).*

Calculation of neutron cross-section

Neutron diffraction (or elastic neutron scattering) is similar to X-ray diffraction. When the neutron wavelength is comparable to atomic spacing in crystals, it is measured following the *Bragg's law*, the so called epoch-making equation to determine the atomic and/or magnetic structure of a material as below

$$n\lambda = 2d \sin\theta \qquad (1)$$

where n is an integer, λ is the wavelength of the incident neutrons, d the inter-planar distance of the planes, and θ the glancing angle at which the neutrons are reflected. The likelihood of interaction of neutron and nucleus between an incident neutron and a target nuclide, independent of the type of reaction, is expressed with the help of the total cross section σ_T which is simply the sum of the two partial cross sections, namely the scattering and absorption cross sections σ_S and σ_A:

$$\sigma_T = \sigma_S + \sigma_A \qquad (2)$$

The neutron transport equation can be used in a nuclear reactor work mathematically to study the motion of neutrons with material; however, the *Ramsauer Model* [8] is used *qualitatively* to estimate σ_T at an energy E,

$$\sigma_T(E) = 2\pi[(R + \lambda_r(E)]^2 (1 - \alpha \cos\beta) \qquad (3)$$

where R is the effective radius of hit nuclei, the reduced wavelength $\lambda_r(E) = h/\sqrt{2mE}$, $\alpha = 1$ if there is no absorption or $\alpha < 1$ if absorption occurs, β denotes phase change (in passing through the nucleus, see *Eq. 1*).

Whether if an elastic or inelastic scatter occurs is dependent on the speed of the neutron, and it is also dependent on the nucleus it strikes and its neutron cross section. Elastic collisions often transfer kinetic energy between particles. In inelastic scattering, the kinetic energy of an incident particle is not conserved as some of the energy of the incident particle is lost or increased, while the neutron interacts with the nucleus, α, β, γ, and protons may be emitted.

For a single element, the *macroscopic* cross section Σ ( ) can be derived from the *microscopic* cross section σ and total atom number N (cm^{-3}) of the material nuclei; similarly, for a compound the *macroscopic* cross-section Σ_{comp} (cm^{-1}), for each element i, there is a relationship with σ_i and N_i,

$$\Sigma_{comp} = \sigma_i N_i, \qquad N_i = \rho N_a n_i / M \qquad (4)$$

where ρ is the density (g/cm^3), $N_a = 6.022 \times 10^{23}$ is Avogadro's number (atoms/mole), n_i is the number of element i in one molecule, M is molecular weight of the compound (g/mole). The mean free path λ between neutron interactions is deduced as follows

$$\lambda = 1/\Sigma \qquad (5)$$

To account for temperature impact of neutrons on materials, we have the following, where T_0 and T are in Kelvins, σ_0 is the cross-section at T_0.

$$\sigma_T = \sigma_0 (T_0/T)^{1/2} \qquad (6)$$

Neutron flux and neutron fluence

The *neutron flux* is defined as the total length travelled by all neutrons per unit time and volume (n/cm^2·s); the *neutron fluence* is defined as the neutron flux integrated over a certain time period (n/cm^2). For a known or measured cascade flux I_1, and at altitude (pressure) A_1 (in g/cm^2), then flux I_2 can be calculated at altitude A_2 as follows,

$$I_2 = I_1 \exp[(A_1 - A_2)/L_n] \qquad (7)$$

where $L_n = 148$ g/cm^2 is the absorption length for neutron [10]. *Eq. 7* was used to calculate the neutron flux attenuation [2].

RESULTS AND DISCUSSION
Results of Neutron Interaction Cross Sections

Table 1: Neutron Flux (n/cm^2·s)

Category	Flux	Notes
Thunderstorms [9]	0.03-90	0-20MeV
New York, sea level [2]	0.013	3.4× in Denver
Puno, Peru, alt. 3889m [2]	0.117	9× of NYC
Pic-du-Midi, alt. 2885m [2]	0.111	8.5× of NYC
New York, alt. 30k ft [11]	0.35	1-10MeV
Denver, alt. 60k ft [11]	1.3	1-10MeV
$^7Li(p, n)^7Be$ [5]	30	15-60keV
$^7Li(p, n)^7Be$ [6]	9.3×10^3	100MeV
Reactor (32 of 40 yrs.) [12]	$\sim10^{12}$	>1MeV
Stars [12]	10^5-10^{11}	*s-process*
Supernovae [12]	10^{22}	*r-process*

Table 2: Temperature variance $(T_0/T)^{1/2}$ on neutron cross-section σ_T (in units of barn, i.e. 10^{-24} cm^2)

°C	T	$(T_0/T)^{1/2}$
20 (T_0)	293.15	1.0
150	423.15	0.8323
125	398.15	0.8581
0	273.15	1.0360
-40	233.15	1.1213
-55	218.15	1.1592

(Note: 0K = −273.15 °C and 273.16K = 0.01°C.)

Compiled in *Table 1* are neutron flux values detected or estimated at various geological locations (*Eq. 7*), compared with the temperature variation (*Table 2* using *Eq. 6*), the neutron flux at high altitude makes big difference, although the flux also increases with the latitudes from North *0* to North *40* degrees. At high altitudes, SEU induced in SRAM was continuously measured over 12 months [2].

Neutron interaction cross section (*Eqs. 1-3*) can be calculated using *JANIS* (Java-based nuclear information software) programs [13]. The total cross-section of a few CMOS elements commonly found in CMOS gate oxide and transistors are calculated, only silicon $^{28}_{14}Si$ ($^{A}_{Z}Si$, mass number $A=28$, atomic number $Z=14$, and its neutron number $N=A-Z=14$) is shown in *Figure 3*.

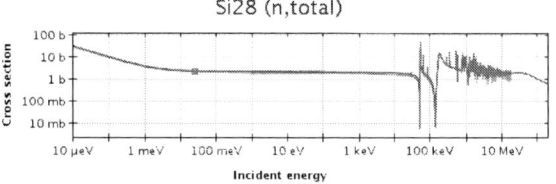

Figure 3: Total neutron cross sections in $^{28}_{14}Si$

The *macroscopic* cross section (*Eq. 4*) and the mean free path (*Eq. 5*) for several compounds are then calculated. For SiO_2 in the gate oxide, its density is 2.27 (g/cm3), M=60.08 (g/mole), using total σ_T for each element at 10MeV ($\sigma_1 = 1.90b$ and $\sigma_2 = 1.32b$), $\Sigma_{comp} = \sigma_i N_i = 0.103 (cm^{-1})$, and $\lambda = 9.709$cm, its one dimensional magnitude (~10^8nm) may indicate that to produce significant SEU at 40 nm scale range, the *neutron fluence* should be at or above 10^8 n/cm^2 order. Similar results at various neutron energies (1keV-100MeV) are also produced as a reference, see *Table 3* and *Table 4*.

Table 3: The macroscopic *cross section and the mean free path of CMOS transistor compounds for 10 MeV neutrons*

Compound	ρ	M	σ_1	σ_2	Σ_{comp}	λ
SiO$_2$	2.27	60.08	1.90	1.32	0.103	9.709
GaAs	5.32	144.6	3.65	3.80	0.165	6.061
SiC	3.16	40.10	1.90	1.18	0.146	6.849
Si	2.329	28.08	1.90	N/A	0.095	10.53

Table 4: The macroscopic *cross section and the mean free path of SiO$_2$ and Si at various neutron energies*

E, MeV	0.001	0.01	0.1	1	10	100
SiO$_2$, Σ_{comp}	0.217	0.213	0.189	0.478	0.103	0.053
SiO$_2$, λ	4.608	4.695	5.291	2.092	9.709	18.87
Si, Σ_{comp}	0.097	0.087	0.050	0.230	0.095	0.055
Si, , λ	10.31	11.49	20.00	4.348	10.53	18.18

Results of Single Event Upsets (SEUs)

SEU can be measured with SER. A few commercial RAM products tested with various neutron fluence (n/cm^2) conditions with reported SER (i.e. FIT) or SEU have been listed to analyze accumulated neutron effect, see *Table 5*.

Table 5: The reported SER (or SEU) values of SRAM products influenced by various neutron fluence conditions

Products	fluence	SER (SEU)	Vendor Ref.
0.6um, 0-11MeV, 14MeV	(5-86)×10^9	0	Xilinx 1998
FPGA 0.35um, 100 MeV	0.3×10^9	(5)	Xilinx 1998
1Gb 130nm@3889m	3.69×10^6	(4000)	[2] 2012
2x1Gb 90nm@2885m	3.5×10^6	(3000)	[2] 2012
16Mb 65nm, 20-250MeV	(2-3)×10^9	838	Cypress 2017
Virtex-6 RAM, 40nm	~1×10^9	83	Xilinx 2009

Discussion

The dominated neutron interactions are by collisions with nuclei, leading to scattering and absorption (capture or fission). The neutron radiation field encompasses secondary particles such as photons, electrons and heavy charged ions that can induce secondary ionizing radiation doses. In

current study, *the energy, fluence, the (macroscopic) cross section* and *the mean free path* are used to evaluate impacts on elements/compounds in CMOS. Calculated results of present work show that temperature has only <20% (*Table 2*) impact with respect to Σ_{comp} and λ that have maximum impact factors of ~2 (*Table 3*) at 10 MeV, while neutron energy has an impact factor of 2-4 (*Table 4*). Published studies (*Table 5*) indicate that without design protection in SRAM, at natural (low flux) low fluence of 10^6 n/cm^2 [2] and experimental (high flux) high fluence of 10^9 n/cm^2 (Cypress 2017 and Xilinx 2009 data), SEU (SER) exhibits a concern.

Reliability is atop critical in high-end applications (ADAS and high-performance computing, HPC). In most advanced application for space flight, electronic components in avionic-critical systems, are exposed to high levels of gamma and neutron radiation for repeated or extended periods of time, they must be designed and hardened (RadHard) to continue to function. Neutrons also have applications in medicine (*e.g.* boron neutron capture therapy, BNCT), security (^{3}He neutron detection at border monitoring and crossings), and materials research (using neutron spectroscopy). In neutron facilities and nuclear events such as in a nuclear disaster, reliable CMOS based electronic equipment resistant to radiation is highly required. In CDM of ESD study on CMOS gates [4], the protection scheme was suggested to be combined with the secondary charged particles from neutrons in future study.

REFERENCES

[1] G. Hubert, L. Artola, D. Regis, "Impact of scaling on the soft error sensitivity of bulk, FDSOI and FinFET technologies due to atmospheric radiation", *Integration, the VLSI Journal*, **50**, 39-47, 2015.

[2] G. Hubert et.al, "Continuous high-altitude measurements of cosmic ray neutrons and SEU/MCU at various locations: correlation and analyses based-on MUSCA SEP3," RADECS 2012 Proceedings.

[3] Microsemi, "Neutron-induced Single Event Upset (SEU)", https://www.microsemi.com/

[4] C.-Z. Chen and D.Y. Hu, "Geometry Effect with Respect to ESD and Radiative Charged Particles in SoC", Semiconductor Technology International Conference (CSTIC,March 12-13, 2017, Shanghai), **IEEE Conference Publications,** http://ieeexplore.ieee.org/document/7919894/

[5] C.-Z. Chen, G. Randers-Pehrson, S.A. Marino, and C.R. Geard, "Development of 'soft' neutron irradiation facility," In *1993 Annual Report* (Center for Radiological Research), P. Kliauga, ed., Columbia University, pp.195-197.

[6] M. Ohlsson, P. Dyreklev, K. Johansson, P. Alfke, "Neutron Single Event Upsets in SRAM-Based FPGAs", Radiation Effects Data Workshop, 1998. IEEE Xplore: 06 August 2002. DOI: 10.1109/REDW.1998.731500

[7] P.Rinard, "Neutron Interactions with Matter," Los Alamos Technical Report, p357-377, 1997.

[8] R.W. Bauer, J.D. Anderson, S.M. Grimes, V.A. Madsen, "Application of Simple *Ramsauer Model* to Neutron Total Cross Sections," *International conference on nuclear data for science and technology*, Trieste (Italy), 19-24 May, 1997.

[9] A. V. Gurevich *et al.*, "Observations of high-energy radiation during thunderstorms at Tien-Shan", Phys. Rev. D, **94**, 023003, 2016.

[10] J.F. Ziegler, "Terrestrial cosmic rays", IBM J. Res. Develop., **40**(1), Jan. 1996.

[11] A. Tabor and E. Normand, "Single Event Upsets in Avionics", IEEE Trans. Nuclear Science, **40**(2): 120-126, 1993.

[12] Neutron flux, https://en.wikipedia.org/wiki/Neutron_flux

[13] JANIS, http://www.oecd-nea.org/janis/

AN IMPROVED S-BOX OF LIGHTWEIGHT BLOCK CIPHER ROADRUNNER FOR HARDWARE OPTIMIZATION

Juhua Liu[1] and Wei Li[2] and Guoqiang Bai[3]*

[1]Institute of Microelectronics, Tsinghua University, Beijing, China
Liujh14@mails.tsinghua.edu.cn
[2] Institute of Microelectronics, Tsinghua University, Beijing, China
[3] Tsinghua National Laboratory for Information Science and Technology
*Corresponding Author's Email: baigq@tsinghua.edu.cn

Abstract

Recently, the need for designing block ciphers targeting on resource constrained device has been repeatedly expressed by professionals. Currently, Roadrunner is one of the most software efficient lightweight ciphers. Its security is provable against linear and differential attacks. In this paper, we design a novel S-Box layer, in order to improve the performance for hardware implementation. Furthermore, the new S-Box layer has eight different 4-bit S-Boxes, which are decomposable. We decompose each 4-bit S-Box by exploiting affine transformations and a single shared quadratic permutation. After being decomposed, these eight 4-bit S-boxes can be merged into one component, sharing the same quadratic permutation. In this way, we can save the area consumption by 51%, compared to the traditional implementation of old S-box layer by using look up tables (LUTs).

Keywords

Lightweight, novel S-Box, hardware implementation, Roadrunner, area-optimization

INTRODUCTION

Over the last few decades, a series of lightweight block ciphers has been proposed, such as Roadrunner[1], Lblock [2], Present [3],and so on. Amongst all these lightweight ciphers, Roadrunner, as proposed in LightSec 2015, is designed to achieve high efficiency in software implementation. At the same time, Roadrunner has a relatively high security margin in contrast to most light weight ciphers [1]. Its security is provable against differential and linear attacks [1].

In Lblock [2], the author used 8 different 4-bit S-Boxes to build up the S-Box layer. Besides, these 8 S-Boxes are chosen carefully to fulfill the following conditions: no fix point, completed, best non linearity, best differential probability, and good algebraic order etc [2]. However, in Roadrunner [1], one single 4-bit S-Box is used eight times to compose the S-Box layer. Since each of these 8 4-bit S-Boxes has the same level of security, apparently, the S-Box layer in Lblock is much safer than the one in Roadrunner. Nevertheless, both of them didn't take the hardware implementation efficiency into account.

In this paper, we proposed a novel S-Box layer to improve the hardware performance of Roadrunner. This S-Box layer consists of eight different optimal 4-bit S-Boxes, which fulfill the natural requirement to be resistant against linear and differential cryptanalysis [4]. According to the notion of linear equivalent S-Boxes [5], these eight S-Boxes belong to 2 different S-Box classes and can be individually decomposed into affine transformations and a single shared quadratic permutation. Thereby this sharable quadratic permutation can be reused in an iterative manner to help many various S-Boxes merged into one component and thus reduce the resource overhead [4].

DESCRIPTION OF ROADRUNNER

Roadrunner is a kind of lightweight block cipher, it relies on a Feistel-type structure, which is shown in Figure 1. The block length of Roadrunner is 64-bit and two key sizes of 80 and 128 bits are supported. 10 rounds are required by 80-bit key and 128-bit key requires 12 rounds. It employs a variant Feistel structure where whitening keys (WK0 and WK1) are used to XOR the left part of the state in the first and last round. There is no swap operation in the final round.The decryption algorithm of Roadrunner is the reverse of encryption procedure where the order of whitening keys, round keys and constants are used reversely [6]. Let $x_0\|x_1\|x_2\|x_3\|x_4\|x_5\|x_6\|x_7$ denote a 64-bit plaintext. The left input data and the right input data of each round F function are $x_0\|x_1\|x_2\|x_3$ and $x_4\|x_5\|x_6\|x_7$ respectively. The function F also takes the Ci as the constant and 96-bit round key. The output of F function will be XORed with $x_4\|x_5\|x_6\|x_7$. F is a 4-round substitution-permutation network (SPN) type function as shown on top right of Figure 1, where SLK is a consecutive application of S-Box layer (S), diffusion layer (L), and key XOR operation (K), as illustrated on the bottom right of Figure 1. The last function S takes the same S-Box in the SLK. After the second SLK function, round constant is XORed with the rightmost byte of the state, which is x3 exactly. For the round $i = 0,1,...,NR-1$, the round constant is $C_i = NR - i$, where NR is the

978-1-5386-5309-8/18 $31.00 © 2018 IEEE

number of rounds, and C_i is represented as 8-bit little endian integer, that is $10 = 00001010$, $11 = 00001011$, etc [6]. Because of the limit of page number, the detail of key schedule and diffusion layer is omitted. Please refer to the original paper[1].

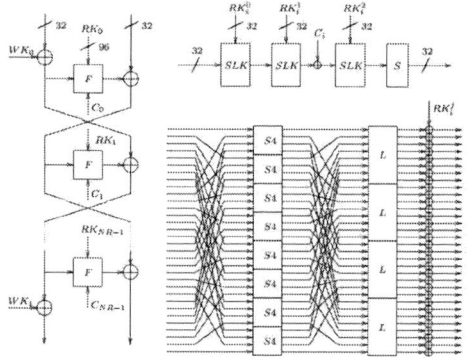

Figure. 1. The algorithmic description of Roadrunner. Feistel structure on left, F function in the top-right corner, and SLK function in the bottom-right corner[1].

S-Box layer

The S-Box layer consists of eight 4×4 S-Boxes in parallel. The S-Box used in Roadrunner is also a 4×4 S-Box, the action of this box in hexadecimal notation is given by Table 1.

Table 1. The actions of S-Box

input	0	1	2	3	4	5	6	7
output	0	8	6	D	5	F	7	C
input	8	9	A	B	C	D	E	F
output	4	E	2	3	9	1	B	A

DECOMPOSITION OF S-BOXES
Optional 4-Bit S-boxes

A 4-bit S-Box is considered as cryptographically optimal, if it fulfills the natural requirement to be resistant against linear and differential cryptanalysis as best as possible [4]. According to the definition in [5], an optimal 4-bit S-Box has the following properties.

Definition 1. Let $S : F2^4 \rightarrow F2^4$ be an S-Box. If $S(\cdot)$ fulfills the following conditions we call $S(\cdot)$ an optimal S-Box:

1. $S(\cdot)$ is a bijection
2. $Lin(S(\cdot)) = 8$
3. $Diff(S(\cdot)) = 4$

The values $Lin(S(\cdot))$ and $Diff(S(\cdot))$ are measures for the resilience of an S-Box against linear attacks and differential attacks, respectively.

Linear Equivalence

The authors of [5] categorize all possible optimal 4-bit S-Boxes into 16 classes of non-equivalent optimal S-Boxes. If an affine transformation is applied on the S-Box, resistance against linear attacks $Lin(S(\cdot))$ of an S-Box and the resistance against differential attacks $Diff(S(\cdot))$ of an S-Box doesn't change. This linear equivalence between two optimal S-Boxes can be expressed formally by Definition 2 [4]:

Definition 2. Two S-Boxes $S(\cdot)$, $S'(\cdot)$ are linear equivalent if there exist two 4×4-bit invertible matrices A, B and two 4-bit vector c, d such that
$$S(\cdot) = A(S'(Bx+c)+d), \quad \forall x \in 0,1,...,15.$$

Based on Definition 2 two linear equivalent S-Boxes $S(\cdot)$, $S'(\cdot)$ with the same properties can be transformed into each other and thus belong to the same categorization of S-Box class [4]. The affine 4×4-bit transformation matrices will be noted in a hexadecimal form from now on:

$$A = \begin{pmatrix} 1 & 0 & 1 & 0 \\ 0 & 1 & 0 & 0 \\ 1 & 0 & 0 & 0 \\ 1 & 0 & 1 & 1 \end{pmatrix} = \begin{pmatrix} a \\ 4 \\ 8 \\ b \end{pmatrix} = \begin{pmatrix} 0 & x & b & 8 & 4 & a \end{pmatrix} \quad (1)$$

Decomposition of Optimal 4-Bit S-boxes

The reason to decompose 4-bit S-Box into a combination of quadratic and linear Boolean functions is to reuse the sharable quadratic permutation for the hardware implementation of the S-Box layer containing 4-bit S-Boxes from different S-Box classes [4]. A composition of an optimal 4-bit S-Box is given by the following definition.

Definition 3. If a vectorial Boolean function such as a permutation $S(\cdot)$ can be written as a composition of several lower degree vectorial Boolean functions $f_1(\cdot)$, $f_2(\cdot)$, ... , $f_n(\cdot)$ i.e $S(\cdot) = f_n(...f_2(f_1(\cdot)))$, then $f_1(\cdot)$, $f_2(\cdot)$, ... , $f_n(\cdot)$ is called the decomposition of $S(\cdot)$. If the order of Boolean function is 2 degree, then wo call this S-box(permutation) quadratic permutation. After all these 4-bit S-Boxes have been decomposed into different affine transformation and the same quadratic permutation, we can reuse the sharable quadratic permutation to construct a 32-bit permutation. Actually, all 4-bit S-Boxes of the S-Box classes 0,1,2,4,5,7,8,13 out of the 16 classes defined in [5] are sharable and form a cluster of S-Box classes referred to as group A_{16}.

Decomposition of all S-boxes of A16 with One Quadratic Permutation

The same quadratic permutation can be used for sharing the 4-bit permutation to construct a 32-bit S-Box layer. In paper [4], a quadratic permutation is selected to reuse in an iterative manner to reduce resource overhead. An arbitrary S-Box can be constructed by exploiting a consecutive execution of an appropriate quadratic permutation G' so that $S(\cdot) = G'(G'(\cdot))$. Paper [4] came up with a special decomposition structure in order to further generalize the idea of sharing a basic quadratic 4-bit permutation as a non-linear mapping core for sharable S-Boxes of A_{16}:

$$S(\cdot) = f_n(\ldots f_2(f_1(\cdot))) = M_n \cdot G(M_{n-1}G(\ldots M_1(G(\cdot)))) \quad (2)$$

where $M_n, \ldots, M_0$ are invertible matrices and G is a quadratic, sharable permutation of A_{16}. So each S-Box of the group A_{16} can be composed by a sequence of other S-Boxes, which means that each S-Box of the group A_{16} can be constructed by the core permutation G and the affine transformation, by using appropriate core permutation G listed in Table 2. Paper [4] gives the construction of each S-Box belonging to one of the classes within the group A_{16} by using the decomposition description of Eq.(2) with an appropriate transposition matrix M^{cls} and G as given in Table 3 [4].

Table 2 Core Permutation G

x	0	1	2	3	4	5	6	7
$G(x)$	0	4	1	5	2	15	11	6
x	8	9	10	11	12	13	14	15
$G(x)$	8	12	9	13	14	3	7	10

Table 3 S-Box representation of A_{16} using core G.

S-Box class	Transformation matrix	Transformation to the S-Box S(x)
0	$M^{cls0} = 0x1249$	$S(\cdot) = G(M^{cls0}G(\cdot))$
1	$M^{cls1} = 0x1248$	$S(\cdot) = G(M^{cls1}G(\cdot))$

NEW S-BOX LAYER FOR ROADRUNNER

Primarily in order to reduce the resource overhead, in this section we propose a new S-Box layer for Roadrunner. We suggest to apply these 2 most simply constructed S-Box classes, which are class 0 and class 1. Then we choose 4 sets of satisfactory parameter A, B, c, d for each class by random to construct the new S-Box layer. All the parameters for affine transformation form class 0 and 1 are given in the appendix. In this way, this new S-Box layer is composed by eight different 4-bit S-Boxes and sharing the same core permutation G. By employing the proposed decomposition structure in Table 3, the core permutation G can be reused and these eight 4-bit S-Boxes can be merged into one component.

HARDWARE IMPLEMENTATION AND RESULTS

Figure 2 depicts the hardware circuit design of the proposed merged S-Box layer. Eight different affine transformations and one same quadratic permutation are used. Because the G permutation should be used twice for producing one 4-bit S-Box, we put eight 4-bit multiplexers in front of it. And to store intermediate data, we used eight 4-bit registers. We note that the M^{cls1} is unit matrix and thus, the matrix multiplication of M^{cls1} can be described as simple wires in hardware directly. In this way, the

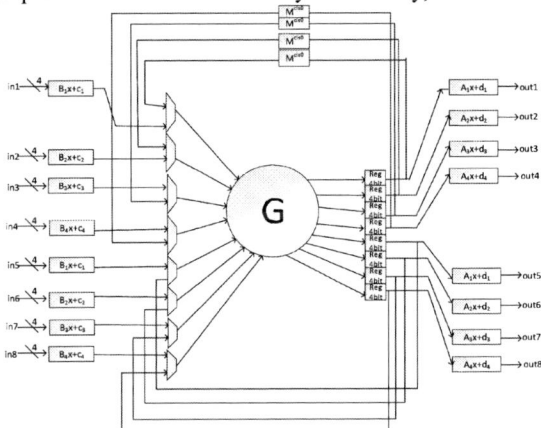

Figure 2 Design of the merged Roadrunner S-Box layer implementation.

quadratic permutation is reused eight times so that a bunch of area resource is saved. Besides, these affine transformations are just linear calculation, which won't cost much area. One might argue that in such a solution, additional registers are required for the storage of temporary state. However, this is not the case, since the registers are already in part of the cipher because of the round based operation. Therefore, we won't list the area consumption for registers in implementation results. Please also note that we didn't give the control logic circuit, because what we are concerned about is the core computation and the control logic can be merged into the bigger design when the whole cipher is implemented. For comparison reasons, we also gave the result of implementation of the S-box layer using LUTs.

We implemented the proposed design in Verilog-DHL and synthesized it on $0.18um$ CMOS technology to check its hardware complexity. In order to compare the area requirements independently it is common to state the area as gate equivalents(GE). One GE is equivalent to the area which is required by the two-input NAND gate with the lowest driving strength of the corresponding technology. The area in GE is derived by dividing the area in um^2 um2 by the area of a two-input NAND gate .

Specifically, in the above implementation, the area requirement is mostly occupied by G permutation, which costs 13.6 GE. Each affine transformation costs around 3.3 GE. Multiplexers also cost a big share of area resource. one 1-bit alternative multiplexer costs 2.6 GE. Here, we need eight 4-bit alternative multiplexers, which in total are 83.2 GE. Table 4 presents the area requirement of proposed S-Box layer. A comparison with the straight implementation using the traditional way of LUTs follows in Table 5.

Table 4 Area requirement of proposed S-Box layer

module	GE	module	GE
G	13.6	1-bit multiplier	2.6
$B_1 x + c_1$	3.3	$A_1 x + d_1$	3.3
$B_2 x + c_2$	3.3	$A_2 x + d_2$	2.6
$B_3 x + c_3$	3.3	$A_3 x + d_3$	3.3
$B_4 x + c_4$	3.3	$A_4 x + d_4$	2.6
		sum	124

Table 5 Area requirement using LUTs

module	GE	module	GE
S_1	23.6	S_5	21.2
S_2	25.2	S_6	31.6
S_3	28.6	S_7	26.9
S_4	26.9	S_8	21.9
sum	251		

The proposed design costs 124 GE, while the LUTs consumes 251 GE. By comparison, we can save up to 51% area resource, through using the proposed method to implement the new S-box layer. This decreased size could help for area limited hardware implementation, e.g., smart cards and to allow more copies of the S-box for parallelism or pipelining of Roadrunner [7].

CONCLUSION

In this paper, we have designed, synthesized and implemented a novel S-box layer consisting of 8 different S-boxes These 8 S-boxes can be decomposed into different affine transformations and a sharable quadratic permutation. By reusing the same quadratic permutation, we proposed a new circuit architecture instead of conventional LUTs. And the hardware implementation of this proposed design requires about 124 GE on $0.18um$ technology. Furthermore, for comparison reasons, we also implement the new S-box layer using LUTs, which costs 251 GE. Hence the consumption is reduced by up to 51% in the S-box layer. The implementation result on ASIC confirms that the new proposed S-box layer design is efficient and economic in hardware and suitable for low resource applications. In the current state, we only give a possibility for the composition of the S-box layer. Noticing that there are 8 different S-box classes and thousands of combination for affine transformations in paper [4], which makes designing other S-box layers possible. How to design new S-box layer for different application situations are left for future work.

ACKNOWLEDGMENT

This work is supported by the National Natural Science Foundation of China (Grants 61472208) and National Key Basic Research Program of China(Grant 2013CB338004).

REFERENCES

[1] A. Baysal and S. Şahin, "Roadrunner: A small and fast bitslice block cipher for low cost 8-bit processors," in International Workshop on Lightweight Cryptography for Security and Privacy. Springer, 2015, pp. 58–76.

[2] W. Wu and L. Zhang, "Lblock: a lightweight block cipher," in Applied Cryptography and Network Security. Springer, 2011, pp. 327–344.

[3] A. Bogdanov, L. R. Knudsen, G. Leander, C. Paar, A. Poschmann, M. J. Robshaw, Y. Seurin, and C. Vikkelsoe, "Present: An ultra-lightweight block cipher," in CHES, vol. 4727. Springer, 2007, pp. 450–466.

[4] S. Kutzner, P. H. Nguyen, A. Poschmann, and M. Stöttinger, "Minimizing s-boxes in hardware by utilizing linear transformations," in International Conference on Cryptology in Africa. Springer, 2014, pp. 235–250.

[5] G. Leander and A. Poschmann, "On the classification of 4 bit s-boxes," Arithmetic of Finite Fields, pp. 159–176, 2007.

[6] J. Liu, G. Bai, and X. Wu, "Efficient hardware implementation of roadrunner for lightweight application," in Trustcom/BigDataSE/I SPA, 2016 IEEE. IEEE, 2016, pp. 224–227.

[7] D. Canright, "A very compact s-box for aes," in International Workshop on Cryptographic Hardware and Embedded Systems. Springer, 2005, pp. 441–455.

AFFINE TRANSFORMATION PARAMETERS

$$A_1 = \begin{pmatrix} 1 & 0 & 0 & 1 \\ 0 & 1 & 0 & 0 \\ 0 & 0 & 1 & 0 \\ 0 & 0 & 0 & 1 \end{pmatrix}, B_1 = \begin{pmatrix} 1 & 0 & 1 & 0 \\ 0 & 1 & 0 & 0 \\ 0 & 0 & 1 & 0 \\ 0 & 0 & 0 & 1 \end{pmatrix}, c_1 = (0\ 0\ 0\ 0), d_1 = (0\ 0\ 1\ 0), \quad (3)$$

$$A_2 = \begin{pmatrix} 1 & 0 & 0 & 0 \\ 0 & 1 & 0 & 1 \\ 0 & 0 & 1 & 0 \\ 0 & 0 & 0 & 1 \end{pmatrix}, B_2 = \begin{pmatrix} 1 & 0 & 0 & 0 \\ 0 & 1 & 1 & 0 \\ 0 & 0 & 1 & 0 \\ 0 & 0 & 0 & 1 \end{pmatrix}, c_2 = (0\ 0\ 1\ 0), d_2 = (0\ 1\ 0\ 0), \quad (4)$$

$$A_3 = \begin{pmatrix} 1 & 0 & 0 & 0 \\ 0 & 1 & 0 & 0 \\ 0 & 0 & 1 & 1 \\ 0 & 0 & 0 & 1 \end{pmatrix}, B_3 = \begin{pmatrix} 1 & 0 & 0 & 0 \\ 0 & 1 & 0 & 0 \\ 0 & 1 & 1 & 0 \\ 0 & 0 & 0 & 1 \end{pmatrix}, c_3 = (0\ 1\ 0\ 0), d_3 = (1\ 0\ 0\ 0), \quad (5)$$

$$A_4 = \begin{pmatrix} 1 & 0 & 0 & 0 \\ 0 & 1 & 0 & 0 \\ 0 & 0 & 1 & 0 \\ 0 & 0 & 1 & 1 \end{pmatrix}, B_4 = \begin{pmatrix} 1 & 0 & 0 & 0 \\ 0 & 1 & 0 & 0 \\ 0 & 0 & 1 & 0 \\ 0 & 1 & 0 & 1 \end{pmatrix}, c_4 = (1\ 0\ 0\ 0), d_4 = (0\ 0\ 0\ 1). \quad (6)$$

CERAMIC INTERCONNECT BRIDGE FOR HETEROGENEOUS INTEGRATION AND MULTIPLE CHIP PACKAGING

Boping Wu, *Senior Member, IEEE*
Huawei Research, Beijing, China
bennywu@ieee.org

ABSTRACT

In this paper, we describe the architecture and performance of a fine pitch multiple chip heterogeneous integration solution using ceramic interconnect bridge on organic substrate package. We present the increased IO density and improvement of electrical high-speed performance on signal integrity are achievable through this novel integration scheme, where dense copper routings on small ceramic elements are served as interconnect bridges. The cost and signal attenuation of using ceramic bridge is far better than that of silicon bridge or wafer interposer on substrate.

INTRODUCTION

The performance gain on semiconductor technologies in the speed of Moore's Law scaling is gradually slowing down. Additionally, the CMOS circuit design rules are becoming more complex, and the mask cost of next advancing node is ever increasing. So advancements in the package technologies are required to open another dimension of high level chip integration, in order to meet the future bandwidth, space- and energy-efficient demands of powerful ICT systems [1]-[3].

System integration with high speed data transmission often demands packaging solution having fine line capability in order to deliver the desired performances with better power and signal integrity. However, while wafer fabrication is advancing at a relentless pace, IC substrate technology has not been able to catch up device's fine feature needs at a reasonable cost per functionality unit. This is mainly due to the substrate material poor mechanical stability and equipment limitations of current manufacturing infrastructure. As such, novel IC wafer interposers such as silicon and glass slices have been investigated intensively with their specific material properties [4]-[11].

ARCHITECTURE

We introduce the architecture and performance of an innovative fine pitch multi-chip integration scheme using ceramic interconnect bridge instead of silicon bridge, as shown in Fig. 1. Compared to alternative interposer approaches, ceramic bridge offers competitive price margin, mechanical reliability, high yield and better electrical performance. For example, ceramic bridge could support larger inter-chip gap than that of wafer-level

chip-scale integrated fan-out (InFO) or quilt packaging. The material cost of using ceramic bridge is at least 30% cheaper than silicon material. The electrical loss tangent, also called energy dissipation factor, of ceramic dielectric is 2x smaller than the organic/glass material and 20x smaller than the lossy silicon wafer. The relative permittivity (dielectric constant) of ceramic is usually above 8. By these physical natures, ceramic bridge leads to smaller signal attenuation and narrower trace for the same impedance target. The aggressive z-height thin profile and design freedom of ceramic bridge is also favorable than that of a complete large panel of interposer, or build-up laminate, or package-on-package (PoP) solution [12]-[16]. Therefore, ceramic bridge has positioned itself well enough as a flexible 2.x-D IC packaging solution for multiple chip heterogeneous integration among CPU and DRAM, GPU, FPGA or even communication devices (e.g. wireless and optical transceivers) for various cutting-edge high-end applications. [17]-[20]

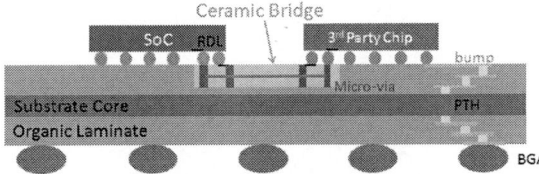

Figure 1: Multiple chip packaging using ceramic interconnect bridge embedded in organic substrate.

TABLE I. KEY ATTRIBUTES OF CERAMIC BRIDGE DESIGN

Features	Specification
Embedded ceramic size	max 10 mm per XY-side
Embedded ceramic thickness	20~60 μm (2~3 layer RDL)
Cu line/space on ceramic	min 3/4 μm
Cu roughness & thickness	< 0.3 μm rms & min 3μm
Ceramic dielectric properties	Dk = 8, Df = 0.0009
Cu conductivity	4.31×10^7 S/m @ 90 °C
Bump dimensions (min.)	10 μm in radius, 30 μm pitch
Micro-via in ceramic (min.)	2 μm in radius, 10 μm pitch
Insertion loss on ceramic (per mm)	0.1dB @ 28Gbps NRZ 0.2dB @ 56Gbps NRZ
Impedance target	50Ω SE / 90Ω DE (<10% tol.)
Latency	< 2 nSec.
Bandwidth density	> 800 Gbps/mm
IO Support	Both serial and parallel bus

Table I concludes the major package attributes

978-1-5386-5309-8/18 $31.00 © 2018 IEEE

including the fine features and superior electrical performance on high-K ceramic material. Ceramic dielectric and dense patterning traces in Fig. 2 provide a significant IO bandwidth increase of package interconnect as well as the reduction of through package-core signal paths, which are the nightmare of impedance mismatch for high speed communications. By incorporating a redistribution layer (RDL) process flow with leeway routings to re-align mixed pitches of assembly bumps, there is a good opportunity for yield improvement and fabrication cycle-time reduction of multiple chip integration.

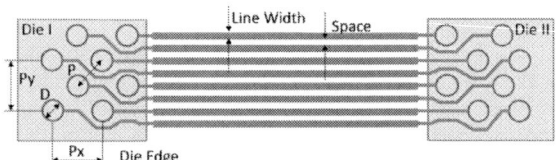

Figure 2: Dense fine line feature on high-K ceramic interconnect.

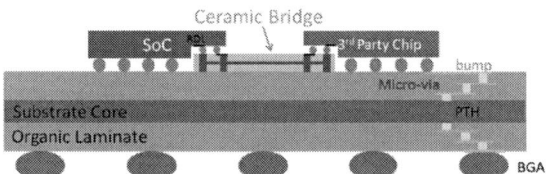

Figure 3: Multi-chip interconnects using ceramic bridge inserted between two chips.

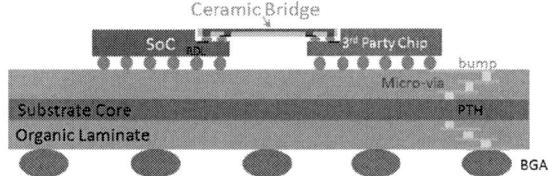

Figure 4: Multi-chip interconnects using ceramic bridge as top quilt between two chips.

Actually, the ceramic bridge does not have to be embedded in the cavity of organic substrate. For connecting multi-chip purpose, it could be either seated in between the inter-chip gap as Fig.3, or suspended in the air like the top quilt in Fig.4. These alternative methods could avoid milling-out process of organic packaging substrate by adding extra efforts on wafer post-processing and assembly precision.

RESULTS AND DISCUSSIONS

From the signal integrity viewpoint, the ceramic technology could provide ultra low loss of signal attenuation as shown in Fig. 5 and minimized noise coupling in Fig. 6. By assigning proper signal-to-ground co-planar inter-lane referencing, we can well control the

mutual/self capacitance. The thickness of the dielectrics also need fine tunings for low loop inductance in the power delivery network (PDN). Thus, this novel packaging architecture could support both serial and parallel signaling busses. It can be generally used in both low voltage driving mode and low current mode.

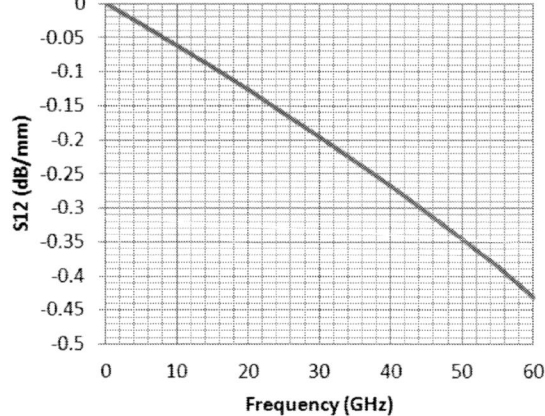

Figure 5: Simulated single-ended insertion loss of Cu trace on ceramic interconnect bridge.

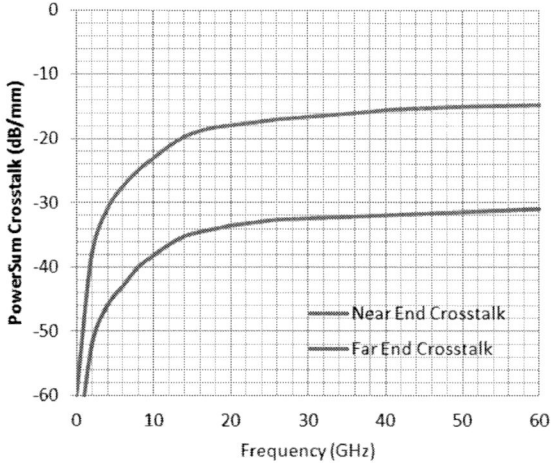

Figure 6: Simulated powersum crosstalk for neighboring traces on ceramic interconnect bridge.

In order to maintain the good electrical performance of the Gb/s digital signaling, the ceramic bridge should allow abundant data passing its entire characteristics while keeping the high resolution of copper interconnects. A full-wave electromagnetic simulation of an optimized 3D layout is given. The design meets the specification targets and outperforms other alternative interconnect technologies. We are now investigating a stitching small input-output (SSIO) interface specifically designed for this kind of ultra short reach and stitching chip links. It provides more configurability compared to the

978-1-5386-5309-8/18 $31.00 © 2018 IEEE

standardized on-package IO. Gladly to see, the low power consumption and low latency is physically achievable in this novel heterogeneous integration scheme. The total IO bandwidth density could overshoot 800 Gbps per mm.

SUMMARY

An innovative multi-chip packaging technology has been developed that extends the physical and electrical performance of traditional organic laminate structure. Ceramic bridge incorporates the advanced feature of fine line and low loss of interconnect dielectrics; altogether becoming the favorable solution choice of heterogeneous integration of multiple wafer chips. In comparison to other approaches of packaging ASIC with different devices, ceramic bridge technology exhibits enhanced electrical signal integrity, superior IO density, aggressive form factor, and intrinsic mechanical reliability. The high yield and design flexibility could also be translated into economic cost benefit.

This novel interconnect bridge method could potentially integrate heterogeneous chips including different lithographic nodes, different IC sizes, different wafer fabs and different chiplet functionalities. In order to revive and enrich the Moore's Law with new level of integration, the packaging fabrication and assembly providers need re-tooling and re-development to catch up with this hybrid interconnect technology, which enables the OSAT service into the sub-micron fine-pitch capability. It will open and expand the emerging market of diversified multi-chip integrations for various as-yet unimagined future applications.

ACKNOWLEDGEMENTS

The author would like to thank managers and colleagues at Huawei Technologies for funding and supporting this work.

REFERENCES

[1] D.L.K. Eng et al., "Heterogeneous packaging of organic electro-optic modulators with RF substrates," IEEE Photon. Tech. Lett., vol.28, no.6, pp.613-616, Mar. 2016.

[2] K.K. Samanta, "Pushing the envelope for heterogeneity: multilayer and 3-D heterogeneous integrations for next generation millimeter- and submillimeter-wave circuits and systems," IEEE Microw. Mag., vol.18, no.2, pp.28-43, Mar. 2017.

[3] D.S. Green et al., "A revolution on the horizon from DARPA: Heterogeneous integration for revolutionary microwave/millimeter-wave circuits" IEEE Microw. Mag., vol.18, no.2, pp.44-59, Mar. 2017.

[4] X. Gu, B. Wu, M. Ritter and L. Tsang, "Efficient full-wave modeling of high density TSVs for 3D integration," in Proc. IEEE Electron. Comp. Tech. Conf. (ECTC), Las Vegas, 2010, pp.663-666.

[5] C.L. Lung et al., "Through-silicon via fault-tolerant clock networks for 3-D ICs," IEEE Trans. CAD of ICs & Syst., vol.32, no.7, pp.1100-1109, July 2013.

[6] U.R. Tida, C. Zhuo, and Y. Shi. "Novel through-silicon-via inductor-based on-chip DC-DC converter designs in 3D ICs," ACM J. Emerg. Technol. Comput. Syst. vol.11, no.2, art.16, Nov. 2014.

[7] N.C. Chen et al., "A novel system in package with fan-out WLP for high speed SERDES application," in Proc. IEEE Electron. Comp. Tech. Conf. (ECTC), Las Vegas, 2016, pp.1495-1501.

[8] T. Lu et al., "Heterogeneous microwave and millimeter-wave system integration using quilt packaging," in Proc. IEEE MTT-S Int. Microw. Symp. (IMS), San Francisco, 2016.

[9] R. Mahajan et al., "Embedded multi-die interconnect bridge (EMIB) - A high density, high bandwidth packaging interconnect," in Proc. IEEE Electron. Comp. Tech. Conf. (ECTC), Las Vegas, 2016, pp.557-565.

[10] B. Wu, "High-bandwidth IC interconnects with silicon interposers and bridges for 3D multi-chip integration and packaging," in Proc. China Semicon. Tech. Int. Conf. (CSTIC), Shanghai, 2017.

[11] S. Jangam et al., "Latency, bandwidth and power benefits of the SuperCHIPS integration scheme," in Proc. IEEE Electron. Comp. Tech. Conf. (ECTC), Orlando, 2017, pp.86-94.

[12] C. Zwenger et al., "Electrical and thermal simulation of SWIFT high-density fan-out PoP technology," in Proc. IEEE Electron. Comp. Tech. Conf. (ECTC), Orlando, 2017, pp.1962-1967.

[13] S.Y. Hou et al., "Wafer-level integration of an advanced logic-memory system through the second-generation CoWoS Technology," IEEE Trans. Electron. Devices, vol.64, no.10, pp.4071-4077, Oct. 2017.

[14] B. Banijamali, S. Ramalingam, N. Kim and C. Wyland, "Ceramics vs. low-CTE organic packaging of TSV silicon interposers," in Proc. IEEE Electron. Comp. Tech. Conf. (ECTC), Lake Buena Vista, 2011, pp.573-576.

[15] C. Lin, N. Wang and J. Tan, "COOL substrate for 2.5D assembly," in Proc. IEEE Electron. Packag. Tech. Conf. (EPTC), Singapore, 2015.

[16] N. Chujo et al., "LTCC package for high-bandwidth logic to memory interconnection," in Proc. IEEE Electrical Design Adv. Pkg. Syst. Symp. (EDAPS), Seoul, 2015, pp.5-8.

[17] S. Kariyazaki, K. Kuboyama, R. Oikawa and T. Funaya, "New signal skew cancellation method for 2 Gbps transmission in glass and organic interposers to achieve 2.5D package employing next generation high bandwidth memory (HBM)," in Proc. Int. 3D Syst. Integr. Conf. (3DIC), Sendai, 2015, pp.TS2.3.1-5.

[18] L. Li et al., "3D SiP with organic interposer for ASIC and memory integration," in Proc. IEEE Electron. Comp. Tech. Conf. (ECTC), Las Vegas, 2016, pp.1445-1450.

[19] Y. Uematsu, N. Ushifusa and H. Onozeki, "Electrical transmission properties of HBM interface on 2.1-D system in package using organic interposer," in Proc. IEEE Electron. Comp. Tech. Conf. (ECTC), Orlando, 2017, pp.1943-1949.

[20] B. Wu, B. Brown and E. Warner, "High-density interconnect board design for wafer-level packaging," IET Electron. Lett., vol.47, no.20, pp.1137-1138, Sept. 2011.

Auto-Generation of Pipelined Hardware Designs for Polar Encoder

Zhiwei Zhong[1,2], Xiaohu You[2], and Chuan Zhang[1,2,*]

[1]Lab of Efficient Architectures for Digital-communication and Signal-processing (LEADS)
[2]National Mobile Communications Research Laboratory, Southeast University, Nanjing, China
Email: {zwzhong, xhyu, chzhang}@seu.edu.cn

Abstract—**This paper presents a general framework for auto-generation of pipelined polar encoder architectures. The proposed framework could be well represented by a general formula. Given arbitrary code length N and the level of parallelism M, the formula could specify the corresponding hardware architecture. We have written a compiler which could read the formula and then automatically generate its register-transfer level (RTL) description suitable for FPGA or ASIC implementation. With this hardware generation system, one could explore the design space and make a trade-off between cost and performance. Our experimental results have demonstrated the efficiency of this auto-generator for polar encoder architectures.**

Index Terms—**Polar encoder, pipelined architecture, hardware auto-generation, high-level synthesis.**

I. INTRODUCTION

Polar code [1], the first channel code which can provably achieve the capacity of the binary-input discrete memory-less channels (BDMCs), has been considered as the recent breakthrough of coding theory. Recently, polar code has been adopted by the enhanced mobile broadband (eMBB) control channels for the 5G NR interface. As pointed out by [1], to achieve a good error-correcting performance of polar code, the code length is expected to be sufficiently long. However, as for polar code, the hardware complexity of fully parallel encoder will be high as the code length increases. Therefore, pipelined architecture should be introduced to reduce the hardware cost. Using folding transformation [2], [3] has proposed both feed-forward and feed-back polar encoder with 2-parallel process-ing; [4] has proposed pipelined polar encoder architecture with 4-parallel processing. Although [4] has claimed that the folding transformation could derive polar encoder with any level of parallelism, the detailed framework is not given.

In synthesizing hardware architectures for an N-bit polar encoder, different level of parallelism leads to different latency, throughput, silicon area and memory cost. Intuitively, the level of parallelism M suitable for an N-bit polar encoder should be $2 \leqslant M \leqslant N/2$, where M is a power of two. Thus, as the code length increases, there will be more choices of M and the design space will be wider. Therefore, it will be exhausting to choose the optimal values of N and M under different hardware constraints.

In order to fulfill the requirements of different applications, a auto-generator which can connivently output polar encoder architecture with given code length N and parallelism M is highly expected. Also, this auto-generator can free the hardware designers from the laborious case designs, bypass the hardware details, and give the design space in a more convenient way. Inspired by a fast Fourier transform (FFT) generator [5] which could automatically generate FFT hardware architecture with arbitrary parallelism and figure out hardware cost, this paper proposes an auto-generation system which could produce polar encoder hardware architecture with arbitrary code length and arbitrary level of parallelism.

The remainder of this paper is organized as follows. In Section II, the brief description of polar encoding is introduced. In Section III, we propose the generation system of polar encoder and an exemplary 32-bit polar encoder with 8-parallel processing. In Section IV, the analysis of the performance of the generation system is given. In Section V, we conclude and remark on the entire paper.

II. PRELIMINARIES

A. Polar Encoder

In polar code encoding, u_0^{N-1} is regarded as the source word and x_0^{N-1} as the codeword. The encoding scheme can be defined by Eq. (1), where G_N and B_N are the generation matrix and the bit-reversal permutation matrix respectively, and $F^{\otimes n}$ is the *Kronecker power* of n with $n = \log_2 N$ and $F = \begin{bmatrix} 1 & 0 \\ 1 & 1 \end{bmatrix}$.

$$x_0^{N-1} = u_0^{N-1} G_N = u_0^{N-1} B_N F^{\otimes n}. \qquad (1)$$

As proved by [3], the data-flow graph (DGF) of polar encoder could be derived from the DFG of FFT processors by replacing all the butterfly modules with XOR-and-PASS modules, and all the twiddle factors with 1's. Therefore, the proposed framework for polar encoder has the potential for implementing the pipelined hardware architecture for FFT by reversing the replacement. An exemplary DGF of an 8-bit polar encoder is shown in Fig. 1. Note that this DFG is similar to the that of an 8-point radix-2 decimation-in-frequency (DIF) FFT processor in the way mentioned above.

III. HARDWARE GENERATION

In this section, we introduce the general pipelined frame-work for polar encoder with arbitrary code length N and arbitrary level of parallelism M. The general framework could be easily denoted by a general formula $F(N, M)$. Then we

---

978-1-5386-5309-8/18 $31.00 © 2018 IEEE

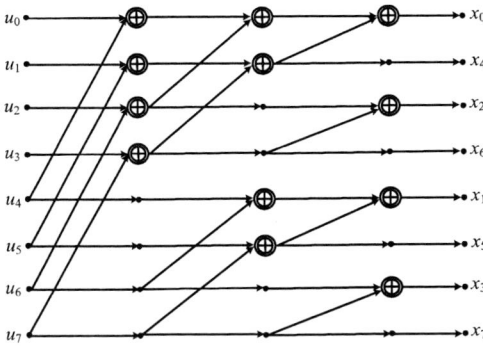

Fig. 1. The data-flow graph of an 8-bit polar encoder.

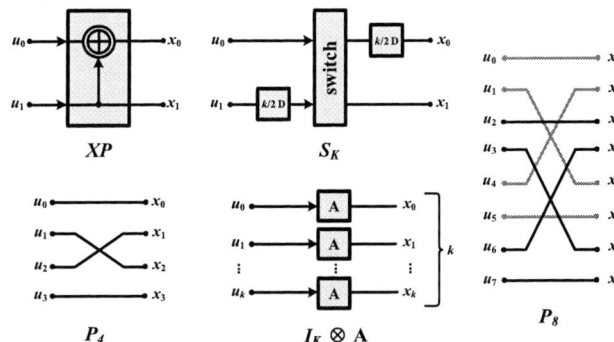

Fig. 3. The symbols and corresponding hardware modules in the formula.

show how to use an algorithm to derive a specific formula $f_{N,M}$ from $F(N, M)$ based on the values of N and M. Finally, a compiler is employed to translate $f_{N,M}$ into RTL description. The hardware generation system is illustrated in Fig. 2.

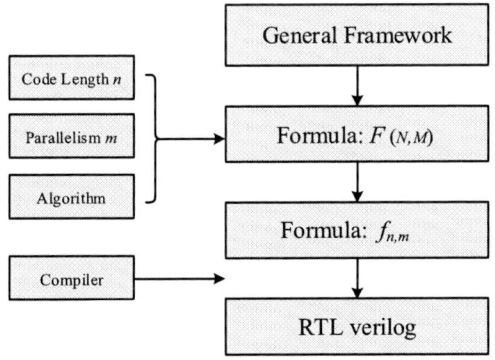

Fig. 2. The hardware generation system for polar encoder.

A. From General Framework to Formula

Consider that the general framework is expected to implement polar encoder with arbitrary code length and arbitrary level of parallelism, the framework should be scalable, i.e., the number of stages and the number of hardware modules in each stage should change with the values of N and M. Such a scalable framework could represented by formula $F(N, M)$ shown in Eq. (2). Here the parameters N and M are powers of 2, and $4 \leqslant M \leqslant N/2$. Before we go into details of $F(N, M)$, we introduce all the symbols that might be used in $F(N, M)$ and $f_{N,M}$, as well as the symbols' corresponding hardware modules. Note that the final hardware implementation of $f_{N,M}$ is the serial connection of the individual modules of different symbols. Fig. 3 illustrates all the exemplary modules, as well as symbols, that might be used in our design, all of which take u_0^{N-1} as input and x_0^{N-1} as output.

Symbol XP represents an XOR-and-PASS module that achieves: $x_0 = u_0 + u_1$ (in GF(2)) and $x_1 = u_1$. The number of inputs of XP is fixed and equals to 2 in our design.

Symbol S_K (K is a power of 2, $K > 1$) represents a switch with $k/2$ delay elements (denoted by D) on each side. A $\log_2 K$-bit counter is needed to control the switch: the value 0 of the most significant bit of the counter infers direct data transfer, and the value 1 infers cross data transfer. The number of inputs of S_K is fixed and equals to 2.

Symbol P_N (N is a power of 2, $N > 2$) denotes the permutation on an N-dimensional vector. The detail function of P_N is illustrated in Algorithm 1. Intuitively, P_N is the duplication of $P_{N/2}$. For example, P_8 could be viewed as partial overlap of two P_4 modules with red wires and black wires respectively.

Symbol $(I_K \otimes A)$ (K is a power of 2, $K > 0$) is a Kronecker product representing K parallel instances of module A, where A is an abstract module and A could be replaced by XP, S_K or P_N. Note that when $K = 1$, $(I_K \otimes A)$ equals to A. Suppose that A has X inputs, the number of inputs of $(I_K \otimes A)$ equals $K \times X$.

The general formula $F(N, M)$ is composed of symbols mentioned above, except that the W in Eq. (2) is a variable module. When deriving $f_{N,M}$ from $F(N, M)$, symbol W should be replaced by P_N or S_K according to its subscript. In Algorithm 2, as the code length and the level of parallelism are given, all the subscripts of each symbol in $F(N, M)$ will be figure out. Then the module $(I \otimes W)$ is replaced by $(I \otimes P)$ or $(I \otimes S)$ based on the value of the subscript of W. Finally, the formula $f_{N,M}$ is determined.

B. Compiler

We have built a compiler in *Python* that takes $f_{N,M}$ as input and automatically connects all the basic modules in $f_{N,M}$ in left-to-right order. Specifically, as we input N and M into $F(N, M)$, the $f_{N,M}$ will be determined and transformed into the register-transfer level (RTL) Verilog by the compiler. The detail of the compiler is beyond the scope of this paper; we only provide a brief introduction here.

978-1-5386-5309-8/18 $31.00 © 2018 IEEE

There are totally three types of basic modules in the formula $f_{N,M}$: the XOR-and-PASS module XP, the switch module S_K, and the permutation module P_N. There are two ways to expand these modules. The first one is to employ the symbol $I_K \otimes$ to layout the duplication of one module in a parallel way. The other one is to change the symbols' subscripts. Therefore, the compiler needs to read each symbol of $f_{N,M}$ from left to right, and recognizes $I_K \otimes$ as well as each symbol's subscript. Then the compiler could determine the specific hardware architecture and print the Verilog files. In fact, the Verilog descriptions of each cascadable module have already been stored in a library separately, what the compiler needs to do is to call these modules sequentially based on $f_{N,M}$. Thus the compiler is of great time efficiency and low complexity.

C. Input and Output Orders

The input and output data of this framework are in regular order. Suppose the input data of $f_{N,M}$ is u_0^{N-1}, since $f_{N,M}$ represents a pipelined architecture, u_0^{N-1} will be divided into N/M M-dimensional vectors $V_{in(i)}$ illustrated in Eq. (3), where $i = 0, 1, ..., (N/M) - 1$. All the data in $V_{in(i)}$ will be entered into the encoder in parallel, and i indicates the sequence of the input vector, i.e., $V_{in(0)}$ is the first set of the input data and the $V_{in(N/M-1)}$ is the last set of the input data. The output data are in bit-reversal order. Specifically, suppose x_0^{N-1} is the theoretical codeword and y_0^{N-1} is in the bit-reversal form of x_0^{N-1}. Then the i-th output vector $V_{out(i)}$ equals to $y_{M \times i}^{(M \times i) + (M-1)}$, where $i = 0, 1, ..., (N/M) - 1$.

$$
\mathbf{V}_{in(i)} = \begin{bmatrix}
u_{(M/2) \times i} \\
u_{(M/2) \times i + (N/2)} \\
u_{(M/2) \times i + 1} \\
u_{(M/2 \times i) + 1 + (N/2)} \\
u_{(M/2) \times i + 2} \\
u_{(M/2 \times i) + 2 + (N/2)} \\
... \\
u_{(M/2) \times i + (M/2) - 1} \\
u_{(M/2) \times i + (M/2) - 1 + (N/2)}
\end{bmatrix}. \tag{3}
$$

For the general framework, the processing latency (clock cycles) is $T_{latency} = (3N/2M) - 1$. The number of XOR gates and delay elements are:

$$
\begin{aligned}
\#XOR &= (M/2) \times \log_2 N; \\
\#MEM &= (3N/2) - M.
\end{aligned} \tag{4}
$$

D. A 32-Bit 8-Parallel Polar Encoder

According to Algorithm 2, given $N = 32$ and $M = 8$, formulas $F(32, 8)$ and $f_{32,8}$ are obtained in Eq. (5) and Eq. (6), respectively. The hardware architecture is illustrated in Fig. 4, which consists of 20 XOR gates and 40 delay elements in accordance with Eq. (4). The architecture could be split in 11 columns, each of which has its relevant symbol under the column. Note that Eq. (6) is actually composed of all the symbols at the bottom of Fig. 4. The order of the input data u ($k = 0, 1, 2, 3$) at the leftmost part of Fig. 4 conforms to the order mentioned above. The output data x is in the bit-reversal order.

Algorithm 1 The Permutation on an N-dimensional Vector

Require: The input vector u_0^{N-1}.
1: **for** $(i = 0; i < N/2; i = i + 2)$ **do**
2:    $x[i] = u[i]$
3: **end for**
4: **for** $(i = N - 1; i > N/2; i = i - 2)$ **do**
5:    $x[i] = u[i]$
6: **end for**
7: **for** $(i = 1; i < N/2; i = i + 2)$ **do**
8:    $x[i] = u[i - 1 + (N/2)]$
9:    $x[i - 1 + (N/2)] = u[i]$
10: **end for**
11: Output x_0^{N-1}.

Algorithm 2 The Generation of Formula $f_{N,M}$

Require: The code length N and the level of Parallelism M ($N = 2^i, M = 2^i, i \geqslant 2, i \in Z, M \leqslant N/2$).
1: Input N and M in to the general formula $F(N, M)$.
2: **if** $(k >= 1, k = 2^i, i \in Z)$ **then**
3:    $(I_{M/2} \otimes W_{1/k}) = (I_k \otimes P_{M/k})$
4: **else**
5:    $(I_{M/2} \otimes W_{1/k}) = (I_{M/2} \otimes S_{1/k})$
6: **end if**
7: $f_{N,M} = F(N, M)$
8: Output the formula $f_{N,M}$.

IV. PERFORMANCE AND COMPLEXITY

Some of the hardware designs derived from the auto-generation system were implemented on the Xilinx Virtex-7 VC709 FPGA platform with Virtex-7 XC7VX690T. All the design examples are of the same code length $N = 1024$, but with different level of parallelism. The synthesis results are illustrated in Table I. From the table, it can be observed that the throughput (T/P) and the number of Slice LUTs and Slice Registers increase as the value of M increases. In an extreme case, the polar encoder with $M = 512$ consumes more Slice LUTs than the polar encoder with $M = 4$ by 5167% but achieves higher throughput by 8710%.

As mentioned in Section III, the value of M conforms to $4 \leqslant M \leqslant N/2$. Then, given the code length N, the generation system could implement $(\log_2 N) - 2$ designs with different M, covering a wide cost/performance trade-off space. Therefore, one could choose the most suitable polar encoder in the design space to fit the application.

V. CONCLUSION

This paper proposes an auto-generation system for the hardware architecture of polar encoder. The system could offer users a wide range of design space so that the users could make a trade-off between cost and performance to best fit their applications. The essence of the generation system lies in the formula-based expression of the general framework for polar encoder that could achieve encoding with arbitrary code length and arbitrary parallelism. This auto-generation can

978-1-5386-5309-8/18 $31.00 © 2018 IEEE

$$(I_{M/2} \otimes XP)(I_{M/4} \otimes P_4) \left\{ \Pi_{i=0}^{\log 2N-3} [(I_{M/2} \otimes W_{N/(2^i M)})(I_{M/2} \otimes XP)] \right\} (I_{M/4} \otimes P_4)(I_{M/2} \otimes S_{N/M})(I_{M/2} \otimes XP) \qquad (2)$$

$$(I_4 \otimes XP)(I_2 \otimes P_4) \left\{ (I_4 \otimes W_4)(I_4 \otimes XP)(I_4 \otimes W_2)(I_4 \otimes XP)W_1(I_4 \otimes XP) \right\} (I_2 \otimes P_4)(I_4 \otimes S_4)(I_4 \otimes XP) \qquad (5)$$

$$(I_4 \otimes XP)(I_2 \otimes P_4) \left\{ (I_4 \otimes S_4)(I_4 \otimes XP)(I_4 \otimes S_2)(I_4 \otimes XP)P_8(I_4 \otimes XP) \right\} (I_2 \otimes P_4)(I_4 \otimes S_4)(I_4 \otimes XP) \qquad (6)$$

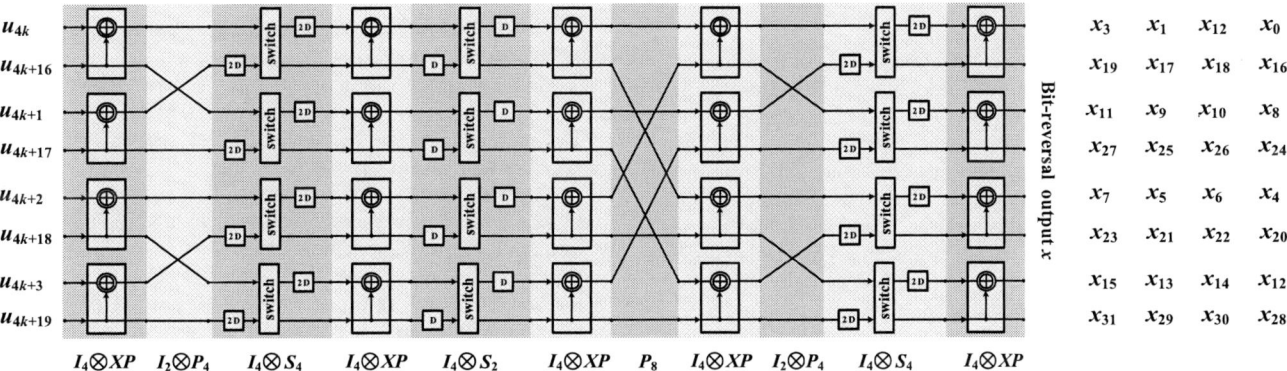

Fig. 4. The hardware architecture of polar encoder with $N = 32$, $M = 8$.

TABLE I

IMPLEMENTATION OF THE HARDWARE DESIGNS DERIVED FROM THE AUTO-GENERATION SYSTEM ON THE XILINX VIRTEX-7 VC709 FPGA PLATFORM WITH VIRTEX-7 XC7VX690T.

N	M	Slice LUTs	Slice Registers	Max freq (MHz)	T/P (Gbps)
1024	4	148	82	519.535	2.07
1024	32	467	312	407.05	13.02
1024	128	1278	845	340.518	43.58
1024	256	1704	1194	348.712	89.27
1024	512	2628	1025	356.223	182.38

help designers to conveniently design polar encoder without touching hardware details. The derivation of design space can further help us to identify the required design.

In this paper, we also introduce the scalable hardware modules associated with the formula, as well as the compiler that could transform the formula into RTL Verilog files. Synthesis results on FPGA have demonstrated the efficiency and the large trade-off space of the auto-generated circuits.

Future work will be directed toward the auto-generation of successive cancellation polar decoder and belief prorogation decoder based on our previous works [6–9], and the design optimization based on the design space.

ACKNOWLEDGEMENT

This work is supported in part by NSFC under grant 61501116, Jiangsu Provincial NSF under grant BK20140636, Huawei HIRP Flagship under grant YB201504, the Fundamental Research Funds for the Central Universities, State Key Laboratory of ASIC & System under grant 2016KF007, ICRI

for MNC, and the Project Sponsored by the SRF for the Returned Overseas Chinese Scholars of MoE.

REFERENCES

[1] E. Arıkan, "Channel polarization: A method for constructing capacity-achieving codes for symmetric binary-input memoryless channels," *IEEE Transactions on Information Theory*, vol. 55, no. 7, pp. 3051–3073, 2009.

[2] K. K. Parhi, C.-Y. Wang, and A. P. Brown, "Synthesis of control circuits in folded pipelined DSP architectures," *IEEE J. Solid-State Circuits*, vol. 27, no. 1, pp. 29–43, 1992.

[3] C. Zhang, J. Yang, X. You, and S. Xu, "Pipelined implementations of polar encoder and feed-back part for SC polar decoder," in *Proc. IEEE International Symposium on Circuits and Systems (ISCAS)*, 2015, pp. 3032–3035.

[4] H. Yoo and I.-C. Park, "Partially parallel encoder architecture for long polar codes," *IEEE Trans. Circuits Syst. II*, vol. 62, no. 3, pp. 306–310, 2015.

[5] P. Milder, F. Franchetti, J. C. Hoe, and M. Püschel, "Computer generation of hardware for linear digital signal processing transforms," *ACM Transactions on Design Automation of Electronic Systems (TODAES)*, vol. 17, no. 2, p. 15, 2012.

[6] C. Zhang, B. Yuan, and K. K. Parhi, "Reduced-latency SC polar decoder architectures," in *Proc. IEEE International Conference on Communications (ICC)*, June 2012, pp. 3471–3475.

[7] C. Zhang and K. Parhi, "Low-latency sequential and overlapped architectures for successive cancellation polar decoder," *IEEE Trans. Signal Process.*, vol. 61, no. 10, pp. 2429–2441, 2013.

[8] C. Zhang and K. K. Parhi, "Latency analysis and architecture design of simplified sc polar decoders," *IEEE Trans. Circuits Syst. II*, vol. 61, no. 2, pp. 115–119, 2014.

[9] J. Yang, C. Zhang, H. Zhou, and X. You, "Pipelined belief propagation polar decoders," in *Proc. IEEE International Symposium on Circuits and Systems (ISCAS)*, May 2016, pp. 413–416.

An aCCELERATOR-AWARE MICROARCHITECTURE SIMULATOR FOR DESIGN SPACE EXPLORATION

Di Gao and Cheng Zhuo*

Key Lab. of Advanced Micro/Nano Electronic Devices & Smart Systems of Zhejiang

College of Information Science & Electronic Engineering

Zhejiang University, Hangzhou, China, 310003

*Corresponding Author's Email: czhuo@zju.edu.cn

ABSTRACT

Specialized hardware accelerator has emerged as an efficient approach to mitigate the issue of power wall. Most accelerator designs focus on accelerator RTL synthesis, and place little focus on the communications between core and accelerator, thereby potentially limiting overall system performance. This paper presents an accelerator-aware micro-architectural simulator that supports accelerator design using high-level language (HLL) description, *e.g.*, C++. The paper also discusses the design of a complete software stack for the simulator, from programming model, user configurability, to power profiler. Designers can then use the tool to conduct case studies and make performance analyses for design space exploration.

INTRODUCTION

With increased transistor density brought by technology scaling, power density continues growing [1]. The so-called "power wall" then leads to the phenomenon of "dark silicon", in which only a small portion of the entire chip is active while the rest are idle or powered off [2]. In order to overcome this challenge, hardware acceleration has emerged as a promising approach, as it delivers up to orders of magnitude performance and energy improvements than general-purpose processors. Research on application of specific accelerator has gained increasing attention [3-5]. The ITRS predicts that there will be hundreds to thousands of dedicated accelerators by 2022 [6].

In other words, the integration of a hardware accelerator has become an essential part of an energy-efficient system. Customized datapath and control logics inside accelerator demand careful trade-offs between hardware complexity and energy efficiency, and hence significant efforts in hardware accelerator modeling. One common approach is to leverage a full-system simulator for CPU and memory hierarchy, and then employ high-level synthesis (HLS) and register-transfer-level (RTL) simulation for accelerator design [7]. However, such an approach demands significant implementation efforts and reconfiguration budgets. Thus, it is highly desired to have an effective and time-saving simulation framework to enable the design space exploration for accelerator design.

This paper proposes a cycle-accurate system-level simulation platform to address the aforementioned issues for accelerator-aware microarchitecture design space exploration. The contributions are summarized as below:

- The platform enables the integration of specialized accelerators using high-level language (HLL) descriptions of their algorithms, and then maps them to accelerator hardware operations, without going through RTL implementation stage.
- Based on the platform we then propose a simulation methodology to combine general-purpose CPU together with customized accelerators, by leveraging a widely used microarchitecture level simulator GEM5 [8]. The methodology is fully supported by a complete software stack, from programming model, user configurability, to power profiler.
- Different from the existing solutions in which power model is built upon time-consuming RTL simulations (*e.g.*, Design Compiler), we develop a light-weighted power model embedded into the simulator, which is trained using McPAT [9] and GEM5.

In order to understand the tradeoff brought by various accelerator design and interaction with host processor for a given application, we can use the proposed platform and methodology to analyze the performance breakdown of both CPU and accelerator. The exploration then enables choices of the optimal solution for higher energy efficiency.

OVERVIEW OF THE ACCELERATOR-AWARE ARCHITECTURE

In this section, we present an overview of the proposed accelerator-aware microarchitecture and the underlying mechanisms for hardware execution.

Accelerator-Aware Architecture

Figure 1 demonstrates an example of an accelerator-aware design, including a sea of customized accelerators compared to a general-purpose computing system. In the figure, accelerators assist the host core to conduct specified tasks with a communication interface, where library-based information for applications is fed into the appropriate accelerator. As shown in Figure 1, L1 cache is shared among core and accelerators, enabling accelerators to access and update data during the invoked stage. Moreover, in order to optimize data access for locality, we utilize local buffer resources inside accelerators for temporary data storage.

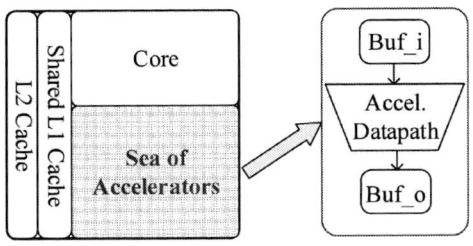

Figure 1: Overview of an accelerator-aware design

Accelerator Execution

Figure 2 describes a hardware execution mechanism for accelerator design in our architecture. The process consists of three stages: (1) CPU invokes the accelerator using the communication interface, and then allocates the appropriate task to a particular accelerator. (2) The application is conducted in the accelerator in terms of customized datapath and control flow. The accelerator itself can organize hardware resources and assign dedicated computing units to complete sub-functions. (3) The application invocation is terminated when the results are returned from the accelerator to shared L1 cache. Then CPU takes the control back.

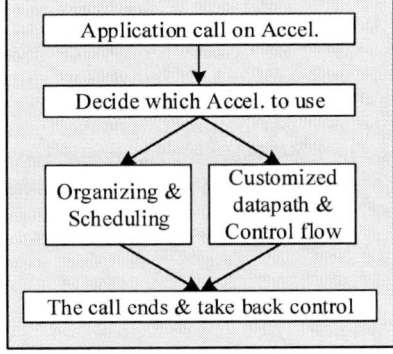

Figure 2: A mechanism for accelerator execution

ACCELERATOR-AWARE SYSTEM-LEVEL SIMULATOR

In this section, we discuss a complete software stack to support the proposed design framework. In order to enable efficient system-level design space exploration, the framework is based on a widely used architectural simulator, GEM5, and capable of boosting an actual OS to run various applications. In the following we will cover the key implementations of our software stack design and simulation flow, which enables easier access and user configurability.

Software Stack Design

GEM5 is an open source platform, which can be compiled into an executable simulation tool in terms of ISAs and architectural modules. We then extend GEM5 to support the integration of desired accelerators, and provide flexible system-level configurations to the CPU model, ISAs, memory hierarchy, and cache coherence protocols. The building and compilation process is shown in Figure 3. The key components inside the software stack design is summarized as following:

- **Accelerator design methodology**. In order to achieve efficient accelerator modeling, we may customize accelerators using high-level language descriptions of their algorithms, *e.g.*, C++, which allows users to easily analyze different algorithms.
- **Programming models**. In order to minimize programming efforts of the proposed accelerator-aware architecture, a library-based programming model is provided. The library is a set of extended instructions for specific functions conducted by the accelerators. The concrete operations of each instruction are then implemented, which leverages the inherent fetching and decoding mechanisms.

We can then implement the proposed software stack in GEM5. All the ISA descriptions can be instantiated into concrete operations controlled by the build scripts.

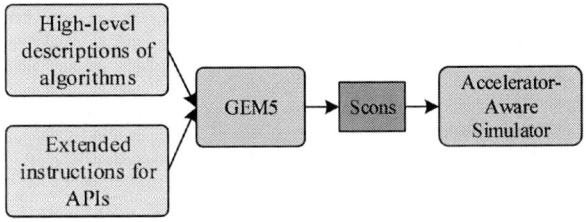

Figure 3: Building and compilation process of the proposed simulator.

Simulation Flow

Based on the proposed software stack design, we implement a complete simulation flow from source code to execution as shown in Figure 4, including the features of user configurability, compiler, and power/performance profiler, *etc*. The flow can be summarized as below:

- **Configuration and compilation:** It compiles C program, along with user specifications into binary, and feeds it to the software stack. An

978-1-5386-5309-8/18 $31.00 © 2018 IEEE   567

accelerator-relevant static library is linked when compiling, to enable OS to identify customized instructions of accelerators.

- **Decoder and executor:** Two executing schemes for instructions are supported: (1) Inherent CPU-centric execution and (2) Extended accelerator-aware instructions.
- **Run-time evaluation and profiler.** A light-weighted profiler is embedded into the GEM5-based simulation platform. The profiler not only calls for GEM5's own profiler, but also leverages the model for power and performance estimation.

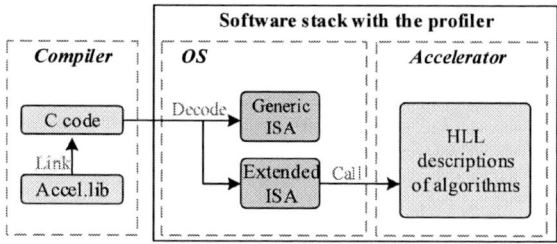

Figure 4: Overall simulation flow of the proposed accelerator-aware simulator

RUN-TIME EVALUATION AND PROFILING

In this section, we present a light-weighted power evaluation profiler embedded into the proposed simulator, to enable fast and run-time power estimation of both core and accelerators.

In order to implement a generic power model, we embed a light-weighted power evaluation profiler into GEM5. The profiler is employed to estimate energy efficiency of various modules, such as CPU, cache and memory. The power model is trained using the given specifications and output data from GEM5, CACTI-3DD, McPAT by running various application traces [10]. Then a linear model between GEM5 performance counter statistics and the reported power for each IP of interest can be extracted. In other words, we use the following equation to evaluate power for each module:

$$Power = w_0 \times s_0 + w_1 \times s_1 ...$$

where w_i's are trained weights and s_i's are performance counter statistics from GEM5. The linear model depends on the microarchitecture details and setup, and is unique to each design. As GEM5 supports breakpoint execution, the profiler can output the power of all the modules, and detect the energy-intensive operations.

In addition to the power profiler of the core, we also need to provide power evaluation for each accelerator by leveraging on the existing power models and accelerator details. It is noted that accelerator is customized and specific to a particular application. Thus, it is not supported by any existing architectural power simulator and hence has to rely on RTL power simulation by translating the design to RTL level descriptions. This inevitably incurs significant run time overhead and conflicts with the desired fast design space exploration purpose of the proposed simulator. Thus, instead of simulating the accelerator power as a whole, we analyze and factorize each accelerator executions to a series of unit operations. The power of each unit operation can be simulated and found from running McPAT on the simple CPU mode of GEM5. Then we can calculate the total power consumption of the accelerator over time by analyzing unit operation power and accelerator executions, which is called sub-profiler for the accelerator.

Thus, with the integration of sub-profilers of accelerators, when an accelerator instruction is triggered, we invoke a sub-profiler for the accelerator to dump its power and performance information. Then combined with the power estimation for core, we can investigate energy efficiency of various algorithms for the accelerator design and optimize the system level energy efficiency.

DESIGN SPACE EXPLORATION

In order to demonstrate the efficiency of the proposed accelerator-aware architecture, we conduct a few experiments with the proposed simulator to explore the energy efficiency. The system to be simulated is based on ARM Cortex A9, including 64K L1 instruction cache and data cache, 1MB L2 cache, 1G main memory, with system clock set to 2 GHz.

We validate the simulator with various applications to ensure the functionality and communications between core and accelerator. Meanwhile, the profiler records the selected performance information and power consumption of different modules over execution time. We choose matrix multiplication as a demonstrative example, which is completed in a customized systolic array based accelerator. Figure 5 illustrates its normalized power (*w.r.t.* maximum accelerator power) over execution time for both core and accelerator.

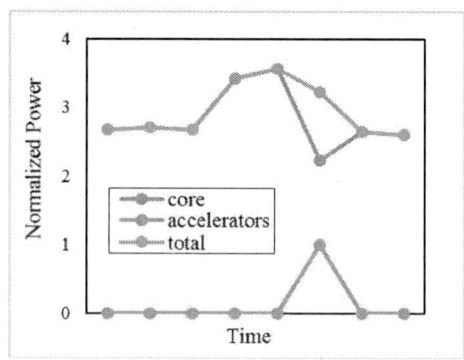

978-1-5386-5309-8/18 $31.00 © 2018 IEEE   568

Figure 5: Power consumption of accelerator and core for a matrix multiplication application

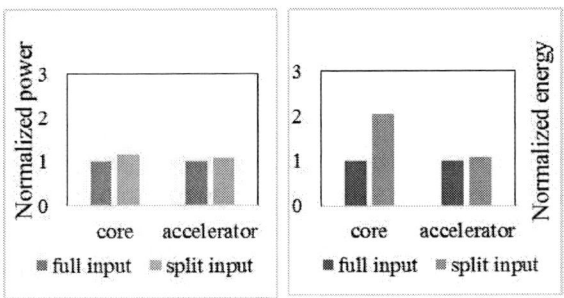

Figure 6: Comparison of power and energy consumption for a matrix multiplication application with different algorithms.

Figure 6 illustrates the normalized power and energy consumption of matrix multiplication application with different algorithms and configurations. This may help architects to achieve better system-level energy efficiency. We demonstrate the power and energy differences for a given application implemented with two algorithms: (1) "Full" algorithm conducts the task with complete input size using the dedicated accelerator; (2) "Split" algorithm divides the application into a few sub-tasks in the host core and then sends each sub-task to the accelerator for computation. For example, "full" algorithm conducts a 32 × 32 matrix multiplication as it is, while "split" algorithm divides the matrix into 16×16 matrices and then merge the results. With the proposed simulator, we can achieve comparisons as depicted in Figure 6 and analyze energy efficiencies of different algorithms.

It is noted that the proposed accelerator design fully relies on the high-level language descriptions, which saves efforts of RTL customization and programming. Combined with the profiler, the simulator can guide the HHL design to achieve an optimal efficiency. Moreover, the framework is aware of the interaction between the accelerator and core, which enables system-level design space exploration.

CONCLUSIONS

In this paper we present a cycle-accurate, system-level simulation platform for accelerator-aware architecture design, which supports accelerator customization with HLL description and run time profiling. The power evaluation profiler embedded into the simulator can collect performance and power statistics for both cores and accelerator. Experimental results show that the proposed simulator can help guide the accelerator design and investigate the potential design space.

ACKNOWLEDGEMENTS

The authors would like to thank the support from NSFC Fund (Grant No. 61601406).

REFERENCES

[1] D. Evans, "The internet of things: how the new evolution of the internet is changing everything", https://www.cisco.com/c/dam/en_us/about/ac79/docs/innov/IoT_IBSG_0411FINAL.pdf, 2011.

[2] H. Esmaeilzadeh, *et. al.*, "Dark silicon and the end of multicore scaling," in *Proc. of ISCA*, pp. 365-376, 2011.

[3] V. Govindaraju, *et. al.*, "Dyser: unifying functionality and parallelism specialization for energy-efficient computing," in *Proc. of MICRO*, pp. 38-51, 2012. .

[4] A. Hansson *et. al.*, "Simulating DRAM controllers for future system architecture exploration," in *Proc. of ISPASS*, pp. 201-210, 2014.

[5] J. Ahn, *et. al.*, "A scalable processing-in-memory accelerator for parallel graph processing," in *Proc. of ISCA*, pp. 105-117, 2015.

[6] A. B. Kahng, "The ITRS design technology and system drivers roadmap: Process and Status," in *Proc. of DAC*, pp. 1-6, 2013.

[7] Y. S. Shao, *et. al.*, "Aladdin: a pre-RTL, power-performance accelerator simulator enabling large design space exploration of customized architectures," in *Proc. of ISCA*, pp. 97-108, 2014.

[8] N. Binkert, *et. al.*, "The gem5 simulator," in *ACM SIGARCH CAN*, vol. 39(2):1–7, 2011

[9] Sheng Li, *et. al.*, "McPAT: An integrated power, area, and timing modeling framework for multicore and manycore architectures," in *Proc. of MICRO*, pp. 469-480, 2009.

[10] C. Zhuo, *et. al.*, "A novel cross-layer framework for early-stage power delivery and architecture co-exploration," in *Proc. of DAC*, pp. 1-6, 2016.

HIGH-SPEED IMPLEMENTATION OF SM2 BASED ON FAST MODULUS INVERSE ALGORITHM

Wei Li[1], Juhua Liu[1], Guoqiang Bai[2]*

1 Institute of Microelectronics, Tsinghua University, Beijing, China
2 National Laboratory for Information Science and Technology, Tsinghua University, Beijing, China
Corresponding author baigq@tsinghua.edu.cn

ABSTRACT

In this paper, we explore the fast modulus inverse algorithm and its implementation. For the first time, we proposed a radix-8 modulus algorithm to speed up the point multiplication in SM2 public key cryptographic algorithm, which is established as the ECC standard of China for commercial applications released by the State Cryptographic Administration of China in December 2010. The critical path delay of our hardware implementation of SM2 is the delay of a one-cycle 256-bit multiplier, which is difficult to get a further reduction. The possibility of further optimization is reducing the number of cycles needed by the binary modulus inverse without changing the critical path delay when converting the Jacob coordinates back to affine coordinates. The radix-8 binary inverse algorithm can reduce the number of cycles significantly by 33.2% on average compared with the radix-4 binary inverse algorithm, which needs 256 cycles at most to complete the conversion.

Keywords—SM2; ECC; radix-8 inverse; high speed; hardware implementation

INTRODUCTION

ECC provides better security compared with other public key cryptographic algorithms, such as RSA, with the same key size [1]. SM2 public key cryptographic algorithm, which is released by the State Cryptography Administration of China in December 2010, is established as the ECC standard of China for commercial applications and is widely used in personal identification, digital signature, finance authentication, etc. SM2 is defined over a Galois field GF(p), where the $p = 2^{256} - 2^{224} - 2^{96} + 2^{64} - 1$, and p is a very special pseudo-Mersenne prime that can be used to compact the process of modulo reduction significantly, more details about SM2 can be found in [2].

Point multiplication, defined as $kP = P + P + \ldots + P$, where $k \in [0, p-1]$, $P \in GF(p)$ and P is the base point, is the most time-consuming operation in SM2 (more than 90%) [3]. It consists of repeated point addition (PADD) and doubling (PDBL) operations of base point P while performing the point multiplication. However, algorithms to perform PADD and PDBL differ by the choice of coordinate system, leading a different quantity of arithmetic operations, as listed in TABLE I. (A→Affine, P→Projective, J→Jacob, C→Chudnovsky, I→Inversion, M→Multiplication, S→Square)

TABLE I OPERATION COUNTS OF PADD AND PDBL

PADD		PDBL	
$A+A \to A$	$1I, 2M, 1S$	$2A \to A$	$1I, 2M, 2S$
$P+P \to P$	$12M, 2S$	$2P \to P$	$7M, 3S$
$J+J \to J$	$12M, 4S$	$2J \to J$	$4M, 4S$
$C+C \to C$	$11M, 3S$	$2C \to C$	$5M, 4S$
$J+A \to J$	$8M, 3S$		
$C+A \to C$	$8M, 3S$		

As shown above that $2J \to J$ for the PDBL and $J+A \to J$, mixed Jacobian-affine coordinate, for the PADD is the best choice to get a high-speed architecture. For a further clarification, we give the detailed formulas of PADD and PDBL operation in TABLE II.

TABLE II PADD AND PDBL FORMULAS

$P_3 \leftarrow P_1 + P_2$
$X_3 = (Y_2 Z_1^3 - Y_1)^2 - (X_2 Z_1^2 - X_1)^2 (X_1 + X_2 Z_1^2)$
$Y_3 = (Y_2 Z_1^3 - Y_1)(X_1 (X_2 Z_1^2 - X_1)^2 - X_3) - Y_1 (X_2 Z_1^2 - X_1)^3$
$Z_3 = (X_2 Z_1^2 - X_1) Z_1$

$P_3 \leftarrow 2P_1$
$X_3 = (3X_1^2 + aZ_1^4)^2 - 8X_1 Y_1^2$
$Y_3 = (3X_1^2 + aZ_1^4)(4X_1 Y_1^2 - X_3) - 8Y_1^4$
$Z_3 = 2Y_1 Z_1$

Once the point multiplication is done, we have to transfer the results to standard affine coordinate system form Jacobian coordinate system, in which an inversion is necessary.

HARDWARE ARCHITECTURE

This section we explore the high-speed hardware implementation of SM2, specifically we will give the hardware architecture of the point multiplication algorithm, as shown in Fig 1. In this architecture we input the base point $P = (X_2, Y_2, Z_2)$ then compute the kP and return the result in affine coordinate (x, y).

Multiplier: A 256-bit multiplier and a two-stage pipeline are introduced to our architecture to get a best trade-off between the number of cycles needed to complete the point multiplication and the minimum critical path of the datapath. In this case, the critical path of this hardware architecture is the delay of the 256-bit one-cycle multiplier [4].

Reduction: A two stage pipeline is applied to the modular-multiplication and a new reduction module is designed to best accelerate the point multiplication. To get a more balanced pipeline, we combined the modular

978-1-5386-5309-8/18 $31.00 © 2018 IEEE

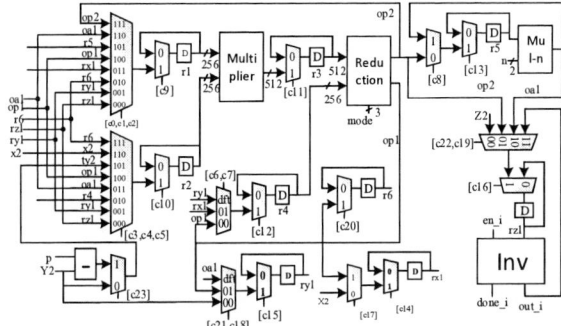

Fig.1 hardware implementation architecture of point multiplication of SM2 algorithm. A 256-bit multiplier and fast modulus algorithm (Inv) are used.

addition and the modular abstraction together with the modular reduction to obtain a new reduction module, which is controlled by signal *mode*, and the specific operation according to the value of *mode* is shown in TABLE III.

Mul-n: This module is designed to compute the 2,3,8 multiple of its input according to the selection signal n.

Inv: The module ***Inv*** coverts the result in Jacob coordinates to affine coordinates.

TABLE III OPERATIONS OF THE NEW REDUCTION MODULE

mode	corresponding operations
000	$out_1 = (in_1 \% p - in_2) \% p, out_2 = (in_1 \% p + in_2) \% p$
001	$out_1 = (in_2 - in_1 \% p) \% p, out_2 = (in_1 \% p + in_2) \% p$
010	$out_1 = (in_1 \% p - in_2) \% p, out_2 = (1.5 in_2 \% p - in_1) \% p$
011	$out_1 = in_1 \% p, out_2 = in_1 \% p$
100	$out_1 = 3 in_1 \% p, out_2 = 3 in_1 \% p$
101	$out_1 = 2 in_1 \% p, out_2 = 2 in_1 \% p$

BINARY MODULUS INVERSE ALGORITHM

As depicted in the Fig 1, the inversion module ***Inv*** that coverts the Jacob coordinates to affine coordinates is another bottleneck of the proposed architecture [5]. So, to get a further optimization on the number of cycles, we proposed a radix-8 binary inverse algorithm. As the best of our knowledge, the inverse of $a, a \in [1, p-1]$, is defined by $b = a^{-1} \bmod p$, $b \in [1, p-1]$, where $a \cdot b = 1 \bmod p$. Generally, we compute the $a^{-1} \bmod p$ as follow:

$$a \cdot x_1 = u \bmod p \qquad (1)$$
$$a \cdot x_2 = v \bmod p \qquad (2)$$

We get the inversion a^{-1} whenever $u = 1$ ($a^{-1} = x_1$) or $v = 1$ ($a^{-1} = x_2$) by right shifting u or v (binary representation), or right shifting after subtracting u from v, if $v > u$ ($u - v$, if $u > v$) on the condition of keeping the two equations Eq(1) and Eq(2) always being true during the whole inversion process. Conventional binary radix-2 modulus inverse algorithm deals with one bit once a clock

TABLE IV
FAST RADIX-4 MODULUS INVERSE ALGORITHM

Radix-4 Modulus Inverse Algorithm on *GF(p)*

Input: *prime* $p, a \in [1, p-1]$

Output: $a^{-1} \bmod p$

1. $u \leftarrow a, v \leftarrow p, x_1 \leftarrow 1, x_2 \leftarrow 0$

2. *if* $v > 0$, *repeatedly execute*

   2.1 $c = u[1:0], d = v[1:0]$

   2.2 *if* $c = 2'b00$, *execute*

      $u \leftarrow u/4, \; x_1 \leftarrow x_1/4$

   2.3 *else if* $d = 2'b00$, *execute*

      $v \leftarrow v/4, \; x_2 \leftarrow x_2/4$

   2.4 *else if* $c = d$, *execute*

      2.4.1 *if* $u > v, u \leftarrow (u-v)/4, \; x_1 \leftarrow (x_1 - x_2)/4$

      2.4.2 *else*, $v \leftarrow (v-u)/4, \; x_2 \leftarrow (x_2 - x_1)/4$

   2.5 *else if* $c[1:0] = 2'b10$, *execute*

      2.5.1 *if* $u/2 > v, u \leftarrow (u/2-v)/2, \; x_1 \leftarrow (x_1/2 - x_2)/2$

      2.5.2 *else*, $v \leftarrow (v - u/2)/2, \; x_2 \leftarrow (x_2 - x_1/2)/2$

   2.6 *else if* $d[1:0] = 2'b10$, *execute*

      2.6.1 *if* $u > v/2, u \leftarrow (u - v/2)/2, \; x_1 \leftarrow (x_1 - x_2/2)/2$

      2.6.2 *else*, $v \leftarrow (v/2 - u)/2, \; x_2 \leftarrow (x_2/2 - x_1)/2$

   2.7 *else if* $u \geq v$, *execute*

      $u \leftarrow (u-v)/2, \; x_1 \leftarrow (x_1 - x_2)/2$

      $v \leftarrow (v-u)/2, \; x_2 \leftarrow (x_2 - x_1)/2$

3. *return* x_1

cycle and radix-4 modulus inverse algorithm, as shown in TABLE IV, deals with two bits once a clock cycle (the LSBs of u and v), and they need 512 clock cycles and 256 clock cycles at most respectively. However, radix-8 modulus inverse algorithm deals with three bits once a clock cycle and only needs 171 clock cycles on average, which means 85 clocks are saved. The details of the algorithm are given by TABLE V below. The focus of the algorithm is to reduce u or v as quickly as possible until one of them is equal to 1, which means the process is over.

Unfortunately, when both of u and v are odd numbers, we can't reduce u or v by simply right shifting. But it's worth noting that when $c + d = 3'b000$, we can reduce the $c + d$ with 3 bits by right shifting once a clock cycle and give the result to the maximum of c and d. Furthermore, when $(c - d) = 3'b100$ or $(d - c) = 3'b100$ we can execute $u \leftarrow (u-v)/4, x_1 \leftarrow (x_1 - x_2)/4$, if $u > v$ to guarantee at least one of c and d is reduced by more than two bits. We give all the cases when c and d are odd numbers in TABLEVI. As we know that deciding the cases that $c + d = 3'b000$ and $\| c - d \| = 3'b100$ is an easy task, so

978-1-5386-5309-8/18 $31.00 © 2018 IEEE

TABLE V
FAST RADIX-8 MODULUS INVERSE ALGORITHM

Radix-8 Modulus Inverse Algorithm on *GF(p)*

Input: prime $p, a \in [1, p-1]$

Output: $a^{-1} \bmod p$

1. $u \leftarrow a, v \leftarrow p, x_1 \leftarrow 1, x_2 \leftarrow 0$

2. if $v > 0$, repeatedly execute

   2.1 $c = u[2:0], d = v[2:0]$

   2.2 if $c = 3'b000$, execute

      $u \leftarrow u/8, \ x_1 \leftarrow x_1/8$

   2.3 else if $d = 3'b000$, execute

      $v \leftarrow v/8, \ x_2 \leftarrow x_2/8$

   2.4 else if $c[1:0] = 2'b00$, execute

      $u \leftarrow u/4, \ x_1 \leftarrow x_1/4$

   2.5 else if $d[1:0] = 2'b00$, execute

      $v \leftarrow v/4, \ x_2 \leftarrow x_2/4$

   2.6 else if $c = d$, execute

      2.6.1 if $u > v, u \leftarrow (u-v)/8, \ x_1 \leftarrow (x_1 - x_2)/8$

      2.6.2 else, $v \leftarrow (v-u)/8, \ x_2 \leftarrow (x_2 - x_1)/8$

   2.7 else if $c[1:0] = d[1:0]$, execute

      2.7.1 if $u > v, u \leftarrow (u-v)/4, \ x_1 \leftarrow (x_1 - x_2)/4$

      2.7.2 else, $v \leftarrow (v-u)/4, \ x_2 \leftarrow (x_2 - x_1)/4$

   2.8 else if $c[0] = 0$, execute

      2.8.1 if $u/2 > v, \ u \leftarrow [(u/2)-v]/2, x_1 \leftarrow [(x_1/2) - x_2]/2$

      2.8.2 else, $u \leftarrow [v-(u/2)]/2, x_1 \leftarrow [x_2 - (x_1/2)]/2$

   2.9 else if $d[0] = 0$, execute

      2.9.1 if $v/2 > u, v \leftarrow [(v/2)-u]/2, x_2 \leftarrow [(x_2/2) - x_1]/2$

      2.9.2 else, $v \leftarrow [u-(v/2)]/2, x_2 \leftarrow [x_1 - (x_2/2)]/2$

   2.10 else if $c + d = 3'b000$

      $\max[u,v] \leftarrow (u+v)/8, x_1/x_2 \leftarrow (x_1 + x_2)/8$

   2.11 else if $u > v$, execute

      2.11.1 if $(c-d) \| (d-c) = 3'b100$

         $u \leftarrow (u-v)/4, \ x_1 \leftarrow (x_1 - x_2)/4$

      2.11.2 else $u \leftarrow (u+v)/4, \ x_1 \leftarrow (x_1 + x_2)/4$

   2.12 else,

      2.12.1 if $(c-d) \| (d-c) = 3'b100$

         $v \leftarrow (v-u)/4, \ x_2 \leftarrow (x_2 - x_1)/4$

      2.12.2 else $v \leftarrow (v+u)/4, x_1 \leftarrow (x_1 + x_2)/4$

3. return x_1

the delay of the ***Inv*** module will not increase so much compared with Radix-2 and Radix-4 inverse algorithm, which ensure the critical path of the whole point multiplication hardware architecture unchanged.

TABLE VI
THE CASES WHEN c AND d ARE ODD NUMBERS ($u > v$)

c	d	operations	c	d	operations
001	011	"+"then">>2"	101	001	"-"then">>2"
001	101	"-"then">>2"	101	011	"+"then">>3"
001	111	"+"then">>3"	101	111	"+"then">>2"
011	001	"+"then">>2"	111	001	"+"then">>3"
011	101	"+"then">>3"	111	011	"-"then">>2"
011	111	"-"then">>2"	111	101	"+"then">>2"

CONCLUSION

In the hardware implementation, we use the basic $0.13um$ CMOS standard cell library to evaluate the performance of our design. For a clear comparison, we give the clock cycles needed to complete the binary inverse in *GF(p)* by radix-2 and radix-4, as well as radix-8 algorithm proposed in this paper, in TABLE VII below.

TABLE VII
CLOCK CYCLES NEEDED TO COMPLETE INVERSE

	Radix-2	Radix-4	Radix-8
Cycles	512	256	171
Percentage in SM2	15.4%	7.7%	5.1%

Compared with radix-2 inverse algorithm, radix-4 inverse algorithm reduces the number of cycles significantly by 100% on average (256 cycles are saved) and improved the performance of SM2 by 7.7%. Similarly, compared with radix-4 inverse algorithm, radix-8 inverse algorithm reduces the number of cycles significantly by 33.2% on average (85 cycles are saved) and improved the performance of SM2 by 2.6%.

ACKNOWLEDGEMENTS

This work is supported by the National Natural Science Foundation of China (Grants 61472208) and National Key Basic Research Program of China (Grant 2013CB338004).

REFERENCES

[1] N. Kobliz, Elliptic curve cryptosystem, Mathematics of Computation 48 (1987) 203-209.

[2] State Cryptography Administration of China: Public Key Cryptographic Algorithm SM2 Based on Elliptic Curves, http://www.oscca.gov.cn/UpFile/2010122214822692.pdf

[3] Tim Guneysu, Christof Paar, Ultra High Performance ECC over NIST Primes on Commercial FPGAs, in: Cryptographic Hardware and Embedded Systems (CHES), LNCS 5154, 2008, pp. 62C78.

[4] Zhenwei Zhao, Guoqiang Bai. Exploring the Speed Limit of SM2. 2014 IEEE 3rd International Conference on Cloud Computing and Intelligence Systems. IEEE, 2014. 456-460.

[5] Zhenwei Zhao, Guoqiang Bai. Ultra High-Speed SM2 ASIC Implementation. The 13th IEEE Internatinal Conference on Trust,Security and Privacy in Computig and Communicatins. IEEE, 2014. 182-188.

[6] Patrick Longa, Catherine Gebotys, Efficient Techniques for High-Speed Elliptic Curve Cryptography, in: Cryptographic Hardware and Embedded Systems (CHES), LNCS 6225, 2010, pp. 80C94.

TECHNOLOGY MEDIATED TUTORIAL ON RISC-V CPU CORE IMPLEMENTATION AND SIGN-OFF USING REVOLUTIONARY EDA MANAGEMENT SYSTEM (EMS) - VSDFLOW

Kunal Promode Ghosh[1]*, Anagha K Ghosh[1]

[1]VLSI System Design Corporation Pvt. Ltd., Bangalore 560037, India

*Corresponding Author's Email: kunalpghosh@gmail.com

ABSTRACT

VSDFLOW (VLSI System Design Flow) is a 'plug and play (PnP)' EDA management system, built for chip designers to implement their ideas and convert to GDSII. 'plug and play (PnP)' refers to switching between any EDA tools, for e.g. user can plug Cadence Genus for synthesis, Synopsys ICC for PNR and Tempus for sign-off STA. The output report will provide a QOR of entire design, which forms the starting point for design analysis. At this moment, it has been tested on medium sized designs like picorv32 - a RISC-V CPU core that implements RV32I instruction set, using 180nm PDK's from OSU and 45nm PDK's from Nangate. 'vsdflow' was also tested on designs like ARM Cortex-M0, which is available as an evaluation IP on ARM website

INTRODUCTION

vsdflow

In recent years, there has been a steep increase in licensed and open-source EDA tool usage, especially, among students and educational institutes, looking for a solution to innovate and implement their ideas. This toolset generally consists of RTL synthesizer, a place-and-route engine and static timing analysis engine. To obtain the best results or compare best between them, the design needs to be going through many different tool-set which again becomes time consuming and tedious process to take the data from one tool to other. 'vsdflow' provides a plug and play (PnP) solution to use entire EDA toolset under single umbrella, where user needs to provide design in RTL language, select necessary tools and 'vsdflow' will generate GDSII and performance metrics of design. This framework enables user to focus on improvising quality of design ideas, while leaving the implementation to 'vsdflow'. 'vsdflow' automates the complete process of comparison and obtaining the performance of design by just plugging in the best tool. As a prototype, currently 'vsdflow' uses mix and match of partial and fully open-source EDA tool-set shown in below Figure 1.

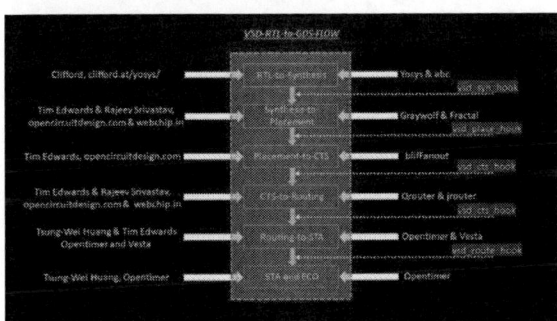

Figure 1: vsdflow framework

Unique features

'vsdflow' is designed in such a flexible way, that user can pitch-in at any point of flow and some of top features are enumerated below

• 'vsdflow' allows user to feed design details and constraints in generalized format in csv file, which is a standard practice in industries, and achieve desired results, there-by enabling user to focus on crafting the best requirements for design. See below Figure 2.

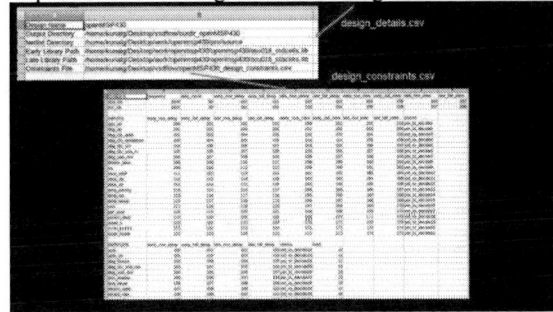

Figure 2: Inputs to 'vsdflow' in csv format

• 'vsdflow' has built-in proprietary free commands, which can be used in a stand-alone fashion on 'vsdflow' terminal, which saves time for now, by knowing only one command to run all tools. An example of proprietary-free command of 'vsdsynth' ('vsdsynth' is a synthesis+STA section of 'vsdflow') is shown in below Figure 3.

978-1-5386-5309-8/18 $31.00 © 2018 IEEE

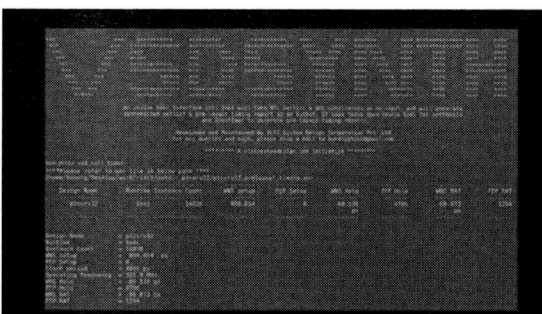

Figure 3: Demo of 'vsdflow' proprietary-free command

Toolset

Currently, 'vsdflow' has used mix and match of partial and fully open-source tools for prototyping, details of which is enumerated below. This concept can be extended to conventional industry standard EDA tools as well.

• **Yosys** –Yosys is a free and opensource software, which performs verilog RTL synthesis with complete Verilog-2005 support.

• **Placer**– 'vsdflow' uses 2 different Placers, viz. Graywolf (fully open-source) and Fractal (partial open-source). **Graywolf** is a fully open-source placement tool, formerly known as Timberwolf, which is currently maintained by Ruben Undheim in a much-streamlined form. **Fractal** is a commercial placement engine works exclusively with webchip.in platform.

• **blifFanout**–blifFanout tool is open source and does buffer tree synthesis. It works in conjunction with Graywolf

• **Router** – 'vsdflow' uses 2 different Routers, viz. Qrouter (fully open-source) and Jrouter (partial open-source). **Qrouter** is a maze router, based on standard Lee Maze router algorithm. **Jrouter** is a commercial signal router. It is at par with industry standard routers and is developed by James Ho of Jspeed Technologies

• **Static timing analysis (STA)**– 'vsdflow' uses 2 different fully open-source STA tools, **Opentimer** and **Vesta**. Opentimer is a high- performance academic timing analysis tool and supports important features like PBA, CPPR and many more. Vesta is an open-source timing analysis tool and comes a part of qflow package

RESULTS USING VSDFLOW

To generate a performance output of the same design with different EDA tools, we need a EDA management system (EMS), which uses a mix and match of above mentioned tools and generate comparison results, there-by making the decision process simpler for tool selection. 'vsdflow' is one such revolutionary EMS which takes inputs from users in a csv format and generates performance metrics output from set of tools selected by user. Comparison metrics between different combination

of tools can be generated using multiple parallel runs of 'vsdflow' with a different set of tool selection by user

'vsdflow' was tested on picorv32 – a RISC-V CPU core which implements RV32I instruction set, and on E31- a RISC-V CPU core which implements RV32IMAC.

RISC-V is a free and open-source RISC instruction set architecture and was originally developed in Computer Science division of EECS Department at University of California, Berkeley. The below Figure 4. correctly describes RISC-V architecture and its implementation flow

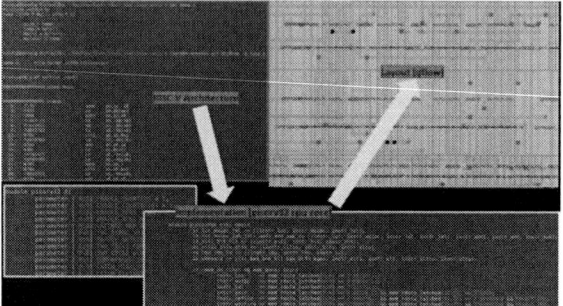

Figure 4: Snippet of RISC-V architecture and its physical design implementation flow

As per 'vsdflow' concept mentioned, it can use a mix and match of tools, so picorv32 was tested using 4 different flows mentioned below with 180nm PDK's from OSU and 45nm PDK's from Nangate. (4 flow with mix n match of different tools) –

• **Flow1** – RTL and Clock Tree Synthesis using **Yosys** and **blifFanout**, followed by placement using **Graywolf** and routing using **Qrouter**. Performance metrics displayed in below **Table I**

• **Flow2** – Synthesis using **Yosys**, followed by placement using **Fractal**, routing and buffer tree synthesis using **jrouter**. Performance metrics displayed in below Table II Comparing results from below Table I and Table II, it can be easily deduced, that **Flow2** has superior performance in terms of runtime and area, but is partially open-source, while **Flow1** is fully open-source and has comparable performance with **Flow2**.

TABLE I.   PERFORMANCE METRICS FOR PICORV32 USING FLOW1

Picorv32	Technology node 180nm, Operating Frequency = 378MHz, RAM = 1GB		
	Runtime	*Inst. count*	*Area um²*
Synthesis	1min	13215	-
Placement	36min	14828	1197 X 876
Routing	39min	14828	1197 X 876

This comparison between 2 flows enables user to decide on which Flow to go for.

TABLE II.   PERFORMANCE METRICS FOR PICORV32 USING FLOW2

Picorv32	Technology node 180nm, Operating Frequency = 378MHz RAM = 1GB		
	Runtime	*Inst. count*	*Area um²*
Synthesis	1min	11956	
Placement	9min	11956	814 X 814
Routing	4min	13826	814 X 814

• **Flow3** – This experiment was performed to find the capacity of 'vsdflow' and E31 RISC-V CPU core was used for this flow with 45nm PDK's from Nangate (3 RAM's implemented separately). Synthesis was done using **Yosys**, followed by placement using **Fractal** and routing using **jrouter**. Performance metrics are displayed in below Table III

TABLE III.   PERFORMANCE METRICS FOR E31 USING FLOW3

E31	Technology node 45nm, RAM = 1.2GB (~10GB including RAM's)		
	Runtime	*Inst. count*	*Area um²*
Synthesis	48min	1073716 (incl. RAMS)	
Placement	25min	94244 (excl. RAM'S)	543 X 543
Routing	10min	94244 (excl. RAM'S)	543 X 543

• **Flow4** – This experiment, where Flow is like Flow1, was performed to identify the capacity of fully open-source Flow1. Flow4 uses an example system of MCU from ARM, which contains a single cortex-m0 processor, internal program memory, SRAM data memory and boot-loader. The evaluation RTL is available on ARM website and performance metrics are displayed in below Table IV.

TABLE IV.   PERFORMANCE METRICS FOR CORTEX-M0 USING FLOW4

cortex-m0	Technology node 180nm, Operating Frequency = 157MHz RAM = 1GB		
	Runtime	*Inst. count*	*Area um²*
Synthesis	2min	27435	
Placement	72min	29479	1696 X 1246
Routing	111min	29479	1696 X 1246

FORTHCOMING RESEARCH

'vsdflow' envisions to add a lot more features and below list enumerates a few of them -
• For designs, with huge instance count post-synthesis (>100k), 'vsdflow' has built-in hierarchical approach using combination of above mentioned partially open-source tools. The same approach needs to be incorporated in 'vsdflow' using commercial EDA tools
• 'vsdflow' is working towards to provide more accurate post-layout timing results with back-annotated parasitics included for fully and partially open-source tool-set.
• 'vsdflow' is enabling an automated timing ECO framework, which will perform timing ECO's using commercial and open-source EDA tools, and provide a comparison performance metrics report post-ECO. This feature will enable user to seamlessly decide and switch between best of all worlds

REFERENCE

Refer to the reference section of this paper, for contributions to Fractal placer and jrouter [1], Graywolf placer and Qrouter [2], Opentimer and vesta, an open-source static timing analysis engine [2] & [3], Yosys synthesis tool [4], picorv32 CPU core [4], E31 CPU core [5] and arm cortex-m0 [6]

Please look for updates on 'vsdflow' on our website at www.vlsisystemdesign.com. If you have any questions regarding 'vsdflow' and its usage, please do not hesitate to contact Kunal at kunalpghosh@gmail.com

ACKNOWLEDGEMENTS

'vsdflow' would not have been possible without the constant contributions and effort by many people.

Rajeev Srivastava & Aditya Pratap Singh [1] have been very generous and helpful towards contributing Fractal placer and Jrouter tools for rigorous testing 'vsdflow'

Tim Edwards [2] continuous and timely support has been helpful in 'vsdflow' inception and making it available for community usage

And finally, really appreciate Tsung-Wei Huang's [3] prompt help on fine tuning Opentimer based on community feedback. He plans to upgrade tool to support advanced timing models

REFERENCES

[1] Rajeev Srivastava & Aditya Pratap Singh, Creators of VLSI App called 'webchip.in', an Open digital circuit design platform.
[2] Tim Edwards, Founder of opencircuitdesign.com, a repository for suite of open-source EDA tools.
[3] Tsung-Wei Huang, Developer of Opentimer, an open-source high-performance timing analysis tool
[4] Clifford Wolf, Developer of Yosys, a framework for verilog RTL synthesis and Owner of picorv32 – a size-optimized RISC-V
[5] SiFive, provider of E31 RISC-V CPU core evaluation RTL.
[6] ARM developer, provider of Cortex-M0 CPU core.

978-1-5386-5309-8/18 $31.00 © 2018 IEEE

An Improved Leakage-Driven Runtime Decap Modulation Algorithm for Microprocessors

Leilei Wang and Pingqiang Zhou

School of Information Science and Technology, ShanghaiTech University, Shanghai, China

Corresponding Author's Email: zhoupq@shanghaitech.edu.cn

Abstract—The leakage power of decaps is a significant portion of the total chip power. Previous work [1] proposed a method to dynamically modulate the decap at runtime to save leakage power consumed by the unused decap. However, it did not consider the fact that the pre-placed decap at chip design stage actually has limited capacity, thus the estimated capacitance by the method [1] may exceed the decap's capacity. In this paper, we address this issue, and propose a method to improve the prior work to make it feasible. The effectiveness of our idea is tested on four benchmarks.

I. INTRODUCTION

Power is delivered from the board-level voltage regulator to each on-chip transistor by a power delivery network (PDN) as shown in Fig. 1(a). The parasitics in the PDN, together with the temporal power variations in the switching circuit blocks, result in transient voltage noise in the power grid, which may impair the performance and reliability of a chip. The on-chip decoupling capacitors (decaps) in a processor chip can serve as temporary power reservoirs, and provide the needed power to its nearby large switching circuit blocks to suppress the supply noise, and thus increase the reliability of the power grid [2]. Unfortunately, the deliberately-added decap can occupy more than 20% of the total chip area in high-end processors and its leakage can contribute to more than 20% of the total power consumption [3].

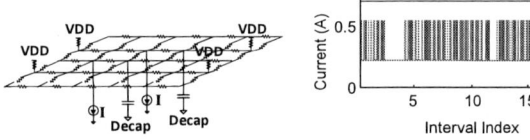

(a) The model of power grid for a given chip, with VDD sources, decaps and circuit block loads.

(b) The simulated power profile of one circuit block in a processor chip. One time interval = 10,000 cycles.

Fig. 1. The model of a power grid and one sample of the current trace for the current loads in the power grid.

Previous work in [1] proposed a method to adapt the amount of "on" capacitance of each pre-placed decap dynamically to the varying need of the power loads (see Fig. 1(b)) at runtime, thus to reduce the decap leakage consumed by the unused "on" capacitance. Although it has been shown that the idea in [1] can effectively reduce the decap leakage power, it ignores one important fact that the amount of the pre-allocated decaps distributed across the chip is limited, and when estimated amount of needed "on" capacitance by the method in [1] is larger than the pre-allocated capacitance, the method in [1] is unfeasible.

In this paper, we improve the method proposed in [1], and address the aforementioned issue. In Section II, we present our proposed method. Then we show the results in Section III, which is followed by the conclusion in Section IV.

II. IMPROVED DECAP MODULATION APPROACH

Generally speaking, if one decap could not afford the demand charge due to limited capacity, its voltage drop will be larger than its surrounding power grid nodes. As a result, its insufficient amount of charge will have to be amortized among this decap's nearby VDD pins and other decaps with enough capacity. Fig. 2(a) shows one example. Decap1 is insufficient,

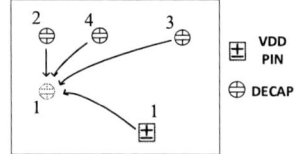

(a) Insufficient decap1 and its surrounding potential helpers.

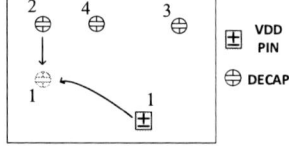

(b) Decap1 is helped only by VDD1 and decap2, because decap3 is remote and decap4 itself is insufficient.

Fig. 2. A simple example showing an insufficient decap and its potential helpers.

so after turning on all its available capacitance, it has to borrow the remaining charge from nearby VDD pin VDD1 and three decaps decap2, decap3 and decap4.

Assume the amount of charge form each helper to this insufficient decap is proportional to the equivalent conductance between the helper and the insufficient decap [1], we can calculate the amortized charge for each decap helper, which is transformed to the extra "on" capacitance of the decap helper. The equivalent conductance matrix $\mathbf{A}$ for the example shown in Fig. 2(a) is given in Fig. 3, where the absolute value of each non-diagonal element represents the equivalent conductance between a decap/vdd and another decap/vdd.

	Vdd1	Decap1	Decap2	Decap3	Decap4
Vdd1	0.4905	−0.1626	−0.0514	−0.1203	−0.1562
Decap1	−0.1626	0.7035	−0.3051	−0.0203	−0.2155
Decap2	−0.0514	−0.3051	0.7824	−0.1277	−0.2982
Decap3	−0.1203	−0.0203	−0.1277	0.4121	−0.1438
Decap4	−0.1562	−0.2155	−0.2982	−0.1438	0.8137

Fig. 3. Matrix $\mathbf{A}$ of the example in Fig. 2(a).

978-1-5386-5309-8/18 $31.00 © 2018 IEEE

A. How to Find the Equivalent Conductance Matrix A?

Given an RC power grid with N nodes, for the calculation of amortized charge, we can approximate it as a purely resistive grid whose system equation can be expressed as [4]

$$GV = J, \quad G \in R^{N \times N}, V \in R^N, J \in R^N. \quad (1)$$

where G is the conductance matrix, V is the voltage vector and J is the vector of current sources. Among the N nodes, we treat the nodes that the M VDD sources and K decaps connected at as port nodes, while other nodes as internal nodes. Then we can apply the macromodeling approach [5] to have

$$\begin{bmatrix} G_{11} & G_{12} \\ G_{12}^T & G_{22} \end{bmatrix} \begin{bmatrix} \mathbf{V}_{int} \\ \mathbf{V}_{port} \end{bmatrix} = \begin{bmatrix} J_{int} \\ J_{port} + I_{port} \end{bmatrix} \quad (2)$$

where

- G_{11} is the conductance matrix of the internal nodes,
- G_{12} is the conductance matrix between the internal nodes and the port nodes,
- G_{22} is the conductance matrix of the port nodes,
- V_{int} and V_{port} are the voltage vectors of the internal and port nodes respectively,
- J_{int} and J_{port} are the current sources connected at the internal nodes and ports respectively.
- I_{port} is the vector of currents flowing into the ports.

The Eq. (2) can be finally transformed into

$$I_{port} = \underbrace{(G_{22} - G_{12}^T G_{11}^{-1} G_{12})}_{\text{Port admittance matrix A}} V_{port} + (G_{12}^T G_{11}^{-1} J_{int} - J_{port})$$
$$(3)$$

where $A = G_{22} - G_{12}^T G_{11}^{-1} G_{12}$ is the equivalent conductance matrix among the port nodes, which is exactly the matrix we want as shown in Fig. 3.

B. Amortization Charge among the Helpers

After obtaining the matrix A, we can start calculating the amortized charge for each decap helper, i.e., its extra required "on" capacitance.

Although all the surrounding decaps or VDDs of an insufficient decap could be the potential helpers, the following two types of decaps/VDDs can be removed from the helper list:

- *The remote helpers.* In reality, a remote helper can only provide very tiny charge to the insufficient decap, because they have a large equivalent impedance to the decap, so we can ignore such remote helpers in the calculation. For instance, in Fig. 2(b) since the equivalent conductance between decap1 and decap3 (0.0203 in Fig. 3) is less than 5% of the self-conductance of decap1 (i.e., the diagonal element , 0.7035), we can assume that the contribution of charge from decap3 is zero, and correspondingly, we can set its equivalent conductance to be zero, and thus achieve a new matrix $\hat{A}$ shown in Fig. 4.

  We can use a list S_k to record all the effective helpers for one insufficient decap k. For the example shown in Fig. 2(b), S_k for decap1 is {Vdd1, decap2 and decap4}.

- *The surronuding insufficient decaps.* All the insufficient decaps have to turn "on" all its available capacitance and are unable to help the other insufficient decaps. We use

	Vdd1	Decap1	Decap2	Decap3	Decap4
Vdd1	0.4905	−0.1626	−0.0514	−0.1203	−0.1562
Decap1	−0.1626	0.7035	−0.3051	0	−0.2155
Decap2	−0.0514	−0.3051	0.7824	−0.1277	−0.2982
Decap3	−0.1203	0	−0.1277	0.4121	−0.1438
Decap4	−0.1562	−0.2155	−0.2982	−0.1438	0.8137

Fig. 4. Matrix $\hat{A}$, obtained by setting the remote helper's conductance to be zero in matrix A, as shown in Fig. 3.

a list C_{inMax} to record such decaps. For instance, in Fig. 2(b), decap4 is insufficient, it would be put into C_{inMax}. Notice that the list C_{inMax} is dynamically updated as we process each insufficient decap in the iteration (see Section II-C).

As a result, the effective helpers for decap k would be those in S_k, but not in C_{inMax}. For an effective helper $j \in S_k, j \notin C_{inMax}$, the extra amount of "on" capacitance it must provide to help decap k, $\triangle C'_j$, is

$$\triangle C'_j = \frac{a_{jk} \cdot (C_k - C_{k,max})}{\sum_{j \in S_k, j \notin C_{inMax}} a_{jk}} \quad (4)$$

where

- a_{jk} is the equivalent conductance between the helper j and decap k.
- C_k is the required "on" capacitance of decap k estimated by the method in [1].
- $C_{k,max}$ is the maximum available capacitance of pre-placed decap k.

For example, in Fig. 2(b) the extra capacitance provided by decap2 to help decap1 is $\triangle C'_2 = \frac{0.3051 \cdot (C_1 - C_{1,max})}{0.3051 + 0.1626}$.

C. Overall Solution and Discussion

The complete improved modulation approach is shown in Algorithm 1. We maintain a list C_{toHelp} to record all the insufficient decaps. After the pre-calculation of required "on" capacitance for each decap by the method in [1], we put the indexes of insufficient decaps into list C_{toHelp} (line $4 \sim 9$). Then for each insufficient decap k in C_{toHelp}, we estimate the amount of charge from its helpers with (4), until C_{toHelp} is empty (line $10 \sim 18$). Once the required amount of capacitance is determined for each decap, we can use the idea of Gated-decap [6] to set the right "on" portion of decaps across the chip.

The algorithm complexity is $O(K^2)$ theoretically, where K is the total number of decaps in the chip. Although new insufficient decaps may appear during the borrowing process in the iteration (line 16), the algorithm will terminate within limited iterations. This is because we know that though several decaps may not have enough available capacitance, in a local chip area, there should be enough decap resource to use, which is ensured at the design step of the processor chip.

Regarding the overhead of physical implementation of our proposed approach, the Gated-decap used in the proposed method (also in [1]) would bring about 5.36% area overhead of the decaps, while the energy overhead is negligible if an appropriate interval of modulation is adopted [6].

978-1-5386-5309-8/18 $31.00 © 2018 IEEE

Algorithm 1 The Improved Online Decap Modulation Approach

1: Obtain matrix $\hat{A}$. (see (3) and Fig. 4) (off-line step)
2: **while** a chip block switches **do**
3:     Create new lists $C_{\_inMax}$ and $C_{\_toHelp}$
4:     **for** $j = 1$ to K **do**
5:         Calculate C_j by the method in [1]
6:         **if** $C_j > C_{j,max}$ **then**
7:            add decap k to $C_{\_inMax}$, add k to $C_{\_toHelp}$
8:         **end if**
9:     **end for**
10:     **while** $C_{\_toHelp}$ is not empty **do**
11:         Pop a decap k from $C_{\_toHelp}$
12:         Find S_k of decap k (see Section II-B)
13:         **for** each $j \in S_k, j \notin C_{\_inMax}$ **do**
14:            re-calculate $C_j = C_j + \triangle C_j'$ with (4)
15:            **if** $C_j > C_{j,max}$ **then**
16:                Add decap j to $C_{\_inMax}$, add j to $C_{\_toHelp}$
17:            **end if**
18:         **end for**
19:     **end while**
20:     Set the "on" capacitance of each decap
21: **end while**

III. EXPERIMENTAL RESULTS

We test our method in four benchmarks. The power traces are obtained by running the PARSEC 2.1 benchmarks [7] on simulators MacPAT [8] and GEM5 [9], and the power grids are adapted from the IBM Power Grids [10].

Fig. 5 shows that modulation result of one insufficient decap in Benchmark1. The demand capacitance estimated by method [1] is shown in Fig. 5(a). But since this decap has a maximum available capacitance of 5nF, it has to be modulated as shown in Fig. 5(b). Using our method, the

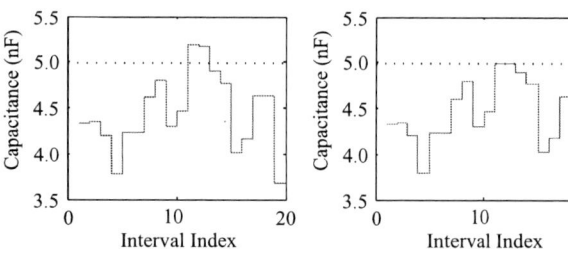

(a) The demand capacitance of one decap in Benchmark1.

(b) With maximum capacitance 5nF, this decap could only be modulated under 5nF.

Fig. 5. The estimated and the modulated capacitance of one insufficient decap in Benchmark1.

insufficient amount of this decap, 186pF, is amortized among its surrounding helpers as shown in Fig. 6. Note that part of the charge is provided by the nearby Vdds, whose numbers were not shown in the figure.

Table I shows the maximum available capacitance (**MaxCap**) and the estimated required capacitance (**EstCap**) of the worst insufficient decaps in all the four benchmarks.

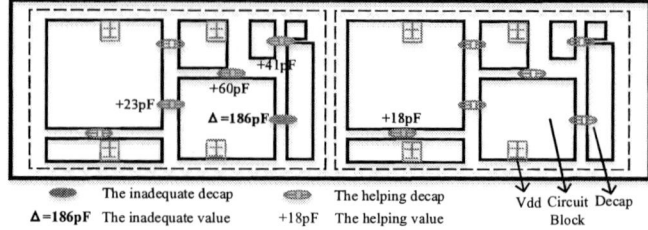

The inadequate decap   The helping decap   Vdd Circuit Decap Block
Δ =186pF The inadequate value   +18pF The helping value

Fig. 6. The amortized charge (more specifically, capacitance) among the helpers in Benchmark1 for one insufficient decap in a two-core processor chip.

The Lowest voltage shows the minimum voltage of all the power grid nodes after using the proposed decap modulation method. From Table I we can see that our method is effective in modulating the insufficient decaps for all the benchmarks.

TABLE I
RESULTS OF ALL THE FOUR BENCHMARKS. THE VDD IS 1V AND WE SET 10% OF VDD AS THE ALLOWED VOLTAGE DROP IN THE EXPERIMENTS.

#Benchmark	Insufficient decap amount		Lowest voltage (V)
	MaxCap (nF)	EstCap (nF)	
Benchmark1	5.000	5.186	0.9039
Benchmark2	3.500	3.606	0.9086
Benchmark3	3.000	3.167	0.9185
Benchmark4	4.000	4.327	0.9123

IV. CONCLUSION

In this paper, we improve our prior work [1] by proposing a method to modulate the decaps for large function blocks in a processor chip, under the practical constraint that each pre-placed decap has limited capacity. Results on several benchmarks show that our method is feasible. In the future, we will test our method on more benchmarks of heterogeneous chips.

REFERENCES

[1] L. Wang and P. Zhou, "Leakage power reduction in multicore chips via online decap modulation," in *CSTIC*, 2016, pp. 1–3.
[2] P. Zhou *et al.*, "Congestion-aware power grid optimization for 3D circuits using MIM and CMOS decoupling capacitors," in *ASP-DAC*, 2009, pp. 179–184.
[3] J. Gu *et al.*, "Design and implementation of active decoupling capacitor circuits for power supply regulation in digital ics," *TVLSI*, vol. 17, no. 2, pp. 292–301, Feb. 2009.
[4] C.-W. Ho *et al.*, "The modified nodal approach to network analysis," *TCAS*, vol. 22, no. 6, pp. 504–509, Jun. 1975.
[5] M. Zhao *et al.*, "Hierarchical analysis of power distribution networks," in *DAC*, 2000, pp. 150–155.
[6] Y. Chen *et al.*, "Gated decap: Gate leakage control of on-chip decoupling capacitors in scaled technologies," *TVLSI*, vol. 17, no. 12, pp. 1749–1752, Dec. 2009.
[7] C. Bienia, "Benchmarking modern multiprocessors," Ph.D. dissertation, Princeton, NJ, USA, 2011.
[8] S. Li *et al.*, "McPAT: An integrated power, area, and timing modeling framework for multicore and manycore architectures," in *Micro*, 2009, pp. 469–480.
[9] N. Binkert *et al.*, "The gem5 simulator," *SIGARCH Comput. Archit. News*, vol. 39, no. 2, pp. 1–7, Aug. 2011.
[10] "IBM power grid benchmarks," available at http://dropzone.tamu.edu/~pli/PGBench/.

978-1-5386-5309-8/18 $31.00 © 2018 IEEE

IMPLEMENTATION OF HEART RATE DETECTION ALGORITHM BASED ON A LOW POWER CHIP

Yingying Wu, Hao chen, Zhongmin Lin, Zirong Chen, Xin-an Wang*

The Key Lab of Integrated Microsystems Peking University Shenzhen Graduate School,
Shenzhen 518055
* Email:wangxa@pkusz.edu.cn

ABSTRACT

This paper presents a low power electrocardiogram (ECG) signal processing application specific integrated circuit (ASIC) chip for real-time and reliable detection of heart rate (HR). At a sampling frequency of 250 Hz, the analog front-end (AFE) accurately senses and digitizes the raw ECG signal, which is then sent to the digital signal processing (DSP). In DSP, after preprocessing, the R-peak detection and RR interval cumulative deviation detection, data will be transmitted to terminal equipment via the universal asynchronous receiver/transmitter (UART) transfer protocol. The terminal will receive instantaneous heart rate (IHR) and mean heart rate (MHR). Under 0.18um standard CMOS process, the area and power are 0.3 mm^2 and 10.9uW, respectively.

INTRODUCTION

The incidence rate of heart disease is on the rise, and real-time monitoring of HR is the most effective way of preventing heart disease. The existing HR monitoring devices can process photoplethysmography (PPG) signal or ECG signal to get HR. However, the performance of PPG analysis method degrades when patients are moving, which increases the complexity of HR extraction algorithm [1]. Most of the ECG signal processing algorithms adopt software algorithms, which have high power consumption and poor endurance, but the single chip designed in this paper can meet the requirements of real-time, high accuracy and low power consumption simultaneously.

Since the R-peak detection is the basis of ECG signal analysis, the structure of most ECG signal processing chips is preprocessing raw signal, then detecting R-peaks, finally detecting heart rate or evaluating heart rate variability (HRV) [2-4]. Although the structure is similar, the detection algorithms of R-peaks and hardware implementations in different chips are totally different. [2] uses adaptive threshold detector to detect R-peaks, which has simple algorithm and low power consumption but poor accuracy. [3-4] adopts wavelet transform to obtain R-peaks, but the algorithm is complex and power consumption is high.

Considering the complexity, accuracy and noise immunity, we realized this low power HR detecting chip. It contains a band-pass filter (BP), envelope extraction,

peak-finding logic and RR interval cumulative deviation check (RRI_check), 4K and 2K on-chip register file (RF), a serial peripheral interface (SPI) and a UART interface. Under the standard CMOS process of 0.18um, the power consumption of the chip is 10.9uW, which meets the needs of real-time monitoring of HR.

CHIP ARCHITECTURE

Figure 1 illustrates the structure of the proposed chip and its upper and lower module, which we called HR monitoring system. AFE samples ECG data and transmits it to the DSP, finally IHR and MHR are sent to the phone to display.

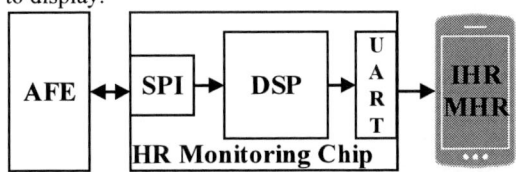

Figure 1: The Architecture of HR monitoring system

The concrete structure of AFE is shown in Figure 2. The sampled 16-bit ECG signal is transmitted to DSP through SPI interface.

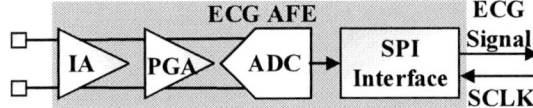

Figure 2: Schematic of the AFE

The proposed chip, as shown in Figure 3, contains 9 modules, and the clock management module and the reset module control other modules. The chip is reset by asynchronous reset and synchronous release, reducing the possibility of reset failure.

In DSP, BP preprocesses the ECG signal to remove ECG noises; Envelope firstly performs first order differential processing to remove the low frequency signal, then squares it to further enhance the R-peaks and reduce the high P waves and the T waves, finally filters the data; Peak-finding uses Hilbert filter firstly, then the adaptive threshold method is used to detect zero crossing point and RR interval [5], which can reduce the false detection caused by noise; RRI_check redetects RR interval based on cumulative deviation to remove the wrong RR interval, correct accidental error detection caused by insertion and eliminate the obviously error RR interval.

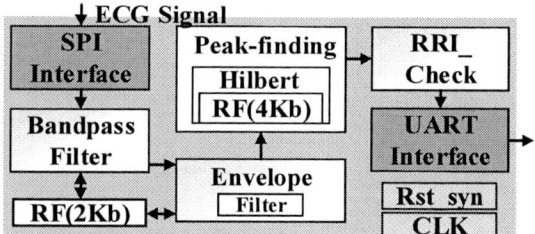

Figure 3: Chip block diagram

HARDWARE IMPLEMENTATION

SPI Interface

Data exchange between the chip and AFE is implemented through the SPI interface. The timing diagram of the interface is shown in figure 4.

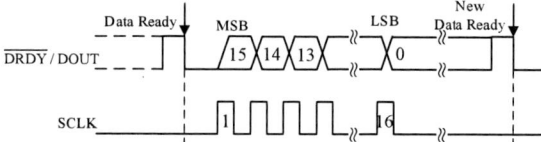

Figure 4: Timing Diagram of SPI interface

AFE informs the chip that the sample has completed through the falling edge of the SPI interface. Then the interface converts the serial ECG data to parallel and sends it to the BP, and outputs the clock line to AFE.

Digital Signal Processing (DSP)

As the data flow shown in Figure 3, the DSP consists of 5 blocks, the most important of which is the filter and limiting.

After receiving the data, we need to filter the 16-bit ECG signal multiple times to remove ECG noises, such as baseline wander, electromyogram (EMG) interface and power frequency interference in the signal, and in Envelope we use a filter to sharpen QRS complex and smooth other components. In Peak-finding we adopt Hilbert filter to transform the peak value of the envelope to zero for the zero point detection, which is easier to achieve than peak detection.

The three filters in the chip have the same structure, all of them are the direct-type of finite impulse response (FIR) filter. The difference equation for the FIR filter is given by:

$$y(n) = \sum_{i=0}^{N-1} h(i)x(n-i) \quad (1)$$

As shown in formula (1), the essence of the direct FIR filter is a multiply-and-accumulate (MAC) operation. Considering the amount of hardware resources that are required, the filter employs a serial structure and reduces

the length. The coefficients are 12-bit fixed-point numbers, which improves the accuracy of the filter. In addition, a folded structure is used to reduce the chip area by only storing half of the coefficients.

In BP, the filter length is 48, the bandwidth is 6 to 18Hz, and the stopband attenuation is 30dB. The filter realizes shift operation by changing the addresses, thereby avoiding the power consumption caused by data-shift and reversal operation in the data delay chain. As shown in Figure 5, a MAC is applied to time division multiplexing, and multiple clock cycles are used to multiply and accumulate to achieve filtering, which greatly saves the hardware resources.

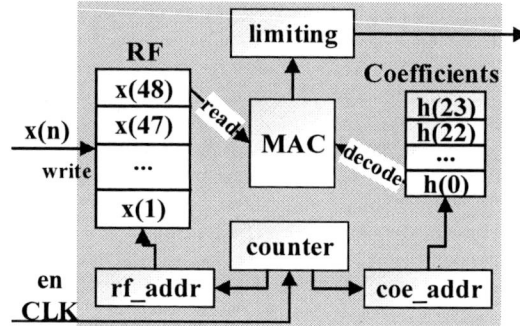

Figure 5: Band-pass filter block structure

In Envelope, the triangular window filter length is two times the width of the QRS complex, and the coefficient set contains 75 values, which is [1,2,3, ..., 37,38,37,36,...,2,1]. In addition, the structure of this filter is the same as that of BP, and the address decoding of the 2Kb RF is shared, which further reduces the total area of RF.

In Peak-finding, considering that the output of Envelope is a low frequency envelope, the Hilbert filter uses the coefficient drop sampling to decrease the number of coefficients from 125 to 50, which greatly reduces the computational complexity and the power consumption without losing quality. The Hilbert filter whose length is 246 has the same structure as BP, and also has a 4Kb RF to store data.

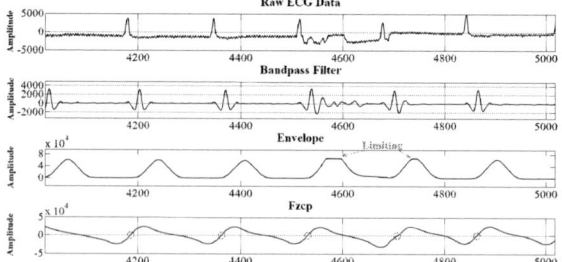

Figure 6: The waveform after each module

Because of the possibility of overflow when the ECG signal is large, we not only increase the intermediate data width of related modules, but also limit the output data in

978-1-5386-5309-8/18 $31.00 © 2018 IEEE   580

each module to 16-bit, that is, when the data exceeds the range of 16 binary number, it is limited to the maximum or the minimum value, which can reduce the amount of calculation as well as the consumption of hardware resources.

The waveform of each module is shown in Figure 6, the points marked by red circles are zero crossing points which correspond to R-peaks. For the waveform of the Envelope is symmetrical, limiting doesn't influence the function of Hilbert filter, which is to change the R-peaks to zero crossing points, so it doesn't affect the accuracy of the chip.

UART Interface

In this block, data is first packaged as the format shown in Figure 6, then the data is sent in bytes through the UART interface, and the number of cycles between adjacent data is fixed.

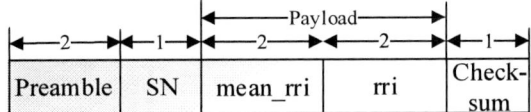

Figure 6: The UART frame format

The 2 bytes preamble ensures that the current data is valid. A sequence number (SN) of 1 byte prevents repeated reception of the same frame data. The 4 bytes of IHR and MHR are used to display the HR on the phone. A checksum of 1 byte can verify whether the received data is correct.

SYSTEM VERIFICATION RESULTS

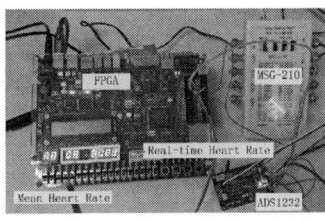 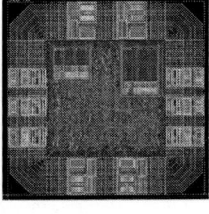

(a)   *(b)*

Figure 8(a): FPGA prototyping system

Figure 8(b): The layout of the chip

To verify the correctness of this hardware implementation of the HR detection algorithm, we build a prototype system using Altera Cyclone II FPGA. As presented in Figure 8(a), the ECG signal is generated by TSTWIT MSG-210. The sensor is ADS1232 from TI for signal sampling. As we can see, the top three numbers in FPGA represent MHR while the latter three represent IHR. It is found that when the signal amplitude is between 0.1mV-50mV, the design can measure the HR accurately.

The layout is shown in Figure 8(b). Table 1shows the comparison of the capabilities of each ECG chip.

TABLE I Performance Comparison

	[2]	[3]	[4]	Our design
Frequency	256Hz	1MHz	1Khz	76.8KHz
Technology	65nm	0.18um	0.35um	0.18um
Voltage	1.0V	1.2V	3V	1.8V
Se(%)	99.58	-	99.90	99.88
+P(%)	99.57	-	99.91	99.30
Power(uW)	0.96	32	13.6	10.9
Area(mm^2)	1.56	23.5	1.2	0.3

As we can see in table I, [2] has low power but Se(sensitivity) and +P(positive prediction) are low,[3-4] has high power, so our design performs best.

SUMMARY

This paper has presented the implementation of the HR detection algorithm based on a low power chip including preprocessing, R-peak detection, RR interval detection and two interfaces. The chip can accurately process ECG signal in real-time, display IHR and MHR on the phone, and the extracted R-peak also provides the basis for the detection of HR disease. This design has achieved the goal and has been validated on the FPGA with good working performance. Under the 0.18um standard CMOS process, the chip area is 0.3 mm^2, and the power consumption is 10.9uW.

ACKNOWLEDGEMENTS

This work is supported by R&D project of Shenzhen Government(Project NO.JCYJ20160229094148396 and NO.JCYJ20170306091821082).

REFERENCES

[1] Krachunov S, Beach C, Casson A J, et al. Energy efficient heart rate sensing using a painted electrode ECG wearable[C]//Global Internet of Things Summit (GIoTS), 2017. IEEE, 2017: 1-6.

[2] Deepu C J, Zhang X, Heng C H, et al. A 3-Lead ECG-on-Chip with QRS Detection and Lossless Compression for Wireless Sensors[J]. IEEE Transactions on Circuits & Systems II Express Briefs, 2016, 63(12):1151-1155..

[3] Kim H, Kim S, Van Helleputte N, et al. A configurable and low-power mixed signal SoC for portable ECG monitoring applications [J]. IEEE transactions on biomedical circuits and systems, 2014, 8(2): 257-267.

[4] Min Y J, Kim H K, Kang Y R, et al. Design of Wavelet-Based ECG Detector for Implantable Cardiac Pacemakers[J]. IEEE Transactions on Biomedical Circuits & Systems, 2013, 7(4):426.

[5] Manikandan M S, Soman K P. A novel method for detecting R-peaks in electrocardiogram (ECG) signal [J]. Biomedical Signal Processing and Control, 2012, 7(2): 118-128.

978-1-5386-5309-8/18 $31.00 © 2018 IEEE

Design And Implementation Of A Low-Complexity R-peak Detection Algorithm

Zirong Chen , Zhongmin Lin, Hao chen, Xin-an Wang, Yingying Wu

The Key Lab of Integrated Microsystems Peking University Shenzhen Graduate School, Shenzhen 518055

* Email:wangxa@pkusz.edu.cn

ABSTRACT

Cardiovascular disease, which is one of the most dangerous diseases in modern medicine, has become a threat to people's health. ECG is a kind of information that can reflect the physiological characteristics of human heart. With the development of modern technology, achieving ECG for health monitoring in the wearable hardware has become increasingly popular. The detection of R-peak is very important in ECG signal processing. This paper presents a low complexity r-peak detection algorithm , which is suitable for ASIC implementation. First of all, the algorithm use digital filter and envelope filter to remove the noise which is in the ECG. Then it use the Hilbert transform convert the r-peak to zero-crossing point. At last, it check the zero-crossing point to finish the R-peak detection. The algorithm is using the first-channel of the MIT-BIH arrhythmia database to test, and achieves average detection accuracy of 99.81%, sensitivity of 99.87%.

INTRODUCTION

In order to check the health of the heart, it often need ECG to provide cardiac physiological information. And ECG signal processing depends heavily on R-peaks detection. Characteristic parameters such as heart rate(HR), heart rate variation rate (HRV) all depend on R-peak detection. The detection of QRS complex, P wave, T wave also depend on it. The location of R-peak, QRS complex, P wave, T wave is shown in Figure 1.

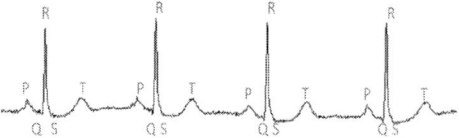

Figure 1: The Location of complexes of ECG

So many methods based on the derivatives [1], wavelet transform [2], empirical mode decomposition(EMD)[3] and neural networks [4] have been developed for the R-peaks detection. The derivatives method takes advantage of the maximum slope value of R-peak, while the slope of other ECG components is small. The R-peak

is detected using the slope threshold varying with the signal [1]. Wavelet transform decomposes the signal into different scales and eliminates noise. It has the characteristics of multi-resolution [2]. However, Wavelet transform has the choice problem of mother wavelet and scales to obtain QRS events. Although the EMD-based approach in [3] can overcome the choice problem of basis function, selection of a set of intrinsic mode functions (IMFs) is very difficult under noisy environments. The neural network is used as an adaptive nonlinear filter, which filters the ECG signal and detects r-peaks[4]. But the neural network needs training, it has very high-complexity and its convergence speed is slow and the generalization is weak.

This paper focuses on how to implement R-peak detection algorithm on hardware. So there are two requirements for the algorithm: (1) high accuracy of R-peak detection. (2) low –complexity. Since the above algorithms are hard to meet these two requirements in hardware implementation, this paper proposes a new R-peak detection algorithm based on envelope filtering and Hilbert transform.

The flow of the algorithm is shown in Figure 2.

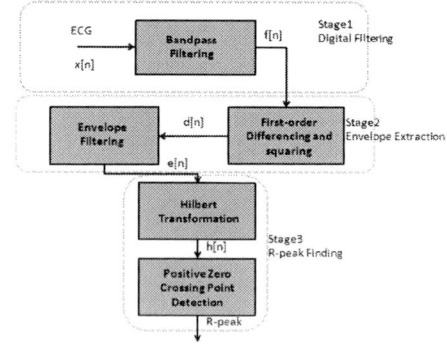

Figure 2: R-peak detection algorithm flow

R-PEAK DETECTION FLOW
Digital filtering

Since the ECG signals collected from the sensors are extremely weak, the noise signals are several orders of magnitude higher than the ECG signals. 90% frequency spectrum of ECG concentrate on 0.25Hz to 45 Hz, the main complex of ECG is QRS wave, its frequency spectrum concentrate on 3Hz to 40 Hz. P wave of frequency spectrum concentrate on 0Hz to 18 Hz. T wave of frequency spectrum concentrate on 0Hz to 8 Hz. The main noise of ECG is as follows: (1) power frequency interference, it is produced by AC network. Its frequency spectrum concentrate on 50Hz and inter-harmonics of 50Hz [5]. (2) base-line drift: baseline drift noise is mainly caused by respiratory, human activities and other factors. Its frequency spectrum concentrate on 0.05-2Hz. (3) EMG interference: EMG disturbance is mainly caused by muscle tension or human activity, which belongs to high frequency interference. Its frequency spectrum concentrate on 10-1000Hz. Its frequency spectrum concentrate on 10-1000Hz. To sum up, in order to remove the influence of noise and the P wave and T wave, a bandpass FIR filter is used in the first stage. The bandpass filter is designed for the bandwidth of 6-18Hz. It can effectively preserve the QRS complexity of ECG and remove interference at the same time. It makes a tradeoff between noise remove effect and computational complexity ,uses 48 order to design the bandpass filter. The figure 3 is the original ECG signal x[n] and the ECG signal f[n] which is filtered. It shows that the band-pass filter has a inhibition on the noise and P wave and T wave.

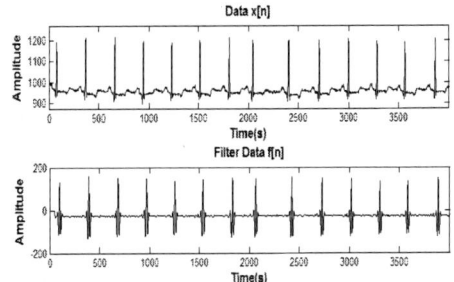

Figure 3 Digital filtering result

Envelope extraction

After filtering the ECG signal, the filtered signal f[n] is given a differential operation and square operation. The formula is :

$$d[n] = (f[n] - f[n-1])^2 \qquad (1)$$

The differential operation and square operation can be regarded as a high pass filter, which can enhance the R-peak of f[n]. And then the envelope is extracted from d[n].Because it can smooth d[n] and get the only one max point from one R-peak. Some algorithms use the rectangle window to extract the envelope. This method is the average calculation of the collected signal, and then extract the R-peak, although this method is simple and convenient in calculation, but the envelope is not sharp, it is

difficult to find the R-peak, and it's not easy for hardware implementation and use more order than triangular window .This paper trade off the hardware cost and the accuracy of R-peak detection and determine to use the triangular window to extract the envelope. The formula is :

$$e[n] = \sum_{i=1}^{55} i \times d[n-i+1] + \sum_{i=1}^{54}(55-i) \times d[n-54-i] \qquad (2)$$

Triangular window allows us to obtain peaks that are sharper than those using rectangular window, and the computation is lower complexity than rectangular windows. At the same time, the triangular window filter can smooth the small noise spikes, and the smoothing effect can be determined by the window order. The higher order filter can remove more noise spikes, but it maybe miss some R-peak of ECG. And the QRS complexes of ECG is about 0.3s, and the triangular window should cover the whole QRS complexes. In summary, for sampling rate is 360Hz,the filter order of 110 is found in this study. Figure 4 shows the signal after the envelope extraction stage.

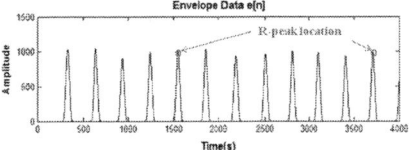

Figure 4 Envelope extraction result

R-peak finding

In order to reduce the computation complexity of finding the R-peak after the envelope extraction, the Hilbert transform is applied to the enveloped signal, and the peak value of e[n] is converted to the positive zero crossing point.

The Hilbert transform is a method often used for signal analysis, and the HT of a real signal x(t) is defined as(3):

$$X(t) = H[x(t)] = \frac{1}{\pi t} \times x(t) = \frac{1}{\pi} \int_{-\infty}^{\infty} \frac{x(\tau)}{t-\tau} d\tau \quad (3)$$

The Hilbert transform can be simplified, the transformation formula can be obtained:

$$H(t) = \frac{t}{1+t^2} \qquad (4)$$

Since the algorithm is implemented on hardware, it is not suitable that the continuous expression of Hilbert transform in time domain. We also need to change (4) to discrete form for hardware implementation. In fact, the discrete Hilbert transform has been proposed by R.N.Bracewell in 1983. In addition, some studies have shown that discrete Hilbert transform can be implemented by FIR filters, and its advantage is low-complexity. So this algorithm uses FIR filter to implement the discrete Hilbert transform. This method can be seemed to add a window to the original signal like the envelope extraction stage. The order of the FIR filter also affects the ability of suppress noise and the accuracy of R-peak detection. It also makes a tradeoff here, chooses 246 as the Hilbert transform filter order. The coefficients are decimated. The expression of Hilbert transform filter is:

$$h[n] = \begin{cases} h[5m-2] = \frac{1}{n} & (0 < m < 26) \\ h[5m-2] = -\frac{1}{n} & (25 < m < 50) \\ 0 & (others) \end{cases} \quad (5)$$

After the Hilbert transform, the h[n] will have a series of positive zero crossing point. These positive zero crossing points represent the location of the R-peak, so it is particularly important to detect the zero crossing point of the real R-peak instead of the zero crossing point caused by noise. The adaptive threshold method is used to detect zero crossing point, which can reduce the false detection caused by noise. Figure 5 shows positive zero crossing point after the Hilbert transformation.

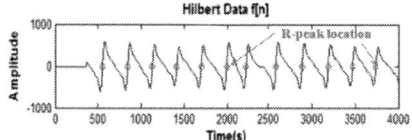

Figure 5 R-peak finding result

RESULT AND DISCUSSION

The R-peak algorithm is evaluated by using several sets of data in MIT-BIH arrhythmia database. We calculated three quantitative results of the proposed R-peak detection algorithm true-positive (TP) when a R-peak is correctly detected by the proposed algorithm, false-negative (FN) when a R-peak is missed, and false-positive (FP) when a noise spike is detected as R-peak. To evaluate the performance of the proposed detection algorithm, the sensitivity (Se), detection accuracy(Acc) can be computed by using the following equations, respectively,

$$Se = \frac{TP}{TP+FN} \times 100\% \quad (6)$$

$$Acc = \frac{TP}{TP+FP} \times 100\% \quad (7)$$

The evaluated result listed in TABLE I , an average sensitivity of 99.87% and an average accuracy of 99.81 % can be obtained. Compared with paper[1][2][3][4], the sensitivity and accuracy of the R-peak detection algorithm presented in this paper are comparable and its complexity is lower.

Data	Num	TP	FN	FP	Sen(%)	Acc(%)
100	2273	2273	0	0	100.00%	100.00%
101	1865	1862	3	6	99.84%	99.68%
102	2187	2187	0	0	100.00%	100.00%
103	2084	2084	0	0	100.00%	100.00%
104	2229	2224	5	8	99.78%	99.64%
105	2572	2567	15	10	99.42%	99.61%
106	2027	2026	1	4	99.95%	99.80%
107	2137	2136	1	5	99.95%	99.77%
total	17374	17359	25	33	99.87%	99.81%

Table 1 Evaluation of algorithm

SUMMARY

This paper presents a R-peak detection algorithm with high accuracy and high sensitivity. It is also low-complexity ,so it can be implemented in ASIC or other hardware. Finally, the MIT-BIH database is used to evaluate the performance of algorithm. From the results, the algorithm has a good performance in accuracy and sensitivity.

ACKNOWLEDGEMENTS

This work is supported by R&D project of Shenzhen Government (Project NO.JSGG20130918140947999 and Project NO.61471011).

References

[1] Wang H M, Lai Y L, Hou M C, et al. A ±6ms-accuracy, 0.68mm 2, and 2.21μW QRS detection ASIC[C]// IEEE International Symposium on Circuits and Systems. IEEE, 2010:1372-1375.

[2] Goodfellow J, Escalona O J, Kodoth V, et al. Denoising and automated R-peak detection in the ECG using Discrete Wavelet Transform[C]// Computing in Cardiology Conference. IEEE, 2017.

[3] Li H, Wang X, Chen L, et al. Denoising and R-Peak Detection of Electrocardiogram Signal Based on EMD and Improved Approximate Envelope[J]. Circuits Systems & Signal Processing, 2014, 33(4):1261-1276.

[4] Sedaaghi M H, Ajami E. Neural-network-based adaptive ECG denoising[J]. Acta Press, 2003.

[5] Van Alsté J A, Schilder T S. Removal of base-line wander and power-line interference from the ECG by an efficient FIR filter with a reduced number of taps [J]. IEEE Trans Biomed Eng. 1985, BME-32(12):1052-1060

AN APPLICATION-LEARNABLE NEUROMORPHIC FREQUENCY SYNTHESIZER WITH RANDOM VECTOR NEURAL NETWORK

Lizhao Gao, Tingbing Ouyang, Chao Zhang, Jiangtao Gu and Bo Wang[*]

Key lab of IMS, School of ECE, Peking University Shenzhen Graduate School

Shenzhen 518055, China

*Corresponding Author's Email: wangbo@pkusz.edu.cn

ABSTRACT

This paper proposes a novel architecture of application-learnable neuromorphic frequency synthesizer (ALNFS). We train a feedforward random vector neural network (RVNN) to obtain weights. Then by shifting trained weights, ALNFS can operate as both Integer-N and Fractional-N synthesizing, with 0 cycle frequency switching. With 50 hidden neurons, the prototype ALNFS achieves -68.71 dBc/Hz Phase Noise at 100 KHz offset of 4 MHz output frequency while power consumption is only 3.59 μW for RVNN.

INTRODUCTION

Current frequency synthesizer architectures are often based on the feedback mechanism of the phase-locked loops (PLLs), which suffer the long locking time. Another synthesizer technology, direct digital synthesis (DDS) [1], can fulfill fast frequency switching due to feedforward architecture. However, its output frequency can merely reach 40% reference frequency with large power consumption.

In this paper, we present a novel application-learnable neuromorphic frequency synthesizer (ALNFS) to fulfill Integer-N and Fractional-N synthesizing, as well as function generator for low power, high resolution and 0 cycle frequency switching. By training the core block of ALNFS with the samples of target output, the ALNFS can be application-learnable.

OUR PROPOSED MODEL

An extremely core and useful property of frequency synthesizing is frequency multiplication, i.e., the generation of an output frequency that is a multiple of the reference input frequency. An excellent frequency synthesizer must ensure not only the stability of output frequency, also fast frequency switching. The current mainstream frequency synthesizer, PLL, can generate stable output frequency, and keep extremely tiny phase error between input and output signal meanwhile. However, it suffers the long locking time (usually hundreds of cycles) due to the feedback loop. Figure 1 depicts the architecture of our proposed learnable neuromorphic frequency synthesizer (ALNFS), which mainly consists of three blocks: pre-processing block (PrePB), neural synthesizing block (NSB) and post-proce-

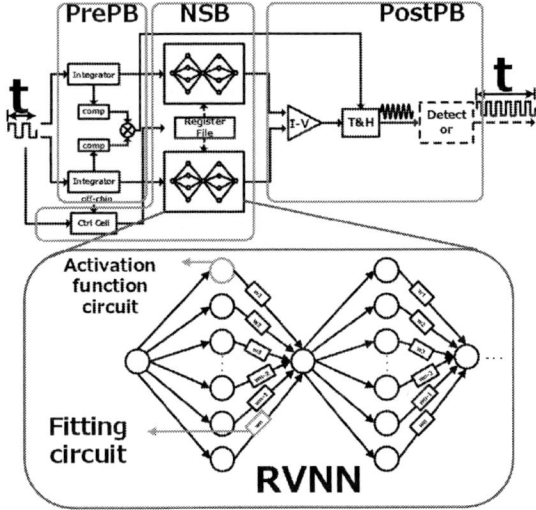

Figure 1: Proposed architecture of application-learnable neuromorphic frequency synthesizer (ALNFS).

ssing block (PostPB). PrePB transforms the reference input (square waveform) to triangle waveform via two integrators, as the input of NSB, which learns the mechanism of 'amplifying' the frequency of reference input by updating the weights of hidden neurons in random vector neural network (RVNN). A register file, totally controlled by signals from off-chip, is needed to store weights or convey the weights into RVNN. Then PostPB transforms the output current of NSB into the voltage by I-to-V amplifier. Also, a Trace and Hold (T&H) circuit is utilized to eliminate glitches caused by RVNNs when shifting the weights. Also, a zero-cross detector can be used to convert it to the square wave for specific applications.

The core of the architecture is RVNN, a single-hidden layer feedforward network that implements the frequency synthesis. Most of the artificial neural networks (ANNs) contain the multi-layer neurons, which may slow the propagation speed. Furthermore, ANNs use back-propagating algorithm to train the enormous parameters which cannot be linearly mapped to analog block precisely. Our proposed RVNN, which draws inspiration from [2], maps the input into high-dimensional feature space

978-1-5386-5309-8/18 $31.00 © 2018 IEEE

randomly and fits the target waveform using norm least-squares solution. For RVNN, more hidden neurons can help enhance the capacity of frequency multiplication but these effects work in limited quantum. To fulfill multiplication as much as possible, we set 2-stage cascade RVNNs (as shown in the blue box of Figure 1). In cascade RVNNs, the first-stage output layer of RVNN is regarded as the second-stage input layer of next RVNN. For instance, 1st-stage RVNN and 2nd-stage multiple M times and N times respectively, so the output frequency is MN times of input frequency totally.

Learning algorithm

We set $[x_1, \dots, x_N]$ as the sample of input of RVNN, $[T_1, \dots, T_N]$ as the target value. For M hidden neurons, the output layer can be described as: $H\beta = T$, where $\beta = [\beta_1, \dots, \beta_M]$ is the weight vector connecting the hidden nodes and the output nodes. H is the output matrix of activation function and described as:

$$H = \begin{bmatrix} \alpha_1 g(\eta_1 \cdot x_1 + b_1) & \dots & \alpha_1 g(\eta_M \cdot x_1 + b_M) \\ \vdots & \ddots & \vdots \\ \alpha_1 g(\eta_1 \cdot x_N + b_1) & \dots & \alpha_M g(\eta_M \cdot x_N + b_M) \end{bmatrix} \quad (1)$$

where $\alpha_j g(\eta_j \cdot x_i + b_j)$ means the tuning curve of jth hidden neuron for i^{th} input. we utilize the smallest norm least-squares solution to calculate β as: $\hat{\beta} = H^\dagger T$ where $H^\dagger$ is the moore-penrose generalized inverse of matrix H.

VLSI implementation of RVNN

As shown in Figure 2, the RVNN consists of two main circuits, namely activation function circuits (hidden neurons) and fitting circuits (weights between hidden layers and neuron layers).

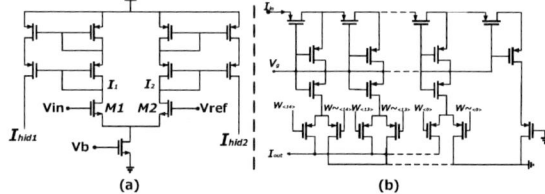

(a)   (b)

Figure 2: Two main circuits of RVNN: (a) activation function circuits (b) fitting circuits

We use the differential pair to implement activation function circuits. The dual tuning curves of activation function circuits are described as below:

$$I_i = I_b / (1 + e^{(2i-3)(\frac{V_{in}}{nU_T} - \frac{V_{ref}}{nU_T})}), \ i=1,2 \quad (2)$$

where I_b is the maximum bias current, V_{in} is the ramp input voltage, V_{ref} is the constant input voltage, and U_T is the thermal voltage. There are two measurements to improve the diversity of sigmoid tuning curve. First, the

gate voltage V_{ref} and V_b of one neuron differ from each other, which is a salient method to achieve a distinct tuning curve for each neuron. Also, the distribution of V_{ref} and V_b adjusted via MATLAB tool, to confirm that tuning curves of activation function circuits are uniformly distributed. Second, Unlike TAB [3] in which each hidden neuron only has one tuning curve, here the current I_1 is copied to $I_{hidden1}$ and current I_2 is copied to $I_{hidden2}$ via a current mirror respectively, which acts as a sigmoid tuning curve for a hidden neuron. The dual output increases the heterogeneity of sigmoid tuning curve, which is in favor of better fitting so that the accuracy of final output can be improved.

We implement the fitting circuit using an R2R circuit with 15-bit switches, the output of which can be described as:

$$I_{out} = \sum_{n=1}^{15} I_{in} W_n / 2^n \quad (3)$$

where $W_n = 1$ when the state of current branch is on while $W_n = 0$ when the state is off. The I_{out} is further routed to currents I_+ or I_-, determined by the polar signal. I_+ joins with I_- to generate the final current that is the final output of RVNN.

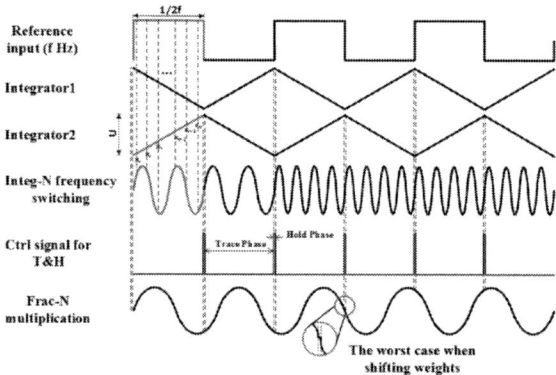

Figure 3: Schematic diagram of process for frequency synthesizing of ALNFS. Waveforms in blue means they are in the training phase

Frequency synthesizing

How can ALNFS 'amplify' the frequency of reference input? As shown in Figure 3, putting the reference input into RVNNs directly violates the uniqueness rule of mapping. So an integrator in front of RVNN converts the reference input into the triangle wave. To eliminate the coupling effect of input triangle signal for RVNN, we parallel two RVNNs and shift the phase of the input of the 2nd RVNN to 180°. The output of both RVNNs is integrated together. We extract the tuning curves of RVNNs and obtain weights with our proposed learning algorithm by virtue of MATLAB. RVNN is a feedforward architecture without locking phase and frequency through a loop. So by shifting weights of RVNN, ALNFS can fulfill 0 cycle frequency switching with continuous-phase for Integ-N frequency synthesis as shown in Figure 3.

Fractional-N frequency synthesis can be implemented similarly. Supposing there is M/N frequency multiplication (M, N are mutually prime), we need 2N groups of weights for each neuron due to shifting weights at the rising and falling edges of the clock, which would bring in glitches that destroy the continuity of output wave. To eliminate the glitch as far as possible, we set up a T&H circuit that divides the period of shifting weights into the holding phase and tracing phase. Two comparators detect slight span of triangle vertex, as the input of an XNOR gate ($\otimes$ in Figure 1), the output of which is taken as the control signal of T&H circuit. T&H circuit may cause discontinuity, especially when sampling steep output as shown in Figure 3. The worst case occurs when the absolute value of the slope of output is greatest, however, too weak to make much difference.

For frequency synthesizing, another crucial performance is frequency resolution, namely the minimum interval between adjacent two output frequencies. In term of ALFNS, theoretically, the resolution can be infinitely small if the capacity of weight dictionary is infinitely large. As for N words register file, we can obtain the desired resolution, $2F_{input}/N$, through reconfiguring the weight dictionary. Furthermore, as an application-learnable architecture, the ALNCS can be trained in the future study as a function generator like [1].

SIMULATION RESULTS

The prototype ALNFS, consists of 50 hidden neurons (including 25 activation function blocks and 50 fitting blocks), is simulated in 0.13um CMOS process by Cadence. Figure 4 shows the actual output of ALNFS for 4 multiplications while the frequency of reference input is 1MHz.

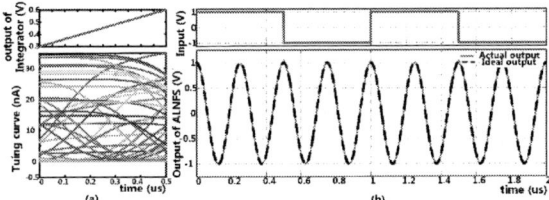

Figure 4: (a) tuning curves of all the neurons in RVNN. (b) Ideal output and actual output of ALNFS at 1MHz reference input and 4 multiplications

Fig. 5 shows the spectrogram of proposed ALNFS. With our proposed method of harmonic eliminating, the effect of triangle signal coupling is alleviated remarkably. The odd harmonics of output decrease 31 dB on average. As is depicted in Fig. 6, there are only flicker noise and white noise in Phase Noise (PN) with corner frequency of 1K Hz. Through our proposed optimization methods, i.e., wide-range and targeted training, the PN of ALNFS decreases from -65.69dBc to -68.71dBc at 100K Hz offset frequency.

With the help of high-speed register file and T&H circuit, ALNFS can fulfill 0 cycle frequency switching with continuous-phase. Overall power consumption of RVNN is 3.59 µW, while most of the power dissipation is the I-V amplifier.

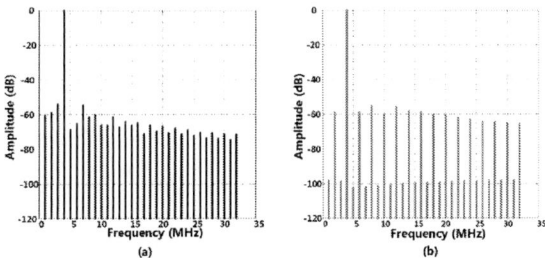

Figure 5: Spectrogram of the prototype ALNFS (a) without, and (b) with the method of harmonic eliminating

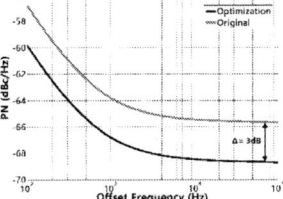

Figure 6: Phase Noise of the prototype ALNFS

CONCLUSION

A novel low power, high resolution, 0 cycle frequency switching neuromorphic frequency synthesizer is presented. We train a feedforward random vector neural block (RVNN) to obtain weights. By shifting trained weights, ALNFS can implement both Integer-N and Fractional-N synthesis, as well 0 cycle frequency switching.

ACKNOWLEDGMENTS.

This work was supported by R&D projects of Shenzhen city (JCYJ20160229094148396) and NSFC (61471011).

REFERENCES

[1] Tierney J, et al., *IEEE Transactions on Audio and Electroacoustics*, 1971, 19(1): 48-57.

[2] Pao Y H, et al., *Neurocomputing*, 1994, 6(2): 163-180.

[3] Thakur C S, et al. *IEEE Transactions on Circuits and Systems I: Regular Papers*, 2016, 63(2): 211-221.

978-1-5386-5309-8/18 $31.00 © 2018 IEEE

EMPOWERING EDGE MINING ON SMARTPHONES WITH RECONFIGURABLE FABRICS

Zeyu Yan, Xiaowei Xu, Guangyu Yu and Hu Yu

School of Optical and Electronic Information, HUST, Wuhan, 430074, China
*Corresponding Author's Email: bryanhu@hust.edu.cn

ABSTRACT

With the prevalence of sensors in smartphones, a large volume of time series is being produced, which contains useful information about personal lifestyles and habitats. Considering privacy and security, the mining of such data is best done on the edge. However, the tight energy constraint imposed by batteries makes this task challenging. Recently many studies focus on edge mining of time series on low-power co-processors for high energy efficiency. However, none of them has explored the opportunity of reconfigurable fabrics, which are well-known energy-efficient and low-cost solutions. In this paper, we propose an efficient smartphone architecture to empower edge mining on smartphones. A configuration lib is associated with the fabrics which can be configured according to the current bottleneck of edge mining tasks. Two widely-adopted algorithms, artificial neural networks (ANNs) and Dynamic Time Warping (DTW) for edge mining, are evaluated. The proposed architecture is implemented on the off-the-shelf smartphones and low-power FPGAs. Experimental results show that compared with smartphones, the proposed architecture can achieve a speedup of 7.8x-21.1x and an energy efficiency improvement of 10.6x-27.1x.

INTRODUCTION

Smartphone has been playing an indispensable role in every aspect of human life. Today's smartphone can have about one dozen of sensors in it. Thus, tens of billions of user-specific data records (or time series) are generated by these smartphones. This huge amount of time series makes it possible to create well-managed healthy lifestyles and can be exploited to discover hidden patterns including frequent activities, classification of physiological data, and clusters of mobile trajectories. Therefore, there is a growing need to mine the value of these personal time series.

Currently, digital signal processor (DSP) and graphical processing unit (GPU) have also been adopted to deal with the time series for better energy efficiency. Georgiev et al. [1] implemented a more complicated audio sensing task on off-the-shelf smartphones. This audio sensing task is well designed and optimized for a low-power DSP co-processor. Meanwhile, a few studies pay attention to efficient data mining on heterogeneous mobile architecture. Prakash et al. [2] achieved high energy efficiency with a static partitioning strategy to execute application kernels across CPU and GPU cores.

However, more efficient reconfigurable fabrics such as Field Programmable Gate Array (FPGA) have not been well explored for smartphones. Reconfigurable fabrics can be configured to a variety of algorithms with high performance, which is very resource-efficient. Furthermore, the energy efficiency of reconfigurable fabrics is much higher than GPUs or DSPs.

In this paper, we propose an efficient smartphone architecture to empower edge mining on smartphones. An algorithm configuration library is associated with the fabrics which can be configured according to the recent bottleneck of edge mining algorithms. Particularly, we implement the widely-used dynamic time warping (DTW) [3] and Artificial Neural Networks (ANNs) [4] for edge mining of time series in the algorithm configuration lib. The results show that compared with smartphones, the proposed architecture has a speedup of 7.8x-21.1x and an energy efficiency improvement of 10.6x-27.1x.

HARDWRAE IMPLEMENTATION

Architecture Overview

As shown in Figure 1, the proposed architecture consists of a commercially available smartphone, an interface and reconfigurable fabrics. The smartphone gets data from inner sensors or external networks, which will be processed by edge mining tasks of time series. The bottleneck computation module analyzes the current edge mining tasks and distributes the computations load to CPU and reconfigurable fabrics. The bottleneck computation module will send the data of computational task to the reconfigurable fabrics and the processing of the data will be accelerated by the accelerator configured according to the configuration lib.

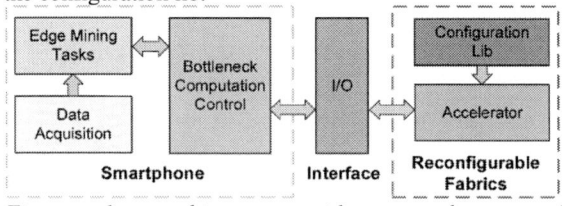

Figure 1: architecture with smartphone and reconfigurable fabrics

ANN Accelerator

As shown as Figure 2, we implement a pipelined ANN accelerator including a Read/Write (R/W) controller, an ANN controller, a processing element (PE) array, a sigmoid memory and a weight memory. The sigmoid module is responsible for the calculation of sigmoid function, and the weight array is used to store the parameters of ANNs. The PE array performs computations in ANNs. The R/W controller handles data reading and writing through the interface.

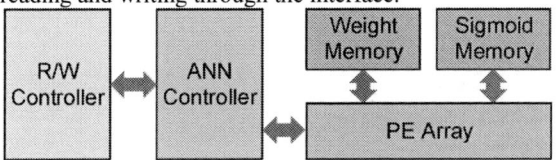

Figure 2: Overall structure of ANN accelerator on reconfigurable fabrics

The PE array module consists of several PEs, and the computing process of PE array is shown in Figure 3. As multiplication is very intense for ANNs and the number of multiplier (MUL) is often limited in FPGAs, a time-division multiplexing strategy for the multipliers is used to compute ANNs with the minimum resource.

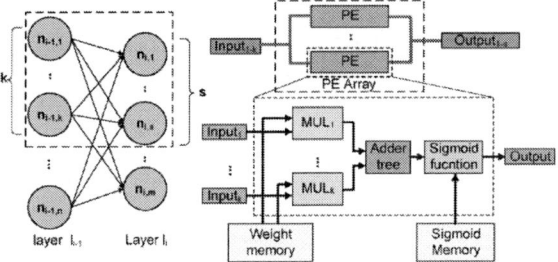

Figure 3: PE structure and its computation processes of ANN

DTW Accelerator

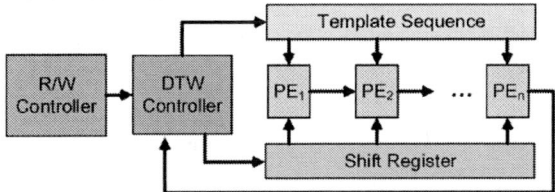

Figure 4: Overall structure of DTW accelerator

As shown in Figure 4, the DTW accelerator processes data streams with no pauses. The length of template sequence is equal to the number of PEs. The Manhattan distance is selected to be the distance function between two data elements. The DTW controller saves the template sequence in registers and there is a one-to-one correspondence between the data in template sequence and PE. The test data streams are sent to a shift register which will be fetched by PEs sequentially.

EXPERIMENTS
Experiment Setup
We select a smartphone, Meizu MX4 for the proposed

TABLE I. HARDWARE IMPLEMENTATIONS FOR ANNS AND DTW

FPGA	Algorithm	Resource Utilization (LEs)	Max frequency
Cyclone IV (EP4CE22)	ANN	7012/22320	56.05 MHz
IGLOO nano (AGLN020)	DTW	5896/6144	27.87 MHz

architecture. The USB2.0 chip FX2LP and two low-power FPGAs: EP4CE22 and AGLN020 are adopted for the experiment. The tested ANN is a 4×4×3 neural network, while the template length and the number of PEs in DTW test are 10, and the resource usage is shown in Table I.

The dataset used for ANNs and DTW are from UCI machine learning repository [5] and UCR time series classification archive [6].

Experiment Result

As shown in Table II, it can be discovered that the power consumption of the smartphone (0.821W-1.066W) is slightly larger than that of our architecture (0.528W-0.776W). However, our architecture has a better performance which is around 8x to that of the smartphone implementation. Thus, our architecture has an overall energy efficient improvement of 10.57x-14.8x compared with the smartphone.

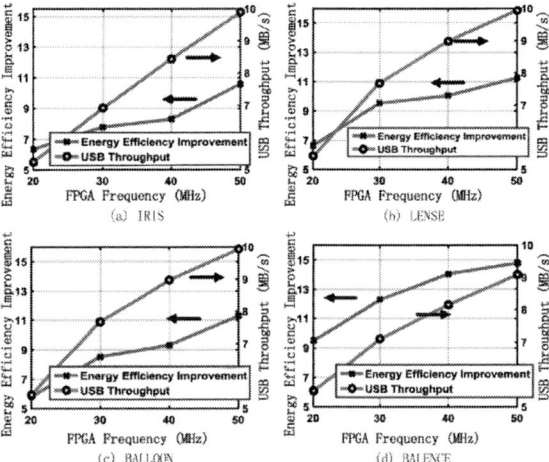

Figure 5: Energy efficient improvement and USB throughput of ANN.

We also discuss the frequency influence of FPGA on power consumptions and USB throughputs. As shown in Figure 5, the throughput of USB and energy efficiency improvement grow as the frequency increases in general. The maximum USB throughput is about 10 MB/s which is below the limit of USB transport capability of the smartphone.

978-1-5386-5309-8/18 $31.00 © 2018 IEEE

TABLE II. SPEEDUP AND ENERGY CONSUMPTION OF ANN

Dataset	Performance (million items/second)		Energy (Watt)		Speedup	Energy Efficiency Improvement
	Smartphone	Our Architecture	Smartphone	Our Architecture		
IRIS	0.171	1.414	0.821	0.642	8.3x	10.6x
LENSE	0.173	1.420	1.066	0.776	8.2x	11.3x
BALANCE	0.167	1.298	1.006	0.528	7.8x	14.8x
BALLONS	0.167	1.419	0.842	0.635	8.5x	11.3x

TABLE III. SPEEDUP AND ENERGY CONSUMPTION OF DTW

Dataset	Performance (million items/second)		Energy (Watt)		Speedup	Energy Efficiency Improvement
	Smartphone	Our Architecture	Smartphone	Our Architecture		
50words	0.087	1.839	0.627	0.490	21.1x	27.1x

As to DTW accelerator, we first perform dimension reduction to preprocess the dataset with the widely-used piecewise constant models [7]. With the model, the template length in the dataset is normalized to 10, as it is equal to the number of PEs in the accelerator. Note that the template length can be changed for specific applications. The performance and energy efficiency are shown in Table III. Just like the results for ANNs, the power consumption of the smartphone (0.627W) is slightly larger than that of our architecture (0.490W), and the speedup and energy efficiency improvement are 21.14x and 27.05x, respectively.

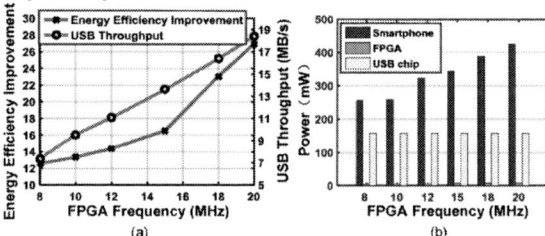

Figure 6: (a) Energy-efficient improvement and USB throughput, and (b) power consumption component of DTW.

Energy efficient improvement and USB throughput with the FPGA frequency are discussed as shown in Figure 6. We can notice that energy efficient improvement and USB throughput are almost positively correlated with the FPGA frequency. However, the throughput of USB is about 18.4MB/s which is close to the USB transport capability (20MB/s), which means we can not get a higher energy efficient improvement by simply increasing the FPGA frequency in our architecture. Meanwhile, the power consumption of FPGA is only a small portion of the whole power consumption of this architecture.

CONCLUSIONS

In this paper, we propose a smartphone architecture, TIGER by embedding a low-power FPGA to existing smartphone architecture for efficient edge mining of time series on smartphones. The results show that compared with smartphones, the smartphone architecture has a speedup of 7.8x-8.5x and an energy efficiency improvement of 10.6x-14.8x for accelerating ANNs, while 21.1x and 27.1x for accelerating DTW algorithm. Meanwhile, experiments show that a better performance can be achieved by adopting a more efficient I/O interface instead of USB

REFERENCES

[1] P. Georgiev, N. D. Lane, K. K. Rachuri, and C. Mascolo. Dsp. ear: Leveraging coprocessor support for continuous audio sensing on smartphones. In Proceedings of the 12th ACM Conference on Embedded Network Sensor Systems, pages 295–309. ACM, 2014.

[2] Prakash, Alok, et al. "Energy-efficient execution of data-parallel applications on heterogeneous mobile platforms." Computer Design (ICCD), 2015 33rd IEEE International Conference on. IEEE, 2015.

[3] G. Cormode. Fundamentals of analyzing and mining data streams. In Tutorial at Workshop on Data Stream Analysis, Caserta, Italy, 2007.

[4] Ronao C A, Cho S B. Human activity recognition with smartphone sensors using deep learning neural networks[J]. Expert Systems with Applications, 2016, 59: 235-244.

[5] A. Asuncion and D. Newman. Uci machine learning repository, 2007.

[6] Y. Chen, E. Keogh, B. Hu, N. Begum, A. Bagnall, A. Mueen, and G. Batista. The ucr time series classification archive. URL www. cs. ucr. edu/~ eamonn/time_series_data, 2015.

[7] E. Keogh, K. Chakrabarti, M. Pazzani, and S. Mehrotra. Locally adaptive dimensionality reduction for indexing large time series databases. ACM Sigmod Record, 30(2):151–162, 2001.

ACCELERATING EARTH MOVERS DISTANCE WITH INSTRUCTION SET EXTENSION FOR IMAGE RETRIEVAL

Guangyu Yu, Xiaowei Xu, Zeyu Yan and Hu Yu

School of Optical and Electronic Information, HUST, Wuhan, 430074, China
*Corresponding Author's Email: bryanhu@hust.edu.cn

ABSTRACT

Image retrieval is one of the most popular applications for computer vision and pattern recognition, in which similarity computation is the computational bottleneck. Earth Movers Distance (EMD) is one of the most popular similarity measure for image retrieval, which has a high time complexity of $O(n^3 log n)$. Recently, with the explosion of image data, EMD acceleration has been emerging. In this paper, we propose an EMD acceleration architecture based on instruction set extensions for image retrieval. The EMD acceleration architecture achieves a speedup of 1.3x-2.2x over software implementations. The main advantage of the proposed architecture over existing hardware accelerations is that it can support larger histograms. Specifically, the number of supported variables in histograms has 1-2 orders of magnitude improvement.

INTRODUCTION

Earth Movers Distance (EMD) is one of the most popular similarity measure for image retrieval. Today, EMD acceleration has been urgent, which is mainly due to the following two reasons. Firstly, the time complexity of EMD is up to , which cannot meet the processing requirement of many realtime applications. Secondly, huge amounts of image data need to be processed more efficiently with the advent of the era of big data.

Recently, studies related to EMD mainly focus on software acceleration. Assent et al. [4] accelerated EMD algorithm in Filter-and-Refine framework and proposed several lower bound methods for EMD. Wichterich et al. [12] presented a new dimension reduction method, and results showed that EMD distances in the derivative space is a lower bound of EMD in the original space.

Another approach [9] [13] for EMD acceleration research focused on EMD simplification for specific applications. Andoni et al. [3] achieved EMD simplification in high dimensional space. Pele et al. [7] simplified EMD calculation by limiting the number of connections in the graphs. Jang et al. [6] and Shirdhonkar et al. [10] proposed linear EMD methods by simplifying EMD, respectively. Andoni et al. [2] proposed a method for accelerating planar EMD. There exists a hardware acceleration work for simplex algorithm, which can support general EMD computation [5]. However, the work can only support 751 variables, which corresponds to only 27 bins for pair-wise EMD computation.

In this paper, we propose an EMD acceleration architecture based on instruction set extensions. Particularly, max-flow min-cost algorithm [1] is adopted which is specific for the EMD problem. With algorithm analysis, we design extended instructions (dedicated hardware) for EMD bottlenecks. To achieve parallel input/output operation, we design a customized memory for the extended instructions. The main advantage of the proposed architecture over existing hardware accelerations is that it can support larger histograms.

ALGORITHM ANALYSIS

Algorithm Process Analysis

The call graph of max-flow min-cost algorithm for EMD is shown in Fig. 1.

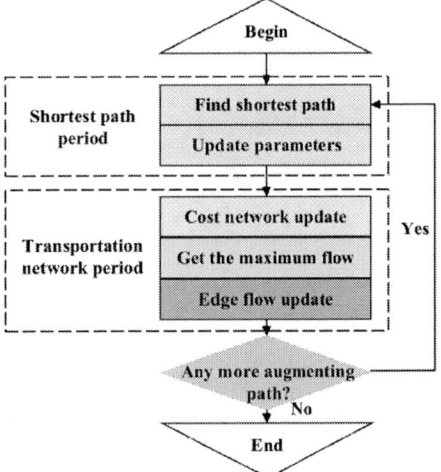

Fig.1. Illustration of max flow min cost algorithm for EMD.

Algorithm Bottleneck Analysis

Based on the analysis of max-flow min-cost algorithm for EMD. We find that finding the shortest cost path and updating the cost of the network operation have high time complexity.

The percentages of runtime of finding shortest cost path and updating network with different experimental configurations are shown in Fig. 2. We find that the two operations of the network account about 75% of the total running time.

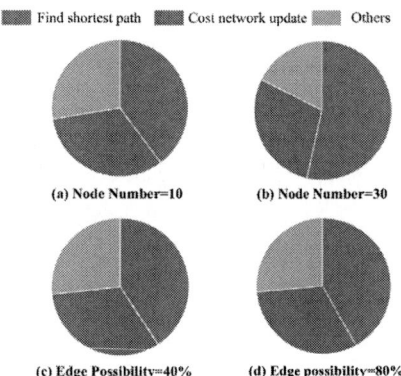

Find shortest path ■ Cost network update ■ Others

(a) Node Number=10   (b) Node Number=30

(c) Edge Possibility=40%   (d) Edge possibility=80%

Fig.2. Illustration of EMD bottleneck.

EMD ACCELERATION ARCHITECTURE
Acceleration Architecture

The EMD acceleration architecture based on instruction set extension is shown in Fig. 3. An on-chip RAM is connected to the extended instructions to get higher throughput. We design three extended instructions: *Initial*, *ExtrMin* and *ReduceCost* instruction. And the three custom instructions are integrated into an extended multi-cycle instruction.

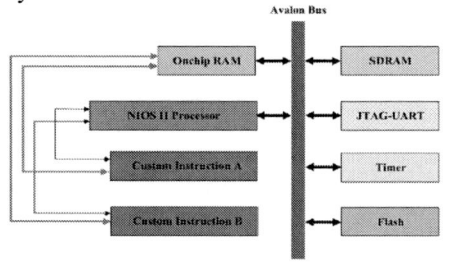

Fig. 3. EMD acceleration architecture based on instruction set extension.

Initialization Instructions Design

Initial instruction mainly realizes the function of loading data to On-chip RAM to store the data in a specific area before calling *ExtrMin* and *ReduceCost* instructions.

Finding the Shortest Cost Path Instruction Design

The hardware implementation of the *ExtrMin* instruction is shown in Fig. 4(a). The corresponding code is shown in Fig. 4(b).

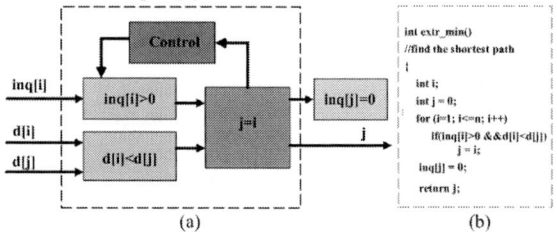

Fig. 4. Instruction design of shortest path operation, ExtrMin.

The main calculation is an iteration operation caused by the *for* operation. Three input data can be read parallel, and all the operation can be fully pipelined. The whole calculation can be processed efficiently and can be processed in one clock cycle.

Cost Network Update Instruction Design

The hardware implementation of cost network update instruction *ReduceCost* as shown in Fig. 5(a). The corresponding code is shown in Fig. 5(b).

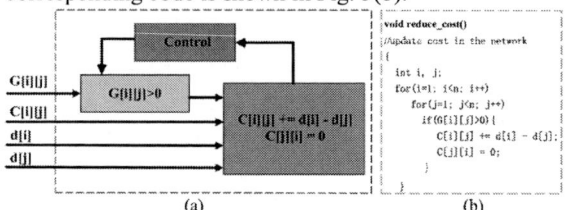

Fig. 5. Instruction design of cost update operation, ReduceCost.

The main calculation is a two-layer iteration operation caused by the *for* operation. Multiple input data can be read in parallel.

EXPERIMENTS
Experiment Setup

We implement the acceleration architecture based on ISE on Altera Nios II processor. Particularly, a FPGA chip EP4CE15 is selected. Hardware design are implemented with Quartus II 13.1, and Nios II processor design are realized with built-in Qsys in Quartus II 13.1. The configuration of the Nios II processor is shown in Table I.

TABLE I.   CONFIGURATION OF THE NIOS II PROCESSOR

Module	Parameter	Value
Nios II processor	Performance	II/f
Onchip memory	Dual Port R/W	Enable
	Data width	32bits
	Memory Size	36000bytes
	R/W Delay	1 cycle
Timer	Delay	1us
	Readable snapshot	Enable
SDRAM controller	Data width	16bits
	Memory Size	32Mbytes
Clock	Frequency	100MHz

Experiment Result

As shown in Fig. 6, the average speedup of *ExtrMin* and *ReduceCost* instruction is about 3.2x and 5.5x compared with software implementation, respectively. The speedup of the overall EMD calculation is only about

1.3x with only *ExtrMin* instruction or *ReduceCost* instruction. When *ExtrMin* and *ReduceCost* instructions are both adopted, the speedup can reach about 2.0x.

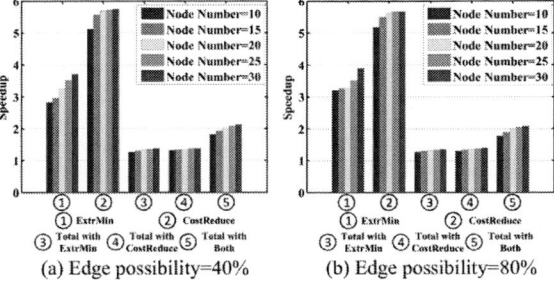

Fig. 6. Speedup with different node number and edge possibility of 40% (a) and 80% (b).

As shown in Fig. 7, the speedup of the number of bins of 30 is significantly greater than the speedup of that of 10. When the number of bins is 30, the speedup reaches 2.2x.

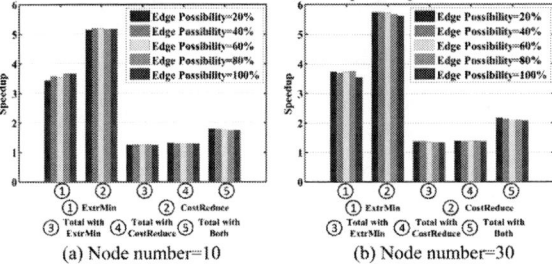

Fig. 7. Speedup with different edge possibility and node number of 10(a) and 30(b).

SUMMARY AND CONCLUSIONS

In this paper, we propose a EMD acceleration architecture based on instruction set extensions. Particularly, max-flow min-cost algorithm [1] is adopted which is specific for the EMD problem. With algorithm analysis, we design extended instructions (dedicated hardware) for EMD bottlenecks. In order to achieve parallel input/output operation, we design a customized memory for the extended instructions. The main advantage of the proposed architecture over existing hardware accelerations is that it can support larger histograms. The ex-perimental results show that the EMD acceleration architecture has a speedup of 1.3x to 2.2x. Specifically, the supported number of bins in histograms has a 1-2 orders of magnitude improvements.

REFERENCES

[1] R. K. Ahuja. Network flows. PhD thesis, TECHNISCHE HOCHSCHULE DARMSTADT, 1988.

[2] A. Andoni, K. D. Ba, P. Indyk, and D. Woodruff. Efficient sketches for earth-mover distance, with applications. Institute of Electrical and Electronics

Engineers, 2010.

[3] A. Andoni, P. Indyk, and R. Krauthgamer. Earth mover distance over high-dimensional spaces. In Proceedings of the nineteenth annual ACM- SIAM symposium on Discrete algorithms, pages 343–352. Society for Industrial and Applied Mathematics, 2008.

[4] I. Assent, A. Wenning, and T. Seidl. Approximation techniques for indexing the earth mover's distance in multimedia databases. In 22nd International Conference on Data Engineering (ICDE'06), pages 11–11. IEEE, 2006.

[5] S. Bayliss, G. A. Constantinides, W. Luk, et al. An fpga implementation of the simplex algorithm. In 2006 IEEE International Conference on Field Programmable Technology, pages 49–56. IEEE, 2006.

[6] M.-H. Jang, S.-W. Kim, C. Faloutsos, and S. Park. A linear-time approximation of the earth mover's distance. In Proceedings of the 20th ACM international conference on Information and knowledge management, pages 505–514. ACM, 2011.

[7] O. Pele and M. Werman. Fast and robust earth mover's distances. In 2009 IEEE 12th International Conference on Computer Vision, pages 460–467. IEEE, 2009.

[8] Y. Rubner, C. Tomasi, and L. J. Guibas. The earth mover's distance as a metric for image retrieval. International journal of computer vision, 40(2):99–121, 2000.

[9] B. E. Ruttenberg and A. K. Singh. Indexing the earth mover's distance using normal distributions. Proceedings of the VLDB Endowment, 5(3):205–216, 2011.

[10] S. Shirdhonkar and D. W. Jacobs. Approximate earth movers distance in linear time. In Computer Vision and Pattern Recognition, 2008. CVPR 2008. IEEE Conference on, pages 1–8. IEEE, 2008.

[11] S. Skiena. Dijkstras algorithm. Implementing Discrete Mathematics: Combinatorics and Graph Theory with Mathematica, Reading, MA: Addison-Wesley, pages 225–227, 1990.

[12] M. Wichterich, I. Assent, P. Kranen, and T. Seidl. Efficient emd-based similarity search in multimedia databases via flexible dimensionality reduction. In Proceedings of the 2008 ACM SIGMOD international conference on Management of data, pages 199–212. ACM, 2008.

[13] J. Xu, Z. Zhang, A. K. Tung, and G. Yu. Efficient and effective similarity search over probabilistic data based on earth mover's distance. Proceedings of the VLDB Endowment, 3(1-2):758–769, 2010.

A UNIVERSAL IMPLEMENTATION OF CARDIOVASCULAR DISEASE SURVEILLANCE BASED ON HRV

Yufan Feng, Ying Zhang, Xiaole Cui, Xin-an Wang[*]

The Key Lab of Integrated Microsystems Peking University Shenzhen Graduate School, Shenzhen 518055

* Email:wangxa@pkusz.edu.cn

ABSTRACT

Heart Rate Variability (HRV) refers to such a quantitative scale measurement, which can reflect the regularity of heart rate changes. Taking two common cardiovascular diseases as an example, this paper obtained 14 eigenvalues by analyzing the HRV data taken from PhysioNet in the time domain and frequency domain through the "normperiod" code. Accordingly, disease classification was carried out using an improved K-Nearest Neighbor (KNN) algorithm, including module of removing boundary, uniform density and distance weighting. Finally, we obtained an accuracy of up to 98% and a sensitivity of up to 100%. The results illustrate that this approach with low computational cost enables surveillance and early warning of a large group of cardiovascular diseases effectively.

INTRODUCTION

A large number of studies over the past decade have fully affirmed that autonomic nervous activity is associated with a variety of diseases, particularly with certain cardiovascular diseases [1]. It is also recognized that HRV is a commonly used quantitative indicator for judging the cardiac autonomic nervous activity, which has very important realistic meaning and research worthiness. Therefore, we consider using the information of HRV in the time domain and frequency domain to analyze and judge the human body health with method of KNN classification algorithm.

The most widely used algorithms of frequency domain analysis are based on Fast-Fourier Transform or autoregressive modeling respectively. However, a necessary requirement for the above two methods is that sampling rate is constant. HRV signals aren't in compliance with it. Interpolation is in need to do that, resulting in high computation complexity and some measurement errors. The algorithm used in this paper is based on the "normperiod" code for unevenly spaced data [2], just overcoming this problem.

In recent years, a variety of improved classification algorithms have been proposed. A weighting approach for KNN algorithm is described in [3], which is combined with the Artificial Bee Colony algorithm to determine the optimal weight. A hybrid approach is proposed for simultaneous feature selection and feature weighting with k value selection of KNN rule based on biogeography based optimization in [4]. Besides, [5] presents a KNN density-based rule for the high dimensional data sets by developing a new cluster.

However, all of the above hybrid algorithms are modified for the general data sets, they can't play a good enough role in HRV classification due to the complexity and preciseness of electrocardiogram data. We optimize the original KNN algorithm with applications to disease surveillance based on HRV accordingly. To be clear, this algorithm includes three configurable modules, which are used for removing boundary, uniform density and distance weighting. In addition, data files used for test are all taken from PhysioNet. Samples are classified on the basis of 14 eigenvalues obtained through analysis in the time and frequency domain. Finally, the experimental results signify that it has higher efficiency with an accuracy of up to 98% and a sensitivity of up to 100%, compared to other methods with 5% improvement at least. It illustrates that the hardware oriented algorithm is of great significance in the field of cardiovascular disease surveillance.

HRV DATA PROCESSING

Data preprocessing

Preprocessing is frequently required before HRV analysis to reduce analysis errors. Due to the limited quantity, data from the 24-hour recording are divided into 5-minute segments. Considering the objectivity and universality of the algorithm, a handful of segments need to be removed, of which the average heart rate is greater than 240 or less than 40. Besides, there are also one or more ectopic intervals leading to analysis errors, which need to be detected and corrected with a threshold filter. The percentage R filter can locate intervals, which is defined by

$$R = \frac{|nn(i) - nn(i-1)|}{nn(i-1)} \tag{1}$$

The threshold is usually a user defined parameter, whose value is given as 20% in this paper. It means that if the percentage that $nn(i)$ changes by more than 20% from $nn(i-1)$, $nn\_pre(i)$ will be corrected as Eq. (2)

$$\begin{cases} nn\_pre(i) = \frac{1}{2}[nn(i-2) + nn(i-3)], R > 20\% \\ nn\_pre(i) = nn(i), R \le 20\% \end{cases} \tag{2}$$

where $nn(i)$ is the i-th NN interval and $nn\_pre(i)$ is the i-th corrected NN interval. Preprocessed time series will be obtained after adding up the new data. Figure 1 illustrates the validity of preprocessing.

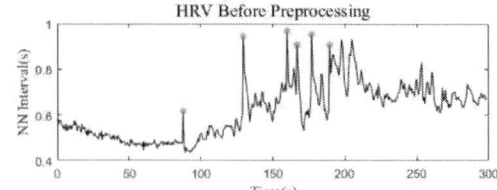

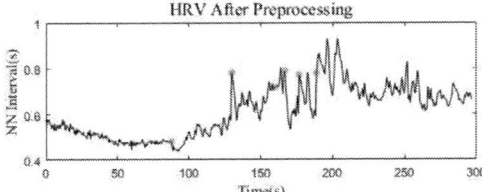

Figure 1: The waveforms before and after preprocessing

Time domain analysis

The traditional analysis of signals is in time domain, often classified as statistical and geometric methods. There are numerous indicators mentioned in the literature, some of which have physiological significances. For example, SDNN reflects complexity of HRV and dynamic balance of the autonomic nervous system. The following is a list of 7 eigenvalues adopted in time domain.

TABLE 1. INDEXES OF TIME DOMAIN

Index	Description
MEANHR	the average heart rate
SDNN	the standard deviation of NN intervals
RMSSD	the root mean square of successive differences of the NN interval series
P50, P20, P10	the ratio of NN intervals that successively differ by more than 50 ms, 20 ms, 10 ms
HRVTI	the HRV triangular index

Frequency domain analysis

Fluctuations in HRV are often thought to be periodic, which can be quantified within the time series by calculating the power spectrum density (PSD). The PSD function can be used to figure out all of the necessary eigenvalues shown as TABLE 2.

TABLE 2. INDEXES OF FREQUENCY DOMAIN

Index	Description
AVLF, ALF, AHF	the absolute spectral power in very low frequency (VLF, 0.003-0.04 Hz), low frequency (LF, 0.04-0.15 Hz) and high frequency (HF, 0.15-0.4 Hz)
ATOTAL	total power in the spectrum
LFHF	the ratio of ALF to AHF
NLF, NHF	the value normalized to total power

[2] puts forward an improved algorithm for spectrum analysis of unevenly spaced data using the "normperiod" code, in which the range and the total number N_P of frequency f_i is given by Eq. (3) and Eq. (4)

$$\frac{1}{4t_r} \leq f_i \leq fratio * \frac{N}{2t_r} \tag{3}$$

$$N_P = \frac{fratio * 4N}{2} \tag{4}$$

where t_r is the record length, N is the number of data points and *fratio* is a user defined parameter that affects the sampling density. Then the frequency step can be calculated by the formulas.

In [2], PSD function *P(T)* at the moment t_i is described by Eq. (5) through fitting a sine curve to data time series using the least-squares method, and the parameter τ is defined by Eq. (6)

$$P(T) = \frac{1}{2\sigma^2} \left\{ \frac{\left[\sum_{j=1}^{N} (x_j - \bar{x}) \cos \frac{2\pi(t_i - \tau)}{T} \right]^2}{\sum_{j=1}^{N} \cos^2 \frac{2\pi(t_i - \tau)}{T}} + \frac{\left[\sum_{j=1}^{N} (x_j - \bar{x}) \sin \frac{2\pi(t_i - \tau)}{T} \right]^2}{\sum_{j=1}^{N} \sin^2 \frac{2\pi(t_i - \tau)}{T}} \right\} \tag{5}$$

$$\tan(\frac{4\pi\tau}{T}) = \frac{\sum_{j=1}^{N} \sin(\frac{4\pi t_j}{T})}{\sum_{j=1}^{N} \cos(\frac{4\pi t_j}{T})} \tag{6}$$

where $\bar{x}$ is the mean of data x_i and σ^2 represents the variance. The spectra of HRV is as shown in Figure 2.

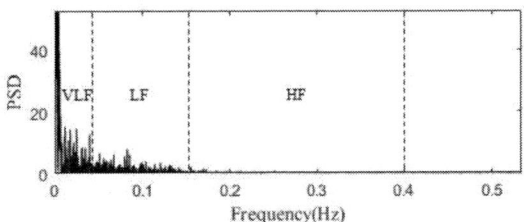

Figure 2: The spectra of HRV

KNN CLASSIFICATION ALGORITHM

According to the results, each sample can be given a 14-dimensional coordinates to be used for KNN classification. Most importantly, the effect heavily depends on the k value, which cannot be greater than square root of the number of training samples. Both too large and too small k value will increase the noise. In order to raise the accuracy, the algorithm has been improved in three configurable modules as following.

Module of removing boundary

The training sets contain a lot of redundant data that increase the storage overhead and computational cost. Some samples that have little to do with classification can be deleted with this module.

The first step is to select the nearest k neighbors from the training sets for each sample. It is obvious that the sample, which has never been used in this process, is on the edge of the feature space. It is almost impossible to become one of k nearest neighbors. These useless samples can be deleted to obtain more effective data sets.

Module of uniform density

Uneven distribution of samples may lead to false judgement, so cutting the data sets in the region of high density is needed pressingly.

978-1-5386-5309-8/18 $31.00 © 2018 IEEE

First, the area is divided into two parts: the inner side and the covered side. Supposing an area A of radius R and center X (one of the samples in the training sets), the condition is that whether all of the samples in the area A are the same kind as X or not. If yes, sample X belongs to the inner side, or else the covered side. Next, considering a parameter $min_{covered}$ for the covered side, the center X of area A, in which the number of samples is greater than $min_{covered}$, is in a high-density region if X belongs to the covered side. If sample Y is in the high-density covered region and belongs to the neighborhood of center X, it says that Y is "high-density" accessible from X as shown in Figure 3. It is the same for parameter min_{inner} and the inner side. Besides, all three parameters above are user defined.

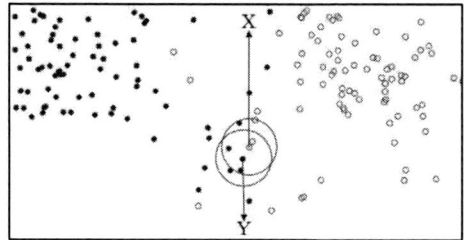

Figure 3: The schematic map of samples in the covered side ($min_{covered}=6$)

After finishing the preparations, all need to do is to locate samples which are "high-density" accessible from the point in the high-density region whether in the inner or covered side and delete them, like Sample Y in Figure 3. This module realizes the function of removing useless points from the inner side and error-prone points from the covered side.

Module of distance weighting

Normally, the closer the point is, the more likely it belongs to the same category. Therefore, a module is put forward so as to assign the heavier weights to the closer neighbors according to Gauss function. Experiments have proved that this method can lead to a better robustness.

DISCUSSION & CONCLUSION

This paper takes Atrial Fibrillation (AF) and Congestive Heart Failure (CHF), focal points in the cardiovascular field, as study objects. Firstly, the atrial fibrillation and normal samples are classified according to the eigenvalues of time domain and frequency domain analysis with k=7. The result is given as classification 1 in TABLE 3. After identifying the patients, it can be further distinguished between patients who are sick or not as classification 2. The algorithm is only equipped with the last module when distinguishing between CHF patients and normal people due to the small size of

dataset, the result of which is given as classification 3. The accuracy (AC) indicates the ratio of real patients to the people who are judged to be ill. The sensitivity (SEN) indicates the percentage of patients who can be detected among all of them.

TABLE 3. RESULT OF ALGORITHM VERIFICATION

NO	Removing boundary	Uniform density	Distance weighting	AC	SEN
1	√	√	√	90%	90%
2	√	√	√	98%	100%
3	--	--	√	92%	91%

Through the correct classification of patients and normal people, electronic medical devices can achieve the surveillance of AF and CHF, which are both closely related to hypertension and coronary heart disease. Then according to classification 2, patients can get different treatments according to the degree of illness.

Our contribution is to propose a novel universal approach based on HRV, which is described with an accuracy of up to 98% and a sensitivity of up to 100%. The performance evaluations confirm the availability of the hybrid algorithm. The HRV data can be classified accurately to judge if a person has some kind of disease or not. In a sense, this method can be widely used with applications to cardiovascular disease surveillance.

ACKNOWLEDGMENTS

This work is supported by R&D project of Shenzhen Government (Project NO. JCYJ20170306091821082 and NO. JCYJ20170306092000960).

REFERENCES

[1] Moorman J R, Lake D E, Griffin M P. *Heart rate characteristics monitoring for neonatal sepsis*[J]. IEEE transactions on bio-medical engineering, 2006, 53(1):126-32.

[2] Pytharouli S I, Stiros S C. *Spectral analysis of unevenly spaced or discontinuous data using the "normperiod" code*[J]. Computers & Structures, 2008, 86(1):190-196.

[3] Yigit H. *A weighting approach for KNN classifier*[C]. International Conference on Electronics, Computer and Computation. IEEE, 2014:228-231.

[4] Kardan A A, Kavian A, Esmaeili A. *Simultaneous feature selection and feature weighting with K selection for KNN classification using BBO algorithm*[C]. Information and Knowledge Technology. IEEE, 2013:349-354.

[5] Tran T N, Wehrens R, Buydens L M C. *Knn density-based clustering for high dimensional multispectral images*[C]. The Workshop on Remote Sensing & Data Fusion Over Urban Areas. IEEE, 2003:147-151.

A Low-Offset Current-Mode CMOS Vertical Hall Sensor Microsystem with Four-Phase Spinning Current Technique

Xingxing Hu[1], Yue Xu[1,2], Jun Xu[1]*

[1]College of electronic and optical engineering & College of microelectronics in Nanjing University of Posts and Telecommunications, Nanjing, Jiangsu Province, China, 210023

[2]National and Local Joint Engineering Laboratory of RF Integration and Micro-assembly Technology, Nanjing 210003, China

*Corresponding Author's Email: yuex@njupt.edu.cn

ABSTRACT

This paper introduces a low-offset and high-sensitivity CMOS vertical Hall sensor microsystem with the integrated front-end. A current-mode four-phase spinning current method and a current negative feedback loop technique are proposed, which enables both the effective elimination of offset and 1/f noise and the great improvement of current sensitivity for CMOS vertical Hall device. Designed by a 0.8-μm high-voltage CMOS technology, the current-mode vertical Hall sensor obtains a residual offset of less than 0.3mT and a high Hall voltage linearity up to 99.9%.

Keywords—Current-mode; CMOS vertical Hall sensor; four-phase spinning current; residual offset.

INTRODUCTION

Due to the low cost, low power consumption, high integration and high reliability, CMOS integrated magnetic Hall sensors have been widely used in industrial control systems, automobiles, intelligent instruments, and consumer electronics for the detection of magnetic field [1,2]. However, CMOS integrated Hall devices, especially the vertical Hall devices have the big disadvantages of low magnetic sensitivity and high offset, which seriously restrict the detection resolution at the low magnetic field applications.

In the traditional voltage-mode Hall sensors, either using constant voltage or current bias, Hall devices with spinning current modulation operation always outputs weak voltage signals in the range of hundreds of millivolt to several microvolt [3].The modulated Hall voltage is then amplified and demodulated back to low frequency Hall signal by signal conditioner, featuring low signal-to-noise ratio (SNR) and high residual offset. Fortunately, the current-mode Hall sensor is a good alternative with respect to the traditional voltage-mode Hall sensor，for the sensitivity in the current-mode is twice larger than that in the voltage-mode [4]. In current-mode Hall sensors, the output is current signal and not voltage [5].

This paper presents a current-mode vertical Hall sensor microsystem with the integrated front-end. A current-mode four-phase spinning current circuit is applied in the symmetrical structure vertical Hall device for the great improvement of current sensitivity and effective cancellation of Hall offset. In addition, a current negative feedback modulation circuit is also proposed to significantly attenuate the offset and 1/f noise of signal conditioner. The vertical Hall sensor microsystem is designed based on 0.8-μm high-voltage CMOS technology. Finally, the advantages of high magnetic sensitivity and low residual offset of the Hall sensor microsystem are validated by simulation.

DESIGN AND SIMULATION OF CURRENT-MODE SWITCHED VERTICAL HALL CIRCUIT

Fig. 1(a) shows the current-mode switched vertical Hall circuit with a four-phase spinning current technique, consisting of 16 NMOSFET transistors and a vertical Hall device (VHD). In this circuit, the four-phase spinning current circuit is controlled by four-phase sequence clock clk_1~clk_4. Its working principle is to perform twice operations of 2-phase spinning current modulation. When clk1 and clk2 are high level respectively, the polarity of the Hall current (I_h) changes periodically, while the offset current (I_{op}) keeps positive polarity as follows.

$$I(0°) = I_H + I_{0P1} \tag{1}$$
$$I(90°) = -I_H + I_{OP2} \tag{2}$$

If a correlated double sampling (CDS) circuit is used, the output current can be given by

$$I(0°) - I(90°) = 2I_H + \triangle I_{OP12}(r) \tag{3}$$

Here, $\triangle I_{OP12}(r)$ is the residual offset produced in the first 2-phase spinning current process.

Similarity, as clk3 and clk4 are high level respectively, the Hall current (I_h) changes its polarity, while the offset current (I_{op}) maintains negative polarity. The output current can also be expressed as

$$I(180°) = I_H - I_{0P3} \tag{4}$$
$$I(270°) = -I_H - I_{OP4} \tag{5}$$
$$I(180°) - I(270°) = 2I_H - \triangle I_{OP34}(r) \tag{6}$$

Here, $\triangle I_{OP34}(r)$ is the residual offset produced in the second 2-phase spinning current process.

Obviously, the offset current (I_{op}) is subtracted in view of (3) and (6). If an adder follows the CDS demodulation circuit, the residual offset can be further reduced. The diagram of output Hall current and offset current corresponding to the four-phase sequence clocks is illustrated in Fig. 1(b).

The vertical Hall device applies an optimized 3-contacts 4-folded symmetric structure for reduced offset, as illustrated in Fig.2 [7]. On the one hand, this VHD has a four-axis symmetrical structure, so it can obtain low initial offset. On the other hand, during the four-phase spinning current operation, the current flowing in the device can always keep symmetrical and the average offset at the four spinning current states is very small. Thus, it is suitable for the four-phase spinning current circuit to achieve low residual offset.

978-1-5386-5309-8/18 $31.00 © 2018 IEEE

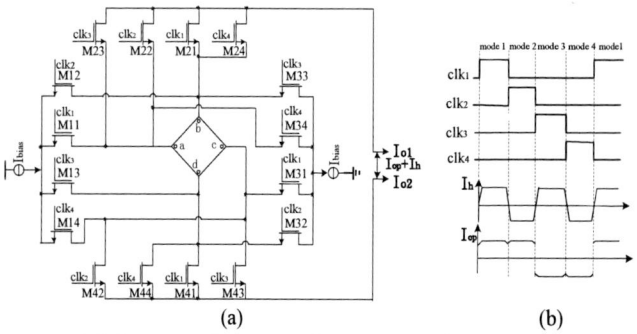

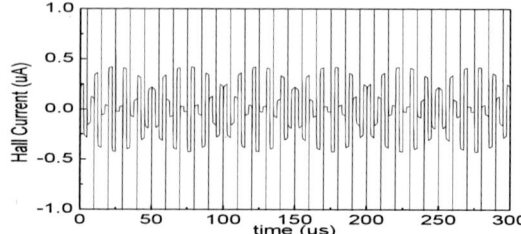

Fig.1.(a) Schematic diagram of current-mode switched vertical Hall circuit with a four-phase spinning current technique; (b)diagram of output Hall current and offset current corresponding to the four-phase sequence clocks.

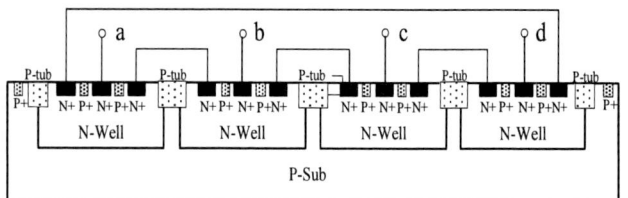

Fig.2. Structure of 3-contacts 4-folded VHD

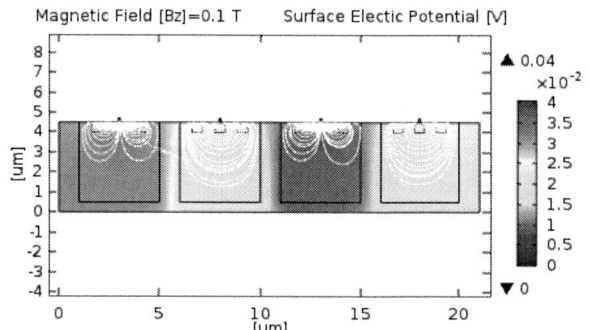

Fig.3.COMSOL simulation of vertical 3-contact four-folded Hall device.

The vertical 3-contact four-folded Hall device was simulated in COMSOL environment. Fig. 3 shows surface electrical distribution and current stream line in the vertical Hall sensor when a large magnetic field (0.1T) and a nominal bias current of 100 µA are applied. The result of differential output current (Hall current) shows the current-mode sensitivity is about 5.4 %T^{-1} which is double the voltage-mode sensitivity.

The transient simulation was done based on 0.8-µm high-voltage CMOS technology. In the circuit simulation, an eight-resistor Verilog-A simulation model of vertical Hall device is implemented in Cadence environment with the current sensitivity of 5.4 %T^{-1}. In addition, the offset is artificially produced by unbalanced Wheatstone bridge resistance in the simulation model. Fig.4 shows the transient simulation waveform of the current-mode four-phase spinning current circuit. The supply voltage is 5V, and the frequency of four-phase sequence clocks clk$_1$- clk$_4$is 100 kHz. The Hall device bias current is 100 µA, and the frequency and amplitude of applied magnetic field signal are 10kHz and 50mT, respectively. As can be seen, during the four-phase spinning

current process, the output differential Hall current changes its polarity periodically and its amplitude correctly reflects the change of magnetic field.

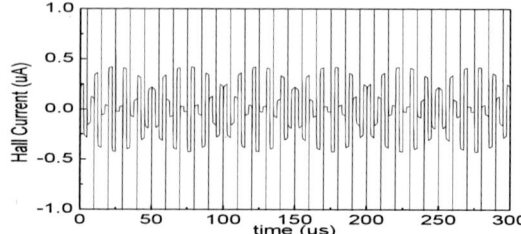

Fig.4. Transient output of current-mode four-phase spinning current circuit.

Design and Simulation of Signal conditioner

Fig.5 shows the schematic of the front-end CMOS vertical Hall sensor microsystem. Firstly, the current-mode four-phase spinning current circuit outputs the differential Hall current and offset current. After the I-V conversion, the Hall voltage and offset voltage are fed to the differential-differential amplifier (DDA) for voltage amplification. After the two times peak-to peak detection by correlated double sampling (CDS) circuit, the output signals corresponding to the two 2-phase spinning current operations are sampled and hold, respectively. As a result, the Hall offset produced in the 2-phase spinning current operation is removed. After that, the two offset signals are averaged to cancel the residual offset by the adder function, whose output voltage is 4 times Hall voltage. The negative feedback loop is used to eliminate the offset components generated after DDA amplification. Finally, by the low-pass filtering, the linear output Hall voltage proportional to magnetic field is obtained.

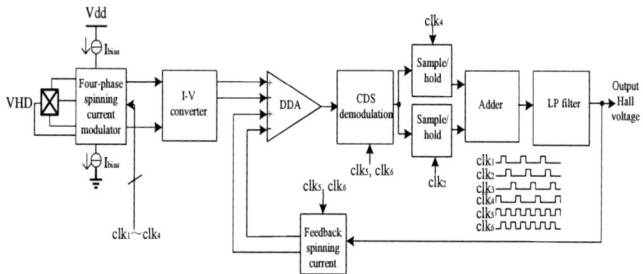

Fig.5. Hall sensor Microsystem block diagram.

Fig.6 illustrates the transient simulation result of signal conditioner. Here, the frequency and amplitude of input sinusoidal magnetic field signal are 10 KHz and 50mT, respectively. Fig. 6(a) shows transient output voltage waveform of DDA, where input Hall voltage is amplified and the offset and 1/f noise in signal conditioner loop is attenuated. Fig. 6(b) shows the output waveform of negative feedback modulation circuit, which has same frequency as the spinning current and amplitude proportion to output Hall voltage. Fig. 6(c) shows the final output Hall voltage of low-pass filtering at B=50mT. It can be seen that the output static point voltage is 2.5V, and the output waveform has same variation with the magnetic field. The simulation indicates that the signal conditioner can work properly and effectively eliminate the offset of Hall device and signal conditioner.

978-1-5386-5309-8/18 $31.00 © 2018 IEEE

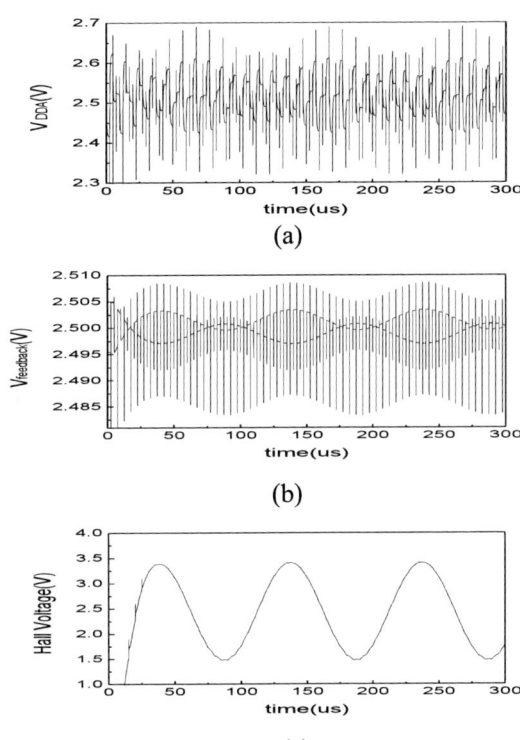

(a)

(b)

(c)

Fig.6. (a) Transient output voltage waveform of DDA. (b) Transient output waveform of feedback modulation circuit. (c) Transient output waveform of Hall voltage after low-pass filter at B=50mT.

Fig.7 shows the output Hall voltage versus amplitude of magnetic field when the field frequency is10kHz and the offset is5mV. It is obvious that the Hall output voltage is linearly changed with the sinusoidal magnetic field and the linearity reaches 99.9% in the magnetic field range from 5mT to 100mT.

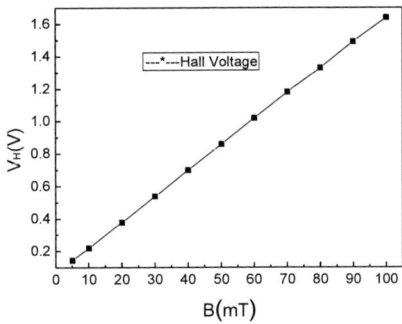

Fig.7. Hall voltage verves magnetic field.

Table I summarizes the key performance parameters of the vertical Hall sensor microsystem. When the supply voltage is 5V, the maximum output peak-to-peak voltage is up to 3.3V, close to the supply voltage and the minimum output peak voltage is about 0.3V. The residual offset is less than 0.3mT, suggesting the strong ability to eliminate offset originating from the Hall device and electronics. In addition, the vertical Hall sensor microsystem achieves a high linearity of 99.9% in the magnetic field range from 5mT to 100mT and the power dissipation is 31.3 mW.

TABLE I
KEY PERFORMANCE PARAMETERS OF THE VERTICAL HALL SENSOR MICROSYSTEM.

Parameters	Value
Supply voltage	5 V
Residual offset	0.3 mT
Linearity	99.9%
Bandwidth	30 kHz
Power dissipation	31.3 mW
Dynamic output voltage	0.3-3.3V

CONCLUSIONS

A front-end CMOS vertical Hall sensor with current-mode four-phase spinning current technique is presented. The current-mode four-phase spinning current circuit not only can significantly improve the current sensitivity of the vertical Hall device, but also can effectively remove the device offset and 1/f noise. Moreover, a current negative feedback loop technique is proposed to further eliminate the offset and 1/f noise of signal conditioner. By means of 0.8-μm high-voltage CMOS technology, the simulation results reveal that the residual offset of the Hall sensor is less than 0.3mT and the output Hall voltage linearity is up to 99.9%. The front-end Hall microsystem exhibits the outstanding advantages of low residual offset and high sensitivity and high linearity.

ACKNOWLEDGMENTS

This work is sponsored by the National Natural Science Foundation of China (No. 61571235) and QingLan Project of Jiangsu Province.

REFERENCES

[1] A. Ajbl, M. Pastre, and M. Kayal, "A fully integrated Hall sensor microsystem for contactless current measurement," IEEE Sensors Journal,vol. 13, pp. 2271–2278, June, 2013.

[2] C. Schott, R. Racz, A. Manco, and N. Simonne, "Cmos single-chip electronic compass with microcontroller," Solid-State Circuits, IEEE Journal of Soild-State Circuits, vol. 42, pp. 2923–2933, December, 2007.

[3] Hadi Heidari, Edoardo Bonizzoni, Umberto Gatti, and Franco Maloberti, "A CMOS Current-Mode Magnetic Hall Sensor With Integrated Front-End," IEEE Transactions on Circuits and Systems, vol. 62,pp, 1270-1278, May , 2015.

[4] Hadi Heidari, Umberto Gatti, and Franco Maloberti, "Sensitivity Characteristics of Horizontal and Vertical Hall Sensors in the Voltage- and Current-Mode," Ph.D. Research in Microelectronics and Electronics. IEEE,7251302,September, 2015.

[5] H. Heidari, E. Bonizzoni, U. Gatti, and F. Maloberti, "A current-
mode CMOS integrated microsystem for current spinning magnetic hall sensors," International Symposium on Circuits and Systems .IEEE, 6865226, July,2014.

[6] Hadi Heidari, Edoardo Bonizzoni, Umberto Gatti, Franco Maloberti, Ravinder Dahiya, "Optimal Geometry of CMOS Voltage-Mode and Current-Mode Vertical Magnetic Hall Sensors," Sensors IEEE, vol.2, pp.1-4,November, 2015,.

[7] Hadi Heidari, Edoardo Bonizzoni, Umberto Gatti, Franco Maloberti, and Ravinder Dahiya, "CMOS Vertical Hall Magnetic Sensors on Flexible Substrate," IEEE Sensors Journal, vol. 16, pp.8736-8743, December, 2016.

A NOVEL GAS SENSOR SIGNAL DRIFT ADJUSTMENT METHOD BASED ON CONTROLLED MEASUREMENT

Chang-Yong Chiu[1]*, and Zhun Zhang[2]

[1]School of Communication & Information Engineering, Shanghai University, Shanghai, China
[2]College of Optoelectronic Engineering, Shenzhen University, Shenzhen, China
*Corresponding Author's Email: qiuchangyong@sina.com

ABSTRACT

Gas sensors of semiconductor types are more or less affected by external environmental factors change, which could inevitably lead to signal drift of measurements. In this study, we propose a novel method to adjust sensor signal drift. The system design includes: (1) real-time feedback of remote environmental factors by wireless sensor networks; (2) a controlled measurement setup used for simulating changeable remote environmental factors; (3) drift adjustment of remote sensor signal based on controlled sensor signal. We process extracted sensitivity feature vector of three cases with Naïve Bayes classifier for comparison, and acquired experimental results demonstrate that our drift adjustment method can effectively improve recognition rate at the presence of sensor signal drift. Especially, this design is appropriate for remote control sensor applications under changeable environmental factors.

INTRODUCTION

Gas detection and measurement have been meaningful and significant to industrial field and daily life. However, gas sensor signal drift problem exists widely in actual measurement. Signal drift error can certainly influence the subsequent data processing results. Therefore, it is necessary to conduct signal drift adjustment in the stage of sensor data preprocessing. Semiconductor gas sensor, electrochemical gas sensor, and conductive polymer sensor all need to take into account signal drift problems which could greatly affect recognition rate. Unexpected disturbances may come from changeable environmental factors such as temperature, relative humidity, gas pressure, gas flow rate, even electric and magnetic fields, light, acoustic waves and so on. Each of these environmental factors could interact with MOX (metal oxide) gas sensor due to multiple effects of semiconductor [1]. Typically, temperature and relative humidity are primary environmental factors, however, under extreme cases, barometric pressure also need to be considered since gas volume concentration (in ppm unit) can change as a result of atmospheric pressure change. In addition, gas flow rate has an influence on sensor response [2], thus it cannot be neglected for concentration measurement of moving gas.

The gas sensor signal drift problem has troubled gas sensor applications over time. To overcome such a shortcoming, various signal drift compensation methods have been reported. To suppress drift to some extent and improve the performance of networks, Zhang et al. introduced the response of temperature sensor into artificial neural networks and eliminated the misjudgments [3]. Vergara et al. adopted an ensemble method based on support vector machines which uses a weighted combination of classifiers trained at different points of time to cope with sensor drift [4]. Liu et al. proposed a novel ensemble method with dynamic weights based on fitting [5] to predict classifier weights by fitting functions obtained from training, and a pre-aging process to improve the stability of the classifier ensemble [6]. Wiezbicki et al. developed a model that uses an ensemble of weighted neural networks to compensate the drift and weighted input data according to their recentness during classifier training [7]. Ziyatdinov et al. proposed a new multivariate method based on common principal component analysis (CPCA) for drift compensation, whose novelty lies in computing the drift direction explicitly as a variance common for all the odors classes [8], and applied Orthogonal Signal Correction (OSC) to drift compensation [9]. Aliaghasarghamish et al. employed an appropriate recursive least squares (RLS) fuzzy model for modeling of the system and a quantitative analysis of the responses in different conditions and used the results of the analysis for the partial compensation of the drift caused errors [10].

The above drift compensation methods mainly deal with applying a better algorithm or optimizing a model or combining several different algorithms or taking into account dynamic and ambient factors, whereas our paper does not mainly focus on classifier algorithm or classifier model aspect but on the aspect of signal measurement. As noted by Vergara et al. [4], sensor drift can be attributed to two kinds: one is called 'real-drift' such as aging and poisoning, the other is measurement system drift which is our main study target. Post-processing of drifted sensor signal for feature extraction is evidently important due to the reason that its accuracy can greatly determine final classification accuracy. Imagine a case, if sensor response of gas type A has drifted to be very approximate to that of gas type B, even a well-trained model would wrongly predict it as gas type B. Therefore, environmental factors should be deemed as input variables of classifiers. However, training work is an onerous and time-consuming task in that a huge database of gas measurement samples needs to be built. Thus, we propose a drift adjustment method in the main part of this paper to save the laborious job. The rest of this paper is organized as follows: we first set forth signal drift analysis and adjustment method, and then describe system design of a controlled measurement based setup; at last, we use the proposed method to adjust drift of measured sensor signal and show the advantage by experimental results.

978-1-5386-5309-8/18 $31.00 © 2018 IEEE

SIGNAL DRIFT ANALYSIS AND ADJUSTMENT METHOD

Gas sensor signal drift could make a well-trained model inaccurate or unreliable for gas type or concentration recognition. The feature extraction is mostly based on sensitivities of gas sensors. Normally, we have a gas sensor sensitivity definition expressed as below,

$$S = (R_{Air} - R_{Gas})/R_{Air} \qquad (1)$$

where R_{Air} is gas sensor's resistance in clean air, and R_{Gas} is gas sensor's resistance in measured gas. Obviously, S is a value within the range $[0, 1)$ due to decrease of sensor's resistance in measured gas. At the presence of sensor signal drift (caused by environmental factors change), even for the same gas type with same ppm concentration, R_{Gas} could change to R_{Gas}', thereby S calculated by equation (1) changes to another value S', resulting in recognition error if not take notice of that R_{Air} could change to R_{Air}', thus ultimate recognition rate is lowered.

Since gas sensor sensitivity is mostly used as the input feature for classifiers, what we do is to adjust gas sensor signal drift as much before feature extraction. Here we have one assumption that environmental factors' contribution to sensor signal drift is additive, namely, with the same deviation from sensor signal under standard case, as shown below,

$$R_{Air} - R_{Air}' = R_{Gas} - R_{Gas}' \qquad (2)$$

where R_{Gas}' and R_{Air}' are unknown variables that all need to be measured.

Since we generally train gas classifiers under standard case before classification, to reuse the well-trained model, we need to convert it to standard case by equation (2). Then we can deduce an adjusted gas sensitivity definition with additive drift,

$$\begin{aligned} S &= (R_{Air} - (R_{Gas}' + (R_{Gas} - R_{Gas}')))/R_{Air} \\ &= (R_{Air} - (R_{Gas}' + (R_{Air} - R_{Air}')))/R_{Air} \\ &= (R_{Air}' - R_{Gas}')/R_{Air} \end{aligned} \qquad (3)$$

Also we can have another assumption that environmental factors' contribution to sensor signal drift is proportional, namely, with the same ratio to sensor signal under standard case, as shown below,

$$R_{Air}/R_{Air}' = R_{Gas}/R_{Gas}' \qquad (4)$$

Similarly, we need to convert it to standard case by equation (4). Then we can derive an adjusted gas sensor sensitivity definition with proportional drift,

$$\begin{aligned} S &= (R_{Air} - (R_{Gas}' \cdot \frac{R_{Air}}{R_{Air}'}))/R_{Air} \\ &= (R_{Air} \cdot R_{Air}' - (R_{Gas}' \cdot R_{Air}))/(R_{Air} \cdot R_{Air}') \\ &= (R_{Air}' - R_{Gas}')/R_{Air}' \end{aligned} \qquad (5)$$

Now we have two formulas (3) and (5) for gas sensor signal drift adjustment. As denoted above, in addition to measuring R_{Gas}', we also need to measure R_{Air}', which is gas sensor's resistance in clean air (considering changed environmental factors, namely, baseline state with drift).

SYSTEM DESIGN OF MEASUREMENT MODULES

To test and verify that if gas sensor sensitivity calculated by the two formulas is able to achieve better recognition accuracy, we design a setup to measure R_{Gas}' and R_{Air}'. The setup consists of two measurement modules (*remote measurement module* and *controlled*

measurement module, which are used for measuring R_{Gas}' and R_{Air}' respectively) and *signal processing module*. Two measurement modules communicate each other through ZigBee based wireless sensor networks, as shown in Figure 1. Signal processing module can connect the two with ZigBee node.

The design of ZigBee node is based on TI CC2530F256 SoC, which is low-power, highly integrated RF transceiver for 2.4-GHz IEEE 802.15.4 ZigBee applications [11]. Its excellent receiver sensitivity and robustness to interference ensure effective wireless connectivity to low-rate sensors and control devices [12]. Under obstacle-free test field of outdoors, its transmission range of single-hop can be up to 250m, while that of multi-hop can be up to 1000m. To ensure data validity of data transmission and receiving, CRC checksum is used for frame error control in both TX and RX modes. Especially, peer-to-peer mode has lower error rate of approximately 1%, compared to broadcast mode. Furthermore, functionality of ACK frame is useful for data re-transmission mechanism against frame loss.

A. Remote measurement module

The module includes: temperature/humidity sensor, ZigBee node, half-opened gas chamber with uniformly dotted holes which are used for keeping gas flow relatively stable, environmental factors measurement circuit which is used for signal-conditioning and measuring temperature/humidity sensor, and sensor signal measurement circuit which is used for gas sensor.

Environmental factors are fed back to *controlled measurement module* to simulate changeable remote environmental factors for sensor baseline state measurement, and gas sensor response data of gas analyte is sent to *signal processing module* for sensor signal drift adjustment.

B. Controlled measurement module

This module includes: thermostat (heater/cooler), humidity controller (humidifier/dehumidifier), ZigBee node, gas chamber which is used for keeping gas concentration stable, MFC (Mass Flow Controller) which is used for accurate gas sampling, gas valve, gas sensor, and sensor signal measurement circuit.

It receives data from *remote measurement module* and parses it into remote environmental factors feedback, then thermostat and humidity controller can control heater/cooler and humidifier/dehumidifier to regulate gas chamber's temperature and humidity accordingly. PID controlled constant temperature and humidity system is introduced to keep environmental factors stable and follow even slight change, which is an automated program control process to simulate remote environmental factors. In practical use, clean air is drawn into gas chamber for measuring R_{Air}' value. In addition, it can generate sensor data in standard condition and extract feature by equation (1) for training classifiers.

Since temperature and relative humidity are main environmental factors contributing to sensor signal drift, only thermostat and humidity controller are used; if there is some other environmental factor which impacts sensor drift, it should be measured and simulated similarly.

C. Signal processing module

This module receives sensor response signal of gas analyte from *remote measurement module* in real time and simulated sensor baseline state signal under remote environmental factors from *controlled measurement module*, respectively. Sensor signal drift adjustment work is done here to train classifier model or predict classification result (gas type and concentration) from it. In Figure 1, this module is a separated module connecting the two modules with ZigBee node. Furthermore, it can also be integrated to either of them.

There can be more than one *remote measurement module*, but only one *controlled measurement module*. If environmental factors of *remote measurement module* are stable, it only needs to measure R_{Air}' once. Gas sensor or gas sensor array should be calibrated to the same status.

To recover sensor to baseline state, clean air is needed. However, in polluted atmosphere, clean air is unavailable or not easy to obtain [13]. The system design can avoid this issue by adopting controlled measurement. In general, controlled measurement makes relatively long recovery time wait unwanted, saving the sensor recovery process of *remote measurement module*.

Without controlled measurement, the difficulties lie in two sides: 1) It is a burdensome task to measure data points of all possible cases, especially for multidimensional environmental factors. It is nearly impossible to prepare all data points for drift compensation, due to the reason that sensor's data specification cannot provide full range sensor drift correction curve. For multidimensional environmental factors, sensor drift correction equation is a multivariate function that can be expressed as below,

$$D(i, x; w1, w2, \ldots, wn) =$$
$$V(i, x; w1, w2, \ldots, wn) - V(i, x; W1, W2, \ldots, Wn) \quad (6)$$

where i is input gas type, x is input gas concentration, w1,w2,…,wn are environmental factors under changeable condition, W1,W2,…,Wn are environmental factors under standard condition, V is measured value. Even if we ignore i and x variables' influence on D and V, n-dimensional environmental factors can bring huge data points to be measured. 2) For unknown or changeable environmental factors, frequent training to update classifier model may be a solution to sensor drift problem, however, it is not practicable in actual use since training with field data is a troublesome job.

Especially for longtime monitoring system, the environmental factors vary frequently, thus controlled measurement is particularly required for effective drift correction. Remote control sensor applications (e.g. gases detection) based on WSN technology [14] can separate human from hazardous or harmful gases for safety; mobile robots equipped with *remote measurement module* can communicate with other modules by ZigBee networks.

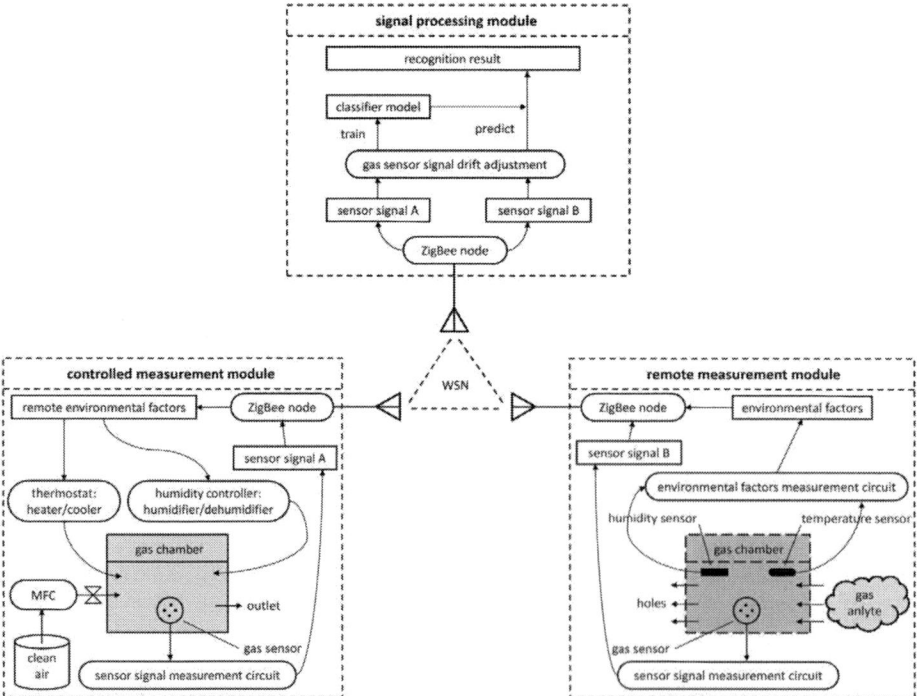

Figure 1. Schematic diagram of controlled measurement based drift adjustment system.

DATA PROCESSING AND EXPERIMENTAL RESULTS

We get experimental data from methane gas of specific concentration value (2400 ppm). The measured resistance values of eight gas sensors are shown in Table 1, including baseline state / response state under standard environmental factors (here we select the normal case of room temperature of 25 ℃ and RH of 45%) and drifted environmental factors (temperature of 19 ℃ and RH of 94%). We then calculate sensitivities for feature extraction, including three cases: 1. No drift adjustment; 2. Additive drift adjustment; 3. Proportional drift adjustment. Sensitivities of eight sensor dimensions under standard environmental factors are compared with that of three cases under drifted environmental factors, as shown in Figure 2.

TABLE I

Typical sensor resistance values in kΩ (measured under standard environmental factors and drifted environmental factors, respectively)

Sensor dimension	Standard baseline state	Standard response state	Drifted baseline state	Drifted response state
1	97.26	72.41	90.86	62.94
2	46.31	17.25	53.15	21.43
3	49.28	36.28	49.80	39.05
4	76.63	33.32	77.76	34.07
5	37.26	22.96	31.46	21.31
6	35.66	12.38	29.99	9.53
7	40.42	24.13	37.26	21.99
8	30.46	22.49	29.06	19.77

We employ Naïve Bayes classifier [15, 16] as the prediction method. It uses Gaussian probability model for gas type or concentration prediction. Usually, PCA is needed to reduce sensor data dimensions, making each two sensor dimensions uncorrelated. For the case of Gaussian distribution, each two uncorrelated dimensions are independent. It uses Gaussian probability density function as the distribution, thus each gas sensor response (namely, sensitivity feature) influences recognition results independently. Naïve Bayes classifier has an assumption that all feature dimensions are independent of each other, obviously it can be met. Therefore, the joint probability can be calculated by multiplication of each sub-feature's probability.

e.g., we have N kinds of gas types or concentrations (namely, g_1, g_2, ..., g_N) to discriminate, we can use a gas sensor array which consists of 8 sensors, thus each input gas kind has 8 sensor responses (namely, sensitivity feature vector $[s_1, s_2, s_3, s_4, s_5, s_6, s_7, s_8]$). Naïve Bayes classifier can find the most likely one by calculating its joint probability. The basic deduction process is as

follows,

$$p(g \mid [s_1\,s_2\,s_3\,s_4\,s_5\,s_6\,s_7\,s_8]) =$$
$$p([s_1\,s_2\,s_3\,s_4\,s_5\,s_6\,s_7\,s_8] \mid g)\,p(g) \,/\, p([s_1\,s_2\,s_3\,s_4\,s_5\,s_6\,s_7\,s_8]) \quad (7)$$

$p([s_1\,s_2\,s_3\,s_4\,s_5\,s_6\,s_7\,s_8])$ can be ignored due to the reason that it is the same to all kinds, and $p(g)$ is deemed as equal probability to all kinds since each kind's probability is unknown in advance. Therefore, the problem is formulated as the calculation of maximum $p([s_1\,s_2\,s_3\,s_4\,s_5\,s_6\,s_7\,s_8] \mid g)$, and due to the independence of all feature dimensions, it can be expressed as below,

$$p([s_1\,s_2\,s_3\,s_4\,s_5\,s_6\,s_7\,s_8] \mid g) = \prod_{i=1}^{8} p(s_i \mid g) \quad (8)$$

Given σ value of 0.25 which is 1/4 of full sensitivity range, then we can plug sensitivity feature vector $[s_1, s_2, s_3, s_4, s_5, s_6, s_7, s_8]$ into equation (8) to acquire joint probability of each gas kind, the one with maximum joint probability is the best target candidate, namely, the result gas kind. $p(s_i \mid g)$ (i=1,…,8) is Gaussian distribution which has probability density function as $p(s_i \mid g) = \frac{1}{\sqrt{2\pi}\sigma} \exp\left(-\frac{(s_i - S_i)^2}{2\sigma^2}\right)$. Thus, to solve $[s_1\,s_2\,s_3\,s_4\,s_5\,s_6\,s_7\,s_8] = \arg\max_s p([s_1\,s_2\,s_3\,s_4\,s_5\,s_6\,s_7\,s_8] \mid g)$ is equivalent to solve

$$[s_1\,s_2\,s_3\,s_4\,s_5\,s_6\,s_7\,s_8] = \arg\min_s \sum_{i=1}^{8}(s_i - S_i)^2 \quad (9)$$

where $[s_1, s_2, s_3, s_4, s_5, s_6, s_7, s_8]$ and $[S_1, S_2, S_3, S_4, S_5, S_6, S_7, S_8]$ are the sensitivity feature vectors of input gas kind and target gas kind, respectively. Here we use measured data in Table 1 to compare three cases with standard case for the same gas kind at the presence of sensor signal drift, calculated results of $\sum_{i=1}^{8}(s_i - S_i)^2$ for three cases are 0.0400, 0.0276, and 0.0140, respectively. According to equation (9), it shows that compensating gas sensor signal drift with the two drift adjustment methods can achieve better recognition accuracy than that without drift adjustment, and proportional drift adjustment method is better than additive drift adjustment method here.

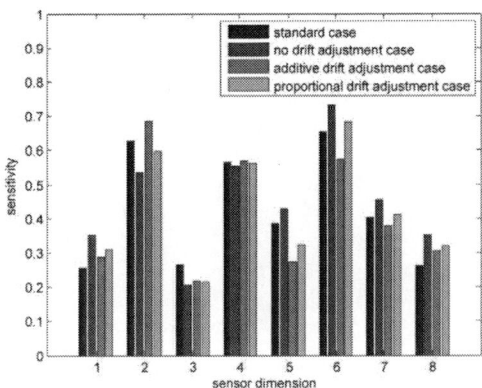

Figure 2. Sensitivity comparison of eight sensor dimensions for standard case and three cases under drifted environmental factors.

CONCLUSION AND PROSPECT

After signal drift adjustment, sensor data is more appropriate for subsequent training process and classification process (pattern recognition or machine learning). Higher recognition rate can be achieved by this method as a result of effective drift adjustment. It is relatively easy to apply this method. Moreover, data obtained from controlled measurement can be used for optimizing a specific data processing algorithm.

The controlled measurement based drift adjustment design is also suitable for other sensor types such as electrochemical gas sensor [17], piezoresistive pressure sensor [18], altimeter senor [19] and so on. The lightweight and cost-effective design of remote measurement module can be carried by mobile robot to perform gas detection task under open environment; while ZigBee technology based wireless sensor networks is effective for self-organized network and real-time data transmission, and ensures on-line task of remote-controlled robot.

REFERENCES

[1] Sze, Simon M.. "Physics of Semiconductor Devices." Physics Today 23.6 (1970):75-75.

[2] Liang, Hong Zeng, et al. "Influence of Gas Flow on Response of Semiconductor Gas Sensor." Instrument Technique & Sensor (2012).

[3] Min, Zhang. "Research on the Temperature Drift Suppressing of Gas Sensor Array." Chinese Journal of Sensors & Actuators 20.6(2007):1237-1239.

[4] Vergara, Alexander, et al. "Chemical gas sensor drift compensation using classifier ensembles." Sensors & Actuators B Chemical 166-167.9(2012):320-329.

[5] Liu, H., and Z. Tang. "Metal oxide gas sensor drift compensation using a dynamic classifier ensemble based on fitting." Sensors 13.7(2013):9160.

[6] Liu, Hang, R. Chu, and Z. Tang. "Metal Oxide Gas Sensor Drift Compensation Using a Two-Dimensional Classifier Ensemble." Sensors 15.5(2015):10180-10193.

[7] Wiezbicki, Thiago, and E. P. Ribeiro. "Sensor drift compensation using weighted neural networks." IEEE

Conference on Evolving and Adaptive Intelligent Systems (2016):92-97.

[8] Ziyatdinov, A., et al. "Drift compensation of gas sensor array data by common principal component analysis." Sensors & Actuators B Chemical 146.2(2010):460-465.

[9] Padilla, M., et al. "Drift compensation of gas sensor array data by Orthogonal Signal Correction." Chemometrics & Intelligent Laboratory Systems 100.1(2010):28-35.

[10] Aliaghasarghamish, M, and S. Ebrahimi. "Recursive least squares fuzzy modeling of chemoresistive gas sensors for drift compensation." International Symposium on Innovations in Intelligent Systems & Applications (2011):1-5.

[11] CC2530 Datasheet, http://www.alldatasheet.com/datasheet-pdf/pdf/461731/TI1/CC2530.html

[12] Adams, Jon T. "ZigBee Wireless Technology and the IEEE 802.15.4 Radio-Enabling Simple Wireless." Texas Wireless Symposium (2008).

[13] Romain, Anne-Claude, and J. Nicolas. "Long Term Stability Of Metal Oxide-Based Gas Sensors For E-nose Environmental Applications: an overview." Olfaction & Electronic Nose: International Symposium on Olfaction & Electronic Nose. American Institute of Physics (2009):443-445.

[14] Somov, Andrey, et al. "Compact Low Power Wireless Gas Sensor Node With Thermo Compensation for Ubiquitous Deployment." IEEE Transactions on Industrial Informatics 11.6(2017):1660-1670.

[15] Zhang, Harry. "The Optimality of Naive Bayes." Seventeenth International Florida Artificial Intelligence Research Society Conference, Miami Beach, Florida, Usa 2004.

[16] Rish, I. "An empirical study of the naive Bayes classifier." Journal of Universal Computer Science 1.2(2001):127.

[17] Zhang, Xiao Jun, M. L. Zhang, and L. I. Xiao-Hui. "Temperature Compensation of the CO Electrochemical Gas Sensor Based on RBF Neural Network." Chinese Journal of Sensors & Actuators 22.1(2009):11-14.

[18] Zhang, Yanhua, et al. "Research on method of temperature compensation for piezoresistive pressure sensor." Electronic Measurement Technology (2017).

[19] Liansheng, H., et al. "Altimeter Sensor with Temperature Compensated Coefficients in UAV." Microcomputer Information 24.10(2008):177-176.

A

Abe Tamotsu	2-19,
Agopian PDG	1-41,
An Xia	1-47,
Anju Eisuke	1-9,
Arabhavi Akshay M	1-1,

B

Bae Kyoungchul	7-9,
Bai Dongshun	7-3,
Bai Guoqiang	9-4, 9-17,
Bai Song	2-41, 3-52,
Bai Xiaolin	7-6,
Baick I	6-23,
Banares Berwin	4-15,
Bao Lin	1-62,
Batllo Francois	5-5,
Bei Duohui	6-29,
Bersuker Gennadi	6-8,
Bömmels Juergen	4-40,
Bolognesi CR	1-1,
Bordallo C	1-41,
Briggs Basoene	4-40,
Brost Bert	8-31,
Brune Jan	7-22,
Buergi Daniel	7-20,

C

Cai Hua	2-28,
Cai Kun	6-2,
Caio Yimao	1-62,
Cao Hui	2-30,
Cao Xinpei	7-12,
Cao Zhijun	1-76,
Cao Zigui	4-19,
Chambers Andrew A	4-26,
Chan Yi Peng	1-64,
Chang CC	7-23,
Chang Edward Yi	8-29,
Chang Juyeon (Jay)	5-12,
Chao Taifong	5-14,
Chen Chen	1-60,
Chen Cheng	1-61,
Chen Chung-Fu	5-22,
Chen Dihu	9-11,
Chen Guodong	3-55,
Chen Hao	9-3, 9-2,

Chen Hualun	1-26,
Chen Hunglin	6-1, 6-2, 6-12,
Chen Jack	4-31,
Chen Kuang-Chao	5-22,
Chen Kuang-Wei	5-22,
Chen Qiaoli	3-15,
Chen Xingjian	3-26,
Chen Xiuguo	2-20,
Chen Yao-Tsong	7-17,
Chen Yen-Ting	5-22,
Chen Zirong	9-2, 9-3,
Chen ZJ	8-9,
Chen Zong-Cyan	7-17,
Chena CZ	9-31,
Cheng Jie	5-23,
Cheng Xie	2-9,
Cheng Xingming	7-5, 7-8,
Chi Min-Hwa	1-24, 1-64,
Chiu Chang-Yong	9-18,
Chiu Tzu-Yin	1-64,
Cho Jiyoon	7-9,
Choi Jinu	7-12,
Choi YC	6-23,
Chung Joo Young	7-11,
Chung KW	7-23,
Chung Stream	7-17,
Claeys C	1-41,
Claeys Cor	1-55,
Clauberg Horst	7-20,
Collaert N	1-41,
Collaert Nadine	1-55,
Cui Liang	7-5,
Cui Xiaole	9-10,

D

Dai Johnny	6-16,
Den Hai	2-18,
Deng Hai	2-30,
Ding P	4-33,
Ding Peijun	4-25,
Ding Tongguo	4-22,
Dong Lisong	2-11,
Dong Tianhua	1-64,
Dong Yunhe	3-57,
Dou Tao	6-29,
Duan M	6-33,
Duan Qingxi	1-39, 1-77,
Duan Wenting	1-26,

Duan Xiansheng	7-7,
Duan Yangyang	7-8,

E

Elliston Bob	3-13,
Endoh Tetsuo	1-54,
Eneman Geert	1-55,
Erickson Stuart	7-10,

F

Fan Rongwei	6-1, 6-2, 6-12,
Fan RX	1-30,
Fang Qiang	5-14,
Fang Yichen	1-62,
Fang Zhiquan	1-26, 1-27, 2-10,
Fang Zhou	2-32,
Fang Ziquan	3-14,
Fann Daniel	7-23,
Feng Danniel	1-65,
Feng Ke	2-37,
Feng Ming	4-31,
Feng Yufan	9-10,
Feng Yulin	1-58,
Fernandez Cyril-Patrick	6-31,
Findleton Kevin	4-15,
Flaim Tony	7-3,
Fu Yali	3-13, 3-18,
Fuh Yiin-Kuen	4-39,

G

Gan Zhifeng	2-34,
Gao Bin	1-32, 1-49,
Gao Di	9-24,
Gao Jiangfeng	1-76,
Gao Junbo	7-12,
Gao Lizhao	9-7,
Gao Rui	1-13,
Gao Siyang	1-47,
Geng Weidong	6-4,
Goglia Peter	3-36,
Gong Enbin	9-16,
Gong Hanmo	2-28,
Gosh Anagha K	9-32,
Gosh Kunal Promode	9-32,
Gu Jiangtao	9-7,
Gu Jie	1-76,
Gu Xiaofang	6-1, 6-2,
Gu Zeguo	1-49,

Guo Eric	3-40,
Guo Fangfang	7-6, 7-7,
Guo Lingyi	1-61,
Guo Wei William	5-1,
Guoping Mao	2-22,

H

Hahn Sookap	5-16,
Haigermoser Christian	3-21,
Han Qiuhua	1-34,
Hanyu Takahiro	1-54,
Hao Peilin	1-47,
Hao Peng	6-8,
Hao Xiucheng	9-16,
Hashimoto Masanori	9-22,
He Liang	1-55,
He Mingxiao	9-16,
He Weiming	
He Xiaobin	1-76, 2-46,
He Yanghua	6-14,
He Yonggen	4-29,
He Youfeng	4-29,
Hegde Hari	3-36,
Heinrich Kathleen	8-17,
Heo Yumi	2-15,
Heyns Marc	1-55,
Hiromoto Masayuki	9-23,
Hori Tsukasa	2-19,
Hoshii T	8-24,
Hou Chaozhao	1-76,
Hou Wenrong	4-31,
Hou Zhaozhao	1-74,
Hsieh Yu-Lin	4-39,
Hsu Chien-Pin Sherman	3-27,
Hsu Heng tung	8-29,
Hsu Hisang-Hua	8-29,
Hsu Michael	7-15,
Hsu Po-Chun	1-55,
Hu David Y	9-22,
Hu Jun	1-26,
Hu Linhui	6-4,
Hu Meili	1-31,
Hu Xingxing	9-12,
Hu Zhigao	6-8,
Huang Chi-Yi	8-29,
Huang Chong	4-22,
Huang Haigou	5-14,
Huang Hung-Jui	4-39,

Huang Qin	1-40,
Huang Qianqan	1-60, 1-61,
Huang Ran	1-28,
Huang Ru	1-39, 1-47, 1-60, 1-61, 1-62, 1-77,
Hunag YH	8-18,
Huang Yahui	3-55, 3-56, 3-57,

I

Inoue Hideo	2-23,
Iwai Hiroshi	8-24, 8-29,

J

Jensen Jim	6-31,
Jeong Seulgi	2-14,
Jeong T	6-23,
Ji H	4-33,
Ji Shiliang	1-34,
Ji X	1-63,
Ji Z	6-33,
Ji Zhigang	1-13,
Jia Yutao	
Jiao Mingjie	3-13, 3-18,
Jiang Hao	1-65, 2-12,
Jiang Qifeng	1-76,
Jiang Xiaojuan	7-5,
Jiang Yulei	1-64,
Jiang Zhongwei	3-55, 3-56, 3-57,
Jin Lan	1-64, 4-29,
Jin M	6-23,
Jin YF	8-18,
Jung Hyenil	2-15,
Jung Kyunghag	7-16,

K

Kakushima K	8-24,
Kamrath Michael	5-5,
Kan X	8-9,
Kang Jian	1-62,
Kau Li-Han	4-39,
Kawaguchi Mark	3-21,
Kawasuji Yasufumi	2-19,
Kim Cheilkyu	6-16,
Kim Jaesung	7-16,
Kim Jinhyung	2-14, 2-15,
Kim Junghwa	7-9,
Kim H	6-23,
Kim Hon Jin	5-14,
Kim Kyun	7-11,

Kim MH	3-18,
Kim Min Sung	7-16,
Kim Sangkyun	7-9, 7-16,
Kim Seunghyun	2-15,
Kim Sung Hwan	2-15,
Kim Yoona	2-14,
Kim Yunjun	2-15,
Kodama Takeshi	2-19,
Koh Jeongdeog (Jeidy)	5-12,
Koh Leong Tee	1-64,
Koli Dinesh	5-14,
Kryman Tim	6-16,
Kunert Bernadette	1-55,
Kwon Kihyeok	7-11,

L

Langenickel Joern	8-17,
Langer Robert	1-55,
Lau John H	7-21,
Lau W S	4-27, 4-28,
Lee Cheng-Kuei	
Lee Chien-Chieh	4-39,
Lee Da-Eun	
Lee Donghwan	7-9, 7-11, 7-16,
Lee Junwoo	7-9,
Lee Ken	7-16,
Lee Tammy	3-18,
Lee Yi-Ting	
Leighton Jamie	5-20,
Li Alan	3-40,
Li Bin	8-21,
Li Cheng	
Li Ding	8-20, 8-21,
Li Eric	4-31,
Li Fenglian	6-29,
Li Jing	1-8,
Li Jinghong	6-25,
Li Jingxian	1-77,
Li Junfeng	1-76,
Li Junjie	1-76, 2-46,
Li Lanxia	7-5,
Li Lidong	1-77,
Li Ming	1-47,
Li Mingming	1-48, 6-9, 6-19,
Li Neil	4-31,
Li Tomi T	4-39,
Li Wei	6-20, 9-4, 9-17,
Li Xuan	3-52,

Li Xuemiao	2-18,
Li Yang	5-23,
Li Yawei	6-8,
Li Yi	1-58,
Li Yudong	1-76,
Li Yue	1-58,
Li Zhiguo	4-22,
Liang Peipei	2-34,
Liao Hualin	1-37, 9-16
Liao Miao	1-22,
Lim Jaebum	2-15,
Lim Sanghak	2-14, 2-15,
Lin Irving	7-23,
Lin Jiantao	7-7,
Lin Ming-Xhi	6-4,
Lin X	1-63,
Lin Yi Shih	6-4,
Lin Yueh-Chin	8-29,
Liu Alex	7-23,
Liu Chang	2-32,
Liu Chunling	4-6,
Liu Donghua	1-26, 1-27, 2-9, 2-10, 3-14, 4-7,
Liu Hao	7-6, 7-7,
Liu Hongjie	7-5, 7-8,
Jiu JL	8-18,
Liu Jianqiang	4-31,
Liu Junhua	1-37, 9-4, 9-17,
Liu Mengyang Elaine	6-3,
Liu Mike	5-20,
Liu Min	5-18,
Liu Shenjian	3-26,
Liu Shiyuan	2-12, 2-20,
Liu Wei	3-58,
Liu Weijie	4-1,
Liu Xuan	2-37,
Liu Xiao	7-3,
Liu Xiaoyan	3-58,
Liu Yiqun	4-29,
Liu Zewen	3-15,
Liu Zexue	1-37,
Liu Zhongyang	9-5,
Liu Ziqian Javaer	6-3,
Liu Zuoyue	6-1,
Long Yin	6-1, 6-12,
Lu Jui-Lin	1-64,
Lu Max	3-40,
Lu Xinchun	5-23,
Lu Yen Pin	5-22,
Lu Yihong	1-76,

Lu Ying Emily	5-1,
Lu Yuanfang	5-14,
Lube Michael	6-14,
Luo Chen	6-8,
Luo Jie	3-23,
Luo Jun	1-65,
Luo Kun	1-76,
Luo Lan	9-11,
Luo Shaoheng	9-5,
Luo Yijie	5-18,
Luoh Tuung	5-22,
Lv Zhu	1-60,

M

Ma Bingxin	4-1,
Ma Guiying	1-21, 1-22,
Ma Shawming	3-13, 3-18,
Ma Shuying	7-15,
Mair Robin	6-16,
Mallik Arindam	4-40,
Marti Diego	1-1,
Martin Joerg	8-17,
Martino JA	1-41,
Maslyk Dan	7-12,
Masuda Yutake	9-22,
Matsumoto Hiroshi	2-23,
Meng Lingkuan	1-76,
Miao JZ	1-30,
Mizoguchi Hakaru	2-19,
Moebius Martin	8-17,
Mols Yves	1-55,
Müller Dirk	7-22,
Mui David	3-21,
Mukundhan Priya	6-16,
Muneta I	8-24,
Muneta Iriya	1-9,

N

Nagatami Kenichi	6-10,
Nakarai Hiroaki	2-19,
Nappa Daria	6-14,
Natsui Masanori	1-54,
Nevarez Hector	6-14,
Ni Qiliang	6-1, 6-2, 6-12,
Ni Tuqiang	3-26,
Nie Miao	3-56,
Nowak Krzysztof M	2-19,

O

Ohno Hideo	1-54,
Oishi Kazuki	9-23,
Ostinelli Olivier	1-1,
Otoshi Kenji	2-23,
Otto Thomas	8-17,
Ouyang Tingbing	9-7,
Oyama Kenichi	3-7,

P

Pae S	6-23,
Pätzel Rainer	7-22,
Pan Yu	1-76,
Park J	6-23,
Park Kwansoon	6-16,
Park Yongyeop	7-9,
Park Yushin	2-15,
Peng Bo	7-1,
Peng Fei	2-16,
Peng Yufei	1-28,
Pey Kinleong	6-8,
Phan-Vu Hai-Au	3-13,
Pilloux Yannick	3-36,
Pristauz Hugo	7-19,
Pu Shanon	6-4,
Puligadda Rama	7-3,

Q

Qian He	1-32, 1-49,
Qian Hongtao HT	6-3,
Qian Jun	1-28,
Qian Wensheng	1-26, 1-27, 3-14, 4-7,
Qin H	4-33,
Qin Schafer	3-40,
Quan Wei	1-1,

R

Ren Qian	2-28,
Ren Zhexuan	6-15,
Rooyackers R	1-41,
Rosa Mike	5-20,
Ruicheng Ran	2-22,

S

Saitou takashi	2-19,
Saleem Mohamed	4-15,
Saman Bander	1-42,
Sasong HC	6-23,
Sato Takashi	9-23,
Saupe Ray	8-17,
Schlichtemann Ulf	9-21,
Sekine Naoki	7-7,
Shen JK	1-20, 1-30, 4-6,
Shen Manhua	3-52,
Shen Yiijiang	2-16,
Shen Zhengkun	9-16,
Shi Irene	3-40,
Shi L	1-30,
Shi X	4-33,
Shi Xuejie	1-22,
Shin H	6-23,
Shin Seungwook	2-15, 6-23,
Shintani Michihiro	9-23,
Shiraishi Yutaka	2-19,
Shu Xiaowei	4-31,
Sima Ge	7-14,
Simoen E	1-41,
Simoen Eddy	1-55,
Soethoudt Job	4-40,
Song Kunshan	4-29,
Song Lijun	1-31,
Song Yang	4-23,
Soumagne Georg	2-19,
Su Buchun	4-19,
Su Tao	9-11,
Sun Andrew	7-12,
Sun Chang	1-28,
Sun Feiyang	8-21,
Sun Hao	1-22,
Sun L	8-9,
Sun Litao	6-8,
Sun Peng	7-14,
Sun Tao	6-8,
Sun Xianhao	1-39,
Sun Yan	4-19,

T

Tan Andrew	7-20,
Tan Wei	7-5, 7-8,
Tanaka Hiroshi	2-19,

Tanaka Yusuke	7-2,
Tang Ling	1-22,
Tao Jiajia	1-18,
Tataurova Yulia	5-18,
Tatsunokuchi Suguru	8-24,
Tay Yuchi	5-22,
Tian Eric	3-40,
Tökei Zsolt	4-40,
Tomg Johnson	5-22,
Tong Zheyuan	6-29,
Treibergs Valts	8-31,
Trichur Ram	7-3,
Tsai Stan	5-14,
Tsutsui K	8-24,
Tu Hailing	1-76,

V

Vaniapura Vijay	3-13,
Versluijs Janko	4-40,

W

Wakabayashi Hitoshi	1-9,
Waldron Niamh	1-55,
Wang Bo	9-7,
Wang Chanfeng	4-23,
Wang Chaolun	6-8,
Wang Chenxu	2-18,
Wang Chuangang	1-37,
Wang Fang	1-58,
Wang Fei	1-20,
Wang Frank	3-18,
Wang Gangning	6-4,
Wang Guilei	1-76,
Wang Hougong	4-25,
Wang JY	8-18,
Wang Jianing	6-15,
Wang Jinfeng	5-12,
Wanf Jing	3-56,
Wang Kai	6-1, 6-12,
Wang Ken	3-13,
Wang Lei	6-7,
Wang Leilei	9-15,
Wang Kuanmao	4-25,
Wang MJ	8-18,
Wang Peng-Fei	6-4,
Wang Qian	7-6,

Wang Qingpeng	2-41,
Wang Qingpeng	6-17,
Wang Runsheng	1-60, 6-8,
Wang Songmiao	6-17,
Wang Teng	7-15,
Wang Weitung	5-18,
Wang Wen	1-65,
Wang Wenwu	1-76,
Wang Xiaofeng	6-20,
Wang Xiaorong	4-31,
Wang Xingjie	4-6,
Wang Xin'an	9-2, 9-3, 9-10,
Wang Yansheng	1-28,
Wang Yi	3-13, 3-18,
Wang Yifan	3-15,
Wang YP	1-30,
Wang Zhen	7-5, 7-8,
Wang Zhongwei	1-62,
Wang Zhou	2-37,
Watanabe Itaru	7-2,
Watanabe Yukio	2-19,
Wei Feng	1-76,
Wei Yayi	2-11,
Wei Yunqing	1-58,
Wei Zhengying	1-28,
Weiss Alexander	8-17,
Wilson J	4-40,
Wong Ying-Chieh	8-29,
Wu Boping	9-30,
Wu Hauqiang	1-32, 1-49,
Wu Jenny	5-1,
Wu Jie	7-20,
Wu Lei	1-31,
Wu Liang	1-64,
Wu Lixiao	5-16,
Wu Liying	6-20,
Wu Qi	7-3,
Wu Qiang	2-34, 2-36, 2-37,
Wu Qiong	4-29,
Wu Tianhan	7-6,
Wu Xing	6-8,
Wu Yi	3-58,
Wu Yingying	9-2, 9-3,
Wu Yuan	1-48, 6-19,
Wu Zhenhua	I-74, 1-76,
Wu Zhong	8-20,

X

Xia Miao	2-41,
Xia W	4-33,
Xia Wei	4-25,
Xiang Jinjuan	1-76,
Xiao Jammy	3-18,
Xiao Zhiyi	7-15,
Xie Jinjin	6-18,
Xiang Wei	1-26,
Xiong Wenjuan	1-76,
Xu Feng	1-49,
Xu Gaobo	1-76,
Xu Jian	7-14,
Xu Jie	4-22,
Xu Jintong	1-62,
Xu Jun	9-12,
Xu Kung	6-17,
Xu Liying	1-39,
Xu Renren	1-76,
Xu Wei	6-17,
Xu WeiJia	1-21,
Xu Xiaojun	6-9, 6-18,
Xu Xiaowei	9-5, 9-8, 9-9,
Xu Yue	8-21, 8-20, 9-12,
Xu Yun	1-35, 1-40, 1-48, 6-9, 6-18,
Xu Zhaozhao	1-26, 2-10, 4-7,
Xue Sophia	3-40,

Y

Yamada Tsuyoshi	2-19,
Yamashita Hiroshi	2-23,
Yamazaki Taku	2-19,
Yan Changfeng	5-16,
Yan Feng	1-63,
Yan Jiang	1-76,
Yan Jun	5-18,
Yan JL	8-9,
Yan Jin	4-35,
Yan Zeyu	9-8, 9-9,
Yanagida Tatsuya	2-19,
Yang Dongxu	2-39,
Yang FH	8-9,
Yang Libo	1-61,
Yang Lin-Wuu	5-22,
Yang Q	1-63,
Yang Sunyoung	2-15,

Yang Ta-Hone	5-22,
Yang Yongsheng	1-18, 1-22,
Yang Yuchao	1-39, 1-62, 1-77,
Yang Zhiyong	2-12,
Yao Jiaxin	I-74, 1-76,
Yao Jing-Neng	8-29,
Yao L	1-30,
Yao Liming	3-55,
Yao Peng	1-32,
Yao Yao	6-4,
Yauw Oranna	7-20,
Ye Kang	4-23,
Ye Le	1-61, 9-16,
Ye Lei	2-39,
Ye Tianchun	2-11,
Ye Zhiyuan	6-4,
Yin Binfeng	6-11,
Yin Huaxiang	I-74, 1-76, 2-46,
Yoon Jiah	7-9,
You Xiaohu	9-35,
Yousong Sun	2-22,
Yu Daquan	7-15,
Yu Guangyu	9-8, 9-9,
Yu Hu	9-8, 9-9,
Yu Jeong Yun	2-14, 2-15,
Yu M	8-18,
Yu Tzuchiang	1-18, 1-22,
Yu Yue	7-1,
Yun Huichan	2-14,
Yuan Q	8-9,

Z

Zeng Guangpeng	6-32,
Zeng Honglin	1-64,
Zeng Xinglin	6-16, 6-18,
Zeng Zhimin	1-48, 6-9,
Zha XY	8-18,
Zhang Bingxin	1-47,
Zhang Chao	9-7,
Zhang Chuan	9-35,
Zhang David Wei	2-9,
Zhang Dongping	3-23,
Zhang Haiyang	3-23,
Zhang Hai-Yang	1-34,
Zhang Hao	9-16,
Zhang Hong	7-3,

Zhang J	1-63,
Zhang JF	6-33,
Zhang John H	5-14,
Zhang Jiang Fu	1-13, 6-8,
Zhang Kailiang	1-58,
Zhang Libin	2-11,
Zhang Lingyue	4-19,
Zhang Luyao	9-16,
Zhang Nancy	3-13, 3-18,
Zhang Qingchun	1-18,
Zhang Qingwen	1-48, 6-9, 6-19,
Zhang Qingzhu	1-76, 2-46,
Zhang Teng	1-77,
Zhang W	4-33,
Zhang WD	6-33,
Zhang Wei	4-1,
Zhang Wenqiang	1-32,
Zhang Xing	6-15,
Zhang Xingdi	6-12,
Zhang Yutian	6-9, 6-18,
Zhang Yuxiang	1-35, 1-40, 1-48,
Zhang Z	8-18,
Zhang Zhaozhao	1-76, 2-46,
Zhang Zhun	9-18,
Zhang Zichao	1-58,
Zhao Hongbin	1-76,
Zhao JC	8-9,
Zhao Junhong	1-28,
Zhao Lijun	2-11,
Zhao Tingchen	8-20, 8-21,
Zhao Xiang Fu	6-27,
Zhao Yang	1-60,
Zhao Yongmin	1-65,
Zheng Deguang	7-1,
Zheng Hui	6-11,
Zhen Jinguo	4-25,
Zhong Zhiwei	9-35,
Zhou Falong	1-18,
Zhou Jie	1-18,
Zhou K	6-11,
Zhou Maopeng	7-7,
Zhou Pingqiang	9-15,
Zhou Wei	1-28,
Zhou Zhangyu	1-76,
Zhu C	1-63,
Zhu Jiadi	1-39, 1-61,
Zhu ZY	8-18,

Zhu Zhimin	1-20,
Zhuo Cheng	9-5, 9-24,
Zhuo Jinu	7-2,
Zou Xiao	6-9,
Zou Xiaodong	6-29,

IEEE
445 Hoes Lane
Piscataway, NJ 08854-4141

ISBN 978-1-5386-5309-8